WebElements: the periodic table on the world-wide web

http://www.webelements.com/

Key:

element name **atomic number** element symbol 1997 atomic weight (mean relative mass)

1	2		3	4	5	6	7	8	9	10	11	12	13	14	15	16	17	18
hydrogen 1 H 1.00794(7)																		helium 2 He 4.002602(2)
lithium 3 Li 6.941(2)	beryllium 4 Be 9.012182(3)												boron 5 B 10.811(7)	carbon 6 C 12.0107(8)	nitrogen 7 N 14.00674(7)	oxygen 8 O 15.9994(3)	fluorine 9 F 18.9984032(5)	neon 10 Ne 20.1797(6)
sodium 11 Na 22.989770(2)	magnesium 12 Mg 24.3050(6)												aluminium 13 Al 26.981538(2)	silicon 14 Si 28.0855(3)	phosphorus 15 P 30.973761(2)	sulfur 16 S 32.066(6)	chlorine 17 Cl 35.4527(9)	argon 18 Ar 39.948(1)
potassium 19 K 39.0983(1)	calcium 20 Ca 40.078(4)		scandium 21 Sc 44.955910(8)	titanium 22 Ti 47.867(1)	vanadium 23 V 50.9415(1)	chromium 24 Cr 51.9961(6)	manganese 25 Mn 54.938049(9)	iron 26 Fe 55.845(2)	cobalt 27 Co 58.933200(9)	nickel 28 Ni 58.6934(2)	copper 29 Cu 63.546(3)	zinc 30 Zn 65.39(2)	gallium 31 Ga 69.723(1)	germanium 32 Ge 72.61(2)	arsenic 33 As 74.92160(2)	selenium 34 Se 78.96(3)	bromine 35 Br 79.904(1)	krypton 36 Kr 83.80(1)
rubidium 37 Rb 85.4678(3)	strontium 38 Sr 87.62(1)		yttrium 39 Y 88.90585(2)	zirconium 40 Zr 91.224(2)	niobium 41 Nb 92.90638(2)	molybdenum 42 Mo 95.94(1)	technetium 43 Tc [98.9063]	ruthenium 44 Ru 101.07(2)	rhodium 45 Rh 102.90550(2)	palladium 46 Pd 106.42(1)	silver 47 Ag 107.8682(2)	cadmium 48 Cd 112.411(8)	indium 49 In 114.818(3)	tin 50 Sn 118.710(7)	antimony 51 Sb 121.760(1)	tellurium 52 Te 127.60(3)	iodine 53 I 126.90447(3)	xenon 54 Xe 131.29(2)
caesium 55 Cs 132.90545(2)	barium 56 Ba 137.327(7)	57-70 *	lutetium 71 Lu 174.967(1)	hafnium 72 Hf 178.49(2)	tantalum 73 Ta 180.9479(1)	tungsten 74 W 183.84(1)	rhenium 75 Re 186.207(1)	osmium 76 Os 190.23(3)	iridium 77 Ir 192.217(3)	platinum 78 Pt 195.078(2)	gold 79 Au 196.96655(2)	mercury 80 Hg 200.59(2)	thallium 81 Tl 204.3833(2)	lead 82 Pb 207.2(1)	bismuth 83 Bi 208.98038(2)	polonium 84 Po [208.9824]	astatine 85 At [209.9871]	radon 86 Rn [222.0176]
francium 87 Fr [223.0197]	radium 88 Ra [226.0254]	89-102 **	lawrencium 103 Lr [262.110]	rutherfordium 104 Rf [261.1089]	dubnium 105 Db [262.1144]	seaborgium 106 Sg [263.1186]	bohrium 107 Bh [264.12]	hassium 108 Hs [265.1306]	meitnerium 109 Mt [268]	ununnilium 110 Uun [269]	unununium 111 Uuu [272]	ununbium 112 Uub [277]		ununquadium 114 Uuq [289]		ununhexium 116 Uuh [289]		ununoctium 118 Uuo [293]

*lanthanides	lanthanum 57 La 138.9055(2)	cerium 58 Ce 140.116(1)	praseodymium 59 Pr 140.90765(2)	neodymium 60 Nd 144.24(3)	promethium 61 Pm [144.9127]	samarium 62 Sm 150.36(3)	europium 63 Eu 151.964(1)	gadolinium 64 Gd 157.25(3)	terbium 65 Tb 158.92534(2)	dysprosium 66 Dy 162.50(3)	holmium 67 Ho 164.93032(2)	erbium 68 Er 167.26(3)	thulium 69 Tm 168.93421(2)	ytterbium 70 Yb 173.04(3)
**actinides	actinium 89 Ac [227.0277]	thorium 90 Th 232.0381(1)	protactinium 91 Pa 231.03588(2)	uranium 92 U 238.0289(1)	neptunium 93 Np [237.0482]	plutonium 94 Pu [244.0642]	americium 95 Am [243.0614]	curium 96 Cm [247.0703]	berkelium 97 Bk [247.0703]	californium 98 Cf [251.0796]	einsteinium 99 Es [252.0830]	fermium 100 Fm [257.0951]	mendelevium 101 Md [258.0984]	nobelium 102 No [259.1011]

Symbols and names: the symbols of the elements, their names, and their spellings are those recommended by IUPAC. After some controversy, the names of elements 101-109 are now confirmed: see Pure & Appl. Chem., 1997, 69, 2471–2473. Names have not been proposed as yet for the most recently discovered elements 110–112, 114, 116, and 118 so those used here are IUPAC's temporary systematic names: see Pure & Appl. Chem., 1979, 51, 381–384. In the USA and some other countries, the spellings aluminum and cesium are normal while in the UK and elsewhere the usual spelling is sulphur.
Periodic table organisation: for a justification of the positions of the elements La, Ac, Lu, and Lr in the WebElements periodic table see W.B. Jensen, "The positions of lanthanum (actinium) and lutetium (lawrencium) in the periodic table", J. Chem. Ed., 1982, 59, 634–636.
Group labels: the numeric system (1–18) used here is the current IUPAC convention. For a discussion of this and other common systems see: W.C. Fernelius and W.H. Powell, "Confusion in the periodic table of the elements", J. Chem. Ed., 1982, 59, 504–508.
Atomic weights (mean relative masses): see Pure & Appl. Chem., 1996, 68, 2339–2359. These are the IUPAC 1995 values. Elements for which the atomic weight is contained within square brackets have no stable nuclides and are represented by one of the element's more important isotopes. However, the three elements thorium, protactinium, and uranium do have characteristic terrestrial abundances and these are the values quoted. The last significant figure of each value is considered reliable to ±1 except where a larger uncertainty is given in parentheses.

Matter in Equilibrium

Matter in Equilibrium

Statistical Mechanics and Thermodynamics

SECOND EDITION

R. STEPHEN BERRY
The University of Chicago

STUART A. RICE
The University of Chicago

JOHN ROSS
Stanford University

New York • Oxford
OXFORD UNIVERSITY PRESS
2002

Oxford University Press

Oxford New York
Athens Auckland Bangkok Bogotá Buenos Aires Calcutta
Cape Town Chennai Dar es Salaam Delhi Florence Hong Kong Istanbul
Karachi Kuala Lumpur Madrid Melbourne Mexico City Mumbai
Nairobi Paris São Paulo Shanghai Singapore Taipei Tokyo Toronto Warsaw

and associated companies in
Berlin Ibadan

Published by Oxford University Press, Inc.
198 Madison Avenue, New York, New York 10016
http://www.oup-usa.org

ISBN: 0-19-514749-9

Full Libary of Congress Cataloging-in-Publication Data for this title can be
found under Physical chemistry, 2nd ed. / R. Stephen Berry, Stuart A. Rice,
John Ross, ISBN: 0-19-510589-3.

9 8 7 6 5 4 3 2 1
Printed in the United States of America
on acid-free paper

To our families

Contents

Preface

Soul of the world, inspired by thee,
The jarring seed of matter did agree,
Thou didst the scatter'd atoms bind,
Which, by the laws of true proportion joined
Made up of various parts one perfect harmony

—Text by Nicholas Brady for the *Ode on St. Cecilia's Day, 1692* by Henry Purcell

The first edition of this book was published twenty years ago. Our goal was to introduce an undergraduate student to physical chemistry in a manner that focused attention on the fundamental principles and conceptual structure of the field, following a logical development from the simple to the more complex descriptions of systems and phenomena. We also emphasized how the content, style, expectation of achievement, and even the nature of what is considered explanation, contribute to the definition of physical chemistry as a field of scholarly scientific endeavor, and keep it a vital contemporary discipline.

The current (second) edition of this book is intended to meet the same goal and to also be useful as a text for a first year graduate course in physical chemistry. Our revision of the text has resulted in the removal of errors and improvements in the quality of the arguments used. We have retained an emphasis on correctness and depth of conceptual argument without recourse to advanced mathematics. The basic concepts and their ramifications are presented in ways that establish both their logical structure and their realms of validity. Points that require close scrutiny in some situations are carefully discussed. The applications of physical chemistry connect with forefront problems in science. The additions in this edition are the following: considerable new material on nuclear magnetic resonance, especially on modern methods; expansion of the treatment of the fundamental principles of statistical mechanics and of the treatment of linear irreversible thermodynamics; updating of the descriptions of the equilibrium properties of liquids and solutions and of nonlinear chemical kinetics and electrode processes. Several vignettes covering forefront areas of research in physical chemistry, by leaders in the fields, have been added. Just as when we finished preparing the first edition of this book, even as we complete this edition, we see new results in the scientific journals that we would have incorporated in the text had they appeared in time.

The first edition of this book found use both as a text and as a reference book in physical chemistry. We trust that the second edition will also be used both ways.

Our approach to the teaching of physical chemistry is based on our experience as researchers and educators. In the Preface to the first edition of this book we described that approach in the following words.

Physical chemistry is an empirical science. A science is a set of constructs, called theories, that link fragments of experience into a consistent description of natural phenomena. The adjective "empirical" refers to the common experiences from which the theories grow, that is, to experiments. Simple working hypotheses are guessed by imaginative insight or intuition or luck, usually from a study of experiments. This repetitive interplay in time leads to the formulation of theories that correlate the accumulated experimental information and that can predict new phenomena with accuracy. As scientists we have, throughout our careers, endeavored to combine both experimental and theoretical work. In this book we try to knit these two inseparable parts of the science into a coherent structure that represents accurately the way they interact in physical chemistry today.

Our goal is the presentation of the three major areas of physical chemistry: molecular structure, the equilibrium properties of systems, and the kinetics of transformations of systems. The theoretical foundations of these subjects are, respectively, quantum mechanics, thermodynamics and equilibrium statistical mechanics, and chemical kinetics and kinetic theory. These theories, firmly based on experimental findings, constitute the structure required for the understanding of past accomplishments and a basis for recognition and development of significant new areas in physical chemistry. The presentation of the theories of physical chemistry requires careful discussions at several levels of exposition. Our approach aims toward depth of understanding of fundamentals more than toward breadth of recognition of the multitude of activities that go on under the name of physical chemistry. The organization of the book, with its three principal sections, should make this clear. The mathematical level begins with elementary calculus and rises to the use of simple properties of partial differential equations and the special functions that enter into their solutions. Our intention is to keep the reader's mind on the science rather than on the mathematics, especially at the beginning. This procedure also corresponds to the pattern, followed by many students, of taking physical chemistry and advanced calculus concurrently. Appendices develop the details of the mathematical tools as they are needed.

The text discussion contains more material than can be covered in the traditional one-year physical chemistry sequence; it is designed to fulfill the dual purpose of providing a clear and incisive treatment of fundamental principles at a level accessible to all students while broadening the perspectives and challenging the minds of the best students. Individual instructors will wish to make their own selections of material for inclusion and exclusion, respectively. We have provided guidance on this matter by suggesting, at the end of this Preface, schema for several different types of physical chemistry courses. These schema suggest both how to use the material discussed in a different order than in the text, and how to omit or downplay classes of topics deemed unsuitable to a particular group of students. For example, we have taught the junior-level physical chemistry course both as organized in this text and with the material of Part II preceding that of Part I. To invert the order of Parts II and I, it is only necessary to ask the student to accept the existence of quantized energy levels and a few specific examples of energy level spectra. This has proved quite easy for students, all of whom have heard of these matters in freshman chemistry courses. Students with strong backgrounds will have seen the material in the first chapter and parts of the second chapter. Chapter 10, on intermolecular forces, and Chapter 11, on the structure of solids, contain large blocks of material that could be passed over in a traditionally oriented one-year course. The thermodynamic description of matter can be emphasized and the statistical molecular description de-emphasized, or vice versa, by selection of the relevant sections of Chapters 21 to 26. Chapter 20, dealing with linear irreversible thermodynamics and negative temperature, can easily be omitted if so wished by the instructor. In Part III, the elements of physical kinetics or transport theory are contained in Sections 27.1 to 28.8 and 29.1; the elements of chemical kinetics are contained in Chapter 30. The specialized topics in the remaining sections may be used for more extensive treatments of these subjects, either in a three-semester physical chemistry sequence or in a senior-graduate course on these topics.

At the other end of the scale there is sufficient material in this book for a first-year graduate course for students whose undergraduate preparation in physical chemistry did not emphasize modern aspects of the subject.

At the end of each chapter we have suggested extra reading for the interested student. The book contains about 750 problems. A few of these problems are designed to acquaint the reader with dimensions, units, and simple manipulations. More are intended to develop intellectual skills, to enable students to master the material of the text discussions by the difficult process of thinking through the kinds of questions encountered in the laboratory. Some of the problems are designed to extend the theoretical analysis of the text to special but interesting situations, and some are designed to require the use of a computer for numerical calculation.

Part I opens with a review of the elementary quantities of the atomic world and how they are measured. Many readers will find some or all of the material in Chapter 1 familiar; they may choose to read quickly through this chapter and start their more intensive study with Chapter 2, where we develop the experimental evidence for the quantum structure of matter at the molecular level. In the process, we examine the harmonic oscillator, the primitive model for many kinds of behavior examined later, and the concept of action.

Chapter 3 is more theoretical and mathematical than its predecessors but begins at a level the well-prepared student will find quite elementary. Waves and wave equations are introduced and are used to describe only very simple situations: the states of particles in boxes of various sorts, and of rigid rotators.

Chapter 4 brings us back to the more realistic problems of the quantum-mechanical oscillator and the simplest atom, hydrogen.

In Chapter 5, we further develop the concepts such as orbitals and transitions between quantum states that are introduced for hydrogen in Chapter 4. Chapter 5 treats atoms with more than two electrons, in particular their electronic states and the interactions among the electrons.

Molecules are first introduced in Chapter 6, which is almost analogous to Chapter 4 in that it too goes in depth into the description and behavior of the simplest molecules, H_2^+ and H_2, just as Chapter 4 examined the H atom. The concepts developed in Chapter 6 are than extended to more complex diatomic molecules in Chapter 7. Chapter 6 deals almost exclusively with electronic states; Chapter 7 introduces molecular vibration and rotation and discusses concepts used to correlate and unify our observations regarding diatomic molecules.

Chapter 8 begins with the primitive species H_3^+ and H_3 and then goes on to ideas that begin to be important with three or more nuclei: hybrid orbitals and delocalized molecular orbitals at the level of electronic states, and normal modes of vibration. Larger molecules are discussed in Chapter 9; here we introduce the concepts of chirality and optical activity, and explore some aspects of ligand field theory and the magnetic properties of molecules.

Chapter 10 is the first in which we go beyond the properties of individual molecules. The discussion of intermolecular forces describes the interactions between charge distributions and how one molecule behaves when it collides with another. The material in this chapter is based wholly on the framework of molecular structure but becomes especially useful to us in Parts II and III, where we study the behavior of matter in the aggregate. Part I concludes with another structural aspect of aggregated matter, the structure of solids. Here, we extend the various concepts of bonding previously developed to include the concept of metallic bonding, to describe the structure and states of periodic condensed phases.

Part II is concerned with the equilibrium properties of bulk matter. Our presentation simultaneously develops the statistical molecular theory and the classical thermodynamic theory in a mutually reinforcing fashion. Despite use of this "mixing" of microscopic and macroscopic points of view, no compromise is made with respect to the rigor of classical thermodynamics, and if desired the two points of view can be separated.

In Chapter 12, we begin with a discussion of the zero-th law of thermodynamics and the concept of temperature. By examining the phenomenological bases for the equation of state and the definition of temperature, together with the elements of the kinetic theory of perfect gases, we establish a first connection between macroscopic and microscopic descriptions. The building of bridges between the two classes of description is a principal theme of succeeding chapters.

Chapters 13 and 14 treat the first law of thermodynamics and some of its many applications. Particular care is devoted to the precise definition of work and heat and to how the nature of these quantities exemplifies the differences between the thermodynamic and mechanical descriptions of matter.

Chapter 15 introduces the concept of entropy by way of the microscopic structure of matter. Given only that every sample of matter has an energy-level spectrum, it is shown that there exists a function of the density of states, the entropy, that behaves like a property only of macroscopic variables of the system, despite its definition in terms of the microscopic energy-level spectrum.

Chapter 16 develops the second law of thermodynamics via the classical Clausius and Kelvin principles, and Chapter 17 is devoted to examples of the use of the second law to solve problems of chemical interest. These chapters contain a careful discussion of the nature of irreversibility and its interpretation in terms of thermodynamic and statistical molecular theories.

Chapter 18 introduces, discusses, and gives applications of the third law of thermodynamics.

In Chapter 19 we examine the central problem of describing equilibrium as a function of the external constraints on the system. The thermodynamic theory of open systems is developed and is used to derive the several criteria of equilibrium that are suitable to different external constraints. With this background, the notion of ensemble is introduced, and the classical thermodynamics of equilibrium is related to the development of the grand canonical, canonical, and microcanonical partition functions of statistical mechanics. The theory is illustrated by analyzing the velocity distribution in a perfect gas.

Chapter 20 introduces a new point of view into the analysis—it deals with the description of systems whose properties vary slowly in time, and with the extension of thermodynamics to systems with negative temperature.

Chapters 12 through 19 establish the principles and develop the tools needed to study the bulk properties of matter. This study is carried out systematically in Chapters 21 through 26. Chapter 21 deals with gases, Chapter 22 with solids, Chapter 23 with liquids, Chapter 24 with phase transformations, Chapter 25 with solutions of nonelectrolytes, and Chapter 26 with solutions of electrolytes. In each chapter the thermodynamic theory is developed first, then the statistical molecular theory. Extensive use is made of the details concerning molecular behavior that are provided by computer simulation studies. In addition, the principles developed are illustrated with data from experimental situations wherever that is appropriate.

The simultaneous development of classical and statistical thermodynamics, employed through Part II, is designed to overcome the difficulties associated with the very abstract nature of purely thermodynamic reasoning, and also to illustrate the richness of the phenomena that can arise from molecular interactions. Nevertheless, for any given problem, the generality of the thermodynamic approach is made evident, as are the wealth of detail and dependence on assumed models of the statistical molecular approach.

Part III is concerned with time-dependent processes, especially the approach to equilibrium. The topic of physical kinetics (transport processes) is introduced in Chapter 27 with a discussion of the mechanics of molecular collisions, mostly binary collisions. We present as simply as possible the elements of kinematics and dynamics, including the concept of scattering, which is illustrated with the hard-sphere model.

In Chapter 28 we consider the kinetic theory of gases, beginning with how velocity distribution functions change with time because of collisions. We present an elementary discussion of the time-dependent transport equations and show how they govern a gas's approach to equilibrium. With these results we can discuss fluxes of mass, momentum, and energy, and study the process of effusion and the simple transport properties (diffusion, viscosity, and thermal conduction) in dilute gases. We conclude with a brief treatment of energy exchange processes, and sound propagation and absorption in gases.

The transport properties of dense phases are taken up in Chapter 29. Transport in liquids is approached with a discussion of Brownian motion, leading to the relation of transport coefficients to autocorrelation functions. Brief discussions of transport in solids and of electrical conductivity in electrolyte solutions conclude the chapter.

Chemical kinetics is treated in a manner parallel to physical kinetics, with an elementary development followed by selected advanced applications. We begin in Chapter 30 with a presentation of the mechanics of reactive collisions, including both kinematics and dynamics. The emphasis is again on the simple hard-sphere model. The collision-theory approach is compared with the activated-complex theory, and both theories are used for an analysis of kinetics in gases and solutions. After a brief survey of experimental methods, we discuss complex reactions and provide an elementary discussion of chemical reaction mechanisms.

Chapter 31 is devoted to various advanced topics in chemical kinetics, including the RRKM theory of unimolecular reactions, symmetry rules in chemical reactions, chain reactions, oscillatory reactions, photochemistry, homogeneous and heterogeneous catalysis and the kinetics of electrode reactions.

Our educational and professional associations have obviously influenced the way we wrote this book, as have a few remarkable volumes. In particular, the books by H. Reiss (*Methods of Thermodynamics,* Blaisdell, N.Y., 1965) and F. Reif (*Statistical Physics,* McGraw-Hill, N.Y., 1965) helped to clarify our independently developed presentations.

The current edition of this book draws heavily on the content of the first edition. We owe a continuing debt to all those who contributed to the first edition, in particular to George P. Flynn and Joseph N. Kushick. It is a pleasure to thank the many colleagues who have helped with critiques of the first edition and made suggestions that, we trust, improve the second edition. Amongst these are Stephen Bernasek, John Brauman, Laurie J. Butler, Claude Diou, John Enderby, Michel Fayer, Harold Friedman, Roald Hoffman, George C. Lie, David Oxtoby, Robert Madix, Peter Radziks, Jan Sengers, and Richard Zare. This edition has been enhanced by the contributions of Christopher Chidsey, Robert Eisenberg, R. A. Marcus, H. Schaefer, D. G. Truhlar, M. Vlad, Peter Wolynes, and Achmed Zewail, whose special topic vignettes are distributed throughout the text; we thank them for these contributions. We also thank Andrea Twiss-Brooks and Fritz Whitcomb for the Appendix describing modern methods of searching the scientific literature, and Ron Friedman and Ron Duchovic who have provided new problems that augment the set of problems carried over from the first edition.

To the readers, we say that we hope that you will have as much delight in using and creating physical chemistry as we have.

R. Stephen Berry
Stuart A. Rice
John Ross

Suggested Topic Selections for Various Courses

The following table illustrates possible choices of subject matter for a variety of physical chemistry courses.

Chapter	One-Year Course (initial emphasis on macroscopic approach)	One-Year Course (initial emphasis on microscopic approach)	Three-Semester Course	Senior or Graduate Course
1	Start with Chapters 12 through 26, with emphasis placed on the thermodynamic analysis. Keep only such elements of the statistical molecular analysis as fits with the instructional goal. Then return to Chapter 2 and follow the outline for the one-year course with initial emphasis on microscopics, going rapidly to Section 3.3	Start with Chapter 1, setting pace as fast as students can absorb Chapters 1 and 2, and Sections 3.1 and 3.2. Touch lightly or omit Section 3.10 and Appendix 3A	All, as introduction	Omit
2			All	Read quickly
3			All	Read Sections 3.1 and 3.2 quickly
4		All	All	All
5		All, but Section 5.5 may be omitted for very well-prepared students	All	Touch lightly on Section 5.5
6		All	All	All
7		Section 7.8 may be dropped	All	All
8		All	All	All
9		Omit Sections 9.6 and 9.7	All	All
10		Omit	All	All
11		Sections 11.1 through 11.5, 11.8, and 11.10	All	All
12		All	All	All
13		All	All	All
14		Omit Section 14.6. Only touch lightly on Sections 14.9, 14.10	All	All
15		Omit optional Section 15.1	Omit optional Section 15.1	All
16		Omit optional Sections and Appendix 16A	Omit Appendix 16A	All
17		All	All	All
18		All	All	All
19		Omit Section 19.3	All	All
20		Omit	Touch lightly on Sections 20.1–20.4, keep 20.5	All
21		Omit Appendices 21A, 21B, 21C	All	All
22		Omit Section 22.5 and Appendix 22A	All	All
23		Omit Section 23.4	All	All
24		Omit Section 24.3. Touch lightly on Section 24.4	All	All
25		Omit latter half of Section 25.8. Touch lightly on Section 25.9	All	All
26		Omit Section 26.6	All	All
27		27.1–4	All	All
28		28.1, 4, 6, 7, 8	28.9, 10 optional	All
29		29.1	29.2–6	All
30		Omit 30.5, 6, 9	All	All
31		As time permits Assign as special topics	31.3, 6–11 and as time permits	All

Glossary of Symbols

Symbol	Meaning	Where used
A	Helmholtz free energy	general
	action	Part I; Eq. 2.25
	area	general
	frequency factor in rate coefficient	Part III; Eq. 30.42
A	operator for a classical property A	Part I; Section 3.4
$A(X)$	electron affinity of X	Eq. 14.46
A_∞	high-pressure frequency factor for unimolecular reactions	Part III; Eq. 30.174
$\mathscr{A}$	amplitude of perturbation	Part III; Eq. 30.127
a	molar Helmholtz free energy	Part II; general
	ionic radius	Eq. 26.69
	radius of Brownian particle	Eq. 29.53
$\mathbf{a}$	acceleration	general
a_i	activity of component i	Parts II and III; general
a_0	Bohr radius	general
$a_\pm$	mean ionic activity	Eq. 26.8
$\mathbf{a}, \mathbf{b}, \mathbf{c}$	primitive vectors of a crystal	p. 315 ff
a, b, c	primitive distances in a crystal	p. 315 ff
a_x, a_y, a_z	x-, y-, z-components of acceleration	general
B	second virial coefficient of a gas	Part II; Eq. 21.17
	second virial coefficient in the concentration expansion of the osmotic pressure of a solution	Part II; Eq. 25.127
	magnitude of the magnetic field strength **B**	general
$\mathbf{B}$	magnetic field strength	general
B_e	rotational constant for a rigid rotor	Eq. 7.21
B_v	rotational constant for a molecule in the vibrational state designated by quantum number v	Eq. 7.21
B_2^*	second virial coefficient in liquid solution	Eq. 25.196
$B(R)$	bridge function in theory of liquids	Eq. 23.53
b	impact parameter	general
	excluded volume term for the van der Waals equation of state	Table 21.2
	excluded volume in a lattice model of a mixture	Eq. 25.193
C_p; C_v	heat capacity at constant pressure; heat capacity at constant volume	general
C_σ	heat capacity along the liquid-vapor coexistence line	Eq. 24.19
C	the Coulomb; SI unit of charge	Table 1.1
C_2	symmetry operator for 180° rotation	Section 31.6
$C(R)$	direct correlation function of a liquid	Eq. 23.24
C	third virial coefficient of a gas	Eq. 21.17
	third virial coefficient in the concentration expansion of the osmotic pressure of a solution	Eq. 25.127
C_d	differential capacitance of electrolyte-electrode double layer	Eq. 26.132
c	speed of light	general
	speed of sound	Section 28.10
c^*	"equivalent" concentration	Eq. 29.103
c_i	concentration of component i	general

Symbol	Meaning	Where used
c_{ij}	coefficient of the jth basis orbital in the expression for the ith molecular orbital	Part I; Eq. 8.17
C^S_{ijkl}	second order adiabatic elastic constant of a crystal	Eq. 22.18
C^T_{ijkl}	second order isothermal elastic constant of a crystal	Eq. 22.18
c_p; c_v	molar heat capacity at constant pressure; molar heat capacity at constant volume	general
c^x_v	contribution of x (for example, vibration) to the constant volume molar heat capacity	Eq. 22.44
D	dielectric constant ($\mathcal{D}$ in Section 30.8)	Eq. 26.58
	Debye unit	Eq. 4.29 ff
	(self-) diffusion coefficient	Table 28.1
D_k	(mutual) diffusion coefficient of component k	Eq. 20.15; App. 16A
D_0	dissociation energy measured from the zero point energy level	general
D_{0i}	energy required to dissociate a molecule of species i into atoms, measured from the zero point level	general
D_e	dissociation energy measured from the minimum in the potential well (well depth)	general
$D(w, w^*)$	distance between points w and w^* in phase space	p. 423
d	spacing between lattice planes	Fig. 11.8
	hard-sphere diameter	Part III; Fig. 27.7
$d(\mathbf{r}_2, \theta_{12} \mid \mathbf{r}_1)$	conditional probability density for finding electron 2 at $\mathbf{r}_2$, θ_{12} when electron 1 is at $\mathbf{r}_1$	Part I; Eq. 5.17
E	energy	general
E	operator for energy	Part I; Eq. 3.32
$\mathbf{E}$	electric field strength	general; Eq. 1.8
$E(R)$	eigenvalue of the Schrödinger equation based on the Hamiltonian $H_{\text{elec}}(r, R)$ for fixed nuclei	Part I; Eq. 6.3
E_{rot}	rotational energy; an eigenvalue of the Schrödinger equation for rotational energy of a molecule	Eq. 7.19
E_{vib}	vibrational energy; an eigenvalue of the Schrödinger equation for vibrational energy of a molecule	Eq. 7.6
E_a	activation energy	Part III; Eq. 30.42
E_v	energy for formation of a vacancy	Section 29.5
E_∞	high-pressure activation energy for a unimolecular reaction	Part III; Eq. 30.174
ΔE_{reac}	energy change in a reaction	general
ΔE^0_0	net change in energy in conversion of reactants to products at $T = 0$	Eq. 21.152
ΔE^*_f, ΔE^*_r	barrier height in forward and reverse directions of reaction	Part III; Fig. 30.6
$\Delta E^\ddagger$	energy of activation (energy difference between activated complex and reactants)	Part III; Section 30.6
$\epsilon_i(R)$	energy of ith molecular orbital	Part I, esp. Chapter 7, ff
e	elementary electronic charge	general; Table 1.1
$\mathcal{E}$	electromotive force	Eq. 26.38
	electric potential	Table 13.1
$\mathcal{E}^0$	standard emf	Eq. 26.37
$\mathcal{E}_{1/2}$	half-cell potential	Part II; Eq. 25.566
F	structure factor	Eq. 11.28
	magnitude of the force $\mathbf{F}$	general
	tension	Part II; Table 13.1
$\mathbf{F}$	force	general
F_{bonding}	magnitude of the force bonding two nuclei	Part I; Eq. 6.8
$\mathcal{F}$	The Faraday constant	general
f	number of degrees of freedom of the macroscopic variables in a multicomponent multiphase system	Part II; Chapter 24
	flux density of a beam	Eq. 2.13; Eq. 30.86
$f(\mathbf{r}, \mathbf{v}, t)$	velocity and space distribution function; number of molecules per unit "volume" in phase space at time t	Part III; Section 28.1
$f(\theta, E)$	quantum mechanical scattering amplitude	Eq. 27.73
$f(\mathbf{r}, t)$	density of molecules in configuration space	Part III; Eq. 28.14
$f(\mathbf{v}, t)$	density of molecules in velocity space	Part III; Eq. 28.2

Symbol	Meaning	Where used
$f(\mathbf{r}, \mathbf{v}, t)$, $f(\mathbf{r}, t)$, $f(\mathbf{v}, t)$	probability densities corresponding to $f(\mathbf{r}, \mathbf{v}, t)$, $f(\mathbf{r}, t)$, and $f(\mathbf{v}, t)$, $f(\mathbf{v}, t) \equiv 1/N\, f(\mathbf{r}, \mathbf{v}, t)$, etc. Note that $f(\mathbf{v}, t)$ corresponds to the $f(\mathbf{v})$ of Section 19.7	Part III; Eqs. 28.4 ff
$f(\mathbf{v})$	equilibrium velocity distribution function	Part II; Eq. 12.8
$f(v)$	equilibrium speed distribution function	Part II; Eq. 12.10
$f(\omega)$	distribution of orientations of a molecule	Eq. 24.96
$f(t)$	randomly fluctuating force on a Brownian particle	Eq. 29.38
f_i^0	fugacity of pure component i	Eq. 21.14
f_i	fugacity of component i in a mixture	Eq. 21.37
f_s	atomic scattering factor	Part I; Eq. 11.26
$f(\epsilon)$	one-particle energy distribution function	general; Eq. 11.56
$f(j, t)$	number of oscillators in quantum state j at time t	Part III; Eq. 28.15
G	Gibbs free energy	general
G_{ij}	Integrated correlation function in Kirkwood–Buff Theory of mixtures	Eq. 25.189
$G(k)$	density of momentum states	Part I; Eq. 11.52
g	molar Gibbs free energy	Part II; general
	acceleration due to gravity	general; Eq. 1.22
g_m	mean molar Gibbs free energy of a mixture	Chapter 25
g_k	degeneracy of the level with energy E_k; the number of states with energy in the range $E_k \le E \le E_k + dE$	Eq. 19.115
g_1	osmotic coefficient	Part II; Eq. 25.125
g_s	Landé magnetic g-factor	Part I; Eq. 5.7
g_{eff}	effective magnetic g-factor	Part I; Eq. 9.15
$g(\epsilon)$	density of one-electron energy states	Eq. 11.54
$g_2(\mathbf{R}_1, \mathbf{R}_2)$, $g_2(R)$	pair correlation function of a liquid	Eq. 11.2; Eq. 23.15
$g_3(\mathbf{R}_1, \mathbf{R}_2, \mathbf{R}_3)$	triplet correlation function of a liquid	Eq. 23.23
$g(t)$	autocorrelation function	Part III; Eq. 29.62
$g_v(t)$	velocity autocorrelation function	Part III; Eq. 29.74
H	enthalpy	general
H	Hamiltonian operator	general; Eq. 3.62
H_{AB}	integral of the Hamiltonian over the orbital functions φ_A^* and φ_B	Part I; Eq. 6.37
H_{HF}	Hartree–Fock Hamiltonian operator	Part I; Eq. 8.16
H_{nuc}	Hamiltonian operator for motion of the nuclei of a molecule in the effective potential of the electrons $E(R)$	Eq. 6.1
H_{elec}	Hamiltonian operator for electronic energy and nuclear potential energy	Eq. 6.1
$H_n(z)$	nth Hermite polynomial	Eq. 4.9
$\mathbf{H}$	magnetic field strength	Table 13.1
$\hat{\mathsf{H}}$	Hamiltonian operator	p. 448
h	Planck's constant $\hbar \equiv h/2\pi$	general
	molar enthalpy	general
h, k, l	Miller indices	Part I; Eq. 11.3
$h(R)$	correlation function in theory of liquids	Eq. 26.89
h_1	$1s$ hydrogenic orbital on atom 1	Part I; p. 360
h_m	mean molar enthalpy of a mixture	Chapter 25
Δh_{fus}	molar heat of fusion	general
Δh_{vap}	molar heat of vaporization	general
Δh_{subl}	molar heat of sublimation	general
$\bar{h}_i$	partial molar enthalpy of component i	general
$\Delta\bar{h}_i \equiv \bar{h}_i$	partial molar enthalpy of component i	Section 14.6
ΔH	heat of formation	general
ΔH^0	standard heat of formation per mole of product	general
$\Delta H^0_{\text{atom'n}}$	enthalpy change on atomization of one mole of a compound under standard state conditions	Section 14.8

Symbol	Meaning	Where used
$\Delta H_{\text{sol'n}}$	integral heat of solution	Section 14.6
ΔH_{∞}	enthalpy change when one mole of solute is added to an effectively infinite amount of solvent	Section 14.6
ΔH_{dil}	integral heat of dilution	Section 14.6
I	intensity of a beam	Eq. 2.8
	electric current	general
	ionic strength	Eq. 26.49
	moment of inertia of diatomic molecule	Eq. 3.129
	nuclear spin quantum number	Eq. 18.15
I_x, I_y, I_z	principal moments of inertia of a polyatomic	Eq. 3.173
I_{IR}	moment of inertia for internal rotation	Eq. 21.129
I_0	intensity of incident light	Eq. 31.22
$I(\mathrm{M})$	ionization potential of M	Eq. 14.46
$\mathbf{i}, \mathbf{j}, \mathbf{k}$	unit vectors in the x-, y-, z-directions	Eq. 3.54
i	net current in an electrolytic cell	Section 31.13
$\mathcal{I}(x, t)$	intensity of a wave at the point x at time t	Eq. 3.47
$\mathbf{J}$	total angular momentum	general; Eq. 5.18
J	rotational quantum number	general; Eq. 7.19
	cell partition function	Part II; Eq. 25.207
J	Joules	general
J	Coulomb integral over atoms A and B	Eq. 6.40
$J(x, y, \ldots /u, v, \ldots)$	Jacobian of the transformation from variables $x, y, \ldots$ to variables $u, v, \ldots$	Eq. 11.18
$\mathbf{J}^Q(\mathbf{r}, t)$	vector flux of Q at $\mathbf{r}$, t	Part II; Eq. 20.2
$\mathbf{J}^m(\mathbf{r}, t)$	mass flux at $\mathbf{r}$, t	Part II; Eq. 20.6
$\mathrm{J}_k^m(\mathbf{r}, t)$	mass flux of component k, relative to the center of mass, at $\mathbf{r}$, t	Part II; Eq. 20.13
$\mathbf{J}^p$	momentum flux	Part II; Eq. 20.21
$\mathbf{J}_{ij}^p$	component ij of the momentum flux tensor	Part II; Eq. 20.23
j	rotational quantum number (usually J in Parts I and II)	Part III; general
$\mathbf{j}^r$	electric current density	Part II; Eq. 29.87
K	Kolmogorov entropy	p. 460
$K(j\,\vert\,j')$	Transition probability for $j' \to j$	Eq. 15C.2
$K(T)$	equilibrium constant for a gas phase reaction (independent of pressure)	Eq. 21.64
K_{p}	equilibrium constant in terms of component pressures	Eq. 21.75
K_V	equilibrium constant in terms of component concentrations	Eq. 21.88
K_x	equilibrium constant in terms of component mole fractions	Eq. 21.86
$K(T, p, \text{solvent})$	equilibrium constant for a reaction in a condensed phase	Eq. 25.131
K_a	equilibrium constant in terms of component activities	Eq. 30.82
K_w	ion product of water ($\equiv [H_3O^+][OH^-]$)	Eq. 31.104
$K_V^{\ddagger}$	value of K_V for formation of an activated complex ($K^{\ddagger}$, Eq. 30.79, is the corresponding equilibrium constant in terms of activities)	Part III; Eq. 30.44
K_{AB}	exchange integral over atoms A and B (sometimes K)	Eq. 6.40
K_b	boiling point constant	Part II; Eq. 25.111
K_f	freezing point constant	Part II; Eq. 25.114
k	force constant of a harmonic oscillator	Eq. 2.32
	wave number ($\equiv 2\pi/\lambda$)	Eq. 3.3
	rate coefficient of a chemical reaction	Part III; general
$\mathbf{k}$	wave vector	general
	unit normal to a surface	general
k_B	Boltzmann's constant	general
$k_2(T, p, \text{solvent})$	Henry's law constant	Part II; Eq. 25.60
k_α	force constant for the αth normal mode of vibration of a molecule	Part I; Eq. 8.38

Symbol	Meaning	Where used
k_a, k_b	rate coefficients for acid and base catalysis	Part II; Eq. 31.92
k_D	diffusion-controlled limit of rate coefficient	Part III; Eq. 30.103
k_f, k_r	rate coefficients for forward and reverse reactions	Part III; general
k_{uni}	effective unimolecular rate coefficient	Part III; Eq. 30.172
k_∞	high-pressure limit of k_{uni}	Part III; Eq. 30.174
k_1, k_2, . . .	rate coefficients for steps 1, 2, . . . in a reaction mechanism (k_{-1} is the coefficient for the step inverse to step 1)	Part III; general
$\mathbf{L}$	orbital angular momentum	Eq. 3.151
L	magnitude of angular momentum $\mathbf{L}$	Eq. 2.31
L^2	operator for the square of the angular momentum $\mathbf{L}$	Eq. 3.176
L_z	operator for the z-component of the angular momentum $\mathbf{L}$	Eq. 3.176
$\mathbf{L}_i$	orbital angular momentum of the ith electron	Eq. 5.20
L_{ik}	Onsager linear transport coefficient	Eq. 20.91
l	length	general
	mean free path	Part III; Eq. 28.65
	orbital angular momentum quantum number	Eq. 3.162
M	Madelung constant	Eq. 11.41
	total mass (in a system of two or more particles)	general
$M(T)$	chemical potential along the critical isochore	Part II; Eq. 24.67
M_i	molecular weight of component i	Eq. 14.14
$\mathbf{M}$	magnetic moment	Table 13.1
M_L	quantum number for the z-component of the total orbital angular momentum $\mathbf{L}$	Eq. 5.22
M_S	quantum number for the z-component of the total spin angular momentum $\mathbf{S}$	Eq. 5.22
m	mass	general
m_e, m_n, m_p	mass of the electron, neutron, proton	Table 1.1
m_{eff}	effective mass	Eq. 11.95
m_i	molality of component i	general
$m_\pm$	mean ionic molality	Eq. 26.8
m_1	molality of pure solvent	Part II; Eq. 25.137
m	quantum number of angular momentum for a particle in a circular box, or for the z-component of angular momentum of an electron	Eq. 3.142; Eq. 3.161
	z-component of any angular momentum	
m_s	spin quantum number; the z-component of the spin angular momentum of an electron	Eq. 5.2
m_{li}, m_{si}	quantum number for the z-components of the orbital and spin angular momentum of electron i	Eq. 5.22
N	number of molecules in a system	general
N_A	Avogadro's number	general
$\mathcal{N}$	number of replicas in an ensemble	Part II; Chapter 15
$N(E_n)$	number of states with energy E_n	Part I; Eq. 4.17
N_v	number of vacancies in a lattice	Eq. 29.86
n	number density	general
	number of moles	general
	integer	Eq. 2.53
	a quantum number	general
n_i	number of moles of component i	general
	principal quantum number for the ith one-particle quantum state	Part I; Eq. 2.10
n_i^*	effective principal quantum number for the ith one electron quantum state	Part I; Eq. 2.12
$n(v)$	number density of molecules with speed v	Part III; Section 27.5
$P(x)$	probability distribution of x	general
$\mathcal{P}(x)$	probability density of x	general
$\mathcal{P}$	excess pressure in a perturbation	Part III; Eq. 28.125

Symbol	Meaning	Where used
P_{12}	operator indicating permutation of particles 1 and 2	Eq. 6.53
p	pressure	general
	magnitude of momentum **p**	general
	probability aftereffect factor	Part II; Eq. 16.A.12
	molecular orbital bond order	Part I; Eq. 9.4
	reaction probability factor	Part III; Eq. 30.27
$\mathfrak{p}$	unit of pressure	Part II; Eq. 21.12
$\mathbf{p}$	momentum	general
p	momentum operator	general; Eq. 3.36
p_θ, p_ϕ	angular momentum	general; Eq. 3.128
p_ϕ	angular momentum operator	general; Eq. 3.138
p_{app}	applied pressure	general
p_σ	vapor pressure of a liquid along the liquid-vapor coexistence line	Eq. 23.1
p^*	reduced pressure (with respect to pair potential parameters)	Eq. 21.178
$\tilde{p}$	reduced pressure (with respect to the critical pressure)	Chapter 21
p, q	probabilities of positive and negative displacements in Brownian-motion model	Part III; Section 29.2
	numbers of protons on acid and proton-acceptor sites on conjugate base	Part III; Eq. 31.106
Q	electric charge	general
	ratio of activities in an electrochemical-cell reaction	Part III; Section 31.13
Q_α	αth normal coordinate	Eq. 8.34
Q_N	canonical partition function for N particles in a volume V at temperature T	general; Eq. 19.131
q	heat absorbed by a system	general
	spatial coordinate	Part I; Eq. 3.98
	electric charge	Eq. 1.5
$\mathbf{q}$	quadrupole moment	Part I; Eq. 10.6
	heat flux vector	Part II; Eq. 20.64
$q_i(R)$	charge density around ion i	Part II; Eq. 26.74
q_1	single-molecule partition function per unit volume	Parts II and III; Eq. 21.148; Eq. 30.54
R	gas constant	general
	interatomic or intermolecular distance	general
	electrical resistance	Eq. 29.101
R_e	internuclear distance for which $E(R)$ is a minimum	general
R_H	Rydberg constant for hydrogen	Eq. 2.7
R_∞	Rydberg constant for a hydrogenic atom with infinite nuclear mass	Eq. 2.69
$R_{nl}(r)$	solution of the radial wave equation for the hydrogen atom	Eq. 4.18
R^*	reduced distance (with respect to pair potential parameters)	Eq. 26.90
$\mathfrak{R}$	excess density in a perturbation	Part III; Eq. 28.126
r	radial distance; radial polar coordinate	general
$\mathbf{r}$	position vector	general
r^*	critical distance for a diffusion-controlled reaction	Part III; Eq. 30.94
r_{kl}	distance between electrons k and l	Eq. 8.14
r_{Kk}	distance between nucleus K and electron k	Eq. 8.14
r_{max}	distance at which the electronic probability density has its maximum	Eq. 4.23
S	entropy	general
	screening constant	Part I; Eq. 2.6
$\mathbf{S}$	spin angular momentum	general; Eq. 5.1
$\mathbf{S}_i$	spin angular momentum of the ith electron	general; Eq. 5.20
S_{AB}	overlap integral over atoms A and B	general; Eq. 6.30
s	molar entropy	general

Symbol	Meaning	Where used
	number of internal oscillators in a molecule	Part III; Section 31.1
s_m	mean molar entropy of a mixture	Chapter 25
$\bar{s}_i$	partial molar entropy of component i	
$\mathbf{s}$	scattering vector	Eq. 23.16
T	absolute temperature	general
T	kinetic energy operator	general
T_e	kinetic energy operator of a molecule when $R = R_e$	Part I; p. 151
T_{vap}, T_{fus}	temperature of vaporization, fusion	general
T^*	reduced temperature (with respect to pair potential parameters)	Eq. 21.178
$\tilde{T}$	reduced temperature (with respect to the critical temperature)	Chapter 21
$\mathcal{T}$	excess temperature in a perturbation	Part III; Eq. 28.127
t	time	general
	centigrade (Celsius) temperature	general
t^*	limiting zero-pressure temperature on the centigrade scale	Eq. 12.34
$t_{1/2}$	half-life for a first-order reaction	Part III; Eq. 30.8
U	internal energy	general
	potential energy of an ionic crystal	Part I; Eq. 11.39
u	molar internal energy	general
	magnitude of the mass velocity (local average velocity)	Part III; Section 28.10
	mobility of a charged particle	Part III; Eq. 29.57
$\mathbf{u}$	velocity	App. 2A
u_m	mean molar internal energy of a mixture	general
$u_2(R)$	pair interaction potential	Part II; general
$u_3(\mathbf{R}_1, \mathbf{R}_2, \mathbf{R}_3)$	potential energy of three molecules	Part II; Eq. 21.159
V	volume	general
	electric potential difference	general
$V(r)$, $V(R)$	potential energy as a function of r, R	general
V	potential energy operator	general, Eq. 3.40b
V^*	reduced volume (with respect to pair potential parameters)	Eq. 21.178
$\tilde{V}$	reduced volume (with respect to the critical volume)	Chapter 21
$\mathbf{V}$	center-of-mass velocity	Eq. 27.14
$\mathcal{V}$	volume of subregion in phase space	Eq. 15A.2
$\mathbf{v}(\mathbf{r}, t)$	local fluid velocity of $\mathbf{r}$, t	Eq. 20.6
$\mathcal{V}_N$	total potential energy of N molecules	Eq. 21.157
$V_{\text{ind}}(R)$	induction energy of interaction between charge distributions	Eq. 10.16
V_f	free volume in solution	Part III; Eq. 30.76
v	molar volume	general
	quantum number for vibrational energy	general
	volume per unit mass	Chapter 20
	speed (magnitude of velocity $\mathbf{v}$)	general
$\bar{v}_i$	partial molar volume of component i	Eq. 25.24
v_m	mean molar volume of a mixture	Eq. 25.17
v_m^E	excess volume per mean mole of mixture	Eq. 25.18
v_x, v_y, v_z	x-, y-, z-components of velocity $\mathbf{v}$	general
$v_{\text{eff}}(k)$	effective velocity	Part I; Eq. 11.94
$\mathcal{W}$	difference between reversible and irreversible heat transfers	Eq. 20.116
w	work done on a system (sometimes W)	general
	parameter in regular solution model	Part II; Eq. 25.29
w_{ad}	adiabatic work	general
w_2	pair potential of average force in a liquid	Eq. 23.18

Symbol	Meaning	Where used
X	intensive generalized force	Eq. 13.13
$\mathbf{X}_i$	force associated with rate of variation of entropy	Eq. 20.89
x, y, z	Cartesian coordinates	general
x_i	mole fraction of component i	general
Y	extensive generalized displacement	Eq. 13.13
$Y_{lm}(\theta, \phi)$	spherical harmonic	Eq. 3.166
y_i	mole fraction of component i in the vapor over a liquid (two-phase, two-component system)	Eq. 25.20
Z	atomic number	general
Z_i	number of elementary charges on ion i	general
Z_N	configurational partition function for N molecules in a volume V at temperature T	Eq. 21.C.5
Z_{12}	collision frequency between species 1 and 2	Eq. 28.17
z	number of electrons transferred in an electrochemical reaction	Chapters 26, 31
α	coefficient of thermal expansion	general
	polarizability	general
	diagonal element of the energy matrix of the Hückel Hamiltonian	Part I; Eq. 9.6
	critical exponent for the temperature dependence of the heat capacity near the critical temperature	Part II; Eq. 24.36
	attenuation constant for sound waves	Part III; Eq. 28.149
	proportionality constant in Hinshelwood-Lindemann mechanism	Part III; Eq. 31.9
	net number of chain carriers produced in a branching step	Part III; Eq. 31.50
	coefficient of thermoelectric potential	Probs. 20.15, 20.16; p. 545
$\alpha(1), \beta(1)$	spin functions for electron 1, with $m_s = \frac{1}{2}$ and $-\frac{1}{2}$, respectively	general; Eq. 6.47
β	$1/k_BT$	general
	nearest-neighbor off-diagonal element of the energy matrix of the Hückel Hamiltonian	Part I; Eq. 9.7
	critical exponent for the temperature dependence of the density near the critical temperature along the coexistence line	Part II; Eq. 24.33
γ	gyromagnetic ratio, magnetic moment per unit of angular momentum	Part I; Eq. 8.52 ff
	ratio of heat capacities, c_p/c_v	general
	critical exponent for the temperature dependence of the compressibility near the critical temperature along the isochore $\rho = \rho_c$, $T > T_c$	Part II; Eq. 24.37
$\gamma_i, \gamma_i', \gamma_i''$	activity coefficient of component i, mole fraction scale, molality scale, molarity scale, respectively	general; Eq. 25.1; Eq. 25.136; Eq. 25.139
$\gamma_\pm$	mean ionic activity coefficient	general; Eq. 26.8
γ_V	$(\partial p/\partial T)_V$	Eq. 23.6
γ_G	Grüneisen constant	Eq. 22.26
$\Gamma(E, V, N)$	total number of quantum states with energy less than E for given N, V	Eq. 15.12
$\Gamma(x)$	Gamma function of x	Eq. 21.170
$\Gamma_n, \Gamma_m, \Gamma_E$	flux densities for numbers of molecules, mass, and energy crossing a plane	Part III; Eqs. 28.65 ff
$\mathbf{\Gamma}_n, \mathbf{\Gamma}_m, \mathbf{\Gamma}_E$	flux density vectors corresponding to Γ_n, Γ_m, and Γ_E ($\Gamma_n = \mathbf{k} \cdot \mathbf{\Gamma}_n$, etc.)	Eqs. 28.67 ff
$\mathbf{\Gamma}_{mv}$	flux density for momentum: pressure tensor (or stress tensor)	Part III; Eq. 28.75
δ	quantum defect	Part I; Eq. 2.12
	critical exponent for the density dependence of the pressure near the critical density along the critical isotherm	Part II; Eq. 24.35
	range of concentration gradient near an electrode	Part III; Section 31.13
δ_{ij}	Kronecker delta	general
$\delta(x)$	delta function	general
Δ	volume of a lattice cell	Eq. 25.191
Δ_{r-1}	determinant of the derivatives $(\partial^2 g_m/\partial x_i \partial x_j)$, $I, j = 2, \ldots, r$	Eq. 25.169
$\Delta_3(1, 2, 3)$	excess of the energy of three molecules over the sum of the pair interaction energies	Eq. 21.159
ϵ	energy parameter (well depth) in the intramolecular potential	general; Eq. 10.23
ε	energy per molecule	general

Symbol	Meaning	Where used
ϵ_0	permittivity of free space	general; Eq. 1.5
ϵ_F	Fermi energy	Eq. 11.58
ϵ_{n_1,n_2,n_3}	energy of a molecule in a state with quantum numbers n_1, n_2, n_3	general
ε_{ijk}	Levi–Civita symbol	Eq. 20.A.1
ζ	spin-orbit coupling constant	Eq. 5.24
	bulk viscosity	Eq. 20.46
	friction coefficient	Eq. 29.38
η	shear viscosity	general; Eq. 1.21
	ground state degeneracy	Part II; Eq. 18.14
	critical exponent for the distance dependence of the pair correlation function near the critical temperature	Part II; Eq. 24.102
	overpotential	Part III; Section 31.13
	efficiency of a heat engine	Eq. 120.132
	information entropy	p. 460
	packing fraction	Eq. 123.40
η_{ij}	rotation independent strain parameter	Eq. 22.5
θ	gas scale temperature	Eq. 12.29
	empirical temperature	Chapter 12
	Lindemann ratio	Eq. 24.98
	angular polar coordinate	general
	fraction of surface sites covered by adsorbed molecules	Part III; Eq. 31.111
θ^*	gas scale temperature in the limit of zero gas pressure	Eq. 12.31
$\Theta(\theta)$	factor of the wave function, for a particle in a spherically symmetric potential which contains the θ dependence of that function	Eq. 3.155
Θ_{rot}, Θ_{vib}, Θ_{elec}	characteristic temperature for rotational, vibrational, and electronic excitation, respectively	Eq. 21.134; Eq. 21.138; Eq. 21.141
Θ_E, Θ_D	characteristic temperature of the Einstein and Debye models of a solid, respectively	Chapter 22
κ	Debye screening length	Part II; Eq. 26.67
	transmission coefficient in activated-complex theory	Part II; Eq. 30.49
κ_T	isothermal compressibility	general
κ_S	adiabatic compressibility	general
κ_f, κ_r	theoretical rate coefficients for forward and reverse reactions	Part III; Eq. 30.34
λ	wavelength	general
	quantum number for orbital angular momentum of an electron along the axis of a linear molecule	general; Eq. 6.43
	absolute activity	Part II; Eq. 21.A.5
	thermal conductivity	general
	progress variable	Eq. 19.95
	length parameter in theory of diffusion-controlled reactions	Part III; 30.101
	reorganization energy	Section 31.13; Eq. 30
	Lyapunov exponent	Eq. 15A.8
λ_{ij}	parameter in the Born–Mayer repulsion between ions i and j	Eq. 14.33
	strain in a crystal	Eq. 22.2
Λ	quantum number for total orbital angular momentum along the axis of a linear molecule	general
	molar conductance	Part III; Eq. 29.108
μ	reduced mass	general; Eq. 2.74
$\boldsymbol{\mu}$, $\boldsymbol{\mu}_e$	electric dipole moment	Eq. 4.29
μ_0	magnetic permittivity of free space	general; Chapter 18
$\boldsymbol{\mu}_l$	orbital magnetic moment	Eq. 5.4
μ_N	nuclear magneton	Eq. 5.9
μ_B	Bohr magneton	Eq. 5.6

Symbol	Meaning	Where used
$\boldsymbol{\mu}_m$	magnetic dipole moment	Eq. 5.3
$\boldsymbol{\mu}_s$	spin magnetic dipole moment	Eq. 5.7
$\boldsymbol{\mu}_{n'n}$	transition dipole moment between states characterized by quantum numbers n' and n	Eq. 4.33
μ_i	chemical potential of component i	general
μ_i^0	chemical potential of the reference state of component i (the reference state is the pure liquid or solid)	general; Eq. 25.5
$\mu_i^{\ominus}$	chemical potential of the reference state of component i (the reference state is a hypothetical substance having the same properties as does the solute at infinite dilution)	general; Eq. 25.23
μ_i^E	excess chemical potential of component i	Eq. 25.13
μ_{JT}	Joule–Thomson coefficient	Prob. 21.7
ν	frequency in cycles per second (Hz)	general
	number of degrees of freedom of a system	Part II; Chapter 15
$\tilde{\nu}$	wavenumber	general
$\tilde{\nu}_e$	frequency factor in the power series expansion of E_{vib} as a function of the quantum number υ (approximately equal to the fundamental vibration frequency in cm^{-1})	Eq. 7.10
$\tilde{\nu}_e x_e$	second coefficient in the power series expansion of E_{vib} as a function of the quantum number υ	Eq. 7.10
ν_i	stoichiometric coefficient of the ith species in a chemical reaction	general
ν_+, ν_-	numbers of positive and negative ions in a salt	Eq. 26.6
ξ	distance scaled to dimensionless units	Part I; Eq. 3.65
	extent of reaction ($\dot{\xi}$, rate of reaction)	Part III; Eq. 30.2
	scale factor in the geometric mean combining law for mixed species pair interaction	Part II; Eq. 25.211
	correlation length in the critical region of a fluid	Part II; Eq. 24.102
ξ_i	ith normal coordinate	Part II; Eq. 24.84
Ξ	grand partition function	general; Eq. 19.126
π_g, π_u	designations of orbitals of a linear molecule with quantum numbers $\lambda = 1$ and with even and odd parity, respectively	general; Fig. 6.8
π	osmotic pressure	general; Eq. 25.121
Π	designation of an electronic state of a linear molecule for which $\Lambda = 1$	general; p. 203
	Peltier coefficient	Eq. 20.95
ρ	mass density	general
	Hammett reaction parameter	Part III; Eq. 30.112
	scale length in Born–Mayer repulsion between ions	Eq. 14.34
$\tilde{\rho}$	reduced density (with respect to the critical density)	Chapter 21
$\rho(\mathbf{r}, t)$	local mass density	Part III; Eq. 28.114
$\rho_Q(\mathbf{r}, t)$	local density of Q at point $\mathbf{r}$ and time t	Eq. 20.1
$\rho_k(\mathbf{r}, t)$	mass density of component k at $\mathbf{r}$, t	Eq. 20.12
ρ_ε	local density of potential energy in a fluid	Eq. 20.64
$\hat{\rho}$	von Neumann density operator	Eq. 15B.6
ρ_{mn}	element of von Neumann density operator	Eq. 15B.4
σ	total scattering cross section	general; Eq. 2.9
	size parameter in the pair interaction between molecules	Eq. 10.23
	electrical conductivity	Eq. 29.86
	time interval	Part III; Eq. 29.74
	Hammett substitution parameter	Part III; Eq. 30.112
	symmetry number	Eq. 21.111
	surface tension	Table 13.1
$\boldsymbol{\sigma}$	stress tensor	Part II; Eq. 20.43
σ_g, σ_u	designations of orbitals of a linear molecule with quantum number $\lambda = 0$ and with even and odd parity, respectively	general, Fig. 6.8

Symbol	Meaning	Where used
σ_v	symmetry operator for reflection through a plane	Section 31.5
$\sigma(v)$	total scattering cross section	Part III; Eq. 27.62
$\sigma(v, \alpha, \beta)$	differential scattering cross section in laboratory coordinates	Eq. 27.53
$\sigma(v, \chi)$	differential scattering cross section, in relative (center-of-mass) coordinates	Eq. 23.54
$\sigma_R(v)$	cross section for reactive scattering	Eq. 30.32 ff
Σ	designation for an electronic state of a linear molecule for which $\Lambda = 0$	general; p. 165
τ	period	general; Eq. 2.38
	relaxation time	general; Eq. 28.111
	effective lifetime of photoexcited state	Part III; Eq. 31.27
	scattering volume	Section 27.5
τ_f	mean free time	Fig. 28.3
ϕ	practical osmotic coefficient	Fig. 26.20
	angle; azimuthal polar coordinate	general
	work function for electron emission	Part II; Eq. 26.55
	phase shift	Part III; Eq. 28.158
	electric potential	Eq. 30.89
ϕ_A	volume fraction of component A	Eq. 25.20
φ_A, φ_B	atomic orbitals centered on nuclei A, B	Eq. 6.19
Φ	quantum yield for a photochemical reaction	Eq. 31.26
Φ_i	electrostatic potential acting on ion i	Eq. 26.61
Φ_A, Φ_B	wave function for molecules A, B	Eq. 11.35
$\Phi(\phi)$	factor of the wave function for a particle in a spherically symmetric potential which contains the dependence of that function on the angle ϕ	Eq. 3.155
χ	magnetic susceptibility	Footnote 4, p. 118
	relative scattering angle	Fig. 27.3, 27.4
$\boldsymbol{\chi}$	susceptibility tensor of an anisotropic crystal	p. 598
χ_i	ith product orbital	Section 31.6
ψ	wave function	general
	electrostatic potential	Part II; general
	general transport property	Part III
$\psi(x)$	spatial part of the wave amplitude along the x-direction	Eq. 3.59
$\psi(1, 2)$	molecular orbital wave function for electrons 1 and 2	Eq. 6.46
$\psi_{\text{elec}}(r, R)$	solution of the electronic wave equation for fixed nuclei (separated by R)	Eq. 6.3
$\psi_{\text{nucl}}(R)$	factor of the molecular wave function describing rotation and vibration	Eq. 6.2
ψ_{MO}	molecular orbital wave function	Section 6.6; Eq. 6.46
ψ_{VB}	valence bond wave function	Section 6.8; Eq. 6.72
Ψ	wave amplitude	Eq. 3.14
Ψ^*	complex conjugate of Ψ	general
ω	angular frequency ($\equiv 2\pi\nu$)	general
	small element of volume	Part II; Chapter 15
$\boldsymbol{\omega}$	angular velocity	general
	set of angles defining the orientation of a molecule	Part II; Eq. 24.96
$\omega(E)$	density of states at energy E	general
$\bar{\omega}_{sl}$	degeneracy of the sth state of molecule 1	Eq. 21.100
Ω	solid angle	general; Eq. 2.15
$\Omega(E, V, N)$	number of quantum states in the energy range between E and $E + dE$ f or given V, N	general; Eq. 15.12
Z	rate of entropy creation	Eq. 20.85

Symbol	Meaning	Where used
Commonly Used Superscripts and Subscripts		
0	Generally used to denote a standard state, or a change in which both initial and final states are standard states.	Part II
0	Denotes pure component.	
θ	Denotes the hypothetical reference state defined to be a pure substance which has the same properties as does the solute at infinite dilution.	
*	Denotes the perfect gas reference state, or the limiting value of the gas property when the pressure approaches zero.	
E	Denotes, for a mixture, the excess value of a thermodynamic function over that expected for an ideal mixture, and, for a pure component, the residual value of a thermodynamic function over that expected for a perfect gas at the same temperature and density (the latter is used only in Chapter 23).	
$\bar{z}_i$	Denotes partial molar "zee" of component *i*.	
i	component *i*	
V,p,T, . . .	constant volume, pressure, temperature, . . .	
σ	along the coexistence curve	
0	ground state	
m	mean molar value for a mixture	
±	mean ionic quantity	
reac,fus vap,subl inv	value of the thermodynamic function for the process indicated	
′	value (of velocity, impact parameter, etc.) after collision	Part III; Eq. 27.1
′ ″	Designates quantum numbers in the upper and lower states of a transition	Eq. 7.28
*	complex conjugate	general
	excited state	Section 5.3
	unusual configuration	Section 5.3
	value at the maximum of probability distribution	Eq. 15.49
	value corresponding to top of potential energy barrier	Eq. 30.22
‡	value for activated complex	Part III; Eq. 30.44

Matter in Equilibrium

PART II

Matter in Equilibrium: Statistical Mechanics and Thermodynamics

In Part I we treated the detailed structure of microscopic systems: individual atoms and molecules and their interactions. Now we shall consider the macroscopic systems of our common experience.

We have seen that an exact description of matter requires the use of quantum mechanics. The solution of the Schrödinger equation (with appropriate boundary conditions) yields as complete a description of a system as it is possible to obtain, consisting of the wave function and the energy levels. In principle all properties of a system, whatever its size, can be calculated with this information. In practice, however, as we saw throughout Part I, this method is tractable only for systems with a few particles or degrees of freedom. The hydrogen atom, with only two particles, can be treated exactly; the hydrogen molecule presents considerably more difficulty, with even the most extensive calculations giving at best only a nearly exact solution. Larger atoms and molecules require the use of approximate methods. The situation is still worse when more than one molecule is involved: Even two interacting argon atoms form a "complex system" that can be treated only with relatively crude approximations. What, then, can one do with a system containing a *mole* of argon atoms?

Consider such a system. A mole of argon is not a large quantity by macroscopic standards: It occupies about 22 liters at room temperature and atmospheric pressure, and weighs a mere 40 g. Yet this paltry amount of gas contains approximately 6×10^{23} argon atoms, each with a nucleus and 18 electrons. Even if we neglect the electrons and consider each atom as a mass point, we have 1.8×10^{24} translational degrees of freedom. There is no conceivable way in which we could solve the Schrödinger equation (or the classical equations of motion) for that many particles. And even if we could somehow obtain such a solution, we would have no way of accumulating and using the enormous quantity of information involved.[1] We must content ourselves with a less than complete description of a macroscopic system; fortunately, such an incomplete description is satisfactory for most purposes.

How much information can we really use? A complete description of our gas in terms of classical mechanics (which is sufficiently accurate to describe the translational motion) would require knowledge of the positions and velocities of all the particles. One reduced level of analysis is that using only the *distributions* of position and velocity, that is, the relative numbers of particles with positions and velocities falling within given ranges; one can also define such distributions for other microscopic properties of a system, such as energy, dipole moment, and so on. The subject of *statistical mechanics* (also known as *kinetic theory,* from its concern with the motions of molecules) is the proper formulation of such distributions, the calculation of various average quantities, and the connection of these averages with the properties of macroscopic systems. For example, the average kinetic energy of the molecules in a gas is found to be identifiable with the macroscopic quantity we call

[1] For systems of moderate size, containing up to a few thousand molecules, it is possible to solve the classical equations of motion with high-speed computers. Such calculations yield the trajectories of all the particles in the system. This information is not itself immediately useful for the interpretation of the macroscopic properties of the system. To make a connection between trajectories and macroscopic properties, new concepts must be introduced. These new concepts include the definition of the kind of averaging over trajectories that is appropriate for the calculation of macroscopic properties and the definitions of the relationships between the distributions of the coordinates and momenta of subsets of molecules and macroscopic properties. Calculations of macroscopic properties from computer simulation data are, for most properties, sufficiently insensitive to the number of molecules in the sample that the analysis generates information useful in understanding much larger systems.

temperature. Statistical mechanics is so called because its models are those of mechanics (whether classical or quantum); it is in obtaining the averages that additional assumptions must be made, primarily with regard to the relative likelihood of distributions of various kinds.

A still greater reduction in the level of description is found in a theory based *only* on macroscopic properties. The properties referred to are of various kinds: some, such as temperature and pressure, are familiar to all (and thus all the more needful of careful analysis); others, such as internal energy and entropy, must be newly defined. The subject of *thermodynamics* is the proper formulation of the relationships among these quantities, relationships that can be derived from a small set of principles called the *laws of thermodynamics.* The formulations of these laws are derived from, and are precise representations of, the essential features of the observed behavior of matter at the macroscopic level. Since these laws are statements of observation at the macroscopic level, they are independent of any assumptions about the microscopic structure of matter, which could be either atomic or continuous as far as the principles of thermodynamics are concerned. Thermodynamics was in fact largely developed before the atomic nature of matter was established, and can be expounded axiomatically in complete isolation from any microscopic considerations; but since we do have our microscopic theory, it would be foolish not to use it. Statistical mechanics is the bridge that links mechanics with thermodynamics. As already noted, one can identify the averages of statistical mechanics with the macroscopic variables of thermodynamics, and such juxtapositions can often serve to illuminate our understanding of both theories. We shall thus undertake a parallel development of the principles of both statistical mechanics and thermodynamics.

The macroscopic properties of matter can be classified into two broad categories, equilibrium and nonequilibrium. In an isolated system at equilibrium the macroscopic properties (thermodynamic variables or statistical averages) are time-independent. In a nonequilibrium system one observes spontaneous and irreversible natural processes, such as flows of matter or heat. Part II of this text is devoted to matter in equilibrium, with the nonequilibrium behavior reserved for Part III. The first several chapters of Part II cover the general principles of our two theories and some of their simpler applications; in the remaining chapters we study the specific properties of the various classes of macroscopic systems (gases, liquids, solids, solutions, etc.).

CHAPTER 12

The Perfect Gas at Equilibrium and the Concept of Temperature

We begin the study of the properties of bulk matter with the simplest type of matter: the dilute gas. It is simple because its molecules are far apart and hardly interact with one another, so that the macroscopic properties are directly related to those of the individual molecules. As indicated in the introduction to Part II, we approach the same properties from the viewpoints of statistical mechanics and thermodynamics.

In both cases we deal in this chapter not with a real gas but with an abstraction known as a "perfect gas" (also called an "ideal gas"). In the microscopic theory the perfect gas is a model with *no* interaction between the molecules. We consider the properties of a collection of such molecules, in particular the pressure they exert on the walls of a container, and develop an equation of state connecting the gas's pressure, volume, and energy.

In thermodynamics the perfect gas is an abstraction from the limiting behavior of real gases at low pressure. Here the equation of state is experimental, as embodied in the familiar ideal gas law, and involves the nonmechanical concept of temperature. We shall devote an extensive discussion to the meaning of temperature, which is much less simple than one might imagine.

Finally, comparing the two equations of state, we shall build our first bridge between microscopic and macroscopic properties. The two equations have the same form if temperature is proportional to average molecular kinetic energy. By this correspondence we can say that the "perfect gases" of statistical mechanics and thermodynamics are the same, so that the results of each theory can be used to throw light upon the other.

12.1 The Perfect Gas: Definition and Elementary Model

According to the *kinetic hypothesis,* the individual molecules of which matter is composed are continually in motion, regardless of whether or not the piece of matter as a whole is moving. These individual motions take place in all directions and with a variety of speeds, so that, as far as gross motion is concerned, the contributions of individual molecules tend to cancel. From the macroscopic viewpoint, however, the molecular motion has two major consequences: (1) The kinetic energy of the molecules contributes to the internal energy of the material; and (2) the impacts of the moving molecules on the container wall contribute to the pressure exerted by the material on its surroundings. Although the molecular motion is also presumed to determine all other properties of the medium, for our present purposes it is sufficient for a dilute gas to consider only the internal energy and the pressure.

We must first define what we mean by a dilute gas. We assume for simplicity that the molecules of the gas are monatomic (such as He, Ne, or Ar), to avoid having to consider the relative motions of the atoms in a polyatomic molecule. We also assume that the molecules are independent and exert no forces on one another, except during occasional binary collisions—collisions in which only two molecules come sufficiently close together to exert a significant force on each other. In an *elastic collision* the translational momentum and kinetic energy of the pair of molecules are conserved. For monatomic molecules we can safely neglect any internal energy changes, e.g., those due to electronic excitation, and regard any collision as elastic. A gas that satisfies all these conditions will be called a *perfect gas.* The model is a fairly good approximation for most real gases at moderate pressures (as in the atmosphere), and such gases can be called "dilute."

We make the additional assumption that the gas is macroscopically at *equilibrium.* Equilibrium, as we use the term here, means that the properties of the system do not change measurably with time. This is a purely macroscopic concept, since the positions and velocities of individual molecules are continually changing at all times; yet experiment shows that macroscopic properties such as pressure and temperature can and do become time-independent in a real gas. Suppose that we have a quantity of dilute gas in a container of fixed volume, which we can isolate in such a way that no interaction of any kind occurs between the gas and the rest of the universe.[1] No matter what the initial state of the gas—even if all the gas is initially in one corner of the container, with some of it hotter than the rest—we find that in time a state is reached in which the properties of the gas are uniform throughout the container and thereafter do not vary measurably. We shall see in this chapter that such quantities as pressure and temperature can be related to averages over the molecular velocities. Although these averages also change whenever a molecular collision occurs, the number of molecules is so enormous that these changes can be—and at equilibrium must be—negligibly small fluctuations in the total quantities. Thus, we shall assume that at equilibrium the velocity distribution—the relative numbers of molecules in different velocity ranges—remains invariant with time.[2]

Additional properties of the equilibrium state can be deduced from the observation that, if there is no external field acting on the gas, the gas becomes uniform throughout its container (see Footnote 1). Measurements of such properties as density and temperature then give the same results if performed at any point within the gas. The gas is also found to be at rest relative to the container, with no macroscopic flows or currents: Any flows that may have been initially present die away with time. What are the consequences of these conditions for the equilibrium distribution of molecular velocities and positions? Clearly the density of gas molecules must be the same in every volume element large enough to perform a measurement on, that is, in every volume element containing very many molecules. The kinetic energy per unit volume (which we shall see is proportional to temperature) must also be the same in every such volume element; for practical purposes, this is possible only if the distribution of velocities among the molecules is the same everywhere. If the gas as a whole is at rest, the average velocity of all the molecules must be zero,

$$\langle \mathbf{v} \rangle \equiv \frac{1}{N}\sum_{i=1}^{N} \mathbf{v}_i = 0 \quad \text{and thus} \quad \sum_{i=1}^{N} \mathbf{v}_i = 0. \tag{12.1}$$

(Remember that velocity is a vector, so that velocities in opposite directions cancel one another; obviously the average molecular speed is not zero.) Furthermore, at any point in the gas there must be equal numbers of molecules traveling in every possible direction, since any imbalance would constitute a net flow of gas; this must hold not just overall but within each range of speeds, if the velocity distribution is to remain the same from point to point. All of the preceding can be summarized as follows: In a perfect gas at equilibrium the distributions of molecular velocities and positions must be *homogeneous* (uniform in space) and *isotropic* (uniform in all directions).

A word here about the types of "densities" we shall use. The ordinary density ρ, or *mass density*, is simply the mass per unit volume. The *number density* n is the number of molecules per unit volume. If there are N molecules, each of mass m, in the volume V, we have

$$n = \frac{N}{V} \quad \text{and} \quad \rho = nm. \tag{12.2}$$

One can also define a kinetic energy density (kinetic energy per unit volume) and other quantities of the same type.

Having defined our terms, let us consider first a very elementary description of the relationship between the pressure exerted by a gas and the kinetic energy of its molecules. The argument is essentially one of dimensional analysis, but yields a result of the right order of magnitude. Suppose that a perfect gas is contained in a cubical box with edge length l, and define a Cartesian coordinate system with axes normal to the walls. Imagine that the walls of the box have the property of reflecting molecules elastically; that is, when a molecule strikes a face of the box, its motion perpendicular to that face is reversed and its motion parallel to that face is unaltered (see Fig. 12.1). Thus, if a molecule initially has a velocity $\mathbf{v}$ with components v_x, v_y, v_z, its velocity at any later time must always be some combination of the components $\pm v_x$, $\pm v_y$, $\pm v_z$. Let S be a face of the box normal to the z axis, with area l^2. Between two successive collisions with S, a molecule must move to the opposite face (which is a distance l away from S) and back. Since the molecule's velocity component perpendicular to S is v_z, the time interval between successive collisions with S is $2l/v_z$, and the frequency of these collisions is $v_z/2l$. At each collision the molecule undergoes a change of momentum in the z direction of amount $2mv_z$, from $+mv_z$ to $-mv_z$; an equal and opposite momentum is transferred to the wall. We conclude that the

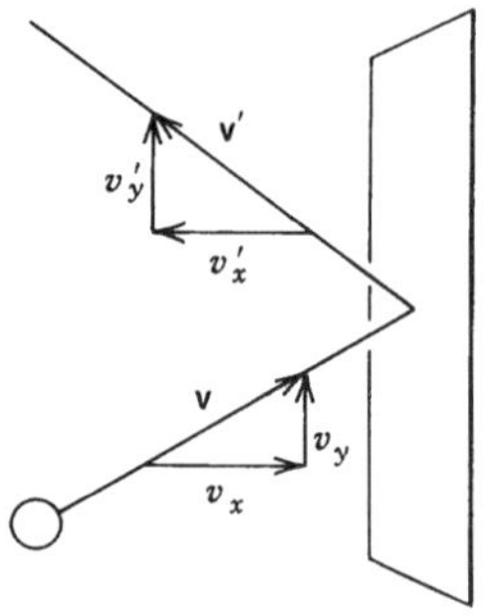

Figure 12.1 Elastic reflection of a particle from the wall of a box. The initial velocity is $\mathbf{v}$, with components v_x perpendicular to the wall and v_y parallel to the wall. After reflection the velocity is $\mathbf{v}'$, with components $v_x' = -v_x$ and $v_y' = v_y$.

[1] We cannot actually isolate a gas from gravity, and the Earth's gravitational field produces a density gradient with height, but this lack of uniformity is ordinarily too small to measure over laboratory distances.
[2] Because the binary collisions are infrequent, a molecule can move a long distance along a linear trajectory without interruption. For the purpose of calculating the properties of the gas at equilibrium, the collisions can be neglected. However, the existence of collisions permits us to describe qualitatively how a gas arrives at an equilibrium state.

total change per unit time in the z component of a single molecule's momentum due to collisions with S is

$$\left[\frac{d(mv_z)}{dt}\right]_S = 2mv_z \cdot \frac{v_z}{2l} = \frac{mv_z^2}{l}. \tag{12.3}$$

We now assume the simplest possible velocity distribution. We imagine that all the molecules have the same velocity components $\pm v_x$, $\pm v_y$, $\pm v_z$ along the x, y, z axes. Then, if there are N molecules in the volume $V = l^3$, the total change in momentum per unit time arising from all collisions with S is

$$\left[\sum_{i=1}^{N} \frac{d(mv_z)_i}{dt}\right]_S = \frac{Nmv_z^2}{l} = \frac{nVmv_z^2}{l} = nml^2v_z^2, \tag{12.4}$$

where n is the number density. Now, by definition, the pressure on a wall is equal to the force exerted per unit area. In turn, by Newton's second law, the force exerted is equal to the rate of change of momentum. Hence, if p is the pressure exerted by the gas, pl^2 (the total force on S) is the rate at which momentum is exchanged by collisions with S, and we have

$$p = nmv_z^2. \tag{12.5}$$

But we know that the pressure in a gas at equilibrium is the same in all directions (Pascal's law), so we must have

$$v_x^2 = v_y^2 = v_z^2 = \tfrac{1}{3}(v_x^2 + v_y^2 + v_z^2) = \tfrac{1}{3}v^2, \tag{12.6}$$

where v is the common speed of all the molecules. We finally obtain

$$p = \tfrac{1}{3}nmv^2. \tag{12.7}$$

Since $mv^2/2$ is the translational kinetic energy of a single molecule and n is the number per unit volume, we conclude that the pressure exerted by the gas equals two-thirds of the kinetic energy density.

12.2 The Perfect Gas: A General Relation between Pressure and Energy

The arguments just given are too restrictive. In general, molecules are not reflected elastically from a vessel wall (which is, after all, also composed of molecules); the molecular speeds are not all the same, but are distributed over a wide range; the directions of motion are random; and the vessel need not be cubical. We now proceed to give a more general derivation of the relationships between the distribution of molecular velocities in a gas and the pressure and energy of the gas. We shall see that for many purposes we need not know the detailed form of the velocity distribution, but only some of its properties.

Consider a perfect gas at equilibrium, with N molecules in a volume V. We expect that the molecules will move with different velocities. Let the number of molecules per unit volume with velocities between $\mathbf{v}$ and $\mathbf{v} + d\mathbf{v}$ (that is, with x components between v_x and $v_x + dx_x$, etc.) be denoted as $f(\mathbf{v})\, d\mathbf{v}$. The function $f(\mathbf{v})$ is known as the *velocity distribution function,* and it plays an important role in the kinetic theory of gases. Because the total kinetic energy of all N molecules is finite, there must be some maximum speed of molecular motion that cannot be exceeded. Thus, we expect that $f(\mathbf{v}) \to 0$ as $|\mathbf{v}| \to \infty$. Furthermore, since $f(\mathbf{v})\, d\mathbf{v}$ is the number of molecules per unit volume with velocities in a given range $d\mathbf{v}$, if we sum over all possible velocity ranges we must obtain the *total* number of molecules per unit volume, which is just the number density n. This last condition can be formalized by writing the integral

$$\int_{\mathbf{v}} f(\mathbf{v})\, d\mathbf{v} = \int_{-\infty}^{\infty}\int_{-\infty}^{\infty}\int_{-\infty}^{\infty} f(\mathbf{v})\, dv_x\, dv_y\, dv_z = n. \tag{12.8}$$

Now, we have already specified that the molecular motion at equilibrium is isotropic—that equal numbers of molecules in any given speed interval must travel in every direction. Thus $f(\mathbf{v})$ must be independent of the direction of $\mathbf{v}$ and can be a function only of its magnitude, the speed v. It is therefore useful to introduce a new distribution function: We define the number of molecules per unit volume with speeds between v and $v + dv$ as $f(v)\, dv$. Suppose that we represent the molecular velocities by points in a "velocity space," a Cartesian space whose coordinates are v_x, v_y, v_z; in this space $d\mathbf{v}$ ($=dv_x\, dv_y\, dv_z$) is a volume element, and v is a radial coordinate. The molecules with speeds between v and dv then occupy a spherical shell of radius v and thickness dv; the formulas of Appendix 2A show that the "volume" of this shell is $4\pi v^2\, dv$. Thus, the relation between the two distribution functions must be

$$f(v) = 4\pi v^2 f(\mathbf{v}). \tag{12.9}$$

(See Section 28.1 for a more systematic proof.) Although we shall not at this time derive the form of $f(v)$, the results of such a derivation (cf. Chapter 19) are illustrated in Fig. 12.2. Recall also that at equilibrium the gas is homogeneous, each

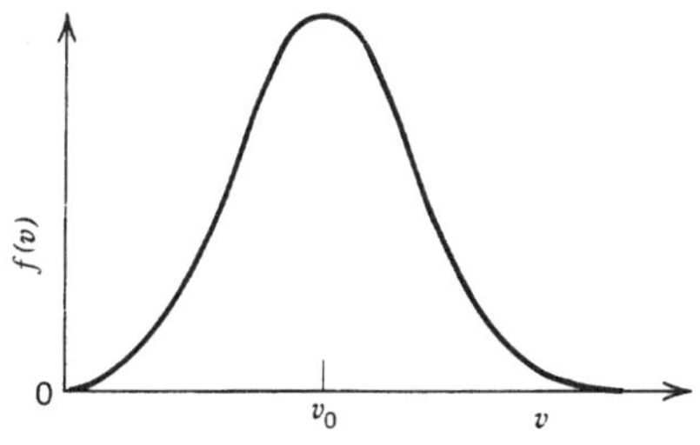

Figure 12.2 The distribution of molecular speeds, $f(v)$. The speed v_0, which is the most probable, is the *mode* of the distribution.

volume element being the same as every other volume element, so $f(v)$ must be independent of position within the gas. A remarkable feature of the function $f(v)$, which will be demonstrated in Chapter 21, is that it is the same for all substances for which the energy of interaction of the molecules is independent of their velocities.

We can now formally calculate the internal energy of the gas arising from the molecular motion. A molecule with mass m and speed v has kinetic energy $mv^2/2$. To obtain the total kinetic energy of the gas, we multiply the kinetic energy of a single molecule with speed v by the number of molecules per unit volume with speeds between v and $v + dv$, then sum contributions from all possible speeds and from all possible volume elements in the total volume. For the perfect monatomic gas (neglecting the internal structure of the atoms) the kinetic energy is the same as the total internal energy, which we shall call U. This internal energy is thus

$$U = \int_V \int_{v=0}^{\infty} \frac{mv^2}{2} f(v)\, dv\, dV. \tag{12.10}$$

But since $f(v)$ must be independent of position, Eq. 12.10 can be immediately integrated over volume to give

$$U = \frac{mV}{2} \int_0^{\infty} v^2 f(v)\, dv \tag{12.11}$$

for the internal energy of the perfect monatomic gas. Other types of gases (which we consider in Chapter 21) may contain polyatomic molecules, or may have significant interactions between molecules. In these cases there are also contributions to the internal energy U from molecular vibration and rotation, or from the intermolecular potential energy.

We wish now to obtain the relationship between internal energy and pressure, corresponding to Eq. 12.7 in the simple model. As we mentioned before, the pressure on the container wall is the force exerted by the gas per unit area of wall or, equivalently, the rate at which momentum is transferred to a unit area of wall. But we need not restrict our definition to the walls: By the assumptions of homogeneity and isotropy, the pressure at equilibrium must be the same anywhere within the gas (this again is equivalent to Pascal's law). Picture a plane surface S fixed at an arbitrary position within the gas; we define a coordinate system with the z axis normal to S (Fig. 12.3). We assume that gas molecules collide elastically with S, in the same way as shown in Fig. 12.1. Thus whenever a molecule whose z component of momentum is mv_z strikes S, an amount of momentum $2mv_z$ is transferred to S. The pressure on S is the total momentum so transmitted per unit time per unit area of S, and at equilibrium must be the same on both sides of S.

Now suppose that the real surface S is replaced by an imaginary plane in the same position. The molecules that would rebound from a real surface are just those that in this model cross S. But a molecule that would transfer

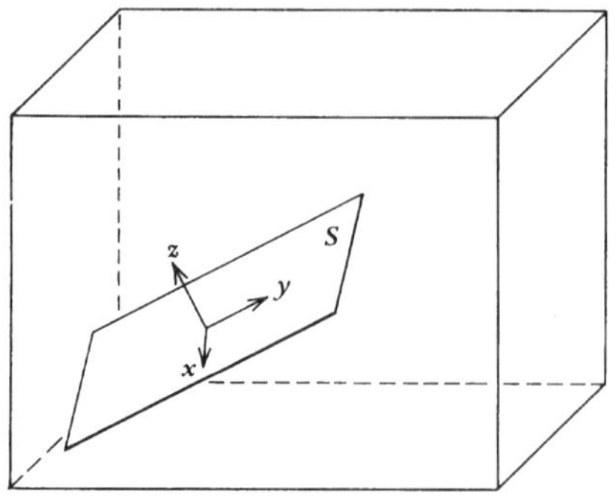

Figure 12.3 The plane surface S is fixed at an arbitrary position in a container of gas. The xyz coordinate system is defined so that the z axis is normal to S.

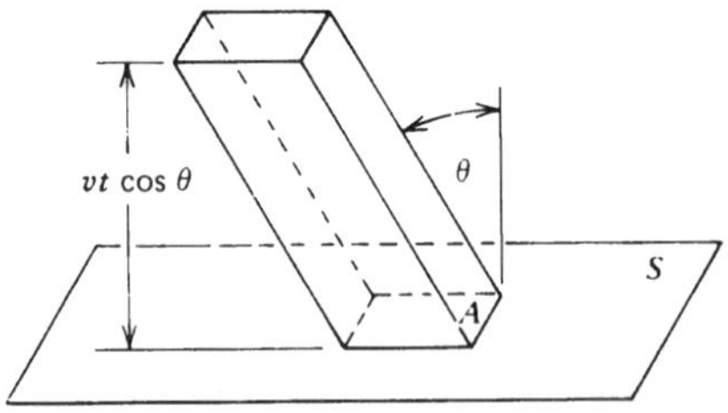

Figure 12.4 All the molecules within the inclined prism that have velocity **v** in the direction θ will strike the area A within time t. The volume of the prism is $A\,vt \cos\,\theta$.

z-momentum $2mv_z$ to the real S carries only mv_z (half as much) across the imaginary S. We *define* the pressure to be the same in both cases, thus obtaining a definition independent of the presence of a wall: The pressure normal to an imaginary plane S is *twice* the momentum transported across S per unit time per unit area of S, by molecules coming from one side of S (since the two sides of a real surface are different areas).

How do we evaluate this rate of momentum transport? The molecules we shall consider are those crossing S in the positive z direction. Let θ be the angle between a given molecule's trajectory and the z axis (Fig. 12.4). We define $F(\theta, v)\, d\theta\, dv$ as the number of molecules with speeds between v and $v + dv$ that cross S per unit area per unit time at angles between θ and $\theta + d\theta$. Each such molecule has a momentum of magnitude mv, with z component $mv \cos\,\theta$. Only the z component of momentum contributes to the pressure. Because of the gas's isotropy, the other components add to give a null result; that is, for a given range of θ there are equal numbers of molecules with x components $+mv_x$ and $-mv_x$, or with y components $+mv_y$ and $-mv_y$, so that the net transport of these components across S is zero. Thus the contribution to the pressure from the molecules in the ranges dv, $d\theta$ is the number per unit time per unit area of such molecules multiplied by twice the z component of momentum transported per molecule, or

$$dp = 2mv \cos\,\theta F(\theta, v)\, d\theta\, dv. \tag{12.12}$$

Integration over all possible values of θ and v—that is, over all molecules crossing S in the positive z direction—gives a total pressure of

$$p=\int_{v=0}^{\infty}\int_{\theta=0}^{\pi/2} 2mv\cos\theta F(\theta,v)\,d\theta\,dv \tag{12.13}$$

(note the limits of integration for θ).

To use Eq. 12.13 we must evaluate $F(\theta, v)$. The expression $F(\theta, v)\,d\theta\,dv$ may be thought of as the product of (1) the number of molecules per unit volume with speeds between v and $v + dv$ moving at angles between θ and $\theta + d\theta$, which we call $f(\theta, v)\,d\theta\,dv$, and (2) the volume occupied by all such molecules capable of crossing unit area of S in unit time. The latter volume is easily evaluated: As shown in Fig. 12.4, it is simply a prism inclined at an angle θ from the normal to its base. For base area A and time t, the volume of such a prism is $Avt\cos\theta$; since our calculation refers to unit area and unit time, the volume we require is $v\cos\theta$. We can thus write

$$F(\theta,v)\,d\theta\,dv=v\cos\theta f(\theta,v)\,d\theta\,dv. \tag{12.14}$$

Remember that $f(v)\,dv$ is the total number of molecules per unit volume with speeds between v and $v + dv$, or $f(\theta, v)$ integrated over all values of θ. Since the molecular motion is isotropic, $f(\theta, v)\,d\theta\,dv$ must bear the same relation to $f(v)\,dv$ that the range of solid angles between θ and $\theta + d\theta$ bears to 4π (cf. Appendix 2A). By Eq. 2B.7, the solid angle between θ and $\theta + d\theta$ is $2\pi\sin\theta\,d\theta$, so we have

$$\frac{f(\theta,v)\,d\theta\,dv}{f(v)}=\frac{2\pi\sin\theta\,d\theta}{4\pi}=\frac{1}{2}\sin\theta\,d\theta. \tag{12.15}$$

Substituting Eqs. 12.14 and 12.15 into Eq. 12.13, we find the pressure to be

$$\begin{aligned}p&=\int_{v=0}^{\infty}\int_{\theta=0}^{\pi/2}(2mv\cos\theta)(v\cos\theta)\left(\frac{1}{2}\sin\theta\,d\theta\right)f(v)\,dv\\&=m\int_0^{\infty}v^2 f(v)\,dv\int_0^{\pi/2}\cos^2\theta\sin\theta\,d\theta\\&=\frac{m}{3}\int_0^{\infty}v^2 f(v)\,dv.\end{aligned} \tag{12.16}$$

But, comparing this result with Eq. 12.11, we obtain for a perfect monatomic gas

$$p=\frac{2}{3}\frac{U}{V}, \tag{12.17}$$

where U/V is the internal energy density.

Thus, the hypotheses we have introduced lead to the conclusion that the pressure in a perfect monatomic gas is just equal to two-thirds of the internal energy density. This result is the same as that obtained in Eq. 12.7, from analysis of our original, very crude model. Note that we have been able to obtain this interesting relationship without specifically calculating $f(v)$. Subsequent calculations, especially in Part III of this book, will require knowledge of the form of $f(v)$, which we shall derive in a subsequent section. Later in the present chapter (Section 12.6) we shall examine critically the analysis just given, but first we begin our study of the macroscopic theory.

12.3 Some Comments about Thermodynamics

As we explained in the introduction to Part II, thermodynamics deals with relationships among the macroscopic properties of matter—more specifically, among a particular class of macroscopic variables that we must define. These relationships can be derived from the *laws of thermodynamics,* which are inferred from observations of the behavior of bulk matter. The laws of thermodynamics are analogous to the principles of quantum mechanics (Chapter 3), in the sense that both are inferred (guessed, if you like) from a wide variety of observations, and ultimately justified by the agreement of the conclusions drawn from them with experiment; and both are extraordinarily fruitful in the range of such conclusions that one can draw. Thermodynamics in fact provides an accurate and useful description of the equilibrium properties of any piece of matter containing a very large number of molecules, and it leads to an interpretation of, and a method of calculating the changes associated with, spontaneous processes. Thermodynamics goes beyond mechanics in providing an interpretation of "the arrow of time"—specifically, why irreversible processes occur.

The methods ordinarily used to formulate the laws of thermodynamics make no assumptions about the microscopic structure of matter. An internally consistent and logically complete theory can be constructed without reference to the existence of atoms and molecules. Naturally, the specific properties of real matter must appear in the theory, in the form of material constants such as heat capacities or compressibilities, but thermodynamics concerns itself only with the relations among these quantities, not with why they have the values they do; to answer the latter question one must resort to microscopic theories of the properties and interactions of molecules. By considering the same quantities from the viewpoints of both microscopic and macroscopic theories, one can gain significant insights into the macroscopic consequences of molecular behavior—and the microscopic implications of macroscopic measurements.

As a concrete example of how the theories can be interrelated on various levels, the pressure of a real (imperfect) gas can be expressed empirically as a power series in the density n; the coefficients of the powers of n are determined from experimental data. Thermodynamics gives the relationships between pressure and other macroscopic quantities (internal energy, entropy, etc.), and thus enables one to calculate these quantities in terms of the coefficients in the density expansion of the pressure. All this is on the macroscopic level; what of the microscopic approach? Here we begin with the concept of a potential energy of interaction between molecules, the $V(R)$ described in Chapter 10. Using the methods of statistical mechanics, one can average $V(R)$ over all the molecular pairs in a gas, over all the triples of molecules, and so on, obtaining average quantities that can be identified with the coefficients in the density expansion of the pressure. Combining the two theories gives one a complete chain of connections between the molecular interactions and the thermodynamic properties. In principle one could obtain $V(R)$, and by extension the macroscopic properties, by solving the Schrödinger equation for the molecular interaction. In practice, as we know, this is extraordinarily difficult, and one ordinarily works in the opposite direction, deducing $V(R)$ from macroscopic measurements of the pressure or other properties, or more directly from molecular scattering.

But we have a long way to go before we can draw such connections as these; they will be developed in Chapter 21. Let us now get down to actually developing the principles of thermodynamics, on what must initially be a much more abstract level. We begin by examining some basic notions, starting with definitions of what it is that thermodynamics studies.

A *system* is that part of the physical world which is under consideration. It may have fixed or movable boundaries of arbitrary shape and may contain matter, radiation, or both. All the rest of the universe is defined to be the *surroundings.* A system is said to be *open* if matter can be exchanged between the system and the surroundings. A *closed system,* which does not exchange matter, may still be able to exchange energy with its surroundings. An *isolated system* has no interactions of any kind with its surroundings.

To describe the properties of a system, one needs to be able to measure certain of its attributes. In thermodynamics the system is described by a set of macroscopic variables or "coordinates," the determination of which requires only measurements over regions containing very many molecules, carried out over periods long enough for very many molecular interactions, and involving energies very large compared with individual quanta. The thermodynamic properties of the system are assumed to be entirely described by the set of macroscopic coordinates. To anticipate a bit, the coordinates to which we refer are most often the temperature, pressure, and volume. When a macroscopic coordinate is fixed in value by the boundary conditions defining the system, there is said to be a *constraint* on the system. There can be other forms of constraints, such as one allowing no deformation of the system. Many of these, and the roles they play in thermodynamics, will be discussed later, in Chapter 19.

The classical theory of thermodynamics deals with the properties of systems at *equilibrium.* This has essentially the meaning already outlined in our treatment of the perfect gas: A thermodynamic system is in equilibrium when none of its thermodynamic properties change with time at a measurable rate. (This should not be taken to include a *steady state,* in which matter or energy flows steadily through the system but the properties measured at any given point are time-independent. An equilibrium state is a state of rest.) The equilibrium state of a system depends on the constraints placed on it, which must be specified. In general, there is only one true equilibrium state for a given set of constraints. However, many systems have states that show no change with time, and thus can be regarded for practical purposes as equilibrium states, even though they are not the true equilibrium state. An example is a room temperature mixture of H_2 and O_2, for which the reaction $H_2 + \frac{1}{2}O_2 \rightarrow H_2O$ is so slow (in the absence of a catalyst) that the mixture can be considered nonreactive for most purposes. Such a condition is called a *metastable equilibrium.* Thermodynamics can also be used to describe some of the properties of systems not at equilibrium; this aspect of the theory is still under active development.

The thermodynamic state of a system is assumed to be uniquely determined when a complete set of independent macroscopic coordinates is given. But what constitutes a "complete" set of thermodynamic coordinates? Given that the mass of the system is known, they are just those variables that must be independently specified for one to reproduce a given thermodynamic state, *as determined by experiment.* It should be noted that the number of thermodynamic coordinates necessary to specify the state of a system is very small (frequently, for a system with fixed mass, the minimum number is only two), whereas the total number of microscopic coordinates defining the same system is of the order of the number of atoms present.

The number of variables required for a complete set differs from system to system. Let us concentrate on the properties of a *fluid,* a form of matter that assumes the shape of its container (i.e., a gas or liquid). In general, the shape is found to be of no importance in determining the bulk properties of the fluid. Experiment shows that, for a given mass of pure fluid (fluid containing only a single chemical species) in the absence of external fields, the specification of the pressure and volume uniquely defines the thermodynamic state; for a fluid mixture one must also specify the composition.[3] For example, given the pressure and volume of a certain mass of oxygen, one finds that all equal masses of oxygen whose pressures and volumes are adjusted to be the same are in identical thermodynamic states. That is, for the given mass of oxygen there is a unique relationship among pressure, volume, and any other thermodynamic

[3] Of course, should one be concerned with, say, the electric or magnetic properties of a system, other variables must also be used. There are also certain anomalous cases, the best known being that of liquid water just above its freezing point: The density increases from 0 to 4°C, then decreases again, so that, say, $p = 1$ atm and $v = 1.000060$ cm^3/g can correspond to either 2°C or 6°C. Thus in this region p, v do not constitute a complete set of variables (but p, T do).

variable; this relationship can be expressed by an empirical *equation of state,*

$$f(p,V,X)=0, \tag{12.18}$$

where X is any third variable. In the next section we shall consider the properties of one such variable, the quantity we call temperature.

12.4 Temperature and the Zero-th Law of Thermodynamics

We wish to describe the properties of matter, beginning with the dilute gas, from a thermodynamic point of view. One concept important in this description is the thermodynamic coordinate known as *temperature,* the nature of which we must now clarify. Our goal is to formulate a quantitative definition of temperature which is free of ambiguities associated with intuitive notions. Although everyone is familiar with the qualitative notion that temperature is associated with the "hotness" of a body we can easily fool ourselves in this regard: For example, try placing both your hands in a bowl of tap water after one has been in ice water and the other in very hot water. This experiment shows that we cannot trust our simple intuitive or physiological reactions to provide a quantitative measure of temperature.[4]

We shall develop our definition of temperature using only the variables required to define the macroscopic state of a system and how those variables are related when particular experiments are carried out. Consider again a mass of pure fluid, the thermodynamic state of which depends on the pressure and volume. We can define a temperature θ by writing an equation of state in the form

$$f(p,\rho,\theta)=0, \tag{12.19}$$

where we have replaced the volume by the density ρ to make the equation independent of the particular mass of fluid.[5] We emphasize that Eq. 12.19 is to be interpreted as a *definition* of θ: We choose some function of pressure and density and call it the temperature. How do we decide what function to choose? Well, certainly θ should be well behaved (finite, continuous, and single-valued); it ought to bear some relationship to the intuitive notion of temperature (increasing θ should make a system "hotter"); and it should be mathematically convenient to use. But there are still infinitely many functions that satisfy these conditions. For example, we shall see that for an ideal gas a convenient temperature function is $\theta = p/\rho$; but $(p/\rho)^2$ or $\log(p/\rho)$, or either of these (or many other combinations of the same variables) times a constant multiplier, would do equally well. We can express this generality by "solving" Eq. 12.19 for temperature in the form

$$g(\theta)=\varphi(p,\rho), \tag{12.20}$$

where $g(\theta)$ is any of the alternative possible temperature functions and $\varphi(p, \rho)$ is the function of pressure and density that defines it. Which of the functions we choose as *the* temperature is strictly a matter of convention, as we shall see in the next section. But first we must introduce some concepts that apply to all temperature scales.

We begin by defining the properties of two idealized walls. Consider two systems 1 and 2, initially isolated and separately at equilibrium. Let the two systems initially contain the same pure fluid at different temperatures—that is, have values of the pressure and density such that, for a given choice of temperature function (Eq. 12.20), we have

$$\varphi(p_1,\rho_1)\neq\varphi(p_2,\rho_2). \tag{12.21}$$

Suppose now that the isolation of the two systems from each other is reduced to their separation by a rigid barrier B, both remaining isolated from the rest of the universe (Fig. 12.5). If the thermodynamic states of the two systems subsequently remain unchanged from the initial states, then the barrier B is called an *adiabatic wall.* In less precise language, an adiabatic wall is a perfect insulator across which heat cannot flow ("adiabatic" means uncrossable), so that the temperatures of the two systems remain different. We are all familiar with real approximations to adiabatic walls—for example, the silver-coated, evacuated, double glass walls of a Thermos bottle, or a thick, rigid sheet of asbestos. Clearly, the concept of a perfect adiabatic wall is an abstraction, but one that is linked to the observed behavior of a number of materials. Obviously, an isolated system must be surrounded by adiabatic walls.

There is another—and more likely—outcome of the experiment. Suppose again that the two systems, initially isolated and at different temperatures, are separated from one another only by a rigid barrier B. The usual result is that

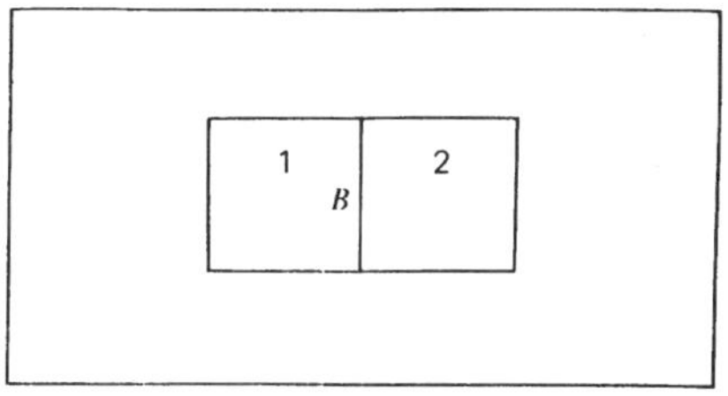

Figure 12.5 The two systems labeled 1 and 2 are separated by a rigid wall B. Both systems are isolated from the remainder of the universe.

[4] Despite the different sensations generated by tap water after exposure of one's hands to hot and cold water, one can train oneself to identify the temperature of an object to within about 1°C, touching the object to be tested to the inside of the forearm or wrist.

[5] This can be done only for those properties, such as temperature, that are in fact independent of the total mass. We shall discuss this distinction in Section 13.4, on intensive and extensive variables.

the thermodynamic states of the two systems change with time, eventually approaching a limit in which they are time-independent. Such a barrier is called a *diathermal wall,* and in the final time-independent state the two systems are said to be in *thermal equilibrium* with each other. In ordinary language, a diathermal wall (*dia,* "through," + *therme,* "heat") allows heat to pass through until the two systems are at the same temperature. All real material walls are diathermal, although in some cases the rate at which heat can cross is very small, so that the approach to thermal equilibrium can be very slow.

The definitions of adiabatic and diathermal walls are not limited to fluid systems, nor to systems of the same chemical composition. In general, the thermodynamic states of the systems may be defined by more variables than p and ρ, but the principle is the same: If the states of two otherwise isolated systems change when they are separated only by a rigid barrier B, then B is diathermal; if both states remain unchanged, then B is adiabatic (except for the trivial case that the systems are in thermal equilibrium to begin with). The requirement that the barrier be rigid is necessary to prevent the performance of work, which we shall discuss in the next chapter.

We said that the original isolated systems were separately in equilibrium, the equilibrium defined by the constraint of isolation. When the constraint of isolation is replaced by the constraint of contact through a diathermal wall, the system moves towards, and ultimately reaches, a new equilibrium state. The overall system is initially out of equilibrium with respect to the new constraint, but eventually approaches the equilibrium state. The way in which this state is reached is immaterial at present; only the final equilibrium itself is of concern to us. Experiment shows that, once two systems have reached thermal equilibrium with each other, one cannot change the state of system 1 without simultaneously changing the state of system 2. Thus the overall system at thermal equilibrium is not described by separate equations of state like Eq. 12.18, but by some single relationship involving the independent coordinates of both systems. For simplicity we shall restrict ourselves again to fluid systems, for which this single relationship has the form

$$F_{12}(p_1, \rho_1, p_2, \rho_2) = 0. \qquad (12.22)$$

We need not know the precise form of the function F_{12}, which will depend on the nature of the two fluids and their individual equations of state.

At this point it is appropriate to introduce the first postulate of thermodynamic theory, which is known as the *zero-th law of thermodynamics.*[6] It is called "zero-th" because, although it was formulated after the first, second, and third laws, it is logically prior to them in a rigorous development of the subject. The zero-th law is stated in the form:

Two systems, each separately in thermal equilibrium with a third system, are in thermal equilibrium with each other.

Like the other laws of thermodynamics, this statement is derived from the results of observations of many systems. It clearly embodies everyday experience: For example, if each of two systems is separately found to "have the same temperature" with respect to, say, a mercury-in-glass thermometer, we do not expect to observe any change when the two systems are brought into contact with each other. But why do we even need to state what seems so obvious a fact? Thus far we have only examined how to define a temperature for a particular system, by some such means as Eq. 12.19. But what we wish to obtain is a universal definition of temperature applicable to all systems. For such a universal temperature to exist, there must be some property that all systems "at the same temperature" (i.e., in thermal equilibrium with one another) have in common. Since we cannot actually investigate all systems, for logical completeness we must *assume* that this is true; the zero-th law has been chosen as the simplest way of formulating this assumption. We shall now see just how the zero-th law leads to a unique definition of temperature.

Consider three fluids, 1, 2, and 3, for each of which p and ρ are the only independent variables. When fluids 1 and 2 are in mutual thermal equilibrium, their thermodynamic states are related by an equation of the form 12.22, whereas for equilibrium between fluids 1 and 3 we have the equation

$$F_{13}(p_1, \rho_1, p_3, \rho_3) = 0. \qquad (12.23)$$

Equations 12.22 and 12.23 both contain the variables p_1 and ρ_1. In principle, each of these equations can be solved for, say, the pressure of fluid 1. As a result of solving these two equations, p_1 can be expressed as some function w_{12} of the variables ρ_1, p_2, ρ_2, or as some function w_{13} of the variables ρ_1, p_3, ρ_3:

$$p_1 = w_{12}(\rho_1, p_2, \rho_2) \quad \text{or} \quad p_1 = w_{13}(\rho_1, p_3, \rho_3). \qquad (12.24)$$

But since p_1 is the same physical quantity in both of these equations, we can immediately write

$$w_{12}(\rho_1, p_2, \rho_2) = w_{13}(\rho_1, p_3, \rho_3). \qquad (12.25)$$

Equation 12.25 is an alternative representation of the conditions placed on the states of systems 1, 2, and 3 by the requirements that 1 and 2 be in mutual equilibrium and that 1 and 3 be in mutual equilibrium.

We now introduce the new element required by the zero-th law. According to this law, if fluids 2 and 3 are separately in thermal equilibrium with fluid 1, they are also in thermal equilibrium with each other. Thus, the relationship between the states of fluids 2 and 3 must be completely described by an equation of the form

$$F_{23}(p_2, \rho_2, p_3, \rho_3) = 0. \qquad (12.26)$$

[6] The first formulation of the zero-th law of thermodynamics was presented by H. von Helmholtz, Crelle's Journal **97,** 159 (1884).

But Eq. 12.25 contains the density ρ_1, whereas Eq. 12.26 does not. If the two equations are to hold simultaneously, as the zero-th law requires, the variable ρ_1 must drop out of Eq. 12.25. That is, the functions w_{12} and w_{13} must be of such a form that we can write a new equation

$$w_2(p_2, \rho_2) = w_3(p_3, \rho_3), \tag{12.27}$$

in which each of the functions w_2 and w_3 depends on the state of a single system only. By an obvious extension of the argument we obtain the more general equation

$$w_1(p_1, \rho_1) = w_2(p_2, \rho_2) = w_3(p_3, \rho_3), \tag{12.28}$$

which can be further extended to any number of fluids in mutual thermal equilibrium. Note that the $w_i\ (p_i,\ \rho_i)$ may be quite different functions of pressure and density for different fluids; it is only their numerical values that must be the same at equilibrium.

Example: These abstract arguments are clarified by a concrete example. Consider the ideal gas, a hypothetical system that obeys the equation of state $pV = nRT$; we shall explain this equation in the next section, but for now take it as a given. If we put ideal gases 1 and 2 in thermal equilibrium and determined the relationship between their states, the form of Eq. 12.22 that would obtain is

$$F_{12}(p_1, \rho_1, p_2, \rho_2) \equiv \frac{p_1 M_1}{\rho_1} - \frac{p_2 M_2}{\rho_2} = 0, \tag{12.22$'$}$$

where M is the molecular weight. Similarly, we have for ideal gases 1 and 3,

$$F_{13}(p_1, \rho_1, p_3, \rho_3) \equiv \frac{p_1 M_1}{\rho_1} - \frac{p_3 M_3}{\rho_3} = 0. \tag{12.23$'$}$$

Solving both equations for p_1, we obtain

$$p_1 = w_{12}(\rho_1, p_2, \rho_2) \equiv \frac{\rho_1 p_2 M_2}{\rho_2 M_1},$$
$$p_1 = w_{13}(\rho_1, p_3, \rho_3) \equiv \frac{\rho_1 p_3 M_3}{\rho_3 M_1}, \tag{12.24$'$}$$

and thus,

$$\frac{\rho_1 p_2 M_2}{\rho_2 M_1} = \frac{\rho_1 p_3 M_3}{\rho_3 M_1}. \tag{12.25$'$}$$

The zero-th law tells us that we must also have

$$F_{23}(p_2, \rho_2, p_3, \rho_3) \equiv \frac{p_2 M_2}{\rho_2} - \frac{p_3 M_3}{\rho_3} = 0, \tag{12.26$'$}$$

which does not contain ρ_1. By the reasoning in the text, this means that ρ_1 must drop out of Eq. 12.25′, as it obviously does, giving

$$w_2(p_2, \rho_2) \equiv \frac{p_2 M_2}{\rho_2} = w_3(p_3, \rho_3) \equiv \frac{p_3 M_3}{\rho_3} \tag{12.27$'$}$$

By extension the function pM/ρ has the same value for all ideal gases in thermal equilibrium; in the next section we shall see the significance of this fact in defining a temperature scale.

We have shown that, given the zero-th law, there must exist a set of functions w_i, each depending only on the state of the ith fluid, which all have the same value for any number of fluids in thermal equilibrium. For this to be true, the w_i must all depend on some single property that all fluids in thermal equilibrium have in common, regardless of their natures. This single property is defined to be their common temperature. Furthermore, the argument is not restricted to fluids: If an arbitrary system is described by the set of independent variables $X_1, Y_1, Z_1, \ldots$, then the F_{12} of Eq. 12.22 becomes $F_{12}\ (X_1, Y_1, Z_1, \ldots, X_2, X_2, Z_2, \ldots)$ and the rest of the derivation proceeds along exactly the same lines. Thus it is possible to define a common temperature for all systems of whatever kind in thermal equilibrium.

The functions $w_i(p_i,\ \rho_i)$ are clearly equivalent to the $\varphi(p,\ \rho)$ of Eq. 12.20, which says that for any fluid one can define a temperature function $g(\theta)$ on the basis of that particular fluid's equation of state. What we are now saying is that each such $g(\theta)$ is a function of the property that all systems in thermal equilibrium have in common. This tells us how to obtain a universal temperature scale: We choose a particular system as a standard, select one of that system's possible temperature functions $g(\theta)$, and define the numerical value of this $g(\theta)$ as the temperature θ; the same numerical value of θ is assigned to any other system in thermal equilibrium with our standard system. Such a standard system is an example of what we call a *thermometer,* and the function used to define a temperature is called a *thermometric property;* in the next section we shall see how such a system and property are chosen.

12.5 Empirical Temperature: The Perfect Gas Temperature Scale

The arguments of the preceding section seem a long way from our intuitive understanding of "temperature." What is the relationship between these abstract considerations and a usable temperature scale?

Any system property that varies with temperature in a well-behaved way can be used to define a temperature scale. The choice among the various possible thermometric properties is arbitrary and based essentially on convenience. One important requirement is that significant changes in the thermometric property be accurately measurable in a small system (a thermometer should be small enough to come into thermal equilibrium with a system without changing the system's temperature significantly). The best-known thermometric property is the volume of a liquid, most commonly mercury; the volume changes are magnified by being measured in a narrow tube attached to a larger reservoir. The first thermometers invented were of this *liquid-expansion* type, with temperature scales defined as linear functions of the liquid column's length. Among the other thermometric properties in common use are the *electrical resistance* of a fine metal wire (usually platinum); the electromotive force generated by a *thermocouple* (the junction between two wires of different metals); and the intensity of the radiation emitted by a hot body, assumed to obey the black-body radiation law (Section 2.6), as measured with a *pyrometer.* But none of these properties is actually used as the basis of our temperature scale; rather, instruments of all these types are calibrated with respect to the scale we shall now describe.

A particularly elegant and useful temperature is that based on the low-pressure properties of a very dilute gas, to which we now finally return. Since the easily measured properties of a gas are its volume (or density) and pressure, we seek a convenient way of relating temperature to these properties, as in Eq. 12.20. The most straightforward procedure is to define the temperature as proportional to one of these properties, with the other held constant. The two alternatives lead to constant-volume and constant-pressure gas thermometers, and instruments of both types in fact exist; but as we shall see, both temperatures converge to the same low-pressure limit.

Suppose that we measure the pressure of a fixed volume of gas in thermal contact with an equilibrium mixture of ice, liquid water, and water vapor (i.e., water at its *triple point*) and call this pressure p_3. We use the pressure p of the *same* volume of gas under other conditions as a thermometric property, defining a constant-volume temperature scale by the equation

$$\theta(p) = \frac{p}{p_3}\theta_3, \quad (V = \text{constant}), \tag{12.29}$$

where θ_3 is an arbitrary constant, the temperature of the triple point of water. By convention (for reasons that will become apparent later) we set the value of θ_3 equal to 273.16 K, the "K" standing for the temperature unit, the *kelvin* (formerly *degree Kelvin,* °K). Alternatively, we can measure the volume V_3 of a given mass of gas at the triple-point temperature and gas pressure p, then use the volume at the same pressure to define the constant-pressure temperature scale

$$\theta(V) = \frac{V}{V_3}\theta_3, \quad (p = \text{constant}). \tag{12.30}$$

The empirical temperature scales defined in Eqs. 12.29 and 12.30 are functions of both the mass of gas in the thermometer and the nature of the gas. Experimentally, however, as one lowers the quantity of gas (and thus the value of p) in a constant-volume thermometer or the constant pressure p in a constant-pressure thermometer, both temperatures are found to approach the same limiting value,

$$\lim_{p\to 0}\theta(p) = \lim_{V\to\infty}\theta(V) = \theta^*, \tag{12.31}$$

which is, furthermore, the same for all gases. This limiting process is illustrated in Fig. 12.6. The limit θ^* is called the *perfect gas temperature.*

Although the elaborate limiting procedure described here is not itself convenient for ordinary temperature measurements, it can be used as a primary standard to calibrate practical thermometers of the various types we have mentioned. The *International Practical Temperature Scale of 1968* (IPTS–68) is a set of reproducible fixed points[7] and interpolation formulas that have been reliably measured in terms of the perfect gas temperature scale. The IPTS–68 is defined in terms of a platinum-resistance thermometer from 13.81 K to 903.89 K; of a platinum/platinum–rhodium thermocouple from 903.89 K to 1337.58 K; and of Planck's black-body radiation law above 1337.58 K. These can in turn be used to calibrate thermometers for ordinary measurements, so that a perfect gas temperature can be determined, even if indirectly, for any system of interest.

Equation 12.31 is a direct consequence of the experimental observations embodied in Boyle's law and Gay–Lussac's law (the latter of which is often called Charles's law), which we can now formulate as follows: If the temperature θ of a mass of gas is held fixed and its pressure p is lowered, then the product pV/n, where V/n is the volume per mole (cf. Section 1.6), approaches for all gases the same limiting value,

$$\lim_{p\to 0} pV/n = \beta(\theta^*) \tag{12.32}$$

where the common limit $\beta(\theta^*)$ is a function of the temperature θ^* only (Boyle's law). Furthermore, the limits obtained at two different temperatures are related by the equation

$$\frac{\beta(\theta_1^*)}{\beta(\theta_2^*)} = \frac{\theta_1^*}{\theta_2^*}, \tag{12.33}$$

---

[7] The fixed points of the IPTS–68 are as follows ("boiling point" and "freezing point" imply a pressure of 1 atm):

13.81K (–259.34°C) triple point of equilibrium hydrogen (see Chapter 21);
17.042 K (–256.108°C) liquid–vapor equilibrium point of equilibrium hydrogen at 33330.6 N/m² pressure;
20.28 K (–252.87°C) boiling point of equilibrium hydrogen;
27.102 K (–246.048°C) boiling point of neon;
54.361 K (–218.789°C) triple point of oxygen;
90.188 K (–182.962°C) boiling point of oxygen;
273.16 K (0.01°C) triple point of water;
373.15 K (100.00°C) boiling point of water;
692.73 K (419.58°C) freezing point of zinc;
1235.08 K (961.93°C) freezing point of silver;
1337.58 K (1064.43°C) freezing point of gold.

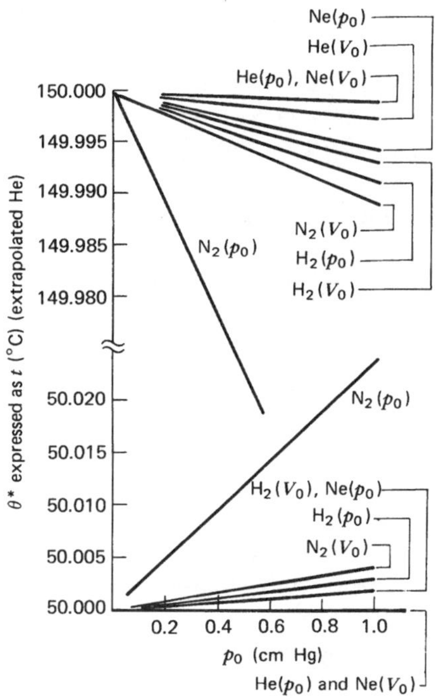

Figure 12.6 As the pressure is lowered in a constant-volume thermometer [$\theta(p)$], or a constant-pressure thermometer [$\theta(V)$], both empirical temperatures approach a common value, namely, the perfect gas temperature θ^*. In the figure p_0 and V_0 are, respectively, the pressure and volume of the gas in the thermometer at the ice point. From F. H. Crawford, *Heat, Thermodynamics and Statistical Physics,* Harcourt Brace Jovanovich, New York, 1963.

where θ^* is the perfect gas temperature defined in Eq. 12.31; that is, $\beta(\theta^*)$ is directly proportional to θ^* (Gay–Lussac's law).

The temperature scale was formerly defined in terms of two fixed points, the temperatures of melting ice and boiling water (each under a pressure of 1 atm). The *Celsius* (*centigrade*) scale was obtained by dividing the interval into 100 degrees, the melting-ice and boiling-water points being designated as 0°C and 100°C, respectively. For a constant-volume gas thermometer, then, one could define (in the perfect gas limit) the Celsius temperature

$$t^* = \lim_{p_0 \to 0} \frac{100(p - p_0)}{p_{100} - p_0}, \quad (V = \text{constant}). \tag{12.34}$$

At that time the absolute temperature θ^* was defined by the equation

$$\theta^* = t^* + \theta_0^* \tag{12.35}$$

where θ_0^*, the ice-point temperature, had to be chosen such that Eq. 12.33 would be satisfied; the experimental value was about 273.15°, corresponding to 273.16° for the triple point θ_3^*. To eliminate the experimental uncertainty, an international agreement in 1954 made θ_3^* = 273.16 K by *definition,* as indicated above. The melting point of ice, θ_0^*, is then still 273.15 K, but the boiling point of water changes from 373.15 to 373.146 K. Since the Celsius degree is now defined as equal to the kelvin, Eq. 12.35 still holds as a definition of the Celsius temperature t^*, but Eq. 12.34 is no longer exactly valid.[8]

We shall show in Chapter 16 that it is possible to establish a thermodynamic temperature scale independent of the properties of any real or hypothetical thermometric substance. However, it will also be shown that this thermodynamic temperature T is in fact equal to the perfect gas temperature θ^*. To simplify our notation, we shall anticipate this result by regularly using T to designate the perfect gas temperature.

We may as well also say something here about the units of pressure. The SI unit (adopted in 1971) is the *pascal* (Pa), or newton per square meter; closely related are the dyne per square centimeter (=0.1 Pa) and the *bar* (≡10^5 Pa). However, most scientific measurements are still expressed in terms of the *atmosphere* (atm), which has the defined value

$$1 \text{ atm} \equiv 1.01325 \times 10^5 \text{ Pa},$$

chosen to fall within the range of actual atmospheric pressures near sea level. We shall see in subsequent chapters that thermodynamic quantities are commonly defined in relation to a standard state at 1 atm. For lower pressures one commonly uses the *torr,* defined as $\frac{1}{760}$ atm; this is for practical purposes the same as the conventional "millimeter of mercury" (mm Hg).[9]

We can now combine Eqs. 12.32 and 12.33 to obtain the familiar perfect gas equation of state,[10]

$$pV = nRT \tag{12.36}$$

where n is the number of moles of gas in the volume V. The proportionality constant R is known as the (*universal*) *gas constant,* and can be evaluated by measuring the limit

$$R = \lim_{p \to 0} \frac{pV}{nT}, \tag{12.37}$$

[8] The 1967 General Conference on Weights and Measures adopted the name "kelvin," symbol "K," for the unit of thermodynamic temperature, with "degree Celsius," symbol "°C," as an alternative unit for a temperature *interval.* Note the omission of the word "degree" and the degree symbol from the kelvin scale.

[9] To be precise, the "millimeter of mercury" is defined as the pressure exerted by a 1-mm column of a fluid with density 13.5951 g/cm^3 (equal to that of mercury at 0°C) at a place where the acceleration of gravity is 9.80665 m/s^2. For the record we note that the torr is named after Torricelli, who was a prominent Renaissance scientist.

[10] This is also and perhaps more commonly known as the *ideal gas law.* One can make a formal distinction between a *perfect* gas, which has the microscopic properties described in Section 12.1, and an *ideal* gas, one which satisfies Eq. 12.36. But to the extent that both have the same limiting properties, we can identify the two and call both perfect gases.

which is the same for all real gases. It has the value

$$R = 0.0820568 \text{ liter atm / mol K}$$
$$= 8.31441 \text{ J / mol K}$$

Division by Avogadro's number (Section 1.6) gives *Boltzmann's constant,*

$$k_B \equiv \frac{R}{N_A} = 1.38066 \times 10^{-23} \text{ J / K}, \tag{12.38}$$

which is simply the gas constant per molecule. The perfect gas equation of state can thus also be written as

$$pV = Nk_BT, \tag{12.39}$$

where N is the number of molecules in the volume V. These equations can be used to give us a thermodynamic definition of a perfect gas—a hypothetical gas that at nonzero pressures obeys the equation of state, Eq. 12.36, that real gases obey only in the limit $p \to 0$. (However, we shall see in the next chapter that this does not suffice for a complete definition of the perfect gas.)

A final note: Given the value of R, one can determine the molar mass ("molecular weight") M of any gas by measuring T, p, and the mass density ρ. From data taken at one temperature and a series of pressures, one makes the extrapolation

$$M = RT \lim_{p \to 0} \frac{\rho}{p}. \tag{12.40}$$

Although Eq. 12.40 is still used to determine molecular weights in some cases, determinations using a mass spectrometer (Section 1.5) are much more accurate and at least as convenient.

12.6 Comparison of the Microscopic and Macroscopic Approaches

The preceding sections have illustrated some of the differences in approach between the microscopic and macroscopic theories. Let us now try to analyze these differences. The questions we wish to answer include: What are the essential features of each approach? Which arguments are general and which specific to a particular model? And how are our two definitions of the "perfect gas" related? We begin with the last of these questions.

Let us consider just what we mean when we speak of a perfect gas. In the microscopic theory a perfect gas is defined entirely in terms of the properties of the constituent molecules, in general as a gas in which there are no significant interactions between molecules. Our treatment in Sections 12.1 and 12.2 was even more restrictive than this, being limited to monatomic molecules that collide elastically with the container walls. It is thus not clear to what extent our results are valid for polyatomic molecules, or for molecules that do interact with one another. This illustrates a major limitation of statistical mechanics, that many of its results are applicable only for a particular model of molecular properties. The perfect gas model can in fact be extended to polyatomic molecules with a more sophisticated treatment, but any such extension must be worked through explicitly.

Contrast this with the definition of the perfect gas used in thermodynamics,[11] an abstraction based on the measured properties of real gases in the zero-pressure limit. No assumptions about the properties—or even the existence—of molecules are required; in particular, it does not matter whether the gas molecules are monatomic or polyatomic. Although neither "perfect gas" has the same properties as any real gas, the thermodynamic definition is at least tied directly to experiment.

How are the two definitions related? Consider the equations of state obtained in the two theories. The final result of our kinetic theory analysis was a relationship, Eq. 12.17, among the pressure, volume, and internal energy—the latter meaning in this case the total kinetic energy of the molecules in a monatomic gas. Although the pressure and volume can be measured directly, we have no direct way of determining the molecular kinetic energy. In contrast, the macroscopic perfect gas equation of state, Eq. 12.36, involves only directly measurable variables. Now we take the step joining the two theories: We *assume* that both refer to the same gas. Both equations of state involve the pressure, which is presumably the same physical quantity in each case. If both describe the same gas, then we can equate the two results for the pressure:

$$p = \frac{2}{3}\frac{U}{V} = \frac{nRT}{V}. \tag{12.41}$$

We thus obtain the important result

$$U = \tfrac{3}{2}nRT = \tfrac{3}{2}Nk_BT; \tag{12.42}$$

that is, the molecular kinetic energy per mole is proportional to the absolute temperature. As we have derived this result it applies only to monatomic gases, but we shall later show that it is generally true, even for gases with intermolecular forces.

What is important about Eq. 12.42 is that it constitutes our first bridge between a microscopically defined quantity

[11] As mentioned earlier, we have not yet given the full definition. In thermodynamics a perfect gas is one that (1) obeys Eq. 12.36 as its equation of state, and (2) has an internal energy that depends on temperature only. The meaning of the second part of the definition must wait until the next chapter.

(the molecular kinetic energy) and a thermodynamic variable (the temperature). But how do we justify the assumption that led us to this bridge, the assumption that both "perfect gases" are the same? The simplest answer is that they were designed to be the same. In effect the microscopic perfect gas is a model devised to mimic the properties of the thermodynamic perfect gas, an attempt to answer the question, "What kind of molecular model would give macroscopic behavior corresponding to the perfect gas equation of state?" We have shown that the model in question does this, *if we identify the molecular kinetic energy with the temperature.* Here this identity may seem arbitrary, but we shall see that the same identification (or its equivalent) can be made for many other types of systems, in each case giving correspondence between a properly designed microscopic model and the observed behavior of a macroscopic system. As always, our assumptions are justified by their agreement with experiment.

The approach outlined here can be extended to other systems. We have already mentioned that for a given mass of pure fluid any two of the variables p, V, T constitute a complete set of independent variables. (Remember that this is an experimental fact.) In short, all fluids, no matter how complex their microscopic structure, have an equation of state of the form $f(p, V, T) = 0$; the only example we have yet examined is Eq. 12.36 for the perfect gas case. On the other hand, any parallel equation of state developed from a molecular model must involve p, V, and the molecular energy in some form; in general, one may have to consider the intramolecular (vibrational, rotational, electronic) and intermolecular energies as well as the kinetic energy. Since the two approaches deal with different sets of variables, it is always necessary to establish a bridge of some kind if we wish to apply both kinds of analysis to the same real system. Equation 12.36 is an elementary example of such a bridge, applicable only to a single kind of system; in later chapters we shall develop bridges with more general validity.

Note that we have not actually done any thermodynamics in this chapter, except for introducing the zero-th law and the temperature concept. What we have talked about instead is various ways of obtaining an equation of state: by actual measurement in a real system, by extrapolation from measurement to a hypothetical system (such as the perfect gas), or from a microscopic model via a bridge between two sets of variables. Thermodynamics deals rather with the relationships between equations of state (however determined), other measurable quantities, and functions such as the internal energy and entropy which we have yet to define. These relationships are completely general. For example, the rule that $(\partial U/\partial V)_T = T(\partial p/\partial T)_V - p$ is true for all systems, no matter what their equations of state, but the equation of state of any specific system can be substituted in this rule to obtain $(\partial U/\partial V)_T$ for that system. The existence of such thermodynamic relationships makes it possible to derive a great deal of information about any system from a relatively small amount of data or from a relatively simple molecular model and bridge.

Thus far it would appear that the advantages of generality lie all with thermodynamics. Although this is largely true, the very generality of thermodynamics often precludes it from making specific predictions about particular systems. For example, experiment shows that the equations of state of all gases approach the perfect gas law in the limit $p \to 0$, and thermodynamics can be used to calculate the behavior of any other property in the same limit. But thermodynamics alone cannot tell us at what pressure measurable deviations from the perfect gas law will occur; for this we must rely on experiment. In contrast, a simple modification of the molecular theory to take into account the interactions between molecules enables us to predict how great the deviations from the perfect gas law will be at any pressure—for a particular model of the intermolecular forces. The choice of model is, of course, still based on experiment, but the experiments used in this case may be much farther removed from the quantity we seek (cf. Section 10.2); and in principle, at least, one can calculate directly the intermolecular forces by quantum mechanics. To sum up, then: Thermodynamics establishes general relationships whose validity transcends the nature of particular molecular models, whereas molecular theory is capable of model-based predictions outside the scope of thermodynamics.

As we said in the introduction to Part II, statistical mechanics and thermodynamics are complementary tools that can be used to study the properties of matter. It is always important to bear in mind the advantages and disadvantages of the two theories. Nevertheless, it should already be clear that an analysis using all the available tools is likely to provide us with greater understanding and more incisive descriptions than an analysis limited to only one approach.

• FURTHER READING

Bailyn, M., *A Survey of Thermodynamics* (AIP Press, New York, 1994), Chapter 1.

Brush, S. G., *Statistical Physics and the Atomic Theory of Matter* (Princeton University Press, Princeton, 1983), Chapter 1.

Kennard, E. H., *Kinetic Theory of Gases* (McGraw-Hill, New York, 1938), Chapter 1.

Kestin, J., A *Course in Thermodynamics,* Vol. I (Blaisdell, Waltham, Mass., 1965), Chapters 2 and 3.

McCullough, J. P., and Scott, D. W. (Eds.), *Experimental Thermodynamics,* Vol. I (Plenum Press, New York, 1968), Chapter 2.

Swindells, J. F. (Ed.), *Temperature,* NBS Special Publication 300, Vol. 2 (1968).

Zemansky, M. W., *Heat and Thermodynamics,* 4th ed. (McGraw-Hill, New York, 1957), Chapter 1.

• PROBLEMS

1. Explain how momentum and energy are conserved even when a gas molecule sticks to a wall after impact.

2. What happens to the kinetic energy of a body when it undergoes an inelastic collision with another body?

3. A delicate galvanometer mirror, if placed in a gas, may be observed to undergo chaotic oscillation. What is the cause of this oscillation?

4. If the number of molecules with speeds between v and $v + dv$ is given by

$$\text{constant} \times 4\pi v^2\, dv,$$

how many molecules have kinetic energies between ϵ and $\epsilon + d\epsilon$?

5. The rate of effusion of a gas through a very small hole from a container to an evacuated space is proportional to the average speed of the molecules of the gas. Explain why this is so.

6. Using the results of Problem 5, devise a method for determining the molar weight of a perfect gas.

7. Suppose that all the atoms in a monatomic gas move with the same speed v, but that the directions of motion are randomly distributed. Compute the average number of atoms striking an element of wall area dA per second, and the average momentum imparted per second to dA in terms of N, V, v, and dA. (*Hint:* Average over all directions of incidence.)

8. Suppose that in a gas the number of atoms with speed between v and $v + dv$ is

$$2N\left(1 - \frac{v}{v_m}\right)\frac{dv}{v_m}$$

for $v < v_m$, where v_m is a constant. As in Problem 7, the directions of motion are randomly distributed. Calculate the average number of molecules striking dA per second and the average momentum imparted to dA per second, in terms of N, V, v_m, and dA.

9. Show that for the gases described in Problems 7 and 8,

$$p = \frac{2}{3}\frac{U}{V}.$$

10. Let $n(v)$ be the average number of molecules per unit volume with speed less than v. Then $n(v)$ increases monotonically as v increases, starting from $n(0) = 0$ and approaching $n(\infty) = n$ as $v \to \infty$.
    (a) Write an expression for the average number of molecules per unit volume with speed less than v that have velocities directed in the solid angle $d\Omega = 2\pi \sin\theta\, d\theta$.
    (b) Write an expression of the average number of molecules per unit volume that have velocities directed between θ and $\theta + d\theta$ and have speed in the range v to $v + dv$.
    (c) From the result of part (b), relate the function $n(v)$ to $f(v)$ introduced in Eq. 12.9.

11. If the average number of molecules per unit volume with velocity between $v_x + dv_x$, $v_y + dv_y$, $v_z + dv_z$ is

$$e^{-\frac{1}{2}cv^2}\, dv_x\, dv_y\, dv_z\,,$$

calculate $n(v)$ defined in the preceding problem. Assume that n, the average number density, is known.

12. Suppose that a box is divided by a thin rigid wall. On one side of the container is gas characterized by number density n_1, mass per molecule m_1, and coefficient c_1 in the exponential of the distribution of Problem 11. On the other side of the container is gas characterized by, respectively, n_2, m_2, and c_2. What condition must be satisfied if the forces exerted on the wall by the gases on the two sides are to balance?

13. The densities of gaseous trimethylamine [$(CH_3)_3N$] and ammonia (NH_3) as a function of pressure at 0°C are given in the following table:

$(CH_3)_3N$		NH_3	
p (atm)	**ρ (g/liter)**	**p (atm)**	**ρ (g/liter)**
0.20	0.5336	0.333	0.25461
0.40	1.0790	0.667	0.38293
0.60	1.6363	0.500	0.51182
0.80	2.2054	1.000	0.77169

Determine the molar weights of trimethylamine and ammonia. If we take, on the chemical scale, the atomic weights of H and C to be 1.008 and 12.010, compute the atomic weight of N from both data sets.

14. From the following data, determine the zero of the perfect gas temperature scale (i.e., the temperature at which a hypothetical perfect gas would have zero volume).

THERMAL EXPANSION OF N_2 AT CONSTANT VOLUME

p (torr) at 0°C	**$\frac{p_{100} - p_0}{p_0}$ Relative Increase in Pressure from 0°C to 100°C**
333.6	0.36653
449.8	0.36667
599.9	0.36687
750.4	0.36707
998.9	0.36741
5,000	0.3734
10,000	0.3809
20,000	0.3956
30,000	0.4101
40,000	0.4247
50,000	0.4396

15. From the following data, show that the coefficient of thermal expansion,

$$\alpha \equiv \frac{1}{V}\left(\frac{\partial V}{\partial T}\right)_p,$$

is the same for all three gases in the limit as the pressure tends to zero.

He		H_2		N_2	
p (torr)	**$\alpha \times 10^6$(K^{-1})**	**p (torr)**	**$\alpha \times 10^6$ (K^{-1})**	**p (torr)**	**$\alpha \times 10^6$ (K^{-1})**
504.8	3658.9	508.2	3660.2	511.4	3667.9
520.5	3660.3	1095.3	3659.0	1105.3	3674.2
760.1	3659.1				
1102.9	3658.2				
1116.5	3658.1				

16. Use the following data to show that the perfect gas law is valid only in the limit as $p \rightarrow 0$. In this table, V^+ is 1 unit when the temperature is 0°C and $p = 1$ atm.

BEHAVIOR OF N_2 AT 0°C		BEHAVIOR OF C_2H_2 AT 20°C	
p (atm)	**V^+**	**p (atm)**	**V^+**
0.33333	3.00087	1	1
0.66667	1.50016	31.6	0.02892
1.00000	1.00000[a]	45.8	0.01705
19.0215	0.052190	84.2	0.00474
23.7629	0.041712	110.5	0.00411
28.4968	0.034735	176.0	0.00365
37.9526	0.026014	282.2	0.00333
50.000	0.019692	398.7	0.00313
52.2160	0.018856		
100.	0.009846		
200.	0.005181		
400.	0.0031392		
600.	0.0025380		
800.	0.0022449		
1000.	0.0020641		

[a] Reference volume, defined to be 1 unit at 0°C and 1 atm.

17. The accepted value of the molar volume of a perfect gas at 273.15 K and 1 atm is 22.4138 liters. Calculate the volume occupied by 1.0000 mol of each of the following gases at 273.15 K and 1 atm from the densities in grams per liter.

Gas	Density (g/liter)	Gas	Density (g/liter)
He	0.1785	Rn	9.73
Ne	0.9002	O_2	1.4290
Ar	1.7824	O_3	2.144
Kr	3.708	F_2	1.695
Xe	5.851	Cl_2	3.214

18. A constant-volume gas thermometer is used to measure a series of reference temperatures. The limiting values of the ratios of pressure at T to pressure at the triple point of water are found to be 0.33017 at the boiling point of oxygen, 0.85770 at the melting point of mercury, and 1.79788 at the boiling point of naphthalene. Calculate the perfect gas temperatures corresponding to these reference points.

19. A Pt-resistance thermometer has a resistance of 9.81 ohms at 0°C, 13.65 ohms at 100°C, and 21.00 ohms at 300°C. Is the thermometer linear over this range? If it is assumed to be linear between 0°C and 100°C, will a temperature of 50°C deduced from a resistance reading of 11.73 ohms be higher or lower than the true temperature?

20. Let X be a general thermometric property of a suitable substance. Following the arguments used in the text, show how a temperature scale may be set up using the property X. What conditions must the property X satisfy?

21. Suppose that for some gas

$$(pV)_0 = A_0 + B_0 p_0,$$
$$(pV)_{100} = A_{100} + B_{100} p_{100}$$
$$(pV)_\theta = A_\theta + B_\theta p_\theta,$$

where the numbers B_0p_0, $B_{100}p_{100}$, $B_\theta p_\theta$ are all small relative to A_0, A_{100}, A_θ, respectively. The numbers A_0, and so on, refer to the limiting perfect gas values of the respective pV products. If we define a thermometric scale in terms of the given pV terms, there will be a difference between temperatures on the defined scale and on the perfect gas scale. If

$$\Delta\theta = \theta^* - \theta_{\text{gas}}$$

is that difference in temperature, work out the correction to the temperature (i.e., the difference between the real gas and perfect gas temperatures) when the real gas is used in a constant-volume thermometer.

22. Assume that the electromotive force produced by a given thermocouple, with one junction at 0°C and the other at θ, can be represented by

$$\mathscr{E} = a_1\theta + a_2\theta^2.$$

Taking $\mathscr{E}$ to be the thermometric property, define a Celsius scale in terms of the thermal emfs. Obtain a formula for the correction $\Delta\theta = \theta_{\text{lim}} - \theta_{\mathscr{E}}$, where $\theta_{\mathscr{E}}$ is the value of the temperature calculated from the definition. If $a_1 = 0.20$mV/°C and $a_2 = 5.0 \times 10^{-4}$ mV/(°C)^2, work out the magnitude of the correction at $\theta = 50$°C.

23. It is sometimes stated that there are no collisions between the molecules of a perfect gas. Imagine a volume of a perfect gas in which there is an initial nonequilibrium distribution of molecular velocities. If the molecules can collide elastically with the walls, but not with each other, is there any mechanism by which an equilibrium state can be reached?

24. The molecules of a perfect gas have negligible volume. In a so-called hard-sphere gas, the molecules have nonzero volume and interact with each other through instantaneous (impulsive) collisions. Such a gas has the equation of state

$$p(V-nb)=nRT$$

where the parameter b is proportional to the molecular volume. (a) Find p/RT correct to second order in the molar density. (b) At low density, does the hard-sphere gas approach perfect gas behavior?

CHAPTER 13

The First Law of Thermodynamics

In this chapter we discuss the first law of thermodynamics, historically the first of the great principles that form the basis of thermodynamics. In simple terms, the first law is the law of conservation of energy; however, it cannot be simply identified with the corresponding mechanical principle. As we did in Chapter 12 for temperature, we must carefully define just what "energy," "work," and "heat" mean in thermodynamics.

Following the pattern of Chapter 12, we carry out a parallel development of the theory from the microscopic point of view, where energy can be given its ordinary (quantum) mechanical meaning. We apply both the thermodynamic and the microscopic models to the perfect gas, still the only system simple enough for us to analyze in detail.

In this chapter we also develop several concepts that are of general importance in thermodynamics (rather than being related specifically to the first law). These include the difference between "intensive" and "extensive" variables, the limiting processes called "quasi-static" and "reversible," and the key notion of "constraints." Most of these ideas have already been used implicitly, but for complete understanding we must analyze them further.

13.1 Microscopic and Macroscopic Energy in a Perfect Gas

As remarked above, the first law of thermodynamics is usually interpreted as a statement of the conservation of energy. The concept of energy conservation is well defined in both classical and quantum mechanics. For a single particle moving in a force field, the kinetic energy is $\frac{1}{2}mv^2$, the potential energy is defined as in Eq. 2.46, and their sum is constant throughout the particle's motion. For a system of interacting particles, the total energy of all the particles is conserved, but not the energy of a single particle. This analysis of the energy can be used only as long as we are dealing, at least implicitly, with particles. In thermodynamics, it will be recalled, the concepts used are defined only in terms of macroscopic operations, without reference to the microscopic structure of the substance involved. What, then, does "energy" mean in thermodynamics, where it must be some function of the thermodynamic variables discussed in Section 12.3? What similarities and differences are there between the mechanical and thermodynamic concepts with the same name?

As implied by these questions, there does exist an essential difference between the two concepts. The nature of this difference can be seen from the following considerations: In the mechanical description of a system of particles, it is presumed that the positions and velocities (and thus the energies) of individual particles can be controlled by suitable manipulation of the forces available to the experimenter. The work done in displacing a given particle is then exactly determined by the controlled displacements of that particle and all the other particles present. The uncertainty principle limits the accuracy with which this process can be carried out, but the quantum mechanical and classical descriptions are basically the same: One follows the motion of the individual particles as closely as nature allows. Contrast this with the thermodynamic description of a system, in which only macroscopic variables such as the volume are controlled. If work is done by displacing the system's boundaries, the positions and velocities of the individual molecules change in an uncontrolled manner. Only the *average* changes in various molecular quantities are determined by the changes in the macroscopic variable. It is these unavoidable and uncontrolled molecular motions that cause thermodynamic energy to differ from mechanical energy, thermodynamic work from mechanical work, and so on.

Nevertheless, connections between the thermodynamic and mechanical quantities can be found, in the sense that thermodynamic variables correspond to mechanical variables averaged over large numbers of particles. We have already seen one example of this. The temperature T is a characteristic variable of a thermodynamic system, defined in a completely nonmechanical way. Yet according to Eq. 12.42, in a perfect monatomic gas T is directly proportional to the total molecular kinetic energy U, defined as in Eq. 12.10 by summing over individual particle kinetic energies. We shall obtain many other correspondences of this sort—including, of course, that between thermodynamic and mechanical energy.

Before going on to the thermodynamic development, let us consider a simple molecular model in an attempt to obtain some intuitive understanding of the interplay between macroscopic and microscopic changes in a system. The model we choose is again that of the perfect monatomic gas, but this time we describe it quantum mechanically. Although this system is simpler than any real substance, the conclusions we shall draw are generally valid, as will be demonstrated more comprehensively in Chapter 16.

Consider a set of N noninteracting particles in a cubical box of volume V. Since the walls of the box prevent particles from leaving or entering, this is a model of a closed system. We showed in Chapter 3 that the possible energies of particles in such a box are quantized, with the energy spectrum for any of the independent particles given by Eq. 3.119 as

$$\epsilon_{n_1n_2n_3} = \frac{\pi^2\hbar^2}{2mV^{2/3}}(n_1^2+n_2^2+n_3^2), \quad (n_1, n_2, n_3 = 1, 2, \ldots) \tag{13.1}$$

where n_1, n_2, n_3 are the quantum numbers corresponding to motion along the x, y, z axes of the cube. If the box is of macroscopic size, the spacings between the energy levels are very small. The form of the energy spectrum is the same for each of the particles, because of our assumption that they do not interact. Therefore the total energy of the N particles in the box is simply the number of particles in each quantum state multiplied by the energy of that state, summed over all states,

$$E = \sum_{n_1,n_2,n_3} N_{n_1n_2n_3}\epsilon_{n_1n_2n_3}, \tag{13.2}$$

where $N_{n_1n_2n_3}$ is the number of particles in the level with energy $\epsilon_{n_1n_2n_3}$.

Now suppose that the box containing the particles is changed in size, but in such a way as to remain a cube. If the volume change dV is small relative to V, each of the energy levels of Eq. 13.1 changes by the amount

$$\begin{aligned} d\epsilon_{n_1n_2n_3} &= -\frac{\pi^2\hbar^2}{3mV^{3/3}}(n_1^2+n_2^2+n_3^2)\,dV \\ &= -\frac{2\epsilon_{n_1n_2n_3}}{3V}\,dV. \end{aligned} \tag{13.3}$$

Thus if the volume is decreased (dV negative), the energies of all the allowed levels are increased. Given that the microscopic energy spectrum changes when a macroscopic coordinate is altered, what happens to the energy of the system of particles? The answer depends on how the change is made.

In general, a perturbing force such as the motion of the walls of our box will induce transitions between the energy levels of a system. But suppose that the perturbation is carried out very slowly. By the adiabatic principle of quantum mechanics (Section 6.5), we know that a sufficiently slow perturbation cannot induce transitions. More specifically, one can show that transitions between states m and n are unlikely if the rate of change of the Hamiltonian with time is much smaller than $|E_m - E_n|/\tau_{mn}$, where the uncertainty principle shows that the characteristic time τ_{mn} is of the order of $\hbar/|E_m - E_n|$. In principle, one can change the volume of the box slowly enough that no transitions take place; such a change is what we call an *adiabatic perturbation.* Then, whatever the initial distribution of particles over the energy levels, there must still be the same number of particles $N_{n_1n_2n_3}$ in each level after the change as there were before. Given that the $N_{n_1n_2n_3}$ are constant, combining Eqs. 13.2 and 13.3 tells us that the change in the total energy of the system of particles is

$$\begin{aligned} dE &= \sum_{n_1,n_2,n_3} N_{n_1n_2n_3}\,d\epsilon_{n_1n_2n_3} \\ &= -\frac{2}{3V}\left[\sum_{n_1,n_2,n_3} N_{n_1n_2n_3}\epsilon_{n_1n_2n_3}\right] dV = -\frac{2E}{3V} \end{aligned} \tag{13.4}$$

when the volume is changed very slowly. We shall see that this adiabatic perturbation in the quantum mechanical sense corresponds to what is called a *reversible adiabatic process* in thermodynamics.

How is the energy change of Eq. 13.4 exhibited in our model system? The only energy the particles in the box have is kinetic energy. If the energy levels $\epsilon_{n_1n_2n_3}$ increase while the occupation numbers $N_{n_1n_2n_3}$ remain unchanged, then clearly the average kinetic energy per particle must increase, as must the total kinetic energy of the system of N particles. But according to Eq. 12.42, this in turn implies an increase in the temperature of the gas. We have thus deduced from purely molecular considerations that in the (reversible) adiabatic compression of a perfect gas the temperature must increase.

But our model is more restrictive than we need. Equation 13.4 applies strictly only to a gas of point-mass particles in a box with perfectly smooth walls. In any real gas the collisions of molecules with the walls and with each other will not be perfectly elastic, and transitions between energy levels will inevitably take place, even if the walls are not moving at all. Yet in either our model or a pure real gas, the molecules are indistinguishable from one another. All one can hope to know about the gas is how many molecules there are in the various energy levels, not which molecules occupy a particular level. And if transitions between levels occur in such a way that there is no net change in the occupation of the individual levels, then the effect on macroscopic properties is the same as if there were no transitions at all. To obtain Eq. 13.4 and similar results, we need only assume that the $N_{n_1n_2n_3}$ remain unchanged with time.

Yet clearly this constancy of the energy-level populations can only describe the average behavior of the gas molecules over time. A snapshot of this distribution at some one instant of time will differ from one at some other instant, even if the differences must average out in the long run. Such differences between the instantaneous molecular distribution (of energy, velocity, etc.) and the average distribution are referred to as *fluctuations*. Fluctuations of one sort or another necessarily occur in all collections of molecules. When the fluctuations are small—that is, when the deviations from the average populations of the energy levels are small compared to these average populations—the distribution of molecules is for most purposes adequately described by the average values of the occupation numbers $N_{n_1n_2n_3}$. This is the same as saying that a thermodynamic description of the system is valid, since the macroscopic variables of thermodynamics correspond to the average properties of the molecular system. But if the fluctuations are not small, then the ordinary thermodynamic description is not adequate; it must be either supplemented by the use of additional concepts or completely replaced. Further discussion of the importance and magnitude of fluctuations is deferred to Chapter 15.

To obtain results corresponding to those of thermodynamics, then, we ordinarily need to know only the average values of the $N_{n_1n_2n_3}$. Equation 13.4 holds for any process in which these average values remain unchanged and fluctuations can be neglected. We conclude from this equation that, when a box containing N independent particles is compressed adiabatically in the quantum mechanical sense, the total energy of the system of particles increases; when the box is expanded in the same way, the total energy decreases. What are the corresponding energy changes in the surroundings? By the law of conservation of energy, the total energy of the box, the N particles, and the rest of the universe must remain constant. Thus energy in some form must be transferred to or from the particles. One obvious way in which this transfer can take place is in the form of *work*. In order to compress the box, it is in general necessary to exert external forces on it. The product of the force per unit area (the applied pressure p_{app}), the total area of the box, and the (uniform) inward displacement is just the work done on the box by the surroundings; this is the ordinary mechanical definition of work as force times distance. In either mechanics or thermodynamics, work is a form of energy. Thus, changing the volume of the box by dV (compression: $dV < 0$; expansion: $dV > 0$) requires that an amount of energy $-p_{app}\,dV$ be transferred from the surroundings to the box in the form of work. Is this the same as the energy change of the particles inside the box? That depends, of course, on the nature of the box.

The box implied by Eq. 13.1 is of a rather special kind. It has perfectly smooth walls, which affect the particles within it only as boundary surfaces, in that the wave function must vanish at the walls. When the volume of the box is changed, the particles "see" only a change in boundary conditions. In fact, the energies of the particles in such a box can be changed only by moving the walls, that is, only by performing work on the box. Another way to say this is that energy can cross the walls only in the form of work. We shall see that such walls correspond exactly to the adiabatic walls of thermodynamics, which we defined in Section 12.4. If our box has such adiabatic walls, then clearly the increase in energy of the system (box plus N particles) must be exactly equal to the work performed on the box by the surroundings.

How does such a box differ from real boxes? Primarily in that the real boxes are made up of molecules, and these molecules can vibrate. Then it is possible to transfer energy across the walls without moving them. If we impart additional vibrational energy to the molecules on the outside wall, this energy will travel from molecule to molecule until it reaches the inside wall, from which it can be transferred to the gas molecules colliding with that wall. This process, which sounds so complicated, is nothing more than what we call a flow of *heat* across the wall. "Heat," we shall see, is simply a term for energy that crosses the boundary of a closed system in a form other than that of work. A wall that allows such a process to take place corresponds to the diathermal wall of Section 12.4. The problem we raised above is more complicated in a box with diathermal walls. Clearly the total energy change of the system (box and particles) must be equal and opposite to that of the surroundings, but how much of the energy is transferred as work and how much as heat? It is questions like this that thermodynamics was devised to answer.

We see in all this another of the differences between mechanics and thermodynamics. In mechanics, if we wish to measure the work performed on a system, we measure the properties of the system itself—basically, that is, we follow the trajectories of particles. But in thermodynamics we measure work in terms of what flows across the boundaries of a system, that is, in terms of the properties of the surroundings. The two kinds of "work" are thus not conceptually the same, and we must introduce a new definition of thermodynamic work; Section 13.3 deals with this definition, which in turn leads directly to the thermodynamic definition of energy.

13.2 Description of Thermodynamic States

To discuss the thermodynamic definitions of energy and work, however, we must first make more precise a number of concepts, beginning with that of thermodynamic state. Our definition of thermodynamic state in Section 12.3 is equivalent to saying that two states are identical if each and every member of a specified set of macroscopic variables (pressure, density, etc.) has the same values in the two states. We have not yet stated just how the relevant macroscopic variables are to be chosen. One must establish by experimentation, for each class of systems, which and how many macroscopic quantities must be fixed, that is, what constraints must be applied, for given properties of the system to be reproduced. Note that each constraint corresponds to fixing one of the macroscopic variables, which implies that the variables chosen must be independently controllable by the experimenter. Only if the latter condition is met can one say that there is an independent set of variables. To sum up: A thermodynamic state is defined uniquely by specifying the values of a complete set of independent variables, each such set of values corresponding to a distinct state of the system. The number and kinds of independent variables needed to describe a given system must be determined by experiment.

There is a convenient and useful way of visualizing thermodynamic states geometrically. In the absence of external fields, a complete set of independent variables for a fixed mass of a pure fluid is constituted by the pressure p and the density ρ (or the specific volume[1] v). Since these determine the temperature by the equation of state, any two of the variables p, v, T form such a set. In terms of the set v, T, then, any thermodynamic state of the given mass of fluid can be represented as a point on the surface defined by the equation

$$p = f_1(v, T) \tag{13.5a}$$

in the three-dimensional space defined by the axes p, v, T; such a surface is shown in Fig. 13.1. Of course, in this space the same surface can be equally described by the equations

$$v = f_2(p, T) \tag{13.5b}$$

or

$$T = f_3(p, v) \tag{13.5c}$$

Each of the three equations 13.5 arises from solving the equation of state for one of the variables p, v, T, the other two then being the independent variables. Because of the equation of state, fixing any two of the variables p, v, T is sufficient to define a state of the system, represented by a point in the three-dimensional p, v, T space.[2]

We can also construct representations of the equation of state in the (p, v), (v, T), or (p, T) plane. For example, for each possible value of p, Eq. 13.5a defines a different curve in the (v, T) plane; a few such curves are illustrated in Fig. 13.2a. Similarly, families of curves can be obtained for each of the other two variables, as shown in Figs. 13.2b and 13.2c. These families of curves are equivalent to the intersections of the various constant-p, constant-v, or constant-T planes with the surface defined by the equation of state in the full (p, v, T) space.

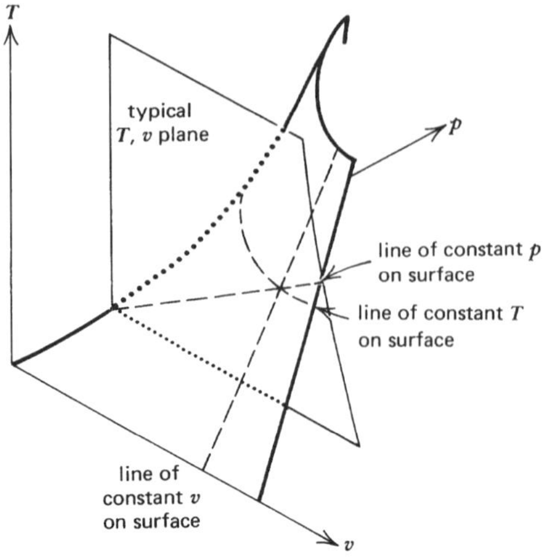

Figure 13.1 Part of a surface representing the equation of state in (p, v, T) space. The particular surface shown is for the perfect gas, with the equation of state $pv = RT$. The lines of constant p, T, and v on the surface are the loci of the intersections of the surface with typical (T, v), (p, v), and (T, p) planes, respectively. A typical (T, v) plane is shown; the others are omitted for clarity.

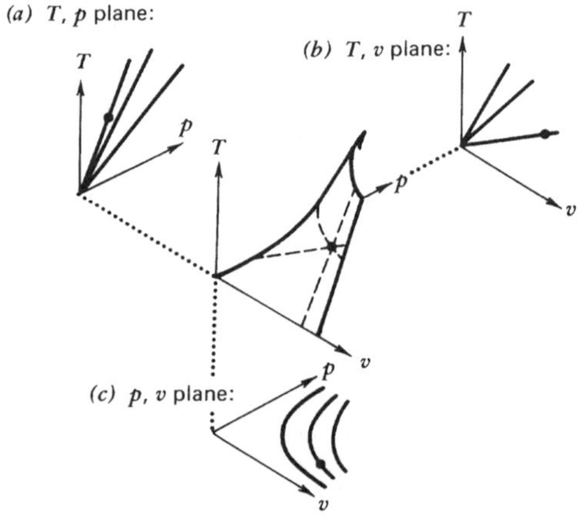

Figure 13.2 Intersections of the perfect gas surface of Fig. 13.1 with planes on which v, p, or T is constant. (a) (T, P) plane—each line in this plane is the locus of the intersection of the p, v, T surface and a plane on which v is constant. (b) (T, v) plane—each line in this plane is the locus of the intersection of the p, v, T surface and a plane on which p is constant. (c) (p, v) plane—each curve (hyperbola) in this plane is the locus of the intersection of the p, v, T surface and a plane on which T is constant.

Every point on the surface defined by Eqs. 13.5 corresponds to a possible equilibrium thermodynamic state of the system, whereas all points in the (p, v, T) space not lying on this surface represent nonexistent states—that is, values of the variables that do not satisfy the equation of state. A change of state corresponds to a shift from one point on the surface to another. Since a change of state is defined simply by the end points (the initial and final thermodynamic states), the same change of state can be carried out by an

[1] That is, the volume per unit mass, which is the reciprocal of the mass density. The term "specific" is generally applied to extensive quantities divided by mass. We shall use exclusively the mole as unit mass, except in Chapter 20, so v will always be the molar volume.

[2] In general, if n independent variables must be specified to determine the thermodynamic state of a system, the corresponding geometric representation requires an $(n + 1)$-dimensional space.

infinite number of different "paths." What we shall later call a *reversible process* is a change of state that can actually be represented by a path lying entirely in the surface, that is, which proceeds through a continuous sequence of equilibrium states; in Section 13.5 we shall examine the properties of such an idealized process.

What about *non*equilibrium states of the system? Strictly speaking, a nonequilibrium state is not represented by a point in the (p, v, T) space at all. The reason is that a system not in equilibrium requires additional variables for a full description of its state. For example, the system may not be homogeneous; then p, v, and T will vary from point to point,[3] and no set of values of these variables can be assigned to the system as a whole. Other possibilities include a system that is spatially homogeneous but that has a nonequilibrium velocity distribution; a system in which a chemical reaction is taking place; or a steady-state system with time-independent properties at any given point, but material flowing in and out. However, any system in which a real change of state is occurring at a measurable rate is necessarily a nonequilibrium system. For example, if a gas is expanded by moving a piston, it takes some time for the resulting pressure change to propagate through the entire gas, and during this time the gas is necessarily inhomogeneous. How, then, can we speak of the "path" of a change of states at all? Fortunately, in many cases a system undergoing a change of state is almost in an equilibrium state, especially when the change of state is slow, and can for some purposes be approximated as being in that equilibrium state.

13.3 The Concept of Work in Thermodynamics

We are now ready to make the thermodynamic concept of "work" more precise. The meaning of the term is similar to that used in mechanics; however, since some of the thermodynamic variables we use do not have direct mechanical analogs, we must discuss several new features of the work process.

Let us recall the mechanical definition of work, introduced in Eqs. 2.23 and 2.24: If an object moves through an infinitesimal displacement $d\mathbf{s}$ under the influence of a force $\mathbf{F}$, the work done on the object is

$$dw = \mathbf{F} \cdot d\mathbf{s} = F_s \, ds, \qquad (13.6)$$

where F_s is the component of $\mathbf{F}$ in the direction of $d\mathbf{s}$. If the object moves through a finite distance, from point 1 to point 2, the total work done upon it is given by the line integral

$$w_{12} = \int_1^2 \mathbf{F} \cdot d\mathbf{s} = \int_1^2 F_s \, ds. \qquad (13.7)$$

Two features of this definition are significant. First, the work depends on the path taken and how the force varies along it; different paths between the same two points entail different amounts of work. Second, work is defined only with respect to motion; the object does not "contain" one amount of work at the beginning of a process, another at its end. Work is a *process variable,* a quantity that passes between systems. We shall see that both these features of mechanical work are also characteristic of thermodynamic work.

There is also an important distinction between mechanical and thermodynamic work to which we now draw attention. In a typical mechanical system, the force acting is derivable from a potential. This will be the case when the following conditions are met:

$$(\partial F_y/\partial x) = (\partial F_x/\partial y), \ (\partial F_z/\partial y) = (\partial F_y/\partial z),$$
$$(\partial F_x/\partial z) = (\partial F_z/\partial x),$$

which is the case when the expression $F_x\,dx + F_y\,dy + F_z\,dz = \mathbf{F} \cdot d\mathbf{s}$ is an exact differential (see Appendix II). When the force is derivable from a potential, the work done in carrying out a displacement is equal to the difference in potential energy between the initial and final points. A mechanical system in which only forces derivable from potentials act is called a conservative system because its energy is conserved. When some or all of the forces acting are not derivable from potentials, the mechanical system is *dissipative.* In the typical thermodynamic system, the forces that act are not derivable from potentials; hence, the work done in carrying out a displacement is not the same as the change in potential energy. However, it is possible, as discussed in Chapter 19, to generalize the definition of potential and to thereby define thermodynamic potentials that are directly related to the amount of work a system is capable of performing under specified conditions.

The simplest (and most commonly considered) type of thermodynamic work is that involved in the expansion or compression of a fluid. Consider a volume of fluid in a container with movable or deformable walls. The pressure the fluid exerts on the walls is what we call p, whereas the pressure applied by the surroundings to the outside of the container is p_{app}. Note that p_{app} need not be a hydrostatic pressure, but can refer to any force exerted on the container, divided by the area to which this force is applied; one example could be a weight placed on a movable piston. In the absence of any constraint fixing the volume (such as rigid walls, or pegs holding a piston in place), at equilibrium we must have $p = p_{\text{app}}$. Suppose now that, from such an initial state of equilibrium, we increase the value of p_{app}. If the walls of the container are free to move, the volume of the fluid must decrease. Consider a small element of the

[3] To speak of values of p, v, T "at a point" implies the existence of *local* equilibrium, that is, the uniformity of the molecular averages to which these quantities correspond over a sufficiently large region (and for a sufficiently long time) for a measurement to be performed.

container's surface, of area dA_i, which moves inward an infinitesimal distance dx_i. The work done on the system by p_{app} is simply the product of the applied force and the distance moved or, for the ith area element,

$$đw_i = F_i\, dx_i = (p_{app})_i\, dA_i\, dx_i = (p_{app})_i\, dV_i \qquad (13.8)$$

where $dV_i \equiv dA_i\, dx_i$ is the change in the volume of the system as a result of the motion of the ith area element; the symbol $đ$ in $đw_i$ will be explained below. Summation over the entire surface of the container gives the total work done on the system for an infinitesimal volume change dV. We now adopt the convention that w is positive when the volume of the system decreases ($dV < 0$); that is, that work is reckoned positive when it is done *on* the system *by* the surroundings. (Later we shall discuss the reasons for this convention.) If we simplify by assuming p_{app} uniform over the surface, this is

$$đw = -p_{app}\, dV. \qquad (13.9)$$

The total work done on the system for a measurable volume change is then just

$$w = -\int_{V_1}^{V_2} p_{app}\, dV, \qquad (13.10)$$

where V_1 and V_2 are, respectively, the initial and final volumes.

We again emphasize that p_{app} is the effective pressure exerted by the surroundings; if work is to be done at a measurable rate, p_{app} cannot be the same as the pressure p within the system. In any case, either p_{app} or p can be varied independently of the volume during the process: p_{app} by adding or removing weights, p by changing the temperature, etc. There is an infinity of such possible variations in p_{app}—that is, of possible paths in the (p_{app}, V) plane—and each such path gives a different value for the integral of Eq. 13.10. This is true even when the paths run between the same initial and final states of the system. The work done on the system is thus incompletely specified if the path by which the process occurs is not specified. In mathematical language we say that $đw$ is an *inexact differential* (cf. Appendix II at the end of the book), in that it is not the differential of some state function $w(p_{app}, V)$; we indicate this by the bar through the d.

When we write a differential in the form $đX$, we imply that the integral $\int_1^2 đX$ between a given pair of end points depends on the path taken. Thus there exists no function X of which $đX$ is the differential, and which could be assigned a unique value in the initial or the final state. Like an object in mechanics, a thermodynamic system in a given state does not "contain" a certain amount of work. To put it another way, work cannot be defined as a function of those variables that determine the thermodynamic state of the system. A variable that can be so expressed is called a *state variable* (or *state function*); examples are p, v, T themselves and (as we shall see) the energy of the system. The change in a state variable between given initial and final states does depend only on those states ($\Delta Y = Y_2 - Y_1$) and is independent of the path followed between them. Remember that in this context "state" refers only to equilibrium states.

Thus far we have considered only the work involved in the expansion or contraction of a fluid (often called "pV work"). As we indicated in Eq. 13.8, this type of work can be interpreted mechanically in a quite straightforward force-times-distance sense. But there are many other possible forms that work can take in thermodynamics, and the relation to mechanical work is often much less clear-cut. To give a single example, the electrical work required to capacitor is found to be

$$w = \int_{Q_1}^{Q_2} \mathscr{E}\, dQ \qquad (13.11)$$

where Q is the charge on the capacitor and $\mathscr{E}$ is the applied potential difference.[4] To include this in the same category as mechanical work and pV work, we need a more general definition of work. We find this in the statement that *the performance of work is equivalent to the lifting of a mass in a gravitational field.*

This statement requires some clarification. Consider the system in Fig. 13.3, in which a fluid expands by lifting a mass atop a frictionless piston. The applied pressure on the piston is the gravitational force mg divided by the piston's area A. As the fluid expands from volume V_1 to volume V_2, the work done on the surroundings is

$$\begin{aligned} -w &= \int_1^2 p_{app}\, dV = \int_1^2 \left(\frac{mg}{A}\right)(A\, dh) = mg\int_1^2 dh \\ &= mg\,\Delta h, \end{aligned} \qquad (13.12)$$

exactly the work done in lifting the mass m a distance Δh. This is, of course, an idealized example, in which the expansion work is applied directly to the lifting of the mass. But imagine next a case of expansion work in which the movable boundary of the system does not move in such a convenient vertical direction. One can still in principal attach to the system walls mechanisms (rods, pulleys, cams, etc.) that will transmit this motion to a vertically moving mass m. If the mechanisms are all ideal (frictionless and weightless), the expansion work can again be completely converted to the work of lifting the mass m. With other kinds of work the process is more complicated. If we discharge the capacitor of Eq. 13.11, for example, we must imagine the resulting cur-

[4] The complementary case of a point charge moving through an electric field has been discussed in Section 1.3; for a charge Q in a field $\mathbf{E}$, the work done on the charge in a displacement $d\mathbf{s}$ is

$$đw = \mathbf{F}\cdot d\mathbf{s} = Q\mathbf{E}\cdot d\mathbf{s} = -Q\, d\mathscr{E}$$

(since electric potential is defined by $d\mathscr{E} = -\mathbf{E}\cdot d\mathbf{s}$, or $\mathbf{E} = -\nabla\mathscr{E}$).

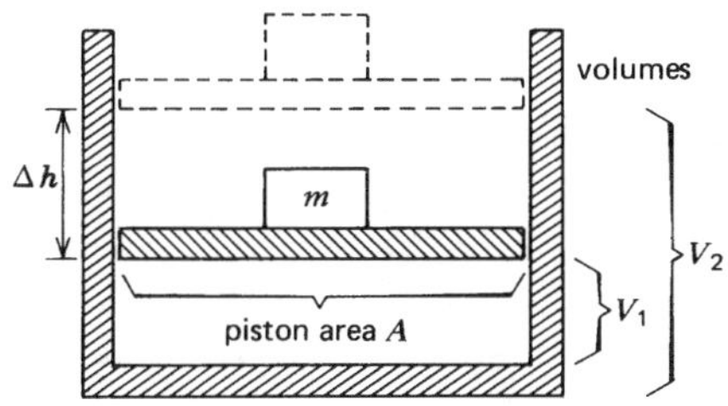

Figure 13.3 Equivalence of work with lifting a mass in a gravitational field. The system is a fluid in a cylindrical container with rigid walls, except for a movable piston atop the container. The piston itself is weightless and frictionless, but is held in place by a mass m. The fluid expands in volume from V_1 to V_2 by upward motion of the piston, thereby lifting the mass m through a vertical distance Δh. (See text for computation of work.)

rent to be operating an ideal electric motor that completely converts the electrical energy to mechanical energy, in turn driving mechanisms of the type already described. For any kind of work one can devise such an idealized process. These processes cannot be carried out perfectly in the real world (where, for example, friction always exists), but for our definition it suffices that they can be described in principle. We can then say that a system performs work on its surroundings if the only effect in the surroundings could be the lifting of a mass. The same reasoning applies when the surroundings perform work on the system, except that the ideal mechanisms must be imagined to be within the system.

The tacit assumption in the above argument is that work is a form of energy, and that various forms of energy can be converted into one another. This is an application of the general physical principle of conservation of energy, which we shall later in this chapter express in thermodynamic terms. The point to note here is that work must have the same dimensions as energy. This is illustrated by the specific examples we have already considered: (force × distance), (pressure × volume), (electric potential × charge) all have the dimensions of energy. There are many other forms of work that may need to be considered in specific chemical problems, and we shall later describe some of these. For now we need only bear in mind that a given system can often perform work (or have work performed on it) in several different ways, and that each case must be examined for its salient features to be determined.

We have stressed the similarity between mechanical and thermodynamic work; just what is the difference? Part of the difference is that thermodynamics can treat a much wider range of variables, but that is not the fundamental point. As we noted at the end of Section 13.1, in mechanics one measures work essentially by observing the trajectories of the particles that make up the system. In thermodynamics, however, we are concerned primarily with what crosses the boundaries of the system. Thermodynamic work is a form of energy that we observe as it enters or leaves a system, identifiable by the fact that it can (in principle) be converted into the lifting of a mass. We do not care what happens to this particular energy before or after it crosses the boundary, only with the crossing process itself. As we shall see, the fundamental question is how much of the total energy flow across the boundary consists of work.

13.4 Intensive and Extensive Variables

Continuing our development of concepts needed for the thermodynamic description of a system, let us now take a closer look at the variables used in thermodynamics. The thermodynamic variables we have used can be classified into two types. Some, such as the temperature or the pressure, are independent of the mass of the system; these are called *intensive variables.* Others, such as the volume or the internal energy, are proportional to the mass of the system when the intensive variables are held constant; these are called *extensive variables.* That is, if we combine two identical systems into one by removing a barrier, the temperature and pressure will remain unchanged, but the final volume and energy will be twice that of either original system. Another way to define the difference is that extensive variables are inherently properties of the system as a whole, whereas intensive variables can be measured at a point within the system. In simple though imprecise terms, the bigger a system is, the more one has of the extensive variables; if a system is divided into subsystems, the extensive properties of the whole system are the sums of those of the subsystems. Intensive and extensive variables play different roles in the description of a system, and we shall see that this difference is important in defining the key concept of the reversible process.

Each intensive variable X has what is called a *conjugate* extensive variable Y, the two being related by a particular work process. Suppose that X has the same value inside and outside a system boundary (e.g., $p = p_{\text{app}}$); the conjugate extensive variable Y is then a quantity such that an infinitesimal change dY does an amount of work

$$đw = X\,dY. \qquad (13.13)$$

Using this relationship, we can see from Eqs. 13.9 and 13.11 that volume is conjugate to pressure, and electric charge is conjugate to potential. There are many other such pairs,[5] some of which are listed in Table 13.1: Note that the product of each such pair of variables must have the dimensions of energy.

Conjugate extensive and intensive variables correspond to the *generalized coordinates* and *generalized forces* of mechanics, respectively. If the potential energy of a mechanical system is expressed as a function of several independent variables q_i (the generalized coordinates), then the generalized force Q_i conjugate to a particular q_i is defined as the

[5] The extensive variable conjugate to temperature is the entropy, which we shall not define until Chapter 16.

Table 13.1 Conjugate Intensive and Extensive Variables and Work

System	Intensive Variable	Extensive Variable	Work Done on System[a]
Fluid (or other system that can expand and contract)	Pressure (p)	Volume (V)	$-\int p\, dV$
Surface film	Surface tension (σ)	Area (A)	$\int \sigma\, dA$
Wire	Tension (F)	Length (l)	$\int F\, dl$
Capacitor	Potential ($\mathscr{E}$)	Charge (Q)	$\int \mathscr{E}\, dQ$
Electrochemical cell	emf ($\mathscr{E}$)	Charge (Q)	$\int \mathscr{E}\, dQ$
Paramagnetic solid	Magnetic field strength ($\mathbf{H}$)	Magnetic moment ($\mathbf{m}$)[b]	$\mu_0 \int \mathbf{H} \cdot d\mathbf{m}$

[a] This column gives the work done on the system only when the intensive variable has the same value inside and outside the boundary (i.e., in a reversible process).
[b] This is the same quantity that we have called $\boldsymbol{\mu}_m$ for a single atom or molecule (Section 9.6).

negative partial derivative of the potential energy with respect to q_i, with all the other coordinates held fixed. Similarly, we shall see that the intensive variable X conjugate to a given extensive variable Y is defined by $X \equiv -(\partial U/\partial Y)$, where U is the internal energy and all other independent extensive variables are held fixed. The differential work in any mechanical process can be expressed as a sum of terms of the form $Q_i\, dq_i$ (where dq_i is a *generalized displacement*), corresponding to Eq. 13.13. In both cases, the product of the two members of a conjugate pair (XY or $Q_i q_i$) must have the dimensions of energy.[6]

In general, the boundaries of a system (or portions thereof) may or may not allow the performance of various kinds of work. For example, expansion work can be performed at a movable boundary, but not at a rigid one; electrical work can be performed at a conducting boundary, but not at an insulating one; and so forth. The performance of a given type of work is always associated with a change in the corresponding extensive variable (cf. Table 13.1) in the system or the surroundings or both: volume changes in expansion work, flows of charge in electrical work, and so on. Whenever a boundary allows the performance of a particular kind of work, including change in the corresponding extensive variables, we can say that the same boundary "transmits" the conjugate intensive variable. By this we mean that at equilibrium the intensive variable in question has the same value on both sides of the boundary—in the system and its surroundings.[7]

Suppose, for example, that part of a system's boundary is displaceable, allowing the performance of expansion work. When such a boundary moves outward, volume changes occur in both the system and the surroundings. The boundary must therefore transmit pressure, the variable conjugate to volume, so that the equilibrium state has the same pressure on both sides of the boundary. And we know that this is indeed true for any movable boundary in mechanical equilibrium, when the forces on each area element must be in balance.

On the other hand, if a given boundary does not allow the performance of a particular kind of work, then in whatever equilibrium state is attained, the corresponding intensive variable need not have the same value in system and surroundings. To continue our example, if part of a system's boundary is rigid (not allowing expansion work), then at equilibrium the pressures inside and outside the boundary need not be the same.

For a system initially at equilibrium to undergo a change of state, its boundaries must be capable of transmitting some intensive variable. That is, any such change of state must involve the passage of either work or heat across the system's boundaries. We have already discussed the various work processes; if none of these is possible, the system can interact with its surroundings only by the flow of heat, that is, by transmission of the intensive variable temperature through a diathermal wall. If neither work nor heat flow is possible, the system is what we have defined as isolated.

13.5 Quasi-static and Reversible Processes

We now return to a consideration of the path by which a change of state takes place. We have earlier pointed out that the change in any state function (volume, internal energy, etc.) depends only on the initial and final states, not on the path between them. Why, then, are we interested in the detailed nature of this path? The answer is that the work and, as we shall see, the heat associated with a given process does depend on the path. The original stimulus for thermodynamics was the study of heat engines, in which the amount of work obtained is clearly all-important. Even when one merely wishes to accomplish a given change of state, one must know *how* best to do it—that is, what routes between the initial and final states are available, and what amounts of work and heat are associated with each route. We shall be concerned with two categories of paths: reversible and irreversible.

Consider, first, a reversible path followed by a system between specified initial and final states. By definition, every point along a reversible path is an equilibrium state of the system, so the sequence of states through which the system passes can be represented by some curve in the kind of space defined in Section 13.2. Both the definition and the representation of a reversible path immediately present a problem, since an equilibrium state is defined as one that

[6] Compare our discussion in Section 3.7 of the *generalized momentum* p_i conjugate to a given generalized coordinate q_i; the product of p_i and q_i, however, must have the dimensions of action (energy × time).
[7] This is not a new result, but merely a restatement of our definition of conjugate variables.

does not change measurably with time. This means that any process occurring at a measurable rate cannot be described by a path in the equilibrium surface. Since any real change of state must occur at a nonzero rate, how do we describe it?

One obvious way to deal with this problem is by a limiting process. Imagine that in repeated experiments a given change of state, say, the expansion of a fluid, is carried out more and more slowly. In the limit we have a process occurring at an infinitesimal rate of change; such a process is called *quasi-static*. As with more rapid processes, a given change of state can be carried out quasi-statically by infinitely many routes: The particular limiting behavior one obtains depends on the conditions, or constraints, placed on the system. Quasi-static processes can be reversible or irreversible; for the case under discussion we restrict attention to reversible quasi-static processes.

The definition of a reversible process requires consideration of how a real process can remain at all times infinitesimally close to equilibrium. The slower the process is carried out, the closer the system can be to an equilibrium state at any given time, the limit being an infinitely slow process in which the system passes through a continuous sequence of equilibrium states.[8] Such a process obviously does have a describable path in the space of equilibrium states, but how does this help us in the case of real processes occurring at measurable rates? The answer is that a very large class of real processes have characteristics (such as the values of work and heat involved) so close to those for the corresponding reversible processes that the latter can be assumed for purposes of calculation. The implication of this assumption, as we shall see later, is that the deviation from equilibrium brought about by the nonzero rate of change tends to zero more rapidly than the rate of change itself, so that during the process the system is always very close to equilibrium. For any given process, however, the validity of this assumption must ultimately be based on experiment.[9]

Consider, now, an irreversible path followed by a system between specified initial and final states. Although the system passes through nonequilibrium states, the work done in traversing that path is calculable, provided p_{app} is known as a function of V.

All this has been very abstract, so let us now give a concrete example. Consider the expansion of a perfect gas from volume V_1 to volume V_2. How can this process be carried out with the gas close to equilibrium at all times? Assume that the expansion is performed by moving a frictionless piston, as in Fig. 13.3, and that initially the external pressure p_{app} equals the internal pressure p. Then decrease p_{app} by a small amount Δp, and the gas will begin to expand. Maintain the relationship $p_{\text{app}} = p - \Delta p$ throughout the expansion (with both p and p_{app} steadily decreasing), and when the desired final volume is reached, set p_{app} again equal to p to stop the expansion. All this is clear enough. Now repeat the process with a smaller value of Δp; since the driving force (equal to Δp times the piston area) is less, the expansion will take place more slowly. In the limit $\Delta p \to 0$, we have an infinitesimal rate of expansion with $p_{\text{app}} = p$ at all times. Since the expansion is so slow, the pressure and temperature remain uniform throughout the gas at all times, and the gas passes through a continuous sequence of equilibrium states. In short, the limit we have described is a reversible expansion of the gas.

To calculate the work performed in such a process, we must specify at least one additional condition, since the state of the gas is described by two independent variables, and we already have the condition $p_{\text{app}} = p$. The simplest case is that of a constant-temperature expansion, which can be performed by giving the system diathermal walls and immersing it in a thermostat. Then, from Eqs. 13.10 and 12.36, we obtain

$$\begin{aligned} w &= -\int_{V_1}^{V_2} p_{\text{app}}\, dV = -\int_{V_1}^{V_2} p\, dV \\ &= -\int_{V_1}^{V_2} \frac{nRT}{V}\, dV = -nRT \ln \frac{V_2}{V_1} \end{aligned} \tag{13.14}$$

for the work performed on the system in the reversible, isothermal expansion of a perfect gas. If we represent a fluid's pressure as a function of volume by a curve in the (p, V) plane, then the magnitude of the work performed in a reversible expansion is simply the area under this curve, as shown in Fig. 13.4. For the isothermal expansion of a perfect gas, the curve AB is part of one of the hyperbolas of

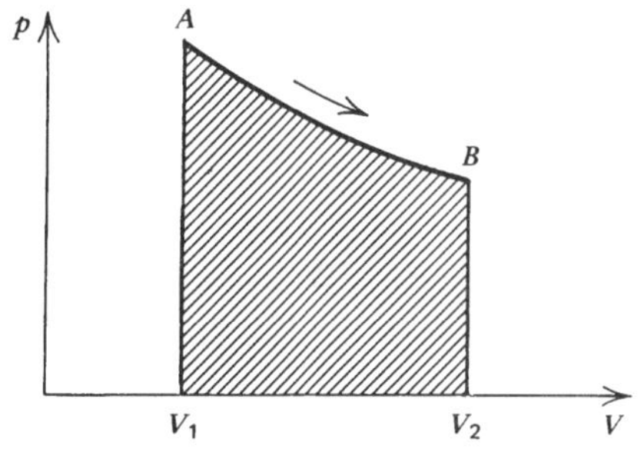

Figure 13.4 Work performed in the reversible expansion of a fluid. The heavy line AB gives the value of p as a function of V during the expansion from V_1 to V_2; the work performed on the fluid is then

$$w = -\int_{V_1}^{V_2} p\, dV,$$

equal to the negative of the area under the curve AB.

[8] Anticipating the discussion of the magnitudes and distributions of fluctuations of the thermodynamic variables of a system presented in Chapter 19, a complementary view of a reversible process can be stated as follows. The values of the thermodynamic variables of a system at equilibrium—e.g., the energy, undergo continuous fluctuations. If the state of a system is changed so slowly that the thermodynamic variables always remains within a standard deviation of their mean values, no single observation can ever show that the state of the system has changed, because the thermodynamic variables are always within the ranges expected for the equilibrium state.

[9] This is true as long as one remains within the bounds of thermodynamics itself. In principle one should be able to determine the limits of validity from a microscopic theory, although this is not always feasible in practice.

Fig. 13.2*c*. Similar graphical representations are easily derived for the other forms of work in Table 13.1. However, Fig. 13.4 is no longer valid if $p_{app} \neq p$, which we shall see means that the process is not reversible.

Note that in the above-described reversible process, (1) expansion work is performed and (2) the intensive variable associated with expansion work, the pressure, has the same value on both sides of the system boundary. This illustrates a general property of reversible processes: The intensive variables corresponding to the types of work being performed (cf. Table 13.1) must be continuous across the boundary at which that work is performed. It is easy to see why this must be so: We pointed out in the last section that at equilibrium an intensive variable has the same value on both sides of a boundary that transmits that variable. But since in a reversible process the system is effectively always in equilibrium, this continuity of intensive variables across the boundary must then also hold at all times. The last column of Table 13.1 therefore gives the work performed on the system in a reversible process of the type specified.

We can now give a complete definition of a reversible process: (1) It takes place at an infinitesimal rate—i.e., it is quasi-static; (2) it passes through a continuous sequence of equilibrium states, and can thus be exactly described by a path in, say, (p, v, T) space; (3) at every point in the process, any intensive variable is continuous across a boundary at which the type of work corresponding to that variable is being performed.

Why is such a process called "reversible"? As already indicated, every point along the path is an equilibrium state. But this means that the direction of the process can be reversed by an infinitesimal change in the external constraints. Suppose, for example, that the process is the reversible expansion of a gas, in which p_{app} must equal (or rather, be infinitesimally less than) p at all times; if p_{app} were made infinitesimally *greater* than p, the gas would contract rather than expand. The direction of a reversible process thus cannot be determined from a description of the instantaneous state of the system. This is an idealization, since all real changes of state are *irreversible processes*—processes during which the system passes through nonequilibrium states, with unbalanced "forces" giving the process a unique direction. We shall have more to say about irreversible processes later.

It is clear that a given reversible path can be traversed in either direction, the details of the process being identical except for the sign of change. Thus if the work done in the forward process is $w_f = -\int_{V_1}^{V_2} p\, dV$, the work done in the corresponding reverse process is $w_r = -\int_{V_2}^{V_1} p\, dV$; since the path is the same, p is the same function of V in both cases, and we must have $w_f = -w_r$. We also have $(\Delta X)_f = -(\Delta X)_r$ for any state function X, but this would be true whether the path were the same or not; however, the equality of forward and reverse work is valid only for reversible processes.

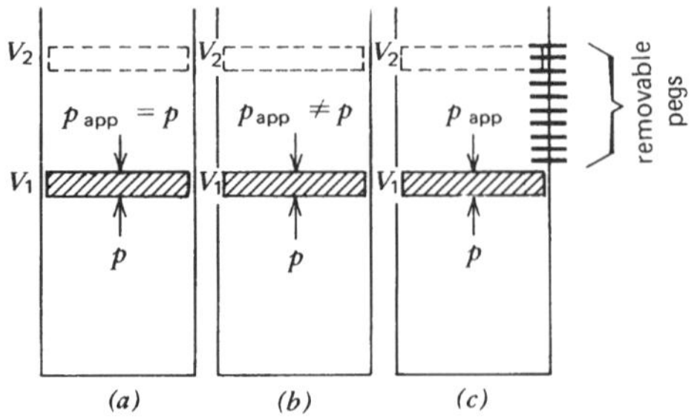

Figure 13.5 Reversible, irreversible, and quasi-static expansion of a fluid. (*a*) Reversible (and thus quasi-static): $p_{app} = p$ at all times, and the expansion proceeds infinitely slowly. (*b*) Irreversible: $p_{app} \neq p$, and the expansion proceeds at a measurable rate. (*c*) Quasi-static but irreversible: See text.

One final point: If a reversible process consists of a continuous sequence of equilibrium states and takes place infinitely slowly, then what is the difference between a system actually in equilibrium and a system undergoing a reversible process? The answer is that true equilibrium is a time-independent state of a system that is exchanging neither work nor heat with its surroundings—a system that either is isolated or might as well be.[10]

Let us now look at some processes that are not reversible. Consider again our earlier example of the expansion of a perfect gas, but this time omit the requirement that $p_{app} = p$. Presumably we are able to set p_{app} to any desired value. In general, the driving force, $A(p - p_{app})$, will be appreciable, and the expansion will occur at a nonzero rate—in short, it will be irreversible (cf. Fig. 13.5*b*). Clearly, Eq. 13.14 no longer gives the work performed, since the substitution $p_{app} = nRT/V$ is not valid unless $p_{app} = p$. Equation 13.10 remains applicable. In computing the work done, however, one must use the actual value of the applied pressure p_{app}. If conditions are such that p_{app} is not actually fixed or measured, calculating the work performed can be a very complicated problem. In particular, the work no longer equals the area under the $p(V)$ curve in Fig. 13.4, but the area under a similar $p_{app}(V)$ curve. It is even possible for *no* work to be performed in an irreversible process, as when a gas expands into a vacuum ($p_{app} = 0$). The nature of irreversible processes requires some careful examination, and we shall return to this problem in Chapter 16.

We have said that reversible processes are only one class of quasi-static processes. This implies that a process can take place at an infinitesimal rate (quasi-statically) and still be irreversible. How can this be? Consider the system in

[10] Anticipating the discussion of Chapter 19, we note that thermodynamic equilibrium is the state that requires the minimum number of variables to specify the distribution functions of any and all of the microscopic properties of the system.

Fig. 13.5*c*, in which a fluid is impeded from expanding by a series of pegs holding the piston in place. If the lowest of these pegs is removed, the fluid will expand irreversibly—but only as far as the next peg; this process can then be repeated over and over. If we imagine more and more pegs installed closer and closer together, the expansion can be carried out as slowly (and apparently continuously) as we like, in the limit quasi-statically. Yet no matter how slowly such a process takes place, it does not become reversible. Rather than passing through a continuous sequence of equilibrium states, the system is in equilibrium only during those moments when the piston is resting against a peg; during the actual expansion steps we presumably have $p_{app} = 0$, and only the pegs hold back the expansion.

A somewhat different type of quasi-static irreversible expansion would be one in which the piston does not move without friction: The work in this case is different from that in a reversible expansion because some of the energy is dissipated as heat. But to demonstrate this we need the first law of thermodynamics, which we are finally ready to introduce.

13.6 The First Law: Internal Energy and Heat

The first law of thermodynamics is an abstraction and generalization of experiment, based on the concepts of work and thermodynamic state. Consider a system enclosed by adiabatic walls, so that its state can be changed only by the performance of work. Since all real materials have some thermal conductivity, no real system can be so enclosed, but real systems exist that closely approximate the ideal one described. Many experiments on such systems lead to the conclusion that the amount of work needed to produce a given change of state is independent of the path. This conclusion is so important that we state it formally:

If the state of an otherwise isolated system is changed from A to B by the performance of work, the amount of work required depends solely on the initial state A and the final state B, and not on the means by which the work is performed, nor on the intermediate stages through which the system passes between the initial and final states.

This is our statement of the first law. The initial and final states referred to must, of course, be equilibrium states.

It is not immediately obvious that this abstract statement is equivalent to the law of conservation of energy. Using it, however, we are at last ready to introduce a thermodynamic definition of energy. In terms of only macroscopic operations, we define the *internal energy U* of a system by the statement:

If an otherwise isolated system is brought from one state to another by performance upon it of an amount of work w_{ad}, the change in the system's internal energy in the process is defined to be an amount ΔU exactly equal to w_{ad}.

(Here w_{ad} stands for "adiabatic work.") Formally, then, for work w_{ad} performed on an otherwise isolated system, we have

$$\Delta U = U_B - U_A = w_{ad}\,, \tag{13.15}$$

where *A* and *B* designate the initial and final (equilibrium) states, respectively. Note that ΔU is positive when work is done *on* the system *by* the surroundings, the condition for which we have set *w* positive by convention. Equation 13.15 holds for both reversible and irreversible work processes.

Two observations must be made here. First, our definition gives only the *change* in internal energy in a given process; it does not establish the zero of energy, which as usual is arbitrary. However, once the internal energy has been fixed arbitrarily at some value U_0 for a given equilibrium state of a system, the value of *U* for every other equilibrium state of that system is uniquely determined. We demonstrate that this is so in Chapter 16, in conjunction with a discussion of the Second Law of Thermodynamics. Second, the first law requires that the work needed to produce a given change of state in an otherwise isolated system be independent of the path. Therefore, ΔU for the same change of state must also be independent of the path, depending only on the initial and final states. In other words, the internal energy *U* is a state function. Like any other state function, *U* must depend only on the small number of macroscopic variables that define the equation of state. Thus we can write for a fluid

$$U = U_1(p,v) = U_2(p,T) = U_3(v,T), \tag{13.16}$$

where the functions U_1, U_2, U_3 differ in form but, since they represent the same state function, must have the same value in any given equilibrium state of the fluid.

As already mentioned, real systems are never enclosed by perfectly adiabatic barriers. Consider now a system whose containing walls are diathermal. When such a system undergoes a given change of state, the work *w* performed upon it is in general different from the work w_{ad} performed when the same change of state occurs in a system bounded by adiabatic walls. (By "the same change of state," remember, we mean a process between the same initial and final equilibrium states.) Whereas by the first law the adiabatic work w_{ad} is independent of the path, the corresponding diathermal work *w* does depend on the path. We can therefore *define* a quantity *q* by the relationship

$$q \equiv w_{ad} - w; \tag{13.17}$$

we call q the *heat* transferred to the system in the diathermal process. Using Eq. 13.15 for the adiabatic process, we can at once write

$$q = \Delta U - w, \tag{13.18a}$$

or, in more common form,

$$\Delta U = q + w. \tag{13.18b}$$

ΔU, the change in a state function, is of course the same in both processes. See Fig. 13.6 for a graphical representation of the relationships among ΔU, w, and q.

Equations 13.18 uniquely define the heat associated with a given process in terms of the work performed and the change of state produced (which determines ΔU). We have mentioned that work performed on the system by the surroundings is taken to be positive. The corresponding convention for heat, embodied in Eq. 13.17, is that heat transferred *to* the system *from* the surroundings is taken as positive. (More on these sign conventions later.) Heat, like thermodynamic work, is a form of energy that we observe only as it crosses the boundaries of a system. In fact, Eqs. 13.18 define q as the amount of (internal) energy that enters a system in a given process in forms other than the performance of work. These equations can thus be taken as a statement of the first law as the law of conservation of energy, in that energy is conserved if both heat and work are taken into account. This was essentially the original formulation of the first law, but in our version the concept of heat is secondary to that of internal energy.

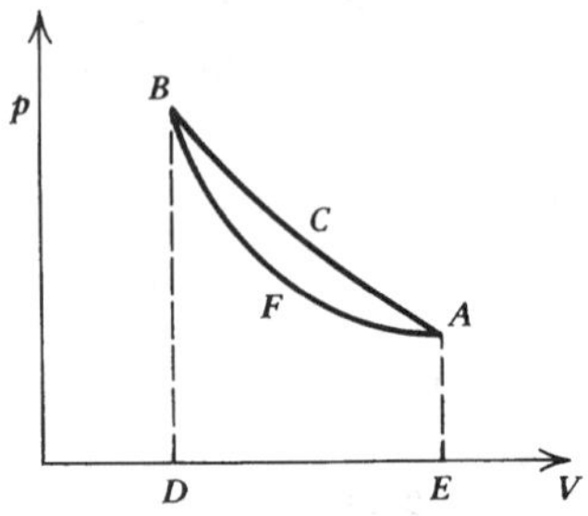

Figure 13.6 Graphical representation of internal energy, work, and heat. Consider a given change of state, in this case the reversible compression of a fluid from state $A(p_1, V_1)$ to state $B(p_2, V_2)$. Assume that ACB is the path followed when this change of state is carried out *adiabatically*. The adiabatic work w_{ad} associated with this change of state is then equal to the area $ACBDEA$, and by Eq. 13.15 this must equal the internal energy change:

$$\Delta U = U_B - U_A = w_{ad} = \text{area } ACBDEA$$

Now consider any other path AFB along which the same change of state can be carried out diathermally. The work performed on the system in the diathermal process is then equal to the area $AFBDEA$, the internal energy change is the same as in the adiabatic process, and the heat q absorbed by the system must equal the area between the two paths:

$$q = w_{ad} - w = \text{area } ACBDEA - \text{area } AFBDEA = \text{area } ACBFA$$

In this case q and w are both positive.

Although the internal energy is a state function, this is clearly true of neither work nor heat. We have already pointed out that the work associated with a given change of state depends on the path; by Eq. 13.18*a* this must also be true of the heat associated with the same change of state. Thus, dU is an exact differential, whereas both work and heat have inexact differentials. In the notation introduced in Section 13.3, we can write

$$dU = đq + đw \tag{13.19}$$

for an infinitesimal change of state. We can also write for a *cyclical* process (one that returns to its starting point)

$$\oint dU = 0 \tag{13.20}$$

where the symbol $\oint$ designates integration over a cyclical path; in Chapter 16 we shall study the relationship between work and heat in such processes. Finally, for a given *reversible* path we showed in the previous section that the work done in corresponding forward and reverse processes is related by $w_f = -w_r$; since we must also have $(\Delta U)_f = -(\Delta U)_r$, we immediately obtain $q_f = -q_r$, again for reversible processes only.

For a system undergoing only reversible processes, we can write Eq. 13.19 in the generalized form

$$dU = đq - \sum_i X_i\, dY_i, \tag{13.21}$$

where X_i, Y_i are the conjugate pairs of intensive and extensive variables involved in the various possible work processes (Table 13.1). We can now see why we can define the intensive variable X_i conjugate to a given extensive variable Y_i by

$$X_i \equiv \left(\frac{\partial U}{\partial Y_i}\right)_{\text{rev,ad},Y_j=\text{const}}. \tag{13.22}$$

where the subscripts identify a reversible adiabatic process in which all other independent extensive variables Y_j are held fixed.[11]

Remember, however, that the first law is not restricted to reversible processes. The adiabatic work defining ΔU is the same for any process, reversible or irreversible, that joins the

[11] We shall see in Chapter 16 that it is not necessary to make a separate assumption of adiabaticity: in a reversible process we have $đq = T\,dS$, where S is the extensive variable entropy.

same initial and final states—provided only that those states are equilibrium states. This is fortunate for the chemist, since a chemical reaction ordinarily takes place irreversibly. Yet, as we shall see in Chapter 14, the first law can indeed be applied to such reactions.

13.7 Some Historical Notes

We have mentioned that the first law was originally formulated in terms of the conservation of energy. One of the cornerstones of modern thermodynamics was thus the demonstration that work could be converted into heat, with all forms of work giving equivalent amounts of heat. It is of interest to sketch some of the background of this principle.

It was Count Rumford who first showed, in 1798, that mechanical work could be continuously converted to heat (by friction, in the boring of cannon). In 1799 Humphry Davy performed a similar experiment: When two pieces of ice were rubbed together, the ice melted.[12] Later experiments made it increasingly clear that heat was in some way a form of energy, where "energy" means roughly the ability to do work. The conservation of energy was already a well-established principle for mechanical systems in the absence of friction (heat). Such systems therefore were called "conservative." It was natural to extend this principle to include heat, but it took some time for the necessary experimental evidence to be obtained.[13] In 1842 J. R. von Mayer first formulated the general principle of the equivalence of different kinds of energy and the conservation of total energy. However, it was primarily the work of J. P. Joule in the next few years that provided the quantitative basis for the law of conservation of energy.

Joule used various ways of producing heat from work, the best known being the rotation of a paddle wheel in a liquid, with the paddle wheel driven by a falling weight. He also measured the heat produced when an electric current passes through a resistance, when bodies are rubbed together, in the expansion and contraction of air, and in other processes. In general, the work done was measured in mechanical or electrical terms, whereas the heat produced was measured by the rise in temperature of a known mass of substance (see Chapter 14). Joule's results established the existence of a consistent proportionality between the mechanical work w dissipated in a system and the heat q produced as a result:

$$w = Jq \tag{13.23}$$

where the quantity J is called the *mechanical equivalent of heat.*[14]

In the older literature J is a conversion factor between work in mechanical units (ergs or joules) and heat in *calories,* one mean 15° calorie ($\text{cal}_{15°}$) being the heat required to raise the temperature of 1 g of water from 14.5 to 15.5°C at atmospheric pressure. In terms of this unit Mayer had estimated a value of J equivalent to 3.6 $\text{J/cal}_{15°}$, whereas Joule obtained (1849) a value of 4.15 $\text{J/cal}_{15°}$; modern measurements give

$$1\ \text{cal}_{15°} = 4.1855\ \text{J}.$$

The modern convention, however, is to define the calorie directly in terms of work, rather than as a function of the properties of water. The definition currently used by chemists, known as the *thermochemical calorie* (cal_{th} or simply cal), is

$$1\ \text{cal}_{\text{th}} \equiv 4.1840\ \text{J},$$

so that 1 $\text{cal}_{15°} = 1.00036\ \text{cal}_{\text{th}}$. (Several other slightly different calories have also been used, so one should be careful when using published data.) The existence of separate energy units for heat and work is superfluous, and the calorie is being phased out of use; however, many of the thermochemical data in the literature are still expressed in terms of calories.

The significance of Eq. 13.23 can be expressed in several ways. One of the most incisive is to regard a thermodynamic system as a reservoir of energy. Then the convertibility of work and heat implies that any energy passing inward across the system's boundaries increases the total amount of energy present, no matter how it enters the system: by performance of mechanical work, transfer of heat, or any other method. This view is equivalent to saying that heat is just energy in transit to or from the system. It is clear that this interpretation is closely associated with the definition of heat given by Eq. 13.17. Furthermore, since heat is defined only for a particular type of energy-transfer process, there is no meaning to the phrase "heat in the system"; this is, of course, the same conclusion we reached earlier for work.

We should say something about the history of the sign conventions for heat and work. In this text, it will be

[12] Recent historical research has revealed that, although Davy's conclusion was valid, his experimental technique was faulty, and the observed melting was caused by heat leaks from the surroundings.

[13] In the early development of thermodynamics, for some time it was supposed that heat is conserved. However, the realization that the generalized quantity we call energy can be given a precise definition, and that it is this abstract quantity which is conserved, is the subtle and brilliant observation that permits formulation of the First Law of Thermodynamics. It required an astounding intellectual advance, generating a conceptual revolution, to realize that "energy" can appear in so many apparently unrelated and qualitatively different forms.

[14] For electrical work Joule found that

$$I^2 Rt = Jq,$$

where I is current, R is resistance, and J has the same value as for mechanical work. The quantity I^2Rt is still known as the "Joule heat" produced by an electric current.

recalled, we use the conventions that w is positive for work done *on* a system, and q is positive for heat absorbed *by* a system. Many textbooks use the opposite sign convention for work, considering work done *by* a system as positive; given this other convention, the first law assumes the form $\Delta U = q - w$. Historically, thermodynamics evolved in large part in response to the stimulus imparted by the introduction of the steam engine and the need to understand the performance of engines. Since engines are constructed for the purpose of performing a useful task, it was natural for work done by an engine to be reckoned positive. This was generalized to the assignment of a positive sign to work done by any system on its surroundings. Today, however, scientists are generally more concerned with the thermodynamic properties of a system itself, particularly its internal energy. The sign of U is fixed by the notion that to do work a system must "expend energy." Thus, U is defined by Eq. 13.15 in such a way as to increase when adiabatic work is done on the system and decrease when the system does adiabatic work on the surroundings, and it now becomes natural to reckon positive that work which *increases* a system's internal energy, that is, work done on the system. Although this convention has received international recommendation, the use of the older convention is still widespread, especially among engineers; one should be careful to note which is used in any thermodynamics text one consults.

The history of the sign convention for heat is more straightforward. The engines with which early thermodynamics dealt were driven by the addition of heat to the system, usually by building a fire under the boiler; thus it was natural to consider heat added to a system as positive. The shift in emphasis from engines to system properties required no change in this convention, since addition of heat to a system also increases its internal energy. In short, both w and q are now defined as positive when they tend to increase the internal energy of the system under consideration.

13.8 Microscopic Interpretation of Internal Energy and Heat

Let us now return to the microscopic level of description. Our thermodynamic definition of a system's internal energy (Section 13.6) is based on adiabatic work processes. "Adiabatic" here is used in the thermodynamic sense, implying that energy crosses the system boundary only in the form of work. How does this correspond to the "adiabatic perturbation" of Section 13.1? The process described there, expansion of a perfect gas, must be thermodynamically adiabatic, since the container walls act only as elastic reflectors for the gas molecules. But the adiabatic perturbation there is also so slow that no transitions take place between energy levels; the gas is thus always in an equilibrium state, and the process is reversible in the thermodynamic sense. We conclude that the adiabatic perturbation of quantum mechanics corresponds to the reversible adiabatic process of thermodynamics. The discrepancy between the two senses of "adiabatic" is unfortunate, but seldom leads to confusion; henceforth we shall ordinarily use the thermodynamic sense only.

Thus, we have a microscopic equivalent of a reversible adiabatic work process, as described for a perfect gas model in Section 13.1. By simply reformulating our definition, we can interpret the change in the system's internal energy as the total change in mechanical energy of the system's constituent particles. For the perfect gas model we saw that a reversible adiabatic compression leads to an increase in the average kinetic energy per particle, but with the distribution of particles over the energy levels remaining unaltered. Using the same model, we can now consider the microscopic interpretation of the heat transferred in nonadiabatic processes.

In general, when a system interacts with its surroundings so that both work and heat are exchanged, both the microscopic energy spectrum and the distribution of particles over the energy levels change. Suppose, for example, that our model perfect gas expands with the temperature fixed. The energy spectrum of the perfect gas, Eq. 13.1, depends only on the dimensions of the box, not on how those dimensions are established; when the box size increases, the energy of each level decreases. We have identified the temperature with the average kinetic energy; for this quantity to remain constant in an expansion, some particles must be promoted to higher energy levels. We showed in Section 13.1 that the average kinetic energy decreases in a reversible adiabatic expansion. For a reversible expansion to the same volume with the average kinetic energy fixed, clearly some energy must be added from the surroundings. It is this added energy, which must be transferred as kinetic energy of the atoms in the box walls, that corresponds to thermodynamic heat.

We have therefore reached these important conclusions: In a reversible adiabatic work process, the distribution of molecules over a system's energy levels remains unchanged; in any other reversible process, the heat transferred across the system boundary is related to the change in the distribution of molecules over energy levels.[15] In an irreversible process, there is in general a change in the molecular distribution whether heat is transferred or not, since the work performed differs from that in the corresponding reversible process. As with the conclusions of Section 13.1, we shall demonstrate later that these statements are valid generally and not merely for the perfect gas model. To quantify the argument, we must first study the second law of thermodynamics and introduce some macroscopic measure of the distribution of molecules. This will be done in Chapters 15 and 16 with the introduction of the concept of entropy.

[15] Of course, the work done in the two processes is also different. The relationship between work done, heat transferred, and changes in the distribution of molecules over a system's energy levels is discussed in Chapter 15.

We can draw one additional macroscopic conclusion from our microscopic perfect gas model. Suppose again that the volume of the box is changed with the temperature held constant. But for a perfect monatomic gas, the only energy attributable to the molecules in the box is their kinetic energy. Since the number of molecules is constant, we conclude that the total energy of the gas does not depend on the box's volume in a constant-temperature process. This result for a quantum mechanical perfect gas is in fact the same as our conclusion in Section 12.6 for a classical perfect gas. By identifying our microscopic model with a macroscopic gas obeying the ideal gas law (an identification that is now confirmed), we obtained Eq. 12.42, stating that

$$U = \tfrac{3}{2} nRT. \tag{13.24}$$

In Eq. 12.42, however, U stood only for the kinetic energy of the molecules; now we can identify U with the thermodynamic internal energy.[16] Given this identification, we can say that the internal energy of a perfect gas depends only on temperature, and is independent of density (volume) or pressure. If one wishes to define a perfect gas without any microscopic assumptions, this must be made part of the definition; we shall consider this point further in Chapter 14.

We can now complete a derivation we began in Section 13.5. By Eq. 13.14, the work performed on a perfect gas in a reversible isothermal expansion is $w = -nRT \ln(V_2/V_1)$. But by the above argument, we must have $\Delta U = 0$ for any isothermal process involving a perfect gas. Substituting in Eqs. 13.18, we immediately obtain

$$q = -w = nRT \ln \frac{V_2}{V_1} \tag{13.25}$$

for the heat absorbed in the reversible isothermal expansion of a perfect gas. The explicit calculation of w and q for other kinds of processes is more complicated; in Chapter 14 we shall see how to treat the general perfect gas case.

One other interesting point: Eq. 13.4 gives the volume dependence of a perfect gas's total energy E in what we can now call a reversible adiabatic expansion. Again identifying E with the thermodynamic internal energy, we obtain

$$dU = -\frac{2U}{3V}\,dV = -\frac{2}{3}\left(\frac{3}{2}nRT\right)\frac{dV}{V} = -nRT\,\frac{dV}{V}$$
$$= -p\,dV, \tag{13.26}$$

where we have substituted Eq. 13.24 and the perfect gas law. The result $dU = -p\,dV$ is, of course, true for the reversible adiabatic expansion of any fluid (since in any adiabatic process $dU = đw$), but here we have derived it microscopically for the perfect gas model.

[16] Remember that the zero of energy is always arbitrary. Here, as in Chapter 12, we have implicitly taken $U = 0$ to correspond to zero kinetic energy of the gas molecules.

13.9 Constraints, Work, and Equilibrium

In Section 12.3 we introduced the notion of constraints, the boundary conditions defining a thermodynamic system. We have made little use of this concept, except in our definition of thermal equilibrium: A new equilibrium is achieved when an adiabatic barrier is replaced by a diathermal one, that is, when a constraint is removed. It is generally true that any change in the constraints imposed on a system changes the equilibrium state of the system. We shall now see that the kind of work done by a system is also defined by the nature of the constraints imposed.

Consider a gas enclosed in a rigid cylinder with adiabatic walls, with a sliding piston dividing the gas into two volumes, V_1 and V_2 (see Fig. 13.7). If the piston is an adiabatic barrier and is fixed in place, then the two subsystems are isolated, and the pressures p_1 and p_2 and temperatures T_1 and T_2 can be completely different; there are four independent variables, two in each subsystem. If the adiabatic barrier is replaced by a fixed diathermal barrier, the two temperatures will equalize but the pressures can still be different; now

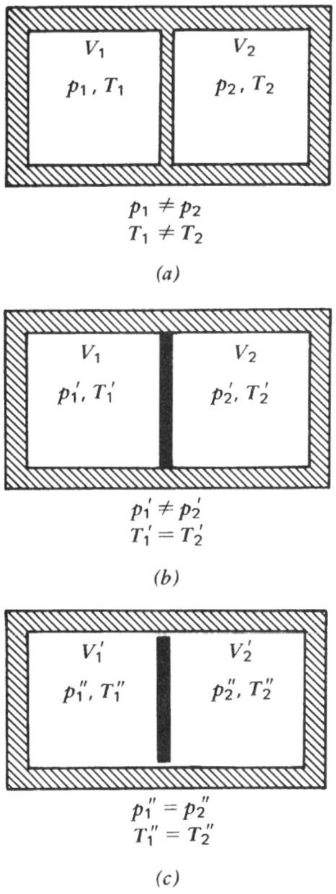

Figure 13.7 Interactions of two gaseous subsystems in a rigid adiabatic container. (*a*) Fixed adiabatic barrier. (*b*) Fixed diathermal barrier. (*c*) Movable diathermal piston.

there are only three independent variables, since $T_1 = T_2$. Finally, if the diathermal piston is allowed to move freely, it will move to a position such that $p_1 = p_2$, while still keeping $T_1 = T_2$; the final system has two independent variables. If in the final process the piston were coupled to an external machine instead of moving freely, external work would be done; assuming that this work process is still adiabatic, the system's internal energy and thus its temperature must decrease. We would still have $T_1 = T_2$, but at a lower value.

What conclusions can we draw about constraints from these and other observations? It is found that:

1. For each constraint imposed on a system, an additional independent variable is required to describe the system. (This can in fact be used as the definition of a constraint.)
2. The kind of equilibrium achieved by a system is defined by the constraints. When a system is initially in equilibrium subject to a given set of constraints, either addition or removal of a constraint will ordinarily generate a new equilibrium state, described by a different set of independent variables.
3. The kind of work performed by a system is determined by the constraints under which it acts. Stated more precisely, whenever a given constraint is removed, the system becomes capable of spontaneously performing a particular kind of work; to restore the system to its original state including the constraint, work must be performed on it. Thus, in the above example, pV work *can* be performed when the constraint of a fixed barrier is removed, and *must* be performed to restore the barrier to its original position. Similarly, electrical work can be performed when a capacitor or electrochemical cell is discharged (i.e., when an electrically insulating barrier is removed).

Another way to express point 2 is that a system is at equilibrium with respect to change in certain variables, with the constraints operationally defining which such variations are allowable. For example, a one-dimensional harmonic oscillator has its equilibrium at a point where the potential energy is a minimum with respect to the displacement x; the system is constrained not to allow displacements in other directions. In Fig. 13.7*c* the system reaches equilibrium with respect to variation of (say) $p_1 - p_2$ and $T_1 - T_2$; but what criterion of equilibrium corresponds to the minimum potential energy of the mechanical system? In Chapter 19, where we expand upon this viewpoint, we shall see that there are many such criteria—in fact, one for each combination of constraints on a system.

For another example of the interplay between equilibrium states and constraints, consider a room-temperature mixture of H_2 and O_2. The system is effectively in metastable equilibrium, since the reaction between the two gases is so slow as to be negligible. We can thus say that there is a constraint against chemical reaction. In this situation one can vary the amount of either gas independently of the other; thus for fixed volume there are three independent variables, which can be chosen from among p, T, $n(H_2)$, and $n(O_2)$. Suppose now that the constraint against chemical reaction is removed, either by adding a suitable catalyst or by raising the temperature. The H_2 and O_2 will react, and a new equilibrium will be attained. Corresponding to the removal of one constraint, there is one less independent variable, since $n(H_2)$ and $n(O_2)$ can no longer be varied independently (they are connected through the equilibrium constant).

In this chapter we have developed the first law of thermodynamics and various associated topics, all in rather general and abstract terms. Chapter 14 will be devoted mainly to applying these concepts to practical calculations. The first law enables us to calculate the change in internal energy (and other state functions) in any real or hypothetical process; it does not tell us anything about whether a given process will occur or not. The latter question is extremely important to chemists (e.g., given a mixture of reactants and products, which way will the reaction go?), and we shall have to come back to it later.

• FURTHER READING

Bailyn, M., *A Survey of Thermodynamics* (AIP Press, New York, 1994), Chapters 2 and 3.
Caldin, E. F., *Introduction to Chemical Thermodynamics* (Oxford University Press, London, 1958), Chapter 2.
Cardwell, D. S. L., *From Watt to Clausius* (Cornell University Press, Ithaca, New York, 1971).
Zemansky, M. W., *Heat and Thermodynamics,* 4th ed. (McGraw-Hill, New York, 1957), Chapters 3 and 4.

• PROBLEMS

1. Consider the system pictured below:

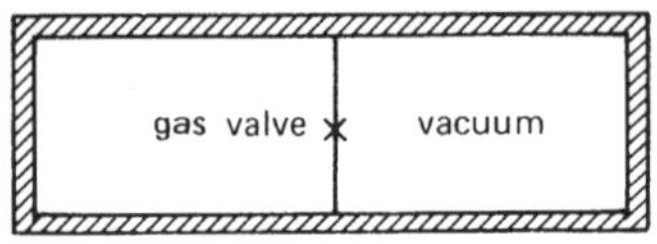

Gas is confined to a subvolume V_1 in an insulated rigid container. The container has an adjoining subvolume V_2, initially evacuated, which can be connected to V_1 by opening a valve. Suppose that the valve is opened and the gas flows out of V_1, finally filling the entire volume $V_1 + V_2$. Calculate the work done by the gas in this expansion, and the change in the internal energy of the gas.

2. A rock of mass M_R, density ρ_R, is dropped from a height h above the bottom of an insulated beaker containing a viscous fluid. Let the density of the fluid be ρ_F and its height in the beaker d. Assume that no fluid splashes out of the beaker when the rock falls in. Calculate the heat generated as a result of dropping the rock into the beaker.

3. An electric current passes through a resistor that is immersed in running water. Let the resistor be the system under investigation.
   (a) Is there a flow of heat into the resistor?
   (b) Is there a flow of heat into the water?
   (c) Is work done on the resistor?
   (d) Assuming the resistor to be unchanging, apply the first law to this process.

4. A current of 1 A flows across a 2 V potential difference for 1 min. If the energy carried by this current were completely converted to work, calculate how high a 100-g mass could be lifted in the earth's gravitational field.

5. Is the product of two intensive variables an intensive variable or an extensive variable? How about the product of an extensive and an intensive variable?

6. The energy, mass, and volume of a system are extensive. How about the molar energy, molar mass, and molar volume? How about the energy and mass densities?

7. In Fig. 13.5*c*, how much work is done by the piston as the gas expands? Assume that the pegs are perfectly rigid.

8. In each of the following cases, an ideal gas expands against a constant external pressure. For each case, decide if the expansion is allowed by the first law, and if so, if the expansion can be carried out reversibly:
   (a) Isothermal (constant temperature) and adiabatic
   (b) Isothermal but not adiabatic
   (c) Neither isothermal nor adiabatic

9. Describe how the apparatus shown in Problem 1 above might be used to determine if a gas behaves ideally.

10. A current of 1 A is passed through a resistance of 1 Ω. How long must the current be maintained to raise the temperature of 250 ml of water from 25°C to 100°C?

11. If the force on a particle depends only on the particle's position (conservative force), show that the work done on the particle in the course of a displacement is independent of path. Mechanical work in a conservative force field is analogous to what kind of thermodynamic work?

12. Evaluate the work integral

$$\int_{0,0}^{1,1} p\,dV$$

along the following paths in the (p, V) plane:
   (a) Straight line connecting 0, 0 and 1, 1
   (b) Parabola connecting 0, 0 and 1, 1 ($p = V^2$)
   (c) Rectangular path $(0, 0 \rightarrow 0, 1 \rightarrow 1, 1)$
   (d) Rectangular path $(0, 0 \rightarrow 1, 0 \rightarrow 1, 1)$

Next, evaluate the integral

$$\int_{0,0}^{1,1} V\,dp$$

along each of the paths (a) through (d); show that the sum of the two integrals is independent of the path of evaluation.

13. Suppose that the isothermal compressibility of a solid, defined by

$$\kappa_T \equiv -\frac{1}{V}\left(\frac{\partial V}{\partial p}\right)_T,$$

is a constant independent of V and T. Write down an equation of state for the solid valid for not too high pressure in terms of κ_T and the volume V_0 ($V = V_0$ when $T = T_0$ and $p = 1$ atm). Calculate the work done in increasing the pressure quasi-statically and isothermally on a solid from p_1 to p_2. Apply this formula to calculate the work done when 10 g of copper is subjected to a pressure increase from 1 atm to 1000 atm at 0°C. For copper we have $\rho = 8.93\text{g/cm}^3$ and $1/\kappa_T = 1.31 \times 10^{12}$ dyn/cm^2 at 0°C.

14. A thin-walled metal bomb of volume V_B contains n mol of a perfect gas at high pressure. Connected to the bomb is a capillary tube and stopcock. When the stopcock is opened slightly, the gas leaks slowly into a cylinder containing a nonleaking frictionless piston. The gas pushes against the piston, and the pressure in the cylinder is maintained at the constant value p_0 by moving the piston.
   (a) Show that, after as much gas as possible has leaked out, an amount of work

$$p_0(V_B - n v_0)$$

   has been done. Here, v_0 is the molar volume of the perfect gas at pressure p_0 and temperature T_0.
   (b) How much work would be done if the gas leaked directly into the atmosphere?

15. During a quasi-static adiabatic expansion of a perfect gas the pressure and volume are related by

$$pV^{\gamma} = \text{constant},$$

where γ is another constant. Show that the work done in expanding from a state (p_i, V_i) to the state (p_f, V_f) is

$$w = \frac{p_f V_f - p_i V_i}{\gamma - 1}$$

If $p_i = 10$ atm, $p_f = 2$ atm, $V_i = 1$ liter, $V_f = 3.16$ liters, and $\gamma = 1.4$, how much work is done?

16. Calculate the coefficient of thermal expansion,

$$\alpha \equiv \frac{1}{V}\left(\frac{\partial V}{\partial T}\right)_p$$

and the isothermal compressibility

$$\kappa_T \equiv -\frac{1}{V}\left(\frac{\partial V}{\partial p}\right)_T,$$

for a perfect gas. Compute α and κ_T numerically in units of K^{-1} and $(\text{torr})^{-1}$ at $T = 273$ K, $p = 1$ atm.

17. Calculate the work done when:
(a) One mole of water freezes, at 0°C, $p = 1$ atm to form 1 mol of ice. The density of ice is 0.917 g/cm^3 at 0°C and that of water is 1.000 g/cm^3 at 0°C.
(b) One mole of water is vaporized at 100°C against an applied pressure of 1 atm. At 100°C the vapor pressure of water is 1 atm. Assume that the vapor is a perfect gas. How much error is made if the volume of the liquid is neglected?

18. If the length of a wire in which there is a tension F is changed from L to $L + dL$, what is the infinitesimal amount of work done?

19. The tension in a wire (length L, cross-sectional area A) is increased quasi-statically and isothermally from F_i to F_f. If the isothermal Young's modulus $Y \equiv (L/A)(\partial F/\partial L)_T$ can be regarded as constant during this process, show that the work done is

$$w = \frac{L}{2AY}(F_f^2 - F_i^2).$$

Assume that the stretching of the wire is so small that one can write $F = k(L - L_0)$, with k a constant.

20. Given that the length L of a thin wire under the action of a stretching force F can be expressed as

$$L = a + bF,$$

what experiments would be necessary to obtain an equation of state?

21. Discuss, in general terms, what operations need to be carried out in order to specify the complete set of macroscopic coordinates needed to specify the thermodynamic state of the system. Consider such variables as size, surface area, gravitational field strength, and any others you think relevant.

22. Suppose that the equation of state of a gas is

$$\left(p + \frac{n^2 a}{V^2}\right)(V - nb) = nRT$$

(the van der Waals equation), where a and b are constants. Show that if the gas is isothermally and reversibly compressed from volume V_1 to volume V_2, the work done is

$$w = -nRT\ln\left(\frac{V_2 - nb}{V_1 - nb}\right) - n^2 a\left(\frac{1}{V_2} - \frac{1}{V_1}\right).$$

23. The isothermal compression of a perfect gas from volume V_1 to volume V_2 requires work equal to

$$w = -nRT\ln\left(\frac{V_2}{V_1}\right).$$

Hence, the work computed in Problem 22 differs from the work of compression of a perfect gas by

$$\Delta w = -nRT\ln\left(\frac{1 - nb/V_2}{1 - nb/V_1}\right) - n^2 a\left(\frac{1}{V_2} - \frac{1}{V_1}\right).$$

Given that the constant a is a measure of intermolecular attraction, and that b is a measure of the volume occupied by the molecules, how do you interpret the above equation for Δw?

24. Suppose that the equation of state of a gas is

$$pv = RT\left(1 + \frac{B}{v}\right),$$

where B can vary with temperature, but not with pressure. Repeat the calculation of Problem 14 and the analysis of Problem 15. How would B be related to the a and b of the van der Waals equation?

The following problems are designed to illustrate, by use of a simple model, the (classical mechanical) kinetic theory interpretations of adiabatic and isothermal work.

25. Imagine that a perfect gas is contained in a cylinder that is closed by a frictionless piston. Suppose that all the molecules have the same speed v and can move only in the direction perpendicular to the piston face. Then the

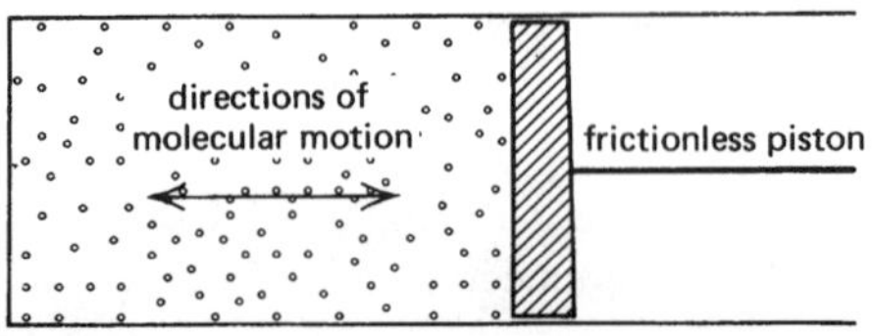

molecular velocity **v** has only two components, namely, $\pm v$. As a result of collisions between the gas molecules and the piston, energy and momentum are exchanged between them. If m is the mass of a gas molecule, and M and v_{p} are the mass and speed of the piston, show that the change in kinetic energy (KE) of a gas molecule as a result of collision with the piston is

$$\Delta\mathrm{KE} = \frac{4mM}{(m+M)^2}\left[\frac{M}{2}v_{\mathrm{p}}^2 - \frac{m}{2}v^2 - \frac{1}{2}(M-m)vv_{\mathrm{p}}\right].$$

26. Suppose that the frictionless piston in the system described in the preceding problem is acted on by a force **F**, which has the consequence that the piston moves with the *constant speed* v_{p}. Furthermore, suppose that energy can be transferred from the piston to the gas molecules, but not to any other part of the system or to the surroundings (adiabatic compression or expansion). Show that if $M >> m$ and $v >> v_{\mathrm{p}}$, the rate of gain of kinetic energy of the gas is

$$\frac{d}{dt}(\mathrm{KE}) = -2Z_c m v_{\mathrm{p}} v,$$

where Z_c is the number of collisions per second of all molecules with the piston. Interpret this result.

27. Show from the result of Problem 26 that the gain of kinetic energy by the gas in some infinitesimal time interval is just the work

$$đw = -p\,dV.$$

28. In Problem 25 we modeled an adiabatic expansion or compression. To model an isothermal expansion or compression, imagine that a frictionless piston is between two perfect gases, and subjected to collisions from both (see diagram). As a result of collisions

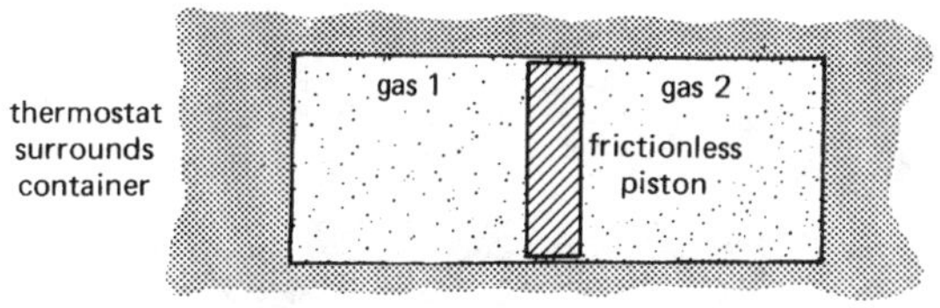

between the piston and the molecules of gases 1 and 2, the piston jiggles back and forth. When it is balanced between the two gases its average velocity, $\langle \mathbf{v}_{\mathrm{p}} \rangle$ is zero, but its mean square velocity $\langle v_{\mathrm{p}}^2 \rangle$, is not zero. The jiggling of the piston is the mechanism by which energy is transferred between gases 1 and 2.

The isothermal compression is imagined to occur as follows. Suppose that there is a very small imbalance of forces on the piston (say, gas 2 pushes on gas 1), so small that the resultant piston velocity is very, very small, indeed sensibly zero. Suppose further that the surroundings of the container are a thermostat, so that the temperatures of gases 1 and 2, and the piston, are fixed throughout the compression. Note that this condition implies that the kinetic energy of the piston cannot change during compression. Show by direct calculation that the kinetic energy of gas 1 is not increased as a result of compression. Use the same simple model for molecular motion as in the preceding problems.

CHAPTER 14

Thermochemistry and Its Applications

We now pause in our development of thermodynamics to see what use can be made of what we have already learned—mainly, the first law. In short, we shall come down from the heights of abstraction and deal for a while with applications. In this chapter we are concerned with the energy and temperature changes that accompany physical and chemical processes of various kinds. The measurement and interpretation of such changes constitutes the science of *thermochemistry.*

To obtain meaningful interpretations of thermochemical measurements, the conditions under which the various processes occur must be carefully defined. We anticipate, for example, that a given chemical reaction will release different amounts of heat when it proceeds under different constraints, since the heat and work associated with any process depend on the path. The first part of the chapter is therefore devoted largely to definitions of the many specific kinds of energy change, and of the standard states relative to which they are defined.

The remainder of the chapter deals with the molecular interpretation of thermochemical data. This is an extremely rich field, in which we finally have the opportunity to combine microscopic and thermodynamic knowledge in a useful way. Although still more can be learned with the help of the second law of thermodynamics, we shall see that the first law alone provides a wealth of fascinating information about such chemically significant topics as bond energies, the shape of molecules, and the stability of oxidation states.

Before starting our exposition, which uses physical processes as examples, we need to point out that there has been a subtle difference of focus in the applications of thermodynamics to physical processes and to biological processes. Traditional thermodynamics grew out of the analysis and optimization of heat engines—machines that convert heat into other forms of energy, especially mechanical energy. For that reason considerable attention is paid to the production of heat in various processes and to the efficiency of the conversion of heat into work. Living systems, for the most part, convert energy from one form to another without going through an intermediate transformation to heat. For example, muscles convert stored chemical energy directly into mechanical energy and membranes maintain concentration gradients without using heat energy; these processes are driven by coupled cycles of chemical reactions. This difference of focus is slowly vanishing as modern technologies such as conversion of sunlight into electrical energy via solar photoelectric cells and conversion of chemical energy into electrical energy in fuel cells stimulate developments in engineering thermodynamics that deal with "direct processes" in which energy is converted from one form to another without an intermediate transformation to heat. At the same time, the interpretation of how living systems function is turning more and more from analyses that deal only with structure to analyses that include processes. These developments have given rise to concepts such as "molecular motors," which have been applied from the organismal and cellular levels all the way to the (macro)molecular level. Indeed, the tools of thermodynamics, the most general of all sciences, are becoming increasingly useful in the analysis of biological function and biological processes.

14.1 Heat Capacity and Enthalpy

We begin by introducing some new thermodynamic functions, including those that are particularly applicable to describing energy transfer, the heat capacities. We have had occasion to mention heat capacity before (as early as Section 2.9), but now we are ready for a precise definition. Let us approach this by way of an idealized thermochemical experiment.

Consider a vessel with adiabatic walls, completely filled with fluid and containing a thermometer and an electric heater. Suppose that the walls are rigid, so that the volume of the vessel is fixed. The system is thus isolated except for the heater. At equilibrium in the initial state the temperature of the fluid is T_1. If some current is passed through the electric heater for a specified time period, a new equilibrium will be established at a higher temperature T_2; we define $\Delta T \equiv T_2 - T_1$. The electrical energy supplied is dissipated as a quantity of heat q. The limiting ratio of q to ΔT as both quantities tend to zero,

$$\lim_{\Delta T \to 0} \left(\frac{q}{\Delta T} \right)_V \equiv C_V , \tag{14.1}$$

is called the *heat capacity at constant volume.*[1] But by the first law of thermodynamics, Eq. 13.18, in the process described we have

$$\Delta U = q_V , \tag{14.2}$$

since no work can be performed by the fluid (i.e., $w_V = 0$). We therefore find that C_V is the rate of change of internal energy with temperature when the system is maintained at constant volume:

$$C_V = \left(\frac{\partial U}{\partial T} \right)_V . \tag{14.3}$$

But C_V is not the only kind of heat capacity. Suppose that we perform the same experiment as before, except that one wall of the vessel is a freely moving (but still adiabatic) piston. The volume of the fluid will no longer be the same after addition of heat. However, we can instead maintain the external pressure p_{app} fixed; if the heat is added so slowly as to approximate a reversible process, the fluid pressure p can also be assumed fixed. The limiting ratio for such a constant-pressure process,

$$\lim_{\Delta T \to 0} \left(\frac{q}{\Delta T} \right)_p \equiv C_p , \tag{14.4}$$

is called the *heat capacity at constant pressure.* (In what follows, pressure-volume work processes should be assumed reversible unless otherwise stated; thus we can consistently take $p = p_{\text{app}}$.)

Although one can define many other heat capacities,[2] C_V and C_p are the only ones with which we now need to concern ourselves. Both C_V and C_p as we have defined them are extensive quantities, proportional to the mass of fluid present. It is customary to refer them to a specific amount of material, usually the mole, in which case they are called *molar heat capacities.* (The term "molar specific heats" is also used, but it is best to restrict "specific heat" to the meaning of heat capacity per gram.) They can be given in such units as joules per degree (°C or K) per mole. The molar heat capacities are state functions characteristic of the particular substance, and we shall see in later chapters how they can be related to the microscopic energy spectrum. For notational consistency we shall use capital letters for extensive functions, for example, U for the total internal energy of an arbitrary mass of substance, and lower case letters for the corresponding molar functions, for example, u for the internal energy per mole. However, because the usage is so common, we also use capital letters for the changes in energy, enthalpy, and so forth per mole of product in a chemical reaction.

When the fluid in the constant-pressure experiment is allowed to expand, work is done by the system on the surroundings ($w_p < 0$). By the first law, adding a given quantity of heat must cause a smaller change in the internal energy of the system when the system does work than when no work is performed. In the constant-volume process, none of the energy added to the system as heat is transferred back to the surroundings as work; it all goes into the translational and internal energy of the molecules. By Eq. 13.18, then, for the same positive value of q we must have $(\Delta U)_p < (\Delta U)_V$. Since internal energy increases with temperature, this suggests that for the same q we should have $(\Delta T)_p < (\Delta T)_V$; when this is true (as it usually[3] is), comparison of Eqs. 14.1–14.4 tells us that $C_p > C_V$.

For a constant-volume process we can rewrite Eq. 14.3 in the differential form

$$(dU)_V = C_V \, dT. \tag{14.5}$$

Does there exist a comparably simple relationship for a constant-pressure process? Using the first law in differential form, Eq. 13.19, we have

$$(dU)_p = C_p \, dT - p \, dV \tag{14.6}$$

for a reversible constant-pressure process in a system that can perform only pV work. The heat added at constant pressure, $C_p \, dT$, thus equals the quantity $(dU + p \, dV)_p$. This suggests that we should introduce a new thermodynamic variable,

$$H \equiv U + pV , \tag{14.7}$$

since we can then write

$$(dH)_p = (dU + p \, dV)_p = C_p \, dT , \tag{14.8}$$

analogous to Eq. 14.5; the analog of Eq. 14.3 is thus

$$C_p = \left(\frac{\partial H}{\partial T} \right)_p . \tag{14.9}$$

[1] As in the partial derivative notation, we use a subscript (here V) to denote a variable that is held constant in the process described.

[2] For example, in a system whose properties depend on the electric field **E**, one might need to use $C_{\mathbf{E}} \equiv \lim_{\Delta T \to 0} (q/\Delta T)_{\mathbf{E}}$.

[3] An exception occurs if the fluid *contracts,* rather than expanding, when heat is added at constant pressure ($q_p > 0$, $w_p > 0$); then we have $(\Delta U)_p > (\Delta U)_V$ and thus $C_p < C_V$. This anomaly occurs in liquid water between 0°C and 4°C.

The function H is called the *enthalpy* of the system. Since U and pV are both unique functions of the thermodynamic state of the system, H must also be a state function; it has the dimensions of energy, and its zero has the same arbitrariness as that of U.

The introduction of the function H may seem arbitrary or even whimsical. This is not so, however. When the system is a simple fluid that can perform only pV work, the natural variables defining the internal energy, those in terms of which the function has the simplest form, are T and V. But many processes are more conveniently carried out at constant pressure than at constant volume,[4] and are thus described more compactly in terms of enthalpy changes than in terms of energy changes. For example, the heat transferred in a constant-pressure process is just $(\Delta H)_p = q_p$.[5] The passage from Eq. 14.6 to Eq. 14.8 is best regarded as a change of independent variable from V to p. If the same change of variable is carried out for a constant-volume process, in which $(\Delta H)_V = q_V$, we obtain

$$(dH)_V = (dU)_V + [d(pV)]_V = C_V\,dT + V\,dp. \tag{14.10}$$

In Section 14.3 we shall see how equations such as these can be used to obtain ΔH for arbitrary changes of state.

The apparatus actually used in experimental thermochemistry is not very different from the idealized models already described. One uses an insulated vessel, called a *calorimeter*, in which heat is delivered to a sample under controlled conditions, usually constant pressure or constant volume. Although our discussion so far has been in terms of a fluid sample, similar measurements can of course be made on solids. The science of calorimetry has advanced to the point where very small energy changes and very small amounts of a substance can be studied. We shall not go into the fascinating experimental details; it is sufficient to note that particular calorimetric studies can involve a wide range of subjects, e.g., sophisticated analyses of the temperature scale coupled with development of special techniques to reach very low or very high temperatures[6] and the use of a bolometer (a form of calorimeter) to measure the intensity of a molecular beam.

The immediate results of calorimetric measurements are values of the heat capacities. Given a sufficient amount of such data, one can obtain C_p and C_V as functions of temperature and pressure (or density) over the whole range of interest. The various equations above can then be integrated to determine ΔU or ΔH for any process; we shall give examples of such calculations in the following sections. Note, however, that to obtain $(\Delta U)_p$ or $(\Delta H)_V$ one must determine $p\Delta V$ or $V\Delta p$, respectively, for the process in question; this requires either measurement or a knowledge of the fluid's equation of state.

Table 14.1 gives data on the heat capacities of various substances. For the same reason that H is more useful than U, most such tables give values of C_p rather than C_V.

For the same substance, C_p and C_V are of course related to one another. One can easily find the relation between the two for the perfect gas. Given Eq. 12.42, $U = \frac{3}{2}nRT$, we can see that $(\partial U/\partial T)_p = (\partial U/\partial T)_V$; we thus have for the perfect gas

$$\begin{aligned} C_p - C_V &= \left(\frac{\partial H}{\partial T}\right)_p - \left(\frac{\partial U}{\partial T}\right)_V \\ &= \left[\frac{\partial (U+pV)}{\partial T}\right]_p - \left(\frac{\partial U}{\partial T}\right)_p \\ &= \left[\frac{\partial (pV)}{\partial T}\right]_p = \left[\frac{\partial (nRT)}{\partial T}\right]_p = nR, \end{aligned} \tag{14.11}$$

where we have substituted $pV = nRT$. That is, the molar heat capacities of a perfect gas should differ by the rather sizable quantity R, or about 8.3 J/K mol. The difference is much less in a solid or liquid, where the molar volume is only about 0.001 as great as in the gas; as a result, pV is only a small fraction of nRT, so that H and U, and thus C_p and C_V, are nearly the same. Later, in Chapter 17, we shall derive an equation giving the relationship between C_p and C_V for the general case.

Some interesting regularities can be obtained from the data in Table 14.1. For example, near room temperature we have $c_p \approx \frac{5}{2}R$ (20.79 J/K mol) for monatomic gases and $c_p \approx \frac{7}{2}R$ (29.10 J/K mol) for diatomic gases; in Chapter 21 we shall see how these values can be obtained from microscopic theory. In general, the value of c_p tends to increase with the number of atoms in a compound or, more precisely, with the number of vibrational modes in the molecule. In most solid metals we have to a good approximation $c_p \approx 3R$ (25 J/K mol). This is the Law of Dulong and Petit, which we introduced in Section 2.9; it can also be derived theoretically, as we shall see in Chapter 22. Similarly, for monatomic-ion salts such as the alkali halides, we find $c_p \approx 6R$ (50 J/K mol) or $3R$ per ion.

The temperature dependence of the heat capacity is shown for a variety of substances in Fig. 14.1. Note that for

---

4 An obvious example is any process open to the atmosphere.

5 It is important to note that $(\Delta H)_p = q_p$ in a constant pressure process whether that process is reversible or irreversible. To see that this is so, imagine a chemical reaction carried out in a cylinder with diathermal walls and closed by a movable piston. Let the fixed external pressure acting on the piston be p_{app}. Before the reaction starts, the system consists of only reactants at pressure p_i, and after it is complete the pressure of the products is p_f. We require that $p_{app} = p_i = p_f$. During the reaction, the system pressure will vary and the piston will move, always against the constant external pressure p_{app}. Since $\Delta H = \Delta U + \Delta(pV)$, where p is the pressure of the system, we can write $\Delta H = q + w + (p_fV_f - p_iV_i) = q + w + p_{app}(V_f - V_i)$. But, by definition, $w = -\int p_{app}\,dV = -p_{app}(V_f - V_i)$ so $\Delta H = q$ at the constant applied pressure p_{app}.

6 See J. P. McCullough and D. W. Scott (Eds.), *Experimental Thermodynamics,* Vol. I (Plenum Press, New York, 1968); B. LeNeindre and B. Vodar (Eds.), *Experimental Thermodynamics,* Vol. II (Butterworths, London, 1975); *Precision Measurement and Calibration: Temperature* (N.B.S. Special Publication 300, Vol. 2, 1968).

Table 14.1 Heat Capacities of Some Substances at 25°C and 1 atm

Substance	c_p (J/K mol)	Substance	c_p (J/K mol)
GASES:		SOLIDS:	
He	20.79	C (graphite)	8.53
Ar	20.79	C (diamond)	6.12
Xe	20.79	S (rhombic)	22.60
H_2	28.83	S (monoclinic)	23.64
N_2	29.12	I_2	54.44
O_2	29.36	Li	23.64
Cl_2	33.84	K	29.51
HCl	29.12	Cs	31.4
CO	29.15	Ca	26.28
H_2S	34.22	Ba	26.36
NH_3	35.52	Ti	25.00
CH_4	35.79	Hf	25.52
CO_2	37.13	Cr	23.35
SO_2	39.87	W	24.08
PCl_3	72.05	Fe	25.23
C_2H_6	52.70	Ni	26.05
n-C_4H_{10}	98.78	Pt	26.57
		Cu	24.47
LIQUIDS:		Au	25.38
Hg	27.98	Al	24.34
Br_2	75.71	Pb	26.82
H_2O	75.15	U	27.45
CS_2	75.65	NaCl	49.69
N_2H_4	98.93	NaOH	59.45
CCl_4	131.7	$NaNO_3$	93.05
H_2SO_4	138.9	LiF	41.90
C_6H_6	136.1	CsI	51.87
n-C_6H_{14}	195.0	AgCl	50.78
C_2H_5OH	111.4	$BaSO_4$	101.8
$(C_2H_5)_2O$	171.1	$K_4Fe(CN)_6$	335.9
CH_3COOH	123.4	CBr_4	128.0
		SiO_2 (quartz)	44.43
		B_2O_3	62.97
		V_2O_5	127.3
		$C_{10}H_8$ (naphthalene)	165.7
		n-$C_{33}H_{68}$	912

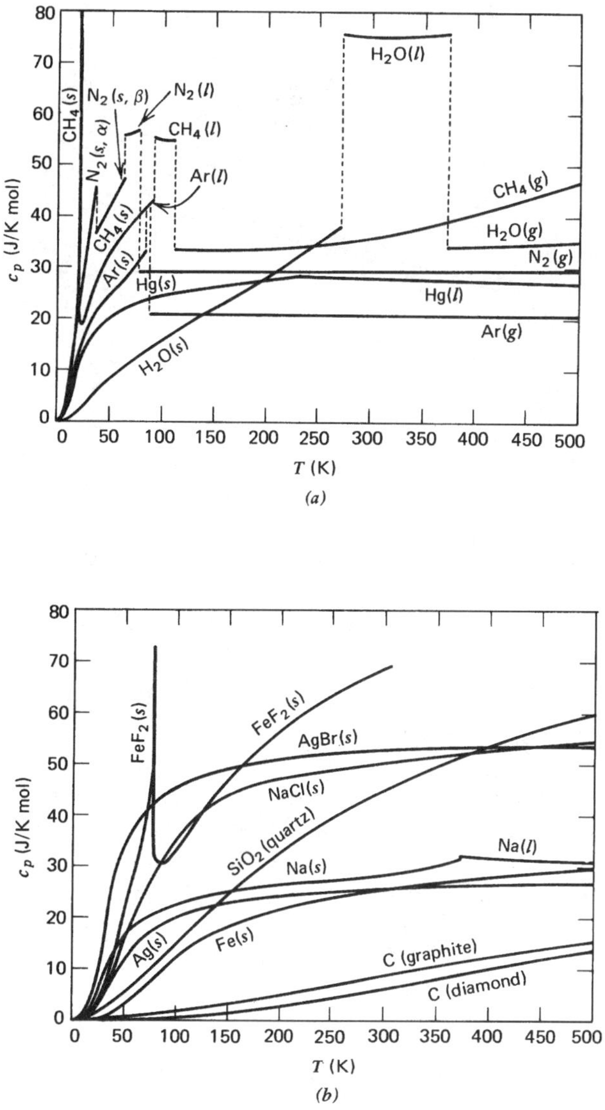

Figure 14.1 Temperature dependence of the heat capacity for various substances. (*a*) Room-temperature gases and liquids. (*b*) Room-temperature solids.

all substances c_p vanishes at absolute zero, rising in most cases to a "plateau" (often the Dulong–Petit value) at higher temperatures. Discontinuities occur at the melting and boiling points, with the liquid usually having a higher heat capacity than the other phases. Some solid–solid phase transitions are also shown: an ordinary discontinuity in N_2, but sharp peaks in CH_4 and FeF_2; the latter are called "lambda points," from the shape of the $c_p(T)$ curve, and represent second-order phase transitions (see Chapter 24). Within a given phase, the temperature dependence is usually well described by an equation of the form $c_p = a + bT + cT^{-2}$, where b and c are typically of the order of 10^{-3} J/K^2 mol and 10^5 J/K mol, respectively. The pressure dependence of the heat capacity is slight in the condensed phases,[7] but can be appreciable for the gas, as shown in Fig. 14.2.

[7] In liquids, c_p typically decreases by about 10% in 2000–3000 atm, then increases more slowly; in metals there is an almost negligible decrease, of the order of 10^{-8}/atm.

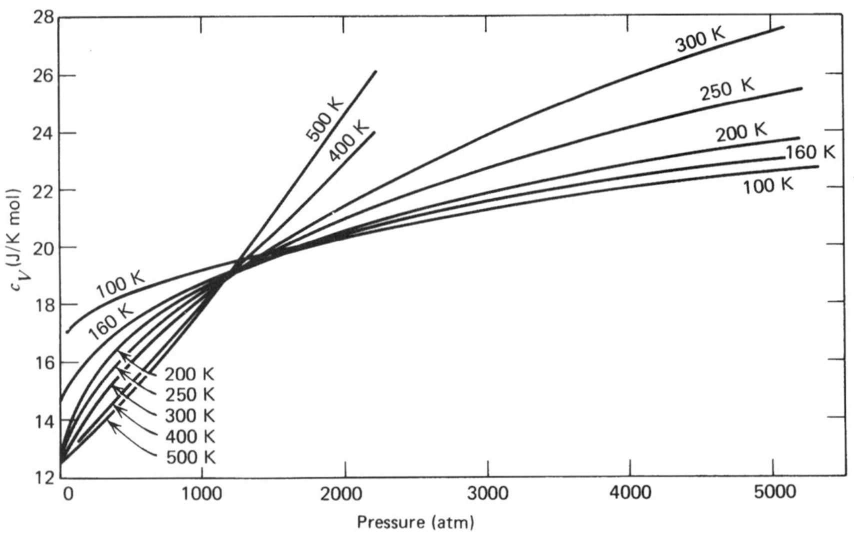

Figure 14.2 Constant-volume heat capacity of gaseous argon as a function of pressure and temperature. From F. Din (Ed.), *Thermodynamic Functions of Gases,* Vol. 2 (Butterworths, London, 1962).

14.2 Energy and Enthalpy Changes in Chemical Reactions

One of the most important applications of the first law of thermodynamics is to the study of the energy changes which accompany chemical reactions. Suppose that we consider a general chemical reaction

$$aA + bB + \cdots = lL + mM + \cdots. \quad (14.12)$$

For the present purposes, this is more conveniently written as an algebraic equation with all the terms on one side,

$$lL + mM + \cdots - aA - bB - \cdots = \sum_{i=1}^{r} \nu_i X_i = 0, \quad (14.13)$$

where the symbols X_i stand for the r species A, B, . . . , and the ν_i are the stoichiometric coefficients a, b, ... By convention we always subtract reactants from products, so the ν_i must be positive for the products and negative for the reactants. To make this more specific, consider the reaction

$$\tfrac{1}{2}H_2 + \tfrac{1}{2}Cl_2 = HCl,$$

for which we have $\nu_{HCl} = 1$, $\nu_{H_2} = -\frac{1}{2}$, $\nu_{Cl_2} = -\frac{1}{2}$ in the notation of Eq. 14.13.

How do we go from an equation in terms of chemical formulas to one involving the thermodynamic properties of substances? Consider the process by which we balance a chemical equation. What we actually do is apply the principle that the atoms of each element are conserved in chemical processes. But mass is also conserved, if we neglect relativistic effects.[8] We can therefore write, for the change of mass in a reaction,

$$\Delta M = \sum_{i=1}^{r} \nu_i M_i = 0, \quad (14.14)$$

where M_i is the molar mass (molecular weight) of substance i.

Now consider the internal energy and the enthalpy. They are state functions, and thus defined for any thermodynamic state of a system. Furthermore, they are extensive functions, so that we can define an energy or enthalpy per mole. We can therefore write equations similar to Eq. 14.14 for the total changes in internal energy and enthalpy in a reaction. These changes will not in general be zero, unless the system is isolated. We have

$$\Delta U = \sum_{i=1}^{r} \nu_i u_i \quad (14.15a)$$

and

$$\Delta H = \sum_{i=1}^{r} \nu_i h_i. \quad (14.15b)$$

If u_i and h_i are the internal energy and enthalpy per mole of component i, respectively, then the ΔU and ΔH defined by

[8] Einstein's fundamental relation of equivalence between mass and energy, $E = mc^2$ (where c is the speed of light), implies that there must be a mass difference between reactants and products if there is an energy change in a reaction. For typical chemical reaction energies (say, 40 kJ/mol), however, the mass change is less than 10^{-9} g/mol, and can be safely neglected.

these equations refer to 1 mol of the stoichiometric reaction. For example, if the reaction is $\frac{1}{2}H_2 + \frac{1}{2}Cl_2 = HCl$, then ΔU is the energy change when 0.5 mol each of H_2 and Cl_2 react to form 1 mol of HCl.

But what are the states for which u_i and h_i are defined? To specify ΔU and ΔH for a reaction, we must clearly know not only the stoichiometric equation but the initial states of the reactants and the final states of the products. By "state," as usual, we mean the specification of the state of aggregation (solid, liquid, or gas) and the temperature and pressure; u_i and h_i in Eqs. 14.15 refer to the initial state for the reactants and to the final state for the products. It is simplest to treat a reaction that occurs entirely at the same T and p, but obviously we cannot expect all real-life reactions to be this simple. A common situation, however, is a reaction occurring in a thermostat subject to atmospheric pressure.

An additional complication is the fact that changes in energy and enthalpy occur even when substances are mixed without reaction. Since in practice chemical processes are usually directed to the production of pure substances, it is customary to take the initial state as consisting of pure reactants isolated from one another, and the final state as isolated pure products. One can then imagine the overall reaction as a three-step process:

1. Take the pure reactants, in their initial equilibrium states, and physically mix them under reaction conditions.
2. Allow the reaction to proceed.
3. Isolate the individual products from the reaction mixture and bring them to their final equilibrium states.

This is not how a reaction ordinarily takes place, of course; reaction begins while step 1 is still under way, and it is often useful to start step 3 before the reaction is over. But for state functions such as U and H, the total change must be the same for this idealized path as in reality, as long as the initial and final states are the same. If an experimenter wishes to determine the energy change in step 2 alone, then, he must be careful to account for the changes corresponding to steps 1 and 3. Step 2 is of particular interest because it is likely to correspond to the model of any microscopic theory of reaction.

Once we have defined exactly what process we are considering, what is it that Eqs. 14.15 tell us? The first of these equations is in fact a statement of the conservation of energy: The total internal energy of the reactants plus the energy absorbed by the system (ΔU) must equal the total internal energy of the products. That is, the difference between the total energies of the reactants and the products gives just the amount of energy that must be absorbed or released for the reaction to occur under the given conditions. If we have a table of the internal energy per mole as a function of T and p for each of the species involved, we can obtain ΔU directly by subtraction. The same is true of the enthalpy change, which is more commonly used, and expressions like Eqs. 14.15 can also be written for other extensive state functions.

The algebraic properties of chemical equations have led us to Eqs. 14.14 and 14.15. Since a set of chemical equations can be thought of as a set of mass balances, it follows that different equations can be combined by addition and subtraction to obtain any desired mass balance. It does not matter whether or not the final equation represents a real chemical reaction, that is, one observed in the laboratory. For an example of this process, consider the several reactions

$$\begin{aligned} &(1) \qquad C + O_2 = CO_2, \\ &(2) \qquad H_2 + \tfrac{1}{2}O_2 = H_2O, \\ &(3) \qquad C_2H_6 + \tfrac{7}{2}O_2 = 2CO_2 + 3H_2O; \end{aligned}$$

if we algebraically add 2(Eq. 1) + 3(Eq. 2) – Eq. 3, multiplying each equation by the indicated numerical factor, we obtain the composite reaction

$$(4) \qquad 2C + 3H_2 = C_2H_6.$$

But what good is such manipulation? The answer is that it is equally valid for extensive quantities other than mass. Until now we have thought of a substance's chemical formula as representing the substance itself, or in stoichiometric contexts a mole of the substance; but it can equally well represent the internal energy, enthalpy, or any other extensive property of a mole of the substance. This makes possible a tremendous reduction in empiricism, since ΔU, ΔH, and other such quantities can be obtained for reactions without actually performing them.

Let us examine this. The enthalpy balance for Eq. 1 above, namely Eq. 14.15*b*, can be written as

$$h_C + h_{O_2} = h_{CO_2} - \Delta H,$$

or, in the more usual shorthand,

$$C + O_2 = CO_2 - \Delta H;$$

similarly for the other equations, and with either ΔH or ΔU. Since Eqs. 14.15 are of the same form as the mass balance, Eq. 14.14, the internal energy and enthalpy changes for Eq. 4 can be computed by adding those for Eqs. 1–3 in the same way as we added the equations themselves:

$$\begin{aligned} \Delta U(\text{Eq. 4}) &= 2\Delta U(\text{Eq. 1}) + 3\Delta U(\text{Eq. 2}) - \Delta U(\text{Eq. 3}), \\ \Delta H(\text{Eq. 4}) &= 2\Delta H(\text{Eq. 1}) + 3\Delta H(\text{Eq. 2}) - \Delta H(\text{Eq. 3}). \end{aligned}$$

These equations are examples of *Hess's law,* which states that the internal energy and enthalpy changes of such composite reactions are additive, provided that all the reactions are carried out under the same conditions (i.e., at the same temperature and pressure). Like Eqs. 14.15 themselves,

Hess's law is a direct consequence of the first law. Observe now what we have shown: If ΔH (or ΔU) has been measured for reactions 1, 2, and 3, the above equations enable us to calculate ΔH (or ΔU) for reaction 4 without ever actually observing that reaction. We thus begin to see the practical value of all the formal structure and laws of thermodynamics, in that they enable us to extend our knowledge beyond what has been measured directly. To do this efficiently, however, it is not enough to organize our knowledge in terms of specific chemical reactions. As we shall see in Section 14.4, a far greater reduction in empiricism is obtained when thermodynamic measurements are analyzed in terms of individual substances.

Hess's law is a consequence of the fact that U and H are state functions, and therefore that ΔU and ΔH are independent of path. It does not matter whether the reaction sequence 2(Eq. 1) + 3(Eq. 2) – Eq. 3 is an actual pathway or even a physically possible pathway for reaction 4. As long as the net reactants and products of the sequence are, respectively, identical in chemical composition and physical state to the reactants and products of the overall reaction, Hess's law is applicable.

Here is a numerical example of the application of Hess's law. Suppose that for the following reactions we know ΔH (often called the *heat of reaction,* because at constant pressure it is the same as q), all at 25°C and 1 atm:

$$(5) \qquad PbO(s)+S(s)+\tfrac{3}{2}O_2(g)=PbSO_4(s), \quad \Delta H=-39.56 \text{ kJ / mol};$$

$$(6) \qquad PbO(s)+H_2SO_4\cdot 5H_2O(l)=PbSO_4(s)+6H_2O(l), \quad \Delta H=-5.57 \text{ kJ / mol};$$

$$(7) \qquad SO_3(g)+6H_2O(l)=H_2SO_4\cdot 5H_2O(l), \quad \Delta H=-9.82 \text{ kJ / mol};$$

where the symbols in parentheses following the chemical formulas indicate that the substances are in the solid (s), liquid (l), or gaseous (g) state. What is the heat of reaction for

$$(8) \qquad S(s)+\tfrac{3}{2}O_2(g)=SO_3(g)$$

at 25°C and 1 atm? In this case, since the product SO_3 is formed directly from the elements S and O_2, the enthalpy change is known as the *enthalpy of formation* (*heat of formation*), ΔH_f, of $SO_3(g)$. We can obtain Eq. 8 by simply subtracting Eqs. 6 and 7 from Eq. 5; the heat of formation of $SO_3(g)$ at 25°C and 1 atm is thus

$$\begin{aligned}\Delta H_f(SO_3,g) &\equiv \Delta H(\text{Eq. } 8)\\ &= \Delta H(\text{Eq. } 5)-\Delta H(\text{Eq. } 6)-\Delta H(\text{Eq. } 7)\\ &= [-39.56-(-5.57)-(-9.82)] \text{ kJ / mol}\\ &= -24.17 \text{ kJ / mol.}\end{aligned}$$

Note how easy the analysis becomes when the chemical formulas are regarded as algebraic symbols, each representing a mole of the corresponding substance. Ordinary algebraic manipulations can then be used to simplify complicated sets of equations.

Because of the sign convention for reactions embodied in Eqs. 14.15, a negative value of ΔU means that energy is liberated in a reaction. This energy usually appears mainly as heat; specifically, when only expansion work is possible, the heat liberated is $-\Delta U$ at constant volume and $-\Delta H$ at constant pressure. In the above example, heat is liberated in forming $SO_3(g)$ from the elements at 1 atm. But in general work other than pressure-volume work may also be performed, and under some circumstances this can be the dominant effect. Indeed, both internal combustion engines and electrochemical cells are designed to maximize the amount of work obtained from a chemical reaction.

In the preceding examples we have considered only reactions in which all the products and reactants are at the same temperature and pressure. Of course, this will not always be true, especially if large amounts of energy are liberated. In practice it is often impossible to keep T and p under control during a reaction (and sometimes a change in T or p is exactly what one wants to produce). What one can do to keep the calculation tractable is to assume that the reaction itself occurs at some constant T and p, and to consider the temperature and pressure changes as separate steps in the overall process. This will still give the right values of ΔU and ΔH, as long as one does not change the initial and final states. But to do this one must know how to obtain the internal energy and enthalpy changes accompanying physical processes such as the expansion of a fluid, or a phase change from liquid to gas. The next section is devoted to this subject.

14.3 Thermochemistry of Physical Processes

Let us consider first the changes in internal energy and enthalpy when a fluid, originally at equilibrium in the state p_1, V_1, T_1, is brought to the new equilibrium state p_2, V_2, T_2. Since both U and H are state functions, all paths connecting the initial and final states must give the same values of ΔU and ΔH. For convenience we can then imagine the change of state to be carried out in two steps, as shown in Fig. 14.3a. First, the fluid is heated (or cooled) at constant volume V_1 until the temperature T_2 is reached. Second, with the temperature T_2 held fixed in a thermostat, the fluid is expanded (or contracted) until the volume reaches V_2; since only two of the variables p, V, T are independent, this must also bring the pressure to p_2. In short, we split the overall change into an *isochoric* (constant-volume) process and an *isothermal* (constant-temperature) process. We are thus considering V and T as our independent variables, and we shall see that this choice is particularly convenient for calculating ΔU.

If the internal energy U is a function of the independent variables T and V, then by Eq. II.4 in Appendix II at the back

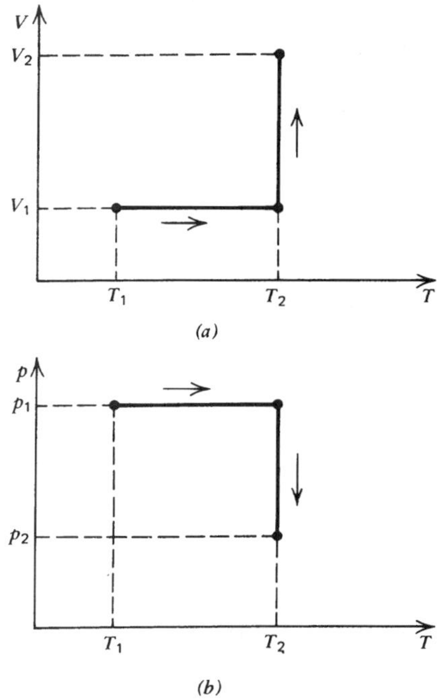

Figure 14.3 Two-step analysis of the change of state of a fluid: (*a*) constant-volume process followed by constant-temperature process, for calculating ΔU; (*b*) constant-pressure process followed by constant-temperature process, for calculating ΔH. In both cases the change is from the initial state p_1, V_1, T_1 to the final state p_2, V_2, T_2.

of the book we can write any infinitesimal change in internal energy as

$$dU = \left(\frac{\partial U}{\partial T}\right)_V dT + \left(\frac{\partial U}{\partial V}\right)_T dV, \qquad (14.16)$$

where dT and dV are the corresponding changes in T and V. To obtain ΔU for the entire process, we must integrate this equation over the path in Fig. 14.3a. In the first step the volume is constant, so the entire contribution to ΔU comes from the first term on the right-hand side of Eq. 14.16; in the second step the temperature is constant, so the entire contribution to ΔU comes from the second term on the right-hand side. Thus, for any change of state in a fluid, we can write

$$\Delta U \equiv U(T_2, V_2) - U(T_1, V_1)$$
$$= \int_{T_1,V_1}^{T_2,V_1} \left(\frac{\partial U}{\partial T}\right)_V dT + \int_{T_2,V_1}^{T_2,V_2} \left(\frac{\partial U}{\partial V}\right)_T dV. \qquad (14.17)$$

Now, by Eq. 14.3 the first term is simply $\int_{T_2,\,V_1}^{T_2,\,V_1} C_V\, dT$. To evaluate the second term, we must know the value of the derivative $(\partial U/\partial V)_T$ for the fluid; for the present we use without proof the relationship

$$\left(\frac{\partial U}{\partial V}\right)_T = T\left(\frac{\partial p}{\partial T}\right)_V - p, \qquad (14.18)$$

which we shall derive later (see Eqs. 17.3–17.17). Substituting in Eq. 14.17, we obtain for the total internal energy change,

$$\Delta U = \int_{T_2,V_1}^{T_2,V_1} C_V(T,V)\,dT$$
$$+ \int_{T_2,V_1}^{T_2,V_2} \left\{ T\left[\frac{\partial p(T,V)}{\partial T}\right]_V - p(T,V) \right\} dV. \qquad (14.19)$$

We do not derive Eq. 14.18 at this time only because the second law of thermodynamics is needed for its proof. But why should we make such a substitution at all? The answer is that $(\partial U/\partial V)_T$ cannot be measured directly in the laboratory, whereas *p, V,* and *T* can. These variables (and the heat capacities) are the primary measurements from which *U* and other thermodynamic functions are computed; that is why we call the combined values of *p, V, T* the equation of state. In other words, we have put our equation for ΔU in terms of directly measurable quantities only. We shall shortly need another substitution of the same type as Eq. 14.18,

$$\left(\frac{\partial H}{\partial p}\right)_T = V - T\left(\frac{\partial V}{\partial T}\right)_p, \qquad (14.20)$$

which will be derived as Eq. 17.50.

To obtain the enthalpy change for a general change of state in a fluid, it is most convenient to consider a different two-step process, that shown in Fig. 14.3*b*. In this case we combine an *isobaric* (constant-pressure) heating step with an isothermal pressure change, using *p* and *T* as independent variables. In analogy to Eq. 14.16, we have

$$dH = \left(\frac{\partial H}{\partial T}\right)_p dT + \left(\frac{\partial H}{\partial p}\right)_T dp. \qquad (14.21)$$

Each term again vanishes along one step of our path, and integration gives for the total enthalpy change

$$\Delta H \equiv H(T_2, p_2) - H(T_1, p_1)$$
$$= \int_{T_1,p_1}^{T_2,p_1} \left(\frac{\partial H}{\partial T}\right)_p dT + \int_{T_2,p_1}^{T_2,p_2} \left(\frac{\partial H}{\partial p}\right)_T dp$$
$$= \int_{T_1,p_1}^{T_2,p_1} C_p\,dT$$
$$+ \int_{T_2,p_1}^{T_2,p_2} \left\{ V(T,p) - T\left(\frac{\partial V(T,p)}{\partial T}\right)_p \right\} dp, \qquad (14.22)$$

in which we have substituted Eqs. 14.9 and 14.20.

Inspection of our equations for ΔU and ΔH reveals an important point. The changes in internal energy and enthalpy (and thus *U* and *H* themselves) can be thought of as having two components, one a function of temperature only (at some given *V* or *p*) and the other describing the change with *V* or *p* at a given temperature. Now consider the case of the perfect gas. Substituting $pV = nRT$ into Eqs. 14.18 and 14.20, we see that $(\partial U/\partial V)_T$ and $(\partial H/\partial p)_T$ both become zero; that is, U and H for the perfect gas are both functions of temperature only.[9] The two components just mentioned can thus be considered as a perfect gas term and a correction term for real fluids; this concept will be useful when we consider the properties of real gases (Chapter 21).

The preceding argument has been concerned with the case that the system can only do pV work. The relationship between changes in internal energy and heat transferred and work done is valid for all kinds of work (Eq. 13.21). If we separate the work done into two categories—pV work and all other work—we can write

$$dU = đq + đw_{pV} + đw_X \tag{13.21a}$$

and then

$$dH = (đq)_p + đw_X. \tag{13.21b}$$

This relationship provides the basis for interpreting enthalpy changes when several kinds of work are done by the system—for example, the work done by an electrochemical cell (see Chapter 26).

We are now ready to apply our results to calculation of the internal energy and enthalpy changes in chemical reactions as functions of temperature and pressure. Here is an example of the kind of question one must know how to answer: Suppose that we know the heat of reaction for

$$SO_2(g) + \tfrac{1}{2}O_2(g) = SO_3(g)$$

at 291 K and 1 atm; what is the heat of reaction at 873 K and 1 atm? Now, the "heat of reaction" is just the difference between the enthalpies of products and reactants. Once ΔH is known at one T and p, it can be easily obtained at any other T and p by applying Eq. 14.22 to each of the substances involved. Substituting in Eq. 14.15b, we find the general expression for the enthalpy of reaction as a function of T and p:

$$\begin{aligned}\Delta H_{\text{reac}}(T_2,p_2) &= \sum_i \nu_i h_i(T_2,p_2)\\ &= \Delta H_{\text{reac}}(T_1,p_1)\\ &\quad + \sum_i \nu_i \left\{ \int_{T_1,p_1}^{T_2,p_1} c_{p,i}\,dT \right.\\ &\quad \left. + \int_{T_2,p_1}^{T_2,p_2} \left[\upsilon_i - T\left(\frac{\partial \upsilon_i}{\partial T}\right)_p \right] dp \right\},\end{aligned} \tag{14.23}$$

where $c_{p,i}$ and υ_i are, respectively, the molar heat capacity and molar volume of the ith substance in the pure state.[10] Similarly, from Eqs. 14.15a and 14.19 we obtain for the internal energy change as a function of T and p

$$\begin{aligned}\Delta U_{\text{reac}}(T_2,p_2) &= \sum_i \nu_i u_i(T_2,p_2)\\ &= \Delta U_{\text{reac}}(T_1,p_1)\\ &\quad + \sum_i \nu_i \left\{ \int_{T_1,\upsilon_{1i}}^{T_2,\upsilon_{1i}} c_{V,i}\,dT \right.\\ &\quad \left. + \int_{T_2,\upsilon_{1i}}^{T_2,\upsilon_{2i}} c_{V,i}\left[T\left(\frac{\partial p_i}{\partial T}\right)_V - p_i \right] d\upsilon \right\},\end{aligned} \tag{14.24}$$

where υ_{1i} and υ_{2i} are the initial and final molar volumes corresponding to the states T_1, p_1 and T_2, p_2 for the ith pure substance, in which $p_i = p(T, \upsilon_i)$. Thus to compute the heat of reaction at T_2, p_2 we need to know in general (1) the heat of reaction at some T_1, p_1, (2) the heat capacities of all compounds involved, as functions of T and p, and (3) the equations of state of all substances involved. For perfect gases, of course, the pressure integrals vanish, because $\upsilon_i = T(\partial \upsilon_i/\partial T)_p$. In reactions involving real fluids, however, the pressure term may make an important contribution to the heat of reaction.

Let us now return to the specific example given above and actually carry out the calculations of Eq. 14.23. Since p_1 and p_2 are the same, the pressure integrals drop out (in the vicinity of 1 atm the perfect gas approximation would be valid anyway). The other data that we need are

$$\begin{aligned}\Delta H(291\text{ K}, 1\text{ atm}) &= -5351\text{ J/mol},\\ c_p(SO_2) &= 2.72 + 4.10\times10^{-4}\,T - 4.89\times10^{4}\,T^{-2},\\ c_p(O_2) &= 1.80 + 1.93\times10^{-4}\,T - 2.1\times10^{4}\,T^{-2},\\ c_p(SO_3) &= 3.02 + 1.53\times10^{-3}\,T,\end{aligned}$$

(T in K, c_p's in J/mol K). Substitution and integration over T then gives

$$\begin{aligned}&\Delta H(873\text{ K}, 1\text{ atm})\\ &= \Delta H(291\text{ K}, 1\text{ atm})\\ &\quad + \int_{291}^{873} \left[c_p(SO_3) - c_p(SO_2) - \tfrac{1}{2}c_p(O_2) \right] dT\\ &= -5351 + \int_{291}^{873} [-0.60 + 1.02\times10^{-3}\,T + 5.94\times10^{4}\,T^{-2}]\,dT\\ &= -5219\text{ J/mol}.\end{aligned}$$

---

[9] As we shall see in Chapter 16, this is not a *consequence* of Eqs. 14.18 and 14.20, but one of the *assumptions* necessary to derive them. We have already reached this conclusion for U from microscopic considerations, in Section 13.5; the result for H follows immediately, because for the perfect gas $H = U + nRT$.

[10] These are in general not the same as the values of c_p and υ assigned to each substance in the reaction mixture (partial molar quantities), because the physical process of mixing also involves energy and enthalpy changes—steps 1 and 3 on p. 393. But we need not consider these effects at this time (cf. Section 14.5), because we have defined ΔH_{reac} as the enthalpy change from pure reactants to pure products.

The equations we have derived in this section are generally valid for fluids or for any system whose state is completely defined by two of the variables p, V, T. The paths we have chosen to connect p_1, V_1, T_1 with p_2, V_2, T_2 were selected for the simplicity of the resulting equations. Other paths are of course possible, and we shall consider some of these later. For all our paths, however, we have assumed that the integrals are continuous—that is, that C_p, U, and so on vary continuously with p, V, and T for all substances. This is clearly not true if a phase change such as freezing or vaporization occurs in the course of the process. We must therefore consider the thermochemistry of phase changes.

14.4 Introduction to Phase Changes

We begin with some definitions. A *phase* is a part of a system, physically distinct, macroscopically homogeneous, and of fixed or variable composition; it is mechanically separable from the rest of the system. Another way to say this is that a phase is a region within which all the intensive variables vary continuously, whereas at least some of them have discontinuities at the borders between phases. For example, ice and water in equilibrium form a two-phase system: The ice and water are mechanically separable (say by sieving), each is internally uniform, and there is a boundary between them at which the density changes discontinuously. It is customary to consider all the ice as a single phase, no matter how many pieces may be in contact with the water. By a *phase change* we mean simply a process that brings a system from a state characterized as phase A to another state characterized as phase B. The transition from water to ice (or vice versa) is an example of a phase change.

Although phase changes in systems with several components can be quite complicated, we are here concerned primarily with one-component systems. For any pure substance, a (p, T) diagram shows the various phases (solid, liquid, gas) as distinct regions separated by sharp boundaries; an example is given in Fig. 14.4. A process that involves a phase change corresponds to a path crossing one of these boundaries. The phase change itself occurs entirely at the point where the path crosses the boundary, and thus at constant temperature and pressure. (Later we shall see why this is so.) In all ordinary phase changes (freezing, vaporization, etc.) there is a difference in density between the two phases involved; whenever this is true, it is found experimentally that a quantity of heat is absorbed or released in the process. This amount of heat is called the *latent heat* of the phase change. Because the pressure is constant, the latent heat is also equal to ΔH for the process, whereas ΔU, of course, equals $\Delta H - p\Delta V$.

To illustrate these concepts, suppose that our system is pure water and our path is the isobar $p = 1$ atm. We start with a piece of ice at some temperature below 0°C and gradually add heat. The ice increases in temperature and expands slightly until 0°C is reached. At this point the ice transforms continuously to liquid water, at constant temperature and pressure, with 19.05 J absorbed for each gram of ice melted. Water is unusual in that it expands on freezing, the liquid being 0.083 g/cm^3 denser than the solid at 0°C. After all the ice is transformed to liquid, adding more heat will raise the temperature again; the liquid contracts slightly up to 4°C and then again expands. These phenomena are illustrated in Fig. 14.5.

Latent heats of fusion (melting) and vaporization are listed in Tables 14.2 and 14.3, respectively, along with the

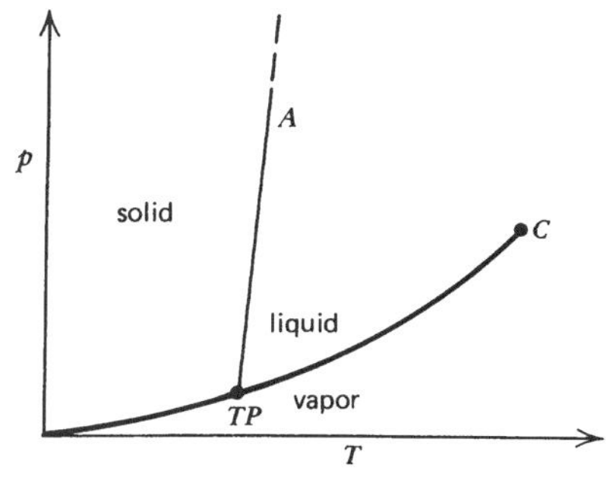

Figure 14.4 A typical phase diagram for a simple substance. The lines represent the locus of pressure and temperature for which two phases can coexist, for example liquid and vapor along the line TP-C. The points TP and C are, respectively, the triple point and the critical point for the substance (see Chapter 24).

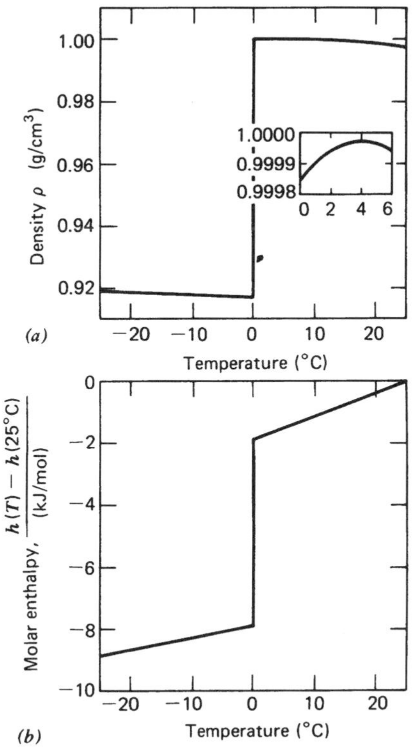

Figure 14.5 Solid-liquid phase transition for water at 1 atm: (*a*) density; (*b*) molar enthalpy (relative to 25°C). The inset in (*a*) shows the anomalous density behavior between 0 and 4°C. Note in (*b*) that the heat capacity $[c_p = (\partial h/\partial T)_p]$ of liquid water is approximately twice that of ice.

Table 14.2 Latent Heats of Fusion of Some Substances

Substance	T_{fus} (K)	Δh_{fus} (kJ/mol)	($\Delta h_{fus}/T_{fus}$) (J/K mol)
Ne	24.55	0.335	13.6
Ar	83.85	1.176	14.0
H_2	13.95	0.117	8.39
O_2	54.34	1.4448	8.19
Cl_2	172	6.406	37.2
HF	189.8	3.928	20.70
HCl	159.0	1.992	12.53
CO	68.1	0.836	12.3
H_2O	273.15	6.007	22.0
NH_3	195.42	5.65	28.9
CH_4	90.6	0.938	10.4
CCl_4	250.3	2.5	10.0
C_6H_6	278.7	9.837	35.30
C_2H_5OH	158.5	5.02	31.7
$(C_2H_5)_2O$	157.0	6.90	44.0
Li	453.7	3.01	6.64
Ca	1123	8.66	7.71
Ba	983	7.66	7.79
Ag	1234.43	11.27	9.13
NaCl	1073	28.8	26.8
LiF	1121	26.4	23.6

Table 14.3 Latent Heats of Vaporization of Some Substances

Substance	T_{vap} (K)	Δh_{vap} (kJ/mol)	$\Delta h_{vap}/T_{vap}$ (J/K mol)
Ne	27.07	1.760	65.02
Ar	87.28	6.519	74.69
H_2	20.39	0.904	44.34
O_2	90.15	6.819	75.64
Cl_2	235.11	20.41	86.81
HF	292.61	7.489	25.59
HCl	188.13	16.15	85.84
CO	81.65	6.04	73.97
H_2O	373.15	40.66	109.0
NH_3	239.74	23.35	97.40
CH_4	111.5	8.180	73.40
CCl_4	349.84	30.0	85.75
C_6H_6	353.25	30.76	87.08
C_2H_5OH	351.55	38.74	110.2
$(C_2H_5)_2O$	307.8	26.60	86.43
Li	1590	148.1	93.14
Ca	1760.3	150.0	85.22
Ba	1910	150.9	79.00
Ag	2485	254.0	102.2
NaCl	1734	170	98.0
LiF	1954	213.3	109.2

corresponding (1-atm) melting and boiling points. Note that the heat of vaporization is normally much larger than the heat of fusion. As can be seen from the table, the ratio of the molar heat of fusion to the melting point, $\Delta h_{fus}/T_{fus}$, is of the order of R (8.3 J/K mol) for simple substances, including metals, but tends to increase with molecular size. For vaporization, the value of $\Delta h_{vap}/T_{vap}$ is close to 85 J/K mol for a wide variety of substances; this generalization is known as *Trouton's rule.*[11] (We shall see in Chapter 17 that $\Delta H/T$ for a phase change is equal to the *entropy* of transition.) Although the data in these tables are all for 1 atm, it should be kept in mind that the transition temperatures and to a lesser extent the latent heats vary with pressure. These and other details of phase equilibria will be discussed at length in Chapter 24.

What concerns us here, however, is the thermochemistry of phase transitions. In general, whenever the path a system follows from state p_1, T_1 to state p_2, T_2 crosses a phase boundary, the latent heat of the transition must be included in ΔH for the process. Thus, consider the above example of the 1-atm ice $\rightarrow$ water transition: The molar enthalpy change, which would be simply $\int_{T_1}^{T_2} c_p\, dT$ in the absence of a phase transition, must in fact be calculated as

$$h(l,T_2)-h(s,T_1)=\int_{T_1}^{0°\mathrm{C}} c_p(s)\, dT+\Delta h_{fus}+\int_{0°\mathrm{C}}^{T_2} c_p(l)\, dT .$$

Clearly, in any other problem involving enthalpy changes, such as the variation of heats of reaction with temperature, one must similarly take account of possible phase changes; in general, one simply adds to Eq. 14.23 a term $\nu_i(\Delta h_{trans})_i$ for each transition along the path of integration.

14.5 Standard States

It should be clear by now that the change in internal energy or enthalpy for a given process measures only the relative properties of the initial and final states. Not only do absolute values of energy and enthalpy not need to be specified, it is meaningless even to speak of "absolute" values, because the zero of energy is arbitrary. It is convenient, however, to have definite reference points to which the energy and enthalpy of substances can be related, because the values so defined can then be tabulated for substances rather than reactions. (This is much more efficient for storing information, because there are of the order of N^2 possible reactions between N substances.) The choice of such reference states is arbitrary. Once such choices have been made, however, they must be used consistently throughout a given set of calculations.

[11] A much more accurate rule is due to Hildebrand. It states that $\Delta h_{vap}/T_{vap}$ is the same for all liquids when the vaporization is carried out at temperatures for which the vapor densities are the same. The value of $\Delta h_{vap}/T_{vap}$ is 93 J/K mol when the vapor density is 2.02×10^{-2} mol/l.

As we shall see, there are many kinds of reference states, usually called *standard states.* For pure substances, however, it is conventional to choose standard states in the following way:

1. The standard state of a chemical substance at a given temperature is taken to be a state of the pure compound or element at that temperature and a pressure of 1 atm.[12] If no temperature is specified, 25°C is to be assumed.
2. Each element in its most stable form at 25°C (298.15 K) and 1 atm is assigned an enthalpy of zero.[13]

One can define a standard state of a given substance in any of its possible physical forms (phases), even one that is not ordinarily stable at the given temperature and pressure: water as liquid or vapor, carbon as graphite or diamond, and so on. If no phase is specified, "standard state" refers to the most stable form under the given conditions. Thermodynamic quantities referring to a standard state are designated by the superscript ^{0}. Thus, the *standard heat of reaction,* the value of ΔH for a reaction in which all the reactants and products are in standard states at the same temperature is designated as ΔH^0 or ΔH_T^0 (where T is the temperature).

We mentioned in Section 14.2 that ΔH for the formation of a compound from its elements is called the compound's heat (enthalpy) of formation. We can now rigorously define the *standard heat of* formation (ΔH_f^0) as ΔH^0 for the reaction in which the compound is formed from its elements in their most stable forms. The values of ΔH_f^0 have been tabulated for a very large number of compounds, usually at 25°C; a sampling is included in Table 14.4. Given our specification of the standard enthalpy of elements, it is clear that for any species at 25°C the standard enthalpy H^0 has the same value as $\Delta H_f^0(25°\text{C})$; to obtain ΔH_f^0 or H_0 at other temperatures one must apply Eq. 14.23. Since the same elements appear on both sides of any chemical reaction equation, it is obvious that the standard enthalpy change in any reaction can be obtained directly from the standard heats of formation of all the species involved:

$$\Delta H^0_{\text{reac}} = \sum_{i=1}^{r} \nu_i h_i^0 = \sum_{i=1}^{r} \nu_1 (h_i^0 - h^0_{i,\text{elements}})$$

$$= \sum_{i=1}^{r} \nu_i \Delta H_f^0(i). \qquad (14.25)$$

Table 14.4 Standard Heats of Formation at 25°C

Substance	ΔH_f^0 (kJ/mol)	Substance	ΔH_f^0 (kJ/mol)
$AlCl_3$	−695.4	NO	90.37
NH_3	−46.19	NO_2	33.85
$BaCl_2$	−860.1	N_2O_4	9.67
$BaSO_4$	−1465	O_3	142
$BeCl_2$	−511.7	PCl_3	−306.4
$BCl_3(l)$	−418.4	PCl_5	−398.9
HBr	−36.2	KBr	−392.2
CaF_2	−1215	KCNS	−203.4
C(diamond)	1.9	KOH	−425.85
CO_2	−393.5	NaCl	−411.0
HCl	−92.3	SiO_2(quartz)	−859.4
CuCl	−136	AgCl	−127.0
$CuCl_2$	−206	NaBr	−359.9
$H_2O(l)$	−285.9	$NaHCO_3$	−947.7
$H_2O(g)$	−241.8	Na_3PO_4	−1925
$FeCl_3$	−405	$Sr(NO_3)_2$	−975.9
$FeSO_4$	−922.6	S(monoclinic)	0.30
$PbSO_4$	−918.4	SF_6	−1100
LiF	−612.1	H_2S	−20.15
$MgCl_2$	−641.8	SO_2	−296.9
Hg_2Cl_2	−264.9	UF_6	−2110
$HgCl_2$	−230	$ZnCl_2$	−415.9

A table of ΔH_f^0 thus gives us ΔH^0_{reac} for all reactions involving only the tabulated substances at the same temperature.

Compounds for which ΔH_f^0 is negative are usually but not always stable with respect to their elements, since energy is released on their formation. A compound with positive ΔH_f^0 can be formed from the elements only when energy is added to the reactive system, and such compounds tend to be unstable. These statements have been hedged because absolute stability is not determined solely by the enthalpy, as we shall see in Chapter 19. Nevertheless, they are correct for nearly all cases in which the magnitude of ΔH_f^0 is large.

Heats of formation are thus very useful, but how are they obtained? For some few compounds one can carry out the formation reaction from the elements directly in a calorimeter. In the many cases where this is not easily feasible, one can obtain ΔH_f^0 by applying Hess's law. In particular, it is usually easy to measure a compound's heat of combustion, after which one can use the heats of formation of the combustion products to calculate that of the original compound. This is especially convenient for organic compounds, whose principal combustion products are CO_2 and H_2O. Although the method of calculation is identical with that described in Section 14.2, the technique is sufficiently important to justify another worked example.

[12] For gases the standard state actually used is that of an equivalent perfect gas at 1 atm (cf. Chapter 21), but in most cases this makes very little difference in thermodynamic quantities.

[13] One could make this assumption for the internal energy or some other energy function, but thermochemical tables and calculations are usually given in terms of the enthalpy. Note that this convention is not suitable in the discussion of nuclear reactions, where transmutation of the elements can occur.

Suppose, then, that we wish to determine the standard heat of formation of gaseous propane (C_3H_8) at 25°C, the formation reaction being

$$3\text{C(graphite)} + 4\text{H}_2(g) = \text{C}_3\text{H}_8(g).$$

The combustion of propane is in turn described by the reaction

$$(1) \qquad \text{C}_3\text{H}_8(g) + 5\text{O}_2(g) = 3\text{CO}_2(g) + 4\text{H}_2\text{O}(g),$$

for which the measured value of ΔH^0 (25°C) is –2044.0 kJ/mol. We can also write the formation reactions

$$(2) \qquad \text{C(graphite)} + \text{O}_2(g) = \text{CO}_2(g),$$
$$\Delta H^0 = \Delta H_f^0(\text{CO}_2, g) = -393.5 \text{ kJ/mol},$$

and

$$(3) \qquad \text{H}_2(g) + \tfrac{1}{2}\text{O}_2(g) = \text{H}_2\text{O}(g),$$
$$\Delta\text{H}^0 = \Delta\text{H}_f^0(\text{H}_2\text{O}, g) = -241.8 \text{ kJ/mol},$$

with the 25°C heats of formation obtained from Table 14.4. Now, the formation reaction for $C_3H_8(g)$ can be constructed by taking the sum 3(Eq. 2) + 4(Eq. 3) – (Eq. 1). By Hess's law, the standard heat of formation of $C_3H_8(g)$ at 25°C must then be

$$\begin{aligned}
&\Delta H_f^0(\text{C}_3\text{H}_8, g) \\
&= 3\Delta H_f^0(\text{CO}_2, g) + 4\Delta H_f^0(\text{H}_2\text{O}, g) - \Delta H^0_{\text{comb}}(\text{C}_3\text{H}_8, g) \\
&= 3(-393.5) + 4(-241.8) - (-2044.0) \\
&= -103.7 \text{ kJ/mol}.
\end{aligned}$$

The use of standard states for pure reactants and products is thus quite straightforward. However, many reactions take place entirely in solution, and one often wants to know the heat of reaction without making additional measurements of the heats of mixing and separation—steps 1 and 3 of the analysis on p. 393. This can be done if one develops for species in solution standard states that do not require referring back to the pure components, together with rules for handling the dependence of enthalpy on concentration. The latter point is important, since (as previously mentioned) there are energy changes even when a solution is formed in the absence of reaction; furthermore, these changes depend on the solvent,[14] so that any definition of standard states must include specification of the solvent.

A variety of reference states are in use for solutions, the choice depending on whether the species in question is solvent or solute, whether or not a solute is ionized, and what types of calculations one wishes to carry out. However, there are two principal conventions. The standard state of a solvent is customarily taken to be the pure solvent at the same temperature and pressure. The treatment of solutes is more complicated. The usual standard state is defined to have the limiting properties of infinitely dilute solutions. The rigorous definition of this state involves concepts we have not yet introduced, and is thus deferred to Chapter 25.[15] What we can consider at this time is the nature of the enthalpy changes that can occur in solution, a subject to which we devote the next section.

14.6 Thermochemistry of Solutions

We are concerned here primarily with the enthalpy changes that occur when substances are mixed to form a solution. These can be referred to by the general term *heats of solution,*[16] but several different quantities are in use to describe these phenomena. In this section we shall define these quantities, say something about their molecular interpretation, and discuss some of the problems involved in calculating heats of reaction in solution.

By "heat of solution" one most often means the *integral heat of solution,* the total change in the system's enthalpy when a given amount of solute is dissolved in a specified amount of solvent at some particular temperature and pressure. For example, dissolving 1 mol (36.5 g) of gaseous HCl in 10 mol (180 g) of liquid water at 25°C and 1 atm releases 69.3 kJ of heat; this process can be represented by the equation

$$\text{HCl}(g) + 10\text{H}_2\text{O}(l) = \text{HCl}(5.55\ m),$$
$$\Delta H_{\text{sol'n}}(\text{HCl}, 5.55\ m) = -69.3 \text{ kJ/mol HCl},$$

where $\Delta H_{\text{sol'n}}$ designates the integral heat of solution and 5.55 *m* is the *molal* concentration (moles per kilogram of solvent) of the HCl. (Since there are 55.5 mol of H_2O to the kilogram, the molality of an aqueous solution is simply $55.5n_2/n_1$, where n_2/n_1 is the solute/solvent mole ratio.) It is useful to write the change of state in a solution process explicitly, as we have done in the above equation; the equation is balanced with respect to mass since "HCl(5.55 *m*)" designates a solution containing 1 mol of HCl and 10 mol of H_2O.[17] In this section all quantities such as $\Delta H_{\text{sol'n}}$ refer to

[14] The usual convention is to denote as *solvent* that component of a solution present in largest concentration. Aqueous solutions are in such common use that an unspecified solvent can usually be assumed to be water.

[15] See also Section 21.2 for the treatment of gaseous mixtures, and Chapter 26 for the modifications necessary for electrolytic solutions.

[16] When two pure liquids are mixed to form a solution, the enthalpy change is commonly called the *heat of mixing.* Although this is formally identical with the heat of solution, the two terms are usually distinguished; often the units used are also different.

[17] Strictly, one should write "HCl(*aq,* 5.55 *m*)," where *aq* identifies an aqueous solution, but our notation is complicated enough already, and water as solvent can usually be assumed.

25°C and a pressure of 1 atmosphere, unless otherwise explicitly stated.

If the solvent is not directly involved in a chemical reaction with the solute, it is convenient to set the enthalpy of the pure liquid solvent (at the temperature and pressure of the solution) equal to zero. In this way we avoid having to manipulate very large and unwieldy numbers for the heats of formation of solvents, which would in any case cancel out when we calculated heats of reaction in solution. Given this convention, we can easily define a heat of formation of any solution for which we know the integral heat of solution. Thus, in the above example we have, using Table 14.4,

$$\Delta H_f(\text{HCl, 5.55 } m) = \Delta H_f^0(\text{HCl, } g) + \Delta H_{\text{sol'n}}(\text{HCl, 5.55 } m)$$
$$= (-92.1) + (-69.3)$$
$$= -161.4 \text{ kJ / mol HCl.}$$

Later in this section we shall see what to do when the solvent is involved in the reaction.

When additional solvent is added to an already existing solution, resulting in a solution of lower concentration, the total change in the system's enthalpy is called the *integral heat of dilution.* For example, adding another 10 mol of water to our 1 mol of 5.55 *m* HCl would release an additional 2.5 kJ of heat:

$$\text{HCl}(5.55\ m) + 10\text{H}_2\text{O}(l) = \text{HCl}(2.78\ m),$$
$$\Delta H_{\text{dil}}(\text{HCl}, 5.55\ m \rightarrow 2.78\ m) = -2.5 \text{ kJ / mol HCl.}$$

The heat of dilution is simply the difference between two integral heats of solution, so that

$$\Delta H_{\text{sol'n}}(\text{HCl}, 2.78\ m)$$
$$= \Delta H_{\text{sol'n}}(\text{HCl}, 5.55\ m) + \Delta H_{\text{dil}}(\text{HCl}, 5.55 \rightarrow 2.78\ m)$$
$$= (-69.3) + (-2.5) = -71.8 \text{ kJ / mol HCl.}$$

The heat of solution for aqueous HCl is plotted as a function of concentration in Fig 14.6, from which it can be seen that the heat of dilution falls off rapidly with increasing dilution. The integral heat of solution is thus fairly independent of concentration in dilute solutions,[18] and in fact reaches a limit for infinite dilution, which we shall discuss shortly.

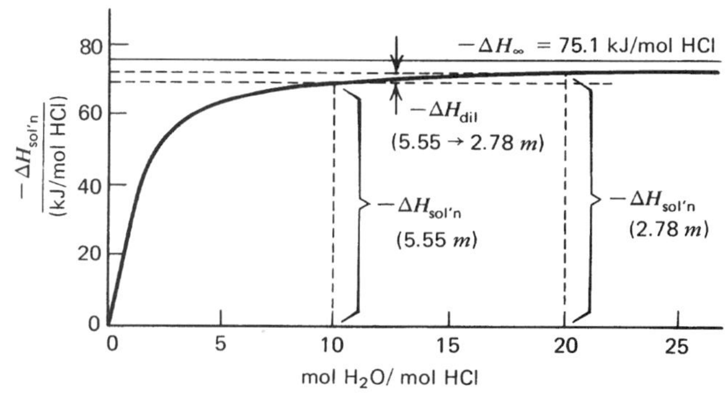

Figure 14.6 Heat of solution of HCl in water.

One other quantity sometimes used is the *differential heat of solution,* $\Delta\bar{h}$. This is the enthalpy change when a small quantity of solute is added to so large a quantity of solution that the concentration can be assumed unchanged by the addition, that is,

$$\Delta\bar{h}_i \equiv \left(\frac{\partial H}{\partial n_i}\right)_{T,p,n_j\,(j\neq i)} = \left(\frac{\partial \Delta H_{\text{sol'n}}}{\partial n_i}\right)_{T,p,n_j\,(j\neq i)}, \tag{14.26}$$

where n_i is the number of moles of solute i present, and j refers to all other components of the solution.[19] In binary solutions there is a simple relationship between the differential and integral heats of solution: At constant T and p, the integral heat of solution is a function of the amounts of the two components, so one can write

$$d(\Delta H_{\text{sol'n}}) = \left(\frac{\partial \Delta H_{\text{sol'n}}}{\partial n_1}\right)_{T,p,n_2} dn_1 + \left(\frac{\partial \Delta H_{\text{sol'n}}}{\partial n_2}\right)_{T,p,n_1} dn_2 = \Delta\bar{h}_1\, dn_1 + \Delta\bar{h}_2\, dn_2\,. \tag{14.27}$$

If we integrate this equation at constant composition, i.e., keeping the ratio n_1/n_2 constant, the $\Delta\bar{h}_i$'s will also remain constant, and we obtain

$$\Delta H_{\text{sol'n}} = \Delta\bar{h}_1\, n_1 + \Delta\bar{h}_2\, n_2\,. \tag{14.28}$$

If we designate the solvent and solute as 1 and 2, respectively, then $\Delta\bar{h}_1$, the solvent's differential heat of solution, is simply the slope of a curve like that in Fig. 14.6. For our previous example one can calculate

$$\Delta\bar{h}_2(\text{HCl}, 5.55\ m) = -64.0 \text{ kJ / mol HCl,}$$
$$\Delta\bar{h}_2(\text{HCl}, 2.78\ m) = -69.4 \text{ kJ / mol HCl.}$$

In the limit of infinite dilution $\Delta\bar{h}_1$ vanishes and $\Delta\bar{h}_2$ becomes equal to the integral heat of solution per mole of solute.

Why is it that enthalpy changes occur when a solution is formed or diluted? This effect is a result of differences in molecular interactions. Consider for simplicity a liquid–liquid solution. If the energies of molecular interaction

[18] For comparison, ordinary "concentrated HCl" is about 16 *m*, or 3.4 mol H_2O/mol HCl.

[19] The $\Delta\bar{h}_i$ defined here is the same as $\bar{h}_i$, the partial molar enthalpy of component *i* (see Chapter 19).

were the same in the two pure liquids A and B, then in the A–B mixture a molecule would find all local environments to have the same energy, independent of composition. In such a case there would be no energy change upon mixing. In reality, however, different molecules do interact in different ways (cf. Chapter 10). The heat of solution is therefore a rough measure of the difference in interaction energies between A–A pairs, B–B pairs, and A–B pairs. In a solution where the interactions between solute molecules differ from those between solute and solvent molecules, the addition of more solvent (dilution) changes the relative amounts of the two interactions and thus alters the total interaction energy.

Let us examine the solution process more closely. Clearly the average distance between solute molecules in a solution is greater than in the pure solid or liquid states of the same solute. In a dilute solution the solute molecules are so far apart that each is surrounded by a nearest-neighbor shell of solvent molecules. In a nonelectrolyte solution the nearest-neighbor interactions are dominant in determining the energy. Let the average (attractive) interaction energy between nearest neighbors be ϵ_{ij}[20] and the number of nearest neighbors be z; then the (integral) solution energy per solute molecule in dilute solution will be

$$\Delta\epsilon_{\text{sol'n}} = z\left[\epsilon_{12} - \tfrac{1}{2}(\epsilon_{11} + \epsilon_{22})\right]. \tag{14.29}$$

This is a very crude model, since in general z may be different in pure solute, pure solvent, and solution. The sign of $\Delta\epsilon_{\text{sol'n}}$ depends on the values of ϵ_{12}, ϵ_{11}, and ϵ_{22}. If $\Delta\epsilon_{\text{sol'n}}$ (and thus $\Delta U_{\text{sol'n}}$) is negative, as is often the case, energy (heat) will be released upon solution; if $\Delta\epsilon_{\text{sol'n}}$ is positive, energy must be added to form the solution. What we have said of the internal energy changes will largely be applicable to the enthalpy changes also, since the change in pV is usually quite small.[21] The simple model outlined here can be greatly refined, as we shall see in Chapter 25. The situation is somewhat more complicated in electrolyte solutions, where there are at least two kinds of solute species interacting by long-range Coulomb forces.

When a solute is dissolved in a sufficiently large amount of solvent, adding more solvent will produce no detectable change in the enthalpy. A solution of this type is said to be *infinitely dilute* in the solute. Of course, different solutions may appear to become infinitely dilute at different concentrations, depending on their particular properties; a solution may also appear to be infinitely dilute with respect to some properties but not with respect to others. To return again to our earlier example, dissolving 1 mol of gaseous HCl in an effectively infinite amount of water, at 25°C and 1 atm, releases a total of 75.1 kJ of heat, as shown in Fig. 14.6; we can represent this by

$$\mathrm{HCl}(g) + \infty\mathrm{H_2O}(l) = \mathrm{HCl}(aq),$$
$$\Delta H_\infty \mathrm{HCl} = -75.1 \text{ kJ / mol HCl},$$

where ΔH_∞ is the enthalpy of infinite dilution (the dilute-solution limit of $\Delta H_{\text{sol'n}}$) and *aq* used alone denotes the state of infinite dilution. Clearly, the enthalpy change when 5.55 *m* HCl is infinitely diluted is just

$$\begin{aligned}&\Delta H_{\text{dil}}(\mathrm{HCl}, 5.55\ m \to 0\ m)\\ &= \Delta H_\infty(\mathrm{HCl}) - \Delta H_{\text{sol'n}}(\mathrm{HCl}, 5.55\ m)\\ &= -75.1 - (-69.3) = -5.8 \text{ kJ / mol HCl}.\end{aligned}$$

Thus far we have discussed only the physical process of solution; let us now consider heats of reaction in solution. We shall illustrate some of the principles involved with ionic reactions in aqueous solution, but the same principles can be applied to reactions in any solvent. We examine first the reactions of ions in (effectively) infinitely dilute solutions. More specifically, we shall indicate how thermochemical studies of such solutions can be used in constructing an ionic model of their structure.

When one mixes aqueous solutions of such salts as LiCl, $CsNO_3$, and $CaCl_2$, no precipitate forms, and at infinite dilution the heat of mixing goes to zero. On the other hand, mixing solutions of $AgNO_3$ and NaCl yields a precipitate, and a nonzero heat of mixing, even when the initial solutions are "infinitely dilute" with respect to their heats of solution. (We assume, of course, that the solutions are not so dilute as to prevent formation of the precipitate; this is a fairly safe assumption here, because the solubility of AgCl or $BaSO_4$ is only about 10^{-5} *m*.) Furthermore, it is found that mixing infinitely dilute solutions of $AgNO_3$ and *any* metal chloride releases the *same* amount of heat, 65.5 kJ/mol of AgCl(*s*) formed. Similarly, mixing infinitely dilute solutions of $Ba(NO_3)_2$ and *any* metal sulfate releases 19.2 kJ/mol of $BaSO_4(s)$ formed. Finally, mixing infinitely dilute solutions of an Arrhenius base (NaOH, KOH) and an Arrhenius acid (HCl, HNO_3) gives the same heat of mixing in all combinations, releasing 55.9 kJ/mol of H_2O formed.

These and many other facts are consistent with the Arrhenius model, in which the species present in dilute aqueous solution are the ions resulting from dissociation of the parent salt, acid, or base. When two infinitely dilute solutions are mixed, provided that the ions undergo no reaction, the environment of each ion remains essentially unchanged (solvent molecules only) and there is no heat of mixing. But suppose that there is a reaction of ions to form a precipitate, water, or some other undissociated compound. In this case ions are removed partially from the solution, and there must be an enthalpy change. In an infinitely dilute aqueous solution the ions are on the average so far apart that each inter-

[20] ϵ_{11}, ϵ_{22} refer to the interaction of like molecules and ϵ_{12} refers to the solvent–solute interaction.

[21] A solution can form spontaneously even when $\Delta H_{\text{sol'n}}$ is positive, because of the entropy of mixing; see Chapter 25.

acts only with the surrounding water molecules.[22] It is therefore to be expected that the enthalpy changes associated with precipitation, acid-base neutralization, and so on, will depend only on the ions removed from solution, and not at all on their original partner ions. We can thus write the reactions described above as

$$Ag^+(aq)+Cl^-(aq)=AgCl(s),$$
$$\Delta H^{\theta}_{reac}=-65.5 \text{ kJ/mol};$$
$$Ba^{2+}(aq)+SO_4^{2-}(aq)=BaSO_4(s),$$
$$\Delta H^{\theta}_{reac}=-19.2 \text{ kJ/mol};$$
$$H^+(aq)+OH^-(aq)=H_2O(l),$$
$$\Delta H^{\theta}_{reac}=-55.9 \text{ kJ/mol}.$$

The superscript $^{\theta}$ is used to denote the state of infinite dilution.

We have seen the advantages of analyzing heats of reaction into contributions from individual reactants and products. Can one apply the same method in ionic solutions, splitting the heat effects into separate contributions from the individual ions present? In other words, can one obtain a set of standard ionic heats of formation, analogous to Table 14.4 for neutral compounds? Consider an infinitely dilute aqueous solution of NaCl, which in our model is a mixed solution of Na^+ and Cl^-. We should therefore have

$$\Delta H^{\theta}_f(NaCl,aq)=\Delta H^{\theta}_f(Na^+,aq)+\Delta H^{\theta}_f(Cl^-,aq).$$

Given our convention that the solvent has zero enthalpy, thermochemical measurements yield

$$\Delta H^{\theta}_f(NaCl,aq)=\Delta H^0_f(NaCl,s)+\Delta H_{\infty}(NaCl)$$
$$=-411.0+3.9=-407.1 \text{ kJ/mol}.$$

But how is this quantity to be divided into contributions from Na^+ and Cl^-? The same problem exists in any other ionic solution: By the condition of electroneutrality, in any macroscopic solution there must be equal amounts of positive and negative charge. We can therefore never isolate and study ions of one species only. The best we can do is to measure the total enthalpy of a set of ions of opposite signs whose charges add up to zero.

It is nevertheless possible to construct a useful set of standard ionic enthalpies. To do so, however, we must make an additional arbitrary definition. If a value is arbitrarily assigned to the enthalpy of any single ion, a consistent set of values can be obtained for all other ions. The convention actually adopted is the following:

The enthalpy of formation of the H^+ ion in infinitely dilute aqueous solution is defined to be zero at all temperatures and pressures.

Let us see how this convention is used to construct a table of standard ionic enthalpies. If $\Delta H^{\theta}_f(H^+, aq)$ is zero by definition, then $\Delta H^{\theta}_f(Cl^-, aq)$ must be the same as $\Delta H^{\theta}_f(HCl, aq)$; from the data already given this is

$$\Delta H^{\theta}_f(Cl^-,aq)=\Delta H^{\theta}_f(HCl,aq)$$
$$=\Delta H^0_j(HCl,g)+\Delta H_{\infty}(HCl)$$
$$=-92.3+(-75.1)$$
$$=-167.4 \text{ kJ/mol}.$$

Combining this in turn with the data for NaCl, we have

$$\Delta H^{\theta}_f(Na^+,aq)=\Delta H^{\theta}_f(NaCl,aq)-\Delta H^{\theta}_j(Cl^-,aq)$$
$$=-407.1-(-167.4)$$
$$=-239.7 \text{ kJ/mol},$$

and so forth for other ions. In Table 14.5 we display the standard enthalpies of formation for a number of ions, deduced in the manner just described. Note that by combining such

Table 14.5 Standard Enthalpies of Formation of Ions in Infinitely Dilute Aqueous Solution at 25°C and 1 atm

Ion	ΔH^{θ}_f(kJ/mol)	Ion	ΔH^{θ}_f(kJ/mol)
H^+	0 (convention)	OH^-	–229.94
Li^+	–278.4	F^-	–329.1
Na^+	–239.7	Cl^-	–167.46
Cs^+	–248	HS^-	–17.7
Cu^+	51.9	I^-	–55.94
K^+	–251.21	SO_3^{2-}	–624.3
Ag^+	105.9	SO_4^{2-}	–907.51
NH_4^+	–132.8		
Ba^{2+}	–538.06		
Ca^{2+}	–542.96		
Cu^{2+}	64.39		
Fe^{2+}	–87.9		
Pb^{2+}	1.6		
Mg^{2+}	–461.96		
Mn^{2+}	–219		
Ni^{2+}	–64.0		
Zn^{2+}	–152.4		
Sr^{2+}	–545.51		
Al^{3+}	5470.33		
Cr^{3+}	5480		
Fe^{3+}	–47.7		

[22] Obviously, ions must come together if any reaction is to take place, but in a solution that is not literally infinitely dilute this can be brought about by random motion. The same principle accounts for reactions in dilute gases. However, these are kinetic effects (see Part III), and the thermodynamic properties depend only on the average distances between particles.

ionic values one can obtain the heat of formation of any infinitely dilute electrolyte solution. For this or any other electrically neutral collection of ions, the effects of the arbitrary choice of zero cancel out.

Heats of reaction in ionic solutions can be calculated in the usual way, by taking the difference between the heats of formation of products and reactants. As already noted, a special problem occurs when the solvent takes part in the reaction. Consider the neutralization reaction between infinitely dilute hydrogen and hydroxyl ions,

$$H^+(aq) + OH^-(aq) = H_2O(l).$$

The water formed in this reaction cannot be treated as having zero heat of formation (our usual convention for the solvent), since it does not appear on both sides of the equation. We must therefore use the standard heat of formation of liquid water from the elements to calculate

$$\begin{aligned}\Delta H^{\ominus}_{\text{reac}} &= \Delta H_f^0(H_2O,l) - \Delta H_f^{\ominus}(H^+,aq) - \Delta H_f^{\ominus}(OH^-,aq)\\ &= -285.84 - 0 - (-229.94) = -55.90 \text{ kJ / mol},\end{aligned}$$

which agrees with the experimental value given earlier.

Reactions (or other processes) at ordinary ionic concentrations are more complicated to describe than at infinite dilution. The complication arises because the interactions (and thus the heats of formation) of the ions vary with their concentrations, which are of course changed by reaction. Indeed, the properties of a given ion are affected by the concentrations of other ions as well. The general theory of electrolyte solutions, a quite complex subject, will be surveyed in Chapter 26.

Regardless of these difficulties, one can of course measure the enthalpy change for any given process, as long as all the relevant concentrations are specified. In interpreting such data, one finds again a special complication for reactions involving the solvent. For the acid-base neutralization reaction in infinitely dilute solution, the addition of 1 mol of water per mole of reaction does not alter the enthalpies of the nonreacting ions; the formation of water as a product is the sole source of the enthalpy change. But at ordinary concentrations one must also consider a dilution effect. If we mix stoichiometrically equivalent amounts of two solutions, one containing x mol of water per mole of HCl and the other containing y mol of water per mole of NaOH, then after neutralization the solution will contain $x + y + 1$ mol of water per mole of NaCl. The total enthalpy change measured thus includes both the heat of reaction and the heat of dilution of Na^+ and Cl^- ions from their hypothetical initial concentration (only $x + y$ mol of water present) to the final concentration after reaction ($x + y + 1$ mol of water present). Formally, one can separate these steps, imagining the overall process to occur in such a way that the mole of water produced by the chemical reaction is removed from the solution, after which an extra mole of "solvent water" is added. This hypothetical sequence of processes is equivalent to the actual reaction, and has the advantage of helping to keep our conventions consistent. As already pointed out, we must say that $\Delta H_f^0 = -285.84$ kJ/mol for the chemically formed water, even though by convention we set ΔH_f^0 equal to zero for the nonreacting solvent water.

14.7 Molecular Interpretation of Physical Processes

In the remainder of this chapter we deal with a variety of phenomena on the molecular level about which information can be derived from thermochemical measurements. The discussion here must be regarded as preliminary, since the second law of thermodynamics is needed for a detailed analysis of these phenomena. We thus give here only those conclusions that can be drawn from the facts already introduced. To begin with, we examine in this section some of the simple physical processes and phase changes discussed in Sections 14.3 and 14.4.

Consider first the vaporization of a liquid. By virtue of the differences in density and thus in average distance between molecules, the molecular interactions are large in a liquid and small in a gas. "Large" and "small" here can be defined with respect to the average kinetic energy in the dilute gas, $\frac{3}{2}RT$ per mole or $\frac{3}{2}k_BT$ per molecule. Thus the heat of vaporization should be a very rough measure of the average intermolecular potential energy in the liquid. By the empirical Hildebrand's rule, we in fact have $\Delta h_{\text{vap}} \approx 11RT_{\text{vap}}$ per mole, of which $RT_{\text{vap}} \approx p\upsilon_{\text{gas}}$ can be attributed to the expansion work and the remainder to the potential energy contribution; the latter is thus about $10k_BT_{\text{vap}}$ per molecule. If we compare intermolecular potential parameters (see Table 21.13) with boiling points (Table 14.3), we find that ϵ, the depth of the potential well, is typically in the range $1 - 2k_BT_{\text{vap}}$ per pair of molecules. The discrepancy between these two figures is due to the fact that each molecule in a liquid has many near neighbors.

We noted in Section 14.4 that the heat of fusion is normally much less than the heat of vaporization. We can now see that this is to be expected, since the density difference between a solid and a liquid is relatively small. In terms of the argument just outlined, neither the expansion work nor the potential energy contribution to Δh_{fus} is large. We shall see in Chapter 24 that the entropy of fusion $\Delta h_{\text{fus}}/T_{\text{fus}}$ is more significant in interpreting microscopic phenomena.

The change in volume of a substance under external pressure is described in terms of the *isothermal compressibility,*

$$\kappa_T \equiv -\frac{1}{V}\left(\frac{\partial V}{\partial p}\right)_T; \tag{14.30}$$

typical values of κ_T are 10^{-4} atm^{-1} for a liquid, 10^{-6} atm^{-1} for a solid, and $1/p$ for a perfect gas. For all mechanically stable substances the volume decreases with increasing pressure, so that κ_T is always positive.[23] The work performed on a substance in compressing it isothermally and reversibly is thus

$$w = -\int_{V_1}^{V_2} p\,dV = \int_{p_1}^{p_2} \kappa_T\, pV\,dp \qquad (14.31)$$

which must also be positive. One might therefore expect $(\partial U/\partial p)_T$ to be positive; yet measurements show $(\partial U/\partial p)_T < 0$ for nearly all fluids (water below 4°C is an exception) up to pressures of the order of 10^4 atm. This is not unreasonable, since the quantity of work involved is small[24] and we have no estimate of the heat given off. The molecular interpretation is simple: In a fluid the average distance between near-neighbor molecules is greater than the potential energy minimum, so compression reduces the total energy; it takes a considerable pressure to get past the minimum. In solids, on the other hand, the average distance is near the potential energy minimum, where $dV(R)/dR = 0$; thus we expect $(\partial U/\partial p)_T$ to be close to zero, and κ_T to be small. But all these arguments are oversimplified, since the main effect of very high pressure is *within* molecules: One can show from the virial theorem that the electronic kinetic energy changes much more than the total energy.

Considerable care must be used in developing molecular arguments of the type employed here. For example, it must require work to compress even a perfect gas; substituting in Eqs. 14.30 and 14.31, we obtain the usual $w = nRT\ln(p_2/p_1)$. Yet we have defined a perfect gas as one for which $(\partial U/\partial p)_T = 0$; thus, the work of compression must be exactly cancelled by the heat given off. In a real fluid the presence of molecular interactions makes the internal energy a function of pressure. In the perfect gas there are no molecular interactions, and the microscopic effect of compression is simply to shift the gas's spectrum of energy levels, as we have already described in Section 13.1.

14.8 Bond Energies

The interpretation of thermochemical data in terms of molecular structures is an important aspect of chemistry. By the methods already described one can obtain fairly accurate values of the total energy of a given species, but there are many problems in interpreting these energies in terms of molecular parameters. One of these problems, to which we have often alluded in Part I, is that the energies involved in chemical bonding are usually small differences of large quantities. The accuracy of quantum mechanical calculations of the properties of small molecules, with only a few atoms and a few electrons, is now good enough to generate a molecular interpretation of their properties. By "properties" we mean such things as electronic structure, energy levels, bond lengths and angles, vibrational frequencies, etc.; by "accurate" we mean reproducing or even predicting the experimental values of these properties. However, the accuracy of quantum mechanical calculations for large molecules with many atoms and electrons is, as of the date of this textbook, often only marginally sufficient or still insufficient for the prediction of their properties or for the construction of a molecular interpretation of thermochemical measurements. Nevertheless, the tools available are effective enough to allow us to rationalize those properties. Although the methodology and accuracy of computational quantum chemistry is steadily improving, the effort required to calculate the properties of a large molecule is considerable. Thus it is still necessary to find methods of bridging the gap between quantum mechanical concepts and macroscopic measurements. Among the most useful bridges are the chemical concepts of bonding and group identity.

The basis of this approach is quite simple: In many molecules the electrons can be described with considerable accuracy as localized in bonds of specific types. Furthermore, a given type of bond has characteristic properties (length, energy, etc.) which are much the same whenever it is found—that is, which are "transferable" from molecule to molecule. For example, the C—C and C—H bonds have essentially the same properties in all saturated aliphatic hydrocarbons (CH_4, C_2H_6, C_3H_8, . . .); we can thus assume that all the bonds in these molecules are approximately "local" in character. When the bonding is well localized, it is often possible to recognize whole groups of atoms that have the same properties in different compounds: the methyl (—CH_3) group, the —OH group in alcohols, and so on.

Under favorable circumstances, then, a molecule may be viewed as an assemblage of localized bonds or groups, each with an identity sufficiently well defined for its properties to persist in different molecules. The theoretical justification for this model has been sketched in Section 9.2, in terms of localized bond orbitals; here we are concerned with its consequences. The model is not completely accurate: In some cases (NO_3^-, benzene, etc.) the bonding electrons are essentially delocalized over the entire molecule, whereas in other molecules the interactions between bonds or groups may be significant, especially the so-called steric effects. Some of these exceptions have also been discussed in Chapter 9, and in the next section we shall look at them from a thermochemical point of view. But the assumption of localized bonds and groups with transferable properties suffices to explain a wide range of chemical phenomena.

To formulate in detail such an explanation, one must assign a consistent set of properties to the various types of chemical bonds. The property with which we are most concerned in this chapter is of course the bond energy. Let us see how one can establish a standard set of bond energies or enthalpies.

[23] If it were negative, the substance would spontaneously contract to a state of lower energy.

[24] If $\kappa_T = 10^{-4}$ and $v = 50$ cm^3/mol (reasonable values for a liquid), the work required for compression to 100 atm is about 150 J/mol; this is small compared to RT, which at 25°C equals 2478 J/mol.

From measured heats of reaction, as already described, one can calculate the total energy or enthalpy for each of the various reactant and product species. But the total energy of, say, liquid water is the sum of intramolecular (bonding) energies and intermolecular energies. We can eliminate the latter by considering reactions in the dilute gaseous phase. Consider, for example, the dissociation of a diatomic molecule:

$$N_2(g) = 2N(g), \quad \Delta H^0_{298} = 941.69 \text{ kJ/mol}.$$

Since the standard enthalpy of $N_2(g)$ is defined to be zero, the enthalpy of formation of atomic nitrogen is just half the enthalpy of dissociation, or 470.84 kJ/mol. More to the point, the enthalpy of dissociation of N_2 is another name for the enthalpy of the N—N bond in N_2. Thus, here we have a direct measurement of a bond enthalpy, and similar measurements can be carried out for other gaseous diatomic molecules. (The "measurements" are usually spectroscopic: cf. Section 7.1.) But how can we extend this method to molecules with more than one bond?

To begin with, we must define our terms. By the *bond energy* we mean the energy required to separate an isolated (i.e., gaseous) molecule into two fragments—atoms or polyatomic groups—at infinite separation from each other, with the molecule and the fragments all in their ground states. In thermodynamic terms, the ground-state restriction means the reaction must be carried out at 0 K, so that the bond energy is ΔU_0 for the dissociation reaction. At all temperatures above absolute zero, however, some fraction of the molecules and fragments will be in excited states. The thermodynamic energy of each species at a given temperature is the average energy of the molecules of that species at a given temperature. Thus the measured value of ΔU_T at any $T > 0$ K contains a temperature-dependent component due to excited states; to obtain the bond energy as we have defined it, one must extrapolate the data to absolute zero.

An additional complication is that one usually measures ΔH rather than ΔU. The *standard enthalpy of a bond* is the measured enthalpy required to break the bond in the gaseous state, usually at 25°C (298.15 K) and 1 atm. This differs from what we have called the bond energy[25] in two respects: the difference between enthalpy and internal energy, and the temperature-dependent contribution just described. If we assume that the reactants and products are all perfect gases, then for any reaction we have

$$\begin{aligned}\Delta H - \Delta U &= \sum (pV)_{\text{products}} - \sum (pV)_{\text{reactants}} \\ &= (\Delta n)RT; \end{aligned} \tag{14.32}$$

Δn, the change in the total number of moles, is of course unity for a bond-dissociation reaction. We thus have

$$\begin{aligned}\Delta H^0_{298} - \Delta U^0_{298} &= (1)(8.31441 \text{ J/mol K})(298.15 \text{ K}) \\ &= 2.4789 \text{ kJ/mol}\end{aligned}$$

for a perfect gas dissociation at 25°C; this is a relatively small fraction of the total dissociation energy (ΔU or ΔH), which is typically several hundred kilojoules per mole. The other component, $\Delta U_T - \Delta U_0$, can be calculated by Eq. 14.24 if one knows the heat capacities of the species involved (cf. Section 21.5); at 25°C, where there is little vibrational excitation, it should be no more than 10 kJ/mol. Thus, the absolute-zero bond energy and the 25°C bond enthalpy usually differ by no more than a few percent.

For diatomic molecules one can define the bond energy or dissociation energy unambiguously and rigorously, and obtain it directly from spectroscopic data. For polyatomic molecules we now also have a rigorous definition of the bond energy, but actually evaluating it is less easy. The motions of the nuclei in a polyatomic molecule are strongly coupled, and one cannot readily extract from the vibrational spectrum a limiting energy corresponding to the breaking of a single bond. Why not then proceed straightforwardly by breaking the molecule in two at the bond in question and measuring the energy required? One way to do this is by bombarding molecules with high-energy electrons until bonds are broken, and then analyzing the masses and energies of the fragments by mass spectrometry or other techniques. Yet even this method is often not feasible, especially for large molecules: The molecules may not exist in the gaseous phase, competing reactions may occur, the bond of interest may not be the one that breaks, the fragments may not be stable, and so on. In practice one must often assign values to bond energies by some averaging method; we shall now look at some methods of this type.

For simple molecules of the form AX_n (cf. Section 9.1) in which there exist only A—X bonds, one can give a rigorous definition of the *average* bond energy, which is simply $1/n$ of the energy required for the reaction

$$AX_n(g) = A(g) + nX(g).$$

To see the significance of this quantity, consider the CH_4 molecule, which is simple enough for the reaction

$$CH_4(g) = CH_3(g) + H(g), \quad \Delta H^0_{298} = 430 \text{ kJ/mol}$$

to be studied directly. By the definition above, 430 kJ/mol is precisely what we mean by the standard enthalpy of the C—H bond in methane. But one can also carry out the subsequent reactions

$$\begin{aligned} CH_3(g) &= CH_2(g) + H(g), & \Delta H^0_{298} &= 473 \text{ kJ/mol}; \\ CH_2(g) &= CH(g) + H(g), & \Delta H^0_{298} &= 422 \text{ kJ/mol}; \\ CH(g) &= C(g) + H(g), & \Delta H^0_{298} &= 339 \text{ kJ/mol}. \end{aligned}$$

[25] What we refer to as standard enthalpies of bonds are often called bond energies in the literature, and the latter term may be used loosely to cover both concepts.

The reason these energies differ from one another is that after each reaction the remaining polyatomic fragment undergoes electronic and nuclear rearrangements that make its bonds somewhat different from what they were in the original molecule. For calculations of the structures of these fragments, the individual bond dissociation energies are required. Yet we know that in the CH_4 molecule all four C—H bonds are equivalent. Thus we are interested in the average energy per bond, which is just one-fourth the sum of the previous four quantities, or 416 kJ/mol. We hypothesize that this is the energy associated with a "typical" C—H bond; is this hypothesis valid for other molecules?

Consider the ethane (C_2H_6) molecule, which has six identical C—H bonds and one central C—C bond. Can we obtain a value for the C—C bond energy without actually measuring it? The enthalpy of complete dissociation (atomization) for C_2H_6 is 2829 kJ/mol; this can be obtained from the ΔH_f^0 values for $C_2H_6(g)$, $C(g)$, and $H(g)$. If the average energy of the C—H bonds were the same as in CH_4, 416 kJ/mol, we could estimate the energy of the C—C bond to be 2829 – 6(416) = 333 kJ/mol. The dissociation energy of this bond has in fact been measured and found to be 352 kJ/mol, so the C—H bonds in C_2H_6 actually have an average energy of 413 kJ/mol. This is a remarkably good match, and indicates that the properties of the C—H bond are *almost* transferable from CH_4 to C_2H_6-. Similar calculations show that the same is true for other saturated hydrocarbons.

By combining data on a large number of molecules, one can obtain average values for the energies of many common types of bonds. Some of these average values are listed in Table 14.6, in terms of both 25°C standard bond enthalpies and absolute-zero bond energies. Although the average energies cannot take the place of measurements on individual molecules, they can be very useful for making predictions in the absence of such measurements. For example, suppose that we wish to compute the 25°C standard heat of formation of gaseous *n*-pentane, which has the structure

$$CH_3CH_2CH_2CH_2CH_3,$$

with four C—C bonds and 12 C—H bonds. We can thus estimate the standard heat of atomization,

$$CH_5H_{12}(g) = 5C(g) + 12H(g),$$

at 25°C to be

$$\begin{aligned}\Delta H^0_{\text{atom'n}}(C_5H_{12}, g) &= 4\Delta H^0_{298}(\text{C—C}) \\ &\quad + 12\Delta H^0_{298}(\text{C—H}) \\ &= 4(342) + 12(416) \\ &= 6360 \text{ kJ/mol},\end{aligned}$$

with bond enthalpies from Table 14.6. The heat of formation of $C_5H_{12}(g)$ is of course the enthalpy change in the reaction

Table 14.6 Some Average Bond Energies and Enthalpies

Bond	ΔU_0(kJ/mol)	ΔH^0_{298}(kJ/mol)
H–H	432.0	435.9
C–C	337	342
C=C	607	613
C≡C	828	845
N–N	155	85
N≡N (in N_2)	941.7	945.4
O–O	142	139
O=O (in O_2)	493.6	498.3
F–F (in F_2)	154.8	158.0
Cl–Cl (in Cl_2)	239.7	243.3
Br–Br (in Br_2)	190.2	192.9
I–I (in I_2)	149.0	151.2
C–H	411	416
N–H	386	354
O–H	458	463
F–H (in HF)	565	568.2
Cl–H (in HCl)	428.0	432.0
Br–H (in HBr)	362.3	366.1
I–H (in HI)	294.6	298.3
Si–H	318	326
S–H	364	339
C–O		343
C=O		707
C–N		293
C≡N		879
C–Cl	326	328

$$5C(\text{graphite}) + 6H_2(g) = C_5H_{12}(g).$$

Applying Hess's law, one can easily show that

$$\begin{aligned}H_f^0(C_5H_{12}, g) &= 5\Delta H_f^0(C, g) + 12\Delta H_f^0(H, g) \\ &\quad - \Delta H^0_{\text{atom'n}}(C_5H_{12}, g) \\ &= 5\Delta H^0_{\text{subl}}(\text{C, graphite}) \\ &\quad + 6\Delta H^0_{298}(\text{H—H}) \\ &\quad - \Delta H^0_{\text{atom'n}}(C_5H_{12}, g) \\ &= 5(716.7) + 6(435.9) - 6360 \\ &= -161 \text{ kJ/mol},\end{aligned}$$

where we have used the measured heat of sublimation of carbon and the heat of atomization just calculated. The value of $\Delta H_f^0(C_5H_{12}, g)$ obtained by more direct thermochemical means (from the heat of combustion) is –146.4 kJ/mol. Our estimate from average bond energies is thus only moderately accurate, but is close enough to be useful if the actual value had not been measured. Note that in calculations of this type the heats of formation of gaseous atoms are crucial. Where the standard state is a gas of diatomic molecules (H_2, O_2, Cl_2, etc.), we have simply

$$\Delta H_f^0(\mathrm{X},g)=\tfrac{1}{2}D_0(\mathrm{X}_2,g)=\Delta H^0(\mathrm{X—X}).$$

A key quantity for organic chemistry is ΔH_f^0 (C,g), the heat of sublimation of graphite; measurements of this quantity were uncertain for a long time, but the value 716.7 kJ/mol now seems established. The heat of sublimation of carbon is *defined* as the enthalpy change of the process

$$\mathrm{C(graphite)} \rightarrow \mathrm{C}(g),$$

where the gaseous carbon is composed of single atoms. In fact, over a very wide range of temperature, carbon vapor consists of C, C_2, C_3, C_5 and still higher polymers. The existence of these polymers and the lack of knowledge of their bond energies made the determination of the heat of sublimation of carbon a very difficult problem for many years.

We can see that average bond energies must be used with caution. The model of localized bonds with transferable properties is only an approximation, after all, and more valid for some molecules than others. The properties of a particular bond always depend to some extent on its environment in the molecule. Some data illustrating this point are given in Table 14.7; it can be seen that some of the variations are quite large. (See also Table 9.2.) The effects here are of various kinds: change in electronegativity (Cl_3C— versus H_3C—), partial double-bond character (HCN, C_2N_2 butadiene), resonance stabilization of fragments 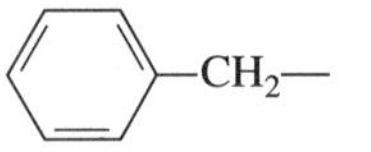 steric effects (the $(CH_3)_3C$—compounds), and so on; many of these phenomena have been discussed in Part I. When only the nearby atoms have a significant effect, one can still obtain average values for the energy of a particular bond in a given environment, or alternatively for the contribution of a particular group of atoms to the molecular energy. For example, the correlation of Benson and Buss states that —CH_3 and —CH_2— groups in a hydrocarbon contribute –42.2 and –20.7 kJ/mol, respectively, to ΔH_f^0 at 25°C; this gives for *n*-pentane

$$\begin{aligned}\Delta H_f^0(\mathrm{C_5H_{12}},g) &= 2\Delta H_f^0(\mathrm{—CH_3})+3\Delta H_f^0(\mathrm{—CH_2—})\\ &= 2(-42.2)+3(-20.7)\\ &= -146.5\ \mathrm{kJ/mol},\end{aligned}$$

which is a vast improvement over our previous calculation. There are many empirical schemes of this sort, all involving lengthy tables of group energies for the various types of molecular environments. Such calculations are naturally more accurate than those using Table 14.6, although no method is completely satisfactory in complicated molecules.

Within limits, then, we can accept as useful approximations the additivity of group or bond energies and their transferability from compound to compound. It is thus possible to estimate the energy change in a chemical reaction in terms of the groups separated and joined, or the bonds broken and formed. In this way estimates can be made for processes on which no experimental data are available. The technique is particularly useful for studying reaction mechanisms, which often involve unstable intermediates of unknown energy, or for predicting the properties of compounds one wishes to synthesize. Although at best one can obtain only approximate values by such calculations, they are often the only methods possible; how useful they are depends on the accuracy required to solve the problem at hand.

Table 14.7 Effects of Molecular Environments on Some Bond Energies

OH bonds	ΔH^0_{298}(kJ/mol)	CH bonds	ΔH^0_{298}(kJ/mol)	CC bonds	ΔH^0_{298}(kJ/mol)
HO–H	498.7	CH_3–H	431.8	CH_3–CH_3	368
HOO–H	374.5	CCl_3–H	377	CF_3–CF_3	406
CH_3O–H	439	CF_3–H	444	CH_3CH_2–CH_3	356
CH_3CH_2O–H	435	N≡C–H	540	$(CH_3)_3C$–CH_3	335
$CH_3C(=O)O$—H	469	$HOCH_2$–H	385	$CH_3C(=O)$—CH_3	335
		CH_3CH_2—H	410		
		$(CH_3)_3$–H	385	N≡C–C≡N	603
		$HC(=O)$—H	364	HC≡C–CH_3	490
		$CH_3C(=O)$—H	360	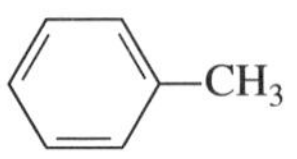	427
		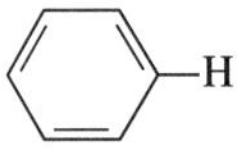	469	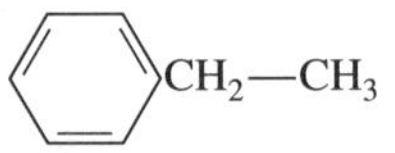	301
		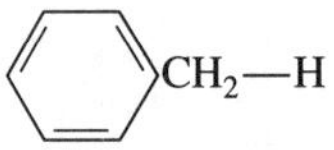	356		

14.9 Some Energy Effects in Molecular Structures

The preceding section was based on the assumption that a molecule is an assemblage of localized bonds or groups with additive and transferable properties. The variations in Table 14.7 are among the evidence showing that this assumption is not completely valid. Yet this failure is in itself useful, in that the deviations from bond additivity can provide information about molecular structure. We have given some examples of this in Chapter 9; here we shall review these and discuss other cases in which quantum mechanical calculations can be combined with thermochemical data.

To begin with, we can now see what is meant by the "resonance energy" in such compounds as benzene (Section 9.2). Suppose that we consider the benzene molecule to have one of the Kekulé structures,

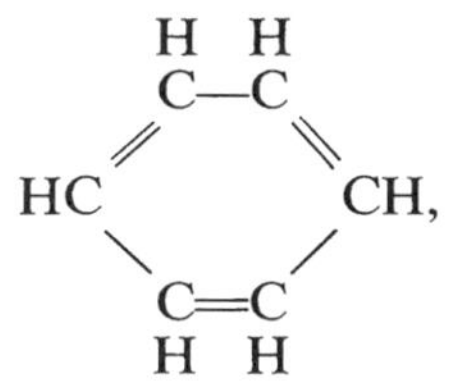

with three C—C and three C═C bonds. If we use the average bond energies in Table 14.6 in the way we did for *n*-pentane, we calculate a standard heat of formation of +247 kJ/mol for $C_6H_6(g)$; yet the experimental value of ΔH_f^0 is +82.93 kJ/mol. The discrepancy of 164 kJ/mol (1.70 eV/molecule) is defined as the resonance energy, which must be accounted for by any theory of the molecule's structure. As we explained in Section 9.2, the resonance energy is essentially the energy of delocalization of the π electrons. Similar effects are found in many other compounds, especially those whose nominal structure contains alternating (*conjugated*) multiple bonds. These include the whole range of aromatic ring compounds (cf. Fig. 9.6), graphite, and some catenated compounds: 1,3-butadiene ($CH_2{=}CH{-}CH{=}CH_2$) has a value of ΔH_f^0 53.6 kJ/mol lower than the nonconjugated 1,2-isomer ($CH_2{=}C{=}CH{-}CH_3$). Delocalization also occurs with nonconjugated multiple bonds, making the adjacent single bonds stronger than usual (such as the C—H bond in HCN, cf. Table 14.7).

Another area that can be investigated thermochemically is the effect of various kinds of strain within molecules, which we have discussed in Section 9.4. One example we mentioned was the cyclopropane molecule,

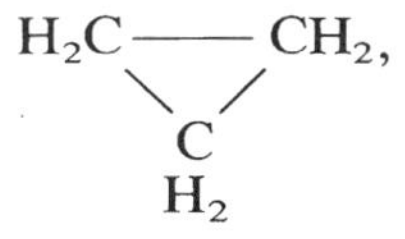

in which the C—C—C angles are constrained to be 60° rather than the normal (tetrahedral) 109°28′. Recall (Chapter 9) that the electron distribution in this molecule could be interpreted to imply that the C—C bonds are best imagined as bent, rather than straight. As a result, its ΔH_f^0 is 120 kJ/mol higher than Table 14.7 would predict; this discrepancy is called the *strain energy.* Another type of strain occurs when atoms not directly bonded to one another are close enough together to interact by van der Waals-type forces, as in branched-chain hydrocarbons, where large side chains can get in each other's way. In all such cases the strain forces bonds to be stretched or bent away from their equilibrium positions, with a resulting increase in energy. Calculations using spectroscopic data to obtain the energies of stretching or twisting bonds have given good agreement with experimental strain energies in many cases.

In Section 9.4 we also discussed the problem of hindered rotation in such molecules as ethane, as illustrated in Fig. 9.12. In ethane the three potential energy minima are identical, so there are three equally likely and indistinguishable conformations for the equilibrium form of this molecule. In a less symmetric molecule, such as 1,2-dichloroethane or cyclohexane (Fig. 9.13), one minimum (the ground state) is slightly lower than the others, but at ordinary temperatures an appreciable fraction of molecules are in the higher-energy conformations. The measured heat of formation must then give a weighted average of the energies of the conformational isomers (*conformers*). This effect can be allowed for by the methods of statistical mechanics, as we shall show in Chapter 21.

In the remainder of this section we consider one of the most fundamental applications of thermochemical data to molecular structure, the estimation of electronic correlation energy. The correlation energy was defined in Part I as the difference between the best possible self-consistent-field calculation of the energy (i.e., the Hartree–Fock energy) and the actual molecular energy. It is not possible to obtain an analytic solution of the Schrödinger equation for any system with more than one electron, but the development of high-speed computers and efficient methods of numerical analysis has permitted the calculation of the Hartree–Fock energies for many molecules. It is also possible to accurately calculate the correlation energy of a small molecule, but it is still very difficult to calculate the correlation energy of a large molecule.

Why should we be concerned about the correlation energy, which is a very small fraction of the total molecular electronic energy? This total energy (defined relative to zero when all the electrons and nuclei are infinitely far from one another) is not really of much interest to us. As chemists we are interested primarily in the energy changes in reactions, which are usually much smaller quantities. If we try to calculate these using our "fairly accurate" theory, we run into the familiar difficulty in determining small differences of large numbers. Consider the following: The Hartree–Fock estimates of the ground-state energies of the N atom and N_2 molecule are, respectively, −1480.25 eV and −2965.69 eV, giving a calculated value of 5.20 eV for the energy of the reaction

$$N_2(g) = 2N(g);$$

the (spectroscopically) measured energies of N and N_2 are –1485.62 eV and –2981.03 eV, with an experimental dissociation energy of 9.79 eV (9446 kJ/mol). The difference between the two sets of values is due to the neglect of correlation energy, the change in which thus accounts for half the dissociation energy. The change in correlation energy is large in this case because the reaction results in the unpairing of electrons; the correlation energy contribution to the energy change in a reaction is not as large (relatively) when the number of electron pairs in the reactants and the products is the same.

We consider first how the correlation energy can be analyzed using ideas from molecular orbital theory and thermochemical data. The correlation energy in a molecule has two components, due to interactions between electrons in the same atom (intraatomic) and in different atoms (interatomic), the first of which should be larger. It will be recalled that in electronic structure calculations a molecular orbital (wave function) is commonly represented as a weighted sum of atomic orbitals. In the separated atoms the correlation energy, obtained from the discrepancy between theory and experiment, can to a good approximation be partitioned into components from each pair of orbitals. It is thus plausible to argue that the intraatomic part of the molecular correlation energy is the sum of these atomic orbital components, with each atomic orbital having the same weighting as in the wave function. Calculations of this kind are now possible for any molecule made up of elements in the first row of the periodic table. This leaves only the interatomic correlation energy to be determined.

The interatomic correlation energy is of essentially the same nature as the van der Waals attraction between separate molecules, which we described in Chapter 10. Such interactions are chemically significant even at intermolecular distances, and thus certainly cannot be neglected within a molecule; an intermediate example is the strain energy mentioned earlier in this section. The change in interatomic correlation energy in a reaction is presumably the residual discrepancy with experiment after one has accounted for all the other energy changes by the methods sketched above. In Fig. 14.7 are shown the results of such calculations for the complete hydrogenation[26] of a number of compounds. Each of these compounds has one central bond between two first-row atoms, the only other bonds being to H atoms; the bulk of the interatomic correlation energy should be that between the two central atoms, and should decrease with increasing distance between them.

The data confirm this hypothesis, the energy falling off approximately as the inverse fourth power of the distance compared with the R^{-6} dependence of the intermolecular van der Waals attraction. On the assumption that the curve of Fig. 14.7 gives the correlation energy between first-row atoms as a function of the distance between them, one can use bond-length data to calculate the change in correlation energy in other reactions. Combining this with all the other calculated values, one obtains the "theoretical" heats of reaction of Table 14.8, which differ from the experimental values by an average of only 9.4 kJ/mol.

At least for simple molecules, then, one can calculate heats of reaction in almost perfect agreement with experimental values, using thermochemical data to estimate the interatomic correlation energy and quantum mechanics to calculate all other components of the energy. This is an interesting demonstration of the power of thermochemistry.

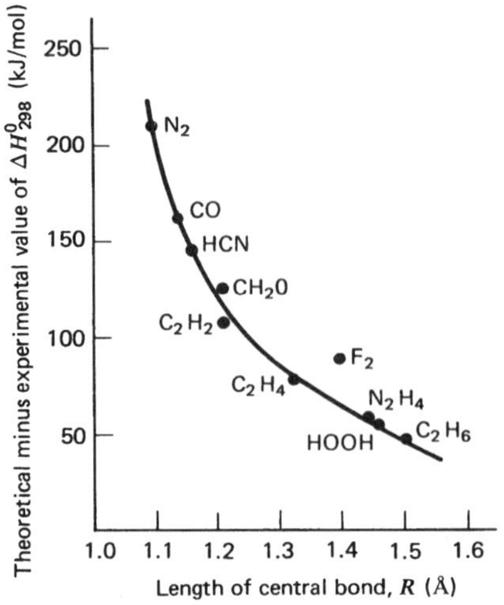

Figure 14.7 Discrepancy between theoretical and experimental values for heat of complete hydrogenation of first-row compounds, presumably due to the change in interatomic correlation energy. (The "central bond" in each case is that between two atoms from the set C, N, O, F.) From L. C. Snyder and H. Basch, *J. Am. Chem. Soc.* **91,** 2189 (1969).

Table 14.8 Theoretical Heats of Reaction[a]

Reaction	**ΔH^0_{298}, Theory (kJ/mol)**	**ΔH^0_{298}, Experiment (kJ/mol)**
$H_2 + C_2H_6 \rightarrow 2CH_4$	–65.7	–64.9
$H_2 + H_2O_2 \rightarrow 2H_2$OABLE	–346.0	–348.1
$2H_2 + CH_2O \rightarrow CH_4 + H_2O$	–214.2	–200.8
$3H + HCN \rightarrow CH_4 + NH_3$	–251.5	–251.0
$3H_2 + C_2H_5N \rightarrow 2CH_4 + NH_3$	–311.3	–321.7
$3H_2 + C_2H_4O \rightarrow 2CH_4 + H_2O$	–328.0	–340.2
$4H_2$ + C3$H_4 \rightarrow 3CH_4$	–487.4	–502.9
$4H_2 + CO_2 \rightarrow CH_4 + 2H_2O$	–144.3	–164.8

[a] The interatomic correlation energy has been estimated from bond-length data and Fig. 14.7. This energy is combined with other calculated values to obtain the theoretical ΔH. Experimental measurements of ΔH^0 at 298 K are also shown.

[26] That is, the reaction to form only simple hydrides, in these cases CH_4, NH_3, H_2O, and HF. For example, the reaction for formaldehyde is

$$CH_2O + 2H_2 = CH_4 + H_2O.$$

Table 14.9 Bond Dissociation Energy Changes for $CH_n \rightarrow CH_{n-1} + H$ at 0 K*

Molecule	Observed (kJ/mol)	Hartree–Fock (kJ/mol)	With Correlation (kJ/mol)
CH_4	471	356	477
CH_3	491	372	493
CH_2 ($^3B_1$)	442	414	437
($^1A_1$)	410		405
CH ($^2\Pi$)	352	238	350
($^4\Sigma^-$)	281		270

* From W. J. Hehre, L. Radom, P. V. R. Schleyer, and J. A. Pople, *Ab Initio Molecular Orbital Theory*, John Wiley & Sons, New York (1986).

We now examine the accuracy of full quantum mechanical calculations of the energy changes in reactions. Consider the reactions $CH_n \rightarrow CH_{n-1} + H$, which we discussed earlier in this section. Table 14.9 displays a comparison between the observed bond dissociation energy changes (corrected to 0 K), the Hartree–Fock estimates of those energies, and the calculated values of those energies including the effects of electron correlation. It is seen that for small molecules, such as involved in these reactions, the calculated and observed energy changes agree, on average, to within 5 kJ/mol.

14.10 Lattice Energies of Ionic Crystals

In Chapter 11 we described the structure and bonding of crystalline solids, but for the most part left their energies for later consideration. Most of our treatment of the thermodynamics of solids is postponed until Chapter 22, but one topic is particularly appropriate at this time. This is the lattice energy of ionic crystals, which can be determined thermochemically, and which can be applied to obtain such data of chemical interest as electron affinities and the relative stability of oxidation states.

Our model for ionic crystals is that introduced in Sections 11.8 and 11.9: We assume the structural units to be discrete positive and negative ions, interacting primarily by Coulomb (electrostatic) forces and relatively short-ranged repulsive forces. We shall consider in this section only crystals made up of monatomic ions, to avoid the complication contributed by internal degrees of freedom and a multitude of atomic interactions. Our initial concern is the lattice energy or cohesive energy, which thermodynamically equals ΔU for a reaction such as

$$\mathrm{MX}(s) = \mathrm{M}^+(g) + \mathrm{X}^-(g);$$

it is thus the negative of the crystal's energy relative to zero for infinitely separated ions.

To begin with, we assume our crystal to be at a temperature of 0 K,[27] and we neglect the zero-point energy of the lattice vibrations. On these assumptions the ions have no kinetic energy, so that the lattice energy is simply a sum of interionic potential energy terms. Assuming the Born–Mayer pair potential of Eq. 11.38,

$$u(R_{ij}) = \left(\frac{Z_i Z_j e^2}{4\pi\epsilon_0 R_{ij}}\right) + \lambda_{ij} e^{-R_{ij}/\rho}, \qquad (14.33)$$

and making the approximation that all the λ_{ij} have the same value λ_{+-}, we derived the total potential energy of the crystal in the form of Eqs. 11.44 and 11.46. We combine these equations as

$$\begin{aligned} U_0 &= \frac{NMZ_+Z_-e^2}{4\pi\epsilon_0 R_0} + zN\lambda_{+-}e^{-R_0/\rho} \\ &= \frac{NMZ_+Z_-e^2}{4\pi\epsilon_0 R_0}\left(1 - \frac{\rho}{R_0}\right), \end{aligned} \qquad (14.34)$$

where N is the number of "molecules" (formula units) in the crystal, M is the Madelung constant, Z_+ and Z_- are the magnitudes of the ionic charges, R_0 is the equilibrium nearest-neighbor distance, z is (approximately) equal to the number of nearest neighbors, and ρ and λ_{+-} are the empirical constants we wish to determine. Designating the energy in Eq. 14.34 as U_0 (the internal energy at 0 K) is simply a convenient choice of energy zero. We could not carry the calculation further in Chapter 11, since we had no way of determining ρ; now, however, we can use thermodynamics to relate ρ to measurable quantities.

To begin with, we again use (still without proof) the relation we introduced as Eq. 14.18. Differentiation with respect to V yields[28]

$$\begin{aligned} \left(\frac{\partial^2 U}{\partial V^2}\right)_T &= T\left[\frac{\partial}{\partial V}\left(\frac{\partial p}{\partial T}\right)_V\right]_T - \left(\frac{\partial p}{\partial V}\right)_T \\ &= \frac{T}{V}\left[\frac{\partial}{\partial T}V\left(\frac{\partial \rho}{\partial V}\right)_T\right]_V - \frac{1}{V}\left[V\left(\frac{\partial p}{\partial V}\right)_T\right] \\ &= \frac{T}{V}\left[\frac{\partial}{\partial T}\left(-\frac{1}{\kappa_T}\right)\right]_V - \frac{1}{V}\left(-\frac{1}{\kappa_T}\right) \\ &= \frac{T}{V\kappa_T^2}\left(\frac{\partial \kappa_T}{\partial T}\right)_V + \frac{1}{V\kappa_T}, \end{aligned} \qquad (14.35)$$

where κ_T is the isothermal compressibility of Eq. 14.30. So far this is a completely general thermodynamic result, valid

[27] See Section 22.2 for the temperature dependence of crystal energies.
[28] We use Eq. II.3 in Appendix II at the back of the book to equate

$$\left[\frac{\partial}{\partial V}\left(\frac{\partial p}{\partial T}\right)_V\right]_T = \left[\frac{\partial}{\partial T}\left(\frac{\partial p}{\partial V}\right)_T\right]_V.$$

for any substance at any temperature. For a solid at 0 K, however, the first term vanishes and we have simply

$$\left(\frac{\partial^2 U}{\partial V^2}\right)_{T=0} = \frac{1}{V\kappa_0}, \tag{14.36}$$

where κ_0 is $\kappa_T(T=0)$.

Now we obtain from our crystal-structure model a result matching Eq. 14.36. The volume of the crystal must be

$$V = KNR_0^3, \tag{14.37}$$

where K is a constant depending on the lattice (in the rock salt lattice, for example, the volume is $2R_0^3$ per ion pair, so we have $K = 2$). We have, in general,

$$\left(\frac{\partial U}{\partial V}\right)_T = \left(\frac{\partial U}{\partial R_0}\right)_T \left(\frac{\partial R_0}{\partial V}\right)_T, \tag{14.38}$$

$$\begin{aligned}\left(\frac{\partial^2 U}{\partial V^2}\right)_T &= \left(\frac{\partial U}{\partial R_0}\right)_T \left(\frac{\partial^2 R_0}{\partial V^2}\right)_T + \left(\frac{\partial R_0}{\partial V}\right)_T \left[\frac{\partial}{\partial V}\left(\frac{\partial U}{\partial R_0}\right)_T\right]_T \\ &= \left(\frac{\partial U}{\partial R_0}\right)_T \left(\frac{\partial^2 R_0}{\partial V^2}\right)_T + \left(\frac{\partial^2 U}{\partial R_0^2}\right)_T \left[\left(\frac{\partial R_0}{\partial V}\right)_T\right]^2.\end{aligned} \tag{14.39}$$

Now, since R_0 is defined as an equilibrium value, it corresponds to a minimum of the potential energy, and in our model at 0 K also a minimum of the internal energy. Thus $(\partial U/\partial R_0)_{T=0}$ vanishes, and we have

$$\begin{aligned}\left(\frac{\partial^2 U}{\partial V^2}\right)_{T=0} &= \left(\frac{\partial^2 U}{\partial R_0^2}\right)_{T=0} \left[\left(\frac{\partial R_0}{\partial V}\right)_T\right]^2 \\ &= \frac{1}{(3KNR_0^2)^2}\left(\frac{\partial^2 U}{\partial R_0^2}\right)_{T=0}.\end{aligned} \tag{14.40}$$

Setting this result equal to Eq. 14.36 gives

$$\left(\frac{\partial^2 U}{\partial R_0^2}\right)_{T=0} = \frac{(3KNR_0^2)^2}{(KNR_0^3)\kappa_0} = \frac{9KNR_0}{\kappa_0}. \tag{14.41}$$

We can now evaluate the parameters ρ and λ_{+-}, the assumed common value of λ_{ij}. Differentiating Eq. 14.34 twice and substituting Eq. 11.45 to clear the exponential, we obtain

$$\begin{aligned}\left(\frac{\partial^2 U}{\partial R_0^2}\right)_{T=0} &= -\frac{NMZ_+Z_-e^2}{2\pi\epsilon_0 R_0^3} + \frac{zN\lambda_{+-}}{\rho^2}e^{-R_0/\rho} \\ &= -\frac{NMZ_+Z_-e^2}{2\pi\epsilon_0 R_0^3}\left(1-\frac{R_0}{2\rho}\right).\end{aligned} \tag{14.42}$$

Setting this result equal to Eq. 14.41 gives

$$-\frac{NMZ_+Z_-e^2}{2\pi\epsilon_0 R_0^3}\left(1-\frac{R_0}{2\rho}\right) = \frac{9KNR_0}{\kappa_0}, \tag{14.43}$$

which rearranges to

$$\frac{R_0}{\rho} = \frac{36\pi\epsilon_0 KR_0^4}{MZ_+Z_-e^2\kappa_0} + 2, \tag{14.44}$$

with all the quantities on the right-hand side known. Once the value of ρ is known, one can also obtain λ_{+-} by rearranging Eq. 11.46 to read

$$\lambda_{+-C} = \frac{MZ_+Z_-e^2\rho}{4\pi\epsilon_0 zR_0^2}e^{-R_0/\rho}. \tag{14.45}$$

Inserting in Eq. 14.44 the experimental value of κ_0 and the appropriate lattice parameters, one can evaluate ρ/R_0, and thus U_0 by Eq. 14.34.

Let us try a sample calculation of the lattice energy of NaCl at 0 K. We have $M = 1.7476$, $K = 2$, $z = 6$ for the rock salt lattice, and experimental values of R_0 = 2.820 Å, $\kappa_T(25°\text{C}) = 4.02 \times 10^{-6}\ \text{atm}^{-1}$ for NaCl. If we assume the same value for κ_0, substitution in Eq. 14.44 gives

$$\begin{aligned}\frac{R_0}{\rho} &= \frac{(36\pi)(8.854\times10^{-12}\ \text{C}^2/\text{N m}^2)(2)(2.820\times10^{-10}\ \text{m})^4}{(1.7476)(1^2)(1.6022\times10^{-19}\ \text{C})^2(4.02\times10^{-6}\ \text{atm}^{-1})} \\ &\quad\times\left(\frac{1.01325\times10^5\ \text{N/m}^2}{\text{atm}}\right)+2 \\ &= 7.11+2 = 9.11,\end{aligned}$$

or ρ = 0.310 Å. Eq. 14.34 then gives for the energy of the crystal

$$\begin{aligned}u_0 &= \frac{(6.022\times10^{23}/\text{mol})(1.7476)(1^2)(1.6022\times10^{-19}\ \text{C})^2}{(4\pi)(8.854\times10^{-12}\ \text{C}^2/\text{N m}^2)(2.820\times10^{-10}\ \text{m})} \\ &\quad\left(1-\frac{1}{9.11}\right) \\ &= -766.5\ \text{kJ/mol},\end{aligned}$$

of which the Coulomb and repulsive terms are –861 and +94.5 kJ/mol, respectively. This is to be compared with an experimental value of –784.9 kJ/mol; the lattice energy as we have defined it is the negative of this quantity. The corresponding value of λ_{+-} from Eq. 14.45 is 2.37×10^{-16} J (1.48 keV). With these values of ρ and λ_{+-} we can calculate the pair potential, Eq. 14.33, as a function of distance; the results are plotted in Fig. 14.8.

Recall that this is only an approximate calculation, since we deliberately omitted several corrections to obtain equations in a simple form. These corrections include the zero-point energy of the lattice, the interionic van der Waals attraction, and the repulsive force between nonnearest neighbors; also, the value of κ_T must be corrected from 25°C

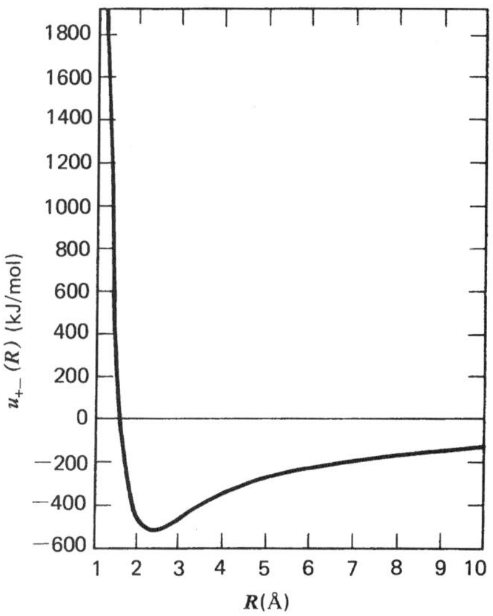

Figure 14.8 The Born–Mayer pair potential for the ionic crystal NaCl.

to 0 K. The inclusion of these effects is straightforward but tedious.[29] When one takes account of them all, the energy of NaCl is found to be –773 kJ/mol. This is an improvement over the simple calculation, but even the latter gave quite good agreement with experiment. At either level of approximation, the results are not really full theoretical evaluations, since experimental data for each crystal must be inserted into the calculations. Nevertheless, the model is useful to the extent that the results obtained for various properties are consistent. The foremost of these properties is the lattice energy. We shall now describe how its experimental value is determined.

Consider a crystal of an alkali halide, with formula MX. It can be synthesized from the elements by two different pathways, as indicated schematically in the following diagram, known as a *Born–Haber cycle:*

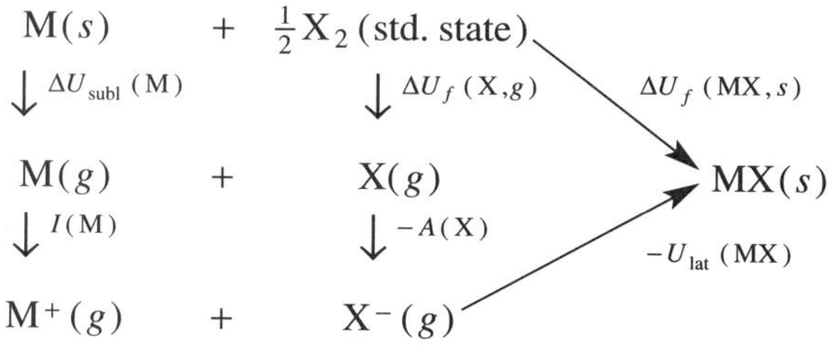

Next to each arrow is written the value of ΔU for the process in question: ΔU_f is the internal energy of formation, and U_{lat} the lattice energy, of the MX crystal; the other quantities are the sublimation energy ΔU_{subl}, the ionization potential I, and the electron affinity A. Note that ΔU_f is the energy change for formation from the elements in their standard states, whereas U_{lat} is $-\Delta U$ for formation from isolated ions. One usually measures ΔH rather than ΔU, but one can be converted to the other ($H = U + pV$) by assuming perfect gas behavior and neglecting pV for solids, so that $H_{\text{gas}} = U_{\text{gas}} + nRT$, $H_{\text{solid}} = U_{\text{solid}}$; we then have, per mole of alkali halide (if X_2 is gaseous), $\Delta u_{\text{subl}} = \Delta h_{\text{subl}} - RT$, $\Delta U_f(\text{X}, g) = \Delta H_f(\text{X}, g) - \frac{1}{2}RT$, and $\Delta U_f(\text{MX}, s) = \Delta H_f(\text{MX}, s) + \frac{1}{2}RT$, to quite high accuracy. The measurements of these various quantities are of course made at temperatures well above 0 K, but one can convert the results to 0 K by applying Eq. 14.19 or 14.22, using the techniques of Chapters 21 and 22 to calculate the heat capacities of gases and solids, respectively.

The point of all this is that, by the first law, the energy change for a given process must be independent of the path. In this case, the energy of formation of MX(s) directly from the elements must equal the sum of the energy changes over the path by way of separated atoms and ions. Rearranging to solve for the lattice energy, we obtain

$$\begin{aligned} u_{\text{lat}}(\text{MX}) &= -\Delta U_f(\text{MX},s) + \Delta u_{\text{subl}}(\text{M}) \\ &\quad + \Delta U_f(\text{X},g) + I(\text{M}) - A(\text{X}) \\ &= -\Delta H_f(\text{MX},s) + \Delta h_{\text{subl}}(\text{M}) \\ &\quad + \Delta H_f(\text{X},g) \\ &\quad - 2RT + I(\text{M}) - A(\text{X}). \end{aligned} \tag{14.46}$$

If all the quantities on the right-hand side are known from thermochemical or spectroscopic measurements, this equation can be used to calculate the lattice energy. If the thermochemical data are obtained at temperature T, the lattice energy at 0 K is

$$\begin{aligned} u_{\text{lat},0}(\text{MX}) &= -u_0(\text{MX}) \\ &= -\Delta H_{f,0}(\text{MX},s) + \Delta h_{\text{subl},0}(\text{M}) \\ &\quad + \Delta H_{f,0}(\text{X},g) + I(\text{M}) - A(\text{X}) \\ &= u_{\text{lat},T}(\text{MX}) + 2RT + \delta(T), \end{aligned} \tag{14.47}$$

where

$$\begin{aligned} \delta(T) &\equiv \int_0^T [c_p(\text{MX},s) - c_p(\text{M},g) - c_p(\text{X},g)]\, dT \\ &= \int_0^T c_p(\text{MX},s)\, dT - 5RT, \end{aligned} \tag{14.48}$$

since c_p for monatomic gases is $\frac{5}{2}R$ (Chapter 21).[30]

In Table 14.10 we compare the values of u_0 obtained by the above method with those calculated by summing the interionic forces. It can be seen that the agreement with the complete theory is quite good, while even the simplified

---

[29] The zero-point and van der Waals energies for NaCl are +5.9 and –24 kJ/mol, respectively. The corrected value of the repulsive energy is 104 kJ/mol (9.5 kJ/mol from nonnearest neighbors).

[30] We assume that neutral atoms and their ions have the same heat capacities (at temperatures too low for excited electronic states to be important), so that I and A are independent of temperature.

Table 14.10 Measured and Calculated Lattice Energies of Alkali Halides at 0 K (kJ/mol)[a]

DATA FOR ELEMENTS

M	Δh_{subl}(M) (298 K)	$I(M)$	X	$\Delta H_f(X, g)$ (298 K)	A(X)
Li	160.7	520.5	F	79.1	332.6
Na	108.4	495.4	Cl	120.9	348.5
K	90.0	418.4	Br	112.1	324.7
Rb	81.6	402.9	I	106.7	295.4
Cs	78.2	375.3			

DATA FOR COMPOUNDS

MX	$-\Delta H_f$(MX, s) (298 K)	$\int_0^{298} c_p$(MX, s) dT	$-u_0$ (meas)[b]	$-u_0$ (calc) (simple theory)[c]	$-u_0$ (calc) (full theory)[d]
LiF	609.6	6.3	1030.9	1060.6	1032.6
NaF	570.3	7.9	915.9	920.1	915.0
KF	562.7	9.6	814.6	807.5	813.4
RbF	551.5	10.5	780.3	764.8	777.8
CsF	565.3	11.7	764.4	730.9	747.7
LiCl	401.7	7.5	850.2	821.3	845.2
NaCl	410.9	10.5	784.9	761.5	777.8
KCl	436.0	11.7	715.9	693.3	708.8
RbCl	432.6	12.6	689.5	665.7	686.2
CsCl[e]	447.3	13.8	674.5	542.2	652.3
LiBr	350.2	8.4	814.6	771.5	797.9
NaBr	359.8	11.7	750.2	722.6	739.3
KBr	392.0	12.6	687.8	662.3	679.5
RbBr	391.2	13.4	664.0	635.5	659.0
CsBr[e]	408.8	14.6	651.9	600.0	632.2
LiI	271.1	10.0	761.1	707.5	739.7
NaI	287.9	13.0	703.3	666.5	692.0
KI	327.6	13.4	648.1	620.1	640.2
RbI	330.5	13.8	627.6	598.7	622.2
CsI[e]	351.0	14.6	618.0	565.7	601.2

[a] Based on D. Cubicciotti, *J. Chem. Phys.* **31,** 1646 (1959); **33,** 1579 (1960); **34,** 2189 (1961).
[b] $-u_0 = -\Delta H_{f,T}(\text{MX}, s) + \Delta h_{\text{subl},T}(\text{M}) + \Delta H_{f,T}(\text{X}, g) + I(\text{M}) - A(\text{X}) + \int_0^T c_p(\text{MX}, s)\, dT - 5RT$; at 298 K, $5RT = 12.55$ kJ/mol.
[c] From Eqs. 14.34 and 14.44, using values of κ_T at 298 K.
[d] Including zero-point, van der Waals, and complete repulsive energies, and corrected for temperature dependence of κ_T.
[e] These crystals have the CsCl lattice, with eight nearest neighbors; all others tabulated have the NaCl lattice.

calculations are adequate for qualitative interpretation. The thermochemical calculation can of course be rearranged to determine any of the quantities in Eq. 14.46 from a knowledge of all the others (including the theoretical lattice energy). This was in fact the method first used to determine the electron affinities of the halogens; it is only in recent years that direct measurements of electron affinities have become sufficiently accurate for the process to be reversed. The data of Table 14.10 give average values of 336.4, 357.3, 336.8, and 307.9 kJ/mol for the A's of F, Cl, Br, and I; all are somewhat higher than the experimental values, showing the existence of a systematic error in even the complete ionic model. This is to be expected, since the real ions in a crystal are not purely spherical charge distributions interacting by simple Coulomb forces, but must be distorted in some way by their neighbors. In addition, the repulsive interaction must be more complicated than the simple Born–Mayer potential. Nevertheless, the ionic model is good enough for interpreting most behavior of chemical interest, and one can anticipate fairly well when it is likely to be in error.

To finish this section, let us now look at one example of how the ionic model can be applied to describe chemical behavior: the relative stabilities of solid metal halides. It is found that for high cation oxidation states the most stable halide is usually the fluoride, that is, that the reaction

$$\text{MX}_{n+1}(s) \rightarrow \text{MX}_n(s) + \tfrac{1}{2}\text{X}_2 \text{ (std. state)}$$

is least likely for X = F (followed by Cl, Br, I in the usual order). To cite some examples, CuI_2 is unstable with respect to CuI at room temperature, whereas all other cupric halides are stable under the same conditions; MnF_4 is stable, but no other tetrahalide of Mn is. To see why this is so, let us begin by setting up a Born–Haber cycle, this time listing the enthalpy change with each step:

$$
\begin{array}{ccc}
MX_{n+1}(s) & \xrightarrow{\qquad \Delta H_{\text{reac}} \qquad} & MX_n(s) \quad + \quad \tfrac{1}{2}X_2\,(\text{std. state}) \\
\downarrow\ u_{\text{lat}}(MX_{n+1})+(n+2)RT & & \downarrow\ u_{\text{lat}}(MX_n)+(n+1)RT \qquad \downarrow\ \Delta H_f(X,g) \\
M^{(n+1)+}(g)+(n+1)X^-(g) & \xleftarrow[\ I(M^{n+})-A(X)\ ]{} & M^{n+}(g)+nX^-(g) \ + \ X(g)
\end{array}
$$

We immediately obtain

$$
\begin{aligned}
\Delta H_{\text{reac}} = {} & u_{\text{lat}}(MX_{n+1})-u_{\text{lat}}(MX_n)-\Delta H_f(X,g) \\
& -I(M^{n+})+A(X)+RT,
\end{aligned}
\tag{14.49}
$$

which should become more negative as X goes from F to I. The data for X atoms, which can be obtained from Table 14.9, are in the right direction:

X	F	Cl	Br	I
$-\Delta H_f(X,g)+A(X)$ (kJ / mol)	253.6	277.6	212.5	188.7

However, we must also estimate the effect of varying X on the lattice energy difference between MX_{n+1} and MX_n.

We already know from Table 14.10 that the simple model of Eq. 14.34 gives qualitatively accurate results for u_0, and we can simplify this even further without serious loss of accuracy. For one thing, the value of ρ does not vary greatly from one crystal to another: The average of the values obtained by Eq. 14.44 for the alkali halides is 0.328 Å, and only one of 20 values differs from this average by more than 0.050 Å. Since this range corresponds to a variation of only ±1.5% in u_0, we can assume that the average value is generally valid. A further approximation comes from Table 11.4, which shows that M/ν (where M is the Madelung constant and ν is the number of ions per "molecule") is roughly constant for a wide range of coordination numbers. To a very rough approximation, then, setting $M = 0.83\nu$ and $\rho = 0.33$ Å, we have

$$
\begin{aligned}
\frac{U_0}{N} &\approx \frac{0.83\nu Z_+ Z_- e^2}{4\pi\epsilon^0 R_0}\left(1-\frac{0.33\ \text{Å}}{R_0\,(\text{Å})}\right) \\
&\approx \left(\frac{1150\ \text{kJ}}{\text{mol}}\right)\frac{\nu Z_+ Z_-}{R_0}\left(1-\frac{0.33\,\text{Å}}{R_0\,(\text{Å})}\right).
\end{aligned}
\tag{14.50}
$$

The value of Eq. 14.50 is that we can apply it without knowing the lattice type, since in general MX_{n+1} and MX_n may have different lattices. The repulsive energy is only a small fraction of the total lattice energy, so the principal effect of such a change of lattice should be through the Coulomb energy, which will be appreciably affected by a change in ν. Although it is very approximate, Eq. 14.50 should therefore serve to indicate a trend. Specifically, the lattice energy terms of Eq. 14.49 become roughly

$$
\begin{aligned}
& u_{\text{lat}}(MX_{n+1})-u_{\text{lat}}(MX_n) \\
& \approx [u_0(MX_n)]_{\text{Coulomb}}-[u_0(MX_{n+1})]_{\text{Coulomb}} \\
& \approx \frac{1150(n+1)n(-1)}{r(M^{n+})+r(X^-)}\left[1-\frac{0.33}{r(M^{n+})+r(X^-)}\right] \\
& \quad -\frac{1150(n+2)(n+1)(-1)}{r(M^{(n+1)+})+r(X^-)}\left[1-\frac{0.33}{(M^{(n+1)+})+r(X^-)}\right],
\end{aligned}
\tag{14.51}
$$

where we introduce ionic radii such that $R_0 = r_+ + r_-$. Next we differentiate with respect to $r(X^-)$, holding all other quantities constant:

$$
\begin{aligned}
& \frac{\partial}{\partial r(X^-)}[u_{\text{lat}}(MX_{n+1})-u_{\text{lat}}(MX_n)] \\
& \approx \frac{1150n(n+1)}{[r(M^{n+})+r(X^-)]^2}\left[1-\frac{0.66}{r(M^{n+1})+r(X^-)}\right] \\
& \quad -\frac{1150(n+2)(n+1)}{[r(M^{(n+1)+})+r(X^-)]^2}\left[1-\frac{0.66}{r(M^{(n+1)+})+r(X^-)}\right].
\end{aligned}
\tag{14.52}
$$

The above expression would be negative even if $r(M^{n+}) = r(M^{(n+1)+})$, and the fact that in general $r(M^{n+}) > r(M^{(n+1)+})$ makes it even more negative; for $r_+ + r_- \approx 3$ Å, the derivative is roughly

$$
\begin{aligned}
& \frac{1150}{9}\left(1-\frac{0.66}{3}\right)[n(n+1)-(n+1)(n+2)] \\
& \approx -200(n+1)\frac{\text{kJ / mol}}{\text{Å}}.
\end{aligned}
$$

Since $r(X^-)$ increases as we go from F^- (1.3 Å) to I^- (2.2 Å), we have established that ΔH_{reac} of Eq. 14.49 becomes more negative with this change. This confirms the experimentally observed order of stability.

Many other relative stabilities of compounds can be similarly explained by crude calculations like those we have carried out here. On a slightly more sophisticated level, one can calculate the relative stabilities of the different lattices a given salt might occupy. This also yields results generally in agreement with experiment, i.e., predicting that the actual lattice is most stable. Much more could be said about the applications of lattice energy calculations, but it is about time for us to terminate this long chapter.

• FURTHER READING

Lewis, G. N., and Randall, M., *Thermodynamics,* 2nd ed., revised by K. S. Pitzer and L. Brewer (McGraw-Hill, New York, 1961), Chapters 5 and 6, and pp. 373–392. For more advanced topics see Pitzer, K. S., *Thermodynamics,* 3d ed. (McGraw-Hill, New York, 1995).

There are many specialized data sources. Some commonly used ones are the following:

Din, F. (Ed.), *Thermodynamic Functions of Gases,* Vols. 1, 2, 3 (Butterworths, London, 1962).

Benson, S. W., *Thermochemical Kinetics,* 2nd ed. (Wiley, New York, 1976), Chapter 2 and Appendices 1–23.

JANAF Thermochemical Tables, 3d ed., J. Phys. Chem. Ref. Data 14, Supplement #1 (1985).

NBS Tables of Chemical Thermodynamic Properties, J. Phys. Chem. Ref. Data 11, Supplement #2 (1982).

Stull, D. R., Westrum, E. F., and Sinke, G. S., *The Chemical Thermodynamics of Organic Compounds* (Wiley, New York, 1969).

Zwolinski, B. J., and Wilhoit, R. C., *Vapor Pressures and Heats of Vaporization of Hydrocarbons and Related Compounds* (Thermodynamics Research Center Data Distribution Office, College Station, Texas, 1971).

Very detailed tables of thermodynamic properties of many individual substances are available in the form of circulars from the National Institute of Standards and Technology (NIST) or the Commission on Thermodynamics of the International Union of Pure and Applied Chemistry (IUPAC). One example is the following:

Angus, S., and de Reuch, K. M. (Eds.), *International Thermodynamic Tables of the Fluid State ^4He* (Pergamon Press, Elmsford, New York, 1977).

• PROBLEMS

1. When 250 g of water at 30°C is poured into a copper calorimeter of mass 3000 g at 0°C, a final temperature of 11.90°C is observed. Taking the specific heat of water to be 4.2 J/g K, calculate the average specific heat of copper.

2. Suppose that the heat capacity of a calorimeter is 913.8 J/K at 23°C. Further, suppose that a piece of metal of mass 35 g, which had previously been heated to 100°C, is immersed in the calorimeter. When equilibrium is achieved the temperature of the calorimeter is observed to have changed from 22.45 to 23.50°C. What is the mean heat capacity of the unknown metal in units of J/g K?

3. Calculate the amount of heat required to increase the temperature of 10^{-3} mol of Ag from 0°C to 900°C at constant pressure, if

$$c_p(\text{Ag}) = 21.3 + 8.54\times10^{-3}\,T + 1.51\times10^{5}\,T^{-2}\ \text{J/K mol.}$$

4. Calculate the heat evolved when 100 g of Au is cooled at constant pressure from 1000 to 0°C, if

$$c_p(\text{Au}) = 23.7 + 5.19\times10^{-3}\,T\ \text{J/K mol.}$$

5. The heat capacity of gaseous CO_2 is

$$c_p = 29.3 + 3.0\times10^{-2}\,T - 7.78\times10^{-6}\,T^2\ \text{J/K mol}$$

in the limit of zero pressure. Calculate the amount of heat needed to raise the temperature of 200 g of gaseous CO_2 from 27 to 227°C at constant pressure. Calculate the amount of heat required to change the temperature by the same amount at constant volume. Assume that CO_2 is a perfect gas.

6. Assume that N_2 behaves as a perfect gas with an average specific heat (at constant pressure) of 1.02 J/g K. Calculate the changes in enthalpy and internal energy when 1 mol of N_2 is heated from 0 to 110°C at 2 atm.

7. Calculate the heat absorbed when 3.1 mol of a perfect gas with c_p = 20.9 J/K mol is isothermally expanded from 1 atm to a volume of 100 liters at 15°C.

8. At a constant temperature of 100°C a perfect gas is transferred reversibly from a pressure of 170 torr and a volume of 4 liters to a pressure of 1 atm. Calculate the work of compression and the amount of heat evolved.

9. Suppose that 1.5 mol of a perfect gas is at a pressure of 5 atm at 0°C. After expansion at constant pressure the gas has a volume of 15 liters. Calculate the work of expansion of the gas and the amount of heat absorbed.

10. One mole of a perfect gas, initially with a volume of 20 liters at 200°C, expands adiabatically against a constant pressure of 1 atm until the pressure of the gas drops to 1 atm. Calculate the work done during this process. Note that this is an irreversible expansion.

11. A combustion experiment is performed by burning a specimen in oxygen in a calorimeter bomb of constant volume. As a result of the reaction the temperature of the water bath surrounding the bomb rises. If the mixture of specimen and oxygen is regarded as the system:
(a) Has heat been transferred?
(b) Has work been done?
(c) What is the sign of ΔU?

12. One mole of a gas obeys the equation of state

$$p(v-b)=RT,$$

where b is a constant. The internal energy of this gas is

$$u=CT,$$

where C is a constant. Calculate the heat capacities c_p and c_V.

13. Derive the equations

(a) $$đq = C\,d_V T + \left[\left(\frac{\partial U}{\partial V}\right)_T + p\right] dV,$$

(b) $$C_p = C_V + \left[\left(\frac{\partial U}{\partial V}\right)_T + p\right] V\alpha,$$

(c) $$đq = C_V\,dT + \left[\left(\frac{\partial U}{\partial V}\right)_T + p\right] dV,$$

where

$$\alpha = \frac{1}{V}\left(\frac{\partial V}{\partial T}\right)_p.$$

14. For water vapor between $\theta = 0°C$ and $\theta = 650°C$, the following formula is approximately valid:

$$c_p = 36.1 + 0.008\theta + 3.0\times 10^{-8}\,\theta^2 \text{ J / K mol}.$$

If $c_p - c_V = R$ determine the increase in internal energy of water when heated from 50 to 650°C.

15. Calculate the changes in internal energy and enthalpy by any two different paths for the following systems and changes of state:

(a) A gas for which the equation of state is

$$p = \frac{RT}{v}\left(1 + \frac{B}{v} + \frac{C}{v^2}\right),$$

where B and C depend on the temperature but not on the volume. The change of state is from (p_1, v_1, T_1) to (p_2, v_2, T_2). Assume c_V is independent of temperature.

(b) A solid for which the equation of state is

$$v = v_0 - Ap,$$

and for which the specific heat at constant volume has the form

$$c_V = C.$$

In the above, A, C, and v_0 are to be considered constants independent of the temperature and volume. As in (a), the change of state is from p_1, v_1, T_1, to p_2, v_2, T_2.

(c) Gaseous CH_4, for which the following data have been obtained:

v (cm^3/mol)	640	640	640
θ (°C)	0	100	200
p (atm)	32.297	46.474	60.486

v (cm^3/mol)	320	320	320
θ (°C)	0	100	200
p (atm)	60.13	91.24	121.81

The change of state is from 30 atm, 0°C to 500 atm, 200°C. *Hint:* Fit the data to the analytic form $pv = RT(1 + B/v)$ and use 100°C temperature intervals to find the temperature dependence of B. Assume c_V is independent of temperature.

16. Calculate the enthalpy change at 298 K and 1 atm for the reaction

$$NaNO_3(s) + H_2SO_4(l) \rightarrow NaHSO_4(s) + HNO_3(g)$$

from the following heats of formation (at 298 K):

$$NaNO_3(s),\quad \Delta H_f^0 = -465.26 \text{ kJ / mol}$$
$$H_2SO_4(l),\quad \Delta H_f^0 = -807.09 \text{ kJ / mol}$$
$$NaHSO_4(s),\quad \Delta H_f^0 = -1118.8 \text{ kJ / mol}$$
$$HNO_3(g),\quad \Delta H_f^0 = -143.93 \text{ kJ / mol}$$

17. Given the following heats of reaction:

$$C_2H_4(g) + 3O_2(g) \rightarrow 2CO_2(g) + 2H_2O(l),\quad \Delta H_{298}^0 = -1411 \text{ kJ / mol}$$

$$H_2(g) + \tfrac{1}{2}O_2(g) \rightarrow H_2O(l),\quad \Delta H_{298}^0 = -286 \text{ kJ / mol}$$

$$C_2H_6(g) + \tfrac{7}{2}O_2(g) \rightarrow 2CO_2(g) + 3H_2O(l),\quad \Delta H_{298}^0 = -1560 \text{ kJ / mol}$$

calculate the standard heat of the reaction

$$C_2H_4(g) + H_2(g) \rightarrow C_2H_6(g)$$

at 298 K.

18. Given

$$C_3H_6(g) + H_2(g) \rightarrow C_3H_8(g), \qquad \Delta H^0_{298} = -1246 \text{ kJ / mol}$$

$$C_3H_8(g) + 5O_2(g) \rightarrow 3CO_2(g) + 4H_2O(l), \qquad \Delta H^0_{298} = -2220 \text{ kJ / mol}$$

$$C(s) + O_2(g) \rightarrow CO_2(g) \qquad \Delta H^0_{298} = -393 \text{ kJ / mol}$$

$$H_2(g) + \tfrac{1}{2}O_2(g) \rightarrow H_2O(l), \qquad \Delta H^0_{298} = -286 \text{ kJ / mol}$$

calculate the standard heats of reaction for

$$C_3H_6(g) + \tfrac{9}{2}O_2(g) \rightarrow 3CO_2(g) + 3H_2O(l)$$

and

$$3C(s) + 3H_2(g) \rightarrow C_3H_6(g)$$

at 298 K.

19. (a) Given

$$Sn(s) + O_2(g) \rightarrow SnO_2(s), \qquad \Delta H^0_{298} = -580.82 \text{ kJ / mol}$$

$$SnO(s) + \tfrac{1}{2}O_2(g) \rightarrow SnO_2(s), \qquad \Delta H^0_{298} = -294.85 \text{ kJ / mol}$$

calculate the standard heat of formation of SnO(*s*).
(b) Given

$$3UO_2(s) + O_2(g) \rightarrow U_3O_8(s), \qquad \Delta H^0_{298} = -318.0 \text{ kJ / mol}$$

$$3U(s) + 4O_2(g) \rightarrow U_3O_8(s), \qquad \Delta H^0_{298} = -3571 \text{ kJ / mol}$$

calculate the standard heat of formation of $UO_2(s)$.

20. (a) On the basis of the energy evolved per gram, which of the following offers the best possibility as a rocket fuel? Assume that gaseous O_2 is the oxidant.
(a) CH_4
(b) C_2H_6
(c) C_3H_8
(d) B_2H_6
(e) NH_3
(f) N_2H_4
(b) Repeat the calculation of part (a), but with gaseous F_2 as the oxidant. Is there any change in your conclusions? Why?

21. Calculate the heat of reaction for

$$2MgO(s) + Si(s) \rightarrow SiO_2(s) + 2Mg(g)$$

at 1000 K. At 298 K the heats of formation of MgO(*s*) and $SiO_2(s)$ are –601.83 kJ/mol and –859.4 kJ/mol, respectively. The heat of vaporization of Mg is 132 kJ/mol at 1393 K. The heat capacities of the substances involved in the reaction are (in J/K mol)

$$\begin{aligned} &MgO(s), && c_p = 45.44 + 5.008 \times 10^{-3}T - 8.732 \times 10^5 T^{-2} \\ &Si(s), && c_p = 24.0 + 2.582 \times 10^{-3}T - 4.226 \times 10^5 T^{-2} \\ &SiO_2(s), && c_p = 45.48 + 36.45 \times 10^{-3}T - 10.09 \times 10^5 T^{-2} \\ &Mg(g), && c_p = 20.79 \\ &Mg(s), && c_p = 24.39 \end{aligned}$$

22. Find ΔH for the following processes:
(a) $KCl(s) \rightarrow KCl(aq)$
(b) Precipitation of AgCl from aqueous solution
(c) Precipitation of PbS from aqueous solution
(d) $HSO_4^-(aq) \rightarrow H^+(aq) + SO_4^{2-}(aq)$

23. Given the standard heats of formation at 25°C,

$$\begin{aligned} &Na_2SO_4(aq), && \Delta H_f^{\ominus} = -1387 \text{ kJ / mol} \\ &Na_2SO_4 \cdot 10H_2O(s), && \Delta H_f^0 = -4324 \text{ kJ / mol} \\ &H_2O(l), && \Delta H_f^0 = -286 \text{ kJ / mol} \end{aligned}$$

calculate the heat absorbed in the process

$$Na_2SO_4 \cdot 10H_2O(s) \rightarrow Na_2SO_4(aq).$$

24. One mole of gaseous SO_2 is dissolved in an infinitely dilute solution containing 2 mol of NaOH [i.e., 2NaOH(*aq*)] at 25°C and 1 atm. Calculate the enthalpy change for this process from

$$\begin{aligned} &OH^-(aq), && \Delta H_f^{\ominus} = -230 \text{ kJ / mol} \\ &SO_3^{2-}(aq), && \Delta H_f^{\ominus} = -624 \text{ kJ / mol} \\ &SO_2(g), && \Delta H_f^0 = -297 \text{ kJ / mol} \\ &H_2O(l), && \Delta H_f^0 = -286 \text{ kJ / mol} \end{aligned}$$

25. When 1 mole of S atoms in rhombic sulfur is dissolved in an infinitely large volume of chloroform at 18°C, 2678J of heat is evolved. When 1 mole of S atoms as monoclinic sulfur is dissolved in an infinitely large volume of chloroform at 18°C, 2343J of heat is evolved. What information can be derived from these data?

26. At 7°C the heat of solution of solid acetic acid in an infinite volume of water is 8910 J/mol, whereas the heat of solution of liquid acetic acid under the same conditions is –1670 J/mol. Calculate the heat of fusion of acetic acid at 7°C.

27. Calculate the maximum temperature and pressure that could be produced by the explosion within a constant-volume bomb of 1 mol of H_2, $\frac{1}{2}$ mol of O_2, and 1 mol of N_2 at 100 torr total pressure and 25°C. Assume that the water vapor formed does not dissociate.

28. One mole of ethyl alcohol is boiled off at 78.1°C and 1 atm with the absorption of 858 J/g. What would you expect the heat of vaporization to be for evaporation into a perfect vacuum?

29. The latent heat of vaporization of water is 40,660 J/mol at 100°C. What part of this enthalpy change is spent on the work of expanding the water vapor?

30. A constant-volume insulated container, initially at 25°C, contains 0.1 mol of $CO(g)$ and 0.05 mol of $O_2(g)$. The gases explode, and 0.1 mol of $CO_2(g)$ is produced. The heat capacity of the container and its contents is 10^3 J/K. Assume that the gases are perfect, and that the standard heats of formation at 25°C of $CO(g)$ and $CO_2(g)$ are –110.5 kJ/mol, and –393.5 kJ/mol, respectively. Find ΔU, ΔT, and ΔH for the system.

31. Use the HF-SCF method along with a 6–31G(d,p) basis set to compute the OH bond dissociation energy, D_e, for the ground electronic state of the reaction

$$\text{OH} \rightarrow \text{O} + \text{H}.$$

Notice the effects of electron correlation by comparing the computed value to the experimental value (corrected for zero-point energy effects) of 448 kJ/mol. A valuable reference is given in Table 14.9. See Problems 5.26 and 7.27 for basis set descriptions and software packages.

32. Repeat Problem 14.31 calculating the bond dissociation energy of H_2. Compare your calculated value to the experimental value (corrected for zero-point energy effects) of 456 kJ/mol.

33. According to Trouton's rule (Section 14.4), the molar entropy of vaporization is approximately 85 J/(K mol) for many substances. Statistically analyze the data in Table 14.3 to determine the effectiveness of Trouton's rule. Which of the substances given in Table 14.3 do not fall within one standard deviation of the mean value? Explain why Trouton's rule fails for these substances.

CHAPTER 15

The Concept of Entropy: Relationship to the Energy-Level Spectrum of a System

In Chapters 12, 13, and 14 we introduced two nonmechanical concepts that differentiate thermodynamics from mechanics, namely, temperature and heat, and we showed how two other concepts from mechanics, work and energy, can be extended to apply to the description of processes in which only gross parameters, such as pressure and volume, are controlled. We also showed how the law of conservation of energy could be used to define heat and work in terms of energy changes, leading to the formulation of the first law of thermodynamics. Finally, we found a very fruitful application of the first law in the subject of thermochemistry.

Now, although the first law of thermodynamics restricts the class of possible processes a system can undergo to the subset that conserves energy, it does not further distinguish among these processes. It does not tell us which energy-conserving changes will, and which will not, actually occur. The following examples illustrate situations where the first law alone is inadequate for the purpose of predicting the behavior of the system.

1. Imagine a box divided into two compartments, one filled with one gas, the other with a different gas. Suppose, for simplicity, that each is a perfect gas, and that the pressure and temperature are the same in the two compartments. If a valve is opened between the two compartments, it is *always* observed that the gases mix. Conversely, if the two compartments are initially filled with the same gas mixture, it is *never* observed that demixing occurs. Yet, given that the gases are perfect, and the temperature is the same in the two compartments, mixing and demixing are processes that do not change the energy of the gases. Why does the mixing process always occur and the demixing process never occur?

2. Imagine a bomb with rigid adiabatic walls filled with two reactants, say, H_2 and O_2. The first law of thermodynamics tells us that the energy of the system is conserved, since no work can be done (rigid walls) and no heat can be transferred (adiabatic walls). Yet we know that an infinitesimal perturbation, such as an internal spark, initiates the exothermic reaction

$$H_2 + \tfrac{1}{2}O_2 = H_2O.$$

How can we predict whether or not this reaction, and others, will occur under specified conditions, that is, whether the chemical equilibrium lies on the product or reactant side?

3. The zero-th law of thermodynamics is based on an essential characteristic of macroscopic systems, namely, that two systems in contact through a diathermal wall, and otherwise isolated, always eventually reach a state of thermal equilibrium. Thus, an isolated copper bar, initially in a state with uniform temperature, never spontaneously undergoes a change to a state with different temperatures at the two ends; and conversely, if prepared with a difference in temperatures at the ends, the bar always approaches the state with uniform temperature. Yet, given that the bar is isolated and can do no work on the surroundings, both processes conserve energy.

The first law alone is also inadequate for the purpose of predicting the direction of a process from knowledge of the energy change in that process. For example, both the mixing of water with concentrated sulfuric acid to form a homogeneous solution and the mixing of water with ammonium chloride to form a homogeneous solution are spontaneous processes. Yet in the former case heat is liberated ($\Delta H < 0$) and in the latter case heat is absorbed ($\Delta H > 0$).

These commonplace illustrations show that we need to develop some criteria for the occurrence of spontaneous processes and for the equilibrium that exists under given constraints. Those criteria cannot be derived from the first law of thermodynamics alone.

The new concept needed for the development of criteria for equilibrium in various situations is embodied in the definition of a thermodynamic function, called *entropy,* and in the formulation of the *second law of thermodynamics,* which defines the relationship between possible states of equilibrium and the entropy of the system. We shall find that the entropy plays the role of a potential function for a macroscopic system in the sense that derivatives of the entropy with respect to parameters of constraint define the generalized forces acting on the system, and the extremal properties of the entropy determine whether or not an equilibrium is stable. We shall also find that the change in entropy in a constant energy process uniquely characterizes whether or not that process can occur spontaneously and that, in general, the change in entropy distinguishes between reversible and irreversible processes.

In this chapter, and the next, we study the second law of thermodynamics from two points of view. First, in this chapter we develop further the notion, introduced in Chapter 13, that there is a relationship between the redistribution of molecules over the energy levels of the system and the heat transferred to or from the system in a reversible thermodynamic process. Our study will require consideration of how much information about the microscopic states of a system of many molecules is contained in the specification of only a few macroscopic parameters, for example, the total energy, mass, and volume. We shall see that a macroscopic system with nonzero energy has an energy spectrum with enormously large degeneracies. That is, specifying that the energy of a macroscopic system lies in the range E to $E + dE$, with $dE/E << 1$, only very mildly constrains the possible distribution of molecules over the energy-level spectrum, there being a huge number of distributions with the same energy. The essential idea in the statistical mechanical description of equilibrium states is to find the most probable distribution of molecules over the energy levels, given values of a few macroscopic parameters, and then to relate other macroscopic properties of the system to averages of microscopic quantities over this most probable distribution. We shall show that in the usual cases the most probable distribution of molecules over the energy levels is very sharply peaked. Furthermore, we shall show that macroscopic properties are sufficiently insensitive to many of the details of the energy-level spectrum of the system that complete knowledge of that spectrum is not needed to establish the general features of the macroscopic behavior.

After constructing the microscopic interpretation of entropy, in Chapter 16 we shall study the classical formulation of the second law of thermodynamics, and of the introduction of the entropy. The link between the two formulations, one microscopic and one macroscopic, provides an illuminating interpretation of the nature of thermodynamic equilibrium.

The development of the second law of thermodynamics and the relationship between the microscopic and macroscopic concepts of entropy are two of the greatest intellectual triumphs of science. A careful study of the contents of this chapter and the next will be rewarding, in that the material presented will lay the foundations for detailed analyses of the chemical and physical properties of matter.

15.1 The Relationship between Average Properties and Molecular Motion in an *N*-Molecule System: Time Averages and Ensemble Averages[1]

Thus far we have accepted as valid the identification of a macroscopic property with the average of a mechanical quantity related to the molecular motion in a system. For example, in the discussion of the kinetic theory of the perfect gas, the average force per unit area exerted on a plane surface by the gas molecules or, equivalently, twice the average momentum transported across a plane per unit area per unit time, was identified with the macroscopic pressure. Three questions immediately arise when we examine this identification. First, why do we use an averaging process? Second, which of several possible averaging processes is the correct one? Third, do different averaging processes lead to the same description of the system?

The answer to the first question has been discussed in several places, but is important enough to be repeated here. A thermodynamic description of a system containing many molecules is characterized by the use of a small number of macroscopic variables, whereas a complete microscopic mechanical description requires a vast number of variables. It is very clear, from the nature of the different analyses, that if a thermodynamic description is to be consistent with a microscopic description, some grouping together, or averaging, or systematic ignoring of microscopic variables must be an inherent part of the connection between the two theories. The thermodynamic description of a system is inevitably coarser than the microscopic description.

The answers to the second and third questions require deeper analysis of how measurements are made and how systems to be studied are prepared. In studying the kinetic theory of the perfect gas we assumed that the system, a container of volume V with N gas molecules, was isolated and in a state of equilibrium. Then the pressure exerted by the gas molecules was calculated by (1) treating all molecules as independent particles, (2) deducing the momentum transported by one molecule across a plane per unit time per unit area, and (3) summing up contributions to the total momentum transport per unit time per unit area from all molecules moving in the appropriate direction. To execute step 3, use was made of the fact that in a gas at equilibrium the velocity distribution function, $f(\mathbf{v})$, is independent of time.

[1] Further information about the nature of ensemble averages can be found in Appendix 15A.

The procedure just described is a calculation of the statistical average of the momentum transport per unit time per unit area. The use of a time-independent velocity distribution function for this purpose implies that averaging the momentum transport for a typical molecular velocity over the distribution of velocities is equivalent to averaging over time the total momentum transported per unit time by all molecules crossing the plane. It is also equivalent to following the trajectory of a *single* molecule for a very long time and averaging the momentum it transports across the plane in successive crossings. All of these procedures, of course, refer to momentum transport in one direction across the plane.

Let us now examine what these equivalences mean. In general, we must expect that the successive crossings of a plane by different molecules are not evenly spaced in time, and that the instantaneous rate of momentum transport varies in time. But in a state of equilibrium the average rate of transport of momentum across a plane by the molecules must be time-independent. Therefore the time average of the rate of transport of momentum, namely, the sum of all momentum transfers in some long interval $\mathcal{T}$ divided by $\mathcal{T}$, must be independent of $\mathcal{T}$ provided that $\mathcal{T}$ is large enough. When $\mathcal{T}$ is small, the differences in the intervals between crossings are comparable to $\mathcal{T}$, so that the rate of momentum transport fluctuates on the time scale of $\mathcal{T}$. On the other hand, when $\mathcal{T}$ is large, differences between the intervals between crossings are small relative to $\mathcal{T}$, the total momentum transported across the plane becomes proportional to $\mathcal{T}$, hence the time average rate of momentum transport becomes independent of $\mathcal{T}$.

Suppose, now, that we imagine following the trajectory of one molecule in the gas. This molecule occasionally collides with other molecules and with the walls as it travels back and forth across the volume V. If we follow the trajectory long enough, we expect that, by virtue of the infrequent collisions mentioned, the molecule we are following will approach arbitrarily close to every point inside the volume V, and have at some time or other a velocity arbitrarily close to any selected velocity consistent with conservation of energy. This expectation is a special case of the *quasi-ergodic hypothesis,* of which we shall say more later. When this hypothesis is valid, and it is generally expected to be valid, we see that if we wait long enough, the trajectory of the molecule we are following will intersect any reference plane with all possible angles and velocities. The distribution of these angles and velocities will be the same as for the collection of molecules with distribution of velocities $f(\mathbf{v})$ transferring momentum across the same plane at one instant. Then, over a sufficiently long time, the distribution of velocities along any one trajectory must approach the time-independent form $f(\mathbf{v})$, thereby rendering averages over a single trajectory equivalent to averages over $f(\mathbf{v})$.

In the case of the dilute gas, collisions between molecules are infrequent and it is possible, in practice, to follow a sufficient length of trajectory to permit the time average of a mechanical quantity to be computed. For example, for the purpose of calculating momentum transport across a reference plane in the gas, collisions between the gas molecules can be neglected. Then the distance a molecule travels in time t simply grows linearly with t. Thus, summing up all contributions to the momentum transport, dividing by t, and letting t become indefinitely large introduces no change into the calculation made in Chapter 12.

If, on the other hand, collisions are very frequent, as in a dense gas or a liquid, a molecule is rarely if ever out of the force field of some other molecule. In these cases the trajectory of a molecule becomes very complicated and, for all practical purposes, impossible to compute.[2] Since we cannot compute directly the time average of a molecular property without knowledge of the molecular trajectories, a different procedure must be used to relate averages of molecular properties with macroscopic properties.

Our study of the perfect gas suggests the procedure needed. We have noted that in the equilibrium state, for a perfect gas, an average of a mechanical quantity over a trajectory is equivalent to an average of the same quantity with respect to the time-independent velocity distribution function. The proper generalization of this equivalence, applicable to any system, was suggested by Gibbs, and independently by Einstein. The idea is to replace the isolated mechanical system and the required time averaging with a scheme of calculation to which they are equivalent. We have already noted many times that the macroscopic description of a system requires only a few gross variables, whereas the microscopic description requires many variables. This implies that, if our knowledge of a system is represented by a few macroscopic parameters, there must exist many different sets of microscopic variables consistent with what is known. Gibbs proposed that, for purposes of calculation, the isolated system under study be replaced by a collection of systems, each having the same macroscopic properties, with the total collection encompassing all possible distributions of the molecular variables consistent with the given macroscopic properties. For example, if all we knew about a perfect gas were the total energy of the N molecules in a box of volume V, Gibbs's proposal would call for imagining a collection of boxes, each with N molecules, each of volume V, and within each of which the distributions of molecular positions and velocities could be anything, provided only that in each box the total energy of the N molecules equaled the originally specified value. Any one of the boxes which represents one of the possible microscopic states is called a *replica* system, and the entire collection of boxes an *ensemble.*

---

[2] The trajectories of individual particles have been followed in computer simulations of molecular motion in systems with $10^3 - 10^4$ particles. Such calculations are lengthy and the results (trajectories) not amenable to simple analytical representation. At present, the interpretation of the results of computer simulations of molecular dynamics rests on concepts introduced in the statistical theory of matter.

Given that there can be many different sets of molecular velocities and positions consistent with our knowledge of a few macroscopic parameters, we can introduce another kind of average, different from a time average. Suppose that we compute the value of some mechanical quantity, say, the kinetic energy per molecule. If we sum up the kinetic energies of the molecules in all of the replicas, and divide by the total number of molecules in the ensemble, we obtain the ensemble average of the kinetic energy per molecule. A similar definition of ensemble average holds for other quantities.[3] Note that the time average would have been obtained by following the trajectories in one isolated replica system for an infinite period of time. It is now postulated that the average of any variable over the distributions of positions and velocities in the ensemble of replica systems (the ensemble average) is equal to the time average of the same variable in any one of the isolated replica systems. If this postulate is valid, time averages, which are difficult to compute, can be replaced by averages over an ensemble, which are easy to compute.

The postulate stated, relating time and ensemble averages, is a direct consequence of what is known as the quasi-ergodic hypothesis. To understand the importance of the quasi-ergodic hypothesis, and for many other reasons as well, we introduce a geometric visualization of the molecular dynamics; for convenience we use the language of classical mechanics. Imagine a many-dimensional coordinate system, the axes of which represent all the independent components of momentum and position of all N molecules in a volume V. If the molecules have no internal structure, and can therefore be treated as point particles, there will be $6N$ such axes. These axes represent the $3N$ position coordinates, $x_1, y_1, z_1, \ldots, x_N, y_N, z_N$, and the $3N$ momentum components $p_{x1}, p_{y1}, p_{z1}, \ldots, p_{xN}, p_{yN}, p_{zN}$. This space is called the *phase space* of the N-molecule system. It is usually assumed that this phase space is a Cartesian space with the distance between two points given by

$$D(w, w^*) = \left[\sum_{j=1}^{2N} (w_j - w_j^*)^2 \right]^{1/2};$$

$$w_j, w_j^* = x_1, y_1, z_1, \ldots, p_{xN}, p_{yN}, p_{zN}.$$

For any given value of the energy the possible values of the momenta and coordinates define a surface in the phase space. For example, for a perfect gas the energy depends only on the particle momenta, $E = (p_{x1}^2 + \cdots + p_{zN}^2)/2m$, which is the equation for the surface of a $3N$-dimensional sphere. When the molecules interact, the energy depends on their instantaneous positions, and the form of the surface is more complicated.

Any fully specified set of values for all the coordinates and momenta corresponding to energy E can be represented as a point on the energy surface in the phase space of the system. Since the energy of an isolated system cannot change, the trajectory the system follows can be represented as a locus of points on the energy surface in phase space. Indeed, it is convenient to think of the evolution in time of an isolated system, after preparation in some initial state, as following a path on the energy surface.

The difference between time averaging and ensemble averaging can be visualized as follows. For any single isolated system, with given E, V, N, the evolution in time is represented as a single path on the energy surface in the system phase space. If we assert that the macroscopic properties of the system are related to the time averages of the corresponding microscopic properties, we are, in effect, following the path representing the mechanical state of the system on the energy surface and then averaging over the sampling of coordinates and momenta along that path. The ensemble average is quite different. Remember that the instantaneous state of a single isolated system is represented by one point on the energy surface, and its evolution by one path. Remember also that the replica systems that make up an ensemble all have the same values of E, V, N, hence the system phase spaces and energy surfaces of all replicas are the same. We can then choose one system phase space, and plot, on one energy surface, representative points for each of the replicas in the ensemble. Each of these points corresponds to a system with the same values of E, V, N, but with different specific instantaneous values of the molecular momenta and coordinates. Clearly, there are so many mechanical states consistent with given E, V, N that it is reasonable to expect the representative points to densely cover the energy surface. The identification of a macroscopic property of the system with the ensemble average of the corresponding microscopic property means adding up the contributions characteristic of each of the representative points on the energy surface and dividing by their number. Unless the path of a single system on the energy surface covers it in a way similar to that achieved by the definition of the ensemble, time averaging and ensemble averaging appear to be different. Despite this apparent difference there are conditions under which the two averages are equivalent. We discuss these conditions in the following paragraphs.

We shall adopt the ensemble average as the fundamental concept of our description of equilibrium. That is, we adopt as *a definition* that a *macroscopic property of a system is to be identified with the ensemble average of the corresponding microscopic dynamical property.* We shall not seek further justification for this identification; we take it as an irreducible concept of the theoretical description. The ultimate justification for our procedure will be based on the comparison of theory and observation. As we shall see, the molecular theory based on ensemble averages is a very successful description of the equilibrium state.

[3] In general, the ensemble average of a quantity is the sum over all replicas of its value in each replica, divided by the total number of replicas.

Ergodic theory, at least as applied to statistical mechanics, is devoted to the study of the equivalence of time and ensemble averages for the properties of a system. The ergodic hypothesis, originally advanced by Boltzmann, is now known to be incorrect. Boltzmann postulated that, if only one waited long enough, the path representing the evolution of a system would pass *through* every point on the energy surface. That this cannot occur in general is a consequence of two facts: First, the equations of mechanics have unique solutions, which implies that the path representing the evolution of a single system cannot cross itself. Second, given the uniqueness of the path, an essentially one-dimensional trajectory can never "fill" an energy surface of dimensionality greater than 1.

The quasi-ergodic hypothesis is a weaker statement about the trajectory on the energy surface. It is based on the *quasi-ergodic theorem* proved by G. Birkhoff in 1931. Briefly, Birkhoff proved that if one cannot draw a boundary on the energy surface such that the trajectory of the representative point lies entirely on one side of the boundary (which implies that the energy surface cannot be divided into sections that have the property that the trajectory of the representative point is confined to lie in only one section) then the trajectory will pass *arbitrarily close to* every point on the energy surface.[4]

The difficulty with applying this theorem, which is valid only in classical mechanics, is that it has not generally been possible to demonstrate whether or not the energy surface corresponding to some given system can be divided as specified. In the absence of such a demonstration we cannot be certain of the equivalence of time and ensemble averages. When the equivalence is postulated to exist, we use the term quasi-ergodic hypothesis.

There are still other problems with attempts to base a description of equilibrium on ergodic theory. The examples cited earlier, namely, mixing of gases, equalization of temperature, and chemical reaction, are typical of macroscopic irreversible processes. Moreover, the idea that the character of an equilibrium depends on constraints, and changes when the constraints are changed, contains the notion of an approach to equilibrium. Clearly, the microscopic theory should at least have the characteristic that, whatever the mechanical state initially, the state to which the system evolves should be time-independent. In this sense the condition of quasi-ergodicity seems necessary but not sufficient. That is, a system may be both quasi-ergodic and quasi-periodic.[5] An example of this behavior is provided by a system of two harmonic oscillators with incommensurable frequencies (ratio of frequencies an irrational number). For this case one can establish that the energy surface is uniformly covered by the trajectory of the representative point, yet any initial state recurs with arbitrary precision. Other more general examples of the same behavior exist.

In the example just given, if two trajectories start from points close together on the energy surface, they remain close together everywhere; that is, the trajectories move apart only smoothly and slowly, the distance between them growing proportional to t. The property needed to avoid reversibility must generate more turbulent trajectories than does quasi-ergodicity. This more general property is called *mixing.* When a system is said to be mixing, two trajectories that start from nearby points on the energy surface diverge from one another at an exponential rate. In these cases the trajectory looks chaotic, and recurrence of the initial state does not occur. Mixing implies quasi-ergodicity, but the reverse is not true. As might be expected, proving that a given system has the mixing property cannot usually be carried out, so its invocation must be considered a hypothesis in most applications.

There are several recent important developments that bear on the foundations of statistical mechanics. First, Sinai has proven that for a system consisting of $N \geq 2$ hard spheres in a box, the only conserved mechanical quantity is the energy, and that the system is mixing and quasi-ergodic. Second, some very general analytical studies and some detailed numerical analyses of the solutions of the equations of motion of modeled systems lead to the conclusion that it is generally to be expected that the trajectory of a representative point becomes very irregular, and uniformly fills the energy surface. In fact, it is now known that a general mechanical system will have, in some energy regions, motion with the characteristic feature that the separation between two trajectories started with infinitesimally different initial conditions grows linearly with increasing time and, in other energy regions, motion with the characteristic feature that the separation between two trajectories started with infinitesimally different initial conditions grows exponentially with increasing time. The latter region is said to exhibit *dynamical chaos.* The numerical studies suggest that in the limit of a system of "thermodynamic size," $N \approx 10^{23}$, for all the systems of interest to us it is satisfactory to describe the equilibrium properties in terms of ensemble averages.

There are three important aspects of the use of ensemble averages in place of time averages not yet mentioned. First, the general dynamical properties of a system, such as the mixing property, have been established only within the framework of classical mechanics. At least to date, quantum ergodic theory is much less well developed. In contrast, the use of ensemble averages to describe equilibrium properties is equally valid in classical and quantum mechanics. Second, the simplest definition of an equilibrium state, namely, that its properties are invariant in time, and the identification

---

[4] This is equivalent to treating the energy as the only constant of the motion.

[5] By quasi-periodic we mean that any initial state recurs, although not necessarily with perfect regularity. A distinction is important here. In any closed mechanical system the initial state will recur to arbitrary precision, given enough time (Poincaré recurrence time). In a quasi-periodic system the initial state recurs *before* the energy surface is uniformly covered, so there is not an irreversible approach to equilibrium.

of macroscopic properties with the averages of microscopic dynamical properties over an infinite time period, both neglect the existence of fluctuations about the equilibrium state.[6] Yet fluctuations do occur, and have observable consequences. Indeed, macroscopic equilibrium is a concept that cannot be understood fully if fluctuations are not considered. We shall see that the use of ensemble averages to represent equilibrium properties does permit analysis of the properties of fluctuations. Third, the kind of equilibrium that exists—that is, the properties of the equilibrium state—depends on the kinds of constraints imposed on the system. It is easy to take this into account when computing ensemble averages, whereas the procedure based on general dynamics is restricted to describing equilibrium in an isolated system with fixed volume and number of particles.

In the preceding discussion we introduced ensemble averaging as a substitute for time averaging. In fact, the use of ensembles and ensemble averaging to relate a macroscopic observable to the corresponding microscopic quantity is far more than a substitute for exact mechanical calculations. Some of the important ways in which the introduction of ensembles influences our description of a system—for example, the concept of "state of a system"—are discussed in Appendices 15A and 15B.

15.2 Ensembles and Probability Distributions

In the preceding section we commented on the relationship between time averaging and ensemble averaging in the kinetic theory of matter. To understand fully the nature of the averaging procedures used, and their implications, it is necessary to examine more thoroughly the notions of an ensemble and a probability distribution. The manner in which these concepts enter our interpretation of the behavior of matter will become clear in succeeding sections of this chapter.

Probability theory is a useful tool for the description of events in which there is a multitude of possible outcomes. We are all familiar with coin-tossing experiments in which each toss can be either "heads" or "tails." Although it is not possible to state with certainty whether any given toss will be "heads" or "tails," it is possible to state with good accuracy that one-half of a large number of tosses will be "heads" if the coin is not loaded. In fact, although having equal numbers of "heads" and "tails" is the most likely outcome of a large number of tosses of an unloaded coin, other possibilities occur with nonzero frequency. We expect that the larger the number of tosses, the more accurate will be the expectation of equal numbers of "heads" and "tails."

There is a familiar feature of elementary probability theory embodied in the expectations stated above, namely, the definition of probability of a specified result as the ratio of the number of events with the required outcome to the total number of events, the so-called relative frequency of the required outcome. In the coin-tossing experiment the probability that any one toss of the coin will result in "heads" is simply the ratio of the number of possible "heads" to the totality of possible outcomes of the toss, namely, "heads" plus "tails," or $\frac{1}{2}$. The probability that out of any n tosses, n_1 will be heads is $[n!/n_1!(n - n_1)!](1/2)^n$, since there are $n!/n_1!(n-n_1)!$ ways in which n tosses of the coin can give n_1 "heads" irrespective of order, and there are 2^n possible sequences of "heads" and "tails" resulting from n tosses of the coin. The formula quoted illustrates how, given that each toss of the coin is regarded as independent of all others, the probability of an event requiring n tosses is constructed by multiplying together two factors. One is the elementary probabilities per toss for a given sequence $[(1/2)^{n_1}(1/2)^{n-n_1} = (1/2)^n$ for any specified sequence of n_1 "heads" and $n - n_1$ "tails," the same for each such sequence]. The other is a combinatorial factor counting the multiplicity of ways a sequence can yield the required event. The formula also illustrates how deviations from the most probable outcome decrease as the length of the sequence of tosses increases. For example, the probability that 2 of 4 tosses will be "heads" is $\frac{6}{16}$ and that 1 of 4 will be "heads" is $\frac{4}{16}$. Clearly, out of 4 tosses even the most probable event, namely, 2 "heads," is not overwhelmingly more probable than other events. In contrast, the probability that 200 out of 400 tosses will be "heads" is about 5×10^{22} times the probability that 100 out of 400 tosses will be "heads."

To show how similar simple probabilistic ideas can be applied to the description of molecular behavior, consider the following question. Given that the molecules of a gas are in motion, the number in any fixed element of volume must vary from moment to moment. Should we expect there to be considerable variation in the gas density when we examine different volume elements, centered at different points in the container? Consider a perfect gas of N molecules in a volume V. Now focus attention on some volume element ω inside the container. Since ω/V is the fraction of total volume occupied by ω, and the gas is uniformly distributed, the probability of finding any selected molecule in ω is just ω/V, and that of finding all N molecules in ω is just $(\omega/V)^N$. If ω is very small, say, with linear dimensions comparable to the average distance between molecules, we must expect the number of molecules in ω to usually be zero or one, and occasionally two, but in any event fluctuating markedly as molecules move in and out of ω along their trajectories. If ω is large, say $V/10$, we expect the number of molecules in ω to be $(\omega/V)N$, and subject only to small fluctuations, since in this case the number of molecules crossing the boundaries of ω at any instant is small compared to the number in ω. Both limiting cases are described by the probability distribution $P(N_1)$, which gives the probability

[6] This statement refers to an infinite time average of the properties of one replica system.

of finding N_1 molecules in ω and $N - N_1$ in $V - \omega$. This is just $[N!/N_1!(N - N_1)!] \times (\omega/V)^{N_1}(1 - \omega/V)^{N-N_1}$, since there are $N!/N_1!(N - N_1)!$ ways of selecting N_1 molecules out of N to place in ω. It is easily shown, given $P(N_1)$ and the methods described later in this section, that the relative deviation in the number of molecules in ω decreases as ω increases, specifically as $(N\omega/V)^{-1/2}$. Therefore, for a perfect gas in equilibrium, the relative deviation in the number of molecules in a small macroscopic volume, say, 1 mm^3, is of the order 10^{-8} at ordinary temperature and pressure. Despite the incessant molecular motion, the macroscopic density scarcely varies at all from point to point in the gas.

As is clear from the example given, the form of probability theory we shall use is based on generalizations of the elementary notions used to describe the coin tossing experiment. In particular, two points merit careful consideration:

1. The relationship between the definition of probability distribution and the class of events that is to be described, that is, the ensemble in which the probability is defined, and
2. How probability theory is related to the fundamental classical or quantum mechanical description of a system (we shall use the quantum mechanical description).

The notions of probability distribution and ensemble are closely related. In the simplest possible terms, as already used in Section 15.1, an ensemble is merely a collection. The nature of the collection, the set of possible states for the objects composing the collection, and other relevant details must be specified before it is possible to use the ensemble to define the probability of finding some one of the objects in a particular condition. As in the elementary theory, the probability of finding a particular characteristic is defined simply as the fraction of all objects in the ensemble having that characteristic. Clearly, this probability depends very much on the nature of the ensemble. For example, the question, "What is the probability of finding a molecule in a sample of perfect gas moving with speed v?", is ambiguous until the macroscopic state of the gas, hence also of the ensemble representing the microscopic states of the gas, is specified. It is evident that it makes a significant difference whether the gas has a temperature T_1 or T_2, whether it is at rest or moving, and so forth. However, once the ensemble is defined, the probability defined in that ensemble is uniquely specified.

As already mentioned, the general name assigned to one of the objects that compose an ensemble is a replica. Implicit in this choice of name is the idea that the objects composing the ensemble are to be considered identical within the constraints imposed by the specifications defining the ensemble. Thus, specifying only that an ensemble consists of all pennies does not require any statement about whether the individual pennies lie "head" up or "tail" up. Similarly, specifying only that an ensemble consists of containers of gas, having rigid adiabatic walls, and each having the same density of molecules (temperature unspecified), does not require any statement about whether or not the pressure of the gas is the same in each container.

The fact that there is an intimate connection between the external constraints defining the state of a system and the nature of the corresponding ensemble of replicas is of fundamental importance to our analysis. We shall see how this relationship can be used to establish a correspondence between the microscopic and macroscopic analyses of the behavior of matter.

Finally, it is generally assumed that an ensemble of replicas is so large that the hypothetical limit $\mathcal{N} \to \infty$ may be taken, where $\mathcal{N}$ is the number of replicas in the ensemble. The use of this abstraction will enable us to consider the probability distribution to be, in many instances, a continuous function. For convenience, in those cases where the probability distribution is a function of a continuous variable, we introduce the notion of *probability density*. The reason for introducing this concept is the following. For a variable that takes on only discrete values, it is meaningful to ask for the probability that some one value is attained out of the set of all possible values. On the other hand, for a variable that takes on a continuous range of values, say, between a and b, the set of all possible values is infinite in size, and the only meaningful question is "What is the probability of finding a value between x_1 and x_2, given that the range of x is such that $a \le x \le b$?" This probability must now be expressed as an integral over a "probability density," $\mathcal{P}(x)$:

$$P(x_1 \le x \le x_2) = \int_{x_1}^{x_2} \mathcal{P}(x)\, dx. \tag{15.1a}$$

If $x_2 = x_1 + dx$,

$$P(x_1 \le x \le x_1 + dx) = \mathcal{P}(x)\, dx. \tag{15.1b}$$

Note that $\mathcal{P}(x)$ must be multiplied by dx to yield an actual probability and that, in the general case, although $P(x)$ is dimensionless, $\mathcal{P}(x)$ is not (it has the dimensions of $1/x$). It is possible, under certain conditions, to reduce a description in terms of a continuous probability distribution to one in terms of a discrete probability distribution, where the number of possible values of the variable of interest becomes countable. This is done by dividing the accessible range of the variable, say, $a \le x \le b$, into arbitrarily small equal intervals of fixed size, Δ. Each such interval can be labeled by some index i, and the value of x for this entire interval is denoted as x_i. The probability of finding the variable in this range is $P(x_i)$, where clearly

$$P(x_i) = \mathcal{P}(x_i)\Delta. \tag{15.2}$$

The resolution of a continuous probability distribution into a discrete probability distribution is valid only if the variation of $\mathcal{P}(x)$ over the interval Δ is sufficiently small, for all intervals, that such variation as does exist can be neglected.[7]

Let x be a variable that can assume only the discrete values $x_1, \ldots, x_N$, with probabilities $P(x_1), \ldots, P(x_N)$. The probabilities $P(x_i)$ may be used to define the *average value* of x (also called the *mean value* or the *expectation value*) by

$$\langle x \rangle \equiv \frac{x_1 P(x_1) + x_2 P(x_2) + \cdots + x_N P(x_N)}{P(x_1) + P(x_2) + \cdots + P(x_N)} = \frac{\sum_i x_i P(x_i)}{\sum_i P(x_i)}. \tag{15.3}$$

Equation 15.3 is just the definition of an arithmetic average, in which each value of the variable is weighted by the $P(x_i)$ with which that value appears (ratio of the number of occurrences of x_i to total number of all possible values of x). More generally, if $f(x)$ is a function of the variable x, the mean value of the function is

$$\langle f(x) \rangle \equiv \frac{\sum_i f(x_i) P(x_i)}{\sum_i P(x_i)}. \tag{15.4}$$

The average values $\langle x^n \rangle$ are called the *moments* of $P(x)$. For the case that x is a continuous variable and $f(x)$ is a continuous function, an obvious extension of Eqs. 15.3 and 15.4 leads to

$$\langle x \rangle \equiv \frac{\int x \mathcal{P}(x)\, dx}{\int \mathcal{P}(x)\, dx}, \tag{15.5}$$

$$\langle f(x) \rangle \equiv \frac{\int f(x) \mathcal{P}(x)\, dx}{\int \mathcal{P}(x)\, dx}, \tag{15.6}$$

using Eq. 15.1. We can also find the simple relations

$$\langle f(x) + g(x) \rangle = \langle f(x) \rangle + \langle g(x) \rangle, \tag{15.7}$$

$$\langle cf(x) \rangle = c \langle f(x) \rangle, \tag{15.8}$$

where c in Eq. 15.8 is a constant. Note that in Eqs. 15.5 and 15.6 the range of integration is the same as the range of x.

[7] The reduction of a continuous probability distribution to a discrete one can be thought of as the use of a histogram, or bar graph, to approximate a smooth curve.

Just as it was convenient to normalize the wave function of a system, it is also convenient to normalize a probability distribution. Let us require that a variable x must have some value between the end points of its defining interval. Then, by our definition of probability as the fraction of events with the required outcome, the total probability that x has a value in the stated range, $\Sigma_i P(x_i)$ or $\int \mathcal{P}(x)\, dx$, must be equal to unity. It is frequently convenient, however, to deal with relative probabilities of some kind. For such applications $\int \mathcal{P}(x)\, dx$ can have any convenient constant value, since this constant cancels out of ratios such as appear in Eqs. 15.3–15.6. In general, then, $\Sigma_i P(x_i)$ or $\int \mathcal{P}(x)\, dx$ is equal to some constant, known as the normalization constant. When the probability is normalized to unity, the denominators of Eqs. 15.3, 15.4, 15.5, and 15.6 are replaced by unity.

A measure of the spread of the probability density distribution $\mathcal{P}(x)$ is given by the difference $\langle x^2 \rangle - \langle x \rangle^2$. Except for the trivial case that $\mathcal{P}(x)$ is unity for some value of x (or range dx about x) and zero for all other values,[8] it is necessarily true that

$$\langle x^2 \rangle > \langle x \rangle^2. \tag{15.9}$$

Equation 15.9 can be understood as follows. In Fig. 15.1*a* is depicted a typical probability density distribution. The important feature of this curve is that it has some width; that is, the distribution of possible values of x is not confined to a single value. The average value of x^2, which is proportional to $\int x^2 \mathcal{P}(x)\, dx$, weights various values of x differently than does the average of x, which is proportional to $\int x \mathcal{P}(x)\, dx$. In the special case that $\mathcal{P}(x)$ is symmetric about $x = 0$ (see Fig. 15.1*b*), the average value of x is zero. However, since it is always true that $x^2 \geq 0$, clearly $\langle x^2 \rangle$ cannot be zero. Equation 15.9 summarizes these observations for a general probability distribution.[9]

[8] This exceptional case is not of interest, since there is no distribution of possible values of x.

[9] To demonstrate that Eq. 15.9 is correct, we define

$$\Delta x \equiv x - \langle x \rangle$$

and calculate the value of $\langle (\Delta x)^2 \rangle$ from

$$\langle (\Delta x)^2 \rangle = \int (x - \langle x \rangle)^2 \mathcal{P}(x)\, dx.$$

Now, we have

$$\begin{aligned} \langle (x - \langle x \rangle)^2 \rangle &= \langle x^2 - 2x\langle x \rangle + \langle x \rangle^2 \rangle \\ &= \langle x^2 \rangle - 2\langle x \rangle \langle x \rangle + \langle x \rangle^2 \\ &= \langle x^2 \rangle - \langle x \rangle^2. \end{aligned}$$

But $(\Delta x)^2$ is always positive or zero, so the right-hand side of the last line above must also be always positive or zero; that is,

$$\langle x^2 \rangle \geq \langle x \rangle^2.$$

Clearly, only when $\langle (\Delta x)^2 \rangle$ vanishes does the equality sign hold; since this can occur only if every value of x is equal to the mean value, Eq. 15.9 is established.

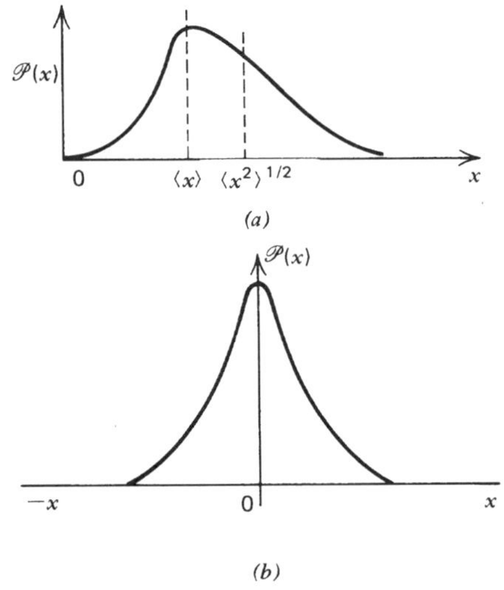

Figure 15.1 (*a*) A typical probability density distribution with $\langle x \rangle \neq 0$. (*b*) A typical symmetric probability density distribution with $\langle x \rangle = 0$.

Equation 15.9 has another important implication. Because a probability distribution describes a range of possible values of the variable, there must be some nonzero chance of finding values of the variable that differ from the average value of the variable. The difference between some particular value of a variable and its average value is called a *fluctuation.* A given fluctuation is more or less probable insofar as the deviation from the average is small or large compared with the spread of $\mathcal{P}(x)$. We have already referred to the existence of fluctuations when a system with many particles is described by only macroscopic variables (see Section 13.1). We shall see later that fluctuations in the values of macroscopic variables are an essential property of matter in bulk, and that the nature of the possible fluctuations around the equilibrium state is related to the properties of that state.

There is one other property of probability distributions that we shall use extensively in later chapters. Suppose that the variable x is distributed as $\mathcal{P}_1(x)$ and the variable y as $\mathcal{P}_2(y)$, where $\mathcal{P}_1(x)$ and $\mathcal{P}_2(y)$ are probability density distributions. If the probability of finding x in the range x to $x + dx$ is independent of the value of y, and vice versa, then the probability of finding x in the range x to $x + dx$ and y in the range y to $y + dy$ simultaneously is the *joint probability distribution*

$$\mathcal{P}(x,y)\,dx\,dy = \mathcal{P}_1(x)\mathcal{P}_2(y)\,dx\,dy. \qquad (15.10)$$

In general, the probability of occurrence of a set of values of independent variables is simply the product of the probabilities of occurrence of the values of each of the variables. When a set of variables is not independent, Eq. 15.10 or its n-variable extension is not valid. The set of variables is then said to be *correlated.* The extent of correlation of the variables can be formulated in a variety of ways. Two convenient formulations of the probability of finding x in the range x to $x + dx$ and y in the range y to $y + dy$ when x and y are not independent are given by

$$\mathcal{P}(x,y)\,dx\,dy = \mathcal{P}_1(x)\mathcal{P}_2(y)g(x,y)\,dx\,dy, \qquad (15.11a)$$

$$\mathcal{P}(x,y)\,dx\,dy = \mathcal{P}_1(x)\mathcal{P}_2(y)[1+h(x,y)]\,dx\,dy. \qquad (15.11b)$$

Both $g(x, y)$ and $h(x, y)$ are called *correlation functions.* Each is used in the molecular theory; the choice is one of convenience. Note that when x and y are independent, we have $g(x, y) = 1$ but $h(x, y) = 0$.

Most of the ideas described in the preceding paragraphs have been used, in an intuitive fashion, in Chapter 12. There the distribution of molecules with respect to speed was introduced and used to describe the properties of the perfect gas of noninteracting molecules. Note, however, that the speed distribution function $f(v)$, which is a probability density, was normalized to the number density n, and not to unity. As an example of the relationship displayed in Eq. 15.9, we observe that at equilibrium the average velocity of a molecule in the gas is zero (because all directions are equally probable), whereas the root-mean-square velocity, $\langle v^2 \rangle^{1/2}$, is not zero. In fact, the mean square of the velocity defines the average kinetic energy per molecule, which in turn is related directly to the temperature of the gas.

As a final comment, we note that knowledge of the full probability distribution implies complete information about the distribution of values of the variables of interest. Although it is often possible to characterize a system by the mean values, mean square values, and so on, of some small set of variables, it is not possible to determine uniquely the full probability distribution from a knowledge of only a few of its moments.

15.3 Some Properties of a System with Many Degrees of Freedom: Elements of the Statistical Theory of Matter at Equilibrium

Given the concepts of ensemble and probability distribution, we are in possession of the tools necessary for the description of the most important of the equilibrium statistical molecular properties of matter. Consider some sample of matter, containing a large number of molecules, N, in the volume V. Suppose the energy of the system lies in the range E to $E + dE$, with $dE/E \ll 1$. From the general theory of quantum mechanics, discussed in earlier chapters of this text, it is clear that there is one feature common to all kinds of matter. This feature is, simply, the existence of an energy spectrum. For our present purposes it does not matter how closely the energy levels of the system are spaced or what the spectral distribution of energy levels is like. What is important, however, is that in any macroscopic system there exists an enormous number of quantum states for given constraints, say, E, V, N. The degeneracy of the system is exceedingly large. Of course, there are also many different distributions of the molecules over the energy levels of the system consistent with the given E, V, N.

In principle, the exact mechanical behavior of the system is determined once the energies and wave functions for all the molecules are given. However, since we do not know how to solve the equations of motion for N interacting molecules, this information is not available for a macroscopic system. Indeed, it is not even of particular interest to seek a full solution of the N-molecule mechanical problem, at least insofar as the description of macroscopic equilibrium is concerned. This is so because the $\Omega(E, V, N)$ states with energy between E and $E + dE$ will correspond to a range of values of, for example, molecular coordinates and velocities. Although each solution of the N-molecule Schrödinger equation will correspond to one, or a small number, of the $\Omega(E, V, N)$ states, others arise from different initial conditions on the many microscopic coordinates that are macroscopically indistinguishable. To obtain the wave functions for all the $\Omega(E, V, N)$ levels, we must obtain many solutions of the Schrödinger equation, sampling a considerable range of initial conditions on the microscopic coordinates. This is, in fact, unnecessary for our purposes, because the properties of macroscopic equilibrium states are determined by $\Omega(E, V, N)$, and not by the detailed behavior of the wave functions for the levels comprising $\Omega(E, V, N)$.[10]

The fundamental postulate of equilibrium statistical mechanics can be stated as follows:

All possible quantum states of an isolated system consistent with a given set of macroscopic parameters of constraint are to be considered as equally probable.

This postulate of equal *a priori* probability cannot be obtained from more fundamental arguments. Although eminently reasonable, and consistent with the laws of mechanics, the postulate stated must be recognized as a fundamental assumption in the development of the statistical theory—an assumption that can only be indirectly verified *a posteriori* by the success of theoretical calculations in interpreting and reproducing the results of observation.

To take advantage of the fundamental postulate of equal *a priori* probabilities, we consider not one system at a time, but rather an ensemble of systems. The ensemble is constructed by collecting a very large number of replica systems, each of which is defined by the same set of macroscopic parameters, in the present case the set E, V, N. As noted in Section 15.2, those properties of the system not defined by the macroscopic constraints are free to vary, provided only that the macroscopic constraints are never violated. In this way we may suppose that all possible quantum states consistent with the definition of the ensemble are represented among the replica systems. Note that the relative frequency definition of probability in an ensemble is in agreement with the assumption of equal *a priori* probability for all quantum states consistent with given E, V, N, for the chance that a random choice from the ensemble will lead to the selection of any one of the replicas is the same, namely, $\mathcal{N}^{-1}$, for each of the $\mathcal{N}$ replicas, and each replica is in one of the quantum states consistent with the given E, V, N.

We must now ascertain which constraints are most suitable for the definition of an ensemble. For the case of interest to us, namely, equilibrium, the macroscopic properties of an isolated system are independent of time. This observation implies that in an isolated system at equilibrium the distribution of molecules over the energy levels is also independent of time, hence a function of only the constants of the motion (see Chapter 3). The most general constants of the motion are the total energy, total linear momentum, and total angular momentum of the system. It is just because each replica system is considered to be isolated that energy, linear momentum, and angular momentum are conserved (no energy may be transferred across the boundaries of an isolated system, and no forces act on an isolated system). Given the constraints that define the ensemble, it is possible to define an appropriate probability distribution. Often only one of the constants of motion is used, and because of its central importance it is usually the total energy of the system that is chosen to define the ensemble. In this case, even though other constants of motion exist, they are unspecified and, just as in the case of the microscopic distribution, are only required to have values consistent with the assigned total energy.

Consider, now, an ensemble of replica systems each with the same volume V and number of molecules N. We complete the specification of the ensemble by requiring that the energy be between E and $E + dE$. Suppose that there are $\Gamma(E, V, N)$ quantum states with energy less than E for this system. The number of quantum states with energy between E and $E + dE$ is defined by the difference

$$\Omega(E, dE, V, N) = \Gamma(E + dE, V, N) - \Gamma(E, V, N). \quad (15.12)$$

The function $\Omega(E, dE, V, N)$ contains information about the degeneracy of the energy-level spectrum of the system, and the dependence of that spectrum on the values of parameters such as V. In general, the parametric dependence of $\Omega(E, dE, V, N)$ on V reflects the influence of boundary conditions on the energy-level spectrum of the isolated system. To simplify the notation, in the following we omit explicit display of the dependence of Ω on dE. We shall see that in the limit

[10] Although $\Omega(E, V, N)$ depends on dE, for the sake of simplicity this dependence will not be displayed explicitly; we show later that for the purposes of obtaining thermodynamic functions the dependence on dE is not important.

that the system is very large the dependence of Ω on dE is insignificantly small and, consequently, does not contribute to the thermodynamic properties of the system.

This last point is so important that it is necessary to expand on such a terse statement. The Schrödinger equation is a second-order partial differential equation and, depending on the form of the potential, the magnitude of the energy, the shape of the container, and so on, has many possible solutions. The selection of the set of solutions appropriate to a given problem is accomplished by applying the correct boundary conditions. (See Chapter 3.) It is because the boundary conditions depend on the nature of the enclosure (the boundary) that the wave functions and allowed energy levels are functionally dependent on the volume of the system. Consider, for example, a free particle enclosed in a box with infinitely repulsive walls. The solutions to the Schrödinger equation for a free particle are of the forms sin kx and cos kx, and form a set continuous in the parameter k. Thus before the application of the boundary conditions there is a continuum of possible solutions. However, when it is recognized that the wave function must be zero at an infinitely repulsive boundary, the values that k may take on are limited, as are the energies. In the absence of bounding walls, there is an infinite number of allowed states in the range from E to $E + dE$, but when the particle is confined to a volume V the number of available states in this range becomes finite and *dependent on the value of V*. The change from an infinite number to a finite number of allowed solutions corresponds to a change in the spectrum of energy levels from a continuum to a discrete set.

How does the fact that a macroscopic system contains a very large number of molecules lead to simplifications in the analysis? Suppose that the energy scale is divided into equal intervals of magnitude dE, with $dE << E$. For a large system even an interval dE that is small by macroscopic standards contains a very large number of energy levels. The number of states between E and $E + dE$, namely, $\Omega(E)$, depends on the magnitude of dE, but there is a wide range over which dE may vary and for which the linear relation

$$\Omega(E)=\left(\frac{\partial\Gamma}{\partial E}\right)dE=\omega(E)\,dE \qquad (15.13)$$

is valid. The function $\omega(E)$ measures the number of states per unit energy interval, and is a characteristic property of the system; $\omega(E)$ is known as the *density of states* of the system. Because of our assertion that $\omega(E)$ is characteristic of a system, it is of obvious interest to examine, even roughly, how sensitive $\Omega(E)$ and $\omega(E)$ are to the energy E of a system of macroscopic size. (See Chapter 11.)

The purpose of the following argument is to demonstrate that $\Omega(E)$ is insensitive to the precise nature of the energy spectrum of the system, provided that the system is large enough. In this limit we shall see that $\Omega(E)$ has the characteristics of a macroscopic parameter even though it is defined in terms of the microscopic energy spectrum. Further, because $\Omega(E)$ has the characteristics of a macroscopic parameter, it plays the role of connecting the statistical molecular description with the thermodynamic description of a system. The argument used to establish these properties of $\Omega(E)$ involves a mixture of dimensional analysis and order of magnitude estimation. It will be seen that the argument contains several self-consistency checks that validate the conclusions reached concerning the properties of $\Omega(E)$.

Let the system with energy E be described by ν quantum numbers, and let the energy per degree of freedom (one for each quantum number) be approximated by

$$\epsilon=\frac{E}{\nu}. \qquad (15.14)$$

For example, a perfect gas composed of N monatomic molecules has $3N$ degrees of freedom, three for each molecule. Then, using Eq. 12.42, we find that the approximation 15.14 is equivalent to writing, for a perfect gas,

$$\epsilon_{\text{perfect gas}}\approx\tfrac{1}{2}k_BT. \qquad (15.15)$$

Equation 15.14 has much deeper significance than is apparent from the way it has been introduced. We shall later see that one of the characteristics of a system of many particles described by classical mechanics is that the total energy is divided among the degrees of freedom in such a way that the average energy per degree of freedom is just $\frac{1}{2}k_BT$. In a system described by quantum mechanics, the same equipartition of energy occurs when k_BT is large relative to the energy-level spacings of the system, but not when k_BT is smaller than or equal to the spacings between the energy levels. For the present we regard Eq. 15.14 as an approximation, and use it only to obtain an order-of-magnitude estimate of $\Omega(E)$ for an arbitrary system with ν degrees of freedom.

Consider again the total number of quantum states with energy less than E. We consider the energy to be separable into contributions from the ν degrees of freedom. Of course, the total energy of the system may be distributed over the ν degrees of freedom in many different ways, and each of these possible distributions contributes a state to $\Gamma(E)$. Let E_i be the energy in the ith degree of freedom and $\Gamma_i(E_i)$ the number of states of the ith degree of freedom with energy less than E_i. Then $\prod_{i=1}^{\nu}\Gamma_i(E_i)$ is the total number of states with energy less than E for a particular subdivision of the energy. The total number of states of the system with energy less than E when all possible subdivisions of E are accounted for is

$$\Gamma(E)=\sum_{\{E_i\}}\prod_{i=1}^{\nu}\Gamma_i(E_i);\qquad E=\sum_i E_i. \qquad (15.16a)$$

Note that the summation over $\{E_i\}$ is over all possible ways of subdividing the total energy subject to the constraint of conservation of energy. For the purpose of making a crude estimate of $\Gamma(E)$ we approximate it by requiring that each of the E_i be just the average energy ϵ. This corresponds to choosing only one term in the sum over all possible subdivisions of the energy, the one which has a uniform distribution of energy over the degrees of freedom. Then we obtain

$$\Gamma(E) \approx \prod_{i=1}^{\nu} \Gamma_i(\epsilon) \tag{15.16b}$$

$$\approx [\Gamma_1(\epsilon)]^{\nu}. \tag{15.16c}$$

Equation 15.16*b* is only a crude estimate of $\Gamma(E)$, but it is sufficiently accurate for our purposes. The right-hand side of Eq. 15.16*c* represents still another order-of-magnitude estimate, this time replacing the product of the numbers of states for all degrees of freedom by the νth power of the number of states corresponding to one degree of freedom. The assumption inherent in this estimate is that the ν degrees of freedom are sufficiently alike that some reasonable average behavior can be ascribed to them, at least for the purposes of this argument. We now note that expansion of the right-hand side of Eq. 15.12 in a Taylor's series, followed by substitution of Eq. 15.16*c*, leads to

$$\Omega(E) = \frac{\partial \Gamma}{\partial E}\, dE \tag{15.17}$$

$$\approx \frac{\partial}{\partial E}\{[\Gamma_1(\epsilon)]^{\nu}\}\, dE \tag{15.18}$$

$$= \nu \Gamma_1^{\nu-1} \frac{\partial \Gamma_1}{\partial(\nu\epsilon)}\, dE = \Gamma_1^{\nu-1} \frac{\partial \Gamma_1}{\partial \epsilon}\, dE, \tag{15.19}$$

where we have used $E = \nu\epsilon$, from Eq. 15.14. Now, $\Gamma_1(\epsilon)$ must increase as E increases. Therefore, when E increases, $\Gamma_1(\epsilon)$ also increases; and because ν is so very large for a macroscopic system, $\Omega(E)$ is a very rapidly increasing function of E. The magnitude of the rate of increase can be appreciated if we rewrite Eq. 15.19 in the form

$$\ln \Omega(E) = (\nu - 1) \ln \Gamma_1(\epsilon) + \ln\left(\frac{\partial \Gamma_1}{\partial \epsilon}\, dE\right). \tag{15.20}$$

As already mentioned, the energy range dE is very much larger than the separation between the energy levels of the system. It is not important just how much larger, because the term $\ln[(\partial\Gamma_1/\partial\epsilon)\, dE]$ in Eq. 15.20 is negligibly small relative to the term $(\nu - 1) \ln \Gamma_1(\epsilon)$. To show this, let us suppose that dE were as large as ϵ. The order of magnitude of the derivative $(\partial\Gamma_1(\epsilon)/\partial\epsilon)$ is Γ_1/ϵ, an estimate that comes from assuming that $\Gamma_1(\epsilon)$ is proportional to ϵ. In fact, studies of simple models, such as the harmonic oscillator or the particle in a box, show that $\Gamma_1(\epsilon)$ increases as some small power of ϵ. For the cases mentioned,

$$[\Gamma_1(\epsilon)]_{\text{harmonic oscillator}} \propto \epsilon,$$

$$[\Gamma_1(\epsilon)]_{\text{particle in a box}} \propto \epsilon^{1/2}.$$

In general, then, $\ln[(\Gamma_1(\epsilon)\epsilon]$ will be a factor ν smaller than $(\nu - 1) \ln \Gamma_1(\epsilon)$, and ν is of the order of 10^{24}. Even if our estimates of $\partial\Gamma_1/\partial\epsilon$ and dE were both off by factors of ν $\ln[(\partial\Gamma_1/\partial\epsilon)\, dE]$ would still be negligibly small relative to $\nu \ln \Gamma_1$, because the logarithm of some small power of ν is a small multiple of $\ln \nu$, which for $\nu = 10^{24}$ means a small multiple of 55. Thus, even multiplying $\Gamma_1(\epsilon)$ by the very large factor ν, or some power of ν, introduces terms in the logarithm that are negligibly small relative to ν itself. An estimate of the magnitude of the remaining term in Eq. 15.20, namely, $(\nu - 1) \ln \Gamma_1(\epsilon)$, can be made as follows: If Δ is the average spacing between levels, then $\Gamma_1(\epsilon)$ must be of the order of magnitude of ϵ/Δ, and $\epsilon/\Delta >> 1$ when ϵ is far from the ground-state energy. Then $\nu \ln \Gamma_1(\epsilon)$ is of the order of magnitude of ν. We conclude that the relation

$$\ln \Omega(E) \approx \nu \ln \Gamma_1(\epsilon) \tag{15.21}$$

implies that $\ln \Omega(E)$ is of the order of magnitude of ν or larger, provided that the energy E is well above the energy of the ground state of the system. Note that ν is so large that the result cited is insensitive to the rate at which $\Gamma_1(\epsilon)$ increases with increasing ϵ. That is, it is immaterial whether $\Gamma_1(\epsilon)$ increases as ϵ^2, ϵ^3, $\epsilon^{1/2}$, or any other small power of ϵ, since our argument is unaffected if there is a coefficient of order 2, 3, $\frac{1}{2}$, and so on, in the relation 15.21.

15.4 The Influence of Constraints on the Density of States

We are now able to examine the role of constraints in defining the density of states, $\omega(E)$, or the number of states, $\Omega(E)$, descriptive of an isolated system with energy between E and $E + dE$. It is convenient to begin with a simple example. Consider, as a model of a perfect gas, a cubical box of volume $V = a^3$ containing N independent point particles. It was shown in Chapter 3 that the wave function for a single particle in a cubical box is

$$\psi_j = A \sin\frac{n_{jx}\pi x}{a} \sin\frac{n_{jy}\pi y}{a} \sin\frac{n_{jz}\pi z}{a}; \quad \begin{matrix} 0 < x < a \\ 0 < y < a \\ 0 < z < a \end{matrix} \tag{15.22}$$

where $j = 1, \ldots, N$ labels the particle, A is a normalization constant, and the particle quantum numbers n_{jx}, n_{jy}, n_{jz} satisfy

$$n_{jx}^2 + n_{jy}^2 + n_{jz}^2 = \frac{2m\epsilon_j}{\pi^2\hbar^2} a^2, \tag{15.23}$$

with ϵ the energy of particle j. The values of n_{jx}, n_{jy}, n_{jz} are restricted to the positive integers since the substitutions $n_{jx} \to -n_{jx}$, $n_{jy} \to -n_{jy}$, $n_{jz} \to -n_{jz}$ do not lead to new linearly independent wave functions.[11] The total energy of the N particles in the box is

$$E = \sum_{j=1}^{N} \epsilon_j, \tag{15.24}$$

which we rewrite in the form

$$\sum_{j=1}^{N} (n_{jx}^2 + n_{jy}^2 + n_{jz}^2) = \frac{2mE}{\pi^2 \hbar^2} a^2. \tag{15.25}$$

To calculate the total number of possible states with energy less than E, we proceed as follows. Imagine a $3N$-dimensional space defined by the orthogonal axes n_{jl} ($j = 1, \ldots, N$; $l = x, y, z$). Since every n_{jl} must be a positive integer, each allowed state is represented by a point in the positive orthant[12] $n_{jl} \geq 0$, all j and l. Furthermore, since Eq. 15.25 defines a hyperspherical surface with radius $R_s = (2mEa^2/\pi^2\hbar^2)^{1/2}$, the total number of states with energy less than E is represented by the points with integer coordinates lying between the positive axes n_{jl} (all j and l) and the hyperspherical energy surface. When the volume of the box is large, the spacing between adjacent energy levels is very small; then the number of points with integer coordinates is very well approximated by the volume of the region they occupy. The volume of a $3N$-dimensional sphere can be shown to be

$$\frac{\pi^{3N/2}}{(3N/2)!} R_s^{3N} = \frac{\pi^{3N/2}}{(3N/2)!} \left(\frac{2mEa^2}{\pi^2 \hbar^2} \right)^{3N/2},$$

and the positive orthant is $(\frac{1}{2})^{3N}$ of the total volume. Therefore,

$$\Gamma(E) = \frac{1}{(3N/2)!} \left(\frac{mEa^2}{2\pi\hbar^2} \right)^{3N/2}. \tag{15.26}$$

Finally, using Eq. 15.13, the number of states between E and $E + dE$ is

$$\Omega(E) = \left[\frac{\partial \Gamma(E)}{\partial E} \right] dE$$

$$= \frac{E^{3N/2-1}}{((3N/2)-1)!} \left(\frac{ma^2}{2\pi\hbar^2} \right)^{3N/2} dE. \tag{15.27}$$

Given Eq. 15.13, the density of states $\omega(E)$ is just the coefficient of dE in Eq. 15.27. Note that $\Omega(E)$ is proportional to V^N, and that V enters, once for each of the N particles, by virtue of the boundary conditions on the wave function.[13] These boundary conditions define the nature of the energy spectrum.

For the example given, only V influences the boundary conditions on the Schrödinger equation, and thereby the spectrum of energy levels. In more general cases there are several macroscopic variables that influence the energy-level spectrum. Suppose that the macroscopic state of a system is defined by specifying the values of a set of variables, say,

$$y_1 = \alpha_1, y_2 = \alpha_2, \ldots, y_n = \alpha_n.$$

Given these values for the macroscopic variables, the set of possible solutions to the Schrödinger equation is restricted in analytical form, and there is a corresponding dependence of the energy-level spectrum on $y_1 = \alpha_1$, $y_2 = \alpha_2, \ldots, y_n = \alpha_n$. We define

$$\Omega \equiv \Omega(E, y_1, \ldots, y_n) \tag{15.28}$$

to be the number of states accessible to the system in the range from E to $E + dE$ when each macroscopic parameter of constraint y_i is in the range from y_i to $y_i + dy_i$.

What general remarks can be made about the dependence of $\Omega(E, y_1, \ldots, y_n)$ on the parameters of constraint $y_1, \ldots, y_n$? First note that when a constraint is removed,

$$\Omega(E, y_1, \ldots, y_{n-1}) > \Omega(E, y_1, \ldots, y_n), \tag{15.29}$$

since removal of a boundary condition always allows a larger range of possible solutions to the Schrödinger equation.

The inequality 15.29 is descriptive of the difference between initial and final states of a system when the number of constraints is reduced. It is far more common, however, for initial and final states of a system to differ by virtue of having different values for the parameters of constraint, the number remaining the same. How does $\Omega(E, y)$ change in such cases? A simple example is, again, illustrative of the behavior to be expected. Consider N particles in a box. For this system the parameter of constraint is the specified value of the volume of the box. The number of states between E and $E + dE$, displayed in Eq. 15.27, is proportional to V^N. Therefore, increasing V increases $\Omega(E, V, N)$ and decreasing V decreases $\Omega(E, V, N)$. In general, $\Omega(E, y)$ can either increase or decrease, depending on how the parameters of constraint $y_1, \ldots, y_n$ are changed between initial and final states. (See Chapter 3, problem 9.)

[11] Note that $\sin(-x) = -\sin x$.

[12] The positive orthant is the generalization of the positive octant of a three-dimensional space. The positive octant is defined by $x > 0$, $y > 0$, $z > 0$.

[13] $\psi_j(x, y, z)$ vanishes at $x = 0$ and a, $y = 0$ and a, $z = 0$ and a, for every $j = 1, \ldots, N$.

To systematize the description of changes in $\Omega(E, y)$ induced by changes in the values of the parameters of constraint, *we adopt the definition that removal or relaxation of the parameters of constraint always leads to $\Omega_f(E, y) > \Omega_i(E, y)$.*

The combination of these observations and the definition lead to the following: Suppose that an isolated system, at equilibrium with a set of constraints, has Ω_i equally accessible states. Suppose, further, as a result of an external change involving no work done on the system, some constraints are removed or relaxed. A new system is thereby created for which, at equilibrium, there exist Ω_f equally accessible states, with Ω_f and Ω_i related by

$$\Omega_f \geq \Omega_i . \tag{15.30}$$

The preceding remarks bring us to a study of the central point in the statistical molecular theory. Consider an ensemble of systems. We start from a state that is in equilibrium with respect to all constraints acting, and in which all the Ω_i accessible states are equally probable. Now let one constraint be removed or relaxed. At first, none of the replicas of the ensemble will be in states excluded by the original constraints. But in the final equilibrium state, because of the change in the parameters of constraint, there are Ω_f states, and our fundamental postulate requires that these be occupied with equal probability. Thus the state of the system immediately following removal or relaxation of a constraint, one in which only Ω_i/Ω_f of the Ω_f accessible states are occupied, cannot be an equilibrium state of the system. The specific way in which Ω increases is determined by the detailed molecular dynamics associated with the way in which the constraint is changed. Although we cannot give general rules describing how the final equilibrium distribution is achieved, it is possible to assert that either removing or relaxing some of the constraints defining the state of an isolated system causes the system to change until a new equilibrium state, consistent with the remaining constraints, is achieved.

Let us now consider the problem inverse to that posed in the last paragraph. Suppose that, in an ensemble of systems, the final equilibrium has been reached following the relaxation of some constraints.[14] Depending on the nature of the constraints, they may or may not be easy to restore, but in general the constraints and the system cannot both be restored to the initial condition without work being done on the system. Imposition of a constraint without work being done on a system leaves $\Omega(E)$ essentially unchanged.[15] To illustrate this statement we again take as an example the

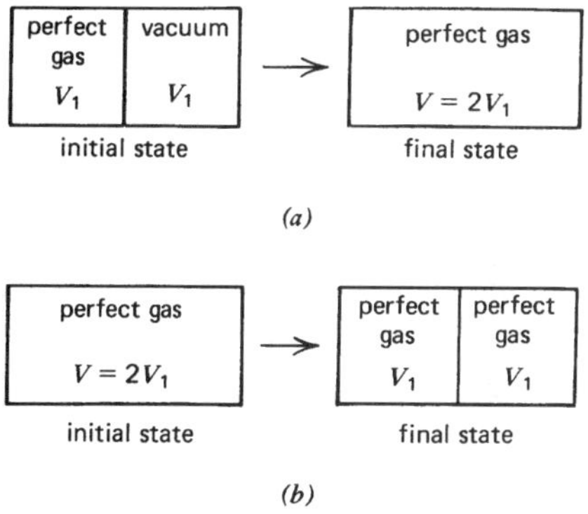

Figure 15.2 Schematic illustration demonstrating that restoration of a previously relaxed constraint without execution of work does not regenerate the initial state of the system.

particle-in-a-box model of a perfect gas. Suppose that the initial state is one in which, as shown in the schematic diagram of Fig. 15.2*a,* perfect gas occupies one-half of the volume of an isolated system because of a rigid barrier that separates the system into two equal volumes. The other half of the system is evacuated. Imagine, further, that the rigid barrier can be removed at will by expenditure of a vanishingly small amount of work. Of course this can be true only in a limiting sense, but it is practical in the laboratory to remove and add barriers in a fashion that approaches the theoretical limit of no expenditure of work. We have already seen that for the perfect gas Ω increases when V increases, so that when the barrier is removed and the gas expands to fill the volume $V = 2V_1$,

$$\Omega_f \geq \Omega_i \qquad \text{(process 15.2}a\text{)}.$$

Consider now what happens when the barrier is reinserted after process 15.2*a* is complete. A sketch of the situation is shown in Fig. 15.2*b.* Clearly, the final state achieved in process 15.2*b* is different from the initial state of process 15.2*a,* since the insertion of the barrier merely divides the box into two parts but leaves gas in each part. How is Ω_f related to Ω_i for process 15.2*b*? From Eq. 15.27 we have that

$$\Omega(E,V,N) = CV^N \left(\frac{E}{N}\right)^{3N/2} dE \tag{15.31}$$

where C is a constant independent of E and V. To obtain Eq. 15.31 we have used the Stirling approximation to $N!$, namely, $N! \approx N^N \exp(-N)$. In the initial state of process 15.2*b,* each particle can be anywhere in the volume V, whereas in the final state $N/2$ particles each are in the left- and right-hand volumes $V/2$. The energy per particle, E/N, is the same in the initial and final states. Since $N/2$ particles each can be placed in the left- and right-hand volumes in $N!/[(N/2)!(N/2)!]$ ways, each of which leads to an equivalent

[14] From here on we use the term relaxation of constraints to include the possibility of reduction of the number of constraints (removal).
[15] The word "essentially" means to terms of macroscopic significance.

state, the use of Eq. 15.31 to describe the initial and final states of process 15.2*b* leads to

$$\Omega_f(E,V,N)$$
$$= C\frac{N!}{(N/2)!(N/2)!}\left(\frac{V}{2}\right)^{N/2}$$
$$\left(\frac{E/2}{N/2}\right)^{3N/4}\left(\frac{V}{2}\right)^{N/2}\left(\frac{E/2}{N/2}\right)^{3N/4} dE \qquad (15.32)$$
$$= CV^N\left(\frac{E}{N}\right)^{3N/2} dE \qquad (15.33)$$

where we again have used the Stirling approximation to $N!$. Since Ω_i is given by Eq. 15.31 we conclude that

$$\Omega_f = \Omega_i \qquad \text{(process 15.2}b\text{)}.$$

Thus, as asserted earlier, imposition of a constraint without work being done on a system leaves $\Omega(E)$ essentially unchanged.

From the macroscopic point of view, to restore the initial state of process 15.2*a*, the gas must be swept out of one of the compartments. One way in which this could be achieved is by compressing the gas with a sliding wall, thereby creating a full and an empty compartment, each of volume V_1. But in order to achieve this compression, work must be done on the system, and if the system were thermally isolated the final temperature of the gas would be different from the initial temperature.

Other schemes to recreate the initial state of process 15.2*a* may be imagined, but it is found after examination of all possible methods of restoring the constraint that if $\Omega_f > \Omega_i$, a simple restoration of the geometry of the system without doing work or allowing energy to cross the boundaries cannot re-create the initial state of the system. Once the molecules are distributed over the Ω_f accessible states, a simple reimposition of boundary conditions cannot lead to a spontaneous rearrangement wherein some of the Ω_f states are vacated and the molecules occupy only the restricted subset of states Ω_i. It is also clear that relaxation of further constraints on the system cannot lead to a restoration of the initial condition, since in each such relaxation even more states become accessible to the system.

We now categorize some of the ways in which a system may change following relaxation of a constraint, so as to lay the groundwork for comparing the statistical molecular and thermodynamic descriptions of reversible and irreversible processes. Consider some process occurring in an isolated system, carrying the system to a final state. If, for an isolated system, the final state is such that imposition or relaxation of constraints without the requirement of external work cannot re-create the initial state, the process undergone by which the final state was achieved is called an *irreversible process*. On the other hand, if the imposition or relaxation of constraints without the requirement of external work on the isolated system can re-create the initial state, the process undergone is said to have been a *reversible process*. In summary, for processes at constant energy and constant total volume:

1. If some of the constraints defining an isolated system are relaxed, $\Omega_f \geq \Omega_i$.
2. If $\Omega_f = \Omega_i$, the systems of the ensemble are distributed over the same accessible states before and after relaxation of the constraint. The system remains in equilibrium at all stages of the process, and the transition $i \to f$ is reversible.
3. If $\Omega_f > \Omega_i$, the distribution of initial systems is over a smaller set of states than is the distribution of final systems. In the transition $i \to f$ equilibrium does not prevail through all stages of the process, and the process is irreversible.

Statements 2 and 3 merit further comment. In Chapter 13 we defined a reversible process as one in which the intensive variables are continuous across the boundary of the system. This definition refers to the case that the system is in contact with some reservoirs. What is meant by a reversible process in an isolated system, with *E*, *V*, and *N* specified? Consider, as an example, the internally divided system sketched in Fig. 15.3. Subsystems 1 and 2 are gases separated by a movable piston on which weights rest. A pulley arrangement provides a means of removing a weight and depositing it inside the system on a ledge that is at the same height. Therefore, removing a weight from the piston and transferring it to the ledge requires no net work. Accordingly, the pulleys and string are merely a device for

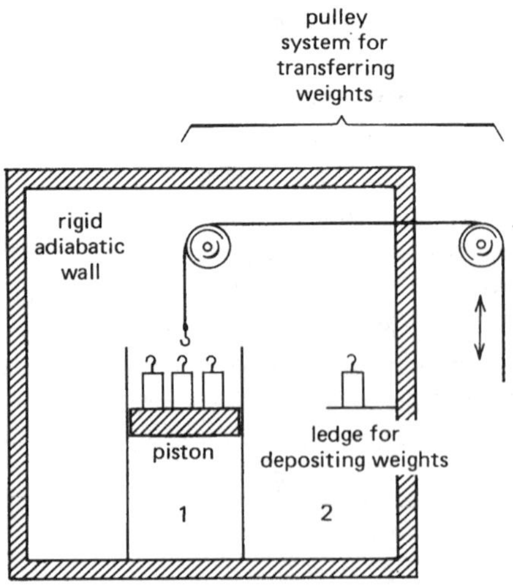

Figure 15.3 Schematic illustration of how a reversible process in an isolated system can be made to occur.

changing a parameter of constraint inside an otherwise isolated system. The external walls are rigid and adiabatic. Provided that the weights transferred inside the system are infinitesimally small, the pressure will be continuous across the piston along the path of the expansion of subsystem 1 and the corresponding compression of subsystem 2. Therefore this path describes a reversible process. Note that in this example subsystem 2 plays the role of surroundings for subsystem 1.

In general, the thermodynamic definition of a reversible process requires that every point on the path be infinitesimally close to an equilibrium state of the system, and that everywhere along the path the relevant intensive variables be continuous across the boundaries of the system. These conditions imply that the extensive variables of the system must be infinitesimally close in value in two states that are infinitesimally close, that is, everywhere along the path of a reversible process. Then, returning to statement 2 and using the language of the statistical molecular theory, we say that neighboring points along the path of a reversible process must have $\Omega_f = \Omega_i$, since it is the extensive variables that are usually the parameters of constraint and that determine the boundary conditions defining Ω.

15.5 The Entropy: A Potential Function for the Equilibrium State

In this section we study some of the properties of the number of accessible states Ω for a composite system over the parts of which the total energy may be distributed. It will be shown that the most probable distribution of energy over the parts of the system leads to a condition that is equivalent to the equality of temperatures between systems in thermal equilibrium. On the basis of this result a new function, the *entropy,* will be introduced and shown to be related to the work and internal energy as defined by the first law of thermodynamics. Thus defined, the entropy will provide a bridge enabling us to relate the macroscopic and microscopic descriptions of a system.

To carry through the program described, it is necessary to specify carefully the nature of the ensemble describing a composite system. Consider two macroscopic systems, A and B. Let E_A and E_B be the energies of the two systems.[16] Further, let $\Omega_A(E_A)$ and $\Omega_B(E_B)$ be, respectively, the numbers of states between E_A and $E_A + dE_A$ and between E_B and $E_B + dE_B$. N_A and N_B, the numbers of A and B molecules, are fixed. The ensemble to be considered is specified by the requirement that the sum of the energies of the two systems A and B, $E_T \equiv E_A + E_B$, is a constant. In the ensemble of pairs of systems, different replica pairs have a different distribution of the energy E_T between the systems A and B. In contrast, the volumes V_A and V_B, and any other necessary parameters of constraint, are separately specified for systems A and B, and hence for each replica pair. For simplicity we refer to the ensemble as being defined by the value of E_T.

In this ensemble of composite systems, if system A of a replica pair has energy E_A, then system B must have energy $E_T - E_A$. Now, when system A has energy E_A, it can be in any one of $\Omega_A(E_A)$ states, while simultaneously system B can be in any one of $\Omega_B(E_T - E_A)$ states. Since every possible state of A can be combined with every possible state of B, the total number of distinct states accessible to the combined system $A + B$, corresponding to energy E_T and subdivision $E_A = E_T - E_B$, is simply

$$\Omega_T(E_T, E_A) = \Omega_A(E_A)\Omega_B(E_T - E_A). \tag{15.34}$$

In the ensemble of pairs that we are examining, E_T is fixed. Therefore, $\Omega_T(E_T, E_A)$ varies because E_A may vary, and $\Omega_T(E_T, E_A)$ may be thought of as a function of the one variable E_A. This is a valid simplification provided we remember that it is only true in the ensemble of pairs having the same total energy E_T.

We now proceed by finding the contributions to $\Omega_T(E_A) = \Omega_T(E_T, E_A)$ from different subdivisions of the energy between A and B. The probability of finding $A + B$ in a configuration with the energy of A between E_A and $E_A + dE_A$ is just the ratio

$$\frac{\Omega_T(E_A)}{\sum_{E_A} \Omega_T(E_A)},$$

where $\Omega_T(E_A)$ is the total number of states accessible to the combined system for a given E_A, and the sum is taken over all possible values of E_A. For convenience, let

$$\sum_{E_A} \Omega_T(E_A) \equiv C^{-1}. \tag{15.35}$$

Using Eqs. 15.34 and 15.35, we can represent the probability of finding the combined system $A + B$ in a state with A having energy between E_A and $E_A + dE_A$ as

$$P(E_A) = C\Omega_A(E_A)\Omega_B(E_T - E_A). \tag{15.36}$$

As E_A increases, $\Omega_A(E_A)$ increases but $\Omega_B(E_T - E_A)$ decreases. Thus, the product displayed in Eq. 15.36 has a maximum for some value of the energy, say E_A^*. Because $\Omega_A(E_A)$ and $\Omega_B(E_T - E_A)$ are very rapidly varying functions

[16] This statement is short for: System A has energy between E_A and $E_A + dE_A$, etc.

of the energy, this maximum is extremely sharp.[17] To locate the position of the maximum, the most probable state of the combined systems, we regard Ω_A and Ω_B as continuous functions and solve

$$\left(\frac{\partial P(E_A)}{\partial (E_A)}\right)_{V,N} = 0. \tag{15.37}$$

Carrying out this operation, with $P(E_A)$ given by Eq. 15.36, leads to

$$C\Omega_B(E_T - E_A)\left(\frac{\partial \Omega_A}{\partial E_A}\right)_{V,N} + C\Omega_A(E_A)\left(\frac{\partial \Omega_B}{\partial E_A}\right)_{V,N} = 0 \tag{15.38}$$

or, after division by $P(E_A)$ and use of $E_A = E_T - E_B$,

$$\left(\frac{\partial \ln \Omega_A}{\partial E_A}\right)_{V,N} = \left(\frac{\partial \ln \Omega_B}{\partial E_B}\right)_{V,N} \tag{15.39}$$

[17] We can illustrate how a sharp maximum develops in the function $P(E_A)$ by examining the simpler function $[4x(1-x)]^n$ in the range $0 \le x \le 1$. This function is the product of two factors, $(2x)^n$ and $(2-2x)^n$, which respectively increase and decrease over this range, thus corresponding to our Ω_A and Ω_B. Clearly, as n increases (corresponding to increasing the number of available states in our model), the increase and decrease of the two factors become steeper and steeper. Whatever the value of n, however, the product of these factors has a maximum value of unity at $x = \frac{1}{2}$. The peak of this function becomes steadily narrower as n increases, sharpening to an infinitely narrow spike in the limit $n \to \infty$. This behavior is illustrated in the accompanying graph.

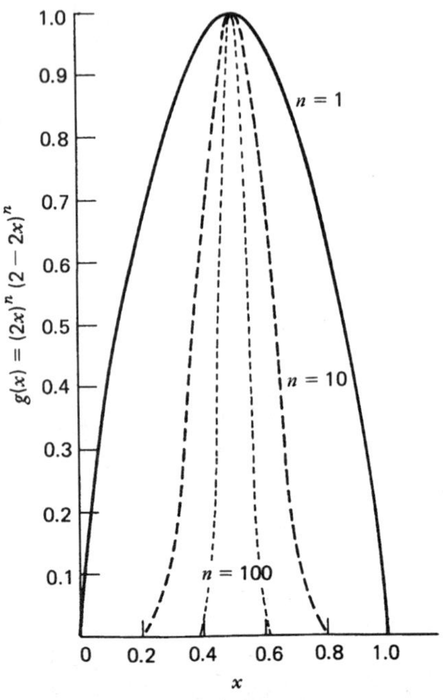

If we define

$$\beta \equiv \left(\frac{\partial \ln \Omega}{\partial E}\right)_{V,N}, \tag{15.40}$$

we see that in the most probable state of two systems over which a given amount of energy is distributed,

$$\beta_A = \beta_B. \tag{15.41}$$

This result is reminiscent of the equality of temperature between two systems in thermal equilibrium. Indeed, we shall see later that we can make the identification $\beta = (k_BT)^{-1}$, where k_B is the Boltzmann constant. The function defined by

$$S \equiv k_B \ln \Omega \tag{15.42}$$

is called the *entropy.* The condition of maximum probability in terms of the entropy, obtained by introducing S in Eq. 15.36, is then

$$(S_A + S_B)_{E_T,V,N} = \text{maximum} \tag{15.43}$$

for the system we have discussed.

We now identify the most probable state of the ensemble of replica pairs with the macroscopic state of thermal equilibrium. As will be shown below, this most probable state is overwhelmingly more likely to occur than are states having even slightly different distributions of energy between the pairs. Thus, for the given macroscopic constraints of constant total volume, total energy, and total mass, a pair of replicas that instantaneously has an energy distribution different from the most probable distribution, as a result of a temporary perturbation, is far more likely to move toward the most probable state than away from it. Another way of saying this is to note that Eq. 15.43 defines a stable equilibrium.

The motivation leading to Eq. 15.42 should now be clear, for we have argued that when a parameter of constraint is relaxed and a system undergoes a change to a new equilibrium state, $\Omega_f > \Omega_i$. Therefore the entropy function can be used to describe the direction in which a system will change upon relaxation of a parameter of constraint, that is, which energy-conserving changes will, and which will not, occur.

Equation 15.43 is of such fundamental importance that we wish to make the following supplementary observations:

1. The derivation given shows that if the ensemble of replica pairs is defined by the requirement that the sum of the energies of two systems is constant, then the most probable distribution of energy in a pair has the feature that $\beta_A = \beta_B$. It is important to note that in this ensemble the systems A and B are not in contact, and there is no flow of energy between them. Thus the energies E_A and E_B of any pair of replica systems are, in our ensemble, constants of the motion and time-independent. Nevertheless, the constraint that $E_A = E_T - E_B$ is sufficient to ensure that in the most probable distribution of energy between A and B the condition $\beta_A = \beta_B$ is satisfied. We emphasize this point because it shows that the most probable distribution of energy is a feature of the equilibrium state that is completely independent of the nature of the approach to equilibrium. Our arguments give no clue whatsoever as to the rate or the mechanism of approach to equilibrium. It is logical to assume, just because the state in which $\beta_A = \beta_B$ represents the overwhelmingly most probable state of a system for which $E_T = E_A + E_B$, that if A and B were in contact, and if energy did then flow between A and B, that the average direction of energy transfer must be such as to establish the state in which $\beta_A = \beta_B$.

2. Although the number of states in the energy range from E to $E + dE$ depends on the size of dE, the entropy function is insensitive to the size of dE, and β is completely independent of dE. To show these two features of the entropy function we rewrite Eq. 15.42, making use of Eq. 15.13, in the form

$$S \equiv k_B \ln\Omega = k_B \ln(\omega\, dE). \qquad (15.44)$$

Clearly, since dE is considered to be a fixed energy interval, we must have

$$\beta \equiv \left(\frac{\partial \ln\Omega}{\partial E}\right)_{V,N} = \left(\frac{\partial \ln\omega}{\partial E}\right)_{V,N}, \qquad (15.45)$$

so that β is independent of dE. Also, if ΔE is some different subdivision of the energy scale than that defined by dE, an entropy S_Δ can be defined by

$$S_\Delta \equiv k_B \ln\Omega_\Delta(E), \qquad (15.46)$$

where $\Omega_\Delta(E)$ is the number of the accessible states in the energy range from E to $E + \Delta E$. The difference between the entropies defined by Eqs. 15.42 and 15.46 is just

$$S - S_\Delta = k_B \ln\left(\frac{dE}{\Delta E}\right), \qquad (15.47)$$

and can be shown to be negligibly small by the following argument: Near the ground-state energy of a system there are few states available and $\Gamma_1(\epsilon)$, the number of quantum states of energy less than ϵ corresponding to one degree of freedom, is unity or some small integer. Provided the energy of the system is not too close to that of the ground state, $\Gamma_1(\epsilon)$ greatly exceeds unity. Then, not too close to the ground state, Eq. 15.21 implies that the entropy of the system is of the order of magnitude of νk_B, in the sense that $\ln \Gamma_1$ is some positive number, larger than unity, and very small compared to ν. On the other hand, if ΔE were larger or smaller than dE by a factor of ν, the term $\ln(dE/\Delta E)$ only ranges between $-\ln\nu$ and $+\ln\nu$. Since $\ln \nu << \nu$ when ν is a large number, the difference $S - S_\Delta$ is negligibly small relative to S. We thus have

$$S = S_\Delta \qquad (15.48)$$

with negligible error. The fact that Eq. 15.21 is only a crude estimate does not influence the strength of the argument given.

3. We can approximate the shape of the probability distribution $P(E_A)$ in the vicinity of the maximum as follows: For small displacements from the most probable distribution, we can write the following Taylor series expansions of $\ln \Omega_A(E_A)$ and $\ln \Omega_B(E_B)$:

$$\begin{aligned}\ln\Omega_A(E_A) &= \ln\Omega_A(E_A^*) + \left(\frac{\partial \ln\Omega_A}{\partial E_A}\right)_{V,N}(E_A - E_A^*)\\ &+ \frac{1}{2}\left(\frac{\partial^2 \ln\Omega_A}{\partial E_A{}^2}\right)_{V,N}(E_A - E_A^*)^2 + \cdots, \end{aligned} \qquad (15.49)$$

$$\begin{aligned}\ln\Omega_B(E_B) &= \ln\Omega_B(E_B^*) + \left(\frac{\partial \ln\Omega_B}{\partial E_B}\right)_{V,N}(E_B - E_B^*)\\ &+ \frac{1}{2}\left(\frac{\partial^2 \ln\Omega_B}{\partial E_B{}^2}\right)_{V,N}(E_B - E_B^*)^2 + \cdots, \end{aligned} \qquad (15.50)$$

where $E_A{}^*$ and $E_B{}^*$ are the values of E_A and E_B for which the probability distribution has its maximum value. It will be recalled that

$$E_T = E_A + E_B = E_A^* + E_B^*,$$

so that

$$\begin{aligned} E_B^* &= E_T - E_A^*,\\ E_A - E_A^* &= E_B^* - E_B, \end{aligned}$$

because of the conservation of energy. Adding Eqs. 15.49 and 15.50 yields (see Eq. 15.36)

$$\begin{aligned}\ln P(E_A) &= \ln[C\Omega_A(E_A)\Omega_B(E_B)]\\ &= \ln[C\Omega_A(E_A^*)\Omega_B(E_B^*)]\\ &\quad + (\beta_A - \beta_B)(E_A - E_A^*)\\ &\quad + \frac{1}{2}\left[\left(\frac{\partial^2 \ln\Omega_A}{\partial E_A^2}\right)_{V,N} + \left(\frac{\partial^2 \ln\Omega_B}{\partial E_B^2}\right)_{V,N}\right]\\ &\quad (E_A - E_A^*)^2. \end{aligned} \qquad (15.51)$$

At the maximum of the probability distribution we have $\beta_A = \beta_B$ by Eq 15.41, and the linear term in Eq. 15.51

vanishes. Equation 15.51 then shows that $P(E_A)$ is the product of two terms (aside from the constant C),

$$F_A(E_A)=\Omega_A(E_A^*)\exp\left[\frac{1}{2}\left(\frac{\partial^2 \ln\Omega_A}{\partial E_A^2}\right)_{V,N}(E_A-E_A^*)^2\right],$$
$$F_B(E_B)=\Omega_B(E_B^*)\exp\left[\frac{1}{2}\left(\frac{\partial^2 \ln\Omega_B}{\partial E_B^2}\right)_{V,N}(E_B-E_B^*)^2\right], \quad (15.52)$$

where the substitution $E_A - E_A^* = E_B^* - E_B$ has been made because the systems are coupled by the constraint $E_T = E_A + E_B$. Now $(E_A - E_A^*)^2$ is always positive, and $\Omega_A(E_A)$ is an increasing function of E_A. Also, $(E_B - E_B^*)^2$ is always positive and $\Omega_B(E_B)$ is an increasing function of E_B and thus a decreasing function of E_A. For the product function $P(E_A)$ to have a maximum at E_A^*, E_B^*, the extrema of $F_A(E_A)$ and $F_B(E_B)$ at these points must also be maxima rather than minima. It is thus necessary that the coefficients of $(E_A - E_A^*)^2$ and $(E_B - E_B^*)^2$ be negative, that is, that the second derivatives $(\partial^2 \ln \Omega_A/\partial E_A^2)$ and $(\partial^2 \ln \Omega_B/\partial E_B^2)$ be negative. To suggest why this is the case, we recall that $\Omega(E)$ is proportional to $[\Gamma_1(\epsilon)]^\nu$, from Eq. 15.21, in connection with which we have argued that $\Gamma_1(\epsilon)$ increases as some small power of ϵ. Thus, at least crudely, $\Omega(E)$ is proportional to $\epsilon_1^\nu = (E/\nu)^\nu$, where it is to be understood that the precise exponent may differ from ν by a factor that is of the order of magnitude of unity. Clearly, this error in our estimate of $\Omega(E)$ cannot influence the sign of the second derivative $[\partial^2 \ln \Omega/\partial E^2]$. Using the cited proportionality, we obtain

$$\left(\frac{\partial \ln\Omega}{\partial E}\right)_{V,N} \propto \frac{\nu}{E}, \quad (15.53)$$
$$\left(\frac{\partial^2 \ln\Omega}{\partial E^2}\right)_{V,N} \propto -\frac{\nu}{E^2}. \quad (15.54)$$

Since ν is positive and E^2 is positive, $-(\nu/E^2)$ must be negative, hence the functions displayed in Eq. 15.52 have maxima.

The first derivative of $\ln \Omega$ with respect to E is also of interest. Combining 15.53 and 15.40, we find

$$\frac{E}{\nu} \propto \frac{1}{\beta}, \quad (15.55)$$

which is a form of the principle of equipartition of energy. Equation 15.55 implies that the average energy per degree of freedom is constant and proportional to β^{-1}, or, with the identification to be made later, $\beta^{-1} = k_BT$, to the temperature. The estimate given for $\Omega(E)$ suffices to determine the orders of magnitude and the signs of derivatives and of functions, but is not sufficiently precise to be a complete justification for the equipartition principle. We shall return to that subject later in this text.

4. We now can show that the probability distribution for the energy is extremely sharply peaked about the energies E_A^* and E_B^*. Indeed, the distribution is so sharply peaked that for many purposes the distribution may be replaced by the value of $P(E_A)$ at the maximum. The combination of Eqs. 15.36 and 15.52, followed by the use of the estimate in 15.54, shows that the arguments of the exponential function in Eq. 15.52 are proportional to

$$-\nu\frac{(E-E^*)^2}{E^{*2}}.$$

Then the energy for which the probability has fallen to $1/e$ of the probability of the maximum is given by

$$-\nu\frac{(E-E^*)^2}{E^{*2}}=-1$$

or

$$\frac{E-E^*}{E^*}=\frac{1}{\sqrt{\nu}}. \quad (15.56)$$

For $\nu \approx 10^{24}$, Eq. 15.56 yields $(E - E^*)/E^* \approx 10^{-12}$, and we conclude that in a system of macroscopic size the probability of finding an energy deviating by any substantial amount from the most probable energy is vanishingly small.

5. The probability distribution is so sharply peaked that the entropy can also be defined in terms of $\Gamma(E)$, the total number of states with energy less than E, the result being negligibly different from Eq. 15.42. That is, because of the very sharp maximum in the energy distribution, practically all the accessible states lie within the range $\Delta E^* = E^*/\sqrt{\nu}$ of E^*. Thus, Eq. 15.13 becomes

$$\Gamma(E)\approx\omega(E^*)\Delta E^*, \quad (15.57)$$

and thereby

$$\ln\Gamma(E)=\ln\Omega(E)+\ln\left(\frac{\Delta E^*}{dE}\right), \quad (15.58)$$

so that

$$S_\Gamma = k_B\ln\Gamma(E) \quad (15.59)$$

differs negligibly from the entropy defined in Eq. 15.42 by application of the same arguments used to establish the equivalence displayed in Eq. 15.48.

There are several properties of the entropy function that can be used to provide a still closer link between the microscopic and macroscopic descriptions of matter. Consider two isolated systems, A and B. At equilibrium in their isolated conditions their energies and entropies are E_A^i, E_B^i and S_A^i, S_B^i, respectively. Imagine that a third system, R, is brought into contact with system A. The system R could be, for example, a container of gas. When the contact between R and A is through a rigid diathermal wall, expansion of the system R will extract energy from A while the volume of A remains fixed. Conversely, compression of R will lead to addition of energy to A, at constant volume of A. We shall use system R to transfer energy between A and B without allowing A and B to be in contact.

We have already learned that the most probable state of an ensemble of replicas of A and B has the properties that

$$\beta_A^f = \beta_B^f \tag{15.60}$$

and

$$E_A^{f*} + E_B^{f*} = E_A^i + E_B^i, \tag{15.61}$$

where, as before, the asterisk refers to the most probable state. We now examine further the implications of the identification of the most probable distribution of energy in the ensemble of replica systems A and B with the equilibrium state that is achieved when macroscopic systems A and B are brought into contact through a diathermal wall. Since contact through a diathermal wall implies energy transfer, we must examine how the properties of A, B, and R are related.

Let the system R, in diathermal contact with A, be expanded[18] so as to remove energy

$$-q_A = E_A^{f*} - E_A^i \tag{15.62}$$

from A. Now let R be isolated from A and placed in diathermal contact with system B, and let R be compressed until energy

$$q_B = E_B^{f*} - E_B^i \tag{15.63}$$

is added to B. As usual, q_A and q_B are defined as positive for energy added to the system in question. At the end of this sequence of processes, because of the conservation of energy (see Eq. 15.61), we note that

$$q_A + q_B = 0, \tag{15.64}$$

and that the system R has undergone no net change in state. Of course, because of the energy changes in systems A and B the entropies of A and B have been changed. From the fundamental relation between the entropy and the number of accessible states we conclude that

$$S(E_A^{f*}) + S(E_B^{f*}) \geq S(E_A^i) + S(E_B^i). \tag{15.65}$$

Equation 15.64 is interesting. Because the volumes of systems A and B have been maintained constant during the diathermal interactions with R, no work is done on or by either A or B. Then, in thermodynamic terminology, the energy changes in A and B are due entirely to heat transfers. It is this observation that motivated the choice of symbol q in Eqs. 15.62 and 15.63. The energy conservation condition, as stated in Eq. 15.64, then says that the heat given off by system A is absorbed by system B.

Further information of interest can be obtained by supposing that system A is very, very much smaller than system B, so that

$$E_A^* << E_B^*. \tag{15.66}$$

Of course, under these conditions system B has many, many more degrees of freedom than does system A. For system B the rate of change of β_B with the energy E_B, namely, $(\partial\beta_B/\partial E_B)_{V,N}$, is of the order of magnitude[19] of $(-\beta_B/E_B^*)$. On the other hand, the maximum amount of energy that can be removed from system A is E_A^*. If system A and system B are brought to a condition of thermal equilibrium, say, by means of the third system R, then because $q_A = -q_B$, it is found that

$$\left|\left(\frac{\partial \beta_B}{\partial E_B}\right)_{V,N} q_B\right| \approx \beta_B \frac{E_A^*}{E_B^*} << \beta_B. \tag{15.67}$$

The importance of Eq. 15.67 is the following: It implies that the characteristic parameter of system B, namely, β_B, remains virtually unchanged no matter how much heat B absorbs from A. When condition 15.67 is satisfied for every energy transfer, system B is considered a reservoir for system A. Thus, if B is a reservoir and condition 15.67 is valid, β_B can be taken as constant throughout any sequence of energy transfers from A to B. Also, and perhaps most important, Eq. 15.67 allows us to establish a direct connection between the amount of energy transferred and the entropy change of the reservoir. After the reservoir has absorbed heat q_B from system A, with $\beta_A = \beta_B$, we can write for the change in the number of states accessible to the reservoir (system B)

$$\begin{aligned}\ln\Omega_B(E_B + q_B) - \ln\Omega_B(E_B) \\ = \left(\frac{\partial \ln\Omega_B}{\partial E_B}\right)_{V,N} q_B + \cdots \\ = \beta_B q_B.\end{aligned} \tag{15.68}$$

[18] The process can also be run in the opposite direction.

[19] In the crude approximation previously used, correct only as to order of magnitude, combining Eqs. 15.53 and 15.54 with Eq. 15.40 yields

$$\beta \propto \frac{\nu}{E},$$

$$\left(\frac{\partial \beta}{\partial E}\right)_{V,N} \propto -\frac{\nu}{E^2};$$

hence,

$$\left(\frac{\partial \beta}{\partial E}\right)_{V,N} \approx -\frac{\beta}{E}.$$

But the left-hand side of Eq. 15.68 is just the entropy change of the reservoir, so that we reach an important conclusion: If a heat reservoir absorbs heat q_B at constant β_B, then the resultant change in the reservoir's entropy is

$$\Delta S_B = k_B(\beta_B q_B) = \frac{q_B}{T_B}, \tag{15.69}$$

with the final form again using the congruence $\beta_B = (k_B T_B)^{-1}$. We have now established a connection between the macroscopic variable heat and the entropy function that has been defined in terms of the microscopic energy spectrum of the system. A similar relation holds for any system that, at constant β, absorbs an infinitesimal amount of heat $đq$ from some other system. Provided that $đq << E$, where E is the energy of the system under study, we have[20]

$$dS = k_B \beta\, dq = \frac{đq}{T}. \tag{15.70}$$

What are the properties of the entropy function when more general interactions between systems are permitted? To answer this question we must first determine the dependence of the variation of the density of states on the variation of the external parameters of constraint defining the system.

Consider, first, the case that the parameter of constraint is the volume V, and let $E_i(V)$ be the energy of the ith state in the range E to $E + dE$. When V is changed by dV, $E_i(V)$ changes by $(\partial E_i(V)/\partial V)\, dV$. It is to be expected that the energies of different states change by different amounts. Now, the number of states between E and $E + dE$ corresponding to some value of V *and having the property that*

$$-(Y + dY) < \left(\frac{\partial E_i}{\partial V}\right)_{\Omega,N} < -Y \tag{15.71}$$

is denoted[21] $\Omega_Y(E, V, N)$. Since each of the levels corresponds to some range from Y to $Y + dY$, it is necessary that

$$\Omega(E,V,N) = \sum_Y \Omega_Y(E,V,N), \tag{15.72}$$

with the summation over all values of Y. Note that the derivative in Eq. 15.71 is defined with Ω and N held constant. If there are N_i particles with energy E_i, holding Ω fixed implies holding all the N_i fixed, and by virtue of the definition of the entropy (Eq. 15.42), also holding the entropy constant.

Given the definition of $\Omega_Y(E, V, N)$, we can calculate how many states are altered when $V \to V + dV$ such that their energies change from being less than E to being more than E. Note that those states for which $(\partial E_i/\partial V)_{\Omega,N} = -Y$ undergo an energy change of $-Y\,dV$. Then the total number of states whose energy is changed from a value less than E to a value more than E is simply equal to the product of the number of states per unit energy corresponding to the given values of V and Y, namely, $\Omega_Y(E, V, N)/dE$, and the energy change, $-Y\,dV$, summed over all possible values of Y—that is, to

$$-\sum_Y \frac{\Omega_Y(E,V,N)}{dE} Y\,dV.$$

Consider now the total number of states between E and $E + dE$ irrespective of the value of Y. When $V \to V + dV$, the number of states in this energy range changes by

$$\begin{aligned}
&\Omega(E,V+dV,N) - \Omega(E,V,N) \\
&= \left(\frac{\partial \Omega(E,V,N)}{\partial V}\right)_{E,N} dV \\
&= \sum_Y \frac{\Omega_Y(E+dE,V,N)}{dE} Y\,dV - \sum_Y \frac{\Omega_Y(E,V,N)}{dE} Y\,dV \\
&= \sum_Y \left(\frac{\partial \Omega_Y(E,V,N)}{\partial E}\right)_{V,N} Y\,dV.
\end{aligned} \tag{15.73}$$

The second equality of Eq. 15.73 is simply the difference between the net number of states that leave the energy range between E and $E + dE$ by having their energy changed from a value less than $E + dE$ to a value more than $E + dE$, and the net number of states that enter this energy range by having their energy changed from a value less than E to a value more than E. But the mean value of Y over all states, each state weighted with equal probability, is

$$\langle Y \rangle = \frac{1}{\Omega(E,V,N)} \sum_Y Y \Omega_Y(E,V,N), \tag{15.74}$$

so that Eq. 15.73 may be written in the form

$$\begin{aligned}
\left(\frac{\partial \Omega}{\partial V}\right)_{E,N} &= \frac{\partial}{\partial E}(\langle Y \rangle \Omega)_{V,N} \\
&= \left(\frac{\partial \Omega}{\partial E}\right)_{\langle Y \rangle,V,N} \langle Y \rangle + \left(\frac{\partial \langle Y \rangle}{\partial E}\right)_{\Omega,V,N} \Omega,
\end{aligned} \tag{15.75}$$

because

$$\langle Y \rangle \Omega(E,V,N) = \sum_Y Y \Omega_Y(E,V,N). \tag{15.76}$$

If Eq. 15.75 is divided by Ω, we find that

$$\left(\frac{\partial \ln \Omega}{\partial V}\right)_{E,N} = \left(\frac{\partial \ln \Omega}{\partial E}\right)_{\langle Y \rangle,V,N} \langle Y \rangle + \left(\frac{\partial \langle Y \rangle}{\partial E}\right)_{\Omega,V,N}. \tag{15.77}$$

[20] The preceding argument shows that Eqs. 15.69 and 15.70 are valid only for reversible heat transfers at the constant temperatures T_B and T, respectively.

[21] We choose to insert the minus sign in the definition of Y for later convenience.

For a macroscopic system the term $(\partial\langle Y\rangle/\partial E)$ is negligible relative to $(\partial \ln \Omega/\partial E)\langle Y\rangle$, since $(\partial\langle Y\rangle/\partial E)$ is of the order of magnitude of $(\langle Y\rangle/E)$ whereas $(\partial \ln \Omega/\partial E)\langle Y\rangle$ is of the order of magnitude of $(\nu/E)\langle Y\rangle$, which differ by the factor[22] $\nu \approx 10^{24}$. In fact, the larger the system, the more negligible does the second term on the right-hand side of Eq. 15.78 become. We therefore can write

$$\left(\frac{\partial \ln \Omega}{\partial V}\right)_{E,N} = \left(\frac{\partial \ln \Omega}{\partial E}\right)_{\langle Y\rangle,V,N} \langle Y\rangle, \tag{15.78}$$

or, by Eq. 15.45,

$$\left(\frac{\partial \ln \Omega}{\partial V}\right)_{E,N} = \beta\langle Y\rangle. \tag{15.79}$$

Since by our definition of Y we have

$$\langle Y\rangle = -\left\langle\left(\frac{\partial E(V)}{\partial V}\right)_{\Omega,N}\right\rangle, \tag{15.80}$$

it can be seen that $\langle Y\rangle$ plays the role of a generalized force complementary to the volume V. In this case the generalized force is the pressure, as we shall shortly demonstrate.

Consider, as an example, a gas of N noninteracting particles in a box of volume V. We have already shown that (see Eqs. 15.27 and 15.31)

$$\Omega(E) \propto V^N E^{(3N/2)-1} dE \tag{15.81}$$

with the proportionality constant independent of E and V. Then

$$\left(\frac{\partial \ln \Omega}{\partial V}\right)_{E,N} = \frac{N}{V} \tag{15.82}$$

because the proportionality constant cancels out of the expression for the derivative. We now must evaluate $\langle(\partial E/\partial V)_{\Omega,N}\rangle$. We have already noted that for constant Ω the number of particles per level is a constant. Then for our system of independent particles

$$E = \sum_i N_i E_i, \tag{15.83}$$

from which we have

$$\left\langle\left(\frac{\partial E}{\partial V}\right)_{\Omega,N}\right\rangle = \left\langle\sum_i N_i\left(\frac{\partial E_i}{\partial V}\right)_{N_i}\right\rangle = -\frac{2}{3}\frac{\langle E\rangle}{V}, \tag{15.84}$$

where the extreme right-hand side of Eq. 15.84 follows from use of Eq. 15.23.[23] By combination of Eqs. 15.82, 15.84, 15.79, and 15.80, we find that the average energy is

$$\langle E\rangle = \frac{3N}{2\beta}. \tag{15.85}$$

But for a perfect gas, from Eq. 12.42, and identification of the ensemble average energy with the average energy of Chapter 12,

$$\langle E\rangle = \tfrac{3}{2}Nk_BT, \tag{15.86}$$

so that we can make the congruence

$$\beta = \frac{1}{k_BT}, \tag{15.87}$$

as already used in Eq. 15.69. Furthermore, from the relation

$$pV = \tfrac{2}{3}\langle E\rangle \tag{15.88}$$

derived in Chapter 12, we find after use of Eq. 15.84 that

$$p = -\left\langle\left(\frac{\partial E}{\partial V}\right)_{\Omega,N}\right\rangle, \tag{15.89}$$

as asserted above.[24]

The preceding argument can be applied to calculate the change in $\Omega(E, y_1, \ldots, y_n)$ arising from changes in any (or several) parameters of constraint y_i. The algebraic manipulations become more tedious, but the ideas used and the procedure followed are the same. Clearly, when there are n parameters of constraint, $\Omega(E, y_1, \ldots, y_n)$ defines n generalized forces by relations of the form

$$\left(\frac{\partial \ln \Omega}{\partial y_\alpha}\right)_{E,N,y_{i\neq\alpha}} = \beta\langle Y_\alpha\rangle. \tag{15.90}$$

By a generalized force is meant simply that if the change of parameter of constraint dy is thought of as a "displacement," then $\langle Y\rangle\, dy$ has the dimensions of work. Furthermore, as discussed in Chapter 3, the general definition of force in

---

[22] To the same, crude, approximation we have used before (cf. discussion after Eq. 15.20), we assume $\langle Y\rangle$ is proportional to E and use Eq. 15.53 for $(\partial \ln \Omega/\partial E)_{V,N}$.

[23] In Eq. 15.23, ϵ_j refers to the energy of particle j, whereas in Eq. 15.83, E_i refers to the energy of level i. Of course, $\Sigma_{j=1}^{N}\epsilon_j = \Sigma_{i=1}^{\infty}N_iE_i$.

[24] It is easy to verify that $(\partial\langle Y\rangle/\partial E)_{\Omega,V,N}$ is very small compared to $(\partial \ln \Omega/\partial E)_{\langle Y\rangle,V,N}\langle Y\rangle$ as stated after Eq. 15.77. Using Eq. 15.88, and setting $\langle Y\rangle = p$, we see that

$$\left(\frac{\partial p}{\partial E}\right)_{\Omega,V,N} = \frac{2}{3V},$$

which is, indeed, negligible relative to

$$\left(\frac{\partial \ln \Omega}{\partial V}\right)_{E,N} = \frac{N}{V}.$$

mechanics is as the negative derivative of the energy with respect to a coordinate.

The notion of generalized force is quite useful. In mechanics the energy is a potential function that defines both the force acting on the system and whether or not the equilibrium state is stable. We have shown in this section that the rate of change of entropy with the parameter of constraint also defines the generalized force acting on the system. By analogy with mechanics, we expect the entropy to determine whether or not the equilibrium state is stable. We shall show, in Chapter 19, that this expectation is correct.

We now examine the nature of the changes in entropy corresponding to an interaction between two systems, *A* and *B*, in which both work and energy transfers are allowed. We shall use arguments similar to those already outlined. Let the two systems interact through a third system *R; A* and *B* are otherwise isolated. Further, let *A* and *B* exchange both energy and work through the use of *R*. This can be accomplished by allowing V_A and V_B to vary (keeping $V_A + V_B$ constant) so that *A* does work on *R* and then *R* does work on *B*, or vice versa. Clearly, for other kinds of work processes equivalent conditions can be defined. As before, since $E_T = E_A + E_B$ is fixed, the energy E_B is determined if E_A is known. Further, the parameters of constraint $V_A, y_1^A, \ldots, y_{n-1}^A$ defining system *A* are, at equilibrium, some function of the parameters of constraint $V_B, y_1^B, \ldots, y_{n-1}^B$ defining system *B*. For example, $V_T = V_A + V_B$ and, since V_T is a constant, a knowledge of V_A fixes V_B. Analogous relations hold for $y_1^A, \ldots, y_{n-1}^A$ and $y_1^B, \ldots, y_{n-1}^B$. In the combined system the total number of states is a function of the energy *E* and the parameters $V, y_1, \ldots, y_{n-1}$. As before, $\Omega_T(E, V, y_1, \ldots, y_{n-1})$ has a very sharp maximum for some values $E = E^*$, $V = V^*$, and $y_\alpha = y_\alpha^*$. Also as before, the mean values of E, V, and $y_1, \ldots, y_{n-1}$ differ negligibly from the most probable values.

Consider now a process by which the system *A*, through interaction with *B* via the intermediary *R*, is changed from an equilibrium state characterized by $E, V, y_1, \ldots, y_{n-1}$, to another equilibrium state characterized by $E + dE$, $V + dV$, $y_1 + dy_1, \ldots, y_{n-1} + dy_{n-1}$. What is the change in the entropy? We can write, using Eqs. 15.45 and 15.79,

$$d\ln\Omega = \left(\frac{\partial \ln\Omega}{\partial E}\right)_{N,V,y} dE + \sum_{\alpha=1}^{n-1}\left(\frac{\partial \ln\Omega}{\partial y_\alpha}\right)_{E,V,N} dy_\alpha + \left(\frac{\partial \ln\Omega}{\partial V}\right)_{E,N,y} dV \qquad (15.91)$$

$$= \beta\left(dE + \sum_{\alpha=1}^{n-1}\langle Y_\alpha\rangle dy_\alpha + p\,dV\right). \qquad (15.92)$$

But, by our definition of a generalized force, the product $\langle Y_\alpha\rangle\, dy_\alpha$ is the work done in changing the parameter of constraint y_α by dy_α, just as $p\,dV$ is the work done in changing V to $V + dV$. Using the same convention as in Chapter 13, we reckon work done on the system as positive, and work done by the system as negative. With this convention we obtain[25]

$$đw = -\sum_{\alpha=1}^{n-1}\langle Y_\alpha\rangle\, dy_\alpha - p\,dV. \qquad (15.93)$$

Combining Eqs. 15.93 and 15.92, we find that

$$d\ln\Omega = \beta[dE - đw] \equiv \beta\, đq. \qquad (15.94)$$

In Eq. 15.94 the relationship

$$đq = dE - đw$$

has been introduced. By this introduction of $đq$ in terms of dE and $đw$, we use the first law of thermodynamics to provide a bridge between the macroscopic definition of heat and the number of states. Indeed, substituting Eq. 15.70 into Eq. 15.94, we are led to

$$đq = \frac{dS}{k_B\beta} = dE - đw, \qquad (15.95)$$

a relationship between the entropy and macroscopic functions.

Before using the entropy function to study the macroscopic properties of matter, we turn to another way of introducing the entropy concept, again in terms of the constraints defining the state of a macroscopic system, but now employing only thermodynamic considerations.

[25] It was to obtain the correct sign in Eq. 15.93 that the negative sign was introduced in the definition of *Y*.

APPENDIX 15A

Comments on Ensemble Theory

The goal of the statistical theory of matter is the provision of a microscopic interpretation of the observed behavior of macroscopic matter. The description of the macroscopic behavior of matter is based on the assumptions that matter and energy are continuous; hence, the natural mathematical representation of observable quantities is by means of continuous functions of the space coordinates and the time. The character of a macroscopic observable in a particular experimental realization, namely its dependence on space and time, is typically determined by a partial differential equation with specified boundary conditions. For example, suppose a fluid with temperature T_0, flowing in the x direction with constant velocity υ, enters (at $x = 0$) a straight circular cylinder with radius R, whose wall is maintained at the constant temperature T_1. The stationary state temperature distribution in the liquid is a solution of the equation

$$\upsilon \frac{\partial T}{\partial x} = k\left(\frac{\partial^2 T}{\partial r^2} + \frac{1}{r}\frac{\partial T}{\partial r} + \frac{\partial^2 T}{\partial x^2} \right) \tag{15A.1}$$

with boundary conditions $T = T_1$ at $r = R$ all $x > 0$; $T = T_0$ at $x = 0$ all $r < R$. The coefficient k is the thermal conductivity of the liquid. In contrast, the microscopic description of matter takes account of the fundamental discontinuities that exist on the atomic length scale and describes the motion of each particle by the basic laws of dynamics. To construct a bridge between observed macroscopic behavior and microscopic dynamics we need a set of correspondence rules for associating, in a unique fashion, macroscopic observables with microscopic dynamical functions. This set of correspondence rules cannot be provided by mechanics for, even were we able to solve the initial value problem for a very large number of interacting particles, the solution would not be in a form that could be used to answer questions about macroscopic behavior because those questions often use nonmechanical concepts to describe the macroscopic properties, e.g., temperature. The correspondence rules we need connect the *concepts* of the macroscopic description of matter with dynamical functions; because *nonmechanical concepts* are involved, the rules of correspondence must come from outside mechanics.

The macroscopic description of the temperature distribution in the example cited above is considered to be deterministic in the sense that every time the experiment is repeated under identical conditions, the same result will be found. But from the microscopic point of view, specifying the temperature distribution in the liquid gives very little information about the particle dynamics; there are an enormous number of molecular configurations (wave functions when quantum mechanics is the basis for the molecular description, particle positions and momenta when classical mechanics is used), which are compatible with the few macroscopic constraints (the boundary conditions and the fluid velocity), and the repetition of the experiment must sample a great variety of these molecular configurations. Since the observed behavior is the same for all of those molecular configurations compatible with the macroscopic constraints, they must in some sense be equivalent and therefore should be treated as such.

The preceding description, by use of the words *identical conditions,* seems to imply that the macroscopic constraints can be defined with infinite precision. Of course, this is not so. We must recognize that the macroscopic constraints are defined by the values of the macroscopic variables used to characterize the system and that these values are always subject to some uncertainty. Although the uncertainty in the value of a macroscopic variable can be made very small relative to the magnitude of the variable itself, the magnitude of the uncertainty is very large on the scale of molecular energy, distance, etc. Consider, for example, the determination that a macroscopic system has energy between E and $E + \Delta E$ with $\Delta E/E << 1$. It is always the case that although ΔE can be made very small on the macroscopic scale, it is enormously larger than the energy per particle. We are then logically compelled to recognize that all of the molecular configurations compatible with the macroscopic constraints *and their uncertainties* must be treated as equivalent.

One way of accounting for the relationship between microscopic dynamics and macroscopic behavior is to attach a weight to each microscopic configuration at $t = 0$. For example, we can assign weight zero to every configuration inconsistent with the macroscopic constraints and an equal nonzero weight to every configuration consistent with the macroscopic constraints. To use this assignment of weights, we must introduce another postulate, namely, we *define* the macroscopic observable (e.g., the energy) to be the average of the corresponding microscopic quantity over all of the configurations of the dynamical system, properly weighted. This definition guarantees that if the system is in any microscopic configuration consistent with the macroscopic constraints, the same values for the macroscopic observables will be obtained. The prescription just given can predict the average outcome of a large number of experiments performed under identical macroscopic conditions, but it cannot predict, and we cannot exclude, specific fluctuations from the average outcome. The theory based on that prescription is, however, capable of describing some of the statistical properties of fluctuations, e.g., their average square values.

It is very important to observe that the argument just given for weighting microscopic configurations posits a definition of the state of the system that differs from that used in mechanics and from that used in thermodynamics. In the former, the state is defined by the wave function (or the positions and momenta of all the particles), and in the latter, it is defined by the constraints determined by the values of a few macroscopic variables. In the statistical description of the many particle system the state is represented by a collection of wave functions (or a collection of points in phase space), each function (or phase point) weighted in a specified fashion. The collection of weighted functions (phase points) defines an ensemble, and we identify the observable value of a dynamical function with the ensemble average of the corresponding microscopic dynamical function.

The definition and description of the equilibrium state is of central importance in the statistical theory of matter. To illustrate the problems encountered in developing a suitable definition of statistical equilibrium consider, first, the macroscopic description of an equilibrium state. Thermodynamics defines an equilibrium state as that state toward which an isolated system evolves. To determine if a system is in a state of thermodynamic equilibrium, its isolation must be broken and the stability with respect to various "displacements" examined (a detailed analysis is presented in Chapter 19). In practical measurements, an equilibrium state with respect to specified constraints is identified when the macroscopic properties of the system are shown to be time independent. However, this definition cannot be used in a many particle dynamical theory because, when molecular configurations are examined, such a state cannot be recognized; the molecular motion incessantly generates fluctuations. The heart of the problem of defining an equilibrium state using many body dynamical theory concerns how to account for fluctuations.

The requirement that the thermodynamic and the many particle dynamical theory definitions of the equilibrium state be consistent implies that the concept of equilibrium introduced should not depend explicitly on the existence of fluctuations yet should implicitly include all possible fluctuations of the system. If we were to construct the statistical dynamic definition of equilibrium without the introduction of ensemble averaging—that is, deal with the behavior of a single system—there are only two possibilities for that definition. These are (1) either a selected microscopic state is defined to be the equilibrium state, or (2) the fluctuations are eliminated by taking average values. We reject the first of these possible definitions because fluctuations in the system then appear as deviations from equilibrium rather than as characteristics of equilibrium. The second possible definition suffers from two difficulties. First, if the equilibrium properties of the system are defined as time averages taken over an infinite time, we must confront the problem that real measurements are always taken over only a finite time; hence, a careful analysis of the relationships between the duration of the observation and the microscopic and macroscopic relaxation times is required. Second, computation of time averages requires, in principle, complete solution of the N-particle equations of motion. This problem is bypassed by the use of the quasi-ergodic hypothesis, which asserts the equality of time averages and ensemble averages, but which introduces ensemble averaging as a mathematical device rather than as a fundamental aspect of the statistical molecular description.

We can overcome the difficulties cited in a different and more perceptive way by adopting Gibbs' point of view. Instead of trying to define the equilibrium state in terms of the observations that can be made on it, we consider as well the means used to prepare that state of the system. The method of preparation defines a Gibbs ensemble.

The ideas used by Gibbs are based on the distinction between an ideal experiment and a statistical experiment. An *ideal experiment* is defined to be one that yields perfectly reproducible results. This implies that the conditions defining the experiment can be so completely controlled that every replication of the experiment yields the same result. In contrast, we define a *trial* to be a nonreproducible experiment. We call an experiment a trial when even the most careful preparation of the system and control of the experimental conditions that can be achieved cannot guarantee that the result of the measurement will be the same for every replication. A *statistical experiment* is carried out by repeating a trial many times. The set of replications of the trial defines the ensemble. Then, if A is any one of the possible results of a trial, with $N(A)$ the number of replications in the ensemble yielding the result A, and $\mathcal{N}$ is the total number of replications defining the ensemble, $N(A)/\mathcal{N}$ is the relative frequency of the result A in this ensemble. For infinite statistical experiments—that is, those consisting of an infinite number of trials—the relative frequencies of the results are postulated to be perfectly reproducible, provided the experimental conditions in the individual trials are identical to within the limits of our ability to control them. With this assumption, we can interpret the relative frequency of a possible trial result A in an infinite statistical experiment as a physical property of that trial, which we define to be its physical probability.

The concept of physical probability is used to construct a Gibbs ensemble with the following argument. A trial conducted with a dynamical system can be thought of as having three stages: preparation, development, and observation. The preparation process cannot be controlled sufficiently to generate a unique dynamical state of the system, else we would have a reproducible experiment instead of a nonreproducible trial. Similarly, the observation cannot be accurate enough to uniquely determine the dynamical state of the system, else it could be incorporated into an improved preparation process that would generate a unique dynamical state and eliminate the need for statistical analysis. The development stage occupies the interval during which the consequences of the preparation stage evolve. We now define an *incomplete trial* as one where the system has been prepared but not observed. The "results" of incomplete trials are the unknown dynamical states that determine the evolution of the development stage, and we use probability theory to characterize these results even though they are unobservable (by construction). We define a *Gibbs ensemble* as the set of systems in various dynamical states generated by many replications of an incomplete trial; the members of the Gibbs ensemble differ only in their dynamical states. Since a trial is initiated by specifying the macroscopic constraints, this definition is equivalent to that in which only macroscopic constraints define the replicas of the ensemble, and every dynamical state consistent with those macroscopic constraints is represented in the ensemble. The statement "this system is in equilibrium" is taken to mean that if a Gibbs ensemble is prepared in the same way as the system was prepared, then that Gibbs ensemble will be in equilibrium.

Although we have introduced the ensemble average definition of the microscopic dynamics-macroscopic observable correspondence as a postulate not needing further justification, it is worthwhile examining how reasonable this postulate is.[26] We examine this issue using the classical dynamical description of a many particle system, since it is much better developed than is the corresponding quantum dynamical description. Fundamental support for the postulate comes from the observation that the trajectories of almost all mechanical systems are extremely unstable with respect to small changes in the initial conditions. This is a consequence of the *mixing* property, which is conjectured to be a common feature of the dynamics of many particle systems. We give a qualitative analysis of mixing below.

Consider a subregion γ in phase space, which we associate with the extent (volume)

$$\mathcal{V}(\gamma) = \int_{\gamma} \prod_{i} d\mathbf{x}_i \, d\mathbf{p}_i = \int_{\gamma} d\Gamma_i . \tag{15A.2}$$

We now show that $\mathcal{V}(\gamma)$ remains constant as the system evolves (Liouville's theorem). The argument that leads to this conclusion starts with the observation that the evolution of the system is represented by a path on the energy surface and that there is only one trajectory through a given point on that surface. If this were not the case, an initial condition could specify two different solutions to the equation of motion, which is inconsistent with the fact that the solution to the equation of motion, for given initial conditions, is unique. Then a phase point outside of $\mathcal{V}(\gamma_0)$ cannot get inside $\mathcal{V}(\gamma_t)$ unless it cuts the boundary of $\mathcal{V}(\gamma_\tau)$ at some time $0 \le \tau \le t$. But the boundary is also defined by the system dynamics, which implies that each point on the boundary could be on the trajectory of some system point. Since no two system point trajectories can cross, an outside phase point cannot get in $\mathcal{V}(\gamma)$, and an inside phase point cannot get out, which then leads to $\mathcal{V}(\gamma_0) = \mathcal{V}(\gamma_t)$. In other words, the volume of any subregion in phase space is a dynamical invariant.

The coupling of the invariance of $\mathcal{V}(\gamma)$ and the extreme instability of system trajectories has an interesting consequence. There is a class of dynamical systems that have local properties that give rise to asymptotically stochastic behavior, the signature of which is system trajectories that diverge from one another with an exponential rate. A system that exhibits exponentially diverging trajectories has the mixing property mentioned above. For such a system, as the phase points contained in a volume element of phase space evolve as dictated by the equations of motion, the volume element contorts, stretching and contracting in orthogonal directions at exponential rates, yet preserving the magnitude of $\mathcal{V}(\gamma_t)$. This is the behavior that underlies mixing, as can be illustrated with the following argument that, for simplicity, we restrict to the case of a finite phase space. Consider, in addition to the volume $\mathcal{V}(\gamma_0)$, some other fixed volume in phase space, say $\mathcal{V}(G)$. We imagine holding $\mathcal{V}(G)$ fixed while $\mathcal{V}(\gamma_0) \to \mathcal{V}(\gamma_t)$, in the process spreading out over the phase space and eventually intersecting $\mathcal{V}(G)$. Let $\gamma_t \cap G$ be the region of intersection of γ_t and G and $\mathcal{V}(\gamma_t \cap G)$ be its phase space volume. As the spreading continues and $\mathcal{V}(\gamma_t)$ intersects $\mathcal{V}(G)$ many times, we expect to find that $\mathcal{V}(\gamma_t \cap G)/\mathcal{V}(G)$ will approach $\mathcal{V}(\gamma_t)/\mathcal{V}$ (total phase space). But $\mathcal{V}(\gamma_0) = \mathcal{V}(\gamma_t)$. Then, if we take the finite volume of the total phase space to be unity, we say that a system is dynamically mixing if $\mathcal{V}(\gamma_t \cap G)/\mathcal{V}(G) = \mathcal{V}(\gamma_0)$ in the limit as $t \to \infty$. Mixing implies ergodicity, the latter being a much weaker property of the system dynamics. The converse is not true: Ergodicity does not imply mixing. If a system is dynamically mixing, the initial distribution of phase points in the ensemble will spread out over the entire accessible phase space with the macroscopic consequence that the system reaches a state of equilibrium.

It is worth exploring a few more qualitative aspects of the relationship between the general character of the dynamics of a system, ergodicity, mixing, and the approach to equilibrium. As before, we consider a point in phase space, $\Gamma(t) \equiv (\mathbf{x}, \mathbf{p})$. The evolution of the system is described by following the evolution of $\Gamma(t)$, which we represent in the form

$$\Gamma(t+t') = \hat{\mathbf{T}}(t')\Gamma(t). \tag{15A.3}$$

[26] For advanced treatments of this and related subjects see Gaspard, P., *Chaos, Scattering and Statistical Mechanics* (Cambridge University Press, Cambridge, 1998), and Gutzwiller, M. C., *Chaos in Classical and Quantum Mechanics* (Springer-Verlag, New York, 1990).

Then the evolution of an arbitrary function of $\Gamma(t)$ can be represented in the form

$$f(\Gamma, t+t') = f(\hat{\mathbf{T}}(t')\Gamma(t)). \tag{15A.4}$$

The motion in the system is ergodic if

$$\begin{aligned}\lim_{T\to\infty} \frac{1}{T}\int_t^{t+T} dt' f(\Gamma(t')) &= \lim_{T\to\infty} \frac{1}{T}\int_0^T dt f(\hat{\mathbf{T}}(t)\Gamma(0)) \\ &= \int d\Gamma f(\Gamma) \equiv \langle f \rangle,\end{aligned} \tag{15A.5}$$

and the equality is independent of t. It is understood that Eq. 15A.5 holds for typical trajectories in the system; there may exist a set of measure zero of trajectories for which this equality is invalid.

Consider now the correlation in the values of the functions $f(t + T)$ and $g(t)$, defined by

$$C(f,g) \equiv \langle f(t+T)g(t)\rangle - \langle f\rangle\langle g\rangle. \tag{15A.6}$$

A system is mixing if

$$\lim_{T\to\infty} C(f,g) = 0. \tag{15A.7}$$

As already noted, mixing is associated with extreme instability in the dynamics of the system, one consequence of which is that the distance between two trajectories initially close together at t grows exponentially with the interval T as T increases. Put another way, arbitrarily small differences in initial conditions result in arbitrarily large deviations between two trajectories (see Section 16.3). If the phase space of the system is finite in at least one dimension, this instability cannot separate the trajectories by more than some characteristic dimension of the phase space, and the trajectories begin to interweave and jumble. But because the volume of any domain surrounding a phase point $\Gamma(t)$ is preserved as t increases (Liouville's theorem), an expansion of the volume element in one coordinate must be compensated by a contraction in one or more other coordinates. Consequently, if we examine the evolution of some compact neighborhood of a phase point from time t to time $t + T$, a picture like Fig. 15.4 is generated. The difference between the way an ergodic trajectory covers phase space and the way a mixing trajectory covers phase space is caricatured in Fig. 15.5. In the former case the trajectory covers the phase space systematically. In the latter case, the trajectory covers the phase space in time T and in the process subdivides it into cells of some average volume. Then in time (approximately) $2T$, the covering of the phase space is repeated in a fashion such that the average volume of the cells generated is halved and so on for successive coverings of the phase space.

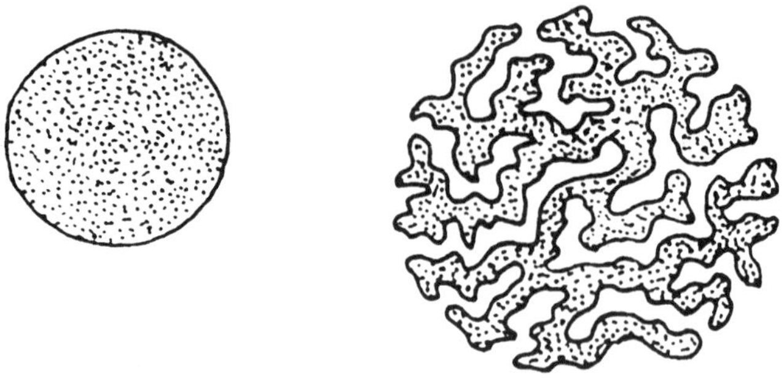

Figure 15.4 Spreading of a volume element in phase space when the dynamics is mixing.

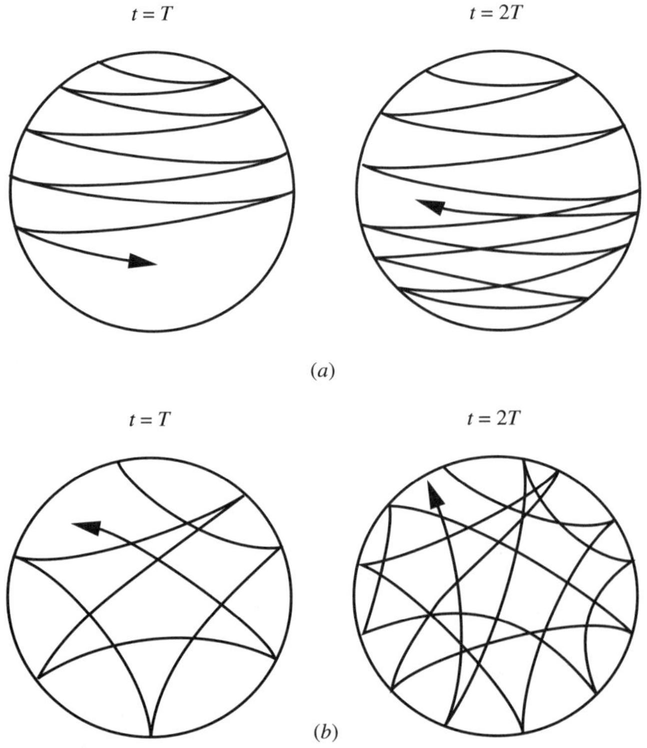

Figure 15.5 Schematic representation of the way the energy surface is covered by a trajectory when (*a*) the dynamics is ergodic and (*b*) the dynamics is mixing.

The formal definition of the local instability that generates the exponential separation of two trajectories that are initially only infinitesimally displaced from one another is that there exists a direction in which the distance between the trajectories grows as

$$D(t) = D(0)\exp(\lambda t), \tag{15A.8}$$

where λ, known as the Lyapunov exponent, is a function of the coordinates of a point in phase space. It can be shown that the relationship between Eq. 15A.8 and the condition expressed by Eq. 15A.7 is

$$C(t) \propto \exp(-\langle \lambda t \rangle), \tag{15A.9}$$

where $\langle \lambda \rangle$ is the inverse of the correlation time. Eqs. 15A.8 and 15A.9 define a relationship between the dynamics of a system and the onset of statistical behavior—i.e., when the motion becomes mixing (see also the discussion in Section 16.3). Of course, to actually use Eqs. 15A.8 and 15A.9, one must find the points in phase space for which the motion is locally unstable. The determination of the onset of chaotic motion, which we shall not discuss, involves many technical difficulties. A simple and intuitively appealing approach[27] assumes that only neighboring trajectories that diverge in the direction perpendicular to the flow in phase space contribute to chaotic motion; the necessary condition for such divergence of neighboring trajectories is the existence of a region where the curvature of the potential energy surface perpendicular to the trajectory is negative.

Finally, we note that although dynamical mixing is much more important for the evolution of a many-particle system to an equilibrium state than is ergodicity, it does not provide a necessary and sufficient justification for the use of ensemble averaging in statistical mechanics. For example, the existence of dynamical mixing provides no information about fluctuations in the equilbrium state.

---

[27] R. Kosloff and S. A. Rice, *J. Chem. Phys.* **74,** 1947 (1981). See also: J. W. Duff and P. Brumer, *J. Chem. Phys.* **65,** 3566 (1977); J. W. Duff and P. Brumer, *J. Chem. Phys.* **67,** 4898 (1977); M. Toda, *Phys. Lett.* **A48,** 335 (1974); C. Cerjan and W. P. Reinhardt, *J. Chem. Phys.* **71,** 1819 (1979).

APPENDIX 15B

$\Omega(E)$ as a System Descriptor

In this appendix, we connect the arguments advanced in Appendix 15A and the choice of $\Omega(E)$ to characterize the properties of a many-particle system (Section 15.3). If we use quantum mechanics to describe the system, the preceding ideas are implemented by introducing the probability w_k of finding the system in some state $\Psi^{(k)}(\mathbf{x})$, where $\Psi^{(k)}(\mathbf{x})$ is one of the many possible wave functions consistent with the macroscopic constraints. We require that $w_k \geq 0$ for all k and that $\Sigma_k w_k = 1$. Let each of these wave functions be represented by an expansion in the orthonormal functions $\{\phi_m\}$:

$$\Psi^{(k)}(\mathbf{x}) = \sum_m c_m^{(k)} \phi_m(\mathbf{x}). \tag{15B.1}$$

The expectation value of some dynamical operator $\hat{\mathbf{f}}$ in the state $\Psi^{(k)}(\mathbf{x})$ is then

$$\begin{aligned} \langle \hat{\mathbf{f}} \rangle^{(k)} &= \int d\mathbf{x} \Psi^{*(k)}(\mathbf{x}) \hat{\mathbf{f}} \Psi^{(k)}(\mathbf{x}) \\ &= \sum_m \sum_n c_m^{*(k)} c_n^{(k)} f_{mn}; \\ f_{mn} &= \int d\mathbf{x} \phi_m^*(\mathbf{x}) \phi_n(\mathbf{x}), \end{aligned} \tag{15B.2}$$

and the observable value, which we define to be the ensemble average, is obtained by averaging the expectation values for the states compatible with the external constraints, including the assigned probabilities for being in those states:

$$\begin{aligned} \langle\langle \hat{\mathbf{f}} \rangle\rangle &= \sum_k w_k \langle \hat{\mathbf{f}} \rangle^{(k)} \\ &= \sum_k w_k \sum_m \sum_n c_m^{*(k)} c_n^{(k)} f_{mn}. \end{aligned} \tag{15B.3}$$

If we introduce

$$\rho_{mn} = \sum_k c_n^{(k)} c_m^{*(k)} \tag{15B.4}$$

(note the order of the indices), we can write

$$\langle\langle \hat{\mathbf{f}} \rangle\rangle = \sum_m \sum_n f_{mn} \rho_{nm}. \tag{15B.5}$$

The ρ_{mn} defined by (15B.4) can be considered to be the matrix elements of an operator $\hat{\rho}$, called the von Neumann density operator, or density matrix. As defined, the density matrix is normalized to unity since

$$Tr(\hat{\rho}) \equiv \sum_m \rho_{mm} = \sum_k \sum_m w_k c_m^{*(k)} c_m^{(k)} = \sum_k w_k = 1, \tag{15B.6}$$

with $\sum_m c_m^{*(k)} c_m^{(k)} = 1$ by the orthonormality of the basis functions.

Since the observable properties of a system in an equilibrium state are, by definition, independent of the time, we infer that when a system is in an equilibrium state, the density matrix can be a function of only the constants of the motion. It is usually the case for a macroscopic system that the only relevant constant of the motion is the energy, so we can write $\hat{\rho} = \hat{\rho}(\hat{\mathsf{H}})$. Then, if we have a representation in which the system Hamiltonian is diagonal, the density matrix is also diagonal. In that representation the matrix elements of $\hat{\rho}$ can be written in the form

$$\rho_{mn} = \Omega^{-1} a_m \delta_{mn}, \tag{15B.7}$$

where $\delta_{mn} = 1$ when $m = n$ and zero otherwise. Clearly, $\Omega^{-1}\alpha_m \geq 0$, since this number is just the probability of finding the system in state m.

Suppose what we know about a system is derived from the measurement of some observable quantity, corresponding to the dynamical operator $\hat{\mathbf{f}}$, which is not necessarily a constant of the motion. If the eigenfunctions of the dynamical operator $\hat{\mathbf{f}}$ are $\{u_n\}$, and the corresponding eigenvalues are f_{nn},

$$\hat{\mathbf{f}} u_n(\mathbf{x}) = f_{nn}(\mathbf{x}). \tag{15B.8}$$

The $\{u_n\}$ are a complete orthonormal set of functions so any state of the system can be represented by

$$\Psi^{(k)} = \sum_n c_n^{(k)} u_n(\mathbf{x}) = \sum_n r_n^{(k)} e^{i\phi_n} u_n(\mathbf{x}), \tag{15B.9}$$

where we have written the probability amplitude for finding the state n as the product of a magnitude and a phase factor. The measurement of the observable $\langle\langle \hat{\mathbf{f}} \rangle\rangle$ will, in general, tell us that the state of the system is equally well represented by a group of g_α neighboring states that have eigenvalues within the uncertainty in our measurement. In the Gibbs ensemble, which we use to represent the system,

$$\rho_{mn} = \frac{1}{\mathcal{N}} \sum_{k=1}^{\mathcal{N}} c_n^{(k)} c_m^{*(k)} = \frac{1}{\mathcal{N}} \sum_{k=1}^{\mathcal{N}} (r_n^{(k)})^2 , \quad (15B.10)$$

and $\rho_{nn} = 0$ if n is not in the group of g_α neighboring states. However, if $n \in g_\alpha$, we know nothing about the value of ρ_{nn}. The most plausible assumption we can make is that all the ρ_{nn} are the same for all states $n \in g_\alpha$.

We now consider the off-diagonal elements ρ_{nm}, which, in general, are nonzero; these have the form

$$\begin{aligned} \rho_{mn} &= \frac{1}{\mathcal{N}} \sum_{k=1}^{\mathcal{N}} c_n^{(k)} c_m^{*(k)} \\ &= \frac{1}{\mathcal{N}} \sum_{k=1}^{\mathcal{N}} r_n^{(k)} r_m^{(k)} e^{i(\phi_n^{(k)} - \phi_m^{(k)})} , \end{aligned} \quad (15B.11)$$

which displays an explicit dependence on the phases of the expansion coefficients $\{c_n^{(k)}\}$. However, a measurement of $\langle\langle \hat{\mathbf{f}} \rangle\rangle$, no matter how accurate, cannot give any information about the phases $\{\phi_n^{(k)}\}$ and is the same no matter what values the phases take. To be consistent with this fundamental property of quantum measurements, we take the replicas of the ensemble to be independent and noninteracting. In each replica the system is in one of the g_α states consistent with the observable value of $\langle\langle \hat{\mathbf{f}} \rangle\rangle$, and the ensemble average is the *incoherent* sum over the contributions from all of the replicas. With this condition on the ensemble average the contribution of the off-diagonal elements of the density matrix to the observable value is zero (because in Eq. 15A.13, the off-diagonal terms with different phases contribute positive and negative terms that sum to zero). The arguments concerning the diagonal and off-diagonal elements of the density matrix can be summarized in the form:

POSTULATE. *The construction of an appropriate representative ensemble, to correspond to the knowledge available from a measurement of the observable value of a dynamical operator* $\hat{\mathbf{f}}$ *for some system, is obtained by assigning equal a priori probabilities and random a priori phases to those states of the system that are consistent with the observed value* $\langle\langle \hat{\mathbf{f}} \rangle\rangle$.

Since in each replica the system is in one of the g_α states consistent with the observable value of $\langle\langle \hat{\mathbf{f}} \rangle\rangle$, Eq. 15B.7 can be written in the form

$$\rho_{mm} = \Omega^{-1} a_m = g_\alpha . \quad (15B.12)$$

Then the hypothesis of equal a priori probability of occurrence of the accessible states requires that

$$\begin{aligned} a_m &= 1 \qquad \text{if } E \le E_m \le E + \Delta E \\ a_m &= 0 \qquad \text{otherwise,} \end{aligned} \quad (15B.13)$$

where E_m is the energy of state m. The normalization of the density matrix, namely,

$$Tr(\hat{\rho}) = \sum_m \rho_{mm} = \sum_m \Omega^{-1} a_m = \Omega^{-1} \sum_{E \le E_m \le E + \Delta E} 1 = 1, \quad (15B.14)$$

requires that

$$\Omega = \sum_{E \le E_m \le E + \Delta E} 1 . \quad (15B.15)$$

Thus, Ω is the number of states with energy between E and $E + \Delta E$ (see Section 15.3).

APPENDIX 15C

The Master Equation

The definition of a state of a system using Gibbs' prescription for the construction of an ensemble implies that its accessible phase space is divided into cells with nonzero volume, with each cell representing a possible macroscopic state of the system. The phase space cells are constructed as follows. If $G(\Gamma)$ is some dynamical function, then any uncertainty in G leads to a range Δg for $G(\Gamma)$. The scale of all values of g accessible to $G(\Gamma)$ may then be subdivided into intervals of size Δg. Each such interval corresponds to a slice of Γ space determined by $g_j < G(\Gamma) < g_{j+1}$, and a macroscopic measurement of G merely leads to a determination of which slice of Γ space contains the phase point for the relevant microscopic state of the system. If there are a number n of macroscopic observables, $G^{(r)}(\Gamma)$, the observation of all such quantities, requires the phase point to lie in that region of Γ space common to all the separate slices $g_j^{(r)} < G^{(r)}(\Gamma) < g_{j+1}^{(r)}$. We take these overlap regions to define our phase cells, each of which has volume

$$\mathscr{V}(j) = \int_{g_j^{(r)} < G(\Gamma) < g_{j+1}^{(r)}} d\Gamma_j \tag{15C.1}$$

Although a complete determination of all the $G^{(r)}(\Gamma)$ fixes the phase cell in which the representative point for the system must lie, and thereby the macroscopic state of the system, there is a very large number of microscopic states compatible with each macroscopic state defined in this fashion. Consider the motion of a phase point, which is in some selected phase cell at t_0. From each such point inside a particular phase cell, there originates a trajectory. The individual trajectories corresponding to the large number of phase points in the cell leave the cell in all possible directions. Moreover, because of the nonzero extension of the phase cells, the phase points jump back and forth between cells in a superficially random fashion, and because of the intercession of interactions on a scale much finer than the extension of the phase cell, a phase point in a particular cell at t_0 may be found in any of the nearby cells for $t > t_0$.

The divergence of the trajectories of the phase points leaving a particular phase space cell resembles the trajectories of particles diffusing away from a source of concentration. Accepting this analogy, we can analyze the motion of a phase point as if that motion is random. If at t_0 the system is in phase space cell j', then the transition probability for finding it in cell j a time s later is $K(j\,|\,j';s)$. This transition probability must satisfy the conditions

$$\begin{aligned} &K(j|j';s=0) = \delta_{jj'}, \\ &K(j|j';s) \geq 0; \quad s \geq 0, \\ &\sum_j K(j|j';s) = 1. \end{aligned} \tag{15C.2}$$

The first line of Eq. 15C.2 states that the system starts in phase space cell j' at time t_0, the second line states that all transition probabilities are positive, and the third line states that a system that starts in phase space cell j' at time t_0 must be found in some other cell (possibly including j') at time $t_0 + s$. The transition probability can be interpreted as a conditional probability or as the fraction of the population of cell j' which passes into cell j in the time interval s when the starting distribution of phase points arose from an ensemble which had at time t_0 constant density within the cell j' and zero density everywhere else in phase space.

When the motion of the phase point is considered to be a random process, the transition probability also satisfies the conservation condition

$$K(j|j';s) = \sum_{j''} K(j|j'';s-\tau)K(j''|j';\tau), \tag{15C.3}$$

which is called the Chapman–Kolmogorov equation. This equation states that the transition probability $K(j\,|\,j';s)$ is independent of any choice of path from j' to j, so that we can imagine the path from j' to j to be constructed from the superposition of all paths from j' to intermediate phase space cells j'', followed by paths from all the phase space cells j'' to the phase space cell j. Put another way, the Chapman–Kolmogorov equation implies that if the phase points that arrive in cell j'' at times $s - \tau$ are redistributed with constant density in j'' and the random process allowed to continue, the final distribution of phase points is unaffected by the intermediate redistributions. In the limit of small time intervals Δt we can use the expansion

$$K(j|j';\Delta t) = \delta_{jj'}\left[1 - \Delta t \sum_{j''} a_{j''j'}\right] + a_{jj'}\Delta t, \tag{15C.4}$$

where $a_{jj'}$ is the transition probability per unit time for transfer of a phase point from cell j' to cell j and Δt must be chosen to be small enough that the macroscopic state of the system does not change much but long enough that the microscopic variables undergo large changes. The substitution of Eq. 15C.4 in Eq. 15C.3 leads to

$$\frac{\partial K(j|j';t)}{\partial t} = \sum_{j''} [a_{jj''} K(j''|j';t) - a_{j''j} K(j|j'';t)]. \quad (15C.5)$$

We now change terminology slightly and refer to a system represented by a phase point in the phase space cell j' as being in state j'. Let $w_j(t)$ be the probability at time t that the representative point of the system is in state j. We have

$$w_j(t) = \sum_{j'} K(j|j';t) w^0_{j'} \quad (15C.6)$$

where $w^0_j = w_j(t = 0)$. If both the left and right hand sides of (15C.5) are multiplied by $w^0_{j'}$ and the resulting equation summed over j', we find

$$\frac{\partial w_j(t)}{\partial t} = \sum_{j''} (a_{jj''} w_{j''} - a_{j''j} w_j), \quad (15C.7)$$

which is known as the *master equation.* Equation 15C.7 makes the simple statement that w_j is increased by all the transitions from states j' to j and is decreased by all transitions away from the state j to other states j'.

The master equation, which is very often used to describe the evolution of a system, has two very interesting properties. First, the assumption embodied in Eq. 15C.6 defines a direction of time and thereby imposes a condition on the system dynamics. To see that this is so we introduce the notation $K(j\,|\,j';t) = K_{jj'}$ and treat Eq. 15C.6 as an algebraic equation. The inverse of that equation then reads

$$w^0_j = \sum_{j'} K^{-1}_{jj'} w_{j'}. \quad (15C.8)$$

Substitution for w_j from Eq. 15C.6 then yields the consistency condition

$$\sum_{j'} K^{-1}_{jj'} K_{j'k} = \delta_{jk}. \quad (15C.9)$$

For $j \neq k$ it is clear that not all of the $K^{-1}_{jj'}$ can be positive, since all of the $K_{j'k}$ are. Thus, Eq. 15C.8 is not an equation relating w^0_j to $w_{j'}$ by a set of transition probabilities, which implies that there is a unique direction in time defined by the relationship defined by Eq. 15C.2. That definition of the direction of time has the consequence that the entropy, defined by (see Chapter 19)

$$S \equiv -\sum_j w_j \ln w_j \quad (15C.10)$$

is an increasing function of time. This conclusion follows from the observation that for any positive x and y, $F(x, y) = x \ln x - x \ln y - x + y \geq 0$. Then $w^0_j(\ln w^0_{j'} - \ln w_j) + w_j - w^0_{j'} \geq 0$, and we have

$$\begin{aligned}
&\sum_j w^0_j \ln w^0_j - \sum_j w_j \ln w_j \\
&\quad = \sum_j w^0_j \ln w^0_j - \sum_{j,j'} K_{jj'} w^0_{j'} \ln w_j \\
&\quad \geq \sum_j w^0_j \ln w^0_j - \sum_{j,j'} K_{jj'} [w^0_{j'} \ln w^0_{j'} - (w^0_{j'} - w_j)] \\
&\quad \geq 0.
\end{aligned} \quad (15C.11)$$

This result provides a different view of the conditions for irreversible transformations stated in Section 15.4.

Second, when equilibrium is reached, the occupation probabilities w_j are time independent so that, from Eq. 15C.7,

$$\sum_{j''} (a_{jj''} w_{j''} - a_{j''j} w_j) = 0 \quad (15C.12)$$

from which we find

$$\frac{w_j}{w_{j''}} = \frac{a_{jj''}}{a_{j''j}}, \quad (15C.13)$$

which is an example of the principle of detailed balance (discussed in Part III of this book).

Equation 15C.3 is an assumption about the random nature of the flow of phase points between the coarse-grained cells defined by Eq. 15C.1. Taken together, Eq. 15C.3 and the assumption displayed as Eq. 15C.6 lead to the master equation, a direction for the flow of time and an entropy function that increases as time increases. These assumptions are, therefore, worth further examination. First, we note that the initial distribution of points in the cell in phase space need not be uniform, since even subregions of a given cell are characterized by trajectories that leave in all possible directions, provided only that the subregion is not too small. The details of the initial distribution of phase points are irrelevant, provided only that the initial distribution is not freakish. One could, in principle, trace back those phase points in j that came from j' in time t, and then construct a density distribution in a subregion of j', such that only these special points are allowed in the subregion. Then the phase points that leave the specially constructed subregion are always found in cell

j at time t, and the assumption that the motion of the phase point is random is not valid. However, the construction of such a freak initial distribution of phase points requires knowledge of the trajectories of all the individual phase points, equivalent to knowledge of the coupled mechanical behavior of all of the molecules in the system. These trajectories are very complicated and the phase points that end in j are initially spread all over j' so that a complex disentanglement is necessary. Insofar as our macroscopic description of the system generates phase space cells of nonzero volume, the needed information concerning the trajectories is never available and the freak ensemble cannot be purposely constructed. Nevertheless, from the point of view of mechanics such freak distributions are possible, and we must therefore explicitly exclude them when Eq. 15C.3 and 15C.6 are postulated to describe the motion of a point in phase space. The Chapman–Kolmogorov equation must, therefore, be understood to mean that if one starts at t_0 from a distibution of points that is uniform in cell j' and zero elsewhere in the phase space, then when $t > t_0$ each phase space cell j will contain some points. In addition, the distribution of points arriving in j'' from j' is assumed to be uncorrelated with the distribution of those points in j'' that will pass into any particular other phase space cell at some still later time. It is this assumption of lack of correlation that so greatly simplifies the description of the motion of representative points in phase space.

The possibility of satisfying the requirements on the distribution of representative points in a phase space cell depends on the nature of the description of the macroscopic system. The phase space cells cannot be too small, or the assumption that the trajectory exhibits random motion will be incorrect; this implies that the set of macroscopic variables specifying the state of the system cannot be too large. On the other hand, if the phase space cells are too large, corresponding to very few macroscopic observables, the representative points will often remain in one phase space cell for a relatively long time, or move rarely from cell to cell, and then the assumption that the trajectory of the phase point is random cannot be used. The set of macroscopic variables used to describe the evolution of a system must be sufficiently large that the future behavior of all members of the set is predictable from a knowledge of the initial values to within the same accuracy as the latter are defined.

• FURTHER READING

Andrews, F. C., *Equilibrium Statistical Mechanics,* 2nd ed. (Wiley, New York, 1975), Chapters 2–6.

Percival, I., and Richards, D., *Introduction to Dynamics* (Cambridge University Press, Cambridge, 1982).

Reif, F., *Statistical and Thermal Physics* (McGraw-Hill, New York, 1965), Chapters 2 and 3.

Zaslavsky, G. M., *Chaos in Dynamic Systems* (Harwood Academic Publishers, Chur, Switzerland, 1985), Chapters 1, 2, and 12.

• PROBLEMS

1. A coin is tossed five times. What is the probability that:
(a) Head is up all five times?
(b) Head is up three times?
(c) The sequence of head and tail is H T H T H?

2. Two dice are thrown. What is the probability of obtaining a total of 7? What is the probability of obtaining 8? What is the probability of obtaining 7 or 8?

3. Two cards are drawn from a deck of well-shuffled cards. What is the probability that both cards drawn are aces?

4. Two cards are drawn from a well-shuffled deck of cards, the first card being returned before the second is taken. What is the probability that both the extracted cards belong to one suit?

5. Consider a box of volume V containing N molecules. What is the probability of finding all the molecules in a portion of the container having volume $\frac{1}{2}V$? If $N = 10^{23}$, what is the numerical value of this probability? Assume that the molecules in the box are free of all interactions, that is, that the equation of state is that of a perfect gas.

6. A volume V contains N_A molecules of type A and N_B molecules of type B. A valve is opened and M molecules flow out. What is the probability that among the M molecules there are m_A of species A and m_B of species B?

7. Suppose that a volume V is mentally subdivided into M subvolumes. Let there be N molecules in V. What is the probability that some one subvolume will contain N' molecules?

8. Suppose that a volume V is subdivided into two compartments. One of these contains N'_A A molecules and N'_B B molecules; the other contains N''_A A molecules and N''_B B molecules. A valve between the compartments is opened and closed and one molecule transferred from the side marked (′) to the side marked (″). After a long time the valve is again opened and closed and one molecule transferred from side (″) to side (′). What is the probability that the molecule transferred in both steps is an A molecule?

9. Consider a perfect gas flowing down a tube with constant average flow velocity $\mathbf{u}$. Describe in simple terms how the distribution of molecular velocities differs in this case from that of a stationary perfect gas. If the temperatures of a flowing and a stationary gas are the same, show in a simple diagram how the respective velocity distributions are related.

10. Consider an ensemble of pairs of systems, A and B, coupled together by a frictionless sliding wall that separates the volume V into V_A and V_B. Each replica in the ensemble has different values of V_A and V_B, but $V_A + V_B = V =$ constant for all replicas. The temperatures in V_A and V_B are the same. Let V_A and V_B each contain 1 mol of a perfect gas.

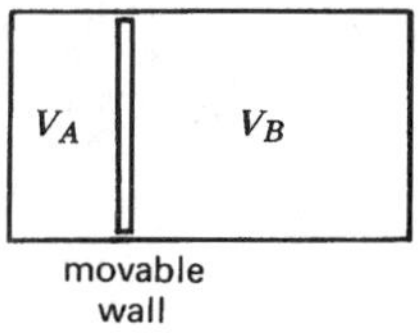

(a) What is the average value of the pressure difference $p_A - p_B$?
(b) What is the average value of $p_A - p_B$ if the ensemble consisted of the same replicas, but with the wall dividing V_A from V_B fixed in position in each replica?
(c) What is the average value of $p_A - p_B$ if the replicas are so constructed that $V_A < V_B$ in all cases?
(d) Does $\langle p_A - p_B \rangle = \langle (p_A - p_B)^2 \rangle^{1/2}$ in cases (a), (b), and (c)?

11. One simple model of a solid treats the solid as if it were N independent harmonic oscillators of the same frequency ν. Consider such a set of N oscillators. The energy of the N oscillators is

$$E = \tfrac{1}{2} Nh\nu + Mh\nu.$$

The first term is the zero-point energy of the system, and the second term is the energy of M quanta of vibration.
(a) Calculate the number of states that have energy E.
(b) The energy of the system is increased by just $h\nu$, and thereby the number of states with energy E is changed. How large is this change?
(c) In this system, does the number of states with energy E increase indefinitely as E increases? Explain your answer.

12. Suppose that N independent particles can each be in only one of the two energy levels $\pm\Delta$.
(a) How many states of the system have energy

$$E = M\Delta,$$

where M is a number between N and $-N$?
(b) How is the number of states with energy E changed when we replace M by $M + 1$?
(c) Discuss how the number of states with energy E varies with M as M changes from very small values to the value N.
(d) How does your answer to part (c) differ from your answer to part (c) of Problem 11? Why should there be a difference in behavior of these two systems?

13. Calculate the entropy corresponding to the system of Problem 11. Express the total energy E in terms of the temperature T.

14. Carry out the calculations of Problem 13, but now for the system of Problem 12.

15. In Problems 13 and 14 the results gave relations involving the energy of the system and the temperature T. From these calculate the specific heats at constant volume corresponding to the two systems.

16. Make a plot of $E/Nh\nu$ versus $k_BT/h\nu$ for the system of Problem 11 and of $E/N\Delta$ versus k_BT/Δ for the system of Problem 12. Also make plots of C_V versus $k_BT/h\nu$ and k_BT/Δ for these two systems. Why are the curves of different shape? How do these shapes reflect the qualitative features of the energy spectra of the two systems?

17. Show that, given the general properties of $\Omega(E, V, N)$, the temperature of a system at equilibrium must be positive.

18. Consider a perfect monatomic gas, consisting of N particles with total energy E, in thermal equilibrium in a box of volume $V = a^3$. From Equation 15.27, show that the probability that a given gas molecule will have energy ϵ is proportional to

$$e^{-\epsilon/k_BT}.$$

Hint: Use the fact that $E/N = \frac{3}{2} k_BT$. The expansion $\ln(1 + x) = x$, valid when $|x| << 1$, will be found useful.

19. Evaluate $\Omega(E, V, N)$ for a gas of N "hard-sphere" molecules of radius r_0. The molecules are impenetrable and do not attract each other.

20. At 0 K, a perfect crystal of CO has molecules that are all lined up but that can either point "up" or "down." Calculate Ω and S at 0 K for 1 mol of a perfect crystal of CO under the assumption that half the molecules point up and the other half point down. Use Stirling's approximation to obtain a numerical value for S.

CHAPTER 16

The Second Law of Thermodynamics: The Macroscopic Concept of Entropy

In Chapter 15 we showed that the entropy,

$$S \equiv k_B \ln \Omega(E, V, N), \qquad (16.1)$$

defined with respect to the number of quantum states of a system between the energies E and $E + dE$, is a function that describes a macroscopic property of the system. We should expect, then, that the entropy of a system can be defined solely in terms of its macroscopic behavior, and without any reference to the system's energy-level spectrum or other microscopic properties. In fact, the entropy function was first introduced into thermodynamic theory in this way, and the definition of Eq. 16.1 was proposed later to provide a microscopic interpretation. This chapter is devoted to the analysis of those aspects of the behavior of macroscopic systems that lead to the second law of thermodynamics, and from the second law to the concept of entropy.

16.1 The Second Law of Thermodynamics

The second law of thermodynamics is a statement about the behavior of real macroscopic systems; it must be considered one of the fundamental principles on which the macroscopic description of matter is based. As is true of all of the laws of thermodynamics, the second law is an abstraction from experiment. It is a formulation of an aspect of macroscopic behavior that need not be derived from, or proved by, appeal to other considerations. Using the statistical molecular theory of matter, we can interpret the meaning and origin of the second law, but that is not a proof.

There are a number of equivalent statements of the second law of thermodynamics. Although we shall not be concerned with proofs of equivalence, it is of interest to set down a few of the superficially different formulations that can be given. The statements most directly representative of experience are the following:

1. *Clausius statement:* It is impossible to devise a continuously cycling engine (the system) that transfers heat from a colder to a hotter body without doing work on the system.
2. *Kelvin statement:* It is impossible to devise a continuously cycling engine that produces no effect other than the extraction of heat from a reservoir at one temperature and the performance of an equal amount of mechanical work.
3. *Caratheodory statement:* In the neighborhood of every equilibrium state of a closed system, there are states that cannot be reached from the first state along any adiabatic path by any spontaneous process or reversible limit of a spontaneous process.

Note that, unlike the first law of thermodynamics, statements 1, 2, and 3 claim the impossibility of something happening. This "statement of impossibility" is a fundamental feature of the second law.

In both the Clausius and Kelvin statements of the second law, an essential role is assigned to cyclic processes. Originally, cyclic processes were singled out for attention because of interest in the use of engines to transform heat into work. By design, an engine is required to repeat exactly some process over and over so that the amount of heat that can be transformed into work is unlimited. More generally, a cyclic process returns a system to its initial state, hence leaves the internal energy of the system unchanged. Implicitly, then, the second law imposes conditions that distinguish between those energy-conserving processes that can occur spontaneously and those that cannot.

It is important to observe that it is relatively simple to invent procedures that transfer heat from a colder to a hotter body, but not in a cyclic process and not without doing work on the system. Consider, for example, a perfect gas contained in a volume V_1 that is in thermal contact with a body at some temperature, say, T_1. Let the gas be expanded isothermally to occupy a larger volume V_e, thereby extracting heat from the body with which it is in contact. A consequence of that isothermal expansion is that the gas does work on the surroundings in the amount $-RT_1 \ln(V_e/V_1)$. Now let the container of gas be isolated and then adiabatically compressed until its volume is the original volume V_1; the pressure of the gas is now p_c. In this process work is done by the surroundings on the system in the amount $(p_cV_1 - p_eV_e)/(\gamma - 1)$ (see Problem 13.15). As a result of the adiabatic compression, the temperature of the isolated gas is increased to $T_c > T_1$. Finally, let the container of gas be brought into thermal contact with a new body, at a temperature T_2 such that $T_c > T_2 > T_1$. Because $T_c > T_2$, heat will be transferred from the gas to the second body. Thus, in the sequence of steps envisaged, we have transferred heat from a colder body to a hotter body. However, at the end of the sequence of steps the state of the gas is different from its initial state, hence the process is clearly not cyclic. Moreover, the magnitude of the adiabatic compression work done on the system to raise its temperature to T_c is greater than the magnitude of the isothermal expansion work done by the system when it expanded to volume V_e. This is easily demonstrated by using the fact that the pressure and volume of a perfect gas subjected to an adiabatic compression or expansion satisfy the relation $pV^{\gamma} = constant$ (see Problem 13.15). Using this relation we find that the adiabatic compression work is equal to $R(T_c - T_1)/(\gamma - 1)$ and the isothermal expansion work is equal to $-(RT_1/(\gamma - 1))\ln(T_c/T_1)$. The magnitudes of the two kinds of work then satisfy $|w_{\mathrm{comp}}| > |w_{\mathrm{exp}}|$ because $T_c - T_1 > T_1 \ln(T_c/T_1)$ for all $T_c > T_1$. Thus, the heat transfer described is not prohibited by the second law. In general, it is found that no cyclic process can be devised that violates statements 1 or 2.

The Caratheodory statement of the second law, 3, is less obviously an abstraction from experiment than are statements 1 and 2. To illustrate how statement 3 is derived, we consider again how the internal energy of a system is determined. It will be recalled that if a system is enclosed by an adiabatic envelope, the state of the system can be altered only by moving part of the envelope, and not by any other external action. Is it possible to reach *all* states of the system by adiabatic means starting from some arbitrary reference state? In other words, are there thermodynamic states of the system that are inaccessible by use of the mechanical processes that define U? The answer to this question comes from experiment. It is concluded that:

1. From any given state of a system, there are other states that cannot be reached by *particular* adiabatic paths. For example, if a paddle wheel is turned in an adiabatically enclosed fluid, the work dissipated in frictional heat can only increase the temperature of the fluid. A state of lower temperature cannot be reached by this particular adiabatic path.
2. It is always possible to connect two thermodynamic states, one of which is inaccessible to the other by an adiabatic work path, by either:
   (a) reversing the direction of some possible adiabatic work path, that is, considering an expansion process in place of a compression process, or
   (b) connecting the two states via separate adiabatic work paths to an identical third state.

Conclusion 1 is clearly related to the Caratheodory statement of the second law. Conclusion 2, in both parts, provides assurance that the internal energy of every possible state of a system can be determined, relative to some reference state, by means of only adiabatic work processes, despite the difficulties posed by the inaccessibility referred to in conclusion 1. It is not implied that the adiabatic work processes used need be of only one kind.

As stated, these conclusions mask a subtle feature of the definition of the internal energy. Specifically, Conclusion 2 *assumes* that if X′ and X″ are states of a system such that X″ cannot be reached from X′ by an adiabatic work path, it is nevertheless true that $w_{\mathrm{ad}}(\mathrm{X}' \to \mathrm{X}'') = -w_{\mathrm{ad}}(\mathrm{X}'' \to \mathrm{X}')$. This assumption can be avoided if we augment our definition of the properties of an adiabatically enclosed system as follows:

Lemma: Given any state X(U, V) of an adiabatically enclosed system, every state of the continuous sequence of states X′(U′, V′), where U′ > U, can be reached from X(U, V). It is necessary that V (and/or any other deformation coordinate used to define the state of the system) have the same value in X(U, V) and X′(U′, V′).

The stated Lemma implies that if X′ and X″ are arbitrarily prescribed states of an adiabatically enclosed system such that transitions X′ → X″ are impossible, then there exist transitions X″ → X′. To prove this statement we proceed as follows. Let $\mathrm{X}'(U', V')$ and $\mathrm{X}''(U'', V'')$ be the states chosen, and let X′ be the initial state. We now imagine changing V' quasistatically until it reaches the value V''. Whatever the value of the energy U^* of the state $\mathrm{X}^*(U^*, V'')$ that is reached by this process, the Lemma requires that $U^* > U''$. If this were not the case, X″ could be reached from X* (and therefore also from X′). If, then, we now take X″ as the initial state we may first make the transition X″ → X* at constant volume and, assuming that the quasistatic processes used are also reversible, we can then make the transition from X* to X′ by reversing the processes that lead from X′ to X*.

The introduction of the Lemma cited leads to a unique definition of the internal energy that has the property $\oint dU = 0$, so that $w_{ad}(X' \to X'') = -w(X'' \to X')$. In particular, the relationship between the values of the adiabatic work in the forward and reverse processes that connect X′ and X″ does not have to be assumed. Then $w_{ad} = -\Delta U$ becomes an operational definition of U for all the states of the system and for all ΔU.

To this point our discussion of the several formulations of the second law makes no reference to the entropy of a system. We now show that the Clausius and Kelvin statements imply both the existence of an entropy function and the Caratheodory statement of the second law.

16.2 The Existence of an Entropy Function for Reversible Processes[1]

We consider, first, how the Clausius and Kelvin statements of the second law can be translated into a mathematical formulation that defines the entropy and its fundamental properties. In this section we analyze reversible processes; the complementary analysis of irreversible processes is deferred to the next section.

Consider a one-component, one-phase closed system. Let the states of this system be displayed on the *T, V* diagram shown in Fig. 16.1. Suppose that the state of the system is transformed from 1 to 2 by a reversible adiabatic process, and from 2 to 2′ by a reversible isothermal process. We seek to close the path in the *T, V* plane by some process that transforms the state of the system from 2′ to 1. *Can we find a reversible adiabatic path connecting* 2′ *to* 1? We proceed by assuming that such a path exists and deducing a contradiction that, *a posteriori,* requires rejection of the assumption.

For any closed cycle in the *T, V* plane,

$$\Delta U_{\text{cycle}} = 0, \tag{16.2}$$

whereupon

$$w = -q. \tag{16.3}$$

In the particular cycle cited, the path 1 → 2 is a reversible adiabatic along which

$$q_{1\to 2} = 0, \tag{16.4}$$

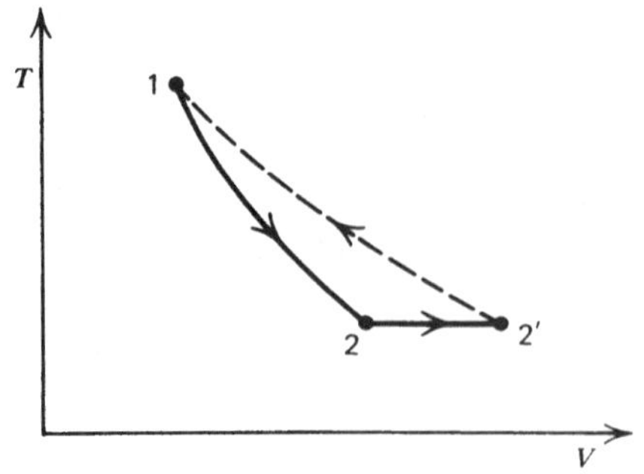

Figure 16.1 Existence of an entropy function for reversible processes: If 1 → 2 by a reversible adiabatic process, and 2 → 2′ by a reversible isothermal process, the path 2′ → 1 cannot be a reversible adiabatic path (see text).

the path 2 → 2′ is a reversible isotherm along which the system expands so that

$$q_{2\to 2'} > 0, \tag{16.5}$$

and the path 2′ → 1 is assumed to be a reversible adiabatic for which

$$q_{2'\to 1} = 0. \tag{16.6}$$

Combining Eq. 16.3 with Eqs. 16.4, 16.5, and 16.6, we conclude that in the cycle described

$$w < 0. \tag{16.7}$$

Combination of Eqs. 16.3, 16.5, and 16.7 shows that the cycle described converts a quantity of heat drawn from a reservoir at one temperature into an equal amount of work,[2] which is a violation of the Kelvin statement of the second law. We conclude that the path connecting states 2′ and 1 cannot be a reversible adiabatic.

The argument just given is valid for every isothermal path between 2 and 2′. Therefore it must be the case that only one reversible adiabatic path passes through each point of the *T, V* space. Put another way, reversible adiabatics in the *T, V* space cannot cross.

The relationship between the behavior of reversible adiabatics just described and the Caratheodory statement of the second law follows from a similar argument. Consider again a one-component, one-phase system. All states of the system that can be reached from state 1 by isothermal transformation lie on the isotherm labeled T_1 in Fig. 16.2; those that can be similarly reached starting from state 2 lie on the isotherm labeled T_2. Through (T_1, V_1), representing state 1 of the system, there can pass one and only one reversible adiabatic; it is shown intersecting the isotherm T_2 at (T_2, V_2).

[1] This argument follows an analysis given by J. G. Kirkwood and I. Oppenheim.

[2] Recall that work done by the system is reckoned negative.

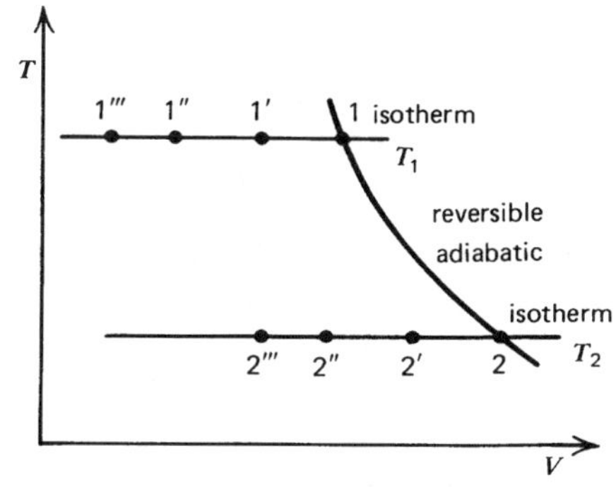

Figure 16.2 Diagram illustrating the behavior of reversible adiabatic paths in the T, V plane and the Caratheodory statement of the second law of thermodynamics.

Other reversible adiabatics, each unique, pass through (T_1, V_1'), (T_1, V_1''), ... By virtue of the uniqueness of the adiabatics, no adiabatic path through 1 can intersect isotherm T_2 at $2'$, $2''$, ... Given that there exist unique adiabatics between $1'$ and $2'$, $1''$ and $2''$, . . . , the points 2, $2'$, $2''$, . . . cannot coincide because reversible adiabatics cannot cross. Although all points on the isotherm T_2 lying to the left of point 2 can be reached via reversible adiabatic paths from points on isotherm T_1 lying to the left of point 1, all points in the T, V plane lying to the left of the reversible adiabatic through point 1 are inaccessible via adiabatic paths from point 1.

If the states of the system depend on more variables than just T and V, the preceding argument still holds for any plane section parallel to the T axis. We conclude that there is a region of space surrounding point 1, which contains states that cannot be reached from 1 along an adiabatic path. This statement will be recognized as equivalent to the Caratheodory formulation of the second law.

We now exploit the uniqueness of the reversible adiabatic through a selected point in the T, V plane in the following fashion. Because of this uniqueness every reversible adiabatic curve can be represented by a function $T = T(V)$ or $f(T, V) = \text{constant}$. Let $S(T, V)$ be such a function, the entropy, which is constant everywhere along the reversible adiabatic through (T, V):

$$S(T,V) = \text{constant along the reversible adiabatic path through } (T,V). \tag{16.8}$$

Clearly, then,

$$dS(T,V) = 0 \text{ along the reversible adiabatic path through } (T,V). \tag{16.9}$$

The prescription given for the entropy $S(T, V)$ does not specify its detailed form, but its existence is ensured by the uniqueness of the reversible adiabatic. Moreover, because

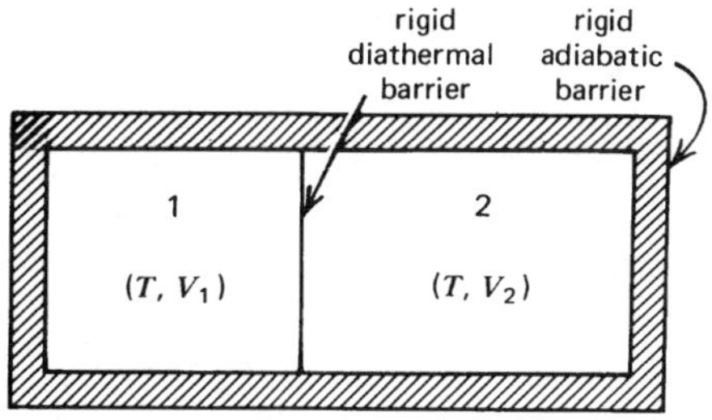

Figure 16.3 Schematic diagram illustrating the relationship between $\Theta(T, V)$ and reversible heat transfers (see text).

reversible adiabatics cannot cross, they must be monotonic functions of T. To proceed further, we write

$$dS(T,V) = \Theta(T,V)\, đq_{\text{rev}}. \tag{16.10}$$

For an adiabatic process $đq_{\text{rev}} = 0$, hence Eq. 16.10 reduces to Eq. 16.9. For a general reversible process $đq_{\text{rev}} \neq 0$, and Eq. 16.10 is an extension of Eq. 16.9. It will be recalled that $đq_{\text{rev}}$ depends on the path of the process that leads to the heat transfer. On the other hand, since $S(T, V)$ describes the unique reversible adiabatic through (T, V), it must be a function of the state of the system, and $dS(T, V)$ must be independent of the path by which the change is effected. Then the function $\Theta(T, V)$ in Eq. 16.10 plays the role of an *integrating factor*.[3]

To determine $\Theta(T, V)$ we examine the behavior of the system shown in Fig. 16.3. A rigid thermally isolated container is divided internally into volumes V_1 and V_2 by a rigid diathermal barrier. At equilibrium the temperatures of the two compartments are the same. Suppose that an infinitesimal amount of heat $đq_{\text{rev}}$ is transferred reversibly from one compartment to the other. Then, because the total system is isolated,

$$đq_{1(\text{rev})} = -đq_{2(\text{rev})}, \tag{16.11}$$

$$dS = 0 = dS_1 + dS_2, \tag{16.12}$$

with Eq. 16.12 following from Eq. 16.9. By substitution of Eqs. 16.10 and 16.11 in Eq. 16.12, we find

$$[\Theta_1(T,V_1) - \Theta_2(T,V_2)]\, đq_{1(\text{rev})} = 0. \tag{16.13}$$

---

[3] An integrating factor is defined as follows: Let $x_1, x_2, \ldots$, be the independent variables in the linear differential form $X_1\,dx_1 + X_2\,dx_2 + \cdots$. The function $\Theta(x_1, x_2, \ldots)$ is an integrating factor for this linear differential form if the product

$$\Theta(x_1,x_2,\ldots)X_1\,dx_1 + \Theta(x_1,x_2,\ldots)X_2\,dx_2 + \cdots = df(x_1,x_2,\ldots)$$

is the exact differential of some function $f(x_1, x_2, \ldots)$. It is also possible to make a function into an exact differential by adding an integrating term, and this method sometimes has advantages with respect to the multiplicative form adopted here.

Since the argument given holds for arbitrary subdivision into V_1 and V_2, and $đq_{1(\text{rev})} \neq 0$, we conclude that

$$\Theta_1(T, V_1) = \Theta_2(T, V_2) = \Theta(T). \tag{16.14}$$

In other words, the integrating factor Θ depends only on the temperature.

The preceding analysis is a special case of the class of reasoning used to establish the character of the equilibrium state (cf. Chapter 19). Here we have tested the equilibrium state by asking how the system responds when an amount of heat $đq_{\text{rev}}$ is transferred between subvolumes while maintaining the equality of temperature characteristic of the original equilibrium. Later we shall reverse the argument; that is, we shall regard Eq. 16.12 as the fundamental relation and deduce the conditions relating intensive variables to one another in the original equilibrium state.

We can take the evaluation of Θ one step further. Our analysis of the behavior of the system shown in Fig. 16.3 did not in any way depend on the nature of the substances in compartments 1 and 2. Accordingly, $\Theta(T)$ must be a universal function of the temperature whose value does not depend on the particular properties of any substance. It is legitimate, then, to choose some very simple substance, evaluate dS for a reversible heat transfer $đq_{\text{rev}}$, and so find $\Theta(T)$. The result will be generally valid because $\Theta(T)$ does not depend on the nature of the substance. To carry out this program we choose the ideal gas, for which

$$pV = nRT, \tag{16.15}$$

$$\left(\frac{\partial U}{\partial V}\right)_T = 0. \tag{16.16}$$

The first law expression for a change in internal energy,

$$dU = đq_{\text{rev}} - p\,dV = C_V\,dT, \tag{16.17}$$

can be rewritten to read (for the ideal gas)

$$đq_{\text{rev}} = C_V\,dT + \frac{nRT}{V}\,dV. \tag{16.18}$$

Using Eqs. 16.10 and 16.14 with Eq. 16.18 leads to

$$dS = \Theta(T)\,đq_{\text{rev}} = \Theta(T)C_V(T)\,dT + \Theta(T)\frac{nRT}{V}\,dV. \tag{16.19}$$

But $S(T, V)$ is a function of state, and dS is an exact differential, hence

$$\left[\frac{\partial}{\partial V}\Theta(T)C_V(T)\right]_T = \left\{\frac{\partial}{\partial T}\left[\Theta(T)\frac{nRT}{V}\right]\right\}_V. \tag{16.20}$$

We now note that

$$\left\{\frac{\partial}{\partial V}\Theta(T)C_V(T)\right\}_T = \Theta(T)\left(\frac{\partial C_V}{\partial V}\right)_T = 0 \tag{16.21}$$

by virtue of Eq. 16.16. Therefore

$$\left\{\frac{\partial}{\partial T}\left[\Theta(T)\frac{nRT}{V}\right]\right\}_V = \frac{nR}{V}\frac{\partial}{\partial T}[T\Theta(T)]_V = 0 \tag{16.22}$$

by virtue of Eqs. 16.20 and 16.21. We conclude that

$$T\Theta(T) = \text{constant}. \tag{16.23}$$

Any function $\Theta(T)$ that satisfies Eq. 16.23 is an integrating factor.[4] We choose the simplest form, namely,

$$\Theta(T) = \frac{1}{T}. \tag{16.24}$$

This is a good point to pause and examine the implications of our study of $\Theta(T, V)$. One of the most important consequences of the second law is the introduction of a thermodynamic scale of temperature independent of the nature of any particular substance or class of substances. Until now, we have used the ideal gas temperature scale, but the fact that $\Theta(T)$, the integrating factor in Eq. 16.10, is independent of substance implies the existence of an absolute temperature scale. The demonstration that $T\Theta(T)$ is a constant, where T is measured on the ideal gas scale, identifies the absolute temperature with the ideal gas temperature to within a scale factor that, for convenience, we take to be unity. An examination of the arguments used shows that a key role is played by the fact that there is a unique reversible adiabatic through the point (T, V) representing the state of a

---

[4] Equation 16.23 can also be deduced from Eq. 16.19 by considering the change of state $(T_1, V_1) \rightarrow (T_2, V_2)$ shown below.

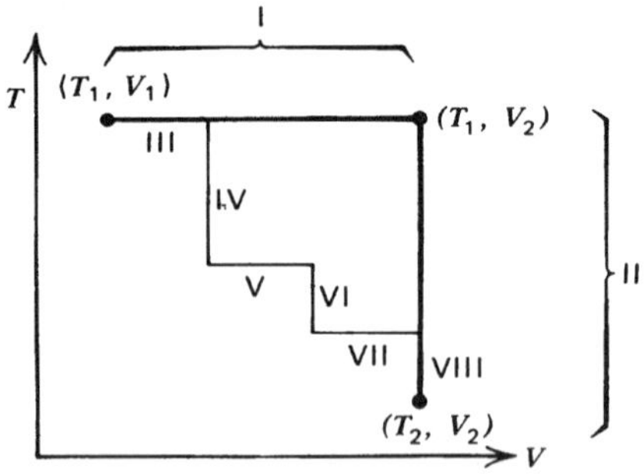

Since S is a state function, the integral of Eq. 16.19 between (T_1, V_1) and (T_2, V_2) is independent of the path. Therefore

$$\int_{\text{I}} \Theta(T)\frac{nRT}{V}\,dV + \int_{\text{II}} \Theta(T)C_V(T)\,dT$$

$$= \int_{\text{III+V+VII}} \Theta(T)\frac{nRT}{V}\,dV + \int_{\text{IV+VI+VIII}} \Theta(T)C_V(T)\,dT = \Delta S.$$

Now note that the integrands of the temperature integrals above are independent of the volume for an ideal gas. The integral over path II must therefore be equal to the integral over IV+VI+VIII. Then, to maintain the same value of ΔS, it must be true that the volume integral over I equals the volume integral over III + V + VII. In other words, the integrands of the volume integrals must be independent of T. Equation 16.23 follows immediately.

system. It follows from the existence of this unique reversible adiabatic that $S(T, V)$ is a function of the state of the system and not of how that state was reached. Therefore we have established that, in an arbitrary reversible process,

$$dS = \frac{đq_{\text{rev}}}{T}, \tag{16.25a}$$

$$S_2 - S_1 = \int_1^2 \frac{đq_{\text{rev}}}{T}, \tag{16.25b}$$

where T is the absolute temperature. For an arbitrary cyclic reversible process,

$$\Delta S_{\text{cycle}} = 0 = \oint \frac{đq_{\text{rev}}}{T}. \tag{16.26}$$

In general, to compute the entropy difference between two states of the system the integral 16.25*b* must be taken along a reversible path. It will be recalled that changes in energy are defined by measuring the work done in adiabatic processes. Similarly, changes in entropy are defined, in principle, by measuring the heats absorbed in reversible processes. Even though a reversible process is only the limiting behavior expected as the rate of a real process tends to zero, and therefore is difficult to achieve experimentally, the heat absorbed in a reversible process is readily computed if enough is known about the properties of the substance involved. Examples of calculations of this type will be given in Chapter 17.

16.3 Irreversible Processes: The Second-Law Interpretation

We now take up the study of an important consequence of the second law. We shall show that for any change which takes place spontaneously in an isolated system the inequality

$$\Delta S > 0 \tag{16.27}$$

must hold. Since it is always observed that a macroscopic system prepared in an arbitrary state approaches a state of equilibrium, and it is never observed that a macroscopic system spontaneously reverts to some original nonequilibrium state even if energy is conserved, expression 16.27 is a statement about the irreversibility of naturally occurring processes.

Before discussing the thermodynamic description of irreversible processes, it is necessary to examine briefly how irreversibility is accounted for by the molecular theory of matter. In fact, the observation of a ubiquitous trend toward an equilibrium state pertains only to systems containing a very large number of particles. This limiting condition is what we imply when we use the word macroscopic. If the system under examination contains only a few particles—for example, an atom—then time-symmetric (reversible) behavior is observed. In principle, since we believe that the equations of quantum mechanics describe all systems,[5] one might expect all systems to show time-reversible behavior. This is both a correct and a subtle observation. A complete description of a macroscopic system, using either quantum mechanics or classical mechanics, does have the mathematical property of time reversibility. That complete description includes, when quantum mechanics is used, the many-particle wave function and its dependence on all coordinates and the boundary conditions; when classical mechanics is used it includes all particle coordinates, positional and velocity correlations and so on. But the variables needed to describe the properties of a macroscopic system are very different from those needed to provide a complete description of a microscopic system; the former uses only a few variables, not all of which are in one-to-one correspondence with mechanical equivalents—for example, the entropy. Furthermore, the thermodynamic, the ensemble theory, and the microscopic definitions of the equilibrium state are somewhat different (see Appendix 15A), although all of the definitions are consistent with one another.

The details of the analysis are very complex, but one can easily grasp the basic idea in the description of the evolution of a many body system from an arbitrary initial state to the equilibrium state. The analysis of the approach to equilibrium in a many-body system is much further developed, and is easier to visualize, when the system is described by classical mechanics, so we use the language of classical mechanics in the following discussion.

Suppose that the initial state of an isolated many-body system can be characterized by some distribution of positions and momenta that deviates from the equilibrium distribution; we denote the collection of initial positions and momenta by Γ_0. As the system evolves in time many collisions between molecules occur. The trajectory of any one molecule includes collisions with other molecules, occasional recollisions with molecules previously encountered, and so forth. The particle trajectories are extremely sensitive to the initial conditions. If we consider an initial condition infinitesimally displaced from Γ_0, say by $\delta\Gamma$, the trajectories that evolve deviate from those that start from Γ_0 at an exponential rate. Consider, as an example, a gas of hard spheres (billiard balls). We show in Fig. 16.4 a sketch of successive collisions between hard spheres. The trajectories shown represent the position of the center of one hard sphere, call it 1, as it scatters off the others, which we denote 2, 3, ... The surfaces at which the trajectory of particle 1 changes direction are each positioned one sphere diameter from the centers of particles 2, 3, ... The trajectories followed by the centers of particles 2, 3, . . . are not shown. The key characteristic of hard sphere collisions is that they are defocusing in the sense that the angle between two trajectories that scatter from the same particle is magnified by the collision. This feature is illustrated in

[5] We omit discussion of relativistic effects and the creation and annihilation of particles.

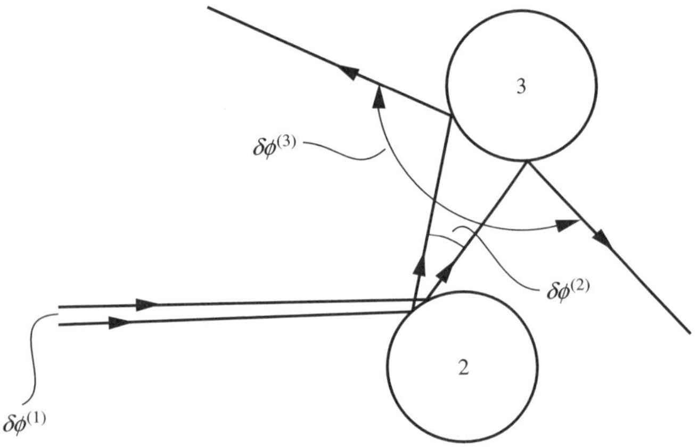

Figure 16.4 Schematic diagram of the scattering of particle 1 from particles 2, 3, . . . , showing the local instability with respect to a small change in initial conditions.

Fig. 16.4, which shows the trajectories corresponding to two initial positions for particle 1, both of which include scattering from particle 2. We assume these initial positions differ very little. Let the angle between the trajectories, after scattering from particle 2, be $\delta\phi^{(2)}$. After a second collision, say with particle 3, the angle between these trajectories is increased to $\delta\phi^{(3)} \sim (2R/\sigma)\delta\varphi^{(2)}$, with σ the hard sphere diameter and R the separation between scatterers 2 and 3. Successive scattering events further increase the deviation between the trajectories that were initially close together. We can estimate the extent of this defocusing by assuming that it is sufficiently accurate to take the distance between scattering events to be the average separation of the hard spheres, $\langle R\rangle$. Then $\delta\phi^{(n)} \approx (2\langle R\rangle/\sigma)\delta\phi^{(n-1)} = (2\langle R\rangle/\sigma)^2\delta\phi^{(n-2)} = \ldots = K^n\delta\phi^{(1)} = \exp(n\ln K)\delta\phi^{(1)}$. Because of the extreme instability of the system trajectories to small changes in the initial conditions, the system dynamics has mixing character and, after only a few collisions, the trajectories become unpredictable (see Appendix 15A). Then only the most probable behavior of the system, or the behavior of the expectation values of dynamical variables, can be predicted.

Given that the solutions to the equations of motion are unique, what is meant by the statement that the trajectories become unpredictable? To answer this question, we must understand what prediction means—that is, how much information is required to accurately define a trajectory. Suppose a system has only two states. We define the information needed to specify the state of the system to be one bit (one binary digit) since the answer to one question—namely, is the system in state 1?—fully specifies its state. Proceeding in this fashion, if a system has N states, and the jth state is known to occur with probability P_j, we measure the amount of information needed to specify the state of the system by the quantity

$$\eta = -\sum_{j=1} P_j \log_2 P_j \,,$$

which is called the *information entropy.* If any $P_j = 1$ we find $\eta = 0$, because a measurement then cannot tell us anything not already known; if all the P_j are equal—namely $1/N$—we find $\eta = \log_2 N$, because we have no knowledge of the state of the system; hence, its determination gives the maximal possible increase in our knowledge.

To apply this analysis to the system of interest to us, imagine that the phase space is divided into N hypercubes with side length ϵ. We also imagine that the time scale is divided into intervals τ. We will now describe a trajectory by its passage through the hypercubes at successive specified times. Let $P(j_0, \ldots, j_N)$ be the joint probability that the phase point of the system lies in cell j_0 at time $t_0 = 0, \ldots,$ and cell j_N at time $t_N = N\tau$. Then the information necessary to specifiy the trajectory to precision ϵ at each of the specified times, assuming that only the probabilities $P(j_0, \ldots, j_N)$ are known, is

$$\eta_N \equiv -\sum_{j_0,\ldots,j_N} P(j_0,\ldots,j_N)\log_2 P(j_0,\ldots,j_N).$$

It then follows that $\eta_N - \eta_{N-1}$ is the additional amount of information needed to specify whether the trajectory, having passed through cells j_1 to j_{N-1}, passes through cell j_N. The Kolmogorov entropy is defined by the limiting process

$$K = \lim_{\tau\to0}\lim_{\epsilon\to0}\lim_{N\to\infty}\frac{1}{N\tau}\sum_{n=0}^{N-1}(\eta_{n+1} - \eta_n)$$
$$= -\lim_{\tau\to0}\lim_{\epsilon\to0}\lim_{N\to\infty}\frac{1}{N\tau}\sum_{n=0}^{N-1} P(j_0,\ldots,j_N)\log_2 P(j_0,\ldots,j_N)$$

When the Kolmogorov entropy is positive, the joint probability for a trajectory to pass through a specified sequence of cells in phase space decreases exponentially as the length of the sequence increases—i.e., as time increases. Of course, as that joint probability decreases, the motion of the system becomes increasingly unpredictable. In general, the time scale on which the motion becomes unpredictable is $1/\ln K$.

The Kolmogorov entropy is also a measure of the instability of the system to changes in initial conditions. When the separation, $D(t)$, between two initially close trajectories grows exponentially in time, $D(t) = D(0)\exp(\lambda t)$, with the coefficient λ known as the Lyapunov exponent; it measures the instability of a trajectory with respect to changes in initial conditions. For exponentially diverging trajectories, the Kolmogorov entropy is related to the Lyapunov exponent by $K = \lambda$. Returning now to the example sketched in Fig. 16.4, we see that the angular separation of the initially close trajectories grows exponentially with collision number and that, taking n as a surrogate for time, K is the Kolmogorov entropy. In the general case, the evolution of the phase space volume element containing the phase points corresponding to different initial conditions that satisfy the macroscopic constraints on the system is accompanied by exponential stretching in some directions and exponential contraction in other directions (Appendix 15A). The exponential stretching of this phase space volume element in some directions implies the existence of positive Lyapunov exponents for the corresponding motions. It can be shown that the Kolmogorov entropy is the sum of the positive Lyapunov exponents for the system.

The preceding arguments are consistent with mechanical reversibility as expressed, for example, in the Poincaré recurrence theorem. This general theorem of mechanics states that an isolated system will achieve a state with the individual positions and momenta of the particles arbitrarily close to what they were in the initial state, if only enough time is allowed to pass. That is, a trajectory started from a particular phase space point will return to an arbitrary neighborhood of that point if one waits long enough. It is readily shown that the time scale on which this recurrence appears, even for systems that are small (for example, a chain of 10 oscillators), is much, much longer than the estimated lifetime of the universe. Furthermore, the Poincaré recurrence theorem does not address the consequences of the extreme sensitivity of the system dynamics to changes in the initial conditions; it refers only to the return of a particular trajectory to a neighborhood of its intial position. Our use of the many-body dynamics to calculate the macroscopic properties of a system rests on the recognition that there are many phase points consistent with the specified macroscopic constraints. This ensemble theory definition of the state of the system, together with the extreme sensitivity of the system trajectories to changes in the initial conditions, means that there is no inconsistency between recurrence of a particular system trajectory and the overall mixing character of the system dynamics. Put another way, if the initial state of a system is represented by the cloud of phase points that are consistent with the macroscopic constraints, that is a set of phase points defined with arbitrary but limited precision, the evolution of the system is determined by bundles of trajectories. In that case it can be shown that the only time evolution consistent with the definition of the initial state is an irreversible approach to an equilibrium state that is independent of the initial state and is dependent only on the temperature, volume, and other macroscopic state variables. The thermodynamic equilibrium state is the end of the evolution because virtually all possible states of the system are close to the equilibrium state. That this statement is true was shown in Chapter 15.

The corresponding quantum mechanical analysis of the approach to equilibrium is much less developed. As of the date of this textbook, the development of a deep understanding of the quantum dynamics of a system that displays classical dynamical chaos, the definition of a quantum analogue of the classical Kolomogorov entropy, the character of the approach to equilibrium, and the nature of the transition to the classical limit of dynamical behavior are all topics of current research.

A particular example of the interplay between reversibility and irreversibility will be found in Appendix 16A.

The macroscopic description of irreversible processes is based on relation 16.27 and a few phenomenological laws such as Fourier's law of heat conduction and Fick's law of diffusion. The latter laws, which we shall discuss in Chapter 20, describe the temporal evolution toward equilibrium of the properties of a system not at equilibrium. In contrast, the inequality 16.27, which we discuss now, describes the overall consequences of an irreversible process in terms of the differences between some initial state and the final equilibrium state.

Since inequality 16.27 refers to a change that can take place in an isolated system, we must prove that $\Delta S > 0$ for an irreversible adiabatic process. Consider again Fig. 16.1. Recall that $1 \to 2$ is a reversible adiabatic path and $2 \to 2'$ is a reversible isothermal path. We showed that $2' \to 1$ could not be a reversible adiabatic path. We now investigate whether or not $2' \to 1$ can be an irreversible adiabatic path. There are two possibilities:

1. The path $2 \to 2'$ is a reversible isothermal expansion (the case shown in Fig. 16.1). If the path $2' \to 1$ is irreversible and adiabatic, heat is transferred only along the path between 2 and $2'$. We have, as before, $\Delta U = 0$, $q_{2\to 2'} = -w$, $q_{2\to 2'} > 0$, and $w < 0$, implying extraction of heat from a reservoir at one temperature and performance of an equal amount of work, which violates the Kelvin statement of the second law. Therefore there can be no adiabatic path, reversible or irreversible, connecting $2'$ and 1 in this case.
2. The path $2 \to 2'$ is a reversible isothermal contraction (see Fig. 16.5). In this case $q_{2\to 2'} < 0$ and an irreversible adiabatic connecting $2'$ and 1 does exist. One possible irreversible adiabatic path from $2'$ to 1 is shown in Fig. 16.5. Imagine that $2' \to 2''$ is a reversible adiabatic compression that is complete when the volume of the system is equal to that in state 1, V_1. Then adiabatic work at constant volume, for example, rotating a paddle wheel

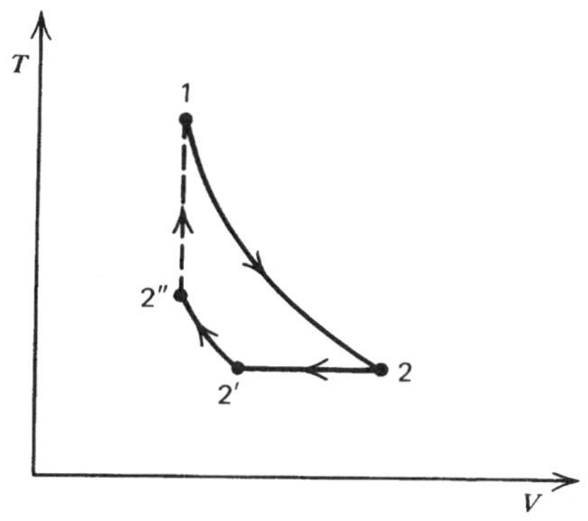

Figure 16.5 Schematic diagram illustrating that $\Delta S > 0$ for an irreversible adiabatic process. The path $1 \to 2$ is a reversible adiabatic, $2 \to 2'$ a reversible isothermal compression, $2' \to 2''$ a reversible adiabatic compression, and $2'' \to 1$ involves constant-volume adiabatic work done on the system (see text).

inside the system, is used to traverse $2'' \to 1$, the amount of work done being just enough to achieve the final temperature desired, namely, T_1. In the case that $V_{2'} < V_1$, the path $2' \to 2''$ can be an adiabatic expansion to V_1 followed by constant-volume adiabatic work sufficient to raise the system's temperature to T_1. Consequently, in the case that $2 \to 2'$ is a compression, there does exist an irreversible adiabatic path connecting $2'$ to 1.

Since the entropy depends only on the state of the system, the difference

$$\Delta S = S_1 - S_{2'} \tag{16.28}$$

is defined even when there is no reversible adiabatic path between the two states. We now rewrite Eq. 16.28 to read

$$\Delta S = (S_1 - S_2) + (S_2 - S_{2'}). \tag{16.29}$$

But

$$S_1 - S_2 = 0 \tag{16.30}$$

because the path $1 \to 2$ is a reversible adiabatic. Then, since

$$S_2 - S_{2'} = \frac{q_{2' \to 2}}{T} > 0, \tag{16.31}$$

we have

$$\Delta S > 0 \tag{16.32}$$

for the irreversible process connecting states $2'$ and 1.

The inequality 16.27 is sufficiently important to warrant another, and more general, demonstration. This time we represent the states of the system on a T, S diagram (see Fig. 16.6). A reversible isotherm is a horizontal line, a reversible adiabatic is a vertical line, and the isotherms and adiabatics meet at right angles. We have already established that the reversible adiabatic through a point representing the state of the system is unique, and that two reversible adiabatics cannot cross. We also know that there is not any adiabatic path connecting state 1 to any state lying to the left of the reversible adiabatic through 1; that is, there is no adiabatic path connecting state 1 to any state with smaller entropy. For a spontaneous adiabatic process connecting two states to occur, the representative point of the final state on the T, S diagram must lie to the right of that for the initial state, or

$$S_{\text{final}} - S_{\text{initial}} \equiv \Delta S > 0. \tag{16.33}$$

Finally, we now demonstrate that

$$\Delta S > \int \frac{đq}{T_{\text{surroundings}}} \tag{16.34}$$

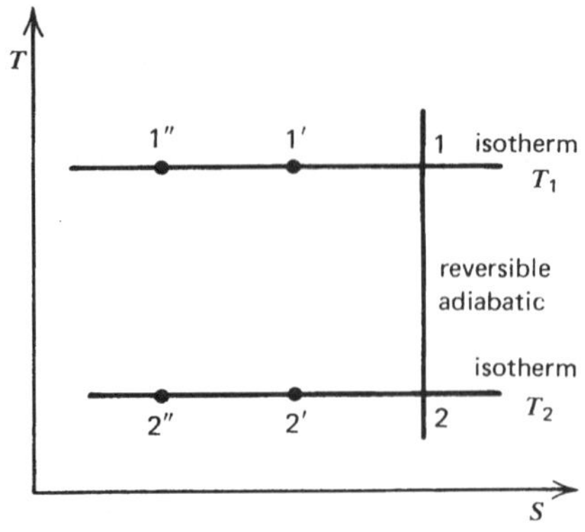

Figure 16.6 Diagram demonstrating that $\Delta S > 0$ for an irreversible process (see text).

for any irreversible process. Consider an isolated compound system consisting of two subsystems. Suppose that some spontaneous process occurs in subsystem A, by virtue of which it undergoes an irreversible change of state. The corresponding change in system B, which we regard as the surroundings of system A, will be assumed to be reversible.[6] An example of such a compound system is shown in Fig. 16.7. Now, since the compound system undergoes a spontaneous process, we have

$$\Delta S_{AB} = \Delta S_A + \Delta S_B > 0, \tag{16.35}$$

which implies that

$$\Delta S_A > -\Delta S_B. \tag{16.36}$$

But because the process in part B is reversible, we can write

$$\Delta S_B = \int \frac{đq}{T_B}, \tag{16.37}$$

which leads directly to

$$\Delta S_A > -\int \frac{đq}{T_B}. \tag{16.38}$$

For the compound isolated system,

$$đq = 0 = đq_A + đq_B, \tag{16.39}$$

[6] The assumption that changes in system B are reversible imposes some restrictions. It is usual to require that B be very much larger than A. It is more important to require that B be sufficiently "well stirred" that no local gradients of temperature, pressure, or composition can develop. The stirring, which can be thought of as mechanical but need not be, homogenizes system B very rapidly on the time scale for changes in system A, hence so far as A is concerned system B is always in a state of internal equilibrium.

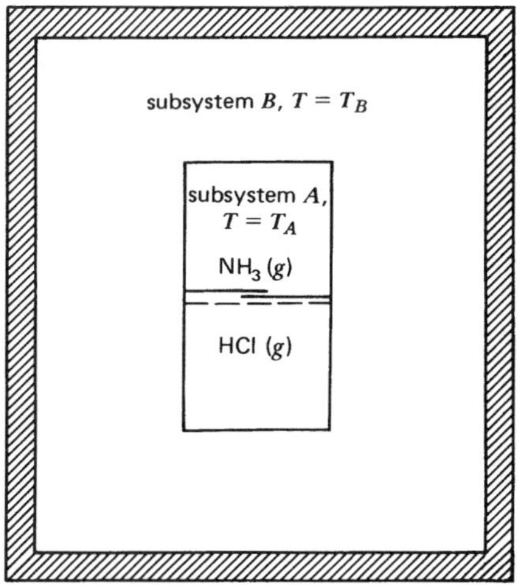

Figure 16.7 Schematic diagram of a compound isolated system that can undergo a spontaneous process in one of its component parts. The outer envelope of B is a rigid adiabatic barrier, whereas A is contained by a rigid diathermal envelope. A is also divided internally by a permeable membrane. A slide covers the permeable membrane so that the two sides of A can be loaded. The slide can be opened without expenditure of work. In the initial state, $NH_3(g)$ is put into one side of A and $HCl(g)$ into the other. Once the slide is opened, the membrane permits passage of both gases, so in the final state the NH_3 and HCl have reacted to form NH_4Cl. Some of the heat released in this reaction is transferred through the diathermal envelope to B.

so that inequality 16.38 can be rewritten

$$\Delta S_A > \int \frac{đq_A}{T_B}, \tag{16.40}$$

and we have proved inequality 16.34.

16.4 The Clausius and Kelvin Statements Revisited

We have shown how the physical statements of the second law given by Clausius and Kelvin lead to the introduction of a new function of state, the entropy, and to a discrimination between reversible and irreversible processes. It is now worthwhile to reverse the argument and use the mathematical formulation of the second law to illuminate the consequences of the physical statements. We shall do this by examining the conversion of heat to work for a system carried through the cyclic processes described in the Clausius and Kelvin statements.

Consider, first, the Kelvin statement of the second law. The cyclic process referred to draws heat from a reservoir at one temperature, say, T_0. For any cyclic process $\Delta S = 0$, so application of inequality 16.40 leads immediately to

$$0 \geq \oint \frac{đq}{T_0}. \tag{16.41}$$

In this cyclic process we also must have $\Delta U = 0$, hence $q = -w$. When the system does work on the surroundings, $w < 0$, requiring that $q > 0$. Since T_0 is positive, inequality 16.41 requires that $q < 0$. We conclude that it is impossible to operate an engine in a cycle so as to produce no effect other than the extraction of heat from a reservoir at one temperature and the performance of an equal amount of work.

Consider, now, the Clausius statement of the second law. We have established that for any process, reversible or irreversible,

$$dS \geq \frac{đq}{T}, \tag{16.42}$$

where T is the absolute temperature of the surroundings. The equality sign holds in relation 16.42 only for the case of a reversible process, in which case the temperatures of the system and surroundings are the same. Clearly, for the same infinitesimal entropy change,

$$\frac{đq_{\text{rev}}}{T} \geq \frac{đq}{T} \tag{16.43}$$

which, when combined with the first law,

$$dU = đq + đw, \tag{16.44}$$

leads to the conclusion

$$đw_{\text{rev}} \leq đw. \tag{16.45}$$

In words, the maximum work extractable from any process is the reversible work.[7] For this reason we proceed by examining the simplest possible reversible cycle for conversion of heat into work. In the design of this cycle it is, of course, necessary to take account of the restrictions imposed by the second law. The Kelvin statement of the second law prohibits continuous conversion of heat into work if the heat is drawn from a source at a single temperature; hence the simplest cycle must use at least two heat sources, say, one at temperature T_1 and the other at T_2 with $T_2 > T_1$. The system undergoing the cycle will then execute two reversible isothermal processes, one at T_1 and the other at T_2. To complete the cycle the two isotherms must be connected by paths along which heat is not transferred, that is, by reversible adiabatics. The combination of isothermal and adiabatic processes just described is known as a (reversible) *Carnot cycle.*

A realization of a Carnot cycle is shown in Fig. 16.8; the working substance need not be specified at this state of the analysis, although the diagram implies that it is a fluid. We

[7] Recall again that we use the convention that work done by the system is reckoned negative.

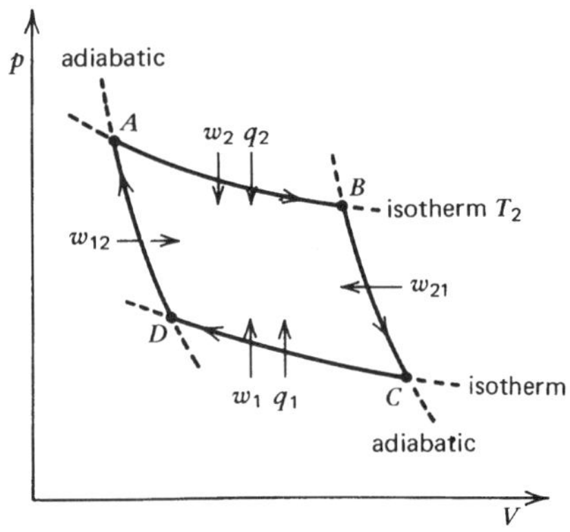

Figure 16.8 Representation of a Carnot cycle for a simple fluid as working substance.

start in state A, with the system at equilibrium in contact with a thermostat at temperature T_2. The steps in the Carnot cycle displayed are as follows:

1. $A \to B$: The system undergoes a reversible isothermal expansion during which an amount of heat q_2 is transferred from the reservoir to the system and an amount of work w_2 is performed by the system in the surroundings, all at temperature T_2.
2. $B \to C$: The system undergoes a reversible adiabatic expansion, the final temperature being $T_1 < T_2$. In carrying out this process the system must be isolated from any thermostat, so that no heat is absorbed or released, but work w_{21} is done by the system in the surroundings.
3. $C \to D$: The system is brought into thermal contact with a reservoir at temperature T_1 and a reversible isothermal compression is carried out. In this process an amount of heat q_1 is transferred from the system to the reservoir and an amount of work w_1 is done by the surroundings on the system.
4. $D \to A$: The system is removed from contact with the thermostat and, by adiabatic compression, restored to state A. In this last reversible process, work w_{12} done by the surroundings on the system.

Recall that we have not made reference to the actual properties of any particular substance. In the simplest possible case the substance would be a fluid with equation of state defined by the independent variables p and V. Each of the reversible transformations defined is then a line in the p, V space. In more general cases the number of independent variables required to define the state of the system is greater than two, and then the isothermal and adiabatic processes described define multidimensional surfaces in the appropriate space.

What are the characteristics of the Carnot cycle we have described? The first important deduction is that the ratio q_2/q_1 is independent of the properties of the substance undergoing the cycle, and therefore is the same for all substances undergoing a Carnot cycle between the temperatures T_1 and T_2. To demonstrate this we note that, by the first law, the change in internal energy for the entire cycle must be zero,

$$\Delta U = q_2 + q_1 + w_2 + w_{21} + w_1 + w_{12} = 0, \qquad (16.46)$$

so that

$$w \equiv w_2 + w_{21} + w_1 + w_{12} = -q_2 - q_1. \qquad (16.47)$$

But since the cyclic process is reversible and heat is transferred only along the isothermal legs, the entropy changes along those legs are

$$\Delta S_2 = \frac{q_2}{T_2}, \qquad (16.48)$$

$$\Delta S_1 = \frac{q_1}{T_1}. \qquad (16.49)$$

For one complete cycle

$$\Delta S = 0 = \frac{q_2}{T_2} + \frac{q_1}{T_1}, \qquad (16.50)$$

or

$$-\frac{q_2}{q_1} = \frac{T_2}{T_1}. \qquad (16.51)$$

Thus, the ratio q_2/q_1 is independent of the nature of the working substance.[8]

The connection of the preceding with the Clausius statement of the second law is made as follows. Since $T_2 > T_1$, Eq. 16.50 implies that

$$|q_2| > |q_1|. \qquad (16.52)$$

But, from Eq. 16.47, if heat is transferred from the reservoir to the system at the lower temperature T_1, then $q_1 > 0$, $-q_2 - q_1 > 0$, hence $w > 0$, and work must be done by the surroundings on the system. In contrast, if heat is transferred from the reservoir to the system at the higher temperature T_2, then $q_2 > 0$, $-q_2 - q_1 < 0$, hence $w < 0$, and work is done by the system on the surroundings. We conclude that no cycle can be devised that produces no effect other than the transfer of heat from a colder to a hotter body.

[8] A heat engine performs work in the surroundings, $w < 0$, when $q_2 > 0$ and $q_1 < 0$. The efficiency of conversion of heat into work by the engine is measured by the ratio $-w/q_2$, namely $\eta = -w/q_2 = 1 - (T_1/T_2)$, which is obtained from the combination of (16.47) and (16.51). A typical heat engine is much less than 100% efficient. For example, if the engine works between 500°C (773 K) and 50°C (323 K), the efficiency of conversion of heat into work, for the Carnot cycle considered above, is 58%. A Carnot engine operating between temperatures T_1 and T_2 is the most efficient of all engines operating between those temperatures.

16.5 The Second Law as an Inequality

It is now convenient to again examine that part of the second law which takes the form of an inequality. In Chapter 15 we argued that the entropy of an isolated system is a maximum subject to the constraints that define its state. To understand better the nature of the inequality 16.27, we must consider how one can carry out a process inverse to one that occurs in an isolated system. We define an *inverse process* as one that, by virtue of heat and work transfers to the surroundings, restores a state of the system from which the initial state could have been reached via a natural process in an isolated system. Since any process that does occur in an isolated system must involve no change in the internal energy, we also require that an inverse process involve no change in the internal energy. Note that, in order to carry out the process inverse to one that occurs in an isolated system, we must break the isolation of that system. Then, by doing work and transferring heat in amounts such that the internal energy of the system is maintained constant, we can restore the system to the state it had while isolated and before alteration of the constraints.

For example, consider the system represented schematically in Fig. 16.9. An isolated box, divided into two equal subvolumes V_1, has perfect gas in one and vacuum in the other. When the diaphragm separating the subvolumes is removed without expenditure of work, the gas spontaneously fills the entire volume $2V_1$. The natural process following removal of the diaphragm is just the irreversible expansion from a volume V_1 to a volume $2V_1$. The inverse process requires compression of the gas from the volume $2V_1$ to the volume V_1, and restoration of the partition. The internal energy of the gas when in V_1 must be the same as when it fills the volume $2V_1$. Since the internal energy of a perfect gas depends only on the temperature, to accomplish the inverse process we couple the system to some machine and reservoir that will compress the perfect gas isothermally. The reversible work required to change the volume from $2V_1$ to V_1 under these conditions is

$$w = -\int_{2V_1}^{V_1} p_{\text{app}}\, dV = -\int_{2V_1}^{V_1} \frac{nRT}{V} dV = nRT \ln 2. \qquad (16.53)$$

But the internal energy of the gas is to be constant throughout this inverse process, hence the first law requires that the system transfer heat to the surroundings in the amount $nRT \ln 2$. Then, using Eq. 16.25*b* for an isothermal process, for which T may be removed from the integrand, we find

$$\Delta S_{\text{inv}} = \frac{q}{T} = -nR \ln 2 < 0. \qquad (16.54)$$

Thus, in the inverse process described, the entropy of the system is decreased.

It is found to be generally true that, if an isolated system is at equilibrium with respect to some set of constraints, any process inverse to one that could have been involved in reaching that equilibrium state is characterized by

$$\Delta S_{\text{inv}} < 0. \qquad (16.55)$$

In terms of the mathematical criteria for a maximum of a function, we state that:

If an isolated system is at equilibrium, then for any infinitesimal variation which is the inverse of a process that could have occurred in an isolated system and during which the original constraints are maintained, it is necessary that

$$\delta S = 0 \qquad \text{(original constraints retained).} \qquad (16.56)$$

This statement is to be interpreted as follows. In the original equilibrium state of the isolated system, the entropy was a maximum with respect to the pertinent constraints. Imagine changing the constraints so as to create a new equilibrium state, one from which the original state could be obtained in a spontaneous process. The entropy of the new equilibrium state must be smaller than the entropy of the original equilibrium state, since the latter can be obtained from the former in an isolated system (definition of spontaneous process). Suppose that the constraints are changed in a way such that every state along the path used is displaced only infinitesimally from the preceding state. Evidently the entropy must decrease, because the new state has lower entropy than the original state. But because the entropy in the original equilibrium state was a maximum with respect

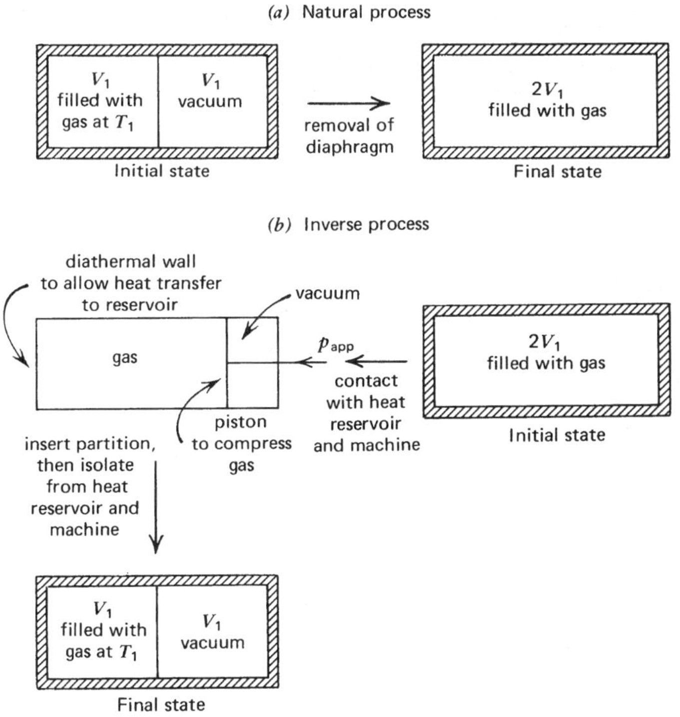

Figure 16.9 Diagram illustrating the nature of an inverse process.

to its constraints, the initial changes in the variables of the system start with zero slope, that is, are connected by the differential equation 16.56. When δS is expressed in terms of changes in variables such as T or V, the coefficients of the differentials in this equation refer only to the initial state and not to the final state. The variation that carries the system from one equilibrium state to another equilibrium state derived from it by an inverse process is called a *virtual variation.* Since the precise nature of the change in the constraints is never invoked (remember that the coefficients in 16.56 refer to the initial state), the changes imposed in defining a virtual variation may be very general, provided only that a true equilibrium state can be defined for the final set of constraints.

16.6 Some Relationships between the Microscopic and Macroscopic Theories

We now examine, briefly, some relationships between the microscopic and macroscopic theories of the entropy.

First, we observe that the thermodynamic temperature is closely related to, and may be considered to measure a fundamental property of, the density of states of the system. Using Eq. 15.42 as the microscopic definition of the entropy, and making the identification

$$\beta = \frac{1}{k_B T}, \tag{16.57}$$

where T is the thermodynamic (absolute) temperature, we see that the definition

$$\beta \equiv \frac{\partial \ln \Omega}{\partial E} \tag{16.58}$$

implies that T is positive. For we have noted that Ω is an increasing function of the energy, so that $(\partial \ln \Omega/\partial E)$ must be positive, and hence $T > 0$.[9] Thus, the macroscopic assignment of the sign of the temperature scale such that the dissipation of energy in an isolated system leads to an increase in temperature is fundamentally related to the fact that Ω increases as E increases. If the opposite macroscopic convention had been adopted, the relation between β and $(\partial \ln \Omega/\partial E)$ would require a minus sign to preserve agreement between the two descriptions. Either convention is internally consistent. What is important is to notice that the direction of change of temperature due to dissipation of energy is connected with the direction in which the number of states of the system increases as a function of the energy.

A second point of interest is that the condition that entropy must increase in a spontaneous process is equivalent to the requirement that heat flow from the system with the greater absolute temperature to the system with the lesser absolute temperature. Using the condition[10]

$$\frac{\partial \ln \Omega_A(E_A^i)}{\partial E}(E_A^f - E_A^i) + \frac{\partial \ln \Omega_B(E_B^i)}{\partial E}(E_B^f - E_B^i) \geq 0, \tag{16.59}$$

derived from Eq. 15.42 and the second law, we find that

$$(\beta_A^i - \beta_B^i)\, q \geq 0, \tag{16.60}$$

[9] The statement is true when all degrees of freedom of the system are accounted for. In some situations it is convenient to focus attention on only one or a few degrees of freedom, and to neglect all others. For example, one might want to ascribe thermodynamic properties to an assembly of atoms, neglecting all states of the system except those associated with two levels, say, the ground level and one excited level of each atom. Alternatively, one might want to ascribe thermodynamic properties to an assembly of atoms, neglecting all states of the system except those associated with the nuclear spin. Suppose that the occupation of the set of states singled out for attention is restricted by the Pauli exclusion principle. Furthermore, suppose that the set of states of interest is effectively decoupled from all other states of the system so that they comprise (nearly) an isolated system. The kind of system described differs from an ordinary one in that there is an upper limit to its possible energy, which occurs when all of the atoms (or spins) are in the highest quantum state. When all of the atoms are in their ground state, or in the highest quantum state, the number of quantum states of the whole system with energy between E and $E + dE$ is very small; e.g., if the ground state is nondegenerate, there is only one state of the system with each atom in its ground state. On the other hand, when half the atoms are excited there are many quantum states with energy between E and $E + dE$ differing only in the assignment of which atoms are excited. Thus $\Omega(E)$ increases as E increases from the ground-state energy, decreases as E approaches the upper bound to the energy of the system (all nuclear spins aligned, all atoms excited, etc.), and therefore has a maximum somewhere in between. In these cases it is possible to define negative "isolated-degrees-of-freedom" temperatures. Such behavior occurs in lasers and in nuclear magnetic resonance experiments.

[10] If two systems, initially with temperatures and T_A^i and T_B^i, are brought into thermal contact through a diathermal wall, there is a transfer of heat, and at thermal equilibrium $T_A^f = T_B^f$. The second law requires that the entropy change for this process be positive,

$$\Delta S_A + \Delta S_B \geq 0.$$

Suppose that T_A^i and T_B^i are only slightly different, and use Eqs. 15.62 and 15.63 to write, for the heats transferred,

$$q_A = E_A^f - E_A^i; \qquad q_B = E_B^f - E_B^i.$$

Then, using the fundamental definition 15.42 and the Taylor series expansions

$$\ln \Omega_A(E_A^f) = \ln \Omega_A(E_A^i) + \frac{\partial \ln \Omega_A(E_A^i)}{\partial E}(E_A^f - E_A^i) + \cdots,$$

$$\ln \Omega_B(E_B^f) = \ln \Omega_B(E_B^i) + \frac{\partial \ln \Omega_B(E_B^i)}{\partial E}(E_B^f - E_B^i) + \cdots,$$

we obtain the expressions

$$\Delta S_A = k_B \ln \left[\frac{\Omega_A(E_A^f)}{\Omega_A(E_A^i)}\right],$$

$$\Delta S_B = k_B \ln \left[\frac{\Omega_B(E_B^f)}{\Omega_B(E_B^i)}\right]$$

for the entropy changes; when added, these give Eq. 16.59 of the text.

so that, if $q > 0$,

$$\beta_A^i \geq \beta_B^i \tag{16.61}$$

or

$$T_A^i \leq T_B^i, \tag{16.62}$$

since $T > 0$. We conclude, as stated, that positive heat is always absorbed by the system with lower temperature, and this consequence of the properties of entropy is also related to the fact that Ω increases as E increases.

Finally, we note that the ensemble probability distribution is very sharply peaked near the mean value of the energy. Indeed, it has been shown in connection with Eq. 15.56 that the fractional width of the distribution in energy is proportional to the inverse square root of the number of degrees of freedom of the system. This fractional width is, therefore, so small for a macroscopic system that one nearly always observes the mean value. Although fluctuations (i.e., states of nonmaximal probability) are a characteristic of the state of equilibrium, a completely macroscopic approach that neglects the existence of states of the system other than the most probable state is satisfactory for most problems. There are some properties of matter, for example, the scattering of light by liquids and gases, that are determined by the fluctuations, and to describe these phenomena states other than the most probable state must be included in the description of the processes involved. The nonzero width of the ensemble probability distribution can also be used to prove that $(\partial T/\partial E) > 0$,[11] which is a stronger condition than $T > 0$, and is an example of a thermodynamic stability condition; that is, no substance can exist for which $(\partial T/\partial E)$ is ever negative or zero.

Because of the importance of entropy in the description of the properties of matter, it is advantageous to gain familiarity with the nature of the entropy changes in various kinds of processes, the magnitudes of the entropy changes, and so on. It is these matters that we shall consider in Chapter 17.

---

[11] Since $\beta \equiv \partial \ln \Omega/\partial E$, and $\beta = 1/k_B T$, we have

$$\frac{\partial \beta}{\partial E} = -\frac{1}{k_B T^2}\frac{\partial T}{\partial E} = \frac{\partial^2 \ln \Omega}{\partial E^2}.$$

But we have shown in Eq. 15.54 that $\partial^2 \ln \Omega/\partial E^2$ is always negative. Then, because T^2 is always positive, it must be true that $(\partial T/\partial E) > 0$.

APPENDIX 16A

Poincaré Recurrence Times and Irreversibility

In 1896 Poincaré proved the theorem: "In a system of material particles under the influence of forces that depend only on the spatial coordinates, a given initial state (i.e., a representative point in phase space) must, in general, recur, not exactly, but to any desired degree of accuracy, infinitely often, provided that the system always remains in the same finite part of the phase space." The time between two consecutive repetitions is called a *Poincaré recurrence time.* Clearly, the length of a recurrence time depends on how precisely the recurrence condition is specified. A simple but crude estimate can be obtained in the following way. Consider a dilute gas in which each molecule undergoes about 10^{10} collisions per second (Chapter 28). If the number density is 10^{24} molecules per cubic meter, the number of collisions per second in a sample of 1 cm^3 volume is $Z_T = 10^{28}$. If we assume that each collision induces a transition between two states of the system, and note that the original state need not recur until the system has passed through all other states, then the average recurrence time, Θ, is given by

$$\Theta \approx \frac{\Omega}{Z_T}, \tag{16A.1}$$

where Ω is the total number of states with energy between E and $E + dE$. Ω is easily estimated from the entropy: $\Omega = \exp(S/k_B) \approx 10^{10^{18}}$. Thus,

$$\Theta \approx \frac{10^{10^{18}}}{10^{28}} \approx 10^{10^{18}} \text{ s}. \tag{16A.2}$$

This is to be compared with the "age" of the universe, which is of order 10^{17} *s*. We see immediately that the recurrence time is greater than this by about 10^{18} orders of magnitude.

Boltzmann made a much more detailed calculation by estimating the volume of phase space available to the system, and dividing it into small cells whose dimensions are of the order of magnitude to which a recurrence is specified. In the system discussed in the previous paragraph, the average separation of neighboring molecules is 10^{-8} m. Boltzmann supposed that the initial state is one in which all the molecules move with velocities of 500 m/s, and that the initial state is reproduced if the same molecules are in the same positions to within 10^{-9} m (10% of the separation) with the same velocities to within 1 m/s. Since the gas is dilute, recurrence of the initial state can be discussed by considering the recurrence of velocity and configurational distributions separately. The volume of the velocity space may be calculated in the following way: The subspaces for each molecular velocity are considered in succession and integrations are carried out over each subspace subject to the condition that the total kinetic energy is fixed. Thus, the first molecule may have a speed v_1 anywhere in the range from zero to $a = 500 \times 10^9$ m/s,[12] the second anywhere in the range from zero to $(a^2 - v_1^2)^{1/2}$, and so on. Since the length of a side of a cell in the velocity space is 1 m/s, it may be treated as a dimensionless number, and the calculated volume of the velocity space will be given as the number of cells Γ_v. We have, from the preceding,

$$\begin{aligned}\Gamma_v &= (4\pi)^{n-1} \int_0^a dv_1 v_1^2 \int_0^{(a^2-v_1^2)^{1/2}} dv_2 v_2^2 \cdots \int_0^{(a^2-v_1^2-v_2^2-\cdots-v_{n-2})^{1/2}} dv_{n-1} v_{n-1}^2 \\ &= \left\{ \frac{\pi^{(3n-3)/2}}{2\cdot 3\cdot 4\cdot \cdots \cdot [3(n-1)/2]} \right\} a^{3(n-1)}; \quad n \text{ odd}, \\ &= \left\{ \frac{2(2\pi)^{(3n-4)/2}}{3\cdot 5\cdot 7\cdot \cdots \cdot [3(n-1)/2]} \right\} a^{3(n-1)}; \quad n \text{ even}.\end{aligned} \tag{16A.3}$$

The number of ways of arranging n labeled points in m cells is $[m!/(m - n)!]$, so the number of configurational cells, Ω_{conf}, is

$$\Omega_{\text{conf}} = \frac{m!}{(m-n)!}. \tag{16A.4}$$

If the system passes through all possible states before returning to its initial state, the recurrence time is

$$\Theta = \frac{\Gamma_v \Omega_{\text{conf}}}{Z_T}. \tag{16A.5}$$

[12] This value of a exceeds the velocity of light, but we ignore all relativistic effects.

The value of Θ is easily estimated for the very large numbers in the example. One finds that[13]

$$\Gamma_v \approx 10^{9n}, \qquad \Omega_{\text{conf}} \approx m^n,$$

so that with $a = 5 \times 10^{11}$ m/s, $n = 10^{18}$ s, and $m = 10^{21}$ (for 1 cm^3),

$$\Theta \approx 10^{10^{19}} \text{ s}, \tag{16A.6}$$

which, although enormously different from 16A.2, nevertheless bears a similar relation to the age of the universe.

The thermodynamic meaning of irreversibility is different from that employed in the study of Poincaré recurrences, because we observe a system only in a gross way. The thermodynamic state of a system corresponds to a large number of trajectories in phase space, as mentioned in Section 16.3. During a Poincaré cycle the trajectory of the point in phase space will pass more or less close to every point in the space, depending on how precisely we define the recurrence, so that a given thermodynamic state will in general recur many times, depending on what fraction of the phase space is occupied by the points corresponding to that state. In general, then, even large fluctuations from a thermodynamic norm will recur, and it is with the frequency of these events that a discussion of irreversibility is concerned.

The problem of fluctuations and their recurrence was analyzed by Smoluchowski[14] in terms of what he called the *probability aftereffect:* Given the state α of a system at some instant, what can we say about α at a time τ later? Consider, for definiteness, that the state of the system is specified by the instantaneous number of particles, ν, in a volume ω. Obviously, when τ is much shorter than the period of the fluctuations of ν, we shall expect to find ν almost unchanged, whereas for very large τ we shall expect the value of ν to be independent of the initial value. For intermediate times we must take into account the speed of the fluctuations explicitly. An example considered by Smoluchowski was the number of colloid particles in a small volume defined within a much larger volume. This number was observed at successive instants at intervals of time τ. In a typical sequence of counts the average number was ≈ 2.0, but frequently as many as 5 particles were found. Such sequences, containing frequent large fluctuations from the mean, show none of the expected behavior of an irreversible process, and yet we shall see that they are compatible with the macroscopic irreversible law of diffusion (see Chapter 20).

Consider a system of colloid particles in equilibrium.[15] If the concentration is not too high, the probability that a particle will enter or leave the volume is not influenced by the number in the volume or near it outside. The probability distribution for the number ν will then be a Poisson distribution

$$P(\nu) = \frac{e^{-\gamma}\gamma^{\nu}}{\nu!}, \tag{16A.7}$$

with

$$\langle \nu \rangle = \gamma \tag{16A.8}$$

and

$$\langle \nu^2 \rangle = \gamma^2 + \gamma. \tag{16A.9}$$

This distribution can be regarded as the limit of a binomial distribution as the total number of particles N in the whole system, and its volume V, tend to infinity while (N/V) remains constant: The probability that any particular particle is to be found in the observed volume ω is (ω/V) if all points in V are equiprobable. Thus, the probability that any ν are found in V is

$$P^{(N)}(\nu) = \frac{N!}{\nu!(N-\nu)!}\left(\frac{\omega}{V}\right)^{\nu}\left(1-\frac{\omega}{V}\right)^{N-\nu}. \tag{16A.10}$$

We now let $N, V \to \infty$ while $N/V = n$, the average density. Then Eq. 16A.10 becomes

$$P^{(N)}(\nu) = \lim_{N\to\infty} \frac{N!}{\nu!(N-\nu)!}\left(\frac{\gamma}{N}\right)^{\nu}\left(1-\frac{\gamma}{N}\right)^{N-\nu},$$

where $\gamma = n\omega$. Noting that

$$\frac{N!}{\nu!(N-\nu)!} \to N^{\nu}, \qquad \left(1-\frac{\gamma}{N}\right)^{N-\nu} \to e^{-\gamma}$$

for large N, we see that

$$\lim_{N\to\infty} P^{(N)}(\nu) = P(\nu). \tag{16A.11}$$

To discuss the speed of fluctuations, we must calculate the probability that m particles will be observed within ω a

---

13
$$\frac{m!}{(m-n)!} \approx \frac{m^m e^{-m}}{(m-n)^{m-n} e^{-(m-n)}} \approx m^n \quad \text{for } m >> n,$$

$$\left\{\frac{\pi^{(3n-3)/2}}{2\cdot 3\cdot 4\cdot \cdots \cdot [3(n-1)/2]}\right\} a^{3n-3}$$

$$\approx \frac{\pi^{3n/2} a^{3n}}{(3n/2)!} \approx \left[\left(\frac{2e}{n}\right)^{1/2} a\right]^{3n} \approx (10^3)^{3n}.$$

14 M. V. Smoluchowski, *Physik. Z.* **17**, 557, 585, (1916).

15 We discuss colloid particles as a surrogate for molecules in a dilute gas because experiments have been performed on colloidal suspensions. Think of the colloidal system as a dilute gas of massive particles.

time τ after ν were found [called the *transition probability* and denoted $K(\nu|m)$]. We also define the probability aftereffect factor p that a particle initially in ω will have left it during τ. In the present example, individual particles are assumed to obey the macroscopic law of diffusion. The probability that a particle, initially at $\mathbf{r}_1$, will have diffused to $\mathbf{r}_2$ in a time τ is (see Chapter 20)

$$(2\pi D\tau)^{-3/2}\exp\left[-\frac{(\mathbf{r}_1-\mathbf{r}_2)^2}{4D\tau}\right],$$

where D is the diffusion coefficient. The value of p is found by integrating this probability over values of $\mathbf{r}_1$ inside ω and values of $\mathbf{r}_2$ outside ω. This definition of p again requires both the assumptions made in deriving $P(\nu)$.

Now, let $A_i^{(\nu)}$ denote the probability that if ν particles are initially observed, some i will have left ω after a time τ. $A_i^{(\nu)}$ is the binomial distribution

$$A_i^{(\nu)}=\frac{\nu!}{i!(\nu-i)!}p^i(1-p)^{\nu-1}. \tag{16A.12}$$

Also, let E_j denote the probability that j particles will enter ω during τ. By assumption, E_j cannot depend on ν, and since the *a priori* probabilities of entry and exit must be the same, we have

$$\begin{aligned}E_j&=\langle A_j^{(\nu)}\rangle\\&=\sum_{\nu=j}^{\infty}A_j^{(\nu)}P(\nu)\\&=\frac{(\gamma p)^j e^{-\gamma p}}{j!},\end{aligned} \tag{16A.13}$$

which is a Poisson distribution with the mean γp. Consequently, $K(\nu|\nu+k)$ is given by

$$K(\nu|\nu+k)=\sum_{i=0}^{\nu}A_i^{(\nu)}E_{i+k}, \tag{16A.14a}$$

and also

$$K(\nu|\nu-k)=\sum_{i=k}^{\nu}A_i^{(\nu)}E_{i-k}. \tag{16A.14b}$$

In Eqs. 16A.14, i cannot be less than k because E_j is not defined for $j<0$. We see that the transition probability is in general given by

$$K(\nu|m)=\sum_{x+y=m}P_1^{(\nu)}(x)P_2(y), \tag{16A.15}$$

where the summation is over all values of x and y consistent with $x+y=m$. This form of distribution is called the sum or convolution of the two component distributions $P_1^{(\nu)}$ and P_2. It is easy to show that the mean and mean-square deviations of m for given ν are the sums of the corresponding moments of $P_1^{(\nu)}$ and P_2:

$$\begin{aligned}\langle x\rangle&=\nu(1-p); &\langle(\Delta x)^2\rangle&=\nu p(1-p);\\\langle y\rangle&=\gamma p; &\langle(\Delta y)^2\rangle&=\gamma p.\end{aligned} \tag{16A.16}$$

Let a subscript 1 on $\langle\cdot\ \cdot\ \cdot\rangle$ denote an average conditional on the given value of ν. Then

$$\langle(m-\nu)\rangle_1=\langle m\rangle_1-\gamma=(\gamma-\nu)p \tag{16A.17}$$

and

$$\langle(m-\nu)^2\rangle_1=[(\gamma-\nu)^2-\gamma]p^2+(\gamma+\nu)p. \tag{16A.18}$$

Equation 16A.17 is important in that it shows that the average tendency is to move toward the mean. This is one of the observed characteristics of irreversible processes.

Upon averaging these results over all values of ν, we find

$$\langle\langle(m-\nu)\rangle_1\rangle=0,\quad\text{since}\quad\langle\nu\rangle=\gamma, \tag{16A.19}$$

and

$$\langle\langle(m-\nu)^2\rangle_1\rangle=2\gamma p. \tag{16A.20}$$

Equation 16A.19 is, of course, to be expected; it simply states the fact that the mean values of two consecutive observations should be the same. This result follows from the initial assumption that the colloid system is in equilibrium.

Both Eqs. 16A.19 and 16A.20 provide a means of testing the theory of the probability aftereffect. Elegant and thorough experiments performed by Svedberg and by Westgren[16] test Smoluchowski's theory. We shall discuss briefly the results of Westgren's experiments, which were obtained with an experimental arrangement designed specifically with the theory in mind. Westgren set up his experiment in such a way that a volume of fluid with linear dimensions of a few micrometers, and containing colloidal gold particles of radius approximately 5 μm, was viewed in an ultramicroscope. The number of particles in view was counted at successive instants. Westgren performed experiments with different concentrations of particles, different sizes of volume ω, and different time intervals. Each experiment consisted of about 1500 observations. A sample of 114 consec-

[16] T. Svedberg, *Z. Physik. Chem.* **77,** 147 (1911); A. Westgren, *Arkiv Matematik, Astronomi och Fysik* **11,** Nos. 8, 14 (1916); **13,** No. 14 (1918).

Table 16A.1 Experimental Verification of the Poisson Distribution

ν	**Obs.**	**Calc.**
0	15	15
1	29	31
2	30	31
3	25	21
4	10	10
5	4	4
6	1	1

utive observations for which the time interval $\tau = 0.81$ s is the following:

02313200112132311235121423432311101021
43232422353413253233602235413244234230
32111122100123210000112123234312021011.

The average value of ν for this sample is $\gamma = 2.018$. The theoretical numbers of occurrences of different values of ν for a Poisson distribution of this average are shown in Table 16A.1. The agreement is seen to be good. The experimental arrangements under which the sample of observations quoted was obtained correspond to a value of the probability aftereffect factor p of 0.613 (from integrating the diffusion expression described above). From Eq. 16A.20 and the sample, the observed value is 0.582. From the theory of errors we expect a fractional variation in possible values of $\sim(114)^{-1/2} \approx 9\%$, and the observed value is well within this limit.

Despite the fact that regression is very much in evidence, theory shows, and observation confirms, that time symmetry is maintained. Let $H(\nu, \nu + k)$ denote the probability that the pair of values $(\nu, \nu + k)$ is observed consecutively. Then it can be shown that

$$\begin{aligned} H(\nu,\nu+k) &= P(\nu)K(\nu|\nu+k) \\ &= P(\nu+k)K(\nu+k|\nu) \\ &= H(\nu+k,\nu), \end{aligned} \tag{16A.21}$$

which is achieved by the rearrangement of factors between the full expressions for $P(\nu)$ and $K(\nu \mid \nu + k)$. This result can be extended to show that the probability of observing the pair (ν, m) at instants separated by $s\tau$ is also symmetrical. Thus, we conclude that the sequence of observations is statistically unaltered by time reversal, and this is borne out fairly well by the sample of observations displayed, considering its small size, as shown in Table 16A.2. Nevertheless, it is still true that if $\nu > \gamma$, the probability of the pair (ν, m) is larger for $m < \nu$ than for $m > \nu$.

Table 16A.2 Number of Occurrences of Pairs (ν, m) for Interval τ, in the Sample Quoted (Top Rows); Calculated Value from Eq. 16A.21 (Bottom Rows)

	m					
ν	**0**	**1**	**2**	**3**	**4**	**5**
0	4	5	4	1		
	4	5	3	1		
1	5	9	8	3	2	
	5	10	8	4	2	
2	2	8	3	14	2	1
	3	8	9	6	3	1
3	1	4	11	1	4	3
	1	4	6	5	3	1
4		2	4	3	1	
		2	3	3	2	1
5		1		2	1	
			1	1	1	

We have seen how a state different from the average tends to regress toward the average, and now come to the question of how soon this state may recur.

Let T_ν be the average lifetime of a state ν, and let Θ_ν be the average recurrence time of that state; T_ν and Θ_ν are related in a simple way. If, in any long time $\mathcal{T}$, the system spends intervals $t_1, t_2, \ldots, t_N$ (N very large) in the state ν, then

$$P(\nu) = \lim_{N\to\infty} \frac{1}{T} \sum_{i=1}^{N} t_i, \tag{16A.22}$$

and

$$T_\nu = \lim_{N\to\infty} \frac{1}{N} \sum_{i=1}^{N} t_i \approx \frac{\mathcal{T}}{N} P(\nu); \qquad \mathcal{T}, N \text{ large.} \tag{16A.23}$$

The time spent not in state ν is

$$\mathcal{T} - \sum_{i=1}^{N} t_i,$$

and therefore,

$$\frac{1}{\mathcal{T}}\left(\mathcal{T} - \sum_{i=1}^{N} t_i\right) = 1 - P(\nu). \tag{16A.24}$$

The average recurrence time of state ν is, therefore, from Eq. 16A.24,

$$\Theta_\nu = \frac{1}{N}\left(\mathcal{T} - \sum_{i=1}^{N} t_i\right),$$

since the state ν has recurred N times. Hence,

$$\Theta_\nu = \frac{1-P(\nu)}{P(\nu)} T_\nu . \qquad (16A.25)$$

We now calculate T_ν following Smoluchowski's work. Since $K(\nu|\nu)$ is the probability that the state ν will be observed on two successive occasions, the probability that the state ν will be observed on $(k-1)$ successive occasions and not on the kth is

$$\phi_\nu(k\tau) = [K(\nu|\nu)]^{k-1}[1-K(\nu|\nu)]. \qquad (16A.26)$$

We can define the average lifetime in a natural way, as

$$\begin{aligned} T_\nu &= \sum_{i=1}^{\infty} k\tau\phi_\nu(k\tau) \\ &= \tau\left[\sum_{k=1}^{\infty} k(K(\nu|\nu))^{k-1}\right][1-K(\nu|\nu)] \\ &= \tau[1-K(\nu|\nu)]^{-1}, \end{aligned} \qquad (16A.27)$$

where we have used

$$\frac{d}{dK}(1-K)^{-1} = (1-K)^{-2} = \sum_{k=1}^{\infty} kK^{k-1}.$$

The observed average lifetimes and recurrence times for the sample of data quoted are shown in Table 16A.3; agreement between theory and experiment is again satisfactory.

So far we have only discussed the case of intermittent observations. In order to study the relation between the recurrence of mechanical and thermodynamic states, we now extend the theory to the case of continuous observations. We must require that the numbers involved be so large that we may ignore the discrete nature of the process and treat it as continuous. We consider this extension in the limit $\tau \to 0$. Then we must expect $p(\tau) \to 0$ and $K(\nu|\nu) \to 1$. From the formula for $K(\nu|\nu)$ we therefore select the leading term ($i = 0$):

$$\begin{aligned} K(\nu|\nu) &= e^{-\gamma p}(1-p)^\nu + \mathcal{O}(p^2) \\ &= 1-(\nu+\gamma)p. \end{aligned} \qquad (16A.28)$$

Hence, the mean lifetime becomes

$$T_\nu = \frac{\tau}{(\nu+\gamma)p}, \qquad (16A.29)$$

and in order that this should tend to a finite limit we must have $p(\tau) \to p_0\tau$ for small τ. This is easily seen to be the case if one considers the example of the number of gas molecules in a given volume. As $\tau \to 0$ we may ignore the effects of collisions and treat the relevant part of the trajectories of the molecules as linear. Then the number leaving the volume element in a time τ is equal to the number striking the surface in that time and is obviously proportional to τ.

In the limit of large numbers the Poisson distribution tends to a Gaussian distribution near its peak:[17]

$$P(\nu) = (2\pi\gamma)^{-1/2} \exp\left[-\frac{(\nu-\gamma)^2}{2\gamma}\right]; \qquad |\nu-\gamma| << \gamma. \qquad (16A.30)$$

Equation 16A.25 then becomes, with Eq. 16A.29,

$$\Theta_\nu \approx \frac{\omega}{\sigma}\left(\frac{m}{k_BT}\right)^{1/2} \gamma^{-1/2} \exp\left[\frac{(\nu-\gamma)^2}{2\gamma}\right], \qquad (16A.31)$$

where ω and σ are the volume and surface area of the cell considered, m is the mass of a gas molecule, and T is the absolute temperature. Values of Θ calculated for a 1% deviation in the density of oxygen in a sphere of radius a at 300 K are shown in Table 16A.4. It is at once clear that for

Table 16A.3 Average Lifetime T_ν and Recurrence Time Θ_ν of the State ν for the Sample Quoted: Observed Values (Top Rows); Calculated Values (Bottom Rows)

ν	T_ν/τ	Θ_ν/τ
0	1.5	11.67
	1.41	9.20
1	1.45	5.20
	1.49	4.06
2	1.125	3.92
	1.42	3.83
3	1.08	4.91
	1.31	5.88
4	1.33	10.
	1.21	11.97
5	1	14.67
	1.13	25.98

[17] This is easily seen by writing

$$\begin{aligned} \ln P(\nu) &= \nu \ln \gamma - \gamma - \ln \nu! \\ &= \nu \ln \gamma - \gamma - (\nu+\tfrac{1}{2}) \ln \nu + \nu - \tfrac{1}{2} \ln 2\pi + \mathcal{O}(\nu^{-1}), \end{aligned}$$

where we have used Stirling's formula for log $\nu!$. If

$$\delta = \nu - \gamma$$

and we restrict ν so that $(\delta/\gamma) << 1$, then

$$\begin{aligned} \ln P(\nu) &= -(\gamma+\delta+\tfrac{1}{2}) \ln\left(1+\frac{\delta}{\gamma}\right) + \delta - \tfrac{1}{2} \ln 2\pi\gamma + \mathcal{O}(\nu^{-1}) \\ &\approx -\frac{\delta^2}{2\gamma} - \tfrac{1}{2} \ln 2\pi\gamma, \end{aligned}$$

whence

$$P(\nu) = (2\pi\gamma)^{-1/2} \exp\left[-\frac{(\nu-\gamma)^2}{2\gamma}\right]; \qquad |\nu-\gamma| << \gamma.$$

Table 16A.4 Recurrence Times for a 1% Deviation in the Density of Oxygen in a Sphere of Radius a at 300 K ($\gamma/\omega = 3 \times 10^{25}$ m^{-3})

a (μm)	10^{-2}	5×10^{-7}	3×10^{-7}	2.5×10^{-7}	10^{-7}
Θ (s)	$10^{10^{14}}$	10^{68}	10^{6}	1	10^{-11}

volumes just at the limit of unaided visual perception, even quite small fluctuations have large recurrence times, so that diffusion is for all intents and purposes an irreversible process. On the other hand, for volumes just at the limit of microscopic vision (~0.25 μm), large fluctuations occur rapidly, and there is no question of irreversibility in the usual sense. Nevertheless, we have seen from a study of such fluctuations that the results are, on average, in accordance with the macroscopic theory of diffusion.

In this appendix we have been concerned with only one gross variable, the number density. However small the number of particles considered, it has been possible to discern both the time symmetry of the sequence of occupation numbers and the regression toward the mean. If the initial state is sufficiently far from the mean, then its recurrence time is extremely long. Alternatively, the recurrence time of a fixed fractional deviation from the mean becomes extremely long as the mean itself increases. The case of the number of gas molecules within a spherical volume discussed above is a striking example of this. It is clear, then, that on the gross level of ordinary observation, when the numbers of particles considered are always large enough that local thermodynamic variables may be defined, a significant fluctuation has so long a recurrence time that for practical purposes it decays irreversibly to equilibrium. Given an ensemble of replica systems, in each of which a prescribed volume element initially contains, say, an excess over the average number of particles, a distribution function for the number of particles instantaneously in the volume can be defined. This distribution function approaches its equilibrium form in a time of the order of that taken by a typical fluctuation to decay. The distribution function remains in its equilibrium form, and does not revert to its initial form if the system is undisturbed. The probability of a fluctuation the size of the initial value, as calculated from the equilibrium distribution function, is then related directly to its recurrence time. We then say that the distribution function has approached equilibrium irreversibly.[18]

• FURTHER READING

Beattie, J. A., and Oppenheim, I., *Principles of Thermodynamics* (Elsevier Scientific Publishing Co., Amsterdam, 1979), Chapter 8.

Epstein, P. S., *Textbook of Thermodynamics* (Wiley, New York, 1937), Chapter 4.

Kestin, J., *A Course in Thermodynamics,* Vol. 1 (Blaisdell, New York, 1966), Chapters 9 and 10.

Kirkwood, J. G., and Oppenheim, I., *Chemical Thermodynamics* (McGraw Hill Book Co., New York, 1961), Chapter 4.

Pippard, A. B., *Classical Thermodynamics* (Cambridge University Press, London, 1957), Chapter 4.

Reiss, H., *Methods of Thermodynamics* (Blaisdell, New York, 1965), Chapter 4.

Some texts dealing with the role of dynamical chaos in the evolution of a system, with more detail than presented in Section 16.3, are:

Casati, G., and Chirikov, B., *Quantum Chaos* (Cambridge University Press, Cambridge, 1995), a collection of papers dealing with general aspects of, and the existence of, quantum chaos.

Gaspard, P., *Chaos, Scattering and Statistical Mechanics* (Oxford University Press, Oxford, 1998).

Ozorio de Almeida, A. M., *Hamiltonian Systems:Chaos and Quantization* (Cambridge University Press, Cambridge, 1988).

Zaslavsky, G. M., *Chaos in Dynamical Systems* (Harwood Academic Publishers, New York, 1985), Chapters 1 and 2 (classical systems), and 9, 10 and 12 (quantum systems).

• PROBLEMS

1. Consider a liquid (at temperature T) in an insulated container. Let the depth of the liquid be d and its density ρ. Above the liquid, at a height h, is suspended a rock of volume v_R and mass M_R. The rock is suddenly dropped into the liquid through a small hole in the top of the insulating container. The rock comes to rest at the bottom of the container. Assume that the mass of the liquid is very much larger than the mass of the rock. What is the entropy change of the liquid?

2. A hot meteorite falls (velocity 200 km/h) into the Atlantic Ocean. The meteorite was originally at a temperature of 1000°C, weighed 1000 g, and had heat capacity of 0.82 J/Kg. If the ocean temperature is 15°C, calculate the change in entropy of the universe as a result of this process.

3. An electric current of 10 A is maintained for 1 s in a resistor of 25 Ω while the temperature of the resistor is kept constant at 27°C.
(a) What is the entropy change of the resistor?
(b) What is the entropy change of the resistor plus surroundings?
The same current is maintained for the same time in the same resistor, but now it is thermally insulated. If the resistor has a mass of 10 g and $C_p = 0.836$ J/Kg,
(c) What is the entropy change of the resistor?
(d) What is the entropy change of the resistor plus surroundings?

4. A mass of some conducting substance is placed in an evacuated container. The walls of the container are

[18] This does not mean that the system is not reversible, in the sense that reversal of all velocities at any time t after the initial ensemble is set up at 0 leads again to the initial state at $2t$.

maintained at the constant temperature 500°C. Show that if the conducting substance is heated from 50 to 150°C by passage of an electric current from an external generator, the process undergone is irreversible.

5. A mass of steam at 100°C is irreversibly condensed by contact with a heat sink at temperature T_0. In this process the latent heat of vaporization is extracted from the steam, and its entropy decreased. Does this process violate the second law of thermodynamics? Why?

6. Take a gas with equation of state

$$p(v-b)=R\theta,$$

with c_V a function of θ only, through a Carnot cycle, and show that

$$T=\theta,$$

where T is the thermodynamic temperature and θ the empirical gas scale temperature.

7. The coefficient of volume expansion of water is negative for $0°\text{C} < t < 4°\text{C}$. Show that over this temperature interval water is cooled by adiabatic compression. Compare this result with the adiabatic compression of a perfect gas.

8. Consider an adiabatic process and an isothermal process, both represented in pV space. Under what conditions would the adiabatic path coincide with the isothermal path?

9. By considering water to be the working fluid in a Carnot cycle, show that the temperature $t = 4°\text{C}$ cannot be reached by cooling (heating) in an adiabatic expansion. Assume the temperature at which the density of water is a maximum is independent of the pressure.

10. Consider the first-law expression for a change in internal energy of a simple fluid:

$$dU = đq - p\,dV.$$

Along a constant-volume path $(đq)_V = C_V\,dT$. Along a path with $đq = 0$ we have $dU = -p\,dV$. Thus, we can write

$$\Delta U = \int_{V_1T_1}^{V_1T_2} C_V(V_1,T)\,dT - \left(\int p_{\text{app}}\,dV\right)_{đq=0},$$

but we also know that

$$\Delta U = \int_{V_1T_1}^{V_1T_2} C_V(V_1,T)\,dT + \int_{V_1T_2}^{V_2T_2}\left(\frac{\partial U}{\partial V}\right)_{T_2} dV.$$

Explain the differences in these two expressions for ΔU. If the isochoric (constant-volume) heating steps are the same, can the final states reached be the same? How is your answer related to the Caratheodory statement of the second law of thermodynamics?

11. If $p\,dV$ were an exact differential for a given substance, could that substance be used as the working fluid in the cylinder of a heat engine undergoing a cyclic process? Explain your answer.

12. Consider a gas for which $U(T)$ is a monotonic increasing function of T, and is independent of V and p. Let this gas be reversibly isothermally compressed. Show that the final state reached in this compression cannot be reached by a reversible adiabatic path starting at the initial state.

13. (a) A sample of gas containing n mol is heated reversibly from T_1 to T_2 at constant volume. Another sample of n mol of the same gas is heated reversibly from T_1 to T_2 at constant pressure. For which process is ΔS larger? Why?
(b) A sample of n mol of Ar is heated reversibly from T_1 to T_2 at constant volume. A sample of n mol of Br_2 is heated reversibly from T_1 to T_2 at constant volume. For which gas is ΔS larger? Why?
(c) A sample of gas containing n mol is expanded from (p_1, V_1) to (p_2, V_2) in a reversible isothermal process. Another sample of n mol of the same gas is expanded from (p_1, V_1) to (p_2, V_2') in a reversible isentropic process. Which is the larger, V_2 or V_2'? Why? Draw a graph illustrating your answer.

14. Consider the change of state of a perfect gas $(p_1, T_1) \rightarrow (p_2, T_2)$. Devise two or more reversible paths between states 1 and 2 and show that the heat absorbed by the gas is different for each path, but that the entropy change is the same.

15. Consider 1 mol of a perfect gas which, at 300 K, occupies a volume V_1 = 10 liters. Suppose that this gas undergoes an isothermal reversible expansion to a volume V_2 = 100 liters.
(a) Calculate the change in entropy of the gas.
(b) Calculate the change in entropy of the surroundings. Demonstrate, by an explicit calculation, that if the isothermal change in volume $V_1 \rightarrow V_2$ is effected by expansion against zero pressure, the expansion is irreversible.

16. Describe qualitatively reversible paths by means of which changes in entropy in the following processes could be calculated.
(a) Superheated steam at 150°C and 2 atm pressure is brought in contact with ice at 0°C and 1 atm pres-

sure, the quantities being chosen so that the final state of the system is only liquid water at 50°C and 1 atm.

(b) A chunk of iron at 500°C is dropped into liquid water at 100°C and 1 atm, steam being evolved at 100°C and 1 atm.

17. Prove that when heat is added to a single phase of any substance, the temperature must rise.

18. Let a closed system be specified in such a fashion that it can only undergo reversible isothermal changes. Show that q for any such change is independent of path.

19. Prove that an isothermal path cannot intersect a reversible adiabatic path twice.

20. Consider a Carnot cycle with 1 mol of a monatomic ideal gas as the working fluid. Calculate w, q, ΔU, and ΔS for each step of the cycle as well as for the entire cycle. Assume that the cycle operates between temperatures T_1 and T_2 and that the volumes at points A, B, C, and D are V_A, V_B, V_C, and V_D, respectively (see Fig. 16.8).

21. Repeat the above problem with 1 mol of the gas described in Problem 12 of Chapter 14 as the working fluid.

22. Calculate the entropy change of 3 mol of an ideal liquid when it is isothermally expanded from 10 liters to 100 liters. By what factor does Ω change during the expansion?

23. In the Carnot cycle shown in Fig. 16.8, heat q_2 is absorbed in the isothermal expansion $A \rightarrow B$ and heat q_1 is absorbed in the isothermal compression $C \rightarrow D$. Is it possible for q_1 to be zero? How is your answer related to the Kelvin statement of the second law?

24. Consider an isolated system consisting of two subvolumes containing perfect gases separated by a frictionless piston (see diagram). As in Problem 25 of

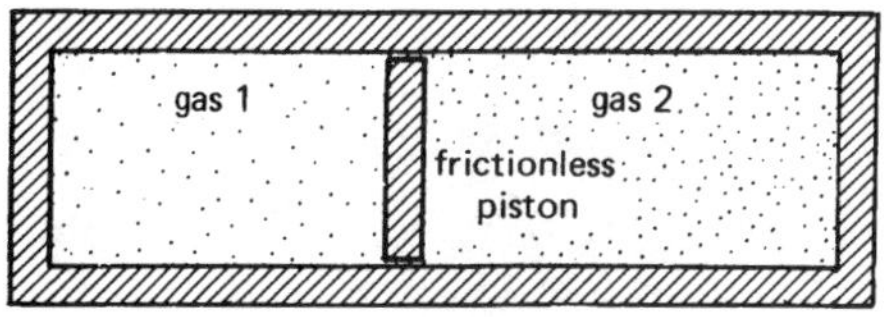

Chapter 13, we imagine that the molecules of gases 1 and 2 have, respectively, the constant speeds v_1 and v_2 and can move only in the direction perpendicular to the piston faces. Because of collisions between the molecules of the two gases and the piston, the piston jiggles back and forth, and thereby transmits energy between the two gases. Suppose that in the initial state the temperatures of gases 1 and 2 are not equal, so that $m_1 v_1^2/2 \neq m_2 v_2^2/2$. Further, assume that any imbalance in forces on the two sides of the piston is so small that the average velocity of the piston, v_p, is sensibly zero, and that any change in kinetic energy of the piston resulting from its motion is vanishingly small. Show that the rate of gain of kinetic energy by gas 1 is

$$\frac{d}{dt}(KE) = \frac{4 Z_{c_1} Z_{c_2} m_1 m_2}{M(Z_{c_1} m_1 + Z_{c_2} m_2)} \left(\tfrac{1}{2} m_2 v_2^2 - \tfrac{1}{2} m_1 v_1^2 \right),$$

where Z_{c_1} and Z_{c_2} are the numbers of collisions per second of all the molecules of gases 1 and 2 with the piston, respectively.

25. The result of the preceding problem shows that the direction of energy flow is from the gas with greater kinetic energy per particle to that with lower kinetic energy per particle. This is, of course, the expected macroscopic result: Heat flows from higher to lower temperature. However, the result in Problem 24 was obtained from a microscopic model using a time-reversible equation of motion. Explain how an assumption used in the model leads to the result obtained, and what that assumption means.

CHAPTER 17

Some Applications of the Second Law of Thermodynamics

In this chapter we examine some more consequences of the second law of thermodynamics by computing the entropy changes accompanying a variety of processes. For simplicity we restrict our attention to the case of a one-component closed system, the state of which is defined by specification of any two of the three variables p, V, T.

17.1 Choice of Independent Variables

The combined first and second laws require that the differential change in internal energy be

$$dU = T\,dS - p\,dV. \tag{17.1}$$

Equation 17.1 is valid for both reversible and irreversible processes since it relates changes in the state functions S, U, and V. The evaluation of dS for an irreversible process requires that we find (invent) a reversible path between the initial and final states, but dS is an exact differential, and the relationship expressed in Eq. 17.1 is unaffected by the nature of the process.

Equation 17.1 suggests that interpretation of the energy change in a process is particularly straightforward if the change can be effected by altering the volume with the entropy held fixed, and then altering the entropy with the volume held fixed. Then, for fixed S the slope of U versus V is just the pressure, and for fixed V the slope of U versus S is just the temperature. The simplicity of this interpretation of the rates of change of U with respect to changes of S and V, that is, their identification with variables from the set p, V, T, leads one to classify S and V as the "natural" independent variables for the function U. For any thermodynamic state function, we define the natural variables as those that allow the determination of *all* other thermodynamic properties by differentiation. This definition provides the motivation to introduce a transformation of variables such that a function $y(x)$ of the independent variable x is rewritten as a function in which the derivative of $y(x)$ at x is the independent variable.

Why should one want to find the natural variables of a thermodynamic state function? Laboratory experiments are usually conducted under conditions that correspond to holding p and T fixed, or sometimes to holding V and T fixed. It is certainly possible to compute the change in U with respect to changes in p and T, or with respect to changes in other pairs of independent variables. However the resultant forms are more complicated than Eq. 17.1 in the sense that the coefficients multiplying the changes in independent variables are not simply the derivatives of U with respect to the variables chosen; rather, they are combinations of functions related to properties of the system. For example, choosing T and V as independent variables for U leads to

$$dU = \left(\frac{\partial U}{\partial T}\right)_V dT + \left(\frac{\partial U}{\partial V}\right)_T dV \tag{17.2}$$

$$= C_V\,dT + \left[T\left(\frac{\partial S}{\partial V}\right) - p\right]_T dV, \tag{17.3}$$

since

$$\left(\frac{\partial U}{\partial V}\right)_T = T\left(\frac{\partial S}{\partial V}\right)_T - p \tag{17.4}$$

from Eq. 17.1. It will be shown soon (see Eqs. 17.16 and 17.17) that $(\partial S/\partial V)_T = (\partial p/\partial T)_V$, so that the coefficient of dV in Eq. 17.3 can be expressed in terms of the measurable quantities p, V, T. Even so, the rate of change of U with respect to changes in V is determined by the balance between p and $T(\partial p/\partial T)_V$, which is clearly not as "simple" as when S and V are chosen as the independent variables.

This illustration provides motivation for the introduction of new thermodynamic functions defined so as to simplify the calculation of heat and work exchanges for each particular choice of independent variables. We shall later see that these new thermodynamic functions have another important property, namely, they also play the role of potential functions for the transitions among equilibrium states defined by constraints imposed by different independent variables.

A general method for rewriting a function of one set of independent variables as an equivalent function of other independent variables is the *Legendre transformation,* which we now describe for the case of one variable. Figure 17.1*a* shows a function $y(x)$ plotted with respect to the variable x. Suppose that, for some reason, we wish to use as independent variable not x but rather the slope of $y(x)$ at x, namely, $y'(x)$. Complete specification of the tangent line $y'(x)$ requires knowledge of its intercept with the y axis, say, $i[y'(x)]$, since both slope and intercept are required to define a straight line uniquely. As shown in Fig. 17.1*b*, if the slope and intercept are known at every point of $y(x)$, the information available is equivalent to knowledge of $y(x)$. Therefore, the curve $y(x)$ in Fig. 17.1 can also described by

$$i(y') = y - xy'(x), \tag{17.5}$$

since

$$y'(x) = \frac{y - i(y')}{x}. \tag{17.6}$$

The function $i(y')$ is the Legendre transformation of y; $i(y')$ is equivalent to $y(x)$ but treats y' as the independent variable in place of x. The generalization of Eq. 17.5 to the case that there is more than one independent variable is straightforward, the result being

$$i(y') = y - \sum_{j=1}^{n} \left(\frac{\partial y}{\partial x_j} \right)_{x_i \neq x_j} x_j \,, \tag{17.7}$$

where the x_j's are the independent variables defining the function $y(x_1, \ldots, x_n)$.

We now apply the Legendre transformation to convert $U(S, V)$ and $H(S, p)$ to equivalent functions in terms of other pairs of independent variables. Consider again choosing T and V as the independent variables. We seek a Legendre transformation of $U(S, V)$ to an equivalent function. Following Eq. 17.5, we find that the required function is

$$\begin{aligned} A(T,V) &= U - \left(\frac{\partial U}{\partial S} \right)_V S \\ &= U - TS, \end{aligned} \tag{17.8}$$

using

$$T = \left(\frac{\partial U}{\partial S} \right)_V \tag{17.9}$$

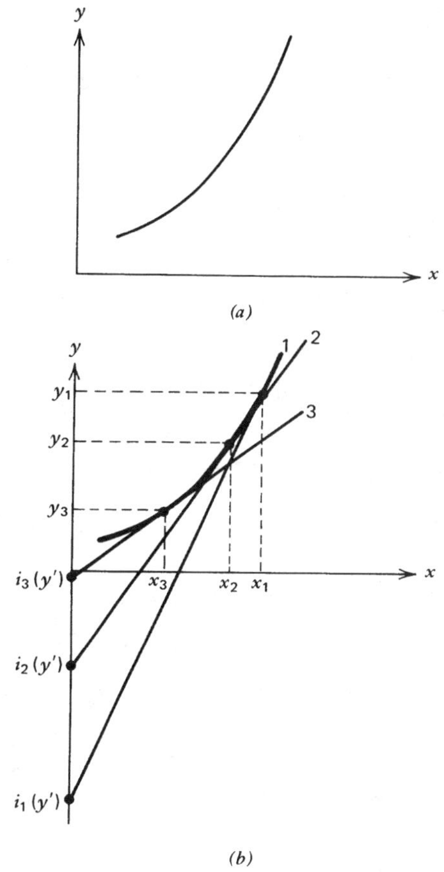

Figure 17.1 Diagram illustrating a Legendre transformation (see text).

from Eq. 17.1. The quantity $A(T, V)$ is known as the *Helmholtz free energy;* the reason for the name "free energy" will be made apparent in the next section. Next consider choosing T and p as the independent variables. In this case we make a Legendre transformation of $H(S, p)$. Recall that $H \equiv U + pV$. Again following Eq. 17.5, we find that the required function is

$$\begin{aligned} G(T,p) &= H - \left(\frac{\partial H}{\partial S} \right)_p S \\ &= H - TS, \end{aligned} \tag{17.10}$$

using

$$T = \left(\frac{\partial H}{\partial S} \right)_p \tag{17.11}$$

from the differential relation

$$\begin{aligned} dH &= dU + p\,dV + V\,dp \\ &= T\,dS + V\,dp. \end{aligned} \tag{17.12}$$

The quantity $G(T, p)$ is known as the *Gibbs free energy.* Both $A(T, V)$ and $G(T, p)$ are, obviously, functions of state.

In summary, for any process described in terms of:

1. Changes in the independent variables S and V,

$$dU = T\,dS - p\,dV. \tag{17.1}$$

$$\left(\frac{\partial T}{\partial V}\right)_S = -\left(\frac{\partial p}{\partial S}\right)_V = \left(\frac{\partial^2 U}{\partial S\,\partial V}\right); \tag{17.13}$$

2. Changes in the independent variables S and p,

$$dH = T\,dS + V\,dp, \tag{17.14}$$

$$\left(\frac{\partial T}{\partial p}\right)_S = \left(\frac{\partial V}{\partial S}\right)_p = \left(\frac{\partial^2 H}{\partial S\,\partial p}\right); \tag{17.15}$$

3. Changes in the independent variables T and V,

$$dA = dU - T\,dS - S\,dT = -S\,dT - p\,dV, \tag{17.16}$$

$$\left(\frac{\partial S}{\partial V}\right)_T = \left(\frac{\partial p}{\partial T}\right)_V = \left(\frac{\partial^2 A}{\partial V\,\partial T}\right); \tag{17.17}$$

4. Changes in the independent variables T and p,

$$dG = dH - T\,dS - S\,dT = -S\,dT + V\,dp, \tag{17.18}$$

$$-\left(\frac{\partial S}{\partial p}\right)_T = \left(\frac{\partial V}{\partial T}\right)_p = \left(\frac{\partial^2 G}{\partial p\,\partial T}\right). \tag{17.19}$$

Equations 17.13, 17.15, 17.17, and 17.19 are known as *Maxwell's relations.* They play an important role in practical thermodynamics, since they relate quantities of interest to the measurable variables p, V, T.

17.2 The Available Work Concept

We showed, in Chapter 13, that the work done in a process depends on the path used to connect the initial and final states. There are some special cases for which the work done depends only on the change in state; for example, in an adiabatic work process, $w = \Delta U$. We shall now discuss the limitations imposed by the second law on the conversion of heat into work.

Consider a closed system in contact with a single heat reservoir whose temperature is T_R. Suppose that the system undergoes the change of state $1 \rightarrow 2$. What is the maximum work that can be accomplished by the system? Other systems in the surroundings can be used to assist the change $1 \rightarrow 2$ provided that they are carried around cyclic processes, hence are in the same states before and after $1 \rightarrow 2$.

For any process $1 \rightarrow 2$ in the system, the first law requires that the work done and the internal energy change of the system be related by

$$w = \Delta U - q, \tag{17.20a}$$

or in differential form that

$$đw = dU - đq \tag{17.20b}$$

for each infinitesimal segment of the path $1 \rightarrow 2$. Consider the case that the heat transfer to the system in any infinitesimal segment of the path, $đq$, is reversible. To make this possible, we arrange to interpose a reversible engine between the system and the reservoir, so that the heat transferred to the system comes from the engine, and only indirectly from the reservoir. This engine works between the temperatures T_R and T, where T is the temperature along that part of the path having heat transfer $đq$. Because heat is transferred via an engine, work is done by the engine on the system during the transfer. For the step in which heat $đq$ is transferred to the system at temperature T, the work done by the reversible engine is

$$đw_R = \left(\frac{T - T_R}{T}\right) đq. \tag{17.21}$$

For a reversible transfer of heat the entropy change of the system is $dS = đq/T$, since T is the temperature of the heat source. In general, if the engine supplies heat at T, then for the system

$$dS \geq \frac{đq}{T}. \tag{17.22a}$$

Suppose a different engine is used for each infinitesimal segment of the path $1 \rightarrow 2$. Then

$$\Delta S \geq \int_1^2 \frac{đq}{T}, \tag{17.22b}$$

where T is the heat source temperature, and differs from T_R by virtue of the engine used to transfer the heat. Using the fact that T_R is a constant, we multiply 17.22*b* by T_R and rearrange to find

$$T_R\,\Delta S - T_R \int_1^2 \frac{đq}{T} \geq 0, \tag{17.23}$$

which, after combination with 17.20*a*, assumes the form

$$w \geq \Delta U - T_R \Delta S - \int_1^2 \left(\frac{T - T_R}{T}\right) đq. \tag{17.24}$$

Note that the total work done on the system by the reversible engines that altogether deliver heat q to the system is

$$w_R = \int_1^2 \left(\frac{T - T_R}{T}\right) đq, \tag{17.25}$$

so that the total work done by the system and by the engines that serve to transfer heat to the system when it undergoes the change of state $1 \to 2$ is

$$w + w_R \geq \Delta U - T_R \Delta S. \tag{17.26}$$

If all the processes involved are reversible, the equality sign holds in 17.26.

Recall that ΔU and ΔS depend only on the system's change of state, $1 \to 2$, and that T_R is fixed. Then if all the processes involved are reversible, $w + w_R$ is the maximum possible work that can be done by the system and engines for the system change of state $1 \to 2$:

$$w_{\max} = \Delta U - T_R \Delta S. \tag{17.27}$$

Since the right-hand side of Eq. 17.27 defines a minimum, why is the left-hand side labeled $w_{\max}$? The point is that work done by the system is a negative quantity, so Eq. 17.27 defines the most negative amount of work done for the system change of state $1 \to 2$, and this is just the maximum amount of work done by the system. Moreover, all reversible processes that bring about the same change of state of the system must produce the same maximum work, since $w_{\max}$ depends only on T_R, ΔU, and ΔS. It must be kept in mind that $w_{\max}$ is a composite work, since it includes the work done by the reversible Carnot engines that effect the heat transfer from the reservoir to the system.

Equation 17.27 can be used to relate $w_{\max}$ to changes in the Helmholtz and Gibbs free energies of a system. Indeed, the following cases are of particular interest.

1. Suppose that the volume of the system is held fixed, hence it cannot do any expansion work. The maximum nonexpansion work that can be done, $w_{x,\max}$, is, from Eq. 17.27,

$$w_{x,\max} = \Delta U - T_R \Delta S; \qquad V \text{ fixed.} \tag{17.28}$$

2. Suppose that the pressure of the system is held fixed, so the expansion work it does is $-p\Delta V$. The maximum nonexpansion work that can be done is, from Eq. 17.27,

$$\begin{aligned} w_{x,\max} &= \Delta U - p\Delta V - T_R \Delta S \\ &= \Delta H - T_R \Delta S; \qquad p \text{ fixed.} \end{aligned} \tag{17.29}$$

3. In cases 1 and 2 no restriction was made concerning the initial and final temperatures of the system. Suppose that the initial and final temperatures of the system are equal to T_R. Then Eq. 17.28 becomes

$$\begin{aligned} &w_{x,\max} = \Delta A; \\ &V \text{ fixed, initial and final temperatures } T_R, \end{aligned} \tag{17.30}$$

and Eq. 17.29 becomes

$$\begin{aligned} &w_{x,\max} = \Delta G; \\ &p \text{ fixed, initial and final temperatures } T_R. \end{aligned} \tag{17.31}$$

In both Eqs. 17.30 and 17.31 the temperature of the system must be T_R in states 1 and 2, but can vary along the path $1 \to 2$.

The reason for the nomenclature "free energy" can now be made clear. Although in an adiabatic process the entire energy change in a system is converted to work, under other conditions less than the entire energy change can be converted to work. The amount *unavailable* for conversion is just $T_R\Delta S$ for a change ΔS in the system's entropy. The differences, $\Delta U - T_R\Delta S$ or $\Delta H - T_R\Delta S$, are then "free energies," in the sense that if all processes used are reversible and the initial and final temperatures of the system are T_R, complete conversion of the free energies to work is possible. Therefore, $w_{x,\max}$ can be thought of as the amount of work available from a reversible change of state in a system coupled to reversible changes in its surroundings. As indicated in the title of this section, it is sometimes called the *available work* or *availability.*

17.3 Entropy Changes in Reversible Processes

We shall now examine the nature of the temperature dependence of the entropy of a one-component system and then, using this information, we shall calculate the entropy change accompanying a general change of thermodynamic state $1 \to 2$.

It is convenient to begin with the combined first and second laws for an infinitesimal general process, namely, Eq. 17.1. From Eq. 17.1 we find

$$\left(\frac{\partial U}{\partial T}\right)_V \equiv C_V = T\left(\frac{\partial S}{\partial T}\right)_V; \tag{17.32}$$

hence C_V/T measures the rate of change of entropy with temperature in a constant-volume process. Suppose, now, that a one-component system undergoes the change of state $1 \to 2$. Suppose, further, that the state of the system is fully defined by any two of the three variables p, V, T. Since the entropy is a function of state, all paths between states 1 and

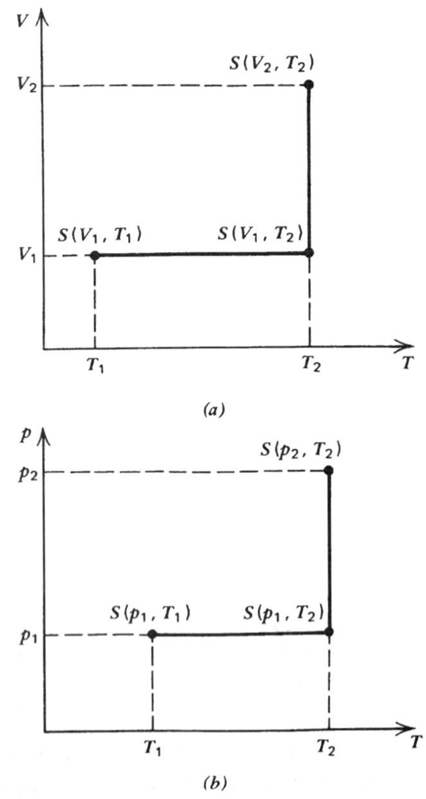

Figure 17.2 Paths along which the entropy change corresponding to the changes in state (*a*) $(V_1, T_1) \to (V_2, T_2)$ (*b*) $(p_1, T_1) \to (p_2, T_2)$ may be computed.

2 lead to the same entropy change. Consider first a path consisting of an isochoric temperature change followed by an isothermal volume change (see Fig. 17.2*a*). Along this path we regard the entropy as a function of the independent variables V and T. Thus, a differential change in the entropy can be written as

$$dS = \left(\frac{\partial S}{\partial T}\right)_V dT + \left(\frac{\partial S}{\partial V}\right)_T dV \tag{17.33}$$

or

$$dS = \frac{C_V}{T} dT + \left(\frac{\partial S}{\partial V}\right)_T dV. \tag{17.34}$$

But, as shown in Eq. 17.17,

$$\left(\frac{\partial S}{\partial V}\right)_T = \left(\frac{\partial p}{\partial T}\right)_V, \tag{17.17}$$

so that we can put Eq. 17.34 in the form

$$dS = \frac{C_V}{T} dT + \left(\frac{\partial p}{\partial T}\right)_V dV. \tag{17.35}$$

We then find the entropy difference between states 1 and 2, calculated along the path described, to be

$$\Delta S \equiv S_2 - S_1 = \int_{V_1,T_1}^{V_1,T_2} \frac{C_V}{T} dT + \int_{T_2,V_1}^{T_2,V_2} \left(\frac{\partial p}{\partial T}\right)_V dV. \tag{17.36}$$

The advantage of the form displayed in Eq. 17.36 is that it permits calculation of the entropy change accompanying the process $1 \to 2$ in terms of the equation of state [from which we find $(\partial p/\partial T)_V$] and the heat capacity as a function of T at some one volume, that is, in terms of directly measured quantities. Provided that the change of state is always in a one-phase region of the substance, Eq. 17.36 is a general result applicable to all one-component systems.

The general result embodied in Eq. 17.36 is made more meaningful when we consider the simple case of the perfect gas. In this instance we obtain

$$\left(\frac{\partial p}{\partial T}\right)_V = \left[\frac{\partial}{\partial T}\left(\frac{nRT}{V}\right)\right]_V = \frac{nR}{V}. \tag{17.37}$$

Now assume that the heat capacity per mole at constant volume, C_V, is independent of temperature (true for a monatomic gas at temperatures low enough that there is no electronic excitation), and insert Eq. 17.37 into 17.36. The result is, for n mol of perfect gas,

$$\Delta S_{\text{perfect gas}} = nc_V \ln \frac{T_2}{T_1} + nR \ln \frac{V_2}{V_1}. \tag{17.38}$$

It is important to observe how the properties of the equation of state are used to reduce the general equation 17.36 to the specific equation 17.38. Note also how the entropy increases, in general, as the system's temperature and/or volume increases.

Referring back to the discussion of Section 15.4, we easily recognize that in the isothermal expansion of a perfect gas $(\Omega_2/\Omega_1) > 1$, so that the corresponding entropy change should be positive. When the energy of the perfect gas is increased at constant volume we also have $(\Omega_2/\Omega_1) > 1$, and therefore an entropy increase. Equation 17.38 represents the quantitative macroscopic description of the entropy changes accompanying these very processes.

In Chapter 14 we showed that, corresponding to the same change of state $1 \to 2$, the internal energy change is

$$\Delta U \equiv U_2 - U_1 = \int_{V_1,T_1}^{V_1,T_2} C_V \, dT + \int_{V_1,T_2}^{V_2,T_2} \left(\frac{\partial U}{\partial V}\right)_{T_2} dV \tag{17.39}$$

or, using Eqs. 17.3 and 17.17,

$$\Delta U = \int_{V_1,T_1}^{V_1,T_2} C_V \, dT + \int_{V_1,T_2}^{V_2,T_2} \left[T\left(\frac{\partial p}{\partial T}\right)_V - p\right] dV. \tag{17.40}$$

As might be expected, Eq. 17.40 takes a particularly simple form for the perfect gas. Then we have

$$T\left(\frac{\partial p}{\partial T}\right)_V - p = p - p = 0, \tag{17.41}$$

whereupon, for a perfect gas,

$$\Delta U_{\text{perfect gas}} = \int_{V_1,T_1}^{V_1,T_2} C_V\, dT = C_V\,(T_2 - T_1). \tag{17.42}$$

The general formula for the energy change, Eq. 17.40, is valid for any one-component, one-phase system characterized by *p, V, T*. In contrast, Eqs. 17.41 and 17.42 are merely statements of internal consistency. It was stated in Chapter 16 that the specification of the properties of a perfect gas requires, as a supplement to the equation of state, the condition

$$\left(\frac{\partial U}{\partial V}\right)_{T,\text{ perfect gas}} = 0. \tag{17.43}$$

In fact, Eq. 17.43 is a definition necessary for the identification $T_{\text{absolute}} = T_{\text{perfect gas}}$. Equations 17.41 and 17.42 are merely reexpressions of this condition. Stated again, the only kind of energy a perfect gas has is the kinetic energy of the constituent molecules, so that U can only be a function of the temperature and the number of molecules in the sample of gas.

The entropy change represented by Eq. 17.36 displays the result of choosing V and T as the independent variables. Of course, we could have chosen p and T as the independent variables; depending on the experimental arrangement, the choice of one or the other of the pairs (V, T) or (p, T) may lead to greater convenience in interpretation of measurements. If p and T are chosen as the independent variables describing the state of the system and we consider an isobaric temperature change from T_1 to T_2, followed by an isothermal pressure change from p_1 to p_2, as in Fig. 17.2*b*, Eq. 17.33 is replaced by

$$\begin{aligned} dS &= \left(\frac{\partial S}{\partial T}\right)_p dT + \left(\frac{\partial S}{\partial p}\right)_T dp \\ &= \frac{C_p}{T}\, dT + \left(\frac{\partial S}{\partial P}\right)_T dp. \end{aligned} \tag{17.44}$$

Here the relation

$$C_p = T\left(\frac{\partial S}{\partial T}\right)_P \tag{17.45}$$

follows from the definition of the enthalpy. The analog of Eq. 17.36 therefore becomes, using Eq. 17.19,

$$\Delta S \equiv S_2 - S_1 = \int_{p_1,T_1}^{p_1,T_2} \frac{C_p}{T}\, dT - \int_{p_1,T_2}^{p_2,T_2} \left(\frac{\partial V}{\partial T}\right)_p dp. \tag{17.46}$$

Provided that the change of state is always in a one-phase region of the substance, Eq. 17.46 is also a general result.[1]

Again, the simple case of the perfect gas is of interest. In this case, for n mol of gas, we have

$$\left(\frac{\partial V}{\partial T}\right)_p = \left[\frac{\partial}{\partial T}\left(\frac{nRT}{p}\right)\right]_p = \frac{nR}{p}. \tag{17.47}$$

Proceeding as before, if we assume C_p to be a constant independent of the temperature (true for monatomic gases under the same conditions that C_V is constant), Eq. 17.47 takes the form

$$\Delta S_{\text{perfect gas}} = nc_p \ln\frac{T_2}{T_1} - nR\ln\frac{p_2}{p_1}. \tag{17.48}$$

Note how the entropy change increases as the ratio of final to initial pressures decreases.

Finally, to round out our arguments, we consider the calculation of the enthalpy change accompanying the change of state of the system under examination. Starting from Eq. 17.12, the analog of Eq. 17.39 is found to be

$$\Delta H \equiv H_2 - H_1 = \int_{p_1,T_1}^{p_1,T_2} C_p\, dT + \int_{p_1,T_2}^{p_2,T_2} \left(\frac{\partial H}{\partial p}\right)_T dp \tag{17.49}$$

---

[1] As is to be expected from their definitions, the difference between C_p and C_V for a given substance depends on the equation of state of that substance. A general relation for $C_p - C_V$ is easily obtained as follows. By definition,

$$C_p \equiv \left(\frac{\partial H}{\partial T}\right)_p = \left(\frac{\partial U}{\partial T}\right)_p + p\left(\frac{\partial V}{\partial T}\right)_p.$$

To evaluate $(\partial U/\partial T)_p$, we write

$$dU = \left(\frac{\partial U}{\partial T}\right)_V dT + \left(\frac{\partial U}{\partial V}\right)_T dV,$$

from which we find

$$\left(\frac{\partial U}{\partial T}\right)_p = C_V + \left(\frac{\partial U}{\partial V}\right)_T \left(\frac{\partial V}{\partial T}\right)_p,$$

or

$$C_p - C_V = \left[p + \left(\frac{\partial U}{\partial V}\right)_T\right]\left(\frac{\partial V}{\partial T}\right)_p.$$

But from Eqs. 17.1 and 17.17,

$$\left(\frac{\partial U}{\partial V}\right)_T = T\left(\frac{\partial S}{\partial V}\right)_T - p = T\left(\frac{\partial p}{\partial T}\right)_p - p,$$

so that

$$C_p - C_V = T\left(\frac{\partial p}{\partial T}\right)_V \left(\frac{\partial V}{\partial T}\right)_p = -T\frac{(\partial V/\partial T)_p}{(\partial V/\partial p)_T} = \frac{\alpha^2 VT}{\kappa_T},$$

where we have used the identity $(\partial p/\partial T)_V(\partial V/\partial p)_T(\partial T/\partial V)_p = -1$ and the definitions $\alpha \equiv (1/V)(\partial V/\partial T)_p$ and $\kappa_T \equiv -(1/V)(\partial V/\partial p)_T$.

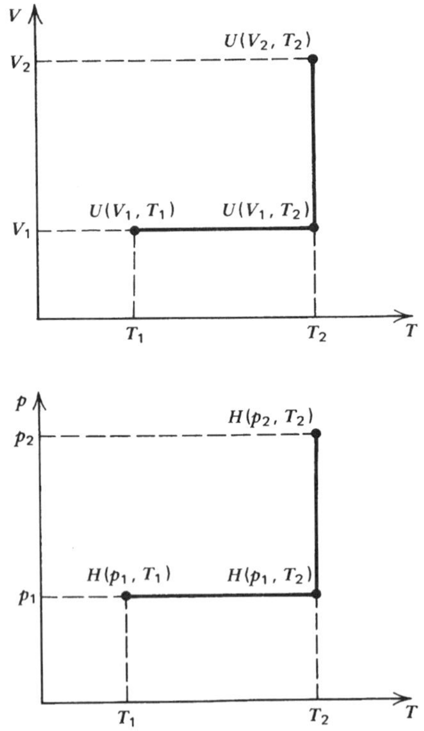

Figure 17.3 Paths along which the internal energy and enthalpy changes corresponding to the changes of state $(V_1, T_1) \to (V_2, T_2)$ and $(p_1, T_1) \to (p_2, T_2)$ may be computed.

(see the second part of Fig. 17.3). Of course, we must express the coefficient $(\partial H/\partial p)_T$ in terms of the equation of state of the substance. From Eqs. 17.12 and 17.19,

$$\left(\frac{\partial H}{\partial p}\right)_T = T\left(\frac{\partial S}{\partial p}\right)_T + V$$
$$= -T\left(\frac{\partial V}{\partial T}\right)_p + V, \qquad (17.50)$$

a result we anticipated in Chapter 14. Finally, we obtain for ΔH the relation

$$\Delta H = \int_{p_1,T_1}^{p_1,T_2} C_p\, dT + \int_{p_1,T_2}^{p_2,T_2} \left[V - T\left(\frac{\partial V}{\partial T}\right)_p\right] dp. \qquad (17.51)$$

Equation 17.51 is particularly simple for the perfect gas, because the internal energy and enthalpy differ only by the term pV, which in a perfect gas is just nRT. Since $U_{\text{perfect gas}}$ depends only on T, we conclude that $H_{\text{perfect gas}}$ must likewise depend only on T. Substituting the perfect gas relation

$$T\left(\frac{\partial V}{\partial T}\right)_p = \frac{nRT}{p} = V \qquad (17.52)$$

into Eq. 17.51 and again assuming C_p to be independent of T, we find as the analog of Eq. 17.42 that

$$\Delta H_{\text{perfect gas}} = C_p(T_2 - T_1). \qquad (17.53)$$

17.4 Entropy Changes in Irreversible Processes

We return, now, to the discussion of the nature of irreversible processes and the connection between irreversibility and the entropy function. We first expand the discussion of Eq. 16.45. Consider the work done by a fluid in an irreversible process. In general, for a fluid, we can write

$$-w = \int_{\text{path}} p_{\text{app}}\, dV, \qquad (17.54)$$

and the entire difference between w_{rev} and w_{irrev} lies in the path of integration of Eq. 17.54 and the consequent relationship between the applied pressure and the system pressure. To compute the work done, in the representation displayed in Eq. 17.54, the pressure to be entered in the integrand is always the external pressure acting on a system. It is only in the limiting case of a reversible process that the difference between the external pressure acting on a system and the pressure exerted on the container by the system is negligible. In that limiting case the external pressure may be replaced by the pressure determined by the equation of state of the system. In contrast, in an irreversible work process the external pressure and the pressure of the system differ by a nonzero amount and work is done at a nonzero rate. For an expansion process this observation implies that $-w_{\text{irrev}} < -w_{\text{rev}}$ because $p_{\text{app}} < p$. (Remember that the work is an algebraic quantity with a sign, and that work done by the system is reckoned negative.) Correspondingly, for a compression process the magnitude of the irreversible work required for the given compression exceeds the magnitude of the reversible work required for the same compression, because in this case $p_{\text{app}} > p$. Taking into account the sign, we always have $-w_{\text{irrev}} < -w_{\text{rev}}$. Furthermore, if we consider all possible reversible and irreversible changes of state between the same end points, the first law of thermodynamics implies that, because the internal energy is a function of state and its change thus independent of the path followed, the reversible change of state absorbs a different amount of heat from all possible irreversible changes of state. This can be seen directly from a consideration of the internal energy changes in the two classes of processes, for if the end points are the same we have

$$\Delta U = q_{\text{rev}} + w_{\text{rev}}$$
$$= q_{\text{irrev}} + w_{\text{irrev}}, \qquad (17.55)$$

and, since

$$-w_{\text{rev}} > -w_{\text{irrev}} \quad \text{for both compression and expansion} \tag{17.56}$$

subtraction gives

$$q_{\text{rev}} > q_{\text{irrev}} \quad \text{for both compression and expansion} \tag{17.57}$$

Although thermodynamic theory clearly distinguishes between reversible and irreversible processes and between their consequences, the source of the irreversibility of a process taking place at a nonzero rate cannot be ascertained from thermodynamics. Rather, to obtain a deeper understanding of the origins of irreversibility in any given process we must appeal to some form of kinetic theory analysis.

The simplest and most obvious example of how conducting a process at a nonzero rate leads to dissipation is provided by a system consisting of a block sliding on a horizontal slab. If there were no friction, lateral displacement of the block on the slab would require no work. However, because of the existence of friction, in order to move the block with constant velocity **v**, a constant force, proportional to **v**, must be applied to the block. The work done in moving the block a distance d is transformed into heat. Since the force required to maintain the velocity **v** is proportional to **v**, any motion of the block with nonzero velocity dissipates work as heat, hence is irreversible. For this example it is readily shown that the rate at which heat is generated is proportional to the square of the velocity of the block.[2] Therefore, the rate at which heat is generated approaches zero more rapidly than the velocity of the block approaches zero, and the translation of the block on the slab approaches reversibility as its velocity approaches zero.

A similar macroscopic description can be given for slow processes in a liquid system. If the velocity of the fluid is not the same everywhere, there are velocity gradients present. Moreover, there is friction between two infinitesimal adjacent layers of liquid moving with different velocities; work must be done against this friction to maintain any nonzero velocity gradient. It is again found that in the limit of very small fluid velocity the rate of dissipation of work into heat approaches zero more rapidly than does the velocity.

[2] The force required to maintain a block of mass M sliding with constant velocity **v** is directed parallel to **v** and just matches the force arising from friction, which is antiparallel to **v**, say, $-\zeta\mathbf{v}$. The work done by the constant force $\zeta\mathbf{v}$ does not increase the kinetic energy of the block, which is fixed at $\frac{1}{2}Mv^2$ for constant speed v. The rate at which work is done on the system (considered as the block plus the slab) is simply ζv, the magnitude of the external force, times v, the distance traveled per unit of time, or ζv^2. Since the kinetic energy of the block is constant, this must be equal to the rate at which work of displacement is converted to heat.

In both the cases described, if the velocity is infinitesimally small, the rate of dissipation of work as heat is an order smaller (e.g., if the velocity is **v**, the rate at which work is dissipated is proportional to v^2). It is therefore meaningful to say that processes in these systems are irreversible if they proceed at nonzero rates, but that they approach reversibility as the rates of the processes approach zero.

In the case of thermal processes—for example, the transfer of heat down a nonzero temperature gradient—it is hard to find a macroscopic mechanical interpretation of the observed irreversibility. Indeed, in cases of this type the macroscopic concept of irreversibility rests on the interpretation of the entropy change in the process, as we now illustrate.

Consider two equal masses of a perfect gas, one maintained at T_{1i} and the other at T_{2i}. Suppose that the containers have rigid walls, so that when the two containers are brought into thermal contact their volumes are maintained constant. Under these conditions no work is done by the gases as they achieve the equilibrium temperature T_f. However, heat is transferred between the two systems by the irreversible process of thermal conduction; the heat transfer persists until the temperatures of the two systems are equal. Since we have specified that the volume is constant and that the masses of gas are equal, the equilibrium temperature is determined by the heat transferred according to the equation

$$C_V = (T_{1i} - T_f) = C_V(T_f - T_{2i}), \tag{17.58}$$

where we have assumed C_V to be independent of T. From Eq. 17.58 we find that

$$T_f = \tfrac{1}{2}(T_{1i} + T_{2i}). \tag{17.59}$$

But as a result of this process the entropies of the two masses of perfect gas change by the amounts

$$\Delta S_1 = C_V \ln\frac{T_f}{T_{1i}},$$
$$\Delta S_2 = C_V \ln\frac{T_f}{T_{2i}}, \tag{17.60}$$

so that the total entropy change is

$$\Delta S = \Delta S_1 + \Delta S_2 = C_V \ln\frac{T_f^2}{T_{1i}T_{2i}}. \tag{17.61}$$

Now, we must have $(T_{2i} - T_{1i})^2 \geq 0$ irrespective of whether T_{1i} or T_{2i} is the greater. By expansion, this becomes

$$(T_{2i} - T_{1i})^2 = (T_{2i})^2 - 2(T_{1i}T_{2i}) + (T_{1i})^2 \geq 0. \tag{17.62}$$

If we now add $4T_{1i}T_{2i}$ to each side of Eq. 17.62 and collect terms, we find that

$$T_f^2 = \tfrac{1}{4}(T_{1i} + T_{2i})^2 \geq T_{1i}T_{2i}, \qquad (17.63)$$

so that ΔS as defined by Eq. 17.61 is always positive if $T_{1i} \neq T_{2i}$. This is, of course, just the thermodynamic characterization of an irreversible process in an isolated system, namely, $\Delta S > 0$. On the other hand, if $T_{1i} = T_{2i}$, heat may be transferred reversibly, and there is no change in total entropy because the increase in entropy of one mass of gas is exactly balanced by the decrease in entropy of the other mass of gas.

17.5 Entropy Changes in Phase Transitions

There are two other applications of the second law that have great importance, namely, to phase changes and to chemical reactions. We defer discussion of the entropy changes in chemical reactions to the next chapter. In Chapter 14 we mentioned that associated with a phase change, such as liquid → gas, there is a latent heat of transition. Moreover, since the enthalpy change occurs at constant temperature, and since equilibrium prevails between the two phases for any relative proportions of the phases so long as the temperature is maintained at the transition temperature, the entropy change accompanying the phase change is just

$$\Delta S_{\text{transition}} = \frac{\Delta H_{\text{transition}}}{T_{\text{transition}}}. \qquad (17.64)$$

What are the magnitudes of typical entropies of fusion and vaporization, and how can they be interpreted? From the data in Table 14.3, we can deduce that the molar entropy of vaporization is of the order of 80 J/K mol for many substances. In terms of the statistical molecular theory of matter, this implies that there are many more states accessible to the gas than to the liquid for a given energy range. This is easily understandable, at least qualitatively. Suppose that we approximate the liquid by a perfect gas of the same density as the liquid. Clearly this approximation is very crude, but it will suffice to give some idea of the important role that the volume per molecule plays in determining the entropy of a fluid. Now for many liquids the molar volume is of the order of 50 cm^3/mol, whereas at the normal boiling point the molar volume of the vapor is usually of the order of magnitude of $3–5 \times 10^4$ cm^3/mol. Using our naive approximation we see that the principal change occurring in the vaporization is the almost 1000-fold increase in the volume per molecule. The isothermal expansion of a perfect gas through such a range of volume leads to a change of entropy per mole of $R \ln(1000) = 6.9R$, or 57 J/K mol. Thus this model suggests that the entropy of vaporization of a liquid has a large contribution arising from the volume change between liquid and gas. Note, however, that there also appears to be a substantial contribution to the entropy of vaporization from the change in the intermolecular force field. That this is so can be verified independently of the crude perfect gas model for the expansion part of the entropy change. Substances that are known to be dimerized in the liquid, such as NO, and substances known to be extensively hydrogen bonded in the liquid, such as H_2O, and that vaporize to give unassociated gases, have large entropies of vaporization. Typical values of the entropies of vaporization of these substances cluster around 13–14R, or ~108–116 J/K mol.

It is possible to give a similar simple picture of the enthalpy change on vaporization. At the boiling point the density of the vapor is usually sufficiently low that for the present purposes it can be assumed to be a perfect gas. Then the enthalpy of the vapor is determined only by the kinetic energy of the molecules, whereas the enthalpy of the liquid contains an important contribution from the intermolecular interactions. Since vaporization is an isothermal process, accepting this interpretation implies that the enthalpy of vaporization measures the work required to free a molecule in the liquid from its neighbors, thereby enabling the volume per molecule to increase. In loose terms, the enthalpy of vaporization measures the work necessary to overcome the attractions between molecules so as to be able to expand the fluid. It must be emphasized that this pictorial description is quantitatively crude, and the dense perfect gas is a very poor model of a liquid.

In general the entropies of fusion of substances, which can be obtained from Table 14.2, range over a considerable scale. For very simple solids, such as Ar, Kr, and Xe, the entropies of fusion per mole are close to R, an amount that again may be interpreted as determined mainly by the change in freedom of motion of a molecule, part of which is reflected in the change in volume per molecule. Since the change of total volume per molecule is small (of the order of 10% of the volume) for the solid → liquid transition, what is important is the change in effective volume per molecule. By effective volume we mean the volume available to the center of mass of the molecule. Because of repulsion between molecules this is less than the volume per molecule; it also changes more on fusion of the solid. For some polyatomic molecules the entropy of fusion is very large. For example, for SO_2 the molar entropy of fusion is 37.4 J/K mol. In most cases the large entropy of fusion of a polyatomic molecular solid reflects a change in the freedom of molecular rotation, since in many solids rotation is hindered or even not possible, whereas in most liquids molecular rotation is almost free. The latter deduction is also supported by a comparison of the entropies of vaporization of polyatomic molecular liquids and, say, liquid Ar.

In the preceding remarks we have used the concept of freedom of motion of a molecule, or volume available per molecule, rather loosely. The reason for our choice of analogy will become apparent later in this book. For the present it is sufficient to realize that we have given only an imperfect analogy so as to invest the observed entropy changes with some simple molecular significance, and that a precise description requires a careful analysis of molecular interactions, the real energy spectrum, and other details characteristic of a particular substance.

It is pertinent to mention briefly the computation of entropy changes that accompany chemical reactions. Clearly, the method to be used must be analogous to that introduced in Chapter 14 for computing the energy and enthalpy changes in a reaction. However, the consideration of entropy changes raises interesting questions concerning standard states, and we therefore defer detailed consideration of this topic until the end of the next chapter.

FURTHER READING

Bailyn, M., *A Survey of Thermodynamics* (AIP Press, New York, 1994), Chapters 3 and 4.

Kestin, J., *A Course in Thermodynamics,* Vol. I (Blaisdell, Waltham, Mass., 1965), Chapters 11–13.

Lewis, G. N., and Randall, M., *Thermodynamics,* 2nd ed. revised by K. S. Pitzer and L. Brewer (McGraw-Hill, New York, 1961), Chapters 7–11.

Pitzer, K. S., *Thermodynamics,* 3d ed. (McGraw Hill, Inc., New York, 1995), Chapters 2, 3, and 4.

PROBLEMS

1. Using the equation of state of the perfect gas, show that Eqs. 17.38 and 17.48 are equivalent. Hint: Start by showing that for a perfect gas with temperature independent heat capacities, $C_p = C_V + nR$.

2. Show that for a change of state $(p_1,T_1) \to (p_2,T_2)$

$$\Delta H_{\text{perfect gas}} = \int_{T_1}^{T_2} C_p\,dT.$$

3. Suppose that a perfect gas is taken from an initial state p_1, V_1, T_1 to some other state p_2, V_2, T_2 (with $V_2 > V_1$ and $T_2 > T_1$), by the following reversible processes:

(a) Adiabatic compression until $T = T_2$ followed by isothermal expansion until $V = V_2$;

(b) Isobaric cooling until $T = T_1$ followed by adiabatic compression until $T = T_2$ and then isothermal expansion until $V = V_2$;

Show that the entropy changes in processes (a) and (b) are equal, and that they are also equal to the entropy change shown in Eqs. 17.48 and 17.38.

4. The constant-volume molar heat capacity of a solid may be approximated by

$$C_V = \frac{12\pi^4}{5} R \left(\frac{T}{\Theta_D}\right)^3$$

if $T \ll \Theta_D$. The constant Θ_D is different for each solid. For Al, $\Theta_D = 398$ K. What is the entropy change of a 5-g mass of Al when it is heated at constant volume from 4 K to 176 K?

5. Using the data of Tables 14.2 and 14.3, calculate entropies of fusion and vaporization for a number of compounds. Include some that hydrogen bond in the condensed phase, such as H_2O and HF. Interpret your calculations in terms of simple models.

6. Consider a space in which T and S are taken as the independent variables. Show that an isochoric process is represented in this space by a curve with slope T/C_V, and that an isobaric process is represented by a curve with slope T/C_p.

7. Calculate the entropy increase when 2 mol of NH_3 are heated at constant pressure from 300 to 400 K. Assume that

$$c_p^{NH_3} = 28.0 + 26.3 \times 10^{-3}\,T\ \text{J/K mol}.$$

8. Derive general expressions for the changes in H and S corresponding to the change of state $p_1, T_1 \to p_2, T_2$ for a gas with equation of state

$$p = \frac{RT}{v-b} - \frac{a}{v^2},$$

with b and a constants.

9. Calculate the changes in U, H, and S when 1 mol of liquid water at 0°C and 1 atm is converted to steam at 200°C and 3 atm.

Data: c_p(liquid) = 75 J/K mol

c_p(steam) = $36.9 - 7.9 \times 10^{-3}T + 9.2 \times 10^{-6}T^2$ J/K mol

Δh_{vap}(100°C, 1 atm) = 40.6 kJ/mol

Assume the steam to be a perfect gas and the liquid to have temperature independent density and c_p.

10. (a) Consider 1 mol of supercooled water at –10°C. Calculate the entropy change of the water and of the surroundings when the supercooled water freezes at –10°C and 1 atm.

Data: c_p(ice) = 38 J/K mol

c_p(water) = 75 J/K mol

Δh_{fus}(0°C) = 6026 J/mol

Consider c_p(ice) and c_p(water) to be independent of temperature.

(b) Is the freezing of supercooled water at –10°C a reversible or an irreversible process? Justify your answer quantitatively.

11. Show that

(a) $$\left(\frac{\partial T}{\partial V}\right)_U = \frac{p - T(\partial p/\partial T)_V}{C_V} = -\frac{T^2}{C_V}\left(\frac{\partial (p/T)}{\partial T}\right)_V.$$

(b) $$\left(\frac{\partial T}{\partial p}\right)_H = -\frac{V - T(\partial V/\partial T)_p}{C_p} = \frac{T^2}{C_p}\left(\frac{\partial (V/T)}{\partial T}\right)_p.$$

12. From the following data, estimate the molar entropy of C_6H_6 imagined as a perfect gas at 25°C and 1 atm. At 25°C the vapor pressure of C_6H_6 is 95.13 torr and its heat of vaporization is 33850 J/mol. The heat of fusion is 9866 J/mol at the freezing point, which is 5.53°C. The average value of $c_p(l)$ between 5.53 and 25°C is 134.0 J/Kmol, and the entropy of the solid at 5.53°C is 128.8 J/K mol.

13. Derive the Gibbs–Helmholtz equation,

$$\left(\frac{\partial (G/T)}{\partial T}\right)_p = -\frac{H}{T^2}.$$

14. The heat capacity of solid iodine between 0°C and the melting temperature 113.6°C is represented by the equation (with t in °C)

$$c_p(l) = 54.68 + 13.4\times 10^{-4}(t-25)^2 \text{ J/K mol}.$$

The molar heat of fusion is 15650 J/mol at the melting point. The entropy of solid iodine is 117 J/K mol at 25°C. What is the entropy of liquid iodine at the melting point?

15. At thermal equilibrium between two parts of an isolated system, $\beta_1 = \beta_2 = \beta$. Given the expression for β (Eq. 15.45) and the facts that for each part,

$$\Omega \propto U^{3N/2} V^N,$$

$$U = \tfrac{3}{2} N k_B T \qquad \text{(valid for ideal gases)},$$

(a) Show that $\beta = 1/k_B T$.
(b) For fixed V, show that $dU = T\,dS$.
(c) Add the known work term $-p\,dV$ to dU of (b), and show that for an ideal gas

$$\left(\frac{\partial S}{\partial V}\right)_U = \left(\frac{\partial S}{\partial V}\right)_T = \frac{p}{T}.$$

(d) Show that, in general,

$$\left(\frac{\partial S}{\partial V}\right)_T = \frac{p}{T} + \frac{1}{T}\left(\frac{\partial U}{\partial V}\right)_T.$$

(e) Consider an isolated system for which

$$\left(\frac{\partial U}{\partial V}\right)_T = 0.$$

Prove that if the system is divided into two parts by a movable partition at equilibrium

$$\gamma_1 = \gamma_2,$$

where

$$\gamma \equiv \left(\frac{\partial \ln \Omega}{\partial V}\right)_U$$

for each system. Finally, using all of your results, prove

$$pV = Nk_BT.$$

16. Evaluate $(\partial U/\partial V)_T$ and $(\partial H/\partial p)_T$ for 1 mol of a van der Waals gas:

$$\left(p + \frac{a}{v^2}\right)(v-b) = RT.$$

17. Calculate $C_p - C_V$ for a van der Waals gas. (See Problem 13 in Chapter 14.)

18. The net work done in an irreversible cyclic process is 50 J. Can you say anything about the net heat absorption?

19. Evaluate ΔA and w for the expansion of 1 mol of ideal gas at 300 K from 10 atm pressure to 1 atm:
(a) Reversibly.
(b) By free expansion into a vacuum.
(c) Against constant external pressure of 1 atm.

20. Evaluate $(\partial S/\partial V)_T$ for 1 mol of a van der Waals gas. For a given isothermal expansion, will ΔS be greater for an ideal gas or for a van der Waals gas?

21. Consider a manufacturing process that involves, because of the change from initial to final states, changes in internal energy, volume, and entropy of a system by ΔU, ΔV, ΔS. We need not specify what the system is. Suppose that the input of energy to the system is accomplished by condensation of steam, so that heat q is transferred to the system at temperature T_S. During the manufacturing process the surroundings absorb heat q_0 and are maintained at the constant temperature T_0, and pressure p_0. Show that if the manufacturing process is reversible,

$$q - \frac{T_S}{T_S - T_0}(\Delta U + p_0\Delta V - T_0\Delta S) = 0.$$

Show that if the process is irreversible,

$$q - \frac{T_S}{T_S - T_0}(\Delta U + p_0 \Delta V - T_0 \Delta S) > 0,$$

and the amount by which q exceeds the second term represents heat input that is wasted and is unavailable for work.

22. Show that

(a) $$\left(\frac{\partial S}{\partial p}\right)_V \left(\frac{\partial T}{\partial V}\right)_p - \left(\frac{\partial T}{\partial p}\right)_V \left(\frac{\partial S}{\partial V}\right)_p = -1.$$

(b) $$\left(\frac{\partial H}{\partial V}\right)_T = -V^2 \left(\frac{\partial p}{\partial T}\right)_V \left(\frac{\partial (T/V)}{\partial V}\right)_p.$$

23. Calculate the entropy of Pb vapor at 25°C and 1 atm from the following data. Between 900 and 1600 K the vapor pressure of Pb (in torr) is given by

$$\log p = 7.822 - \frac{9854}{T}.$$

The gas contains only Pb atoms. At 25°C the molar entropy of Pb(*s*) is 64.98 J/K mol. Below 900 K the average value of $c_p(s)$ is 28.0 J/K mol, of $c_p(l)$ is 26.4 J/K mol, and of $c_p(g)$ is 20.8 J/K mol. The melting point of Pb at 1 atm is 600 K, and the entropy of fusion is 7.95 J/K mol.

24. Measurements of the properties of some gaseous substance give

$$\left(\frac{\partial v}{\partial T}\right)_p = \frac{R}{p} + \frac{a}{T^2},$$

$$\left(\frac{\partial v}{\partial p}\right)_T = -Tf(p),$$

where a is a constant and $f(p)$ depends only on the pressure. It is also found that

$$\lim_{p \to 0} c_p = c_p \text{ for monoatomic ideal gas} = \tfrac{5}{2}R.$$

Show that:

(a) $f(p) = \dfrac{R}{p^2}$.

(b) $pv = RT - \dfrac{ap}{T}$.

(c) $c_p = \dfrac{2ap}{T^2} + \dfrac{5}{2}R$.

25. The molecular theory of the perfect gas leads to the relations

$$U = an^{5/3}V^{-2/3}e^{2S/3nR},$$

$$A = \frac{3}{2}nRT \ln \frac{2aen^{2/3}}{3RTV^{2/3}},$$

where a is a constant characteristic of the gas, e is the base of natural logarithms, and n is the number of moles. Show that both of these equations yield

$$pV = nRT.$$

26. The molar Gibbs free energy of some gas is

$$g(T, p) = a + bT + cT \ln T + RT \ln p - \left(\frac{d}{T^3}\right)p,$$

where a, b, c, and d are constants. Find the equation of state of this gas.

CHAPTER 18

The Third Law of Thermodynamics

Thus far we have examined some aspects of the thermodynamic description of a system, how that description is related to its energy-level spectrum, and how changes in thermodynamic functions are measured. With respect to the last point, we have shown that with the equation of state and either C_p or C_V as a function of T at one value of p or V, respectively, we can calculate all changes in the thermodynamic functions of a one-component system. For example, for the change of state $p_1, T_1 \rightarrow p_2, T_2$ entirely in a one phase region we have (see Eqs. 17.51 and 17.46)

$$\Delta H = \int_{p_1, T_1}^{p_1, T_2} C_p \, dT + \int_{p_1, T_2}^{p_2, T_2} \left[V - T\left(\frac{\partial V}{\partial T}\right)_p \right] dp, \tag{18.1}$$

$$\Delta S = \int_{p_1, T_1}^{p_1, T_2} \frac{C_p}{T} \, dT - \int_{p_1, T_2}^{p_2, T_2} \left(\frac{\partial V}{\partial T}\right)_p dp. \tag{18.2}$$

Consider, now, an isothermal process in some system, say, at temperature T_2. The Gibbs free-energy change in this process is

$$\Delta G = \Delta H - T_2 \Delta S \tag{18.3}$$

$$= \Delta H + T_2 \left(\frac{\partial \Delta G}{\partial T}\right)_p, \tag{18.4}$$

where Eq. 18.4 follows from Eq. 17.18. Equation 18.4 is known as the *Gibbs–Helmholtz equation;* we shall say more about it, and use it extensively, later.

We shall show, in Chapters 19 and 21, that the equilibrium composition of a reacting mixture

$$aA + bB + \cdots = lL + mM + \cdots$$

is determined by the difference in Gibbs free energies per mole of the products and reactants, $\Delta G = \Sigma_i \nu_i g_i$. Accordingly, it is of interest to be able to calculate the change in Gibbs free energy for the transformation of reactants to products from thermal measurements alone. If that is possible, reaction equilibria can be predicted, and the necessity for sometimes difficult measurements of the composition of reacting mixtures can be avoided. To see what is involved in calculating ΔG from thermal data, we imagine a reaction to be carried out at constant temperature T_2 and constant pressure p_2. Then

$$\Delta G = \Delta H_{T_1} - T_2 \Delta S_{T_1} + \int_{T_1}^{T_2} \Delta C_p \, dT - T_2 \int_{T_1}^{T_2} \frac{\Delta C_p}{T} dT, \tag{18.5}$$

where the subscript T_1 refers to evaluation of ΔH and ΔS at that temperature. Note that ΔH_{T_1} and ΔS_{T_1} play the roles of constants of integration; their values must be known in order that ΔG may be computed. If we choose for the reference temperature $T_1 = 0$, the absolute zero. Then we can write

$$\Delta G = \Delta H_0 - T_2 \Delta S_0 + \int_0^T \Delta C_p \, dT - T_2 \int_0^T \frac{\Delta C_p}{T} \, dT. \tag{18.6}$$

Equation 18.6 will be valid only if ΔH_0, ΔS_0, and the integrands of the two integrals displayed, do not diverge as $T_1 \rightarrow 0$. This condition imposes a restriction on the temperature dependence of the specific heat near the absolute zero. In addition, Eq. 18.6 still contains the limiting changes in enthalpy and entropy at $T_1 = 0$. The possibility that ΔG can be calculated from thermal data alone depends, finally, on the behavior of ΔS_{T_1} and ΔC_p as $T_1 \rightarrow 0$.

We now ask if the changes of entropy and of Gibbs free energy observed to occur show any special pattern as $T \rightarrow 0$. This question was addressed by Nernst in the early years of the twentieth century. He noticed that experimental data for the enthalpy and the Gibbs free energy changes in reactions appeared to show

$$\lim_{T \rightarrow 0} \Delta G \rightarrow \Delta H. \tag{18.7}$$

The limited data then available did not extend to very low temperatures, but the trend in the data was obvious. If relation 18.7 is in fact valid, and $\Delta G = \Delta H$ when $T = 0$ it must be true that

$$\lim_{T\to 0} \Delta S = 0. \tag{18.8}$$

About the same time, as the technology for obtaining low temperatures improved, measurements of C_p were extended to lower and lower temperatures. The then available data strongly suggested that

$$\lim_{T\to 0} C_p = 0. \tag{18.9}$$

Using these observations, Nernst conjectured that the following limiting laws are generally valid for isothermal processes:

$$\begin{aligned} &\lim_{T\to 0} \Delta G \to \Delta H,\\ &\lim_{T\to 0}\left(\frac{\partial \Delta G}{\partial T}\right)_p = \left(\frac{\partial \Delta H}{\partial T}\right)_p,\\ &\left.\begin{aligned}\lim_{T\to 0} \Delta S &= 0,\\ \lim_{T\to 0} \Delta C_p &= 0,\end{aligned}\right\} \text{ independent of } p \text{ and any other external variable.}\end{aligned} \tag{18.10}$$

The limiting laws (18.10) represent the original formulation of the *Nernst heat theorem.*

Numerous more recent investigations have substantiated Nernst's conjecture and shown its limitations. From the point of view of practical thermodynamics, the Nernst heat theorem, or the *third law of thermodynamics* as it has come to be known, is of great importance. We shall see that there are several equivalent statements of the third law of thermodynamics one of which, due to Planck, asserts that the entropy of any system vanishes in the state for which $T = 0$. Exploitation of the latter form of the third law of thermodynamics permits calculation of an absolute entropy and, as we shall see later, equilibrium constants for heterogeneous reactions from thermochemical measurements alone. On the other hand, there remain significant theoretical questions concerning the formulation and use of the third law.

Rather than follow the historical development, we shall open our discussion of the third law by examining the information available from the microscopic spectrum of states of a system. The definition of the entropy in terms of the number of accessible states in the range from E to $E + dE$ has implications with respect to the properties of the entropy function in the limit as the thermodynamic temperature tends to zero. We shall explore these implications, then set up a scale of absolute values for the entropy and examine the nature of entropy changes in chemical reactions.

18.1 The Magnitude of the Entropy at $T = 0$

As a preliminary step to the discussion of the limiting value of the entropy of a single substance at $T = 0$, we consider how $\ln \Omega(E)$ varies with E. A plot of $\ln \Omega(E)$ against E has slope $\partial \ln \Omega(E)/\partial E$, which by definition is equal to $1/k_BT$ (Eq. 15.40). Thus for energies close to the ground-state energy, the slope of $\ln \Omega(E)$ versus E is very large, and it decreases continuously as E increases. Furthermore, the curve of $\ln \Omega(E)$ versus E has curvature $\partial^2 \ln \Omega(E)/\partial^2 E$ which, in terms of temperature, is $-(1/k_BT^2)(\partial T/\partial E)$. At the end of Chapter 16 we demonstrated that $\partial T/\partial E$ is positive; therefore the plot of $\Omega(E)$ versus E curves toward the E axis. By combining these two properties we can construct the schematic diagram shown in Fig. 18.1.

Let the ground-state energy of a system be E_0. Usually, corresponding to the energy E_0 there exists only one state, or at most a small number of states, of the system. Thus, $\Omega(E)$ approaches unity, or at most some small number, in the limit $E \to E_0$. Suppose that $\Omega(E_0) = 1$. Then the entropy $k_B \ln \Omega(E_0)$ vanishes. Even if $\Omega(E_0) = \eta$, where η is a small number, the entropy $k_B \ln \eta$ is negligibly small relative to the entropy when $E >> E_0$; the latter entropy is of the order of magnitude of $k_B\nu$, where ν is the number of degrees of freedom of the system (see Eq. 15.21).

We have already indicated that

$$\frac{\partial T}{\partial E} > 0. \tag{18.11}$$

This is just another way of saying that the specific heat of a one-phase system is always positive or, in different words, that adding energy to a system in a one-phase region can only increase its temperature. Conversely, removing energy from a one-phase system can only decrease its temperature. Since the lowest possible energy the system can have is E_0, from Eq. 18.11 and the requirement that T be positive we deduce that

$$\lim_{E\to E_0} T = 0; \tag{18.12}$$

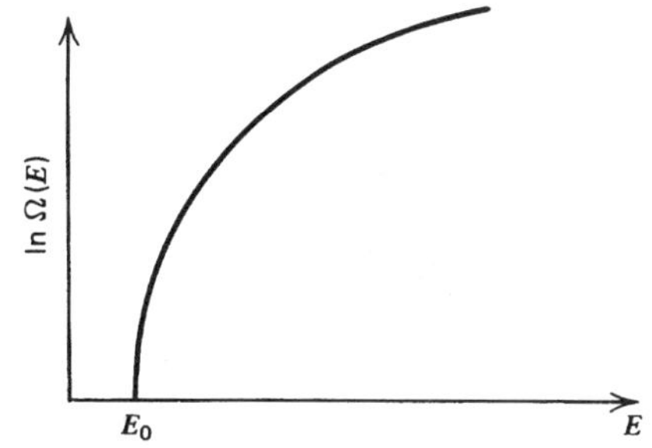

Figure 18.1 Schematic plot of the dependence of $\ln \Omega(E)$ on E.

in words, in the limit that the energy of the system approaches the ground-state energy, the temperature of the system approaches zero. This property is usually stated the other way around because it is normal to use the temperature as the independent variable. We therefore cite the complementary relation to Eq. 18.12, namely,

$$\lim_{T\to 0} E = E_0\,; \tag{18.13}$$

in words, in the limit that the temperature approaches zero, the energy of the system approaches the ground-state energy. Thus, by combining Eq. 18.13 with the assumption that $\Omega(E_0)$ is a small number, we are led to the deduction that

$$\lim_{T\to 0} k_B \ln \eta \approx 0, \tag{18.14}$$

which is to be interpreted to say that in the limit that the temperature approaches zero, the entropy of the system also effectively approaches zero.

Under what conditions is the assumption that $\Omega(E_0) = \eta$ a valid representation of the states of the system? Clearly, this depends on the nature of the system under examination.

Consider, first, the case that the ground state of the system is not degenerate. A good example of a system with this characteristic is He with isotopic mass 4, since the He^4 nucleus has zero spin, and the closed shell electron configuration also has zero spin. The heat capacity as a function of temperature for He^4, in the low temperature region, is shown in Fig. 18.2. The λ shaped maximum in the heat capacity at about 2.2 K is the signature of a phase transition between ordinary liquid He^4 and superfluid liquid He^4. Theoretical studies of the low temperature superfluid phase predict that its entropy depends on the temperature as $S \propto T^{3/2}$, hence its heat capacity is also proportional to $T^{3/2}$. This prediction is verified by the data displayed in Fig. 18.2, which show that as $T \to 0$ the heat capacity of the low temperature phase rapidly approaches zero, in agreement with the third law of thermodynamics.

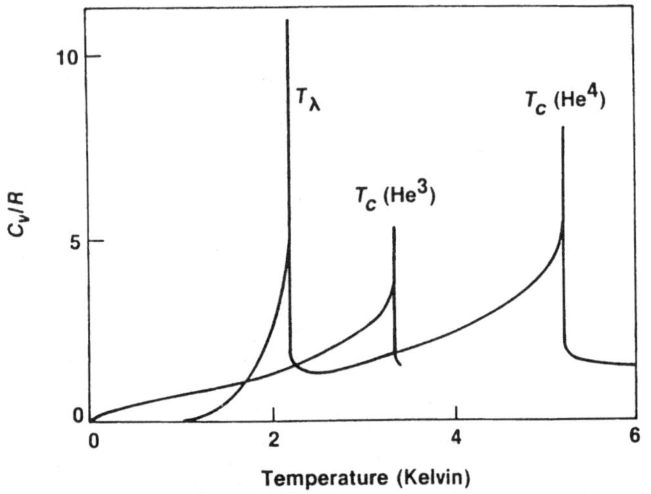

Figure 18.2 The constant volume heat capacity of He^3 and of He^4 at their respective critical densities (see Chapter 24).

Consider, now, the case that the true ground state of a system is nondegenerate but that there exists a set of states near to the absolute ground state of the system that span an energy range small compared with the thermal energy k_BT, where T is the lowest temperature to which the system can be brought. Then the number of accessible states remains very large unless the system is brought to a temperature T_0 such that k_BT_0 becomes small compared to the energy spacing of this set of states. Examples of this behavior are numerous. One well-studied case is the set of energy levels corresponding to the possible orientations of N nuclear spins. Now, the nuclear spin–spin interaction is very weak. Yet if the temperature were made low enough, this interaction would become large relative to the thermal energy, and a system of nuclear spins would collapse into a well-defined ground state. However, in such systems it is typically the case that even for a temperature of the order of 1 K the thermal energy available is large relative to the spacings of the spin-state energy levels. To illustrate the consequences of the resultant effective degeneracy of states, consider the limiting case when the interactions between the atoms of the sample are independent of the nuclear spin and there is no external magnetic field; then all allowed orientations of the nuclear spin are of equal energy. If the nuclear spin quantum number is I, there are $2I + 1$ states per nucleus, of equal energy; and for a sample of N spins, $\Omega(E_0)$ is just $(2I + 1)^N$. The limiting value of the entropy then becomes

$$\lim_{T\to T_0} S = k_B \ln(2I+1)^N = Nk_B \ln(2I+1) = S_0\,. \tag{18.15}$$

Of course the spin–spin interactions are never exactly zero, but for most nonmagnetic solids they are so small that, as indicated above, it is usually a considerable task to achieve a temperature low enough that the thermal energy is less than the energy spacing of the stationary states of nuclear spin. For systems of this type Eq. 18.14 must be replaced by

$$\lim_{T\to T_0} S = S_0\,, \tag{18.16}$$

where T_0 indicates some very low temperature, but one large enough that the thermal energy exceeds the energy difference between the states, and S_0 is a constant. Since the spins are already distributed over all accessible states at T_0, an increase in the temperature does not affect the states of spin orientation, and S_0 is thus independent of the exact value of T_0.

There is another limiting case of interest: If one set of energy levels is separated from another set by an energy very large relative to the thermal energy, the second set of levels can be ignored for many practical purposes. For example, the spacings of the energy levels of all nuclei are such that at all temperatures of ordinary interest for chemical purposes, say, less than 100,000 K, nuclei remain in their ground states throughout all reactions and changes of state. Insofar as nuclear excitation is concerned, any temperature up to 100,000 K is effectively zero.

We now consider what can be learned about the limiting properties of the entropy solely from thermodynamic arguments. Suppose that a system is constrained to occupy a constant volume. Then the entropy change accompanying a change in temperature $T_1 \to T_2$ is

$$\Delta S \equiv S_2 - S_1 - \int_{T_1}^{T_2} \frac{C_V}{T}\, dT. \qquad (18.17)$$

Now consider the limit $T_1 \to 0$. We have

$$\lim_{T_1 \to 0} S_2(T_2) = S_1(0) + \int_0^{T_2} \frac{C_V}{T}\, dT. \qquad (18.18)$$

If the entropy is to be finite at any temperature T, it is necessary that $S_1(0)$ be finite and that $C_V(T)$ approach zero as T approaches zero; C_V must approach zero at least as some positive power of T.[1] In symbolic form, we have

$$\lim_{T \to 0} C_V(T) = 0. \qquad (18.19)$$

Note that the vanishing of the specific heat is a deduction dependent only on the assumption that the entropy function remains finite at $T = 0$, and not on any specific value of the function. If we consider a system constrained to be at constant pressure during the change of state in which $T_1 \to T_2$, reasoning similar to that just outlined leads to the conclusion

$$\lim_{T \to 0} C_p(T) = 0. \qquad (18.20)$$

Thus, specific heats C_V and C_p both vanish in the limit of zero temperature.

Although the results cited thus far follow from the simple argument that the entropy must be finite, many more far-reaching results can be derived from the following postulate, originally due to Nernst:[2]

In any system in internal equilibrium undergoing an isothermal process between two states, the entropy change of the process approaches zero as the temperature of the system approaches zero.

This is one form of the third law of thermodynamics.

The restriction to states of internal equilibrium is important. Frequently, during the approach to $T = 0$, a system develops internal constraints that prevent the achievement of internal equilibrium. These internal constraints are not under the control of the experimenter, and they prevent the achievement of equilibrium with respect to the applied and controlled external constraints. For example, a glass is unstable with respect to crystallization into a solid.[3] The glass is prevented from crystallizing, on the usual time scales of interest, by the extremely large viscosity of the medium, which acts to inhibit atomic motion. Usually internal constraints change slowly in time; for example, ordinary silica glass crystallizes (devitrifies) at room temperature on a time scale of many centuries. Since the internal constraints described are not under the control of the experimenter, it cannot be guaranteed that they will remain fixed during some process in which the externally controlled constraints are varied; the alteration of the internal constraints will, in general, lead to an irreversible process and hence an entropy change. On the other hand, if the internal constraints remain fixed, and if a reversible process can be devised to connect two states of a system with the same internal constraints, then the third law as stated above is valid.

The processes that lead to internal constraints proceed at a variety of rates. It is therefore possible for a system to develop internal constraints with respect to one property, or one degree of freedom, and not another, depending on the method by which the system is prepared. A glass is out of equilibrium with respect to the atomic arrangement that is most favorable, but given the atomic geometry of the glass the motions of the atoms are typical of equilibrium. As the atomic geometry changes very slowly in time, so also do the motions of the atoms. In this particular case the nonequilibrium geometry is created by cooling a molten substance (or mixture) more rapidly than atomic rearrangement to the crystalline form can occur. On the time scales of usual experimental interest, the glass we have just described does not appear to change, and hence its properties can be adequately described thermodynamically provided that the special

---

[1] Unless C_V approaches zero as some positive power of T, the integral term in Eq. 18.18 diverges at the lower limit. For example, were C_V to be of the form

$$C_V = A + BT + CT^2 + \cdots,$$

then substitution into Eq. 18.18 and integration would lead to terms of the form

$$A \ln T\Big|_0^T + BT\Big|_0^T + \tfrac{1}{2}CT^2\Big|_0^T + \cdots.$$

The logarithmic term diverges at $T = 0$; the other terms do not. If C_V is not restricted to depend on integral powers of T, the same argument shows that so long as

$$C_V = DT^m,$$

with $m > 0$, the entropy defined by Eq. 18.18 remains finite at $T = 0$.

[2] As stated, this is not the original form of the Nernst postulate. That form was: *The entropy change in any isothermal process approaches zero as the temperature at which the process occurs approaches zero* (see Eq. 18.10). This form is now known to be inadequate because of the occurrence of long-lived nonequilibrium states, such as glasses.

[3] We choose a glass as an example of a system not at complete equilibrium. Other examples could be cited.

nature of the uncontrolled internal constraints is recognized. Other systems may change too rapidly for such characterization; each case must be considered separately.

If two glasses are compared it is found, in general, that cooling under different conditions leads to the generation of different internal constraints. If the internal constraints in one glass remain fixed, the entropy change in an isothermal process in that glass will approach zero as the temperature approaches zero. However, the entropy difference between the two glasses, which depends on two sets of uncontrolled internal constraints, will not necessarily approach zero as the temperature approaches zero.

We may summarize the preceding interpretation of the third law by saying that, if a particular set of constraints on a system remains fixed through an isothermal process, then, whether these constraints are controlled or uncontrolled, the entropy change in that process tends to zero as the temperature approaches zero.

What are the consequences of the third law? Consider the derivative $(\partial S/\partial p)_T$, which measures the rate of change of entropy with respect to an isothermal pressure change. From the third law we have, for a single substance,

$$\lim_{T\to 0}\left(\frac{\partial S}{\partial p}\right)_T = 0; \tag{18.21}$$

combining this with the expression

$$\left(\frac{\partial S}{\partial p}\right)_T = -\left(\frac{\partial V}{\partial T}\right)_p \tag{18.22}$$

(see Eq. 17.19), we obtain

$$\lim_{T\to 0}\left(\frac{\partial V}{\partial T}\right)_p = 0. \tag{18.23}$$

In other words, the coefficient of thermal expansion,

$$\alpha \equiv \frac{1}{V}\left(\frac{\partial V}{\partial T}\right)_p, \tag{18.24}$$

must vanish in the limit as the temperature approaches zero.[4]

A similar argument can be given to show that, since by the third law $(\partial S/\partial V)_T$ must be zero in the limit of zero temperature, it is necessary that

$$\lim_{T\to 0}\left(\frac{\partial p}{\partial T}\right)_V = 0. \tag{18.25}$$

The preceding arguments suggest a simple scheme for defining absolute entropies. The statistical theory leads to the assignment $S = 0$ to a system for which $\Omega(E) = 1$. Suppose that we cannot achieve a temperature low enough to observe temperature variations in the nonzero residual entropy S_0 corresponding to differential excitation of closely spaced energy levels. It is then feasible to adopt a practical scale in which the entropy of a substance is assigned the value zero at the "practical" zero of temperature (i.e., $T = T_0$). It must be remembered, however, that in adopting this practical scale we must restrict the temperature range considered to be such that the smallest thermal energy ($k_B T_0$) is much greater than the energy spacing of any possible quasi-degenerate states of the system. Since in the consideration of any entropy changes the constant S_0 will cancel out, we can effectively neglect it. In the practical scale to be discussed, S_0 is assigned the value zero.

A different statement of the third law is as follows:

The entropy of any system vanishes in the state for which

$$T \equiv \left(\frac{\partial U}{\partial S}\right)_V = 0.$$

This form, due to Planck, is the usual statement of the third law of thermodynamics, and must be interpreted, together with the practical zero of the temperature scale and the nature of internal constraints and internal equilibrium, in the manner indicated. It is generally believed that if a sufficiently low temperature were achieved and equilibrium established at that temperature, then the Planck statement would be found to be valid. But in the absence of a means to achieve low enough temperatures, and in the absence of knowledge as to whether or not there exist sets of internal states of the system with exceedingly small energy spacing, only the practical scale of absolute entropies can be operationally defined.

In the practical scale of absolute entropies we define the *calorimetric entropy* by

$$S(T,p) = \left(\int_{T_0}^{T_f}\frac{C_p}{T}dT\right)_{\text{crystal}} + \frac{\Delta H_{\text{fus}}}{T_f} + \left(\int_{T_f}^{T_b}\frac{C_p}{T}dT\right)_{\text{liquid}} + \frac{\Delta H_{\text{vap}}}{T_b} + \left(\int_{T_b}^{T}\frac{C_p}{T}dT\right)_{\text{vapor}} \tag{18.26}$$

at the temperature T and pressure p. Clearly, if at (T, p) the substance is a solid, the entropies of fusion and vaporiza-

[4] Note that, accepting the third law as correct, a classical perfect gas cannot exist at $T = 0$, for the coefficient of thermal expansion of a classical perfect gas is

$$\alpha = \frac{1}{T},$$

which obviously does not approach zero as T approaches zero. Similarly, a gas with specific heat independent of temperature near $T = 0$ cannot exist. The quantum properties of matter are of dominant importance near $T = 0$, and the failures just cited arise from the inapplicability of classical mechanics to a system with energy very close to the ground-state energy E_0. Near $T = 0$ or $E = E_0$, a perfect quantum gas (defined by absence of interparticle potential energy) has an equation of state different from the classical equation of state; the new equation of state, and also the specific heat of the gas, satisfy the third law. An example illustrating this behavior is described in Section 22.5.

tion and the entropy changes in the liquid and gaseous phases, all included in the general Eq. 18.26, are to be omitted. In Table 18.1 are shown the practical absolute entropies of a number of substances. Note how large the entropy of a gas is relative to the entropy of a liquid or solid, and also how the entropy increases as the molecule becomes more complicated.

The scheme used to calculate the entropy change in a chemical reaction is identical with that described in Chapter 14 for the calculation of enthalpy changes. Thus, if s_i is the entropy per mole of substance i, the reaction

$$\nu_A A + \nu_B B + \cdots = \nu_M M + \nu_N N + \cdots \tag{18.27}$$

results in an entropy change of

$$\Delta S = \sum_{i=1}^{n} \nu_i s_i . \tag{18.28}$$

Again, as in the calculation of enthalpy changes, it is convenient to choose a reference state for the entropy. We denote the molar entropy of substance i in the reference state by s_i^0. It is usual to choose $T = 298.15$ K, $p = 1$ atm as the conditions defining the reference state. Thus, when the reaction

$$\tfrac{1}{2}\mathrm{H_2}(g) + \tfrac{1}{2}\mathrm{Cl_2}(g) = \mathrm{HCl}(g) \tag{18.29}$$

is carried out at 298.15 K and 1 atm, the entropy change is (see Table 18.1)

$$\begin{aligned}\Delta S &= 186.678 - \tfrac{1}{2}(130.587) - \tfrac{1}{2}(222.949) \\ &= 9.910\ \mathrm{J/K}.\end{aligned} \tag{18.30}$$

The calculation of the entropy change in a reaction under conditions differing from the standard state is carried out in a fashion parallel to the corresponding enthalpy calculation described in Chapter 14. Using the defining relation, Eq. 18.28, we can express the entropy of each substance participating in the reaction in terms of its entropy in the standard state, its specific heat, and its equation of state. Since the procedure followed is similar to that already described in Chapter 14, we shall not provide an explicit illustration here; examples will be found in the problems.

Having worked out procedures for calculating both enthalpy and entropy changes in chemical reactions and other processes, we may ask why the calculations are of interest. We shall see in Chapter 19 that the nature of the equilibrium in a thermodynamic system is defined by the competition between enthalpy (or energy) changes and entropy changes. The precise character of this competition is defined by the constraints on the system. Thus, we shall show that knowledge of the energy and entropy changes for a given thermodynamic system and process enables us to predict the properties of the equilibrium state and its changes as the constraints vary.

Table 18.1 Standard Entropies of Some Substances ($T = 298.15$ K, $p = 1$ atm)

Substance	S^0 (J/K mol)	Substance	S^0 (J/K mol)
Ag(s)	42.702	P(s, white)	44.4
Al(s)	28.321	P(s, red)	63.2
Ba(s)	66.9	Pb(s)	64.81
Be(s)	9.54	S(g)	167.716
Br(g)	174.912	S(s, monoclinic)	32.55
Br_2(g)	245.34	Sn(s, white)	44.8
Br_2(l)	152.3	Sn(s, gray)	51.5
C(s, diamond)	2.377	Zn(s)	41.21
C(s, graphite)	5.694	AgF(s)	84
Cl(g)	165.088	AgCl(s)	96.11
Cl_2(g)	222.949	AgBr(s)	107.11
Cu(s)	33.30	AgI(s)	114.2
F(g)	158.645	Al_2O_3(s)	50.986
F_2(g)	203.3	$BaCl_2$(s)	125
H(g)	114.612	BeO(s)	14.10
H_2(g)	130.587	CH_4(g)	202.50
Hg(l)	77.4	C_2N_2(g)	242.09
I(g)	180.682	CO(g)	197.907
I_2(g)	260.580	CO_2(g)	213.639
I_2(s)	116.7	CS_2(g)	237.82
K(s)	63.856	CuO(s)	42.7
Mg(s)	32.51	$CuSO_4$(s)	113.4
N(g)	153.197	$CuSO_4 \cdot 5H_2O$(s)	305.4
N_2(g)	191.489	D_2O(g)	198.234
Na(s)	51.417	D_2O(l)	75.990
O(g)	144.218	H_2O(g)	188.724
O_2(g)	205.029	H_2O(l)	69.940
O_3(g)	237.7	HF(g)	173.51
HCl(g)	186.678	$SnCl_4$(l)	258.6
HBr(g)	198.476	$ZnCl_2$(s)	43.9
HI(g)	206.330	$ZnSO_4$(s)	124.7
H_2S(g)	205.64	CH_4(g)	186.19
H_2SO_4(l)	156.86	C_2H_6(g)	229.49
KF(s)	66.57	C_3H_8(g)	269.91
KCl(s)	82.68	n-C_4H_{10}(g)	310.12
KBr(s)	96.44	n-C_5H_{12}(g)	348.95
KI(s)	104.35	n-C_6H_{14}	388.40
KNO_3(s)	132.93	C_2H_2(g)	200.819
K_2SO_4(s)	175.7	C_2H_4(g)	219.45
MgO(s)	26.8	C_6H_6(g)	269.20
$MgSO_4$(s)	95.4	CH_3OH(l)	126.8
NH_3(g)	195.51	CH_3OH(g)	237.7
NH_4Cl(s)	94.6	C_2H_5OH(l)	160.7
NO(g)	210.618	C_2H_5OH(g)	282.0
NO_2(g)	211.17	CH_3COOH(l)	159.8
N_2O_4(g)	304.30	CH_3COOH(g)	293.3
PCl_3(g)	311.67	CF_4(g)	262.3
PCl_5(g)	352.7	CCl_4(g)	309.41
PH_3(g)	210.0	CBr_4(g)	358.2
PbS(s)	91.2	CHF_3(g)	223.84
$PbSO_4$(s)	147.3	$CHCl_3$(g)	296.48
SO_2(g)	248.53	$CHBr_3$(g)	331.29
SO_3(g)	256.23	CH_3F(g)	223.01
$SiCl_4$(g)	330.79	CH_3Cl(g)	234.18
SiH_4(g)	204.14	CH_3Br(g)	245.77

18.2 The Unattainability of Absolute Zero

We cannot leave the study of the third law without briefly commenting on the *unattainability of absolute zero.* Indeed, to some the statement that the absolute zero cannot be attained in any system using any sequence of processes with a finite number of steps *is* the third law of thermodynamics!

Suppose that we consider some system which is adequately described by two independent variables. Ordinarily—for example, for a fluid—the independent variables chosen would be V and T, or p and V. However, for our present purposes it is convenient to choose as the independent variables T and S. Then any other property of the system is expressed as a function of T and S, for example, $V = V(T, S)$. Consider some arbitrary property of the system, $Z = Z(T, S)$. In Fig. 18.3*a* are represented two lines along which Z has the constant values Z_1 and Z_2, respectively. The third law requires that in any isothermal process, such as that along the lines P_1P_2 or $P_1'P_2'$, the entropy change must tend to zero as T tends to zero. Therefore the curves of constant Z must come together at the origin where $T = 0$ and $S = 0$. Now suppose that we have available some process, based on the property Z, which can be used to reduce the temperature of the system. For example, the magnetic moments of the ions in a paramagnetic salt can be aligned in a strong magnetic field. Clearly, when the magnetic field is strong enough to align the moments, the thermal energy is small compared to the energy of interaction of an ionic moment with the field. If in the absence of the magnetic field the moments were random in spatial orientation, alignment must result in a reduction of the entropy of the salt. If the magnetization is carried out isothermally, the reduction of entropy must lead to a transfer of heat from the salt to the surroundings. If the magnetic field is then removed adiabatically, randomization of the moments must lead to a decrease in temperature, because no energy can be transferred to the salt. This method of attaining low temperatures, known as *adiabatic demagnetization,* was suggested independently by Debye (1926) and Giauque (1927). The hypothetical process based on Z to which we referred could be adiabatic demagnetization if the system were a paramagnetic salt, or any other process appropriate to the nature of the system; for our present purposes we need only hypothesize that a suitable process exists.

We imagine that it is possible to change Z isothermally from Z_2 to Z_1 along the path P_1P_2 of Fig. 18.3*b*, and then to change Z from Z_1 to Z_2 along the adiabatic path P_2P_3. If the system is a paramagnetic salt, Z and Z_2 correspond to zero magnetic field and large magnetic field, respectively. As a result of the process described, the temperature of the system has been reduced from T_1 to T_3 (see diagram). But, because the two curves merge at $T = 0$, it is not possible to reach $T = 0$ by the process described in a finite number of steps. As can be easily seen, each successive drop in temperature, corresponding to another cycle of operation, is smaller than the preceding one, and the limit $T = 0$ is approached only as the number of cycles of operation becomes indefinitely large. We therefore conclude that the principle of the unattainability of absolute zero is equivalent to the statement of the third law already given.[5] We leave it as an exercise for the reader to prove that if the third law were not valid, that is, if the isothermal entropy change in a process did not vanish as $T \to 0$, then absolute zero could be attained in a finite series of steps of some process or processes.

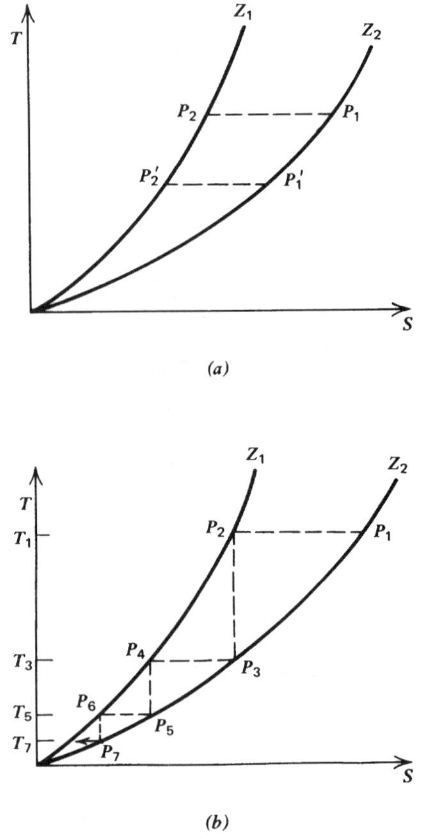

Figure 18.3 (*a*) Curves of constant Z in the (T,S) plane near $T = 0$. These are determined by the intersection of the surface $Z = Z(T,S)$ with the planes $Z = Z_1$ and $Z = Z_2$. (*b*) A hypothetical sequence of operations, based on the use of the variation of property Z, to reduce the temperature of a system toward $T = 0$.

[5] It is more correct to say that the Nernst heat theorem implies the unattainability of absolute zero, but that the converse implication is not valid; the unattainability principle can be shown to lead to the Nernst heat theorem only under restrictive conditions. There is, then, a residual inequivalence between these two formulations of the behavior of systems as $T \to 0$.

18.3 Experimental Verification of the Third Law

Although we opened this chapter by pointing out that experimental studies led to the postulation of the Nernst heat theorem, our analysis has presented the third law as a postulate that is made plausible by the statistical molecular theory of matter. In this section we return to the empirical side and examine a few experimental data related to verification and demonstration of the limitations of the third law.

Consider first the limiting behavior of C_p as $T \to 0$. In Fig. 18.4 are shown the values of C_p as a function of T for several substances. In every case it is clear that the natural extrapolation of the observations corresponds to $C_p \to 0$ as $T \to 0$. We shall see later, in Chapter 22, that for a nonconducting simple solid such as NaCl

$$\lim_{T\to 0} C_p \propto T^3, \tag{18.31}$$

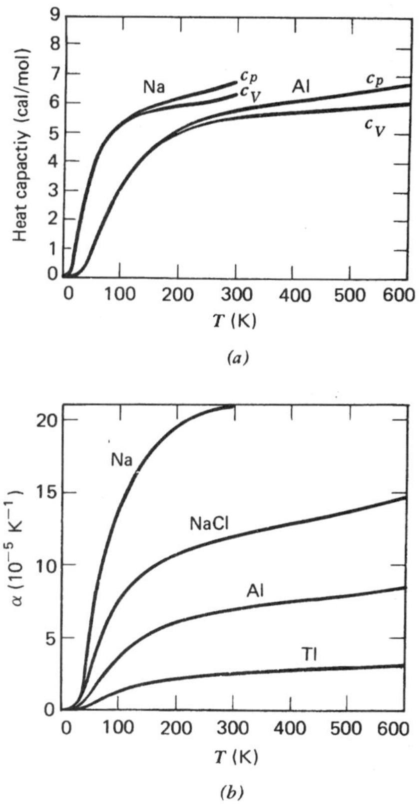

Figure 18.4 (*a*) Typical behavior of c_p and c_V as functions of temperature for two substances. (*b*) Experimental curves of the thermal expansion coefficient α as a function of T at $p = 0$; α is not necessarily positive, but $\alpha \to 0$ as $T \to 0$. Notable examples of negative thermal expansion are Ge and Si, where α is negative at low T, then becomes positive and continues to increase as T increases. In addition, for anisotropic crystals the thermal expansion may be quite different along the different crystallographic directions, as for example along the c and a axes for Zn and Cd. From D. C. Wallace, *Thermodynamics of Crystals* (Wiley, New York, 1972).

whereas for a simple metal

$$\lim_{T\to 0} C_p \propto \gamma T + \beta T^3. \tag{18.32}$$

Both relations 18.31 and 18.32 satisfy Eq. 18.20 (see also footnote 1 following Eq. 18.18). Even when the low-temperature behavior of the specific heat of the solid is more complex, as shown in Fig. 18.5 for $KCr(SO_4)_2 \cdot 12H_2O$, it is still true that $C_p \to 0$ as $T \to 0$.

It will be shown, in Chapter 21, that the thermodynamic functions of a gas can be calculated from a knowledge of the properties of an isolated molecule. For example, for the case of a diatomic molecule we must know the mass, the moment of inertia, the vibration frequency, and the degeneracy of the electronic ground state. From these we compute the so-called *spectroscopic entropy*. It can be compared with the absolute calorimetric entropy obtained from Eq. 18.26 with $T_0 = 0$. Some typical values are shown in Table 18.2. It is easily seen that for most of the examples cited the residual entropy S_0 (Eq. 18.16) is zero. This is not the case for CO, which provides an example of freezing in of metastable configurations as $T \to 0$. Since CO has a very small dipole moment and is almost symmetric, an individual CO molecule can be distributed in the crystal

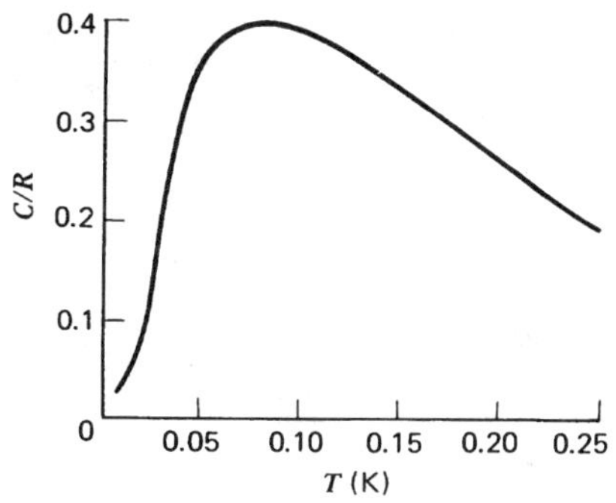

Figure 18.5 Specific heat of potassium chrome alum in units of R. Adapted from B. Bleaney, *Proc. Roy. Soc.* (*Lond.*) **A204,** 216 (1950).

Table 18.2 Comparison of Calorimetric and Spectroscopic Entropies of Some Diatomic Gases

Substance	Calorimetric Entropy (J/K mol)	Spectroscopic Entropy (J/K mol)	Residual Entropy (J/K mol)
N_2	192.0	191.6	0
O_2	205.4	205.1	0
HCl	186.2	186.8	0
HBr	199.2	198.7	0
HI	207.1	206.7	0
CO	193.3	198.0	$5.76 \equiv R \ln 2$

with two orientations with respect to a fixed reference frame. These two orientations, CO · · · CO · · · CO · · · and CO · · · OC · · · OC · · · ,[6] have nearly the same energy. Presumably, if complete equilibrium is achieved the ordered structure CO · · · CO · · · CO is obtained. Since it requires energy to rotate a CO molecule in the force field of its neighbors, a disordered configuration can be frozen in if equilibrium is not established before T drops to where k_BT is less than the energy of reorientation. Assuming that the CO molecules are randomly distributed with respect to orientation, and that all the many disordered distributions have about the same energy, there is an effective ground-state degeneracy of the crystal of 2^N,[7] hence for CO,

$$S_0 = Nk_B \ln 2 = 5.763 \text{ J / K mol}.$$

This value of S_0 is in good agreement with the observed difference between the calorimetric and spectroscopic entropies of CO.

We shall not comment further on the experimental verification of the third law. It is sufficient to say that in addition to the examples cited there are many others that verify its accuracy and utility.

FURTHER READING

Bailyn, M., *A Survey of Thermodynamics* (AIP Press, New York, 1994), Chapter 8.

Caldin, E. F., *An Introduction to Chemical Thermodynamics* (Oxford University Press, Oxford, 1958), Chapter 8.

Epstein, P. S., *Textbook of Thermodynamics* (Wiley, New York, 1937), Chapters 13–15.

Pitzer, K. S., *Thermodynamics,* 3d ed. (McGraw Hill, Inc., New York, 1995), Chapter 6.

Wilson, A. H., *Thermodynamics and Statistical Mechanics* (Cambridge University Press, Cambridge, 1957), Chapter 7.

PROBLEMS

1. Show that Eq. 18.25 is valid.

2. Suppose that the equation of state of a solid is

$$pv + G(v) = \Gamma u,$$

where $G(v)$ is a function of volume only, Γ is a constant, and u is the internal energy per mole. Prove that c_v approaches zero as T approaches zero.

3. At $T = 0$, $(\partial S/\partial V)_T = 0$, and also

$$\left[\frac{\partial}{\partial V}\left(\frac{\partial S}{\partial V}\right)_T\right]_{T=0} = 0.$$

Using these facts, prove that

$$\lim_{T\to 0}\left[\frac{\partial(1/\kappa_T)}{\partial T}\right]_V = 0.$$

4. Calculate the standard entropy changes in the following reactions:
   (a) $CH_4(g) + 2H_2S(g) = CS_2(g) + 4H_2(g)$,
   (b) $CO_2(g) + H_2S(g) = OCS(g) + H_2O(g)$,
   (c) $2SO_2(g) + O_2(g) = 2SO_3(g)$,
   (d) $C_2H_4(g) + H_2(g) = C_2H_6(g)$.

5. Calculate the entropy change in the reaction

$$H_2(g) + \tfrac{1}{2}O_2(g) = H_2O(g)$$

at 1000 K and 2 atm:
   (a) Assuming the equations of state of all substances to be $pv = RT$,
   (b) Assuming the equations of state of all substances to be

$$\left(p + \frac{a}{v^2}\right)(v - b) = RT$$

Data:

H_2	$c_p = 2.87 + 1.17 \times 10^{-3}T - 0.92 \times^{-6}T^2$ J/K mol
O_2	$c_p = 36.16 + 0.845 \times 10^{-3}T - 4.310 \times 10^{-5}T^{-2}$ J/K mol
H_2O	$c_p = 30.1 + 11.3 \times 10^{-3}T$ J/K mol
H_2	$a = 0.244 \times 10^{-6}$cm^6 atm/mol^2
O_2	$a = 1.36 \times 10^{-6}$cm^6 atm/mol^2
H_2O	$a = 5.46 \times 10^{-6}$cm^6 atm/mol^2
H_2	$b = 26.6$ cm^3/mol
O_2	$b = 31.8$ cm^3/mol
H_2O	$b = 30.5$ cm^3/mol

6. According to the Debye model of crystals (see Chapter 22), the heat capacity of crystals at low temperatures is given by $C_V = AT^3$, where A is a constant for any crystal sample. Show that $C_p - C_V$ at low temperature is proportional to T^7.

7. Calculate the calorimetric entropy of Na_2SO_4 at 100 K from the following data:

[6] This notation is not intended to represent the actual molecular distribution, but only to suggest that there is an ordered structure (symbolized by CO · · · CO · · · CO · · ·) and many disordered structures (symbolized by CO · · · OC · · · OC · · ·), where the CO can have two orientations at each lattice site.

[7] Since each molecule has two equivalent orientations.

T	c_p (J/K mol)
13.74	0.715
16.25	1.197
20.43	2.619
27.73	6.757
41.11	18.18
52.72	29.42
68.15	43.85
82.96	55.56
95.71	64.14

8. Calculate the entropy of gaseous nitromethane at 298.15 K from the following data:

T	c_p (J/K mol)
15	3.72
20	8.66
30	19.20
40	28.87
50	35.69
60	40.84
70	44.77
80	47.89
90	50.63
100	52.80
120	56.74
140	60.46
160	64.06
180	67.74
200	71.46
220	75.23
240	78.99
260	104.64
280	105.31
300	106.06

The melting point of the solid is 244.7 K with $\Delta h_{\text{fus}} =$ 9703 J/mol. The vapor pressure at 298.15 K is 36.66 mm Hg, and the heat of vaporization at that temperature is $\Delta h_{\text{vap}} = 38{,}271$ J/mol.

9. Which of the following solids do you expect to have nontrivial residual entropies at, say, 1 K? Explain your conclusion.
(a) O_2
(b) H_2O
(c) $AgCl_{0.5}Br_{0.5}$
(d) NaCl

10. Prove that if the isothermal entropy change in a process did not vanish as $T \to 0$, then absolute zero could be attained in a finite series of steps of some process or processes.

11. Three forms of phosphine are known; β-phosphine exists from 0 K to 49.43 K, at which temperature it transforms to α-phosphine, and γ-phosphine exists from 0 K to 30.29 K, at which temperature it also transforms to α-phosphine. Using the data tabulated below [from C. C. Stephenson and W. F. Giauque, *J. Chem. Phys.* **5,** 149 (1937)], devise a test of the third law of thermodynamics based on the computation of the entropy of α-phosphine at 49.43 K by two different routes.

T	c_p (β-phospine) (J/K mol)	T	c_p (γ-phospine) (J/K mol)	T	c_p (α-phospine) (J/K mol)
15.80	4.502	17.71	8.314	31.34	34.31
18.63	6.535	20.58	11.049	32.22	36.53
21.71	8.745	23.65	14.36	33.46	39.25
25.04	11.263	25.55	16.70	34.21	42.51
28.62	13.807	27.62	20.20	34.84	44.14
33.31	17.44	29.05	23.41	35.33	47.49
37.13	20.44	29.29	24.30	35.66	abrupt decrease
41.12	23.31	30.29	transition		
45.10	26.42		$\Delta h = 82.0$ J/mol	36.12	40.21
47.85	29.21			40.87	41.17
49.43	transition			44.17	42.59
	$\Delta h = 777$ J/mol			47.52	43.97
				51.14	45.48

12. The magnetic field on a solid with very low concentration of paramagnetic ions is increased from zero to **H**. The extra work required to generate the field **H** by virtue of the presence of 1 mol of magnetically polarizable material instead of vacuum is (see Table 13.1)

$$\Delta w = \mu_0 \int \mathbf{H} \cdot d\mathrm{M},$$

or, in differential form,

$$đw = \mu_0 \mathbf{H} \cdot d\mathrm{M},$$

where **M** is the molar magnetic polarization. Suppose that the volume of the solid is independent of magnetic field, so that no pV work is done as the field is changed. Show that the following are valid:

$$T\,ds = c_{\mathbf{M}}\,dT - \mu_0 T \left(\frac{\partial \mathbf{H}}{\partial T}\right)_{\mathbf{M}} d\mathbf{M},$$

$$T\,ds = c_{\mathbf{H}}\,dT + \mu_0 T \left(\frac{\partial \mathbf{M}}{\partial T}\right)_{\mathbf{H}} d\mathbf{H}.$$

Here μ_0 is the magnetic permeability of free space, $c_{\mathbf{M}}$ is the molar heat capacity at constant magnetization, and $c_{\mathbf{H}}$ is the molar heat capacity at constant field strength.

13. When the field on a paramagnetic solid is held constant and the temperature is varied, it is found that an increase in temperature always leads to a decrease in magnetization **M**. Show, using this fact, that in an isothermal reversible change of magnetic field strength, heat is rejected by the solid when **H** is increased and absorbed when **H** is decreased.

14. Suppose that the magnetic field strength acting on a paramagnetic solid is adiabatically and reversibly changed. If the temperature change produced by this adiabatic process, ΔT, is very small compared to T, and $c_{\mathbf{H}}$ can be regarded as independent of T over that range, show that

$$\Delta T = -\frac{\mu_0 T}{c_{\mathbf{H}}} \int_{\mathbf{H}_1}^{\mathbf{H}_2} \left(\frac{\partial \mathbf{M}}{\partial T} \right)_{\mathbf{H}} d\mathbf{H}.$$

What is the sign of ΔT for an adiabatic decrease of **H**?

15. The magnetization of gadolinium sulfate satisfies Curie's law,

$$\mathbf{M} = \frac{\mathscr{C}_M}{T} \mathbf{H},$$

where $\mathscr{C}_M$, the Curie constant, has the value 780 erg K/mol oersted^2 in cgs emu. In this system of units energy is measured in ergs, **H** in oersteds, and $\mu_0 = 1$. Given that the molar specific heat of gadolinium sulfate at constant field strength is

$$c_{\mathbf{H}} = (2.66 \times 10^7 + 7.80 H^2) T^{-2} \text{erg/K mol}$$

for $T < 15$ K, calculate the final temperature achieved if 1 mol is equilibrated at 15 K in a field of 10,000 oersteds and then the field is adiabatically and reversibly reduced to zero.

16. Very low temperatures cannot be easily measured; for example, gas thermometry fails. Suppose that a "magnetic temperature" is defined by

$$T^* \equiv \frac{\mathscr{C}_M}{\chi_M}.$$

$\mathscr{C}_M$ is the Curie constant (see Problem 15), and χ_M, the molar magnetic susceptibility, is defined by

$$\chi_M \equiv \frac{\mathbf{M}}{\mathbf{H}}.$$

Suppose also that a series of demagnetization curves from different fields and starting temperatures is used to calculate the entropy as a function of T^* at zero field and an apparent heat capacity defined by

$$C' = T^* \left(\frac{\partial S}{\partial T^*} \right)_{\mathbf{H}=0}.$$

In a separate series of experiments, by introduction of heat by absorption of γ-rays, a different apparent heat capacity is measured:

$$C'' = \lim_{\Delta T^* \to 0} \left(\frac{q}{\Delta T^*} \right)_{\mathbf{H}=0}.$$

Show that the thermodynamic scale of temperature is related to the magnetic scale by

$$T = T^* \frac{C''}{C'}.$$

CHAPTER 19

The Nature of the Equilibrium State

Thus far we have discussed how an equilibrium state of matter can be described in terms of macroscopic variables and thermodynamic functions and how these thermodynamic functions can be evaluated if the equation of state and heat capacities are known. We have also examined the nature of the relationship between the entropy and the energy spectrum of the system. We now investigate the nature of the equilibrium state in more detail. Our studies will be of several kinds. First, we examine what the first and second laws imply is characteristic of the equilibrium state, and what relations between thermodynamic variables are imposed by the condition that an equilibrium state be stable. Second, using the general characteristics of the equilibrium state thus obtained, we introduce several new relationships among thermodynamic functions, defined macroscopically, and the energy spectrum of the system. These new relationships act as bridges between the macroscopic and microscopic descriptions of a system; they supplement the fundamental definition $S \equiv k_B \ln \Omega$. Finally, we demonstrate how chemical reaction equilibria are related to the thermodynamic properties of the reactants and products.

19.1 Properties of the Equilibrium State of a Pure Substance

To investigate the properties of the equilibrium state of a system we must understand how the thermodynamic functions of the system depend on its mass. We begin with an analysis of the first law including mass transfer from one part of the system to another. Consider a closed system with two parts, I and II. As shown in Fig. 19.1, in part I, with volume V, there are n mol of a pure substance at temperature T and pressure p; in part II there are dn mol of the same substance at the same temperature and pressure, and occupying volume $v\, dn$, where v is the molar volume. Let the boundary between part I and the surroundings be deformable in such a way that, as V varies, the pressure of the substance inside never differs more than infinitesimally from the external pressure. Finally, let there be an insulated piston connected to II so that external work can be done on the dn mol of substance in that part of the system. The boundary between parts I and II is permeable to matter; it may be thought of as an imaginary boundary, or it may be real.

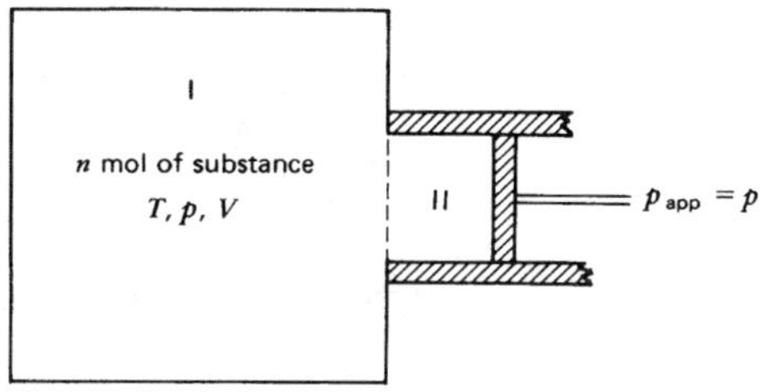

Figure 19.1 Schematic representation of a system with two parts. In part I, which has volume V, are n mol of substance at T and p. In part II are dn mol of the same substance, also at T and p. The piston shown permits the volume of part II to be decreased to zero. The dashed line indicates the boundary between parts I and II. Heat can be transferred to part I, but not to part II, and part I can do work on the surroundings.

We consider two states of the total system. In the initial state of the system the dn mol of substance in part II are outside the envelope defining the volume of part I. The internal energy of the total closed system is then[1]

$$nu + u\, dn,$$

where u is the molar internal energy of the substance. In the final state of the system the dn mol of substance in part II have been isothermally and reversibly driven at constant

[1] We take advantage of the fact that the state of a one-component system is defined by T and p, so that two samples of the same substance with the same T and p have the same internal energy per mole.

pressure p into part I, whose volume thereby changes from V to $V + dV$. In this change of state the surroundings coupled to the piston do work $pv\,dn$ on the system, the expansion of part I does work $+p\,dV$ on the surroundings, and heat $đq$ is transferred to part I from the surroundings. Let the internal energy in the final state be denoted by

$$nu + dU.$$

The first law gives, for the internal energy change of the total system in the process described,

$$dU - u\,dn = đq - p\,dV + pv\,dn. \tag{19.1}$$

Using the definition of the molar enthalpy,

$$h \equiv u + pv, \tag{19.2}$$

we find for the internal energy change of part I,

$$dU = đq + h\,dn - p\,dV, \tag{19.3}$$

which is the form taken by the first law when mass transfer is permitted.

Consider now the form taken by the second law when mass transfer is permitted. In the initial state of the system described above the entropy is

$$ns + s\,dn,$$

where s is the molar entropy. In the final state the entropy is denoted by

$$ns + dS.$$

As a result of the process by which the dn mol of substance are added to part I, there are entropy changes associated with the mass transfer, the work done, and the heat transferred. Since the dn mol of substance were added to part I isothermally and reversibly, the change in entropy of the total system, $dS - s\,dn$, must be

$$dS - s\,dn = \frac{đq}{T} = \frac{dU}{T} - \frac{h}{T}\,dn + \frac{p}{T}\,dV. \tag{19.4}$$

Using the definition of the molar Gibbs free energy,

$$g \equiv h - Ts, \tag{19.5}$$

we find that the generalization of Eq. 17.1 to include mass transfer is

$$dU = T\,dS - p\,dV + g\,dn. \tag{19.6}$$

The molar Gibbs free energy of a pure substance is also called the *chemical potential* of that substance, and is usually designated as μ. A more general definition of the chemical potential is given in Section 19.4.

Having obtained Eq. 19.6, we have the tools necessary to examine some of the most important characteristics of the equilibrium state of a one-component system. We consider, as a concrete example, the equilibrium properties of a one-component isolated system containing two phases. A liquid in contact with solid (e.g., ice and water) or a liquid in contact with gas (e.g., water and water vapor), enclosed in a rigid container with adiabatic walls, are simple particular cases of this situation.

The starting point in our analysis is the second law of thermodynamics. As discussed in Chapter 16, in an isolated system at equilibrium the entropy has the maximum value possible. This fact was expressed, formally, in the two conditions:

1. $\Delta S_{\text{inv}} < 0$ for a finite process with the original constraints on the system retained, (19.7*a*)
2. $\delta S \le 0$ for an infinitesimal process with the original constraints on the system retained. The inequality refers to situations in which virtual displacements in only one direction are allowed. (19.7*b*)

The character of an inverse process was defined in Section 16.5. Note that the equality sign in condition 19.7*b* implies that the entropy is an extremum with respect to positive and negative virtual displacements, and therefore the change of entropy in any such displacement starts with zero slope. The inequality sign implies that even though the entropy is an extremum, the change in entropy with the allowed unidirectional virtual displacement need not start with zero slope. We now rephrase the conditions 19.7 to make the role of the constraints that determine equilibrium more explicit.

The most general class of isolated systems is that for which only the mass and energy are specified. In such cases it is possible, for example, for the volume to change throughout the natural process that leads to establishment of the equilibrium state. Since the energy of an isolated system cannot change, Gibbs stated the most general criterion for equilibrium of such a system in the form:

For the equilibrium of any isolated system it is necessary and sufficient that, in all possible variations of the state of the system that do not alter its energy, the variation of its entropy shall either vanish or be negative.

Gibbs also stated the criterion for equilibrium in the form:

For the equilibrium of any isolated system it is necessary and sufficient that, in all possible variations of the state of the system that do not alter its entropy, the variation of its energy shall either vanish or be positive.

These statements, when suitably interpreted, include the possibilities of existence of stable, neutral, and metastable equilibria (see Section 19.3). When translated into formal mathematical statements, they read[2]

$$(\Delta S)_U \leq 0, \tag{19.8}$$

$$(\Delta U)_S \geq 0. \tag{19.9}$$

In keeping with the discussion of inverse processes in Chapter 16, the possible variations considered must not alter the constraints that determined the original equilibrium state. Aside from meeting this requirement, the possible variations include all changes in state produced by removing the system from isolation and operating on it with various processes. The new equilibrium states generated have the property that all will evolve to the original equilibrium state via natural processes when only the constraints that determine that original state are retained.

For some classes of isolated systems auxiliary constraints are specified—for example, that the volume of the system is fixed, or that there is a given subdivision of the fixed volume into subvolumes. Conditions 19.8 and 19.9 remain valid, the new constraints merely adding extra restrictions on the processes that can be used to generate different equilibrium states.

To understand how statement 19.8 is used, we suppose a system to be described by m variables. We represent the way in which the l constraints acting on the isolated system define its equilibrium state by a set of equations of the form

$$\begin{aligned} f_1(y_1,\ldots,y_m) &= 0, \\ f_l(y_1,\ldots,y_m) &= 0, \end{aligned} \tag{19.10}$$

where the y_i are the variables of the system. For example, in a one-component, two-phase isolated system the volumes of the two phases, V_1 and V_2, are subject to the constraint $V = V_1 + V_2$. When the isolation of the system is broken and various processes are used to drive it to a new equilibrium state, the values of the y_i may change. However, possible changes in the y_i are restricted by Eqs. 19.10, since in the inverse process the constraints on the isolated system are retained. From Eqs. 19.10 we find that the conditions which changes in the y_i must satisfy are

$$\begin{aligned} df_1 &= 0 = \frac{\partial f_1}{\partial y_1} dy_1 + \frac{\partial f_1}{\partial y_2} dy_2 + \cdots + \frac{\partial f_1}{\partial y_m} dy_m, \\ df_2 &= 0 = \frac{\partial f_2}{\partial y_1} dy_1 + \frac{\partial f_2}{\partial y_2} dy_2 + \cdots + \frac{\partial f_2}{\partial y_m} dy_m, \\ &\vdots \\ df_l &= 0 = \frac{\partial f_l}{\partial y_1} dy_1 + \frac{\partial f_l}{\partial y_2} dy_2 + \cdots + \frac{\partial f_l}{\partial y_m} dy_m. \end{aligned} \tag{19.11}$$

[2] The equivalence of these statements implies that if an isolated system is at equilibrium, all possible variations in its state must satisfy *both* Eqs. 19.8 and 19.9. It is not possible that some variations in its state satisfy 19.8 but not 19.9, or 19.9 but not 19.8.

Variations of the state of the system that satisfy Eqs. 19.11 are what we call virtual variations.

We now proceed to the analysis of the one-component, two-phase isolated system. By the definition of an isolated system:

1. The internal energy is constant;
2. The total volume is constant;
3. The total mass of the one component in the system is constant.

Our interpretation of Eq. 19.7*b* requires possible variations of the system, when it is no longer isolated, to satisfy

$$(\delta S)_{U,V,n} \leq 0, \tag{19.12}$$

where the subscripts denote the constraints that are retained. Let the superscripts (1) and (2) refer to the two phases in the system. Using Eqs. 19.10, we formulate the conditions of constraint in the form

$$\begin{aligned} U &= U^{(1)} + U^{(2)} = \text{constant}, \\ V &= V^{(1)} + V^{(2)} = \text{constant}, \\ n &= n^{(1)} + n^{(2)} = \text{constant}. \end{aligned} \tag{19.13}$$

Then the possible variations of the variables $U^{(1)}$, $U^{(2)}$, $V^{(1)}$, $V^{(2)}$, $n^{(1)}$, and $n^{(2)}$ are connected, as shown in Eqs. 19.11, by the relations

$$\begin{aligned} dU &= 0 = dU^{(1)} + dU^{(2)}, \\ dV &= 0 = dV^{(1)} + dV^{(2)}, \\ dn &= 0 = dn^{(1)} + dn^{(2)}. \end{aligned} \tag{19.14}$$

But, by the definition of an inverse process, the system remains in equilibrium throughout the variation in state imposed by the several processes operating. The combination of the first and second laws including mass transfer between parts of a closed system, Eq. 19.6, becomes in this case

$$T^{(1)}\, dS^{(1)} \leq dU^{(1)} + p^{(1)}\, dV^{(1)} - g^{(1)}\, dn^{(1)}, \tag{19.15a}$$

$$T^{(2)}\, dS^{(2)} \leq dU^{(2)} + p^{(2)}\, dV^{(2)} - g^{(2)}\, dn^{(2)}. \tag{19.15b}$$

The total entropy change is the sum of the entropy changes of phases 1 and 2; that is,

$$dS = dS^{(1)} + dS^{(2)}. \tag{19.16}$$

Equations 19.15*a* and 19.15*b* can easily be transformed to provide an explicit form of Eq. 19.16. After division of both sides of Eq. 19.15*a* by $T^{(1)}$, and of Eq. 19.15*b* by $T^{(2)}$, substitution into Eq. 19.16 yields

$$\begin{aligned} dS \leq \frac{dU^{(1)}}{T^{(1)}} + \frac{dU^{(2)}}{T^{(2)}} + \frac{p^{(1)}}{T^{(1)}}\, dV^{(1)} + \frac{p^{(2)}}{T^{(2)}}\, dV^{(2)} \\ - \frac{g^{(1)}}{T^{(1)}}\, dn^{(1)} - \frac{g^{(2)}}{T^{(2)}}\, dn^{(2)}. \end{aligned} \tag{19.17}$$

Equations 19.14 provide a set of relationships between the variables appearing in Eq. 19.17. Using Eqs. 19.14 to write $dU^{(1)} = -dU^{(2)}$, $dV^{(1)} = -dV^{(2)}$, $dn^{(1)} = -dn^{(2)}$, we reduce Eq. 19.17 to the form

$$dS \le \left[\frac{1}{T^{(1)}} - \frac{1}{T^{(2)}}\right] dU^{(1)} + \left[\frac{p^{(1)}}{T^{(1)}} - \frac{p^{(2)}}{T^{(2)}}\right] dV^{(1)} - \left[\frac{g^{(1)}}{T^{(1)}} - \frac{g^{(2)}}{T^{(2)}}\right] dn^{(1)}. \qquad (19.18)$$

Consider, first, the case that $dU^{(1)}$, . . . can be positive or negative. Then the equality sign describes the variation, and along the particular constrained path defined by Eq. 19.13 we set dS in Eq. 19.18 equal to zero. Now, the conditions of constraint, namely, Eqs. 19.10 or 19.13, are independent. Thus, the variations in Eqs. 19.11 or 19.14 are also independent. Then, if the right-hand side of Eq. 19.18 is to be equal to zero, as required by Eq. 19.12, it is necessary that the coefficients of $dU^{(1)}$, $dV^{(1)}$, and $dn^{(1)}$ separately vanish. We thereby conclude that

$$T^{(1)} = T^{(2)}, \qquad (19.19a)$$

$$p^{(1)} = p^{(2)}, \qquad (19.19b)$$

$$g^{(1)} = g^{(2)}. \qquad (19.19c)$$

To obtain Eqs. 19.19*b* and 19.19*c*, Eq. 19.19*a* has been used. The set of conditions 19.19*a*–19.19*c* must be characteristic of the original equilibrium state. It is important to observe that the various processes that lead to a change of state of the system need not be specified, and therefore the detailed nature of the new equilibrium state to which the system is displaced also need not be specified. This result is a consequence of the fact that Eqs. 19.11 and 19.14 are differential equations representing how the variables describing a system may change when a specified set of constraints is retained. Note that the retained constraints refer to the original isolated system and that there is one equilibrium condition for each constraint imposed. Provided that the final state is only infinitesimally displaced from the initial state, the conditions 19.19*a*–19.19*c* make no demands on the final state.

Consider, now, the case where one or more of $dV^{(1)}$, . . . can vary in only one direction. This situation could not arise in the case of equilibrium between two phases, as described above. But in a different case it could arise, for example, if the container holding the system had an internal movable partition that could not pass stops in one direction but could move unopposed in the other direction (giving, say, $dV^{(1)} > 0$ only). The equilibrium condition now requires use of the inequality in Eq. 19.18. Clearly, if $dV^{(1)} > 0$ only, then

$$\frac{p^{(1)}}{T^{(1)}} - \frac{p^{(2)}}{T^{(2)}} < 0$$

in order to satisfy the requirement $dS < 0$. The implication of this condition is that if energy can flow in both directions so $T^{(1)} = T^{(2)}$, then $p^{(1)} < p^{(2)}$. Successive application of this argument leads to the following conditions for the inverse process:

$$\text{if } dU^{(1)} > 0 \text{ only,} \quad T^{(1)} > T^{(2)}; \qquad (19.19d)$$

$$\text{if } dV^{(1)} > 0 \text{ only,} \quad \frac{p^{(1)}}{T^{(1)}} < \frac{p^{(2)}}{T^{(2)}}; \qquad (19.19e)$$

$$\text{if } dn^{(1)} > 0 \text{ only,} \quad \frac{g^{(1)}}{T^{(1)}} > \frac{g^{(2)}}{T^{(2)}}. \qquad (19.19f)$$

In the inverse process to which Eq. 19.18 refers, the system is driven from the original equilibrium state to a new equilibrium state. If the extra constraints that lead to this change of state are relaxed, the system returns to the original state, and an equation just like 19.18 except for reversal of the sign of the inequality describes the entropy change. Clearly, for the case that $dS > 0$ the inequalities in Eqs. 19.19*d*, 19.19*e* and 19.19*f* are reversed. We conclude that in a spontaneous process

$$\text{if } dU^{(1)} > 0 \text{ only,} \quad T^{(1)} < T^{(2)}; \qquad (19.19g)$$

$$\text{if } dV^{(1)} > 0 \text{ only,} \quad \frac{p^{(1)}}{T^{(1)}} > \frac{p^{(2)}}{T^{(2)}}; \qquad (19.19h)$$

$$\text{if } dn^{(1)} > 0 \text{ only,} \quad \frac{g^{(1)}}{T^{(1)}} < \frac{g^{(2)}}{T^{(2)}}. \qquad (19.19i)$$

In words, in the spontaneous approach to an equilibrium state energy flows from higher to lower temperature, volume from higher to lower pressure, and matter from higher to lower molar Gibbs free energy.

We now return to the more common case described by the equality in Eq. 19.18.

It may seem disappointing that two self-evident conditions result from our analysis, namely, the equality of temperatures and pressures in the two phases of the one-component system. However, condition 19.19*c*, the equality of the chemical potentials of the two phases, is by no means self-evident from earlier considerations. It is true that in the case considered, because the system remains in equilibrium everywhere along the path of the variation, the definitions of thermal and mechanical equilibrium lead directly to Eqs. 19.19*a* and 19.19*b*. We have carried through the analysis in this fashion because the technique used is generally valid; in situations where there exist complicated constraints, and where thermal or mechanical equilibrium may be modified by adiabatic walls, or rigid walls, it is not always so easy to write out the equivalents to Eqs. 19.19*a*–19.19*c*.[3]

---

[3] For example, consider the equilibrium between a pure liquid substance *a* and a liquid mixture containing *a* and other substances *b, c,* . . . . When the two are separated by a rigid disthermal wall that permits *a* to pass but not *b, c, d,* . . . , thermal equilibrium requires that $T^{(1)} = T^{(2)}$. However, determination of the pressures $p^{(1)}$ and $p^{(2)}$ requires use of the methods described, since the pressures are dependent on the permeability of the wall to component *a*. This phenomenon, known as *osmotic* equilibrium, will be studied in Chapter 25.

19.2 Alternative Descriptions of the Equilibrium State for Different External Constraints

The Gibbs criteria, Eqs. 19.8 and 19.9, describe equilibrium in an isolated system. But most of the systems of interest to us, both in nature and in the laboratory, are not isolated. For example, it is common to find systems constrained to have fixed temperature and pressure because of contacts with reservoirs. Clearly, Eqs. 19.8 and 19.9 are not applicable to the equilibria found in these cases. What relations can be used to characterize equilibrium states in systems in contact with their surroundings? This section is devoted to answering that question.

It is convenient to begin by remarking that condition 19.8 can be thought of as a test for the existence of equilibrium in an isolated system. When an isolated system is so tested, that is, when an inverse process is used to drive the system to a new equilibrium state, it is always found that work must be done on the system. Put in formal terms, the inverse process retains the constraint that the system's energy is fixed, so the work necessary to generate the new equilibrium state is

$$w_{\text{inv}} = -\int T\,dS. \tag{19.20}$$

Since T is positive, the integral in Eq. 19.20 can be evaluated by use of the mean value theorem. The result is

$$w_{\text{inv}} = -\bar{T}\int dS = -\bar{T}\Delta S_{\text{inv}}, \tag{19.21}$$

where $\bar{T}$ is some temperature in the range over which T varies during the change of state driven by the inverse process. But the second law of thermodynamics requires that $\Delta S_{\text{inv}} < 0$ if the isolated system is at equilibrium. Therefore,

$$w_{\text{inv}} > 0. \tag{19.22}$$

Alternatively, an inverse process can be used to test the existence of equilibrium in an isolated system via condition 19.9. In this case the retained constraint requires that the entropy of the system be constant, hence only reversible adiabatic paths can be employed to drive it to a new equilibrium state. In place of Eq. 19.20 we now have

$$w^S_{\text{inv}} = \int dU, \tag{19.23}$$

so that, using condition 19.9, the work done on the system is

$$w^S_{\text{inv}} > 0. \tag{19.24}$$

The superscript S is used in Eqs. 19.23 and 19.24 to indicate that in this inverse process the entropy of the system is fixed; a corresponding notation would replace w_{inv} in Eqs. 19.20–19.22 with w^U_{inv}. In general, when the particular retained constraint is C we write w^C_{inv}.

We have already noted that if the constraints added by an inverse process are removed, an isolated system will spontaneously revert to its original state of equilibrium. This spontaneous change of state could, if properly coupled to the surroundings, be used to do work. If the coupling is such as to retain the constraints on the original isolated system, and the work is done reversibly, the maximum work that can be done by the system in the reversion to the original equilibrium state is $-w_{\text{inv}}$.

Consider, now, a system that is not isolated. In this case the nature of the equilibrium state is determined by the kinds of reservoirs with which the system is in contact. To find the criteria that describe this equilibrium state, we examine the maximum work that can be done by the system by virtue of a change in its state. In analogy with our discussion of the behavior of isolated systems, we shall identify this maximum work with the negative of the work done on the system in an appropriately constrained inverse process, $-w^C_{\text{inv}}$. The constraints retained in the inverse process are the same as those used to define the equilibrium state of the system studied.

The systems of interest to us are closed, and the influence on them of contact with reservoirs is to fix intensive variables, such as T or p, at values determined by the reservoirs. We represent this as follows. The system of interest, call it I, is imbedded in a very large medium whose temperature and pressure are T_0 and p_0, respectively. This surrounding medium is contained by a rigid adiabatic envelope, and it is assumed to be so large that any change in state in system I does not alter T_0 or p_0 perceptibly. System I plus surroundings, as defined, constitute a composite system. We now calculate the greatest amount of work the composite system can do on its surroundings when system I undergoes the change in state $1 \rightarrow 2$.

It was shown in Chapter 17 that the maximum work that can be done by a system in contact with a single heat reservoir at temperature T_0 is

$$w_{\max} = \Delta U - T_0\Delta S. \tag{19.25}$$

But in the present case system I is imbedded in a medium that has the fixed pressure p_0. Therefore, if system I expands by ΔV, it must do work $-p_0\Delta V$ to push back its surroundings, and this work cannot be done on the surroundings of the composite system. We conclude that the maximum work that the composite system can do when system I undergoes the change of state $1 \rightarrow 2$ is

$$w'_{\max} = \Delta U + p_0\Delta V - T_0\Delta S, \tag{19.26}$$

which depends only on the change of state $1 \rightarrow 2$ and p_0, T_0 established by contact with the surrounding reservoir.

It is important to notice that when system I is in internal equilibrium and has temperature T_0 and pressure p_0,

$w'_{max} = 0$. This follows from the properties of the Gibbs free energy G. We showed in Chapter 17 that for an infinitesimal reversible process in a closed system

$$dG = -S\,dT + V\,dp, \tag{19.27}$$

so that $dG = 0$ for any reversible process during which p and T are fixed. But $G = U + pV - TS$, so that for a process at fixed p and T,

$$\Delta G = 0 = \Delta U + p\Delta V - T\Delta S, \tag{19.28}$$

which, when compared with Eq. 19.26, gives the result $w'_{max} = 0$.

It is proper, then, to interpret w'_{max} as the greatest possible work that system I can do on the environment of the composite system when the change of state $1 \to 2$ involves removal of a constraint that prevented achievement of internal equilibrium, that is, prevented achievement of equilibrium with respect to *only* the constraints defined by contacts between system I and the surrounding reservoir.

We now turn the argument around. Suppose that our system is at equilibrium while in contact with a reservoir that fixes the pressure and temperature at p_0 and T_0. To test that equilibrium we subject the system to an inverse process during which the constraints $T = T_0$, $p = p_0$ are maintained *and* a reversible work process is used to drive the system to a new state. The new state is in equilibrium with respect to the old constraints and any new ones added by the inverse process. The new state generated is unstable with respect to the original equilibrium state when only the constraints that determine that original state are retained. Hence, if the extra constraints associated with the reversible process that generated the new state are removed, a spontaneous change occurs and the original equilibrium state is regenerated. With the identification

$$w^C_{inv} = -w'_{max} \tag{19.29}$$

and the definition of a spontaneous process,

$$w'_{max} < 0, \tag{19.30}$$

we conclude that

$$w^C_{inv} > 0. \tag{19.31}$$

Combining Eqs. 19.31, 19.29, and 19.26, we have

$$(\Delta U + p_0\Delta V - T_0\Delta S)_{inv} > 0. \tag{19.32}$$

Since Eq. 19.32 must be satisfied for all possible variations of the state of the system that retain the original constraints,

$$(\delta U + p_0\delta V - T_0\delta S)_{p_0 T_0} = 0 \tag{19.33}$$

defines a minimum of the Gibbs free energy of the system.

We have derived Eqs. 19.32 and 19.33 for a particular choice of retained constraints, namely, $p = p_0$, $T = T_0$. Examination of the argument shows that the term $p_0\Delta V$ represents the amount of work associated with the retained constraints. The generalizations of Eqs. 19.32 and 19.33 to the description of equilibrium determined by an arbitrary set of constraints are

$$\left(\Delta U - w^C - \int T\,dS\right)_{inv} > 0, \tag{19.34}$$

$$(\delta U - \delta w - T\,\delta S)_C = 0, \tag{19.35}$$

where the superscript and subscript C's refer to the retained constraints. The function that is minimized is different for each choice of retained constraints.

Eq. 19.35 defines a different extremum for each set of retained constraints. To determine whether these extrema are maxima or minima we return to the fundamental equation 19.34. We already know that S is a maximum when the energy, volume, and mass are fixed. This follows from Eq. 19.34 since, under these conditions,

$$-\int T\,dS > 0$$

or

$$\Delta S < 0. \tag{19.36a}$$

Suppose that the system is defined under conditions such that the entropy, volume, and mass are fixed. Then Eq. 19.34 gives

$$\Delta U > 0, \tag{19.36b}$$

or, in words, the internal energy must be a minimum. Similarly, using Eq. 19.34 and the constraints S, p, n constant, T, V, n constant, T, p, n constant, we deduce that, respectively,

$$\Delta U - w^C > 0, \tag{19.36c}$$

$$\Delta U - \int T\,dS > 0, \tag{19.36d}$$

$$\Delta U - w^C - \int T\,dS > 0, \tag{19.36e}$$

which correspond to the enthalpy, Helmholtz free energy, and Gibbs free energy each being a minimum under the appropriate conditions. All these results are summarized in Table 19.1.

Suppose that the system under study can only do expansion work. The following criteria of equilibrium can be obtained from Eq. 19.35 and the inequality 19.34:

1. The system is defined under conditions of constant energy, volume, and mass:

$$(\delta S)_{U,V,n} \le 0; \tag{19.37}$$

2. The system is defined under conditions of constant entropy, volume, and mass:

$$(\delta U)_{S,V,n} \ge 0; \tag{19.38}$$

3. The system is defined under conditions of constant entropy, pressure, and mass:

$$(\delta U)_{S,p,n} + p(\delta V)_{S,p,n} \ge 0 \tag{19.39}$$

or

$$(\delta H)_{S,p,n} \ge 0; \tag{19.40}$$

4. The system is defined under conditions of constant temperature, volume, and mass:

$$(\delta U)_{T,V,n} - T(\delta S)_{T,V,n} \ge 0 \tag{19.41}$$

or

$$(\delta A)_{T,V,n} \ge 0; \tag{19.42}$$

5. The system is defined under conditions of constant temperature, pressure, and mass:

$$(\delta U)_{T,p,n} + p(\delta V)_{T,p,n} - T(\delta S)_{T,p,n} \ge 0 \tag{19.43}$$

or

$$(\delta G)_{T,p,n} \ge 0. \tag{19.44}$$

We now see that the Helmholtz and Gibbs free energies, A and G, introduced in Chapter 17 to help evaluate derivatives of the entropy function, play a fundamental role in describing the equilibrium state of a system; A is an extremum (a minimum) when the equilibrium is defined under the constraints of constant temperature, volume, and mass, and G is an extremum (a minimum) when the equilibrium is defined under the very common constraints of constant pressure, temperature, and mass.

At this point it is also convenient to tabulate some of the more important relations that are useful in manipulating the various thermodynamic functions. This tabulation is given in Table 19.2; the derivations of most of the relations should be obvious.

Table 19.1 Summary of Criteria for Stable Equilibrium

Defined Constraints	Function That Is an Extremum	Type of Extremum	For Virtual Variation
U, V, n constant	S	Maximum	$\Delta S_{\text{inv}} < 0$
S, V, n constant	U	Minimum	$\Delta U_{\text{inv}} > 0$
S, p, n constant	$H \equiv U + pV$	Minimum	$\Delta H_{\text{inv}} > 0$
T, V, n constant	$A \equiv U - TS$	Minimum	$\Delta A_{\text{inv}} > 0$
T, p, n constant	$G \equiv H - TS$	Minimum	$\Delta G_{\text{inv}} > 0$

Table 19.2 Summary of Some Useful Relations for a One-Component System

Function	Exact Differential Form[a]	Useful Relations	Cross-Differentiation Identities
U	$dU = T\,dS - p\,dV + \mu\,dn$	$C_V = T\left(\frac{\partial S}{\partial T}\right)_{V,n}$; $p = -\left(\frac{\partial U}{\partial V}\right)_{S,n}$; $T = \left(\frac{\partial U}{\partial S}\right)_{V,n}$; $\mu = \left(\frac{\partial U}{\partial n}\right)_{S,V}$	$\left(\frac{\partial T}{\partial V}\right)_{S,n} = -\left(\frac{\partial p}{\partial S}\right)_{V,n}$
H	$dH = T\,ds + V\,dp + \mu\,dn$	$C_p = T\left(\frac{\partial S}{\partial T}\right)_{p,n}$; $V = \left(\frac{\partial H}{\partial p}\right)_{S,n}$; $T = \left(\frac{\partial H}{\partial S}\right)_{p,n}$; $\mu = \left(\frac{\partial H}{\partial n}\right)_{S,p}$	$\left(\frac{\partial T}{\partial p}\right)_{S,n} = \left(\frac{\partial V}{\partial S}\right)_{p,n}$
$A \equiv U - TS$	$dA = -S\,dT - p\,dV + \mu\,dn$	$S = -\left(\frac{\partial A}{\partial T}\right)_{V,n}$; $p = -\left(\frac{\partial A}{\partial V}\right)_{T,n}$; $\mu = \left(\frac{\partial A}{\partial n}\right)_{T,V}$	$\left(\frac{\partial S}{\partial V}\right)_{T,n} = \left(\frac{\partial p}{\partial T}\right)_{V,n}$
$G \equiv H - TS$	$dG = -S\,dt + V\,dp + \mu\,dn$	$S = -\left(\frac{\partial G}{\partial T}\right)_{p,n}$; $V = \left(\frac{\partial G}{\partial p}\right)_{T,n}$; $\mu = \left(\frac{\partial G}{\partial n}\right)_{T,p}$	$\left(\frac{\partial S}{\partial p}\right)_{T,n} = -\left(\frac{\partial V}{\partial T}\right)_{p,n}$

[a] In this table we use μ for the Gibbs free energy per mole (chemical potential), previously denoted by g. For a pure component we have $g = G/n = \mu$. As will be seen in the next section, a more general definition, valid in a multicomponent system, is $\mu_i \equiv (\partial G/\partial n_i)_{T,p,n_j}$, where n_j $(j \ne i)$ is the number of moles of j in the system.

19.3 The Stability of the Equilibrium State of a One-Component System

There still remains the problem of the kind of equilibrium defined by the differential equation 19.35. That is, Eq. 19.35 is a necessary condition for an extremum, but is not in itself sufficient to determine if the equilibrium is stable.

In the discussion leading to Eq. 19.34, the arguments used assumed a case of stable equilibrium. That assumption entered through the use of the inequality 19.7. (The corresponding inequalities 19.36*c*, 19.36*d*, and 19.36*e* also refer to a stable equilibrium.) The importance of 19.35 is that it is a differential equation that can, in principle, be solved. The possible solutions describe the relationships between thermodynamic functions required by the retained constraints in the vicinity of an extremum of the entropy. Such relationships cannot be obtained only from the inequality 19.7. Of the many possible solutions to the differential equation, we select those corresponding to the case of stable equilibrium by using the inequality 19.7 as an extra condition—a generalized boundary condition, as it were.

In general, states of equilibrium (i.e., states for which Eq. 19.35 is satisfied) can be divided into four categories:

1. A system is in a state of *stable equilibrium* if, after displacement to a new state and release of the constraint causing the displacement, it always returns to the original state;
2. A system is in a state of *neutral equilibrium* if states that can be reached by virtual variation at constant energy, volume, and mass have the same entropy;
3. A system is in a state of *unstable equilibrium* if, after displacement to a new state and release of the constraint causing the displacement, it never returns to the original state;
4. A system is in a state of *metastable equilibrium* if displacements to nearby states satisfy condition 1, but displacements to distant states satisfy condition 3.

Schematic representations of the change of entropy for these cases are shown in Fig. 19.2*a*, and for the energy (at constant entropy, volume, and mass) in Fig. 19.2*b*.

All of the preceding discussion on the properties of the equilibrium state assumed that we were dealing with case 1. Of course, for many purposes—for example, if the variations of interest all lead to states in the neighborhood of the original state—cases 4 and 1 are equivalent. This is true even if there exist variations that demonstrate the metastability of the original state.

This categorization of equilibrium states can be connected with the discussion in Chapter 15 on the reversibility and irreversibility of processes that follow the removal of a constraint. In an equilibrium of type 1 the removal of the constraint causing the virtual variation leads to an irreversible approach to the original state. This is also true in case 4 for states close to the original state. On the other hand, if removal of the constraint results in a reversible process, then the state of equilibrium is neutral. This statement is clearly equivalent to that in Chapter 15, where a reversible process was shown to occur only if $\Omega_i = \Omega_f$. As might be anticipated, unstable equilibria do not actually occur; this will later be shown to be related to the properties of the fluctuations in the system.

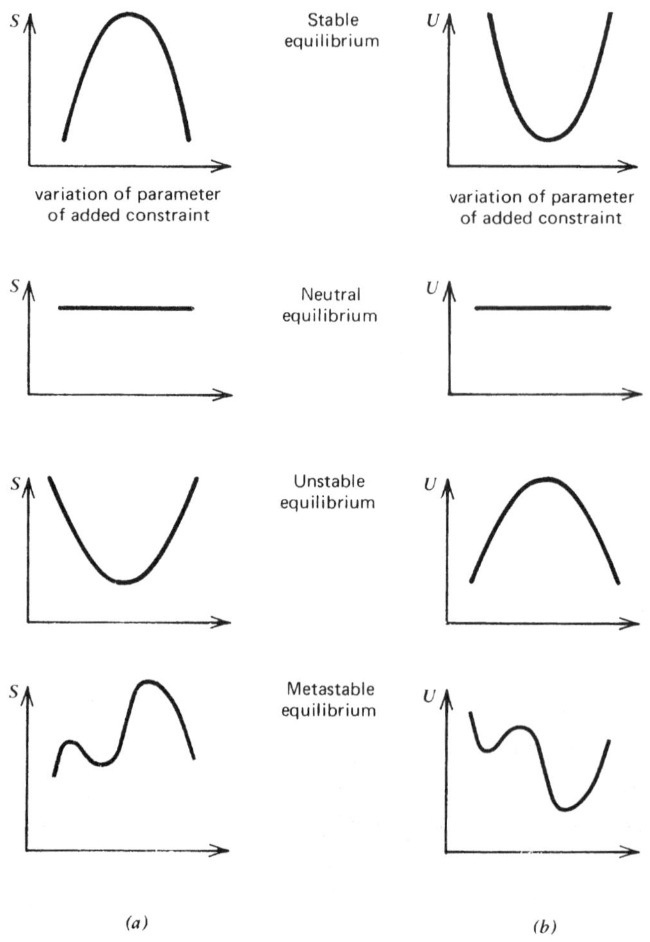

Figure 19.2 Schematic representations of the variations of S and U with change of added constraint for the four classes of equilibrium.

If we suppose that a system is in a state of stable equilibrium, satisfying the appropriate one of Eqs. 19.36*a*–19.36*e*, the properties the system may have are restricted. Certainly we must expect, for a stable equilibrium, the following common-sense observations to be valid:

1. An increase in the external pressure at constant temperature and mass must lead to a decrease in the volume of the system.
2. The addition of heat (at constant pressure or volume and mass) must lead to an increase in the temperature of the system.

To examine further the consequences of the definition of a stable equilibrium we study the virtual variation depicted in

Fig. 19.3. We shall use Eq. 19.36*b* to describe the virtual variation. We insert a partition at the midpoint of a rigid container of volume $2V$ containing some substance. Initially, each of the new subvolumes has the same volume V, entropy S, and number of moles n, and the system is still in stable equilibrium. We then produce a virtual variation by moving the partition to a point where the two parts of the system have volumes $V + \Delta V$ and $V - \Delta V$, and entropies $S + \Delta S$ and $S - \Delta S$, respectively. Suppose that ΔV and ΔS are, in this particular variation, sufficiently small that the internal energies of the two parts of the system can be accurately represented by

$$\begin{aligned} &U(S+\Delta S, V+\Delta V, n) \\ &= U(S, V, n) + \left(\frac{\partial U}{\partial S}\right)_{V,n}(\Delta S) + \left(\frac{\partial U}{\partial V}\right)_{S,n}(\Delta V) \\ &+ \frac{1}{2}\left[\left(\frac{\partial^2 U}{\partial S^2}\right)_{V,n}(\Delta S)^2 + 2\left(\frac{\partial}{\partial V}\left(\frac{\partial U}{\partial S}\right)_{V,n}\right)_{S,n}(\Delta V\,\Delta S)\right. \\ &\left. + \left(\frac{\partial^2 U}{\partial V^2}\right)_{S,n}(\Delta V)\right] \end{aligned} \tag{19.45a}$$

and

$$\begin{aligned} &U(S-\Delta S, V-\Delta V, n) \\ &= U(S, V, n) - \left(\frac{\partial U}{\partial S}\right)_{V,n}(\Delta S) - \left(\frac{\partial U}{\partial V}\right)_{S,n}(\Delta V) \\ &+ \frac{1}{2}\left[\left(\frac{\partial^2 U}{\partial S^2}\right)_{V,n}(\Delta S)^2 + 2\left(\frac{\partial}{\partial V}\left(\frac{\partial U}{\partial S}\right)_{V,n}\right)_{S,n}(\Delta V\,\Delta S)\right. \\ &\left. + \left(\frac{\partial^2 U}{\partial V^2}\right)_{S,n}(\Delta V)^2\right], \end{aligned} \tag{19.45b}$$

respectively. For a stable equilibrium it is then necessary that

$$\begin{aligned} \Delta U = {} & U(S+\Delta S, V+\Delta V, n) + U(S-\Delta S, V-\Delta V, n) \\ & - 2U(S, V, n) > 0. \end{aligned} \tag{19.46}$$

A combination of Eqs. 19.46, 19.45*a*, and 19.45*b* shows that the stability condition must be

$$\begin{aligned} &\left(\frac{\partial^2 U}{\partial S^2}\right)_{V,n}(\Delta S)^2 + 2\left(\frac{\partial}{\partial V}\left(\frac{\partial U}{\partial S}\right)_{V,n}\right)_{S,n}(\Delta V\,\Delta S) \\ &+ \left(\frac{\partial^2 U}{\partial V^2}\right)_{S,n}(\Delta V)^2 > 0. \end{aligned} \tag{19.47}$$

The left-hand side of the inequality in 19.47 is a quadratic form in the variables ΔS and ΔV. For that quadratic form to be positive for all possible values of ΔS and ΔV, it is necessary and sufficient that

$$\left(\frac{\partial^2 U}{\partial S^2}\right)_{V,n} > 0,$$

$$\left(\frac{\partial^2 U}{\partial S^2}\right)_{V,n}\left(\frac{\partial^2 U}{\partial V^2}\right)_{S,n} - \left(\frac{\partial}{\partial V}\left(\frac{\partial U}{\partial S}\right)_{V,n}\right)_{S,n}^2 > 0. \tag{19.48}$$

Then, using the combined first and second laws and the definitions of C_p, C_V, the isothermal compressibility κ_T, and the coefficient of thermal expansion α, we find that[4]

$$\left(\frac{\partial^2 U}{\partial S^2}\right)_{V,n} = \frac{T}{C_V}, \tag{19.49a}$$

$$\left(\frac{\partial^2 U}{\partial V^2}\right)_{S,n} = \frac{C_p}{V\kappa_T C_V}. \tag{19.49b}$$

$$\left(\frac{\partial}{\partial V}\left(\frac{\partial U}{\partial S}\right)_{V,n}\right)_{S,n} = -\frac{\alpha T}{\kappa_T C_V}. \tag{19.49c}$$

---

[4] The evaluation of $(\partial^2 U/\partial S^2)_{V,n}$, etc., in terms of measurable quantities, is an example of the utility of thermodynamic manipulations. The formulas given are derived as follows:

Eq. 19.49*a*: $\left(\frac{\partial^2 U}{\partial S^2}\right)_{V,n} = \left(\frac{\partial}{\partial S}\left(\frac{\partial U}{\partial S}\right)_{V,n}\right)_{V,n} = \left(\frac{\partial T}{\partial S}\right)_{V,n} = \frac{T}{C_V}.$

Eq. 19.49*b*: $\left(\frac{\partial^2 U}{\partial V^2}\right)_{S,n} = \left(\frac{\partial}{\partial V}\left(\frac{\partial U}{\partial V}\right)_{S,n}\right)_{S,n} = -\left(\frac{\partial p}{\partial V}\right)_{S,n}$

$= \left(\frac{\partial S}{\partial V}\right)_{p,n} \Big/ \left(\frac{\partial S}{\partial p}\right)_{V,n}$ using $\left(\frac{\partial p}{\partial V}\right)_{S,n}\left(\frac{\partial S}{\partial p}\right)_{V,n}\left(\frac{\partial V}{\partial S}\right)_{p,n} = -1,$

$= -\left(\frac{\partial p}{\partial T}\right)_{S,n} \Big/ \left(\frac{\partial V}{\partial T}\right)_{S,n}$ using the cross-differentiation identities,

$= -\dfrac{\left(\frac{\partial S}{\partial T}\right)_{p,n} \Big/ \left(\frac{\partial S}{\partial p}\right)_{T,n}}{\left(\frac{\partial S}{\partial T}\right)_{V,n} \Big/ \left(\frac{\partial S}{\partial V}\right)_{T,n}}$ using $\begin{cases} \left(\frac{\partial S}{\partial p}\right)_{T,n}\left(\frac{\partial p}{\partial T}\right)_{S,n}\left(\frac{\partial T}{\partial S}\right)_{p,n} = -1 \\ \left(\frac{\partial S}{\partial V}\right)_{T,n}\left(\frac{\partial V}{\partial T}\right)_{S,n}\left(\frac{\partial T}{\partial S}\right)_{V,n} = -1 \end{cases}$

$= -\dfrac{C_p \Big/ -\left(\frac{\partial V}{\partial T}\right)_{p,n}}{C_V \Big/ \left(\frac{\partial p}{\partial T}\right)_{V,n}}$ using the cross-differentiation identities,

$= \dfrac{C_p}{V\kappa_T C_V}$ using $\left(\frac{\partial V}{\partial T}\right)_{p,n}\left(\frac{\partial T}{\partial p}\right)_{V,n}\left(\frac{\partial p}{\partial V}\right)_{T,n} = -1.$

Eq. 19.49*c*: $\left(\frac{\partial}{\partial S}\left(\frac{\partial U}{\partial V}\right)_{S,n}\right)_{V,n} = \left(\frac{\partial}{\partial V}\left(\frac{\partial U}{\partial S}\right)_{V,n}\right)_{S,n} = \left(\frac{\partial T}{\partial V}\right)_{S,n}$

$= -\left(\frac{\partial S}{\partial V}\right)_{T,n} \Big/ \left(\frac{\partial S}{\partial T}\right)_{V,n}$ using $\left(\frac{\partial T}{\partial V}\right)_{S,n}\left(\frac{\partial V}{\partial S}\right)_{T,n}\left(\frac{\partial S}{\partial T}\right)_{V,n} = -1,$

$= -\left(\frac{\partial p}{\partial T}\right)_{V,n}\frac{T}{C_V}$ using the cross-differentiation identities,

$= T\left(\frac{\partial V}{\partial T}\right)_{p,n} \Big/ C_V\left(\frac{\partial V}{\partial p}\right)_{T,n}$ using $\left(\frac{\partial p}{\partial T}\right)_{V,n}\left(\frac{\partial V}{\partial p}\right)_{T,n}\left(\frac{\partial T}{\partial V}\right)_{p,n} = -1,$

$= -\dfrac{\alpha T}{\kappa_T C_V}.$

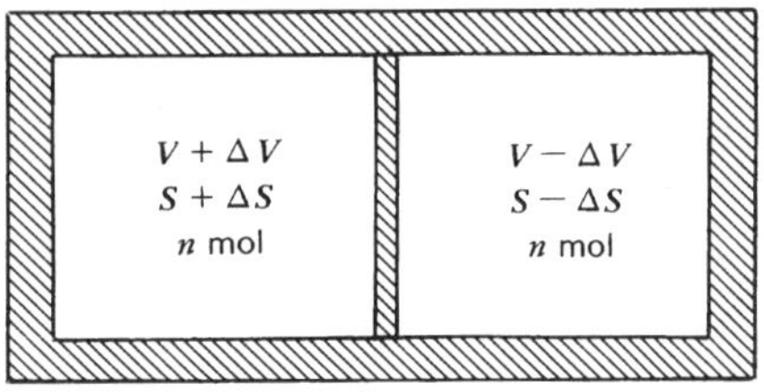

Figure 19.3 A system in a state generated by a virtual variation, for the retained constraints U = constant, V = constant, n = constant. The virtual variation is generated by moving the partition from an initial position in which $\Delta V = 0$ and $\Delta S = 0$ to the position shown.

Thus, the first of conditions 19.48 is equivalent to requiring that

$$\frac{T}{C_V} > 0 \quad \text{or} \quad C_V > 0, \tag{19.50a}$$

and therefore, since $C_p > C_V$,

$$C_p > 0. \tag{19.50b}$$

The second part of conditions 19.48 becomes

$$\frac{T}{C_V}\left(\frac{C_p}{V\kappa_T C_V}\right) - \frac{\alpha^2 T^2}{\kappa_T^2 C_V^2} > 0, \tag{19.50c}$$

implying that

$$\kappa_T > \frac{\alpha^2 TV}{C_p} > 0. \tag{19.50d}$$

The inequalities 19.50*b* and 19.50*d* are equivalent to the "common-sense" requirements for stability mentioned earlier in this section. By choosing different sets of constraints, still other conditions can be derived, for example, that the adiabatic compressibility $\kappa_S \equiv -(1/V)(\partial V/\partial p)_S$ must be positive. The inequalities derived are known collectively as *stability conditions,* since no stable one-phase, one-component system can exist for which they are not valid. This formulation of the properties of a stable equilibrium may be summarized in the following statement, known as *Le Chatelier's principle:*

It is a feature of stable equilibrium that spontaneous processes induced by a deviation from the initial state proceed in such a direction as to restore the system to the initial state.

19.4 The Equilibrium State in a Multicomponent System

Most systems of chemical interest contain more than one component. We must therefore extend our study of the properties of the equilibrium state to examine how, if at all, the presence of several components influences the deductions we have made in earlier sections of this chapter. As in Section 19.1, our first task is to learn how to write the first and second laws when it is possible to transfer some component from one part of a closed system to another.

One of the key concepts in the thermodynamic theory of multicomponent systems is that of the *semipermeable barrier.* A semipermeable barrier is one that permits the passage of one substance but not others. In the idealized thermodynamic concept the selectivity of this barrier is assumed perfect: it offers infinite resistance to the passage of all substances except one. Semipermeable barriers are very common, although real barriers are rarely perfectly selective. For example, hot Pd will permit the passage of gaseous H_2, but is impermeable to other gases; the walls of most living cells are permeable or partially permeable to some but not all organic and inorganic compounds (they are often completely impermeable to the macromolecules synthesized in the cell, while allowing the passage of the small molecules from which the macromolecules are synthesized); liquid films are permeable to gases that dissolve in the liquid, and resist the passage of gases only slightly soluble in the liquid. The thermodynamic concept is, as usual, an abstraction based on the observed behavior of real barriers.

Suppose that the system depicted in Fig. 19.1 is replaced by that in Fig. 19.4. The two systems differ only in that the latter has semipermeable barriers, one for each component, across which the mass of the substances added to the mixture must pass. Across each rigid semipermeable membrane there exists a pressure difference, since there is only one component on one side and r components on the other side. Let the pressures of each of the pure components, each in equilibrium across its own semipermeable membrane with the mixture, be $p_1, p_2, \ldots, p_r$, respectively. We define v_i,

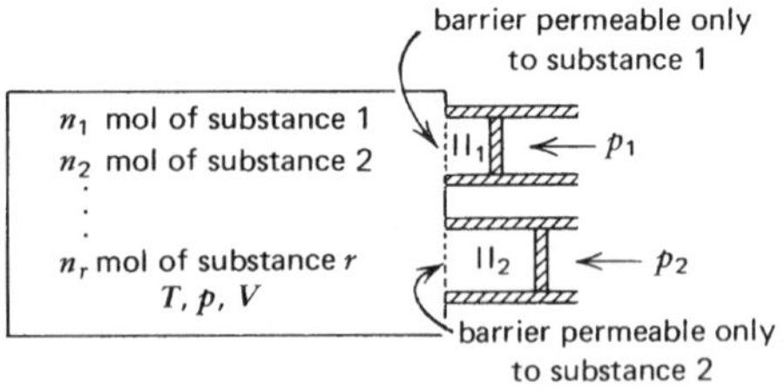

Figure 19.4 Schematic representation of an r-component system with $r + 1$ parts. In part I, which has volume V, are n_1 mol of substance 1, n_2 mol of substance 2, . . . , n_r moles of substance r, at T and p. In part II_1 are dn_1 mol of substance 1, at T and whatever pressure p_1 represents equilibrium across the semipermeable membrane. In general, this pressure is different from p. In part II_2 are dn_2 mol of substance 2, at T and whatever pressure p_2 represents equilibrium across the semipermeable membrane. There are r semipermeable membranes, each permitting only one of the components to enter part I; only two are shown in the diagram. The semipermeable membranes define the boundary of part I. Heat may be transferred from the surroundings to part I, but not to parts $II_1, II_2, \ldots, II_r$. Also, part I may do work on the surroundings.

u_i, s_i as the molar volume, internal energy, and entropy of the pure component in part II_i. We now proceed, exactly as in Section 19.1, to evaluate the change in internal energy resulting from the transfer of dn_i mol of substance i from part II_i into part I of the system. As before, heat $đq_i$ is transferred to part I from the surroundings, and the expansion of part I does work $(p\,dV)_i$ on the surroundings. We find, corresponding to Eq. 19.1, the internal energy change of the total system

$$dU_i - u_i\,dn_i = đq_i - (p\,dV)_i + p_i v_i\,dn_i. \qquad (19.51)$$

Using the definition

$$h_i \equiv u_i + p_i v_i \qquad (19.52)$$

we obtain, in analogy with Eq. 19.3, for the internal energy change of part I

$$dU_i = đq_i + h_i\,dn_i - (p\,dV)_i. \qquad (19.53)$$

In Eqs. 19.51–19.53, the subscript i refers to that contribution arising from the transfer of dn_i mol of i. Continuing as before, we finally obtain for the total change in internal energy of part I

$$dU = T\,dS - p\,dV + \sum_{i=1}^{r} \mu_i\,dn_i, \qquad (19.54)$$

with the definition

$$\mu_i \equiv h_i - Ts_i. \qquad (19.55)$$

Equation 19.55 is understood to apply to the substance i in equilibrium with the mixture across the semipermeable barrier that permits passage of i, that is, to substance i in part II_i. Equation 19.54 is the multicomponent-system generalization of Eq. 19.6.

Equations 19.54 and 19.55 are of such fundamental importance that we now give an alternative derivation that emphasizes a different view.

Suppose that the size of a system is increased by a constant factor ξ, for example, by increasing V to ξV, $n_1, \ldots, n_r$, to $\xi n_1, \ldots, \xi n_r$, and S to ξS. Since U is an extensive function, increasing the system size by the factor ξ must also increase U by the same factor. In mathematical form this reads

$$U(\xi S, \xi V, \xi n_1, \ldots, \xi n_r) = \xi U(S, V, n_1, \ldots, n_r). \qquad (19.56)$$

An alternative mathematical representation of the properties of an extensive function is derived from examining the rate of change of the function with the "size parameter" ξ. The differentiation of Eq. 19.56 with respect to ξ gives the simple result

$$\left[\frac{\partial U}{\partial(\xi S)}\right]_{V,n_j} \frac{\partial(\xi S)}{\partial \xi} + \left[\frac{\partial U}{\partial(\xi V)}\right]_{S,n_j} \frac{\partial(\xi V)}{\partial \xi} + \sum_{j=1}^{r}\left[\frac{\partial U}{\partial(\xi n_j)}\right]_{V,S,n_{i\neq j}} \frac{\partial(\xi n_j)}{\partial \xi} = U, \qquad (19.57)$$

or, after evaluation of $\partial(\xi S)/\partial\xi$, $\partial(\xi V)/\partial\xi$, and $\partial(\xi n_j)/\partial\xi$,

$$\left[\frac{\partial U}{\partial(\xi S)}\right]_{V,n_j} S + \left[\frac{\partial U}{\partial(\xi V)}\right]_{S,n_j} V + \sum_{j=1}^{r}\left[\frac{\partial U}{\partial(\xi n_j)}\right]_{V,S,n_{j\neq i}} n_j = U. \qquad (19.58)$$

But by the definition of an extensive function, Eqs. 19.56, 19.57, and 19.58 must be valid for all values of ξ. For the particular case $\xi = 1$, then, we have

$$\left(\frac{\partial U}{\partial S}\right)_{V,n_j} S + \left(\frac{\partial U}{\partial V}\right)_{S,n_j} V + \sum_{j=1}^{r}\left(\frac{\partial U}{\partial n_j}\right)_{V,S,n_{i\neq j}} n_j = U, \qquad (19.59)$$

from which we obtain

$$TS - pV + \sum_{j=1}^{r} \mu_j n_j = U, \qquad (19.60)$$

with the new definition

$$\mu_j \equiv \left(\frac{\partial U}{\partial n_j}\right)_{S,V,n_{i\neq j}}; \qquad (19.61)$$

μ_j is the chemical potential of component j. We shall see that this is the same as the function introduced in Eq. 19.55. In Eq. 19.60 we have assumed that a variation of U with S or with V, when the composition is fixed throughout the variation, must lead to the same results as do the corresponding variations in a one-component system. Therefore, we have made the identifications

$$\left(\frac{\partial U}{\partial S}\right)_{V,n_j} = T, \qquad \left(\frac{\partial U}{\partial V}\right)_{S,n_j} = -p.$$

From Eq. 19.60 and the definitions of the enthalpy, Helmholtz free energy, and Gibbs free energy, it follows that

$$H = U + pV = TS + \sum_{j=1}^{r} n_j \mu_j, \qquad (19.62)$$

$$A = U - TS = -pV + \sum_{j=1}^{r} n_j \mu_j, \qquad (19.63)$$

$$G = U - TS + pV = \sum_{j=1}^{r} n_j \mu_j. \qquad (19.64)$$

The chemical potential is particularly important in those common situations in which equilibria are defined at constant temperature and pressure, since under these constraints G is a minimum, and, by Eq. 19.64, G is simply related to the several μ_j.

From Eq. 19.60 it also follows that

$$dU = T\,dS + S\,dT - p\,dV - V\,dp + \sum_{j=1}^{r} \mu_j\,dn_j + \sum_{j=1}^{r} n_j\,d\mu_j. \tag{19.65}$$

But the internal energy is a function of the variables S, V, and all the n_j. Therefore, a differential change in U must have the form

$$dU = \left(\frac{\partial U}{\partial S}\right)_{V,n_j} dS + \left(\frac{\partial U}{\partial V}\right)_{S,n_j} dV + \sum_{j=1}^{r} \left(\frac{\partial U}{\partial n_j}\right)_{S,V,n_{i\neq j}} dn_j, \tag{19.66}$$

or

$$dU = T\,dS - p\,dV + \sum_{j=1}^{r} \mu_j\,dn_j. \tag{19.67}$$

Note that Eq. 19.67 is the same as Eq. 19.54. Comparison of Eqs. 19.65 and 19.67 shows that we must have

$$S\,dT - V\,dp + \sum_{j=1}^{r} n_j\,d\mu_j = 0, \tag{19.68}$$

which is known as the *Gibbs–Duhem equation.* Equation 19.68 plays the role of a restriction on the possible coupled variations of the intensive variables T, p, and μ_j.

Consider now the representation of G given in Eq. 19.64. We note that in a one-component system

$$G = n\mu, \tag{19.69}$$

or

$$g = h - Ts = \frac{G}{n} = \mu. \tag{19.70}$$

The usual notation for the chemical potential is μ, and we henceforth adopt this symbol. For a mixture, it follows from Eq. 19.67 that (see Table 19.2)

$$dH = T\,dS + V\,dp + \sum_{j=1}^{r} \mu_j\,dn_j, \tag{19.71}$$

$$dA = -S\,dT - p\,dV + \sum_{j=1}^{r} \mu_j\,dn_j, \tag{19.72}$$

$$dG = -S\,dT + V\,dp + \sum_{j=1}^{r} \mu_j\,dn_j. \tag{19.73}$$

But from Eq. 19.64 we have

$$dG = \sum_{j=1}^{r} n_j\,d\mu_j + \sum_{j=1}^{r} \mu_j\,dn_j, \tag{19.74}$$

and it follows that

$$\sum_{j=1}^{r} n_j\,d\mu_j = \sum_{j=1}^{r} n_j \left[\left(\frac{\partial \mu_j}{\partial T}\right)_{p,n_j} dT + \left(\frac{\partial \mu_j}{\partial p}\right)_{T,n_j} dp\right]. \tag{19.75}$$

Comparison of Eqs. 19.73, 19.74, and 19.75 shows that

$$-S = \sum_{j=1}^{r} n_j \left(\frac{\partial \mu_j}{\partial T}\right)_{p,n_j} = -\sum_{j=1}^{r} n_j \bar{s}_j, \tag{19.76}$$

$$V = \sum_{j=1}^{r} n_j \left(\frac{\partial \mu_j}{\partial p}\right)_{T,n_j} = \sum_{j=1}^{r} n_j \bar{v}_j. \tag{19.77}$$

In a multi-component system $\bar{s}_j$ and $\bar{v}_j$ are called the *partial molar* entropy and volume. For a one-component system $\bar{s}$ and $\bar{v}$ become the molar entropy and molar volume.

Finally, from Eqs. 19.71–19.73 we have the relations, equivalent to Eq. 19.61,

$$\mu_j = \left(\frac{\partial H}{\partial n_j}\right)_{S,p,n_{i\neq j}} = \left(\frac{\partial A}{\partial n_j}\right)_{T,V,n_{i\neq j}} = \left(\frac{\partial G}{\partial n_j}\right)_{T,p,n_{i\neq j}}; \tag{19.78}$$

we thus obtain

$$\mu_j = \left[\frac{\partial}{\partial n_j}(H - TS)\right]_{T,p,n_{i\neq j}} = \bar{h}_j - T\bar{s}_j, \tag{19.79}$$

using the general definition

$$\bar{z}_j \equiv \left(\frac{\partial Z}{\partial n_j}\right)_{T,p,n_{i\neq j}}. \tag{19.80}$$

We are now able to determine what restrictions are placed on the properties of a multicomponent system by the condition that it be in a state of stable equilibrium. Clearly, for any virtual variation in which the composition remains the same at all points in the system, the restrictions derived for a one-component system are valid, for, in this case, the homogeneous mixture behaves like a single substance. When the composition varies, Eq. 19.68 defines a restriction on the possible coupled variations of the intensive variables T, p, μ_i in a fluid system. Thus, at constant temperature and pressure, any change in the composition of a system leads to changes in the chemical potentials such that

$$\left(\sum_{i=1}^{r} n_i \, d\mu_i\right)_{T,p} = 0. \tag{19.81}$$

In a two-component system, addition of dn_1 mol of component 1 requires that

$$n_1 \left(\frac{\partial \mu_1}{\partial n_1}\right)_{T,p,n_2} = -n_2 \left(\frac{\partial \mu_2}{\partial n_1}\right)_{T,p,n_2}. \tag{19.82}$$

Since n_1 and n_2 are positive, the signs of $(\partial\mu_1/\partial n_1)_{T,p,n_2}$ and $(\partial\mu_2/\partial n_1)_{T,p,n_2}$ must be opposite. The extension to an r-component system is straightforward. To determine the sign of $(\partial\mu_1/\partial n_1)_{T,p,n_2}$ we construct a virtual variation for a two-component system defined to be at equilibrium at some T and p. Let the system consist of a deformable container filled with a mixture of $2n_1$ mol of 1 and $2n_2$ mol of 2 (see Fig. 19.5). Imagine that a semipermeable barrier (passing 1 but not 2) is inserted, dividing the container so that there are n_1 mol of 1 and n_2 mol of 2 on each side. Now let the barrier be displaced from the position in which it was inserted. In this displacement component 1 flows freely through the barrier, but however much of component 2 was on each side of the barrier remains there. We assume that during the displacement the temperature and pressure are maintained constant by exchange of heat and work with the surroundings. After the semipermeable barrier is displaced through some distance, an impermeable barrier is slipped in alongside it, and the combined barriers are displaced back to a position dividing the volume equally. In this second displacement the composition of the mixture on each side of the barrier remains fixed. Again we assume that T and p are maintained constant by exchange of heat and work with the surroundings. As a result of the operations described, we have generated a virtual variation in which, although T and p are the same, the composition of the mixture is different on the two sides of the barrier. Because the original state was one of stable equilibrium, in the virtual variation described it is necessary that

$$\Delta G > 0. \tag{19.83}$$

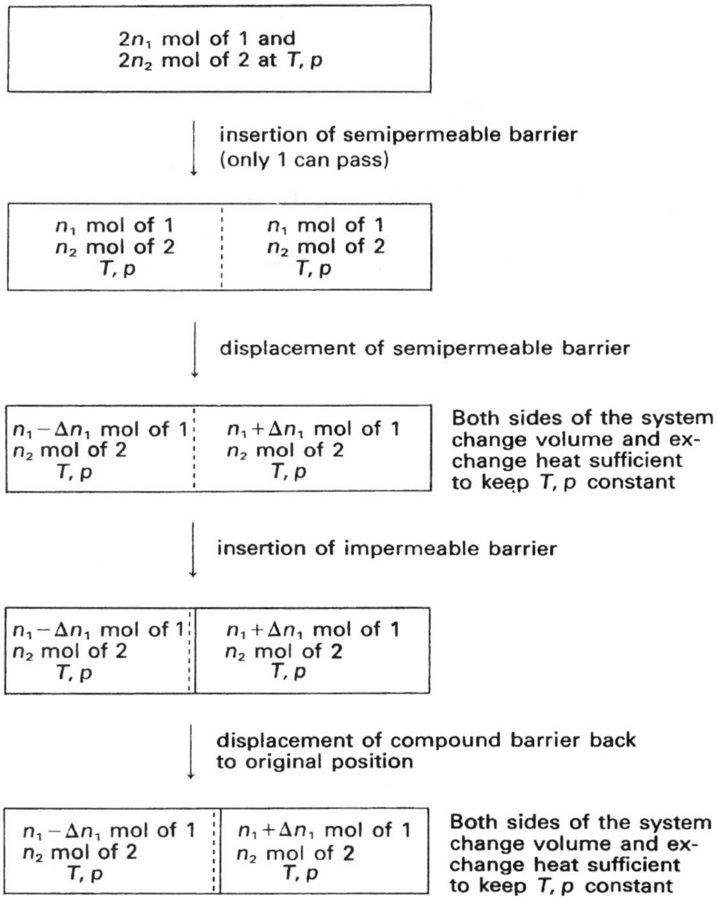

Figure 19.5 A virtual variation at constant T and p that generates a state with a composition difference from an originally homogeneous mixture.

Proceeding as before, we write for the Gibbs free energies on the two sides of the barrier $G(T,p,n_1 + \Delta n_1, n_2)$ and $G(T,p,n_1 - \Delta n_1, n_2)$. Then we must have

$$\Delta G \equiv G(T,p,n_1+\Delta n_1,n_2)+G(T,p,n_1-\Delta n_1,n_2) - 2G(T,p,n_1,n_2)>0. \tag{19.84}$$

Assuming Δn_1 to be small enough that a Taylor series expansion of G may be truncated after the first nonvanishing term, we find that

$$\Delta G = \left(\frac{\partial^2 G}{\partial n_1{}^2}\right)_{T,p,n_2} > 0. \tag{19.85}$$

Using the definition of the chemical potential, we see that Eq. 19.85 is equivalent to

$$\left(\frac{\partial \mu_1}{\partial n_1}\right)_{T,p,n_2} > 0, \tag{19.86}$$

and hence, by Eq. 19.82, to

$$\left(\frac{\partial \mu_2}{\partial n_1}\right)_{T,p,n_2} < 0. \tag{19.87}$$

The stability condition displayed in the inequality 19.85 states that the addition of component i at constant temperature, pressure, and mass of other components must lead to an increase in the chemical potential of component i.

The conditions 19.86 and 19.87 may be used to provide useful information about the chemical potential. Experiments show that $n_2(\partial\mu_2/\partial n_1)_{T,p,n_2}$ is negative, as required, and approaches a constant nonzero value, say, $-B$, for small values of the ratio n_1/n_2. Thus, as $n_1 \to 0$ since $n_2(\partial\mu_2/\partial n_1)_{T,p,n_2}$ becomes constant, $n_1(\partial\mu_1/\partial n_1)_{T,p,n_2}$ must approach the same constant value. But in order that

$$\lim_{n_1\to 0} n_1\left(\frac{\partial \mu_1}{\partial n_1}\right)_{T,p,n_2}$$

be nonzero, it is necessary that

$$\lim_{n_1\to 0}\left(\frac{\partial \mu_1}{\partial n_1}\right)_{T,p,n_2} = \infty. \tag{19.88}$$

Let

$$\lim_{n_1\to 0} n_2\left(\frac{\partial \mu_1}{\partial n_1}\right)_{T,p,n_2} = -B. \tag{19.89}$$

Then for small n_1, Eq. 19.82 gives

$$\left(\frac{\partial \mu_2}{\partial n_1}\right)_{T,p,n_2} = -\frac{B}{n_2} = -\frac{n_1}{n_2}\left(\frac{\partial \mu_1}{\partial n_1}\right)_{T,p,n_2}, \tag{19.90}$$

or

$$\left(\frac{\partial \mu_1}{\partial n_1}\right)_{T,p,n_2} = \frac{B}{n_1}. \tag{19.91}$$

The integration of Eq. 19.91 gives for the chemical potential of component 1 the simple formula

$$\mu_1 = B\ln n_1 + f(T,p,n_2), \tag{19.92}$$

where $f(T, p,n_2)$ is an arbitrary function of its arguments. But the chemical potential is an intensive function and cannot depend on the number of moles of 1 or 2, only on some function of their ratio. We note that for the pure component μ_1 can only be a function of T and p. We conclude that Eq. 19.92 should be rewritten as

$$\mu_1 = B\ln\frac{n_1}{n_1+n_2} + \mu_1^0(T,p). \tag{19.93}$$

In the pure fluid $n_1/(n_1 + n_2) \to 1$, and $\mu_1^0(T, p)$ is identified as its chemical potential. The constant B may depend on T and p; we shall later show that it has the value RT.

Equation 19.93 plays a fundamental role in chemical thermodynamics, since it gives an explicit representation of the change in Gibbs free energy with composition. We shall see in the next section how this information can be used to describe chemical equilibrium in a reacting mixture.

19.5 Chemical Equilibrium

As a first step in the description of chemical reactions, we must deduce the consequences of the condition stated in Eq. 19.37 when applied to chemical equilibrium.

Consider a closed homogeneous system in which only one chemical reaction occurs. The stoichiometry of the reaction can be expressed in the form

$$\sum_{i=1}^{r} \nu_i \mathrm{X}_i = 0, \tag{19.94}$$

where the ν_i are the stoichiometric coefficients and each X_i represents a mole of the ith component. Because of the conservation of mass, the changes in the amounts of the components are all linked through Eq. 19.94, and we can write

$$\frac{dn_1}{\nu_1} = \frac{dn_2}{\nu_2} = \cdots = \frac{dn_r}{\nu_r} = d\lambda, \tag{19.95}$$

where dn_i is the change in number of moles of component i. As before, the ν_i are positive for products and negative for reactants. The quantity λ is known as the *progress variable* of the reaction. Clearly, if $d\lambda/dt = 0$, no reaction is in progress, and λ (measured relative to some arbitrary zero) gives the equilibrium composition after the reaction is complete. Given Eq. 19.95, Eq. 19.38 can be used to define the nature of the equilibrium state. For a closed homogeneous system in which the masses of the several species are allowed to change, we have from Eq. 19.54, using Eq. 19.95, the condition

$$(dU)_{S,V} = \sum_{i=1}^{r} \mu_i \, dn_i = \sum_{i=1}^{r} \nu_i \mu_i \, d\lambda. \tag{19.96}$$

The extremum defined by Eq. 19.96 must be a minimum (see Eq. 19.36*b*), so that for sufficiently large displacements from equilibrium we have

$$\left[\left(\sum_{i=1}^{r} \nu_i \mu_i \right) \Delta\lambda \right]_{S,V} > 0. \tag{19.97}$$

Clearly $\Delta\lambda \equiv \lambda - \lambda_{\text{equil}}$ can be positive or negative, since there can be a mixture with either more or less product than at equilibrium. Thus,

$$\begin{aligned} \Delta\lambda > 0 \quad &\text{implies} \quad \left(\sum_{i=1}^{r} \nu_i \mu_i \right) > 0, \\ \Delta\lambda < 0 \quad &\text{implies} \quad \left(\sum_{i=1}^{r} \nu_i \mu_i \right) < 0, \end{aligned} \tag{19.98}$$

the two relationships holding on opposite sides of equilibrium; in the equilibrium state, then, we must have

$$\sum_{i=1}^{r} \nu_i \mu_i = 0. \tag{19.99}$$

Equation 19.99 states a necessary relationship between the chemical potentials of reacting species when equilibrium is attained.

We can now use Eqs. 19.99 and 19.93 to describe the equilibrium composition in a reacting mixture. First note that the generalization of Eq. 19.93 to a mixture with *r* components is

$$\mu_1 = B \ln \frac{n_1}{n_1 + \cdots + n_r} + \mu_1^0(T, p). \tag{19.100}$$

The quantity

$$x_1 \equiv \frac{n_1}{n_1 + \cdots + n_r} \tag{19.101}$$

is known as the *mole fraction* of component 1. Using an equation of the form of 19.100 for each component, we can write Eq. 19.99 in the form

$$\sum_{i=1}^{r} \nu_i [B \ln x_i + \mu_i^0(T, p)] = 0, \tag{19.102}$$

or

$$-\frac{1}{B} \sum_{i=1}^{r} \nu_i \mu_i^0(T, p) = \sum_{i=1}^{r} \ln x_i^{\nu_i}. \tag{19.103}$$

The left-hand side of this equation depends only on *T* and *p*, the right-hand side only on the equilibrium concentrations of each component. The entire equation implies that as the concentration of component *i* is changed at constant *T* and *p*, the concentrations of all other components must change so as to give the same value for $\Sigma \ln x_i^{\nu_i}$, since the sum must be independent of concentration by its equality with $-(1/B)\Sigma \nu_i \mu_i^0(T, p)$. As we have defined the chemical potential, each μ_i^0 is the molar Gibbs free energy of the *i*th pure component. The sum on the left-hand side of Eq. 19.103 is, therefore, just the difference in Gibbs free energy between products and reactants per mole of reaction (see also Chapter 14). Using $B = RT$, as will be shown later, and defining

$$\ln K(T, p) \equiv \sum_{i=1}^{r} \ln x_i^{\nu_i}, \tag{19.104}$$

we deduce that

$$\sum_{i=1}^{r} \nu_i \mu_i^0 = \Delta G_{\text{reac}} = -RT \ln K(T, p). \tag{19.105}$$

$K(T, p)$ is known as the *equilibrium constant*; it plays a fundamental role in the description of all chemical equilibria.

As derived, Eq. 19.92 is valid only when n_1/n_2 is small; the assumption that this form for the concentration dependence is correct at all concentrations is not, in general, valid. However, the corrections required do not alter our qualitative conclusions. Similarly, Eq. 19.104 requires corrections when Eqs. 19.92 or 19.100 do, but the qualitative conclusion remains valid that some combination of the concentrations can be found which does not itself vary with concentration, but only with *T* and *p*. These topics will be discussed in greater detail in later chapters.

19.6 Thermodynamic Weight: Further Connections between Thermodynamics and Microscopic Structure

As the last general topic in this chapter we examine the relationship between the various ways of describing an equilibrium state and possible new bridges between thermodynamic variables and the microscopic structure of a system. In Chapter 15 we introduced the fundamental definition

$$S \equiv k_B \ln \Omega. \tag{19.106}$$

This definition of the entropy in terms of the number of available states can be interpreted in a different way by reversing the roles of S and Ω. We thus rearrange Eq. 19.106 to read

$$\Omega = e^{S/k_B}. \tag{19.107}$$

In this form we can say that the entropy plays the role of a *thermodynamic weight.* That is, if we treat S as the independent variable, the entropy weighs most heavily those microscopic configurations for which $\Omega(E)$ is largest.

The ensemble we considered in defining $\Omega(E)$ is characterized by the statement that each of the isolated replicas has the same values of E, V, and N. This characterization is strikingly similar to that defining a macroscopic state in which S is a maximum at equilibrium. Can the other criteria which define equilibrium states be used to provide formulas analogous to 19.106? This is the problem we now seek to solve.

A hint to the nature of relationships analogous to 19.107 is obtained by transforming the observation just made into the following statement: The entropy is the thermodynamic weight for a microscopic configuration in an ensemble defined by a set of constraints that, when applied to a macroscopic system, lead to a maximum in the entropy. An obvious generalization of this observation is as follows: In an ensemble defined by some set of constraints, the probability of occurrence of a microscopic configuration is proportional to the positive exponential of that thermodynamic function which is a maximum in the corresponding macroscopic system. As noted in Table 19.1, $-A$ is a maximum and A is a minimum in a macroscopic system constrained to have constant V, T, and N. Thus, if an ensemble is defined by the requirements that V, T, and N be constant, we guess that the probability of occurrence of a microscopic configuration is proportional to $\exp(-A/k_BT)$. It will be shown below that this guess is correct.

Our analysis has two parts. First, we examine the characteristic features of the most probable state of an ensemble in which the energy and composition of the replica systems differ from replica to replica. Second, we establish the relationships between the probability of occurrence of a microscopic configuration, the energy spectrum of a replica system, and thermodynamic functions.

We consider an ensemble similar to that studied in Section 15.5. Let each replica in the ensemble consist of a pair of systems, A and B. The volumes V_A and V_B, and any other necessary constraints, are separately specified for systems A and B, and hence for each replica pair. For simplicity we shall assume that V_A and V_B have the common value V. The replica pair is further defined by specifying the total energy $E_T = E_A + E_B$, and the total number of molecules of each species present,

$$\begin{aligned} N_{T1} &= N_{A1} + N_{B1}, \\ N_{T2} &= N_{A2} + N_{B2}, \\ &\vdots \\ N_{Tr} &= N_{Ar} + N_{Br}. \end{aligned} \tag{19.108}$$

The replica pair is considered to be an isolated system, and the ensemble of replica pairs is also an isolated system. Furthermore, each of the members of any replica pair is isolated. However, because only the total energy and total numbers of molecules in the pair are specified, different replica pairs have different subdivisions of the energy between A and B, and different numbers of molecules of each species in A and B. Thus, in the ensemble of replica pairs all possible energy distributions and composition distributions are represented. Our problem is to find the most probable distribution of energy and composition.

Let $\Omega_A(E_A, N_{A1}, \ldots, N_{Ar})$ and $\Omega_B(E_B, N_{B1}, \ldots, N_{Br})$ be, respectively, the numbers of states between E_A and $E_A + dE_A$ in a system with numbers of molecules between N_{A1} and $N_{A1} + dN_{A1}, \ldots, N_{Ar}$ and $N_{Ar} + dN_{Ar}$, and between E_B and $E_B + dE_B$ in a system with numbers of molecules between N_{B1} and $N_{B1} + dN_{B1}, \ldots, N_{Br}$ and $N_{Br} + dN_{Br}$. The total number of states accessible to the replica pair with energy E_T, composition $N_{T1}, \ldots, N_{Tr}$, and subdivisions

$$\begin{aligned} E_A &= E_T - E_B, \\ N_{A1} &= N_{T1} - N_{B1}, \\ &\vdots \\ N_{Ar} &= N_{Tr} - N_{Br} \end{aligned} \tag{19.109}$$

is (see Eq. 15.34)

$$\begin{aligned} &\Omega(E_T, E_A, \{N_{Ti}\}, \{N_{Ai}\}) \\ &\quad = \Omega_A(E_A, \{N_{Ai}\})\Omega_B(E_T - E_A, \{N_{Ti} - N_{Ai}\}). \end{aligned} \tag{19.110}$$

To find the contributions to Ω_T from different distributions of energy and numbers of molecules between A and B we proceed as in Chapter 15. The key observation is that Ω_A increases and Ω_B decreases when E_A and $N_{A1}, \ldots, N_{Ar}$) increase. Thus, the product $\Omega_A\Omega_B$ has a maximum for some partition of the energy and composition. Since the composition variables are independent and the energy may be varied at fixed composition, this maximum is located by setting equal to zero, separately, the partial derivatives of Ω_T with respect to E_A and each of the N_{Ai}. It is found, by means similar to the derivation of Eq. 15.41, that the most probable distribution of energy and composition in the ensemble of replica pairs is characterized by the equalities

$$\begin{aligned} T_A &= T_B, \\ \mu_{A1} &= \mu_{B1}, \\ &\vdots \\ \mu_{Ar} &= \mu_{Br}, \end{aligned} \tag{19.111}$$

where

$$\frac{1}{k_BT_A} \equiv \left(\frac{\partial \ln \Omega_A}{\partial E_A}\right)_{V,N_{Ai}} \qquad (i = 1, \ldots, r), \tag{19.112}$$

$$\frac{\mu_{Ai}}{k_BT_A} \equiv -\left(\frac{\partial \ln \Omega_A}{\partial N_{Ai}}\right)_{V,E_A,N_{Ai}} \qquad (j \neq i = 1, \ldots, r). \tag{19.113}$$

The usual identification of the entropy with $k_B \ln \Omega$, combined with the general thermodynamic formulas of Eqs. 19.71–19.73, shows that μ_{Ai} is the chemical potential of component i in system A. In other words, the most probable state in an ensemble of replica pairs between which energy and matter may be distributed is characterized by equality between A and B of the temperature and the chemical potentials of all species. We identify the most probable state in the ensemble described with the equilibrium state of the corresponding macroscopic system.

The preceding analysis can be generalized. Let each replica consist of s isolated systems, each with fixed volume V. If only the total energy and total number of molecules of each species in the s systems are specified, different replicas can have different energy and composition distributions. It can then be shown that the most probable state in the ensemble is characterized by equality of temperature and equality of chemical potentials of all species between all s systems in the replica.

To determine the probability of finding a system with a particular energy and composition, we consider an ensemble of $\mathcal{N}$ replica systems, each with volume V. Each of the replica systems is isolated; however, we specify only the total energy and total numbers of molecules in the ensemble. We can now ask what is the most probable state of the ensemble itself, as opposed to a single replica. Imagine that the entire ensemble is itself a replica system in a super-ensemble of ensembles. We then have a situation exactly like that in the previous paragraph, and the corresponding conclusion can be drawn. The most probable state in the super-ensemble; that is, the state in which the largest number of ensembles should be found, is one in which there is equality of temperature and of the chemical potentials of all species between all replicas in a given ensemble.

Let the energy and numbers of molecules of species $1, \ldots, r$ in a given replica be E and $N_1, \ldots N_r$. We divide the energy spectrum of a replica into levels of width dE, and label the levels with the index k. There are, for each replica, some possible energy levels in the range $E_k \le E \le E_k + dE$; the number of such states of course depends on the composition of the replica. We denote by $\Omega(m)$ the number of quantum states available to the entire ensemble when exactly m replicas have the specific composition $N_1, \ldots, N_r$ and an energy in the range $E_k \le E \le E_k + dE$.[5] Since all these replicas have the same composition, they also have the same number of states between E_k and $E_k + dE$; we denote this number of states by g_k.

We regard the entire ensemble as an isolated system and compute $\Omega(m)$ for the purpose of calculating the entropy of the ensemble. By the definition of the energy range E_k to $E_k + dE$, each of the g_k states in that range is equally likely to be populated. Therefore, since the first replica of the m with energy between E_k and $E_k + dE$ can be in any of the g_k states, the second in any of the g_k states, and so on, there are g_k^m ways in which the m replicas can be distributed over the g_k states. Furthermore, since any of the $\mathcal{N}$ replicas in the ensemble is equally likely to belong to the subset m, the first selected can be any of $\mathcal{N}$, the second any of $\mathcal{N} - 1, \ldots,$ the mth any of $\mathcal{N} - m + 1$ replicas. However, since it is immaterial in what order the m replicas are selected, we must divide by $m!$, the number of equivalent ways of selecting the same m replicas. Thus, there are

$$\frac{\mathcal{N}(\mathcal{N}-1)(\mathcal{N}-2)\cdots(\mathcal{N}-m+1)}{m!} = \frac{\mathcal{N}!}{m!(\mathcal{N}-m)!} \tag{19.114}$$

different ways in which m of the $\mathcal{N}$ replicas can be selected. Finally, since each distribution of the m replicas over the g_k states can be combined with each of the possible choices of m replicas out of $\mathcal{N}$, the number of quantum states available to the ensemble with the constraint that exactly m replicas have the specified energy and composition can be written

$$\Omega(m) = \frac{\mathcal{N}!}{m!(\mathcal{N}-m)!} g_k{}^m \Omega_R(\mathcal{N}-m), \tag{19.115}$$

where $\Omega_R(\mathcal{N} - m)$ is the number of states available to the remaining $\mathcal{N} - m$ replicas of the ensemble. The quantity $\Omega_R(\mathcal{N} - m)$ must be understood to be evaluated with the ensemble considered as an isolated system, that is, with the total number of molecules of each species and the total energy and volume held constant.

We have already identified the most probable state of the ensemble with the equilibrium state of the corresponding macroscopic system. We have also stated that in the most probable state of the ensemble the systems have equal temperatures and chemical potentials of all species. As in the cases considered in Chapter 15, the distribution of energy and composition is so sharply peaked about the most probable distribution that the entropy of the ensemble can be defined in terms of the properties of the most probable state. The arguments justifying these remarks can be developed systematically using the techniques of Chapter 15.

Of course, since the ensemble is an isolated system, the entropy of the most probable state of the ensemble must be a maximum. The ensemble may be subjected to a virtual variation, as defined previously, by transferring one of the m replicas from the subset in the g_k states and adding it to the remainder of the system. Because m is a large number and the total energy, volume, and composition of the ensemble are fixed, this variation must not alter the entropy of the ensemble. We thus assume that

$$\delta S(m) = S(m-1) - S(m) = 0; \tag{19.116}$$

[5] As usual, the ensemble of replicas is assumed to be so large that, for every possible composition, there are enough replicas to give a distribution over all E_k of interest. We treat E and N differently because the possible values of N (the integers) are known independent of E, whereas obtaining E for a given N would require solution of a quantum mechanical problem.

we can expand $\delta S(m)$, by using Eq. 19.115, into

$$\begin{aligned}\delta S(m) &= k_B \ln m! - k_B \ln(m-1)! + k_B \ln(\mathcal{N}-m)! \\ &\quad - k_B \ln(\mathcal{N}-m+1)! - k_B \ln g_k - S_R(\mathcal{N}-m) \\ &\quad + S_R(\mathcal{N}-m+1) \\ &= k_B \ln \frac{m}{(\mathcal{N}-m+1)g_k} + [S_R(\mathcal{N}-m+1) \\ &\quad - S_R(\mathcal{N}-m)] \\ &= 0, \end{aligned} \tag{19.117}$$

where $S_R \equiv k_B \ln \Omega_R$ is the entropy of the $\mathcal{N} - m$ replicas of the "remainder." But the expression $[S_R(\mathcal{N} - m + 1) - S_R(\mathcal{N} - m)]$ is just the increase in entropy of the remainder of the ensemble when one of the m replicas with energy in the range $E_k \le E \le E_k + dE$, volume V, and composition $N_1, \ldots, N_r$ is added to it. We must now evaluate this quantity.

Suppose that $\mathcal{N} >> m >> 1$, and consider the limit $\mathcal{N} \to \infty$, in conformity with our initial discussion of the nature of ensembles in Chapter 15. Then we have

$$\begin{aligned}\delta S_R &= S_R(\mathcal{N}-m+1) - S_R(\mathcal{N}-m) \\ &= \Delta V_R \left(\frac{\partial S_R}{\partial V}\right)_{E,N} + \Delta E_R \left(\frac{\partial S_R}{\partial E}\right)_{V,N} \\ &\quad + \sum_{i=1}^{r} \Delta N_{Ri} \left(\frac{\partial S_R}{\partial N_i}\right)_{E,V,N_{j\neq i}} \quad . \end{aligned} \tag{19.118}$$

The use of derivatives is possible because, as $\mathcal{N} \to \infty$, the relative change in S_R arising from addition or subtraction of a single replica becomes infinitesimally small. Also, the changes ΔV_R, ΔE_R, and ΔN_{Ri} are just

$$\begin{aligned}\Delta E_R &= E_k, \\ \Delta V_R &= V, \\ \Delta N_{Ri} &= N_i, \end{aligned} \tag{19.119}$$

since just one replica is transferred in the virtual variation. A combination of Eqs. 19.118 and 19.119 leads to

$$\begin{aligned}\delta S_R &= V \left(\frac{\partial S_R}{\partial V}\right)_{E,N} + E_k \left(\frac{\partial S_R}{\partial E}\right)_{V,N} \\ &\quad + \sum_{i=1}^{r} N_i \left(\frac{\partial S_R}{\partial N_i}\right)_{E,V,N_{i\neq 1}} \\ &= \frac{pV}{T} + \frac{E_k}{T} - \frac{\sum_{i=1}^{r} N_i \mu_i}{T} \end{aligned} \tag{19.120}$$

where the derivatives have been evaluated on the basis of Eq. 19.54. Note that here μ_i is the chemical potential per molecule.

By our definition of an ensemble, the probability of finding a preselected state of a system is the ratio of the number of replicas in that state to the total number of replica systems in the ensemble; that is, if we select at random one replica system out of the whole ensemble of systems, the probability that it is one of the set m is just $(m/\mathcal{N})$. This ratio is the total probability of finding a system with volume V, energy in the range $E_k \le E \le E_k + dE$, and composition $N_1, \ldots, N_r$. With the fundamental hypothesis that each of the g_k states of this energy and composition has equal *a priori* probability, the probability of finding the system in a specified one of the g_k states is just $(m/g_k\mathcal{N})$. We now define the limit

$$P_{k,N}(T,V,\mu_1,\ldots,\mu_r \equiv \lim_{\substack{\mathcal{N}\to\infty \\ \left(\frac{m}{\mathcal{N}}\right)=\text{constant}}} \left(\frac{m}{\mathcal{N}-m+1}\right), \tag{19.121}$$

which becomes, with the use of Eqs. 19.117 and 19.120,

$$P_{k,N}(T,V,\mu_1,\ldots,\mu_r) \equiv g_k \exp\left(-\frac{E_k + pV - \sum_{i=1}^{r} N_i\mu_i}{k_B T}\right). \tag{19.122}$$

Since $\mathcal{N} >> m$, we see that $P_{k,N}$ is the probability $(m/\mathcal{N})$ of finding the system in any one of the g_k states with $E_k \le E \le E_k + dE$ and composition $N_1, \ldots, N_r$.

Given the probability 19.122, how can we use it to calculate the thermodynamic properties of a system from the energy spectrum of the system? Because $P_{k,N}$ is the probability of finding a system in the state specified by E_k and $N_1, \ldots, N_r$, and because the system must be in some one of all the possible states, the sum of $P_{k,N}$ over all possible states must be unity. That is, the probability is normalized to unity. Thus we can write

$$\sum_{N_1=0}^{\infty}\sum_{N_2=0}^{\infty}\cdots\sum_{N_r=0}^{\infty}\sum_{k} P_{k,N}(T,V,\mu_1,\ldots,\mu_r) = 1. \tag{19.123}$$

We again emphasize that the reason the sums in Eq. 19.122 contain the composition variables $N_1, \ldots, N_r$ is that the energy of a replica depends on its composition, and we must sum $P_{k,N}$ over all possible states, even those of very low probability where the composition deviates greatly from the most probable composition. Similarly, the sum over all energy level includes states of low probability. Equation 19.123 now provides a link between thermodynamic variables and the energy spectrum. For, using Eq. 19.122, we obtain

$$\begin{aligned}1 = \exp\left(\frac{-pV}{k_B T}\right) \sum_{N_1,\ldots,N_r=0}^{\infty} \exp\left(\sum_i \frac{N_i\mu_i}{k_B T}\right) \sum_k g_k \\ \exp\left[\frac{-E_k(V,N_1,\ldots,N_r)}{k_B T}\right], \end{aligned} \tag{19.124}$$

or

$$\frac{pV}{k_B T} = \ln \Xi, \tag{19.125}$$

where

$$\Xi \equiv \sum_{N_1,\ldots,N_r=0}^{\infty} \exp\left(\sum_i \frac{N_i \mu_i}{k_B T}\right) \sum_k g_k \exp\left[\frac{-E_k(V, N_1, \ldots, N_r)}{k_B T}\right]. \tag{19.126}$$

The quantity Ξ is called the *grand partition function,* and the ensemble to which it corresponds is called the *grand canonical ensemble.* Despite the formidable appearance of Eq. 19.126, we shall see that it is easy and convenient to use.

It should be noted, in confirmation of the conjecture made at the beginning of this section, that in a macroscopic system defined to have constant T, V, and $\mu_1, \ldots, \mu_r$ the quantity pV is a maximum.

A different ensemble corresponds to a macroscopic system defined to have constant T, V, and $N_1, \ldots, N_r$. In this case each replica in the ensemble has the same composition. We can imagine generating such an ensemble by selecting from the grand canonical ensemble described above only those replicas that each have the specified composition $N_1, \ldots, N_r$. As usual, we assume that the number of replicas thus obtained is very large. The subset of replicas defined to have the same volume and composition is now considered to be an isolated system. As such it constitutes an ensemble, called the *canonical ensemble.*

As generated, the canonical ensemble consists of a collection of isolated systems with the same composition but variable energy. Alternatively, the ensemble could have been constructed from scratch by assembling a collection of systems of the same composition and specifying only the total energy of the ensemble. The distribution of energy amongst the isolated replica systems is not specified—hence all possible energies are represented in the ensemble. Then, whichever way the canonical ensemble has been constructed, the most probable state of the ensemble is characterized by equality of the temperatures of all replica systems.

The probability of finding a system in a state of specified energy in a canonical ensemble is obtained from Eq. 19.122 by picking only those terms with the required values of $N_1, \ldots, N_r$, that is, only those terms corresponding to the specified uniform composition of all the replicas. Because a replica system must be in some one of the possible energy states, the normalization condition equivalent to Eq. 19.123 becomes

$$\sum_k P_k^{CE}(T, V, N_1, \ldots, N_r) = 1 \tag{19.127}$$

or

$$\exp\left(\frac{-pV}{k_B T}\right) \exp\left(\sum_i \frac{N_i \mu_i}{k_B T}\right) \sum_k g_k \exp\left(\frac{-E_k}{k_B T}\right) = 1. \tag{19.128}$$

But using Eqs. 19.63 and 19.64, we find that

$$A = -pV + \sum_{i=1}^{r} N_i \mu_i. \tag{19.129}$$

Combining Eqs. 19.128 and 19.129, we find that

$$\frac{A}{k_B T} = -\ln Q_{N_1,\ldots,N_r}, \tag{19.130}$$

where

$$Q_{N_1,\ldots,N_r} \equiv \sum_k g_k \exp\left(\frac{-E_k}{k_B T}\right) \tag{19.131}$$

is called the *canonical partition function.*

We again note, in support of our interpretation of thermodynamic weight, that in a macroscopic system constrained to have constant T, V, and $N_1, \ldots, N_r$ the function $-A$ is a maximum.

There is still another ensemble that can be generated from a subset of the systems making up the grand canonical ensemble. Suppose that we choose from the set of all replicas in the grand canonical ensemble that set which both has uniform concentration $N_1, \ldots, N_r$ and has energy in the range $E_k \le E \le E_k + dE$. This is, of course, just the set of m systems originally described in our derivation of Eq. 19.122. The m isolated systems constitute an ensemble known as the *microcanonical ensemble.* Since all m replica systems have the same composition and energy, the normalization condition equivalent to Eq. 19.123 becomes

$$P_k^{MCE}(E, V, N_1, \ldots, N_r) = 1, \tag{19.132}$$

or

$$\exp\left(\frac{-pV}{k_B T}\right) \exp\left(\sum_i \frac{N_i \mu_i}{k_B T}\right) g_k \exp\left(\frac{-E_k}{k_B T}\right) = 1. \tag{19.133}$$

Using Eq. 19.60 we recognize that

$$S = \frac{E_k}{T} + \frac{pV}{T} - \sum_{i=1}^{r} \frac{N_i \mu_i}{T}, \tag{19.134}$$

so that we find, by substitution,

$$S = k_B \ln g_k. \tag{19.135}$$

Of course, in the microcanonical ensemble we have defined, g_k is the same as the function usually denoted as $\Omega(E_k)$.

In all the preceding discussion we have emphasized that the replica systems constituting an ensemble are individually isolated. It is, nevertheless, a property of the most probable state of the ensemble that each replica has the same value of the intensive variable conjugate to an extensive quantity that may be distributed, for example, the temperature when only the total energy of the ensemble is fixed. Thus, these ensembles mimic the corresponding macroscopic situations in which systems in contact can transfer energy and/or mass. However, had energy and/or mass transfer been permitted between replica systems of the ensemble, no replica system would have had mechanical constants of motion, and our description in terms of well-defined energy states would have been inappropriate. Indeed, measurement of an intensive parameter necessarily imposes some uncertainty on the conjugate extensive quantity, since the "meter" must be open to exchange of the extensive quantity with the system whose properties are being measured (e.g., measurement of pressure requires "transfer" of volume, of temperature requires transfer of energy, etc.). Although it is often convenient to think of the ensembles described as if energy and/or mass can be transferred between replicas, the consequences of this interpretation and the properties of the ensemble as a model of a physical situation must always be remembered.

We close this section with a brief discussion of alternative representations of the entropy. For any system, the canonical partition function determines the Helmholtz free energy via Eq. 19.130, and the entropy of the system can then be obtained from the derivative of the Helmholtz free energy with respect to temperature at constant volume. An analogous procedure, using thermodynamic relationships, can be used to obtain the entropy from the grand partition function, Eq. 19.126. An alternative procedure for calculating the entropy starts with the differential form of the First Law of Thermodynamics

$$dU = T\,dS - p\,dV, \tag{19.136}$$

which we can rewrite to read

$$\begin{aligned} dU &= \sum_k E_k\, dP_k^{CE} + \sum_k P_k^{CE}\, dE_k \\ &= -k_B T\, d\left(\sum_k P_k^{CE} \ln P_k^{CE}\right) - p\,dV \end{aligned} \tag{19.137}$$

using

$$P_k^{CE}(T,V,N_1,\dots,N_r) = \frac{e^{-E_k/k_BT}}{Q_{N_1,\dots,N_r}} \tag{19.138}$$

and

$$\begin{aligned} U &= \sum_k E_k P_k^{CE} = -T^2\left(\frac{\partial}{\partial T}\left(\frac{A}{T}\right)\right)_V \\ &= k_B T^2\left(\frac{\partial \ln Q_{N_1,\dots,N_r}}{\partial T}\right)V. \end{aligned} \tag{19.139}$$

Comparison of Eqs. 19.137 and 19.136 yields

$$dS = -k_B d\left(\sum_k P_k^{CE} \ln P_k^{CE}\right), \tag{19.140}$$

which can be integrated to

$$S = -k_B \sum_k P_k^{CE} \ln P_k^{CE}. \tag{19.141}$$

The constant of integration that results from the transformation of Eqs. 19.140–19.141 is set equal to zero to guarantee agreement with the Third Law of Thermodynamics.

Similarly, using the same argument, it is possible to show that, starting with the probability distribution in a grand canonical ensemble,

$$S = -k_B \sum_{k,N_1,\dots,N_r} P_{k,N}(T,V,\mu_1,\dots\mu_r) \ln P_{k,N}(T,V,\mu_1,\dots,\mu_r). \tag{19.142}$$

Finally, we note that Eq. 19.135, the entropy defined in the microcanonical ensemble, can be written in the same form as Eqs. 19.141 and 19.142 when it is noted that the probability of finding a particular one of the g_k states with energy in the range $E_k \le E \le E_k + dE$ is $P_k^{MCE} = 1/g_k$ and that $\Sigma_k P_k^{MCE} = 1$, so that

$$\begin{aligned} S = k_B \ln g_k &= -k_B \sum_k \frac{1}{g_k} \ln \frac{1}{g_k} \\ &= -k_B \sum_k P_k^{MCE} \ln P_k^{MCE}. \end{aligned} \tag{19.143}$$

Eqs. 19.141, 19.142, and 19.143 clearly display how the entropy of a system is related to the distribution of systems with respect to the parameters that define the equilibrium state for that system.

19.7 An Application of the Canonical Ensemble: The Distribution of Molecular Speeds in a Perfect Gas

In Chapter 12 we introduced, for the purpose of describing the properties of a perfect gas, a function representing the distribution of molecular speeds. At that point we did not have available the tools necessary to determine the form of the distribution function. We close this chapter by showing how the representation of a perfect gas by a canonical ensemble permits calculation of the speed distribution function.

In using a canonical ensemble to represent a perfect gas, we must decide what the appropriate replica system is. By definition, in a perfect gas collisions between molecules may be neglected. Then each molecule can be considered to be "essentially" isolated; that is, those collisions that occur provide a mechanism for the transfer of energy and maintenance of equilibrium, but they are so infrequent that the energy spectrum of a molecule is unaltered from what it would be in the absence of collisions, and the energy of a molecule is constant for periods of time many orders of magnitude larger than the duration of a collision. A canonical ensemble that is a model of the perfect gas can then be constructed by regarding each molecule in the volume V to be an isolated replica system, and the collection of N molecules in the volume V to be an ensemble. Clearly, because the (infrequent) collisions do lead to transfer of energy between molecules, the ensemble representing the gas must allow for unequal distribution of the total energy of the gas amongst the molecules. Thus, the representative ensemble is chosen to be a canonical ensemble. A microcanonical ensemble, in which each replica has the same energy, would be inappropriate, as would a grand canonical ensemble, in which each replica has different composition.

We take as a model of the perfect gas a set of N independent molecules in a cubic box of volume V. The energy of one molecule in the box is

$$\epsilon_{n_1 n_2 n_3} = \frac{\pi^2 \hbar^2}{2mV^{2/3}}(n_1^2 + n_2^2 + n_3^2), \tag{19.144}$$

or

$$\epsilon_n = \frac{\pi^2 \hbar^2 n^2}{2mV^{2/3}}, \tag{19.145}$$

where $n^2 = n_1^2 + n_2^2 + n_3^2$. In the canonical ensemble we are using, ϵ_n is the energy of one replica system. Now the probability of finding a replica (molecule) with energy ϵ_n is

$$P_n^{CE}(\epsilon_n) = \frac{g_n \exp(-\epsilon_n / k_B T)}{\sum_n g_n \exp(-\epsilon_n / k_B T)}, \tag{19.146}$$

which is obtained simply by dividing Eq. 19.122 by the normalization condition 19.127. To evaluate Eq. 19.146 we must first compute g_n, the number of quantum states with energy in the range $\epsilon_n \le \epsilon \le \epsilon_n + d\epsilon$. In a macroscopic box the spacings between the energy levels of the spectrum are very small, and at ordinary temperatures n_1, n_2, and n_3 are very large. In what follows we shall find it convenient to take advantage of this fact and treat n_1, n_2, and n_3 as continuous variables.

Let n_1, n_2, and n_3 be three perpendicular axes defining a space. Each state of the molecule is defined by the triplet of numbers n_1, n_2, n_3; hence the energy of each state can be represented as a point in the space described. Equation 19.144, which gives the energy as a function of n_1, n_2, and n_3, describes a spherical surface. Since n_1, n_2, and n_3 can change only by integer values, there is one energy state per unit volume of n_1, n_2, n_3 space. Then the number of states with ϵ in the range $\epsilon_n \le \epsilon \le \epsilon_n + d\epsilon$ can be computed from the volume of the spherical shell of radius n and thickness dn.[6] Actually, since n_1, n_2, and n_3 must be positive, only one-eighth the volume of this shell may be counted, namely, that eighth lying in the positive octant of the space. Thus, the volume required is

$$g_n = \tfrac{1}{8} \cdot 4\pi n^2 \, dn. \tag{19.147}$$

The denominator of Eq. 19.146 can be evaluated by replacing the summation by an integration, treating n as a continuous variable. The result is

$$\begin{aligned} \sum_n g_n \exp\left(\frac{-\epsilon_n}{k_B T}\right) &= \frac{\pi}{2}\int_0^\infty n^2 \exp\left(\frac{-\pi^2\hbar^2 n^2}{2mV^{2/3}k_B T}\right) dn \\ &= \left(\frac{m k_B T}{2\pi\hbar^2}\right)^{3/2} V. \end{aligned} \tag{19.148}$$

We now convert the variable from n to v, the speed of a molecule. To do so we note that a molecule has only kinetic energy and that Eq. 19.145 can be rewritten to read

$$n^2 = \frac{2mV^{2/3}}{\pi^2\hbar^2}\epsilon = \frac{2mV^{2/3}}{\pi^2\hbar^2}\left(\frac{1}{2}mv^2\right). \tag{19.149}$$

From Eqs. 19.147 and 19.149, we easily find that

$$g_n = \frac{\pi}{2} n^2 \, dn = \frac{m^3 V v^2 \, dv}{2\pi^2\hbar^3}. \tag{19.150}$$

The substitution of Eqs. 19.150 and 19.148 into Eq. 19.146 gives, for the distribution of molecular speeds,[7]

$$\begin{aligned} P_v^{CE} &= f(v)\, dv \\ &= 4\pi v^2 \left(\frac{m}{2\pi k_B T}\right)^{3/2} \exp\left(\frac{-mv^2}{2k_B T}\right) dv. \end{aligned} \tag{19.151}$$

Equation 19.151, known as the *Maxwell speed distribution,* is one of the most fundamental relations of chemistry and physics. The distribution of molecular speeds plays an important role in determining the rates of chemical reactions and the thermodynamic properties of matter. We shall later see that if the interaction energy between a pair of molecules is independent of the molecular velocity, then Eq. 19.151 is equally valid for condensed phases and imperfect gases.

[6] Note that $d\epsilon$ is proportional to dn.

[7] Note that the "$f(v)$" in Chapter 12 is the number of molecules per unit volume with speeds between v and $v + dv$, while here $f(v)\,dv$ is the fraction of molecules with speeds between v and $v + dv$.

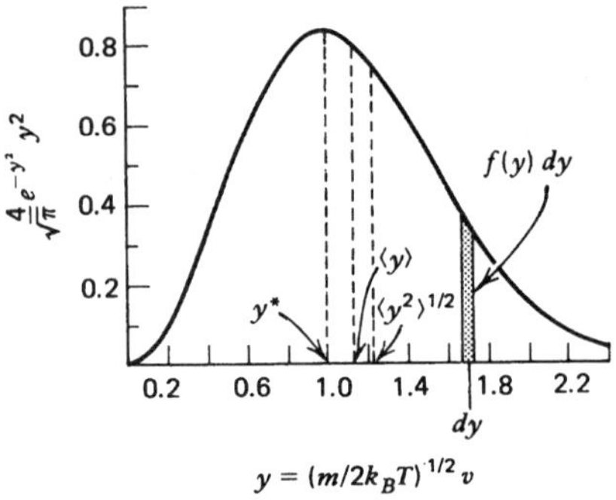

Figure 19.6 The Maxwell speed distribution. In the plot we use the reduced variable

$$y=(m/2k_BT)^{1/2}\,v.$$

In terms of y, the distribution function reads

$$f(y)\,dy=\frac{4}{\sqrt{\pi}}e^{-y^2}\,y^2\,dy.$$

From F. H. Crawford, *Heat, Thermodynamics and Statistical Physics* (Harcourt Brace Jovanovich, New York, 1963).

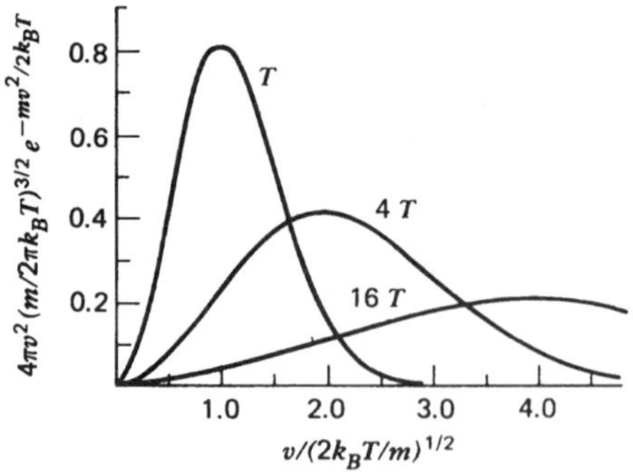

Figure 19.7 The Maxwell speed distribution for several temperatures. The unit along the horizontal axis is the most probable speed for the curve at the extreme left. The other two curves are for the temperatures $4T$ and $16T$, respectively. From F. H. Crawford, *Heat, Thermodynamics and Statistical Physics* (Harcourt Brace Jovanovich, New York, 1963).

The Maxwell distribution of speeds is so important that we digress to study a few of its features. Note that this distribution predicts there are no molecules with either zero or infinite speed. In between these extremes the distribution function has a maximum (see Fig. 19.6). The most probable speed, obtained by setting $df(v)/dv$ equal to zero, is

$$v^*=\left(\frac{2k_BT}{m}\right)^{1/2}, \tag{19.152}$$

which increases as the temperature increases, as does the breadth of the distribution (see Fig. 19.7).

Since the speed distribution function is not symmetric about the most probable speed, the mean speed $\langle v\rangle$ and the root-mean-square speed $\langle v^2\rangle^{1/2}$ differ from v^*. We have, by the definition of average values,

$$\langle v\rangle\equiv\frac{\int vf(v)\,dv}{\int f(v)\,dv}=\left(\frac{8k_BT}{\pi m}\right)^{1/2}, \tag{19.153}$$

$$\langle v^2\rangle\equiv\frac{\int v^2f(v)\,dv}{\int f(v)\,dv}=\left(\frac{3k_BT}{m}\right). \tag{19.154}$$

Thus,

$$\langle v\rangle=\frac{2}{\pi^{1/2}}v^*=1.128v^*, \tag{19.155}$$

$$\langle v^2\rangle^{1/2}=\left(\frac{3}{2}\right)^{1/2}v^*=1.224v^*. \tag{19.156}$$

These two average speeds are, respectively, 13% and 22% greater than the most probable speed.

Finally, we notice that the average kinetic energy of a molecule is

$$\frac{1}{2}m\langle v^2\rangle=\frac{3k_BT}{2}. \tag{19.157}$$

Recalling that the velocity distribution in a perfect gas must be isotropic, we have

$$\langle v^2\rangle=\langle v_x^2+v_y^2+v_z^2\rangle=\langle v_x^2\rangle+\langle v_y^2\rangle+\langle v_z^2\rangle \tag{19.158}$$

with

$$\langle v_x^2\rangle=\langle v_y^2\rangle=\langle v_z^2\rangle. \tag{19.159}$$

The average kinetic energy per degree of freedom is thus,

$$\frac{m}{2}\langle v_x^2\rangle=\frac{m}{2}\langle v_y^2\rangle=\frac{m}{2}\langle v_z^2\rangle=\frac{k_BT}{2}. \tag{19.160}$$

This result is one example of the *principle of equipartition of energy.* This principle states that the average energy associated with each degree of freedom in the mechanical motion of the system that is described by a squared term in the mechanical energy is the same, and equal to $k_BT/2$ per molecule. The theorem is valid only for classical mechanics, and only for squared terms in the energy expression. At high temperature, when the thermal energy exceeds all energy spacings in the spectrum, a real quantum mechanical system approaches the behavior of the corresponding classical system, so that the equipartition theorem is a useful general description of many systems of interest to chemists.

FURTHER READING

Balian, R., *From Microphysics to Macrophysics* (Springer-Verlag, Berlin, 1991), Volume I, Chapters 5 and 6.

Hill, T. L., *Statistical Mechanics* (McGraw Hill Book Co., New York, 1956), Chapters 1, 2, 3, and 4.

Kestin, J., *A Course in Thermodynamics,* Vol. II (Blaisdell, Waltham, 1968), Chapter 14.

Kirkwood, J. G., and Oppenheim, I., *Chemical Thermodynamics* (McGraw-Hill, New York, 1961), Chapter 6.

Mayer, J. E., *Equilibrium Statistical Mechanics* (Pergamon Press, Elmsford, N.Y., 1968), Chapters 1 and 8.

McQuarrie, D. A., *Statistical Mechanics* (Harper & Row, New York, 1976), Chapters 2 and 3.

Münster, A., *Statistical Thermodynamics* (Springer-Verlag, Berlin, 1969), Volume I, Chapters 1, 2, and 3.

Reiss, H., *Methods of Thermodynamics* (Blaisdell, Waltham, Mass., 1965), Chapters 7 and 8.

PROBLEMS

1. A perfect gas is contained in a cylinder fitted with a piston. The cylinder is insulated; hence the system is initially isolated, and any changes of state of the gas when isolation is "broken" are adiabatic. Let the pressure acting on the gas be p_0. Consider adding to the system just described new constraints that might change V or T, while retaining the constraint $p_{\text{app}} = p_0$. Show *by direct evaluation* that under the original conditions the entropy of the system was a maximum.

2. Consider a solid in equilibrium with its own vapor across an unconstrained interface. Let the vapor pressure be p_0 at temperature T_0. Suppose that a barrier is interposed between the solid and the vapor. The barrier is diathermal and permeable to molecules. If the barrier is rigid, so that the solid can be compressed (or fixed in volume as T increases), is equilibrium between solid and vapor at p_0 and T_0 possible? What differences do you expect between the system with and without the barrier as T changes?

3. Consider a chemical reaction involving perfect gaseous reactants and perfect gaseous products. Let the volume of the products at 1 atm be less than the volume of the reactants at 1 atm and the same temperature. If a closed container with a reaction mixture is compressed at constant temperature, how do you expect the equilibrium concentrations of the components to change?

4. If the reaction described in Problem 3 is constrained to:
(a) Take place in a fixed volume at constant temperature,
(b) Take place under constant external pressure and temperature in a vessel of variable volume,
(c) Take place at fixed volume in a thermally isolated container,

do you expect the equilibrium concentrations of the components to be the same? How would they change in (b) relative to (a)? In (c) relative to (a)? What are the most convenient thermodynamic descriptions of the equilibria under conditions (a), (b), and (c)?

5. Write down expressions for the average energy of a system represented by (a) a canonical ensemble and (b) a grand canonical ensemble. Show that

$$\langle U\rangle^{CE} = -\left(\frac{\partial \ln Q_N}{\partial(1/k_BT)}\right)_V,$$

$$\langle U\rangle^{GCE} = -\left(\frac{\partial \ln \Xi}{\partial(1/k_BT)}\right)_{\mu/T,V}.$$

6. Show that for a system represented by a grand canonical ensemble the average density, $\langle N\rangle/V$, is given by

$$\frac{\langle N\rangle}{V} = \frac{1}{V}\left(\frac{\partial \ln \Xi}{\partial(\mu/k_BT)}\right)_{T,V}.$$

7. Show that for a system represented by (a) a canonical ensemble and (b) a grand canonical ensemble the pressure is given by

$$p^{CE} = k_BT\left(\frac{\partial \ln Q_N}{\partial V}\right)_T,$$

$$p^{GCE} = k_BT\left(\frac{\partial \ln \Xi}{\partial V}\right)_{T,\mu}.$$

8. Show that in a perfect gas the distribution of kinetic energies has the form

$$f(\epsilon) = \frac{2}{\pi^{1/2}(k_BT)^{3/2}}\epsilon^{1/2}\exp\left(\frac{-\epsilon}{k_BT}\right).$$

9. Calculate the number of molecules in 1 mol of He at 1000 K that have components of velocity in the positive x direction of less than 10^4 cm/s and 10^5 cm/s. *Hint:* The answer can be expressed in terms of an integral that can be related to

$$\operatorname{erf}(x) \equiv \frac{2}{\sqrt{\pi}}\int_0^x e^{-x^2}\,dx,$$

known as the *error function.* Tables of numerical values of this function are available (see, for example, the *Handbook of Physics and Chemistry*). Cast your answer into a form such that you can use these tables for the numerical work.

10. Calculate the number of molecules in 1 mol of He at 1000 K that have speeds greater than 10^4 cm/s and 10^5 cm/s.

11. Set up the formulas that give the fraction of molecules in a perfect gas that have speeds between $\frac{1}{4}$ and 2 times the most probable speed. Evaluate this fraction for He at 298 K and 1 atm.

12. Evaluate the root-mean-square speed at 300 K for a molecule of UF_6 and a hydrogen atom.

13. Obtain molecular speed distributions, analogous to Eq. 19.143, for a gas whose molecules are constrained to move in two dimensions only, and also for a one-dimensional gas.

14. Equation 19.143 and the above problem deal with distributions of the magnitudes of molecular velocities. The magnitude can, of course, take on only non-negative values: $0 \leq v < \infty$. Now consider a gas whose molecules can move only along the x axis. Let v_x be the magnitude and direction of a molecule's velocity, so that v_x has the range $-\infty < v_x < \infty$. Obtain the distribution of v_x, $P(v_x)$, in terms of the one-dimensional speed distribution of the above problem. [Hint: For any v_x, one should have $P(v_x) = P(-v_x)$.]

15. Generalize your expression for $P(v_x)$ in the above problem to obtain an equation for the three-dimensional vector velocity distribution, $P(v_x, v_y, v_z)$. (Hint: v_x, v_y, and v_z should be independent and equivalent to one another.) If you now convert $P(v_x, v_y, v_z)$ to spherical coordinates and integrate over the angular coordinates, you should be able to rederive Eq. 19.151.

16. Check, by direct calculation, the validity of the Euler theorem for the homogeneous function

$$F_n = x^a y^b z^{n-a-b}.$$

The Euler theorem states that if F is a homogeneous function of its variables of degree n, say, $F(x^n, y^n, \ldots)$, then

$$x\frac{\partial F}{\partial x} + y\frac{\partial F}{\partial y} + \cdots = nF(x, y, \ldots)$$

(see Eqs. 19.56 and 19.57 of the text).

17. The standard Gibbs free energy change at 25°C for the dissociation of 1 mol of iodine vapor is 140.3 kJ. What fraction of iodine vapor at 25°C exists in the atomic state at (a) 1 torr pressure, (b) 10^{-6} torr, (c) 10^{-12} torr? Assume ideality. (760 torr = 1 atm.)

18. Verify Eqs. 19.152–19.154.

19. According to the principle of equipartition, what should C_V be for the following (ideal) gases: Ar, N_2, CO_2, H_2O?

20. The most fundamental condition that must be met for a laser to operate between single-particle energy ϵ_i and single-particle energy ϵ_j (with $\epsilon_j > \epsilon_i$) is that the population be "inverted," that is, the number of molecules with energy ϵ_j must exceed the number of molecules with energy ϵ_i.

(a) If the population inversion is to be produced by heating to some temperature T, what condition must the degeneracies of the energy levels satisfy?

(b) Can thermally induced laser action exist between vibrational states of a diatomic molecule?

(c) Given the fact that vibrational laser action has been produced in the states of product HCl generated in the reaction $H_2 + Cl_2 \rightarrow 2HCl$, can these product states be in thermal equilibrium?

21. The magnetic susceptibility of a linear molecule in a magnetic field is quantized and can have only the values

$$\chi_m = \chi_0 m^2, \qquad m = 1, 0, -1$$

where χ_0 is a constant. The energy of the molecule in a magnetic field H_0 is

$$W_m = \tfrac{1}{2} H_0^2 \chi_m .$$

(a) Considering only the magnetic field effect on the molecule, write a formal expression for the canonical partition function in terms of W_m.

(b) Write a formal expression for $\langle \chi \rangle$, the average value of the susceptibility.

(c) Write an expression for $\langle \chi \rangle$ in terms of H_0, k_B, T, and the partition function Q.

(d) Assuming that $W_m << k_BT$, obtain an explicit expression for $\langle \chi \rangle$ correct to H_0^2.

22. Show that the Helmholtz free energy is given by

$$A = E^2 \frac{\dfrac{\partial}{\partial E}\left(\dfrac{\ln \Omega}{E}\right)}{\dfrac{\partial \ln \Omega}{\partial E}},$$

where Ω is the microcanonical partition function. Evaluate A for 1 mol of an ideal gas with energy E and volume V.

23. The three levels of a molecule are

$$\epsilon_n = n\gamma,$$

where γ is a constant and $n = -1, 0, 1$. The degeneracies g_n are $g_{-1} = 1$, $g_0 = 2$, and $g_1 = 1$. A system of N such molecules is at equilibrium.

(a) Find the average energy at temperature T.

(b) Evaluate C_V if $\gamma << k_BT$.

24. For the reaction

$$\text{C(graphite)} + 2\text{H}_2\,(\text{g}) \rightarrow \text{CH}_4\,(\text{g}),$$

the enthalpy change at 600°C is –88.052 kJ/mol. The third-law entropies (in J/K mol) at 600°C and 1 atm are 20 for graphite, 163 for H_2, and 237 for CH_4.
(a) Calculate the equilibrium constant at 600°C.
(b) Assume that ΔH for the reaction does not vary significantly and find the equilibrium constant at 750°C. *Hint:* Use the Gibbs–Helmholtz equation.
(c) What conditions of temperature and pressure favor the yield of CH_4?

25. We showed in Chapter 15 that the distribution of states in a system at equilibrium is extremely sharply peaked, so sharply peaked that for many purposes the distribution can be replaced by its value at the maximum. Show that when there is effectively only one term of importance in the sum defining

$$\Delta \equiv \sum_V \sum_k g_k \exp\left(\frac{-E_k}{k_B T}\right) \exp\left(\frac{-pV}{k_B T}\right),$$

then

$$G = -k_B T \ln \Delta.$$

Δ is called the *isothermal–isobaric partition function;* it is defined in the isothermal–isobaric ensemble in which each replica has the same number of molecules, the total energy and total volume of the ensemble of replicas are fixed, but the energy and volume of an individual replica can differ from those of another replica.

26. Show that

$$P_k^{CE} = \frac{g_k \exp\left(\frac{-E_k}{k_B T}\right)}{\sum_k g_k \exp\left(\frac{-E_k}{k_B T}\right)},$$

and then that

$$S = -k_B \sum_k P_k^{CE} \ln P_k^{CE}.$$

27. Show that fluctuations of the number of molecules in a volume V at temperature T and with chemical potential μ satisfy

$$\langle N^2 \rangle - \langle N \rangle^2 = \frac{\langle N \rangle^2 k_B T \kappa_T}{V}.$$

Show that the numerical value of the relative fluctuation in number of molecules in V is very small. Hint: Use

$$\langle N \rangle = k_B T \left(\frac{\partial \ln \Xi}{\partial \mu}\right)_{V,T},$$

$$\langle N^2 \rangle = \frac{k_B T}{\Xi} \frac{\partial}{\partial \mu} (\langle N \rangle \Xi).$$

28. Suppose that as a result of some approximate calculation, the following canonical partition function is found for a model of a gas:

$$Q_N = \frac{1}{N!} \left(\frac{2\pi m k_B T}{h^2}\right)^{3N/2} (V - Nb)^N \exp\left(\frac{aN^2}{V k_B T}\right)$$

with a and b constants determined by the model. What is the equation of state of the model gas?

29. At a temperature of 298.15 K, calculate the fraction of argon atoms whose speeds are within 10% (i.e., ± 10%) of the mean speed $\langle v \rangle$. Perform the calculation two ways: (a) Estimate the fraction by assuming that $f(v)$ is a constant over the range of speeds and is equal to $f(\langle v \rangle)$; (b) Compute an accurate value for this fraction of atoms by using a computer-based mathematical package such as MAPLE, MATHEMATICA, MACSYMA, or DERIVE to evaluate the required integral.

30. Calculate the number of molecules in 1.00 mole of O_2 gas at 298.15 K that have speeds between the mean speed $\langle v \rangle$ and the root-mean-square-speed $\langle v^2 \rangle^{1/2}$. Use a computer-based mathematical package such as MAPLE, MATHEMATICA, MACSYMA, or DERIVE to evaluate the required integral.

31. Calculate the fraction of H_2 gas molecules at 298.15 K that have speeds greater than or equal to the most probable speed v^*. Use a computer-based mathematical package such as MAPLE, MATHEMATICA, MACSYMA, or DERIVE to evaluate the required integral.

32. The expression for the mean-square-speed is given in equation (19.154) and the relation between the root-mean-square-speed and the most probable speed is given in equation (19.156). Derive an expression for the mean-cube-speed $\langle v^3 \rangle$ (the third moment of the probability distribution of speeds) and determine the relationship between $\langle v^3 \rangle^{1/3}$ and v^*. Finally, derive the general result for the nth moment $\langle v^n \rangle$ and its nth root. Use a computer-based mathematical package such as MAPLE, MATHEMATICA, MACSYMA, or DERIVE to evaluate the required integrals.

CHAPTER 20

An Extension of Thermodynamics to the Description of Nonequilibrium Processes

In the preceding chapters we have developed the foundations of the thermodynamic and statistical molecular descriptions of matter at equilibrium. In doing so we took an uncompromisingly rigid view of how the thermodynamic functions were to be defined—at every stage of the analysis we chose methods that were valid for the case that equilibrium prevailed. A consequence of the approach we have used is that the applicability of the thermodynamic relations derived is restricted to the description of systems at equilibrium. For example, we defined the concept of temperature by use of the zeroth law of thermodynamics, which explicitly requires that a state of thermal equilibrium be established between the thermometer and the system under investigation, including internal equilibrium of each. It would appear, then, that we could not meaningfully talk about the temperature of the reacting gases in a rocket nozzle. Yet it is possible, by a variety of well-defined procedures, to measure the temperature of that reacting, flowing mixture. Moreover, with suitable precautions as to the interpretation, the temperature so measured gives valuable information about the system probed. Numerous other examples of the validity of the measurement of temperatures of systems out of equilibrium can be given. We conclude that it should be possible to extend the definitions of the thermodynamic functions, previously introduced only for matter at equilibrium, to the description of matter not at equilibrium.

There are two key concepts in the extension of thermodynamic arguments to systems not at equilibrium, both related to the notion of separation of time scales. The simpler of the two is the following: If the rate at which a macroscopic inhomogeneity in a system relaxes toward uniformity is small compared to the rate of establishment of equilibrium between the molecular degrees of freedom, local values of thermodynamic functions can be defined. It is then natural to assume that the relations between local values of the thermodynamic functions are the same as those in the state of complete equilibrium (*hypothesis of local equilibrium*). For example, if a pulse of heat is generated at some point in a fluid,[1] a spatially nonuniform distribution of temperature is created. We can speak of such a temperature distribution because the rate at which collisions between molecules establish a canonical distribution of kinetic energy is very large relative to the rate at which collisions transfer energy over a macroscopic distance. The kinetic energy in an element of volume close to the point of injection is higher than elsewhere, hence so is the temperature. We expect there to be, correspondingly, a distribution of density in the fluid, since an increase in local temperature implies more energetic collisions between molecules, or a larger amplitude of motion, both of which establish a local increase in the volume per molecule. Provided that the time scales for establishment of the local canonical distribution of kinetic energy and the local volume per molecule are both short compared to the time required for relaxation to the uniform state, the local temperature and density will be related by the ordinary macroscopic equation of state. This situation occurs fairly commonly—for example, in chemical reactors, distillation columns, and nonuniform fluids such as the oceans and the atmosphere. When macroscopic thermodynamic variables such as temperature describe small but still macroscopic

[1] This can be done by passing a pulse of current through a small coil of resistance wire immersed in the fluid.

regions, we speak of "local thermal equilibrium." The validity of the local thermal equilibrium supposition makes the description of the behavior of a nonequilibrium system far more tractable than if it requires specification of nonequilibrium microscopic distribution functions. The latter topic will be pursued further in Chapters 28 and 29.

The second key concept relates to the differences in rates of approach to equilibrium of different degrees of freedom. For example, at room temperature it takes 10^3 to 10^4 collisions to transfer vibrational to translational energy, 10 to 10^2 collisions to transfer rotational to translational energy, whereas almost every collision transfers translational energy between molecules. It is possible, then, by suitable excitation of a system and aging, to have some degrees of freedom at equilibrium and others not. Furthermore, even though it may take $\sim 10^4$ collisions on average to transfer vibrational energy to translational energy, it usually takes only about 10 to 10^2 to establish a canonical distribution of vibrational energy within the vibrational degrees of freedom. When this situation prevails it is possible to assign different temperatures to different degrees of freedom. Unlike the spatial inhomogeneity example described above, the relationships between the defined thermodynamic functions are now, in general, not the same as in the state of equilibrium. Macroscopic concepts can still be used to describe the system, but where their validity requires complete internal equilibrium (e.g., the relationship between compressibility and difference in specific heats), they are inapplicable. Of course, if there is a canonical distribution of energy within the levels of one degree of freedom, macroscopic concepts related only to that degree of freedom can be used, provided that the relaxation time criteria previously mentioned are met (e.g., a relation between vibrational entropy and vibrational temperature is valid if the vibrational energy distribution is canonical). Note that our definition of equilibrium with respect to a degree of freedom requires establishment of a canonical distribution of energy in that degree of freedom. If this is not the case, thermodynamic functions cannot be defined for that degree of freedom.

The most fully developed extension of thermodynamic analysis to the description of nonequilibrium phenomena is in its application to hydrodynamic problems. It is usual to proceed by assuming that the local equilibrium hypothesis is valid. We shall use a variant of the usual procedure and show that, starting with the equations that describe fluid motion, the Fourier law of heat conduction implies that locally defined thermodynamic functions satisfy the same relationships as at equilibrium.

20.1 General Form of the Equations of Continuity

To describe a flowing fluid we must use variables whose values are local in space and time. Thus, instead of having state variables that are defined over an entire system, we now regard the pressure, density, and energy to be functions of position within the system, **r**, and time, *t*. At any particular space-time point these functions are assumed to exist.

In fluid mechanics, when we refer to a point in space we mean a region of space that is small by macroscopic standards. Just as in thermodynamics, however, the molecular nature of matter is ignored. A "point" inside a volume of fluid is therefore sufficiently large that it contains a great many molecules. In other words, matter is considered to be continuous.

The starting point of fluid mechanics is a consideration of conservation laws. The usual conditions of conservation of mass, momentum, and energy must be expressed in a manner that reflects the assumed continuous nature of matter. In this format, the conservation laws are often called *equations of continuity.*

The developments of the equations of continuity for the various conserved quantities in a fluid follow a similar line of reasoning. Let Q represent any conserved quantity. Consider an arbitrary closed fixed volume V inside the fluid. At each point in V the local density of Q, $\rho_Q(\mathbf{r}, t)$, is defined. The total amount of Q contained in V at time t is given by

$$Q(t) = \int_V d\mathbf{r}\rho_Q(\mathbf{r},t). \tag{20.1}$$

Because Q is a conserved quantity, the only way $Q(t)$ can vary with time is if Q passes across the boundary surface S that defines V. This surface can be divided into infinitesimal sections, with each section labelled by a vector $d\mathbf{s}$ whose magnitude is equal to the area of the section and whose direction is perpendicular to the section and pointing outward from V (see Fig. 20.1). At each point in V there is a vector flux density $\mathbf{J}^Q(\mathbf{r}, t)$ associated with Q. This vector points in the direction of the flow of Q and has the units of Q per unit area per unit time. The total amount of Q *leaving* V per unit time must equal the surface integral of $\mathbf{J}^Q(\mathbf{r}, t)$ over S. We can therefore write

$$\frac{dQ}{dt} = -\int_S d\mathbf{s}\cdot\mathbf{J}^Q(\mathbf{r},t) \tag{20.2}$$

The minus sign in Eq. 20.2 is required as a result of the convention in assigning the direction of $d\mathbf{s}$.

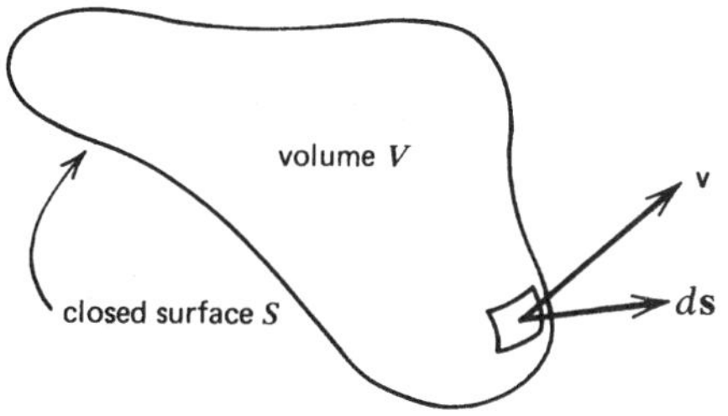

Figure 20.1 An elementary area $d\mathbf{s}$ on the surface S enclosing the volume V, and the mass velocity vector **v**. The shape of the volume V is arbitrary.

The surface integral in Eq. 20.2 can be converted to a volume integral if we make use of the Gauss divergence theorem:

$$\int_S d\mathbf{s} \cdot \mathbf{J}^Q(\mathbf{r},t) = \int_V d\mathbf{r}\, \nabla \cdot \mathbf{J}^Q(\mathbf{r},t). \tag{20.3}$$

From Eq. 20.1 we also have

$$\frac{dQ}{dt} = \int_V d\mathbf{r} \frac{\partial}{\partial t} \rho_Q(\mathbf{r},t). \tag{20.4}$$

Because V is an arbitrary volume, the combination of Eqs. 20.2, 20.3, and 20.4 yields

$$\frac{\partial}{\partial t} \rho_Q(\mathbf{r},t) = -\nabla \cdot \mathbf{J}^Q(\mathbf{r},t), \tag{20.5}$$

which is the general form of the equations of continuity. In order that Eq. 20.5 be useful, expressions for the flux $\mathbf{J}^Q$ must be introduced, and the particular format each of these expressions takes will vary with the property Q under consideration. In the following sections we shall discuss the detailed forms of Eq. 20.5 for mass, momentum, and energy.

20.2 Conservation of Mass and the Diffusion Equation

If $Q(t)$ represents the mass contained in an arbitrary volume of a fluid, Q can change only when an actual flow of matter occurs across the boundary surface of the volume. The mass flux $\mathbf{J}^m(\mathbf{r}, t)$ is then given by

$$\mathbf{J}^m(\mathbf{r},t) = \rho(\mathbf{r},t)\mathbf{v}(\mathbf{r},t), \tag{20.6}$$

where ρ is the local mass density and $\mathbf{v}$ is the local fluid velocity. If we substitute Eq. 20.6 in Eq. 20.5 and use the rule for differentiation of a product,

$$\nabla \cdot (\rho \mathbf{v}) = \rho \nabla \cdot \mathbf{v} + \mathbf{v} \cdot \nabla \rho, \tag{20.7}$$

we find

$$\frac{\partial \rho}{\partial t} + \mathbf{v} \cdot \nabla \rho + \rho \nabla \cdot \mathbf{v} = 0 \tag{20.8}$$

as the local condition expressing conservation of mass.

We can write Eq. 20.8 in a slightly different form if we consider the total time derivative of the density. This is

$$\begin{aligned} \frac{d}{dt}\rho(\mathbf{r},t) &= \frac{\partial \rho}{\partial t} + \sum_i \frac{\partial \rho}{\partial x_i} \frac{dx_i}{dt} \\ &= \frac{\partial \rho}{\partial t} + \sum_i \frac{\partial \rho}{\partial x_i} v_i \\ &= \frac{\partial \rho}{\partial t} + \mathbf{v} \cdot \nabla \rho. \end{aligned} \tag{20.9}$$

Here, the index i runs over the three Cartesian coordinates and x_i is the ith component of the position vector $\mathbf{r}$. Thus, Eq. 20.8 is equivalent to

$$\frac{d\rho}{dt} + \rho \nabla \cdot \mathbf{v} = 0. \tag{20.10}$$

The total derivative $d\rho/dt$ is to be interpreted as the time rate of change of the density of an element of the fluid as it moves in space. By contrast, the partial derivative $\partial\rho/\partial t$ represents the rate of change of the density of the fluid at a fixed point in space.

In the case of incompressible fluids (a very good approximation for most liquids), Eq. 20.10 takes on a particularly simple form. The density is then constant, and the equation of continuity reduces to

$$\nabla \cdot \mathbf{v} = 0. \tag{20.11}$$

So far, the equation of continuity is still not in a useful form, because it is a differential equation for ρ in terms of an unknown velocity field $\mathbf{v}(\mathbf{r}, t)$. An additional relationship between ρ and $\mathbf{v}$ is needed so that $\mathbf{v}$ can be eliminated from Eq. 20.8. To illustrate this point, suppose that the fluid under consideration is a mixture in which there are not any chemical reactions. Then the same arguments concerning conservation of mass can be applied to the rate of change of ρ_k, the mass per unit volume of component k, as to the total mass density. If $\mathbf{v}_k$ is the velocity of component k, the analog of Eq. 20.8 is

$$\frac{\partial \rho_k}{\partial t} = -\nabla \cdot (\rho_k \mathbf{v}_k). \tag{20.12}$$

The rate of change of the total density of the fluid is still given by Eq. 20.8 provided that $\mathbf{v} = \Sigma_k \rho_k \mathbf{v}_k / \rho$ is the velocity of the center of mass. Let

$$\mathbf{J}_k^m = \rho_k(\mathbf{v}_k - \mathbf{v}) \tag{20.13}$$

be the flux of k relative to that carried along by motion of the center of mass. Substitution of Eq. 20.13 into Eq. 20.12 and use of Eq. 20.9 lead to

$$\frac{d\rho_k}{dt} = -\rho_k \nabla \cdot \mathbf{v} - \nabla \cdot \mathbf{J}_k^m. \tag{20.14}$$

We now close Eq. 20.14 by introducing the phenomenological relation known as *Fick's law.*

In the absence of convection currents, and if spatial gradients in the density of component k are small, it is reasonable to assume that the flux of k is everywhere proportional to its density gradient. Fick's law expresses this proportionality and introduces a *diffusion coefficient* D_k by the relation

$$\mathbf{J}_k^m(\mathbf{r},t) = -D_k \nabla \rho_k(\mathbf{r},t). \tag{20.15}$$

The minus sign in Eq. 20.15 expresses the observation that flow occurs from a region of high density to a region of low density (D_k is positive). The substitution of Eq. 20.15 in Eq. 20.12 yields, for a fluid at rest ($\mathbf{v} = 0$), a partial differential equation called the *diffusion equation:*

$$\frac{\partial \rho_k}{\partial t} = D_k \nabla^2 \rho_k . \tag{20.16}$$

For given initial conditions and boundary conditions the diffusion equation can be solved for the spatial dependence of the density of k as a function of time. For example, suppose that at $t = 0$ a mass m_k of component k is concentrated at the origin. The solution to Eq. 20.16 for this condition shows that, at time t, component k is distributed in space according to

$$\rho_k(\mathbf{r},t) = \frac{m_k}{8(\pi D_k t)^{3/2}} \exp\left(\frac{-r^2}{4 D_k t}\right). \tag{20.17}$$

Note that, since the diffusion is isotropic, ρ_k depends only on r, the magnitude of the displacement vector.

Clearly, as time passes the mixture becomes uniform. A schematic representation of how the solution approaches uniform composition is shown in Fig. 20.2.

Equation 20.17 can also be used to calculate the average distance that a "particle" of a diffusing fluid travels in a given time. The mean-square displacement from the origin at time t is found to be

$$\langle r_k^2 \rangle = 6 D_k t. \tag{20.18}$$

The square-root-of-time dependence of the distance traveled is characteristic of diffusive motion, and Eq. 20.18 can be considered a practical definition of the diffusion coefficient.

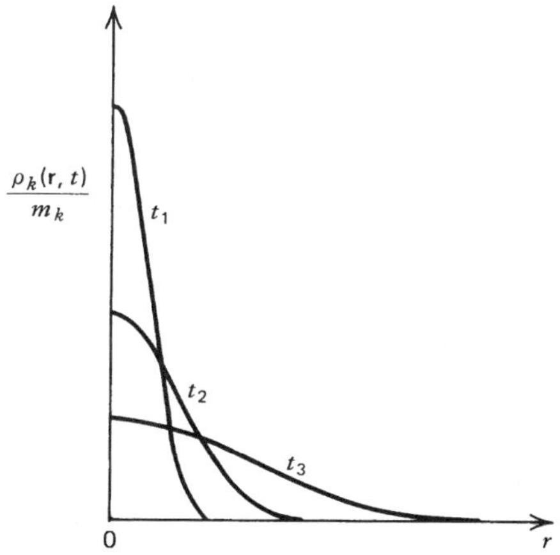

Figure 20.2 Schematic representation of the concentration profile as a function of time for the diffusion of a substance originally concentrated at the origin of the coordinates. The times t_1, t_2, and t_3 are ordered such that $t_1 < t_2 < t_3$.

Of course, Fick's law is not the only way of transforming the equation of mass continuity into a useful form. Under other circumstances there may be mass flux terms due to convection, or gradients in other quantities such as the temperature. The principles involved in the use of other relationships for closure are the same as those described, although details of the analysis will necessarily be different.

20.3 Conservation of Momentum and the Navier–Stokes Equation

The development of the equation of continuity (Eq. 20.5) for momentum is somewhat more complicated than for mass. First, we must realize that momentum is a vector quantity. The flux conjugate to the scalar mass is a vector; given that momentum is a vector quantity, what is the nature of the momentum flux? To answer this question we select an arbitrary fixed coordinate system and focus on a particular component of the local momentum density. Since the ith component of the momentum of a small region of the fluid with mass dm is

$$dp_i(\mathbf{r},t) = v_i(\mathbf{r},t)\, dm(\mathbf{r},t), \tag{20.19}$$

the momentum density has components of the form

$$\rho_{p_i} = \rho(\mathbf{r},t) v_i(\mathbf{r},t). \tag{20.20}$$

The ith component of the momentum density must satisfy an equation of continuity, so we can write

$$\frac{\partial}{\partial t}[\rho(\mathbf{r},t) v_i(\mathbf{r},t)] = -\nabla \cdot \mathbf{J}^{p_i} . \tag{20.21}$$

Here, $\mathbf{J}^{p_i}$ is the vector flux for the ith component of the momentum. We can write the divergence of $\mathbf{J}^{p_i}$ in component form:

$$\nabla \cdot \mathbf{J}^{p_i} = \sum_{j=1}^{3} \frac{\partial}{\partial x_j} J_j^{p_i} \tag{20.22}$$

A typical component of $\mathbf{J}^{p_i}$, say $J_y^{p_i}$, is the flow of i component of momentum in the y direction. Then, taking account of all three components of the momentum, we see that nine quantities of the form $J_j^{p_i}$ will appear in Eq. 20.22, since the two indices i and j both range from 1 to 3. We now change our notation slightly to emphasize the existence of two indices in the momentum flux. Henceforth we write

$$J_{ji}^{p} = J_j^{p_i} . \tag{20.23}$$

The nine quantities J_{ji}^{p} are the components of a new type of quantity, called a tensor, which is seen to be an extension of the vector concept. More formally, the momentum flux is an example of a second-rank tensor, so named because two indices are required to specify its components in a particular coordinate system. An ordinary vector is formally a first-rank tensor; the tensor concept can be generalized to an arbitrary number of indices.

We shall represent the momentum flux tensor by the symbol $\mathbf{J}^p$. The divergence of $\mathbf{J}^p$ is a vector whose ith component is (see Eq. 20.22)

$$(\nabla \cdot \mathbf{J}^p)_i = \sum_{j=1}^{3} \frac{\partial}{\partial x_j} J^p_{ji}. \tag{20.24}$$

We can therefore write Eq. 20.21 without reference to components in a particular coordinate system:

$$\frac{\partial}{\partial t}[\rho(\mathbf{r},t)\mathbf{v}(\mathbf{r},t)] = -\nabla \cdot \mathbf{J}^p. \tag{20.25}$$

What is the form of the momentum flux tensor? We can start to answer this question by carrying out the differentiation indicated on the left-hand side of Eq. 20.25:

$$\frac{\partial}{\partial t}[\rho(\mathbf{r},t)\mathbf{v}(\mathbf{r},t)] = \mathbf{v}\frac{\partial \rho}{\partial t} + \rho\frac{\partial \mathbf{v}}{\partial t}. \tag{20.26}$$

If we substitute Eq. 20.8 for $\partial\rho/\partial t$ in the above equation, we find that

$$\mathbf{v}\frac{\partial \rho}{\partial t} = -\mathbf{v}[\nabla \cdot (\rho\mathbf{v})]. \tag{20.27}$$

The other term on the right-hand side of Eq. 20.26 contains $\partial\mathbf{v}/\partial t$, the acceleration of the fluid at ($\mathbf{r}$, t). This acceleration has two contributions, one stemming from the pressure within the fluid, and the other resulting from *viscosity.*

For the moment let us suppose that the fluid is nonviscous. (Fluids with no viscosity are sometimes called *ideal fluids.*) If we again consider the fluid volume shown in Fig. 20.1, we can write the force acting on an element $d\mathbf{s}$ of the surface as

$$d\mathbf{f} = -p\, d\mathbf{s}, \tag{20.28}$$

where p is the pressure on the surface element. The force is defined to be positive if it points toward the interior of the volume. The total force acting on the fluid volume is then

$$\mathbf{f} = -\int_S d\mathbf{s}\, p(\mathbf{r},t). \tag{20.29}$$

The surface integral can be converted to a volume integral in a manner similar to Eq. 20.3:

$$\mathbf{f} = -\int_V d\mathbf{r}\, \nabla p(\mathbf{r},t). \tag{20.30}$$

We interpret Eq. 20.30 by saying that a force per unit volume equal to $-\nabla p$ acts on any volume element of the fluid. We equate the force per unit volume to the mass per unit volume times the total acceleration of the volume element:

$$\rho\frac{d\mathbf{v}}{dt} = -\nabla p. \tag{20.31}$$

The total derivative $d\mathbf{v}/dt$ can be expanded in partial derivatives in a manner analogous to Eq. 20.9, and the result is

$$\rho\frac{\partial \mathbf{v}}{\partial t} + \rho\mathbf{v}\cdot\nabla\mathbf{v} = -\nabla p. \tag{20.32}$$

Note that the gradient $\nabla\mathbf{v}$ is a tensor whose components have the form

$$(\nabla\mathbf{v})_{ij} = \frac{\partial v_j}{\partial x_i}. \tag{20.33}$$

The dot product $\mathbf{v}\cdot\nabla\mathbf{v}$ is then a vector whose jth component is

$$(\mathbf{v}\cdot\nabla\mathbf{v})_j = \sum_i v_i \frac{\partial v_j}{\partial x_i}. \tag{20.34}$$

The analogy between Eq. 20.34 and the operations in Eq. 20.9 should be apparent.

Equation 20.32 is known as *Euler's equation.* If we use it and Eq. 20.27 in Eq. 20.26, we find that

$$\begin{aligned}\frac{\partial}{\partial t}(\rho\mathbf{v}) &= -\mathbf{v}[\nabla\cdot(\rho\mathbf{v})] - \rho\mathbf{v}\cdot\nabla\mathbf{v} - \nabla p\\ &= -\nabla\cdot(\rho\mathbf{v}\mathbf{v}) - \nabla p,\end{aligned} \tag{20.35}$$

which can be checked by calculating the individual components. The components of the tensor $\rho\mathbf{v}_i\mathbf{v}_i$ have the form $\rho v_i v_j$. The divergence of this tensor has components

$$\begin{aligned}(\nabla\cdot\rho\mathbf{v}\mathbf{v})_j &= \sum_i \frac{\partial}{\partial x_i}(\rho v_i v_j)\\ &= v_j\sum_i\frac{\partial}{\partial x_i}(\rho v_i) + \rho\sum_i v_i\frac{\partial v_j}{\partial x_i},\end{aligned} \tag{20.36}$$

so that

$$\nabla\cdot\rho\mathbf{v}\mathbf{v} = \mathbf{v}(\nabla\cdot\rho\mathbf{v}) + \rho\mathbf{v}\cdot\nabla\mathbf{v} \tag{20.37}$$

as in Eq. 20.35.

It is convenient to rewrite the pressure gradient in Eq. 20.35 in a somewhat different form. We define the unit tensor $\mathbf{1}$ as one whose diagonal elements are all unity and whose off-diagonal elements are all zero; the components of $\boldsymbol{1}$ are often denoted by the Kronecker symbol

$$\delta_{ij} = \begin{cases} 1 \text{ if } i = j \\ 0 \text{ if } i \neq j \end{cases}. \tag{20.38}$$

The jth component of the pressure gradient can then be written

$$(\nabla p)_j = \frac{\partial p}{\partial x_j} = \sum_i \frac{\partial}{\partial x_i}(p\delta_{ij}), \tag{20.39}$$

or

$$\nabla p = \nabla \cdot (p\boldsymbol{1}). \tag{20.40}$$

We can now write Eq. 20.35:

$$\frac{\partial}{\partial t}(\rho \mathbf{v}) = -\nabla \cdot (\rho \mathbf{v}\mathbf{v} + p\boldsymbol{1}). \tag{20.41}$$

Comparison with Eq. 20.25 yields (finally!) the momentum flux tensor for a nonviscous fluid:

$$\mathbf{J}^p = \rho \mathbf{v}\mathbf{v} + p\boldsymbol{1}. \tag{20.42}$$

The physical origin of the two terms in $\boldsymbol{J}^p$ is easy to see. The first term, $\rho \mathbf{v}\mathbf{v}$, represents momentum that is transferred from point to point by an actual flow of fluid. Note that $\rho \mathbf{v}\mathbf{v}$ is the local momentum density, $\rho \mathbf{v}$, times the local velocity. This contribution to the momentum flux is in direct analogy to Eq. 20.6 for the mass flux.

In contrast to mass transfer, momentum can be transferred by means other than fluid flow. We express these other contributions to the momentum flux by introducing the *stress tensor,* whose definition is

$$\boldsymbol{\sigma} = -(\mathbf{J}^p - \rho \mathbf{v}\mathbf{v}). \tag{20.43}$$

For nonviscous fluids we have already shown that the stress tensor is equal to $-p\boldsymbol{1}$. This quantity represents momentum that can be transferred from point to point by means of the application of pressure. When viscosity is present, the stress tensor takes on a somewhat more complicated form.

What do we mean when we say that a fluid is viscous? Suppose that a region of fluid is contained between two flat parallel plates, as shown in Fig. 20.3. The fluid is supposed to be initially at rest, and one of the plates is then set in uniform motion in its plane while the other plate is held fixed. The moving plate drags a thin layer of fluid along with it, while the layer of fluid adjacent to the stationary plate remains stationary. A velocity gradient has thus been set up in the fluid, and it points perpendicular to the two plates.

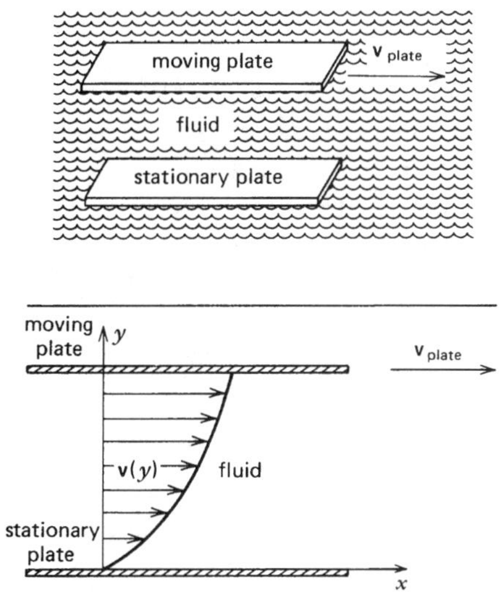

Figure 20.3 Schematic representation of the fluid velocity profile between a moving plate and a stationary plate.

Now, if the fluid is viscous, momentum will be transferred in the direction opposite to the velocity gradient, so that the stationary plate will feel a force in the direction of the moving plate's motion. This transfer of momentum is behind our intuitive notion of viscosity.

If the velocity gradients in the fluid are small, it is reasonable to assume that the momentum flux due to viscosity is proportional to the local velocity gradient. This assumption is analogous to Eq. 20.15 in our discussion of mass diffusion. Therefore, there must be a term in $\mathbf{J}^p$ that is proportional to $-\nabla \mathbf{v}$. In Appendix 20A we show that the momentum flux tensor must be symmetric, which means that in any particular coordinate system,

$$J^p_{ij} = J^p_{ji}. \tag{20.44}$$

Specifically, we demonstrate that every term in J^p_{ij} of the form $\partial v_j/\partial x_i$ must be balanced by a term $\partial v_i/\partial x_j$. The velocity gradient must therefore occur in the symmetrized form $(\nabla \mathbf{v})^s$, where

$$(\nabla \mathbf{v})^s_{ij} = \frac{\partial v_j}{\partial x_i} + \frac{\partial v_i}{\partial x_j}. \tag{20.45}$$

In an isotropic fluid, the only other possible contribution to the momentum flux tensor components is $(\nabla \cdot \mathbf{v})\delta_{ij}$, which is both symmetric and proportional to elements of the velocity gradient. It is customary to combine these various terms in a certain way, and to define the stress tensor to be

$$\boldsymbol{\sigma} = -p\boldsymbol{1} + \eta[(\nabla \mathbf{v})^s - \tfrac{2}{3}(\nabla \cdot \mathbf{v})\boldsymbol{1}] + \zeta(\nabla \cdot \mathbf{v})\boldsymbol{1}. \tag{20.46}$$

The stress tensor contains the pressure term and a combination of two viscosity terms; the coefficients η and ζ are called the coefficients of shear and bulk viscosity, respectively. The *shear viscosity* is a measure of the friction experienced in a shear flow, that is, one like that generated in the example of moving parallel plates. The *bulk viscosity* is a measure of the friction experienced in a pure expansion or compression flow. If the liquid is composed of molecules with internal structure, any lag in establishment of internal equilibrium amongst the molecular degrees of freedom contributes to ζ.

Equation 20.3 now takes on the form

$$\frac{\partial}{\partial t}(\rho \mathbf{v}) = -\nabla \cdot (\rho \mathbf{v}\mathbf{v} - \boldsymbol{\sigma}). \tag{20.47}$$

If we recall Eq. 20.37, and subtract Eq. 20.27 from Eq. 20.47, we find that

$$\rho \frac{\partial \mathbf{v}}{\partial t} = -\rho \mathbf{v} \cdot \nabla \mathbf{v} + \nabla \cdot \boldsymbol{\sigma} \tag{20.48}$$

or, using Eq. 20.46 in component form,

$$\begin{aligned} \rho\left(\frac{\partial v_i}{\partial t} + \sum_j v_j \frac{\partial v_i}{\partial x_j}\right) &= -\frac{\partial p}{\partial x_i} + \eta \sum_j \left(\frac{\partial}{\partial x_j}\frac{\partial v_j}{\partial x_i} + \frac{\partial}{\partial x_j}\frac{\partial v_i}{\partial x_j}\right) \\ &\quad - \frac{2}{3}\eta \frac{\partial}{\partial x_i}(\nabla \cdot \mathbf{v}) + \zeta \frac{\partial}{\partial x_i}(\nabla \cdot \mathbf{v}). \end{aligned} \tag{20.49}$$

However,

$$\sum_j \frac{\partial}{\partial x_j}\frac{\partial v_j}{\partial x_i} = \frac{\partial}{\partial x_i}(\nabla \cdot \mathbf{v}),$$

and

$$\sum_j \frac{\partial}{\partial x_j}\frac{\partial v_i}{\partial x_j} = (\nabla \cdot \nabla) v_i = \nabla^2 v_i, \tag{20.50}$$

so that

$$\begin{aligned} \rho\left(\frac{\partial v_i}{\partial t} + \sum_j v_j \frac{\partial v_i}{\partial x_j}\right) &= -\frac{\partial p}{\partial x_i} + \eta \nabla^2 v_i + \left(\zeta + \frac{1}{3}\eta\right)\frac{\partial}{\partial x_i}(\nabla \cdot \mathbf{v}). \end{aligned} \tag{20.51}$$

In vector notation, Eq. 20.51 is

$$\rho\left(\frac{\partial \mathbf{v}}{\partial t} + \mathbf{v} \cdot \nabla \mathbf{v}\right) = -\nabla p + \eta \nabla^2 \mathbf{v} + \left(\zeta + \frac{1}{3}\eta\right)\nabla(\nabla \cdot \mathbf{v}). \tag{20.52}$$

Equation 20.52, called the *Navier–Stokes equation,* is a fundamental hydrodynamic equation of motion; it is a direct consequence of the law of conservation of momentum. The particular form of the stress tensor that appears in the Navier–Stokes equation has been obtained under the assumption that the velocity gradient is small, but is found to be very generally valid. Equation 20.48, where $\boldsymbol{\sigma}$ includes the pressure contribution $-p\mathbf{1}$, is always valid. The Navier–Stokes equation has a great many practical applications to problems of fluid flow in restricted volumes, to effects of viscous drag, and so forth. A consideration of these applications would take us quite far afield. However, the statement of the Navier–Stokes equation will be important in a somewhat different connection in the following pages.

We close this section by noting that for incompressible fluids $\nabla \cdot \mathbf{v} = 0$, and the Navier–Stokes equation becomes

$$\frac{\partial \mathbf{v}}{\partial t} + \mathbf{v} \cdot \nabla \mathbf{v} = -\frac{1}{\rho}\nabla p + \frac{\eta}{\rho}\nabla^2 \mathbf{v}. \tag{20.53}$$

20.4 Conservation of Energy and the Second Law of Thermodynamics

We shall now consider the hydrodynamic statement of the law of conservation of energy and some of the consequences of that law. In a sense, we shall come full circle in this section. By introducing a phenomenological assumption in the energy conservation equation similar to Fick's law, and using the concept of local equilibrium, we shall be able to make contact with the second law of thermodynamics.

Until now, our conservation equations have been derived by considering changes in a quantity Q in a closed fixed volume V (see Eq. 20.1). In this section it will be more convenient to think of a closed volume V' that moves along with the fluid. To emphasize this change we write the total amount of Q contained in V' as Q', where

$$Q'(t) = \int_{V'} d\mathbf{r}\, \rho_Q(\mathbf{r}, t). \tag{20.54}$$

The form of Eq. 20.4 must now be changed slightly. Since V' moves with the fluid, there cannot be any flow of matter across the boundary of V'. Therefore we must subtract from the right-hand side of Eq. 20.4 the contribution to dQ/dt resulting from fluid flow. From Eqs. 20.6 and 20.42 we see that such flow contributes a term to the flux of Q that has the general form $\rho_Q \mathbf{v}$. We can therefore rewrite Eq. 20.4 as

$$\frac{dQ'}{dt} = \int_{V'} d\mathbf{r} \frac{\partial}{\partial t}\rho_Q + \int_{S'} d\mathbf{s} \cdot \rho_Q \mathbf{v}. \tag{20.55}$$

Note that we have used Eq. 20.2 to remove from Eq. 20.4 that part of dQ/dt which results from flow across the boundary surface.

Suppose that

$$\rho_Q = \rho(\mathbf{r},t) f_Q(\mathbf{r},t), \tag{20.56}$$

where ρ is the mass density and f_Q is some function of space and time. Equation 20.56 has been satisfied by both the properties discussed in the previous two sections. From Eqs. 20.55 and 20.56 we obtain

$$\frac{dQ'}{dt} = \int_{V'} d\mathbf{r}\left[\frac{\partial}{\partial t}(\rho f_Q) + \nabla\cdot(\rho f_Q \mathbf{v})\right]. \tag{20.57}$$

If we expand the derivatives in the above equation, and use Eq. 20.8, we find that

$$\begin{aligned}\frac{dQ'}{dt} &= \int_{V'} d\mathbf{r}\left(\rho\frac{\partial f_Q}{\partial t} + \rho\mathbf{v}\cdot\nabla f_Q\right)\\ &= \int_{V'} d\mathbf{r}\,\rho\frac{df_Q}{dt},\end{aligned} \tag{20.58}$$

where df_Q/dt is the total derivative discussed in Eq. 20.9.

Equation 20.48 can also be rewritten in terms of a total derivative:

$$\rho\frac{\partial\mathbf{v}}{\partial t} + \rho\mathbf{v}\cdot\nabla\mathbf{v} = \rho\frac{d\mathbf{v}}{dt} = \nabla\cdot\boldsymbol{\sigma}. \tag{20.59}$$

We now let Q' represent the total kinetic energy, so that

$$f_Q = \tfrac{1}{2}v^2. \tag{20.60}$$

If we use Eqs. 20.58 and 20.59, we find that

$$\frac{dQ'}{dt} = \int_{V'} d\mathbf{r}\,\rho\mathbf{v}\cdot\frac{d\mathbf{v}}{dt} = \int_{V'} d\mathbf{r}\,\mathbf{v}\cdot\nabla\cdot\boldsymbol{\sigma}. \tag{20.61}$$

However, from the Gauss divergence theorem,

$$\begin{aligned}\int_{S'} d\mathbf{s}\cdot\mathbf{v}\cdot\boldsymbol{\sigma} &= \int_{V'} d\mathbf{r}\,\nabla\cdot(\mathbf{v}\cdot\boldsymbol{\sigma})\\ &= \int_{V'} d\mathbf{r}\,\mathbf{v}\cdot\nabla\cdot\boldsymbol{\sigma} + \int_{V'} d\mathbf{r}\,\boldsymbol{\sigma}\cdot\nabla\cdot\mathbf{v},\end{aligned} \tag{20.62}$$

so that

$$\begin{aligned}\frac{dQ'}{dt} &= \frac{d}{dt}\int_{V'} d\mathbf{r}\left(\frac{1}{2}\rho v^2\right)\\ &= \int_{S'} d\mathbf{s}\cdot\mathbf{v}\cdot\boldsymbol{\sigma} - \int_{V'} d\mathbf{r}\,\boldsymbol{\sigma}\cdot\nabla\cdot\mathbf{v}\end{aligned} \tag{20.63}$$

Recall that $d\mathbf{s}\cdot\boldsymbol{\sigma}$ is a force, and that the scalar product of force with velocity gives power, or rate of energy change. The surface integral in Eq. 20.63 therefore represents the energy change in V' brought about by the work of pressure and viscous forces.

Let $\rho\varepsilon$ be the local density of potential energy in the fluid. Then the law of conservation of total energy (or the first law of thermodynamics) implies that we can write for the time derivative of the total energy in V'

$$\frac{d}{dt}\int_{V'} d\mathbf{r}\left(\frac{1}{2}\rho v^2 + \rho\varepsilon\right) = \int_{S'} d\mathbf{s}\cdot\mathbf{v}\cdot\boldsymbol{\sigma} - \int_{S'} d\mathbf{s}\cdot\mathbf{q}. \tag{20.64}$$

The first integral on the right of Eq. 20.64 is the work term discussed above. The second integral introduces the *heat flux vector* $\mathbf{q}$. The quantity $-\int_{S'} d\mathbf{s}\cdot\mathbf{q}$ measures the rate at which heat enters V'. The heat flux vector should not be confused with the scalar heat q that has appeared in previous chapters. If V' were fixed in space, Eq. 20.64 would have another term equal to the rate of energy change due to the flow of matter across the boundary surface. We would then write

$$\begin{aligned}&\frac{d}{dt}\int_{V} d\mathbf{r}\left(\tfrac{1}{2}\rho v^2 + \rho\varepsilon\right)\\ &= \int_{S} d\mathbf{s}\cdot\mathbf{v}\cdot\boldsymbol{\sigma} - \int_{S} d\mathbf{s}\cdot\mathbf{q} - \int_{S} d\mathbf{s}\cdot\left(\frac{1}{2}\rho v^2 + \rho\varepsilon\right)\mathbf{v}.\end{aligned} \tag{20.65}$$

If we subtract Eq. 20.63 from Eq. 20.64, we find that

$$\begin{aligned}\frac{d}{dt}\int_{V'} d\mathbf{r}\,\rho\varepsilon &= \int_{V'} d\mathbf{r}\,\rho\frac{d\varepsilon}{dt}\\ &= -\int_{S'} d\mathbf{s}\cdot\mathbf{q} + \int_{V'} d\mathbf{r}\,\boldsymbol{\sigma}\cdot\nabla\cdot\mathbf{v}.\end{aligned} \tag{20.66}$$

In the first step above we have used Eq. 20.58. The surface integral of the heat flux can be converted into a volume integral (divergence theorem again), and the fact that V' is arbitrary means that

$$\rho\frac{d\varepsilon}{dt} + \nabla\cdot\mathbf{q} - \boldsymbol{\sigma}\cdot\nabla\cdot\mathbf{v} = 0. \tag{20.67}$$

Let us define the specific volume υ to be

$$\upsilon = \frac{1}{\rho}. \tag{20.68}$$

Equation 20.10 can then be rewritten as

$$\rho\frac{d\upsilon}{dt} = \nabla\cdot\mathbf{v}. \tag{20.69}$$

If we also define $\boldsymbol{\sigma}'$ to be the viscous part of the stress tensor (see Eq. 20.46),

$$\boldsymbol{\sigma}' \equiv \boldsymbol{\sigma} + p\mathbf{1}, \tag{20.70}$$

we see that

$$\begin{aligned}\boldsymbol{\sigma}\cdot\nabla\cdot\mathbf{v} &= -p\nabla\cdot\mathbf{v} + \boldsymbol{\sigma}'\cdot\nabla\cdot\mathbf{v}\\ &= -p\rho\frac{d\upsilon}{dt} + \boldsymbol{\sigma}'\cdot\nabla\cdot\mathbf{v}.\end{aligned} \tag{20.71}$$

The assumption of local equilibrium implies that ε must be a function of p and v only. We can, therefore, expand the derivative $d\varepsilon/dt$ as follows:

$$\frac{d\varepsilon}{dt}=\frac{\partial \varepsilon}{\partial v}\frac{dv}{dt}+\frac{\partial \varepsilon}{\partial p}\frac{dp}{dt}. \tag{20.72}$$

The substitution of Eqs. 20.72 and 20.71 in Eq. 20.67 yields

$$\rho\left(\frac{\partial \varepsilon}{\partial v}+p\right)\frac{dv}{dt}+\rho\frac{\partial \varepsilon}{\partial p}\frac{dp}{dt}+\nabla\cdot\mathbf{q}-\boldsymbol{\sigma}'\cdot\nabla\cdot\mathbf{v}=0. \tag{20.73}$$

We can now introduce two thermodynamic functions, $\theta(p, v)$ and $\omega(p, v)$, and require that they satisfy the relations

$$\frac{\partial \varepsilon}{\partial v}+p=\theta\frac{\partial \omega}{\partial v} \tag{20.74a}$$

and

$$\frac{\partial \varepsilon}{\partial p}=\theta\frac{\partial \omega}{\partial p}. \tag{20.74b}$$

Equation 20.73 then becomes

$$\rho\theta\frac{d\omega}{dt}+\nabla\cdot\mathbf{q}=\boldsymbol{\sigma}'\cdot\nabla\cdot\mathbf{v}. \tag{20.75}$$

The functions θ and ω seem arbitrary at this point. However, the theory of differential equations can be used to demonstrate that θ and ω are well determined by Eqs. 20.74 if one additional condition is imposed. That condition is that θ satisfy the relation

$$\mathbf{q}=-\lambda\nabla\theta, \tag{20.76}$$

where λ is a positive function of p and v.

Equation 20.76 is reminiscent of another relation between a flux and a gradient. In Fick's law (Eq. 20.15) the assumption was made that the mass flux is proportional to the local density gradient. A similar phenomenological statement can be made about the heat flux. This relation is known as *Fourier's law,* and it asserts that the heat flux is proportional to the temperature gradient. If we identify θ with the temperature, Eq. 20.76 is just a statement of Fourier's law, and the proportionality factor λ is called the *thermal conductivity.*

We can now write

$$\begin{aligned}\nabla\cdot\frac{\mathbf{q}}{\theta}&=\frac{1}{\theta}\nabla\cdot\mathbf{q}-\frac{\mathbf{q}}{\theta^2}\cdot\nabla\theta\\&=\frac{1}{\theta}\nabla\cdot\mathbf{q}+\frac{\lambda}{\theta^2}(\nabla\theta)^2.\end{aligned} \tag{20.77}$$

The use of Eq. 20.77 in Eq. 20.75 yields

$$\rho\frac{d\omega}{dt}+\nabla\cdot\frac{\mathbf{q}}{\theta}=\frac{\lambda}{\theta^2}(\nabla\theta)^2+\boldsymbol{\sigma}'\cdot\nabla\cdot\mathbf{v}. \tag{20.78}$$

For a moment, suppose that $\nabla\theta=0$ and $\nabla\mathbf{v}=0$, so that the fluid is in uniform motion and the temperature is uniform. The right-hand side of Eq. 20.78 is then zero. If we integrate the left-hand side of the equation over V' and use Eq. 20.58 and the divergence theorem, we find that

$$\frac{d}{dt}\int_{V'}d\mathbf{r}\,\rho\omega+\int_{S'}d\mathbf{s}\cdot\frac{\mathbf{q}}{\theta}=0. \tag{20.79}$$

The surface integral is the rate of heat loss from V' divided by the temperature. Equation 20.79 then is nothing more than the second law of thermodynamics for reversible processes! To make the connection, all we must do is to associate $\int_{V'} d\mathbf{r}\,\rho\omega$ with the entropy contained in V'. The function ω is thus the local specific entropy. If Eq. 20.79 is then integrated over time, the familiar statement of the second law results.

More generally, the right-hand side of Eq. 20.78 is not equal to zero. However, since λ is positive, we have $(\lambda/\theta^2)(\nabla\theta)^2>0$. Furthermore, if the components of $\boldsymbol{\sigma}'$ given in Eq. 20.46 are used, it will be seen that $\boldsymbol{\sigma}'\cdot\nabla\cdot\mathbf{v}>0$, so we have the general result

$$\frac{d}{dt}\int_{V'}d\mathbf{r}\,\rho\omega+\int_{S'}d\mathbf{s}\cdot\frac{\mathbf{q}}{\theta}\geq 0, \tag{20.80}$$

in accord with the second law of thermodynamics.

Of course, to derive Eq. 20.80 we have used the phenomenological Fourier's law. The second law of thermodynamics is itself much more general than our development leading to Eq. 20.80. Nevertheless, it has been of considerable interest to demonstrate in this simple example how the equations of hydrodynamics, most often used to describe problems of fluid flow, can be used to make a connection with equilibrium thermodynamics. The key element in the connection is the assumption of local equilibrium.

There are many applications in which the simultaneous solutions to hydrodynamic and thermodynamic equations are required, for example, supersonic flow of air over an airplane body. We shall not discuss any of these interesting analyses—what is important is that a vast array of physical phenomena can be accurately described using the hypothesis of local equilibrium, and thermodynamic concepts are valid for those cases. The local equilibrium hypothesis fails when it is not possible to separate time scales as described earlier. Each situation must be examined separately and the important dynamic processes assessed to establish whether or not separate time scales exist for the several processes involved in the phenomenon being investigated.

20.5 Linear Transport Processes

We now examine briefly the extension of macroscopic thermodynamics to the analysis of linear transport processes. A linear transport process is one in which the flux generated by a nonzero generalized force is linear in that force. Examples of such linear transport processes are diffusion (Eq. 20.15), in which the flux of matter is proportional to the gradient of the concentration, and heat conduction (Eq. 20.76), in which the flux of heat is proportional to the gradient of the temperature; the coefficients in these relations are the diffusion constant and the thermal conductance, respectively. If two or more linear irreversible processes occur in the same system at the same time, they couple to give rise to new effects. For example, the superposition of a flow of heat and an electric current gives rise to *thermoelectricity*—that is, an electric potential generated by a temperature gradient. The new element introduced by the thermodynamic description of linear transport processes is the prediction of a relationship between the coefficients of the "cross components," which in the example cited are the proportionality constants between the electric current and the temperature gradient and between the heat flux and the gradient of the electric potential.

The key assumptions that underlie the thermodynamic analysis of linear transport processes are:

1. The condition of local equilibrium is applicable;
2. At each point in the system the relationships between the local thermodynamic variables are the same as for the case of complete equilibrium, so that, for example,

$$T\,dS = dU + đw - \sum_i \mu_i\,dn_i \tag{20.81}$$

   is valid at each point $\mathbf{r}$;
3. The entropy depends explicitly only on T, U, and μ; the time and spatial dependences of the entropy are generated only by the time and spatial dependences of these variables;
4. The total variation in the entropy of the system as a result of the irreversible process is the sum of the entropy variations at each point in the system;
5. The fluxes are linear homogeneous functions of the forces.

Assumptions 1 and 2 are generalizations to all systems of the results of the analysis of Section 20.4. Assumption 3 implies that at each point $\mathbf{r}$

$$T\frac{dS}{dt} = \frac{dU}{dt} + \frac{đw}{dt} - \sum_i \mu_i \frac{dn_i}{dt}, \tag{20.82}$$

and 4 is equivalent to the statement that the entropy is an extensive quantity. Assumption 5 will be considered in more detail below.

Since the entropy is an extensive quantity we can write for its total value

$$S = \int_V \rho s\,dV, \tag{20.83}$$

and for its rate of change with time

$$\frac{dS}{dt} = \int_V \frac{\partial(\rho s)}{\partial t}\,dV, \tag{20.84}$$

where ρ is the mass density and s the entropy per unit mass. Eq. 20.2 equates the total amount of the conserved quantity Q which leaves the volume V per unit time to the integral the flux $\mathbf{J}^Q(\mathbf{r}, t)$ over the surface A enclosing V. In a system in which an irreversible process occurs, the entropy is not conserved, so to apply Eq. 20.2 to the entropy balance it must be modified to include the production of entropy within V. We thus write

$$\int_V \frac{\partial(\rho s)}{dt} = -\int_A \mathbf{J}^S \cdot d\mathbf{A} + \int_V \Sigma\,dV, \tag{20.85}$$

where Σ is the rate of creation of entropy per unit volume. If d_eS is the change in entropy caused by the interaction of the system with its surroundings and d_iS is the creation of entropy in an irreversible process, we can write

$$\frac{d_eS}{dt} \equiv -\int_A \mathbf{J}^S \cdot d\mathbf{A} \quad \text{and} \quad \frac{d_iS}{dt} \equiv \int_V \Sigma\,dV. \tag{20.86}$$

Consider, now, a system that is in an adiabatic enclosure; the state of this system is described by parameters $\{a_i\}$, $i = 1, \ldots, n$. The values of these parameters when the system is at equilibrium are $\{a_i^0\}$, and their deviations from the equilibrium values will be denoted $\{\alpha_i = a_i - a_i^0\}$. In an equilibrium state, the entropy is a maximum, and all of the $\{\alpha_i\}$ are zero. In a nonequilibrium state, the entropy will be less than that at equilibrium; it can be represented in the form

$$\Delta S = S - S^0 = -\tfrac{1}{2}\sum_{i,k} g_{ik}\alpha_i\alpha_k, \tag{20.87}$$

where the matrix with elements g_{ik} is positive definite. We identify the fluxes in the out-of-equilibrium system with the time derivatives of the parameters $\{\alpha_i = a_i - a_i^0\}$, so that

$$\mathbf{J}^i \equiv \frac{d\alpha_i}{dt} = \dot{\alpha}_i, \tag{20.88}$$

and the forces with the rates of variation of the entropy difference with the parameters $\{\alpha_i = a_i - a_i^0\}$,

$$\mathbf{X}_i \equiv \frac{\partial(\Delta S)}{\partial \alpha_i} = -\sum_{k=1}^{n} g_{ik}\alpha_k. \tag{20.89}$$

Combining Eqs. 20.87 and 20.89 leads to

$$\frac{dS}{dt} = \Sigma = -\sum_{i,k} g_{ik}\,\dot{\alpha}_i\,\alpha_k = \sum_i \mathbf{J}^i \cdot \mathbf{X}_i. \tag{20.90}$$

For all linear transport processes (see assumption 5)

$$\mathbf{J}^i = \sum_{k=1}^{n} L_{ik}\,\mathbf{X}_k, \tag{20.91}$$

where the L_{ik} are phenomenological coefficients. The diagonal terms L_{ii} are the familiar coefficients of thermal conductivity, diffusion, etc., while the off-diagonal terms L_{ik} are the cross-coefficients that describe the coupling that leads to a flux of i in response to a force k. The Onsager reciprocal relations establish that

$$L_{ik} = L_{ki} \qquad (i,k = 1,2,\ldots,n) \tag{20.92}$$

when the forces and fluxes are defined as in Eqs. 20.88 and 20.89. Eq. 20.92 is valid when the L_{ik} are not a function of a magnetic field and/or a rotational velocity. If such functional dependences exist, then Eq. 20.92 must be replaced with $L_{ik}(\mathbf{H},\omega) = -L_{ki}(-\mathbf{H},-\omega)$, where $\mathbf{H}$ is the magnetic field strength and ω is the rotational velocity. We shall not consider these more complicated situations any further. The Onsager reciprocal relations can be obtained from statistical mechanical arguments using the invariance of the equations of motion with respect to variation of the sign of time (microscopic reversibility), the theory of fluctuations and the assumption that the temporal decay of a fluctuation has the same form as does the decay of its macroscopic counterpart. Alternatively, from the point of view of the thermodynamics of linear transport processes, we can take Eqs. 20.91 and 20.92 as new principles that are derived from experimental data.

As an example of the use of the Onsager reciprocal relations, we examine the relationships between heat flow, electric potential difference and temperature gradient when two dissimilar metals are in contact. Consider a system consisting of two wires of dissimilar metals; the wires are joined at their ends, one of which is held at temperature T and the other at temperature $T + dT$ (see Fig. 20.4). It is observed that in the steady state there is an electric potential difference between the two junctions, and there are heat flows in and out of the junctions. The Seebeck effect refers to the situation when the temperature difference is maintained between the junctions but no current I is allowed to flow; in this case the electric potential difference between the junctions can be represented as

$$\mathrm{E} = -\int_{x_0}^{x_0'} \left(\frac{d\phi}{dx}\right)_{I=0} dx \tag{20.93}$$

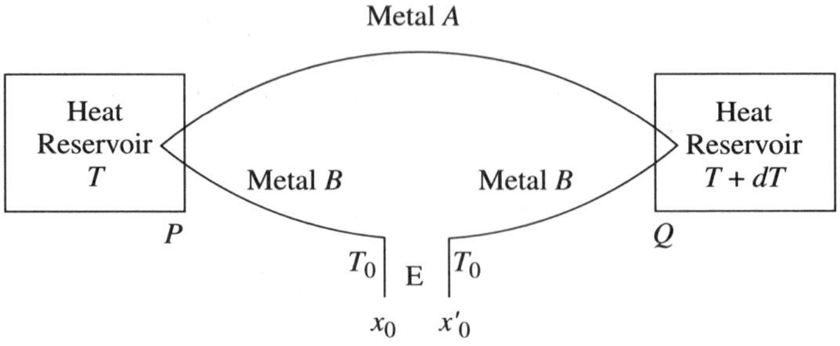

Figure 20.4 Schematic diagram of a thermocouple. The terminals of a potentiometer (to study the Seebeck effect) or a battery (to study the Peltier effect) are attached at the points x_0 and x_0'.

where ϕ is the electrical potential. Its derivative

$$\frac{d\mathrm{E}}{dT} = -\frac{d\phi}{dx}\frac{dx}{dT} = -\frac{d\phi}{dT} \tag{20.94}$$

is called the thermoelectric power. The Peltier effect refers to the situation when the two junctions are kept at the same temperature and a current is passed through the wires. A heat flux J^q, proportional to the current, is absorbed at one junction, and a heat flux $-J^q$ is absorbed at the other junction. The Peltier coefficient, Π, is defined to be

$$\Pi \equiv \left(\frac{J^q}{I}\right)_{dT=0}, \tag{20.95}$$

where I is the current. For this system Eq. 20.91 takes the form

$$\begin{aligned} J^q &= L_{11}\left(-\frac{1}{T}\frac{dT}{dx}\right) + L_{12}\left(-\frac{d\phi}{dx}\right) \\ I &= L_{21}\left(-\frac{1}{T}\frac{dT}{dx}\right) + L_{22}\left(-\frac{d\phi}{dx}\right). \end{aligned} \tag{20.96}$$

The condition defining the Seebeck effect is $I = 0$, whereupon

$$\frac{d\mathrm{E}}{dT} = -\left(\frac{d\phi}{dT}\right)_{I=0} = \frac{L_{21}}{TL_{22}}. \tag{20.97}$$

The condition defining the Peltier coefficient, Eq. 20.95, when substituted into Eq. 20.96, yields

$$\Pi = \left(\frac{J^q}{I}\right)_{dT=0} = \frac{L_{21}}{L_{22}}, \tag{20.98}$$

so that, using the Onsager reciprocal relation

$$\frac{L_{12}}{L_{21}} = 1 \tag{20.99}$$

Table 20.1 Test of Eq. 20.100 (Adapted from D. G. Miller, Chem. Rev. **60,** 15 (1960))

Materials	**Temperature (°C)**	$\frac{\Pi}{T}$ **(expt)**[a]	$\frac{dE}{dT}$ **(predicted)**[a]	$\frac{dE}{dT}$ **(expt)**[a]
Cu/Ni	0	18.6	18.6	20.0
	14	20.2	20.2	20.7
	22	20.5	20.5	22.3
Cu/Pt	0	3.66	3.66	3.67
Cu/Hg	13.9	6.19	6.19	6.19
	56.2	7.36	7.36	7.35
	77.7	7.90	7.90	7.94
	99.7	8.49	8.49	8.54
	132.2	9.28	9.28	9.46
	184.4	10.76	10.76	10.80
Fe/Ni	16	33.1	33.1	31.2
Fe/Hg	18.4	16.72	16.72	16.66
	56.5	16.17	16.17	16.14
	99.6	15.57	15.57	15.42
	131.6	14.89	14.89	14.81
	182.3	13.88	13.88	13.74
Cd/Ni	17	22.1	22.1	22.6
Cd/Pb	0	3.03	3.03	3.02
Zn/Ni	17	22.1	22.1	22.0
Zn/Sn	0	2.65	2.65	2.56
Zn/Bi	0	25.4	25.4	25.1
Bi/Pb	0	–17.5	–17.5	–17.2

[a] Microvolts per degree.

we find

$$\Pi = T\frac{dE}{dT}. \tag{20.100}$$

Equation 20.100 is known as the Kelvin relation. Independent measurements of Π and (dE/dT) can be used to test the Onsager reciprocal relation or, adopting that relation the measurement of Π permits prediction of (dE/dT), which prediction can be tested. Table 20.1 displays experimental data that test the accuracy of the prediction of (dE/dT) from knowledge of Π and use of the Onsager reciprocal relation.

20.6 Negative Temperature

We showed in Chapter 16 that the second law of thermodynamics implied the existence of an absolute temperature scale satisfying the condition $T\Theta(T)$ = constant, with $\Theta(T)$ the integrating factor relating the entropy change to the heat transferred in an infinitesimal reversible process. For our particular choice of absolute temperature scale, T is an everywhere-positive function. This choice of sign has the consequence that a dissipative process in an isolated system leads to an increase in the temperature. We have also shown that the absolute temperature is a measure of the dispersion in energy in a canonical distribution, and that at equilibrium, aside from some minor effects arising from level degeneracy (see Fig. 21.13), the population of energy levels decreases as $\exp(-E/k_BT)$ as the energy increases. We now ask if there are ever circumstances in which it is desirable to extend the domain of the temperature function to allow it to take on negative values.

We note, first, that the introduction of negative absolute temperatures creates an absurdity, for if that negative temperature characterizes a canonical distribution of energy, proportional to $\exp(-E/k_BT)$, the population of the energy levels increases as E increases, and diverges as $E \to \infty$. In all real systems there is no upper bound to the energy scale,[2] although quantum theory does establish the existence of a

[2] If the constituents of the system have internal degrees of freedom, they can be dissociated into fragments with arbitrarily large kinetic energy. If the constituents are elementary particles, these can have arbitrarily large kinetic energy.

lower bound to the energy. We conclude that a negative absolute temperature can never exist in an equilibrium state.

The conclusion just reached is valid for the equilibrium states of a system, but it does not preclude the possibility that nonequilibrium states of the system can be characterized by a negative temperature. Consider a system with an energy-level spectrum that can be meaningfully divided into two groups of levels, the interaction between these two groups being very weak. Moreover, let one of the groups consist of a small number of levels, the highest of which has a bounded energy. The fact that the energy of this subgroup has an upper bound is important. One example of a class of systems with the properties described is a solid containing nuclei with nonzero spin. It is often the case that a canonical distribution of energy in the subset of nuclear spin levels can be achieved very much more rapidly than can a canonical distribution of energy over the combined spin and lattice levels. In the case of LiF, for example, at 300 K the former takes $\sim 10^{-5}$ s and the latter $\sim 10^2 - 10^3$ s.

Suppose that the nuclear spins in some solid are aligned by a magnetic field, and then the magnetic field is reversed so rapidly that the spins cannot follow. The reversed-field situation corresponds to a nonequilibrium state of the system that can persist long enough for many kinds of measurements (~5 min for LiF). Now, at equilibrium in the initial state with, say, magnetic field pointing up, there is a canonical distribution of energy over the levels of the spin system. For a nucleus with spin I, the possible energy levels span the range $-\mu HI$ to μHI, with μ the magnitude of the nuclear magnetic moment and H the magnitude of the magnetic field strength. Note that the most positive energy, μHI, is bounded for any given field **H**. For simplicity, suppose that the nuclear spin is $\frac{1}{2}$, so that the spin energy levels are $-\frac{1}{2}\mu H$ and $\frac{1}{2}\mu H$. At equilibrium the canonical distribution gives, for the ratio of populations of these two levels,

$$\frac{N_+}{N_-} = \exp\left(\frac{\mu H}{k_B T}\right), \tag{20.101}$$

where T is the *common* temperature of the spin and lattice degrees of freedom and N_+ is the number of nuclei with spin parallel to **H**. If, as stated, the spins do not follow the reversal of the magnetic field, then the nonequilibrium state that is created has an inverted population ratio, that is, a higher population in the upper energy level than in the lower energy level. When this inverted distribution is canonical, it can be characterized by a negative temperature. Equation 20.101 then becomes, by definition, for the particular situation described,

$$\frac{N_+}{N_-} = \exp\left[\frac{\mu(-H)}{k_B(-T)}\right]. \tag{20.102}$$

The other degrees of freedom of the system, to which the spins are very weakly coupled, remain at the original temperature, which is positive. As in the case of LiF, because the interaction of the spins with these other degrees of freedom is weak, the spin system can remain in the negative-temperature state described for a considerable period of time, but eventually even the weak interactions are sufficient to drive the system to a new equilibrium state with, as usual, T positive.

Why is the condition that the energy-level spectrum be bounded from above so important? We showed, in Chapter 15, that for all ordinary systems (those with unbounded energy-level spectra), $\Omega(E, V, N)$ and $S \equiv k_B \ln \Omega$ are both everywhere-increasing functions of E. It then follows that the slope $(\partial S/\partial E)_{V,N} = 1/T$ is positive, hence so is T. But if the energy-level spectrum is bounded from above, it is not true that $\Omega(E, V, N)$ is an everywhere-increasing function of E, hence the remainder of the argument fails. Consider, for simplicity, a two-level system with energies ϵ_0 and $2\epsilon_0$. The canonical partition function is $Q_N = (\exp(-\beta\epsilon_0) + \exp(-2\beta\epsilon_0))^N$, from which we find

$$\begin{aligned} A_N &= -k_B T \ln Q_N = -k_B T \ln[e^{-\beta\epsilon_0}(1+e^{-\beta\epsilon_0})]^N \\ &= N\epsilon_0 - Nk_B T \ln(1+e^{-\beta\epsilon_0}), \end{aligned} \tag{20.103}$$

$$\begin{aligned} S_N &= -\left(\frac{\partial A_N}{\partial T}\right)_V \\ &= Nk_B \ln(1+e^{-\beta\epsilon_0}) + Nk_B\beta\epsilon_0 \frac{e^{-\beta\epsilon_0}}{1+e^{-\beta\epsilon_0}}, \end{aligned} \tag{20.104}$$

$$U_N = A_N + TS_N = N\epsilon_0\left(1+\frac{e^{-\beta\epsilon_0}}{1+e^{-\beta\epsilon_0}}\right), \tag{20.105}$$

$$C_V = \left(\frac{\partial U_N}{\partial T}\right)_V = Nk_B(\beta\epsilon_0)^2 \frac{e^{-\beta\epsilon_0}}{(1+e^{-\beta\epsilon_0})^2}. \tag{20.106}$$

Note that S_N and U_N are functions of $\tilde{U} \equiv U_N/N\epsilon_0$. In fact,

$$\tilde{S} \equiv \frac{S_N}{Nk_B} = \ln(1+e^{-\beta\epsilon_0}) + \beta\epsilon_0 \frac{e^{-\beta\epsilon_0}}{1+e^{-\beta\epsilon_0}}, \tag{20.107}$$

$$\tilde{U} = 1 + \frac{e^{-\beta\epsilon_0}}{1+e^{-\beta\epsilon_0}}, \tag{20.108}$$

and we find, using

$$\tilde{U} + \tilde{U}e^{-\beta\epsilon_0} = 1 + 2e^{-\beta\epsilon_0},$$

that

$$\beta\epsilon_0 = \ln\frac{2-\tilde{U}}{\tilde{U}-1}, \tag{20.109}$$

$$\tilde{S} = -(\tilde{U}-1)\ln(\tilde{U}-1) - (2-\tilde{U})\ln(2-\tilde{U}), \tag{20.110}$$

$$\frac{1}{T} \equiv \frac{k_B}{\epsilon_0}\left(\frac{\partial \tilde{S}}{\partial \tilde{U}}\right)_V = \frac{k_B}{\epsilon_0}\ln\frac{2-\tilde{U}}{\tilde{U}-1}. \tag{20.111}$$

Equation 20.111 shows that this system has $T < 0$ when $\tilde{U} > \frac{3}{2}$; that is, it is in a state of negative temperature (see Fig. 20.5). Thus for energies in the range $(3N\epsilon_0/2) < U < 2N\epsilon_0$ in this system there is a population inversion relative to that expected from a canonical distribution with

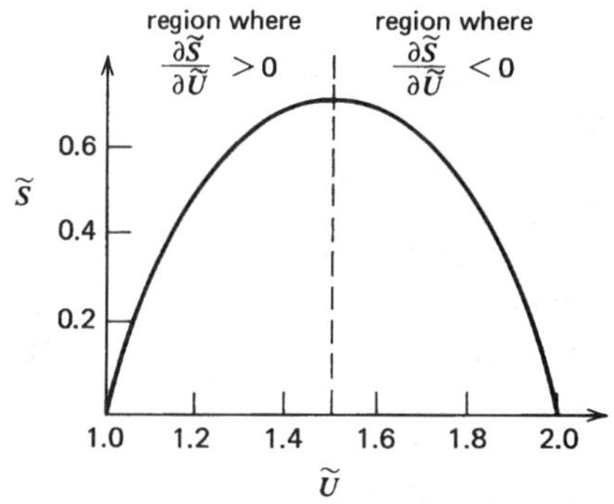

Figure 20.5 $\tilde{S}$ as a function of $\tilde{U}$ for the two-level system described in the text.

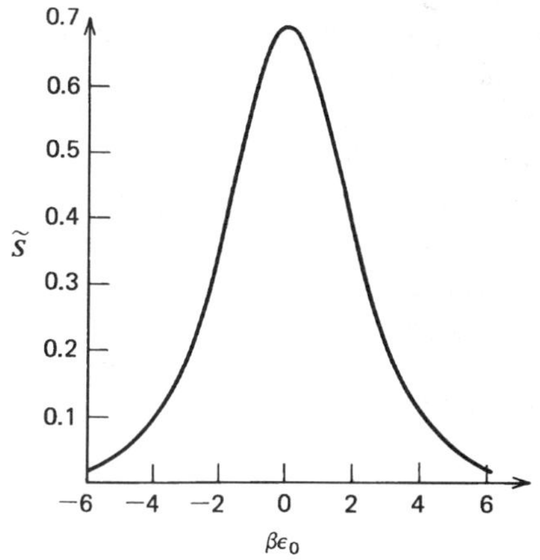

Figure 20.6 $\tilde{S}$ as a function of $1/T$ for the two-level system described in the text.

positive temperature. The point is that as $T \to 0$ only the lowest level of a system is occupied whether the energy-level spectrum is bounded from above or not. But as $T \to \infty$, if the energy-level spectrum is bounded, all levels become equally populated, whereas if the spectrum is unbounded, this does not happen. Using $1/T$ as the variable rather than T, we see that negative temperatures join on smoothly after $1/T$ passes through zero (see Fig. 20.6). In this sense negative temperatures correspond to "hotter" situations than do positive temperatures.

Another way of ascertaining the importance of an upper bound to the energy-level spectrum for the definition of negative temperature is the following. Given the fact that in a system of energy levels bounded from above the populations of the levels approach equality as $T \to \infty$, the heat capacity associated with that system of levels must approach zero as $T \to \infty$ (see Fig. 20.7). Indeed, the heat capacity of such a system tends to zero sufficiently rapidly that only a finite amount of energy is needed to raise its temperature to infinity. Since the coupling of the subsystem under discussion to the lattice degrees of freedom is very weak, this addition of energy can be effected without much influence on the lattice modes. If still more energy is added to the system, it is

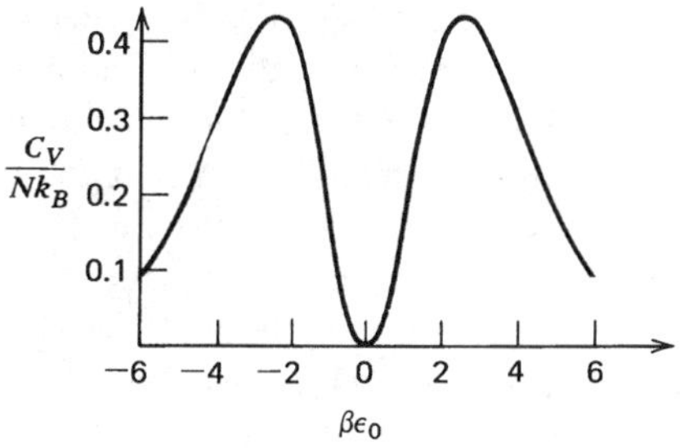

Figure 20.7 C_V as a function of $1/T$ for the two-level system described in the text.

forced into the negative-temperature domain. In our first example the energy was forced to become positive by reversing the magnetic field so rapidly that the nuclear spins could not follow, hence they wound up in positive energy states.

The description of more complex systems than the example given is fundamentally the same. The key ideas are that (1) the energy spectrum of some subset of levels of a system is bounded from above, (2) the degrees of freedom corresponding to that subset of levels interact very weakly with all the others in the system, (3) achievement of a canonical distribution of energy within the subset of levels is much more rapid than is equilibration of energy over all the levels of the system, and (4) a means exists for putting the subsystem corresponding to the bounded energy-level spectrum into a very high energy state. The negative-temperature state created in this process is a transient. If there were no contact with the world of positive temperatures it could persist, but because of that contact it is eventually destroyed. It is important to discuss the existence of negative temperatures as an exception to the rule that T is always positive, but it must be emphasized again that such situations never occur with complete systems in equilibrium. It is only for very special transiently isolatable systems that the concept of negative temperature is useful.

20.7 Thermodynamics of Systems at Negative Absolute Temperature

The arguments of the last section concerning negative absolute temperature were based on an analysis of the microscopic states of a system for which the energy spectrum is bounded above. The key lessons derived from that analysis are the existence of a domain in which the entropy is not an increasing function of the energy and the role of irreversible processes in the transfer of a system from the domain of positive absolute temperature to the domain of negative absolute temperature. These observations can be made the basis for a macroscopic analysis of the conditions for existence of, and the properties of systems with, negative absolute temperature.

Taking a thermodynamic point of view, the only condition necessary for the existence of a negative absolute temperature is that the entropy not be restricted to be a monotone increasing function of the internal energy. Then, any domain in which $(\partial S/\partial U)_x \equiv 1/T < 0$ is one in which the absolute temperature is negative. Although $S(U)$ is an increasing function of U for all systems for which the energy of the system is not bounded above, which is the overwhelming majority of all systems, we have shown that there are circumstances for which it is useful to consider a subset of levels as an isolated system whose energy does have an upper bound, in which case $S(U)$ has the general form displayed in Fig. 20.5. Accepting for the domain $T < 0$ the validity of the second law of thermodynamics in the form that states that the entropy change in a spontaneous process is always positive (see below), the important influence on the thermodynamic properties of a system played by an upper bound to its energy can be illustrated with the curve $S(U)$ displayed in Fig. 20.8. Consider an adiabatic process starting from a point on this curve; several such processes are indicated. The points displayed as $S(U)$ represent stable states of the system. At each energy, the points below the curve represent states of the system that have lower entropy than the stable state, hence are unstable,

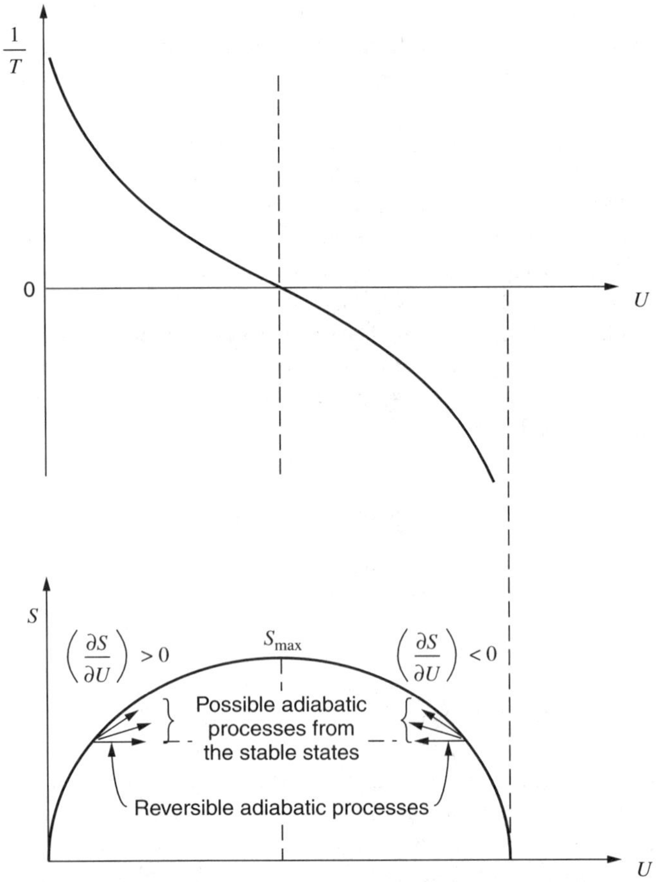

Figure 20.8 Variation of $1/T$ with U and of S with U for the two level system described in the text. Adapted from G. N. Hatsopoulos and J. H. Keenan, Principles of General Thermodynamics, Wiley, New York 1965.

while the points above the curve represent states that cannot exist with the given constraints on the system. The points on the curve $S(U)$ to the left of the maximum represent states with positive absolute temperature, while those to the right of the maximum represent states with negative absolute temperature. Clearly, the only allowed adiabatic processes that can start from a state with positive absolute temperature increase the energy of the system (work is done on the system), whereas the only allowed adiabatic processes that can start from a state with negative absolute temperature decrease the energy of the system (work is done by the system). No adiabatic process can start from the state for which S is a maximum.

We now examine in a little more detail the first, second, and third laws of thermodynamics. We take the definitions of work and heat to be the same for positive and negative absolute temperature, which implies that the first law of thermodynamics has the same form for positive and negative absolute temperature. However, there is a difference between the constraints on the interconversion of heat to work and vice versa in the positive and negative domains of absolute temperature. When $T > 0$ experiments establish that work can be completely converted to heat, but heat cannot be completely converted to work. When $T < 0$ nuclear magnetic resonance experiments establish that heat can be completely converted to work, but work cannot be completely converted to heat.

The form of the second law of thermodynamics which defines the existence of the entropy and the increase of entropy in spontaneous processes is also the same for positive and negative absolute temperature, which can be demonstrated as follows. We write

$$dU = đq + đw \tag{20.112}$$

and define the entropy via the differential relation

$$dS = \frac{đq_{\text{rev}}}{T}. \tag{20.113}$$

Now consider two equilibrium states, denoted 1 and 2, of a system with $T < 0$. Suppose heat $đq_{\text{irr}}$ is transferred in a irreversible process $1 \rightarrow 2$, during which work $đw_{\text{irr}}$ is done. Then

$$đq_{\text{irr}} = dU - đw_{\text{irr}}. \tag{20.114}$$

If the same transition between states of the system is carried out reversibly

$$đq_{\text{rev}} = dU - đw_{\text{rev}}, \tag{20.115}$$

so that, by subtraction,

$$đq_{\text{irr}} - đq_{\text{rev}} = đw_{\text{rev}} - đw_{\text{irr}} \equiv \mathcal{W}. \tag{20.116}$$

The quantity $\mathcal{W}$ must be different from zero, since otherwise the irreversible process $1 \to 2$ can be carried out reversibly in the direction $2 \to 1$, which would imply reversing an irreversible process with no change in the surroundings. The quantity $\mathcal{W}$ cannot be negative, since this would imply that after a cycle $1 \to 2 \to 1$ external work has gone into heating a reservoir without any change in other parts of the universe, which violates the (experimental) condition that when $T < 0$ work cannot be completely converted to heat. However, $\mathcal{W}$ can be positive, since this case corresponds to doing work in the cycle $1 \to 2 \to 1$ at the expense of heat drawn from a reservoir which, when $T < 0$, can be carried out with 100% efficiency. We therefore conclude that

$$đq_{\text{irr}} > đq_{\text{rev}}$$
$$đw_{\text{rev}} > đw_{\text{irr}} \tag{20.117}$$

which, since $đq_{\text{rev}} = T\,dS$, leads to

$$T\,dS < đq_{\text{irr}}. \tag{20.118}$$

In an adiabatic irreversible process $đq_{\text{irr}} = 0$, so in the domain where $T < 0$ we still have

$$dS > 0. \tag{20.119}$$

We conclude that the second law of thermodynamics, in the form that states that the entropy change in a spontaneous process in an isolated system is positive, holds both when $T > 0$ and when $T < 0$. The basic relationship for the thermodynamics of systems in states with $T < 0$ is then

$$T\,dS \le dU - đw. \tag{20.120}$$

We shall see later that the Kelvin statement of the second law needs modification to be applicable to both the domain where $T > 0$ and $T < 0$.

The various formulations of the third law of thermodynamics remain unchanged in the domain $T < 0$ provided that the absolute zeroes of the positive and negative temperature scales are treated the same. However, the physical states with $T = +0$ K and $T = -0$ K are completely different. When the domain is $0 < T < \infty$, the state at $T = 0$ corresponds to an absolute minimum of the energy, whereas when the domain is $-\infty < T < 0$, the state at $T = 0$ corresponds to the absolute maximum in the energy (see Fig. 20.8). A system cannot become colder than $T = +0$ K because no energy can be extracted from it; a system cannot become hotter than $T = -0$ K since it can no longer absorb any energy. The third law of thermodynamics can then be stated as follows:

It is impossible, in a finite number of operations, to cool any system down to $T = +0$ K or to heat any system up to -0 K.

The first part of this statement refers to the domain $0 < T < \infty$ and the second part to the domain $-\infty < T < 0$. The heat capacity vanishes both when $T \to +0$ K and when $T \to -0$ K.

The importance of irreversible processes for effecting the transfer of a system from the domain of positive absolute temperature to the domain of negative absolute temperature becomes evident when we examine the relationship between the absolute temperature scale and an empirical temperature scale, which is determined by some macroscopic variable. Suppose the state of a simple system is determined by T and some macroscopic variable Y. We assume that

$$T = \varphi(\theta) \tag{20.121}$$

where θ is the empirical temperature. Then, using (see Eq. 13.21)

$$đq = dU + X\,dY, \tag{20.122}$$

and (20.113) leads to

$$\frac{\left(\frac{\partial U}{\partial T}\right)_Y dT + \left[\left(\frac{\partial U}{\partial Y}\right)_T + X\right] dY}{T} = \left(\frac{\partial S}{\partial T}\right)_Y dT + \left(\frac{\partial S}{\partial Y}\right)_T dY, \tag{20.123}$$

from which we find

$$\left(\frac{\partial S}{\partial T}\right)_Y = \frac{1}{T}\left(\frac{\partial U}{\partial T}\right)_Y$$
$$\left(\frac{\partial S}{\partial Y}\right)_T = \frac{1}{T}\left[\left(\frac{\partial U}{\partial Y}\right)_T + X\right]$$
$$\frac{\partial}{\partial Y}\left(\frac{1}{T}\left(\frac{\partial U}{\partial T}\right)_Y\right) = \frac{\partial}{\partial Y}\left(\frac{1}{T}\left[\left(\frac{\partial U}{\partial Y}\right)_T + X\right]\right) \tag{20.124}$$

or

$$T\left(\frac{\partial X}{\partial T}\right)_Y = \left(\frac{\partial U}{\partial Y}\right)_T + X. \tag{20.125}$$

In the ordinary case $Y = V$ and $X = p$.

We assume that $T = \varphi(\theta)$ is a single valued function that can be inverted to read $\theta = \psi(T)$ and

$$T\left(\frac{\partial X}{\partial \theta}\right)_Y \frac{d\theta}{dT} = \left(\frac{\partial U}{\partial Y}\right)_\theta + X, \tag{20.126}$$

or

$$\frac{dT}{T} = \frac{\left(\frac{\partial X}{\partial \theta}\right)_Y d\theta}{\left(\frac{\partial U}{\partial Y}\right)_\theta + X}, \tag{20.127}$$

which can be integrated to read

$$\ln \frac{T_2}{T_1} = \int_{\theta_1}^{\theta_2} d\theta \left[\frac{(\partial X/\partial \theta)_Y}{(\partial U/\partial Y)_\theta + X} \right] \equiv \mathcal{T}. \tag{20.128}$$

Therefore, we can write

$$T_2 = T_1 \exp(\mathcal{T}), \tag{20.129}$$

where T_2 and T_1 are the absolute temperatures corresponding to the empirical temperatures θ_2 and θ_1, respectively.

This relation between T_2 and T_1 implies that when a system undergoes a reversible transition from one state to another, say $1 \rightarrow 2$, the absolute temperature cannot change sign—i.e., T is always positive or always negative. The conventional choice $T \geq 0$ implies that $C_Y > 0$, where $C_Y \equiv (\partial U/\partial T)_Y$. In the ordinary case $Y = V$ and $X = p$, so this choice is equivalent to the statement that adding heat to a system constrained to have constant volume always increases the temperature of the system. However, from the relation deduced for $T_2/T_1 = f(\theta_2, \theta_1)$, and without the additional condition just cited, it does not follow that negative absolute temperatures cannot exist (along with positive absolute temperatures). Note that the possibility that negative absolute temperatures can exist in some systems cannot be denied by the third law of thermodynamics, since the unattainability of $T = 0$ only implies the impossibility of a transition through $T = 0$ to the domain of negative absolute temperature. The important point about the sign of the absolute temperature is that it is determined by an additional condition connected with defining which temperature is larger and which is smaller. Moreover, since the unchanging nature of the sign of T is established by considering reversible processes there remains the possibility that the sign of T can change if a system passes from one equilibrium state to another by an irreversible process. The conclusion that T cannot change sign is valid if the states that can be reached from a given state irreversibly can always also be reached reversibly. This is, indeed, true in all ordinary cases. However, states with negative absolute temperature do exist for some special systems, e.g., a simple nuclear spin systems such as described in Section 20.6. As stated there, to achieve this population inversion that transfers the system to the domain of negative absolute temperature, it must be subjected to a rapid irreversible process that transfers energy to the system; the amount of energy transferred must be larger than that corresponding to the transition from $T = 0$ to $T = \infty$. It is impossible to carry out this energy transfer for most systems, since the internal energies of most systems increase indefinitely as $T \rightarrow \infty$. In contrast, if the internal energy of a system approaches an asymptotic value that is finite as $T \rightarrow \infty$, it is possible to place the system in states with negative absolute temperatures.

A convenient graphical visualization of the relationship between states with positive and negative absolute temperature is obtained when we map the temperature line $-\infty < T < \infty$ onto a circle tangent at $T = 0$; this construction is shown in Fig. 20.9. This mapping makes it evident that to pass from the domain where $T > 0$ to the domain where $T < 0$, we must go through $T = \pm\infty$. Therefore, when a system is in a state with $T < 0$, it is "hotter" than at $T = \infty$. Put another way, the range $-\infty < T \leq 0$ lies above $T = \infty$, not below $T = 0$.

As already mentioned several times, in order that a system be capable of being in a state with negative absolute temperature the energy of that system must have a finite limit as $T \rightarrow \infty$. Clearly, the system to be placed in a state with negative absolute temperature must be thermally isolated from any others that do not satisfy the condition on the energy stated. In other words, the time needed to establish equilibrium in the system of interest must be small relative to the time required for an appreciable amount of energy to be lost or acquired from other systems.

Some of the unusual properties of systems in states with $T < 0$ are:

1. When heat $đq_{rev}$ is transferred to such a system its entropy decreases, since $dS = đq_{rev}/T$ and $T < 0$. The molecular interpretation of this observation is that adding heat to a system in a negative temperature state increases the ordering in the system, which is consistent with –0 K being the state with maximum energy.
2. Consider two systems, each in a state with $T < 0$. Let the two system temperatures be T_1 and T_2, and imagine that the systems are placed in contact. If heat $đq_{rev}$ is transferred from system 1 to system 2, the entropy change is

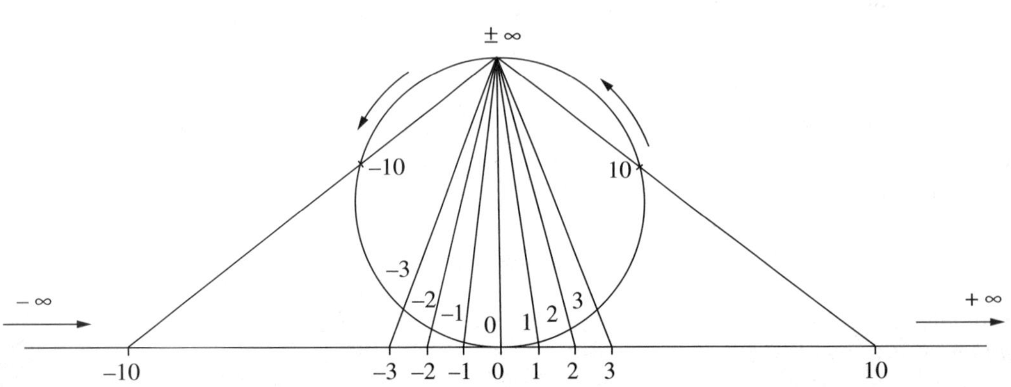

Figure 20.9 Mapping of the temperature line into a circle.

$$-\frac{đq_{\text{rev}}}{T_1}+\frac{đq_{\text{rev}}}{T_2}. \tag{20.130}$$

Since $dS > 0$ for a spontaneous process, the heat will flow from $1 \rightarrow 2$ if $T_1 > T_2$—i.e., T_1 is less negative than is T_2. Therefore, just as in the domain $T > 0$, heat flows from the hotter to the cooler body.

3. When a system is in a state with $T < 0$ it can act as an amplifier, e.g., the use of population inversion in a laser.

4. Consider a heat engine operating via a Carnot cycle (see Section 16.4) between two reservoirs with absolute temperatures T_1 and T_2. These reservoirs deliver heats q_1 and q_2, respectively, to the system, and work w is accomplished per cycle of operation. The second law of thermodynamics requires that the sum of the entropy changes of all systems involved in the process must be greater than or equal to zero. Since the entropy change of the engine carried around one cycle is zero we must have, in general

$$-\frac{q_1}{T_1}-\frac{q_2}{T_2}\geq 0, \tag{20.131}$$

where the value zero characterizes a reversible engine. The efficiency of the heat engine is defined by

$$\eta \equiv \frac{w}{q_1}, \tag{20.132}$$

with q_1 the heat absorbed by the engine from the hot reservoir. Then for a reversible heat engine

$$\eta_{\text{rev}} = 1 - \frac{T_2}{T_1}. \tag{20.133}$$

We now consider three cases for which the reservoir at temperature T_1 is hotter than the reservoir at temperature T_2. For $T_1 \geq T_2$ we find $0 \leq \eta_{\text{rev}} < 1$, so that the efficiency of a reversible heat engine operating between reservoirs with positive absolute temperatures is less than one but greater than or equal to zero. For $0 > T_1 > T_2$ we find $-\infty < \eta_{\text{rev}} < 0$, so that the efficiency of a reversible heat engine operating between reservoirs with negative absolute temperatures is negative. The implication of this deduction is that to operate a reversible heat engine so that it receives heat from the hotter negative temperature reservoir and rejects heat to the colder negative temperature reservoir work must be supplied from the surroundings. Conversely, for a reversible heat engine operating between two reservoirs with negative absolute temperatures to perform work on the surroundings it must receive heat from the colder reservoir and reject it to the hotter reservoir. Finally, for $T_2 > 0 > T_1$ we find $+\infty > \eta_{\text{rev}} > 1$, so that the efficiency of a reversible heat engine operating between a hot reservoir with negative absolute temperature and a cold reservoir with positive absolute temperature is greater than unity.

Consider again the case for which $0 > T_1 > T_2$, and imagine running the engine as a refrigerator which transfers heat from the colder to the hotter reservoir. In this case the engine performs work on the surroundings. Suppose the engine carries out one cycle. Now imagine that the hot and cold reservoirs are placed in thermal contact, say via a thermal switch. Heat will then flow from the hotter to the colder reservoir; we arrange the switch so the amount of heat transferred from the hot to the cold reservoir when in thermal contact is the same as the amount of heat transferred from the cold to the hot reservoir in one cycle of operation of the engine. The combination of the original engine and the once per cycle thermal contact between the reservoirs after each cycle is a new cyclic engine that, without causing any change in the surroundings, does work by conversion of heat drawn from a single (cold) reservoir. This arrangement violates the Kelvin statement of the second law of thermodynamics, which must be amended to read as follows to be applicable for both positive and negative absolute temperature:

Heat can be completely converted into work by taking a medium through a cyclic process if and only if that heat is drawn from a reservoir which has a negative absolute temperature.

Finally, we consider the stability of thermodynamic states when $T < 0$. From the fundamental relation

$$T\,dS < dU + p\,dV, \tag{20.134}$$

we conclude that in an isolated system that has negative absolute temperature the equilibrium state corresponds to a maximum of the entropy. Then, with U and V constant,

$$\begin{aligned}\Delta S &< 0\\ \delta S &= 0\\ \delta^2 S &< 0,\end{aligned} \tag{20.135}$$

characterize the displacements from the equilibrium state. When the system is not isolated we find

$$\begin{aligned}dA &> -S\,dT - p\,dV\\ dG &> -S\,dT + V\,dp\end{aligned} \tag{20.136}$$

so that in a process at constant T and V, in the domain where $T < 0$, the Helmholtz free energy is a maximum at equilibrium, and in a process at constant T and p, in the domain where $T < 0$, the Gibbs free energy is a maximum:

$$\begin{aligned}&\begin{pmatrix}\Delta A < 0\\ \delta A = 0\\ \delta^2 A < 0\end{pmatrix} \text{ at constant } T \text{ and } V \text{ when } T<0\\ &\begin{pmatrix}\Delta G < 0\\ \delta G = 0\\ \delta^2 G < 0\end{pmatrix} \text{ at constant } T \text{ and } p \text{ when } T<0\end{aligned} \tag{20.137}$$

Proceeding as in our discusssion of the stability of the equilibrium state in the domain where $T > 0$, we find that in the domain where $T < 0$

$$\left(\frac{\partial^2 U}{\partial S^2}\right)_{V,n} (\Delta S)^2 + 2\left(\frac{\partial}{\partial V}\left(\frac{\partial U}{\partial S}\right)_{V,n}\right)_{S,n} (\Delta V \Delta S) + \left(\frac{\partial^2 U}{\partial V^2}\right)_{S,n} (\Delta V)^2 < 0 \tag{20.138}$$

Note that the sign of the inequality is the reverse of that for the domain where $T > 0$. This inequality will be satisfied for arbitrary ΔV and ΔS if

$$\text{(a)} \quad \left(\frac{\partial^2 U}{\partial S^2}\right)_{V,n} < 0$$

$$\text{(b)} \quad \left(\frac{\partial^2 U}{\partial S^2}\right)_{V,n} \left(\frac{\partial^2 U}{\partial V^2}\right)_{S,n} - \left(\frac{\partial}{\partial V}\left(\frac{\partial U}{\partial S}\right)_{V,n}\right)_{S,n}^2 < 0. \tag{20.139}$$

Inequality (a) gives

$$\left(\frac{\partial^2 U}{\partial S^2}\right)_{V,n} = \frac{T}{C_V} < 0, \tag{20.140}$$

or $C_V > 0$ when $T < 0$. Inequality (b), after some algebraic manipulation, yields

$$\left(\frac{\partial p}{\partial V}\right)_T < 0. \tag{20.141}$$

We conclude that the conditions determining the stability of thermodynamic states are the same for $T < 0$ and $T > 0$. By choosing U to be a monotone increasing function of T, which is always possible, we ensure that $C_V > 0$, so thermodynamic states with $T < 0$ are stable.

APPENDIX 20A

Symmetry of the Momentum Flux Tensor

In this appendix we show that if the ijth component of the momentum flux tensor has a term of the form $a(\partial v_i/\partial x_j)$, where a is a constant, the term $a(\partial v_j/\partial x_i)$ must also appear. To prove this statement, we note that if the entire volume of fluid under consideration is subjected to uniform rotation, no momentum is transferred within the fluid as a result of viscosity. So, we let $\mathbf{v} = \mathbf{\Omega} \times \mathbf{r}$, where $\mathbf{\Omega}$ is a constant angular velocity, and calculate $\partial v_i/\partial x_j$.

In dealing with vector cross products, it is convenient to introduce the Levi–Civita symbol ε_{ijk}. This quantity requires three indices (it is a third-rank tensor!), each of which ranges from 1 to 3. If $\mathbf{A} = \mathbf{B} \times \mathbf{C}$ is a vector relationship, a little experimentation will convince the reader that

$$A_i = \sum_{j,k} \varepsilon_{ijk} B_j C_k, \tag{20A.1}$$

where ε_{ijk} has the defined values

$$\varepsilon_{ijk} = \begin{cases} 1 \text{ for } (i,j,k)=(1,2,3) \text{ or } (2,3,1) \text{ or } (3,1,2), \\ -1 \text{ for } (i,j,k)=(1,3,2) \text{ or } (2,1,3) \text{ or } (3,2,1), \\ 1 \text{ for } (i,j,k) = \text{all other combinations.} \end{cases} \tag{20A.2}$$

With the use of the Levi–Civita symbol, we see that

$$\begin{aligned} \frac{\partial v_i}{\partial x_j} &= \frac{\partial}{\partial x_j}(\mathbf{\Omega}\times\mathbf{r})_i = \frac{\partial}{\partial x_j}\sum_{k,l}\varepsilon_{ikl}\Omega_k x_l \\ &= \sum_{k,l}\varepsilon_{ikl}\Omega_k \frac{\partial}{\partial x_j} x_l = \sum_{k,l}\varepsilon_{ikl}\Omega_k \delta_{jl} \\ &= \sum_k \varepsilon_{ikj}\Omega_k. \end{aligned} \tag{20A.3}$$

Since the viscous contribution to the stress tensor must be zero, a term like $-\Sigma_k\, \varepsilon_{ikj}\Omega_k$ must also be present. Notice now that

$$\frac{\partial v_j}{\partial x_i} = \sum_k \varepsilon_{jki}\Omega_k, \tag{20A.4}$$

through a sequence of steps analogous to Eq. 20A.4. But by the definition of ε_{ijk},

$$\varepsilon_{ikj} = -\varepsilon_{jki}, \tag{20A.5}$$

so that

$$\frac{\partial v_i}{\partial x_j} + \frac{\partial v_j}{\partial x_i} = 0. \tag{20A.6}$$

Thus, only symmetric combinations like

$$\frac{\partial v_i}{\partial x_j} + \frac{\partial v_j}{\partial x_i}$$

can appear in the momentum flux tensor.

FURTHER READING

Abragam, A., *Principles of Nuclear Magnetism* (Oxford University Press, London, 1961), Chapter 5.

de Groot, S. R., and Mazur, P., *Nonequilibrium Thermodynamics* (North Holland, Amsterdam, 1962).

Fitts, D. D., *Nonequilibrium Thermodynamics* (McGraw-Hill, New York, 1962).

Haase, R., *Thermodynamics of Irreversible Processes* (Dover Publications, New York, 1990).

Kestin, J., A *Course in Thermodynamics,* Vol. II (Blaisdell, Waltham, Mass., 1968), Chapter 24.

Loudon, R., *The Quantum Theory of Light* (Oxford University Press, London, 1973), pp. 33–37.

Nicolis, G., and Prigogine, I., *Self-Organization in Nonequilibrium Systems* (Wiley, New York, 1977).

Prigogine, I., *Introduction to the Thermodynamics of Irreversible Processes* (Charles C. Thomas, Springfield, Ill., 1955).

Ramsey, N. F., *Phys. Rev.* **103,** 20 (1956).

PROBLEMS

1. Verify that Eq. 20.17 is a solution of the diffusion equation.

2. Calculate the integral of $\rho_k(\mathbf{r}, t)$ (Eq. 20.17) over all space. Can you predict the result before doing the integral?

3. Derive Eq. 20.18 from Eq. 20.17.

4. Prove the following vector relations:
(a) $\nabla \cdot (\nabla S) = \nabla^2 S$, ($S$ is scalar);
(b) $\nabla \cdot (S\mathbf{v}) = S\nabla \cdot \mathbf{v} + \mathbf{v} \cdot \nabla S$.

5. Prove the following tensor relations.
(a) $\nabla \cdot (\mathbf{v} \cdot \boldsymbol{T}) = \mathbf{v} \cdot (\nabla \cdot \boldsymbol{T}) + \boldsymbol{T}{:}\nabla \mathbf{v}$.
The double dot product of two tensors is a scalar whose value is $\boldsymbol{T}{:}\boldsymbol{X} = \Sigma_{ij} T_{ij} X_{ji}$.
(b) $\boldsymbol{1}{:}\nabla \mathbf{v} = \nabla \cdot \mathbf{v}$.

6. Use the properties of the Levi–Civita tensor ε_{ijk} to prove the following relations:
(a) $\nabla \times (\nabla \times \mathbf{v}) = \nabla(\nabla \times \mathbf{v}) - \nabla^2 \mathbf{v}$;
(b) $\nabla \cdot (\mathbf{v} \times \mathbf{u}) = \mathbf{u} \cdot \nabla \times \mathbf{v} - \mathbf{v} \cdot \nabla \times \mathbf{u}$;
(c) $\mathbf{v} \times (\nabla \times \mathbf{v}) = \frac{1}{2} \nabla v^2 - \mathbf{v} \cdot \nabla \mathbf{v}$.

7. Use Euler's equation, Eq. 20.32, to derive an equation of continuity expressing the conservation of angular momentum for a nonviscous (ideal) fluid.

8. Let s be the entropy per unit mass of an ideal fluid (a fluid with zero viscosity). For this fluid the equation of motion is

$$\frac{\partial \mathbf{v}}{\partial t} + \mathbf{v} \cdot \nabla \mathbf{v} = -\frac{1}{\rho} \nabla p.$$

Show that for adiabatic motion, in which $ds/dt = 0$,

$$\frac{\partial (\rho s)}{\partial t} + \nabla \cdot (\rho s \mathbf{v}) = 0.$$

9. If the motion of the fluid is isentropic, $s =$ constant. Use the combined first and second laws of thermodynamics and the equation of motion to show that, for an ideal fluid,

$$\frac{\partial \mathbf{v}}{\partial t} + \mathbf{v} \cdot \nabla \mathbf{v} = -\nabla h,$$

where h is the enthalpy per unit mass.

10. A fluid can be in mechanical equilibrium without being in thermal equilibrium. A case of this type arises if a small vertical temperature gradient is imposed on the fluid. Show that such a situation can be maintained (be stable) only if

$$\frac{dT}{dz} > -\frac{gT}{c_p v}\left(\frac{\partial v}{\partial T}\right)_p,$$

where g is the acceleration due to gravity. Interpret this result in simple physical terms. *Hint:* As a result of the temperature gradient, an element of fluid at height z having volume per unit mass $v(p, s)$ tends to rise. Consider this rise to be adiabatic and use the condition of mechanical stability, together with a Taylor series expansion in the entropy difference $(s' - s)$ corresponding to the height difference Δz, to derive the quoted formula.

11. Use Eqs. 20.89 and 20.90 to show that if $X_1, \ldots, X_m$, where $m < n$, are kept constant, the minimum rate of entropy production corresponding to variable forces $X_{m+1}, \ldots, X_n$ occurs when the system is in a steady state with $J_{m+1} = 0, \ldots, J_n = 0$.

12. Consider an isotropic solid across which a temperature gradient is maintained, say, in the x direction. Show that, if thermal expansion can be neglected, at each point in the interior of the solid the rate of change of entropy per unit volume is

$$\rho\left(\frac{\partial S}{\partial t}\right) + \frac{\partial}{\partial x}\left(\frac{\mathbf{J}^q}{T}\right) = \frac{\mathbf{J}^q}{T^2}\left(\frac{\partial T}{\partial x}\right),$$

where

$$\mathbf{J}^q = -\lambda\left(\frac{\partial T}{\partial x}\right).$$

13. Use the second law of thermodynamics and the equation for the rate of change of entropy per unit volume derived in Problem 12 to prove that the coefficient of thermal conductivity must be positive ($\lambda > 0$).

14. In Section 20.15 we examined some aspects of the thermodynamic analysis of thermoelectricity. Recall that (a) across a junction between two different conductors, each one maintained at a different temperature, there is a difference of electrical potential (*Seebeck effect*); (b) when an electric current flows through the junction between two conductors, each maintained at the same temperature, at the junction a quantity of heat is absorbed or released that is proportional to the current (*Peltier effect*); and (c) when an electric current flows in a conductor in which there is also a temperature gradient, in addition to the usual Joule heating (proportional to the resistance and square of the current), there is also heat generated proportional to the temperature gradient and the current (*Thomson effect*). Neglecting volume changes associated with the variation of temperature, for all of these phenomena, the fundamental equation of local equilibrium is

$$T\,dS = dU - (\mu - \mathcal{F}\phi)$$

where $\mathcal{F}$ is the Faraday (the charge on 1 mol of electrons), μ is the ordinary density- and temperature-dependent part of the chemical potential, and ϕ is the electrostatic potential (see Chapter 26).

Use the equations describing conservation of charge and conservation of energy to calculate the rate of change of entropy per unit volume at a point inside the conductor when both a temperature gradient and a potential gradient are present.

15. Let the temperature and potential gradients be along the x axis. The superposition of flows arising from the imposed gradients is represented by

$$\mathbf{J}^q = -(L_{11}/T)(\partial T/\partial x) + L_{12}(\mathbf{E} + (\partial/\partial x)(\mu\mathcal{F})),$$
$$\mathbf{I} = -(L_{21}/T)(\partial T/\partial x) + L_{22}(\mathbf{E} + T(\partial/\partial x)(\mu/T\mathcal{F}))$$

with $L_{12} = L_{21}$ by the Onsager Reciprocal Relation. Solve these equations for $\mathbf{J}^q$ and $\mathbf{E}$, writing the result in the form

$$\mathbf{J}^q = -\lambda(\partial T/\partial x) - \Pi\mathbf{I},$$
$$\mathbf{E} = (\mathbf{I}/\sigma) - \alpha(\partial T/\partial x) - (\partial/\partial x)(\mu/\mathcal{F}),$$

where σ is the electrical conductivity. Find expressions for λ, Π, α, and σ in terms of L_{11}, L_{12}, L_{21}, and L_{22}. Note that Π is the Peltier coefficient and α the coefficient of thermoelectric potential.

16. From the form of the equations for $\mathbf{J}^q$ and $\mathbf{E}$ in terms of λ, Π, σ, and α, show that λ is the thermal conductivity and σ the electrical conductivity. Also show that $\Pi = T\alpha$, which relates the Peltier coefficient Π to the coefficient of thermoelectric potential, α.

17. Consider a thermocouple circuit as shown in the accompanying diagram.

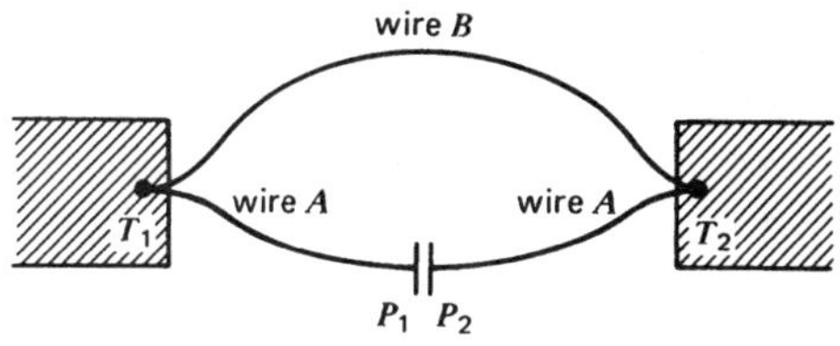

Wires A and B are of different composition, and $T_2 > T_1$. Show that the potential generated across the condenser plates P_1, P_2 is

$$\Delta V = (\alpha_A - \alpha_B)(T_2 - T_1)$$

when α is considered independent of temperature.

18. Consider two wires, A and B, at the same temperature. Let a current I flow through a junction between the wires. Show that the heat released at the junction is $(\Pi_2 - \Pi_1)I$, so that the Peltier heat is proportional to the current passed through the junction.

19. How does the quantum (band) theory of metals explain the Peltier effect, and the relation between the Peltier and Seebeck effects? *Hint:* The Fermi levels of the electrons of the metals in contact are not the same. See Chapter 11.

20. Suppose that a system consists of N independent identifiable molecules and that the energy levels of each molecule are equally spaced:

$$\epsilon_n = n\epsilon_0, \quad n = 1, 2, \ldots \to \infty.$$

Show that the entropy of this system is

$$\frac{S}{Nk_B} = \frac{U}{N\epsilon_0}\ln\left(\frac{U}{N\epsilon_0}\right) - \left(\frac{U}{N\epsilon_0} - 1\right)\ln\left(\frac{U}{N\epsilon_0} - 1\right).$$

Demonstrate that S is a monotonic increasing function of U and that the system cannot have a negative temperature.

21. We have introduced the concept of negative temperature from the point of view of the statistical molecular description of matter. This concept can also be introduced by adding a macroscopic condition to the usual laws of thermodynamics. That condition is as follows: A system can attain a state of negative temperature if and only if its energy approaches a finite limit as $T \to \infty$. Discuss the relationship between this condition and the nature of the energy spectrum for a system that can attain a negative temperature.

CHAPTER 21

The Properties of Pure Gases and Gas Mixtures

In the preceding chapters we have developed the statistical molecular and thermodynamic formalisms required for a systematic description of the properties of matter at equilibrium. In this chapter we make use of these formalisms to analyze the behavior of gases and gas mixtures. Our goal is to answer the following questions:

1. How may observations of the properties of a gas be translated into values of the thermodynamic functions that represent the state of the gas?
2. What is the detailed relationship between the structure of the molecules that compose the gas and the thermodynamic properties of the gas?
3. How do molecular interactions influence the thermodynamic properties of the gas?
4. Which aspects of the molecular structure and the molecular interactions influence chemical equilibrium?

21.1 Thermodynamic Description of a Pure Gas

It is amusing to speculate on what the course of development of thermodynamics and of chemistry might have been if the pressure at the surface of the earth were 1000 atm. At a pressure of 1 atm most gases can be described by equations of state that deviate only slightly from the perfect gas equation of state. Indeed, it is this accident of nature that led to early recognition of the existence of the perfect gas equation of state. At a pressure of 1000 atm the volumes of gases are comparable with the volumes of condensed phases, and it is easier to notice the differences in the properties of substances than to recognize their similarities. If 1000 atm were the "normal" earth pressure, it might have taken much longer to discover the unifying simplicity of the perfect gas concept than it actually did. As it is, the perfect gas concept has served four important functions in the development of chemistry and of thermodynamics:

1. It provided the first method of establishing relative molar weights.
2. It provided a realizable temperature scale that could be identified with the thermodynamic temperature scale.
3. It provides a moderately accurate representation of the properties of real gases near 1 atm.
4. It is a statement of a limiting law of behavior for all gases if only the pressure is low enough.

Functions 1 and 2 are equally well or better executed today without using the properties of gases.[1] On the other hand, function 3 is still useful, and function 4 is still important as an organizing principle when p, V, T measurements on a variety of gases are compared and interpreted.

The properties of real gases may be summarized as follows:

1. At sufficiently low pressure, all gases satisfy the perfect gas equation of state. For simple substances, such as N_2, Ar, or CO, at room temperature, the perfect gas equation of state is accurate to about 1% at pressures less than 15 atm. For more complicated substances, such as H_2O, CH_3OH, or HF, deviations from perfect gas behavior can be observed at lower pressure.
2. At sufficiently low temperature and sufficiently high pressure, all gases condense to a liquid or solid phase. There is a range of temperature and pressure throughout which liquid and gas coexist in equilibrium; there is also a range of temperature and pressure throughout which solid and gas coexist in equilibrium. There is one point at

[1] Relative molar weights are more conveniently and accurately determined by mass spectrometry, and other temperature scales exist that have a well-defined relationship with the thermodynamic scale—for example, those based on the magnetic properties of a paramagnetic salt.

which gas, liquid, and solid coexist at equilibrium (*the triple point*).

3. For each substance there is a temperature, the *critical temperature,* above which it is impossible to condense a gas to a liquid no matter how high the pressure. In a gas–liquid system at equilibrium, the difference in density between the gas and liquid vanishes at the critical temperature. At the point at which the density difference vanishes, the pressure and density are denoted the *critical pressure* and *critical density,* respectively.
4. The properties of many different gaseous chemical substances can be approximately represented by one equation of state if the variables are chosen to be the *reduced temperature,* $\tilde{T} \equiv \dfrac{T}{T_c}$, the *reduced pressure,* $\tilde{p} \equiv \dfrac{p}{p_c}$, and the *reduced volume* or *reduced density,* $\tilde{V} \equiv \dfrac{V}{V_c}$, $\tilde{\rho} \equiv \dfrac{\rho}{\rho_c}$, where T_c, p_c, V_c, and ρ_c are the critical temperature, pressure, volume, and density, respectively. This equation of state is an example of the consequences of the *principle of corresponding states.*

The features of the equation of state of real gases described above are illustrated in Figs. 21.1–21.5.

In this chapter we restrict our attention to the description of the properties of the gas; the nature of the equilibria between solid and gas, liquid and gas, and solid and liquid will be discussed in Chapter 24.

A thermodynamic description of a gas is taken to mean a set of relationships between the derived thermodynamic functions such as U and S, and the measured equation of state and heat capacities. We have already calculated, in Chapters 14 and 17, the energy, enthalpy, and entropy changes accompanying the change of state of a fluid from p_1, V_1, T_1 to p_2, T_2, V_2. Here we just quote the previously obtained results:

$$U(T_2,V_2) = U(T_1,V_1) + \int_{T_1,V_1}^{T_2,V_1} C_V(T,V_1)\,dT + \int_{T_2,V_1}^{T_2,V_2} \left[T_2 \left(\frac{\partial p}{\partial T} \right)_V - p \right] dV, \tag{21.1}$$

$$H(T_2,p_2) = H(T_1,p_1) + \int_{T_1,p_1}^{T_2,p_1} C_p(T,p_1)\,dT + \int_{T_2,p_1}^{T_2,p_2} \left[V - T_2 \left(\frac{\partial V}{\partial T} \right)_p \right] dp, \tag{21.2}$$

$$S(T_2,p_2) = S(T_1,p_1) + \int_{T_1,p_1}^{T_2,p_1} \frac{C_p(T,p_1)}{T}\,dT - \int_{T_2,p_1}^{T_2,p_2} \left(\frac{\partial V}{\partial T} \right)_p dp. \tag{21.3}$$

We now use statement 1 in the summary of properties of gases to write

$$\lim_{p\to 0} \frac{pV}{nRT} = 1, \tag{21.4}$$

as already indicated in Chapter 12. Because Eq. 21.4 is valid for all gases, a convenient reference state from which to measure the thermodynamic properties of a gas is defined by a hypothetical perfect gas composed of the same molecules. Clearly, the properties of this hypothetical perfect gas can be established only in a limiting sense. Each thermodynamic function[2] of a real gas approaches the corresponding function of a perfect gas when the gas density approaches zero, but the rates of approach of the various functions to their limiting forms are different.

In a perfect gas, molecular interactions have negligible influence on the equation of state. Thus, the properties of a perfect gas are determined by the structure of the component molecules, for example, the moments of inertia, vibrational frequencies, electronic energy spectrum. When the hypothetical perfect gas is chosen as the reference state for a given real gas, it follows that the thermodynamic properties of that real gas are the sum of two contributions: a temperature-dependent perfect gas component arising from the distribution of the molecules over the possible internal molecular states, and a density- and temperature-dependent component arising from molecular interactions and the consequent deviations of the real gas equation of state from the perfect gas equation of state.

Let $C_p^*(T)$ denote the limit approached by the constant-pressure heat capacity as the gas pressure approaches zero; $C_p^*(T)$ is a function of temperature only and is not the same

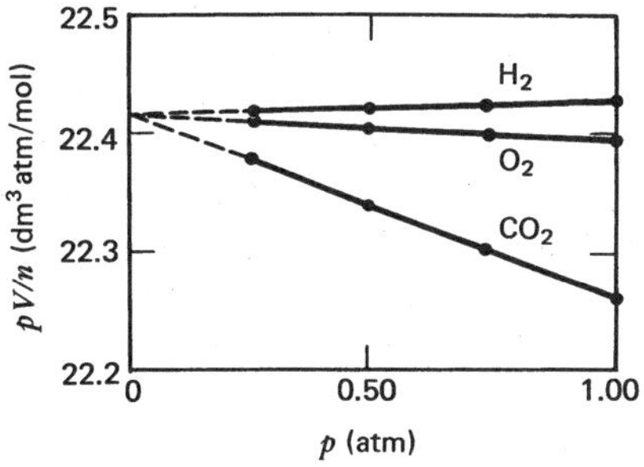

Figure 21.1 The pV product for three gases, showing the universal approach to the ideal gas value as the pressure tends to zero. From R. H. Cole and J. S. Coles, *Physical Principles of Chemistry* (Freeman, San Francisco, 1964).

[2] There is an exception: The Joule–Thomson coefficient $\mu_{JT} \equiv (\partial T/\partial p)_H$ does not approach zero as $n/V \to 0$.

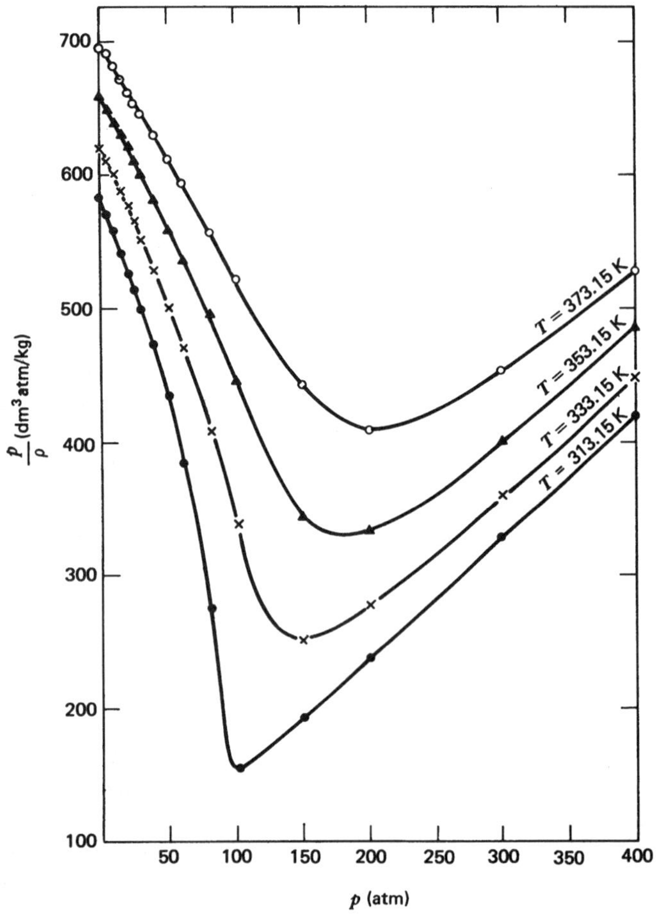

Figure 21.2 Some isotherms of CO_2 for temperatures above the critical temperature.

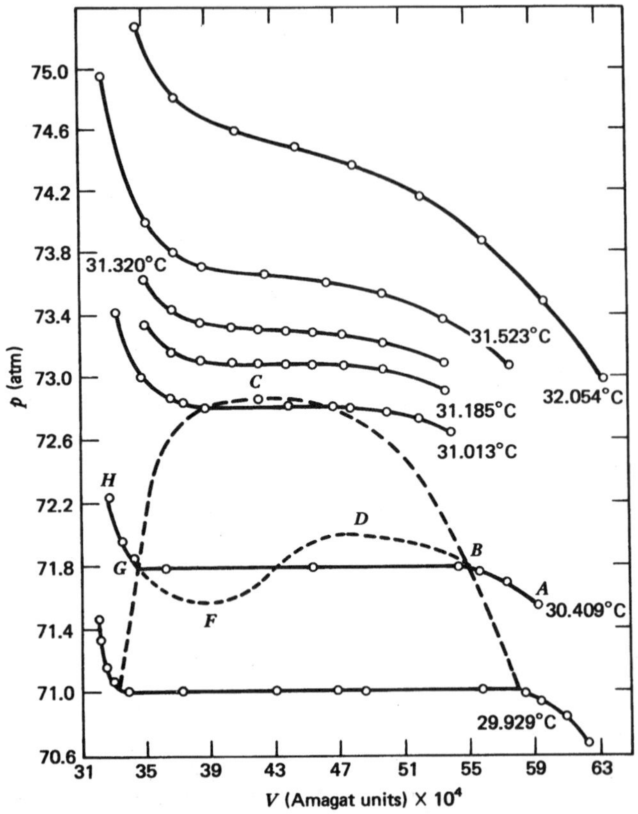

Figure 21.3 Some isotherms of CO_2 in the immediate vicinity of the critical temperature. From E. A. Moelwyn-Hughes, *Physical Chemistry* (Cambridge University Press, Cambridge, 1940); data from A. Michels, B. Blaisse, and C. Michels, *Proc. Roy. Soc.* **A160,** 367 (1937). One Amagat unit of volume for CO_2 is 2.2398×10^4 cm^3/mol.

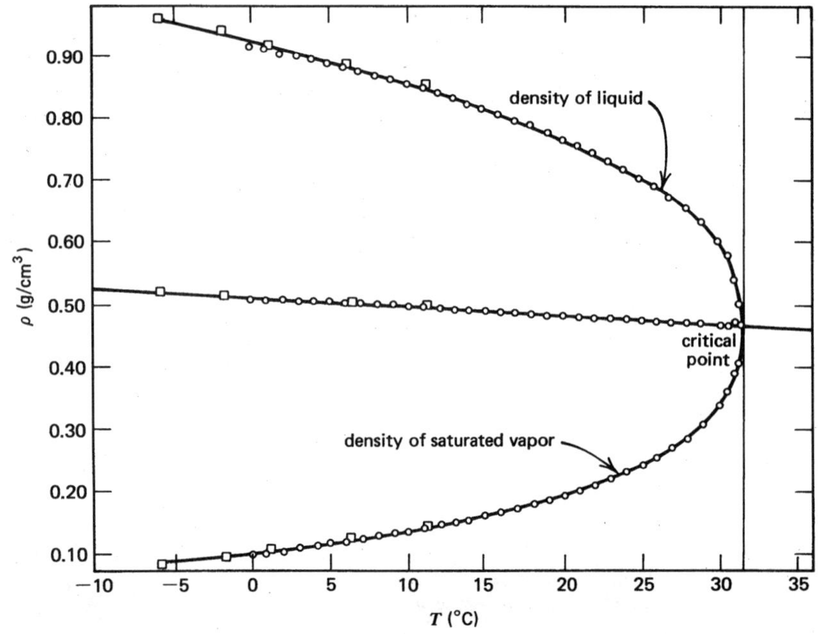

Figure 21.4 The coexistence line for CO_2, showing the approach of the gas and liquid densities to a common value at the critical point. From E. A. Moelwyn-Hughes, *Physical Chemistry* (Cambridge University Press, Cambridge, 1940); data from A. Michels, B. Blaisse, and C. Michels, *Proc. Roy. Soc.* **A160,** 367 (1937).

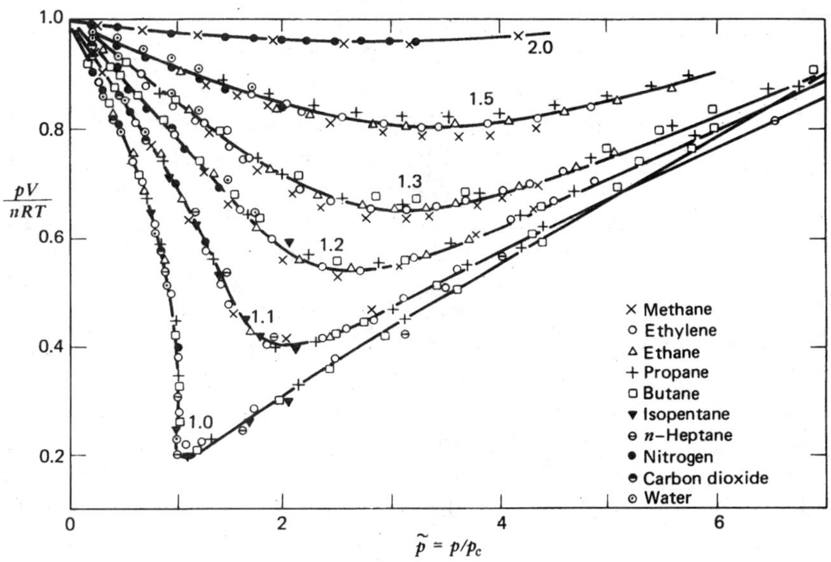

Figure 21.5 The compressibility factor pV/nRT for several gases plotted as a function of the reduced pressure p/p_c. The curves are for different values of T/T_c. From G. J. Su, *Ind. Eng. Chem.* **38,** 803 (1946).

for all gases (see Table 21.1). Let $C_V^*(T)$ be the low pressure limiting value of the constant volume heat capacity and let the low-pressure limits of the internal energy and enthalpy of a gas be denoted by $U^*(T)$ and $H^*(T)$. Using Eq. 21.4, it is possible to relate $C_V^*(T)$ to $C_p^*(T)$ by the equation

$$C_p^*(T) = C_V^*(T) + nR, \tag{21.5}$$

since

$$C_p^*(T) \equiv \lim_{p\to 0}\left(\frac{\partial H}{\partial T}\right)_p = \frac{dH^*}{dT}, \tag{21.6a}$$

$$C_V^*(T) \equiv \lim_{p\to 0}\left(\frac{\partial U}{\partial T}\right)_V = \frac{dU^*}{dT}, \tag{21.6b}$$

and

$$\begin{aligned} H^* &= U^* + \lim_{p\to 0}(pV) \\ &= U^* + nRT. \end{aligned} \tag{21.7}$$

It is $C_p^*(T)$ and $C_V^*(T)$ that determine the "perfect gas" contributions to the thermodynamic functions, namely, those that arise from the distribution of molecules over internal molecular states.

We may now transform the general formulas 21.1, 21.2, and 21.3 into a representation in which the properties of the real gas at T_2, p_2, V_2 are referred to the properties of a

Table 21.1 Limiting Low-Pressure Molar Heat Capacities of Some Gases

Substance	c_p^* (J/K mol)[a]
He	20.786
Ne	20.786
Ar	20.786
Kr	20.786
Xe	20.786
O_2	29.36
H_2	28.87
F_2	31.46
Cl_2	33.93
I_2	36.86
HF	29.08
HCl	29.12
HI	29.16
ClF	32.09
SO_2	39.79
SO_3	50.63
NH_3	35.66
CH_3F	37.45
CH_2F_2	42.84
CHF_3	53.05
$CHClBr_2$	69.12

[a] at 298.15 K

hypothetical perfect gas at T_2, p_2, V_2. We set $T_1 = T_2$ (eliminating the integrals over T), $p_1 = 0$, and $V_1 = \infty$, so that Eqs. 21.1 and 21.2 take the forms

$$U(T_2, V_2) = U^*(T_2) + \int_{T_2,\infty}^{T_2,V_2} \left[T_2 \left(\frac{\partial p}{\partial T} \right)_V - p \right] dV, \quad (21.8)$$

$$H(T_2, p_2) = H^*(T_2) + \int_{T_2,0}^{T_2,p_2} \left[V - T_2 \left(\frac{\partial V}{\partial T} \right)_p \right] dp. \quad (21.9)$$

The reduction of Eq. 21.3 to a form referring the entropy to that of a hypothetical perfect gas is a bit tricky. In the general formulas for $U(T, V)$ and $H(T, p)$, namely, Eqs. 21.1 and 21.2, the second integrals on the right-hand sides vanish if the gas obeys the perfect gas equation of state. On the other hand, the second integral on the right-hand side of Eq. 21.3 has the form

$$\int_{T_2,p_1}^{T_2,p_2} \frac{nR}{p}\, dp$$

for the case of a perfect gas. This integral is undefined when $p_1 \to 0$ because the integrand then becomes indefinitely large. To circumvent this difficulty we must eliminate the divergence of the integral term. It is observed that in a real gas

$$\lim_{p \to 0} \left[\left(\frac{\partial V}{\partial T} \right)_p - \frac{nR}{p} \right] = f(T), \quad (21.10)$$

and that $f(T)$ is bounded. That is, subtraction of the perfect gas value of $(\partial V/\partial T)_p$ from the real gas value eliminates the divergence as $p \to 0$. We exploit this observation by adding and subtracting nR/p in the second integrand of Eq. 21.3, casting the equation into the form

$$S(T_2, p_2) = S(T_1, p_1) + \int_{T_2,p_1}^{T_2,p_1} \frac{C_p(T, p_1)}{T}\, dT$$
$$- \int_{T_2,p_1}^{T_2,p_2} \left[\left(\frac{\partial V}{\partial T} \right)_p - \frac{nR}{p} \right] dp - \int_{T_2,p_1}^{T_2,p_2} \frac{nR}{p}\, dp. \quad (21.11)$$

Let $\mathfrak{p}$ be the unit pressure, that is, the value of the unit in which the pressure is expressed (e.g., 1 atm). Then, letting $p_1 \to 0$, Eq. 21.11 can be rearranged to yield

$$S(T_2, p_2) = S^*(T_2, \mathfrak{p})$$
$$+ \int_{T_2,0}^{T_2,p_2} \left[\frac{nR}{p} - \left(\frac{\partial V}{\partial T} \right)_p \right] dp - nR \ln \frac{p_2}{\mathfrak{p}}, \quad (21.12)$$

where $S^*(T_2, \mathfrak{p})$ is the entropy of a perfect gas at $(T_2, \mathfrak{p})$, defined by

$$S^*(T_2, \mathfrak{p}) = \int_{T_{\text{ref}}}^{T_2} \frac{C_p^*(T)}{T}\, dT + S_0^*(T_{\text{ref}}, \mathfrak{p}), \quad (21.13)$$

and the constant $S_0^*(T_{\text{ref}}, \mathfrak{p})$ depends on the lower limit of integration of Eq. 21.13 (cf. Chapter 18). Note that $p/\mathfrak{p}$ is a dimensionless number with magnitude equal to p in whatever units are chosen: It is important to include $\mathfrak{p}$ explicitly because the logarithm of a dimensional quantity is undefined. Note also that the entropy of a perfect gas depends on the pressure. The integral $\int_{p_1}^{p_2}(nR/p)\, dp$, which diverges as $p_1 \to 0$, is subtracted from both S and S^* to obtain Eq. 21.12.

Equations 21.8, 21.9, and 21.12 for a real gas are more complicated in form than the corresponding equations for a perfect gas. The formulas for the real and perfect gases can be made to look alike as follows. Define the function

$$f \equiv p \exp\left[\frac{1}{RT} \int_0^p \left(v - \frac{RT}{p'} \right) dp' \right], \quad (21.14)$$

called the *fugacity* of the gas. The fugacity has the same units as pressure, and it plays the same role in a real gas as does pressure in a perfect gas. By definition f contains all of the effects of gas imperfection and has the limiting behavior $f \to p$ as $p \to 0$. With the definition 21.14, the chemical potential of a real gas has a dependence on f like that the chemical potential of a perfect gas has on p. From Eqs. 21.9 and 21.12 we have, for a real gas,

$$\mu(T, p) \equiv \frac{G}{n} \equiv \frac{H - TS}{n}$$
$$= h^*(T_{\text{ref}}) - Ts_0^*(T_{\text{ref}}, \mathfrak{p}) + RT \ln \frac{p}{\mathfrak{p}}$$
$$+ \int_{T_{\text{ref}}}^{T} \left(1 - \frac{T}{T'} \right) C_p^*(T')\, dT'$$
$$+ \int_{0,T}^{p,T} \left(v(p') - \frac{RT}{p'} \right) dp'$$
$$= \mu^*(T, \mathfrak{p}) + RT \ln \frac{p}{\mathfrak{p}} + \int_{0,T}^{p,T} \left(v(p') - \frac{RT}{p'} \right) dp'$$
$$= \mu^*(T, \mathfrak{p}) + RT \ln \frac{f}{\mathfrak{p}}, \quad (21.15)$$

where $\mu^*(T, \mathfrak{p})$ is the chemical potential of a perfect gas at T, $\mathfrak{p}$). For a perfect gas Eq. 21.15 reduces to

$$\mu = \mu^*(T, \mathfrak{p}) + RT \ln \frac{p}{\mathfrak{p}} \quad \text{(perfect gas)}. \quad (21.16)$$

The entropy, enthalpy, and other thermodynamic functions of the real gas can be obtained directly from Eq. 21.15 by standard manipulations.

Final reduction of the formulas 21.8, 21.9, and 21.12 requires information about the equation of state of the gas. There have been many suggested generalizations of the perfect gas equation of state, all designed to account somehow for the influence of molecular interactions on the behavior of a gas. In general, the interaction between a pair of molecules is conveniently represented as the superposition of a very short-range repulsion and a somewhat longer-range attraction (see Fig. 21.6). Suppose that only interactions between pairs of molecules are important in the density region under study. If the temperature is high enough that the average kinetic energy of a molecule greatly exceeds the molecular attraction, the repulsion between molecules is most important. Then, since the effective volume in which a molecule may move is reduced by the volume excluded by the repulsion of other molecules, we expect the real gas pressure to exceed the perfect gas pressure at the same density and temperature. Conversely, if the temperature is low enough that the average kinetic energy of a molecule is less than the attractive potential, the pressure of the real gas is less than that of a perfect gas at the same density and temperature. A very convenient form of the equation of state, valid for all gases, is the power series

$$\frac{pV}{nRT} = 1 + \frac{nB}{V} + \frac{n^2C}{V^2} \dots, \qquad (21.17)$$

known as the *virial equation of state.*[3] The coefficients B, C, . . . depend on the temperature but not the density, and are known as *virial coefficients.* The behavior of B and C as a function of temperature is shown in Fig. 21.7: Both B and C are negative at low temperature and positive at high temperature. As will be seen in Section 21.7, the virial coefficients are determined by the intermolecular potential. B represents the effects of interactions between pairs of molecules, C interactions among triplets of molecules, and so forth.

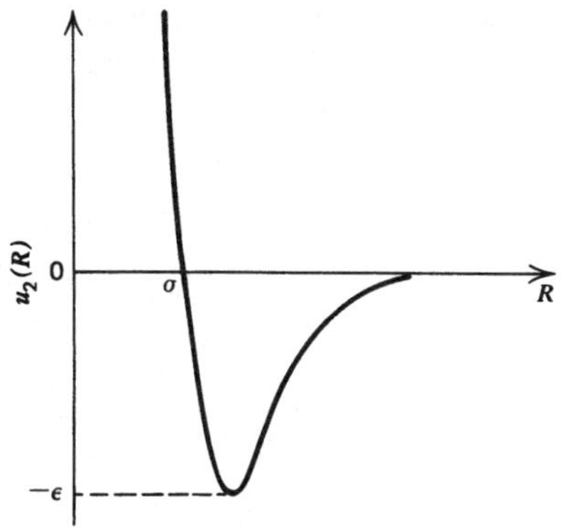

Figure 21.6 A representation of the intermolecular potential between a pair of molecules as a function of the distance between them.

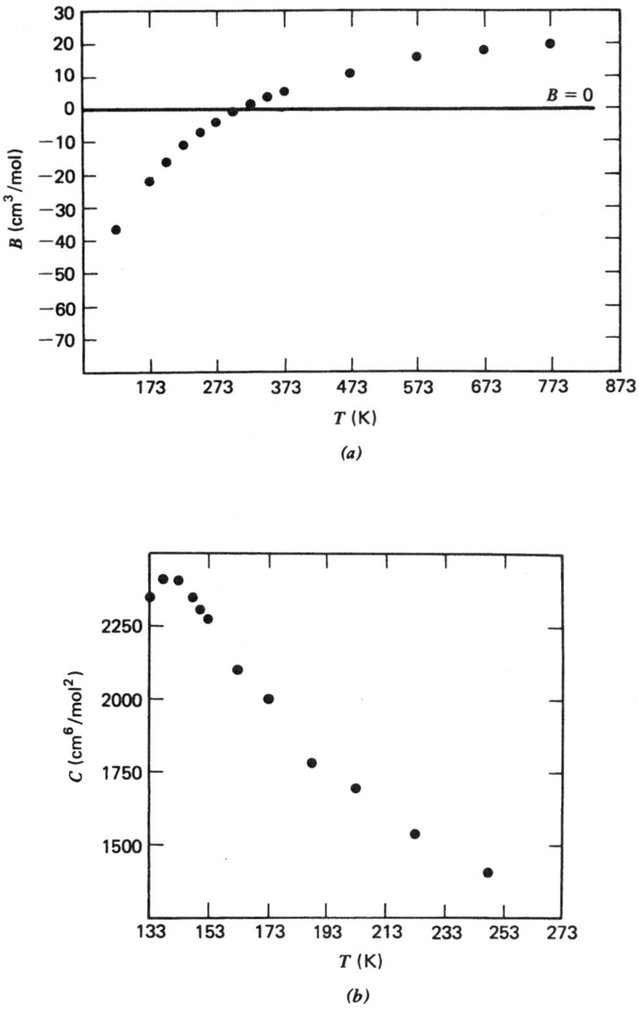

Figure 21.7 The second and third virial coefficients of Ar.

[3] The virial equation of state is sometimes written

$$pV = nRT + nB'p + n^2C'p^2 + \cdot\cdot\cdot\,. \qquad (21.17')$$

This form is less convenient than Eq. 21.17 for comparison with the statistical molecular theory of gases, but is sometimes used in thermodynamic calculations. It may be shown that

$$B = B'$$

but

$$C \neq C'$$

In calculating the enthalpy we have used the form of the virial equation of state with pressure as the independent variable, namely Eq. 21.17′. Thus the approximations to the equation of state that lead, respectively, to Eqs. 21.21 and 21.22 are almost but not quite the same. The approximations differ by quantities that are small relative to the terms retained.

For many purposes it is adequate to keep only the first correction term, B, in the virial equation of state. Then we have

$$T\left(\frac{\partial p}{\partial T}\right)_V - p = \frac{nRT}{V} + \frac{n^2BRT}{V^2} + \frac{n^2RT^2}{V^2}\frac{dB}{dT} - \frac{nRT}{V} - \frac{n^2BRT}{V^2} = \frac{n^2RT^2}{V^2}\frac{dB}{dT}, \tag{21.18}$$

$$V - T\left(\frac{\partial V}{\partial T}\right)_p = nB - nT\frac{dB}{dT}, \tag{21.19}$$

$$\left[\left(\frac{\partial V}{\partial T}\right)_p - \frac{nR}{p}\right] = n\frac{dB}{dT}. \tag{21.20}$$

The use of Eqs. 21.18, 21.19, and 21.20 in, respectively, Eqs. 21.8, 21.9, and 21.12 gives for the internal energy

$$U(T_2, V_2) = U^*(T_2) - n\frac{RT_2^2}{V_2}\frac{dB}{dT}, \tag{21.21}$$

for the enthalpy

$$H(T_2, p_2) = H^*(T_2) + n\left(B - T_2\frac{dB}{dT}\right)p_2, \tag{21.22}$$

and for the entropy

$$S(T_2, p_2) = S^*(T_2, \mathfrak{p}) - nR\ln\frac{p_2}{\mathfrak{p}} - n\frac{dB}{dT}p_2. \tag{21.23}$$

The Gibbs and Helmholtz free energies can be obtained from simple combinations of Eqs. 21.21, 21.22, and 21.23. We leave the calculation of G and A as an exercise for the reader.

Examination of Fig. 21.7 shows that dB/dT is always positive in the temperature range covered, but B is negative at low temperature and positive at high temperature. There is one temperature at which $B = 0$; this temperature is called the *Boyle point.* At and in the neighborhood of the Boyle point the real gas, if described by an equation of state (21.14) retaining no terms beyond B, behaves like a perfect gas and obeys Boyle's law, hence the name. When the temperature is low the average kinetic energy of a molecule is less than the depth of the molecular interaction; in this region of temperature B is negative. When the temperature is high the average kinetic energy of a molecule is greater than the depth of the molecular interaction; in this region of temperature B is positive. Note that if $B < 0$ the pressure of the real gas is less than that of the corresponding perfect gas, and if $B > 0$ the pressure is greater than that of the corresponding perfect gas. Finally, Fig. 21.7 seems to indicate that $dB/dT > 0$ approaches zero as T increases. Theoretical calculations of B predict that it passes through a maximum at high temperature and then decreases very slowly. For Ar that maximum occurs near 3000 K, and for He near 200 K.

A consequence of molecular interaction is that the distribution of molecules in space cannot be completely random. If the average intermolecular attraction exceeds the average molecular kinetic energy, there is greater than random chance of finding molecules closer than the average separation; if intermolecular repulsion exceeds, on the average, the average molecular kinetic energy, there is less than random chance of finding molecules closer than the average separation. We see from Eq. 21.23 that the entropy of the real gas is less than that of the corresponding perfect gas when $dB/dT < 0$, independent of whether B is positive or negative. This is to be expected, since the nonrandom distribution of molecules in space resulting from molecular interaction decreases the number of configurations of gas molecules with equal energy, hence the number of configurations that can be occupied with equal probability, and thus also the entropy of the gas. Equation 21.21 shows that the internal energy of the gas is also less than that of the corresponding perfect gas. Clearly, if molecular attraction leads to an enhancement of the frequency of less than average molecular separations, there is a negative contribution to the internal energy, as found from Eq. 21.21. At very high temperature, where $dB/dT < 0$, the internal energy exceeds that of the corresponding perfect gas, just as expected if repulsive interactions are dominant. Note that because the repulsive interaction between molecules has a much shorter range than the attractive interaction, $B > 0$ only when the average molecular kinetic energy exceeds the molecular attraction by a factor of two or three. For the average spacing of molecules typical of a gas for which Eq. 21.17 is valid, the dominant interaction between molecules is attractive, not repulsive, and $dB/dT > 0$, so that the internal energy of the real gas is less than that of the corresponding perfect gas. In the hypothetical case that the molecular interaction is everywhere repulsive, $dB/dT < 0$, and the energy of the real gas is greater than that of the corresponding perfect gas, in agreement with intuitive considerations.

A similar discussion of the properties of the enthalpy may be based on Eq. 21.22.

The virial equation of state is not the only form used to relate p, V, and T for a gas. Many empirical equations of state have been proposed, but only the virial equation of state has a simple and unambiguous relationship with the statistical molecular theory of gases. Some suggestions to account for the properties of real gases are listed in Table 21.2. None of the early equations of state gives a good representation of pV/nRT over a wide range of temperature and pressure; however, all reproduce qualitatively the dominant features of pV/nRT when corrections to the perfect gas law are small. All of the equations of state listed give an inaccurate description of the critical point and gas–liquid condensation.

Table 21.2 Empirical Gas Equations of State

Name	General Form	Features	Form of Second Virial Coefficient[a]
Early			
van der Waals (1877)	$p=\frac{nRT}{V-nb}-\frac{n^2a}{V^2}$	a, b independent of T, V	$B=b-\frac{a}{RT}$
Dieterici (1899)	$p=\frac{nRT}{V-nb}e^{-na/VRT}$	a, b independent of T, V	$B=b-\frac{a}{RT}$
Berthelot (1907)	$p=\frac{nRT}{V-nb}-\frac{n^2a}{TV^2}$	a, b independent of T, V	$B=b-\frac{a}{RT^2}$
Later			
Keyes (1914)	$p=\frac{nRT}{V-n\delta}-\frac{n^2A}{(V+nl)^2}$	$\delta=\beta e^{-n\alpha/V}$; α, β, l, A independent of T, V	$B=\beta-\frac{A}{RT}$
Beattie–Bridgeman (1927)	$p=\frac{nRT(1-\epsilon)}{V^2}(V+nB)-\frac{n^2A}{V^2}$	$B=B_0\left(1-\frac{nb}{V}\right)$ $A=A_0\left(1-\frac{nA}{V}\right)$ $\epsilon=\frac{nc}{VT^3}$ c, B_0, b, A_0, a independent of T, V	$B=B_0-\frac{A_0}{RT}-\frac{c}{T^3}$
Redlich–Kwong (1949)	$p=\frac{nRT}{V-nb}-\frac{n^2a}{T^{1/2}V(V+nb)}$	a, b independent of T, V	$B=b-\frac{a}{RT^{3/2}}$

[a] The second virial coefficient is obtained, in each case, by expanding the given equation of state in the form of Eq. 21.17.

Later equations of state, also displayed in Table 21.2, are more successful in reproducing the volumetric properties of real gases. Note that the Keyes and Beattie–Bridgeman equations of state require four and five empirical constants, respectively. There are other empirical equations of state that require still more constants.

An example of the reliability of typical empirical equations of state is provided by comparison of the predicted and observed values of the internal energy of an imperfect gas. Eq. 21.8 for the internal energy of an imperfect gas, expressed in molar units, takes the following forms for the van der Waals, Dieterici, Beattie–Bridgeman and Redlich–Kwong equations of state:

van der Waals
$$u=u^*(T)-\frac{a}{v}$$

Dieterici
$$u=u^*(T)-\frac{a}{b}e^{a/bRT}\left[Ei\left(\frac{a}{bRT}\right)-Ei\left(\frac{a}{bRT}\left(1-\frac{b}{v}\right)\right)\right]$$
$$Ei(x)\equiv\int_{-\infty}^{x}\frac{e^t}{t}\,dt$$

Beattie–Bridgeman
$$u=u^*(T)-\left(A_0+\frac{3Rc}{T^2}\right)\frac{1}{v}+\frac{1}{2v^2}\left(aA_0-3\frac{RcB_0}{T^2}\right)+\frac{RcbB_0}{T^2v^3}$$

Redlich–Kwong
$$u=u^*(T)-\frac{3a}{2bT^{1/2}}\ln\left(\frac{v+b}{v}\right)$$

The internal energy per mole of CO_2 predicted by each of these empirical equations of state is compared with experimental data in Fig. 21.8. Clearly, all of the empirical equations of state fail at high pressure.

A common idea in the construction of the typical empirical equation of state (but not the Beattie–Bridgeman equation) is that the pressure can be expressed as the sum of two terms: the kinetic contribution to the pressure corrected for the excluded volume of the molecules (taken to be hard spheres), and the potential energy contribution to the pressure (taken to arise from only the attractive component of the interaction between molecules). That is

$$p=p_{HS}+p_A.$$

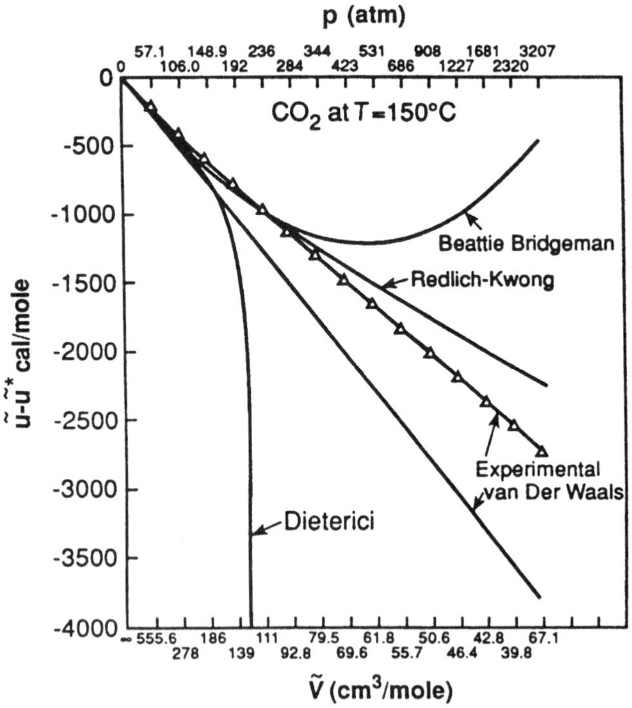

Figure 21.8 Comparison of predicted and measured internal energy of CO_2 using several empirical equations of state. From H. T. Davis, *Statistical Mechanics of Phases, Interfaces and Thin Films*, VCH Publishers, New York, 1996).

The entries in Table 21.2 show that the van der Waals, Dieterici, Berthelot, Keyes, and Redlich–Kwong equations of state all assume that the correction to the kinetic pressure arising from the excluded volume of the molecules has the simple form $p_{HS} = RT/(v - b)$. This particular dependence of the pressure on excluded volume is exact only in the case that the gas is confined to one dimension; it is known to be very inaccurate when the gas is three-dimensional. The results of both theoretical studies and computer simulations of a three-dimensional hard sphere gas (see Chapter 23) show that an excellent approximation to the equation of state, due to Carhahan and Starling, is

$$p_{HS} = \frac{RT}{v}\left(\frac{1+y+y^2-y^3}{(1-y)^3}\right); \qquad y \equiv \frac{b}{4v}.$$

If this form for p_{HS} is used in place of the simple form $RT/(v - b)$, the agreement between predicted and observed properties of the imperfect gas is improved.

At present there are no known analytic formulas capable of reproducing the equation of state of a real gas at all densities and temperatures. Because of its relationship with the molecular theory, the virial equation (21.14) is preferred; from the point of view of experiment, each virial coefficient is a temperature-dependent function to be determined. Thus, the virial equation of state has great mathematical flexibility. Nevertheless, there exists doubt that it can adequately represent the properties of a gas at densities comparable to the liquid density.

21.2 Thermodynamic Description of a Gas Mixture

When considering the properties of a mixture, in addition to the temperature, pressure, and total volume, it is necessary to specify the composition. In general, the composition of a mixture is represented in terms of the masses (in some units) of the several species present. But this simple statement hides an important point. For a one-component fluid any two of the three variables T, p, V determine the third through the equation of state. That is, for a one-component fluid system there are only two independent variables, independent meaning that they can be altered at will and without relationship to each other. Similarly, to specify the composition of a mixture we must use only the set of independent composition variables. For example, suppose that we are given a system containing H_2, I_2, and HI, in the gas phase, at equilibrium. The chemical reaction

$$H_2 + I_2 = 2HI$$

imposes a condition on the way the concentrations of the three molecular species may change relative to one another. Indeed, given the concentrations of H_2 and I_2, and auxiliary thermodynamic information concerning the Gibbs free-energy change in the reaction, the concentration of HI is determined. In this system only the concentrations of H_2 and I_2 (or any two of the set of three concentrations) are independent variables. *In all our considerations we specify the composition of the system in terms of the independently variable concentrations, taking into account the constraints imposed by the conditions of equilibrium.*

Consider a homogeneous gas containing r components. Let this gas be enclosed in one compartment of a container separated into two parts by a semipermeable membrane (see Fig. 21.9). Suppose that compartment I of the container holds pure component i in equilibrium with component i in the gas mixture in compartment II. Suppose also that the semipermeable membrane which allows i to pass is diathermal and therefore permits transfer of energy, but is rigid

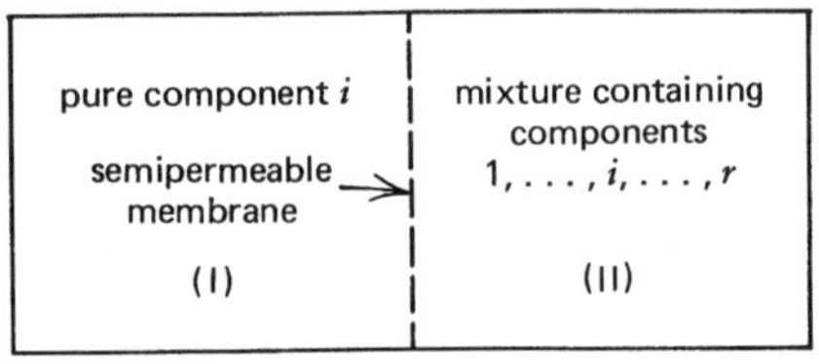

Figure 21.9 Schematic diagram of a box with semipermeable membrane separating pure component i from a mixture of gases containing component i. Only i can pass through the membrane.

(nondeformable) and therefore does not allow performance of work. Then, because the system is at equilibrium, we have

$$\mu_i^{\mathrm{I}} = \mu_i^{\mathrm{II}} = \mu_i, \tag{21.24}$$

$$T^{\mathrm{I}} = T^{\mathrm{II}} = T. \tag{21.25}$$

To define the contribution to the total pressure in the gas mixture made by component *i*, it is necessary to introduce a new extrathermodynamic consideration. If there are $n_1, \ldots, n_r$ mol each of components $1, \ldots, r$ in a mixture, and if

$$x_i \equiv \frac{n_i}{n_1 + n_2 + \cdots + n_r} \tag{21.26}$$

is the mole fraction of component *i*, then it is observed that

$$\lim_{p \to 0} \frac{p_i}{p} = x_i, \tag{21.27}$$

where p_i is the pressure that pure component *i* would exert at the same temperature if its chemical potential were the same as that of component *i* in the mixture. In the perfect gas limit this means that p_i is the contribution to the total pressure from component *i*, called the *partial pressure* of component *i*, and is computed just as if no other gas were present. The mathematical representation of this observation is

$$\lim_{p \to 0} p = \frac{RT}{V}(n_1 + n_2 + \cdots + n_r) = p_1 + p_2 + \cdots p_r. \tag{21.28}$$

Equation 21.28 is a statement of *Dalton's law of partial pressures.*

How do we treat the case of gas mixtures at nonzero pressure? In the pure gas the chemical potential is given by the equation

$$\mu_i^{\mathrm{I}} = \mu_i^*(T, \mathfrak{p}) + RT \ln \frac{p^{\mathrm{I}}}{\mathfrak{p}} + \int_{0,T}^{p^{\mathrm{I}},T} \left[v_i(p') - \frac{RT}{p'} \right] dp'. \tag{21.15}$$

As usual, lowercase letters refer to molar quantities; for example, $v_i(p')$ is the molar volume of pure component *i* at temperature *T* and pressure p'. Consider again the situation depicted in Fig. 21.9. The equilibrium condition Eq. 21.24 and the definition of the partial pressure lead to $p_i^{\mathrm{I}} = p_i^{\mathrm{II}}$ and in the zero-pressure limit $p_i^{\mathrm{I}} \equiv p^{\mathrm{I}} = x_i^{\mathrm{II}} p^{\mathrm{II}}$. Therefore we can write for the chemical potential of component *i* in the mixture

$$\mu_i^{\mathrm{II}} = \mu_i^*(T, \mathfrak{p}) + RT \ln \frac{x_i^{\mathrm{II}} p^{\mathrm{II}}}{\mathfrak{p}} + \int_{0,T}^{p^{\mathrm{II}},T} \left[\bar{v}_i(p') - \frac{RT}{p'} \right] dp', \tag{21.29}$$

where

$$\bar{v}_i \equiv \left(\frac{\partial \mu_i}{\partial p} \right)_{T,n_j} = \frac{\partial}{\partial p} \left(\frac{\partial G}{\partial n_i} \right)_{T,n_{j \neq i}} = \frac{\partial}{\partial n_i} \left(\frac{\partial G}{\partial p} \right)_{T,n_{j \neq i}} = \left(\frac{\partial V}{\partial n_i} \right)_{T,p,n_{j \neq i}} \tag{21.30}$$

is the *partial molar volume* of component *i*.[4]

Just as in the case of the pure gas, evaluation of Eq. 21.29 requires knowledge of the equation of state of the gas. We extend the virial equation 21.17 to read

$$v_m \equiv \frac{V}{\sum_{i=1}^{r} n_i} = \frac{RT}{p} + B(x_1, x_2, \ldots, x_{r-1}, T). \tag{21.31}$$

Note that v_m is an "average molar volume" for the mixture. Since *B* is a measure of the interaction between pairs of molecules, we expect it to be proportional to the frequencies of occurrence of possible pairs of all compositions. The probability that a molecule of species *i* and a molecule of species *j* will, at random, undergo collision is proportional to the mole fraction of *i*, x_i, and the mole fraction of *j*, x_j. Thus we conclude that $B(x_1, \ldots, x_{r-1}, T)$ has a dependence on concentration of the following form:

$$B(x_1, \ldots, x_{r-1}, T) = \sum_{i=1}^{r} \sum_{j=1}^{r} x_i x_j B_{ij}(T). \tag{21.32}$$

Of course, $B_{ij} = B_{ji}$, since the two molecules in an *ij* pair are the same as those in a *ji* pair.

The evaluation of Eq. 21.31 is tedious. For our purposes it suffices to illustrate the evaluation for the specific case of a binary mixture and then to state the general answer. For a binary mixture the equation of state 21.31 becomes

$$\frac{V}{n_1 + n_2} = \frac{RT}{p} + \frac{n_1^2 B_{11} + 2n_1 n_2 B_{12} + n_2^2 B_{22}}{(n_1 + n_2)^2}. \tag{21.31'}$$

---

[4] The final term on the right-hand side of Eq. 21.29 is just the integral of the $(\partial \mu_i / \partial p)_{T,n_j}$ term in

$$d[\mu_i - \mu_i\,(\text{perfect gas})] = \left\{ \frac{\partial}{\partial T} [\mu_i - \mu_i\,(\text{perfect gas})] \right\}_{p,n_j} dT + \left\{ \frac{\partial}{\partial p} [\mu_i - \mu_i\,(\text{perfect gas})] \right\}_{T,n_j} dp.$$

Here μ_i (perfect gas) is given by the first two terms on the right side of Eq. 21.29. Note that for a mixture of perfect gases we have

$$\mu_i\,(\text{mixture};\, T, p) = \mu_i\,(\text{pure } i;\, T, p) + RT \ln x_i.$$

We multiply both sides of Eq. 21.31′ by (n_1+n_2) and use the definition (see Eq. 21.30)

$$\bar{v}_1 \equiv \left(\frac{\partial V}{\partial n_1}\right)_{T,p,n_2} \tag{21.30′}$$

to obtain

$$\bar{v}_1 = \frac{RT}{p} + \frac{n_1^2 B_{11} + 2n_1 n_2 B_{11} + 2n_2^2 B_{12} - n_2^2 B_{22}}{(n_1+n_2)^2}. \tag{21.33}$$

We can rewrite Eq. 21.33 in the form

$$\bar{v}_1 = \frac{RT}{p} + x_1^2(2B_{11} - B_{11}) + x_1 x_2 (2B_{11} - B_{12}) + x_2 x_1 (2B_{12} - B_{21}) + x_2^2 (2B_{12} - B_{22}), \tag{21.34}$$

remembering that $B_{12} = B_{21}$. We have rewritten the equation for $\bar{v}_1$ in the form of Eq. 21.34 to provide an illustration of the general result for an r-component mixture, which we now quote:

$$\bar{v}_1 = \frac{RT}{p} + \sum_{j=1}^{r}\sum_{k=1}^{r} x_j x_k (2B_{ij} - B_{jk}). \tag{21.34′}$$

The substitution of Eq. 21.34′ into Eq. 21.29 leads to

$$\mu_i = \mu_i^*(T) + RT\ln\frac{p x_i}{\mathfrak{p}} + \sum_{j=1}^{r}\sum_{k=1}^{r} x_j x_k (2B_{ij} - B_{jk}) p, \tag{21.35}$$

which is to be compared with

$$\mu_i = \mu_i^*(T) + RT\ln\frac{p}{\mathfrak{p}} + B_{ii} p \tag{21.36}$$

for a pure gas.

To use Eq. 21.36, we need equation-of-state data for the mixture as a function of T, p, and composition. One of the classical goals of chemistry has been the prediction of the properties of mixtures from the properties of pure components. But there is no reason to expect a simple quantitative relation to exist between the intermolecular interactions involving, say, species 1 and 2, and those involving species 1 and 1, or 2 and 2. There are, of course, qualitative relations between these interactions, but all studies have revealed their inadequacy for precise predictions of the behavior of mixtures. Nevertheless, because there are so many possible mixtures, it is useful to have even a crude approximation relating the properties of a mixture to those of its pure components. Lewis and Randall studied this problem and proposed the following rule: We define the fugacity of component i in the mixture by

$$f_i \equiv x_i p \exp\left[\int_0^p \left(\bar{v}_i - \frac{RT}{p'}\right)\frac{dp'}{RT}\right]. \tag{21.37}$$

Then, in a real gas mixture, by analogy with Dalton's law of partial pressures for a mixture of perfect gases, it is assumed that

$$f_i(T,p,x_i) = x_i f_i^0(T,p), \tag{21.38}$$

where the superscript (^{0}) refers to the pure component i. A mixture for which this assumption is valid is known as an *ideal solution.*[5] What does Eq. 21.38 imply? In terms of the fugacity, the chemical potential is given by Eqs. 21.15 and 21.29 as

$$\mu_i^0 = \mu_i^*(T,\mathfrak{p}) + RT\ln\frac{f_i^0}{\mathfrak{p}} \tag{21.15}$$

for a pure gas and

$$\mu_i = \mu_i^*(T,\mathfrak{p}) + RT\ln\frac{f_i}{\mathfrak{p}} \tag{21.39}$$

for a gas mixture. Furthermore, instead of using as a reference state the properties of the perfect gas at unit pressure, we can choose to refer the chemical potential of i in the mixture to the value the chemical potential of pure i has at the same temperature and pressure. This is done simply by subtracting Eq. 21.15 from Eq. 21.39. The result is

$$\mu_i = \mu_i^0 + RT\ln\frac{f_i}{f_i^0}, \tag{21.40}$$

or for an ideal solution

$$\mu_i = \mu_i^0 + RT\ln x_i, \tag{21.41}$$

using Eq. 21.38. Thus, in an ideal solution the chemical potential of a component of the mixture, referred to the pure component as reference, has a very simple form, namely, Eq. 21.41. Note that within the approximation described by Eqs. 21.17 and 21.32 for the equations of state, Eqs. 21.15 and 21.17 assume the forms

$$f_i^0 = p\,\exp\left(\frac{B_{ii} p}{RT}\right) \tag{21.42}$$

[5] For an ideal solution of perfect gases, f_i approaches p_i, the partial pressure of component i, and

$$p_i(T,p,x_i) = x_i p.$$

and

$$f_i = x_i p \exp\left[\frac{\sum_{j=1}^{r}\sum_{k=1}^{r} x_j x_k (2B_{ij} - B_{jk}) p}{RT}\right], \tag{21.43}$$

and that $f_i^0 \to p_i^0$ and $f_i \to p_i$ as $p \to 0$.

One of the most interesting and fundamental differences between a mixture and a pure gas is the following: The entropy of a mixture is greater than the sum of the entropies of the pure components that constitute the mixture. How large is the difference, and what is its origin?

The generalization of Eq. 21.12 to a mixture is obtained from the relation $\bar{s}_i = -(\partial \mu_i/\partial T)_{p,x_j}$ and Eq. 21.29 for μ_i. We find for the partial molar entropy of component i

$$\bar{s}_i = s_i^*(T,\mathfrak{p}) - R\ln\frac{px_i}{\mathfrak{p}} - \int_{0,T}^{p,T}\left[\left(\frac{\partial \bar{v}_i}{\partial T}\right)_{p',n_i} - \frac{R}{p'}\right] dp'. \tag{21.44}$$

Since the entropy is an extensive function, we can write for the total entropies of the mixture and of the separated pure components

$$S^0 = \sum_{i=1}^{r} n_i s_i^0 \quad \text{(pure gases)}, \tag{21.45}$$

$$S^m = \sum_{i=1}^{r} n_i \bar{s}_i \quad \text{(mixture)}.^6 \tag{21.46}$$

The difference between Eq. 21.46 and Eq. 21.45 is the entropy of mixing. This is

$$\begin{aligned}\Delta S_{\text{mix}} &= S^m - S^0 = \sum_{i=1}^{r} n_i(\bar{s}_i - s_i^0) \\ &= -R\sum_{i=1}^{r} n_i \ln x_i \\ &\quad + \sum_{i=1}^{r} n_i \int_{0,T}^{p,T}\left[\left(\frac{\partial v_i}{\partial T}\right)_{p'} - \left(\frac{\partial \bar{v}_i}{\partial T}\right)_{p',n_j}\right] dp'.\end{aligned} \tag{21.47}$$

In the perfect gas-ideal solution limit we must have

$$\left(\frac{\partial \bar{v}_i}{\partial T}\right)_{p,n_j} = \left(\frac{\partial v_i}{\partial T}\right)_p, \tag{21.48}$$

so that

$$\Delta S_{\text{mix}}\,(\text{perfect gases}) = -R\sum_{i=1}^{r} n_i \ln x_i. \tag{21.49}$$

We now undertake an examination of the implications of Eq. 21.49. It has often been argued that Eq. 21.49 is difficult to understand because it assigns the same entropy of mixing to all possible mixtures of species (provided that the gases are perfect), but zero entropy of mixing to the case when the "mixture" is made by adding together portions of the same gas. Such a view, which defines what is known as *Gibbs's paradox,* overlooks the important role played by constraints and possible work processes in a thermodynamic system. From the point of view of molecular theory, the mixing of two different perfect gases with the same thermodynamic properties, Ar and Kr, for example, is fundamentally different from mixing two volumes of Ar, because molecules of the same species are indistinguishable, whereas molecules of different species, even though similar, are distinguishable. Even in the extreme case of two isotopes of the same element, the difference in nuclear composition permits a separation procedure to be devised on the basis of the mass difference between the isotopes. How does this property of indistinguishability of identical molecules enter the thermodynamic analysis? Recall that Dalton's law of partial pressures plays an essential role in the description of the gas mixture. According to this law, a gas of one species, when present in a perfect gas mixture, behaves as if there were no other gas present. Consider a container with two compartments, each containing a different perfect gas, as shown in Fig. 21.10. Let the pressure and temperature be the same on both sides of the partition. Then we have

$$n_a = \frac{pV^{\text{I}}}{RT}, \qquad n_b = \frac{pV^{\text{II}}}{RT}. \tag{21.50}$$

When the partition is removed, the gases mix and fill the entire volume. From Dalton's law, after mixing we must have

$$p = \frac{RT}{V^{\text{I}} + V^{\text{II}}}(n_a + n_b). \tag{21.51}$$

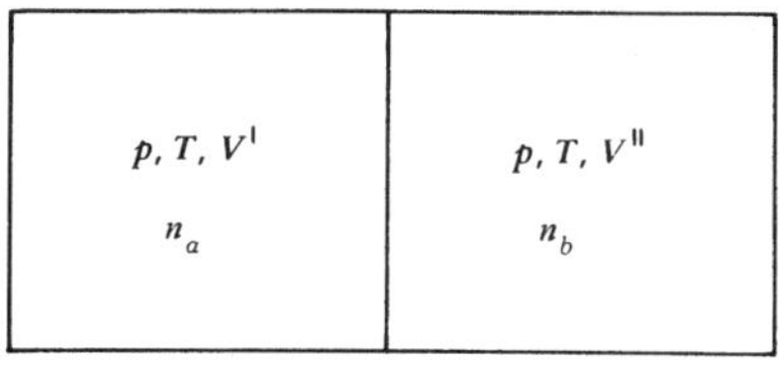

Figure 21.10 Schematic representation of a box separated into two compartments I and II, both at the same temperature and pressure, but containing n_a mol of a and n_b mol of b, respectively.

---

6 See Eq. 19.76.

Now, if the gases behave as if they were individually present in the box, that is, if they are indifferent to each other, the process of mixing may be interpreted as an isothermal expansion of each of the gases. As shown in Chapter 17, the entropy change accompanying the isothermal expansion of a perfect gas is

$$\Delta S = nR\ln\frac{V_2}{V_1}, \tag{21.52}$$

so that in this case we obtain, using Eqs. 21.50 and 21.51,

$$\begin{aligned}\Delta S_{\text{mix}} &= \Delta S_1 + \Delta S_2 = n_a R\ln\frac{V^{\text{I}}+V^{\text{II}}}{V^{\text{I}}} + n_b R\ln\frac{V^{\text{I}}+V^{\text{II}}}{V^{\text{II}}}\\ &= -n_a R\ln\frac{n_a}{n_a+n_b} - n_b R\ln\frac{n_b}{n_a+n_b}\\ &= -n_a R\ln x_a - n_b R\ln x_b,\end{aligned} \tag{21.53}$$

which is, of course, the same as Eq. 21.49. To understand why Eq. 21.53 represents the effects of molecular distinguishability, let us consider the inverse process of separating a gas mixture. We imagine that there is a suitable semipermeable membrane which allows species *a* to pass through, but not species *b*. We insert this membrane at one end of the container holding the gas mixture. The mixture is then compressed using the semipermeable membrane. Let us calculate the work done in this separation process. Since a perfect gas may be compressed to an arbitrarily small volume, we may effect a complete separation of *a* and *b* by starting the semipermeable membrane at the volume $V^{\text{I}} + V^{\text{II}}$, and compressing to a volume 0^+. (This volume will drop out of the final answer, so we need not worry about taking the logarithm of zero. We use 0^+ to denote "essentially" zero volume, in the sense that the expression involves the limit as the volume approaches zero.) For isothermal compression of an ideal gas we have (see Eq. 13.14)

$$w = -nRT\ln\frac{V_2}{V_1}. \tag{21.54}$$

Now in our case the membrane does no work on component *a* in the compression (*a* passes through), but this leaves *a* in a volume $V^{\text{I}} + V^{\text{II}}$ and *b* in a volume 0^+. To restore the initial state we must now compress *a* and expand *b* to their original volumes with a nonpermeable piston. The total work done in this sequence of processes is then

$$\begin{aligned}w &= -\left(n_b RT\ln\frac{0^+}{V^{\text{I}}+V^{\text{II}}} + n_a RT\ln\frac{V^{\text{I}}}{V^{\text{I}}+V^{\text{II}}} + n_b RT\ln\frac{V^{\text{II}}}{0^+}\right)\\ &= -(n_a RT\ln x_a + n_b RT\ln x_b).\end{aligned} \tag{21.55}$$

But an isothermal process in a perfect gas is characterized by $\Delta U = 0$, so that the first law of thermodynamics yields

$$T\int dS = \int p\,dV = -w_{\text{sep'n}}, \tag{21.56}$$

and the entropy change accompanying the separation is

$$\Delta S_{\text{sep'n}} = n_a R\ln x_a + n_b R\ln x_b, \tag{21.57}$$

which is just the negative of the entropy of mixing, as required. It is important to note that the inverse process used depends on the constraint that the membrane be semipermeable. But the very concept of a semipermeable membrane requires the existence of some feature distinguishing the molecules of the mixture. Without the existence of such a distinguishing feature, no "separator" can be envisaged. Since the separation process described is just the inverse of the mixing process, we see that the concept of distinguishability, and its embodiment in Dalton's law of partial pressures, leads to a qualitatively new feature in the thermodynamics of mixtures, namely, the entropy of mixing. At the risk of pedanticism, it is worthwhile again to emphasize that the formulation of the entropy of mixing requires information from outside of thermodynamics, in this case Dalton's law.

One other point concerning the entropy of mixing is important. The distinguishability/indistinguishability of species appears in classical mechanics only insofar as it depends on what we know to distinguish species. For example, if we have no way to measure molecular masses and know nothing about isotopes we would say that all CO_2 molecules are identical and compute the thermodynamic properties accordingly. If we know about isotopes and can measure molecular masses, we can make distinctions among all the various isotopic species of CO_2. At a more fundamental level, however, classical mechanics allows us, in principle, to put labels on every atom and to make them all distinguishable. In quantum mechanics all of the particles of one species, such as all electrons or all ^{4}He atoms, are totally indistinguishable, and the statistical and thermodynamic descriptions of such particles are thereby rendered different from those of classical particles, as we saw in Chapter 6 and will see in more detail in Appendix 21B.

As might be expected, not only the enthalpy but also *V*, *U*, *G*, and *A* can differ in a mixture from the respective sums of their values in the pure component gases. For ease of discussion we now call the difference between the value some property has in the mixture and the mole-fraction-weighted sum of the corresponding properties of the pure components the *excess value* of that property. For example, the excess volume and enthalpy of a mean mole of mixture are

$$\upsilon^E \equiv \frac{V^E}{\sum_{i=1}^{r} n_i} = \upsilon_m - \sum_{i=1}^{r} x_i \upsilon_i^0 \tag{21.58a}$$

and

$$h^E \equiv \frac{H^E}{\sum_{i=1}^{r} n_i} = h_m - \sum_{i=1}^{r} x_i h_i^0. \tag{21.58b}$$

As usual, the quantities with superscript zero refer to the pure component. Note that the excess entropy of a mixture is just what we have called the entropy of mixing.

There are, in general, two contributions to an excess function. One arises from the difference between intermolecular forces in the mixture and the pure components. The second term on the right-hand side of Eq. 21.47 is of this type. The other contribution arises from the distinguishability of different molecular species, that is, from the perfect gas entropy of mixing. The first term on the right-hand side of Eq. 21.47 is of this type. The excess volume and the excess enthalpy of a perfect gas mixture are zero; the only contributions to a nonzero excess enthalpy and excess volume arise from gas imperfection, hence from intermolecular forces. Using arguments similar to those employed in the calculation of the entropy of mixing, it is found that, in general, for any gas mixture,

$$H^E = \int_0^p \left[V^E - T\left(\frac{\partial V^E}{\partial T}\right)_{p',n_j} \right] dp'. \tag{21.59}$$

On the other hand, because of the distinguishability of molecules of different chemical species, the excess entropy (i.e., the entropy of mixing) is not zero for a perfect gas mixture, and therefore the excess Gibbs and Helmholtz free energies of a perfect gas mixture are nonzero.

In summary, for a perfect gas mixture we have

$$s^E \equiv \frac{\Delta S_{\text{mix}}}{\sum_{i=1}^{r} n_i} = -R\sum_{i=1}^{r} x_i \ln x_i, \tag{21.60a}$$

$$h^E = u^E = 0, \tag{21.60b}$$

$$v^E = 0, \tag{21.60c}$$

$$g^E \equiv \frac{\Delta G_{\text{mix}}}{\sum_{i=1}^{r} n_i} = RT\sum_{i=1}^{r} x_i \ln x_i, \tag{21.60d}$$

$$a^E \equiv \frac{\Delta A_{\text{mix}}}{\sum_{i=1}^{r} n_i} = RT\sum_{i=1}^{r} x_i \ln x_i. \tag{21.60e}$$

The thermodynamic functions of a real gas mixture are obtained by combining Eqs. 21.47, 21.59, and the obvious forms for other mixture properties.

It is interesting to consider the properties of an ideal solution, which we defined in Eq. 21.38. For an ideal solution Eqs. 21.41, 19.76, and 21.47 yield

$$\Delta S_{\text{mix}} = \sum_{i=1}^{r} n_i \left[\left(\frac{\partial \mu_i^0}{\partial T}\right)_p - \left(\frac{\partial \mu_i}{\partial T}\right)_{p,n_j} \right]$$

$$= -R\sum_{i=1}^{r} n_i \ln x_i, \tag{21.60a'}$$

the same result obtained in Eq. 21.49 for a mixture of perfect gases. Similarly, it can be shown that the other excess functions for the ideal solution have the same forms as in Eqs. 21.60. An ideal solution is a mixture whose excess functions are the same as those of a perfect gas mixture of the same composition.

21.3 Thermodynamic Description of Gaseous Reactions

With the concepts introduced in Sections 21.1 and 21.2 we can describe the equilibrium composition of a mixture of gases between which chemical reactions occur. It was shown in Chapter 19 that at equilibrium in a mixture of reacting components it is necessary that

$$\sum_{i=1}^{r} \nu_i \mu_i = 0 \tag{21.61}$$

(see Eq. 19.99). We shall now use Eq. 21.61, together with the expression for the chemical potential in a gas mixture, to evaluate the equilibrium constant in a gaseous reaction (see Eq. 19.81). Substitution of Eq. 21.29 in Eq. 21.61 gives

$$\sum_{i=1}^{r} \nu_i \mu_i^* (T,\mathfrak{p}) = -RT\sum_{i=1}^{r} \left[\ln\frac{x_i p}{\mathfrak{p}} + \int_{0,T}^{p,T} \left(\bar{v}_i - \frac{RT}{p'}\right)\frac{dp'}{RT} \right] \nu_i. \tag{21.62}$$

But the left-hand side of this equation is a function of only the temperature. Hence, if the temperature is fixed, the relative composition is also fixed. We write, then, an equilibrium constant $K(T)$, defined by

$$-RT \ln K(T) \equiv \sum_{i=1}^{r} \nu_i \mu_i^* (T,\mathfrak{p}), \tag{21.63}$$

and transform Eq. 21.62 to read

$$K(T) = \prod_{i=1}^{r} \left(\frac{x_i p}{\mathfrak{p}}\right)^{\nu_i} \exp\left[\frac{\nu_i}{RT}\int_{0,T}^{p,T} \left(\bar{v}_i - \frac{RT}{p'}\right) dp' \right] \tag{21.64}$$

$$= \prod_{i=1}^{r} \left(\frac{f_i}{\mathfrak{p}}\right)^{\nu_i}, \tag{21.65}$$

using Eq. 21.37 to introduce the fugacity. In the perfect gas limit we have $f_i \to px_i$, since the integrand in Eq. 21.37 vanishes. If we then define an equilibrium constant in terms of partial pressures,

$$K_p \equiv \prod_{i=1}^{r} \left(\frac{x_i p}{\mathfrak{p}} \right)^{\nu_i}, \tag{21.66}$$

we have $\lim_{p \to 0} K_p = K(T)$.

The sum $\Sigma \nu_i \mu_i^*$ is, by definition, the difference in molar free energies of the pure components in their respective hypothetical perfect gas states at pressure $\mathfrak{p}$. That each μ_i^* is given by this prescription is easily verified by substitution of $p = \mathfrak{p}$ into Eq. 21.15. Thus, Eq. 21.63 may also be written in the form

$$-RT \ln K(T) = \Delta G^*(T, \mathfrak{p}), \tag{21.67}$$

where ΔG^* is the referred-to molar free energy difference between reactants and products. Note that ΔG^* is a function of the temperature but not a function of the pressure. Therefore, the equilibrium constant $K(T)$ depends on T but not on p, whereas K_p is a function of p.

The specific form of the dependence of $K(T)$ on temperature is determined by examining the derivative $d \ln K/dT$. We find that

$$\begin{aligned} -R\frac{d \ln K}{dT} &= \frac{d}{dT}\left(\frac{\Delta G^*}{T} \right) \\ &= -\frac{\Delta G^*}{T^2} + \frac{d\Delta G^*}{dT} \\ &= -\frac{1}{T}\left(\frac{\Delta G^*}{T} + \Delta S^* \right) \\ &= -\frac{\Delta H^*}{T^2}, \end{aligned} \tag{21.68}$$

where we have used the interchange operation

$$\frac{d\Delta G^*}{dT} = \Delta\left(\frac{dG^*}{dT} \right) = -\Delta S^*. \tag{21.69}$$

Thus, the larger the heat of reaction (for a given T), the larger the rate of change of equilibrium constant with temperature.

Even though it is merely a combination of Eqs. 21.67 and 21.69, the relationship

$$\begin{aligned} \Delta S^* &= R\frac{d}{dT}(T \ln K) \\ &= \frac{\Delta H^* - \Delta G^*}{T} \end{aligned} \tag{21.70}$$

is sufficiently important to be worth separate display. Of course, $K(T)$ increases as ΔS^* increases. The extent to which a reaction proceeds is seen to be determined by a balance between the enthalpy required for the chemical transformation and T times the entropy change accompanying the same transformation. If the entropy changes are the same, an exothermic reaction proceeds further to completion than does an endothermic reaction. If the enthalpy changes are the same, a reaction that leads to more product molecules than reactant molecules proceeds further to completion than does a reaction that leads to the same number or fewer product molecules than reactant molecules. In all of these comments we are assuming that "completion" means total conversion of reactants to products. Of course, which chemicals we choose to call reactants and which products is arbitrary; the changes in ΔH^*, ΔS^*, and so on must all be defined consistent with whatever choice is made.

In Eqs. 21.68 to 21.70, ΔH^* and ΔS^* are the differences in enthalpy and entropy between products and reactants in their respective hypothetical perfect gas states. The tabulated thermodynamic properties of gases refer to the hypothetical perfect gas state at 1 atm and 25°C as the standard state of reference. In some cases the equation of state of the gas is not known to sufficient accuracy to distinguish between the actual state of the gas at 1 atm and 25°C and the hypothetical perfect gas state under the same conditions. In fact, for almost all gases there are only very small differences in properties between the actual and hypothetical perfect gas states at 1 atm and 25°C. There is, however, a distinction to be drawn between these states. If the actual state of the gas at 1 atm and 25°C is used as reference state, the equilibrium constant depends very slightly on the pressure because the excess volume of the gas mixture is not exactly zero. Note that the $K(T, p)$ we defined in Section 19.5 was such an equilibrium constant. When the hypothetical perfect gas state is used as reference state, the equilibrium constant is strictly independent of pressure.

Thus, using tables of standard enthalpies, entropies, and Gibbs free energies, the equilibrium composition of a reacting mixture, and the temperature dependence of the composition, may be readily computed. Since data for reactions may be combined to "synthesize" new reactions, the equilibrium properties of reacting mixtures may be predicted from information obtained for other reactions.[7] Of course, because of gas imperfection the equations of state of all components must be known. However, for many applications of the theory of chemical equilibrium, the perfect gas approximation to $K(T)$ is of sufficient accuracy. Indeed, under essentially all conditions, whether some particular species predominates in the reaction mixture can be determined simply from whether or not $\Delta G^* > 0$ or $\Delta G^* < 0$ for the reaction leading to that component. If the

[7] Just as tables of standard enthalpies permit a great economy in the recording of data for the calculation of heats of reaction, tables of standard Gibbs free energies permit a great economy in the recording of data for the calculation of equilibrium constants.

former is the case, the component does not predominate in the reaction mixture; if the latter, it does. If the entropy change in the reaction is small enough that $T\Delta S^* << \Delta H^*$, similar qualitative information can be obtained from the sign of ΔH^*.

There are, however, important qualitative effects that arise from the interactions between the molecules. For example, if terms up to the second virial coefficient suffice for the equation of state, and if the approximation

$$B_{jk} = \tfrac{1}{2}(B_{jj} + B_{kk}) \tag{21.71}$$

is adopted, it is easily shown that

$$\ln \frac{K_p}{K(T)} = -p\frac{\Delta B}{RT}, \tag{21.72}$$

where

$$\Delta B \equiv \sum_{i=1}^{r} \nu_i B_{ii}(T). \tag{21.73}$$

The approximation of Eq. 21.71 was introduced by Lewis and Randall; it is not an exact representation and cannot be justified from the molecular theory of the second virial coefficient. Nevertheless, Eq. 21.73 is found to be reasonably accurate over a range of gas pressure exceeding the range for which Eq. 21.31 is an accurate equation of state. Equation 21.73 is a useful approximation precisely because it eliminates all reference to the thermodynamic properties of the mixture; for these it substitutes a combination of properties of the pure components. Of course, this equivalence is just what the molecular theory denies is correct if the molecular interactions between *jj, kk,* and *jk* pairs are all different.

Now suppose that the temperature is such that $B > 0$ for all the reactants and $B < 0$ for all the products, so that $\Delta B < 0$. Then from Eq. 21.72, we see that $K_p/K(T)$ is greater than unity and increases as p increases, implying that the yield of products is greater than would be obtained in a reaction between perfect gases. When might this situation arise? In general, polar molecules, such as NH_3, attract each other strongly and at low temperature have large negative second virial coefficients. Thus, when nonpolar reactants, say, H_2 and N_2, react to form a polar product, say, NH_3, we anticipate that $\Delta B < 0$. In fact, the synthesis of NH_3 by this reaction is carried out at high pressure to maximize the yield. The effects of nonideality in this and other cases are by no means small. The study of gas imperfection is one of the important parts of the design of an economically viable manufacturing process, and is part of chemical engineering as well as fundamental chemistry.

Before concluding this section, we must point out a subtle assumption used in the preceding arguments. Equation 19.95 implies that the course of a chemical reaction may be specified at any given point by a value of the progress variable. This further implies that it is possible to stop the chemical reaction at an arbitrary point and measure λ. The reason this supposition is made is that it is necessary, at least in principle, to know the thermodynamic properties of the mixture for all possible compositions. If such knowledge cannot be obtained, at least in principle, the operational interpretation of thermodynamic equilibrium in terms of the changes induced by adding or removing constraints becomes meaningless. There are, indeed, some chemical reactions that can be quenched, and for which the supposition made is a valid extrapolation from experience. On the other hand, an uncatalyzed dimerization reaction such as

$$2NO_2 = N_2O_4$$

cannot be rigorously described by the thermodynamic theory because it is not possible to determine the thermodynamic properties of either pure component as a function of T and p from only macroscopic measurements. That is, variation of the pressure or temperature of NO_2 or N_2O_4 is always accompanied by chemical conversion following the stoichiometry shown. Thus it is never possible to determine directly from only macroscopic measurements the properties of either pure N_2O_4 or pure NO_2 as functions of T and p. Despite this limitation, it is very useful to apply the methods of thermodynamic analysis to this class of reactions. In the particular example cited, the concentration of NO_2 can be measured spectroscopically (Eq. 2.9), and its thermodynamic properties then can be calculated, provided the spectrum of NO_2 does not depend on the N_2O_4 concentration or is a known function of the N_2O_4 concentration.

21.4 An Example: The Haber Synthesis of NH_3

Much of the thermodynamic theory discussed in the preceding sections can be illustrated by the analysis of the conditions under which the reaction

$$\tfrac{1}{2}N_2(g) + \tfrac{3}{2}H_2(g) = NH_3(g) \tag{21.74}$$

gives a high yield of NH_3. In the Haber process a mixture of H_2 and N_2 under pressure and in the presence of an iron-based catalyst (usually an iron oxide mixed with aluminum oxide and potassium oxide) is caused to combine at 450–600°C. Extensive investigations of the equilibrium composition of mixtures of H_2 and N_2 under a variety of conditions lead to the data in Table 21.3. These data demonstrate that the formation of NH_3 is favored by low temperature and high pressure. Temperatures below 400°C cannot be used in commercial production because the rate of reaction is too small for economic use of the equipment needed. Temperatures above 600°C lead to yields of NH_3 that are too small for economic exploitation.

Table 21.3 Mole Percent NH_3 in 3/1 H_2/N_2 Equilibrium Mixtures

	p (atm)			
T (°C)	200	300	400	500
400	38.74	47.85	54.87	60.61
450	27.44	35.93	42.91	48.84
500	18.86	26.00	32.25	37.79
550	12.82	18.40	23.55	28.31
600	8.77	12.93	16.94	20.76

Table 21.4 Perfect Gas Equilibrium Constant for the Formation of NH_3

	p (atm)			
T (°C)	200	300	400	500
400	15.897×10^{-3}	18.060×10^{-3}	20.742×10^{-3}	24.065×10^{-3}
450	8.023×10^{-3}	8.985×10^{-3}	10.134×10^{-3}	11.492×10^{-3}
500	4.409×10^{-3}	4.893×10^{-3}	5.408×10^{-3}	6.013×10^{-3}
550	2.598×10^{-3}	2.836×10^{-3}	3.102×10^{-3}	3.392×10^{-3}
600	1.622×10^{-3}	1.751×10^{-3}	1.890×10^{-3}	2.036×10^{-3}

Suppose that the N_2, H_2, and NH_3 form a perfect gas mixture under all the conditions for which data are entered in Table 21.3. Then the equilibrium constant for the formation of NH_3 is

$$K_p = \frac{p_{NH_3}/\mathfrak{p}}{(p_{N_2}/\mathfrak{p})^{1/2}(p_{H_2}/\mathfrak{p})^{3/2}} = \frac{x_{NH_3}}{x_{N_2}^{1/2} x_{H_2}^{3/2}} \cdot \frac{\mathfrak{p}}{p}. \tag{21.75}$$

If, as supposed, N_2, H_2, and NH_3 do form a perfect gas mixture, then the several mole fractions and p will change in compensatory ways, leaving K_p a function of T alone; if not, the value of K_p computed from experimental values of the mole fractions will vary with p.

Consider the datum that at 500°C and 500 atm there is 37.79 mol% NH_3 in the equilibrium reaction mixture. The stoichiometry of the reaction requires that for the formation of y mol of NH_3, $\frac{1}{2}y$ mol of N_2 and $\frac{3}{2}y$ mol of H_2 must be consumed. If we start with a stoichiometric mixture, 0.5 mol of N_2 and 1.5 mol of H_2, we must have at any point

$$x_{NH_3} = \frac{y}{\frac{1}{2}(1-y)+\frac{3}{2}(1-y)+y} = \frac{y}{2-y}. \tag{21.76}$$

Setting $x_{NH_3} = 0.3779$ and solving for y, we find that under the stated conditions there are 0.5485 mol of NH_3 in the equilibrium mixture. Given the value of y, the total numbers of moles of all components are easily computed, and hence also the mole fractions of N_2 and H_2. In this particular case we find that

$$\text{total number of moles} = \frac{1}{2}(1-y)+\frac{3}{2}(1-y)+y = 2-y = 1.4515, \tag{21.77a}$$

$$x_{N_2} = \frac{\frac{1}{2}(1-y)}{2-y} = 0.1555, \tag{21.77b}$$

$$x_{H_2} = \frac{\frac{3}{2}(1-y)}{2-y} = 3x_{N_2} = 0.4666. \tag{21.77c}$$

Thus, the equilibrium constant for formation of NH_3 at 500°C and 500 atm has the value

$$K_p = 6.014 \times 10^{-3}.$$

Values of the perfect gas equilibrium constant obtained in this way are entered in Table 21.4. Clearly, the computed constant varies with pressure and thus cannot be equal to the equilibrium constant $K(T)$ as we previously assumed. We conclude that the perfect gas approximation is not adequate for the description of the N_2-H_2-NH_3 mixture.

As a first step in our further analysis, we now compute the equilibrium constant $K(T)$ for the formation of NH_3, using thermochemical data. The relevant data are

$$NH_3\text{:}\quad \Delta H_f^0 = -46.19 \text{ kJ/mol},$$
$$\Delta G_f^0 = -16.64 \text{ kJ/mol},$$
$$c_p^* = 29.79 + 25.48\times10^{-3}\,T - \frac{1.665\times10^5}{T^2} \text{ J/K mol},$$

$$N_2\text{:}\quad c_p^* = 27.82 + 4.184\times10^{-3}\,T \text{ J/K mol},$$

$$H_2\text{:}\quad c_p^* = 28.66 + 1.17\times10^{-3}\,T - 0.92\times10^{-6}T^2 \text{ J/K mol}.$$

Since the temperature dependence of the equilibrium constant is determined by the relation

$$\frac{d}{dT}\ln K(T) = \frac{\Delta H^*}{RT^2}, \tag{21.78}$$

we have, by integration,

$$\ln\frac{K(T_2)}{K(T_1)} = \int_{T_1}^{T_2} \frac{\Delta H^*}{RT^2}\,dT. \tag{21.79}$$

Of course, ΔH^* depends on the temperature. With the definition of standard state we are using, ΔH^* is the heat of reaction not as it actually takes place, but at a sufficiently low pressure for the gases to behave as perfect gases. Since the enthalpy of a perfect gas is independent of pressure, ΔH^* can be identified with the heat of formation in the (hypothetical) perfect gas state at 1 atm. The temperature dependence of ΔH^* is then computed by the method described in Chapter 14. With a reference temperature 298.15 K, for which ΔH^* becomes the standard heat of formation ΔH_f^0, we reduce Eq. 14.23 to

$$\Delta H^*(T_2) = \Delta H_f^0 + \sum_i \nu_i \int_{298.15}^{T_2} c_p^i(T)\, dT.$$

Given the data cited above, for the reaction under consideration,

$$\begin{aligned}\Delta H^*(T) = {} & \Delta H_f^0 + (29.79 - \frac{1}{2}\times 27.82 \\ & - \frac{3}{2}\times 28.66)(T - 298.15) \\ & + \frac{10^{-3}}{2}(25.48 - \frac{1}{2}\times 4.184 \\ & - \frac{3}{2}\times 1.17)[T^2 - (298.15)^2] \\ & + 1.665\times 10^5\left(\frac{1}{T} - \frac{1}{298.15}\right) \\ & - \frac{3}{2}\times\frac{0.92\times 10^{-6}}{3}[T^3 - (298.15)^3] \\ = {} & \Delta H_f^0 + 6576 - 27.11T + 10.82\times 10^{-3}T^2 \\ & + \frac{1.665\times 10^5}{T} - 0.46\times 10^{-6}T^3. \end{aligned} \tag{21.80}$$

Then Eq. 21.79 becomes

$$\begin{aligned}\ln\frac{K(T_2)}{K(T_1)} = {} & \frac{39614}{R}\left(\frac{1}{T_2} - \frac{1}{298.15}\right) - \frac{27.11}{R}\ln\frac{T_2}{298.15} \\ & + \frac{10.82\times 10^{-3}}{R}(T_2 - 298.15) \\ & - \frac{23.0\times 10^{-8}}{R}[T_2^2 - (298.15)^2] \\ & - \frac{1.99\times 10^4}{R}\left(\frac{1}{T_2^2} - \frac{1}{(298.15)^2}\right). \end{aligned} \tag{21.81}$$

Again for the purpose of illustration we choose $T_2 = 773.15$ K. We find that

$$\ln\frac{K(773.15)}{K(298.15)} = -12.234. \tag{21.82}$$

But from the data given, the equilibrium constant for the formation of NH_3 at 298.15 K is

$$K(298.15) = \exp\left(-\frac{\Delta G_f^0}{RT}\right) = 8.23\times 10^2, \tag{21.83}$$

so that

$$\begin{aligned} K(773.15) &= 8.23\times 10^2 \times 10^{-12.234/2.303} \\ &= 4.00\times 10^2. \end{aligned} \tag{21.84}$$

Thus, at 773.15 K (500°C), the deviation from perfect gas behavior increases K_{NH_3} by 10% at 200 atm and about 50% at 500 atm (see Table 21.4).

We now turn to calculating the influence of gas imperfection on the value of the equilibrium constant K_p. If the equation of state keeping only the second virial coefficient is adequate, we can apply Eqs. 21.72 and 21.73, using the values of B_{ii} in Table 21.5, to obtain

$$\begin{aligned} \frac{K_p}{K(T)} &= \exp\left(-\frac{p\Delta B}{RT}\right) \\ &= 1.21 \text{ at } 773 \text{ K and } 200 \text{ atm}, \\ &= 1.46 \text{ at } 773 \text{ K and } 400 \text{ atm}. \end{aligned} \tag{21.85}$$

The observed values are 1.10 and 1.35, respectively (see the data in Table 21.4).

At higher pressures the limitation of the equation of state to the second virial coefficient term and the use of Eqs. 21.71 and 21.72 result in inaccurate predictions of the equilibrium constants. The limitation imposed by the equation of state can be lifted by (1) using a more accurate equation of state in analytic form, (2) using p, V, T data directly without recourse to an analytic representation (numerical integration), or (3) using the fact that all gases behave almost alike when described by the reduced variables $\tilde{p}$, $\tilde{T}$ and $\tilde{V}$. Figure 21.11 shows the ratio of the fugacity to the pressure as a function of the reduced pressure $\tilde{p} = p/p_c$ at several reduced temperatures $\tilde{T} = T/T_c$. These curves may be used to estimate accurately the fugacities of the gaseous reactants and products when they are pure components. The only way to improve upon the approximation of Eq. 21.71 is to use equation of state data for the gas mixture. Such data are rarely available, so the approximations cited remain useful.

Table 21.5 Second Virial Coefficients of H_2, N_2, and NH_3 (cm^3/mol)

T (K)	B_{H_2}	B_{N_2}	B_{NH_3}
673	15.6	24.2	–38.2
723	15.7	25.5	–29.3
773	15.7	26.6	–23.4
823	15.7	27.6	–19.8
873	15.7	28.3	–16.7

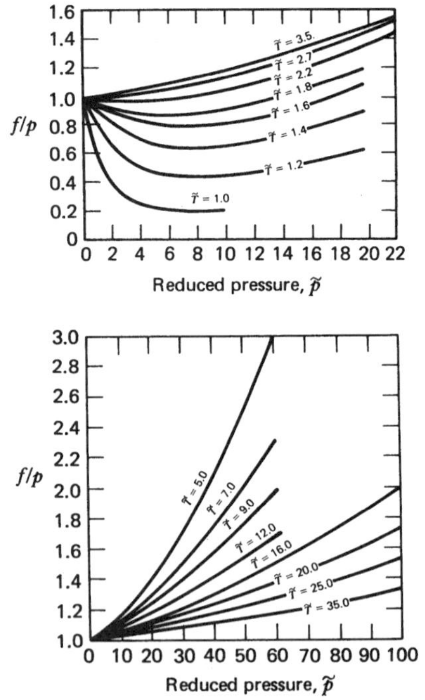

Figure 21.11 The ratio f/p of the fugacity to the pressure in reduced units.

There remain three topics to be mentioned in this section.

1. Occasionally it is useful to define an equilibrium constant of a different form from that we have used. There are two other forms commonly used: One is

$$K_x \equiv \prod_{i=1} x_i^{\nu_i}, \tag{21.86}$$

so that

$$K_x = K_p \left(\frac{p}{\mathfrak{p}}\right)^{-\sum \nu_i} = K_p \left(\frac{p}{\mathfrak{p}}\right)^{-\Delta\nu}, \tag{21.87}$$

where $\Delta\nu$ is the change in the number of moles in the reaction if carried to completion. The other is

$$K_V \equiv \prod_{i=1} \left(\frac{n_i}{V}\right)^{\nu_i}. \tag{21.88}$$

In the perfect gas limit we can set

$$p_i = \frac{n_i RT}{V} \tag{21.89}$$

and obtain

$$K_V = K(T)\left(\frac{RT}{\mathfrak{p}}\right)^{-\Delta\nu}. \tag{21.90}$$

Note that K_V has dimensions, unlike either K_x, K_p, or $K(T)$. From the definitions given, it is clear that even for a perfect gas mixture, K_x depends on the pressure unless $\Delta\nu = 0$; $K(T)$ is independent of pressure by definition, K_p and K_V are independent of pressure for perfect gases only. The equilibrium constants K_x, K_p, and $K(T)$ have the same logarithmic derivative with respect to temperature, namely,

$$\frac{d\ln K(T)}{dT} = \left(\frac{\partial \ln K_x}{\partial T}\right)_p = \frac{\Delta H^*}{RT^2}. \tag{21.91}$$

On the other hand, for K_V we obtain, using Eq. 21.68,

$$\left(\frac{\partial \ln K_V}{\partial T}\right)_p = \frac{d\ln K(T)}{dT} - \sum_i \nu_i \frac{d}{dT}\ln\left(\frac{RT}{\mathfrak{p}}\right)$$
$$= \frac{\sum_i \nu_i (h_i^* - RT)}{RT^2} = \frac{\Delta U^*}{RT^2}, \tag{21.92}$$

where ΔU^* is the heat of reaction at constant volume. The terms $\ln K_V$ and $\ln(RT/\mathfrak{p})$ must be understood to have a formal meaning since, strictly, the logarithm of a quantity with dimensions is undefined. However, their derivatives do have a meaning, since one can always set $d\ln y/dx = d\ln(y/y_0)/dx$, where y_0 is an arbitrary constant with the same units as y.

2. At pressures sufficiently low to set $K_x = K(T)(p/\mathfrak{p})^{-\Delta\nu}$, the pressure dependence of K_x is as follows: If $\Delta\nu < 0$, corresponding to a decrease in volume as the forward reaction proceeds, an increase in pressure favors the formation of more product. The use of K_x rather than $K(T)$ yields the clearest indication of the effect of pressure on reaction yield.

Returning to the synthesis of NH_3, we see that for the formation of NH_3 the perfect gas approximation gives

$$K_x = K(T)\frac{p}{\mathfrak{p}}; \tag{21.93}$$

that is, the formation of NH_3 is promoted by increasing the pressure. In the case of the dissociation reaction

$$N_2O_4 = 2NO_2, \tag{21.94}$$

we can write, treating the system as a perfect gas mixture,

$$K_x = \frac{x_{NO_2}^2}{x_{N_2O_4}} = K(T)\frac{\mathfrak{p}}{p}. \tag{21.95}$$

In this case, with $\Delta\nu > 0$, the forward reaction is promoted by lowering the pressure.

3. It may be advantageous to add an "inert" gas to a reaction mixture, since in that case the yields of various compounds can be altered. If the gases are imperfect, the addition of a gas that does not participate in the reaction can alter the fugacities of all the reacting species. For example, if the added gas has a very strong attraction for one of the reactants, the fugacity of that reactant will be decreased. The

Table 21.6 Mole Percent of NH_3 at Equilibrium in a 3/1 H_2/N_2 Mixture with 20% Inert Gas

	p (atm)			
***T* (°C)**	**200**	**300**	**400**	**500**
400	25.36	31.41	36.14	40.00
450	17.87	23.49	28.14	32.10
500	12.22	16.92	21.05	24.73
550	8.28	11.92	15.31	18.44
600	5.65	8.35	10.97	13.47

effects of an inert gas on the fugacities of the reactants and products are not accounted for by the Lewis and Randall rule. If they are important, analysis of the behavior of the gas mixture must be based on p, V, T measurements of the mixture and cannot be estimated using only pure-component data.

The effect of adding an inert gas is most evident in terms of K_x, since the equilibrium concentration is altered even if the gas mixture is perfect, unless $\Delta\nu = 0$. This happens because the mole fractions of all components are altered by the addition of another species to the mixture, even if that species does not take part in the reaction. For example, in the synthesis of NH_3, if there were initially present a mol of N_2, $3a$ mol of H_2, and b mol of inert gas, and y mol of NH_3 are formed at equilibrium, we find that

$$K_x = \frac{y/(4a+b-y)}{[(a-y/2)/(4a+b-y)]^{1/2}[(3a-3y/2)/(4a+b-y)]^{3/2}}$$
$$= K(T)\frac{p}{\mathfrak{p}}, \tag{21.96}$$

since $\Delta\nu = -1$ for this reaction. Given the values of $K(T)$ and p, Eq. 21.96 may be solved for y. In this case, and indeed in general, the algebra of such solutions is complicated and numerical analysis is to be preferred. The calculated mole percent of NH_3 in an equilibrium mixture initially containing 20 mol% inert gas is shown in Table 21.6. Comparison of Tables 21.3 and 21.6 shows how the mole fraction of NH_3 is reduced by the addition of inert gas to the mixture.

21.5 Statistical Molecular Theory of Gases and Gas Reactions

A. The Partition Function

Earlier in this book, in the discussion of atomic and molecular structure, we examined how the general principles of quantum mechanics restrict the energy spectrum of a system. The statistical interpretation of the thermodynamic properties of a particular system depends, of course, on the energy spectrum of that system.

A particularly interesting situation exists with respect to the requirements imposed on the symmetry properties of the wave function of a system. It is necessary that the wave function representing a system of many identical particles be either symmetric (i.e., have exactly the same value) or antisymmetric (i.e., be multiplied by –1) after two identical particles are interchanged. Any given system is of one or the other type, and the actual number of accessible states within a particular energy range may vary greatly depending on what symmetry the wave function has (see Appendix 21A).

Consider, now, the description of the motion of the center of mass of a molecule, that is, the translational motion. Let $\Lambda \equiv h/(2\pi m k_B T)^{1/2}$ be the *thermal de Broglie wavelength* associated with a molecule. We expect that if Λ is comparable with the average spacing between molecules, we must take into account the effect of diffraction upon the molecular motion. On the other hand, if $\Lambda << (V/N)^{1/3}$, we expect diffraction effects to be negligibly small. Except for He and H_2 the masses of all molecular species are sufficiently large that $\Lambda << (V/N)^{1/3}$ in the temperature ranges in which they are liquid or gaseous. Thus, for liquid Ar (~85 K) quantum effects do not influence the translational motion of the molecules, whereas for liquid Ne (~20 K), which is a borderline case, they amount to only a few percent of the classical motion contribution. On the other hand, in liquid He (~4 K) the effects of quantum mechanical diffraction dominate the behavior of the substance.

The reader should not derive from these paragraphs the impression that quantization of translational motion is rare in nature. The properties of metals, and of nuclei, bear eloquent testimony to the effects of quantization of the energy levels of an electron gas and of a nuclear gas.

There are other symmetry effects, of similar origin, that restrict the number of states of diatomic and polyatomic molecules (see Appendix 21B). Thus, interchange of two identical nuclei, which leads to changes in the electronic, vibrational, and rotational wave functions of a molecule, must satisfy the restriction of being of even or odd parity. We shall see that in the case of H_2 this restriction requires that there be two classes of H_2 molecules—one set of molecules that can have only even rotational states, and the other only odd rotational states. Because of the large spacing of the rotational energy levels of H_2, this particular restriction has marked thermodynamic consequences.

Consider, now, a one-component perfect gas. We shall analyze the properties of this system using the canonical ensemble to represent the relationship between the energy spectrum and thermodynamic properties. Using Eq. 19.131, we write the canonical partition function for the N-molecule system,

$$Q_N = \sum_i g_i \exp\left(-\frac{E_i}{k_B T}\right), \tag{21.97}$$

where E_i is the energy and g_i the number of states in the ith energy level. In a perfect gas there are no interactions between the molecules; hence, it is possible to define the energy of one molecule in meaningful terms. Of course, the total energy of the system, E_i, is just the sum of the energies of the molecules. The energy of any one molecule is denoted by $\epsilon_l(n_x, n_y, n_z, j, \upsilon, \ldots)$ where n_x, n_y, n_z are the quantum numbers specifying translational motion, j is the quantum number specifying rotational motion, υ is the quantum number specifying molecular vibration, the . . . implies other quantum numbers, for example, for electronic excitation, and $l = (1, 2, \ldots, N)$. For a polyatomic molecule j and υ represent several quantum numbers corresponding to the several kinds of rotational and vibrational motion. For simplicity we shall collectively represent the set of all quantum numbers except n_x, n_y, n_z by the symbol s. In a perfect gas the energy of a molecule can be written in the form

$$\epsilon_l(n_x, n_y, n_z, j, \upsilon, \ldots) = \epsilon_l^{\text{trans}}(n_x, n_y, n_z) + \epsilon_l^{\text{int}}(s); \tag{21.98}$$

that is, the translational motion is completely independent of the internal motion of the molecule.

To determine the equation of state from Eq. 21.98, we start by writing the total energy of N molecules as a sum of individual molecule energies:

$$E_i = \sum_{l=1}^{N} [\epsilon_l^{\text{trans}}(n_x, n_y, n_z) + \epsilon_l^{\text{int}}(s)]. \tag{21.99}$$

A selected molecule may be in any one of the possible quantum states provided only that for a given value of E_i the sum of the energies of all the molecules is just E_i. We are, of course, interested in all possible values of E_i (and therefore also in all possible sets of values of the individual energies of the molecules), since the evaluation of Q_N requires that E_i range over all possible values. If molecule l is in an internal state with quantum numbers collectively represented by s, there is a degeneracy[8] $\bar{\omega}_{nsl}$, so that in a gas of independent molecules the overall degeneracy of the state with total energy E_i is

$$g_i \equiv g_i(E_i) = \frac{1}{N!} \sum_{\mathbf{ns}} \prod_{l=1}^{N} \bar{\omega}_{nsl}. \tag{21.100}$$

The factor $(1/N!)$ appears because of the indistinguishability of the molecules—there are $N!$ different ways of assigning N molecules to a specified set of N energies ϵ_i, each of which has the same total energy. The sum $\Sigma_{\mathbf{ns}}$ appears because there are many ways of subdividing internal energy and translational energy that satisfy Eq. 21.99. Each particular subdivision has a degeneracy $(1/N!)\Pi_l\, \bar{\omega}_{nsl}$, but the internal quantum numbers s are, of course, different for each subdivision, so to get the overall degeneracy corresponding to energy E_i we must add up all such contributions. The sum $\Sigma_{\mathbf{ns}}$ is over all sets of values of the internal and translational quantum numbers of the N molecules, but the possible subdivisions of the internal and translational energies of the molecules must satisfy Eq. 21.99. Introducing Eqs. 21.99 and 21.100 into Eq. 21.97, we find that

$$\begin{aligned}\sum_i g_i(E_i)\exp\left(-\frac{E_i}{k_BT}\right) &= \frac{1}{N!}\left\{\sum_{ns1} \bar{\omega}_{ns1} \exp\left[-\frac{\epsilon_1^{\text{trans}}(n1) + \epsilon_1^{\text{int}}(s1)}{k_BT}\right]\right\} \\ &\times \left\{\sum_{ns2} \bar{\omega}_{ns2} \exp\left[-\frac{\epsilon_2^{\text{trans}}(n2) + \epsilon_2^{\text{int}}(s2)}{k_BT}\right]\right\} \\ &\times \cdots \times \left\{\sum_{ns\,N} \bar{\omega}_{ns\,N} \exp\left[-\frac{\epsilon_N^{\text{trans}}(nN) + \epsilon_N^{\text{int}}(sN)}{k_BT}\right]\right\},\end{aligned} \tag{21.101}$$

where the numbers $ns1, \ldots, nsN$ stand for all the quantum numbers specifying the states of molecules $1, \ldots, N$. Equation 21.101 includes the fact that, as stated by Eq. 21.100, any allowed state of one molecule can be combined with any allowed states of all other molecules. Of course every molecule in the gas is the same, and therefore the sum over all the states of a molecule must have the same value for each molecule. Since there is one such sum for each of the N molecules, we have

$$\begin{aligned}\sum_i g_i(E_i)\exp\left(-\frac{E_i}{k_BT}\right) &= \frac{1}{N!}\left\{\sum_{ns1} \bar{\omega}_{ns1} \exp\left[-\frac{\epsilon_1^{\text{trans}}(n1) + \epsilon_1^{\text{int}}(s1)}{k_BT}\right]\right\}^N \\ &= \frac{Q_1^N}{N!},\end{aligned} \tag{21.102}$$

where

$$Q_1 \equiv \sum_{ns1} \bar{\omega}_{ns1} \exp\left[-\frac{\epsilon_1^{\text{trans}}(n1) + \epsilon_1^{\text{int}}(s1)}{k_BT}\right] \tag{21.103}$$

is the partition function of a single-molecule system. From here on we drop the subscript 1 in the summand of Eq. 21.103, since Q_1 is the same for each molecule in the perfect gas.

To determine the precise form of the equation of state we must evaluate Q_1. Consider first the translational motion. We know that

$$\epsilon^{\text{trans}} = \frac{h^2}{8mV^{2/3}}(n_x^2 + n_y^2 + n_z^2) \tag{21.104}$$

[8] We use $\bar{\omega}_{nsl}$ for the degeneracy of the nsth state of molecule l so as to distinguish it from the degeneracy of the N-molecule system, g_i. **ns** indicates summation over all molecules.

for a free particle in a cubical box of volume V. For all cases discussed in this chapter, $h^2/8mV^{2/3}$ is very much smaller than the thermal energy k_BT. This means that the difference in population of two adjacent energy levels is very small, and it is reasonable to replace the sum over energy levels in the definition of Q_1 with an integral over energy levels:

$$\sum_{ns}\varpi_{ns}\exp\left[-\frac{\epsilon^{\text{trans}}+\epsilon^{\text{int}}}{k_BT}\right]=\sum_{n_x,n_y,n_z}\exp\left[-\frac{\epsilon^{\text{trans}}}{k_BT}\right]\times\sum_s\varpi_s^{\text{int}}\exp\left[-\frac{\epsilon^{\text{int}}}{k_BT}\right]=Q_1^TQ_1^{RVE}, \tag{21.105}$$

where

$$Q_1^T\equiv\sum_{n_x,n_y,n_z}\exp\left(-\frac{\epsilon^{\text{trans}}}{k_BT}\right)=\int_0^\infty\int_0^\infty\int_0^\infty dn_x\,dn_y\,dn_z\exp\left[-\frac{h^2(n_x^2+n_y^2+n_z^2)}{8mV^{2/3}k_BT}\right], \tag{21.106}$$

and

$$Q_1^{RVE}\equiv\sum_s\varpi_s^{\text{int}}\exp\left(-\frac{\epsilon^{\text{int}}}{k_BT}\right), \tag{21.107}$$

the superscripts T, RVE referring to the translational and rotation-vibration-electronic (internal) components, respectively. The triple integral defining Q_1^T reduces to products of $\int\exp(-ax^2)\,dx$ and is easily found in a table of integrals. Integration gives

$$Q_1^T=\left(\frac{2\pi mk_BT}{h^2}\right)^{3/2}V=\Lambda^{-3}V. \tag{21.108}$$

What are the effects of internal motion? For simplicity we shall consider the case of a diatomic molecule and temporarily neglect possible electronic excitations (see below). In this case, to a good approximation we can write

$$\epsilon^{\text{int}}=\left(\upsilon+\frac{1}{2}\right)h\nu+J(J+1)\frac{h^2}{8\pi^2I}, \tag{21.109}$$

as deduced earlier (see Eqs. 7.8 and 7.19). In Eq. 21.109, ν is the frequency of vibration and I is the moment of inertia of the diatomic molecule. The lowest energy state ($\upsilon=0$, $J=0$)[9] has energy $h\nu/2$ relative to the point of lowest potential energy for a classical oscillator of frequency ν. This is an example of the consequences of the uncertainty principle, since the molecule cannot be at rest even in its ground state. Let us, as before, take the ground-state energy as the zero of energy. Then we write

$$\epsilon^{\text{int}}-\epsilon_0=\upsilon h\nu+J(J+1)\frac{h^2}{8\pi^2I}. \tag{21.110}$$

Before proceeding, it is instructive to explain why we called Eq. 21.109 an approximation (although a very good one). All molecular vibrations are to some extent anharmonic, and all chemical bonds will stretch to some extent as the speed of rotation of the molecule increases. Thus the internal energy of a molecule contains small anharmonicity corrections [proportional to $(\upsilon+\frac{1}{2})^2$ and higher powers], small centrifugal distortion corrections [proportional to $J^2(J+1)^2$ and higher powers], and a small vibration-rotation interaction correction [whose leading term is proportional to $(\upsilon+\frac{1}{2})\times J(J+1)$]. The latter correction arises as follows: If the bond between the two nuclei in a diatomic molecule stretches both because of centrifugal distortion and by virtue of vibrational excitation in an anharmonic potential, then as the speed of rotation is increased, more anharmonic regions of the interatomic potential are reached. It should be noted that for almost all molecules the corrections cited are small, of the order of magnitude of one part per hundred at about 500°C. Having noted their existence, which is important, we shall not need to refer to them any further in this context.

Given Eq. 21.110, we can divide Q_1^{RV} into its rotational and vibrational components (recall that we are temporarily neglecting electronic excitation, so $Q_1^{RVE}\to Q_1^{RV}$):

$$Q_1^{RV}=\frac{1}{\sigma}\sum_{J=0}^{\infty}(2J+1)\exp\left(-\frac{J(J+1)h^2}{8\pi^2Ik_BT}\right)\sum_{\upsilon=0}^{\infty}\exp\left(-\frac{\upsilon h\nu}{k_BT}\right) \tag{21.111}$$

$$\equiv Q_1^RQ_1^V. \tag{21.112}$$

The factors $(1/\sigma)$ and $(2J+1)$ in Q_1^R will be explained below.[10] We note here only that we have written the degeneracy ϖ_s^{int} as the product $\varpi_s^{\text{int}}=(2J+1)/\sigma$.

Since $\upsilon h\nu$ is always greater than zero, $\exp(-\upsilon h\nu/k_BT)$ is always less than unity, and Q_1^V can be simplified as

$$Q_1^V\equiv\sum_{\upsilon=0}^{\infty}\exp\left(-\frac{\upsilon h\nu}{k_BT}\right)=\sum_{\upsilon=0}^{\infty}\left[\exp\left(-\frac{h\nu}{k_BT}\right)\right]^{\upsilon}=\left[1-\exp\left(-\frac{h\nu}{k_BT}\right)\right]^{-1}. \tag{21.113}$$

To compute Q_1^R, we note that the degeneracy of a rotational state of quantum number J is $(2J+1)$, which has been

---

[9] See Appendix 21B.

[10] See also Appendix 21B.

entered in Eq. 21.111. But for a homonuclear diatomic molecule, such as I_2, Eq. 21.111 counts every distinguishable configuration twice, since a rotation of a homonuclear diatomic molecule by 180° brings the molecule back to a position indistinguishable from the starting position. This is corrected for by the factor σ, which has the value 1 for heteronuclear diatomic molecules, and 2 for homonuclear diatomic molecules.[11] Because for most molecules (but not H_2) the difference in energy between rotational energy levels is very small relative to the thermal energy k_BT, the argument leading to Eq. 21.106 gives here

$$\begin{aligned} Q_1^R &\equiv \frac{1}{\sigma}\sum_{J=0}^{\infty}(2J+1)\exp\left(-\frac{J(J+1)h^2}{8\pi^2 Ik_BT}\right) \\ &= \frac{1}{\sigma}\int_0^{\infty}(2J+1)\exp\left(-\frac{J(J+1)h^2}{8\pi^2 Ik_BT}\right)dJ \\ &= \frac{8\pi^2 Ik_BT}{\sigma h^2} \approx 0.0419\frac{IT}{\sigma}. \end{aligned} \qquad (21.114)$$

This approximation is not valid for the vibrational energy levels, but as seen above, Q_1^V can be evaluated in closed form without the use of any approximation. The numerical form on the right-hand side of Eq. 21.114 is valid when I is given in amu Å^2 and T in K. When Eq. 21.114 is compared with the translational partition function, Eq. 21.108, we see that here the molecular mass is replaced by the moment of inertia, the volume is missing, and the exponent is different (1 instead of $\frac{3}{2}$). That I replaces m is a consequence of the mechanical fact that in rotational motion the moment of inertia plays the same role as does the mass in translational motion. The reason the V is missing is that the space in which rotation occurs is fixed by the center of rotation, and has nothing to do with the volume of real space. That is, rotation about the center of mass constrains the atoms to move on a spherical surface; the area of a sphere of unit radius is 4π. Finally, the difference between the exponents arises because translational space has three dimensions, whereas the rotational space of diatomic molecules has only two dimensions (cf. later discussion of polyatomic molecules).

By combination of Eqs. 21.108, 21.113, and 21.114, and using the relation

$$\frac{A_N}{k_BT} = -\ln Q_N = -N\ln Q_1 - \ln N!, \qquad (21.115)$$

(see Eq. 19.130), we find that

$$\begin{aligned} \frac{A_N}{k_BT} = -N\ln\Bigg\{ & V\left(\frac{2\pi mk_BT}{h^2}\right)^{3/2}\left(\frac{8\pi^2 Ik_BT}{\sigma h^2}\right) \\ & \times\left[1-\exp\left(-\frac{h\nu}{k_BT}\right)\right]^{-1}\Bigg\} + \ln N!. \end{aligned} \qquad (21.116)$$

But from the thermodynamic relation

$$p = -\left(\frac{\partial A}{\partial V}\right)_{T,N} \qquad (21.117)$$

and Eq. 21.116, we find that

$$p = \frac{Nk_BT}{V}, \qquad (21.118)$$

which is just the perfect gas equation of state. The derivation just given shows that, as long as there are no molecular interactions and quantum mechanical diffraction effects may be neglected,[12] the perfect gas equation of state describes equally well gases of monatomic and polyatomic molecules.

The thermodynamic functions of the perfect gas of diatomic molecules are easily obtained from Eq. 21.116. We reduce that equation by using Stirling's approximation,

$$\ln N! - N\ln N - N, \qquad (21.119)$$

which is very accurate for N large, and write

$$\begin{aligned} A_N = -Nk_BT\ln\Bigg\{ & \frac{V}{N}\left(\frac{2\pi mk_BT}{h^2}\right)^{3/2}\left(\frac{8\pi Ik_BT}{\sigma h^2}\right) \\ & \times\left[1-\exp\left(-\frac{h\nu}{k_BT}\right)\right]^{-1}\Bigg\} - Nk_BT. \end{aligned} \qquad (21.120)$$

From the thermodynamic relation

$$S = -\left(\frac{\partial A}{\partial T}\right)_{V,N} \qquad (21.121)$$

we find that

$$\begin{aligned} S_N = Nk_B\Bigg\{ & \frac{7}{2} + \frac{h\nu}{k_BT}\cdot\frac{\exp(-h\nu/k_BT)}{1-\exp(-h\nu/k_BT)} \\ & + \ln\left[\frac{V}{N}\left(\frac{2\pi mk_BT}{h^2}\right)^{3/2}\left(\frac{8\pi^2 Ik_BT}{\sigma h^2}\right)\right] \\ & + \ln\left[\frac{1}{1-\exp(-h\nu/k_BT)}\right]\Bigg\}, \end{aligned} \qquad (21.122)$$

whereas from the thermodynamic relations

$$U = A + TS \qquad (21.123)$$

and

$$H = A + TS + pV \qquad (21.124)$$

[11] See Appendix 21B.

[12] See Appendix 21A.

we find that

$$U_N = \frac{5}{2} Nk_BT + Nh\nu \frac{\exp(-h\nu/k_BT)}{1-\exp(-h\nu/k_BT)} \tag{21.125}$$

and, with Eq. 21.118,

$$H_N = \frac{7}{2} Nk_BT + Nh\nu \frac{\exp(-h\nu/k_BT)}{1-\exp(-h\nu/k_BT)}. \tag{21.126}$$

Before proceeding to a description of the other properties of a perfect gas in terms of the relationships derived, we consider how the analysis must be extended when electronic excitation of the molecules occurs, or when the molecules are polyatomic.

For most molecules the lowest electronic state is nondegenerate, and the next electronic state has an energy so much higher than the groundstate energy that it is not appreciably populated even at temperatures as high as 1000 K. It is easily verified that, if the first electronic excitation requires 1 eV, where $k_BT \approx 0.08$ eV at 1000 K, we find $\exp(-\Delta\epsilon/k_BT) \approx 10^{-6}$. Thus the electronic partition function may be taken to be unity with very small error. There are a few cases in which the first excited electronic state of the molecule is very close in energy to the ground electronic state. For example, the ground state of the F atom is of species ${}^2P_{3/2}$, and the other member of the doublet state, namely, ${}^2P_{1/2}$, is only 0.05 eV higher in energy. The next excited state is several electron volts higher in energy, and at all ordinary temperatures does not contribute to the thermodynamic properties of the gas.

As a result of the preceding remarks, we conclude that it is ordinarily sufficiently accurate to approximate the electronic partition function of a molecule by only the first two terms,

$$Q_1^E = \overline{\omega}_0^E + \overline{\omega}_1^E \exp\left(-\frac{\Delta\epsilon^E}{k_BT}\right), \tag{21.127}$$

where $\overline{\omega}_0^E$ and $\overline{\omega}_1^E$ are the degeneracies of the ground (0) and first excited (1) electronic states, and $\Delta\epsilon^E$ is the excitation energy of the first electronic state. It should be noted, however, that the separation of the electronic partition function from the rotational and vibrational partition functions is not a good approximation if an excited electronic state is appreciably populated. Different electronic states have, in general, different internuclear separations and different vibration frequencies, so that the rotational and vibrational motion of a molecule is not independent of the electronic state of the molecule. Only when the population of the excited electronic state is small, or the excited electronic state (fortuitously) has the same geometry and vibration spectrum as the ground state, is the separation of the partition functions a good approximation.

Consider, now, a perfect gas of polyatomic molecules. Clearly, the only difference between this and a perfect gas of diatomic molecules arises from the greater number of internal degrees of freedom. That is, a polyatomic molecule has more internal vibrations and, for nonlinear molecules, an extra degree of rotational freedom. For a classically excited rigid rotator, that is, one for which the spacing between rotational energy levels is small relative to k_BT, it can be shown that

$$Q_1^R = \frac{8\pi^2(2\pi k_BT)^{3/2}}{\sigma h^3}(I_xI_yI_z)^{1/2}, \tag{21.128}$$

where I_x, I_y, I_z are the principal moments of inertia (see Section 3.12) and σ is the number of equivalent positions of the polyatomic molecule (*symmetry number*). How does Eq. 21.128 differ from Eq. 21.114? Actually, all linear molecules do have a very small moment of inertia about the internuclear axis due to the motion of the electrons (see Section 7.2). For linear molecules $I_x = I_y$, whereas I_z is very small. But the spacing of rotational energy levels is always inversely proportional to the moment of inertia, so it is clear that excitation of rotation about the internuclear axis (corresponding to I_z) is very dffficult. Consequently, because the energy required for excitation is so large, only the ground state of rotation about the internuclear axis is of thermodynamic importance.

When all three moments of inertia are different, there is no simple relationship for the rotational energy levels, although they have been worked out in detail for some molecules. For almost all gases the rotational energy spacing is sufficiently small that Eq. 21.128 is valid, so that these complications need not concern us here. It suffices for our purposes to note that they exist and must be taken into account when low-temperature phenomena are discussed.

There is another kind of molecular motion that can be considered a rotation, viz., internal rotation of one part of a molecule relative to some other part. A molecule exhibiting this kind of motion is ethane, C_2H_6. In general, this motion is not free (see Section 9.4). That is, since the energies of the molecule corresponding to non-equivalent positions are somewhat different, the energy of the molecule must change as the angle of internal rotation changes. In passing from one position of minimum potential energy to another, the molecule must surmount some sort of energy barrier because, as can be seen in Fig. 21.12, this is implicit in the notion of energy minimum. When the energy barrier is very small and the energy-level spacing is small relative to k_BT, the internal rotation is nearly free, and it is found that

$$Q_1^{IR} = \frac{(8\pi^3 I_{IR}k_BT)^{1/2}}{\sigma_{IR}h},$$
$$I_{IR} = \frac{I_1I_2}{I_1+I_2}, \tag{21.129}$$

where I_1 and I_2 are the moments of inertia about the bond axis of the two counterrotating parts of the molecule. As before, σ_{IR} is the symmetry number, here for the internal

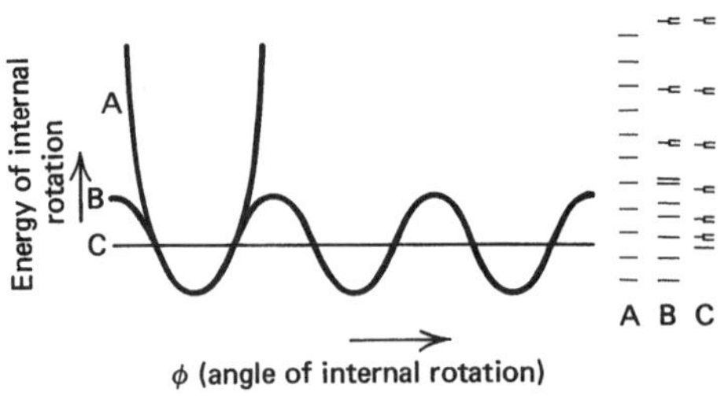

Figure 21.12 Potential functions (on the left) and energy levels (on the right) for a restricted rotator B, a harmonic oscillator A, and a free rotator C. When the barrier is large relative to k_BT, a parabola can be fitted to the bottom of the potential curve and the internal motion becomes a torsional vibration. When the barrier is small relative to k_BT, the internal motion becomes a free rotation. From K. S. Pitzer, *Quantum Chemistry* (Prentice-Hall, Englewood Cliffs, N.J., 1953).

rotation. Where there are several bonds around which there can be classical free rotation, there is a contribution to Q_1 of the form of Eq. 21.129 for each of the bonds. On the other hand, if the barrier to rotation is very large, the separate parts of the molecule oscillate relative to one another in a torsional motion. This is very like a vibration, and can be treated as such. For intermediate barrier heights no general results can be stated—direct calculations are needed.

Finally, a polyatomic molecule has many modes of vibrational motion. Provided that the restoring force acting on any atom of the molecule obeys Hooke's law, the total vibrational motion can be separated into a sum of motions along normal coordinates. The number of normal coordinates is equal to the number of vibrational degrees of freedom, and motion along the normal coordinates is harmonic. If ν_i is the frequency of the ith normal mode of motion, the vibrational partition function is just

$$Q_1^V = \prod_{i=1}^{3n-6}\left[1-\exp\left(-\frac{h\nu_i}{k_BT}\right)\right]^{-1}, \tag{21.130}$$

because:

1. By resolving the motion of the atoms into a sum of motions along independent normal coordinates, we can express the vibrational energy as a sum of the independent energies of motion along the normal coordinates;

2. For an n-atom molecule, using three coordinates to specify each atom's position and subtracting three coordinates for center-of-mass translational motion and three for rotational motion (nonlinear molecule) leaves $3n - 6$ vibrational degrees of freedom (if there is not internal rotation). Thus, the complete partition function for an isolated polyatomic molecule is

$$Q_1 = Q_1^T Q_1^R Q_1^E Q_1^{IR} Q_1^V, \tag{21.131}$$

provided that the energies for the separate motions are, to an adequate approximation, independent of one another.

B. The Heat Capacity of a Perfect Gas of Diatomic Molecules

Given the results of the calculations of the last section, it is easy to compute the heat capacity of a perfect gas and to analyze the contributions to it from translational motion, rotational motion, vibrational motion, and electronic excitation. For simplicity we consider only diatomic molecules.

The key idea in the interpretation of the heat capacity is that of *characteristic temperature.* For each class of molecular motion we consider the ratio of the spacing of the energy levels to k_BT. If the ratio is large, the occupation of the various energy levels corresponding to that class of motion is limited to only the lowest levels, and the corresponding contribution to the heat capacity is very small; if the ratio is small, many levels are occupied and the corresponding contribution to the heat capacity reaches a classical limiting value, the equipartition value. If a characteristic temperature is defined by $\Theta \equiv \Delta\epsilon/k_B$, the former case corresponds to $\Theta/T >> 1$ and the latter to $\Theta/T << 1$.

Using the thermodynamic definition of C_V and Eqs. 21.102 and 21.115,

$$C_V \equiv \left(\frac{\partial U}{\partial T}\right)_{V,N} = T\left(\frac{\partial S}{\partial T}\right)_{V,N} = -T\left(\frac{\partial^2 A}{\partial T^2}\right)_{V,N}$$
$$= Nk_BT\left[\frac{\partial^2}{\partial T^2}(T\ln Q_1)\right]_{V,N}, \tag{21.132}$$

we can compute the contributions to the heat capacity from translational motion, rotational motion, vibrational motion, and electronic excitation, C_V^T, C_V^R, C_V^V, and C_V^E, respectively. Of course, $C_V = C_V^T + C_V^R + C_V^V + C_V^E$.

1. *Translational heat capacity.* For all gases the spacing of translational energy levels is small relative to k_BT, and the heat capacities are

$$C_V^T = \tfrac{3}{2}Nk_B, \qquad C_p^T = \tfrac{5}{2}Nk_B. \tag{21.133}$$

As can be seen from Table 21.7, there is excellent agreement between theory and experiment for the monatomic gases. Equation 21.133 implies that the contribution to the internal energy and enthalpy of a perfect gas from translational motion is linear in the temperature (Fig. 21.13). We shall later see that such behavior is generally characteristic of the case when the spacing of the energy levels is small relative to the thermal energy k_BT, and is not restricted to translational motion.

Table 21.7 Heat Capacities of Monatomic Gases at 298.15 K

Substance	C_V (calc) (J/K mol)	C_V (obs) (J/K mol)	C_p (calc) (J/K mol)	C_p (obs) (J/K mol)
He	12.472	12.519	20.786	20.782
Ar	12.472	12.544	20.786	20.920

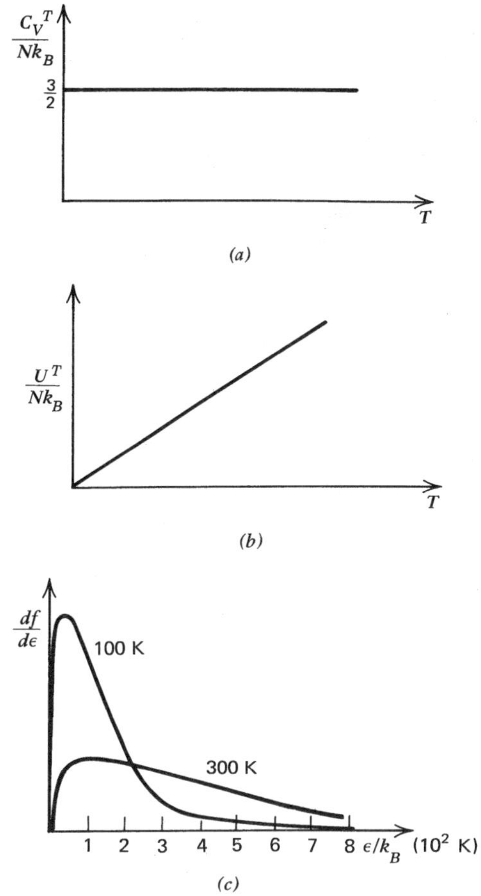

Figure 21.13 (*a*) Translational motion contribution to the heat capacity at constant volume. See Appendix 21A for a description of how C_V is modified near $T = 0$ when the spacing of the translational energy levels becomes comparable to and then smaller than k_BT. (*b*) Translational motion contribution to the internal energy. The slope of this curve is $\frac{3}{2}$. (*c*) Molecular translational energy distribution at two temperatures. To illustrate the relationship between the peak position and the energy scale, the latter is expressed as a temperature,

$$\frac{df_{\text{trans}}}{d\epsilon} = \frac{2\pi N}{(\pi k_B T)^{3/2}} \epsilon^{1/2} \exp\left(\frac{-\epsilon}{k_B T}\right).$$

2. *Rotational heat capacity.* Let

$$\Theta_{\text{rot}} \equiv \frac{h^2}{8\pi^2 I k_B} \tag{21.134}$$

be called the *rotational temperature.* (The reader should compare Eq. 21.134 with the exponent in Eq. 21.111.) Θ_{rot} is small relative to T for most molecules of interest (see Table 21.8). Then the rotational partition function is accurately given by Eq. 21.114, and the corresponding contribution to the heat capacity is

$$C_V^R = Nk_B. \tag{21.135}$$

At sufficiently low temperatures ($\Theta_{\text{rot}} \lesssim 1.5$) the sum in the first line of Eq. 21.114 can no longer be approximated by an integral, and Eq. 21.135 no longer holds Q_1^R and C_V^R must then be calculated numerically, the results for diatomic (or linear polyatomic) molecules being shown in Fig. 21.14. Only in the case of H_2 and its isotopic modifications, however, does this behavior occur in the normal gaseous range.

For H_2, D_2, and T_2 the situation is further complicated by the quantum mechanical symmetry requirements applicable to homonuclear molecules.[13] As described in Section 6.7, the wave function for a system of two identical particles must be either symmetric or antisymmetric to exchange of the particles, depending on whether the particles have integral or half-integral spin, respectively. Since the proton has spin $\frac{1}{2}$, the wave function for H_2 must be antisymmetric to exchange of the nuclei. The translational wave function involves only the molecule's center of mass, and at ordinary temperatures only the ground states for electronic and vibrational motion are occupied, and the corresponding wave functions are symmetric. For the total wave function, which is the product of the various contributions, to be antisymmetric, then, either the rotational or the nuclear spin wave function must be antisymmetric.

As described in Section 3.12, the rotational wave function of a symmetric rotor like H_2 is what is called a spherical harmonic. Since we did not examine these in detail, we must state without proof that the spherical harmonics are symmetric for even values of the rotational quantum number J and antisymmetric for odd J with respect to inversion of the coordinates: $x \to -x$, $y \to -y$, $z \to -z$.[14] Even and odd values of J must thus be associated only with antisymmetric and symmetric nuclear spin wave functions, respectively. In the absence of a catalyst such as activated charcoal, there are virtually no transitions between the symmetric and antisymmetric spin states, so that H_2 is effectively a mixture of two stable species. These are called *orthohydrogen* (symmetric spin, J odd) and *parahydrogen* (antisymmetric spin, J even), abbreviated o-H_2 and p-H_2.

The spacing between nuclear spin energy levels is very small relative to k_BT in a gas, so these states may be regarded as degenerate. If a nucleus has $\overline{\omega}$ degenerate spin states, the homonuclear molecule must have $\overline{\omega}^2$, of which it can be shown that $\overline{\omega}(\overline{\omega}+1)/2$ are symmetric and $\overline{\omega}(\overline{\omega}-1)/2$ are antisymmetric. For H_2 we have $\overline{\omega} = 2$ (spin $+\frac{1}{2}$ and $-\frac{1}{2}$), so that $\overline{\omega}_s = 3$, $\overline{\omega}_a = 1$. The rotational partition function is thus

$$\begin{aligned} Q_1^R = & \sum_{J\,\text{even}} (2J+1)\exp\left(-\frac{J(J+1)\Theta_{\text{rot}}}{T}\right) \\ & +3\sum_{J\,\text{odd}} (2J+1)\exp\left(-\frac{J(J+1)\Theta_{\text{rot}}}{T}\right), \end{aligned} \tag{21.136}$$

[13] See Appendix 21B.

[14] This can be seen for the examples given in Table 3.1 as follows: Inversion takes θ into $\pi - \theta$ and ϕ into $\pi + \phi$, or $\sin\theta \to \sin\theta$, $\cos\theta \to -\cos\theta$, and $\exp(im\phi) \to (-1)^m \exp(im\phi)$. A function $Y_{J,m}(\theta, \phi)$ has even powers of $\cos\theta$ multiplying even powers of $\exp(i\phi)$ if J is even, and even powers of $\cos\theta$ multiplying odd powers of $\exp(i\phi)$ if J is odd. The symmetric or antisymmetric character of the harmonics $Y_{J,m}(\theta, \phi)$ follows immediately.

Table 21.8 Vibrational and Rotational Constants of Some Diatomic Molecules

Substance	Electronic Ground State	$\tilde{\nu}_e$ (cm^{-1})	r_e (Å)	B_e^{a} (cm^{-1})	$\frac{\epsilon_1 - \epsilon_0^{b}}{hc}$ (cm^{-1})	Θ_{vib}^{c} (K)	Θ_{rot}^{d} (K)	Θ_{elec}^{e} (K)
H_2	$^1\Sigma_g^+$	4395.2	0.7417	60.809	91,690	6320	87.5	129,000
HD	$^1\Sigma_g^+$	3817.1	0.7414	45.655	91,711	5500	65.8	129,000
D_2	$^1\Sigma_g^+$	3118.5	0.7416	30.429	91,698	4490	43.8	129,000
HF	$^1\Sigma^+$	4138.5	0.9171	20.939		5990	31.5	
HCl	$^1\Sigma^+$	2989.7	1.2746	10.591	44,000	4330	15.2	61,800
HBr	$^1\Sigma^+$	2649.7	1.4138	8.473	35,000	3820	12.2	49,200
HI	$^1\Sigma^+$	2309.5	1.6041	6.551	27,500	3340	9.4	38,600
N_2	$^1\Sigma_g^+$	2359.6	1.094	2.010	69,290	3390	2.89	97,400
CO	$^1\Sigma^+$	2170.2	1.1282	1.9314	65,075	3120	2.78	91,500
NO	$^2\Pi_{1/2}$	1904.0	1.1508	1.7046	43,966	2745	2.45	61,700
O_2	$^3\Sigma_g^-$	1580.4	1.2074	1.4457	7,918	2278	2.08	11,100
F_2	$^1\Sigma_g^+$	892.1	1.435		34,500	1290		48,500
Cl_2	$^1\Sigma_g^+$	564.9	1.988	0.2438	18,311	814	0.35	25,700
Br_2	$^1\Sigma_g^+$	323.2	2.2836	0.0809	13,814	465	0.12	19,400
I_2	$^1\Sigma_g^+$	214.6	2.6666	0.0374	11,888	309	0.05	16,700
BeO	$^1\Sigma^+$	1487.3	1.3308	1.6510	9,406	2150	2.38	13,200
C_2	$^1\Sigma_g^+$	1855	1.2425	1.8195	717	2678	2.62	1006
LiH	$^1\Sigma^+$	1405.6	1.5953	7.5131	26,516	2035	10.8	37,300
SO	$^3\Sigma^-$	1123.7	1.4933	0.7089	39,356	1625	1.02	55,200
ZnH	$^2\Sigma^+$	1607.6	1.5945	6.6794	23,270	2320	9.60	32,500

[a] $B \equiv h/8\pi^2 I_e c$.

[b] Energy of first electronic excited state relative to the ground state.

[c] $\Theta_{vib} \equiv \dfrac{hc\tilde{\nu}_e}{k_B}$.

[d] $\Theta_{rot} \equiv \dfrac{h^2}{8\pi^2 I_e k_B}$.

[e] $\Theta_{elec} \equiv \dfrac{\epsilon_1 - \epsilon_0}{k_B}$.

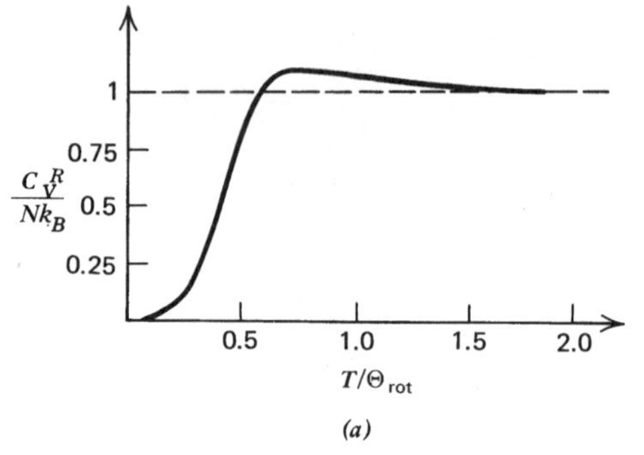

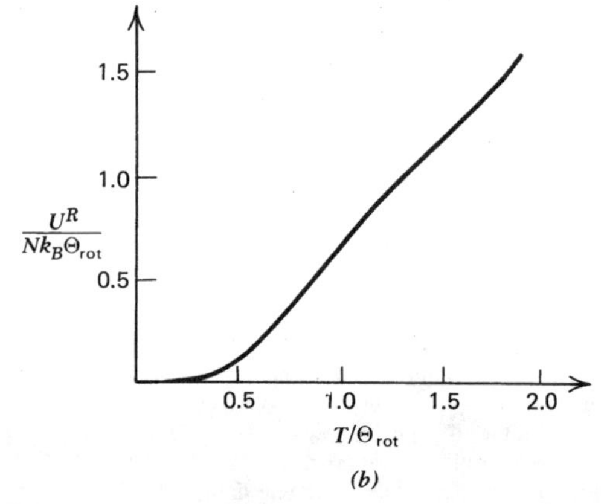

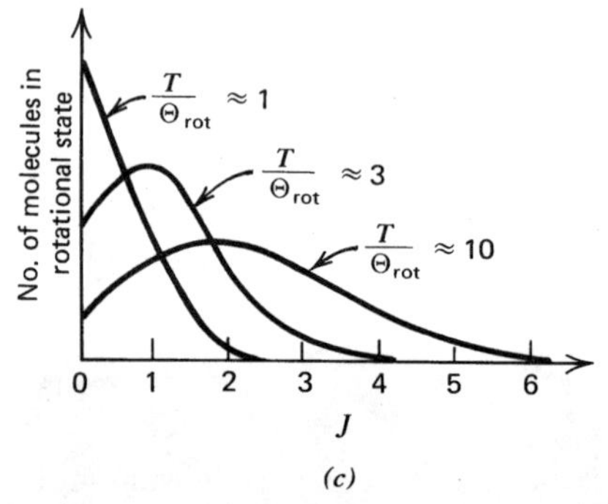

Figure 21.14 (*a*) Rotational motion contribution to the heat capacity at constant volume. (*b*) Rotational motion contribution to the internal energy. (*c*) Molecular rotational population distribution,

$$f_{rot} = \frac{N}{Q_1^R}(2J+1)\exp\left(\frac{-J(J+1)\Theta_{rot}}{T}\right).$$

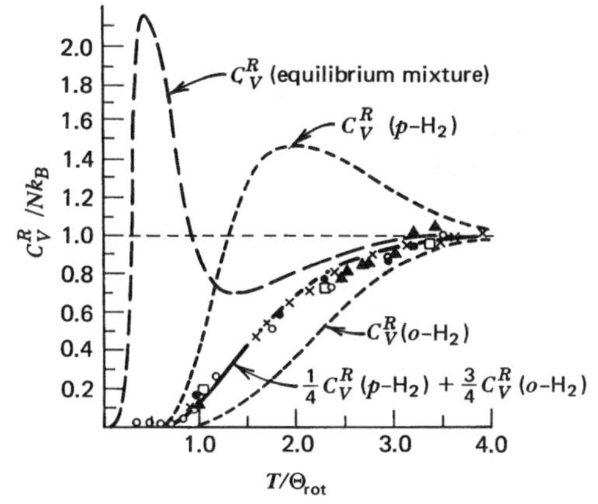

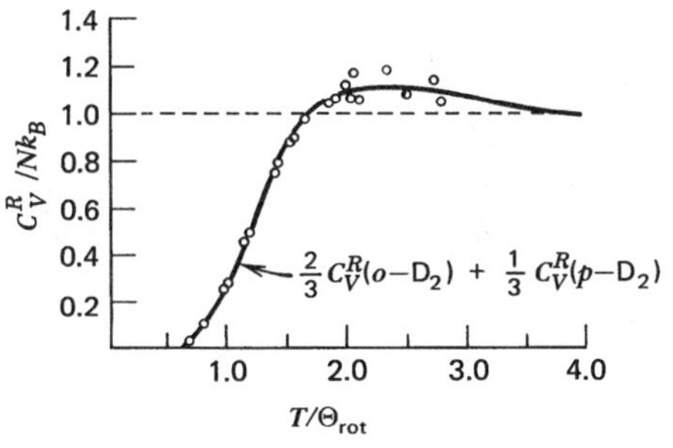

Figure 21.15 Rotational contribution to the constant volume molar heat capacity of H_2 and D_2. For H_2 the species with odd values of J is called ortho, but for D_2 it is the species with even values of J that is called ortho. The measured values of C_V show good agreement with the calculated values.

where $\Theta_{rot} \equiv h^2/8\pi^2 Ik_B$; this equation applies to an equilibrium mixture (e-H_2) of the para and ortho molecules, in proportions given by the two terms displayed, and yields the curve labeled e-H_2 in Fig. 21.15. As explained above, however, this equilibrium does not exist unless catalyzed, and H_2 prepared at room temperature or higher, where the ortho-para ratio reaches a limiting value of 3, will retain this composition (normal or n-H_2) at lower temperatures. The heat capacity of such a mixture is given by

$$C_V^R(n\text{-H}_2) = \tfrac{3}{4}C_V^R(o\text{-H}_2) + \tfrac{1}{4}C_V^R(p\text{-H}_2); \tag{21.137}$$

the ortho and para heat capacities are calculated from the two terms in Eq. 21.136 separately. This result, as shown in Fig. 21.15, does agree with experiment.

Behavior similar to that just described is found for D_2 and T_2. For all other homonuclear diatomic molecules, however, the rotational energy-level spacing is so much smaller than the gas-phase values of k_BT that both sums in the equivalent of Eq. 21.136 can be replaced by integrals. Since the resulting integrals are identical, the differences between ortho and para states vanish under these conditions. The trend toward this high-temperature limit is apparent in Fig. 21.15. Because of the alternation of terms in the sums, the integrals equal one-half the integral in the second line of Eq. 21.114, this being the quantum mechanical explanation of the symmetry factor $\sigma = 2$.

3. *Vibrational heat capacity.* Let

$$\Theta_{vib} \equiv \frac{h\nu}{k_B} = \frac{hc\tilde{\nu}_e}{k_B} \tag{21.138}$$

be called the *vibrational temperature* of the molecule. For essentially all molecules at ordinary temperatures, $\Theta_{vib} >> T$ (see Table 21.8). We then find that

$$\begin{aligned} C_V^V &= Nk_B\left(\frac{h\nu}{k_BT}\right)^2 \frac{\exp(-h\nu/k_BT)}{[1-\exp(-h\nu/k_BT)]^2} \\ &= Nk_B\left(\frac{\Theta_{vib}}{T}\right)^2 \frac{\exp(-\Theta_{vib}/T)}{[1-\exp(-\Theta_{vib}/T)]^2}, \end{aligned} \tag{21.139}$$

and that C_V^V is very small at 298 K for most molecules. It is interesting to examine the behavior of C_V^V at high temperature as another example of the principle of equipartition. As T increases, the heat capacity rises smoothly from zero to

$$\begin{aligned} \lim_{T\to\infty} C_V^V &= \lim_{T\to\infty} Nk_B\left(\frac{\Theta_{vib}}{T}\right)^2 \frac{\exp(-\Theta_{vib}/T)}{[1-\exp(-\Theta_{vib}/T)]^2} \\ &= Nk_B\left(\frac{\Theta_{vib}}{T}\right)^2 \frac{1}{(\Theta_{vib}/T)^2} \\ &= Nk_B. \end{aligned} \tag{21.140}$$

It is interesting that, as shown by a numerical calculation, even when the vibrational heat capacity approaches the value Nk_B per vibration, most of the molecules are still in the $v=0$ and $v=1$ vibrational levels. It is not necessary that the molecules be widely distributed over the available energy levels before the limiting value of C_V^V is achieved. The approach to the limit is shown in Fig. 21.16.

4. *Electronic heat capacity.* Let

$$\Theta_{elec} \equiv \frac{\epsilon_1 - \epsilon_0}{k_B} = \frac{\Delta\epsilon^{elec}}{k_B} \tag{21.141}$$

be the *electronic temperature* of the molecule. As seen from the data in Table 21.8, $\Theta_{elec} >> T$ for almost all molecules. For only a few gases is the electronic heat capacity of any consequence, and even for these only two levels are of importance. From the partition function 21.127, we find

$$C_V^E = Nk_B \frac{\overline{\omega}_0^E\overline{\omega}_1^E \exp(-\Delta\epsilon^{elec}/k_BT)(\Delta\epsilon^{elec}/k_BT)^2}{[(\overline{\omega}_0^E + \overline{\omega}_1^E \exp(-\Delta\epsilon^{elec}/k_BT)]^2}. \tag{21.142}$$

If the next excited electronic state lies at very much higher energy, $\epsilon_2 >> \epsilon_1$, C_V^E ultimately tends to zero as T increases indefinitely (Fig. 21.17). The "few-quantum-states case" is the exception rather than the rule, and even in the example of electronic excitation the limitation to a few states is only apparent—there are other states at high energy, and C_V^E eventually rises again when k_BT becomes comparable to their excitation energies.

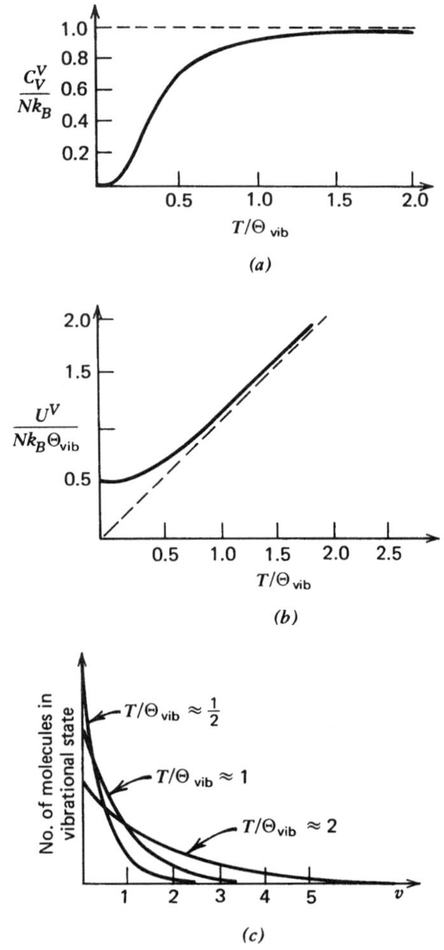

Figure 21.16 (*a*) Vibrational contribution to the constant volume molar heat capacity. (*b*) Vibrational contribution to the internal energy. (*c*) Molecular vibrational population distribution,

$$f_{vib} = \frac{N}{Q_1^V}\exp\left(\frac{-\Theta_{vib}}{T}\right).$$

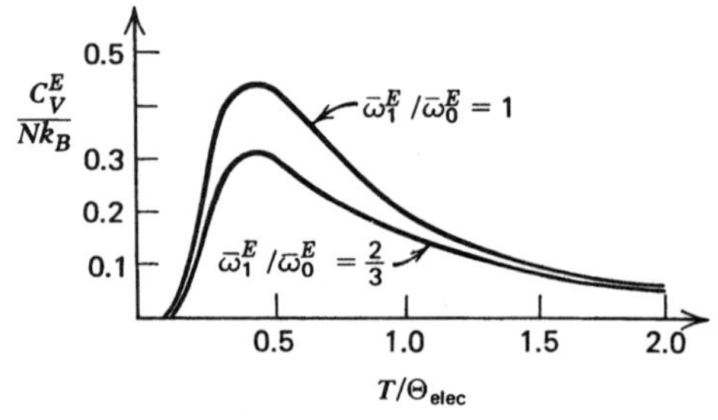

Figure 21.17 Electronic contribution to the molar heat capacity of a gas in the case that only one electronic excited state (labeled 1) is available.

Examination of each of the contributions to the heat capacity we have studied shows that when the relevant energy spacing becomes small compared with the thermal energy k_BT, so that the corresponding characteristic temperature is small relative to T, and when there is an effectively infinite set of quantum states available to the system, each degree of freedom of the system contributes just $Nk_B/2$ to the heat capacity (Fig. 21.18). In the classical theory of statistical mechanics this is known as the *equipartition of energy*. It is found that, in general, for every term in the Hamiltonian of a system that is quadratic (e.g., a kinetic energy term, a harmonic potential energy term kx^2, etc.), the limiting contribution to the heat capacity as $T \to \infty$ is just $Nk_B/2$. When the Hamiltonian contains terms that are not quadratic, the resulting form of C_V is quite different.[15]

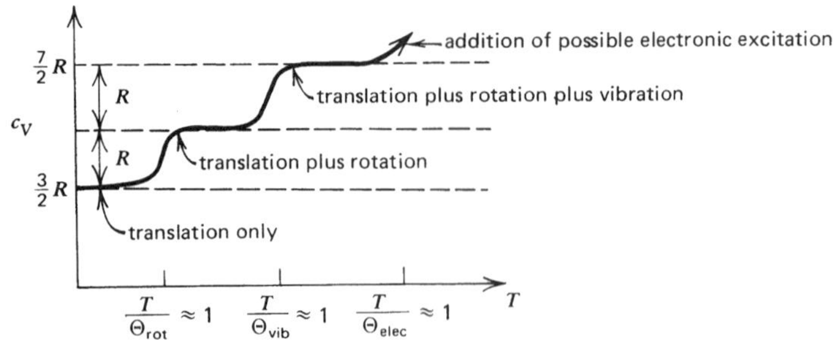

Figure 21.18 Schematic diagram of the variation of c_V with temperature for a perfect gas of diatomic molecules. Note that each increment to c_V becomes important at about its characteristic temperature.

Table 21.9 A Comparison of Computed and Calorimetric Entropies of Some Gases

Substance	S^0_{298} (calc) (J/K mol)	S^0_{298} (obs) (J/K mol)
Ne(*g*)	146.22	146.
Ar(*g*)	154.74	155.
Hcl(*g*)	186.68	186.
HBr(*g*)	198.48	199.
Cl_2(*g*)	222.95	223.1
CO_2(*g*)	213.64	214.
CH_3Br(*g*)	243.	242.

The thermodynamic functions of a perfect gas may be easily computed using the formulas shown. Some comparisons between theory and experiment are given in Table 21.9.

The extension of our analysis to the description of polyatomic molecules is straightforward. Two of the resulting changes are worth noting. First, for a nonlinear molecule a temperature analogous to Θ_{rot} can be defined. When the polyatomic molecule can be treated as a classically excited rigid rotator, we find, using Eq. 21.128, that the rotational contribution to the heat capacity is $\frac{3}{2}Nk_B$, which is larger than for the corresponding linear molecule. Second, for a polyatomic molecule there is a term of the form 21.139 for each normal mode of vibration. Since each normal-mode frequency is different, so is its characteristic temperature.

[15] See, for example, A. Peterlin, *Am. J. Phys.* **28,** 716 (1960).

The temperature dependence of the vibrational contribution to the heat capacity is, then, a superposition of the temperature dependences of the contributions of all the normal vibrations, each with a different value Θ_{vib}/T.

21.6 The Statistical Molecular Theory of the Equilibrium Constant

Given the thermodynamic formulation of the theory of chemical equilibrium in gaseous mixtures, together with the formulas relating the thermodynamic functions of a gas to its molecular properties, it is clear that *a priori* calculations of chemical equilibrium from molecular properties are feasible in principle. If the gas mixture is imperfect, this program also requires a statistical molecular theory of the deviation from perfect gas behavior, for example, a theory of the second and higher virial coefficients. The theory of gas imperfection is discussed in the next section; for the present we consider equilibrium in a perfect gas mixture.

Starting from Eq. 21.115 and using the thermodynamic definition of the chemical potential per molecule,

$$\mu \equiv \left(\frac{\partial A}{\partial N}\right)_{T,V}, \tag{21.143}$$

we easily find that

$$\begin{aligned}\mu &= -k_B T \ln\left(\frac{Q_1^T Q_1^R Q_1^V Q_1^E}{N}\right)\\ &= -k_B T \ln\left(\frac{Q_1^T Q_1^{\text{int}}}{N}\right).\end{aligned} \tag{21.144}$$

In all of the preceding discussion of the computation of the partition function of a molecule we have used as the zero of the energy the lowest accessible quantum state of the molecule. Let the energy of that state, relative to the energy of the molecule when dissociated into atoms that are infinitely far apart, be ϵ_0. Clearly, the constant ϵ_0 differs from one species to another. Now, the partition function weight factor $\exp(-\epsilon_j/k_BT)$ can be evaluated for any choice of the zero of energy. To describe chemical equilibrium between reacting gases we must choose a common zero of energy for all reactant molecules. If we simply write

$$\epsilon_j = (\epsilon_j - \epsilon_0) + \epsilon_0 \tag{21.145}$$

and substitute Eq. 21.145 into the definition of the partition function, we find that

$$\begin{aligned}Q_1 &= \sum_j \varpi_j \exp\left(-\frac{\epsilon_0}{k_BT}\right)\exp\left(-\frac{\epsilon_j-\epsilon_0}{k_BT}\right)\\ &= \exp\left(-\frac{\epsilon_0}{k_BT}\right)\sum_j \varpi_j \exp\left(-\frac{\epsilon_j-\epsilon_0}{k_BT}\right).\end{aligned} \tag{21.146}$$

Since ϵ_0 is measured relative to the dissociation limit, the partition functions for all molecules can be referred to a common zero of energy simply by using the lowest accessible quantum state as the internal zero for each class of motion, and then multiplying by $\exp(-\epsilon_0/k_BT)$. For most molecules the only contributions to ϵ_0 are from zero-point vibrational motion and the total bond energy. There are a few exceptions to this generalization. For example, if o-H_2 is involved in a reaction at temperatures comparable to Θ_{rot}, we must note that its lowest rotational state has $J = 1$, whereas the lowest rotational state of most other molecules may be taken as $J = 0$ with the symmetry number σ used to correct the counting of states.

All of the partition functions studied in Section 21.5 were calculated using the lowest quantum state as reference for the energy. Therefore, we need merely multiply the product $Q_1^T Q_1^R Q_1^V Q_1^E$ by $\exp(-\epsilon_0/k_BT)$, and return to Eq. 21.144, to find that

$$\mu - \epsilon_0 = -k_BT\ln\left(\frac{Q_1^T Q_1^{\text{int}}}{N}\right), \tag{21.147}$$

when calculated for each gas, will refer to a common zero of energy.

We now convert Eq. 21.147 to a form resembling the thermodynamic representation of the chemical potential. Let

$$\begin{aligned}Q_1^T &= \left(\frac{2\pi m k_B T}{h^2}\right)^{3/2} V = q_1^T V,\\ q_1^T &\equiv \left(\frac{2\pi m k_B T}{h^2}\right)^{3/2} \equiv \Lambda^{-3},\end{aligned} \tag{21.148}$$

and use Eq. 21.118 to replace V/N by k_BT/p. Then, multiplying the argument of the logarithm by unity in the form $\mathfrak{p}/\mathfrak{p}$, where $\mathfrak{p}$ is the unit of pressure, and separating the terms of the argument, we find that

$$\begin{aligned}\mu &= \epsilon_0 - k_BT\ln\left(q_1^T Q_1^{\text{int}}\frac{k_BT}{\mathfrak{p}}\right) + k_BT\ln\frac{p}{\mathfrak{p}}\\ &= \mu^*(T,\mathfrak{p}) + k_BT\ln\frac{p}{\mathfrak{p}},\end{aligned} \tag{21.149}$$

where, of course,

$$\mu^*(T,\mathfrak{p}) = \epsilon_0 - k_BT\ln\left(q_1^T Q_1^{\text{int}}\frac{k_BT}{\mathfrak{p}}\right). \tag{21.150}$$

The notation used here is the same as in the thermodynamic theory; μ^* is the chemical potential (Gibbs free energy per molecule) in a perfect gas state at unit pressure (1 atm). Using the condition defining chemical equilibrium, Eq. 21.63, we find that the equilibrium constant is

$$K(T) = \prod_{i=1}^{r}\left(q_{1i}^T Q_{1i}^{\text{int}}\frac{k_BT}{\mathfrak{p}}\right)^{\nu_i}\exp\left(-\frac{\Delta E_0^0}{k_BT}\right), \tag{21.151}$$

where

$$\Delta E_0^0 \equiv \sum_{i=1}^{r} \nu_i \epsilon_{0i} \qquad (21.152)$$

is the net change in energy in the conversion of reactants to products, assuming the components to be perfect gases at $T = 0$ K. Of course, when $T = 0$ we also have $\Delta E_0^0 = \Delta H_0^0$, since for a perfect gas H and U differ by just Nk_BT. The energy change ΔE_0^0 may easily be related to the dissociation energies of the reacting molecules. We imagine the reaction to occur in two steps:

1. Let all the reactant molecules be dissociated into isolated atoms.
2. Let the atoms be reassembled into the product molecules.

If D_{0i} is the energy required to dissociate into atoms a molecule of the ith species in its lowest quantum state, the sum of the energy changes in steps 1 and 2 is

$$\Delta E_0^0 = -\sum_{i=1}^{r} \nu_i D_{0i}. \qquad (21.153)$$

Note again that D_{0i} excludes the zero-point energy of the molecule, and thus does not equal D_{ei}, the dissociation energy measured from the minimum in the potential energy (see Section 7.1).

The form of $\mu^*(T, \mathfrak{p})$ in Eq. 21.150 permits easy conversion from one set of units to another. It is written there for the usual choice of standard state, namely, the perfect gas at 1 atm. Then the factor V/N in Eq. 21.147 (the V from Q_1^T, the N already in the denominator) is just $k_BT/\mathfrak{p}$, as shown in Eq. 21.150. If we choose a standard state with gas concentration of 1 molecule/cm^3, the factor V/N becomes 1 cm^3; if, instead, we choose a standard state with gas concentration of 1 mol/liter, the factor V/N becomes (1000 cm$^3/N_A$). The different standard states are, of course, related to equilibrium constants derived from the standard form $K(T)$. For example, K_V, defined in terms of the concentrations in moles per liter of all the reacting species, is

$$K_V^{\text{perfect gas}} = \exp\left(-\frac{\Delta E_0^0}{k_BT}\right)\prod_{i=1}^{r}(q_{1i}^T Q_{1i}^{\text{int}}/N_A)^{\nu_i}. \qquad (21.154)$$

Other forms of the equilibrium constant may be easily worked out.

Consider, as an example, the reaction

$$\text{MX}(g) = \text{M}^+(g) + \text{X}^-(g). \qquad (21.155)$$

The equilibrium constant for this reaction is, from Eq. 21.151,

$$K(T) = \exp\left(-\frac{\Delta E_0^0}{k_BT}\right) \times \frac{(2\pi m_{\text{M}^+}k_BT/h^2)^{3/2}(k_BT/\mathfrak{p})Q_{\text{M}^+}^{\text{int}}\,(2\pi m_{\text{X}}k_BT/h^2)^{3/2}(k_BT/\mathfrak{p})Q_{\text{X}^-}^{\text{int}}}{(2\pi m_{\text{MX}}k_BT/h^2)^{3/2}(k_BT/\mathfrak{p})Q_{\text{MX}}^{\text{int}}} \qquad (21.156)$$

Three examples of alkali halides illustrate how the various factors of Eq. 21.156 govern the position of equilibrium. Consider LiF, NaCl, and CsI at 3000 K and at 4000 K. The necessary atomic and molecular constants are collected in Table 21.10. The equilibrium constant we shall compute is that for MX ⇌ $\text{M}^+ + \text{X}^-$, so the energy ΔE_0^0 of Eq. 21.153 will represent the dissociation of MX to ions. The zero of energy, the reference state for the dissociation energy of MX, is the ground state of the free atoms, so the partition function for the alkali and halide ions must contain Boltzmann factors to reflect the work done to convert M^0 to $\text{M}^+ + e^-$, and the work released when $\text{X}^0 + e^-$ goes to X^-. The ions are all in nondegenerate ^1S states, and the electronically excited states of both the ions and the molecules can be neglected because they contribute factors of unity (within the uncertainties of the molecular parameters) to the partition functions. Hence, $Q_{\text{M}^+}^{\text{int}}$ and $Q_{\text{X}^-}^{\text{int}}$ are simple exponentials. Assume that the interactions between ions in the gas can be neglected at the temperatures of interest.

Let us take the molecules to be rigid rotors and harmonic oscillators, so we can use Eqs. 21.114 and 21.113 for the rotational and vibrational partition functions, respectively. We must note whether the dissociation energy is D_0, the difference in energy of the vibrationless ground states of the

Table 21.10 Data to Compute Equilibrium Constants for LiF, NaCl, and CsI

	Alkali Atom	Ionization Potential (eV)	
	Li	5.390	
	Na	5.138	
	Cs	3.893	
Halogen Atom	**Electron Affinity (eV)**	**Spin–Orbit Splitting of X^0**	
F	3.400	404 cm^{-1} equiv. to 581 K	
Cl	3.613	881 cm^{-1} equiv. to 1207 K	
I	3.081	7603 cm^{-1} equiv. to 10,936 K	
Molecule	**Moment of Inertia (amu Å^2)**	**Vibration Frequency (cm^{-1})**	**Dissociation Energy D_e (eV)**
LiF	11.66	910.2	5.96
NaCl	77.33	364.6	4.23
CsI	713.7	119.2	3.55

molecule and the free atoms at rest, or D_e, the difference in energy between the bottom of the adiabatic potential energy curve and the dissociation limit; D_0 is smaller than D_e by the zero-point energy, $h\nu/2$, of the harmonic oscillator. The values in Table 21.10 are D_e's, not D_0's, so we must multiply Q_1^V (Eq. 21.113) by $\exp(-h\nu/2k_BT)$ or, equivalently, decrease the dissociation energies by $h\nu/2$. Table 21.11 shows the electronic factors

$$\exp(-\Delta E_0^0 / k_B T) = \exp\{[D_e(\mathrm{MX}) + IP(\mathrm{M}) - EA(\mathrm{X})] / k_B T\},$$

the translational factors $[2\pi m_{\mathrm{M}^+} m_{\mathrm{X}^-} k_B T/(m_{\mathrm{M}^+} + m_{\mathrm{X}^-})h^2]^{3/2}$ $k_BT/\mathfrak{p}$, which we abbreviate $\bar{Q}_1^T$, the rotational factors Q_1^R, and the vibrational factors Q_1^V, as well as the equilibrium constants. Note that 1 atm is equivalent to 1.01325×10^5 N/m^2 or 1.01325×10^6 dyn/cm^2, which is the value we take for our reference pressure $\mathfrak{p}$.

In Table 21.11, note first the magnitudes of the four factors. The electronic term is very small and very sensitive to temperature because ΔE_0^0 is so large. The translational factors taken together as $\bar{Q}_1^T$ give a large and only moderately temperature-sensitive, mass-sensitive number. The rotational partition function, five orders of magnitude smaller, is less sensitive to masses and temperature, and the vibrational partition function is of order unity and the least sensitive term of all. One can look upon the magnitude of $\bar{Q}_1^T$ as the quantitative expression of how it is the increase of entropy that allows most thermal dissociation processes to occur at temperatures far below D_e/k_B.

Table 21.11 Typical Equilibrium Constants $K(T) = \exp(-\Delta E_0^0/k_BT)\bar{Q}_1^T/Q_1^R Q_1^V$ for Three Alkali Halides

Molecule		3000 K	4000 K
LiF	$\exp(-\Delta E_0^0/k_BT)$	4.41×10^{-14}	9.82×10^{-11}
	$\bar{Q}_1^T$	1.45×10^8	2.97×10^8
	Q_1^R	1.47×10^3	1.95×10^3
	Q_1^V	2.272	3.04
	$K(T)$	1.915×10^{-9}	4.82×10^{-6}
NaCl	$\exp(-\Delta E_0^0/k_BT)$	2.15×10^{-10}	5.60×10^{-8}
	$\bar{Q}_1^T$	6.58×10^8	1.35×10^9
	Q_1^R	9.73×10^3	1.297×10^4
	Q_1^V	5.71	7.62
	$K(T)$	2.54×10^{-6}	7.65×10^{-4}
CsI	$\exp(-\Delta E_0^0/k_BT$	4.35×10^{-8}	3.01×10^{-6}
	$\bar{Q}_1^T$	6.605×10^9	1.356×10^{10}
	Q_1^R	8.97×10^4	1.196×10^5
	Q_1^V	17.5	23.3
	$K(T)$	1.83×10^{-4}	1.465×10^{-2}

21.7 The Statistical Molecular Theory of the Real Gas

In this section we turn to an examination of the influence of intermolecular forces on the properties of a gas. It has already been made clear that the qualitative form of the intermolecular potential is such as to involve repulsion between molecules at small pair separations and attraction between molecules at large pair separations. Corresponding to this interaction between the molecules, the energy spectrum of the system is changed from that characteristic of the perfect gas. In many cases the vibrations and rotations of the molecules are little altered by the presence or absence of neighboring molecules and can, then, still be separated from the translational degrees of freedom. Then that part of the energy of the system associated with internal molecular motion is just the sum of the energies for the separate internal motions. There are also phenomena, however, in which the coupling between the internal motion of the molecule and the intermolecular potential is very important, e.g., rate of loss of vibrational energy from a molecule. One must examine each case carefully to determine if the internal degrees of freedom are indeed independent of the intermolecular potential, but we shall assume here that no such coupling exists.

We consider the case that the spacing between the translational energy levels of the gas is very small relative to k_BT. As mentioned previously, this condition is satisfied for all gases. When k_BT is large relative to the energy spacing in a system, and the de Broglie wavelength, $h/(2\pi mk_BT)^{1/2}$, is very small, classical mechanics provides an accurate description of the molecular motion. Given a monatomic gas containing N molecules in the volume V, the classical energy is

$$E_N = \sum_{i=1}^{N} \frac{p_i^2}{2m} + \mathcal{V}_N, \tag{21.157}$$

where $\mathcal{V}_N$ is the intermolecular potential acting between the N molecules in the gas. Now, $\mathcal{V}_N$ is often thought of as arising from the sum of the interactions of all possible pairs of molecules. If this is correct, the implication is that the potential energy of, say, three molecules is exactly equal to the potential energies of the three pairs of molecules into which the triplet can be broken; that is,

$$u_3(1,2,3) = u_2(1,2) + u_2(1,3) + u_2(2,3), \tag{21.158}$$

where u_i is the interaction energy between i molecules. The best available data indicate that Eq. 21.158 is not completely accurate, and that there is an extra interaction between three molecules over and above that given by Eq. 21.158. Thus, we write

$$u_3(1,2,3) = u_2(1,2) + u_2(1,3) + u_2(2,3) + \Delta_3(1,2,3). \tag{21.159}$$

Presumably, corresponding to larger groups of molecules there exist other many-molecule contributions to the energy, but almost nothing is known about these at the present time. For our purposes it is not necessary to imagine $\mathcal{V}_N$ as arising from pair interactions, although this picture is so useful that the reader is urged to use it, bearing in mind the limitations mentioned.

The general theory of imperfect gases is complicated even though the physical ideas on which it is based are simple. If we restrict attention to pressures sufficiently small that only the second-virial-coefficient correction to the perfect gas law is important, simple arguments analogous to those employed in Chapter 12 may be used to calculate B. We shall study these simple arguments in this section; a much more general analysis, using the grand partition function, is given in Appendix 21C.

In Chapter 12 we computed the pressure exerted by a gas of N monatomic particles by examining the rate of momentum transferred across an arbitrary plane anywhere inside the gas. We could have computed the pressure from the change of momentum when molecules are reflected from a plane connected to a force-measuring device. The results of the two computations are the same. The reason is, of course, that if molecules that move across the reference plane in the positive direction (say, to the right) transfer positive normal momentum, then molecules that move across the reference plane in the negative direction transfer negative normal momentum. Hence the effect of those molecules moving in the negative direction on the momentum of the gas lying on the positive side of the reference plane is the same as if they had crossed in the positive direction carrying an equal amount of positive momentum. The net effect, then, is to double the rate of transfer of momentum by the molecules moving in the positive direction, just as if a reflecting wall had been present. We thus find that the pressure and mean square velocity are related by $p = \frac{1}{3}nm\langle v^2\rangle$ (see Eqs. 12.17 and 12.10).[16]

We now assume that the distribution of molecular velocities is unaffected by the intermolecular forces. Provided that $\mathcal{V}_N$ depends only on the molecular positions this is correct, as is verified in Appendix 21C. Then in a dilute gas the rate of transfer of momentum by passage of molecules across a reference plane is the same whether or not there are intermolecular forces. This contribution to the pressure, called the kinetic contribution, is just $p_K = \frac{1}{3}nm\langle v^2\rangle$. Using the Maxwell velocity distribution to evaluate $\langle v^2\rangle$, we find, as expected, $p_K = Nk_BT/V$, that is, the perfect gas value.

In addition to the convective flow of momentum just described, the effect of the intermolecular force between a pair of molecules lying on opposite sides of the reference plane is to transfer momentum across the plane, irrespective of whether or not the molecules actually cross the plane. There is, then, a contribution to the pressure from the intermolecular forces. Suppose that a molecule lies on one side of the reference plane. We choose our coordinates so that the origin lies at the center of mass of this molecule. The probability of finding a particular molecule in the volume element dV a distance R from the molecule at the origin would be dV/V if there were no molecular interaction. However, because of the molecular interaction, it is

$$\frac{dV}{V}\exp\left(-\frac{u_2(R)}{k_BT}\right).$$

The probability of finding any molecule (not a selected molecule) in dV a distance R away is

$$\frac{N\,dV}{V}\exp\left(-\frac{u_2(R)}{k_BT}\right),$$

since there are N molecules in the volume V and 1 is negligibly small relative to N.

As shown in Fig. 21.19, a molecule in dV will exert a force on the molecule at the origin with a component $-(du_2/dR)\cos\theta$ normal to the reference plane, where θ is the angle between the line joining the centers of the two molecules and the normal to the reference plane. On the average there are $(N/V)\,dz$ molecules lying at a distance between z and $z + dz$ from a given unit area of the reference plane.[17] The total normal component of the force exerted by the $n\,dz$ molecules between z and $z + dz$ on all molecules lying on the other side of the reference plane is

$$-n^2\,dz\int\frac{du_2}{dR}\cos\,\theta\exp\left(-\frac{u_2}{k_BT}\right)dV,$$

integrated over the whole space beyond the reference plane. Thus, the contribution to the pressure from intermolecular forces between all pairs of molecules is

$$\begin{aligned} p_{\text{int}} &= -2\pi n^2\int_0^\infty dz\int_z^\infty\frac{du_2}{dR}\exp\left(-\frac{u_2}{k_BT}\right)R^2\,dR \\ &\quad\times\int_0^{\cos^{-1}z/R}\cos\,\theta\sin\,\theta\,d\theta \\ &= -2\pi n^2\int_0^\infty R^2\frac{du_2}{dR}\exp\left(-\frac{u_2}{k_BT}\right)dR \\ &\quad\times\int_0^{\pi/2}\cos\,\theta\sin\,\theta\,d\theta\int_0^{R\cos\theta}dz, \end{aligned} \qquad (21.160)$$

[16] Here n is the number density of molecules.

[17] This statement is valid for the purpose of calculating the second virial coefficient, but in general is not valid. The situation is the same as in the calculation of the first-order perturbation correction to the energy of a state; the correction to the energy is obtained by calculating the expectation value of the difference between the potential energy of the real system and the unperturbed system with respect to the unperturbed wave function. In the case of the gas, the analogue of the unperturbed wave function is $n^2\exp(-u_2/k_BT)$, i.e., the perfect gas distribution of a pair of molecules which interact with the potential energy function u_2; to compute the first order perturbation correction to the pressure the force per unit area generated by the pair potential, $-du_2/dR$, is averaged over the unperturbed distribution of molecules.

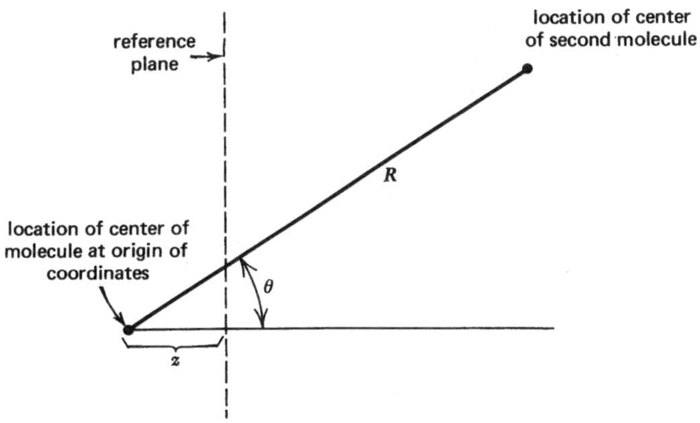

Figure 21.19 Schematic representation of the distances and angle referred to in the derivation of Eq. 21.160.

or

$$p_{\text{int}} = -\frac{2\pi n^2}{3}\int_0^\infty R^3 \frac{du_2}{dR}\exp\left(-\frac{u_2}{k_BT}\right) dR. \qquad (21.161)$$

By requiring that

$$\lim_{R\to\infty} R^3 \frac{du_2}{dR} = 0, \qquad (21.162)$$

and integrating Eq. 21.161 by parts, we find that

$$p_{\text{int}} = 2\pi n^2 k_BT\int_0^\infty R^2\left[1-\exp\left(-\frac{u_2}{k_BT}\right)\right] dR. \qquad (21.163)$$

We now note that p_{int} is proportional to n^2 and, therefore, can be related to the second virial coefficient. In fact, since $p_{\text{int}} = p - Nk_BT/V$, when only the second-virial-coefficient term in the equation of state is significant, $p_{\text{int}} = N^2Bk_BT/N_AV^2$ and we find the simple formulas

$$B = 2\pi N_A\int_0^\infty [1-\exp(-u_2/k_BT)]R^2\, dR \qquad (21.164)$$

and

$$B = -\frac{2\pi N_A}{3k_BT}\int_0^\infty R^3\frac{du_2}{dR}\exp\left(-\frac{u_2}{k_BT}\right) dR, \qquad (21.165)$$

which are related by an integration by parts and the use of Eq. 21.162 (N_A is Avogadro's number). Equation 21.164 gives the more usual representation of B; Eq. 21.165 is often useful for interpretative purposes, since the intermolecular force appears in the integrand and no difference need be taken.

From the above expressions for B it is clear that repulsive interaction between a pair of molecules will result in a positive contribution to B, and attractive interaction in a negative contribution. Thus, if B is found to be positive, we may deduce that the interaction between molecules is predominantly repulsive, and vice versa. This justifies the conclusion we reached by qualitative arguments in Section 21.1.

Since B is determined by the interaction between a pair of molecules, it is easy to generalize Eq. 21.164 or 21.165 to describe the several second virial coefficients of a mixture; we need only replace u_2 by the appropriate potential. The general properties of B_{ij} are in no way different from those of B_{ii}, so the remarks that follow are relevant to the interpretation of the behavior of both pure gases and gas mixtures.

Theoretical calculations of B, based on Eq. 21.164 or 21.165, have been made for many assumed forms of the pair potential u_2. Some insight into the behavior of B as a function of u_2 can be obtained from the following simple models.

1. *The hard-sphere potential.* Suppose that

$$u_2(R) = +\infty \qquad (R \le \sigma),$$
$$u_2(R) = 0 \qquad (R > \sigma). \qquad (21.166)$$

A representation of this potential is presented in Fig. 21.20*a*. It can only be considered an idealized model because no real repulsive force can be infinitely large. However, the hard-sphere potential is useful in illustrating the influence on the equation of state of the volume occupied by the molecules. We note that

$$\exp\left(-\frac{u_2}{k_BT}\right) - 1 = 0 \qquad (R > \sigma),$$
$$\exp\left(-\frac{u_2}{k_BT}\right) - 1 = -1 \qquad (R \le \sigma), \qquad (21.167)$$

for the potential given by Eq. 21.166. Substituting Eq. 21.167 into Eq. 21.164, we find that

$$B = 2\pi N_A\int_0^\sigma R^2\, dR$$
$$= \frac{2\pi N_A\sigma^3}{3}. \qquad (21.168)$$

Since the volume of one hard-sphere molecule is $\pi\sigma^3/6$, B is just four times the volume of one hard sphere. This is just one-half the volume excluded to the center of mass of one hard sphere of diameter σ by another of diameter σ (see Fig. 21.20*b*).

2. *Point centers of repulsion.* Suppose that

$$u_2(R) = \frac{d}{R^\delta} \qquad (\delta > 3). \qquad (21.169)$$

A representation of this potential is presented in Fig. 21.20*c*. The evaluation of Eq. 21.165 with Eq. 21.169 involves nonelementary functions. It may be shown that

$$B = \frac{2\pi N_A}{3}\left(\frac{d}{k_BT}\right)^{3/\delta}\Gamma\left(\frac{\delta-3}{\delta}\right), \qquad (21.170)$$

where $\Gamma(x)$ is called the *gamma function.* Tables of $\Gamma(x)$ are available; for our purposes all that matters is that $\Gamma[(\delta-3)/\delta]$ is a positive number and is not a function of the temperature.

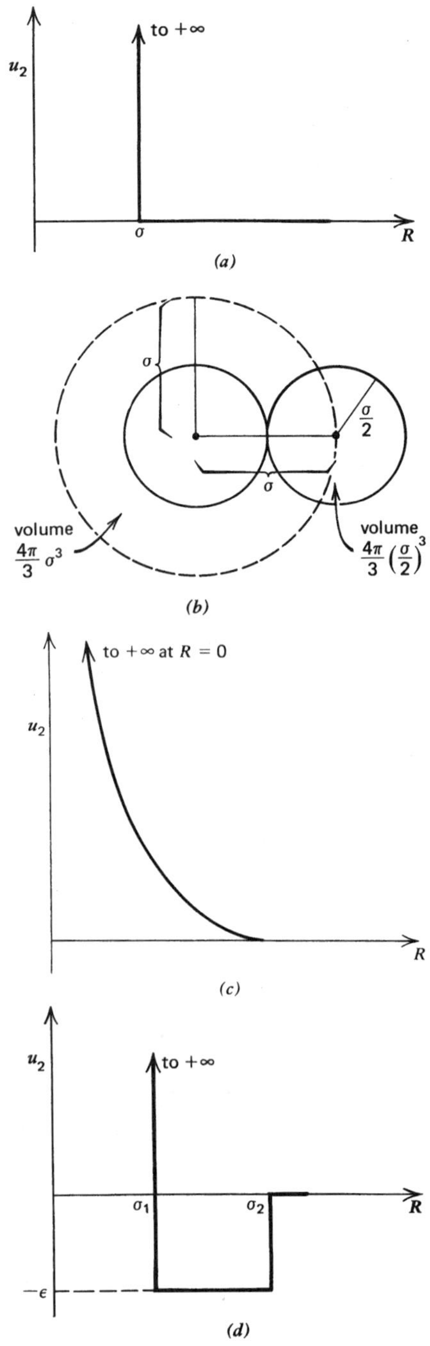

Figure 21.20 (*a*) The potential energy of interaction between two hard-sphere molecules of diameter σ. (*b*) The volume excluded to the center of one hard sphere by another of equal diameter σ. Note that the excluded volume is eight times the molecular volume. (*c*) The potential energy of interaction between two point centers of repulsion. (*d*) The potential energy between two molecules with a square-well interaction.

Equation 21.170 shows that when the intermolecular force is "softer" than that between two hard spheres, the effective excluded volume becomes temperature-dependent. This is the expected result, since as T increases so does the kinetic energy of a molecule, and hence its ability to penetrate closer to the center of mass of the other molecule with which it interacts. Note that for this potential $B > 0$ everywhere and $(dB/dT) < 0$.

3. *The square-well potential.* Some idea of the relative importance of the attractive and repulsive parts of the intermolecular potential can be obtained from examining the potential

$$\begin{aligned} u_2(R) &= +\infty && (R \le \sigma_1), \\ u_2(R) &= -\epsilon && (\sigma_1 < R \le \sigma_2), \\ u_2(R) &= 0 && (\sigma_2 < R), \end{aligned} \tag{21.171}$$

which is shown in Fig. 21.20*d*. From Eq. 21.164 we obtain

$$\begin{aligned} B = 2\pi N_A \Bigg\{ &\int_0^{\sigma_1} \left[1 - \exp\left(-\frac{u_2}{k_B T}\right)\right] R^2\, dR \\ &+ \int_{\sigma_1}^{\sigma_2} \left[1 - \exp\left(-\frac{u_2}{k_B T}\right)\right] R^2\, dR \\ &+ \int_{\sigma_2}^{\infty} \left[1 - \exp\left(-\frac{u_2}{k_B T}\right)\right] R^2\, dR \Bigg\} \\ = \frac{2\pi N_A \sigma_1^3}{3} &+ 2\pi N_A \frac{[1 - \exp(\epsilon/k_B T)](\sigma_2^3 - \sigma_1^3)}{3}. \end{aligned} \tag{21.172}$$

When $k_B T >> \epsilon$, the second term in Eq. 21.172 tends to zero, and B approaches the positive value $2\pi N_A \sigma_1^3/3$, which is one-half the volume excluded to the center of one sphere by another sphere of diameter σ_1 (see Fig. 21.20*b*), the same value as for hard spheres of diameter σ_1. On the other hand, when $\epsilon >> k_B T$, B approaches $-(2\pi N_A/3)(\sigma_2^3 - \sigma_1^3) \times \exp(\epsilon/k_B T)$, which is negative. In a restricted range of temperature, where it is adequate to write $1 + \epsilon/k_B T$ for $\exp(\epsilon/k_B T)$, B simulates in form the second virial coefficient characteristic of a van der Waals gas (see Table 21.1). Note, however, that even for this very simple potential the van der Waals form is not an accurate representation of B over the entire temperature range.

4. *The Lennard–Jones potential.* The potential

$$u_2 = 4\epsilon\left[\left(\frac{\sigma}{R}\right)^{12} - \left(\frac{\sigma}{R}\right)^{6}\right] \tag{21.173}$$

has been used extensively in the theoretical interpretation of the properties of gases. The term proportional to R^{-6} represents the London dispersion interaction between closed-shell atoms and/or molecules (see Chapter 10). The term proportional to R^{-12} is an approximation to the strong short-range repulsion between molecules. As shown in Fig. 21.6, which is in fact a Lennard–Jones potential curve, ϵ measures the depth of the potential well and σ is that point for which $u_2 = 0$.

The calculation of B using Eq. 21.173 for the potential can only be carried out numerically. Some results are listed in Table 21.12 as a function of the variable $T^* = k_B T/\epsilon$. Just as for the square-well potential, B is negative at low temperature and positive at high temperature. Furthermore, just as in the case of the point centers of repulsion, the effective excluded volume of Lennard–Jones molecules decreases with T when repulsive forces are dominant (see Table 21.12 for $T^* > 30$).

Table 21.12 The Second Virial Coefficient Calculated from the Lennard–Jones Potential[a] $T^* \equiv (k_BT/\epsilon)$; $b_0 \equiv (2\pi/3)N_A\sigma^3$; $B(T) \equiv b_0B^*(T^*)$

T^*	B^*	T^*	B^*
0.30	–27.8806	2.00	–0.6276
0.35	–18.7549	2.50	–0.3126
0.40	–13.7988	3.00	–0.1152
0.45	–10.7550	3.50	+0.0190
0.50	–8.7202	4.00	+0.1154
0.55	–7.2741	4.50	+0.1876
0.60	–6.1980	5.00	+0.2433
0.65	–5.3682	10.00	+0.4609
0.70	–4.7100	20.00	+0.5254
0.80	–3.7342	30.00	+0.5269
0.90	–3.0471	50.00	+0.5084
1.00	–2.5381	100.00	+0.4641
1.20	–1.8359	200.00	+0.4114
1.40	–1.3758	300.00	+0.3801
1.60	–1.0519	400.00	+0.3584
1.80	–0.8120		

[a] Adapted from J. O. Hirschfelder, C. F. Curtiss, and R. B. Bird, *Molecular Theory of Gases and Liquids* (Wiley, New York, 1954).

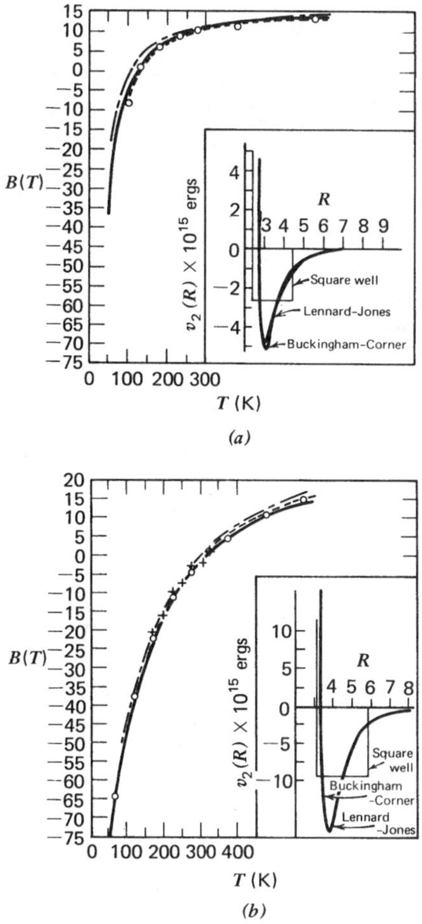

Figure 21.21 (*a*) Second virial coefficients for neon calculated for several molecular models. The potential functions obtained from the experimental $B(T)$ data are also shown. The experimental data are those of L. Holborn and J. Otto, *Z. Physik* **33,** 1 (1925). (*b*) Second virial coefficients for argon calculated for several molecular models. The potential functions obtained from the experimental $B(T)$ data are also shown. The experimental data are those of L. Holborn and J. Otto, *Z. Physik* **33,** 1 (1925), and A. Michels, H. Wijker, and H. Wijker, *Physica,* **15,** 627 (1949). From J. O. Hirschfelder, C. F. Curtiss and R. B. Bird, *Molecular Theory of Gases and Liquids* (Wiley, New York, 1954).

The most direct method for determining the interaction potential between a pair of molecules is to study the scattering of one molecule by another; this subject will be discussed in detail in Chapter 27. The experiments are difficult but are now carried out in many laboratories. For many years, prior to the development and maturation of the molecular beam scattering method, what little we learned about intermolecular forces came from studies of the second virial coefficients of gases and from studies of other gas properties. Although these older methods for inferring the form of the intermolecular potential have been superceded by the molecular beam scattering method, it is still worthwhile to examine what information the former are capable of yielding. The key limitations to the use of, say, the second virial coefficient to infer the intermolecular potential are the lack of a unique inversion procedure and, associated with the lack of that inversion procedure, that B is not very sensitive to the form of the intermolecular potential. In Fig. 21.21 are plotted experimental data and three theoretical curves: B as calculated from the Lennard–Jones potential, from the square-well potential, and from

$$u_2(R) = b\exp\left[-\alpha\left(\frac{R}{R_m}\right)\right] - \left(\frac{C}{R^6} + \frac{C'}{R^8}\right)\exp\left[-4\left(\frac{R_m}{R} - 1\right)^3\right] \qquad (R \le R_m),$$

$$u_2(R) = b\exp\left[-\alpha\left(\frac{R}{R_m}\right)\right] - \left(\frac{C}{R^6} + \frac{C'}{R^8}\right) \qquad (R > R_m),$$

where

$$\begin{aligned} b &= [-\epsilon + (1+\beta)CR_m^{-6}]\exp(\alpha), \\ C &= \epsilon\alpha R_m^6[\alpha(1+\beta) - 6 - 8\beta]^{-1}, \\ C' &= \beta R_m^2 C, \end{aligned} \qquad (21.174)$$

known as the Buckingham–Corner potential. As can be seen, all fit the data reasonably well, so that deducing the form of $u_2(R)$ from measurements of B requires measurements of very high precision over an extended range of temperature. The following general conclusions can be drawn from experiment and theory.

1. In order to reproduce the observed temperature dependence of B, $u_2(R)$ must have a strongly repulsive region (be positive) at small pair separations, and have an attractive region (be negative) at larger pair separations. This form of the intermolecular potential is that expected from fundamental quantum mechanical arguments (Chapter 10), as well as from physical arguments based on the existence and small compressibility of solids and liquids.

2. Because B is negative at low temperatures and positive at high temperatures, B must be zero at some intermediate temperature. The temperature at which $B = 0$ is called the Boyle temperature, denoted T_B. Near T_B a gas behaves nearly like a perfect gas over a limited range of pressure.

3. The intermolecular potential must have a repulsive portion that is "softer" than a hard core; the precise form of the repulsion cannot be fixed from studies of B alone.

4. A comparison between an empirical potential function that fits the molecular beam scattering data and the vibrational spectrum of the Argon dimer (stable at very low temperature) and the best fit Lennard–Jones potential for Ar is shown in Fig. 21.22. These potential functions are very similar but not identical. Note that the well depth of the potential function that fits the scattering data is greater than that for the best fit Lennard–Jones potential and that the curvatures of the attractive components of these potentials are slightly different.

5. Studies of the third virial coefficient establish that the potential energy is not precisely the sum of pair interaction energies, that is, that Δ_3 in Eq. 21.159 is nonzero. This conclusion is supported by a variety of other studies. Aside from the fact that $\Delta_3 \neq 0$, little is known about the functional form. A first order perturbation theory calculation, by Axilrod and Teller, leads to

$$\Delta_3(R_{ij}, R_{jk}, R_{ik}) = \frac{9}{16}\frac{I\alpha^3(1+3\cos\theta_i\cos\theta_j\cos\theta_k)}{(R_{ij}R_{jk}R_{ik})^3}, \tag{21.175}$$

where I is the ionization potential, α is the polarizability, and θ_i, θ_j, and θ_k are the three angles of a triangle whose sides are R_{ij}, R_{jk}, and R_{ki}. Second order and third order corrections to additivity of pair potentials have been reported but are difficult to use. A comparison of the predicted and measured third virial coefficients of Ar is shown in Fig. 21.23. The best available estimates of the correction to additivity of pair potentials come from full, all-electron, quantum theory calculations. These corrections are very important in systems in which hydrogen bonding is a large component of the interaction between molecules.

6. The intermolecular potential between unlike pairs of molecules, even when fitted to the same functional form, does not exactly satisfy any simple additivity rules. For example, if 1 and 2 are the two components of a mixture, it is found that

$$\begin{aligned} \epsilon_{12} &\neq (\epsilon_{11}\epsilon_{22})^{1/2} \\ \sigma_{12} &\neq \tfrac{1}{2}(\sigma_{11}+\sigma_{22}), \end{aligned} \tag{21.176}$$

although the deviations are usually small. The combining rules defined by the equal signs in Eq. 21.176 are what would be expected if the repulsive forces were like those between hard spheres (additivity of radii) and if the attractive forces could be approximated by the crudest version of the London dispersion formula using average ionization energies.

7. There is no fundamental reason to believe that a two-parameter function such as Eq. 21.173 can adequately represent the intermolecular pair potential. Indeed, when the molecule is nonspherical and/or has a nonspherical charge distribution, the intermolecular pair potential is considerably more complex than that represented in Eq. 21.173. In prin-

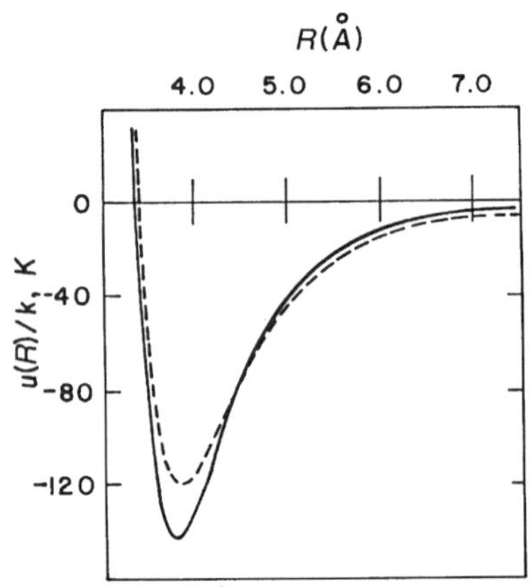

Figure 21.22 Comparison of the Lennard–Jones potential that best fits the second virial coefficient over the temperature range for which data exist and the empirical potential function derived from fitting the molecular beam scattering cross section. From H. T. Davis, *Statistical Mechanics of Phases, Interfaces and Thin Films,* VCH Publishers, New York, 1996.

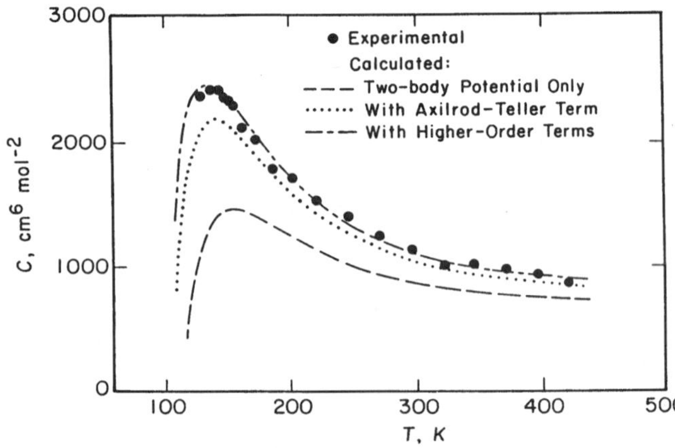

Figure 21.23 Comparison of computed and measured third virial coefficient of Ar, showing the influence of deviations from pairwise additivity of the molecular interaction. From J. M. Prausnitz, R. N. Lichtenthaler, and E. Gomes de Azevado, *Molecular Thermodynamics and Fluid-Phase Equilibrium,* 2nd ed, PTR Prentice Hall, Englewood Cliffs, NJ, 1986.

ciple, the effects of a complex intermolecular pair potential on the thermodynamic properties of the gas are calculable by the general methods already described, but in practice the calculations are very complex, and often only numerical evaluation of the various integrals is possible.

For further details, the reader is referred to the comprehensive monograph, *Molecular Theory of Gases and Liquids,* by J. O. Hirschfelder, C. F. Curtiss, and R. B. Bird (Wiley, New York, 1954), and to the text *Molecular Thermodynamics of Fluid-Phase Equilibrium,* by J. M. Prausnitz, R. N. Lichtenthaler, and E. M. Gomes de Azevado (Prentice Hall, New Jersey, 1986).

As the last topic in this section, we analyze the justification for the observation that the properties of all gases are the same when expressed in the reduced variables $\tilde{T}$, $\tilde{p}$, and $\tilde{V}$. Although a much more general argument can be used, we shall demonstrate the *principle of corresponding states,* as this observation is known, through study of the second virial coefficient. Suppose that the intermolecular potential can be written in the form

$$u_2 = \epsilon f\left(\frac{\sigma}{R}\right), \tag{21.177}$$

where ϵ and σ are constants with the dimensions of energy and distance, respectively. Some values of ϵ and σ for common gases and gas mixtures are presented in Table 21.13 and Table 21.14, respectively. All of the potentials we have thus far considered, except the Buckingham–Corner potential, can be written in the form of Eq. 21.177. Introduction of Eq. 21.177 into Eq. 21.164 leads to the form

$$B = 2\pi N_A \int_0^\infty \left\{1 - \exp\left[-\frac{\epsilon f(\sigma/R)}{k_B T}\right]\right\} R^2\, dR. \tag{21.178}$$

We now define three reduced variables by the equations

$$T^* \equiv \frac{k_B T}{\epsilon}, \tag{21.179a}$$

$$V^* \equiv \frac{V}{N\sigma^3}, \tag{21.179b}$$

$$p^* \equiv \frac{p\sigma^3}{\epsilon}. \tag{21.179c}$$

Introducing Eq. 21.179*a* into Eq. 21.178 leads to

$$\begin{aligned} B &= 2\pi N_A \sigma^3 \int_0^\infty \left\{1 - \exp\left[-\frac{\epsilon f(\sigma/R)}{k_B T}\right]\right\}\left(\frac{R}{\sigma}\right)^2 d\left(\frac{R}{\sigma}\right) \\ &= \frac{2\pi N_A \sigma^3}{3} B^*(T^*). \end{aligned} \tag{21.180}$$

Thus, if the intermolecular potential can be written in the form of Eq. 21.177, the reduced virial coefficient B^* is a function only of T^*. The equation of state

$$p = \frac{N k_B T}{V}\left(1 + \frac{N}{N_A}\frac{B}{V}\right) \tag{21.181}$$

can then be rewritten in the form

$$\begin{aligned} p^* &= \frac{N k_B T}{V}\cdot\frac{\sigma^3}{\epsilon}\left(1 + \frac{N}{N_A}\frac{B}{V}\right) \\ &= \frac{T^*}{V^*}\left(1 + \frac{2\pi}{3}\frac{B^*}{V^*}\right). \end{aligned} \tag{21.182}$$

Equation 21.182 shows that the equations of state of all gases are the same when expressed in the variables T^*, V^*, and p^*. These variables may be related to T_c, V_c, and p_c (see Chapter 24), thereby completing the rationalization of the observed behavior of gases.

Table 21.13 Parameters of the Lennard–Jones Potential

Substance	σ(Å)	ϵ/k_B (K)
He	2.556	10.22
H_2	2.928	37.00
D_2	2.928	37.00
Ne	2.749	35.60
Ar	3.405	119.8
Kr	3.60	171
Xe	4.100	221
O_2	3.58	117.5
CO	3.763	100.2
N_2	3.698	95.05
CH_4	3.817	148.2

Table 21.14 Lennard–Jones Parameters for Mixtures

Pair	σ_{12} (Å)	ϵ_{12}/k_B (K)
He—H_2	2.74	19.4
He—Ne	2.65	19.1
He—N_2	3.13	31.3
He—O_2	3.07	34.7
He—Ar	2.98	61.2
Ne—H_2	2.84	36.3
Ne—N_2	3.23	58.4
Ne—O_2	3.16	64.7
Ne—Ar	3.08	65.9
N_2—O_2	3.64	106
Ar—H_2	3.16	67.2
Ar—O_2	3.49	120

The analysis given may be extended to all the virial coefficients, and to other equivalent general forms of the equation of state. Provided Eq. 21.177 is valid, and that the total potential energy is the sum of pair energies, that is, that Eq. 21.158 is valid, it is found that the law of corresponding states is also valid. Deviations from Eq. 21.177 arise when the molecular force field is not spherical, as must be the case, for example, in hexane. The law of corresponding states can be extended to cover this class of potentials by introduction of a new parameter that measures the asymmetry of the molecular field. As to higher order virial coefficients, or the full equation of state involving contributions from interactions between triplets, quadruplets, . . . of molecules, there can be no doubt that nonadditivity of the pair potential is very important in systems in which hydrogen bonding occurs, e.g., water, and it must be expected that there are nonadditive corrections analagous to those displayed in Eq. 21.159 associated with the nonspherical electron density distribution of most molecules. Of course, one additional parameter cannot define a classification scheme for all forms of molecular interaction. Thus, it is convenient to divide molecules into classes, each class corresponding to a particular deviation of the intermolecular potential from Eqs. 21.158 and 21.177. In that sense, the principle of corresponding states can be applied to complex fluids for which Eq. 21.177 fails.

It is important to recognize the value of correlations in the properties of different species inferred from the principle of corresponding states, and not overemphasize its deficiencies. Despite the existence of deviations from Eqs. 21.159 and 21.177, the law of corresponding states is an organizing principle of great value, since it provides a standard of expected behavior. In the evolution of a science it is often the study of deviations from some well-defined model behavior that leads to deeper understanding of the properties of matter. In this chapter we have seen how deviations from the perfect gas law can be interpreted in terms of the properties of intermolecular forces and, to a lesser extent, how deviations from models based on pair interactions lead to information about many-molecules interactions. Despite the efforts of many scientists over a period of more than 100 years, there is much still to be learned from careful study of the properties of gases.

APPENDIX 21A

Influence of Symmetry of the Wave Function on the Distribution over States: Fermi–Dirac and Bose–Einstein Statistics

We stated in Section 21.5 that the symmetry properties of the wave function of a system impose a restriction on the number of accessible states. In general, the thermodynamic consequences of these symmetry induced restrictions are important only when k_BT is comparable to or smaller than the typical spacing of energy levels. The importance of such quantum mechanical effects is not limited to improvement of the numerical accuracy of prediction of thermodynamic functions. Rather, it is only when the appropriate quantum mechanical description of a system is used that all the general laws of macroscopic behavior are satisfied. For example, under almost all conditions the spacing of translational energy levels in a perfect gas is small compared to k_BT. Then Eq. 21.133 is valid. But as $T \to 0$, so also must $C_V \to 0$, by the third law of thermodynamics. Clearly, Eq. 21.133 must fail when k_BT becomes comparable to the spacing of translational levels, and quantum mechanical effects must then also change the forms of the equation of state and thermodynamic functions such as U or S.

To illustrate the influence on thermodynamic properties of the symmetry of the system's wave function, we examine a perfect gas of point particles. The microscopic state of the gas is defined by the number of particles with momentum **p**, n_p, and the associated kinetic energy

$$\epsilon_p = \frac{p^2}{2m}. \tag{21A.1}$$

For the particle-in-a-box model we have used so often,

$$\mathbf{p} = \frac{2\pi\hbar}{L}\mathbf{n}, \tag{21A.2}$$

where **n** is a vector with integer components and $V = L^3$. Note that the particles are not identified; only the number with momentum **p** is specified.

Suppose that the wave function for the system is antisymmetric under interchange of two particles. This is the case for electrons and other elementary particles with half-integral spin. Then because only two spin states are allowed for each value of **p**,

$$n_p = 0,1, \tag{21A.3}$$

which we recognize as the Pauli exclusion principle discussed in Section 5.2. In contrast, if the wave function for the system is symmetric under interchange of two particles, as is the case for ^4He and elementary particles with integer spin, it is found that any number of particles can occupy a given state, so that

$$n_p = 0,1,2,3,\ldots. \tag{21A.4}$$

Systems for which Eq. 21A.3 is valid are said to obey *Fermi–Dirac statistics,* and those for which Eq. 21A.4 is valid are said to obey *Bose–Einstein statistics.*

To see how Eqs. 21A.3 and 21A.4 influence the thermodynamic properties of the perfect gas, we shall evaluate the grand partition function (Eq. 19.126) for the two cases. By definition,

$$\Xi = \sum_{N\geq 0} \lambda^N Q_N = \sum_{N\geq 0} \sum_{\substack{\{n_p\} \\ \sum_p n_p = N}} \lambda^{\Sigma n_p} \exp\left(-\beta \sum n_p \epsilon_p\right), \tag{21A.5}$$

where $\lambda \equiv \exp(\beta\mu)$ and $\beta \equiv (k_BT)^{-1}$. The second form on the right-hand side of Eq. 21A.5 follows from the facts that for a perfect gas the energy is the sum of individual particle energies, $E = \Sigma_p n_p\epsilon_p$, and the total number of molecules is the sum over all the numbers with all possible values of momentum, $N = \Sigma_p n_p$. The notation $\{n_p\}$, $\Sigma_p n_p = N$ means that we include in the summation every possible set of values for the various occupation numbers n_p subject to the restriction $\Sigma_p n_p = N$. We now note that when the summation over all possible values of N is carried out, each n_p will range over all possible values. Equation 21A.5 can then be rewritten

$$\Xi = \sum_{N\geq 0} \sum_{\substack{\{n_p\} \\ \sum n_p = N}} \prod_p [\lambda \exp(-\beta\epsilon_p)]^{n_p}$$

$$= \sum_{n_0=0}^{n^{\max}} \sum_{n_1=0}^{n^{\max}} \sum_{n_2=0}^{n^{\max}} \cdots \prod_p [\lambda \exp(-\beta\epsilon_p)]^{n_p}$$

$$= \prod_p \left\{ \sum_{n_p=0}^{n^{\max}} [\lambda \exp(-\beta\epsilon_p)]^{n_p} \right\}. \qquad (21A.6)$$

To obtain an explicit form for the equation of state, we must evaluate Eq. 21A.6. As will be seen, this is very easy for the Bose–Einstein (BE) and Fermi–Dirac (FD) cases.

For the BE case, $n_p = 0, 1, 2, \ldots$; hence,

$$\Xi = \prod_p \frac{1}{1-\lambda \exp(-\beta\epsilon_p)}, \qquad (21A.7)$$

since if we define $x \equiv \lambda \exp(-\beta\epsilon_p)$ it is easily recognized that

$$\sum_n x^n = \frac{1}{1-x}$$

when $x < 1$. Indeed, $x < 1$ for the BE perfect gas, since $\exp(-\beta\epsilon_p) < 1$ by virtue of the fact that $\epsilon_p > 0$, and $\lambda = \exp(\beta\mu) < 1$ when $\mu < 0$, which is the case.

For the FD case, $n_p = 0, 1$ only, hence

$$\Xi = \prod_p [1+\lambda \exp(-\beta\epsilon_p)]. \qquad (21A.8)$$

We then find for the equations of state,

$$\frac{PV}{k_B T} = \ln \Xi = -\sum_p \ln[1-\lambda \exp(-\beta\epsilon_p)] \quad \text{(BE)}, \qquad (21A.9a)$$

$$\frac{PV}{k_B T} = \ln \Xi = \sum_p \ln[1-\lambda \exp(-\beta\epsilon_p)] \quad \text{(FD)}. \qquad (21A.9b)$$

In this Appendix, and only here, we use P for the pressure to avoid confusion with the magnitude of the momentum, p. Although explicit and exact, Eqs. 21A.9*a* and 21A.9*b* are not convenient, because they give the equation of state as a function of λ, that is, as a function of the chemical potential. To eliminate λ we calculate, first, the average number of particles in the gas. This is, by definition,

$$\langle N \rangle = \frac{\sum_{N\geq 0} N\lambda^N Q_N}{\Xi} = \lambda \frac{\partial}{\partial \lambda} \ln \Xi, \qquad (21A.10)$$

so that

$$\langle N \rangle = \sum_p \frac{\lambda \exp(-\beta\epsilon_p)}{1-\lambda \exp(-\beta\epsilon_p)} \quad \text{(BE)}, \qquad (21A.11a)$$

$$\langle N \rangle = \sum_p \frac{\lambda \exp(-\beta\epsilon_p)}{1+\lambda \exp(-\beta\epsilon_p)} \quad \text{(FD)}. \qquad (21A.11b)$$

We can also find the average occupation numbers, $\langle n_p \rangle$, from the definition

$$\langle n_p \rangle = \frac{\sum_{N\geq 0} \lambda^{\sum n_p} \sum_{\{n_p\}, \sum n_p = N} n_p \exp(-\beta \sum n_p \epsilon_p)}{\Xi} \qquad (21A.12)$$

$$= -\frac{1}{\beta} \frac{\partial}{\partial \epsilon_p} \ln \Xi, \qquad (21A.13)$$

so that

$$\langle n_p \rangle = \frac{\lambda \exp(-\beta\epsilon_p)}{1-\lambda \exp(-\beta\epsilon_p)} \quad \text{(BE)}, \qquad (21A.14a)$$

$$\langle n_p \rangle = \frac{\lambda \exp(-\beta\epsilon_p)}{1+\lambda \exp(-\beta\epsilon_p)} \quad \text{(FD)}, \qquad (21A.14b)$$

and

$$\langle N \rangle = \sum_p \langle n_p \rangle \qquad (21A.15)$$

as expected.

For a macroscopic system the spacing between adjacent translational energy levels is very small, hence the sum over momentum states can be replaced with an integral over momentum. Using Eq. 21A.2, the conversion reads

$$\sum_p \cdots \to \frac{V}{h^3} \int d\mathbf{p} \cdots$$

and is valid if the summand is finite for all p. This is the case for the perfect Fermi–Dirac gas, so we can write

$$\frac{P}{k_B T} = \frac{4\pi}{h^3} \int_0^\infty dp\, p^2 \ln\left[1+\lambda \exp\left(-\frac{\beta p^2}{2m}\right)\right], \qquad (21A.16)$$

$$n \equiv \frac{\langle N \rangle}{V} = \frac{4\pi}{h^3} \int_0^\infty dp\, p^2 \frac{\lambda}{\lambda + \exp(\beta p^2/2m)}. \qquad (21A.17)$$

In contrast, for the perfect Bose–Einstein gas the summands diverge as $\lambda \to 1$ because the single term corresponding to $p = 0$ diverges. To treat this case we split off the term with $p = 0$ and replace the rest of the sum by an integral. The result is

$$\frac{P}{k_B T} = \frac{4\pi}{h^3} \int_0^\infty dp\, p^2 \ln\left[1+\lambda \exp\left(-\frac{\beta p^2}{2m}\right)\right] - \frac{1}{V}\ln(1-\lambda), \qquad (21A.18)$$

$$n = \frac{4\pi}{h^3}\int_0^\infty dp\, p^2 \frac{\lambda}{\exp(\beta p^2/2m) - \lambda} + \frac{1}{V}\frac{\lambda}{1-\lambda}, \tag{21A.19}$$

where the last terms on the right-hand sides of Eqs. 21A.18 and 21A.19 come from the $p = 0$ terms of Eqs. 21A.9*a* and 21A.14*a*, respectively. The term

$$\frac{\langle n_0 \rangle}{V} = \frac{1}{V}\frac{\lambda}{1-\lambda}$$

becomes significant only when a finite fraction of the particles occupy the level with $p = 0$, and is otherwise infinitesimal, as are the terms of the form $\langle n_p \rangle/V$ for all conditions; $\langle n_p \rangle/V$ is not infinitesimally small when the Bose–Einstein gas condenses.

To calculate P/k_BT as a function of n, the ordinary equation of state, we must eliminate λ between Eqs. 21A.16 and 21A.17 for the FD gas, and between Eqs. 21A.18 and 21A.19 for the BE gas. We shall show how this is done for the FD gas. Consider the behavior of λ as determined from

$$n\Lambda^3 = f_{3/2}(\lambda), \tag{21A.20}$$

$$f_{3/2}(\lambda) \equiv \frac{4}{\sqrt{\pi}}\int_0^\infty dx \frac{\lambda x^2}{e^{x^2} + \lambda}. \tag{21A.21}$$

Equation 21A.20 is obtained from Eq. 21A.17, using the new variable $x^2 \equiv p^2/2mk_BT$ and the definitions of the de Broglie wavelength $\Lambda = (h^2/2\pi mk_BT)^{1/2}$ and of $f_{3/2}(\lambda)$ given by Eq. 21A.21. We now make the expansion

$$\begin{aligned}\frac{\lambda x^2}{e^{x^2} + \lambda} &= \frac{\lambda x^2}{e^{x^2}(1 + \lambda e^{-x^2})}\\ &= \lambda x^2 e^{-x^2}(1 - \lambda e^{-x^2} + \lambda^2 e^{-2x^2} - \cdots),\end{aligned}$$

valid when $\lambda e^{-x^2} < 1$. Then, using the integral

$$\int_0^\infty dx\, x^2 e^{-ax^2} = \frac{1}{4}\left(\frac{\pi}{a^3}\right)^{1/2},$$

we find

$$\begin{aligned}f_{3/2}(\lambda) &= \lambda - \frac{\lambda^2}{2^{3/2}} + \frac{\lambda^3}{3^{3/2}} - \cdots\\ &= \sum_{l=1}^{\infty} \frac{(-1)^{l+1}}{l^{3/2}}\lambda^l.\end{aligned} \tag{21A.22}$$

When $n\Lambda^3 << 1$, corresponding to the average interparticle separation $\mathcal{O}(n^{-3})$ being much larger than Λ,

$$n\Lambda^3 \approx \lambda - \frac{\lambda^2}{2^{3/2}},$$

which may be solved to give

$$\lambda \approx n\Lambda^3 + \frac{1}{2^{3/2}}(n\Lambda^3)^2$$

and the equation of state

$$\begin{aligned}\frac{P}{k_BT} &= \frac{4\pi}{h^3}\int_0^\infty dp\, p^2 \ln\left[1 + \lambda\exp\left(-\frac{\beta p^2}{2m}\right)\right]\\ &= \frac{4\pi}{h^3}\int_0^\infty dp\, p^2\left[\lambda\exp\left(-\frac{\beta p^2}{2m}\right) - \frac{1}{2}\lambda^2\exp\left(-\frac{\beta p^2}{m}\right) + \cdots\right]\\ &= n + \frac{1}{2^{5/2}}n^2\Lambda^3.\end{aligned} \tag{21A.23}$$

Thus, the pressure of a perfect FD gas exceeds that of a perfect gas of particles obeying classical mechanics. This is true for all densities, even though we have shown it only in the low-density limit. The origin of the excess pressure is an effective repulsion between particles generated by the symmetry-imposed requirement that $n_p = 0, 1$ only.

Proceeding in a similar fashion, we find for the perfect BE gas that

$$\frac{P}{k_BT} = n - \frac{1}{2^{5/2}}n^2\Lambda^3, \tag{21A.24}$$

which is smaller than the pressure of a corresponding perfect gas of particles obeying classical mechanics. In this case the symmetry requirements on the system's wave function give rise to an effective attraction. This effective attraction eventually leads to condensation of the BE gas; no such condensation can occur for the perfect FD gas.

Both the FD and BE gases approach classical behavior as $n\Lambda^3 \to 0$. We find from Eqs. 21A.14*a* and 21A.14*b* that

$$\lim_{n\Lambda^3 \to 0}\langle n_p \rangle = n\Lambda^3 \exp(-\beta\epsilon_p), \tag{21A.25}$$

which defines what is sometimes called the *Boltzmann gas.* Then, since $n\Lambda^3 << 1$ implies that $\lambda << 1$,

$$\begin{aligned}\frac{P}{k_BT} &= \frac{4\pi n\Lambda^3}{h^3}\int_0^\infty dp\, p^2 \exp\left(-\frac{\beta p^2}{2m}\right)\\ &= n,\end{aligned} \tag{21A.26}$$

which is the equation of state of the perfect classical gas. We can also evaluate the grand partition function for the perfect classical gas. Using the semiclassical canonical partition function (see Appendix 21C),

$$Q_N = \frac{Q_1^N}{N!}, \tag{21A.27}$$

we write

$$\Xi = \sum_{N\geq 0}\lambda^N Q_N = \sum_{N\geq 0}\lambda^N\frac{Q_1^N}{N!} = e^{\lambda Q_1}. \tag{21A.28}$$

Since

$$Q_1 = V\Lambda^{-3} \tag{21A.29}$$

and

$$\frac{PV}{k_BT} = \lambda Q_1, \tag{21A.30}$$

we find

$$\mu = k_BT \ln \frac{P\Lambda^3}{k_BT} \tag{21A.31}$$

and

$$V = \left(\frac{\partial G}{\partial p}\right)_{T,N} = \frac{Nk_BT}{P} \tag{21A.32}$$

for the perfect classical gas chemical potential and equation of state.

In this appendix we have briefly described what are called *weakly degenerate* FD and BE gases. To show consistency of thermodynamic behavior with the third law we must study strongly degenerate quantum gases, for which $n\Lambda^3 \approx 1$. The FD case, exemplified by electrons in a metal, will be briefly considered in Chapter 22; the BE case, although very important (e.g., for the understanding of superconductivity and superfluidity), will not be considered further in this text. Since 1995 there have been great advances in the experimental and theoretical study of Bose–Einstein condensation, and the properties of the condensate phase. These advances are described very clearly by W. Ketterle, *Physics Today* **52** (12), 30 (1999) and by K. Burnett, M. Edwards, and C. W. Clark, *Physics Today* **52** (12), 37 (1999).

APPENDIX 21B

Symmetry Properties of the Molecular Wave Function: Influence of Nuclear Spin on the Rotational Partition Function

We discussed in Appendix 21A how the symmetry of the wave function for a perfect gas of point particles influences the equation of state. When the perfect gas is composed of molecules that have internal structure, new symmetry properties of the wave function must be considered; these are associated with the effects of permutation of the nuclei within a molecule. Because of their spin, the nuclei of a molecule must behave as bosons (obeying BE statistics) or fermions (obeying FD statistics), as the case may be. The point is that the overall symmetry of the molecular wave function must include the properties of the nuclear wave function when the nuclei are permuted within the molecule. As will be seen, nuclear spin effects, acting through the corresponding wave function symmetry, have an important influence on the system's properties.

Consider, first, a heteronuclear diatomic molecule with nuclei a and b. By virtue of the nuclear spin I, there are several nuclear spin states, namely, $\overline{\omega}^{(n)} = 2I + 1$. Then nucleus a has spin eigenfunctions $\psi_1(a), \ldots, \psi_{\overline{\omega}_a^{(n)}}(a)$ and nucleus b has spin eigenfunctions $\psi_1(b), \ldots, \psi_{\overline{\omega}_b^{(n)}}(b)$. We have, therefore $\overline{\omega}_a^{(n)}\overline{\omega}_b^{(n)}$ combined nuclear eigenfunctions of the type $\psi_r(a)\psi_s(b)$ and, correspondingly, a degeneracy $\overline{\omega}_a^{(n)}\overline{\omega}_b^{(n)}$ in the partition function.

Consider, next, a homonuclear diatomic molecule. Of course each nucleus has the same eigenfunctions and degeneracy $\overline{\omega}^{(n)}$. For the molecule there are

$\frac{1}{2}\overline{\omega}^{(n)}(\overline{\omega}^{(n)} - 1)$
antisymmetric nuclear wave functions of the type
$\psi_r(a)\psi_s(b) - \psi_s(a)\psi_r(b)$,

$\frac{1}{2}\overline{\omega}^{(n)}(\overline{\omega}^{(n)} - 1)$
antisymmetric nuclear wave functions of the type
$\psi_r(a)\psi_s(b) - \psi_s(a)\psi_r(b)$,

$\frac{1}{2}\overline{\omega}^{(n)}(\overline{\omega}^{(n)} - 1)$
antisymmetric nuclear wave functions of the type
$\psi_r(a)\psi_s(b) - \psi_s(a)\psi_r(b)$,

Thus, there are $\frac{1}{2}\overline{\omega}^{(n)}(\overline{\omega}^{(n)} - 1)$ antisymmetric nuclear wave functions and $\frac{1}{2}\overline{\omega}^{(n)}(\overline{\omega}^{(n)} + 1)$ symmetric nuclear wave functions, for a total of $(\overline{\omega}^{(n)})^2$, the same as the number of nuclear eigenfunctions of a heteronuclear diatomic molecule. We now must combine the nuclear wave functions with the rest of the molecular wave function and make the total wave function antisymmetric with respect to exchange of protons and neutrons (both of which obey FD statistics).

What are the symmetry properties of the wave functions corresponding to the non-nuclear molecular degrees of freedom? These are as follows:

1. The eigenfunction descriptive of translation of the center of mass does not contain the coordinates of individual nuclei relative to the center of mass. Therefore the translational eigenfunction is totally symmetric in the nuclear coordinates.

2. The rotational eigenfunctions with even quantum numbers, $J = 0, 2, 4, \ldots$, are symmetric in the nuclear coordinates, and those with odd quantum numbers, $J = 1, 3, 5, \ldots$, are antisymmetric in the nuclear coordinates.

3. The vibrational eigenfunctions are symmetric in the nuclear coordinates (they depend on the displacement from the equilibrium separation).

Since the total molecular eigenfunction must be antisymmetric with respect to proton and neutron permutation, it must be antisymmetric under permutation of the nuclei when they have odd mass numbers, and must be symmetric under permutation of the nuclei when they have even mass numbers.

Consider the case when there is no vibrational excitation of the molecule ($v = 0$). Then, because the vibrational eigenfunction is symmetric in the nuclear coordinates, the following combinations must be considered:

1. Nuclei with odd mass numbers—symmetric rotational wave functions with antisymmetric nuclear wave functions, or antisymmetric rotational wave functions with symmetric nuclear wave functions. Therefore, the terms with even J, including $J = 0$, in the rotational partition function have degeneracy $\frac{1}{2}\overline{\omega}^{(n)}(\overline{\omega}^{(n)} - 1)$ from the nuclear spin states, and those with odd J have degeneracy $\frac{1}{2}\overline{\omega}^{(n)}(\overline{\omega}^{(n)} + 1)$.

2. Nuclei with even mass numbers—symmetric rotational wave functions with symmetric nuclear wave functions, or antisymmetric rotational wave functions with antisymmetric nuclear wave functions. Therefore, the terms with even J, including $J = 0$, in the rotational partition function have degeneracy $\frac{1}{2}\overline{\omega}^{(n)}(\overline{\omega}^{(n)} + 1)$ from the nuclear spin states, and those with odd J have degeneracy $\frac{1}{2}\overline{\omega}^{(n)}(\overline{\omega}^{(n)} - 1)$.

We conclude that the partition function for rigid-body rotation of a diatomic molecule in the vibrational ground state, including the effects of nuclear spin, takes the form

$$\overline{\omega}_a^{(n)}\overline{\omega}_b^{(n)}\sum_{J=0}^{\infty}(2J+1)\exp\left[-\frac{J(J+1)\Theta_{\text{rot}}}{T}\right] \quad \text{(heteronuclear diatomic molecule)}, \tag{21B.1}$$

or

$$\frac{1}{2}\overline{\omega}^{(n)}(\overline{\omega}^{(n)}-1)\sum_{J=0,2,\ldots}^{\infty}(2J+1)\exp\left[-\frac{J(J+1)\Theta_{\text{rot}}}{T}\right] + \frac{1}{2}\overline{\omega}^{(n)}(\overline{\omega}^{(n)}+1)\sum_{J=1,3,\ldots}^{\infty}(2J+1)\exp\left[-\frac{J(J+1)\Theta_{\text{rot}}}{T}\right]$$
$$\text{(homonuclear diatomic molecule, odd nuclear mass numbers)}, \tag{21B.2}$$

or

$$\frac{1}{2}\overline{\omega}^{(n)}(\overline{\omega}^{(n)}+1)\sum_{J=0,2,\ldots}^{\infty}(2J+1)\exp\left[-\frac{J(J+1)\Theta_{\text{rot}}}{T}\right] + \frac{1}{2}\overline{\omega}^{(n)}(\overline{\omega}^{(n)}-1)\sum_{J=1,3,\ldots}^{\infty}(2J+1)\exp\left[-\frac{J(J+1)\Theta_{\text{rot}}}{T}\right]$$
$$\text{(homonuclear diatomic molecule, even nuclear mass numbers)}. \tag{21B.3}$$

Consider now the case when $\Theta_{\text{rot}} << T$. Then, to great accuracy, the sum over even rotational quantum numbers is equal to the sum over odd rotational quantum numbers, because the important terms in the sum involve large values of J. Therefore,

$$\sum_{J=0,2,\ldots}^{\infty}(2J+1)\exp\left[-\frac{J(J+1)\Theta_{\text{rot}}}{T}\right] = \sum_{J=1,3,\ldots}^{\infty}(2J+1)\exp\left[-\frac{J(J+1)\Theta_{\text{rot}}}{T}\right] = \frac{1}{2}\sum_{J=0}^{\infty}(2J+1)\exp\left[-\frac{J(J+1)\Theta_{\text{rot}}}{T}\right]. \tag{21B.4}$$

The factor $\frac{1}{2}$ that arises naturally in this limit is just the reciprocal of the symmetry number of the homonuclear diatomic molecule. From the standpoint of classical mechanics, the appearance of the factor $\frac{1}{2}$ means that configurations generated by rotating the homonuclear diatomic by 180° are to be considered identical, so that the number of distinguishable states is only half the classical total. That is, the classical value of the rotational partition function must be divided by the symmetry number, which is 1 for a heteronuclear diatomic and 2 for a homonuclear diatomic molecule.

The case of polyatomic molecules is similar. Just as for a diatomic molecule, the symmetry of the molecule imposes requirements on the symmetry of the molecular wave function when nuclei are permuted. Consideration must then be given to the construction of appropriate eigenfunctions of nuclear spin and molecular rotation. The effect is, as before, to restrict the rotational quantum states available to a molecule. In the high-temperature limit, which almost always accurately describes a polyatomic molecule, we again use a symmetry number σ to correct the classical partition function. The symmetry number is the number of indistinguishable positions into which the molecule may be carried by rigid body rotation. For example, $\sigma = 12$ for CH_4, since the molecule can be rotated into three indistinguishable positions about any CH axis, and there are four such axes in the molecule. For benzene $\sigma = 12$, because there are six indistinguishable positions obtained by rotation about an axis perpendicular to the ring, and two indistinguishable positions obtained by 180° rotation about an axis through the 1,4 carbon atoms and lying in the plane of the ring. Clearly, for polyatomic molecules with high symmetry σ can be quite large, hence considerably influences the rotational contribution to the Helmholtz free energy.

When $\Theta_{\text{rot}} > T$, no simplification of the combined rotation-nuclear spin partition function is possible, and the full forms of Eq. 21B.2 or 21B.3 must be used to calculate the thermodynamic properties of the diatomic gas. The best-known case where this is necessary is H_2, as described in Section 21.5B.

APPENDIX 21C

The Semiclassical Partition Function; The Equation of State of an Imperfect Gas

In setting up the classical equivalent of the partition functions already introduced, we are led to the following question: How precisely can we define the energy level in a continuous spectrum of states? According to the uncertainty principle, a particle constrained to one dimensional motion cannot have its momentum and position simultaneously specified to better than $\Delta p\, \Delta x \approx h$. Thus, if p and x are chosen as coordinate axes, the point representing a one-dimensional system cannot be located more precisely than within a box with sides Δp and Δx and area h. If there are N molecules moving in a three-dimensional space, we need $3N$ position coordinates and $3N$ momentum coordinates to specify the system. Imagine a space of $6N$ dimensions, by which is meant that the coordinates of one point require the specification of $6N$ numbers. When the $6N$ coordinates are chosen to be the $3N$ position coordinates and the $3N$ momenta, the space is called the *phase space* of the system (cf. Chapter 2). Clearly, in a classical mechanical description, the mechanical state of our system at any instant of time is given by *one* point in the $6N$-dimensional phase space. But if the product of each coordinate and the corresponding momentum is uncertain to the amount $\Delta p_i\, \Delta x_i \approx h$, then from the point of view of quantum mechanics, the finest allowable subdivision of phase space is a cell of hypervolume h^{3N}. According to quantum theory, we cannot distinguish any variation in any function over a hypervolume less than h^{3N}. Then, from Eq. 19.131,

$$Q_N = \sum_j g_j \exp\left(-\frac{E_j}{k_B T}\right), \tag{21C.1}$$

we see that each term in the summation corresponds to

$$\exp\left(-\frac{E_j}{k_B T}\right) = \frac{1}{N!h^{3N}} \int_{E \le E_j \le E+dE} \cdots \int \exp\left[-\frac{1}{k_B T}\left(\sum \frac{p_i^2}{2m} + \mathcal{V}_N\right)\right] dx_1\, dy_1\, dz_1 \cdots dp_{x_N}\, dp_{y_N}\, dp_{z_N}. \tag{21C.2}$$

The factor $N!h^{3N}$ appears because the shell $E \le E_j \le E + dE$ in phase space has volume $N!h^{3N}$. The term h^{3N} arises because each product $dx\, dp_x$ ranges over h, and there are $3N$ such products (all subject to the uncertainty relation). The term $N!$ arises because there are $N!$ permutations of the molecules in the cell among the momenta and positions for which the energy is unchanged. The weighting factor g_j, originally chosen to represent the number of quantum mechanically distinct states with energies in the interval $E \le E_j \le E + dE$, is taken to be unity, because dE is now the smallest distinguishable energy difference. Thus, we have

$$Q_N = \frac{1}{N!h^{3N}} \int \cdots \int \exp\left[-\frac{1}{k_B T}\left(\sum \frac{p_i^2}{2m} + \mathcal{V}_N\right)\right] dx_1 \cdots dp_{z_N}, \tag{21C.3}$$

where the integral is over the entire range of variation of p and x (the entire phase space), for the semiclassical form of the canonical partition function.[18] The range of the space variables x_i is limited by the volume of the system. Although the momenta cannot be infinite if the energy is finite, $\exp(-p^2/2mk_BT)$ tends to zero so rapidly as $p^2 \to \infty$ that no error is incurred by allowing the range of momentum to be infinite.

[18] The form 21C.3 is called the semiclassical partition function because the energy is described by classical mechanics but the factor $N!h^{3N}$ corrects the counting of states for the effects of molecular indistinguishability and the uncertainty principle.

If, as is usually the case, $\mathcal{V}_N$ depends only on the position variables, we obtain

$$Q_N = \frac{1}{N!h^{3N}}\left[\int_{-\infty}^{\infty} \exp\left(-\frac{p^2}{2mk_BT}\right)dp\right]^{3N} \times \int\cdots\int \exp\left(-\frac{\mathcal{V}_N}{k_BT}\right) dx_1\cdots dz_N$$

$$= \left(\frac{2\pi mk_BT}{h^2}\right)^{3N/2} \frac{1}{N!}\int\cdots\int \exp\left(-\frac{\mathcal{V}_N}{k_BT}\right) dx_1\cdots dz_N$$

$$= \left(\frac{Q_1^T}{V}\right)^N \frac{Z_N}{N!}, \qquad (21C.4)$$

where

$$Z_N \equiv \int\cdots\int \exp\left(-\frac{\mathcal{V}_N}{k_BT}\right)dx_1\cdots dz_N \qquad (21C.5)$$

and Q_1^T is given by Eq. 21.108; Z_N is known as the *configurational partition function* for N molecules.

We may now readily obtain the general virial equation of state from the grand partition function. From Eq. 19.125 we find that

$$\frac{pV}{k_BT} = \ln\Xi, \qquad (21C.6)$$

$$\frac{pV}{k_BT} = \ln\left[1+\exp\left(\frac{\mu}{k_BT}\right)Q_1 + \exp\left(\frac{2\mu}{k_BT}\right)Q_2 + \cdots\right], \qquad (21C.7)$$

and by expanding the logarithm,

$$\frac{pV}{k_BT} = \lambda Q_1 + \lambda^2\left(Q_2 - \frac{1}{2}Q_1^2\right)+\cdots, \qquad (21C.8)$$

where $\lambda \equiv \exp(\mu/k_BT)$ and Q_i is the canonical partition function for i molecules in the volume V. We also have, from the definition of average value,

$$\langle N\rangle \equiv \frac{\sum_N N\lambda^N Q_N}{\sum_N \lambda^N Q_N} = \frac{\sum_N \lambda\left(\frac{\partial}{\partial\lambda}\lambda^N Q_N\right)_{T,V}}{\sum_N \lambda^N Q_N} = \left(\frac{\partial\ln\Xi}{\partial\ln\lambda}\right)_{T,V}$$

$$= \lambda Q_1 + \lambda^2(2Q_2 - Q_1^2)+\cdots. \qquad (21C.9)$$

Introducing Eq. 21C.4, we find that Eqs. 21C.8 and 21C.9 become

$$\frac{pV}{k_BT} = (\lambda Q_1^T)Z_1 + (\lambda Q_1^T)^2\frac{Z_2 - Z_1^2}{2}+\cdots \qquad (21C.10)$$

and

$$\langle N\rangle = (\lambda Q_1^T)Z_1 + (\lambda Q_1^T)^2(Z_2 - Z_1^2)+\cdots, \qquad (21C.11)$$

which are to be regarded as two simultaneous equations from which λQ_1^T is to be eliminated. The general problem of solving Eq. 21C.11 for λQ_1^T in terms of $\langle N\rangle$ is algebraically complicated, but if we keep only the terms displayed (up to λ^2), the quadratic equation is easily solved, giving λQ_1^T as an explicit function of $\langle N\rangle$. Thus, subtraction of Eq. 21C.11 from Eq. 21C.10, followed by elimination of λQ_1^T in terms of $\langle N\rangle$, leads to

$$\frac{pV}{k_BT} = \langle N\rangle - \frac{(Z_2 - Z_1^2)\langle N\rangle^2}{2Z_1^2}+\cdots. \qquad (21C.12)$$

But, by definition, we must have

$$Z_1 = \int_V dx\,dy\,dz = V, \qquad (21C.13)$$

since Z_1 refers to one molecule in a volume V. Note that, using Eq. 21C.13, Eq. 21C.12 is already in the form of the virial equation of state:

$$\frac{pV}{\langle N\rangle k_BT} = 1 + \frac{(Z_1^2 - Z_2)}{2V}\frac{\langle N\rangle}{V}+\cdots. \qquad (21C.14)$$

Indeed, if we compare coefficients of the terms in Eqs. 21C.14 and 21.17, we see that

$$B = \frac{N_A}{2V}(Z_1^2 - Z_2)$$

$$= \frac{N_A}{2V}\left[\left(\int dx_1\,dy_1\,dz_1\right)^2 - \int\exp\left(-\frac{u_2}{k_BT}\right) dx_1\,dy_1\,dz_1\,dx_2\,dy_2\,dz_2\right]$$

$$= \frac{N_A}{2V}\iint\left[1-\exp\left(-\frac{u_2}{k_BT}\right)\right]dx_1\cdots dz_2. \qquad (21C.15)$$

To show that B is independent of the volume, we take advantage of the fact that $u_2(1, 2)$ depends only on the distance between the monatomic molecules. Then, introducing the new coordinates

$$X = \tfrac{1}{2}(x_1 + x_2), \quad Y = \tfrac{1}{2}(y_1 + y_2), \quad W = \tfrac{1}{2}(z_1 + z_2),$$

$$x = x_2 - x_1, \quad y = y_2 - y_1, \quad z = z_2 - z_1,$$

$$\mathbf{R} = \mathbf{i}x + \mathbf{j}y + \mathbf{k}z, \qquad (21C.16)$$

we find that

$$B = 2\pi N_A \int_0^\infty \left[1 - \exp\left(-\frac{u_2}{k_B T}\right)\right] R^2\, dR, \qquad (21C.17)$$

since keeping R fixed implies that u_2 is fixed, and the integrand is constant for all possible values of X, Y, W. The integration of the position of the center of mass over V has just the value V because the integrand has the same value everywhere and the total domain of integration is V.

By similar but more complicated manipulations the third virial coefficient for monoatomic species can be shown to have the form

$$C = -\frac{1}{3}\iint d^3R_{12}d^3R_{23}d^3R_{13}$$
$$\{[\exp(-\beta u_{12})-1][\exp(-\beta u_{23})-1][\exp(-\beta u_{13})-1]\}$$
$$-\frac{1}{3}\iint d^3R_{12}d^3R_{23}d^3R_{13}\,[\exp(-\beta(u_{12}+u_{23}+u_{13})]$$
$$[\exp(-\beta\Delta_3)-1] \qquad (21C.18)$$

where $\beta \equiv 1/k_BT$.

• FURTHER READING

Davidson, N., *Statistical Mechanics* (McGraw-Hill, New York, 1962), Chapters 8–11, 13, and 15.
Davis, H. T., *Statistical Mechanics of Phases, Interfaces and Thin Films* (VCH Publishers, New York, 1996).
Hirschfelder, J. O., C. F. Curtiss, and R. B. Bird, *Molecular Theory of Gases and Liquids* (Wiley, New York, 1954), Chapters 3, 4, and 6.
Mason, E. A., and T. H. Spurling, *The Virial Equation of State* (Pergamon Press, Elmsford, N.Y., 1969).
Mayer, J. E., and M. G. Mayer, *Statistical Mechanics,* 2nd ed. (Wiley, New York, 1977), Chapters 7, 8, and 11.
Pitzer, K. S., *Thermodynamics,* 3d ed. (McGraw Hill, Inc., New York, 1995), Chapter 9.
Prausnitz, J. M., R. N. Lichtenthaler and E. Gomes de Azevado, *Molecular Thermodynamics and Fluid-Phase Equilibrium,* 2nd ed. (PTR Prentice Hall, Inc., Englewood Cliffs, N.J., 1986), Chapters 3, 4 and 5.
Rushbrooke, G. S., *Introduction to Statistical Mechanics* (Oxford University Press, 1949), Chapters 6–12, 15, and 16.

• PROBLEMS

1. Plot $B(T)$ versus T for a van der Waals gas. Provide an interpretation for the behavior shown on your graph.

2. For real gases, there is a temperature at which $B(T)$ passes through a maximum. Why is this behavior not shown by a van der Waals gas? Would you expect a maximum value of $B(T)$ for a gas whose molecules interact through the potential shown in Fig. 21.6?

3. Show that at low density the Dieterici equation of state reduces to the van der Waals equation.

4. Write out expressions for the Gibbs and Helmholtz free energies of a real gas correct to terms including the second virial coefficient.

5. Show that if $(\partial U/\partial V)_T = 0$ and $(\partial H/\partial p)_T = 0$, the equation of state must have the form $pV = aT$, where a is a constant independent of p, V, and T.

6. Consider the following experiment (see diagram).

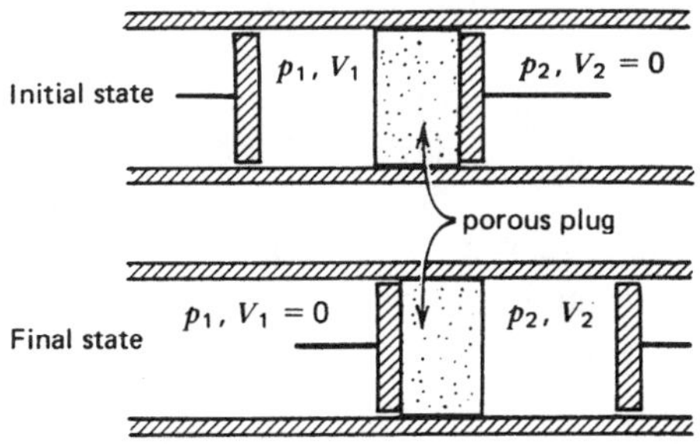

Gas contained in a cylinder is initially present only on one side of a porous plug that divides the cylinder into two volumes. The entire system is thermally isolated from its surroundings. The gas is pushed through the porous plug by the pistons in such a way that both p_1 and p_2 are maintained constant but not equal to one another. The flow is so slow that the gas emerging from the plug is equilibrated at p_2, so p_2 can be taken as uniform throughout the expansion on side 2. Show that for this throttling process

$$h_1 = h_2.$$

7. When the experiment described in Problem 6 is carried out with a real gas as a continuous-flow process that is so slow there is no appreciable kinetic energy associated with the gas flow, the steady-state temperatures on sides 1 and 2 are different. The temperature change is proportional to $\mu_{JT} \equiv (\partial T/\partial p)_H$, called the Joule–Thomson coefficient. Show that

$$\mu_{JT} = \frac{1}{c_p}\left[T\left(\frac{\partial v}{\partial T}\right)_p - v\right] = \frac{v}{c_p}[\alpha T - 1]$$
$$= -\frac{1}{c_p}\left(\frac{\partial h}{\partial p}\right)_T.$$

8. Evaluate μ_{JT} for a real gas correct to terms including the second virial coefficient.

9. At 480°C and 723 torr total pressure, 76% of the HCl in a mixture, which originally contained 49% HCl and 51% O_2 by volume, reacts with oxygen to form chlorine and water. Calculate the equilibrium constant of the reaction with atmospheres as units.

10. Calculate the enthalpy and entropy changes for the reaction

$$I_2 = 2I$$

at 1000 K from the following data:

T(K)	1274	1173	1073	973	872
K(T)	1.678×10^{-1}	4.803×10^{-2}	1.088×10^{-2}	1.801×10^{-3}	1.81×10^{-4}

11. Calculate the thermal de Broglie wavelength for argon at 85 K, neon at 20 K, and helium at 4 K.

12. Consider the ground state and lowest excited electronic state of a molecule. Suppose that both states are nondegenerate and the characteristic electronic temperature of the molecule is 50,000 K. Find the probability that a molecule is in the excited state at 300 K.

13. Why does the electronic heat capacity depend on the difference of electronic energies, rather than their actual values?

14. Calculate a classical (high-temperature) vibrational partition function by approximating Eq. 21.113 by an integral, in a manner analogous to Eq. 21.106 or Eq. 21.114. Show that Eq. 21.140 results directly from the classical partition function.

15. The entropy of CH_3OH at 298.15 K and 1 atm is calculated to be 244.26 J/K mol from spectroscopic data, and found to be 236.94 J/K mol from calorimetric data. Suggest an interpretation of the residual entropy (7.32 J/K mol) in terms of possible properties of the CH_3OH molecule.

16. The entropy of H_2 at 298.15 K and 1 atm is calculated to be 130.67 J/K mol from spectroscopic data, and found to be 124.43 J/K mol from calorimetric data. Offer an explanation for the residual entropy of 6.28 J/K mol in terms of the properties of o-H_2 and p-H_2. *Hint:* o-H_2 rotates in the crystal at the lowest temperature used in the calorimetric measurements.

17. Consider the gas-phase isomerization reaction

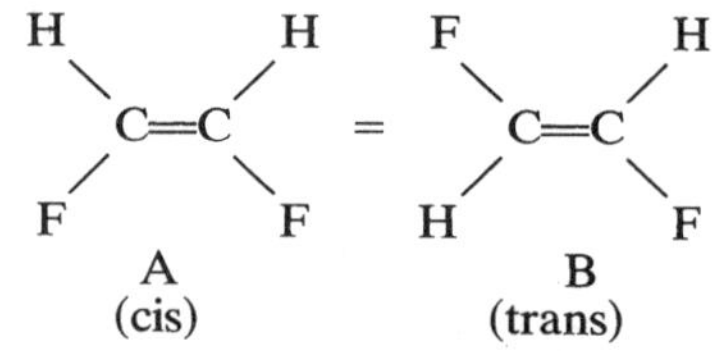

In the presence of a catalyst, A and B are in equilibrium. The energy levels for A and B are

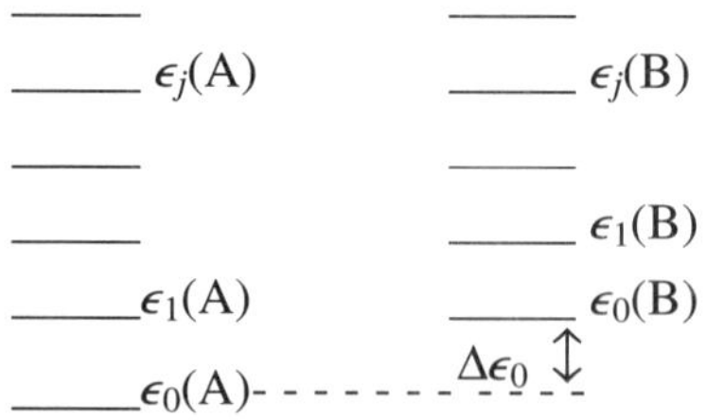

In the absence of the catalyst, the partition function for pure A is

$$Q_A = \sum_j \exp\left[-\frac{\epsilon_j(A)}{k_B T}\right]$$

and similarly, for pure B,

$$Q_B = \sum_j \exp\left[-\frac{\epsilon_j(B)}{k_B T}\right].$$

Now let the catalyst be present. Consider $\epsilon_0(A)$ as the energy zero of all $C_2H_2F_2$ molecules. Let N_0 be the number of $C_2H_2F_2$ molecules in the state with energy $\epsilon_0(A)$ and let $N_j(A)$ be the number of molecules in the state with energy $\epsilon_j(A)$.

(a) Write an expression for $N_j(A)$ in terms of N_0 and $\epsilon_j(A)$. Use this to find the total number of *cis* molecules, N_A.

(b) Express $N_j(B)$ in terms of N_0 and the total number of *trans* molecules, N_B.

(c) Find the equilibrium constant for the reaction.

18. Show that in a canonical ensemble for an ideal gas

$$\langle (E - \langle E \rangle)^2 \rangle = k_B T^2 C_V = \frac{3N}{2}(k_B T)^2,$$

where the brackets $\langle \cdot \; \cdot \; \cdot \rangle$ indicate an ensemble average.

19. Suppose that N molecules of an ideal gas are in volume V. Prove that the probability of there being n molecules in a subvolume v is

$$P(n) = (\lambda Q_1)^n \frac{e^{-\lambda Q_1}}{n!}.$$

Equation 21A.28 will prove helpful. The distribution $P(n)$ is called the *Poisson distribution.* Find the average value of n.

20. The prescription of accounting for molecular indistinguishability by dividing by $N!$ (see Eq. 21.100) is strictly a semiclassical approximation and is valid only under limiting conditions. To see why, consider a system of two particles where each particle can be in either of two nondegenerate energy levels, ϵ_1 and ϵ_2.

(a) If the particles are distinguishable, prove that the canonical partition function Q_2 for the system is

$$Q_2 = Q_1^2,$$

where Q_1 is the single-particle partition function

$$Q_1 = \exp\left(-\frac{\epsilon_1}{k_B T}\right) + \exp\left(-\frac{\epsilon_2}{k_B T}\right).$$

(b) Suppose that the particles obey Fermi–Dirac statistics. Remember that there can be only one possible state of the system, and evaluate Q_2^{FD}.

(c) Suppose that the particles obey Bose–Einstein statistics. Take careful account of the indistinguishability of the particles and evaluate Q_2^{BE}.

(d) Show that the application of the 1/2! correction factor for indistinguishability to Q_1^2 gives neither Q_2^{FD} nor Q_2^{BE} but rather their average:

$$\frac{1}{2!}Q_1^2 = \frac{1}{2}(Q_2^{\mathrm{BE}} + Q_2^{\mathrm{FD}}).$$

(e) Suppose that it were exceedingly unlikely for two particles to have the same energy. Show that under this condition

$$Q_2^{\mathrm{BE}} = Q_2^{\mathrm{FD}} = \frac{1}{2!}Q_1^2.$$

(f) For most systems (but not the one in this problem!) at ordinary temperatures and densities, the number of available molecular energies is vastly greater than the number of particles. Explain why $Q_1^N/N!$ is then a good approximation for the partition function of N indistinguishable particles. Chapter 4 of *Statistical Mechanics* by D. A. McQuarrie (Harper & Row, New York, 1976) contains an excellent discussion of this point.

21. Show that for D_2 gas (nuclear spin = 1), the rotational contribution to the heat capacity is

$$C_V^R = \tfrac{1}{3}C_V^R(p) + \tfrac{2}{3}C_V^R(o),$$

where p (para) refers to odd rotation states and o (ortho) refers to even rotation states.

22. Show that when Q_1 has the form $f(T)V$, the ideal gas equation of state follows.

23. In Section 21.7 it is stated that the probability of finding two molecules in a moderately dense gas separated by a distance R is proportional to $\exp[-u_2(R)/k_B T]$. Sketch this function for various forms of $u_2(R)$ (hard-sphere, square-well, Lennard–Jones) and interpret the features of the graphs.

24. Calculate the second virial coefficient for the potential shown in the accompanying diagram.

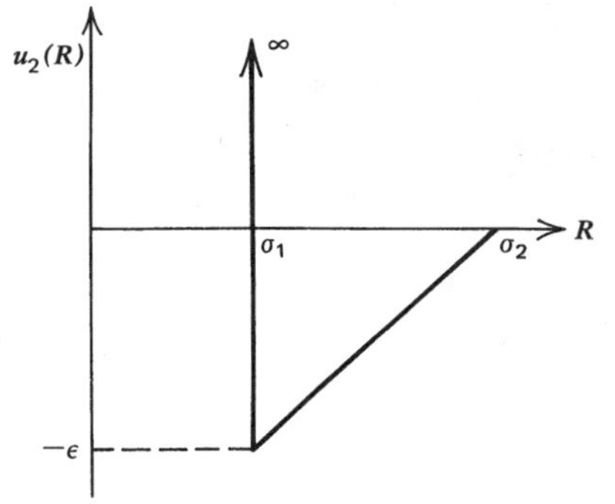

25. A barometric formula allows one to calculate the pressure of a gas in a gravitational field. For an ideal gas, derive the barometric formula

$$p(z) = p(0)\exp\left(-\frac{mgz}{k_B T}\right),$$

where z is the height above sea level, m is the mass of a gas molecule, and g is the acceleration of gravity.

26. In Chapter 19, the Maxwell speed distribution was derived for an ideal gas (Eq. 19.143). Now suppose that the molecules interact through a potential energy function that is independent of velocity. Write the canonical partition function (Eq. 21C.3) and show that the Maxwell distribution applies equally well to a nonideal system.

27. The ^{2}P ground electronic state of the fluorine atom has two fine-structure components, $^2P^{1/2}$ and $^2P^{3/2}$. Use (a) the HF–SCF, (b) the CISD, and (c) the MP2 methods along with a 6–31G(d) basis set to compute the excitation energy $\Delta\epsilon$ between the two fine-structure levels. Then, taking the degeneracies into account, calculate the electronic partition function via equation (21.127) and report the fraction of the population in each level. Finally, compute the electronic heat capacity via equation (21.142). See Problems 5.26 and 7.27 for basis set descriptions and software packages.

28. Repeat Problem 21.27 for the ^3P ground electronic state of the oxygen atom that has three fine-structure components, $^3P_{2,1,0}$. To complete this calculation, generalize equations (21.127) and (21.142) to include a third state.

29. One mole of a gas described by the Dieterici equation of state undergoes a reversible expansion from a volume of 1.00 L to a volume of 5.00 L at a constant temperature of 298.15 K. (a) Calculate the work done during the expansion process; (b) Calculate the change in the entropy of the system during the expansion. Use a computer-based mathematical package to evaluate the required integrals. (*Hint:* Use one of Maxwell's relations given in Section 17.1.)

30. Repeat Problem 21.29 for a gas described by the Redlich–Kwong equation of state.

CHAPTER 22

Thermodynamic Properties of Solids

22.1 Differences between Gases and Condensed Phases

In studying the properties of the gas phase, we have thus far dealt with situations in which the average density of molecules, N/V, is very small relative to the density of closest packing. In a state of closest packing, molecules are as close to one another as is permitted by their size, that is, as is permitted by the repulsive forces that provide a measure of the "diameter" of a molecule. A two-dimensional visualization of the closest-packing configuration is provided by surrounding a penny with six other pennies, the central penny being in contact with each of its neighbors, or by the configuration in which a set of pool balls is arranged before the opening shot of a game. When the density of molecules is small, the average distance between molecules is large relative to the molecular diameter, and it is possible to think of the motion of a molecule as being essentially undisturbed by the presence of other molecules except for occasional binary collisions (even these being ignored in the perfect gas). Such a configuration is similar to that encountered when the set of pool balls is distributed at random over the pool table. Now it is clear that, since molecules have both nonzero size and range of interaction, as the density is increased and correspondingly the average distance between molecules decreased, simultaneous interactions between larger and larger groups of molecules become important. It is also easy to see that when the density becomes so large that the average volume per molecule, $V/N = 1/n$, is only slightly larger than the volume of a molecule, then every molecule must be in continuous interaction with a large number of neighboring molecules. Densities for which this situation occurs are characteristic of the condensed (liquid and solid) phases of matter.

There is another characteristic difference between a gas and a liquid or solid. The observation that in a dilute gas the motion of a molecule is only occasionally disturbed is equivalent to saying that in the gas, on average, the kinetic energy corresponding to a molecule's translational motion is large relative to its potential energy. In the liquid and solid states the translational energy is of the same order as or less than the height of the localized potential energy maxima that separate the positions at which the molecules are located. As a result of this alteration in the relative importance of translational kinetic energy and potential energy, completely new considerations must be invoked in the study of liquids and solids, the most important of which is the concept of structure. As we use the term, the concept of structure includes any spatial arrangement of atoms and molecules that differs from a random distribution; the term is used in the same fashion as in our earlier discussions of molecular structure and crystal structure (see Chapters 9 and 11). Note that we include in the concept of structure the partially regular spatial arrangements such as are characteristic of the liquid phase. We have already discussed, in Chapter 11, some aspects of crystal structure and their relationship to types of chemical bonding. In this chapter we study the thermodynamic properties of solids.

22.2 The Influence of Crystal Symmetry on Macroscopic Properties

In Section 11.2 we described the relationship between space lattices and crystal symmetry. A crystal is a solid whose microscopic properties are periodic in three dimensions. The fundamental structural element of the crystal is the unit cell, from which the entire crystal is produced by repeated displacement by integral multiples of three independent basis vectors. The internal arrangement of atoms in the unit cell (the basis) and the network of points defining the locations of the repeated cell centers (the lattice) together constitute the crystal structure. A crystal may have, in addition to translational symmetry, other symmetries—e.g., overall rotation and combined translation-rotation moves that transform the position and orientation of one molecule in the unit cell into those of another molecule in the unit cell. Although the symmetry of the local environment of a molecule in a crystal varies from point to point, commonly used macroscopic external fields vary very slowly on the scale length of the unit cell, with the consequence that the crystalline solid can be considered a continuous material with uniform anisotropic point properties. The uniform anisotropy accounts for the symmetry properties of the crystal.

The macroscopic properties of crystals can be described and analyzed quantitatively in terms of the different responses of an anisotropic continuum to particular forces. Consider the case that the response, $\mathscr{R}$, of the crystal to an applied field, **F**, is linear in the applied field, i.e., has the form

$$\mathscr{R} = \chi \mathbf{F}.$$

The response $\mathscr{R}$ is a vector, as is the field **F**, and the susceptibility χ is a tensor (see footnote 2 in Chapter 10 for the analagous relation between the induced dipole moment of a nonspherical molecule and the applied field, in which the polarizability is a tensor quantity); in general $\mathscr{R}$ is not parallel to **F**. The components of χ incorporate the symmetry properties of the crystal. The physical properties of the crystal (which are described by the components of χ) must be invariant under the symmetry operations that characterize its structure, since these operations transform the crystal between indistinguishable configurations. Then not all of the components of χ are independent, and for some crystal lattices, some of the components of χ may vanish. The many macroscopic properties of crystals that can be treated using a response function of the form of 22.1 include polarization of a dielectric in an electric field, electrical conduction, thermoelectric phenomena, double refraction and other optical phenomena, crystal elasticity, and piezoelectric phenomena.

The thermodynamic analyses we have developed in previous chapters refer to isotropic materials. For the purpose of developing the equation of state and the related thermodynamic description of an anisotropic solid, we must generalize our previous analyses. The first step in this generalization is consideration of some aspects of the description of the elasticity of a homogeneous anisotropic solid.

A solid that is acted on by external forces is said to be in a state of stress. Consider a volume element inside a stressed solid. The forces that act on this volume element may be divided into two categories, namely, body forces, which act throughout the solid and whose magnitudes are proportional to the volume of the element, and forces exerted on the surface of the volume element by the material surrounding it, and which are proportional to the area of the surface of the volume element. The force per unit area acting on the surface of the volume element is called the stress. The stress exerted on a solid leads to a deformation that alters a volume element in the unstressed solid by displacement and change of shape. The strain at any point in the solid is defined as the rate of change of displacement with distance; it is a dimensionless quantity.

We consider the case that the applied stresses are constant on a given crystal surface and that the resulting strains are uniform throughout the crystal. Consider the strain associated with the change of the solid from some initial configuration to a different final configuration. If the position of a small element of mass in the initial configuration is $\boldsymbol{X}$ and the position of the same element of mass in the final configuration is $\boldsymbol{x}$, when the strain is uniform everywhere we have

$$x_i = \sum_j \lambda_{ij} X_j,$$
$$X_i = \sum_j \xi_{ij} x_j, \tag{22.1}$$

where i and j represent the Cartesian coordinates x, y, and z. For uniform strains λ_{ij} and ξ_{ij}, which can be represented as the derivatives

$$\lambda_{ij} = \frac{\partial x_i}{\partial X_j},$$
$$\xi_{ij} = \frac{\partial X_i}{\partial x_j}, \tag{22.2}$$

are constants. The displacement associated with the strain $\boldsymbol{X} \to \boldsymbol{x}$ is the vector $\boldsymbol{u}$, whose components along the Cartesian axes are $u_i \equiv x_i - X_i$. The displacement gradients are defined by the derivative

$$u_{ij} \equiv \frac{\partial u_i}{\partial X_j}. \tag{22.3}$$

Using the definition of the displacement and Eq. 22.2, we find the following relation between the displacement gradient and the coefficients defining the transformation $\boldsymbol{X} \to \boldsymbol{x}$:

$$\lambda_{ij} = \delta_{ij} + u_{ij}, \tag{22.4}$$

where $\delta_{ij} = 1$ if $i = j$ and 0 if $i \neq j$. A little algebraic manipulation of Eq. 22.1 shows that

$$\sum_i x_i^2 = \sum_{i,j,k} \lambda_{ij}\lambda_{jk} X_j X_k$$
$$= \sum_i X_i^2 + \sum_{i,j}\left(u_{ij} + u_{ji} + \sum_k u_{ki}u_{kj} \right) X_i X_j$$
$$= \sum_i X_i^2 + 2\sum_{i,j} \eta_{ij} X_i X_j$$
$$\eta_{ij} \equiv \frac{1}{2}\left(u_{ij} + u_{ji} + \sum_k u_{ki}u_{kj} \right)$$
$$= \frac{1}{2}\left(\sum_k \lambda_{ki}\lambda_{kj} - \delta_{ij} \right). \tag{22.5}$$

The final configuration of the crystal resulting from the transformation $\boldsymbol{X} \to \boldsymbol{x}$ is completely specified by the coefficients λ_{ij} but not by the coefficients η_{ij}, since the latter involves a symmetric combination of displacement gradients, which has the effect of eliminating information about rotation of the crystal.

The development of a thermodynamic description of a stressed crystal requires evaluation of the work done by the crystal against the applied stresses when its configuration is changed from $\boldsymbol{X}$ to $\boldsymbol{X} + \boldsymbol{dX}$. We consider the case that the crystal is in a state of static equilibrium, which requires that there be no net torque on the crystal. In this case the applied stress is represented by the symmetric tensor σ, the components of which are $\sigma_{ij} = \sigma_{ji}$. We denote the surface of the crystal with configuration $\boldsymbol{X}$ by Σ. The component of force in the ith direction acting on the element of surface $d\Sigma$, due to the applied stress, is

$$f_i = \sum_j \sigma_{ij} \, d\Sigma_j . \tag{22.6}$$

If the displacement gradients corresponding to the strain $\Delta\boldsymbol{X}$ are $\{\Delta u_{ij}\}$, the displacement of the element of surface in the ith direction is

$$\Delta X_i = \sum_k \Delta u_{ik} X_k . \tag{22.7}$$

Then the work done by the crystal against the stress applied to the element of surface $d\Sigma$ is

$$-\sum_i f_i \Delta X_i = -\sum_{i,j,k} \sigma_{ij} \, d\Sigma_j \Delta u_{ik} X_k , \tag{22.8}$$

where we use the common convention in elasticity theory, which assigns a positive value to the stress directed outward from the crystal surface, i.e., positive values of the diagonal elements σ_{ii} are associated with a tensile stress and negative values with a compressive stress. The use of this convention implies that an expansion of the crystal does work on the surroundings, the opposite of the convention that we adopted in the discussion of pressure-volume work done by isotropic systems. The total work done by the crystal against the applied stress is then, using Gauss's theorem to transform a surface integral to a volume integral,

$$\begin{aligned} w &= -\int_\Sigma d\Sigma_j \sum_{i,j,k} \sigma_{ij} \Delta u_{ik} X_k \\ &= -\int_V dV \sum_{i,j} \sigma_{ij} \Delta u_{ij} \\ &= -\sum_{i,j} \sigma_{ij} \Delta u_{ij} V(X), \end{aligned} \tag{22.9}$$

since σ_{ij} and Δu_{ij} are constants. By assumption, $\sigma_{ij} = \sigma_{ji}$, so the antisymmetric part of Δu_{ij} makes no contribution to w. Then, to first order in the displacement gradients, Eq. 22.5 shows that Δu_{ij} can be replaced by $\Delta\eta_{ij}$. We thus find for the differential work done by the crystal against the applied stress

$$đw = -V \sum_{i,j} \sigma_{ij} \, d\eta_{ij} , \tag{22.10}$$

and the differential forms for the internal energy and the Helmholtz free energy are

$$dU = T \, dS + V \sum_{i,j} \sigma_{ij} \, d\eta_{ij} \tag{22.11}$$

and

$$dA = -S \, dT + V \sum_{i,j} \sigma_{ij} \, d\eta_{ij} . \tag{22.12}$$

It follows that

$$\begin{aligned} \sigma_{ij} &= \frac{1}{V}\left(\frac{\partial U}{\partial \eta_{ij}}\right)_{S,\eta'} = \frac{1}{V}\left(\frac{\partial A}{\partial \eta_{ij}}\right)_{T,\eta'} ; \\ T &= \left(\frac{\partial U}{\partial S}\right)_\eta ; \\ S &= -\left(\frac{\partial A}{\partial T}\right)_\eta . \end{aligned} \tag{22.13}$$

The derivatives in Eq. 22.13 with respect to η_{ij} are carried out holding all other η_{kl} constant; the other derivatives are carried out at constant crystal configuration. The derivatives that determine the stresses are evaluated at the initial configuration of the crystal, $\boldsymbol{X}$, corresponding to all $\eta_{ij} = 0$, because in our calculation of the work the stresses were evaluated for the initial crystal configuration (see Eq. 22.9). To calculate the variation of stress with strain, we must evaluate $\sigma_{ij}(\boldsymbol{X} + \Delta\boldsymbol{X}) - \sigma_{ij}(\boldsymbol{X})$. Note that the sign of the definition of the stress as the derivative of the internal energy with respect to strain differs from the definition of the pressure as the derivative of the internal energy with respect to the volume; this difference arises from the already mentioned difference in the convention concerning the sign of the work done by the crystal on its surroundings.

In our previous discussions of the properties of isotropic materials, we have taken the internal energy to be a function of the volume and the entropy. When we consider anisotropic crystals, it is more convenient to take the internal energy to be a function of the crystal configuration and the entropy, so that the dependent variables are the stresses and the temperature. We then write $U = U(\boldsymbol{x}, S)$; the corresponding form for the Helmholtz free energy is $A = A(\boldsymbol{x}, T)$. We require that U be invariant with respect to translation and/or rotation, without deformation, of the crystal. Since the origin of the coordinate system does not enter our description of the crystal, the translational invariance of U is guaranteed. The requirement that U exhibit rotational invariance implies that it (and all other state functions) can depend on the final crystal configuration $\boldsymbol{x}$ only through the initial configuration $\boldsymbol{X}$ and the rotation independent strain parameters η_{ij} (see Eq. 22.5), so that the internal energy can be represented as

$$U(\boldsymbol{x},S)=U(\boldsymbol{X},\eta_{ij},S), \tag{22.14}$$

and the Helmholtz free energy as

$$A(\boldsymbol{x},T)=A(\boldsymbol{X},\eta_{ij},T). \tag{22.15}$$

The expansion of U and A as power series in the strains leads to

$$U(\boldsymbol{X},\eta_{ij},S)=U(\boldsymbol{X},S)+V\sum_{i,j}\sigma_{ij}\eta_{ij}+\frac{V}{2}\sum_{i,j,k,l}C^{S}_{ijkl}\eta_{ij}\eta_{kl}+\cdots \tag{22.16}$$

and

$$A(\boldsymbol{X},\eta_{ij},S)=A(\boldsymbol{X},S)+V\sum_{i,j}\sigma_{ij}\eta_{ij}+\frac{V}{2}\sum_{i,j,k,l}C^{T}_{ijkl}\eta_{ij}\eta_{kl}+\cdots, \tag{22.17}$$

where

$$C^{S}_{ijkl}=\frac{1}{V}\left(\frac{\partial^2 U}{\partial\eta_{ij}\partial\eta_{kl}}\right)_{S,\eta'} \quad\text{and}\quad C^{T}_{ijkl}=\frac{1}{V}\left(\frac{\partial^2 A}{\partial\eta_{ij}\partial\eta_{kl}}\right)_{T,\eta} \tag{22.18}$$

are the adiabatic and isothermal (second order) elastic constants. These adiabatic and isothermal material properties are identified by the superscripts S and T on the C_{ijkl}. Because the strains η_{ij} are symmetric and the elastic constants are second derivatives of U or A with respect to the strains, there are many fewer independent values of the C_{ijkl} than the maximum 81 fourfold combinations of $ijkl$ ($i, j, k, l = x, y, z$). Indeed, because the η_{ij} are symmetric under the exchange of i and j, the C_{ijkl} are symmetric under that exchange as well as under the exchange of the pairs ij and kl of two strains. Then there are only 21 independent values of either the adiabatic or isothermal elastic constants C_{ijkl}. These are conveniently represented using the Voigt notation in which the following correspondence is defined between the pairs of indices ij and kl and new single indices m and n:

ij or kl:	xx	yy	zz	yz	zx	xy	zy	xz	yx
m or n:	1	2	3	4	5	6	7	8	9

Then, by virtue of the symmetry of η_{ij}, we have $C_{11} = C_{xxxx}$, $C_{12} = C_{xxyy}$, . . . , $C_{46} = C_{yzxy}$, . . . , and the nonzero values of C_{mn} can be written in the array

$$\begin{matrix} C_{11} & C_{12} & C_{13} & C_{14} & C_{15} & C_{16} \\ C_{12} & C_{22} & C_{23} & C_{24} & C_{25} & C_{26} \\ C_{13} & C_{23} & C_{33} & C_{34} & C_{35} & C_{36} \\ C_{14} & C_{24} & C_{34} & C_{44} & C_{45} & C_{46} \\ C_{15} & C_{25} & C_{35} & C_{45} & C_{55} & C_{56} \\ C_{16} & C_{26} & C_{36} & C_{46} & C_{56} & C_{66}. \end{matrix}$$

For most crystals the number of elastic constants is reduced further by virtue of the crystal symmetry. For example, a cubic crystal has only three elastic constants. This follows from the observation that if the coordinate axes coincide with the crystal axes, the invariance of the cubic crystal to a rotation of 90° implies that only C_{11}, C_{12}, and C_{44} are nonzero because the sign of η_{ij} for $i \neq j$ cannot affect the value of σ_{ij}; hence, $C_{iiij} = 0$, etc. The full array of elastic constants for a cubic crystal is then

$$\begin{matrix} C_{11} & C_{12} & C_{12} & 0 & 0 & 0 \\ C_{12} & C_{11} & C_{12} & 0 & 0 & 0 \\ C_{12} & C_{12} & C_{11} & 0 & 0 & 0 \\ 0 & 0 & 0 & C_{44} & 0 & 0 \\ 0 & 0 & 0 & 0 & C_{44} & 0 \\ 0 & 0 & 0 & 0 & 0 & C_{44}. \end{matrix}$$

The thermodynamics of crystals is a very rich subject with many interesting applications. We shall consider only three examples: the anisotropy of thermal expansion of a crystal, the description of thermoelastic behavior, and the equation of state.

The constant stress symmetric thermal strain tensor is defined by

$$\alpha_{\sigma;ij}\equiv\left(\frac{\partial\eta_{ij}}{\partial T}\right)_{\sigma} \tag{22.19}$$

and the volume thermal expansion coefficient at constant stress by

$$\alpha_{\sigma}\equiv\frac{1}{V}\left(\frac{\partial V}{\partial T}\right)_{\sigma}. \tag{22.20}$$

To relate α_σ to the $\alpha_{\sigma;ij}$ we use the following identity:

$$\frac{V(\boldsymbol{x})}{V(\boldsymbol{X})}=J=\det[\lambda_{ij}]. \tag{22.21}$$

This identity is a consequence of the transformation from $V(\boldsymbol{x})$ to $V(\boldsymbol{X})$, since, if L_{ij} is the jth Cartesian component of the ith crystal edge when the crystal has configuration $\boldsymbol{X}$, and l_{ij}, the corresponding component when the crystal has configuration $\boldsymbol{x}$, we can write $V(\boldsymbol{x}) = \det[l_{ij}] = \det[(\lambda L)_{ij}] = \det[\lambda_{ij}]V(\boldsymbol{X})$. Then, in configuration $\boldsymbol{X}$,

$$\alpha_{\sigma}=\left(\frac{\partial J}{\partial T}\right)_{\sigma}=\sum_{i,j}\left(\frac{\partial J}{\partial\eta_{ij}}\right)_{\eta'}\left(\frac{\partial\eta_{ij}}{\partial T}\right)_{\sigma}=\sum_{i}\alpha_{\sigma;ii}, \tag{22.22}$$

Table 22.1 Principal Thermal Expansion Coefficients of Some Crystals (Unit = 10^{-6}/°C)

Crystal	Temperature	Symmetry of α_{ij}	α_{11}	α_{22}	α_{33}
Gypsum	313K	$\alpha_{11}\neq\alpha_{22}\neq\alpha_{33}$	1.6	42	29
Aragonite	313K		35	17	10
Zinc	60K	$\alpha_{11}=\alpha_{22}\neq\alpha_{33}$	–2		55
	150K		8		65
	300K		13		64
Calcite	313K		–5.6		25
Copper	298K	$\alpha_{11}=\alpha_{22}=\alpha_{33}$	16		
Diamond	298K		0.89		
Sodium Chloride	298K		40		

using $(\partial J/\partial \eta_{ij})_{\eta'} = \delta_{ij}$. As illustrated by the entries in Table 22.1, the expansion coefficients along different crystal axes can be very different; indeed, some can be negative. Nevertheless, the thermal expansion tensor of a crystal must have the symmetry of the crystal; hence, excluding the possibility of a phase transition, thermal expansion cannot change the symmetry of the crystal.

The thermal expansion of a crystal is only one of the possible thermoelastic responses to changes in the state of stress and the temperature. Indeed, the nine stress components σ_{ij} and the temperature can be taken as independent variables; they then determine the strain components η_{ij} and the entropy. Then we can write for the differential changes in a stress component and in the entropy

$$d\eta_{ij} = \sum_{k,l}\left(\frac{\partial \eta_{ij}}{\partial \sigma_{kl}}\right)_T d\sigma_{kl} + \left(\frac{\partial \eta_{ij}}{\partial T}\right)_\sigma dT$$

$$dS = \sum_{k,l}\left(\frac{\partial S}{\partial \sigma_{kl}}\right)_T d\sigma_{kl} + \left(\frac{\partial S}{\partial T}\right)_\sigma dT. \quad (22.23)$$

The first terms in the differential changes of strain and entropy correspond, respectively, to the isothermal elastic response and the piezocaloric effect (the heat produced as a result of isothermal application of a stress); the second terms correspond, respectively, to the thermal expansion and the heat capacity at constant stress. Consider the generalized free energy

$$\Phi = U - TS + \sum_{i,j}\sigma_{ij}\eta_{ij}, \quad (22.24)$$

from which we find

$$d\Phi = -S\,dT + \sum_{i,j}\eta_{ij}\,d\sigma_{ij}. \quad (22.25)$$

The coefficients of thermal expansion are related to the coefficients of the piezocaloric effect by

$$-\left(\frac{\partial \eta_{ij}}{\partial T}\right)_\sigma = \left(\frac{\partial S}{\partial \sigma_{ij}}\right)_T, \quad (22.26)$$

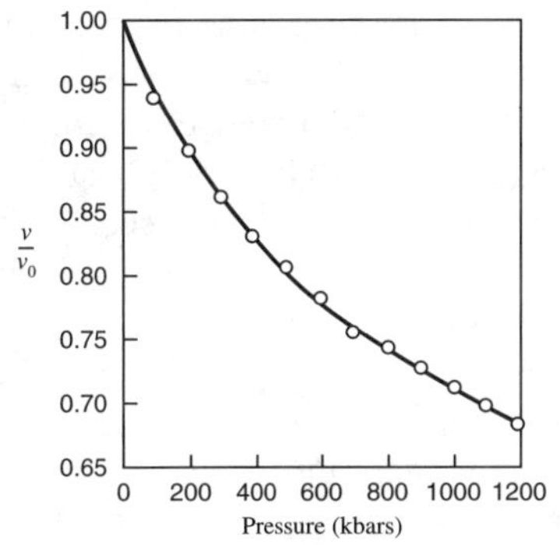

Figure 22.1 The isothermal equation of state of vanadium. The solid line is obtained from Eq. 22.29. The circles represent the experimental data. From O. L. Anderson, *J. Phys. Chem. Solids,* **27,** 547 (1966).

which follows from applying the cross-differentiation identity to (22.25).

In general, the equation of state of a crystal will be a relation, $f(\sigma_{ij};X,\eta_{ij};T) = 0$, between the stress, strain, and temperature for crystal configuration $\boldsymbol{X}$. We shall examine only the simplest case, namely the isothermal relation between pressure and volume. An approximate equation of state, due to Murnaghan, is developed from the following argument. Suppose the isothermal bulk modulus (the inverse of the isothermal compressibility) is a linear function of pressure. That is, we assume that

$$B_T \equiv -V\left(\frac{\partial p}{\partial V}\right)_T = B_T^0 + \left(\frac{\partial B_T^0}{\partial p}\right)_{T,p=0} p$$
$$= B_T^0 + B'p. \quad (22.27)$$

Then, from Eq. 22.27,

$$\frac{dV}{V} = -\frac{dp}{B_T^0 + B'p}, \quad (22.28)$$

which can be integrated to read

$$\ln\left(\frac{V(p,T)}{V_0(T)}\right) = \frac{1}{B'}\ln\left(\frac{B_T^0 + B'p}{B_T^0}\right), \quad (22.29)$$

where $V_0(T) = V(T, p = 0)$. As illustrated by the example in Fig. 22.1, the Murnaghan equation (Eq. 22.29) fits the experimental data for many solids very well.

22.3 Microscopic Theory of the Thermal Properties of Crystalline Solids

In Chapter 11 we focused attention on those properties of the solid state concerned with the nature of the crystalline order and the nature of the bonding. Corresponding to the several broad categories of bonding type, there are characteristic ranges of electrical conductivity, thermal conductivity, strength, and so forth. Of course, the many properties of the various classes of crystalline solids are related to their particular electronic and molecular structure. This very rich subject, which forms the basis for modern electronics technology, will not be considered in this chapter. Instead, we shall examine a few subjects related to the thermal properties of solids, trusting that these are sufficient to whet the student's appetite to learn more.

The oldest aspect of the theory of the solid state is the study of its equation of state and thermodynamic properties. We have briefly discussed the isothermal equation of state of a solid in the last section; we now examine the temperature dependent properties of a solid.

Let us consider first the case of an electrically insulating solid, such as crystalline Ar, NaCl, or diamond. Although the nature of the bonding differs greatly among these crystals, all share the feature that at ordinary temperatures the concentration of free electrons in the solid is negligibly small. Given that the lattice point is the position of minimum potential energy, it is to be expected that the atoms or ions, which have kinetic energy, will oscillate back and forth about their lattice sites. The determination of the modes of motion of a body containing N particles is obviously a very difficult chore, but this analysis may be made, at least in principle, by using the same techniques as described in our discussion of the internal motions of a molecule. Thus, suppose that we regard a perfect crystalline solid as one huge molecule. If there are N atoms in the crystal, there must be $3N - 6$ internal degrees of freedom and six degrees of translational and rotational freedom of the crystal as a rigid body. Clearly, when $N \approx 10^{24}$, the translation and rotation of the entire crystal may be disregarded, and we need consider only the internal degrees of freedom. If the components of the solid are basically monatomic, as they are in Ar, diamond, and NaCl (the several structural units are single particles with only electronic degrees of freedom), then all internal degrees of freedom must correspond to vibrations. If the solid contains polyatomic species, as do crystalline CH_4, N_2 and others, then some of the internal degrees of freedom will correspond to rotation (or torsion) about a lattice site. If the molecular species is very complex, say, $CH_4CH_2CH_2CH_2CH_3$, then there are complicated internal modes that have some rotational and some vibrational character. The moral of these observations is simple—each solid must be examined separately to see if there are any special features that must be considered in the theoretical description.

In the simplest case, when all the internal degrees of freedom may be considered to be vibrations, it is possible to imagine a normal coordinate analysis carried out. The starting point for this analysis is the representation of the potential energy of the crystal as a quadratic form in the displacements of all the particles from defined lattice positions (the structure of the crystal is assumed given). Just as for a polyatomic molecule, a new set of coordinates is found in which the $3N - 6$ vibrational motions are independent; the internal motion of the crystal is then completely described by these normal modes and the corresponding normal frequencies. Let ν_i be the frequency of the ith normal mode. Because, by definition, each normal mode of vibration is independent of all others, the canonical partition function of the crystal is of the form, similar to Eq. 21.113,

$$Q_N^{\text{crystal}} = \prod_{i=1}^{3N-6}\left[1-\exp\left(\frac{h\nu_i}{k_BT}\right)\right]^{-1}. \tag{22.30}$$

Because the crystal contains N atoms, we have labeled the partition function with a subscript N. However, Q_N^{crystal} really corresponds to the single-molecule partition function Q_1 of Chapter 21 because all the atoms are in one crystal.

Now, since $N \approx 10^{24}$, and the spectrum of frequencies is dense, we may consider them to be a continuous distribution in the variable ν. Let $\mathcal{P}(\nu)\,d\nu$ be the fraction of frequencies lying in the range between ν and $\nu + d\nu$. We shall later associate one degree of freedom with each frequency, so that $\mathcal{P}(\nu)$ is just proportional to the density of vibrational states of the crystal. Clearly we have

$$\int_0^\infty \mathcal{P}(\nu)\,d\nu = 1, \tag{22.31}$$

and the partition function can be written in the form

$$\begin{aligned}\ln Q_N^{\text{crystal}} &= -\sum_{i=1}^{3N-6}\ln\left[1-\exp\left(-\frac{h\nu_i}{k_BT}\right)\right]\\ &= -3N\int_0^\infty \mathcal{P}(\nu)\ln\left[1-\exp\left(-\frac{h\nu}{k_BT}\right)\right]d\nu. \end{aligned}\tag{22.32}$$

For convenience we now choose to measure the energy of each normal mode of vibration relative to the energy of its lowest quantum state. As discussed in Section 21.6, to effect this change of energy zero we multiply through by $\exp(-\epsilon_{0i}/k_BT)$ for each normal mode. In the present case $\epsilon_{0i} = \frac{1}{2}h\nu_i$. Proceeding as usual, we then can obtain from Q_N^{crystal} the thermodynamic properties of the crystal. In the case of the perfect crystal we have

$$\begin{aligned}U - E_0 &= k_BT^2\left(\frac{\partial \ln Q_N^{\text{crystal}}}{\partial T}\right)_V\\ &= -3Nk_BT^2\\ &\quad\times\int_0^\infty \mathcal{P}(\nu)\left\{\frac{\partial}{\partial T}\ln\left[1-\exp\left(-\frac{h\nu}{k_BT}\right)\right]\right\}d\nu\\ &= -3Nk_BT^2\\ &\quad\times\int_0^\infty \mathcal{P}(\nu)\left[\frac{-(h\nu/k_BT^2)\exp(-h\nu/k_BT)}{1-\exp(-h\nu/k_BT)}\right]\\ &= 3Nh\int_0^\infty \frac{\nu\mathcal{P}(\nu)}{\exp(h\nu/k_BT)-1}\,d\nu, \end{aligned}\tag{22.33}$$

where E_0 is the energy of the crystal at absolute zero, the sum of all oscillator zero-point energies.

From Eq. 22.33 for $U - E_0$, we find, for the heat capacity at constant volume,

$$C_V \equiv \left(\frac{\partial U}{\partial T}\right)_{V,N}$$
$$= 3Nh\int_0^\infty \nu \mathcal{P}(\nu)\left\{\frac{\partial}{\partial T}\left[\frac{1}{\exp(h\nu/k_BT)-1}\right]\right\} d\nu$$
$$= 3Nk_B\int_0^\infty \left(\frac{h\nu}{k_BT}\right)^2 \mathcal{P}(\nu)\frac{\exp(h\nu/k_BT)}{[\exp(h\nu/k_BT)-1]^2}\, d\nu. \tag{22.34}$$

The other thermodynamic functions and the equation of state can be obtained in the usual manner (cf. Section 21.5).

Equation 22.34 is typical of the representations of the thermodynamic properties of a crystal in terms of a molecular model. All of these representations involve integration over $\mathcal{P}(\nu)\, d\nu$. Clearly, to reduce the general equations to forms suitable for specific calculation, one must determine the frequency spectrum $\mathcal{P}(\nu)$. It is at this point that the description of the thermal properties of a solid becomes difficult. In general, the computation of $\mathcal{P}(\nu)$ is long and tedious, and before discussing its general features it is worthwhile to examine the consequences of assuming particular simple forms for this distribution function.

The theory of the heat capacity of a crystalline solid played an important role in the evolution of the quantum theory. If a solid is modeled as a collection of harmonic oscillators, then the equipartition theorem of classical theory requires that each vibration contribute k_B to the heat capacity. Thus the classical prediction is that the heat capacity of a solid should be equal to $3Nk_B$ and be independent of temperature. At about the turn of the century, measurements of the heat capacities of many solids showed that the classical theory was incorrect; the heat capacity decreased as the temperature decreased and appeared to approach zero as the temperature approached absolute zero. These observations could not be reconciled with the classical theory of crystals. In 1907 Einstein suggested that the motions of the atoms in a crystal are quantized. Then the typical characteristic vibrational temperature of a solid could be larger than room temperature (e.g., diamond), about equal to room temperature (e.g., Al), or smaller than room temperature (e.g., Pb). When the temperature is much higher than the solid's characteristic temperature, the heat capacity becomes $3Nk_B$ (see Section 21.5). When the temperature decreases below the solid's characteristic temperature and ultimately approaches zero, the heat capacity of the solid decreases below $3Nk_B$ and ultimately approaches zero. The Einstein model thereby explains the outstanding qualitative features of the heat capacity-versus-temperature curve of a solid.

The Einstein model of a solid is very simple. It is based on the assumption that all the vibrational frequencies of the solid have the common value ν_E. Then $\mathcal{P}(\nu)\, d\nu$ has the value unity when $\nu = \nu_E$, and zero for all other values of ν. Returning to Eq. 22.34 and inserting the form of $\mathcal{P}(\nu)\, d\nu$ just described, the integrand is nonzero only at the frequency ν_E. Thus for the Einstein model we find that

$$C_V^{\text{Einstein}} = 3Nk_B\left(\frac{h\nu_E}{k_BT}\right)^2 \frac{\exp(h\nu_E/k_BT)}{[\exp(h\nu_E/k_BT)-1]^2}$$
$$= 3Nk_B\left(\frac{\Theta_E}{T}\right)^2 \frac{\exp(\Theta_E/T)}{[\exp(\Theta_E/T)-1]^2}. \tag{22.35}$$

A plot C_V^{Einstein} versus T/Θ_E is shown in Fig. 22.2. The quantity Θ_E is of course defined by

$$\Theta_E = \frac{h\nu_E}{k_B}, \tag{22.36}$$

and is seen to correspond to the characteristic vibrational temperature of a molecule introduced in Section 21.5.

Notwithstanding its general success, the Einstein model has failings that are attributable to the oversimplified form chosen for $\mathcal{P}(\nu)$. For example, as $T \to 0$, the heat capacity approaches zero exponentially. To show this we examine Eq. 22.35, note that $\exp(\Theta_E/T) \to \infty$ as T approaches 0, so that the term unity in the denominator becomes negligibly small relative to $\exp(\Theta_E/T)$, and obtain after division of common factors the simple form

$$\lim_{T\to 0} C_V^{\text{Einstein}} = 3Nk_B\left(\frac{\Theta_E}{T}\right)^2 \exp\left(-\frac{\Theta_E}{T}\right). \tag{22.37}$$

Experiments show that, although the heat capacity approaches zero as T approaches zero, it does so more slowly than Eq. 22.37 predicts. Thus the theory requires improvement.

The assumption that $\mathcal{P}(\nu)\, d\nu$ is unity for $\nu = \nu_E$ and zero for all other values treats the atomic oscillators as if they were independent of one another. Because of the strong interactions between the atoms this is, of course, not true. It is convenient, now, to examine some of the general properties of $\mathcal{P}(\nu)$, without actually calculating this function. In the process, we shall see how simple ideas can be used to improve upon the Einstein model of the crystal.

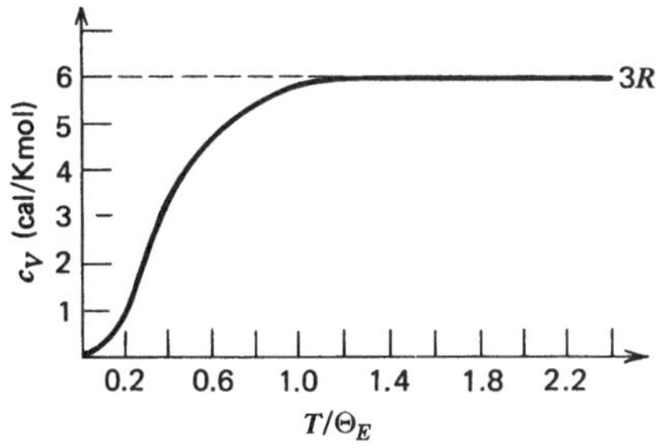

Figure 22.2 The heat capacity corresponding to the Einstein model of a solid [$\mathcal{P}(\nu)\, d\nu = 1$ for $\nu = \nu_E$, $\mathcal{P}(\nu) = 0$ otherwise].

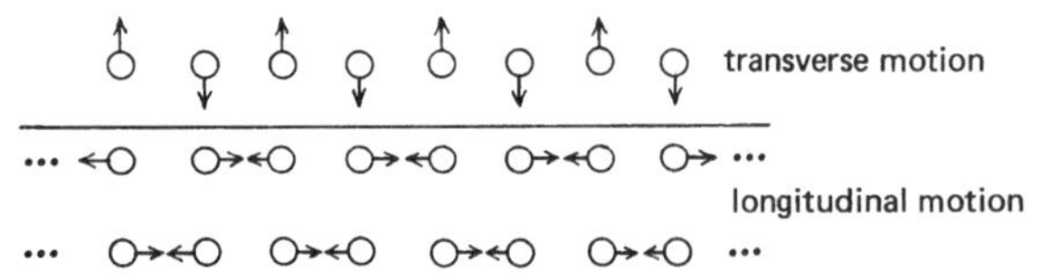

Figure 22.3 Schematic representation of the smallest-wavelength atomic vibration in a one-dimensional crystal.

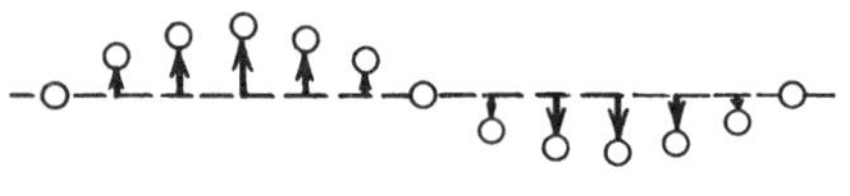

Figure 22.4 Schematic representation of a long-wavelength normal mode in a one-dimensional crystal.

1. In a crystalline medium with atomic structure, and hence a regular lattice spacing, the smallest-wavelength vibration that can occur describes the antiphase motion of two adjacent molecules. By antiphase motion of adjacent molecules we mean that they move in opposite directions either parallel or perpendicular to the line between them, as shown schematically in Fig. 22.3. Why is this the shortest-wavelength motion allowed? Normal modes are defined by the patterns of phased displacements of groups of atoms. Since atoms cannot be closer than one lattice spacing, there can be no simple phase-coherent[1] motion of the atoms with a wavelength less than their separation. Note how a hypothetical continuous medium differs from a real, molecularly coarse medium in this respect. We thus conclude that, for a crystal,

$$\mathcal{P}(\nu) = 0 \qquad \text{for} \qquad \nu > \nu_{\max}, \tag{22.38}$$

where $\nu_{\max}$ is the frequency corresponding to the shortest-wavelength normal mode of the crystal.

2. Just as for a molecule, the total number of degrees of freedom of a crystal is determined by the number of atoms in the crystal. Neglecting the six degrees of freedom corresponding to translation and rotation of the entire crystal, if the particles at the lattice sites are monatomic, the total number of degrees of freedom must be $3N$.

3. Since we have

$$\exp\left(\frac{h\nu}{k_B T}\right) = 1 + \frac{h\nu}{k_B T} + \cdots, \tag{22.39}$$

in the high-temperature limit Eq. 22.34 becomes

$$\begin{aligned} \lim_{T\to\infty} C_V &= 3Nk_B \int_0^\infty \left(\frac{h\nu}{k_B T}\right)^2 \mathcal{P}(\nu)\left[\frac{1}{(h\nu/k_B T)^2}\right] d\nu \\ &= 3Nk_B, \end{aligned} \tag{22.40}$$

where we have used Eq. 22.31. Thus, in the high-temperature limit the contribution to the specific heat from each of the $3N$ modes of vibration is just k_B. The result cited is, again, an example of the equipartition of energy in the limit $T \to \infty$. (See Section 21.5.)

4. Consider now the limit in which the wavelength of the normal vibration is very large relative to the spacing between the particles, such as is shown schematically for a one-dimensional lattice in Fig. 22.4. It is clear that if the wavelength is large enough, so that many atoms are displaced in the normal vibration, the atomic structure of the crystal is of little relevance. In this limit the medium behaves as if there were no atomic structure, that is, as if it were continuous. The vibrations of a continuous medium are well understood. Each possible characteristic vibration of an elastic[2] medium corresponds to a sound wave of a single frequency. That this is so is most easily visualized for the continuous one-dimensional case, which is the violin string. Each possible characteristic vibration of the violin string leads to a sound wave (because the string "drives" the surrounding air) of a single frequency. The superposition of vibrations created when the bow is drawn across the string is the one-dimensional analog of the superposition of vibrations created in an elastic solid when it is struck. Just as an arbitrary displacement of the violin string can be represented as the sum of the displacements corresponding to different characteristic vibrations, so can a displacement in a solid be represented as a sum of normal-mode displacements. It is this possibility of representing a distribution of compressions and expansions in terms of the normal modes of vibration corresponding to sound waves that permits us, in the long-wavelength limit, to associate the distribution of atomic amplitudes with the corresponding continuous density distribution, and thereby with the frequency spectrum of sound waves. Now, it can be shown that for a continuous medium[3]

$$\mathcal{P}(\nu) \propto \nu^2. \tag{22.41}$$

Since $\mathcal{P}(\nu)$ is proportional to the density of vibrational states in the crystal, Eq. 22.12 states that for $\nu \ll \nu_{\max}$ the density of vibrational states is proportional to ν^2. Debye suggested (1912) that the approximation 22.41 be used for the entire frequency distribution, that is, up to $\nu_{\max}$. A sketch of this approximation to the frequency spectrum is displayed in Fig. 22.5. Since the total number of normal modes of

---

[1] By coherent we mean exactly in phase for all time. This includes the notion of exactly out of phase for all time, since "out of phase" just specifies the direction of motion.

[2] By elastic we mean that the force and displacement are related, as they are for a perfect spring, by Hooke's law.

[3] The argument is very similar to that used to derive Eq. 15.25. The numbers of longitudinal and transverse modes with frequencies less than ν in a cube of elastic continuum with edge length L are just the numbers of positive integers that satisfy, respectively,

$$\frac{c_l^2}{L^2}(n_1^2 + n_2^2 + n_3^2) \le \nu^2$$

$$\frac{2c_t^2}{L^2}(n_1^2 + n_2^2 + n_3^2) \le \nu^2$$

where c_l and c_t are the velocities of longitudinal and transverse waves. The remainder of the argument follows that given in Chapter 15.

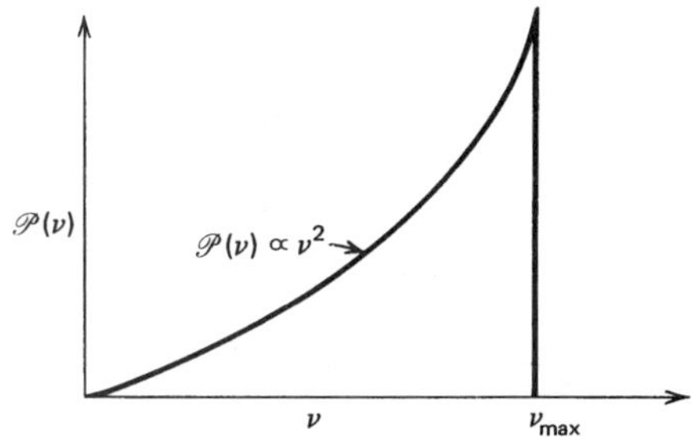

Figure 22.5 Schematic representation of a Debye spectrum.

vibration is $3N$, ν_{max} is fixed by a normalization condition. We now examine the limiting behavior of the heat capacity as $T \to 0$. At low temperature we have $h\nu_{max}/k_BT >> 1$, and only the lowest-frequency vibrations of the solid are thermally excited. It is in this temperature range that setting $\mathcal{P}(\nu) \propto \nu^2$ is a good approximation. If $x \equiv (h\nu/k_BT)$, then Eq. 22.34 is easily transformed to read

$$C_V = 3Nk_B \int_0^{x_{max}} x^2 \mathcal{P}(x) \frac{e^x}{(e^x - 1)^2} \left(\frac{k_BT}{h} \right) dx, \qquad (22.42)$$

so that, if we substitute $\mathcal{P}(x) \propto x^2T^2$,

$$C_V \propto T^3 \int_0^{x_{max}} \frac{x^4 e^x}{(e^x - 1)^2} dx. \qquad (22.43)$$

But if $x_{max} >> 1$, the integrand tends to zero for $x << x_{max}$, and therefore the upper limit of integration may be made infinity without introducing appreciable error. The integral is then just a number, independent of temperature, and we can write[4]

$$\lim_{T \to 0} C_V \propto T^3. \qquad (22.44)$$

The specific value of the proportionality constant depends on the properties of the solid, and need not concern us. What is important is the functional form of the predicted temperature dependence, which is verified for many solids when the temperature is low enough. Of course, Eq. 22.44 depends on the assumption embodied in Eq. 22.41, and that is a valid low-frequency approximation only for three-dimensional systems. Crystals of polymer molecules, or crystals containing strong linear or strong planar interactions, are, in a certain sense, "weakly three-dimensional." For these cases, Eq. 22.41 cannot be used and, correspondingly, a different low-temperature limit is found for C_V.

5. In general, except in the very-long-wavelength region, the distribution of vibrations of a crystal differs considerably from that of the Debye model. A qualitative understanding of the nature of the vibrational spectrum of a crystal can be built up as follows.

[4] The thermodynamic consequences of Eq. 22.15 were discussed in Chapter 18.

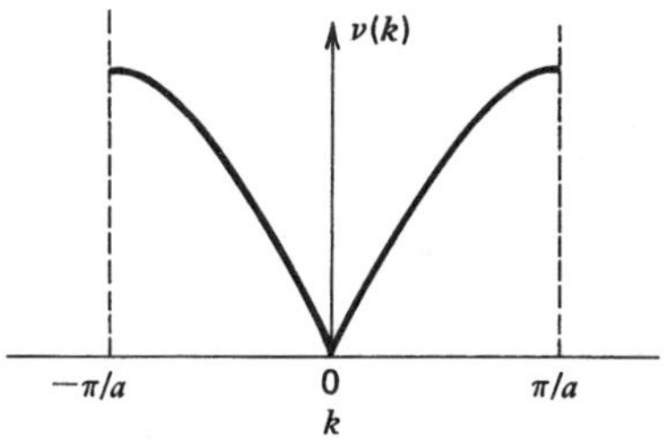

Figure 22.6 Dispersion relation, $\nu(k)$ versus k, for a linear lattice with nearest-neighbor spacing a and only nearest-neighbor interaction. The branch shown corresponds to the longitudinal mode. For this lattice, with one atom per unit cell, this is the longitudinal acoustic mode—note that $\nu(k) \to 0$ as $k \to 0$. For a mode of frequency ν, the time dependence of the vibrational displacement of a lattice point is proportional to $\cos(ka - 2\pi\nu t)$. Note that $\nu(k)$ must be periodic with period $2\pi/a$, since the substitution $k \to k + 2\pi/a$ introduces no change in the displacement pattern. It is therefore customary to restrict k to the range $-\pi/a \le k \le \pi/a$.

Consider the one-dimensional lattice shown in Fig. 22.3. As indicated, there are two kinds of atomic motion: along the chain, say, the z axis (*longitudinal modes*), and perpendicular to the chain, say, along the x and y axes (*transverse modes*). Each class of atomic motion gives rise to a *branch* of the vibrational spectrum. Clearly, in the case illustrated in Fig. 22.3, the transverse vibrations are degenerate (x and y are equivalent). The characteristic vibrations of the linear lattice are described by their frequencies ν_1, ν_2, . . . and by the pattern of displacements of the atoms that correspond to a wave on the lattice in which the displacements of adjacent atoms differ by only a phase factor. Therefore, the pattern of displacements associated with a vibration of frequency ν_i can be described by the wave vector $k_i = 2\pi/\lambda_i$ for that vibration. The dependence of vibrational frequency on k is known as the *dispersion relation.* The vibrations of a crystal have dispersion; that is, they have frequencies that are different for different wavelengths, for the same reason that the symmetric and antisymmetric stretches in the H_2O molecule have different frequencies, namely, interaction between the atoms in different oscillators.[5] In a molecule of n atoms, the $3n - 6$ vibrations are usually well separated in frequency. In contrast, as already pointed out when $\mathcal{P}(\nu)$ was introduced, in a crystal of N atoms the allowed wavelengths are so close together that we can take k to be a continuous variable. The dispersion relation for longitudinal vibrations of a linear lattice with one atom per unit cell, lattice spacing a, is shown in Fig. 22.6.

Consider next a linear lattice with two atoms per unit cell, as shown in Fig. 22.7. In general, the restoring force between the two atoms in one unit cell will be different from

[5] If the O atom of H_2O did not move, and the two OH stretching motions were independent, they would have the same frequency. But the O atom does move when a H atom moves, so the two OH motions are not independent. Also, on stretching the OH there is a contribution to the change in potential energy from their interaction. The net result is the replacement of two independent OH motions with symmetric and antisymmetric combined motions of different frequencies.

Figure 22.7 Schematic representation of a linear lattice with two atoms, A and B, per unit cell.

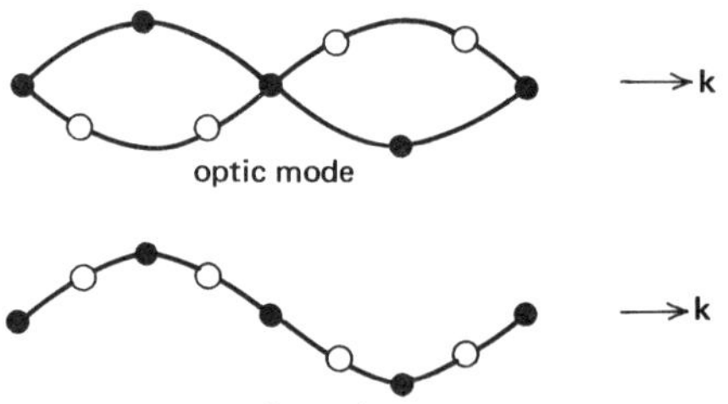

Figure 22.8 Schematic representation of the transverse optic and transverse acoustic modes of a diatomic linear lattice. The atomic displacements are illustrated for a common value of the wave vector (hence also the wavelength).

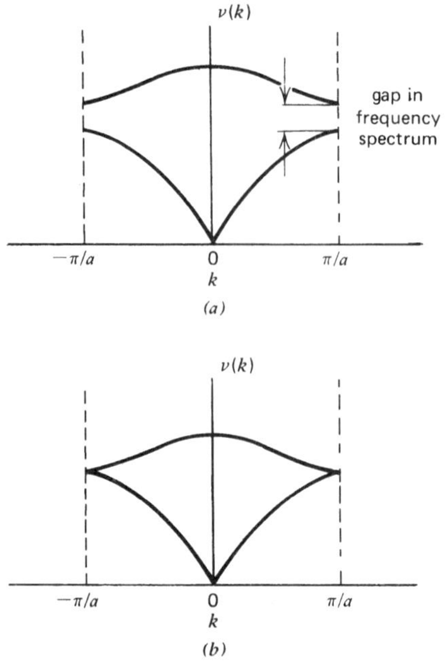

Figure 22.9 Dispersion relations for a linear lattice with two atoms per unit cell. Case (*a*): $m_A \neq m_B$, $a = 2b$. Case (*b*): $m_A = m_B$, $a = 2b$. Nearest-neighbor interactions only in these cases.

that between adjacent atoms in different unit cells, hence the vibrational spectrum is more complicated than that for a linear lattice with only one atom per unit cell. It is easy to distinguish two classes of atomic motions. In one class the unit cells, that is, the pairs of A and B atoms, are displaced with respect to the lattice points. In the other class the atoms vibrate with respect to one another within the unit cells. These internal vibrations can occur, and have a definite phase with respect to the internal vibrations in other unit cells, even when the centers of mass of the unit cells are undisplaced. The lattice with only one atom per unit cell does not have this degree of freedom. Modes involving the displacement of the center of mass of the unit cell are called *acoustic,* whereas those involving internal motions of the unit cell are called *optic* (see Fig. 22.8). The two classes of modes are most easily distinguished in the long-wavelength limit.

It is easy to see that there is a fundamental difference between acoustic and optic modes when their wavelengths become indefinitely large, at $k = 0$. For the case of the acoustic mode, in this limit the centers of mass of all the unit cells move in phase, and there is no longer a restoring force. Consequently, for the acoustic mode we expect that $\nu(k) \to 0$ as $k \to 0$. In contrast, for the optic mode, even if all internal displacements of all unit cells are in phase, there is a nonvanishing restoring force internal to the unit cell. Consequently, for the optic mode we expect that $\nu(k) \to \nu_0 \neq 0$ as $k \to 0$. In fact, ν_0 is usually the highest-frequency vibration of the system. The longitudinal acoustic and optic mode dispersion relations for a linear lattice are shown in Fig. 22.9.

In general, for a three-dimensional lattice, if there is only one atom per unit cell there are three branches in the vibrational spectrum, one corresponding to longitudinal modes and two to transverse modes. These three are acoustic modes and are really purely longitudinal or transverse only for

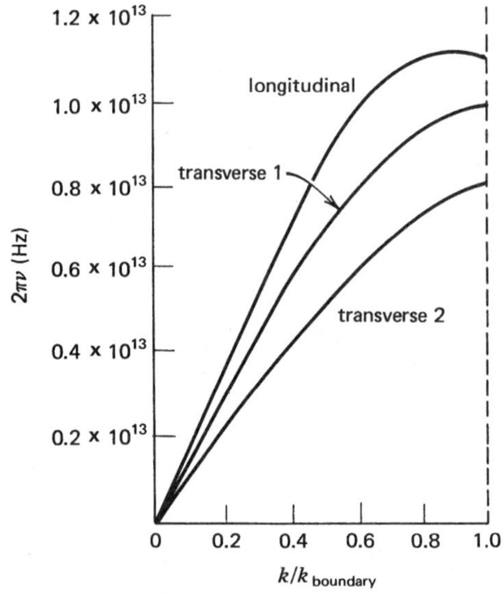

Figure 22.10 Typical dispersion relation in a face-centered cubic crystal in the [210] direction (data are for solid argon). From J. Reissland, *The Physics of Phonons* (Wiley, New York, 1973).

waves propagated along symmetry directions in the crystal structure. When there is more than one atom per unit cell, intracell vibrations give rise to optic branches in the vibrational distribution. If there are s atoms per unit cell, there are $3s - 3$ optic branches in the vibrational spectrum. It is generally the case that for the acoustic modes $\nu(k) \to 0$ as $k \to 0$, whereas for the optic modes $\nu(k) \to \nu_0$ as $k \to 0$. Figure 22.10 shows a set of calculated frequency dispersion curves for crystalline Ar. In this solid there is one atom per unit cell, hence only three acoustic branches to the spec-

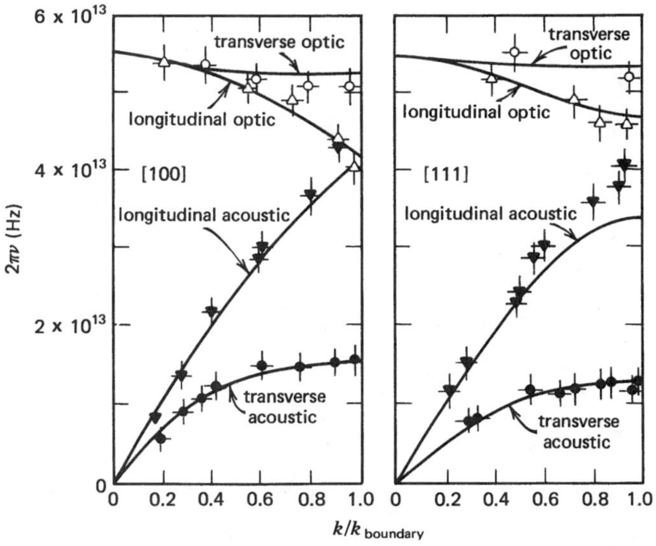

Figure 22.11 Dispersion relations for the [100] and [111] directions in Ge. From W. Cochran, *Proc. Roy. Soc. A* **253,** 260 (1959).

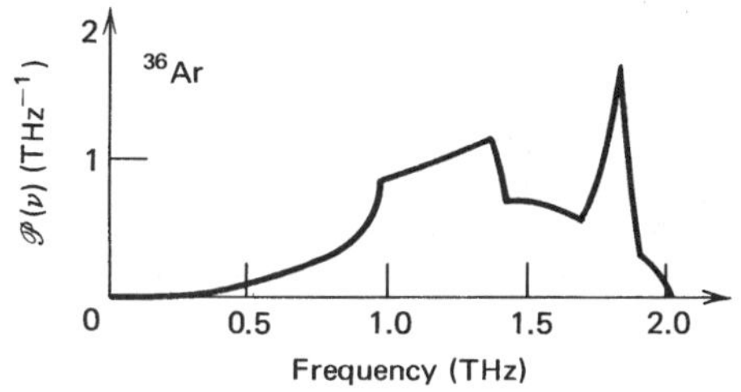

Figure 22.12 Vibrational frequency distribution for crystalline Ar. From D. N. Batchelder, M. F. Collins, B. C. G. Haywood, and R. G. Sidney, *J. Phys. C. Solid State Phys.* **3,** 249 (1970).

trum. Figure 22.11 shows a set of frequency dispersion curves for crystalline Ge. In this case there are two atoms per unit cell, hence three acoustic and three optic branches in the vibrational spectrum.

Finally, it is of interest that there are some points on the frequency dispersion curves where $(\partial \nu/\partial k) = 0$. For example, one such point is the maximum in the longitudinal acoustic branch shown in Fig. 22.10. At the frequencies for which the gradient in the dispersion relation vanishes, there are discontinuities in the frequency derivative of the density of vibrational states, namely, in $\partial \mathcal{P}(\nu)/\partial \nu$. These are called *van Hove singularities,* and there is at least one in the density of vibrational states of a crystal. They show up as kinks in $\mathcal{P}(\nu)$, as seen in Fig. 22.12.

The reason these occur is as follows. Imagine two surfaces of constant ν in k space (see Section 11.11), corresponding to ν and $\nu + d\nu$. These two surfaces enclose a volume dV_k. The volume dV_k can be obtained by multiplying an element of area on one surface, $d\sigma$, by the distance between surfaces, Δk, and integrating the contributions over the entire constant frequency surface. Then we have

$$\mathcal{P}(\nu)\, d\nu = \frac{1}{8\pi^3}\, dV_k = \frac{1}{8\pi^3}\int \Delta k\, d\sigma,$$

where the factor $1/8\pi^3$ is the number of allowed values of k per unit volume (see Eq. 15.25). We now write

$$d\nu = \left|\left(\frac{\partial \nu}{\partial k}\right)\right| \Delta k + \text{higher-order terms},$$

which, on substitution in the equation for $\mathcal{P}(\nu)$, gives

$$\mathcal{P}(\nu) = \frac{1}{8\pi^3}\int \frac{d\sigma}{|(\partial \nu/\partial k)|}.$$

Then, if $(\partial \nu/\partial k)$ is zero, or if it discontinuously changes sign, the next-higher term in the Taylor series must be kept and we expect a cusp of some sort in $\mathcal{P}(\nu)$. In general, $(\partial \nu/\partial k) = 0$ for special values of k because of the symmetry of the crystal; it can change sign discontinuously only at points in k space where ν is degenerate (degeneracy occurs when different ν-versus-k curves cross, as happens for some three-dimensional crystal structures).

We conclude that, in general, $\mathcal{P}(\nu)$ has maxima, discontinuities in $d\mathcal{P}(\nu)/d\nu$, and other features specific to the type of lattice and range of the intermolecular forces.

A comparison of the Debye frequency spectrum (Fig. 22.5) with either of the frequency spectra shown in Fig. 22.12 or Fig. 22.13 shows that it does not reproduce at all well the high-frequency end of the true spectrum. Nevertheless, C_V is not very sensitive to the frequency spectrum, and the Debye model provides a good approximation to the specific heat of a monatomic solid. In Fig. 22.14*b* we see that the specific heats of a variety of solids may be reasonably well described in terms of a characteristic Debye temperature, $\Theta_D \equiv h\nu_{max}/k_B$. Very precise measurements show that Eq. 22.43 does not fit the heat capacity data perfectly; the deviations can be expressed as an empirical variation of Θ_D with temperature which arises from the deviations of the Debye spectrum from the true vibrational spectrum of the solid (Fig. 22.15), and can be accounted for by use of the correct $\mathcal{P}(\nu)$.

It should now be clear that, by using the considerations outlined, one can interpret the thermodynamic properties and equation of state of a crystalline solid in terms of the potential between its constituent atoms or molecules. Of course, the motion of the atoms in a solid can be represented by the harmonic approximation only when the amplitudes of vibration are small. Near the melting point the average amplitude of atomic displacement is large, anharmonic components of the force are important, and all our considerations must be supplemented. Similarly, when the solid is severely compressed or extended, beyond the domain in which Hooke's law is valid, inelastic components of the response to external force must be considered. In the two limits cited, calculation of the equation of state of the solid in terms of the interatomic potential is very dffficult, since the simplifications used in our analysis are no longer adequate. Since many of the most interesting properties of solids are those observed at high temperature or pressure, it is worthwhile to examine some aspects of the influence of anharmonicity on crystal properties.

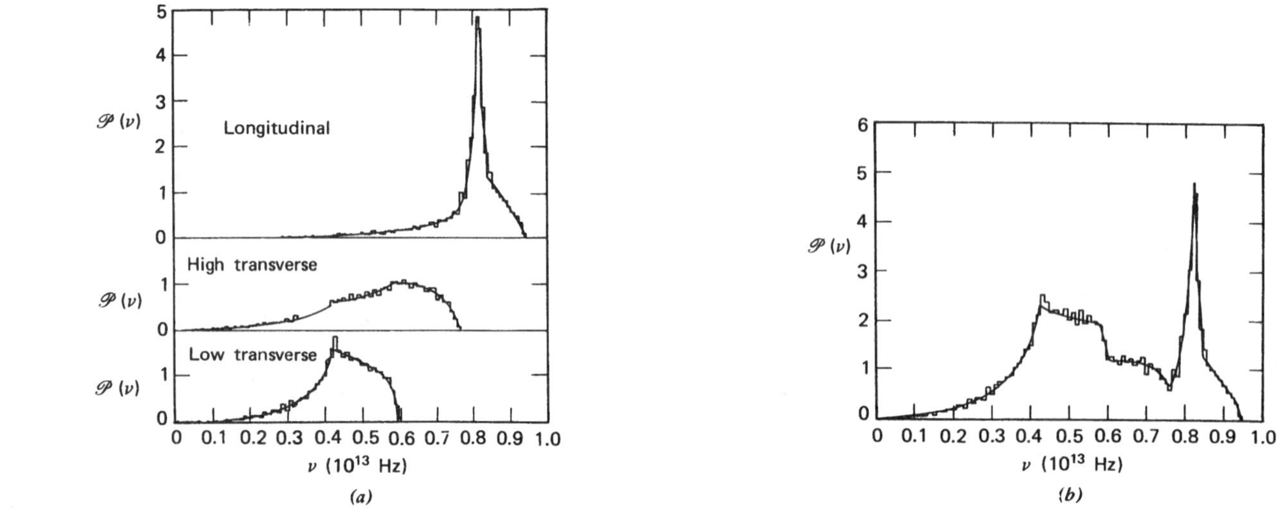

Figure 22.13 (*a*) Vibrational frequency spectra for the three branches of Al. The histograms are obtained from computed eigenfrequencies for 2791 wave vectors. The smooth curves are obtained from the histograms together with the inclusion of the singularities caused by the various critical points. (*b*) Total vibrational frequency spectrum for Al at 300 K. The histogram is obtained from 8373 calculated eigenfrequencies. The smooth curve is obtained from the histogram after inclusion of the singularities arising from the critical points of the various branches. From C. B. Walker, *Phys. Rev.* **103,** 547 (1956).

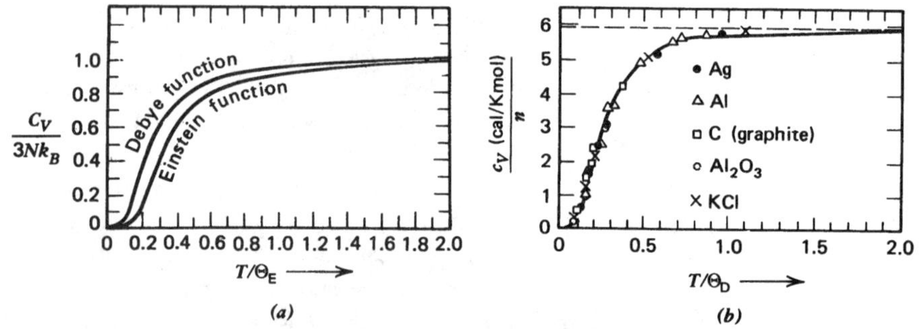

Figure 22.14 (*a*) A comparison of the Einstein and Debye heat capacity curves. The frequency associated with the Einstein curve is equal to the maximum Debye frequency. (*b*) Comparison of the Debye heat capacity curve and the observed heat capacities of a number of simple substances. From F. Seitz, *Modern Theory of Solids* (McGraw-Hill, New York, 1940).

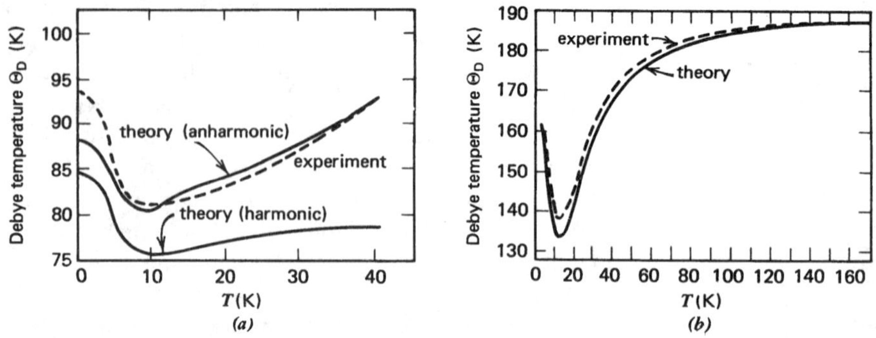

Figure 22.15 Debye temperature Θ_D versus T for (*a*) inert gas solids [J. W. Leech and J. A. Reissland, *Phys. Lett.* **14,** 305 (1965)] and (*b*) NaI [R. A. Cowley, W. Cochran, B. N. Brockhouse, and D. B. Woods, *Phys. Rev.* **131,** 1036 (1963)]. The value of Θ_D plotted is calculated at each temperature by requiring that the Debye model exactly reproduce the observed value of C_V.

22.4 The Contribution of Anharmonicity to the Properties of a Crystal

The potential energy surface of a crystal can be represented by the harmonic approximation only for small-amplitude motion, and this approximation cannot be used to interpret many important crystal properties, such as thermal expansion. Although the general theory of anharmonic crystals is too complex to describe in this text, an idea of the nature of anharmonic contributions to the crystal properties can be obtained from two simple examples.

Consider a one-dimensional anharmonic oscillator that has the Hamiltonian

$$H = \frac{p^2}{2m} + \frac{1}{2}m\omega^2 x^2 + a_3 x^3 + a_4 x^4. \tag{22.45}$$

For illustrative purposes we assume that a_3x^3 and a_4x^4 are small enough relative to the other terms and to k_BT that first-order perturbation theory can be used to evaluate their influence on the oscillator's partition function. For our purposes it is sufficient to evaluate the classical partition function. Just as in quantum mechanical perturbation theory, the first-order correction to the classical partition function is obtained by averaging the perturbation energy over the distribution characteristic of the unperturbed system. We must, then, evaluate

$$\begin{aligned} Q &= \frac{1}{h}\int_{-\infty}^{\infty} dp \exp\left(-\frac{p^2}{2mk_BT}\right)\int_{-\infty}^{\infty} dx \exp\left(-\frac{m\omega^2x^2}{2k_BT}\right) \\ &\quad\times\left(1 - \frac{a_3x^3}{k_BT} - \frac{a_4x^4}{k_BT} + \frac{a_3{}^2x^6}{2(k_BT)^2} + \cdots\right) \\ &= \frac{2\pi k_BT}{h\omega}\left(1 - \frac{3a_4k_BT}{m^2\omega^4} + \frac{15}{2}\frac{a_3{}^2k_BT}{m^3\omega^6} + \cdots\right). \end{aligned} \tag{22.46}$$

As usual, $u = k_BT^2(\partial \ln Q/\partial T)$, from which we find

$$u = k_BT - \left[3\frac{a_4}{(m\omega^2)^2}k_B{}^2 - \frac{15}{2}\frac{a_3{}^2}{(m\omega^2)^3}k_B{}^2\right]T^2, \tag{22.47}$$

and

$$c_V = k_B - \left(6\frac{a_4}{(m\omega^2)^2}k_B{}^2 - \frac{15}{2}\frac{a_3{}^2}{(m\omega^2)^3}k_B{}^2\right)T. \tag{22.48}$$

Of course, the classical harmonic contribution to c_V is just k_B, so that anharmonicity of the type shown leads to an extra contribution to the heat capacity linear in T. The quantum mechanical prediction gives the same temperature dependence.

The results of a detailed calculation of anharmonic contributions to the heat capacities of crystalline Ne, Ar, Kr, and Xe are shown in Table 22.2. Clearly, the prediction that the anharmonicity adds a term to C_V that grows linearly with T is also valid in the general case. When this anharmonic contribution to C_V is accounted for, the observed and predicted variations of Θ_D with temperature are in good agreement (see Fig. 22.15).

A much simpler, yet qualitatively useful, way of accounting for some of the effects of anharmonicity on the properties of a crystal can be based on the following notion. Although thermal expansion of the crystal leads to changes in atomic separations, the displacement of an atom from any position of equilibrium is adequately approximated by a harmonic potential surface. That is, the crystal is harmonic at all temperatures, but the frequencies change with T through a dependence on the volume. To see how this idea is quantified, we return to Eqs. 22.32 and 22.33 and note that if $\mathcal{P}(\nu)$ is the Debye distribution, then the Helmholtz free energy A_{Debye} and the internal energy U_{Debye} are functions of the variable Θ_D/T. We now write

$$\begin{aligned} p_{\text{Debye}} &= -\left(\frac{\partial A_{\text{Debye}}}{\partial V}\right)_T \\ &= -\left(\frac{\partial A_{\text{Debye}}}{\partial \Theta_D}\right)_T\left(\frac{\partial \Theta_D}{\partial V}\right)_T. \end{aligned} \tag{22.49}$$

Table 22.2 Anharmonic Contributions to the Heat Capacity, Showing the Approach to the Linear Law at High Temperatures[a]

T(K)	$\dfrac{c_V^0}{3k_B}$	$\dfrac{\Delta c_{V\,(\text{anh})}}{3k_B}$	$-\dfrac{10^4\,\Delta c_{V\,(\text{anh})}}{3k_BT}$ (K^{-1})
(a) Neon			
4	0.020	– 0.015	35.5
8	0.165	– 0.058	72.5
12	0.368	– 0.123	103
16	0.536	– 0.156	97.5
20	0.655	– 0.175	87.5
24	0.738	– 0.193	80.4
(b) Argon			
10	0.158	– 0.014	14.0
20	0.523	– 0.040	20.0
30	0.727	– 0.050	16.7
40	0.830	– 0.059	14.8
50	0.885	– 0.072	14.4
60	0.918	– 0.085	14.2
70	0.938	– 0.098	14.0
80	0.952	– 0.111	13.9
(c) Krypton			
20	0.647	– 0.021	10.5
40	0.886	– 0.038	9.45
60	0.946	– 0.055	9.09
80	0.969	– 0.072	9.05
100	0.979	– 0.090	9.04
120	0.986	– 0.108	9.03
(d) Xenon			
20	0.699	– 0.014	7.02
40	0.907	– 0.025	6.32
60	0.957	– 0.037	6.16
80	0.975	– 0.049	6.14
100	0.984	– 0.061	6.13
120	0.988	– 0.074	6.13
140	0.991	– 0.086	6.13

[a] c_V^0 is the harmonic part of the heat capacity. The calculations all refer to constant volume corresponding to that at $T = 0$ K. Reproduced, with permission, from J. W. Leech and L. A. Reissland, *J. Phys. Chem. Solid State Phys.* **3**, 997 (1970), copyright The Institute of Physics.

Noting that U_{Debye} is of the form $\Theta_D f(\Theta_D/T)$, and using the change of variables

$$-T\left[\frac{\partial}{\partial T} f\left(\frac{\Theta_D}{T}\right)\right]_{\Theta_D} = \Theta_D\left[\frac{\partial}{\partial \Theta_D} f\left(\frac{\Theta_D}{T}\right)\right]_T, \quad (22.50)$$

we find from the thermodynamic relation

$$U = -T^2\left[\frac{\partial}{\partial T}\left(\frac{A}{T}\right)\right]_V \quad (22.51)$$

that

$$U_{\text{Debye}} = \Theta_D\left(\frac{\partial A_{\text{Debye}}}{\partial \Theta_D}\right)_T, \quad (22.52)$$

since holding V constant is equivalent to holding Θ_D constant. Then, using Eq. 22.49, we have (simplifying the subscript to D)

$$p_D = -\frac{U_D}{\Theta_D}\left(\frac{\partial \Theta_D}{\partial V}\right)_T \quad (22.53)$$

$$= \frac{U_D}{V}\gamma_G, \quad (22.54)$$

where

$$\gamma_G \equiv -\left(\frac{\partial \ln \Theta_D}{\partial \ln V}\right)_T \quad (22.55)$$

is called the Grüneisen constant. Clearly, the Grüneisen constant is a measure of the effects of anharmonicity inasmuch as it represents the rate of change of crystal vibrational frequencies with the volume. Now, differentiating Eq. 22.54 with respect to T, holding V fixed,

$$\left(\frac{\partial p_D}{\partial T}\right)_V = \gamma_G\frac{C_V}{V}, \quad (22.56)$$

and using the identity

$$\left(\frac{\partial p}{\partial T}\right)_V = -\left(\frac{\partial p}{\partial V}\right)_T\left(\frac{\partial V}{\partial T}\right)_p, \quad (22.57)$$

gives

$$\gamma_G = \frac{\alpha V}{\kappa_T C_V}, \quad (22.58)$$

where α and κ_T are, as usual, the coefficients of thermal expansion and isothermal compressibility, respectively. Equation 22.58 provides a means of calculating γ_G and thereby generating an equation of state for the solid (i.e., Eq. 22.54). According to Eq. 22.58, γ_G is temperature-independent to the extent that α, κ_T, and C_V are.

It is generally found that the Grüneisen equation of state gives a reasonable but not exact representation of the properties of a crystal; some values of γ_G are listed in Table 22.3. Given the crudity of the physical ideas, it could not be expected to be better.

Table 22.3 Examples of the Grüneisen Parameter for Cubic Crystals[a]

Crystal	Temperature Range (K)	γ_G Range
Ne	0–25	3.1–3.5
Ar	0–50	2.8–3.1
Kr	0–50	2.75–3.05
Xe	0–50	2.75–3.0
Cu	0–600	1.7–1.95
Ag	0–300	2.7–2.8
Au	0–300	2.3 ± 0.05
Al	0–400	2.3–2.7
Na	0–200	0.9–1.1
K	0–150	1.05–1.2

[a] Reproduced from J. A. Reissland, *The Physics of Phonons* (Wiley, New York, 1973).

22.5 Some Properties of Complex Solids and of Imperfect Solids

The considerations of the preceding two sections suffice to provide an interpretation of the thermodynamic properties of solids composed of monatomic molecules, but are inadequate to interpret the rich spectrum of properties exhibited by solids composed of polyatomic molecules. In this section we mention a few of the new features introduced into the description when the molecules interact with directional forces and are polyatomic.

Consider a crystal composed of polyatomic molecules. In addition to the spectrum of vibrations describing the motions of the centers of mass of the molecules, there are vibrations corresponding to the internal degrees of freedom of the molecules. Given nonzero interactions between the molecules, it is not strictly correct to separate the vibration spectrum of the crystal into intramolecular vibrations and intermolecular vibrations, since all are coupled together and, in principle, of mixed character. In practice, however, the forces within a molecule are so much larger than the forces between molecules that it is an excellent approximation to regard the intramolecular vibrations as being distinct from the intermolecular vibrations, and to assume that the former are only slightly influenced by the latter in the solid. In the simplest cases, then, the vibrational spectrum of a crystal of polyatomic molecules consists of the discrete set of molecular vibrations superimposed on the continuous spectrum arising from center-of-mass motion. Of course, it is also necessary to account for the rotation of the polyatomic molecules. When the rotation is unhindered, the contribution to the energy and other thermodynamic functions is the same as that corresponding to rotation in the dilute gas. When the rotation is hindered, it may be necessary to include in the vibrational spectrum a set of torsional oscillations. Clearly, all the considerations discussed in the study of the internal motion of polyatomic molecules in Chapter 21 are relevant. It must also be remembered that, because of the intermolecular forces in the crystal, the properties of the molecule may be somewhat altered from what they were in the dilute gas. For example, experiments show that the internal vibration frequencies of molecules in the crystalline phase often differ by ~10 cm^{-1} from the corresponding values in the gas phase, even when there is not specific bonding between molecules. If such bonding exists, as when hydrogen bonding occurs, the shifts in frequencies of bonded groups can be much larger (~100 cm^{-1}). Of course, if there is any form of covalent bonding between molecules in the crystal (e.g., in I_2), large alterations in the intramolecular vibrational frequencies are found.

It is easy to construct a useful approximation to the thermodynamic properties of a crystal of polyatomic molecules when hydrogen bonding and covalency can be neglected. Then the spectrum of vibrations may be regarded as the superposition of several Einstein spectra, one each for the several intramolecular vibration frequencies, and a Debye spectrum for the centers of mass. If there exist low-frequency molecular torsional modes, these also may be represented by Einstein terms in the spectrum (see Fig. 22.16). Consequently, a first approximation to the heat capacity of a complex solid is a superposition of Einstein heat capacity terms (one for each internal vibration and each torsion) and a Debye heat capacity term. All other thermodynamic functions may be constructed in the same way.

Thus far in our description of the thermal properties of crystals it has been assumed that the solid state has a perfectly ordered lattice in which the environment of any selected molecule is the same as that of any other molecule in the lattice. This is an idealization; naturally occurring crystalline matter has several different kinds of inherent imperfections.[6] These imperfections make a significant contribution to a number of the properties of a crystal—for example, the strength, coefficient of self-diffusion, magnitude of ionic conductivity (in alkali halides), and so forth. The study of the properties of imperfections is an important

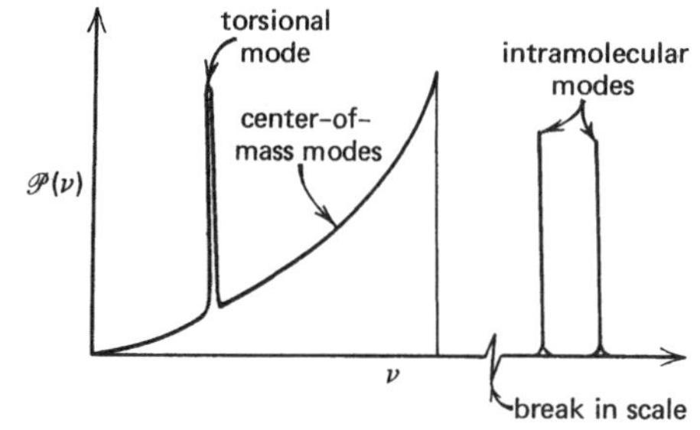

Figure 22.16 Schematic representation of a simple approximation to the frequency spectrum of a polyatomic solid. The torsional mode often has a frequency comparable with those of center-of-mass modes, whereas intramolecular modes usually have much higher frequencies. Note the break in the ν axis.

part of the science of metallurgy, and much of the recent progress in the design of new alloys has arisen from increased understanding of imperfections and how to control them. In this book we shall give only a brief qualitative description of several typical kinds of imperfections.

The first and most obvious deviation from perfect lattice symmetry is the existence of crystal surfaces. Molecules at the surface of a crystal obviously cannot have the same environment as a molecule deep inside the lattice. At the level of thermal properties, the differences in interaction energies and vibrational motions between a molecule in the surface and one in the bulk leads to a *surface free energy.* The magnitude of the surface free energy plays a role in determining the shape of a crystal—a perfect crystal at equilibrium has a shape that minimizes the surface free energy. At the level of electronic structure, the existence of a surface leads to energy levels associated with the surface and differing from those characteristic of the bulk crystal. Such energy levels are important in phenomena as diverse as catalysis and electrical conduction in composite semiconductors. Of course, the ratio of numbers of surface to bulk atoms, $N^{2/3}/N \approx N^{-1/3} \approx 10^{-8}$ in 1 mol, implies that the specific properties of the surface of the crystal need be taken into account only when the phenomena of importance are limited to the surface zone (as in catalysis).

Two common types of crystal imperfections are vacant lattice sites and molecules in interstitial positions. Vacancies occur at random throughout a crystal. In general, the concentration of vacancies is extremely small even close to the melting point of the crystal. We can see why this is so as follows. A vacancy can be imagined as generated by removing a molecule from the bulk of the crystal and placing it on the surface. If each molecule in the crystal interacts with Z neighbors, the energy of sublimation of the crystal is $NZ\epsilon/2$, where ϵ is the strength of the interaction between two molecules at lattice sites. The divisor 2 arises from the need to avoid counting each interaction twice. Now, a molecule on the surface has, approximately, half as many neighbors with which to interact as does one in the

[6] That defects must be present follows from the observation that they represent atomic configurations whose energies differ from that of an ideal perfect crystal by only *finite* amounts, and they are not prohibited from occurring by the constraints that define the equilibrium state.

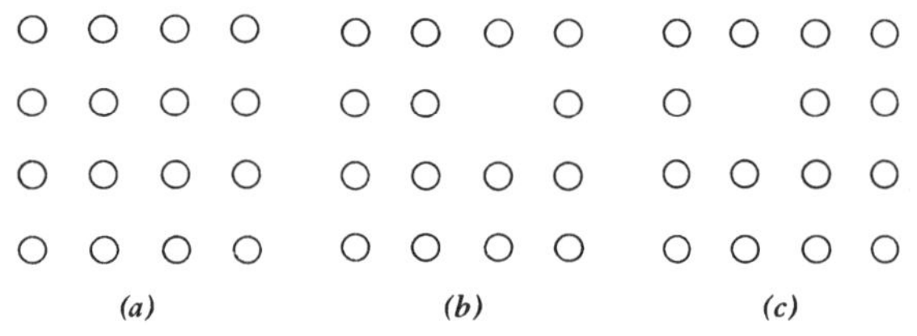

Figure 22.17 Effect of vacancies on atomic displacement in crystal. (*a*) Perfect lattice. (*b*) Lattice with one vacancy. (*c*) Same lattice as in (*b*), but with vacancy moved one column.

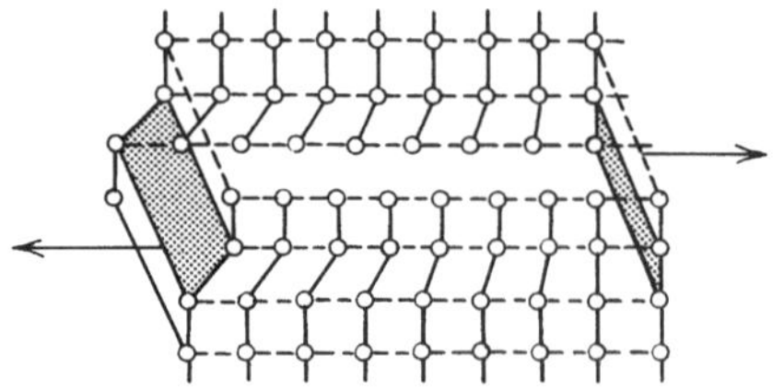

Figure 22.18 Schematic representation of an edge dislocation. The shaded area is the dislocation plane. The arrows indicate the direction of slippage that produces the dislocation.

bulk. Thus the change in energy of interaction on moving one molecule from the bulk to the surface is $NZ\epsilon/2$. To make 1 mol of vacancies, therefore, requires energy about equal to the energy of sublimation. To estimate the entropy of formation of a vacancy, we consider vacancies to be a chemical species and compute the entropy of mixing of atoms and vacancies. This is of the form $-N_{\text{atom}}k_B \ln x_{\text{atom}} - N_{\text{vacancy}}k_B \ln x_{\text{vacancy}}$. Furthermore, because of changes in the local motions of molecules surrounding a vacancy, there is a vibrational contribution to the entropy of formation. Ordinarily the latter contribution is not large compared to the entropy-of-mixing term. The equilibrium concentration of vacancies is determined by the balance between the energy and entropy changes just described. For all solids the energy of sublimation is so large that x_{vacancy} is very small.

Although the concentration of vacancies is small, it has an important influence on the properties of the crystal. When a molecule moves into a vacancy, it leaves a new vacancy behind (see Fig. 22.17). In this way the existence of vacancies permits diffusion of molecules and, in ionic crystals, conduction of an electric current by the ions. Some special electronic states can be associated with vacancies; for example, an electron can be trapped in a negative-ion vacancy in an alkali halide, thereby generating new electronic states in the crystal.

It is possible to estimate the concentration of interstitial molecules by the method outlined above. Because of the strong repulsive forces between molecules when close together, formation of an interstitial is usually much harder than is formation of a vacancy. Clearly, interstitials are easier to form if the molecule is small than if it is large. In some crystals this effect is quite pronounced. For example, AgI is a typical salt and behaves as such below 146°C. In that low-temperature region the conductivity of the salt is small, as is the rate of diffusion of Ag^+ ions. Between 146 and 552°C the crystal has the I^- ions arranged in a body-centered cubic lattice, but the very small Ag^+ ions move so freely through the interstices of this lattice that the electrical conductivity of the solid is very high. On going from 143°C (where the Ag^+ ions are on a lattice) to 146°C, the conductivity increases 5000-fold, but on going from 146 to 540°C it increases only twofold. A more thorough discussion of transport in solids will be found in Section 29.5.

There also occur, mainly in the inorganic oxides but also in other compounds, deviations from perfect stoichiometry. During the preparation of a compound there is a nonzero probability that perfect stoichiometry is not maintained. To maintain overall electrical neutrality in an ionic solid, the "missing" ions must go either on the surface of the crystal or into interstitial positions. Electrical neutrality can also be maintained by substitution of a different ion for the one "missing." A well-studied case of this type is ZnO. It is found, for example, that the conductivity of ZnO is sensitive to the pressure of a surrounding O_2 atmosphere, since at equilibrium that pressure regulates the concentration of O^{2-} defects in the ZnO lattice.

There are also many kinds of crystal imperfections that result when crystal planes are mislocated relative to one another. These are classed as *dislocations;* they influence the plasticity and strength of materials and they can change the local reactivity with selected reagents. A schematic diagram of an edge dislocation is shown in Fig. 22.18.

For further information on the fascinating chemistry of imperfect crystals, the student is referred to *Imperfections in Crystals* by H. G. van Bueren (North Holland, Amsterdam, 1960) and *Theory of Dislocations* by J. P. Hirth and J. Lothe (Wiley, New York, 1982).

22.6 Electronic Heat Capacity of Metals

As a class of materials, metals are distinguished from nonmetals by having a large electrical conductivity. Shortly after the discovery of the electron, it was realized that the conductivity of a metal implied the existence, in its interior, of mobile ("free") electrons. That idea, slightly modified, remains an inherent feature of the modern quantum mechanical description of a metal, aspects of which were discussed in Section 11.11.

In the classical theory of metals the free electrons are treated as a perfect gas. An immediate implication of this supposition is that the N free electrons of a monovalent metal contribute $3Nk_B$ to its heat capacity. Therefore, the heat capacity of a monovalent metal should approach $6Nk_B$ at high temperature, and not vanish at $T = 0$. Of course, neither of these predictions is correct since, as we now know, electrons in a metal cannot be described as a classical gas. In general, the energy spectrum of the electrons in a metal is complicated by strong electron–electron and electron–ion interactions and by the restrictions imposed via the Pauli exclusion principle. For example, the periodic arrangement of the ion cores on a lattice leads to a separation of the allowed energy levels of an electron into bands. Although the detailed structure of these energy bands (the dispersion relation between electron energy and electron momentum) influences the electron contribution to the heat capacity, the effects are usually not large, and we shall not discuss them. For the present it is sufficiently accurate to assume that shielding of the electron–electron and electron–ion interactions is efficient enough that the electrons can be treated as a perfect Fermi–Dirac gas occupying the molar volume of the metal. However, because the mass of the electron is very small, the deviations of the properties of the gas from classical behavior are very large. As we shall show below, the electronic contribution to the heat capacity is proportional to T. Since the lattice motion contribution to the heat capacity is proportional to T^3 at low temperatures (Eq. 22.44), there is a low-temperature regime where the electronic contribution is dominant. Otherwise, for most of the temperature range where the solid metal is stable, the electronic contribution to the heat capacity is very small. The analysis we present was first made in 1928 by Sommerfeld, who applied the theory of the Fermi–Dirac gas to the earlier free-electron theory of metals worked out by Drude and Lorentz.

Consider first the distribution of electrons in a perfect Fermi–Dirac gas at $T = 0$. Clearly, all the electrons occupy the lowest possible energy levels. But for a Fermi–Dirac gas the Pauli exclusion principle prohibits there being more than two electrons per energy level, one with z component of spin angular momentum $+\frac{1}{2}\hbar$, the other with $-\frac{1}{2}\hbar$. The lowest energy of the gas is obtained by filling each energy level with two electrons, hence the energy at $T = 0$ is very different from zero.

Let ϵ_F be the energy of the electrons in the highest quantum state filled at $T = 0$. Since all states of lower energy are filled, and all states with higher energy are empty (see Fig. 22.19), we can obtain the value of ϵ_F as follows. For a perfect Fermi–Dirac gas of electrons, from Eq. 21A.17, we have[7]

$$\langle N\rangle = 2\cdot\frac{4\pi V}{h^3} \times \int_0^\infty dp\, p^2 \frac{\exp(\mu/k_BT)}{\exp(\mu/k_BT)+\exp(-p^2/2mk_BT)}. \tag{22.59}$$

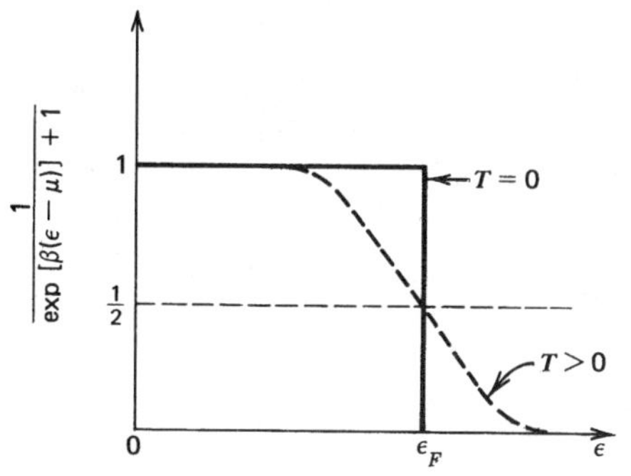

Figure 22.19 The distribution of occupied levels in a perfect Fermi–Dirac gas.

But for $T = 0$, $\exp(-p^2/2mk_BT) = 0$, and the upper limit of integration becomes $p_F = (2m\epsilon_F)^{1/2}$. Therefore,

$$\langle N\rangle = \frac{8\pi V}{3h^3}(2m\epsilon_F)^{3/2}, \tag{22.60}$$

or

$$\epsilon_F = \frac{h^2}{8m}\left(\frac{3n}{\pi}\right)^{2/3}, \tag{22.61}$$

with $n \equiv \langle N\rangle/V$. The quantity ϵ_F is known as the *Fermi energy* of the system. The total energy of the $\langle N\rangle$ electrons at $T = 0$ is

$$\begin{aligned} U_0 &= 2\cdot\frac{4\pi V}{h^3}\int_0^{p_F} dp\, p^2\left(\frac{p^2}{2m}\right) \\ &= \frac{3}{2}\langle N\rangle\epsilon_F\int_0^{\epsilon_F} d\left(\frac{\epsilon}{\epsilon_F}\right)\left(\frac{\epsilon}{\epsilon_F}\right)^{3/2} \\ &= \frac{3}{5}\langle N\rangle\epsilon_F. \end{aligned} \tag{22.62}$$

Note that the average energy per electron at $T = 0$ is $\frac{3}{5}\epsilon_F$, which is a very large value since for typical metals $\epsilon_F \approx 5-10$ eV. As might be expected, the contribution of the electrons to the pressure is large because U_0 is large. In fact, at $T = 0$ we have

$$\begin{aligned} p_0 &= -\left(\frac{\partial U_0}{\partial V}\right)_T = -\frac{3}{5}\langle N\rangle\frac{\partial \epsilon_F}{\partial V} \\ &= \frac{2}{5}n\epsilon_F = \frac{2}{3}\frac{U_0}{V}, \end{aligned} \tag{22.63}$$

which is the same relation between pressure and internal energy density as for a classical perfect gas (Eq. 12.41), but in this case neither p nor U is zero at $T = 0$. Typically, $p_0 \approx 10^5$–10^{16} atm for metals. Clearly, in a stable metal the electron gas pressure must be balanced by forces arising from the electron–ion attraction.

[7] The extra factor of 2 in Eq. 22.30 accounts for the two spin states of an electron.

To calculate the heat capacity of the electron gas, we must first evaluate the energy from

$$U = 2\cdot\frac{4\pi V}{h^3} \times \int_0^\infty dp\, p^2\left(\frac{p^2}{2m}\right)\frac{\exp(\mu/k_BT)}{\exp(\mu/k_BT)+\exp(-p^2/2mk_BT)}$$
$$= \frac{3\langle N\rangle}{2\epsilon_F^{3/2}}\int_0^\infty d\epsilon\frac{\epsilon^{3/2}}{\exp[(\epsilon-\mu)/k_BT]+1}. \tag{22.64}$$

Since $\epsilon_F >> k_BT$, the distribution of electrons differs very little from what it is for $T = 0$. Consequently, U and all other thermodynamic properties of the dense perfect Fermi–Dirac gas can be determined by expanding the integrand about its value for $T = 0$ and integrating term by term. The integral in Eq. 22.64 is of the form

$$I = \int_0^\infty d\epsilon\frac{f(\epsilon)}{\exp[(\epsilon-\mu)/k_BT]+1}. \tag{22.65}$$

Using the expansion procedure mentioned, it is shown in Appendix 22A that

$$I = \int_0^\mu f(\epsilon)\, d\epsilon + \frac{(\pi k_BT)^2}{6}\left(\frac{df}{d\epsilon}\right)_{\epsilon=\mu} + \cdots. \tag{22.66}$$

We calculate the chemical potential of the gas first. Rewriting Eq. 22.59 in the form

$$\langle N\rangle = \frac{3\langle N\rangle}{2\epsilon_F^{3/2}}\int_0^\infty d\epsilon\frac{\epsilon^{1/2}}{\exp[(\epsilon-\mu)/k_BT]+1} \tag{22.67}$$

enables us to apply Eq. 22.65 with $f(\epsilon) = \epsilon^{1/2}$. Then we find

$$1 = \left(\frac{\mu}{\epsilon_F}\right)^{3/2}\left[1+\frac{\pi^2}{8}\left(\frac{k_BT}{\mu}\right)^2+\cdots\right], \tag{22.68}$$

or

$$\mu = \epsilon_F\left[1-\frac{\pi^2}{12}\left(\frac{k_BT}{\mu}\right)^2+\cdots\right] \tag{22.69}$$
$$= \epsilon_F\left[1-\frac{\pi^2}{12}\left(\frac{k_BT}{\epsilon_F}\right)^2+\cdots\right], \tag{22.70}$$

with Eq. 22.70 following from Eq. 22.69 by iterative substitution of the expansion for μ in the term $(k_BT/\mu)^2$.

To evaluate Eq. 22.64 for the energy we set $f(\epsilon) = \epsilon^{3/2}$. Applying Eq. 22.66 again, we find that

$$U = \frac{3}{2}\frac{\langle N\rangle}{\epsilon_F^{3/2}}\left[\frac{2\mu^{5/2}}{5}+\frac{(\pi k_BT)^2}{6}\left(\frac{3\mu^{1/2}}{2}\right)+\cdots\right]$$
$$= \frac{3\langle N\rangle}{5}\left(\frac{\mu}{\epsilon_F}\right)^{3/2}\mu\left[1+\frac{5\pi^2}{8}\left(\frac{k_BT}{\mu}\right)^2+\cdots\right] \tag{22.71}$$
$$= \frac{3\langle N\rangle}{5}\epsilon_F\left[1+\frac{5\pi^2}{12}\left(\frac{k_BT}{\epsilon_F}\right)^2+\cdots\right], \tag{22.72}$$

the last line (Eq. 22.72) following from use of Eq. 22.70 to eliminate μ in favor of ϵ_F. From Eq. 22.72 we immediately obtain the electronic heat capacity,

$$C_V^{\text{el}} = \frac{\pi^2}{2}\langle N\rangle k_B\left(\frac{k_BT}{\epsilon_F}\right), \tag{22.73}$$

which is linear in T as stated earlier. For typical metals, Eq. 22.73 is of the order of 1% of $3Nk_B$ at 300 K, neatly accounting for the qualitative aspects of the observed behavior.

The physical basis for the result displayed in Eq. 22.73 is easily understood. Because of the restrictions imposed by the Pauli exclusion principle, electrons with energy much less than ϵ_F cannot be excited into neighboring states because these are already occupied. The excitation of an electron requires that there be an empty state of proper energy available. Consequently, only those electrons lying within about k_BT of ϵ_F can be excited. The fraction of these is of the order of (k_BT/ϵ_F), so the increase in energy corresponding to thermal excitation is of the order of $N(k_BT/\epsilon_F)(k_BT)$, and the contribution to the heat capacity is proportional to $Nk_B(k_BT/\epsilon_F)$, which is of the form displayed in Eq. 22.73.

It is also clear that Eq. 22.73 satisfies the third law of thermodynamics, whereas the classical value for the electronic heat capacity does not.

APPENDIX 22A

Evaluation of Fermi–Dirac Integrals

This appendix is concerned with evaluation of integrals of the type

$$I=\int_0^\infty d\epsilon \frac{f(\epsilon)}{\exp[(\epsilon-\mu)/k_BT]+1}. \qquad (22A.1)$$

As indicated in Chapter 22, when $T = 0$ the function $\{\exp[(\epsilon - \mu)/k_BT] + 1\}^{-1}$ is a step function, having the value unity for $\epsilon \le \epsilon_F$ and zero for $\epsilon > \epsilon_F$. Moreover, since $\epsilon_F/k_B \approx 5 \times 10^4$ K, the distribution of electrons will be very like that at $T = 0$ for all T of interest. Evaluation of Eq. 22A.1 can then be based on an expansion of the integrand about its value for $T = 0$.

We proceed as follows. Integration of Eq. 22A.1 by parts gives

$$I=\int_0^\infty d\epsilon \frac{\beta \exp[\beta(\epsilon-\mu)]}{\{\exp[\beta(\epsilon-\mu)]+1\}^2}\int_0^\epsilon d\epsilon' f(\epsilon'); \qquad (22A.2)$$

the other terms vanish at the limits of integration. Let $y \equiv \beta(\epsilon - \mu)$ and $F(\epsilon) \equiv \int_0^\epsilon d\epsilon' f(\epsilon')$. Then, expanding $F(y)$ as a Taylor series about $y = 0$, which corresponds to $T = 0$ where $\mu = \epsilon_F$, we find that

$$\begin{aligned} F(y) &= F(0)+\sum_{n=1}^\infty \frac{y^n}{n!}\left(\frac{\partial^n F}{\partial y^n}\right)_{y=0} \\ &= \int_0^\mu d\epsilon\, f(\epsilon)+\sum_{n=1}^\infty \frac{(yk_BT)^n}{n!}\left(\frac{\partial^{n-1} f(\epsilon)}{\partial \epsilon^{n-1}}\right)_{\epsilon=\mu}. \end{aligned} \qquad (22A.3)$$

Substitution of Eq. 22A.3 into Eq. 22A.2 then gives

$$\begin{aligned} I = {} & F(0)\int_0^\infty d\epsilon \frac{\beta \exp[\beta(\epsilon-\mu)]}{\{\exp[\beta(\epsilon-\mu)]+1\}^2} \\ & +\sum_{n=1}^\infty \frac{(k_BT)^n}{n!}\left(\frac{\partial^{n-1} f(\epsilon)}{\partial \epsilon^{n-1}}\right)_{\epsilon=\mu} \\ & \times \int_{y=\beta\mu}^\infty d\epsilon\, y^n \frac{\beta \exp[\beta(\epsilon-\mu)]}{\{\exp[\beta(\epsilon-\mu)]+1\}^2}. \end{aligned} \qquad (22A.4)$$

Consider the first term on the right-hand side of Eq. 22A.4. We have

$$\begin{aligned} \int_0^\infty d\epsilon \frac{\beta \exp[\beta(\epsilon-\mu)]}{\{\exp[\beta(\epsilon-\mu)]+1\}^2} &= \left\{\frac{1}{\exp[\beta(\epsilon-\mu)]+1}\right\}_0^\infty \\ &\approx 1, \end{aligned} \qquad (22A.5)$$

since $\exp(-\beta\mu) << 1$. Next consider the second term on the right-hand side of Eq. 22A.4. We rewrite the term.

$$\frac{\beta \exp[\beta(\epsilon-\mu)]}{\{\exp[\beta(\epsilon-\mu)]+1\}^2}=\frac{\beta}{(e^y+1)(e^{-y}+1)} \qquad (22A.6)$$

to show that it is symmetric in y, and that it approaches zero exponentially as $y \to +\infty$ and $y \to -\infty$. Then the lower limit of the integral can be extended to $-\infty$, since the error made is exponentially small. Since Eq. 22A.6 is symmetric, and the extended range of integration ($\pm\infty$) is symmetric, all those contributions to the second term on the right-hand side of Eq. 22A.4 for which n is odd vanish. Finally, by making the expansion

$$\frac{\beta e^y}{(e^y+1)^2}=\frac{\beta e^{-y}}{(1+e^{-y})^2}=-\beta\sum_{m=1}^\infty (-)^m m e^{-my}, \qquad (22A.7)$$

and integrating term by term, we find that

$$\begin{aligned} \int_{-\infty}^\infty dy \frac{\beta y^n}{(e^y+1)(e^{-y}+1)} &= -2\beta\sum_{m=1}^\infty (-)^m m\int_0^\infty y^n e^{-my}\, dy \\ &= -2\beta n!\sum_{m=1}^\infty \frac{(-)^m}{m^n}. \end{aligned} \qquad (22A.8)$$

The sums in Eq. 22A.8 are known; they have the values

$$\begin{aligned} n=2:&\quad -\sum_{m=1}^\infty \frac{(-)^m}{m^2}=\frac{\pi^2}{12}; \\ n=4:&\quad -\sum_{m=1}^\infty \frac{(-)^m}{m^4}=\frac{7\pi^4}{720}. \\ &\ \vdots \end{aligned} \qquad (22A.9)$$

Combining Eqs. 22A.4, 22A.5, 22A.8, and 22A.9 we finally obtain

$$I = \int_0^{\mu} d\epsilon\, f(\epsilon) + \frac{(\pi k_B T)^2}{6}\left(\frac{\partial f}{\partial \epsilon}\right)_{\epsilon=\mu} + \frac{7}{360}(\pi k_B T)^4 \left(\frac{\partial^3 f}{\partial \epsilon^3}\right)_{\epsilon=\mu} + \ldots, \quad (22A.10)$$

which is the form quoted in the text.

FURTHER READING

Reissland, J. A., *The Physics of Phonons* (Wiley, New York, 1973).
Smith, R. A., *Wave Mechanics of Crystalline Solids* (Wiley, New York, 1961).

More advanced texts:

Born, M., and K. Huang, *Dynamical Theory of Crystal Lattices* (Oxford University Press, London, 1954).
Chaquard, P., *The Anharmonic Crystal* (W. A. Benjamin, Reading, Mass., 1967).
Maradudin, A. A., E. W. Montroll, G. H. Weiss, and I. P. Ipatova, *Theory of Lattice Dynamics in the Harmonic Approximation* (Academic Press, New York, 1971).
Stoneham, A. M., *Theory of Defects in Solids* (Oxford University Press, London, 1975).
van Bueren, H. G., *Imperfections in Crystals* (North Holland, Amsterdam, 1960).
Wallace, D. C., *Thermodynamics of Crystals* (Wiley, New York, 1972).

PROBLEMS

1. Suppose that the interaction between two atoms has the form

$$u_2(R) = -\frac{C_1}{R^n} + \frac{C_2}{R^m}.$$

Show that a necessary condition for $u_2(R)$ to have a minimum is that the repulsive force be of shorter range than the attractive force (i.e., $m > n$).

2. Consider a one-dimensional chain of alternating monovalent cations and anions. Show that the Madelung constant (see Chapter 14) for this linear lattice is 2 ln 2.

3. Suppose that the molar internal energy at T can be written in the form

$$u(v, T) = e_0(v) + e_{zp}(v) + e_{vib}(v, T),$$

where $e_0(v)$ is the static lattice energy, $e_{zp}(v)$ is the zero-point energy, and $e_{vib}(v, T)$ is the energy associated with thermally excited vibrations. For a close-packed cubic crystal in which the interaction is of the Lennard–Jones type, we have

$$\frac{e_0(v)}{4N_A\epsilon} = 6.006\left(\frac{v_0}{v}\right)^4 - 14.454\left(\frac{v_0}{v}\right)^2,$$

where $v_0 = N_A\sigma^3$, and

$$\frac{e_{zp}(v)}{4N_A\epsilon} = 2.291\Lambda^*\left(\frac{v_0}{v}\right)^{7/3}\left[4.283 - 2\left(\frac{v}{v_0}\right)^2\right],$$

where $\Lambda^* \equiv h/\sigma(4m\epsilon)^{1/2}$. Calculate the pressure as a function of volume at $T = 0$ K. From the formula obtained, evaluate v/v_0 as a function of p/p_0 for H_2, for which $\Lambda^* = 1.73$ Å, $v_0 = 15.12$ cm^3/mol, $p_0 \equiv 4\epsilon/\sigma^3 = 203$ atm. Compare your calculations with the measurements of J. W. Stewart and C. A. Swenson, *Phys. Rev.* **94,** 1069 (1954), **97,** 578 (1955).

4. In ionic crystals $e_{zp}(v)$ is usually very small compared with $e_0(v)$ (both defined as in Problem 3). At high pressures the repulsive part of the non-Coulomb interaction and the Coulomb attraction dominate the ion–ion potential. Show that if the repulsive part of the non-Coulomb interaction in an ionic crystal is of the form R^{-n}, then at high pressure

$$p \propto v^{-(n+3)/3}.$$

5. For a one-dimensional chain of particles with mass m, lattice spacing a, and displacement force constant f, show that the dispersion relation is

$$\omega = 2\pi\nu = 2\sqrt{\frac{f}{m}}\sin\left(\frac{ka}{2}\right).$$

6. For a one-dimensional chain with lattice spacing a, the number of standing waves in the wave vector interval dk is $L\,dk/\pi$, where L is the length of the chain and $0 \le k \le \pi/a$. Show that the energy of the chain is

$$U = \frac{2L}{\pi a}\int_0^{\omega_{max}} \frac{\hbar\omega\, d\omega}{[\exp(\hbar\omega/k_BT) - 1](\omega_{max}^2 - \omega^2)^{1/2}},$$

where $\omega_{max}^2 = 4f/m$. Draw the curve of C_V versus T for this lattice for $\hbar\omega_{max}/k_B = 200$ K.

7. The continuum approximation to the linear chain represents the number of vibrational modes in the range $d\omega$ as $L\,d\omega/\pi c_s$, where c_s is the velocity of "sound wave" propagation on the continuous linear chain. Use the Debye method to calculate C_V, for this model and compute C_V versus T for the case $\Theta_D = 200$ K, where $\Theta_D \equiv \hbar\omega'_{max}/k_B$. Compare the two curves of C_V versus T from the calculation of Problem 6 and this one. Note that ω'_{max} here is not the same as ω_{max} in Problem 6.

8. Consider a crystal of N atoms where each atom has one vibrational degree of freedom. Assume that the vibrational energy of each atom can be written as

$$\epsilon = x^2 h\nu,$$

where x can take on all continuous positive values. As in the Einstein treatment, each atom has only one vibrational frequency ν.
(a) Calculate the partition function for this crystal.
(b) Compute the average vibrational energy.
(c) Compute the heat capacity of the crystal.

9. Consider a one-dimensional diatomic lattice such as shown in Fig. 22.7. For this problem let $2b = a$, and let the two masses be M and m. Show that the vibrational frequencies satisfy the dispersion relation

$$\omega^2 = f\left(\frac{1}{m}+\frac{1}{M}\right) \pm f\left[\left(\frac{1}{m}+\frac{1}{M}\right)^2 - \frac{4\sin^2(kb)}{Mm}\right]^{1/2},$$

where f is the Hooke's law force constant for displacement of either atom. Show that the plus sign generates the optical branch and the minus sign the acoustical branch of the spectrum.

10. Discuss the heat capacity of a two-dimensional square lattice with nearest-neighbor separation a using the Debye approximation. The information needed with respect to the dispersion of frequencies can be obtained from the two-dimensional analog of the analysis of footnote 3 on page 604.

11. Show that at low temperature the heat capacity of a one-dimensional chain is proportional to T, and that of a square-planar lattice is proportional to T^2.

12. Show that for an Einstein crystal, when $T >> \Theta_E$,

$$S \approx 3Nk_B\left[1+\ln\left(\frac{k_BT}{\hbar\omega_E}\right)\right].$$

13. Show that for a crystalline solid containing N particles

$$\int_0^\infty [3Nk_B - C_V(T)]\, dT = U(0),$$

where $U(0)$ is the internal energy at $T = 0$ K.

14. The Debye model is an essentially exact treatment of the low-frequency modes of a solid but it does not properly treat the high-frequency modes. Why should this model be a good approximation at low temperature?

15. Prove that as long as there is a maximum vibrational frequency in a crystal, the high-temperature limit of the heat capacity is $3Nk_B$.

16. At 60 K the heat capacity of copper is 8.70 J/K mol. Predict the heat capacity at 4 K using the Einstein and Debye low-temperature limits. Which result agrees better with the experimental value of 5.78×10^{-3} J/K mol?

17. Consider a (hypothetical) perfect lattice containing N atoms. Suppose that we create N_υ vacancies, where $N_\upsilon << N$. Let the energy of creation of a vacancy be w_υ, and the change in vibrational contribution to the entropy per vacancy be ΔS_υ. Show that

$$\frac{N\upsilon}{N} = \exp\left(\frac{\Delta S_\upsilon}{k_B}\right)\exp\left(-\frac{w\upsilon}{k_BT}\right).$$

18. A very simple model can be used to estimate the magnitude of ΔS_υ in Problem 17. Suppose that the crystal is described in the Einstein approximation. Suppose further that the atoms adjacent to a vacancy have frequency of oscillation ν', all others frequency of oscillation ν. Show that

$$\Delta S_\upsilon = 3Zk_B \ln\left(\frac{\nu}{\nu'}\right),$$

where Z is the number of nearest neighbors to a vacancy.

19. Frenkel defects are formed when atoms leave lattice sites and occupy interstitial positions in the lattice. A Frenkel defect, therefore, consists of an interstitial atom and a vacancy. Show that the number of Frenkel defects in equilibrium at T in a crystal is

$$N_{\rm Fr} = (NN_i)^{1/2} \exp\left(\frac{\Delta S_\upsilon}{2k_B}\right)\exp\left(\frac{w_{\rm Fr}}{2k_BT}\right),$$

where N is the number of atoms, N_i is the number of possible interstitial positions, ΔS_υ is the change in vibrational entropy per Frenkel defect, and $w_{\rm Fr}$ is the energy of formation of a Frenkel defect. Note that in chemical language the formation of a Frenkel defect is equivalent to the reaction

occupied lattice site + vacant interstitial site
= vacancy + interstitial.

20. Estimate the number of vacancies per atom in a state of equilibrium in a crystal at $T = 100$ K, 500 K, and 1000 K for the cases that $w_\upsilon = 80$ kJ/mol and $w_\upsilon = 150$ kJ/mol.

21. Suppose that the valence electrons of solid Na form a perfect degenerate Fermi gas. Calculate the average speed of an electron in this gas. What would the corresponding "classical temperature" be? What is the electronic contribution to the beat capacity of Na at 300 K?

22. Suppose that the energy of a free electron "at rest" inside a metal is zero. To escape from the metal, the electron must have an energy of motion perpendicular to the surface of at least ϵ_s. Then if z is the coordinate perpendicular to the surface, the electron must have momentum $p_z > p_{z_0}$ to escape, where $p_{z_0}^2/2m = \epsilon_s$. Show that the emission current density is

$$I = \frac{4\pi emk_B^2}{h^3} T^2 \exp\left(-\frac{\phi}{k_B T}\right),$$

which is known as the Dushman–Richardson equation. Hint: $\phi >> k_B T$ for all metals below the melting point, so only the tail of the Fermi distribution contributes to I.

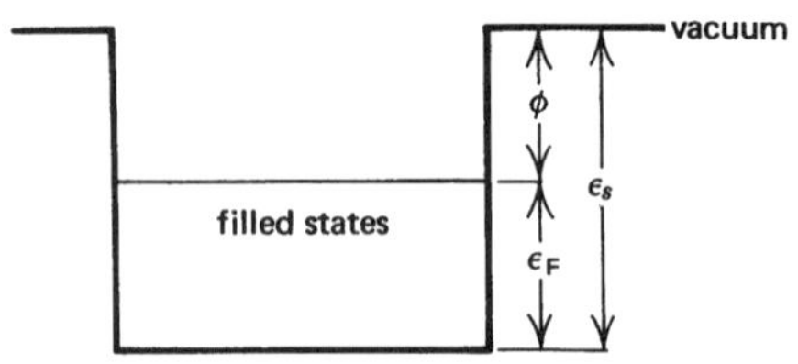

23. At a very high pressure, say, $p > 10^6$ atm, the volumetric properties of a metal are determined by the compressibility of the degenerate electron gas. Indeed, in this pressure range the distinction between solid, liquid, and gas becomes meaningless. Show that for a perfect degenerate electron gas

$$p = \frac{1}{5}\left(\frac{3}{8\pi}\right)^{2/3} \frac{h^2}{m} n^{5/3}.$$

For all elements except hydrogen the ratio of the atomic weight to atomic number is close to 2. (The nucleus has approximately equal numbers of protons and neutrons.) Calculate the pressure corresponding to a mass density of 100 g/cm^3 using an atomic weight/atomic number ratio of 2.

24. A solid contains N distinguishable and noninteracting "magnetic sites." At each site is a magnetic moment $\boldsymbol{\mu}$ which can either point along an applied magnetic field $\mathbf{H}$ or opposite to the field. If N_1 of the moments point along the field, the total energy of the solid is

$$E = -N_1 \boldsymbol{\mu}\cdot\mathbf{H} + (N - N_1)\boldsymbol{\mu}\cdot\mathbf{H}.$$

(Each moment pointing along the field contributes energy $-\boldsymbol{\mu}\cdot\mathbf{H}$, whereas each moment pointing opposite to the field contributes energy $+\boldsymbol{\mu}\cdot\mathbf{H}$.) Suppose that the solid is isolated and has energy E. Compute the entropy S as a function of E, $\mathbf{H}$, $\boldsymbol{\mu}$, and N.

25. Suppose that the solid of Problem 24 is maintained at a constant temperature T. Compute its entropy.

26. Suppose that a perfect free-electron gas has n electrons per unit volume. In the presence of a magnetic field $\mathbf{H}$, let there be n_p with spin magnetic moment parallel to $\mathbf{H}$ and n_a with spin magnetic moment antiparallel to $\mathbf{H}$. The magnetization (magnetic moment per unit volume) is

$$\mathbf{M} = (n_p - n_a)\boldsymbol{\mu}_B$$

where $\boldsymbol{\mu}_B$ is the Bohr magneton (see Chapter 5). Show that if the electron distribution were classical and $\boldsymbol{\mu}_B \cdot \mathbf{H} << k_B T$, then

$$\mathbf{M} = \frac{n\mu_B^2}{k_B T}\mathbf{H} \equiv \chi_p^{\text{cl}}\mathbf{H}.$$

27. The result of Problem 26 disagrees with experimental data for simple metals because the electron distribution *must* be described by Fermi–Dirac statistics. As a result of applying the field $\mathbf{H}$, the distribution of electrons over the energy levels is changed as shown in the accompanying diagram. All levels below ϵ_F are filled, as usual. However, electrons with spin magnetic moment parallel to $\mathbf{H}$ have their energies lowered by $\boldsymbol{\mu}_B \cdot \mathbf{H}$, those antiparallel have it raised by $\boldsymbol{\mu}_B \cdot \mathbf{H}$, where $\boldsymbol{\mu}_B \cdot \mathbf{H} << \epsilon_F$. The energy ϵ is common to both the spin states of the electrons. Show that now

$$\mathbf{M} = \frac{4\pi m}{h^2}\left(\frac{3n}{\pi}\right)^{1/3} \mu_B^2 \mathbf{H}$$
$$\equiv \chi_p \mathbf{H},$$

where $\chi_p \equiv 3n\mu_B^2/2k_B T_F$. Compare this result with χ_p^{cl} of Problem 26.

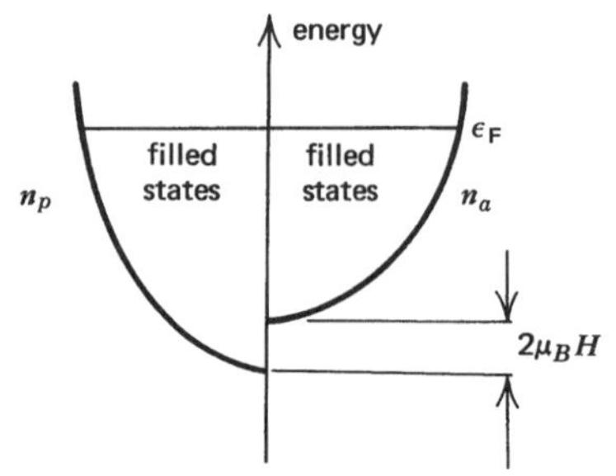

CHAPTER 23

Thermodynamic Properties of Liquids

Our universe contains matter subjected to an enormous range of conditions—from the near-zero temperature and pressure of space to the extremely high temperatures and pressures in the interiors of stars. Liquids exist over only a small part of this range. That domain is, however, of extraordinary importance for technology and for life processes. In this chapter we examine how the thermodynamic properties of a liquid can be understood in terms of the forces between its constituent molecules. The characteristic flow properties of a liquid are associated with nonequilibrium phenomena. These are discussed in Part III of this text.

23.1 Bulk Properties of Liquids

How do the equation of state and the thermal properties of liquids differ from those of gases and solids?

Consider, first, the pressure exerted by a liquid at temperature T. It was shown in Chapter 19 that two parts of a one-component system in contact through a deformable boundary that permits transfer of energy and matter must have the same pressure, temperature, and chemical potential. The boundary separating a liquid and its vapor is of this type, hence the pressure of the vapor must be equal to that of the liquid. We denote the pressure along the line of equilibrium for coexisting liquid and vapor phases by p_σ.

We shall consider details of the nature of the liquid–vapor equilibrium in Chapter 24, but two points, related to the dependence of p_σ on the state of the liquid, deserve mention here. First, in a liquid–vapor system containing only one component, p_σ is a function of only the temperature (see Chapter 24). If, however, a uniform pressure is applied to the liquid by, say, an inert gas that is insoluble in the liquid, the vapor phase consists of two components (the inert gas and the vapor from the liquid), and p_σ is a function of both the temperature and the applied pressure. Indeed, since the chemical potentials of the vapor and liquid are equal,[1]

$$\mu^L = \mu^*(T) + RT \ln \frac{p_\sigma}{\mathfrak{p}}. \tag{23.1}$$

Moreover, since the pressure derivative of the chemical potential of the liquid is given by

$$\left(\frac{\partial \mu^L}{\partial p}\right)_T = v^L, \tag{23.2}$$

we find that the rate of change of vapor pressure with applied pressure is given by

$$RT\left(\frac{\partial \ln p_\sigma}{\partial p}\right)_T = v^L, \tag{23.3}$$

which is often called the *Poynting equation.* The molar volume of a liquid is of the order of 100 cm^3, so at room temperature $(\partial \ln p_\sigma/\partial p)_T$ is small, of the order of 4×10^{-3} atm^{-1}. That is, the vapor pressure changes on the order of 0.5% per atmosphere of added pressure. Second, the temperature dependence of p_σ is determined by the differences in entropies and molar volumes of the vapor and liquid. Consider an alteration of the common temperature and pressure along the liquid–vapor coexistence line. Since $\mu^L = \mu^G$ in both the initial and final states, it is necessary that their changes, $d\mu^L$ and $d\mu^G$, also be equal:

$$\left.\begin{aligned} d\mu^G &= d\mu^L, \\ -s^G\,dT + v^G\,dp_\sigma &= -s^L\,dT + v^L\,dp_\sigma \end{aligned}\right\} \text{along the liquid–vapor coexistence line.} \tag{23.4}$$

[1] Equation 23.1 is written for the case that the vapor is a perfect gas. When gas imperfection is important, the fugacity replaces the pressure in Eqs. 23.1 and 23.3.

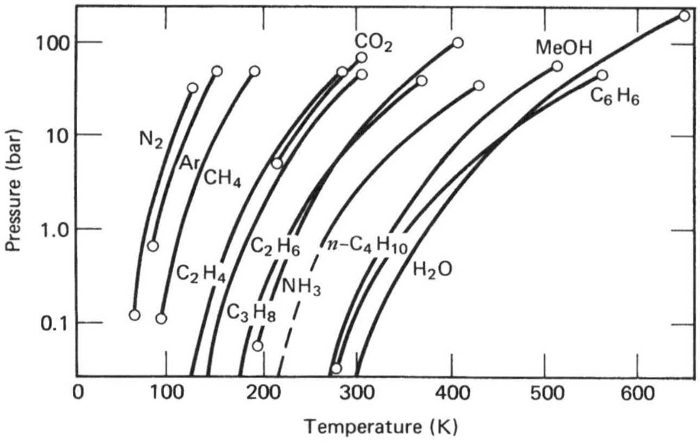

Figure 23.1 The logarithms of the vapor pressures of 12 liquids as functions of the absolute temperature. The circles mark the triple and critical points (see Chapter 24). From J. S. Rowlinson, *Liquids and Liquid Mixtures,* 2nd ed. (Butterworths, London, 1969).

Equation 23.4 is immediately rearranged to

$$\left(\frac{\partial p}{\partial T}\right)_\sigma = \frac{s^G - s^L}{v^G - v^L}, \tag{23.5}$$

which is known as the Clausius–Clapeyron equation. Since $s^G > s^L$ and $v^G >> v^L$, we expect $p_\sigma(T)$ to be a strongly increasing function of the temperature. That this is so is shown in Fig. 23.1. Note that along the liquid–vapor coexistence line the molar volume of the liquid varies.

The formulas used to compute thermodynamic functions are usually expressed in terms of constant-pressure or constant-volume properties of the fluid. Consider then the *thermal pressure coefficient*

$$\gamma_V \equiv \left(\frac{\partial p}{\partial T}\right)_V, \tag{23.6}$$

which measures the rate of change of pressure with temperature at fixed liquid volume. The functional behavior of γ_V determines the volume dependence of the internal energy through (see Eq. 17.40)

$$\left(\frac{\partial u}{\partial v}\right)_T = T\left(\frac{\partial p}{\partial T}\right)_V - p. \tag{23.7}$$

A schematic representation of the dependence of γ_V on temperature is shown in Fig. 23.2*a*. It is observed that γ_V changes smoothly with temperature along the liquid–vapor coexistence line (see Chapter 24). On the right-hand side of Fig. 23.2*a* are lines representing γ_V for different values of v; these lines intersect the coexistence curve at points corresponding to the respective values of the volumes. Note that the slope of γ_V versus T is always small. γ_V can be either

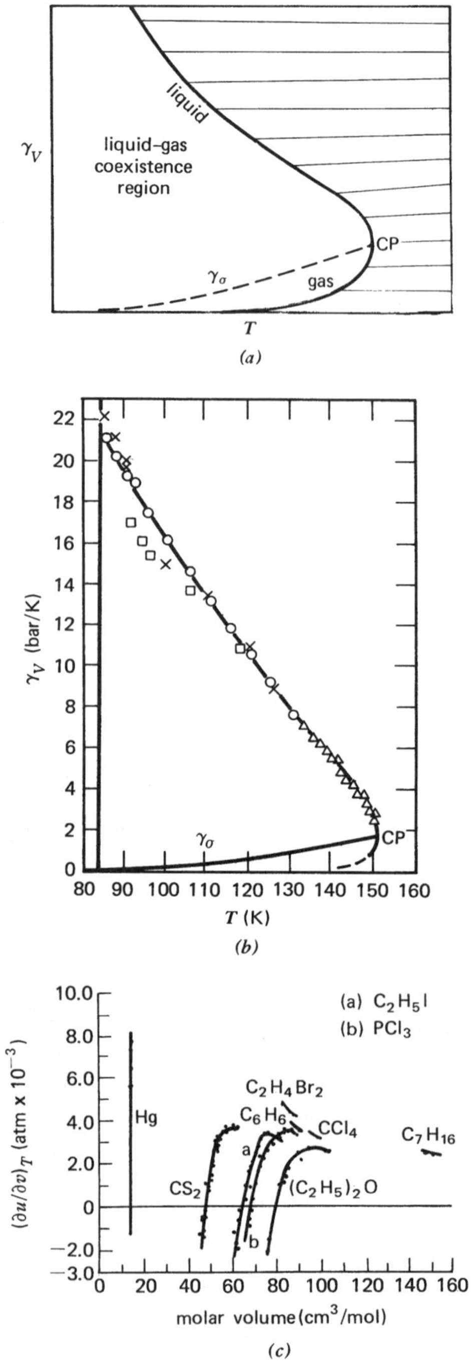

Figure 23.2 (*a*) The variation of the thermal pressure coefficient with temperature for the saturated gas and liquid, and for the homogeneous fluid at temperatures above saturation. (*b*) Thermal pressure coefficient of argon. From J. S. Rowlinson, *Liquids and Liquid Mixtures,* 2nd ed. (Butterworths, London, 1969). (*c*) $(\partial u/\partial v)_T$ as a function of molar volume. From J. H. Hildebrand and R. L. Scott, *Regular Solutions* (Prentice-Hall, Englewood Cliffs, N.J., 1962).

negative or positive, depending on whether the coefficient of thermal expansion is negative or positive, since

$$\gamma_V \equiv \left(\frac{\partial p}{\partial T}\right)_V = -\frac{(\partial v/\partial T)_p}{(\partial v/\partial p)_T} = \frac{\alpha}{\kappa_T}. \tag{23.8}$$

Thermodynamic stability requires that $\kappa_T > 0$ (see Chapter 19), but only that γ_V and α have the same sign. Finally, also note that the vapor pressure curve, defined by $\gamma_\sigma \equiv (\partial p/\partial T)_\sigma$, crosses the two-phase region and cuts the coexistence curve at the critical point, at which $\gamma_\sigma = \gamma_V$. Some values of γ_V along the liquid–vapor coexistence line of argon are shown in Fig. 23.2*b*, and values of $(\partial u/\partial v)_T$ as a function of molar volume for several liquids are shown in Fig. 23.2*c*.

Direct measurements of the p, v, T properties of liquids lead to the following generalizations:

1. The isothermal compressibility κ_T decreases with increasing pressure, the rate of decrease becoming smaller as the pressure increases.
2. The isothermal compressibility increases as the temperature increases, the effect being largest at low pressure and diminishing as the pressure increases.
3. The coefficient of thermal expansion, α, increases as the temperature increases.

A tabulation of data for liquid argon along the liquid–vapor coexistence line is given in Table 23.1.

The values of γ_V discussed and Eq. 23.7 can be used to reach an interesting conclusion. Since T is positive, and α is usually positive, we expect γ_V to be positive. Then at low pressure u must decrease as v decreases. If the pressure is increased a point is reached where the term $[T(\partial p/\partial T)_V - p]$ in Eq. 23.7 becomes zero, and then for larger pressures this term becomes negative. At the pressure where this change takes place, the net internal forces change from attractive to repulsive. In the low-pressure region energy actually flows out of a liquid during an isothermal compression.

There is no generally valid, simple analytic equation of state for a liquid. Just as for the case of imperfect gases, empirical equations of state are used to represent observations in a compact form. The two most widely used are the *Tait equation,*

$$\frac{v_0 - v}{v_0 p} = \frac{A}{B+p}, \tag{23.9}$$

and the *Huddleston equation,*

$$\ln\left(\frac{p v^{2/3}}{v_0^{1/3} - v^{1/3}}\right) = A + B(v_0^{1/3} - v^{1/3}), \tag{23.10}$$

where v_0 is the molar volume at "zero" pressure and A and B are positive constants (different in the two equations). For the high-density region above the critical temperature, a virial expansion (see Chapter 21) with as many as six, seven, or even more terms has sometimes been used to represent pv/RT. As we saw in Chapter 21, the terms in the virial expansion can be related to the interaction of pairs, triplets, . . . of molecules, and thereby each term is given an interpretation in terms of the properties of the molecules of the system. No such interpretation is known, even formally, for the two empirical equations 23.9 and 23.10.

Finally, we note that c_p for a liquid increases as the temperature increases, but c_V is much less sensitive to temperature (see Table 23.1 and Fig. 23.3).

Table 23.1 Some Properties of Liquid Argon[a,b]

T (K)	p_σ (bar)	v (cm³/mol)	α_σ (× 10³/K)	α_p	κ_S (× 10⁻⁴/bar)	κ_T	γ_σ (bar/K)	γ_V	c_σ	c_p (J/K mol)	c_V
83.81[t]	0.68905	28.19	4.37	4.39	0.948	2.033	0.080	21.6	41.8	41.9	19.5
85	0.789	28.34	4.40	4.42	0.970	2.089	0.087	21.2	41.9	42.0	19.5
87.29[b]	1.013	28.66	4.47	4.49	1.022	2.217	0.107	20.3	42.1	42.2	19.4
90	1.336	29.00	4.55	4.58	1.077	2.37	0.133	19.3	42.2	42.4	19.3
95	2.149	29.67	4.75	4.80	1.211	2.74	0.193	17.5	42.4	42.7	19.0
100	3.25	30.38	5.00	5.08	1.363	3.16	0.256	16.1	43.1	43.5	18.7
105	4.73	31.15	5.29	5.41	1.556	3.71	0.333	14.6	44	45	19
110	6.67	32.03	5.7	5.9	1.80	4.47	0.435	13.2	45	46	18
120	12.17	34.25	7.3	7.8	2.49	7.40	0.670	10.5	49	51	17
130	20.28	37.50	10	11	3.99	14.1	0.965	7.9	55	60	17
140	31.67	42.42	16	21	7.59	38	1.32	5.6	70	86	18
150.86[c]	48.98	74.6	∞	∞	∞?	∞	1.8	1.8	∞	∞	∞?

[a] From J. S. Rowlinson, *Liquids and Liquid Mixtures,* 2nd ed. (Butterworths, London, 1969).

[b] $t \equiv$ triple point $b \equiv$ boiling point $c \equiv$ critical point

$$\alpha_\sigma \equiv \frac{1}{V}\left(\frac{\partial V}{\partial T}\right)_\sigma, \quad \alpha_p \equiv \frac{1}{V}\left(\frac{\partial V}{\partial T}\right)_p,$$

$$\kappa_T \equiv -\frac{1}{V}\left(\frac{\partial V}{\partial p}\right)_T, \quad \kappa_S \equiv -\frac{1}{V}\left(\frac{\partial V}{\partial p}\right)_S.$$

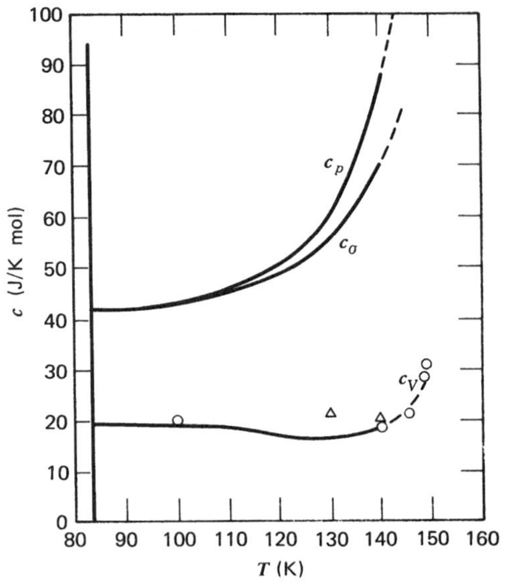

Figure 23.3 The heat capacities of liquid argon. From J. S. Rowlinson, *Liquids and Liquid Mixtures,* 2nd ed. (Butterworths, London, 1969).

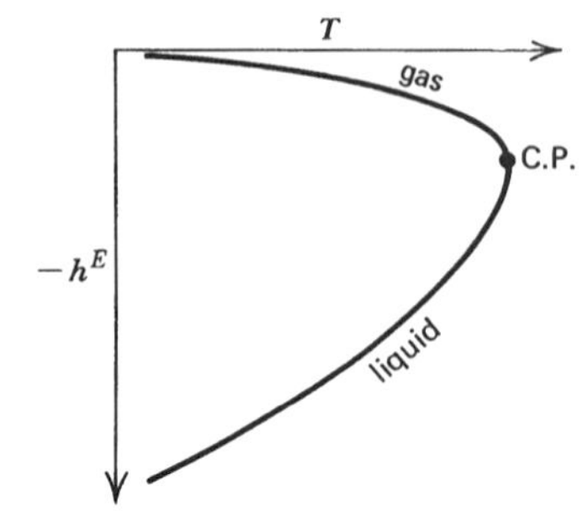

Figure 23.4 Sketch of the variation with temperature of the residual enthalpy of gas and liquid along the liquid–vapor coexistence curve. The difference between the two curves is the latent heat of evaporation. From J. S. Rowlinson, *Liquids and Liquid Mixtures,* 2nd ed. (Butterworths, London, 1969).

Table 23.2 Residual Properties of Liquid Argon[a,b]

T (K)	Δh_{vap}	$h^E + RT$ (J/mol)	u^E	c_p^E (J/K mol)	c_V^E
83.81[t]	6610	−5940	−5940	29.4	7.0
85	6570	−5910	−5910	29.5	7.0
87.29[b]	6517	−5840	−5840	29.7	6.9
90	6440	−5760	−5760	29.9	6.8
95	6290	−5590	−5590	30.2	6.5
100	6110	−5420	−5430	31.2	6.2
105	5900	−5230	−5240	32	6
110	5660	−5030	−5050	34	5
120	5100	−4630	−4670	39	5
130	4350	−4170	−4250	48	5
140	3350	−3610	−3750	74	6
150.86[c]	0	−2030	−2400	∞	∞?

[a] From J. S. Rowlinson, *Liquids and Liquid Mixtures,* 2nd ed. (Butterworths, London, 1969).
[b] $t \equiv$ triple point $b \equiv$ boiling point $c \equiv$ critical point

We showed, in Chapter 17, that the changes in entropy, internal energy, and other thermodynamic functions of a fluid can be calculated from knowledge of the heat capacity as a function of temperature and the equation of state. For a real gas, a convenient reference from which to measure changes is provided by the properties of an ideal gas at the same temperature and pressure. Since interpretation of the properties of the liquid state depends on understanding the role of intermolecular forces, it is convenient again to choose the perfect gas to define the reference state from which changes in thermodynamic properties are measured. The residual value of any function f, denoted f^E, is defined by

$$f^E(T,v) = \int_\infty^v \left[\left(\frac{\partial f}{\partial v}\right)_T - \left(\frac{\partial f}{\partial v}\right)_T^{\text{perfect gas}} \right] dv. \quad (23.11)$$

We can imagine the integral in Eq. 23.11 to be evaluated in two steps. Let the initial state be one with the system at effectively zero pressure, hence effectively infinite volume, and imagine an isothermal compression from $v = \infty$ to v^G, where v^G is the volume of the vapor at pressure p_σ and temperature T along the coexistence line for liquid–vapor equilibrium. Next, isothermally condense the gas to form a liquid with volume v^L, and add the consequent change in f less any change in $f^{\text{perfect gas}}$. For example, the residual enthalpy of a liquid is

$$h^E(T,v) = \int_\infty^v \left[\left(\frac{\partial h}{\partial v}\right)_T - \left(\frac{\partial h}{\partial v}\right)_T^{\text{perfect gas}} \right] dv$$

$$= \int_\infty^v \left[T\left(\frac{\partial p}{\partial T}\right)_V + v\left(\frac{\partial p}{\partial v}\right)_T \right] dv + \Delta h_{vap}. \quad (23.12)$$

A plot of the typical variation of h^E with temperature is shown in Fig. 23.4, and the residual enthalpy and energy of argon are listed in Table 23.2. Note how the heat of vaporization decreases continuously and becomes zero at the critical temperature, above which liquid and gas cannot be distinguished.

An important use for the computed residual thermodynamic properties of a liquid is to assess the existence of correlations in the behavior of widely different types of liquids, for removal of the perfect gas contributions leaves only the influence of intermolecular forces on the thermodynamic functions, and it is differences in these that define the behavior of different liquids. Comparisons of residual properties of liquids are thereby free of differences resulting from peculiarities in molecular structure, internal molecular motion, and the like, provided these are independent of density. The most important tool in the comparison of different liquids is the principle of corresponding states, an example of which was discussed in Chapter 21. This principle is consistent with the assumptions that (1) the internal degrees of freedom of the molecules are independent of the density, (2) the translational motion of the molecules can be described by classical mechanics, (3) the total potential energy of the liquid can be expressed as the product of an energy parameter, ϵ, and a function of distance between molecular centers i and j scaled by the molecular diameter, R_{ij}/σ. When the principle of corresponding states is valid, the thermodynamic functions of the liquid depend only on the reduced temperature $T^* \equiv k_BT/\epsilon$, reduced pressure $p^* \equiv p\sigma^3/\epsilon$, and reduced density $n^* \equiv n\sigma^3$. The principle is valid when the molecular interaction is spherically symmetric, as in simple liquids like argon. In liquids with more complicated molecules, the potential energy must depend on the orientation of the molecules and other factors, so the simple form of the principle of corresponding states is inadequate. Several schemes have been proposed for generalizing the corresponding-states concept. These are essentially empirical in nature, and all have the feature of adding extra parameters with which the observed p, υ, T data can be correlated. The difficulty with such approaches is not inaccuracy—they tend to work rather well as interpolation schemes. Rather, their principal defect is the lack of defined criteria to determine whether or not a given liquid lies outside the domain of applicability of the correlation.

To see what is involved in extending the concept of corresponding states to complicated molecules, consider possible relations between the equations of state of straight-chain hydrocarbons. These constitute an important class of industrial chemicals, and their equations of state have been thoroughly studied. We suppose that the linear hydrocarbons can be thought of as CH_2 groups linked by valence bonds that somewhat alter their properties. In particular, relative to a liquid of unlinked CH_4 molecules, valence bonding in a hydrocarbon chain reduces the number of nearest-neighbor groups with which a given CH_2 can interact at a specified temperature and density [from ~10 in liquid CH_4 to some smaller number in CH_3—$(CH_2)_{n-2}$—CH_3]. Valence bonding also reduces the effective number of degrees of freedom relative to liquid CH_4 with the same number density of C atoms, since the C—C bond has very high stretching frequency relative to the center-of-mass motions in liquid CH_4 and this motion is frozen out at ordinary temperatures (see Section 21.5, where similar phenomena are discussed in terms of the relative sizes of T and the reduced temperatures Θ_{vib} and Θ_{elec}). If these notions are valid, it should be possible to reduce the equations of state of all linear hydrocarbons to a universal form. To test this idea, define the reduced variables

$$\tilde{\tilde{\upsilon}} = \frac{\upsilon}{\upsilon_0(n)},$$
$$\tilde{\tilde{p}} = \frac{p\upsilon_0(n)}{u_0(n)},$$
$$\tilde{\tilde{T}} = \frac{Ts_0(n)}{u_0(n)}, \tag{23.13}$$

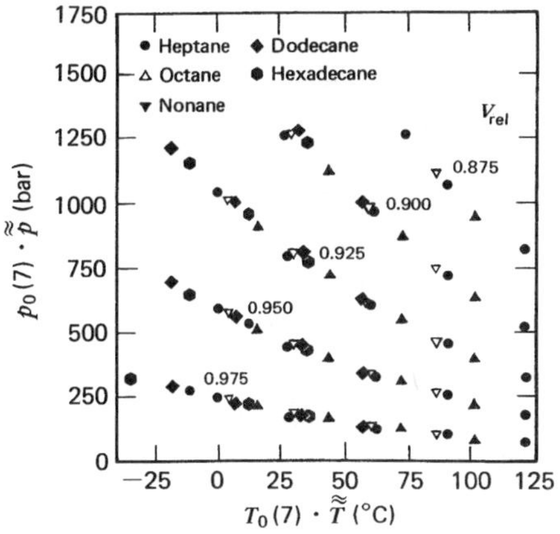

Figure 23.5 Reduced pressure versus reduced temperature curves at constant relative volume for n-heptane, n-octane, n-nonane, n-dodecane, and n-hexadecane. Units have been chosen relative to heptane. From J. Hijmans, *Physica* **27,** 433 (1961).

where $\upsilon_0(n)$, $u_0(n)$, and $s_0(n)$ are the molar volume, intermolecular contribution to the molar internal energy, and intermolecular contribution to the molar entropy of an n-carbon linear hydrocarbon, respectively. If the reduced volume, pressure, and temperature are properly defined, the reduced internal energy, entropy, and other thermodynamic properties should be linear functions of n. This is found to be the case for $n = 1$ to $n = 20$. It is also necessary that relative functions such as

$$V_{rel} \equiv \frac{\tilde{\tilde{\upsilon}}(\tilde{\tilde{p}})}{\tilde{\tilde{\upsilon}}(\tilde{\tilde{p}}=0)} \tag{23.14}$$

be independent of chain length, since all dependence on n is incorporated in the definition of $\tilde{\tilde{\upsilon}}$. Figure 23.5 shows that this expectation is fulfilled.

23.2 The Structure of Liquids

What do we mean by the structure of a liquid? As briefly hinted at the beginning of Chapter 22, we shall use the term structure to describe the arrangement of molecules in space. In the case of a simple liquid, such as Ar, only the relative positions of the centers of the atoms need be specified; in the case of a liquid of polyatomic molecules, for example, H_2O, both the relative positions of the molecular centers of mass and the relative molecular orientations must be specified.

Consider first a simple liquid of monatomic molecules. A representative form for the interaction between a pair of molecules is shown in Fig. 23.6. In liquids of this class, as in most liquids, the average volume per molecule is only

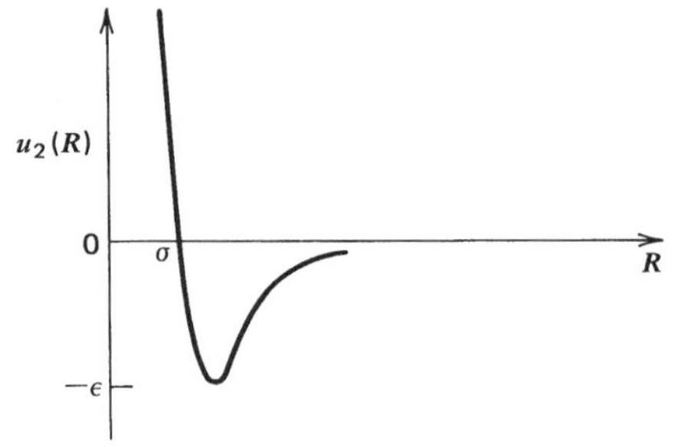

Figure 23.6 A typical molecular pair interaction as a function of internuclear separation.

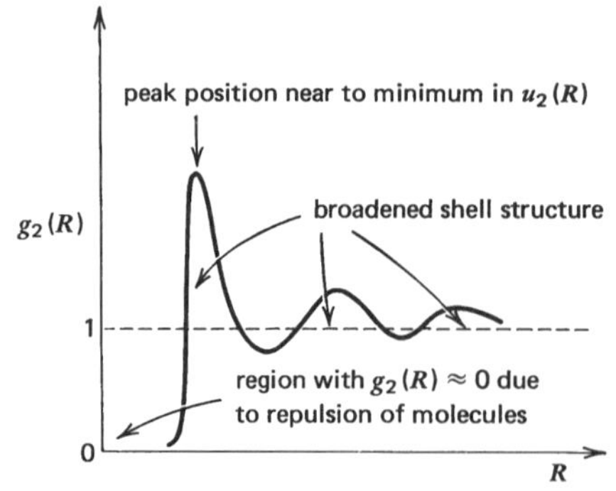

Figure 23.7 A typical pair-correlation function for a simple liquid, for example, Ar.

about 20–30% larger than the volume of a molecule.[2] Also, every molecule is surrounded by from 8 to 12 nearest neighbors, but the very strong repulsion between any pair of molecules prevents the distance between their centers from becoming much less than about one molecular diameter.[3] Put in other words, the probability of finding a pair of molecules separated by much less than σ is vanishingly small. This geometric restriction implies that in a liquid there must be a shell of molecules at a distance slightly greater than σ from any given molecule, because in the high-density condensed phase the average spacing between molecules is only slightly greater than σ. Continuing the argument, the existence of such a shell of molecules virtually prevents there being any other molecules with centers between one and two molecular diameters from the central molecule, and so forth. The shells of molecules so defined are spread over a range of molecular separations, and the rate of exchange of molecules in any one shell with the surrounding fluid is very large. Consequently, beyond a distance of about three molecular diameters from the central molecule the range over which a shell is spread becomes so large, and the corresponding increase in the local density over the average density so small, that all vestigial structure has vanished.

The kind of structure described above may be represented as follows. In general, in a fluid (gas or liquid), a molecule can be found with equal probability in equal-sized volume elements anywhere in the volume. In fact, if the volume element has magnitude ΔV and the total volume is V, the probability of finding a molecule in ΔV is just $\Delta V/V$. *In the absence of intermolecular forces, each molecule moves throughout the entire volume independent of the presence of other molecules.* Hence the probability of finding *one particular* molecule in the volume ΔV at the point $\mathbf{R}_1$, and *another particular* molecule in the volume ΔV at the point $\mathbf{R}_2$, is simply the product $(\Delta V/V)^2$, and is independent of $\mathbf{R}_1$ and $\mathbf{R}_2$. Since there are N molecules in the volume V, the probability of simultaneously finding *any* molecule in ΔV at $\mathbf{R}_1$ and *any other* molecule in ΔV at $\mathbf{R}_2$ is $N(N-1)(\Delta V/V)^2$, and is independent of $\mathbf{R}_1$ and $\mathbf{R}_2$. Then the joint probability density of finding any molecule at $\mathbf{R}_1$ and any other molecule at $\mathbf{R}_2$ is, for noninteracting molecules, $N(N-1)/V^2 = n^2$, and the conditional probability density of finding a molecule a distance R from the center of a given molecule is n. When there are no intermolecular forces, we say that the system has no structure, since the probability of finding a pair of molecules with given separation is independent of that separation.

Consider now a liquid in which molecular interactions exist. Let the probability per unit volume of finding a molecule a distance R from the center of a given molecule be denoted as $ng_2(R)$. Then, for a random distribution, such as is characteristic of noninteracting molecules, we would have $g_2(R) = 1$ everywhere (see preceding paragraph). But for a liquid with molecular interactions, $g_2(R)$ has the form displayed in Fig. 23.7. It is particularly important to notice the overwhelming role played by the repulsive interaction between a pair of molecules in determining the structure of the liquid. The function $g_2(R)$, known as the *pair-correlation function,* may be determined from studies of the intensity of scattered x-rays and thermal neutrons as a function of scattering angle.[4] (See Chapter 11 for the use of pair-correlation functions in the description of solids.)

The determination of the structure of a liquid from x-ray and/or neutron diffraction experiments is based on the same fundamental principle as is the determination of the structure of a crystal. The important point is this: If the wavelength of the x-rays or neutrons is about the same as the average spacing between atoms, there is interference between the secondary waves originating at each of the atoms in the incident beam of radiation. Because the order in the liquid is only of short range, the diffraction pattern from a liquid does not

[2] We take, for present purposes, the volume of a molecule to be defined by $\pi\sigma^3/6$, where σ is the distance at which the interaction between a pair of molecules is zero (see Fig. 23.6). When the separation of a pair of molecules is less than σ, the interaction between them is very strongly repulsive. Indeed, the molecules behave almost like billiard balls with diameter σ.

[3] Although this distance is not fixed, because the distance to which one molecule can penetrate another's force field depends on the kinetic energy available, it is sufficiently well defined to be useful in constructing our pictorial representation.

[4] See Appendix 23A.

have the sharp reflections characteristic of the Bragg spectrum of a solid. In the case of a solid, the reflections occur within very narrow ranges about a small set of angles precisely because phase coherence between the secondary waves originating at each of the atoms is ensured by the long-range order characteristic of the crystal structure. In the liquid the diffraction pattern is diffuse, and there is some scattering at all angles. Indeed, it can be shown[5] that the intensity of x-rays or neutrons scattered without change in energy is, as a function of the scattering angle θ,

$$I(\theta) \propto \frac{1}{2}(1+\cos^2\theta)I_0 \times \left\{1+4\pi n\int_0^\infty [g_2(R)-1]\frac{\sin sR}{sR}R^2\,dR\right\} \qquad (23.15)$$

where

$$s \equiv \frac{4\pi}{\lambda}\sin\frac{\theta}{2} \qquad (23.16)$$

and I_0 is the intensity of the incident radiation of wavelength λ. As written, Eq. 23.15 refers to scattering from monatomic particles; the effects of interference within one atom, because of its internal electronic structure, are incorporated in the proportionality factor. The extension of Eq. 23.15 to describe scattering from a liquid of polyatomic molecules requires two kinds of generalizations. First, account must be taken of the internal structure of the polyatomic molecule. The effects of this structure on the scattering are important, since intramolecular atom–atom separations, especially the intramolecular non-nearest neighbor distances, are comparable in magnitude with intermolecular separations. Second, there are new interference terms corresponding to the addition of the scattered waves from two centers of different atomic number. In principle, the inversion of the scattering data for a polyatomic fluid requires as many independent scattering experiments as there are kinds of pairs of atoms, for example, in H_2O three experiments to get the H–H, H–O, and O–O distribution functions. Rarely are these data available. Instead, approximations are made,[6] such as neglecting the scattering from H atoms, neglecting cross correlations of the X–Y type, and so on. Consequently, inferences concerning the structure of a polyatomic liquid must be examined carefully to filter out errors introduced when observations are reduced using such approximations.

Finally, Eq. 23.15 is valid only for that component of the scattered radiation which has the same wavelength (and thus the same energy) as the incident radiation—this is the case of elastic scattering. It is also possible to study the intensity of the radiation scattered with different wavelength, called inelastic scattering. This is more conveniently done with neutrons than with x-rays, because the change in energy of the scattered radiation relative to the energy of the incident radiation is more easily measured for neutrons. The elastic scattering gives information about the spatial distribution of atoms in the liquid, whereas the inelastic scattering gives information about the dynamics of particle motion. For the present we confine attention to the elastic scattering and the pair-correlation function deduced therefrom.

Systematic studies of the pair-correlation function, $g_2(R)$, as a function of the thermodynamic state of the fluid, show (see Figs. 23.8 and 23.9):

1. The structural features of the liquid are predominantly determined by the liquid density, and vary little with temperature.

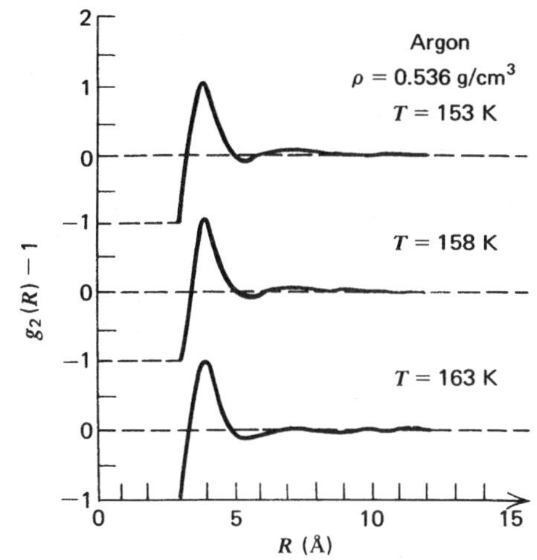

Figure 23.8 Pair-correlation functions of liquid Ar at three different temperatures, but with the same density. From C. J. Pings, in H. N. V. Temperley, J. S. Rowlinson, and S. Rushbrooke (Eds.), *Physics of Simple Fluids* (North Holland, Amsterdam, 1968).

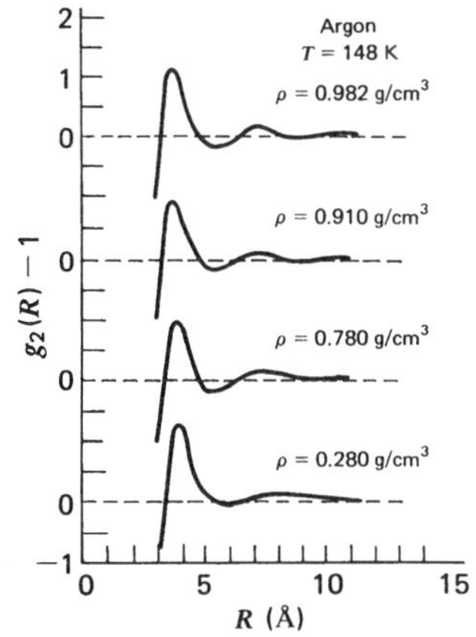

Figure 23.9 Pair-correlation functions of liquid Ar at four different densities, but with the same temperature. From C. J. Pings, in H. N. V. Temperley, J. S. Rowlinson, and S. Rushbrooke (Eds.), *Physics of Simple Fluids* (North Holland, Amsterdam, 1968).

[5] See Appendix 23A.

[6] Different approximations are used in the interpretation of x-ray and neutron diffraction data.

2. The structure of a simple fluid varies smoothly and continuously in going from a high-density (liquid) state to a low-density (gas) state.
3. As the density increases, the ordering in the liquid extends farther and farther from the central molecule.
4. If a coordination number is defined by $Z \equiv 4\pi n \times \int_0^{R_m} g_2(R)R^2\,dR$, where R_m is the first minimum in the integrand,[7] then it is found that Z varies markedly with the thermodynamic state of the fluid.[8] The value of the coordination number is primarily determined by the density of the fluid, as implied by point 1 above. At any given density the fluctuation in the value of Z arising from exchanges of molecular position is of the same order of magnitude as Z itself. That is, the short-range order in the liquid is not static. Molecules readily enter or leave specific shells, and there is a continuous distribution of molecular separations. In this sense the order in the liquid differs fundamentally from that of the solid, where the atoms are almost immobile.
5. In confirmation of a point made in 4 above, computer simulations[9] show that the first peak in the pair-correlation function decays on a time scale of the order of magnitude of 10^{-12} s (see Fig. 23.10).

To understand Fig. 23.10, picture some instantaneous arrangement of atoms of the liquid, and call the time at which that arrangement exists $t = 0$. Select a typical atom as the origin of a coordinate system. At $t = 0$ the average density of atoms located a distance R from the one at the origin is just $ng_2(R)$. Now consider some later time. Because the atoms can move, we expect the spatial arrangement to change. In general, every atom will tend to move away from where it was at $t = 0$. Clearly, close to $t = 0$ the spatial arrangement must be very similar to that at $t = 0$, but as t increases it will resemble that arrangement less and less. Figure 23.10 displays, for several times, the conditional probability, given an atom is at the origin at $t = 0$, that another atom will be a distance R from the origin. For t small, say, 0.5×10^{-12} s, the geometric constraints that lead to the nearest-neighbor peak of $g_2(R)$ dominate; but as t increases, both the probability of finding the original particle at the origin and the probability of finding a different particle at R change. In the long-time limit, an atom can be found with equal probability in all volume elements of equal size, so the peak at $R = \sigma$ decreases and the low average density near $R = 0$ increases, until a uniform density is established.

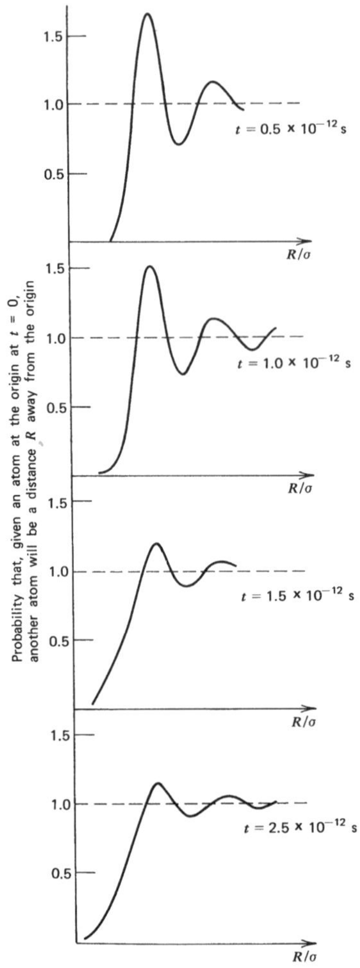

Figure 23.10 The probability that, given an atom is at the origin of coordinates at $t = 0$, a different atom will be a distance R away from the origin, as a function of time. The distance R is measured in units of the size parameter σ of the Lennard–Jones potential. From A. Rahman, *Phys. Rev.* **136A,** 405 (1964).

The preceding comments are valid for simple liquids, in which the constituent particles have spherically symmetric force fields. It is to be expected that a liquid of polyatomic molecules, in which the constituent particles have complicated nonspherical force fields, will have a structure somewhat different from that of a simple liquid. Consider, as an extreme case, liquid water. The water molecule has a dipole moment and higher multipole moments, reflecting the deviation from spherical symmetry of its electronic charge distribution. Of course, the interaction between two polar molecules depends on the angle between their dipole moments, quadrupole moments, and so on. More important, if the molecular charge distribution is not spherical, then neither will be the short-range repulsive forces between molecules. Hence the local packing can deviate considerably from that shown in Fig. 23.9, which is fundamentally like that in a fluid of hard spheres. Finally, and for H_2O most important, water molecules form hydrogen bonds with one another. The hydrogen bond is the dominant interaction between

---

[7] If there is no minimum in the integrand, as happens at high temperature and low density, the concept of coordination number loses all value.

[8] Other definitions of coordination number may be used. However, the qualitative aspects of the deductions cited are not altered by other choices of definition than the one given.

[9] The first computer simulations of this type, carried out by Rahman (1965), were based on solving Newton's equations for 864 molecules interacting with a Lennard–Jones potential corresponding to Ar–Ar. The density chosen was like that in liquid Ar at 85 K.

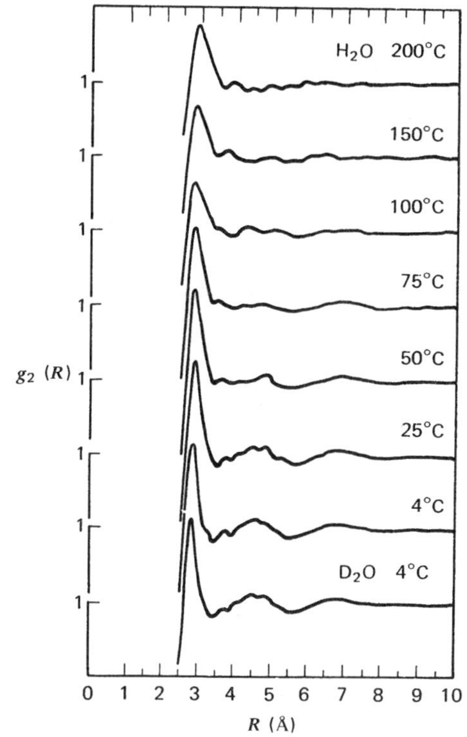

Figure 23.11 Pair-correlation functions for liquid H_2O at various temperatures and for liquid D_2O at 4°C, as determined by Narten *et al.* (1967). Note that the baseline of each curve is one unit above that for the curve below. From A. H. Narten, ORNL-4578 UC4, Oak Ridge National Laboratory, 1970.

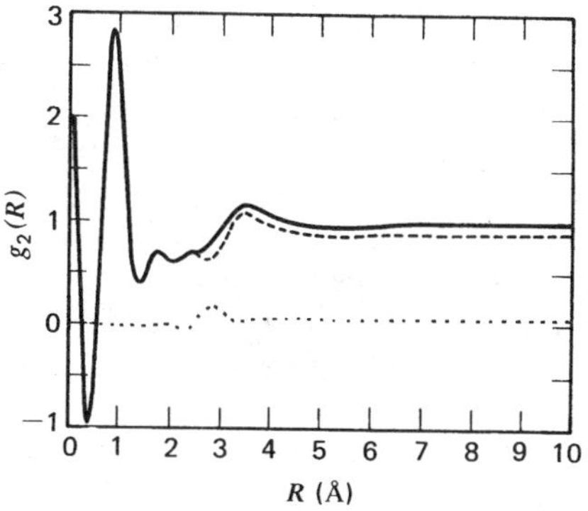

Figure 23.12 Atom pair-correlation functions for liquid D_2O. The solid curve is the weighted sum of contributions from O–O, O–D, and D–D correlations, the dotted curve is the O–O pair-correlation function, and the dashed line is the appropriately weighted difference pair-correlation function. In the difference pair-correlation functions shown, the contributions from O–D and D–D pairs have nearly equal weight. From A. H. Narten, *J. Chem. Phys.* **56,** 5681 (1972).

water molecules; it is strongly directional and has a depth very much larger than k_BT.

Although the electronic charge distribution of a water molecule is not spherical, the deviation is too small to show up in the x-ray diffraction pattern from the liquid. Therefore, the scattering of x-rays from water depends almost completely on only the distribution of molecular centers, that is, on the positions of the O atoms. Figure 23.11 displays O–O pair-correlation functions for liquid H_2O at various temperatures and constant density. These data show that:

1. There is perceptible structure in the local environment of a water molecule to about 8 Å distance from the molecule at the origin.
2. The nearest-neighbor separation is only weakly temperature-dependent at constant density, changing from 2.82 Å at 4°C to 2.94 Å at 200°C.
3. The O–O pair-correlation function is sensibly the same in H_2O and D_2O.
4. The average coordination number in liquid water is sensibly constant at about 4 from 4°C to 200°C (constant density of 1 g/cm^3).

Clearly, water differs appreciably from liquid argon, in which the number of nearest neighbors is about 10, and which is "loosely close-packed." Observations 1 through 4 imply that water is more "open" than is liquid argon, and also that there is tetrahedral coordination in the local structure surrounding a given water molecule. Tetrahedral coordination implies the existence of a correlation between the orientations of water molecules. How far does this orientational correlation extend? This question can be answered by examining neutron scattering from D_2O.

X-rays are scattered by electrons, whereas neutrons are scattered by nuclei. As applied to the study of liquid D_2O, neutron scattering is about equally sensitive to O–D and D–D pair correlations, and rather insensitive to O–O pair correlations. Using the O–O pair-correlation function obtained by x-ray scattering, we can learn, by difference, something about the sum of the O–D and D–D pair correlations, and therefore about orientational correlations in the liquid. The data displayed in Fig. 23.12 show that in liquid D_2O orientational correlations vanish beyond the first shell of four water molecules.

Just as the packing in liquid argon is a blurred and imperfect derivative from the pattern characteristic of closest packing of spheres, so the local packing in liquid water is a blurred and imperfect derivative from some extended hydrogen-bonded network in which each water molecule is coordinated to four neighboring molecules by hydrogen bonds. Even though both are liquids, and macroscopically isotropic, the local structures of Ar and H_2O are very different, reflecting the very different kinds of interaction between molecules in the two systems.

It is worthwhile to examine a few more examples of the relationship between molecular interactions and the structure of a liquid.

1. Suppose that the molecule is very asymmetric, such as a long-chain paraffin hydrocarbon—for example, dodecane. The force field surrounding such a molecule follows the contours of the carbon skeleton, and is also "long and thin." The structure of liquid paraffin hydrocarbons can be interpreted by first considering a collection of randomly oriented hydrocarbon molecules.[10] In this random configuration the density of packing is very low relative to what it would be in an ordered parallel array of molecules. When the density of molecules is low, the total energy of attraction is necessarily small; there are many possible disordered arrangements with the same low energy of attraction. On the other hand, there is only one way in which the molecules can be arranged in the completely ordered array that maximizes the attractive interactions between the molecules. The structure of liquid paraffins is determined by the competition between the tendency toward disorder induced by collisions between molecules and other thermal motions, and the order corresponding to the lowest-energy arrangement, that is, by the competition between entropy and energy terms in the free energy. Thus, the tendency for the molecules to become aligned is stronger the longer the molecular chain, but the ordering is never as complete as in the crystalline phase. As in simple liquids, there is a distribution of possible near-neighbor distances within the constraint of partial parallel alignment enforced by the balance of repulsive and attractive interactions between the molecules.

2. The paraffin hydrocarbons typify a class of substances with asymmetric interactions. What happens when, in addition to these, there are strong local directional interactions, such as are induced by hydrogen bonding? In liquids of the long-chain alcohols the combination of molecular asymmetry and directional hydrogen bonding leads to an accentuation of the features discussed above. It is found that the structure of liquid linear alcohols is in accord with a picture in which the alcohol molecules are linked in chains by hydrogen bonds between hydroxyl groups, while the aliphatic groups remain approximately parallel as in the corresponding hydrocarbons.

3. In liquid sulfur there is evidence that there exists a wide variety of chain and ring molecules, and that the distribution of atoms among chains and rings depends on the temperature. The structure of liquid sulfur is determined by a complex balance between chemical bonding forces and intermolecular forces of the type found in simple liquids.

4. In all the types of liquids thus far described, the range of the interaction is about equal to the molecular diameter. But both molten salts and liquid metals contain charged particles, and the range of the Coulomb interaction between charged particles is very large relative to the particle diameter. Moreover, the electrostatic interaction is so large relative to the thermal energy that it is difficult to separate positive and negative charges. The consequences of these characteristics of the potential are somewhat different in molten salts and in liquid metals.

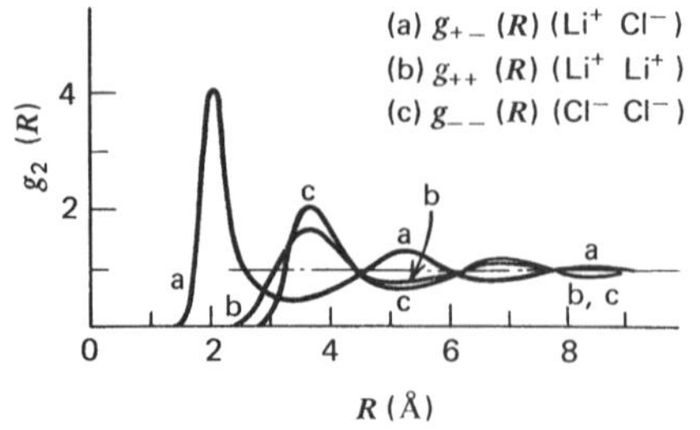

Figure 23.13 Pair-correlation functions from a computer simulation of liquid LiCl, $T = 833$ K, $v = 28.3$ cm^3/mol. From L. Woodcock, *Chem. Phys. Lett.* **10,** 257 (1971).

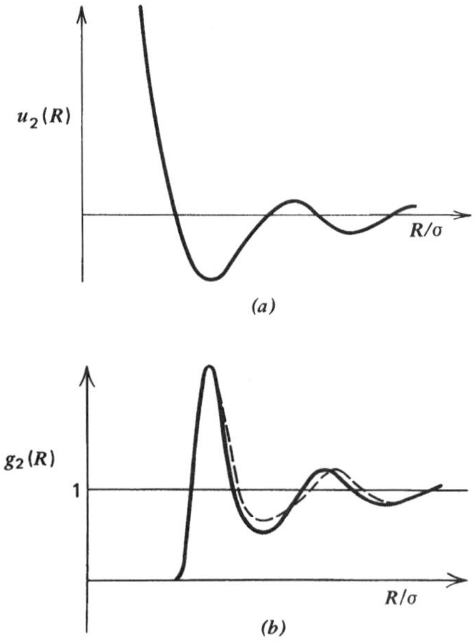

Figure 23.14 (*a*) General shape of the effective pair potential between ion cores in a liquid metal. (*b*) Pair-correlation functions for liquid Ar (- - -) and liquid Cs (—), scaled so that the positions of the first maxima coincide. From A. Paskin, *Adv. Phys.* **16,** 223 (1971).

In molten salts each positive ion is surrounded by negative ions, and vice versa (see Fig. 23.13). The primary effect of the electrostatic interaction between ions is to lower the total energy of the liquid. Given the overall charge alternation of that structure, short-range forces like those in argon then influence the ionic motion and play an important role in determining details of the local ionic environment.

In a liquid metal the two charged species are ions and electrons. Just as in a molecule, the electrons move so much more rapidly than do the ions that the latter move in an effective potential generated by the distribution of electrons that minimizes the energy for each ion–ion separation *R*. Consequently, the effective ion–ion potential is not the Coulomb potential, but rather some relatively short-range screened potential. Moreover, because the electron wavelength and ion–ion separation are comparable, diffraction of the electrons influences the form of the ion–ion potential—it has oscillations (see Fig. 23.14*a*). However, despite the differences between the detailed shapes of the ion–ion interaction and, say, the Ar–Ar interaction, the fact that both are

[10] A visualization of the random configuration is a disorderly heap of matchsticks. In the disorderly heap there are many "holes," and the density of packing is low. The configuration with closest packing (highest density) is with all the matches parallel and touching as in the matchbox.

short-range is enough to ensure that the local structure in a liquid metal is very much like that in a simple liquid (see Fig. 23.14*b*).

23.3 Relationships between the Structure and the Thermodynamic Properties of a Simple Liquid

How can the thermodynamic properties of a liquid be related to its structure? Consider the calculation of the internal energy, and suppose that the total potential energy of molecular interaction can be accurately represented as a sum of pair interactions (see Section 21.7). As usual, we suppose that the pair interactions are independent of the molecular velocities. Given these conditions, the contribution to the internal energy from the kinetic energy of the molecules is $\frac{3}{2}Nk_BT$, just the same as for a perfect gas of monatomic particles. However, in the case of the liquid (or the dense gas) there is an additional contribution to the internal energy from the potential energy of interaction. Since two molecules a distance R apart have a potential energy $u_2(R)$, the average potential energy of the fluid is simply obtained by adding together the potential energy arising from all $\frac{1}{2}N(N-1)$ possible distinct pairs of molecules.[11] The structure of the liquid enters the description when we specify the number of pairs of molecules separated by the distance R. On the average, a given molecule has $4\pi R^2 n g_2(R)\,dR$ neighbors in the range R to $R + dR$. Thus, its average potential energy of interaction with those molecules between R and $R + dR$ is $u_2(R) \cdot 4\pi R^2 n g_2(R)\,dR$, and with all molecules is $\int_0^\infty u_2(R) \cdot 4\pi R^2 n g_2(R)\,dR$. Since there are N molecules in the liquid, to obtain the total energy we multiply the integral expression by N and then divide by 2 to avoid counting each pair interaction twice (once when the molecule is selected as the "given" molecule and once when it is a neighbor of some other "given" molecule). Thus, the internal energy of the liquid is

$$U = \frac{3}{2}Nk_BT + 2\pi Nn\int_0^\infty u_2(R)g_2(R)R^2\,dR. \tag{23.17}$$

Of course, from U as determined by Eq. 23.17 we can obtain the heat capacity and other thermodynamic functions, but such calculations require knowledge of $(\partial g_2(R)/\partial T)_V$, and so on.

Consider now the calculation of the equation of state. The problem to be solved is that of sorting out the effects of collisions with a wall (as in a dilute gas) from the effects of interactions between molecules. The fundamental ideas needed were introduced in Section 21.7 in the calculation of the second virial coefficient. However, before we can use those ideas to calculate the equation of state of a liquid, we must examine what physical picture underlies them.

In Section 21.7 we asserted that the probability density for finding another molecule a distance R from a molecule at the origin of the coordinates is $n\exp(-u_2/k_BT)$. This assertion is correct only when the density of molecules is so low that interactions between all other molecules and the two mentioned can be neglected. Then $n\exp(-u_2/k_BT)$ is seen to be a special case of the probability density for finding a given configuration (separation of a pair of molecules by R) in a canonical ensemble.[12] In a dilute gas the average separation between the molecules is much greater than the range of the potential u_2, and in almost all instances two molecules can be moved toward one another along the line connecting their centers without coming close to any of the other molecules in the gas. Thus at very low density $u_2(R)$ is also just the work required to bring the two molecules to a separation R from a distance large enough that their interaction is vanishingly small. Consider now a liquid. In this case there are many molecules within the range of the pair potential $u_2(R)$, and we cannot isolate any one pair of molecules from interaction with the other molecules. A necessary implication of this observation is that the work required to bring two molecules to a separation R from a distance at which the direct interaction u_2 is vanishingly small, denoted as $w_2(R)$, is different from $u_2(R)$. The point is that in a dense system moving two molecules must result in some rearrangement in the positions of all the other molecules. Thus, when two molecules are brought to a separation R from some initial large separation, there are uncontrolled changes in the rest of the system. Just as in our discussion of the first law of thermodynamics (see Chapter 13), this characteristic behavior of the system of molecules makes w_2 different from simple mechanical work. In general, w_2 is both temperature- and density-dependent; it is known as the *potential of average force,* since the rate of change of w_2 with displacement is the average force acting between the two molecules; for example, $-\partial w_2/\partial x$ is the x component of the average force between the molecule at the origin and the molecule a distance R away.

The generalization of the arguments of Section 21.7 to a dense fluid consists in asserting that the probability density for finding another molecule a distance R from a molecule at the origin of the coordinates is

$$n\exp\left(-\frac{w_2(R)}{k_BT}\right). \tag{23.18}$$

But we have also defined $ng_2(R)$ to be the probability density of finding another molecule a distance R from a molecule at the origin of the coordinates. Comparison of these two definitions establishes the identity

$$g_2(R) \equiv \exp\left(-\frac{w_2(R)}{k_BT}\right). \tag{23.19}$$

[11] It does not matter whether the pair AB is constructed by first choosing A or first choosing B. Hence the $N(N-1)$ possible choices of pairs include each pair, for example, the pair AB, twice; we then divide by 2 to obtain the number of distinct pairs.

[12] See Section 19.6.

A few words about w_2 are appropriate. First, when the density is very low, w_2 must approach the pair-interaction potential u_2, since in the low-density limit two molecules may be moved without influencing any of the other molecules in the system. Second, in a dense system w_2 becomes large and positive for small pair separation [the region where $g_2(R)$ is essentially zero], is negative at a distance corresponding to the first peak in $g_2(R)$, and then oscillates between positive and negative values while decaying nearly to zero after three molecular diameters (see Fig. 23.15). Note that the range of w_2 is longer than that of u_2. This difference arises because w_2 incorporates the effects of rearrangement in the medium, hence of geometric restraints. To illustrate how geometric restraints influence w_2, imagine moving two hard-sphere molecules (see Fig. 23.16) in a dense fluid of hard-sphere molecules. When the two are far apart, collisions with the other molecules are uniformly distributed in all directions, so the net force acting on each vanishes, as does the net force between them. But if the two hard spheres are less than two diameters apart (see Fig. 23.16), other molecules can collide only with the "outside" region, and none can get between them. Thus each molecule is subjected to a net force, that molecule on the left in Fig. 23.16 to a force acting to the right, and the reverse for the other molecule. The net force between the molecules is the source of an effective attraction. We see that even though two isolated hard spheres never attract one another, two hard spheres in a dense fluid of hard spheres do attract one another when $R < 2\sigma$. More generally, we find that in a real fluid w_2 takes on positive and negative values reflecting the balance between the effects of the real potential energy u_2 and the geometric restraints imposed by "impenetrability" of the molecules.

The calculation of the equation of state of a liquid proceeds in the same way as the calculation in Chapter 21. Whenever the potential energy of interaction of the molecules is independent of the molecular velocities, the contribution to the pressure from collisions with the wall, the *kinetic pressure,* is

$$p_K = \frac{Nk_BT}{V}. \tag{23.20}$$

That this is so is easily seen from the arguments preceding Eq. 21.160 or, more formally, from the separation of the par-

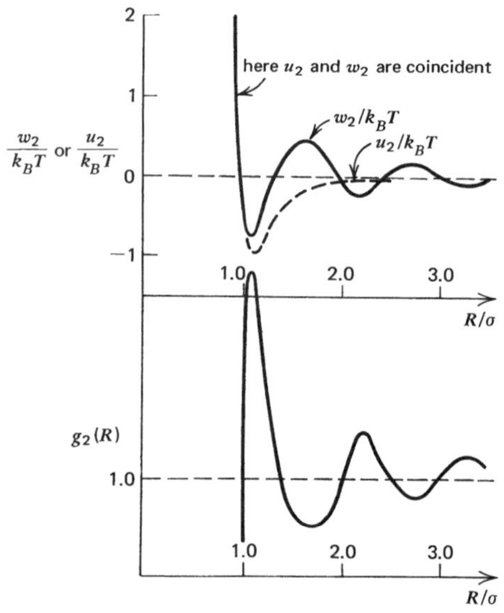

Figure 23.15 The potential of mean force, w_2, and the pair potential, u_2, for a dense fluid. The pair-correlation function g_2 of the fluid to which w_2 and u_2 refer is plotted in the lower part of the figure.

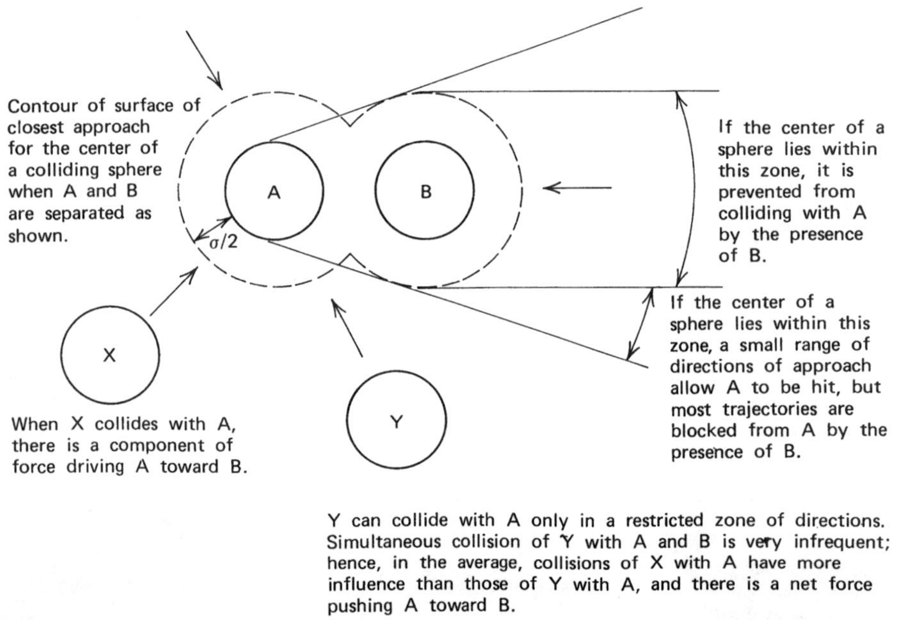

Figure 23.16 Schematic representation of the origin of w_2 for hard spheres. Note that a position and a direction of motion are associated with each sphere.

tition function into translational and configurational contributions, as in Eq. 21C.4. The calculation of the molecular interaction contribution to the pressure starts with the substitution of $n^2 \exp(-w_2/k_BT)$ for $n^2 \exp(-u_2/k_BT)$ in Eq. 21.160, and then proceeds through all the same steps. Of course, the direct force exerted by the molecule a distance R from the one at the origin is still $-(du_2/dR) \cos\theta$. Integrating as before, adding p_K from Eq. 23.20 above, and using Eq. 23.19 to replace $\exp(-w_2/k_BT)$ with $g_2(R)$, we find that the equation of state of the liquid is

$$p = nk_BT - \frac{2\pi}{3} n^2 \int_0^\infty R^3 \frac{du_2}{dR} g_2(R)\, dR. \tag{23.21}$$

From Eqs. 23.17 and 23.21 we can obtain the other thermodynamic functions, at least in principle.[13]

If we consider a realistic intermolecular potential, such as is shown in Fig. 23.6, we see that du_2/dR is negative at small intermolecular separations and positive at large separations, where it very rapidly ($\propto R^{-7}$) falls again to zero. The dominant contribution to the integral in Eq. 23.21 is thus from the region near the molecule at the origin. At high densities there are enough neighbors in the region where $du_2/dR < 0$ to make the integral negative and hence p higher than the perfect gas value, but at moderate densities the contrary is true. This can be seen from the data presented in Fig. 21.5.

The equations we have obtained thus far involve the pair-correlation function $g_2(R)$. To construct an interpretation of liquid properties based entirely on theoretical considerations, it is necessary to relate $g_2(R)$ to the intermolecular potential (as a function of temperature and density, of course). There are various theoretical approaches directed to this end, differing in the use of various approximations, most of which cannot be easily visualized. In Section 23.5 we sketch several of these theoretical approaches. However, before describing those theoretical approaches we shall discuss some of the physical ideas involved and illustrate some of the results obtained.

23.4 The Molecular Theory of Monoatomic Liquids: General Remarks

We have seen that a principal difference between a dilute gas and a liquid can be stated as follows. In a dilute gas a typical molecule is usually outside the force fields of all other molecules and only occasionally in the force field of one other molecule (binary collision), whereas in a liquid a typical molecule is usually within the force fields of, say, 10 nearest-neighbor molecules and is never completely free of the influence of other molecules. One consequence of the multiplicity of simultaneous interactions in the liquid is that the work required to bring a molecule to a distance R from another molecule at the origin of the coordinates is the potential of average force $w_2(R)$, and *not* the pair potential $u_2(R)$. We can think of $w_2(R)$ as the potential of the force acting on two molecules, say 1 and 2, averaged over the possible configurations of the remaining molecules of the liquid. In symbols, the average force acting on molecule 1 at $\mathbf{R}_1$, when molecule 2 is known to be at $\mathbf{R}_2$, is[14] $-\nabla_1 w_2(R_{12})$, where $R_{12} \equiv |\mathbf{R}_1 - \mathbf{R}_2|$ is the distance between the two molecules. Note that by using Eq. 23.19 we can rewrite this average force as $k_BT\nabla_1 \ln g_2(R_{12})$. The average force can be thought of as the resultant of that force due to molecule 2 at $\mathbf{R}_2$ alone and the average force due to all the other particles. But the probability of finding a molecule in the volume element $d\mathbf{R}_3$ at $\mathbf{R}_3$, when it is known that other molecules are located at $\mathbf{R}_1$ and $\mathbf{R}_2$ is, using the argument of Section 23.2,

$$\frac{n^3 g_3(\mathbf{R}_1, \mathbf{R}_2, \mathbf{R}_3)}{n^2 g_2(R_{12})} d\mathbf{R}_3.$$

The denominator of this expression is just the probability density for finding two molecules separated a distance R_{12}, including the effect of intermolecular forces. The numerator is, by definition, the probability density for finding three molecules at positions $\mathbf{R}_1$, $\mathbf{R}_2$, and $\mathbf{R}_3$ multiplied by the volume element $d\mathbf{R}_3$. The *triplet correlation function* $g_3(\mathbf{R}_1, \mathbf{R}_2, \mathbf{R}_3)$ accounts for the effect of intermolecular forces in the medium just as does $g_2(R_{12})$. We can now write for the average force on the molecule at $\mathbf{R}_1$ due to all other molecules except the molecule at $\mathbf{R}_2$ the expression

$$\int d\mathbf{R}_3 \left[-\nabla_1 u_2(R_{13}) \frac{n g_3(\mathbf{R}_1, \mathbf{R}_2, \mathbf{R}_3)}{g_2(R_{12})} \right].$$

Adding the contributions, we have

$$\nabla_1 w_2(R_{12}) = \nabla u_2(R_{12}) + n \int d\mathbf{R}_3 \nabla_1 u_2(R_{13}) \frac{g_3(\mathbf{R}_1, \mathbf{R}_2, \mathbf{R}_3)}{g_2(R_{12})}, \tag{23.22}$$

and using Eq. 23.19 we can rewrite Eq. 23.22 to be an equation for $g_2(R_{12})$:

$$-k_BT\nabla_1 \ln g_2(R_{12}) = \nabla_1 u_2(R_{12}) + n \int d\mathbf{R}_3 \nabla_1 u_2(R_{13}) \frac{g_3(\mathbf{R}_1, \mathbf{R}_2, \mathbf{R}_3)}{g_2(R_{12})}. \tag{23.23}$$

[13] Note that the integral in Eq. 23.21 is not simply proportional to the second virial coefficient B, since $g_2(R)$ itself incorporates density dependence. At low density, however, $w_2(R) \to u_2(R)$ and $g_2(R) \to \exp(-u_2(R)/k_BT)$, and we are led back to Eq. 21.165.

[14] ∇_1 is shorthand notation for differentiation with respect to the position of molecule 1.

If we could solve Eq. 23.23, we would find the pair-correlation function in terms of the pair potential and the temperature and density, with the thermodynamic variables entering via the averaging process. However, to solve Eq. 23.23 we must know the function $g_3(\mathbf{R}_1, \mathbf{R}_2, \mathbf{R}_3)$, and we do not! Several of the theories of simple liquids are based on Eq. 23.23 and a supplementary assumption concerning the form of $g_3(\mathbf{R}_1, \mathbf{R}_2, \mathbf{R}_3)$.

In Eq. 23.23 the terms $\nabla_1 u_2(R_{12})$ and $\nabla_1 u_2(R_{13})$ are "direct" forces, whereas the effect of the averaging in the second term on the right-hand side is to account for "indirect" effects that influence the distribution of molecules. A different way of representing the structure of a liquid also relies on the idea of direct and indirect effects. The idea is that $[g_2(R) - 1]$ measures the total correlation between pairs of particles. To see the motivation for this statement, notice that $[g_2(R) - 1]$ tends to zero as $R \to \infty$, since there is no correlation between the positions of two widely separated molecules. We now *define* $[g_2(R) - 1]$ to be the sum of two contributions. The first contribution, $C(R)$, consists of correlations transmitted directly between the two molecules; we need not specify how this occurs. The second contribution comes from correlations transmitted via a third molecule, and is made up of direct correlation between molecules 1 and 3 multiplied by the total correlation function between molecules 2 and 3 and the probability of finding a third molecule in the volume element $d\mathbf{R}_3$, integrated over all possible positions of molecule 3. Adding these two contributions we are led to the *definition*[15]

$$g_2(R_{12}) - 1 \equiv C(R_{12}) + n \int d\mathbf{R}_3 C(R_{23})[g_2(R_{23}) - 1]. \tag{23.24}$$

If we could solve Eq. 23.24, we would find the pair-correlation function required for the description of the structure of the liquid. But to do so we must know $C(R_{12})$, and we do not! Several of the theories of simple liquids are based on Eq. 23.24 and a supplementary hypothesis concerning the form of $C(R_{12})$.

Equation 23.23 and Eq. 23.24 are exact, yet they cannot be used without extra information that must be obtained from considerations other than those used to derive these equations. We shall examine some aspects of these considerations and the approximate theories of liquid structure to which they lead in Section 23.5. In this section we quote some of the results of approximate theories to illustrate the accuracy with which the structure and thermodynamic properties of a liquid can be predicted.

Before proceeding further in the discussion of the theory of liquid structure, it is appropriate to mention briefly the technique of computer simulation. Although fundamentally a theoretical procedure, it has much in common with experiment, as will become evident from the following description.

There are two computer simulation techniques of importance for the study of liquids. In the method of *molecular dynamics* the classical equations of motion of N interacting particles are solved by step-by-step numerical integration, thereby determining the trajectories of all particles over a suitable time interval. Typical calculations, involving $100 < N < 1000$ particles, use time steps of 10^{-14} s and give the complete history of the system for $10^{-14} < t < 10^{-10}$ s; for some calculations it is necessary to use shorter time steps or to cover longer intervals of elapsed time. The thermodynamic properties are obtained, using the quasi-ergodic hypothesis, from the time average of appropriate molecular properties (see Section 15.1). In the *Monte Carlo* method one generates a sequence of particle configurations in such a way that the probability of a particular configuration, of potential energy $\mathcal{V}_N$, appearing in the sequence is proportional to $\exp(-\mathcal{V}_N/k_BT)$. Then the unweighted average of a molecular property over the configurations of the sequence gives the canonical average of that function (see Section 19.6). Typical calculations involve sequences of 10^6–10^7 configurations. The method of molecular dynamics provides information about the structure and properties of the liquid at equilibrium, and about transport properties and the approach to equilibrium. It is particularly valuable for the study of dynamical properties of single molecules, or groups of molecules, in strong interaction with others. The Monte Carlo method is restricted to the study of equilibrium properties of the system. When properly executed, both methods yield the true properties of the system unbiased by errors arising from approximations.

The molecular dynamics and Monte Carlo methods are similar to experiments in the sense that they generate data which must be analyzed using theoretical concepts introduced by other methods. Unlike real experiments, a computer simulation can test the effects of adding and subtracting particular forces to the intermolecular force, and can provide tests of theoretical ideas by computing the exact behavior of the model used in developing those ideas. It is a powerful addition to the tools available for the study of matter.

Consider what can be learned about $g_2(R)$ from the study of a simple model. We have already noted that the experimental data indicate that $g_2(R)$ is relatively insensitive to the temperature if the density of the liquid is kept fixed. We have also deduced the general dependence of $g_2(R)$ on R from arguments based on the consideration of only the repulsive forces acting between molecules. Suppose that we treat the molecules as if they were billiard balls; that is, suppose that the hard-sphere potential is used to represent the repulsive interaction between molecules, with the effects of the attractive interaction ignored. Clearly this potential cannot lead to a stable liquid because there is no attraction between the molecules. Nevertheless, for fixed density we can examine the extent to which our previous arguments are verified. In particular, we wish to know if $g_2(R)$ for a dense

[15] Equation 23.24 defines the function $C(R_{12})$ in terms of $g_2(R)$; $C(R_{12})$ is called the *direct correlation function.*

fluid of hard spheres behaves as proposed and resembles the distribution of molecules in a real liquid.

The earliest answer to this question was given by Morrell and Hildebrand, who constructed a model of a liquid by placing several hundred gelatin spheres, including a few colored ones, in a vessel containing a boiled gelatin solution of the same density and refractive index. The vessel was shaken and, from photographs of the contents, the distribution of distances between a pair of colored spheres determined. The observed distribution is very much like that observed in a real liquid. Using the Monte Carlo or molecular dynamics computer simulation methods described earlier, a more sophisticated version of the same experiment can be conducted. Typical results are shown in Fig. 23.17*a*. Alternatively, one of the approximate theories we mentioned can be used; typical results of such a calculation are also shown in Fig. 23.17*a*. Aside from the discontinuous fall of $g_2(R)$ to zero for $(R/\sigma) < 1$, the pair-correlation function of a rigid-sphere fluid does show the characteristic features of the distribution of matter in a real liquid. The discontinuous change in $g_2(R)$ at $(R/\sigma) = 1$ is a consequence of the discontinous change in potential at $(R/\sigma) = 1$ that is characteristic of the hard-sphere interaction. For a more realistic, "softer," continuous repulsive potential, the pair-correlation function is substantially the same as for hard spheres everywhere, but does not have the unphysical sudden change in value at $(R/\sigma) = 1$. This is shown in Fig. 23.17*b*. These results for fluids in which there are only repulsive forces, together with the observation that $g_2(R)$ for a rigid-sphere fluid depends only on the fluid's density and not on its temperature, confirm our analysis of the overwhelming importance of repulsive forces in determining the structure of a liquid.

What about the properties of a real liquid? There are, fundamentally, two uncertainties in the theoretical analysis. First, it is now known that the interactions in a liquid cannot be satisfactorily represented as a sum of interactions between pairs of molecules (see Eq. 21.159). Although the contribution to the potential energy by three-body terms is not large, they are required for quantitative prediction of the properties of the liquid. Higher-order, many-body interactions appear to be unimportant for liquid argon, and probably also for other simple liquids. Given an accurate expression for the potential energy, the method of computer simulation provides excellent predictions of the thermodynamic properties of a simple liquid (see Tables 23.3 and 23.4). Indeed, the method of computer simulation, by calculating the properties of the liquid with no simplifications apart from those inherent in the assumed potential, tests the accuracy of a potential energy expression via comparison of the calculated properties with those of the real liquid.

Useful as computer simulations are, theoretical concepts and understanding are usually derived from analytical, albeit approximate, theories of the liquid state. We must acknowledge that the use of approximations in the solution of Eq. 23.23 or 23.24 introduces error in the predicted properties of the liquid—this is the second of the uncertainties in

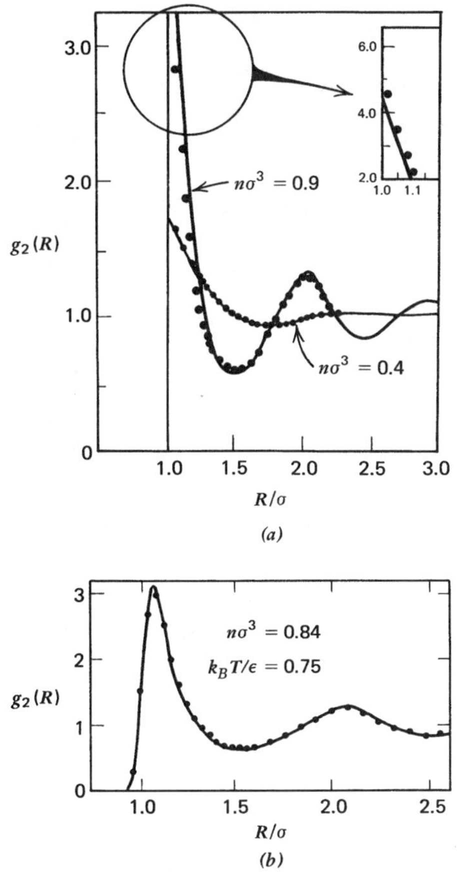

Figure 23.17 (*a*) Pair-correlation functions for a hard-sphere fluid at two densities, $n\sigma^3 = 0.9$ and $n\sigma^3 = 0.4$. • Computer simulation results; ——— PY theory. The PY theory is obtained by making the approximation

$$C(R) = \left[1 - \exp\left(\frac{u_2(R)}{k_BT}\right)\right] g_2(R)$$

and then solving Eq. 23.24 for $g_2(R)$. From R. O. Watts and I. J. McGee, *Liquid State and Chemical Physics* (Wiley, New York, 1976). (*b*) Pair-correlation function for a fluid with purely repulsive pair interactions of the form $u_2 = 4\epsilon[(\sigma/R)^{1/2} - (\sigma/R)^6] + \epsilon$ for $R \le R_m \equiv 2^{1/6}\sigma$, $u_2 = 0$, for $R > R_m$. From J. P. Hansen and I. R. McDonald, *Theory of Simple Liquids* (Academic Press, New York, 1976).

Table 23.3 Contributions to the Internal Energy of Liquid Argon[a]

T (K)	v cm³/mol	$u_1(2b)$[b] cal/mol	$u_1(3b)$[c] cal/mol	$u_1(Q)$[d] cal/mol	u_1(thy)[e] cal/mol	u_1(exp) cal/mol
100.00	27.04	−1525.2	87.1	15.6	−1423	−1432
100.00	29.66	−1393.6	67.9	12.5	−1313	−1324
140.00	30.65	−1284.7	62.8	9.3	−1213	−1209
140.00	41.79	−951.4	39.5	6.4	−906	−922
150.87	70.73	−603.8	26.6	4.6	−573	−591

[a] From J. A. Barker and D. Henderson, *Rev. Mod. Phys.* **48,** 587 (1976).
[b] Pair potential (two-body) contribution.
[c] Triplet potential (three-body) contribution.
[d] Quantum correction (the de Broglie wavelength Λ is, for the temperature listed, large enough to make about a 1% quantum correction to the internal energy necessary).
[e] $u_1(\text{thy}) = u_1(2b) + u_1(3b) + u_1(Q)$.

Table 23.4 Contributions to the Pressure of Liquid Argon[a]

T (K)	v cm^3/mol	$p(2b)$[b] (atm)	$p(3b)$[c] (atm)	$p(Q)$[d] (atm)	p(thy)[e] (atm)	p(exp) (atm)
100.00	27.04	+239.9	364.2	42.2	646	652
100.00	29.66	−148.0	238.8	23.3	116	105
140.00	30.65	+348.9	214.3	16.7	580	583
140.00	41.79	−33.7	49.0	2.7	18	37
150.87	70.73	+34.5	13.2	1.2	49	49

[a] From J. A. Barker and D. Henderson, *Rev. Mod. Phys.* **48,** 587 (1976).
[b] Pair potential (two-body) contribution.
[c] Triplet potential (three-body) contribution.
[d] Quantum correction (the de Broglie wavelength Λ is, for the temperatures listed, large enough to make about a 1% quantum correction to the internal energy necessary).
[e] $p_1(\text{theory}) = p(2b) + p(3b) + p(Q)$.

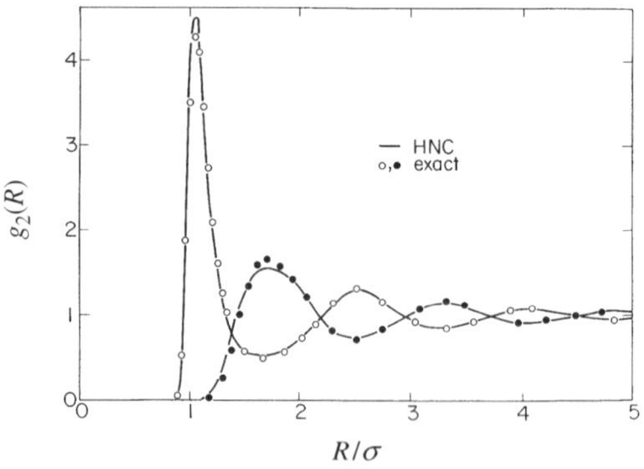

Figure 23.19 Comparison between molecular dynamics simulation data and the HNC prediction of the pair correlation functions for a model univalent molten salt, which mimics molten NaCl. From J. P. Hansen and I. R. McDonald, *Theory of Simple Liquids* (Academic Press, New York, 1976). The open circles are simulation data for the pair correlation function of positive and negative ions, and the filled circles are simulation data for the pair correlation function of like charged ions (the same for positive–positive and negative–negative pairs of ions). The positive and negative ions are assumed to have the same size, i.e., are assigned the same value of σ.

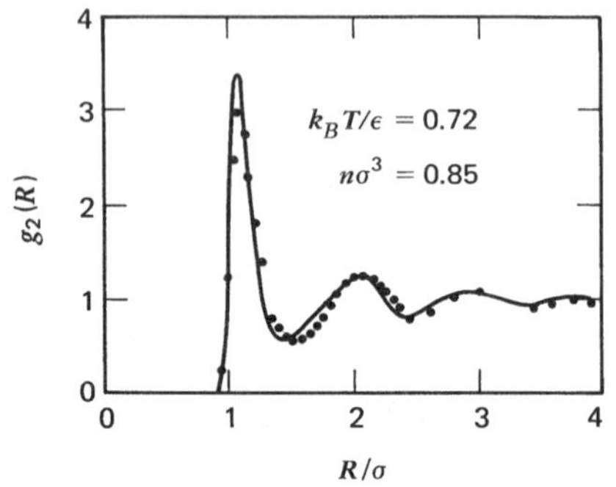

Figure 23.18 The pair-correlation function of the Lennard–Jones 6–12 liquid near its triple point. The points are determined by the method of molecular dynamics, the line is the prediction of the PY theory (see caption to Fig. 23.17*a*). From J. A. Barker and D. Henderson, *Rev. Mod. Phys.* **48,** 587 (1976).

theoretical analysis. Modern theories of the structure of a liquid are actually rather good, although not exact. The error in prediction of the pair-correlation function for a typical liquid density can be ascertained by examination of Fig. 23.17*a,* Fig. 23.18 and Fig. 23.19. Fig. 23.18 displays the pair-correlation function for a liquid with pair interaction of the form of Eq. 21.173, the so-called Lennard–Jones 6–12 potential, and Fig. 23.19 displays the pair-correlation functions for a model molten salt with pair interaction of the form $(e^2/\sigma)[(1/9)(\sigma/R)^9 - (\sigma/R)]$ with e the unit charge. We see that in the three cases illustrated the overall agreement between theory and computer experiment is very good, but it is equally evident that there are small discrepancies with respect to prediction of the height of the first peak of $g_2(R)$ and with respect to the position and shape of the second peak. The thermodynamic functions computed from Eqs. 23.17 and 23.21 are very sensitive to the precise shapes of $u_2(R)$ and $g_2(R)$. Consequently these small discrepancies lead to appreciable errors in the predicted thermodynamic properties of the liquid, particularly the pressure. In addition, because of the form of the approximation used to solve Eq. 23.23 or 23.24, it is usually found that self-consistency is lost, and the pressures calculated in different ways do not have the same value. An example of this behavior is shown in Fig. 23.20. Clearly, even though the available theory is rather good at predicting $g_2(R)$, it must be improved before fully quantitative predictions of the thermodynamic properties of the liquid can be made. That improvement is the subject of much current research, and remains to be accomplished.

The thrust of the preceding argument is that the structure of a liquid is mostly, but not entirely, determined by the repulsion between molecules. The very close resemblance between the pair-correlation functions of a rigid-sphere fluid, a Lennard–Jones 6–12 fluid, and liquid argon suggests the following question. Can we construct an analysis of the properties of a real liquid in which the attraction between molecules acts only as a small perturbation to the distribution induced by the repulsion between molecules? Just as in the application of perturbation theory to problems of quantum mechanical behavior, the success of this approach depends on the suitability of the choice of unperturbed reference system, and on the rate of convergence of the calculated corrections. For example, given what we have learned about the similarity between the pair-correlation functions of a hard-sphere fluid and a real liquid, we might assume that to calculate the influence of attractive interactions on, say, the Helmholtz free energy of the liquid, it is sufficiently accurate to imagine the molecules distributed as in a hard-sphere fluid, and average the attractive potential over the possible molecular separations in a hard-sphere fluid (BH theory).[16] Of course, real molecules do not repel like hard

[16] J. A. Barker and D. Henderson, *J. Chem. Phys.* **47,** 2856, 4714 (1967); *Can J. Phys.* **45,** 1959 (1967).

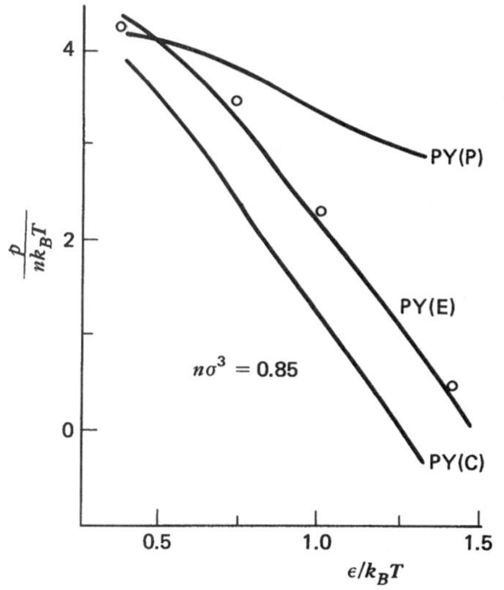

Figure 23.20 The equation of state, p/nk_BT, predicted by PY theory (see the caption to Fig. 23.17*a*). PY(P): Calculated from Eq. 23.21. PY(E): Calculated by first integrating the energy equation 23.17 with respect to temperature to obtain the Helmholtz free energy, then differentiating this Helmholtz free energy with respect to volume to obtain the pressure. PY(C): Calculated from

$$k_BT(\partial n/\partial p)_T = 1 + n\int [g_2(R) - 1]\, d\mathbf{R}$$

and integration with respect to the density to obtain the pressure. From J. A. Barker and D. Henderson, *Rev. Mod. Phys.* **48,** 587 (1976).

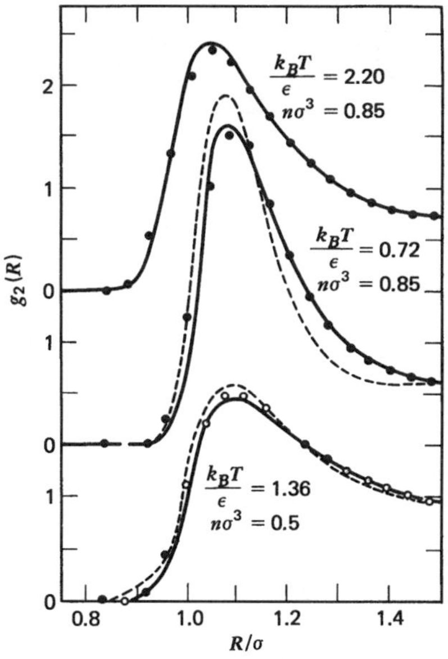

Figure 23.21 Pair-correlation functions of the Lennard–Jones 6–12 liquid. The points give the results of molecular-dynamics calculations, the broken curve the prediction of PY theory, and the solid curve the prediction of BH perturbation theory using the rigid-sphere fluid as reference. From J. A. Barker and D. Henderson, *Rev. Mod. Phys.* **48,** 587 (1976).

spheres; the repulsive energy of real molecules is something like $\epsilon(\sigma/R)^{12}$, and the distance of closest approach depends on the temperature. Furthermore, there can be large fluctuations in the environment of a molecule in a liquid. Assuming that the attractive contribution can be computed using only the hard-sphere distribution of distances is equivalent to replacing the distribution of potential energies of the molecules by an average potential that is the same for every molecule. Both the "softness" of the repulsive interaction and the deviations from average potential behavior can be accounted for approximately. Alternatively, instead of choosing the rigid-sphere fluid as reference and then correcting for the "softness" of the real repulsive interaction, one can divide the molecular interaction into a part containing all the *repulsive forces* and another part containing all the *attractive forces*. The structure of the reference fluid is then defined by the repulsive force component, and the effect of the attractive force component is treated via perturbation theory (WCA theory).[17] The BH form of perturbation theory has the advantage that it uses the same reference system irrespective of the form of the potential, and the properties of that reference system are very well known from computer simulation studies. It has the disadvantage that the perturbation must be carried to second order, or higher, to obtain high accuracy of prediction. The WCA form of perturbation theory has the disadvantage that the reference system is, or can be, different for different forms of the potential. In addition, the properties of the reference system depend on both temperature and density and are less well known than are those of a hard-sphere fluid, whose properties depend only on density. On the other hand, the WCA form of perturbation theory converges very rapidly, and only first-order corrections need be included to obtain very accurate predictions of liquid properties. The two forms of theory are complementary in a useful fashion. Both can be extended to include three-molecule interactions—the term Δ_3 in Eq. 21.159.

To illustrate the use of perturbation theory in the theory of liquids, we show in Fig. 23.21 the region of the peak of the pair-correlation function of a Lennard–Jones 6–12 liquid. Note how the fit to the molecular dynamics calculations improves. A comparison of predicted and (computer) experimental thermodynamic properties of the 6–12 liquid is given in Tables 23.5 and 23.6.

Just as for theories based on Eq. 23.23 or 23.24, application of the BH of WCA perturbation theory to the description of real liquids requires an accurate molecular interaction function. Given that potential energy function, one can expect predictions of the properties of real liquids as satisfactory as we show is possible for the model Lennard–Jones 6–12 liquid.

[17] J. D. Weeks, D. Chandler, and H. C. Andersen, *J. Chem. Phys.* **54,** 5237 (1971); **55,** 5422 (1971).

Table 23.5 Values of the Reduced Helmholtz Free Energy, A_N/Nk_BT, for a Lennard–Jones 6–12 Liquid[a]

k_BT/ϵ	$n\sigma^3$	MD	PY(E)[b]	BH	WCA
2.74	0.60	–0.34	–0.33	–0.33	–0.33
	0.70	+0.01	+0.01	+0.01	+0.01
	0.80	0.43	0.43	0.42	0.41
	0.90	0.93	0.95	0.95	0.92
	1.00	1.59	1.61	1.62	1.56
1.35	0.60	–1.77	–1.75	–1.75	–1.74
	0.70	–1.65	–1.62	–1.63	–1.63
	0.80	–1.41	–1.37	–1.41	–1.41
	0.90	–1.02	–0.99	–1.01	–1.02
	0.95	–0.72	–0.72	–0.72	(–0.74)
1.15	0.60	–2.29	–2.28	–2.30	–2.26
	0.70	–2.25	–2.23	–2.26	–2.24
	0.80	–2.06	–2.06	–2.10	–2.09
	0.90	–1.79	–1.74	–1.76	–1.77
0.75	0.60	–4.24		–4.29	–4.18
	0.70	–4.53	–4.50	–4.28	–4.51
	0.80	–4.69	–4.63	–4.74	–4.69
	0.90		–4.55	–4.67	(–4.62)

[a] From J. A. Barker and D. Henderson, *Rev. Mod. Phys.* **48,** 587 (1976).
[b] PY(E) denotes the results obtained from a calculation using the pair-correlation function derived from the theory of J. K. Percus and G. J. Yevick, *Phys. Rev.* **110,** 1 (1958).

Table 23.6 Values of p/nk_BT for a Lennard–Jones 6–12 Liquid[a]

k_BT/ϵ	$n\sigma^3$	Molecular Dynamics	Monte Carlo	PY(E)	BH	WCA
2.74	0.65	2.22		2.23	2.22	2.18
	0.75	3.05		3.11	3.10	3.04
	0.85	4.38		4.42	4.44	4.30
	0.95	6.15		6.31	6.40	6.10
1.35	0.10	0.72		0.72	0.74	0.77
	0.20	0.50		0.51	0.52	0.53
	0.30	0.35		0.36	0.36	0.31
	0.40	0.26		0.29	0.26	0.17
	0.50	0.30		0.33	0.27	0.18
	0.55	0.41		0.43	0.35	0.27
	0.65	0.80		0.85	0.74	0.71
	0.75	1.73		1.72	1.64	1.64
	0.85	3.37		3.24	3.36	3.28
	0.95	6.32		5.65	6.32	(5.90)
1.00	0.65	–0.25		–0.22	–0.36	–0.50
	0.75	+0.58	0.48	+0.57	+0.53	+0.40
	0.85	2.27	2.23	2.14	2.25	2.20
	0.90	3.50		3.33	3.53	(3.55)
0.72	0.85	0.40	0.25	0.33	0.25	0.26
	0.94		1.60	1.59	1.63	(1.83)

[a] From J. A. Barker and D. Henderson, *Rev. Mod. Phys.* **48,** 587 (1976).

23.5 The Molecular Theory of Monoatomic Liquids: Approximate Analyses

We now examine in more detail the quantitative basis for the aspects of liquid structure discussed in the last section. It is convenient to begin with an analysis that shows that knowledge of the pair correlation function over the density range $(0, n)$ permits calculation of the equation of state without explicit knowledge of the form of the interaction between the molecules. The starting point for this analysis is the relation between the mean number of molecules $\langle N\rangle$ in a volume V in an open system with a specified temperature and chemical potential. Using the definition of the grand partition function (Eq. 19.126), we find

$$\langle N\rangle = \frac{\partial \ln \Xi}{\partial \beta\mu} = \sum_{N=0}^{\infty} \frac{N e^{\beta N\mu} Q_N}{\Xi}, \tag{23.25}$$

with $\beta \equiv 1/k_BT$. At constant temperature, the Gibbs–Duhem equation (19.68) implies that $(nd\mu)_T = dp$, so that the isothermal compressibility of the fluid, $\kappa_T \equiv n^{-1}(\partial n/\partial p)_T$, can be computed from

$$\kappa_T = \left(\frac{\beta V}{\langle N\rangle^2}\right)\frac{\partial\langle N\rangle}{\partial \beta\mu}. \tag{23.26}$$

From the form of the grand partition function we also find that

$$\begin{aligned}\frac{\partial\langle N\rangle}{\partial \beta\mu} &= \sum_{N=0}^{\infty}\frac{N^2 e^{\beta N\mu}Q_N}{\Xi} - \left(\sum_{N=0}^{\infty}\frac{N e^{\beta N\mu}Q_N}{\Xi}\right)^2 \\ &= \langle N^2\rangle - \langle N\rangle^2.\end{aligned} \tag{23.27}$$

By definition

$$\begin{aligned}\langle N\rangle &= \int_V d\mathbf{R}_1 n(\mathbf{R}_1), \\ \langle N(N-1)\rangle &= \int_V d\mathbf{R}_1\, d\mathbf{R}_2 n^{(2)}(\mathbf{R}_1,\mathbf{R}_2),\end{aligned} \tag{23.28}$$

where $n^{(2)}(\mathbf{R}_1, \mathbf{R}_2) = n(\mathbf{R}_1)n(\mathbf{R}_2)g_2(\mathbf{R}_1, \mathbf{R}_2)$ is the two particle distribution function. Combining 23.28 with 23.27 we find

$$\begin{aligned}\frac{\partial\langle N\rangle}{\partial \beta\mu} &= \langle N^2\rangle - \langle N\rangle^2 = \langle N\rangle \\ &\quad + \int_V d\mathbf{R}_1\, d\mathbf{R}_2 n(\mathbf{R}_1)n(\mathbf{R}_2)[g_2(\mathbf{R}_1,\mathbf{R}_2)-1].\end{aligned} \tag{23.29}$$

In a homogeneous liquid of monoatomic species, $n(\mathbf{R}) = \langle N/V\rangle$ is a constant and the pair correlation function $g_2(\mathbf{R}_1, \mathbf{R}_2)$ depends only on the distance $|\mathbf{R}| = |(\mathbf{R}_1 - \mathbf{R}_2)|$.

If we now change coordinates to $\mathbf{R}_1$ and $\mathbf{R}$, integrating 23.29 with respect to d^3R and using 23.26 we find

$$k_B T\left(\frac{\partial n}{\partial p}\right)_T = nk_B T\kappa_T = 1 + n\int d^3R[g_2(R) - 1]. \quad (23.30)$$

In principle, Eq. 23.30 can be integrated to define the pressure as a function of density,

$$p(n) = k_B T\int_0^n \frac{dn'}{1 + n'\int d^3R[g_2(R,n') - 1]}, \quad (23.31)$$

provided the pair correlation function is known as a function of density. Note that, unlike Eq. 23.21, Eq. 23.31 does not have an explicit dependence on the pair potential. The influence of the interaction between the molecules is implicit in Eq. 23.31 through its influence on the pair correlation function. Moreover, Eq. 23.31 is more general than Eq. 23.21 in that it is valid for any one component classical liquid, regardless of whether or not the interaction potential can be decomposed into only pair interactions.

The isothermal compressibility can also be represented in terms of the direct correlation function defined by Eq. 23.24. Multiplying the right and left hand sides of Eq. 23.24 by d^3R and rearranging terms we find

$$\int d^3R[g_2(R) - 1] = \frac{\int d^3RC(R)}{1 - n\int d^3RC(R)}. \quad (23.32)$$

Comparing 23.32 and 23.30 leads to

$$nk_B T\kappa_T = \frac{1}{1 - n\int d^3RC(R)}. \quad (23.33)$$

We now examine several of the theories of the liquid state generated by approximations for either the pair correlation function or the direct correlation function. As pointed out following Eqs. 23.23 and 23.24, some extra information is required before those equations can be solved for the pair correlation function. The general form that information is provided is via a *closure condition,* i.e., a condition that relates g_2 and g_3 to supplement Eq. 23.23, or a condition that relates g_2 and C to supplement Eq. 23.24.

The earliest of the closure conditions is known as the Kirkwood superposition approximation. This approximation states that the potential of mean force defined by

$$w_3(\mathbf{R}_1, \mathbf{R}_2, \mathbf{R}_3) \equiv -k_B T \ln g_3(\mathbf{R}_1, \mathbf{R}_2, \mathbf{R}_3) \quad (23.34)$$

can be represented as the sum of pair potentials of mean force:

$$w_3(\mathbf{R}_1,\mathbf{R}_2,\mathbf{R}_3) = w_2(\mathbf{R}_1,\mathbf{R}_2) + w_2(\mathbf{R}_2,\mathbf{R}_3) + w_2(\mathbf{R}_1,\mathbf{R}_3). \quad (23.35)$$

This approximation resembles the representation of the total potential energy of N molecules as the sum of all the pair interaction energies but is much less accurate because of the neglect of molecular configurations in which there are regions of excluded volume different from the sum of two particle excluded volumes. If Eq. 23.25 is adopted, we can write

$$g_3(\mathbf{R}_1,\mathbf{R}_2,\mathbf{R}_3) = g_2(\mathbf{R}_1,\mathbf{R}_2)g_2(\mathbf{R}_2,\mathbf{R}_3)g_2(\mathbf{R}_1,\mathbf{R}_3), \quad (23.36)$$

and Eq. 23.22 becomes a closed equation for the pair correlation function (the BGYK equation):

$$-k_B T \ln g_2(|\mathbf{R}_1 - \mathbf{R}_2|) = \nabla_1 u_2(|\mathbf{R}_1 - \mathbf{R}_2|) + n\int d\mathbf{R}_3 \nabla u_2(|\mathbf{R}_1 - \mathbf{R}_3|) g_2(|\mathbf{R}_1 - \mathbf{R}_3|) g_2(|\mathbf{R}_2 - \mathbf{R}_3|). \quad (23.37)$$

This integral equation for the pair correlation function must be solved numerically.

A different set of approximate theories of the liquid state uses closures involving the direct correlation function $C(R)$ and the Ornstein–Zernicke equation (Eq. 23.24). It can be shown that away from the critical point of a fluid composed of molecules that interact with a spherically symmetric potential, the direct correlation function has the asymptotic form $C(R) \to -\beta u_2(R)$ as $R \to \infty$. Suppose the pair interaction function in the liquid can be represented in the form

$$u_2(R) = \infty \quad \text{when } R < \sigma$$
$$u_2(R) = u_2^{\text{cont}}(R) \quad \text{when } R > \sigma, \quad (23.38)$$

where $u_2^{\text{cont}}(R)$ is a continuous function of R, and where σ is the impenetrable diameter of the molecule. Then the pair correlation function must vanish for separations smaller than the diameter of the molecule: $g_2(R) = 0$ for $R < \sigma$. Lebowitz and Percus[18] suggested using the asymptotic form of $C(R)$ for all $R > \sigma$ coupled with the condition $g_2(R) = 0$ for $R < \sigma$, i.e.,

$$\begin{aligned} g_2(R) &= 0 && \text{for } R < \sigma \\ g_2(R) &= \tau(R) && \text{for } R > \sigma \\ C(R) &= -\tau(R) && \text{for } R < \sigma \\ C(R) &= -\frac{u_2^{\text{cont}}(R)}{k_B T} && \text{for } R > \sigma. \end{aligned} \quad (23.39)$$

Using this approximation, called the mean spherical approximation (MSA), the Ornstein–Zernicke equation can be

[18] J. L. Lebowitz and J. K. Percus, *Phys. Rev.* **144,** 251 (1966).

solved for several simple cases, included amongst which is the hard sphere fluid. Wertheim[19] showed that

$$C(R) = -(1-\eta)^{-4}\left[(1+2\eta)^2 - 6\eta\left(1+\frac{\eta}{2}\right)^2\left(\frac{R}{\sigma}\right) + \frac{\eta}{2}(1+2\eta)^2\left(\frac{R}{\sigma}\right)^3\right] \quad \text{for } R < \sigma$$

$$C(R) = 0 \quad \text{for } R > \sigma, \tag{23.40}$$

where $\eta = \pi n \sigma^3/6$. The pair correlation function evaluated at the contact distance of the hard spheres is found to be

$$g_2(\sigma) = \frac{1+\frac{\eta}{2}}{(1-\eta)^2}. \tag{23.41}$$

For the hard sphere interaction potential we have

$$-\frac{1}{k_BT}\frac{du_2(R)}{dR}e^{-u_2(R)/k_BT} = \frac{de^{-u_2(R)/k_BT}}{dR} = \delta(R-\sigma),$$

so that Eq. 23.21 for the pressure assumes the form

$$p = nk_BT\left[1 + \frac{2\pi n\sigma^3}{3}g_2(\sigma)\right]. \tag{23.42}$$

Using 23.41 it follows that the equation of state in the MSA is

$$p^V = nk_BT\left[\frac{1+2\eta+3\eta^2}{(1-\eta)^2}\right]. \tag{23.43}$$

If the pressure is computed, instead, from the compressibility relation 23.33, we find

$$p^C = nk_BT\left[\frac{1+\eta+\eta^2}{(1-\eta)^3}\right]. \tag{23.44}$$

The difference between Eqs. 23.43 and 23.44 is a measure of the inconsistency of the MSA. If the assumptions that define the MSA model of the hard sphere fluid were not approximations, the pressures computed from Eqs. 23.21 and 23.33 would be the same. A comparison of the pressure of a hard sphere fluid as a function of density obtained from computer simulation data and computed from Eqs. 23.43 and 23.44 is shown in Fig. 23.20. Overall, for the hard sphere fluid, the equation of state Eq. 23.44 is slightly superior in accuracy to Eq. 23.43.

In general, predictions of fluid properties based on the MSA are more accurate for fluids in which the interaction between particles is strongly repulsive and very short ranged (on the scale of the molecular diameter) than it is for the more typical continuous interaction potential of the Lennard–Jones type (for which the attractive component of the interaction has a range of the order of, or slightly greater than, the molecular diameter). A better approximation, due to Percus and Yevick[20] (PY), relates the direct correlation function and the pair correlation function as follows:

$$C(R) = g_2(R)[1 - e^{u_2(R)/k_BT}]. \tag{23.45}$$

The use of Eq. 23.45 in Eq. 23.24 leads to the following integral equation for the pair correlation function:

$$e^{u_2(R)/k_BT}g_2(R) = 1 + n\int d^3R'[g_2(|\mathbf{R}-\mathbf{R}'|)-1][1-e^{u_2(R')/k_BT}]g_2(R'), \tag{23.46}$$

which must be solved numerically for the general continuous pair potential. In the limiting case of the hard sphere interaction Eq. 23.46 can be solved; in that case, the PY approximation leads to the same equation of state as does the MSA. However, in the general case, the PY approximation and the MSA lead to different equations of state for the liquid.

As might be expected from the observation that when the interaction between particles is of hard sphere form the MSA and the PY approximation lead to the same equation of state, the PY approximation leads to more accurate predictions for liquids in which the pair potential is short ranged than for liquids in which the pair interaction is relatively long ranged. For the latter case a different closure, the hypernetted chain (HNC) approximation, yields superior results. The HNC approximation sets

$$C(R) = g_2(R) - 1 - \ln g_2(R) - \frac{u_2(R)}{k_BT}, \tag{23.47}$$

which, when substituted in Eq. 23.24, then leads to an integral equation for the pair correlation function that reads

$$\ln[g_2(R)e^{u_2(R)/k_BT}] = n\int d^3R'[g_2(|\mathbf{R}-\mathbf{R}'|)-1] \times \left[g_2(R') - 1 - \ln g_2(R') - \frac{u_2(R')}{k_BT}\right], \tag{23.48}$$

which must be solved numerically for the general continuous pair potential.

The predictions of the properties of a liquid using these approximate theories are in good agreement with available data at the high temperature-low density end of the liquid range (but not in the vicinity of the critical point, cf. Chap-

[19] For the case of the hard sphere fluid the mean spherical approximation is equivalent to the Percus–Yevick approximation. It is the Percus–Yevick integral equation which Wertheim solved. See M. S. Wertheim, *Phys. Rev. Lett.* **10**, 321 (1963).

[20] J. K. Percus and G. Y. Yevick, *Phys. Rev.* **110**, 1 (1958).

ter 24) but display serious discrepancies with available data at the high density low temperature end of the liquid range. To illustrate these conclusions, the direct correlation functions and the pair correlation functions of a Lennard–Jones fluid, computed using the MSA, PY and BGYK approximations, are compared with each other in Fig. 23.22 for several densities at the reduced temperature $k_B T/\epsilon = 1.65$, and the pressure and internal energy of a Lennard–Jones fluid at several densities and the reduced temperature 2.74 are displayed in Table 23.7, along with computer simulation data for the same conditions. The data displayed in Fig. 23.22 clearly show that at low density all three approximations are reasonbly accurate. However, as shown by the data entries in Table 23.7, at higher densities the predictions based on these three approximations diverge from each other and from the simulation data, with the PY approximation giving the most accurate predictions. Several other comparisons of the predictions derived from the PY equation and simulation data are displayed in Figs. 23.17–23.20.

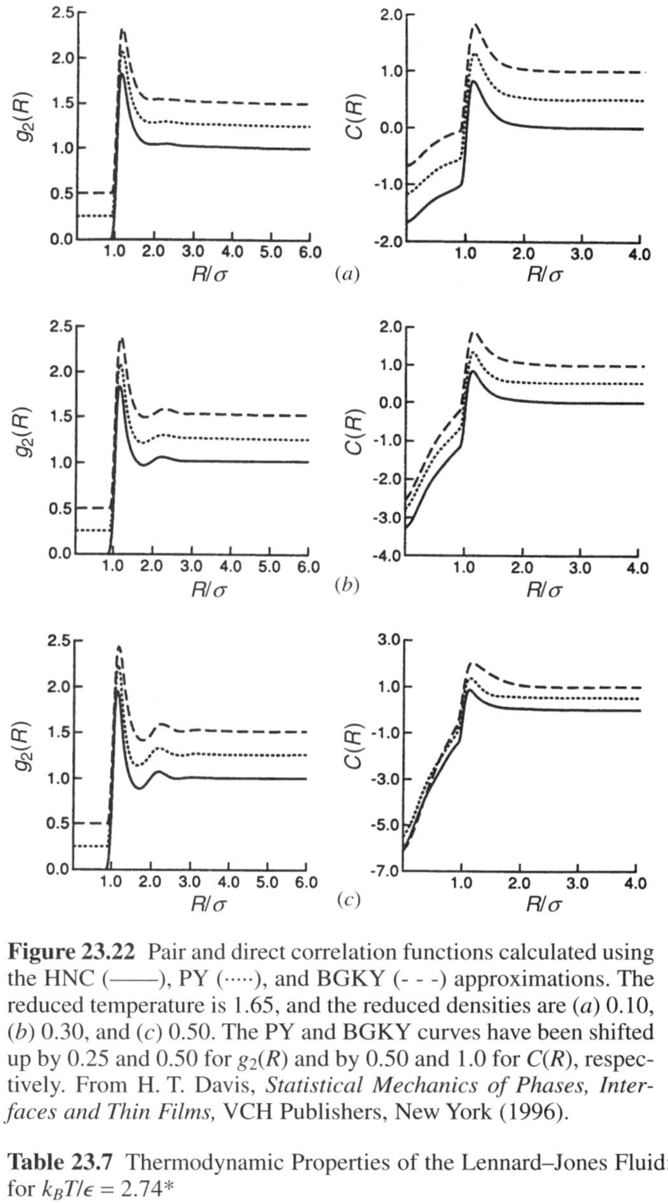

Figure 23.22 Pair and direct correlation functions calculated using the HNC (——), PY (·····), and BGKY (- - -) approximations. The reduced temperature is 1.65, and the reduced densities are (*a*) 0.10, (*b*) 0.30, and (*c*) 0.50. The PY and BGKY curves have been shifted up by 0.25 and 0.50 for $g_2(R)$ and by 0.50 and 1.0 for $C(R)$, respectively. From H. T. Davis, *Statistical Mechanics of Phases, Interfaces and Thin Films,* VCH Publishers, New York (1996).

The BGYK, PY, HNC, and MSA approximations, which we have stated above, can be obtained in several fashions, e.g., the selective summation of terms in the formal representation of the density dependence of the pair distribution function as a power series in the density (an expansion that is analagous to the formal representation of the density dependence of the pressure via the virial expansion described in Appendix 21C), or taking the leading term in the functional Taylor series expansion (see Appendix 23B) of a function of the one particle density $n(\mathbf{R})$ in terms of the difference between the one particle density in the field of another molecule and the average density of the bulk liquid; the different integral equations follow from different choices of the function of the one particle density that is expanded in the functional Taylor series. We now consider how the BGKY, PY, HNC, and MSA theories can be improved by inclusion of some of terms neglected in the derivation of the Eqs. 23.37, 23.46, and 23.48. Although the details of the several analyses will not be discussed in this text, we will sketch some of the ideas used to develop better approximations than those already described.

The improvements to the BGYK, MSA, PY, and HNC approximations involve, in a sense, keeping the largest of the previously neglected terms in the formal representations mentioned above. Arguably the simplest approach is to generate an improvement to the BGYK description of a liquid by correcting the superposition approximation displayed in Eqs. 23.35 and 23.36. Denoting the distances between three particles by r, s, and t, the exact three-particle correlation function can be expressed in the form $g_3(r, s, t) =$

Table 23.7 Thermodynamic Properties of the Lennard–Jones Fluid: Comparison of Simulation Data and Predictions of Various Theories for $k_BT/\epsilon = 2.74$*

	MONTE CARLO		HNC			PY		
$n\sigma^3$	p/nK_BT	$U^E/N\epsilon$	p^c/nk_BT	p^V/nk_BT	$U^E/N\epsilon$	p^c/nk_BT	p^V/nk_BT	$U^E/N\epsilon$
0.3	1.040	−1.783	1.050	1.083	−1.787	1.056	1.070	−1.791
0.4	1.199	−2.371	1.177	1.279	−2.351	1.186	1.234	−2.368
0.55	1.653	−3.207	1.542	1.901	−3.127	1.614	1.722	−3.238
0.7	2.641	−3.902	2.213	3.160	−3.693	2.437	2.646	−3.931
0.8	3.604	−4.281	2.895	4.528	−3.852	3.345	3.603	...
1.0	7.388	−4.180	5.095	9.113	−3.261	6.808	6.670	−4.576

* Adapted from J. P. Hansen and I. R. McDonald, *Theory of Simple Liquids* (Academic Press, New York, 1976).

$g_2(r)g_2(s)g_2(t)\exp[\tau(r, s, t)]$, where the function $\tau(r, s, t)$ corrects for the deviation of the three-particle potential of mean force from the sum of the separate two particle potentials of mean force (see Eq. 23.35). This correction can be represented as the sum of contributions arising from the simultaneous interactions of three selected particles with one other particle, with two other particles, and so forth. The form that each contributing term takes is an integral over the positions of the particles with which the selected three interact, of a product of pair correlation functions, one for each interacting pair of particles. For example, the leading term in the correction to the superposition approximation, involving the simultaneous interaction of the three selected particles (called 1, 2, and 3) with one other (called 4) has the form $\tau(r, s, t) = \int d\mathbf{R}_4 h_{14}h_{24}h_{34}$, where $h(R) \equiv g_2(R) - 1$, and the subscripts label the particular interacting particles between which the distance is varied (e.g., $h_{14} \equiv h(R_{14})$). The next order of correction to the superposition approximation is constructed of terms representing the interaction of three selected particles with two others; for example, one of the seven contributions to this order of correction has the form $\int d\mathbf{R}_4 d\mathbf{R}_5 h_{14}h_{15}h_{24}h_{25}h_{34}h_{45}$, and the other six terms are similar in structure. When these two leading corrections to the three particle correlation function are used in Eq. 23.23, an improved integral equation for the pair correlation function is generated. This integral equation must be solved numerically; it is found that the calculated equations of state of the hard sphere fluid and the Lennard–Jones fluid are in much better agreement with computer simulation data than are calculations based on the original BGYK equation, but there remain discrepancies between the predicted and observed equations of state.[21] The three-particle correlation function determined by the approximation of neglecting all terms except those involving simultaneous interactions with one and two other molecules is in substantial agreement with computer simulation data for that correlation function, i.e., many of the defects of the simple superposition approximation are corrected. However, as with the equation of state, there are also significant differences between the calculated three-particle correlation function and the simulation data[22] in the high density-low temperature region of the liquid domain. Although in principle the correction to the superposition approximation can be carried to arbitrarily high order, in practice the calculations are so tedious it is not possible to go beyond the terms already described.

As mentioned, the PY and HNC approximations can be obtained as, respectively, the leading terms in the functional Taylor series expansion of particular functions of the one particle density in terms of the difference between the one particle density in the field of another molecule and the bulk density. Inclusion of the second order term in the expansion leads, for both the PY and HNC approximations, to expressions for the direct-correlation function that involve the three-particle correlation function. The cumbersome integral equations that result from the use of these second-order approximations for the direct-correlation function in Eq. 23.24 are no longer closed because of the presence of the triplet-correlation function term in the integrand; an approximation similar in spirit but different in detail from that described above is used to complete the closure, thereby generating the PY(2) and HNC(2) integral equations.[23] The predicted properties of a simple liquid obtained from the PY(2) and HNC(2) equations are substantially more accurate than those obtained from the original PY and HNC equations, but there remain discrepancies between the predicted and observed properties of the liquid in the high density-low temperature region (see Table 23.8).

An alternative route to an improved theory of liquids starts by generalizing the conceptual structure of the MSA approximation to apply to the case when the intermolecular pair potential is a continuous function.[24,25] The key new element in the analysis is the division of the pair potential into a reference potential $u_2^{\mathrm{o}}(R)$ and a perturbing potential $u_2^{\Delta}(R)$ according to

$$u_2(R) = u_2^{\mathrm{o}}(R) + u_2^{\Delta}(R). \tag{23.49}$$

If $u_2^{\mathrm{o}}(R)$ is short ranged and repulsive, the structure of the reference liquid is well represented using the PY approximation. On the other hand, the influence of the long ranged perturbation potential on the structure of the liquid is well represented using the HNC approximation. Then

$$C(R) - C^{\mathrm{o}}(R) = g_2(R) - g_2^{\mathrm{o}}(R) - \frac{u_2^{\Delta}(R)}{k_B T} - \ln\left[1 - \frac{g_2(R) - g_2^{\mathrm{o}}(R)}{g_2^{\mathrm{o}}(R)}\right]. \tag{23.50}$$

It is convenient to choose the reference and perturbation potentials by taking for the former the part of $u_2(R)$ where $R < R_m$, and for the latter the part of $u_2(R)$ where $R > R_m$, with R_m the location of the minimum of the potential. This subdivision implies that

$$\begin{aligned} u_2^{\mathrm{o}}(R) &= u_2(R) + \epsilon \qquad && R < R_m \\ u_2^{\mathrm{o}}(R) &= 0 && R > R_m \\ u_2^{\Delta}(R) &= -\epsilon && R < R_m \\ u_2^{\Delta}(R) &= u_2(R) && R > R_m, \end{aligned} \tag{23.51}$$

[21] S. A. Rice and J. Lekner, *J. Chem. Phys.* **42,** 3559 (1965); S. A. Rice and D. Youing, *Discuss. Faraday Soc.* **43,** 16 (1967); D. A. Young and S. A. Rice, *J. Chem. Phys.* **46,** 536 (1967) and **47,** 4228, 5061 (1967).

[22] A. D. J. Haymet, S. A. Rice and W. G. Madden, *J. Chem. Phys.* **74,** 3033 (1981) and **75,** 4696 (1981); W. J. McNeil, W. G. Madden, A. D. J. Haymet and S. A. Rice, *J. Chem. Phys,* **78,** 388 (1983).

[23] L. Verlet, *Physica* **30,** 95 (1964); **31,** 959 (1965).

[24] J. Chihara, *Prog. Theor. Phys.* **50,** 1156 (1973).

[25] W. G. Madden and S. A. Rice, *J. Chem. Phys,* **72,** 4208 (1980).

Table 23.8 Thermodynamic Properties of the Lennard–Jones Fluid: Comparison of Simulation Data and Predictions of Various Theories for $k_BT/\epsilon = 1.35$*

	MONTE CARLO		HNC		PY		PY(2)	
$V\sigma^3$	p/nk_BT	$U^E/N\epsilon$	p^V/nk_BT	$U^E/N\epsilon$	p^V/nk_BT	$U^E/N\epsilon$	p/nk_BT	$U^E/N\epsilon$
0.3	0.352	–2.090	···	···	0.396	–2.18	0.36	–2.17
0.35	0.298	–2.405	···	···	0.376	–2.48	0.30	–2.47
0.4	0.272	–2.747	0.388	–2.79	0.386	–2.77	0.27	–2.77
0.45	0.280	–3.030	0.442	–3.08	0.434	–3.07	0.25	–3.06
0.5	0.303	–3.372	0.557	–3.37	0.532	–3.38	0.30	–3.36
0.55	0.415	–3.704	0.755	–3.68	0.692	–3.69	0.45	–3.66
0.65	0.850	–4.343	1.492	–4.26	1.256	–4.32	1.22	–4.26
0.7	1.166	–4.684	2.086	–4.52	1.689	–4.65	···	···

*Adapted from J. P. Hansen and I. R. McDonald, *Theory of Simple Liquids* (Academic Press, New York, 1976).

where $\epsilon = -u_2(R_m)$. If we now introduce the approximation $g_2(R) = g_2^{\mathrm{o}}(R)$, adopt the PY approximation for $C^{\mathrm{o}}(R)$, and use Eq. 23.50, we obtain

$$C(R)=[1-e^{u_2^{\mathrm{o}}(R)/k_BT}]g_2(R)-\frac{u_2^{\Delta}(R)}{k_BT}. \qquad (23.52)$$

The use of Eq. 23.52 in Eq. 23.24 generates a generalization of the MSA integral equation, sometimes called the soft-sphere MSA equation. Although Eq. 23.52 has been derived using plausible approximations, it can also be formally derived by the method of functional Taylor series expansion.[26] Figure 23.23 displays a comparison of computer simulation data for a high density-low temperature Lennard–Jones liquid with the pair correlation functions predicted using the soft-sphere MSA and PY equations. The agreement displayed in Fig. 23.23 is seen to be very good; however, the agreement between predicted properties and simulation data is not equally good for all properties of the liquid. For example, the liquid-gas coexistence curve (see Chapter 24) of a Lennard–Jones liquid predicted using the soft-sphere MSA approximation deviates from the liquid-gas coexistence curve obtained from computer simulations (see below).

Just as the BGYK, PY, and HNC theories of the pair correlation function can be improved by inclusion of higher order corrections, so too can the soft-sphere MSA theory be improved. The starting point is the exact relation[27]

$$g_2(R)=\exp\left[-\frac{u_2(R)}{k_BT}\right] \times \exp[g_2(R)-1-C(R)]\exp[B(R)], \qquad (23.53)$$

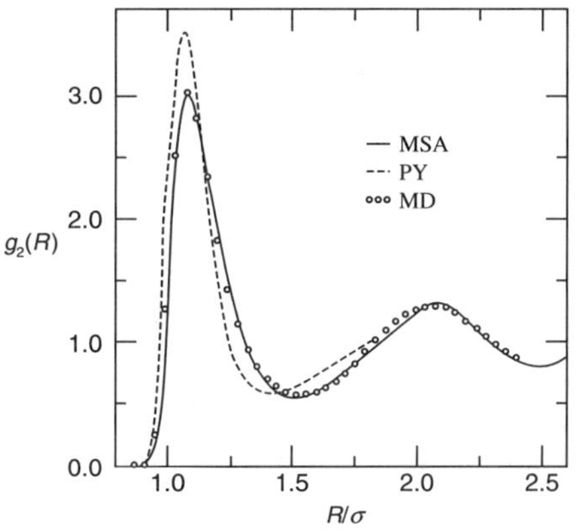

Figure 23.23 Comparison of the pair correlation function for the Lennard–Jones fluid at $k_BT/\epsilon = 0.719$ and $n\sigma^3 = 0.85$ calulated from molecular dynamics simulations, the MSA, and the PY approximation. From W. G. Madden and S. A. Rice, *J. Chem. Phys.* **72,** 4208 (1980).

where $B(R)$ is called the bridge function. Equation 23.53 can be obtained, after considerable algebraic manipulation, from the expansion of the pair correlation function as a power series in the density; $B(R)$ is then represented as a power series in the density with the coefficient of n^{α} determined by a particular subset of all the possible interactions between two selected particles, 1 and 2, and α other particles. Equation 23.53 can also be considered a definition of $B(R)$, in the sense that it is the function that, when multiplied by the first two terms on the right hand side of Eq. 23.53, generates the exact pair correlation function. Note that the HNC approximation is obtained from Eq. 23.53 by setting $B(R) = 0$, and the PY approximation is obtained by setting $B(R) = 0$ followed by linearization of the term $\exp[g_2(R) - 1 - C(R)]$. The exact form of the bridge function is not known for any pair potential, but approximations to it for different pair potentials have been calculated. It is found that the bridge functions for different pair potentials are very similar, so it is reasonable to seek a common functional form for $B(R)$

[26] G. Zareh and J. P. Hansen, *J. Chem. Phys.* **84,** 2336 (1986).
[27] See, for example, S. A. Rice and P. Gray, *The Statistical Mechanics of Simple Liquids,* Wiley, New York (1965), Chapter 2.

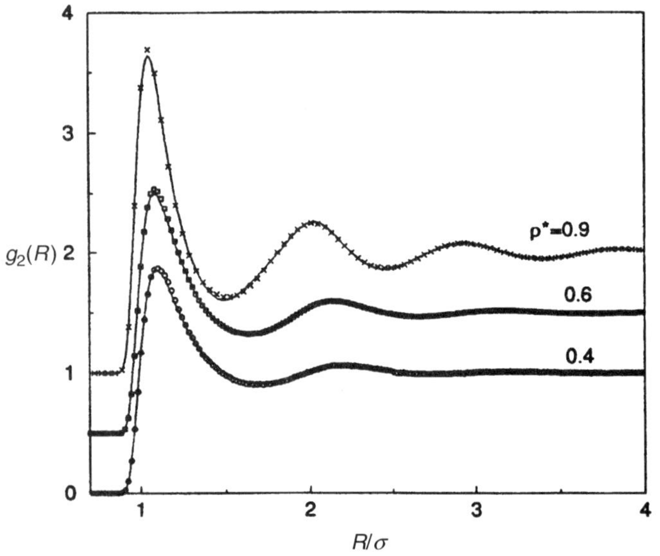

Figure 23.24 The pair correlation function of the Lennard–Jones fluid calculated using Eqs. 23.54 and 23.55. The x's and circles are the results of computer simulations and the lines are calculated from the theory sketched in the text. The curves are displaced by 0.5 for clarity. From D-M. Duh and A. D. J. Haymet, *J. Chem. Phys.* **103,** 2625 (1995).

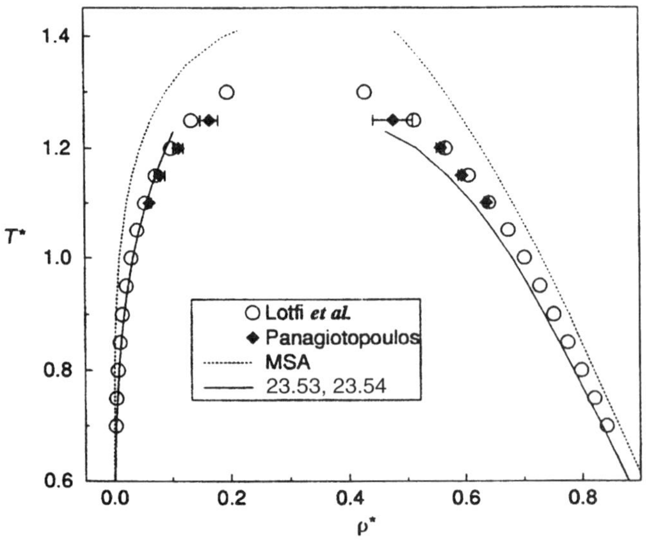

Figure 23.25 The phase diagram of the Lennard–Jones fluid calculated using Eqs. 23.53 and 23.54 (——) and using the MSA (·····) compared with the results obtained from computer simulations. From D-M. Duh and A. D. J. Haymet, *J. Chem. Phys.* **103,** 2625 (1995).

independent of the form of the pair potential. Using both theoretical and semiempirical considerations associated with analysis of the Lennard–Jones fluid and a separation of the pair potential similar to that displayed in 23.51, Duh and Haymet[28] have suggested for this common functional form

$$B(s) = -\frac{s(R)^2}{2\left[1+\left(\frac{5s(R)+11}{7s(R)+9}\right)s(R)\right]},$$

$$s(R) = g_2(R) - 1 - C(R) - \frac{u_2^{\mathrm{A}}(R)}{k_B T}. \tag{23.54}$$

As before, the use of Eq. 23.54 and 23.53 as a closure for the Ornstein–Zernicke equation leads to an integral equation for $g_2(R)$ that must be solved numerically. Several comparisons for different liquid densities of the pair correlation functions calculated from this equation and computer simulation data are shown in Fig. 23.24. The calculated liquid-vapor coexistence curve (see Chapter 24) is compared with that deduced from computer simulation data in Fig. 23.25.

From the preceding considerations we conclude that there exists a reasonably accurate representation of the properties of a simple liquid in terms of the interactions between the molecules in the liquid. In particular, the statistical description of the ordering is fairly accurate. Unfortunately, the kind of theory sketched above does not describe accurately the critical region of a fluid, nor does it lead to an understanding of the molecular origins of freezing. The generalization of the theory of liquids to achieve an accurate description of phase transitions is one of the most challenging of the unsolved problems of the statistical molecular theory of matter.

23.6 The Molecular Theory of Polyatomic Liquids

Since most chemistry is carried out in liquid solution, and the solvents used are always composed of polyatomic molecules, the development of a theory of the behavior of polyatomic liquids is a challenging and important problem for chemists. We have already mentioned in Section 23.2, a variety of ways in which the interactions between polyatomic molecules can influence the structure of a liquid. The comments made in Section 23.2 focused attention on the observations that polyatomic molecules come in all shapes and sizes and that their interactions exhibit a variety of generic and specific characteristics. Consequently, the structures, equations of state and thermodynamic properties of the liquid phases of polyatomic molecules display a broad range of behavior. Noting the complications encountered in predicting the properties of a simple liquid from knowledge of the molecular interaction and the density, it is not surprising that there does not presently exist any comparably accurate analysis of the properties of liquids of polyatomic molecules. In this section we briefly survey the current status of our understanding of the molecular theory of polyatomic liquids.

[28] D-M. Duh and A. D. J. Haymet, *J. Chem. Phys.* **103,** 2625 (1995).

It is convenient to begin our discussion with the observation that x-ray and neutron scattering studies of a polyatomic liquid, which are the principal tools available for the study of liquid structure, usually yield only limited information about that structure. As indicated in the discussion following Eq. 23.15, the angular distribution of x-rays or neutrons scattered from a liquid of polyatomic molecules depends on both the internal structure of the polyatomic molecule and the separation and orientation of the polyatomic molecules in the liquid. Since intramolecular atom–atom separations, especially the intramolecular non-nearest neighbor distances, are comparable in magnitude with intermolecular separations, the angular distribution of the scattered x-rays or neutrons is determined by a weighted sum of the distributions of distances between pairs of atoms within one molecule and in two distinct molecules; the weighting function depends on the species of atoms that form the pair. In principle, the inversion of the scattering data for a liquid of polyatomic molecules requires as many independent scattering experiments as there are kinds of pairs of atoms; for example, in H_2O three experiments to get the H–H, H–O, and O–O distribution functions. By exploiting the change in x-ray scattering cross section with x-ray wavelength and the differences in the neutron scattering cross sections of different isotopes of the same element, it is possible, in principle, to accumulate the required data. In practice, rarely are these data available. Lacking the necessary data, it is common practice to introduce "reasonable" approximations,[29] such as neglecting the scattering of x-rays from H atoms (but H atoms scatter neutrons strongly), neglecting cross correlations of the X–Y type, and so on. As stated in Section 23.2, inferences concerning the structure of a polyatomic liquid must be examined carefully to filter out errors introduced when the x-ray and/or neutron scattering data are reduced using such approximations.

Before sketching the basis for the molecular theory of polyatomic liquids, it is worthwhile to extend the remarks of Section 23.2 with a brief examination of the signatures of intramolecular and intermolecular atom–atom correlations in the measured pair correlation function. The key to identifying those signatures is recognizing that the overall pair correlation function has, as pointed out above, contributions from pairs of atoms in the same molecule and from pairs of atoms in different molecules, and that these contributions have different characteristic length scales. In a simple liquid there is only one length scale, determined by the diameter of a molecule. Consequently, the first peak in the pair correlation function occurs when the separation between molecules is about one molecular diameter. In a liquid of polyatomic molecules, there are several length scales. For example, in a liquid of diatomic molecules, there is one length scale associated with the intramolecular nuclear separation, a second with the smallest separation between atoms in different molecules, and a third with the sum of these. In this case it is expected that the pair correlation function will display three peaks that will be, respectively, narrow (centered at a distance corresponding to the bond length of the diatomic molecule), very similar to the first neighbor peak in a monoatomic liquid (centered at a distance corresponding to the diameter of an atom in the molecule), and a broad shoulder on the large distance side of the first peak (centered at a distance corresponding to the sum of the intramolecular spacing and the diameter of an atom in the molecule). Figure 23.26 displays the pair correlation functions for liquid N_2 and liquid O_2, which show the structure mentioned.[30] The first peak is narrow and tall because the amplitudes of thermal motion of the atoms bound in a molecule are small relative to the width of the distribution of intermolecular distances. The second peak resembles the corresponding feature in the pair correlation function of a simple liquid because the interaction between bound atoms in different molecules is rather similar to that between atoms in a simple liquid. The large distance shoulder in the pair correlation

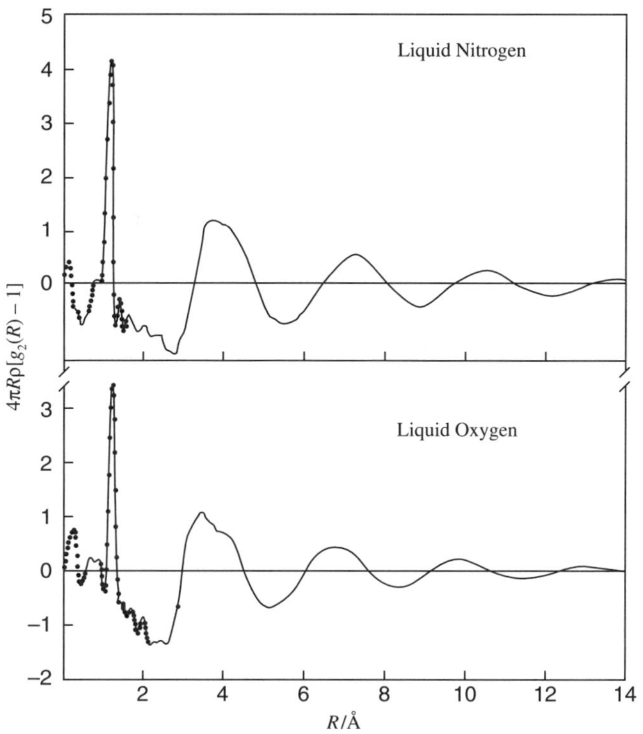

Figure 23.26 The pair correlation functions of liquid N_2 and O_2. From J. C. Dore, G. Walford and D. I. Page, *Mol. Phys.* **29,** 565 (1975).

[29] See, for example, Y. Waseda, *Novel Applications of Anomalous X-Ray Scattering for Structural Characterization of Disordered Materials,* Springer-Verlag, Berlin, 1984; D. L. Price, Marie-Louise Sabounpi, R.-Z. He, T. Egami, and W. S. Howells, *Phys. Rev. B55,* 11249 (1997).

[30] J. C. Dore, G. Walford, and D. I. Page, *Mol. Phys.* **29,** 565 (1975).

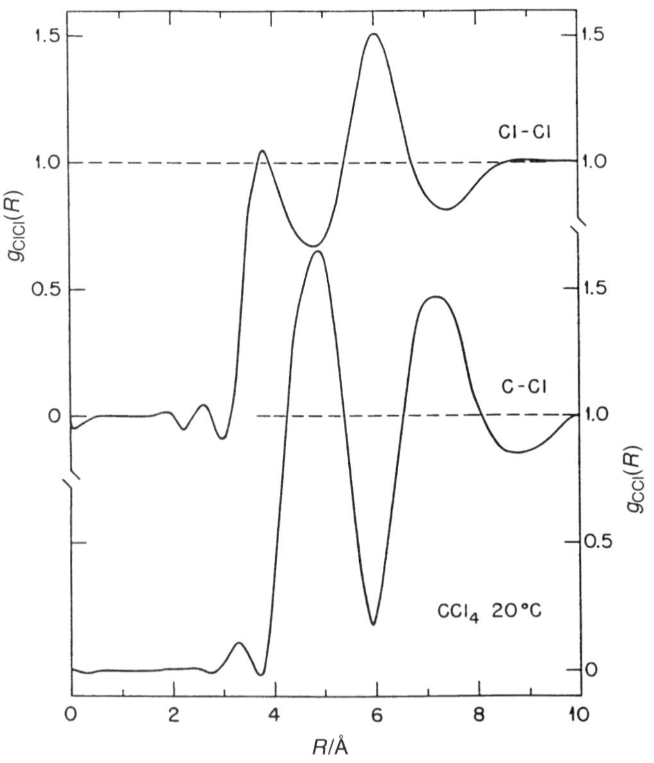

Figure 23.27 The atom–atom pair correlation functions of liquid CCl_4. From A. H. Narten, *J. Chem. Phys.* **65,** 573 (1976).

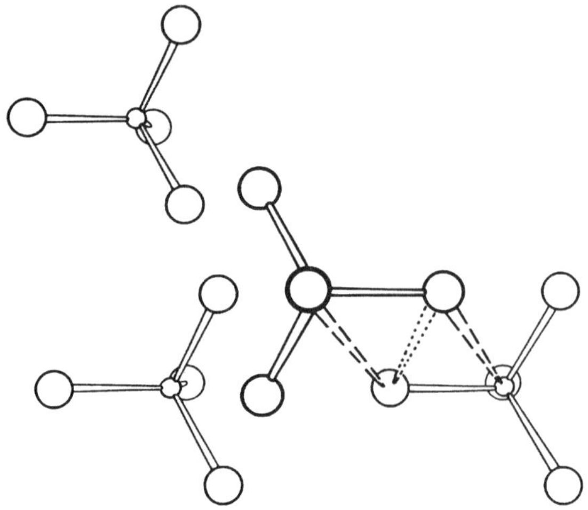

Figure 23.28 Orientational correlation between pairs of neighboring molecules in liquid CCl_4 inferred from the atom-atom pair correlation functions shown in Fig. 23.27. From A. H. Narten, *J. Chem. Phys.* **65,** 573 (1976).

function is very broad because the width of the distribution of molecular orientations further increases the spread of the intermolecular distribution of near neighbor separations. Somewhat larger rigid molecules, e.g., CCl_4, sometimes exhibit orientational correlation of nearest neighbors in the liquid state. Figure 23.27 shows the intermolecular part of the pair correlation function for liquid CCl_4, determined by combining x-ray and neutron diffraction data.[31] The function $g_{ClCl}(R)$ has peaks at 3.8 and 6.0Å, while the function $g_{CCl}(R)$ has peaks at 4.9 and 7.0Å. Note that the locations of the maxima in $g_{CCl}(R)$ coincide with the positions of the minima in $g_{ClCl}(R)$. This observation, coupled with the values of the several atom–atom separations, implies that each chlorine atom in a central molecule is surrounded by alternating shells of chlorine and carbon atoms from neighboring molecules, and that the orientations of neighboring molecules are correlated. Specifically, although liquid CCl_4 is nearly close packed, and the number of nearest neighbors to a molecule at the origin is of order 12, the atom–atom separations in the pair correlation function suggest that four of those nearest neighbors interact in a special way with the central molecule—namely, that each of four molecules that are neighbors to the central molecule has one of its chlorine atoms nestled into one of the four hollows of the valence electron density distribution of the central molecule, and one of the chlorine atoms of the central molecule is nestled in a hollow of a nearest neighbor. A sketch illustrating this molecular packing is shown in Fig. 23.28. Unfortunately, there are very few polyatomic liquids that have been studied as thoroughly as has CCl_4. Usually the only data available concerning the structure of a polyatomic liquid is the total pair correlation function, i.e., the sum of the intramolecular and intermolecular pair correlation functions.

[31] A. H. Narten, *J. Chem. Phys.* **65,** 573 (1976).

The preceding qualitative discussion has, implicitly, introduced one of the key elements of the molecular theory of polyatomic liquids. The interpretation of the CCl_4 pair correlation function sketched above is based on the pairwise interaction of atoms—in particular, the short ranged repulsion between pairs of atoms in different molecules. In general, the intermolecular potential energy of a large class of polyatomic liquids can be represented with good accuracy as the sum of atomic pair interactions or, with somewhat greater generality, the sum of pair interactions between sites in different molecules. In the simplest case the sites coincide with the nuclei of the atoms. The short range repulsive interactions between polyatomic molecules are particularly well described by this representation. However, depending on the molecule, the longer range attractive interactions, and some aspects of the intermediate range repulsive interactions, may require that the interaction sites be different from the positions of the atomic nuclei and, indeed, that they may differ in number from the number of atoms in the molecule. For example, if a diatomic molecule has a large quadrupole moment (see Chapter 10) the minimal model charge distribution requires three interaction sites, e.g., charges (with net charge zero) centered on the atoms and between the atoms. Another case in which atom–atom interactions may not be adequate to represent the intermolecular interaction occurs when the polyatomic molecule is conjugated and the delocalization of the valence electron density contributes to the character of the intermolecular interaction.

The key element in the molecular theory of polyatomic liquids is the generalization of the Ornstein–Zernike equation to account for both intramolecular and intermolecular correlations. For convenience we rewrite Eq. 23.24 in a more compact notation:

$$\begin{aligned} h(R) &= C(R) + n\int dR'C(|R-R'|)h(R') \\ &= C(R) + nC*h(R), \end{aligned} \tag{23.55}$$

where $h(R) \equiv g_2(R) - 1$, and the star in the second line of Eq. 23.55 implies the convolution defined by the integral in the line above. The generalization of Eq. 23.55 for a polyatomic liquid is[32]

$$\begin{aligned} \mathbf{h}(R) = {} & \boldsymbol{\omega}(R)*\mathbf{C}(R)*\boldsymbol{\omega}(R) \\ & + n\boldsymbol{\omega}(R)*\mathbf{C}(R)*\mathbf{h}(R), \end{aligned} \tag{23.56}$$

where $\mathbf{h}(R)$ is the matrix with elements $h_{\alpha\gamma}$, $\mathbf{C}(R)$ is the matrix with elements $C_{\alpha\gamma}$ (each of which is a particular site-site direct correlation function), and $\boldsymbol{\omega}(r)$ is a matrix with elements $\omega_{\alpha\gamma}(r) = \delta(r)$ when $\alpha = \gamma$ and $\omega_{\alpha\gamma}(r) = p_{\alpha\gamma}(r)$ when $\alpha \neq \gamma$, where $p_{\alpha\gamma}$ is the probability density for finding atoms α and γ, in the same molecule, a distance r apart. Equation 23.56 explicitly incorporates intramolecular bonding and its influence on intermolecular correlations. As in the case of simple fluids, a closure relating $\mathbf{C}(R)$ and $\mathbf{h}(R)$ is required to solve Eq. 23.56. Use of the analogue of the Percus–Yevick closure (Eq. 23.45), namely,

$$C_{\alpha\gamma}(R) = [e^{-u_{\alpha\gamma}(R)/k_BT} - 1]e^{u_{\alpha\gamma}(R)/k_BT}g_{\alpha\gamma}(R) \tag{23.57}$$

generates the reference interaction site model (RISM) theory of polyatomic liquids.[32] The generalized Ornstein–Zernicke equation has served as the starting point for several variants of the theory of polyatomic liquids, which we will not discuss.[33]

We consider three tests of the RISM theory. Figure 23.29 displays the calculated pair correlation functions for a model

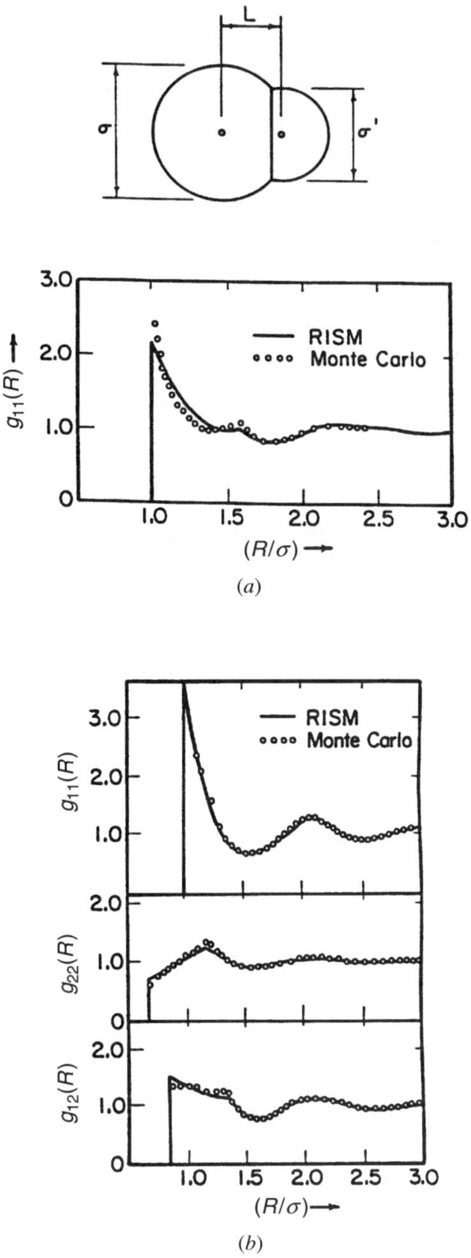

Figure 23.29 Comparison of computer simulations of the pair correlation function for a model liquid of diartomic molecules consisting of fused hard spheres with unequal diameters and the pair correlation function calculated using the RISM approximation. (*a*) $\sigma' = \sigma$ and L = 0.6 σ. (*b*) $\sigma' = 0.675\ \sigma$ and L = 0.346 σ. In both (*a*) and (*b*) the density of the fluid is 0.47 v_0^{-1}, where v_0 is the volume of one hard core molecule. From D. Chandler in *The Liquid State of Matter: Fluids, Simple and Complex,* eds. E. W. Montroll and J. L. Lebowitz, North Holland Publishing Co., Amsterdam, 1982.

[32] D. Chandler in *The Liquid State of Matter: Fluids, Simple and Complex,* eds. E. W. Montroll and J. L. Lebowitz, North Holland Publishing Co., Amsterdam, 1982.

[33] For further details see P. A. Monson and G. P. Morriss, *Adv. Chem. Phys.* **77,** 451 (1990).

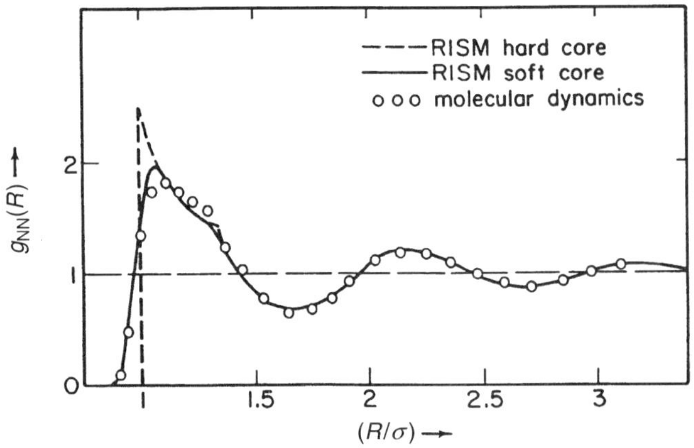

Figure 23.30 The intermolecular pair correlation function for liquid N_2 when $k_BT/\epsilon = 1.83$ and $n\sigma^3 = 0.696$. The line represents the results obtained using the RISM approximation. The circles represent the results of molecular dynamics simulations. From D. Chandler in *The Liquid State of Matter: Fluids, Simple and Complex,* eds. E. W. Montroll and J. L. Lebowitz, North Holland Publishing Co., Amsterdam, 1982.

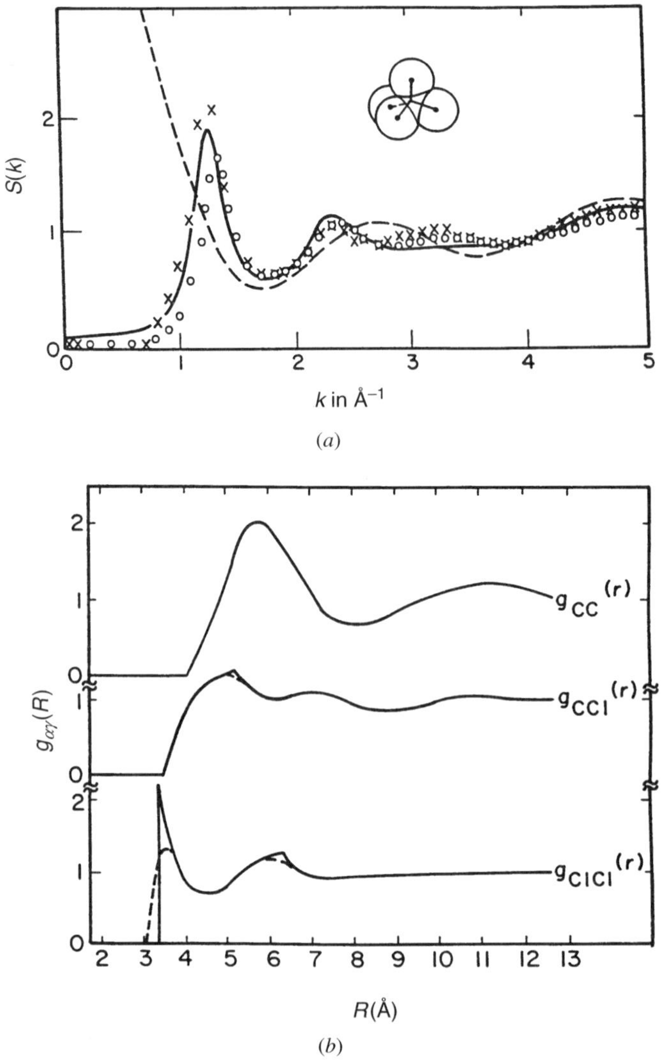

Figure 23.31 (*a*) Comparison between the observed structure function of liquid CCl_4 and that calculated using the RISM approximation with a hard core model of the molecule. (*b*) The various atom-atom intermolecular pair correlation functions in liquid CCl_4 calculated using the RISM approximation. Note the similarity between the results of these calculations and the experimental data shown in Fig. 23.27. From D. Chandler in *The Liquid State of Matter: Fluids, Simple and Complex,* eds. E. W. Montroll and J. L. Lebowitz, North Holland Publishing Co., Amsterdam, 1982.

liquid in which the particles are pairs of fused spheres with unequal diameter; the pair correlation functions obtained from Monte Carlo simulations of the same liquid are also shown. Figure 23.30 displays the calculated pair correlation functions for a model of liquid N_2, along with the results of molecular dynamics simulations of liquid N_2. Finally, Fig. 23.31 displays a comparison of the calculated and measured structure function (effectively the Fourier transform of the pair correlation function) of liquid CCl_4 and the calculated atom–atom pair correlation functions in liquid CCl_4. These examples show that the RISM theory captures the essential elements that determine the structure of a polyatomic liquid.

The ideas just sketched have been extended to treat polymeric liquids,[34] liquids with strong orientational forces (e.g., water),[35] as well as other cases.[36] Although the current form of the theory provides a good representation of the principal features of the structure of polyatomic liquids, there are significant quantitative discrepancies between predicted and observed properties; hence, further development of the theory is necessary.

[34] K. S. Schweitzer and P. Curro, *Adv. Chem. Phys.* **98,** 1 (1997).
[35] D-M. Duh, D. N. Perera, and A. D. J. Haymet, *J. Chem. Phys.* **102,** 3736 (1995).
[36] See, for example, Y. V. Kalyuzhnyi, and P. T. Cummings, *J. Chem. Phys.* **105,** 2011 (1996).

APPENDIX 23A

X-ray Scattering from Liquids: Determination of the Structure of a Liquid

This appendix is concerned with the derivation of Eq. 23.15. Although the analysis will be couched in terms descriptive of x-ray scattering, the same principles can be used to describe electron and neutron scattering.

X-rays are scattered by the electrons in a sample of matter. In the classical picture the very high-frequency x-ray electromagnetic radiation induces an oscillating dipole moment in an atom. This oscillating dipole, in turn, emits radiation at the frequency at which it is driven. The net effect is that some energy is removed from the incident field and scattered (reradiated) in different directions. The quantum mechanical picture is similar, but now the atomic polarization and reradiation is described in terms of the mixture of excited states and ground state induced by the electric field rather than in terms of a classical harmonically driven bound electron.

The geometry of a typical scattering event is shown in Fig. 23A.1.

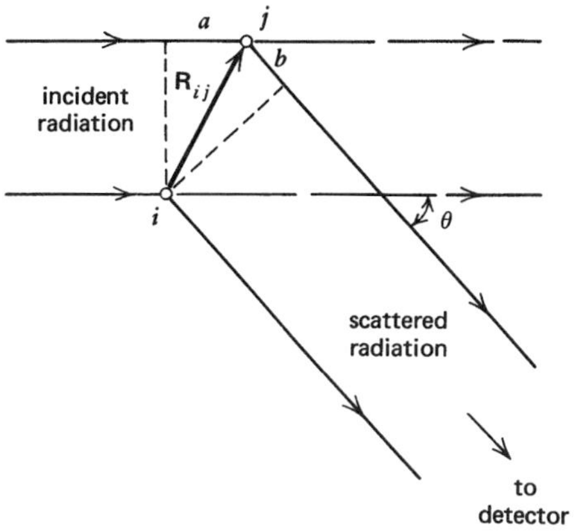

Figure 23A.1 Geometry of a typical scattering event. Particle i is at $\mathbf{R}_i$, particle j at $\mathbf{R}_j$, and their vector separation is $\mathbf{R}_{ij} = \mathbf{R}_j - \mathbf{R}_i$. The path-length difference between waves scattered from i and j is just $a + b$. Since $\mathbf{u}_0$ is a unit vector specifying the direction of the incident radiation, $a = \mathbf{R}_{ij} \cdot \mathbf{u}_0$. Since $\mathbf{u}_s$ is a unit vector specifying the direction of the scattered radiation, and the angle between $\mathbf{u}_s$ and $\mathbf{R}_{ij}$ is greater than 90° for all θ greater than 0°, $b = -\mathbf{R}_{ij} \cdot \mathbf{u}_s$. Here θ is the scattering angle.

Consider first, for simplicity, the scattering of x-rays by N structureless point particles. We suppose the scattering to be so weak that only single scattering events need to be considered. Under these conditions a scattered x-ray photon is not scattered again before it reaches the detector. Furthermore, we suppose that the scattering is elastic. This assumption is an excellent approximation in the case of x-ray photons, since the amount of energy that can be gained from or lost to the translational motion of the atoms is a very small fraction of the x-ray photon energy ($\approx k_BT/h\nu \approx 10^{-2}$ eV/10^4 eV $= 10^{-6}$). Then if the initial radiation has wave vector $\mathbf{k}_0 = (2\pi/\lambda)\mathbf{u}_0$, and the scattered radiation has wave vector $\mathbf{k}_s = (2\pi/\lambda)\mathbf{u}_s$, where $\mathbf{u}_0$ and $\mathbf{u}_s$ are unit vectors specifying the directions of the incident and scattered waves, we have

$$|\mathbf{k}_0| = |\mathbf{k}_s| = \frac{2\pi}{\lambda}. \tag{23A.1}$$

Consider two particles, at $\mathbf{R}_i$ and $\mathbf{R}_j$, in the incident field of radiation (see Fig. 23A.1). Each will act as a secondary emitting center. The extent of the interference between the waves originating at these two centers is determined by the difference in path lengths traversed, since this determines the phase difference between the two waves. Let $\mathbf{R}_{ij} = \mathbf{R}_j - \mathbf{R}_i$ be the vector separation of the molecules. As readily seen from Fig. 23A.1, the extra path length referred to is just $\mathbf{R}_{ij} \cdot \mathbf{u}_0 - \mathbf{R}_{ij} \cdot \mathbf{u}_s$, so that the phase difference is

$$\begin{aligned} \Delta\phi &= \frac{2\pi}{\lambda}(\mathbf{R}_{ij} \cdot \mathbf{u}_0 - \mathbf{R}_{ij} \cdot \mathbf{u}_s) \\ &= -\mathbf{R}_{ij} \cdot (\mathbf{k}_s - \mathbf{k}_0) \\ &= -\mathbf{R}_{ij} \cdot \mathbf{s}. \end{aligned} \tag{23A.2}$$

As shown in Fig. 23A.2, $|\mathbf{s}| = 2|\mathbf{k}_0|\sin(\theta/2)$, where θ is the scattering angle. Let ϕ_j be the phase of the wave scattered from the particle at $\mathbf{R}_j$. If there were no interference between particles, the electric field of this scattered wave would fall off as the reciprocal of the distance from

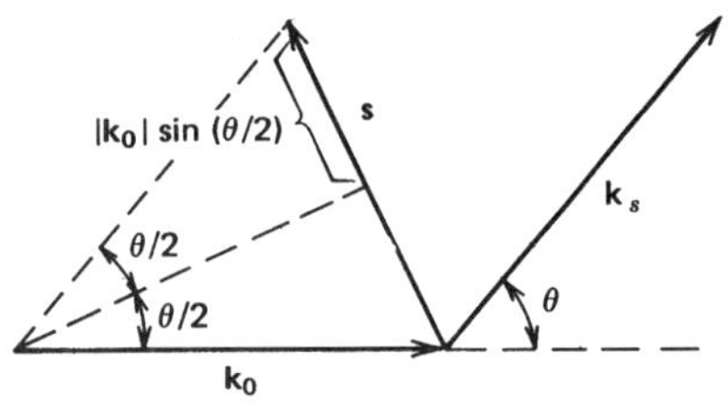

Figure 23A.2 Relations between the incident wave vector $\mathbf{k}_0$, the scattered wave vector $\mathbf{k}_s$, and the difference vector $\mathbf{s} = \mathbf{k}_s - \mathbf{k}_0$. Note that

$$s = |\mathbf{s}| = 2|\mathbf{k}_0|\sin\left(\frac{\theta}{2}\right) = \frac{4\pi}{\lambda}\sin\left(\frac{\theta}{2}\right).$$

particle j. But because of the other particle at $\mathbf{R}_i$, the electric field at some point $\mathbf{R}$ of the scattered wave of amplitude A is

$$\frac{A}{|\mathbf{R}_i - \mathbf{R}_j|}\exp(i\phi_j) = \frac{A}{|\mathbf{R}_i - \mathbf{R}_j|}\exp(i\phi_i)\exp(-\mathbf{s}\cdot\mathbf{R}_{ij}), \quad (23A.3)$$

using $\phi_j = \phi_i + \Delta\phi$ and $\Delta\phi$ from Eq. 23A.2. To obtain the effects of interference between all the scattering centers, we must sum Eq. 23A.3 over all molecules. When the detector is very far from the scattering region, so that $|\mathbf{R}|$ is large compared to all dimensions of the scattering region, the magnitude of the electric field of the scattered radiation is

$$E_s = \frac{A}{R}\exp(i\phi_i)\exp(i\mathbf{s}\cdot\mathbf{R}_i)\sum_{j=1}^{N}\exp(-i\mathbf{s}\cdot\mathbf{R}_j), \quad (23A.4)$$

and its intensity is

$$I_s = \frac{E_s E_s^*}{4\pi} = \frac{A^2}{4\pi R^2}\left[N + \sum_{i\neq j}\exp(-i\mathbf{s}\cdot\mathbf{R}_{ij})\right]. \quad (23A.5)$$

To obtain Eq. 23A.4 we have used the condition $|\mathbf{R}| \gg |\mathbf{R}_j|$, hence to an excellent approximation $|\mathbf{R} - \mathbf{R}_j| \to R$ for all $\mathbf{R}_j$. To obtain Eq. 23A.5 we have separated out the term $i = j$ from $[\Sigma_{j=1}^{N}\exp(-i\mathbf{s}\cdot\mathbf{R}_j)] \times [\Sigma_{i=1}^{N}\exp(-i\mathbf{s}\cdot\mathbf{R}_i)]^* = \Sigma_{i,j}\exp(-i\mathbf{s}\cdot\mathbf{R}_{ij})$.

Clearly, since the intensity of the scattered wave depends on all the values of $\mathbf{R}_{ij}$, it depends on the distribution of pairs of molecules in the scattering region. For the case of interest to us, namely, a uniform liquid, the probability density for finding molecules at $\mathbf{R}_1$ and $\mathbf{R}_2$ is just $n^2 g_2(R_{12})$. Then the sum in Eq. 23A.5 can be replaced by an integral over $d\mathbf{R}_1\, d\mathbf{R}_2$ weighted by the distribution of pairs of molecules:

$$I_s = \frac{A^2}{4\pi R^2}\left\{N + n^2\iint d\mathbf{R}_1\, d\mathbf{R}_2 \exp(-i\mathbf{s}\cdot\mathbf{R}_{12}) + n^2\iint d\mathbf{R}_1\, d\mathbf{R}_2\exp(-i\mathbf{s}\cdot\mathbf{R}_{12})[g_2(R_{12})-1]\right\}. \quad (23A.6)$$

The first term in Eq. 23A.6 is the intensity of scattering from N independent molecules (without interference), the second term is the intensity of scattering when $g_2(R_{12}) = 1$ at the surface of the volume, and the third term is the portion of the scattered intensity that is modulated by interference between the scattering centers. For a macroscopic system the surface scattering term is negligibly small. When the volume element $d\mathbf{R}_1\, d\mathbf{R}_2$ is rewritten in terms of center-of-mass and relative coordinates, the integration over the center of mass gives a factor of V (see Eqs. 21C.16 and 21C.17), which when multiplied by n gives N. After this manipulation we can rewrite Eq. 23A.6 in the form

$$I_s = I_m\left[1 + 4\pi n\int\frac{\sin sR_{12}}{sR_{12}}[g_2(R_{12})-1]R_{12}^2\, dR_{12}\right], \quad (23A.7)$$

where I_m is the intensity of scattering from N independent molecules. Aside from a proportionality factor to be discussed below, this is Eq. 23.15.

X-rays are scattered from electrons and we must not forget that atoms have an internal electronic structure. Inclusion of this internal structure of the atom modifies Eq. 23A.7. In addition, the amplitude of x-rays scattered by electrons depends on the polarization of the incident beam. For an unpolarized beam the intensity contains a factor $(1 + \cos^2\theta)$, where θ is the angle between the scattered and incident radiation. This term appears as a prefactor as shown in Eq. 23.15. Returning to the influence of the structure of the atom, the distribution of electrons about an atom leads to interference effects in the scattered radiation. If $\mathbf{x}_i^{(n)}$ is the position of the nth electron bound to the ith atom, this interference leads to a term of the form

$$F(s) = \left\langle\left|\sum_{i=1}^{Z}\exp[i\mathbf{s}\cdot(\mathbf{x}_i^{(n)} - \mathbf{R}_i)]\right|^2\right\rangle_{QM},$$

where the subscript QM means a quantum mechanical expectation value (the electron distribution is not influenced by the atomic distribution). This so-called *atomic form factor* can be evaluated using atomic wave functions. It too appears as a prefactor in the equation for the scattered intensity (it was omitted from Eq. 23.15; hence, the proportionality sign).

There are serious technical problems associated with inverting Eq. 23.15 to determine $g_2(R)$ from the measured intensity of scattered x-rays, but we shall not discuss that matter here.

APPENDIX 23B

Functional Differentiation

The definition of functional differentiation is a generalization of the definition of the derivative of a function. Recall that an ordinary function of s variables, $f(x_1, x_2, \ldots, x_s)$, is assigned a value by assigning the individual values of the s independent variables. A functional $F[f(x)]$ is a "function of a function" in the sense that the functional associates a real or complex number with each function $f(x)$ defined within a specified domain. A very simple example of a functional is $F[f] = \int_0^1 dx\, f(x)$, which assigns a value to $F[f]$ for every $f(x)$ that is integrable over the interval $0 \le x \le 1$. Consider now the change in the function $f(x_1, x_2, \ldots, x_s)$ induced by small changes in each of the independent variables. We write

$$df = \sum_{i=1}^{s} A_i(x_1, x_2, \ldots, x_s)\, dx_i \tag{23B.1}$$

and define the partial derivative of f with respect to x_i by

$$\left(\frac{\partial f}{\partial x_i}\right)_{x_j} = A_i(x_1, x_2, \ldots, x_s). \tag{23B.2}$$

Following the same procedure, the change in $A_i(x_1, x_2, \ldots, x_s)$ induced by small changes in each of the independent variables is

$$dA_{i'} = \sum_{i=1}^{S} B_{ii'}(x_1, x_2, \ldots x_s)\, dx_{i'}, \tag{23B.3}$$

which leads to the definition of the second partial derivatives of f with respect to x_i and $x_{i'}$:

$$\left(\frac{\partial^2 f}{\partial x_i \partial x_{i'}}\right)_{x_k \neq x_i, x_{i'}} = \left(\frac{\partial A_i}{\partial x_{i'}}\right)_{x_k \neq x_i, x_{i'}} = B_{ii'}(x_1, x_2, \ldots x_s). \tag{23B.4}$$

The definition of a functional derivative is based on a generalization of the argument just given. The variation of $F[f]$ is obtained by changing the value of its argument $f(x_1, x_2, \ldots, x_s)$ by $\delta f(x_1, x_2, \ldots, x_s)$ at every point of the argument $\{x_1, x_2, \ldots x_s\}$ of $f(x_1, x_2, \ldots, x_s)$. The functional $F[f]$ then changes by an amount that is linear in $\delta f(x_1, x_2, \ldots, x_s)$ as $\delta f(x_1, x_2, \ldots, x_s) \to 0$. The analogue of 23B.1 is

$$\delta F = \int d\mathbf{x}\, A[\mathbf{x}; f(\mathbf{x})]\delta f(\mathbf{x}), \tag{23B.5}$$

where $\mathbf{x}$ denotes the set of variables $\{x_1, x_2, \ldots x_s\}$. By analogy with Eq. 23B.2, we define the *functional derivative* of $F[f]$ by

$$\frac{\delta F}{\delta f(\mathbf{x})} = A[\mathbf{x}; f(\mathbf{x})]. \tag{23B.6}$$

Of course, $A[\mathbf{x}; f(\mathbf{x})]$ is still a functional of $f(\mathbf{x})$ and may also depend separately on $\mathbf{x}$. The variation of $A[\mathbf{x}; f(\mathbf{x})]$ is

$$\delta A[\mathbf{x}; f(\mathbf{x})] = \int d\mathbf{x}' B[\mathbf{x}, \mathbf{x}'; f(\mathbf{x}')]\delta f(\mathbf{x}'), \tag{23B.7}$$

from which we define the second functional derivative of $F[f]$ to be

$$\frac{\delta^2 F}{\delta f(\mathbf{x})\delta f(\mathbf{x}')} = \frac{\delta A[\mathbf{x}; f(\mathbf{x})]}{\delta f(\mathbf{x}')} = B[\mathbf{x}, \mathbf{x}'; f]. \tag{23B.8}$$

We note that the functional derivative of $A[\mathbf{x}; f(\mathbf{x})]$ with respect to $\delta f(\mathbf{x}')$ depends on the two "indices" $\mathbf{x}$ and $\mathbf{x}'$, hence can be thought of as a matrix. It is then natural to define the inverse of the derivative $\delta A[\mathbf{x}; f(\mathbf{x})]/\delta f(\mathbf{x}')$ by the corresponding matrix inverse. This definition implies that

$$\int d\mathbf{x}'' \frac{\delta A[\mathbf{x}; f(\mathbf{x})]}{\delta f(\mathbf{x}'')} \frac{\delta f(\mathbf{x}'')}{\delta A[\mathbf{x}'; f(\mathbf{x}')]} = \delta(\mathbf{x} - \mathbf{x}'). \tag{23B.9}$$

Clearly, this procedure can be continued to define higher order functional derivatives of $F[f]$, and so also a functional Taylor series expansion of $F[f]$.

We now use the method of functional differentiation to demonstrate some important properties of a system of N interacting molecules. For simplicity we restrict attention to the case of monoatomic particles in an external field Ψ, which acts independently on each particle. In this case

$\Psi = \Sigma_{i=1}^{N} \psi_i(\mathbf{R}_i)$ depends only on the coordinates of the particles and the total potential energy of the system consists of the sum of the intermolecular interaction energy and the potential energy of each particle in the external field. The partition function of the system can then be considered a functional of the external field. Using Eq. 21A.29 in Eq. 21C.4, we can write the canonical partition function in the form

$$Q_N[\psi] = \frac{1}{N!}\Lambda^{3N} Z_N[\psi], \tag{23B.10}$$

where

$$Z_N[\psi] = \int_V d\mathbf{R}_1 \ldots d\mathbf{R}_N \exp(-\mathcal{V}_N/k_BT)\exp\left(-\sum_{i=1}^{N}\psi_i/k_BT\right). \tag{23B.11}$$

Consider now the variation of $\ln Z_N$ with respect to ψ. We find

$$\begin{aligned}\delta \ln Q_N[\psi] &= \delta \ln Z_N(\psi) \\ &= -\frac{1}{k_BTZ_N[\psi]}\int d\mathbf{R}_1 \ldots d\mathbf{R}_N \left(\sum_{i=1}^{N}\psi_i\right) \\ &\quad \times \exp[-\mathcal{V}_N/k_BT]\exp\left[-\sum_{i=1}^{N}\psi_i/k_BT\right].\end{aligned} \tag{23B.12}$$

Since each term in the integrand of Eq. 23B.12 gives the same contribution, we can write

$$\begin{aligned}\delta \ln Q_N[\psi] &= -\frac{N}{k_BTZ_N[\psi]}\int d\mathbf{R}_1 \ldots d\mathbf{R}_N \delta\psi_1 \\ &\quad \times \exp[-\mathcal{V}_N/k_BT]\exp\left[-\sum_{i=1}^{N}\psi_i/k_BT\right] \\ &= -\frac{1}{k_BT}\int d\mathbf{R}_1\delta\psi_1(\mathbf{R}_1)\times\frac{N}{Z_N[\psi]} \\ &\quad \times \int d\mathbf{R}_2 \ldots d\mathbf{R}_N \exp[-\mathcal{V}_N/k_BT] \\ &\quad \times \exp\left[-\sum_{i=1}^{N}\psi_i/k_BT\right].\end{aligned} \tag{23B.13}$$

The second term on the right hand side of the second line of Eq. 23B.13 is, by definition, the single particle density distribution function in the presence of the external field, since the number density of particles at position $\mathbf{R}$ in the fluid is the expectation value of $\delta(\mathbf{R}_i - \mathbf{R})$. That is,

$$\begin{aligned}n(\mathrm{R}) &= \frac{1}{Z_N[\psi]}\sum_{i=1}^{N}\int d\mathbf{R}_1\, d\mathbf{R}_2 \ldots d\mathbf{R}_N\delta(\mathbf{R}-\mathbf{R}_i) \\ &\quad \times \exp[-\mathcal{V}_N/k_BT]\exp\left[-\sum_{i=1}^{N}\psi_i/k_BT\right] \\ &= \frac{N}{Z_N[\psi]}\int d\mathbf{R}_2 \ldots d\mathbf{R}_N \exp[-\mathcal{V}_N/k_BT] \\ &\quad \times \exp\left[-\sum_{i=1}^{N}\psi_i/k_BT\right],\end{aligned} \tag{23B.14}$$

so that

$$\delta \ln Q_N[\psi] = -\frac{1}{k_BT}\int d\mathbf{R}_1\delta\psi_1(\mathbf{R}_1)n(\mathbf{R}_1;\psi) \tag{23B.15}$$

or

$$-k_BT\frac{\delta \ln Q_N[\psi]}{\delta\psi(\mathbf{R})} = n(\mathbf{R};\psi). \tag{23B.16}$$

In words, the one particle distribution function is the first functional derivative of the Helmholtz free energy with respect to the external field; it measures the response of the Helmholtz free energy of the fluid to a small variation in the fluid Hamiltonian. If the external field is set equal to zero, Eq. 23B.16 becomes

$$-k_BT\left(\frac{\delta \ln Q_N[\psi]}{\delta\psi(\mathbf{R})}\right)_{\psi=0} = n(\mathbf{R}). \tag{23B.17}$$

Using the same procedure again it can be shown that the pair distribution function is directly related to the second functional derivative of the Helmholtz free energy with respect to the external field (equivalent to the first functional derivative of the one particle distribution function with respect to the external field), and so forth for higher order distribution functions. Combining the functional derivative of the one particle distribution function with respect to the external field with its inverse, using Eq. 23B.9, generates the Ornstein–Zernike equation (Eq. 23.24).

The functional Taylor series can be used to derive equations for the pair correlation function. For the external potential that acts on the particles of the fluid, we take the interaction potential of an extra particle (labeled with the index 0) with all of the fluid particles. The extra particle is of the same type as the fluid particles, and we fix its position to be at $\mathbf{R}_0$. Then $\psi = \Sigma_{i=1}^{N} u_2(\mathbf{R}_i - \mathbf{R}_0)$. If the functional $n(\mathbf{R}_1;\psi)\exp[u_2(\mathbf{R}_0 - \mathbf{R}_1)/k_BT]$ is expanded with respect to $n(\mathbf{R};\psi)$ and only the linear term retained, we obtain Eq. 23.45, which, when combined with the Ornstein–Zernike equation yields the Percus–Yevick equation for the pair correlation function. If, instead, the functional $\ln\{n(\mathbf{R};\psi)\exp[u_2(\mathbf{R}_0 - \mathbf{R}_1)/k_BT]\}$ is expanded with respect to $n(\mathbf{R};\psi)$ and only the linear term retained, we obtain Eq. 23.47, which, when combined with the Ornstein–Zernike equation yields the Hypernetted Chain equation for the pair correlation function. If the functional $n(\mathbf{R};\psi)\nabla u_2(\mathbf{R}_1 - \mathbf{R}_1)/k_BT$ is expanded with respect to $\ln\{n(\mathbf{R};\psi)\exp[u_2(\mathbf{R}_0 - \mathbf{R}_1)/k_BT]\}$ and only the linear term retained, we obtain the Yvon–Born–Green equation for the pair correlation function.

FURTHER READING

Barker, J. A., and Henderson, D., *Rev. Mod. Phys.* **40,** 587 (1976).
Ciccotti, G., Frenkel, D., and McDonald, I. R., *Simulations of Liquids and Solids* (North Holland, Amsterdam, 1987).
Croxton, C. A., *Liquid State Physics* (Cambridge University Press, Cambridge, 1974), Chapters 2, 3, and 5.
Gray, C. G., and Gubbins, K. E., *Theory of Molecular Fluids* (Clarendon Press, Oxford, 1984).
Hansen, J. P., and McDonald, I. R., *Theory of Simple Liquids* (Academic Press, New York, 1976), Chapters 1–6.
Hirschfelder, J. O., Curtiss, C. F., and Bird, R. B., *Molecular Theory of Gases and Liquids* (Wiley, New York, 1954), Chapters 4 and 6.
McQuarrie, D. A., *Statistical Mechanics* (Harper & Row, New York, 1976), Chapters 13 and 14.
Marcus, Y., *Introduction to Liquid State Chemistry* (Wiley, New York, 1977), Chapters 1–3.
Montroll, E. W., and Lebowitz, J. L., Eds., *The Liquid State of Matter: Fluids, Simple and Complex* (North Holland Publishing Co., Amsterdam, 1982).
Rice, S. A., and Gray, P., *The Statistical Mechanics of Simple Liquids* (Wiley, New York, 1965), Chapter 2.
Rowlinson, J. S., *Liquids and Liquid Mixtures,* 2nd ed. (Butterworths, London, 1969), Chapters 1, 2, 7, and 8.
Watts, R. O., and McGee, I. J., *Liquid State Chemical Physics* (Wiley, New York, 1976), Chapters 3, 5–10.

PROBLEMS

1. Suppose that the coefficient of thermal expansion of a liquid in equilibrium with its vapor is measured. Since the vapor pressure increases as the temperature increases, along the liquid-vapor equilibrium line one measures

$$\alpha_\sigma \equiv \frac{1}{V}\left(\frac{\partial V}{\partial T}\right)_\sigma$$

and not

$$\alpha \equiv \frac{1}{V}\left(\frac{\partial V}{\partial T}\right)_p,$$

which refers to a process at constant pressure. Show that

$$\alpha = \alpha_\sigma + \kappa_T\left(\frac{\partial p}{\partial T}\right)_\sigma,$$

where, as above, a subscript σ implies that the process is along the liquid–vapor equilibrium line.

2. Derive the following relationships between the several coefficients descriptive of the properties of a liquid:

(a) $\alpha = \kappa_T \gamma_V$

(b) $\gamma_V - \gamma_\sigma = \dfrac{\alpha_\sigma}{\kappa_T}$

(c) $\alpha_\sigma = \alpha\left(1 - \dfrac{\gamma_\sigma}{\gamma_V}\right)$

(d) $\left(\dfrac{\partial H}{\partial T}\right)_\sigma = C_\sigma + V\gamma_\sigma$

(e) $\left(\dfrac{\partial U}{\partial T}\right)_\sigma = C_\sigma + pV\alpha_\sigma$

3. The relative volume of water as a function of pressure at 0°C is given below:

p (atm)	v/v_0
1	1.00000
501	0.97668
1501	0.93924
2001	0.92393
2501	0.91065
3001	0.89869

Calculate the constants A and B in the Tait equation,

$$\frac{v_0 - v}{v_0 p} = \frac{A}{B + p}.$$

4. Calculate the isothermal compressibility for a liquid that is described by the Tait equation. Using the values obtained in Problem 3, compute κ_T at 1, 250, 1000, 2000, and 3000 atm and 0°C for water, and compare these with the observed values, 46, 43, 37, 30, and 25×10^{-12} cm^2/dyn, respectively.

5. The specific volume of liquid Hg as a function of pressure at 0°C is tabulated below:

p (atm)	v(cm^3/g)
1	0.073554
1000	0.073270
2000	0.072993
3000	0.072724
4000	0.072463
5000	0.072213
6000	0.071973
7000	0.071744

Calculate the isothermal compressibility of Hg at 0°C for several pressures in the range of the data given. How do these values compare with those for water? How can you explain the difference found?

6. Suppose that the equation of state of a liquid is of the form

$$p = nk_B T F_1(n) + F_2(n),$$

where $F_1(n)$ and $F_2(n)$ are functions of density but not of temperature. Show that γ_V is a function only of the molar volume, and that the entropy of the liquid is determined solely by $F_1(n)$.

7. Let v_0 be the molar volume of a liquid at $p = 1$ atm and $T = T_0$. Suppose that the equation of state can be represented by

$$v = v_0[1 + a_0(t) + a_1(t)p + a_2(t)p^2 + \cdots],$$

where $t \equiv T - T_0$ and the various a's are functions of temperature alone.

(a) Calculate the coefficients of thermal expansion and isothermal compressibility in terms of a_0, a_1, and a_2.

(b) Show that

$$c_p = c_p^0 - v_0 T\left(\frac{d^2 a_0}{dT^2}p + \frac{1}{2}\frac{d^2 a_1}{dT^2}p^2 + \cdots\right),$$

where c_p^0 is the specific heat at a constant pressure of 1 atm (essentially zero pressure for our purposes).

8. An alternative form for the equation of state of a liquid is

$$p = p_0(T) + b_1(T)\left(\frac{v_0 - v}{v_0}\right) + b_2(T)\left(\frac{v_0 - v}{v_0}\right) + \cdots,$$

where p_0 is the pressure that must be applied to reduce the volume to v_0 from v. Show that the coefficients of this and the equation of state in problem 7 are related by

$$p_0 = -\frac{a_0}{a_1}$$

$$b_1 = -\frac{1}{a_1}\left(1 + \frac{2a_0 a_2}{a_1^2}\right)$$

$$b_2 = -\frac{a_2}{a_1^3}.$$

9. Let c_V^0 be the specific heat at constant volume v_0.

(a) Show that

$$c_V = c_V^0 - v_0 T\left[\frac{d^2 p_0}{dT^2}\left(\frac{v_0 - v}{v_0}\right) + \frac{1}{2}\frac{d^2 b_1}{dT^2}\left(\frac{v_0 - v}{v_0}\right)^2 + \cdots\right],$$

(b) Show that

$$c_p^0 - c_V^0 = \frac{v_0 T}{2a_1}\frac{d^2}{dT^2}(a_0^2).$$

(c) Compare the result (b) with the exact relation

$$c_p - c_V^0 = Tv\frac{\alpha^2}{\kappa_T},$$

using the values of a_0 and a_1 previously obtained, and show that your calculations are internally consistent.

10. Show that the residual entropy of a liquid can be expressed in the form

$$s^E = -\int_v^\infty \left[\left(\frac{\partial p}{\partial T}\right)_v - \frac{R}{v}\right] dv.$$

11. Consider the following model of a one-dimensional liquid. The "molecules" are line segments of length σ; they are constrained to move on a line of length L. In all there are N such molecules on the line, and we imagine that their interaction has the following form:

$$u_2(x) = \infty, \quad x \le \sigma,$$
$$u_2(x) = 0, \quad x > \sigma.$$

At the ends of the line there are impenetrable barriers (hard walls). Show that the equation of state is

$$p(l - \sigma) = k_B T,$$
$$l \equiv \frac{L}{N}.$$

Hint: The molecules cannot pass through one another, so once an ordering is established it must stay the same. There are $N!$ equivalent orderings of the molecules on the line.

12. A simple model of a liquid, called the cell model, is described as follows. Imagine that each monatomic molecule is constrained, because of repulsive forces exerted by neighbors, to move only in the volume v_f. The idea is that the motion is irregular, unlike that in a crystal, but that v_f is of molecular dimensions. Let the canonical partition function be written as

$$Q_N = \left(\frac{2\pi m k_B T}{h^2}\right)^{3N/2} \frac{Z_N}{N!}.$$

The molecules are localized, so we must account for the $N!$ ways of arranging N molecules in N cells.

(a) Show that the equation of state is

$$p = \frac{N k_B T}{v_f}\left(\frac{\partial v_f}{\partial V}\right)_T.$$

(b) Suppose that the volume of the cell in which the molecule moves is $v \equiv V/N$, and the molecule is a hard sphere of diameter σ. Then if the cell is represented as a hollow sphere, the center of the molecule can approach the wall no closer than $\sigma/2$. Show that the equation of state is

$$pv = k_B T\left[1 - \left(\frac{\pi\sigma^3}{6v}\right)^{1/3}\right]^{-1}.$$

Hint:

$$v_f = \frac{4\pi}{3}\left(\frac{d}{2} - \frac{\sigma}{2}\right)^3,$$

where d is the diameter of the cell of volume v.

(c) Take the limit $\sigma \to 0$, but still allowing motion only in the cell of volume v. Compare the Helmholtz free energies and entropies of this limiting model and a perfect gas. What is the nature of the difference?

13. An explicit and simple way of seeing how attractive interactions influence the properties of a liquid is generated from the assumption that the potential energy inside a cell is constant and negative, but rises to infinity at the walls so that a molecule cannot escape from the cell. Suppose that there are 12 nearest neighbors that define a cell, and that $u_2(R)$ is the Lennard–Jones potential,

$$u_2(R) = 4\epsilon\left[\left(\frac{\sigma}{R}\right)^{12} - \left(\frac{\sigma}{R}\right)^{6}\right].$$

Counting only nearest-neighbor interactions, show that the equation of state is

$$pv = k_B T\left[1 - \left(\frac{v_0}{v}\right)^{1/3}\right]^{-1} + 48\epsilon\left[2\left(\frac{v_0}{v}\right)^4 - \left(\frac{v_0}{v}\right)^2\right],$$

with $v_0 \equiv \sigma^3/\sqrt{2}$.

Hint: Use the Lennard–Jones potential to evaluate the potential energy when all molecules are at the cell centers. By hypothesis, this is the value of the constant potential energy in the cell.

14. A more sophisticated version of the cell model improves the calculation of v_f with the following idea. Imagine some reference lattice with the same molecules as the liquid, having static lattice energy E_L. Write the potential energy of the liquid in the form

$$\mathcal{V}_N \equiv (\mathcal{V}_N - E_L) + E_L.$$

Then $\mathcal{V}_N - E_L$ can be regarded as the potential energy of irregular displacement from the reference lattice sites. If the total potential is the sum of pair potentials, show that

$$v_f = 4\pi\int_{\text{cell}} e^{-\beta[\epsilon(r)-\epsilon(0)]} r^2\,dr,$$

where $\epsilon(r) - \epsilon(0)$ is the energy corresponding to displacement r from the cell center.

15. Using the general principles described in the text, and the known properties of a liquid, criticize the ideas used in formulating the cell model of a liquid.

16. Show that the equation of state of a rigid-sphere fluid can be written in the form

$$\frac{p}{nk_BT} = 1 + \frac{2\pi n\sigma^3}{3} g_2(\sigma),$$

where $g_2(\sigma)$ is the value of the pair-correlation function at contact. *Hint:* For rigid spheres, $u_2(R) = 0$ if $R > \sigma$, and $u_2(R) = \infty$ if $R \le \sigma$. Then $\exp[-\beta u_2(R)] = 1$ if $R > \sigma$ and 0 if $R \le \sigma$. Differentiation of $\exp[-\beta u_2(R)]$ with respect to R gives $-\beta(du_2/dR)\exp[-\beta u_2(R)]$, which is unity when $R = \sigma$ and zero everywhere else.

17. The residual chemical potential, internal energy, and entropy of a liquid can be represented by

$$\mu^E = \mu - k_B T \ln\left(\frac{p\Lambda^3}{k_B T}\right),$$

$$u^E = u - \frac{3}{2}k_B T,$$

$$s^E = s - \frac{5}{2}k_B + k_B \ln\left(\frac{p\Lambda^3}{k_B T}\right).$$

Show that

$$Ts^E = u^E - \mu^E + (pv - k_B T).$$

18. In the case of a rigid-sphere fluid, $u^E = 0$. Show that

$$\frac{\mu^E}{k_B T} = -\ln\left(\frac{pv}{k_B T}\right) + \left(\frac{pv}{k_B T} - 1\right) + \int_v^\infty\left(\frac{p}{k_B T} - \frac{1}{v}\right)dv$$

and

$$\frac{s^E}{k_B} = \ln\left(\frac{pv}{k_BT}\right) - \int_v^\infty \left(\frac{p}{k_BT} - \frac{1}{v}\right) dv.$$

Hint: For a rigid-sphere fluid the Helmholtz free energy per molecule is

$$a^0 - a(v) = -\int_v^{v^0} p\, dv = (\mu^0 - k_BT) - (\mu - pv).$$

19. A very good approximate equation of state for the rigid-sphere fluid is the Carnahan–Starling form,

$$\frac{p}{nk_BT} = \frac{1+\eta+\eta^2-\eta^3}{(1-\eta)^3},$$

with $\eta = \pi n\sigma^3/6$. Calculate the residual entropy and chemical potential for a rigid-sphere fluid using this equation of state.

20. Suppose that two atoms interact by a square-well potential (see Fig. 21.19*d*). Consider a gas of such atoms in the low-density region where $u_2(R) \approx w_2(R)$.

(a) Prove that in this low-density limit we have

$$U = \frac{3}{2}Nk_BT - \frac{N^2\epsilon}{V}\exp\left(\frac{\epsilon}{k_BT}\right)\left[\frac{2\pi}{3}(\sigma_2^3 - \sigma_1^3)\right].$$

(b) Sketch $g_2(R)$ versus R for the low-density gas.

(c) How will $g_2(R)$ for the liquid differ from your answer to (b)?

21. Suppose that the potential energy of a liquid is given by

$$\mathcal{V}_N = \sum_{i>j} u_2(R_{ij}) + \sum_{i>j>k} \Delta_3(R_{ij}, R_{ik}, R_{jk})$$

as in Eq. 21.159. Express U in terms of $g_2(R)$ and an analogously defined three-particle correlation function g_3.

22. A general, exact equation of state is the virial equation of state (derived from the virial theorem of mechanics), which can be written

$$pV = Nk_BT + \tfrac{1}{3}\left\langle \sum_{i=1}^{N} \mathbf{R}_i \cdot \mathbf{F}_i \right\rangle,$$

where $\mathbf{R}_i$ and $\mathbf{F}_i$ are the (vector) position of particle i and the total force on particle i due to all the other particles, and the angular brackets indicate an ensemble average. For the case of a fluid containing identical spherical particles, and where the intermolecular potential is pair-additive, derive Eq. 23.21 from the virial equation.

23. Find the value of $4\pi\int_0^\infty g_2(R)R^2\, dR$ for a fluid containing N particles in a volume V.

CHAPTER 24

Phase Equilibria in One-Component Systems

Among the most dramatic of observable phenomena are the sudden alterations in the properties of a substance when it changes from solid to liquid form, or from liquid to gaseous form. In the cases just cited there are discontinuous changes in the density, enthalpy, and entropy of the substance as the transition is traversed. Equally dramatic are the appearance of superconductivity in many metals and of superfluidity in liquid helium, but in these cases the more obvious properties of the substance undergo continuous transformations as the transition is traversed. Why and how do phase transitions occur, and what changes in properties are to be expected for a given kind of transition? It is these questions that we study in this chapter. Wherever possible we shall examine both the macroscopic thermodynamic description and the corresponding statistical molecular interpretation. We shall see that, although it is possible to formulate a convenient and systematic classification of phase transitions by the methods of thermodynamics, the present understanding of the underlying molecular processes is both meager and imperfect.

24.1 General Survey of Phase Equilibria

We have previously cited, as examples of phase transitions, the evaporation of a liquid and the melting of a solid. Recalling the definition of a phase given in Section 14.4, one can easily see that the signatures of the processes mentioned are discontinuities in the material properties of the system on traversal of the transition. That is, when a solid melts, for example, ice $\rightarrow$ water, the process occurs at constant temperature and pressure with a marked change in the density, heat capacity and other properties of the substance. Note that in referring to the solid phase as distinct from the liquid phase, we imply that there exists a well-defined crystal structure in the solid phase, so that the symmetry of the environment of a molecule in the solid, which defines the ordering, is different from that of the environment of a molecule in the liquid. The symmetry of the environment of a molecule in an amorphous solid, such as glass, is the same as that of a molecule in the corresponding liquid; in both cases long-range molecular order is absent. An amorphous solid gradually softens as it is heated and becomes liquid without undergoing a discontinuous change of properties. This behavior can be traced to the fact that the entropies of a crystalline solid and a liquid are different, reflecting largely the difference in ordering between the two phases, whereas a glass has positional disorder very much like that in a liquid. It is then to be expected that there will not be a sharp transition from solid glass to liquid, and that there will be a continuous transformation between these two states. In fact, some of the properties of a substance undergoing the glass $\rightarrow$ liquid transition do change considerably over a small temperature range, but they do not change discontinuously. Thus, glass and liquid are not distinct phases.

A phase transition occurs at a fixed temperature for any specified constant pressure, but the transition temperature changes when the pressure is changed. Also, in the transformation from one phase to another, there is an enthalpy change. Thus, in the absence of external influences, two phases can coexist indefinitely long at the transition temperature corresponding to a specified pressure. However, at the same pressure and temperatures above or below that transition temperature, only one or the other phase is stable. For example, the line representing the pressure at which liquid and vapor coexist in equilibrium, called the vapor pressure curve, divides the (p, T) plane into two regions (see Fig. 24.1), one where the liquid is the stable phase, the other where the gas is the stable phase. The points defining the vapor pressure curve represent states in which an arbitrary quantity of liquid may coexist with an arbitrary quantity of gas. The latter property of a phase

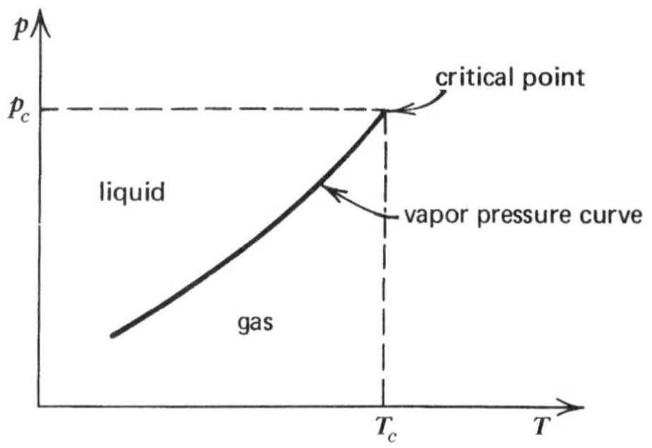

Figure 24.1 Schematic diagram of a vapor pressure curve.

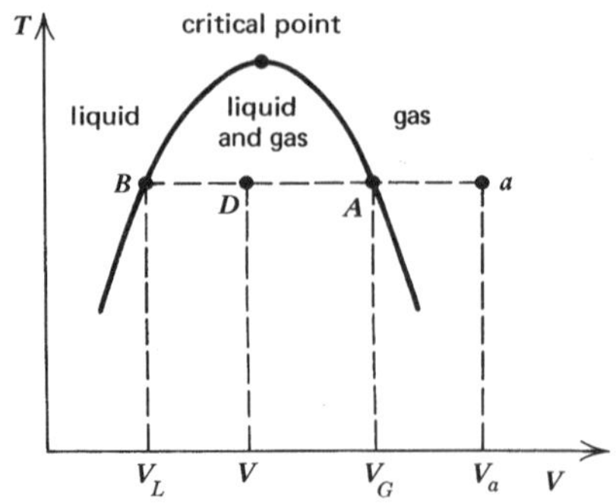

Figure 24.2 Schematic diagram of the coexistence region of liquid–gas equilibrium in the (V, T) plane. The line $aADB$ represents the condensation path described in the text.

transition is easily appreciated when the (V, T) plane is examined (see Fig. 24.2). Suppose that a gas with volume V_a is compressed at constant temperature until condensation begins, which occurs at $V = V_G$. As the system is compressed further, the amount of gas in the system decreases while the amount of liquid increases. When $V = V_L$, the substance is entirely converted to liquid. Consider the point D in Fig. 24.2. Let the fractions of gas and liquid at D be x and $1 - x$. Then the total volume of the system is

$$V = xV_G + (1-x)V_L, \tag{24.1}$$

so that

$$x = \frac{V - V_L}{V_G - V_L}, \tag{24.2}$$

$$1 - x = \frac{V_G - V}{V_G - V_L}. \tag{24.3}$$

The ratio of the amounts of gas and liquid is, from Eqs. 24.2 and 24.3,

$$\frac{x}{1-x} = \frac{V - V_L}{V_G - V_L} = \frac{\text{length } BD}{\text{length } AD}, \tag{24.4}$$

the final form referring to the lines in Fig. 24.2. This result is known as the *lever rule.*

The phase diagram in the (p, V) plane is similar to that in the (T, V) plane, and both differ from the phase diagram in the (T, p) plane. Note that each point on the line AB of Fig. 24.2 represents a different thermodynamic state, while in a (T, p) diagram all these states are represented by a single point on the vapor pressure curve. The region of separation into two coexisting phases is only a line in the (T, p) plane, but occupies an area in the (T, V) and (p, V) planes. This difference results from the requirement that coexisting phases have the same temperature and pressure, although they need not have the same density. These remarks are easily verified by examination of the (p, V, T) surface of a substance, as, for example, that in Fig. 24.3.

As the temperature rises, the pressure at which there is coexistence between liquid and vapor also rises. Corresponding to the increase in vapor pressure of the substance, the difference in densities of the liquid and gaseous phases decreases. There exists a temperature and pressure at which the densities of liquid and gas are the same and the two phases can no longer be distinguished from one another. At higher temperatures or pressures, liquid and gas are everywhere indistinguishable; that is, there is only one phase. Thus, the locus of points of coexistence of liquid and gas in the (p, T) plane terminates at a point T_c, p_c, called the *critical point.* (See Section 21.1.)

In the (T, V) and (p, V) planes the approach to the critical point is manifested by the approach to equality of the densities of the two phases as the temperature of the system increases. (See Fig. 21.4.) The manner in which the pressure approaches the critical pressure in the (p, V) plane, as shown in Fig. 24.4, and the manner in which the critical volume is approached in the (T, V) plane, are very different from that expected from the equations of state listed in Table 21.2. Indeed, the thermodynamic state of the system near the critical point has some very peculiar characteristics. Furthermore, the existence of a critical point in the gas–liquid phase transition strongly suggests that there is no fundamental difference between the liquid and gaseous states. It is in fact possible to pass from the liquid to the gaseous state around the critical point without undergoing any discontinuous change of state. This observation can be made on the basis of a classification of the kinds of transitions in which critical points can be expected to occur. Note that a critical point marks the termination of a line representing the locus of points of coexistence of two phases. Suppose, as in the case of a gas and a liquid, that there are only quantitative differences between the properties of the two phases. The quantitative differences arise because, accompanying the difference in densities, there is a difference in the averaged molecular environments between the gas and liquid. But in both phases the local structure has spherical symmetry,[1]

[1] We neglect here the case of liquid crystals, which can exist in several fluid phases, some of which are anisotropic and only one of which is isotropic. The transitions between the several anisotropic phases of a liquid crystal are analogous to solid–solid transitions from one crystal structure to another and not to the gas–liquid transition. However, the transition between the so-called blue phase III and the isotropic phase is analogous to the gas–liquid transition.

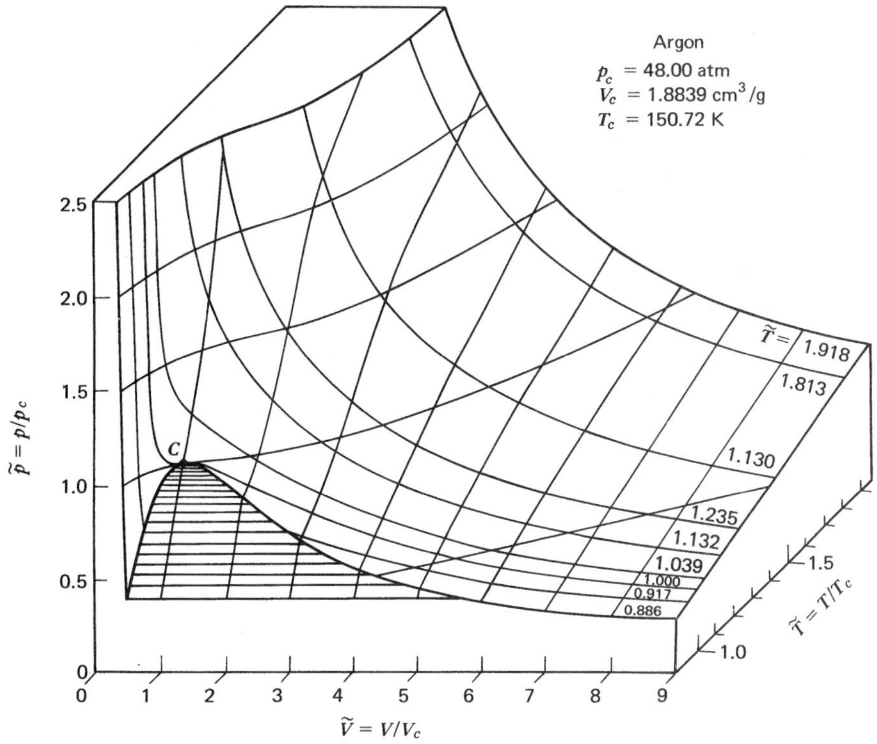

Figure 24.3 Part of the (p, V, T) surface of argon, showing lines of constant p, V, and T.

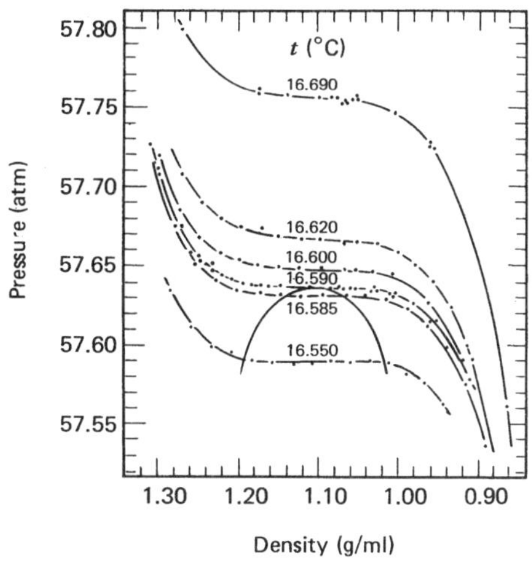

Figure 24.4 Experimental pressure against density isotherms in the critical region of xenon. From H. W. Habgood and W. G. Schneider, *Can. J. Chem.* **32,** 98 (1954).

hence both phases are isotropic. Thus, when the densities of the two phases are made to approach equality, the phases become more and more alike. When the densities are made equal, the two phases merge into one with no change in local symmetry. When there is no difference between the symmetries of two phases, but only quantitative differences in the densities and associated properties, the coexistence line between the phases has a critical point.

On the other hand, if the two phases under consideration have different internal symmetries, as is the case for a liquid and a crystal, they must be regarded as qualitatively different. Adjusting the densities of, for instance, a crystal and a liquid to be equal, if it could be done, would still leave a difference in local symmetries between them. In cases of this type there is no critical point on the coexistence curve representing equilibrium between the two phases. Either the coexistence curve in the (p, T) plane goes to infinity or it terminates by intersecting the coexistence curves of still other phases.

The preceding argument is based on the observation that a particular symmetry property exists or does not exist, and that it belongs to all the molecules of the phase as a whole. A symmetry property can only appear or disappear entirely; it cannot be "partially present." It follows that if we are able to recognize that a symmetry property is present, we can always determine to which of two phases it belongs. Therefore, two phases that have different symmetry properties can, in principle, always be distinguished from one another, and no critical point can terminate their coexistence line.

Although the most widely known phase transitions are melting and vaporization, there exist many other kinds. For example, it is found that at different temperatures and pressures a solid may exist in different crystalline states, each with a specific structure. These different crystalline forms, called crystal modifications, are to be regarded as different phases of the same substance. The different polymorphic crystalline forms usually have different densities; these different modifications can be in equilibrium with one another only along certain lines in the (p, T) plane. In general, when such a coexistence line is crossed, and one crystal form is changed into another, there is absorption (release) of heat. Examples of polymorphism are found in elements (diamond

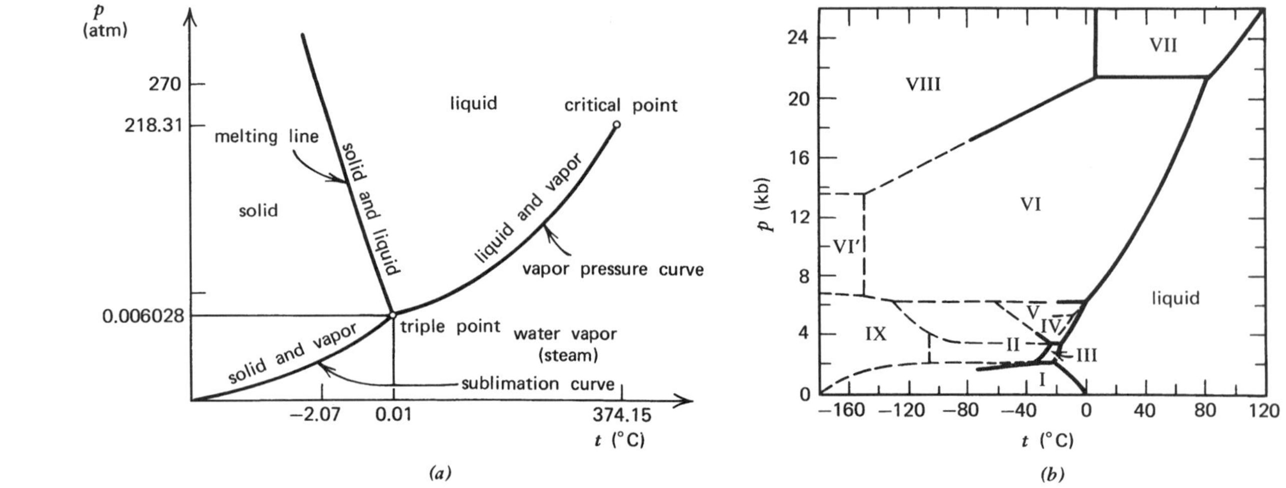

Figure 24.5 Phase diagrams for water. (*a*) Low-pressure ice–water vapor–liquid equilibrium. From J. Kestin, *A Course in Thermodynamics* (Blaisdell, Waltham, Mass., 1966). (*b*) High-pressure equilibria among various forms of ice. From B. Kamb, in E. Whalley, S. Jones, and L. Gold (Eds.), *Physics and Chemistry of Ice* (University of Toronto Press, Toronto, 1973).

and graphite are forms of C) and compounds (quartz, tridymite, and cristobalite are forms of SiO_2). The phase diagram of water, a portion of which is shown in Fig. 24.5, displays many crystal modifications, most of which are stable only at high pressures. In this diagram, ordinary ice is stable in region I.

The liquid–gas, solid–gas, and liquid–solid transitions, and most crystal–crystal transformations, involve a discontinuous change in density across the coexistence line. Along with the discontinuity in density there are also discontinuities in the molar entropy, internal energy, and specific heat. The class of phase transitions just described is known as *transformations of the first* kind (or *first-order*).[2] It is a transformation of this type that connects each of the pairs of phases in both the ordinary-pressure and the high-pressure phase diagrams of water.

It is also possible, in principle, to have a discontinuous change of symmetry type without there being discontinuities in the volume, entropy, or internal energy of the substance. As an example of how this might happen, consider a tetragonal crystal, and suppose that the lengths of the two unequal edges of the unit cell, a and c, change with the temperature at different rates. If c is larger than a at some temperature, and if $(\partial a/\partial T)_p > (\partial c/\partial T)_p$, then as T increases, the lengths of the sides of the unit cell become more nearly equal. Suppose that there is a temperature at which a and c become equal, so that the unit cell is cubic. On further heating all three sides of the cubic unit cell increase in length at the same rate, therefore remaining equal in length. Thus, when $a = c$ and the crystal symmetry has changed from tetragonal to cubic, we have a different crystal modification of the same

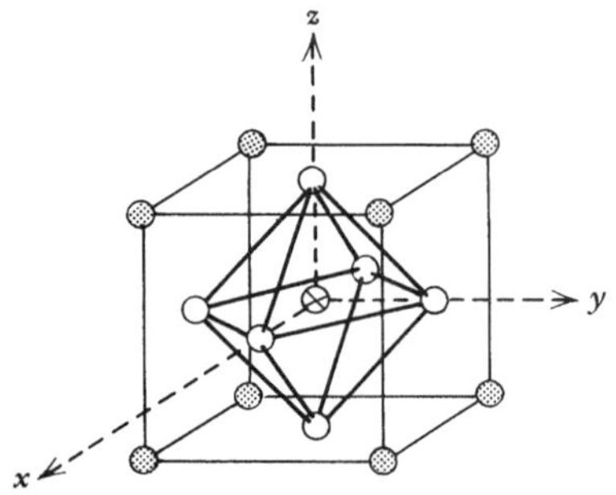

Figure 24.6 Schematic diagram of the unit cell of the cubic perovskite structure with the B ion at the center: ●A, ⊗B, ○O. The general chemical formula is ABO_3.

substance. The characteristic feature of this kind of transition is that there is *not* a discontinuous change in the state of the crystal, because the configuration of the atoms in the crystal changes continuously as T changes. So long as all three sides of the unit cell are equal in length, the phase remains cubic. However, when the temperature drops, the appearance of an infinitesimal difference between a and c converts the phase to tetragonal symmetry. A transition of the type just described is called *a transition of the second* kind (or *second-order*).

The example given, although a possible transformation, is not common. A well-studied class of compounds, the perovskites ABO_3, have the crystal structure shown in Fig. 24.6. Phase transitions between crystal modifications of these compounds involve rotations of the BO_6 octahedra. In $SrTiO_3$ the cubic high-temperature structure undergoes a tetragonal distortion at the transition, corresponding to a rotation of the BO_6 octahedra about a cube axis. In $LaAlO_3$ the cubic high-temperature phase undergoes a trigonal distortion to a rhombohedral structure described by a rotation

[2] Transitions of the *n*th kind are also known as *n*th-order transitions. We shall use both names, choosing one or the other for stylistic reasons.

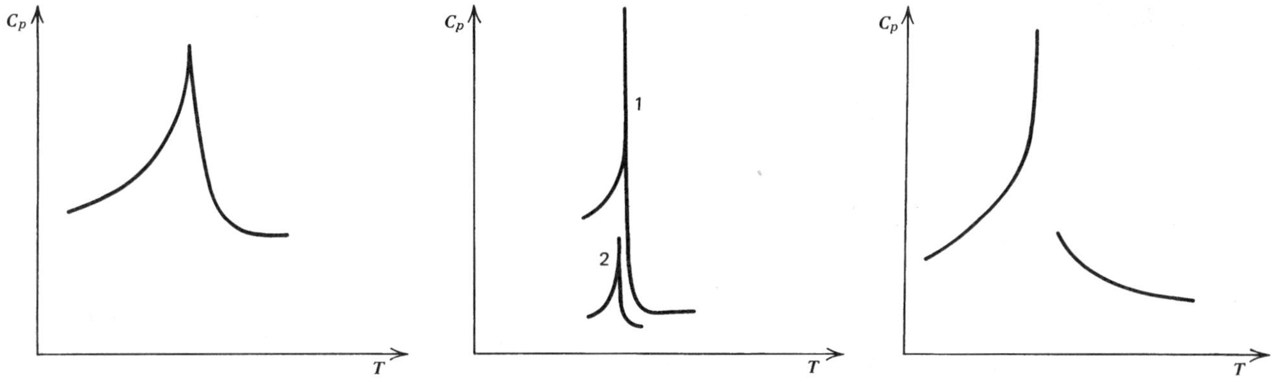

Figure 24.7 Some schematic heat capacity versus temperature curves near transition points. (*a*) Order–disorder transition in a binary alloy. (*b*) Liquid–gas transition; curves 1 and 2 are approximately to scale for argon and helium, respectively. (*c*) Order–disorder transition in a magnet; the curve is for nickel chloride hexahydrate.

of the AlO_6 octahedra about a cube diagonal. Still other kinds of distortions occur in other perovskites.

We have emphasized that in a phase transition of the second kind there is a discontinuous change of symmetry type. Nevertheless, the transition is continuous in the sense that the volume, internal energy, and entropy change continuously from one phase to the other.[3] Thus, such a transition is not accompanied by the absorption or evolution of heat. There are, however, discontinuous changes in the temperature and pressure derivatives of the volume, energy, and entropy when the transition occurs. In the expository model described above, the thermal expansion of the crystal is different in the two phases. In the cubic phase the crystal expands isotropically, whereas in the tetragonal phase the expansion is anisotropic. The symmetry properties of the two crystalline phases enter the problem because the definition of anisotropy requires a nonzero difference between properties along different directions in space. That is, there cannot be a continuous change in the coefficient of thermal expansion across the transition temperature because above the transition temperature $\alpha_{a_1} = \alpha_{a_2} = \alpha_{a_3} = \alpha_c$, whereas below the transition temperature $\alpha_{a_1} = \alpha_{a_2} \neq \alpha_{a_3}$, hence the difference $\alpha_{a_3} - \alpha_c$ must be nonzero at the transition temperature.

The existence of discontinuities in the temperature derivatives of, for example, the internal energy is a principal characteristic of transitions of the second kind. In Fig. 24.7 is displayed a typical variation of the heat capacity with temperature in the vicinity of a transition temperature. It should be noted that the available experimental data suggest that there are very few second-order transitions. Some first-order transitions have such small discontinuities of entropy, volume, and energy that they exhibit major features in common with the hypothesized properties of second-order transitions.

There are many other kinds of transitions, but the examples cited suffice for illustrative purposes. Indeed, as will be seen, in the study of the properties of liquid solutions we deal almost exclusively with first-order phase transitions.

We now turn to an examination of how thermodynamic and statistical molecular analyses can be used to provide quantitative descriptions of the possible equilibria between phases.

24.2 Thermodynamics of Phase Equilibria in One-Component Systems

A. The Clausius–Clapeyron Equation

We have previously shown, in Section 19.1, that at equilibrium between two phases of a one-component system the temperatures, pressures, and chemical potentials of the two phases are equal. In this section we deduce the consequences of this characteristic feature of phase equilibrium.

[3] The usual definition of the order of a transition is in terms of the behavior of derivatives of the Gibbs free energy. An *n*th-order transition is one in which discontinuities appear only in the *n*th and higher derivatives of G with respect to T and p. Thus, the ordinary first-order transitions have discontinuities in

$$V = \left(\frac{\partial G}{\partial p}\right)_T, \qquad S = -\left(\frac{\partial G}{\partial T}\right)_p$$

and thus also in H, U, etc. Second-order transitions have discontinuities in

$$\alpha = \frac{1}{V}\left(\frac{\partial V}{\partial T}\right)_p = \frac{1}{V}\left[\frac{\partial}{\partial T}\left(\frac{\partial G}{\partial p}\right)_T\right]_p,$$

$$\kappa_T = -\frac{1}{V}\left(\frac{\partial V}{\partial p}\right)_T = -\frac{1}{V}\left(\frac{\partial^2 G}{\partial p^2}\right)_T,$$

$$C_V = T\left(\frac{\partial S}{\partial T}\right)_V = -T\left[\frac{\partial}{\partial T}\left(\frac{\partial G}{\partial T}\right)_p\right]_V,$$

$$C_p = T\left(\frac{\partial S}{\partial T}\right)_p = -T\left(\frac{\partial^2 G}{\partial T^2}\right)_p,$$

and so forth. The order of a particular transition must be determined experimentally.

Consider a transition of the first kind, say, the liquid–gas transition. For each temperature and pressure at which the liquid and gas coexist, we must have

$$\mu^L(T,p) = \mu^G(T,p). \tag{24.5}$$

If the equations of state of the two phases are known, Eq. 24.5 may be used directly to obtain the vapor pressure curve by solving for the pressure as a function of the temperature. In general, however, the equations of state of the two coexisting phases are not known. How, then, can Eq. 24.5 be used? From Eq. 24.5 it is possible to derive a differential equation whose solution represents the coexistence curve, in our example the vapor pressure curve. It will be seen that this differential equation is simple in form and permits us to calculate the rate of change of pressure with temperature without complete knowledge of the equations of state of the two phases.

If Eq. 24.5 is valid everywhere along the coexistence curve of the two phases in equilibrium, then it must be true that for a change of T and p along this curve

$$d\mu^L(T,p) = d\mu^G(T,p); \tag{24.6}$$

that is, for the phases to remain in equilibrium a differential change in the chemical potential of one phase must be matched by an equal change in the other phase. But we have in a one-component closed system

$$\begin{aligned} d\mu &= \left(\frac{\partial \mu}{\partial T}\right)_p dT + \left(\frac{\partial \mu}{\partial p}\right)_T dp \\ &= -s\,dT + v\,dp, \end{aligned} \tag{24.7}$$

where, as before, s and v are the entropy and volume per mole. Then

$$-s_L\,dT + v_L\,dp = -s_G\,dT + v_G\,dp, \tag{24.8}$$

or

$$\left(\frac{dp}{dT}\right)_\sigma = \frac{s_L - s_G}{v_L - v_G}. \tag{24.9}$$

Equation 24.9, known as the *Clausius–Clapeyron equation,* describes the rate of change of pressure with temperature along the coexistence curve (which is what the subscript σ denotes). Since

$$s_G - s_L = \Delta s_{\text{vap}} = \frac{\Delta h_{\text{vap}}}{T_{\text{vap}}}, \tag{24.10}$$

we can rewrite Eq. 24.9 in the form

$$\left(\frac{dp}{dT}\right)_\sigma = \frac{\Delta h_{\text{vap}}}{T\Delta v_{\text{vap}}}. \tag{24.11}$$

A similar argument may be used to describe the coexistence line between the liquid and solid phases. The analog of Eq. 24.11 is

$$\left(\frac{dp}{dT}\right)_{\sigma f} = \frac{\Delta h_{\text{fus}}}{T\Delta v_{\text{fus}}}, \tag{24.12}$$

where the subscript *fus* refers to the fusion process, and σf to the solid–liquid coexistence line.

Equations 24.11 and 24.12 cannot be integrated unless we know the temperature and pressure dependencies of the latent heats of vaporization and fusion, and of the volume changes on vaporization and fusion. It is at just this point that information about the equations of state of the two phases is required. Nevertheless, Eqs. 24.11 and 24.12 may be used backwards to determine Δh_{vap} and Δh_{fus}, if the corresponding volume changes and rates of change of pressure with temperature along the coexistence lines are known. Measurements of dp/dT along the coexistence lines are commonly made, and density determinations are often not difficult, so that Eqs. 24.11 and 24.12 are, in fact, widely used to determine Δh_{vap} and Δh_{fus}.

We can also use Eqs. 24.11 and 24.12 in qualitative arguments. For example, since $\Delta h_{\text{vap}} > 0$ always, and $\Delta v_{\text{vap}} > 0$ always, it follows that $(dp/dT)_\sigma > 0$; that is, the boiling temperature always increases when the pressure is increased. Although $\Delta h_{\text{fus}} > 0$ always, Δv_{fus} can be either positive or negative, and correspondingly we have $(dp/dT)_{\sigma f} > 0$ when $\Delta v_{\text{fus}} > 0$ and $(dp/dT)_{\sigma f} < 0$ when $\Delta v_{\text{fus}} < 0$. The most famous example for which $(dp/dT)_{\sigma f} < 0$ is the ice–water transition (cf. Fig. 24.5). It can now be seen that the depression of the freezing point of water with increase of pressure is a consequence of the fact that the density of ice near 0°C is less than the density of water near 0°C. This relationship between the sign of $(dp/dT)_{\sigma f}$ and the sign of Δv_{fus} is not an "explanation" of the phenomenon, but just a description of the interrelationship of macroscopic properties required by the constraint of phase equilibrium.

Suppose that the vaporization of a liquid is carried out at a low enough temperature that the vapor formed is at low pressure. Then, as a first approximation, we can take the equation of state of the vapor to be that of a perfect gas. In this limit it is also true that $v_G >> v_L$, so that $\Delta v_{\text{vap}} \approx v_G$. Equation 24.11 now assumes the approximate form (using $p = RT/v_G$)

$$\left(\frac{dp}{dT}\right)_\sigma = \frac{\Delta h_{\text{vap}}}{Tv_G} = \frac{\Delta h_{\text{vap}}\,p}{RT^2}, \tag{24.13}$$

which is also known as the *Clausius–Clapeyron* equation. Equation 24.13 may be integrated to give

$$\ln\left[\frac{p_2(T_2)}{p_1(T_1)}\right]_\sigma = \int_{T_1}^{T_2} \frac{\Delta h_{\text{vap}}}{RT^2}\,dT. \tag{24.14}$$

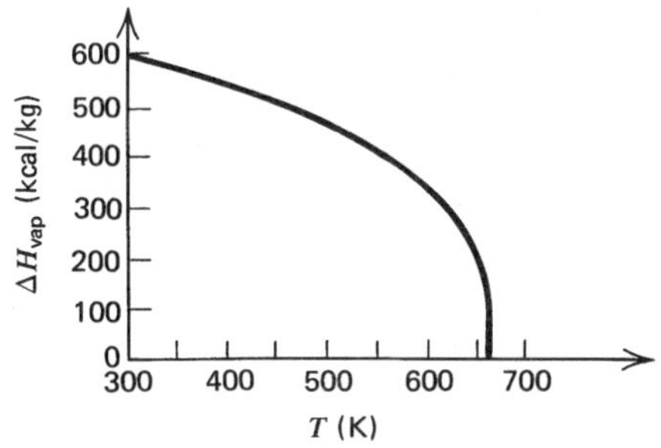

Figure 24.8 The heat of vaporization of water as a function of temperature. From J. Kestin, *A Course in Thermodynamics* (Blaisdell, Waltham, Mass., 1966).

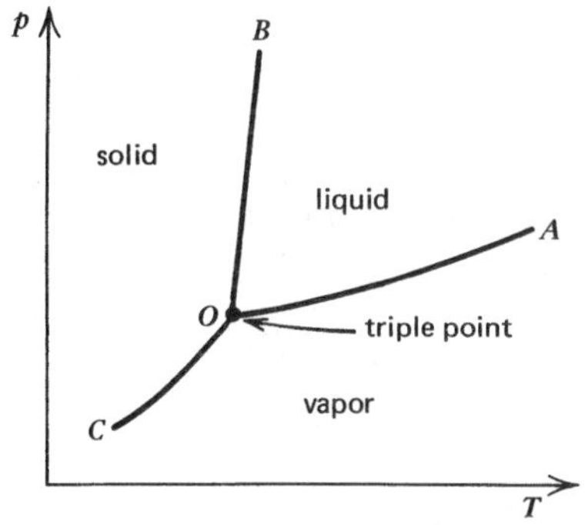

Figure 24.9 Schematic diagram showing the triple point and the equilibrium of solid, liquid, and vapor.

Finally, if the temperature range $T_1 \to T_2$ is sufficiently small that the variation of Δh_{vap} with temperature may be neglected, we obtain

$$\ln\left[\frac{p_2(T_2)}{p_1(T_1)}\right]_\sigma = -\frac{\Delta h_{vap}}{R}\left(\frac{1}{T_2}-\frac{1}{T_1}\right), \tag{24.15}$$

so that a measurement of the pressure of vapor in equilibrium with liquid at two temperatures will permit calculation of Δh_{vap} for the average temperature. Of course, Δh_{vap} does vary with temperature, and indeed becomes zero at the critical temperature[4] (see Fig. 24.8), so that Eq. 24.15 should be regarded as an approximation to Eq. 24.14.

Before considering other properties of the liquid–vapor equilibrium, it is pertinent for us to examine the possibility of equilibrium among three phases of a pure substance. Let the three phases be numbered 1, 2, 3. At equilibrium between the phases we must have

$$\begin{aligned}\mu^{(1)} &= \mu^{(2)},\\ \mu^{(1)} &= \mu^{(3)},\\ \mu^{(2)} &= \mu^{(3)},\end{aligned} \tag{24.16}$$

where the third line may also be considered to be a consequence of the zeroth law of thermodynamics. Corresponding to the three pairs of equilibria, we can find three Clausius–Clapeyron equations,

$$\begin{aligned}\left(\frac{dp_{12}}{dT}\right)_{\sigma_{12}} &= \frac{\Delta h_{12}}{T\Delta v_{12}},\\ \left(\frac{dp_{13}}{dT}\right)_{\sigma_{13}} &= \frac{\Delta h_{13}}{T\Delta v_{13}},\\ \left(\frac{dp_{23}}{dT}\right)_{\sigma_{23}} &= \frac{\Delta h_{23}}{T\Delta v_{23}},\end{aligned} \tag{24.17}$$

[4] At the critical temperature the molar volumes of the two phases are equal, hence Eq. 24.13 is invalid because of the neglect of the volume of the liquid phase.

Table 24.1 Triple Points of Pure H_2O

Phases in Equilibrium	p	T (°C)
Ice I, liquid, vapor	4.579 torr	+0.01
Ice I, liquid, ice III	2,115 kg/cm^2	–22.0
Ice I, ice II, ice III	2,170 kg/cm^2	–34.7
Ice II, ice III, ice V	3,510 kg/cm^2	–24.3
Ice III, liquid, ice V	3,530 kg/cm^2	–17.0
Ice V, liquid, ice VI	6,380 kg/cm^2	+0.16
Ice VI, liquid, ice VII	22,400 kg/cm^2	+81.6

only the first two of which are independent, the third being a consequence of them. Indeed, the third equation in Eq. 24.17 is only a consequence of the zero-th law.

A plot of the pressure of the system versus temperature looks like Fig. 24.9. Each of the three curves in Fig. 24.9 corresponds to one of the three equations of Eq. 24.17. Any point in the (p, T) plane not on one of the lines OA, OB, or OC corresponds to a single phase of the system, and the lines OA, OB, OC to the points of coexistence of pairs of phases. The point O represents a state of the system in which all three phases are in equilibrium; it is called a *triple point.*

The ordinary triple points of most substances lie at pressures that are small relative to atmospheric pressure. When such a substance is heated, at a constant pressure that exceeds the triple-point pressure but is less than the critical pressure, the change of state can be plotted as a horizontal line above the triple point in Fig. 24.9. In this case the crystal melts to a liquid, and then later vaporizes to a gas. Heating the substance at a pressure less than the triple-point pressure leads to a direct conversion of the crystal to vapor. This change of state can be represented by a horizontal line below the triple point in Fig. 24.9. Triple points of various phases of pure water are shown in Table 24.1.

The phase diagram in Fig. 24.9 shows that each of the three phases is stable in a region of the (T, p) plane within which the temperature and the pressure may be varied independently of each other. Any two of the phases can coexist only on a line on the diagram, and along such a line either the temperature or the pressure, but not both, can be considered an independent variable. Finally, all three phases can coexist only at a single point, the triple point, at which the temperature and pressure are both fixed. Thus, if the number of independent variables is called f and the number of coexisting phases is called P, the above observations can be summarized by writing $f = 3 - P$.

The equation $f = 3 - P$ is a simple example of a general and useful relation known as the *Gibbs phase rule.* Consider heterogeneous equilibrium in a system containing P coexisting phases, with C components in each phase. By a *component* we mean a chemical species whose concentration in a phase can be varied independently of all of the other species' concentrations. The number of components is not necessarily equal to the number of chemical species present. For example, if two species are interconvertible through a chemical reaction, and are in chemical equilibrium, their concentrations are not independent of each other. Two species in chemical equilibrium are therefore considered to form a single component. In order to characterize a multicomponent phase completely, we must specify the temperature and pressure, and the mole fractions of each of the C components in the P phases, a total of $C + 2$ variables for that phase. At first sight it appears that to characterize heterogeneous equilibrium in a system containing P coexisting phases with C components in each phase, we must specify $P(C + 2)$ variables. However, these $P(C + 2)$ variables are not all mutually independent. In each phase the mole fractions of the various components must add to unity. This condition means that there are only $C - 1$ independent concentration variables in each of the P phases, and it imposes P constraints on the number of independent variables required to define the system. In addition, the existence of phase equilibrium requires that the temperatures and pressures be equal in all P phases and that the chemical potential of each component be the same in each of the P phases. There are, then, $P - 1$ independent equalities involving the pressures of the phases, $P - 1$ independent equalities involving the temperatures of the phases, and $C(P - 1)$ relationships among the chemical potentials of the components of the phases. To find the number of independent variables describing the system we must subtract from $P(C + 2)$ the number of equations relating these variables. The number of independent variables (also called the number of *degrees of freedom*) is therefore $f = P(C + 2) - P - 2(P - 1) - C(P - 1)$, which yields $f = C - P + 2$. This relation is the Gibbs phase rule; it predicts how many variables are required to describe a system, and it also places a limit on the number of phases that can coexist in a system containing C components. For one-component systems, that limit is three.

We have previously discussed in detail the heat capacity of a one-phase system. When two phases must be kept in equilibrium, the addition of heat to the system is described by neither c_V nor c_p, since the constraint of phase equilibrium is not the same as the constraints of constant volume or constant pressure. We now define a third heat capacity as the limiting ratio of the heat absorbed to the temperature rise generated under the constraint that the system consists of two phases maintained in equilibrium. It is clear that the existence of a nonzero latent heat of transition will influence the magnitude of the heat capacity of a two-phase system.

Consider, for example, a two-phase system consisting of liquid and vapor in equilibrium. The first law of thermodynamics,

$$\text{đ}q = du - \text{đ}w, \tag{24.18}$$

can be used to define the heat capacity of the two-phase system. With the definition cited, we obtain

$$\begin{aligned} c_\sigma &\equiv \left(\frac{\text{đ}q}{dT}\right)_{\text{phase equilibrium}} \\ &= \left[\left(\frac{du}{dT}\right) - \left(\frac{d-w}{dT}\right)\right]_{\text{phase equilibrium}} . \end{aligned} \tag{24.19}$$

For the case of the liquid–vapor transition, the expansion of the system results in the performance of pressure–volume work only, so that we have (per mole)

$$c_\sigma = \left(\frac{du}{dT} + p\frac{dv}{dT}\right)_{\text{vapor pressure curve}} . \tag{24.20}$$

Let $c_{\sigma L}$ and $c_{\sigma G}$ be the heat capacities of saturated liquid and vapor, respectively. Then, using the equalities $p_L = p_G = p_\sigma$ and $T_L = T_G$, both valid because of the constraint of phase equilibrium, we find that

$$\begin{aligned} c_{\sigma L} &= \left(\frac{du_L}{dT}\right)_\sigma + p_\sigma \left(\frac{dv_L}{dT}\right)_\sigma , \\ c_{\sigma G} &= \left(\frac{du_G}{dT}\right)_\sigma + p_\sigma \left(\frac{dv_G}{dT}\right)_\sigma . \end{aligned} \tag{24.21}$$

By subtracting the first of equations 24.21 from the second we obtain

$$c_{\sigma G} - c_{\sigma L} = \left[\frac{d}{dT}(u_G - u_L)\right]_\sigma + p_\sigma \left[\frac{d}{dT}(v_G - v_L)\right]_\sigma . \tag{24.22}$$

By definition, the latent heat of vaporization per mole is

$$\Delta h_{\text{vap}} \equiv h_G - h_L \equiv u_G - u_L + p_\sigma (v_G - v_L), \tag{24.23}$$

so that, by combining Eqs. 24.22 and 24.23, we find

$$\begin{aligned} \frac{d\,\Delta h_{\text{vap}}}{dT} &= \left[\frac{d}{dT}(u_G - u_L)\right]_\sigma + p_\sigma \left[\frac{d}{dT}(v_G - v_L)\right]_\sigma \\ &\quad + (v_G - v_L)_\sigma \frac{dp_\sigma}{dT} \\ &= c_{\sigma G} - c_{\sigma L} + (v_G - v_L)_\sigma \frac{dp_\sigma}{dT} . \end{aligned} \tag{24.24}$$

But the last term on the right-hand side of Eq. 24.24 may be evaluated from the Clausius–Clapeyron equation, because the constraint of phase equilibrium is implicit in Eq. 24.24. Thus, introducing Eq. 24.11 in Eq. 24.24, we obtain

$$c_{\sigma G} - c_{\sigma L} = \frac{d\,\Delta h_{\mathrm{vap}}}{dT} - \frac{\Delta h_{\mathrm{vap}}}{T}. \tag{24.25}$$

If the temperature is low, $c_{\sigma L}$ is almost equal to c_{pL}. With this approximation entered into Eq. 24.25, we find for the heat capacity of the saturated vapor

$$c_{\sigma G} = c_{pL} + \frac{d\,\Delta h_{\mathrm{vap}}}{dT} - \frac{\Delta h_{\mathrm{vap}}}{T}. \tag{24.26}$$

Consider, as an example, water vapor in equilibrium with liquid water at 100°C. Using tables of the thermodynamic properties of water, we find Δh_{vap} = 40.67 kJ/mol, $(d\Delta h_{\mathrm{vap}}/dT) = -47.93$ J/K mol, $c_{pL} = 75.86$ J/K mol. Thus in this case $c_{\sigma G} = -80.91$ J/K mol; the heat capacity of saturated water vapor at 100°C is negative. What does this result mean? If saturated water vapor is compressed adiabatically, its temperature increases with pressure more rapidly than along the coexistence curve between liquid and vapor. Conversely, if saturated water vapor is adiabatically expanded, its temperature decreases with decreasing pressure more rapidly than along the coexistence curve, so some condensation to liquid must occur. In our numerical example, if water vapor at 100°C and 760 torr were adiabatically compressed to 787.6 torr (the vapor pressure at 101°C) its temperature would rise to about 103°C; 80.91 J/mol would then have to be withdrawn at constant pressure to bring the vapor back to the coexistence curve at 101°C. To an external observer this appears as a negative heat capacity of the saturated vapor.

B. The Critical Point

It has been pointed out in Section 21.1 that there exists a critical temperature, T_c, above which gas and liquid cannot be distinguished. The conditions describing phase equilibrium, namely, the equality of chemical potentials and hence of pressures and temperatures between the two phases, in principle permit there to be a critical point between any two coexisting phases. Given the equations of state of the two phases and the constraint of phase equilibrium, we have five equations in six variables:

$$\left.\begin{aligned} p_1 &= p_2 \\ T_1 &= T_2 \\ \mu_1(p_1, T_1) &= \mu_2(p_2, T_2) \end{aligned}\right\}\text{constraint of phase equilibrium,}$$

$$\left.\begin{aligned} p_1 &= p_1(v_1, T_1) \\ p_2 &= p_2(v_2, T_2) \end{aligned}\right\}\text{equations of state.}$$

Thus, five of the six variables p_1, T_1, v_1, p_2, v_2 are determined by these five equations, leaving one relation between the remaining two variables. This is, for example, the equation connecting the vapor pressure and temperature along the coexistence line between vapor and liquid. Then, if the values of v_1 and v_2 become equal for some value of the temperature, the two phases become identical if they have the same symmetry. This temperature is defined to be the critical temperature, and the corresponding values of p and v to be the critical pressure (p_c) and critical volume (v_c), respectively.

It may happen that no temperature can be found for which $v_1 = v_2$ given that $p_1 = p_2$ and $T_1 = T_2$. In this case there will be no critical point on the coexistence line between the two phases. It may also happen that the equality of temperature, pressure, and chemical potential between the two phases does not provide a complete description of the phase equilibrium. For example, in the description of crystal–crystal transitions it is necessary to know the symmetries of the crystal lattices for the two phases. Although the volumes of the two crystals might be the same, unless the lattice structures were also the same the phases would not be identical, and the temperature at which $v_1 = v_2$ would not be a critical point. The role of symmetry properties in defining the properties of transitions of the second kind was described in Section 24.1. Our use of the concept of symmetry is related to the concept of distinguishability. Just as in the description of the entropy of mixing (see Section 21.2), it is distinguishability that determines the qualitative features of the phenomenon described.

We now restrict attention to the study of the gas–liquid critical point and some of the properties of the system in the vicinity of that point. We begin by defining conditions that locate the critical point of a fluid. A study of the isotherms of pressure versus volume, as shown in Fig. 24.4, reveals that one isotherm passes through a point in the $(p,\ v)$ plane at which it is horizontal, and at which it also has a point of inflection (change from positive to negative curvature). These two properties,

$$\left(\frac{\partial p}{\partial v}\right)_{T=T_c} = 0 \quad \text{(horizontal tangent)},$$

$$\left(\frac{\partial^2 p}{\partial v^2}\right)_{T=T_c} = 0 \quad \text{(point of inflection)}, \tag{24.27}$$

define the critical point of the fluid.

What are the properties of a fluid at and near the critical point?

1. Since $(\partial v/\partial p)_T < 0$ for any substance at equilibrium, it follows that

$$\lim_{\substack{T\to T_c \\ v\to v_c}} \left[-\frac{1}{v}\left(\frac{\partial v}{\partial p}\right)_T\right] = \infty; \tag{24.28}$$

hence, at the critical point the isothermal compressibility becomes positive infinite.

2. Suppose that $(\partial u/\partial v)_T$ is finite. Then from

$$\left(\frac{\partial u}{\partial v}\right)_T = T\left(\frac{\partial p}{\partial T}\right)_V - p,$$

it follows that $(\partial p/\partial T)_V$ is finite. But along the gas–liquid coexistence line we have, by Eqs. 24.9 and 24.11,

$$\left(\frac{dp}{dT}\right)_\sigma = \frac{s_G - s_L}{v_G - v_L},$$

so that we may write

$$\lim_{\substack{T\to T_c \\ v\to v_c}} \left(\frac{dp}{dT}\right)_\sigma = \left(\frac{\partial s}{\partial v}\right)_{T=T_c} \tag{24.29}$$

or, using Eq. 17.17,

$$\left(\frac{dp}{dT}\right)_{\sigma,\text{ critical point}} = \left(\frac{\partial p}{\partial T}\right)_{v=v_c}. \tag{24.30}$$

Since $(\partial p/\partial T)_V$ is always finite, $(dp/dT)_{\sigma,T=T_c}$ is finite. Then, using

$$\left(\frac{\partial v}{\partial T}\right)_p = -\left(\frac{\partial v}{\partial p}\right)_T \left(\frac{\partial p}{\partial T}\right)_V,$$

one can easily show that

$$\lim_{\substack{T\to T_c \\ v\to v_c}} \frac{1}{v}\left(\frac{\partial v}{\partial T}\right)_p = \infty. \tag{24.31}$$

In words, the coefficient of thermal expansion at the critical point is positive and infinite.

3. Since the Clausius–Clapeyron equation can be written in the form

$$\Delta h_{\text{vap}} = T(v_G - v_L)\left(\frac{dp}{dT}\right)_\sigma,$$

then at the critical point, when $T = T_c$, $v_G = v_L$, and $(dp/dT)_\sigma$ is finite, we must have

$$\lim_{T\to T_c} \Delta h_{\text{vap}} = 0. \tag{24.32}$$

It can also be shown that $(d\Delta h_{\text{vap}}/dT)_{T=T_c} = -\infty$. A glance at Fig. 24.8 shows that the predicted behavior of Δh_{vap} and $d\Delta h_{\text{vap}}/dT$ as $T \to T_c$ is in agreement with the observed behavior.

The fact that so many thermodynamic properties of a system become either indefinitely large or zero at the critical point is sufficient evidence that the region of the critical point of a fluid has some very peculiar characteristics. Some of the observations that require interpretation follow:

1. As the critical point is approached, the liquid and gas densities converge toward one another as[5]

$$(\rho_L - \rho_G) \propto |T_c - T|^\beta; \qquad \beta = 0.326 \pm 0.002 \tag{24.33}$$

(see Figs. 24.10 and 24.11). Note that to a first approximation $\beta = \frac{1}{3}$, and the coexistence curve near the critical point is cubic. However, this is not true to within the experimental error. Because of the algebraic form of Eq. 24.33, β is called a *critical exponent* (as are δ, α, and γ defined below).

2. The critical point is associated with a striking thermal anomaly. In fact, at a constant density equal to the critical density (see Fig. 24.12),

$$c_V(T) \to \infty \tag{24.34}$$

as $T \to T_c$, from above or below with $v = v_c$.

3. The rates at which various thermodynamic functions approach the critical point are given by

$$(\mu - \mu_c) \propto |\rho - \rho_c|^{\delta-1}(\rho - \rho_c); \quad \delta = 4.80 \pm 0.02, \tag{24.35}$$

$$c_V(T) \propto |T - T_c|^{-\alpha}; \quad \alpha = 0.110 \pm 0.003, \tag{24.36}$$

$$\left(\frac{\partial \rho}{\partial \mu}\right)_T = \rho^2 \kappa_T \propto |T - T_c|^\gamma; \quad \gamma = 1.239 \pm 0.002. \tag{24.37}$$

Arguably the most remarkable feature of the functional relationships displayed in Eqs. 24.33–24.37 is that the values of the exponents are the same for all fluids (and, indeed, for all materials which have short-ranged interactions between the molecules). We will see later that a key element in the modern theory of critical phenomena is the classification of materials in universality classes. Each universality class is defined to have the same spatial dimensionality and the same spin dimensionality, the latter of which refers to the tensorial character of the order parameter (see Section 22.4). For ordinary fluids the spatial dimensionality is 3, and the order parameter is a scalar, so the spin dimensionality is 1. The existence of a universality class for fluids can be attributed to the fact that near the critical point there are fluctuations, e.g., in the density, which have a range much greater

[5] For the present we do not distinguish between approaching the critical point from $T > T_c$ or $T < T_c$. The available experimental evidence suggests that the critical exponents are the same for both approaches to the critical point.

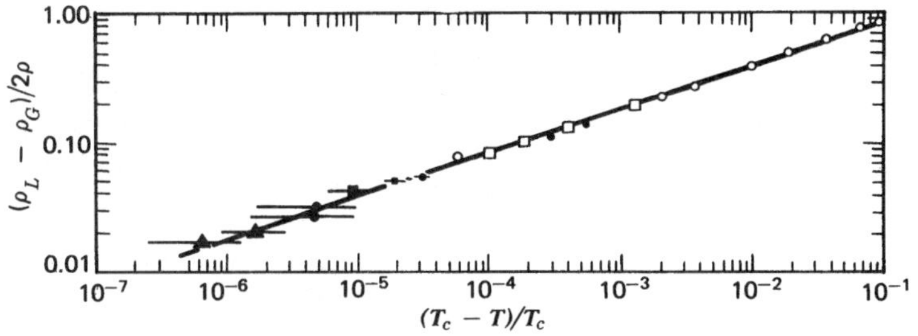

Figure 24.10 The dependence of $\rho_L - \rho_G$ on $T - T_c$ near T_c for CO_2. From P. Heller, *Rep. Prog. Phys.* **30** (II), 731 (1967).

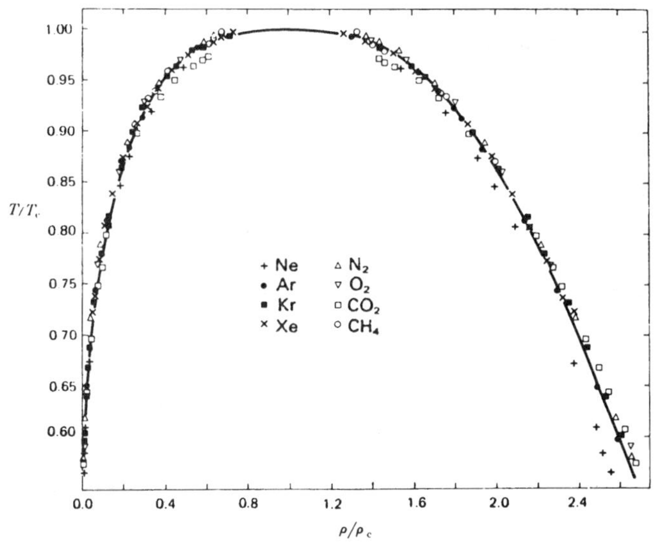

Figure 24.11 The liquid–gas coexistence curve for simple substances. The curve is a composite of data for Ne, Ar, Kr, Xe, N_2, O_2, CO_2, and CH_4. Note the use of the reduced variables T/T_c and ρ/ρ_c. From E. A. Guggenheim, *J. Chem. Phys.* **13,** 253 (1945).

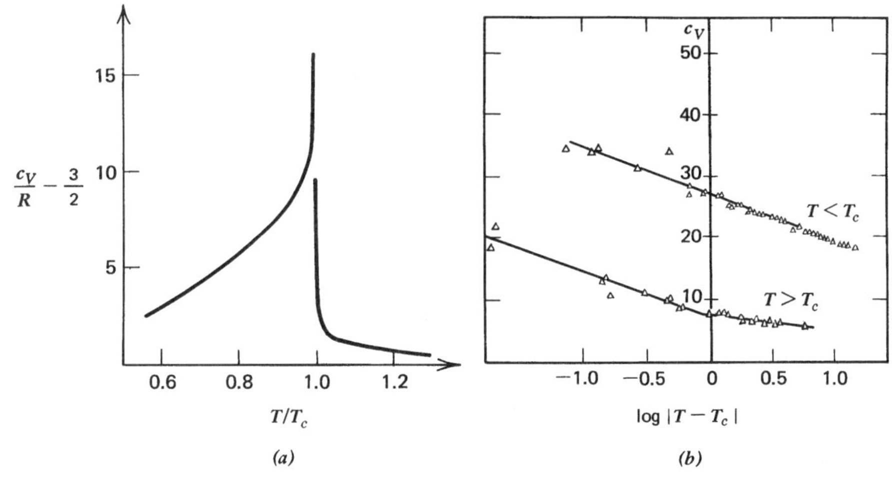

Figure 24.12 (*a*) Schematic diagram of the constant-volume heat capacity of argon near the critical point. (*b*) Variation of the constant-volume heat capacity of argon along the critical isochore. The data show that the variation of c_V near T_c is logarithmic. From P. A. Egelstaff and J. W. Ring [data of M. I. Bagatskil, A. V. Voronel, and B. G. Gusak, *Zh. Eksperim. i Teor. Fiz.* **43,** 728 (1962); *JETP* **16,** 517 (1963)], in H. N. V. Temperley, J. S. Rowlinson, and G. S. Rushbrooke (Eds.), *Physics of Simple Liquids* (North Holland, Amsterdam, 1968).

than the range of the molecular interactions or any other microscopic length scale.

The values of the exponents given in Eqs. 24.33–24.37, which are those currently accepted to be the most accurate, are obtained from so-called renormalization group theory calculations, which we do not discuss in this text, except for a brief sketch of the principal ideas in Section 24.4. The best experimental values of these exponents are in agreement with the theoretical values within experimental accuracy.[6]

[6] For example, two experimental values of the critical exponent β are 0.327 ± 0.009 for SF_6 and 0.311 ± 0.036 for Xe, and two experimental values of α are 0.116 for He^4 and 0.101 for CO_2 (see the compilation in S. C. Greer and M. R. Moldover, *Ann. Rev. Phys. Chem.* **32,** 233 (1981)).

We note that there is no *simple* equation of state of a fluid that predicts the detailed behavior described in Eqs. 24.33–24.37 *and* the behavior far from the critical point, although the nature of the transition in thermodynamic behavior between these domains is understood.

24.3 Phase Transitions Viewed as Responses to Thermodynamic Instabilities

The chemical potential of a pure substance depends only on the pressure and the temperature, the functional forms being different for the solid, liquid, and vapor phases. When the chemical potential of one phase, say, the liquid, becomes less than that of another, say, the vapor, then the latter is unstable with respect to the former. Viewed this way, the condition of phase equilibrium, $\mu^L = \mu^G$ in the particular case mentioned, specifies the boundary of stability between the two phases, and we can think of the phase transition as the system's response to an instability. In this section we pursue this idea.

It is convenient, for the purpose of visualizing the way a thermodynamic instability develops, to represent the equilibrium properties of a one-component system by a surface in a suitable space of variables. Consider a specified mass, say, 1 mol, of a pure substance. Suppose that we know the energy as a function of the entropy and volume, $u(s, v)$. Then, if we construct a three-dimensional space with orthogonal axes u, s, and v, the function $u(s, v)$ is a surface in this space. Each point on this surface is associated with an equilibrium state of the system, and the slopes of $u(s, v)$ at such a point parallel to the s and v axes are simply

$$\left(\frac{\partial u}{\partial s}\right)_v = T, \tag{24.38}$$

$$\left(\frac{\partial u}{\partial v}\right)_s = -p. \tag{24.39}$$

Therefore, any plane parallel to the u, v plane intersects the surface $u(s, v)$ in a curve that has negative slope everywhere (since $p > 0$) and any plane parallel to the u, s plane intersects the surface $u(s, v)$ in a curve that has positive slope everywhere (since $T > 0$). These features of the surface $u(s, v)$ are sketched in Fig 24.13.

The surface $u(s, v)$ has some interesting properties. Consider some point Γ_0 on this surface, and the tangent plane to the surface that passes through it. The equation of that tangent plane is

$$u = u_0 + T_0(s - s_0) - p_0(v - v_0), \tag{24.40}$$

the subscript zero referring to values at Γ_0. Now consider the plane $v = v_0$. This plane is parallel to the (u, s) plane and passes through Γ_0. Moreover, it intersects the plane tangent at Γ_0 in a straight line the equation of which is

$$u = u_0 + T_0(s - s_0) \tag{24.41}$$

(remember that the intersection also lies in the plane for which $v = v_0$), and this straight line intersects the u axis at a value

$$u = u_0 - T_0 s_0 \equiv a_0 \tag{24.42}$$

(see Fig. 24.13). Similarly, the plane $s = s_0$ passes through Γ_0 and intersects the tangent plane at Γ_0 in the straight line

$$u = u_0 - p_0(v - v_0), \tag{24.43}$$

which intercepts the u axis at the point.

$$u = u_0 + p_0 v_0 \equiv h_0 \tag{24.44}$$

(see Fig. 24.13). Finally, the tangent plane at Γ_0 itself intercepts the u axis at

$$u = u_0 - T_0 s_0 + p_0 v_0 \equiv g_0 \tag{24.45}$$

(see Fig. 24.13).

Since a one-component system with fixed mass has definite values of u, s, and v irrespective of the number of phases present, the surface $u(s, v)$ must exist for all the system's equilibrium states. We now examine what that surface looks like when several phases are present (the case of heterogeneous equilibrium).

Consider a one-component system at its triple point. There is some point of the surface $u(s, v)$ representing the

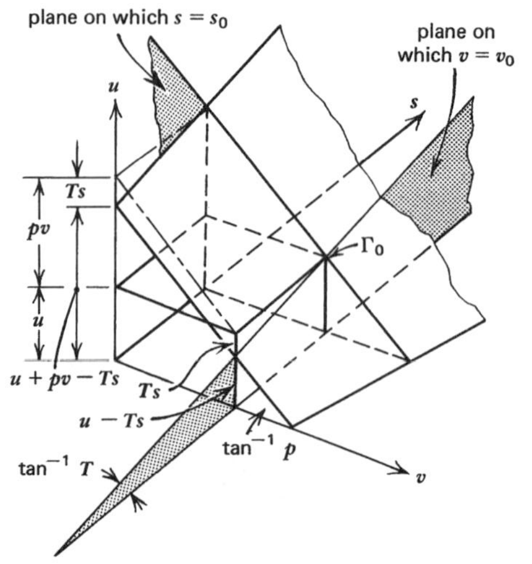

Figure 24.13 The tangent plane to the thermodynamic surface. Drawing courtesy of Prof. J. A. Beattie.

properties of the solid at the triple point, call it Γ_3^S. Similarly, there are points Γ_3^L and Γ_3^G on $u(s, v)$ representing the states of the liquid and the vapor at the triple point. Now construct the three tangent planes at the points Γ_3^S, Γ_3^L, and Γ_3^G. By Eq. 24.45 the intercepts of these tangent planes with the u axis are g_3^S, g_3^L, and g_3^G, respectively. But at the triple point $g_3^S = g_3^L = g_3^G$, since for a one-component system μ is just the Gibbs free energy per mole. Moreover, since $T_S = T_L = T_G$ and $p_S = p_L = p_G$ at the triple point, the slopes of the three tangent planes parallel to the s and v axes are the same at Γ_3^S, Γ_3^L, and Γ_3^G. We conclude that the three tangent planes coincide or, better put, that Γ_3^S, Γ_3^L, and Γ_3^G have a common tangent plane. Therefore, a mixture of solid, liquid, and vapor in equilibrium at the triple point will have its state represented by a point on the so-called *derivative surface,* which is just the region of the common tangent plane to Γ_3^S, Γ_3^L, Γ_3^G lying within the triangle formed by these three points. The position of the point representing the equilibrium mixture of the three phases is determined by

$$\begin{aligned} u &= x_S u_S + x_L u_L + x_G u_G, \\ s &= x_S s_S + x_L s_L + x_G s_G, \\ v &= x_S v_S + x_L v_L + x_G v_G, \end{aligned} \tag{24.46}$$

where x_i is the mole fraction of phase i.

Consider now liquid–vapor equilibrium in a one-component system of fixed mass. A similar argument leads to the surface shown in Fig. 24.14. The derived surface is the set of straight lines tangent simultaneously at Γ_2^L and Γ_2^G representing the coexisting liquid and vapor phases. These two points of tangency on $u(s, v)$ approach one another and eventually coincide at the critical point. Notice that there is a gap in the surface drawn that is spanned by the derived surface. The gap coincides with that region of $u(s, v)$ which would have negative curvature.

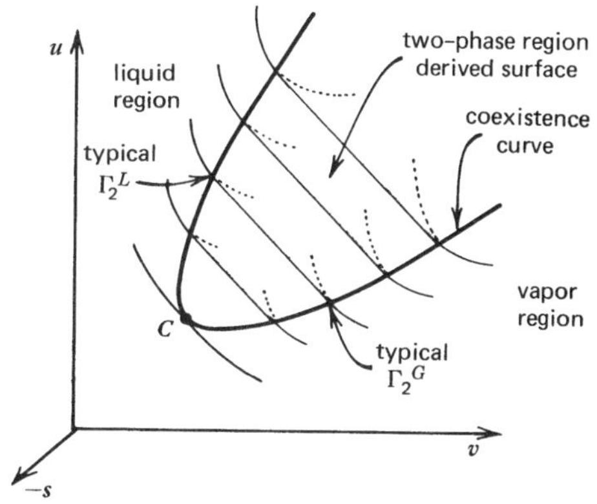

Figure 24.14 Schematic representation of the surface $u(s, v)$ in a region of vapor–liquid equilibrium. The derived surface lies below $u(s, v)$; each line shown in the two-phase region is an example of the common tangent to $u(s, v)$ at points Γ_2^L and Γ_2^G. The locus of the points of tangency defines the coexistence curve. C is the critical point. The dotted portions of $u(s, v)$ show regions of metastability.

To understand how an instability is represented, and the reason for the gap in Fig 24.14, we examine the nature of internal stability in a homogeneous phase. Suppose that we have a homogeneous pure substance isolated in a volume v; its energy and entropy are u and s. We test the stability of this phase with respect to variations that keep the energy and entropy fixed. This is just the situation treated in Chapter 19 (see Fig. 19.4, Eqs. 19.45–19.50). We saw there that for the system to be stable it is necessary and sufficient that

$$\left(\frac{\partial^2 u}{\partial s^2}\right)_v = \frac{T}{c_V} > 0, \tag{24.47a}$$

$$\left(\frac{\partial^2 u}{\partial s^2}\right)_v \left(\frac{\partial^2 u}{\partial v^2}\right)_s - \left(\frac{\partial^2 u}{\partial v \partial s}\right)_{s,v} = \frac{T}{c_V}\left(\frac{c_p}{v\kappa_T c_V}\right) - \frac{\alpha^2 T^2}{\kappa_T^2 c_V^2} > 0, \tag{24.47b}$$

the latter implying that $\kappa_T > \alpha^2 T v / c_p > 0$. At least for infinitesimal displacements, a one-phase system is stable so long as Eqs. 24.47*a* and 24.47*b* are satisfied. However, the stability of the system against the formation of a new phase has not been tested by the class of displacements considered. This is an important observation, as we shall now show.

Consider a region of the surface $u(s, v)$ representing two-phase equilibrium, say, between liquid and vapor. The intersection of this region of $u(s, v)$ and a plane in which s is constant is shown in Fig. 24.15. The straight line that is the common tangent to the points labeled Γ_2^L and Γ_2^G is the intersection of the derived surface and the same plane; it has slope $p = -(\partial u/\partial v)_s$, as required by its definition. In either of the pure phases the pressure is a monotonic function of the volume since, by Eq. 24.47*b*, $v\kappa_T = -(\partial v/\partial p)_T > 0$. This

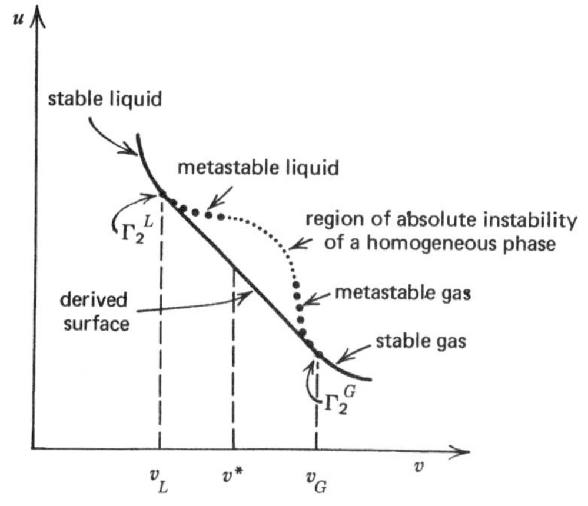

Figure 24.15 Schematic representation of the intersection of the two-phase region of $u(s, v)$ and a plane on which s is constant. The dotted lines are continuations of the homogeneous-phase energies; they represent metastable states of the system. If we attempt to prepare a homogeneous system with the volume v^* shown, it breaks up into liquid and gas, this being the region of absolute instability of a homogeneous phase.

condition can be satisfied in a homogeneous phase beyond the temperature and pressure at which a new phase emerges. Hence, *if we imagine that it is possible to prevent formation of a new phase,* the curve of internal energy versus volume may be continued until Eqs. 24.47*a* and 24.47*b* become invalid. The continuations are shown by dots in Fig 24.15; they stop at the points of inflection where $(\partial^2 u/\partial v^2)_s = 0$. There can be no continuation of u versus v between these points of inflection since in that domain $(\partial^2 u/\partial v^2)_s < 0$ and a homogeneous system is absolutely unstable. Notice that the continuation of $u(s, v)$ for the liquid corresponds to a pressure less than that along the common tangent, and the continuation of $u(s, v)$ for the vapor corresponds to a pressure greater than that along the common tangent. The continuations shown correspond to metastable states of a homogeneous vapor or liquid. They can sometimes be observed. For example, the formation of the clusters of molecules that convert a gas to a liquid can occur slowly enough that the supersaturated vapor (i.e., vapor at a pressure in excess of the vapor pressure of the liquid at the same temperature) can persist long enough to be observed. The supersaturated vapor must be prepared in the absence of the liquid phase, since the homogeneous system is stable only to infinitesimal displacements that do not include the appearance of a new phase. Once the liquid phase appears, the supersaturated vapor cannot persist as it is unstable with respect to that displacement which creates the liquid.

The region of the surface $u(s, v)$ that would have negative curvature is just a gap; this gap is spanned by the derived surface described earlier. Because of the absolute instability of systems whose representative point falls in the gap region of $u(s, v)$, if we were to imagine a hypothetical homogeneous system prepared with volume v^*, with $v_L < v^* < v_G$, it would immediately decompose into liquid and vapor, the amounts of each being determined by Eq. 24.1 (see Fig. 24.15). That is, the energy of the system is lowest along the common tangent from the derived surface, and along this tangent $v = x_L v_L + x_G v_G$.

A convenient way of portraying the relationship between the coexistence curves along which phase equilibria exist, and the limit of absolute stability for a system, is to project the surface $u(s, v)$ onto an (s, v) plane, as shown in Fig. 24.16.

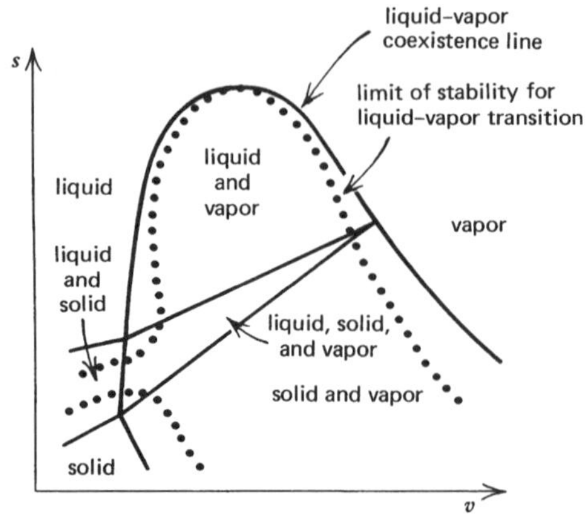

Figure 24.16 Projection of the surface $u(s, v)$ onto an (s, v) plane, showing the several coexistence curves for equilibria involving pairs of phases and also the limit-of-stability curve (spinodal curve) for the liquid–gas transition. The area labels show projections of regions of $u(s, v)$ where the specified phases are stable, and of the derived surfaces where mixtures of phases are stable. Drawing courtesy of Prof. J. A. Beattie.

As an example of the use of stability criteria in discussing phase transitions, consider the van der Waals equation of state,

$$p = \frac{RT}{v - b} - \frac{a}{v^2}, \tag{24.48}$$

the isotherms of which are shown in Fig. 24.17*a*. Clearly, the isotherms for $T < T_c$ have a region with $(\partial v/\partial p)_T > 0$. But the conditions for stability of a homogeneous phase discussed in Chapter 19 and summarized in Eqs. 24.47*a* and 24.47*b* show that unless $(\partial v/\partial p)_T < 0$ a homogeneous system is unstable. Consequently, the van der Waals gas cannot remain one homogeneous phase at all temperatures. Consider a typical isotherm along which condensation occurs. In Fig. 24.17*b* we plot the volume as a function of pressure along the isotherm, and the chemical potential along the same isotherm. From the relation

$$d\mu = -s\, dT + v\, dp, \tag{24.49}$$

we see that

$$\mu(p_2) - \mu(p_1) = \int_{p_1}^{p_2} v(p)\, dp \tag{24.50}$$

along the isotherm. In the unstable region between p_B and p_C, $v(p)$ is triple-valued. For every value of p in this region, two of the values of $v(p)$ lie on curves for which $(\partial v/\partial p)_T < 0$, whereas the third does not. The branches where a homogeneous system would be stable include regions where each is metastable with respect to formation of the other phase. The branches of $v(p)$ where the liquid is stable or metastable are DE' and $E'C$, respectively. The branches of $v(p)$ where the vapor is stable or metastable are AE and EB, respectively. The tie line EE' is the common tangent from the derived surface associated with $u(s, v)$ in the liquid–vapor equilibrium region. The prediction of a third value of $v(p)$ is a major failing of the van der Waals equation of state, hence of its underlying model of molecular interactions.

In the (p, v) plane we now have the following picture of the liquid–gas phase transition: At the pressure p_E the two phases have the same chemical potential. When the pressure is changed isothermally, the locus of equilibrium points lies on either the liquid branch DE' or the gas branch AE. If we start at A in the diagram and make an isothermal compression, the path followed is $AEE'D$, the system being entirely gaseous along AE, entirely liquid along $E'D$, and a mixture of gas and liquid along EE'. (Eqs. 24.2–24.4 give the relative fraction of gas and liquid along EE'.)

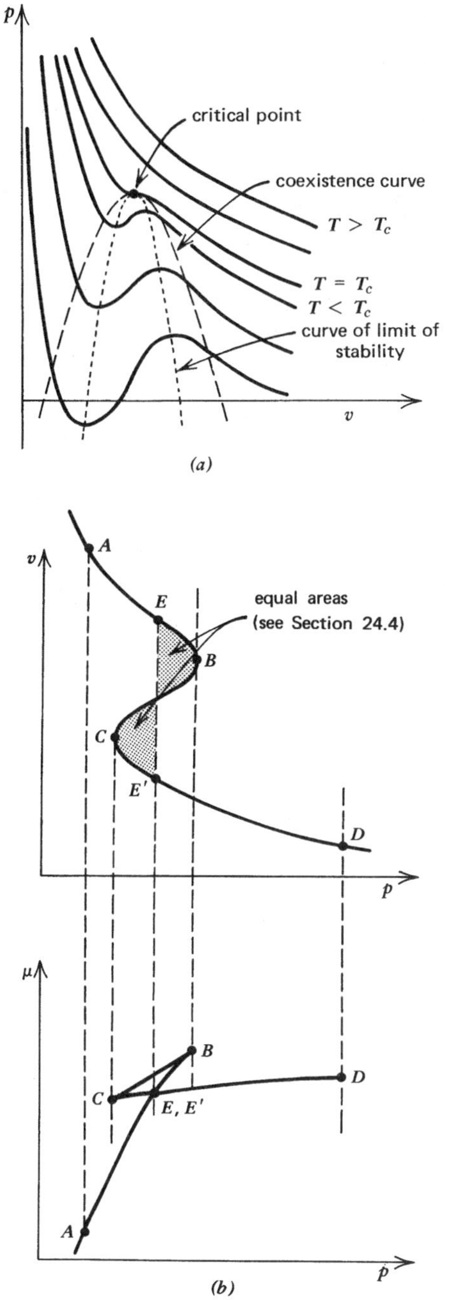

Figure 24.17 (*a*) Schematic representation of isotherms of the van der Waals equation of state. (*b*) Representation of the relation between v, p and μ, p plots for a van der Waals gas.

Thus far we have discussed the stability conditions with respect to conditions away from the critical point. At the critical point we have $(\partial p/\partial v)_T = 0$ and $(\partial^2 p/\partial v^2)_T = 0$, hence the condition $(\partial v/\partial p)_T < 0$ cannot be satisfied, and the stability criteria deduced in Chapter 19 are insufficient to describe the system. If one higher order in the expansion 19.45 is carried, it may be deduced that when $(\partial p/\partial v)_T = 0$, $(\partial^3 p/\partial v^3)_T < 0$ is a stability requirement.

At the critical point the isothermal compressibility becomes indefinitely large. This implies that an isothermal compression of the fluid requires vanishingly little work to be done, which further implies that vanishingly little work need be done to effect a compression or expansion along an arbitrary thermodynamic path. Imagine that the volume of a system at its critical point is divided into subvolumes that can exchange matter and energy with their surroundings. Because $(\partial p/\partial v)_T = 0$, we anticipate that fluctuations in the densities of the subvolumes are large. However, since the average density of the entire system is fixed by the amount of matter in the total volume, subvolumes with higher than average density must be compensated for by the existence of subvolumes with lower than average density. This observation implies that the fluctuations in the densities of the subvolumes cannot be independent, since the constraint that the average density of the total volume is fixed leads to a coupling of fluctuations. That the fluctuations in density at and near the critical point are very large is easily demonstrated experimentally. Corresponding to the fluctuations in density, there exist fluctuations in the fluid's refractive index that lead to scattering of a light beam passed through the fluid. The larger the fluctuations in refractive index, the more intense the scattering becomes. At the critical point the fluid scatters so much of the light that it appears milky and opaque. This phenomenon is known as *critical opalescence.*

Now, it is also true that $(\partial p/\partial v)_T = 0$ in the two-phase region of the liquid–gas transition. By analogy with the argument given above, should there not be large fluctuations in density in the two-phase region? The answer to this question is, basically, yes. In the two-phase region the density fluctuates from point to point just because the system is not homogeneous, and in moving from point to point it is possible to go into and out of each of the two phases present. Thus, along any path in the volume containing the system, the density does fluctuate between that characteristic of the liquid and that characteristic of the gas. Also, during the isothermal condensation process an increase or decrease in volume changes the relative proportions of gas and liquid, since the equilibrium pressure is fixed. In this case, because the pressure is fixed at p_σ, the work done is just $-p_\sigma \Delta v$, where Δv is the volume change. Of course, at the critical point we have

$$\lim_{p_\sigma \to p_c} p_\sigma \Delta v \to 0.$$

It should now be clear that the properties of a fluid at and near the critical point are very strange. Besides the divergences displayed in Eqs. 24.28, 24.31, and 24.33–24.37, it can be seen that, although the vapor pressure line in the (p, T) plane is the locus of points representing vapor–liquid equilibrium in a transition of the first kind, it terminates in a critical point that has some of the characteristics of a transition of the second kind ($\Delta s = 0$, $\Delta v = 0$). Also, the Clapeyron equation defines the slope $(dp/dT)_\sigma$ at all temperatures $T < T_c$, but the ratio $(\Delta s/\Delta v)$ becomes indeterminate when $T = T_c$. As usual, to evaluate $(\Delta s/\Delta v)$ in the limit that numerator and denominator both vanish, we use l'Hospital's rule:

$$\lim_{T \to T_c} \frac{\Delta s}{\Delta v} = \lim_{T \to T_c} \frac{(\partial \Delta s/\partial T)_p}{(\partial \Delta v/\partial T)_p} = \frac{1}{Tv}\left(\frac{c_{pG} - c_{pL}}{\alpha_G - \alpha_L}\right). \quad (24.51)$$

Experiment shows that $(dp/dT)_\sigma$ is finite up to and at the critical point, whereas from Eqs. 24.31 and 24.36 we see that each term in the numerator and denominator of Eq. 24.51 tends to infinity. It is possible that the differences $c_{pG} - c_{pL}$ and $\alpha_G - \alpha_L$ are finite at T_c even though the individual terms are not, but that is not presently known. Proceeding once more to use l'Hospital's rule on the right-hand side of Eq. 24.51, we find another ratio that is indeterminate, this time of the form 0/0. Further application of l'Hospital's rule leads to functions with unknown behavior. Presumably the very strange behavior of the thermodynamic functions at the critical point reflect the extraordinary circumstance that the critical point lies on both the coexistence curve *and* the curve of the limit of stability (spinodal curve), yet is a stable state of the system.

Despite the difficulties just sketched, there are three fruitful applications of thermodynamics to the description of critical phenomena. The first of these provides inequalities that the critical exponents $\alpha, \beta, \gamma, \delta$ of Eqs. 24.33–24.37 must satisfy, the second introduces a hypothesis which leads to the result that the inequalities between critical exponents are reduced to equalities, and the third provides an equation of state in the immediate vicinity of the critical point. We shall discuss an example of each type of analysis.

Consider the pV diagram shown in Fig. 24.18. The critical isotherm is $A'CA$, and BCB' is the liquid–vapor coexistence curve. Imagine that the system executes the cycle $CABDC$. Since no work is done along the legs AB and DC, and $\Delta U_{\text{cycle}} = 0$, we can write for an arbitrary mass of substance

$$q + w = 0 = T_c(S_A - S_C) - \int_C^A p\,dV + \int_{T_c}^{T_1} C_{VG}\,dT$$
$$+ T_1(S_D - S_B) - \int_B^D p\,dV$$
$$+ \int_{T_1}^{T_c} (C_V)_{V=V_c}\,dT, \tag{24.52}$$

where the two-phase heat capacity $(C_V)_{V=V_c}$ is evaluated along the path DC where $V = V_c$. Using the identity

$$S_A - S_C \equiv S_A - S_B + S_B - S_D + S_D - S_C,$$

and Eq. 24.9 in the form[7]

$$S_B - S_D = (V_G - V_c)\left(\frac{dp}{dT}\right)_{T=T_1},$$

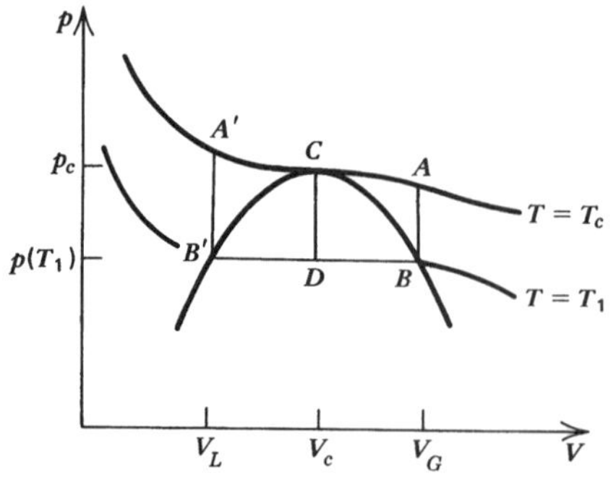

Figure 24.18 Schematic representation of the coexistence curve and isotherms near the critical point.

we can rewrite Eq. 24.52 as

$$\int_{T_1}^{T_c} \frac{T_c - T}{T}(C_V)_{V=V_c}\,dT$$
$$= \int_{T_1}^{T_c} \frac{T_c - T}{T} C_{VG}\,dT + \int_C^A (p_c - p)\,dV$$
$$+ (T_c - T_1)(V_G - V_c)\left[\left(\frac{dp}{dT}\right)_{T=T_1} - \frac{p_c - p(T_1)}{T_c - T_1}\right]. \tag{24.53}$$

By the same reasoning, when the system executes the cycle $A'CDB'A'$,

$$\int_{T_1}^{T_c} \frac{T_c - T}{T}(C_V)_{V=V_c}\,dT$$
$$= \int_{T_1}^{T_c} \frac{T_c - T}{T} C_{VL}\,dT + \int_{A'}^{C} (p - p_c)\,dV$$
$$- (T_c - T_1)(V_c - V_L)\left[\left(\frac{dp}{dT}\right)_{T=T_1} - \frac{p_c - p(T_1)}{T_c - T_1}\right]. \tag{24.54}$$

Now divide Eq. 24.53 by $V_G - V_c$ and Eq. 24.54 by $V_c - V_L$ and add. The result is

$$\left(\frac{1}{V_G - V_c} + \frac{1}{V_c - V_L}\right)\int_{T_1}^{T_c} \frac{T_c - T}{T}(C_V)_{V=V_c}\,dT$$
$$= \frac{1}{V_G - V_c}\int_{T_1}^{T_c} \frac{T_c - T}{T} C_{VG}\,dT + \frac{1}{V_c - V_L}$$
$$\times \int_{T_1}^{T_c} \frac{T_c - T}{T} C_{VL}\,dT + \frac{1}{V_G - V_c}\int_{V_c}^{V_G} (p_c - p)_{T_c}\,dV$$
$$+ \frac{1}{V_c - V_L}\int_{V_L}^{V_c} (p - p_c)_{T_c}\,dV. \tag{24.55}$$

The integral on the left-hand side of Eq. 24.55 follows the path DC. Along this path, $x_L = (V_G - V_c)/(V_G - V_L)$ and $x_G = (V_c - V_L)/(V_G - V_L)$ both vary. In contrast, on the right-hand side of Eq. 24.55 the integrals with respect to temperature

[7] Since $S_D = x_L S_{B'} + x_G S_B$ and $V_c = x_L V_L + x_G V_G$, we have

$$\frac{S_B - S_D}{V_G - V_c} = \frac{(S_B - S_{B'})x_L}{(V_G - V_L)x_L} = \left(\frac{dp}{dT}\right)_{T=T_1}.$$

follow the paths BA and BA', along which V_L and V_G are fixed. We can then rewrite Eq. 24.55 in an illustrative form by explicitly displaying this difference:

$$\int_{T_1}^{T_c} \frac{T_c - T}{T}(C_V)_{V=V_c}\, dT = \int_{T_1}^{T_c} \frac{T_c - T}{T}[x_L(T_1)C_{VL} + x_G(T_1)C_{VG}]\, dT + x_G(T_1)\int_{V_c}^{V_G}(p_c - p)_{T_c}\, dV + x_L(T_1) \times \int_{V_L}^{V_c}(p - p_c)_{T_c}\, dV, \tag{24.56}$$

where $x_G(T_1) = [V_c - V_L(T_1)]/[V_G(T_1) - V_L(T_1)]$ and $x_L(T_1) = [V_G(T_1) - V_c]/[V_G(T_1) - V_L(T_1)]$ are considered fixed values. In the integral on the left-hand side of Eq. 24.56, the amounts of vapor and liquid adjust appropriately as T is varied; but in the integral on the right-hand side of Eq. 24.56, no adjustment is permitted, just as if phase equilibrium were suppressed.

We now note that the first two terms on the right-hand side of Eq. 24.55 are always positive since $T_c > T_1$, $V_G > V_c$, $V_c > V_L$, $C_{VG} > 0$, $C_{VL} > 0$ in the integrals considered. Therefore, removing this positive term from the right-hand side of Eq. 24.55 leaves the inequality

$$\left(\frac{1}{V_G - V_c} + \frac{1}{V_c - V_L}\right)\int_{T_1}^{T_c} \frac{T_c - T}{T}(C_V)_{V=V_c}\, dT > \frac{1}{V_G - V_c}\int_{V_c}^{V_G}(p_c - p)_{T_c}\, dV + \frac{1}{V_c - V_L}\int_{V_L}^{V_c}(p - p_c)_{T_c}\, dV. \tag{24.57}$$

If we assume that[8]

$$|V - V_c| = B(T_c - T)^{\beta}, \tag{24.58}$$

$$p - p_c = -D|V - V_c|^{\delta - 1}(V - V_c), \tag{24.59}$$

$$(C_V)_{V=V_c} = A(T_c - T)^{-\alpha} \tag{24.60}$$

for $T < T_c$ near the critical temperature, we find from Eq. 24.57 that

$$\alpha + \beta(\delta + 1) \geq 2, \tag{24.61}$$

known as *Griffith's inequality.*

Using other thermodynamic arguments it is also possible to show that

$$\alpha + 2\beta + \gamma \geq 2, \tag{24.62}$$

$$\gamma(\delta + 1) \geq (2 - \alpha)(\delta - 1), \tag{24.63}$$

$$\gamma \geq \beta(\delta - 1). \tag{24.64}$$

One of the consequences of the theory that generates the values of the critical exponents α, β, γ, δ displayed in Eqs. 24.33–24.37 is that the thermodynamic inequalities Eqs. 24.61–24.64 become equalities. Indeed, using the numerical values displayed in Eqs. 24.33–24.37, we find

$$\begin{aligned}
&(24.61):\quad \alpha + \beta(\delta + 1) = 2.064 \pm 0.009 = 2,\\
&(24.62):\quad \alpha + 2\beta + \gamma = 2.001 \pm 0.004 = 2,\\
&(24.63):\quad \gamma(\delta + 1) = 7.186 \pm 0.025\\
&\qquad\qquad = (2 - \alpha)(\delta - 1) = 7.182 \pm 0.038,\\
&(24.64):\quad \gamma = 1.239 \pm 0.002\\
&\qquad\qquad = \beta(\delta - 1) = 1.239 \pm 0.007,
\end{aligned}$$

as required by the theory.

A discussion of the theoretical basis for the calculation of critical exponents, the so-called renormalization group theory is out of place in this text, although we do sketch the principal ideas used in that theory in Section 24.4. It is possible, without using renormalization group theory, to see which properties of the equation of state are sufficient to make Eqs. 24.61–24.64 hold with the equality sign by following an argument due to Widom.[9] The key elements in this argument are (1) that the thermodynamic inequalities Eqs. 24.61–24.64 constrain the Helmholtz free energy of the system to be a function of only two variables, and (2) that the Helmholtz free energy can be approximated by a function of one reduced variable.

Let $M(T)$ be the chemical potential along the critical isochore $\rho = \rho_c$. At the critical point, the chemical potential satisfies the conditions[10]

$$\left(\frac{\partial \mu}{\partial \rho}\right)_{T=T_c} = 0, \tag{24.65}$$

$$\left(\frac{\partial^2 \mu}{\partial \rho^2}\right)_{T=T_c} = 0. \tag{24.66}$$

[8] Since we consider only the region very close to the critical point, Eqs. 24.58–24.60 are equivalent to Eqs. 24.33–24.37. For example, $(\rho_L - \rho_G) = \upsilon_L^{-1} - \upsilon_G^{-1} = (\upsilon_G - \upsilon_L)(\upsilon_G \upsilon_L)^{-1}$. Clearly, the variation is all in the difference $\upsilon_G - \upsilon_L$, as $\upsilon_G \upsilon_L = \upsilon_c^2$ to very high precision close to υ_c. Different investigators use Eqs. 24.58–24.60 or 24.33–24.37 as convenient for the analysis in question.

[9] B. Widom, *J. Chem. Phys.* **43**, 3898 (1965).

[10] Starting from

$$d\mu = -s\, dT + \upsilon\, dp = -s\, dT + \rho^{-1}\, dp,$$

we easily find

$$\left(\frac{\partial \mu}{\partial \rho}\right)_T = \frac{1}{\rho}\left(\frac{\partial p}{\partial \rho}\right)_T = -\upsilon^3\left(\frac{\partial p}{\partial \upsilon}\right)_T,$$

and by definition $(\partial p/\partial \upsilon)_{T=T_c} = 0$. A similar argument establishes the validity of Eq. 24.66 at $T = T_c$.

Let the difference in value between $\mu(\rho, T)$ and its value on the critical isochore, $M(T)$, be expressed as a Taylor series expansion in ρ. By virtue of Eqs. 24.65 and 24.66, the first two terms in this series vanish. If $T = \tau(\rho)$ is the equation of the vapor–liquid coexistence curve in the (T, ρ) plane, the first nonvanishing term in the series is

$$\mu(\rho,T)-M(T)=(\rho-\rho_c)[T-\tau(\rho)]\Phi, \tag{24.67}$$

because $(\partial^2\mu/\partial\rho\,\partial T) \neq 0$. The van der Waals equation, and the others described in Chapter 21 (Table 21.1), require that Φ be a constant at the critical point. None of these predicts the correct values of α, β, γ, and δ. That Φ is a constant at the critical point is not, however, a necessary condition on the equation of state. In view of the failure of the classical equations of state in the description of the critical region, B. Widom suggested that the expansion 24.67 still be used to describe that region, but with Φ a function of density and temperature. That function is not known, which is another way of saying that the equation of state in the critical region is not known. But *if Φ is assumed to be a homogeneous function of its variables,*[11] the equation of state can be shown to satisfy Eqs. 24.61–24.64 with the equality signs. This is such a remarkable prediction, and it agrees so well with otherwise puzzling observations, that we give one example of the form of analysis used.

Let

$$z=T-T_c, \tag{24.68}$$

$$y=T_c-\tau(\rho), \tag{24.69}$$

$$\Phi=\Phi(z,y), \tag{24.70}$$

so that Eq. 24.67 becomes

$$\mu(\rho,T)-M(T)=(\rho-\rho_c)(z+y)\Phi(z,y). \tag{24.71}$$

If $\Phi(z, y)$ is a homogeneous function of degree $\gamma - 1$, then

$$\Phi(z,y)=y^{\gamma-1}\Phi\left(\frac{z}{y},1\right) \tag{24.72}$$

$$=z^{\gamma-1}\Phi\left(1,\frac{y}{z}\right),\quad \text{if } z>0. \tag{24.73}$$

$$=(-z)^{\gamma-1}\,\Phi\left(-1,-\frac{y}{z}\right)\quad \text{if } z>0. \tag{24.74}$$

Equation 24.71 determines the chemical potential in the one-phase region, $T > T_c$. From the assumed form

$$T_c-\tau(\rho)=b|\rho-\rho_c|^{1/\beta} \tag{24.75}$$

(see Eqs. 24.33 and 24.71), it follows that:

1. On the critical isochore, where y = 0 and z > 0,

$$\left(\frac{\partial\mu}{\partial\rho}\right)_T=\Phi(1,0)z^{\gamma}. \tag{24.76}$$

2. On the critical isotherm, where $z = 0$,

$$\left(\frac{\partial\mu}{\partial\rho}\right)_T=\left(1+\frac{\gamma}{\beta}\right)\Phi(0,1)y^{\gamma}. \tag{24.77}$$

3. On the coexistence curve, where $z + y = 0$,

$$\left(\frac{\partial\mu}{\partial\rho}\right)_T=\left(\frac{1}{\beta}\right)\Phi(-1,1)y^{\gamma}$$

$$=\left(\frac{1}{\beta}\right)\Phi(-1,1)(-z)^{\gamma}. \tag{24.78}$$

Equations 24.76 and 24.78 show that the compressibility becomes infinite proportional to $|T-T_c|^{-\gamma}$ as the critical point is approached along either the critical isochore or the coexistence curve. The degree-of-homogeneity index γ can, therefore, be identified with the critical index γ introduced in Eq. 24.37. The same exponent γ appears in Eqs. 24.76 and 24.78 because of the assumed homogeneity of the function $\Phi(z, y)$.

Using Eq. 24.72 in Eq. 24.71 gives for $\mu - M(T)$ along the critical isotherm [where $z = 0$ and $M(T_c) \equiv \mu(\rho_c, T_c)$],

$$\mu-\mu(\rho_c,T_c)=\Phi(0,1)(\rho-\rho_c)y^{\gamma}, \tag{24.79}$$

which is just the integrated form of Eq. 24.77. But the coexistence curve itself is of algebraic degree $1/\beta$ (see Eq. 24.75), so that the algebraic degree of the critical isotherm must be[12]

$$\delta=1+\frac{\gamma}{\beta} \tag{24.80}$$

or

$$\gamma=\beta(\delta-1), \tag{24.81}$$

[11] The properties of generalized homogeneous functions are described in Appendix 24A, along with the consequences for the calculation of critical exponents of assuming that the singular part of the Gibbs free energy is a generalized homogeneous function of $(T-T_c)/T$ and p. The use of Eq. 24.67 and Widom's hypothesis concerning Φ separates the chemical potential into an analytic part that describes "background" behavior and a singular part, which is Φ.

[12] The right-hand side of Eq. 24.79 is proportional to $(\rho-\rho_c)^{1+1/\beta}$. Integration of the Gibbs–Duhem equation and use of Eq. 24.59 shows that $\mu-\mu_c \propto (\rho-\rho_c)^{\delta}$, from which Eq. 24.80 follows.

which is just Eq. 24.64 *with an equality sign.* By similar methods of analysis Widom finds the equality sign to hold in the rest of the expressions previously displayed.

The assumption that $\Phi(z, y)$ is a homogeneous function of its variables provides a powerful tool for thermodynamic analysis. We now assume that $\Phi(z, y)$ is a universal function—that is, the same for all substances. A test of this assumption is shown in Fig. 24.19. The test is carried out by rewriting Eq. 24.67 as

$$\Delta\mu(T) = \mu(\rho, T) = M(T)$$

$$= \Delta\rho(\Delta T + b|\Delta\rho|^{1/\beta})|\Delta\rho|^{(\gamma-1)/\beta}\,\Phi\left(\frac{\Delta T}{|\Delta\rho|^{1/\beta}}, 1\right)$$

$$= \Delta\rho|\Delta\rho|^{\gamma/\beta}\,\Psi\left(\frac{\Delta T}{|\Delta\rho|^{1/\beta}}\right), \qquad (24.82)$$

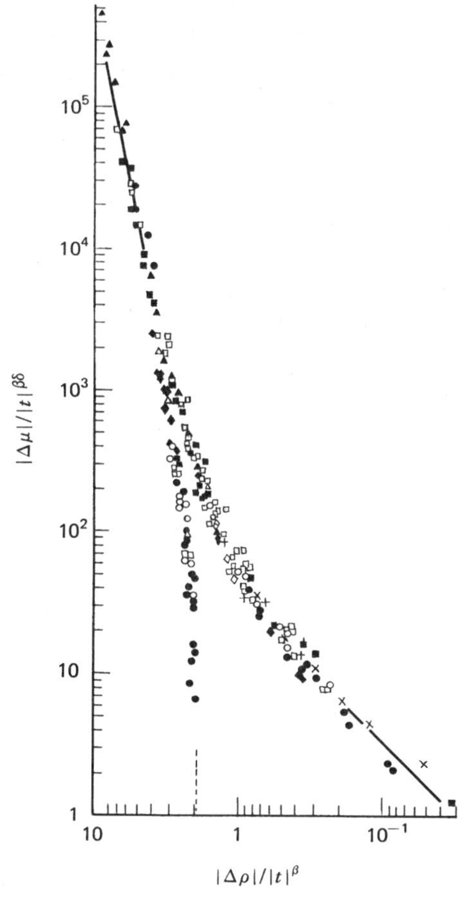

Figure 24.19 A scaled plot of $|\Delta\mu|/|t|^{\beta\delta}$ versus $|\Delta\rho|/|t|^{\beta}$ in the critical regions of N_2O, $CClF_3$, CO_2, SF_6, and Xe. The quantity t is defined as $\Delta T/T_c$. From M. S. Green, M. Vincenti-Missioni, and J. M. H. Levett-Sengers, *Phys. Rev. Lett.* **18,** 1113 (1967).

using

$$z + y = T - \tau(\rho) = \Delta T + b|\Delta\rho|^{1/\beta} \qquad (24.83)$$

from Eqs. 24.68 and 24.69, and

$$\Phi(\Delta T, b|\Delta\rho|^{1/\beta}) = |\Delta\rho|^{(z-1)/\beta}\,\Phi(\Delta T/b|\Delta\rho|^{1/\beta}, 1)$$

from Eq. 24.72. In this form $\Delta\mu(T)$ is expressed as a function Ψ of the single variable $\Delta T/|\Delta\rho|^{1/\beta}$. Using $\delta = 1 + \gamma/\beta$ and dividing leads to

$$\frac{|\Delta\mu(T)|}{|\Delta T|^{\beta\delta}} = \left(\frac{|\Delta\rho|^{1/\beta}}{|\Delta T|}\right)\Psi\left(\frac{\Delta T}{|\Delta\rho|^{1/\beta}}\right), \qquad (24.84)$$

which predicts that $|\Delta\mu(T)|/|\Delta T|^{\beta\delta}$ is a universal function of $\Delta T/|\Delta\rho|^{1/\beta}$. The function has two branches, for $\Delta T > 0$ and $\Delta T < 0$, which become asymptotically identical as $\Delta T \to 0$.

It is seen, from the display in Fig. 24.19, that the experimental data available are in agreement with Widom's hypotheses concerning $\Phi(z, y)$. It must be noted that Eq. 24.84 is accurate only in the immediate vicinity of the critical point. As the distance from the critical point increases, the behavior of the system, for example, the dependence of $V - V_c$ on $T_c - T$, increasingly approaches that predicted by the classical equations of state.

As already noted, the principle of critical point universality is a consequence of (1) the existence of long-range fluctuations in a system near a critical point, and (2) the fact that the range of these fluctuations is much larger than any other microscopic scale length of the system. From the arguments given above, the applicability of this principle appears, at first sight, to be restricted to the immediate vicinity of the critical point. However, the range of the fluctations in a system remains substantially greater than the range of molecular interactions for a considerable region of temperature and density away from the critical point, essentially for all temperatures and densities where any enhancement of the compressibility (beyond that expected from extrapolation of data from the normal fluid range) is observed. This observation has been used as the basis for extending the principle of critical point universality to develop a representation of the equation of state for the entire region in which there is an enhanced compressibility and to characterize how the thermodynamic properties of a fluid cross over from singular nonclassical critical behavior to classical behavior outside the critical region. A detailed discussion of this crossover behavior can be found in the literature.[13] Fig. 24.20 illustrates the fit between theory and experimental data for the constant volume heat capacity of ethane along the critical isochore in a considerable temperature domain centered on the critical temperature.

---

[13] See M. A. Anisimov, S. B. Kislev, J. V. Sengers, and S. Tang, *Physica* **A188,** 487 (1992).

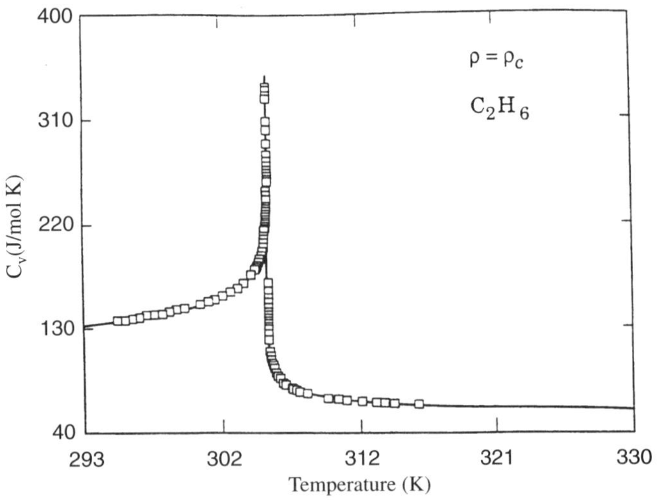

Figure 24.20 Isochoric heat capacity of ethane at the critical density as a function of temperature. The squares identify the experimental data points, and the solid line represents the behavior expected from the thermodynamic theory sketched in the text. From Z. Y. Chen, A. Abbaci, S. Tang, and J. V. Sengers, *Phys. Rev.* **A42,** 4470 (1990).

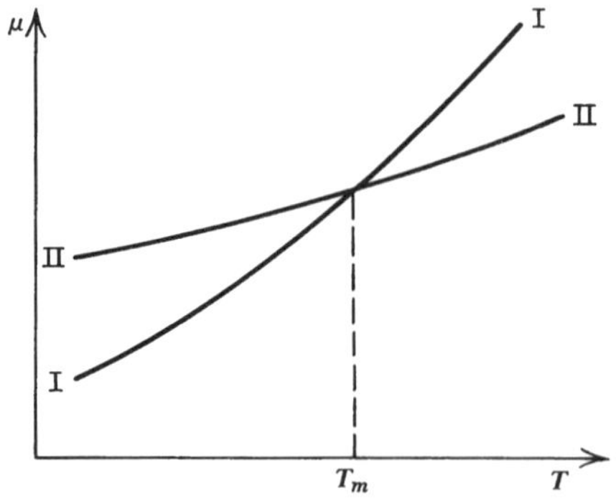

Figure 24.21 Schematic representation of the location of a transition point by comparing the chemical potentials for two models, one for each of the phases. When phases I and II are the solid and liquid, respectively, T_m is the melting point.

24.4 The Statistical Molecular Description of Phase Transitions

We have already remarked that the existence of a phase transition signals the instability of one phase relative to some other phase. In this section we examine how the characteristic features of a phase transition can be related to molecular properties. The molecular theory of phase transitions is complex and only partially developed, so we shall be content to examine the key ideas and some qualitative indications of their implications.

There are three paths that can be followed in constructing a molecular theory of phase transitions. The simplest of these uses directly the thermodynamic condition defining the equilibrium state under a given set of constraints, for example, that the Gibbs free energy must be a minimum for fixed temperature and pressure. Suppose that one molecular model is proposed for phase I, say, the crystal phase, and a different model for phase II, say, the liquid phase. It is, in principle, possible to calculate the Gibbs free energies of phases I and II assuming that they are isolated from one another and also that they are stable over the range of temperature and pressure of interest. The techniques employed in these calculations are just those sketched in Chapters 22 and 23.

Suppose that the pressure is assumed constant and the Gibbs free energies of phases I and II are computed as functions of the temperature. A schematic plot of the results is displayed in Fig. 24.21. As shown in Fig. 24.21, the crystal phase (I) has lower chemical potential for $T < T_m$, whereas the liquid phase (II) has lower chemical potential for $T > T_m$. The temperature T_m can then be identified as the melting point predicted by the statistical molecular model considered.

Models of the type just described have frequently been used to describe phase transitions. However, they are deficient because they do not describe the phase transition itself at all. What the model does describe are the sets of real and hypothetical stable and metastable states of the separate phases. A purely thermodynamic criterion is then used to decide which of the phases is the more stable. No information is obtained about the special character of the distribution of molecules near the transition point, about the presence or absence of special kinds of molecular ordering, or about the "origin" of the phase transition.

An alternative approach is based on the postulation of very simple models that retain only the skeleton of the real properties of a system. These models are specifically designed so that an exact mathematical analysis can be carried to completion. It is then hoped that the general features of the phase transition deduced will be independent of the simplifications of the basic model, indeed, that these features will be universal in the sense of being independent of the nature of the intermolecular potential. Actually, the mathematical description of the statistical mechanics of phase transitions is so difficult that there are very few models for which exact solutions have been obtained. Nevertheless, the results obtained are of great importance, because they show how delicate is the balance between competing factors in the critical region, and also because they are often indicative of behavior not predicted by approximate calculations for more realistic models. Consider, for example, the model known as the *lattice gas.* Imagine that there are $\mathcal{N}$ lattice sites in some regular crystalline array, and that N of these are occupied by particles and $\mathcal{N} - N$ are vacant. Only one particle can be at a given lattice site. Finally, if two particles occupy near-neighbor lattice sites, there is an attractive interaction of magnitude w, whereas particles farther apart do not interact. It is clear that the lattice gas model has some of the features of real matter:

1. The repulsive interaction between a pair of molecules is mimicked by the prohibition of joint occupation of one lattice site by two particles.
2. The attractive interaction between a pair of molecules is mimicked by the attraction between near neighbors, with the range of the potential such that there is zero interaction between pairs separated by more than a near-neighbor distance.
3. Changes in density are represented by changes in the number of vacant lattice sites, $\mathcal{N} - N$. When the density of particles is high, $\mathcal{N} - N$ is small relative to $\mathcal{N}$, and the lattice gas mimics the liquid. When the density of particles is low, $\mathcal{N} - N$ is almost as large as $\mathcal{N}$, and the lattice gas mimics the dilute gas. The major difference between the lattice gas and real liquids or gases, and the principal source of simplification in the mathematical analysis, is the replacement of the continuous space of real systems with the discrete space of a lattice. That is, in real space molecules may be any distance apart, whereas in the lattice gas their separation is determined by the lattice spacing and lattice structure. Clearly, although the lattice gas has some skeletal features in common with real fluids, it also has some bizarre characteristics not found in any real system.

From our point of view the important characteristic of the lattice gas is that, if the interaction between near-neighbor particles is attractive, there is a temperature below which the system separates into two phases, one with large density and the other with small density. The model thereby mimics a two-phase liquid–gas system with a critical temperature. If the near-neighbor interaction is repulsive, particles will tend to occupy alternate lattice sites at low temperature, but without splitting into two phases. In this case there can still be a critical temperature.

With great insight and mathematical dexterity, L. Onsager has evaluated exactly the partition function for a *two-dimensional* infinite square-lattice gas. He obtained the following results:

1. For a half-filled lattice, with an attractive interaction between near neighbors, the system splits into two phases below a critical temperature T_c.
2. The internal energy is continuous at the critical point, but has an infinite slope there. The heat capacity near the critical point diverges at a rate proportional to $\ln |(T - T_c)/T_c|$.
3. Let ρ_G be the density of the gas phase and ρ_L the density of the liquid phase along the coexistence line between liquid and gas. Then for the two-dimensional lattice gas model one has

$$(\rho_L - \rho_G) \propto (T - T_c)^{1/8} \tag{24.85}$$

near the critical point. Also,

$$(p - p_c) \propto (\rho_L - \rho_G)^8 \tag{24.86}$$

and $(\partial p/\partial T)_{V=V_c}$ is continuous at the critical point.

For the moment we merely call the reader's attention to the form of the approach to the critical state. A discussion of the nature of the results is deferred until we have examined the third approach to the study of phase transitions.

This approach also makes use of simple models of the substances of interest. In this case, however, the models are designed to incorporate as many of the real-system properties as is possible without overly complicating the necessary mathematical analysis. There is a fine line dividing models that have qualitative features in common with those of real systems and for which the statistical mechanics *cannot,* at present, be done exactly, from models that are so simple that the corresponding statistical mechanics *can* be done exactly. Since much of our qualitative understanding of phase transitions (but not of critical phenomena) comes from studies of models of the former class, we consider a few examples.

We examine first crystal–crystal transitions. The majority of crystal–crystal transitions are first order, but there also are continuous (second order) transitions which involve a change in symmetry at the transition point. We focus attention on this latter class of transitions, in which the structural change is associated with an instability in some lattice vibration. To be more specific, the frequency of the mode corresponding to the displacement that generates the structural change decreases substantially as the transition temperature is approached from below. This "soft mode" generates a mechanical instability in the sense that displacement of the original atomic configuration toward the new atomic configuration is resisted less and less as the temperature approaches the transition point. Our description of the transition must include this behavior.

Clearly, what is needed is an expression for the thermodynamic properties of the crystal as a function of the displacement that generates the transition. Landau suggested that we find this as follows. Let $\rho(\mathbf{r})$ represent the atomic distribution in the crystal phase with lower symmetry and $\rho_0(\mathbf{r})$ the atomic distribution in the crystal phase with higher symmetry. Then, writing

$$\rho(\mathbf{r}) = \rho_0(\mathbf{r}) + \Delta\rho(\mathbf{r}), \tag{24.87}$$

we have

$$\Delta\rho(\mathbf{r}) = \sum_i c_i \xi_i, \tag{24.88}$$

where the ξ_i are normal coordinates of the high-symmetry lattice. Let η be a normalized linear combination of the coefficients c_i; the normalization is chosen such that η vanishes in the phase with higher symmetry. For example, in one class of transitions η represents the amplitude of the *mean* displacements of the atoms from their high-symmetry lattice positions, which necessarily vanishes in the high-symmetry phase. It is important to note that the symmetry of the crystal changes only when $\eta = 0$; any nonzero value of η generates the same symmetry as for $\eta = 1$. It is only when η vanishes continuously at $T = T_c$ that the transition is second order; if η changes discontinuously at $T = T_c$, the transition is first order. Since the two phases become identical at a second order transition point, the symmetry of the crystal at the transition point must include all the symmetry elements of both high-temperature and low-temperature crystal structures. In the case under consideration, η describes the amplitude of the mean displacements of the atoms from their high-symmetry positions. We assume that the Gibbs free energy per mole of the crystal can be represented as a Taylor series expansion in powers of η:

$$G = G_0 + \alpha\eta + A\eta^2 + B\eta^3 + C\eta^4 + \cdots, \qquad (24.89)$$

where $A, B, C \ldots$ *are functions of* p and T.

Before proceeding it is important to make a few remarks concerning the expansion displayed in Eq. 24.89. Given the character of a second order transition, wherein two phases become identical at the transition point, we must expect the system to exhibit singular behavior of some sort at the transition point. In a first order transition between phases with different symmetry—for example, the melting of a crystal—the chemical potentials of the two phases are equal at the transition point, but at that temperature and pressure the properties of the individual phases are ordinary. Indeed, the chemical potentials of each phase can be extended past the transition point into the regions where the respective phases are metastable. In a second order transition extrapolation of the chemical potentials of each phase past the transition point is not possible, since the Gibbs free energy does not have even a local minimum in that domain. A first order transition between phases with the same symmetry ends at a critical point at which the two phases become identical, similar to the behavior exhibited in a second order transition. As discussed earlier in this chapter, in the vicinity of the critical point that terminates a first order transition, the thermodynamic properties are singular. This observation implies that the expansion displayed in Eq. 24.89 cannot be carried out to arbitrary order and that the expansion coefficients may have singularities as functions of p and T.

Since for fixed p and T the Gibbs free energy is a minimum, we impose the conditions

$$\left(\frac{\partial G}{\partial \eta}\right)_{T,p} = 0, \qquad (24.90)$$

$$\left(\frac{\partial^2 G}{\partial \eta^2}\right)_{T,p} > 0, \qquad (24.91)$$

and immediately find

$$\alpha = 0, \qquad (24.92)$$

so that there is no term linear in η in the expansion of G about G_0. Since the transition occurs at $T = T_c$, and η represents the amplitude of the *mean* displacements of the atoms from their high-symmetry lattice positions, we must have $\eta = 0$ in the high-symmetry phase, say, for $T > T_c$. The behavior of the coefficient A follows from the observations that G must be a minimum both above and below the transition temperature T_c. This requirement implies that $A > 0$ in the phase with $\eta = 0$, $A = 0$ at $T = T_c$, and $A < 0$ in the phase with $\eta \neq 0$:

$$\begin{aligned} A > 0; &\quad T > T_c, \\ A = 0; &\quad T = T_c, \\ A < 0; &\quad T < T_c. \end{aligned} \qquad (24.93)$$

Evaluation of B and C requires another consideration. Starting at some $T > T_c$, $\eta = 0$ and remains zero until $T = T_c$ and then continuously increases as T decreases in the region $T < T_c$. At the transition point, $G(p, T, \eta)$ must be a minimum with respect to variation of η, which implies that (since $A(T_c) = 0$, Eq. 24.93),

$$\begin{aligned} B &= 0, \\ C &> 0. \end{aligned} \qquad (24.94)$$

We consider only the case that $B \equiv 0$ by virtue of the symmetry properties of the crystal. Then from

$$\left(\frac{\partial G}{\partial \eta}\right)_{T,p} = 0 = 2A\eta + 4C\eta^3 \qquad (24.95)$$

we have

$$\eta^2 = -A/2C \qquad (T < T_c) \qquad (24.96)$$

and, of course, $\eta = 0$ if both $A > 0$ and $C > 0$. This is as far as a general macroscopic analysis of the crystal–crystal transition can carry us. To proceed further we must either introduce some hypothesis concerning the pressure and temperature dependencies of the coefficients A and C, or calculate their properties from a statistical molecular theory. We consider just the major ideas from the latter.

As we remarked in Chapter 22, a perfectly harmonic lattice has zero thermal expansion or, put another way, on average the atoms execute thermally excited motion centered on the same lattice sites for all T. Thermal expansion, that is, the change of lattice spacing with change in temperature, arises from anharmonicity of the potential surface. We also saw in Chapter 22 that a simple picture of the effect of thermal expansion could be constructed on the basis that at each T the motion of the atoms was harmonic, but that the centers about which oscillation took place changed as T changed. Then the existence of a crystal–crystal phase transition, which can be characterized as involving a change in positions of atoms, necessarily implies that the potential energy surface is anharmonic with respect to displacements about either set of the crystal lattice sites. Although complete solution of the dynamics of coupled anharmonic oscillators is not possible at present, we can construct a physically plausible picture of the main features to be expected from such an analysis. Suppose that the potential energy is expanded as a power series in the displacement of an atom from a lattice site, but retaining terms that are cubic, quartic, etc., in the displacement. Suppose further that we seek a way of representing the averaged effects of the anharmonic terms in the potential energy, much in the spirit of the analysis of Section 23.3. One such approximate representation is based on the notion that a principal effect of the anharmonicity is to reduce the restoring force returning a displaced atom to its lattice site. To make this clear, consider a one-dimensional example. Suppose that the potential energy of a one-dimensional anharmonic oscillator is $V_2x^2 - V_4x^4$. We represent the average effect of the V_4x^4 term by making the approximation $x^4 \to \langle x^2\rangle x^2$, where the thermally averaged mean-square displacement is evaluated by, say, perturbation theory. In first-order perturbation theory $\langle x^2\rangle$ is determined by weighting x^2 with the distribution of x for a harmonic oscillator; the result is $\langle x^2\rangle = k_BT/V_2$. Substituting this value into the potential energy expression for the anharmonic oscillator gives

$$\begin{aligned} V_2x^2 - V_4x^4 &\approx (V_2 - V_4\langle x^2\rangle)x^2 \\ &\approx \left(V_2 - \frac{k_BTV_4}{V_2}\right)x^2, \end{aligned} \tag{24.97}$$

from which one derives an effective temperature-dependent frequency

$$\omega \propto \left(1 - \frac{k_BTV_4}{V_2^2}\right)^{1/2} V_2^{1/2} \propto \omega_0(1 - aT)^{1/2}, \tag{24.98}$$

which decreases as T increases. This soft mode will, eventually, generate the instability that leads to the phase transition.

In a real crystal the atomic dynamics is much more complicated than the one-dimensional model mentioned. Nevertheless, when the anharmonic contributions to the potential energy are partially averaged, the main result is the same, namely, the prediction of the existence of a soft mode. The precise form of the temperature dependence is related to the nature of correlations near the transition point. If the interactions between particles are very long-range, then $\omega \propto (T_c - T)^{1/2}$, but if they are short-range, then $\omega \propto (T_c - T)^{\beta}$ with $\beta = \frac{1}{3}$. This difference in behavior is similar to that observed for various theoretical models of the gas–liquid critical point, and we shall see later that it has the same origin.

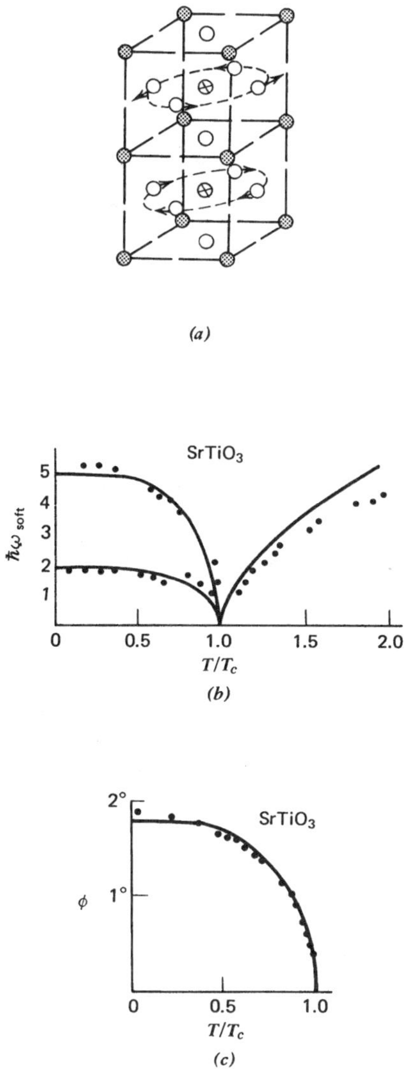

Figure 24.22 (*a*) Soft-mode motion in $SrTiO_3$ transition. (*b*) Soft-mode frequencies as a function of T/T_c both above and below T_c. Note that there are two soft modes in the low-temperature phase. (*c*) Angle of rotation, ϕ, of oxygen octahedron in $SrTiO_3$ transition. From J. F. Scott, *Rev. Mod. Phys.* **46,** 83 (1974).

Does the description just given agree with the experiment? We have already mentioned the existence of second-order transitions in the perovskite structures. Figure 24.22*a* shows the soft mode in $SrTiO_3$, Fig. 24.22*b* the continuous decrease of the vibrational frequency as T_c is approached, and Fig. 24.22*c* the corresponding continuous displacement of the oxygen octahedron. In each case the line represents the predicted behavior. Clearly, there is generally good agreement between theory and observation.

Actually, the behavior observed and to be expected is even richer than so far discussed. The theoretical description sketched is based on the idea that anharmonicity leads to the softening of one or more vibrational modes and eventually to an instability that generates a phase transition. In this analysis the atomic motion is progressively influenced by the

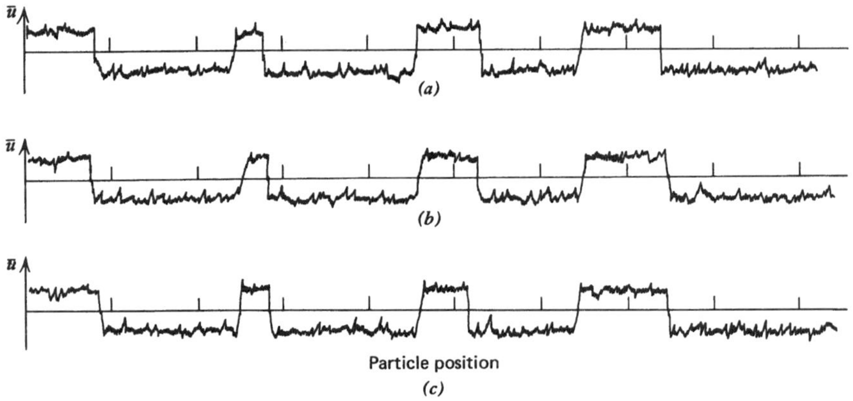

Figure 24.23 Particle displacements along a chain of local double-well potentials. Nearest-neighbor coupling only, Hamiltonian

$$H=\sum_i \frac{p_i^2}{2m}+\sum_i\left[\frac{A}{2}u_i^2+\frac{B}{4}u_i^4+\frac{1}{2}(u_{i+1}-u_i)^2\right].$$

The horizontal axis measures position along the chain (1000 particles long), and the vertical axis the displacement in units of $(|A|/B)^{1/2}$, that is, $\bar{u}=u/u_0$. Curves (*a*), (*b*), and (*c*) are for times $\bar{t}\equiv(|A|/m)^{1/2}$ of 0, 100, and 200 units. The temperature is $\bar{T}=k_BTB/A^2=0.117$. The mean energy per particle is less than the barrier height. From T. R. Koehler, A. R. Bishop, J. A. Krumhansl, and J. R. Schrieffer, *Solid State Comm.* **17,** 1515 (1975).

anharmonicity as T changes. There is theoretical and computer simulation evidence that when the anharmonicity is very large, as when there is a double-well potential for an atom, there are two classes of elementary excitations (see Fig. 24.23). One, already discussed, is quasi-harmonic motion influenced, on average, by anharmonic coupling. The energy required to create one of these excitations is spread throughout the lattice. The other corresponds to a localized large-amplitude displacement, and the energy to create it is relatively localized in the lattice. This large-amplitude displacement can move through the lattice, and has been taken to be representative of "domain walls" that bound clusters of atoms with properties different from their surroundings, for example, a group of atoms in a phase-II configuration surrounded by atoms in a phase-I configuration. We expect such traveling cluster waves to exist in measurable concentration only when their energy of formation is not very large compared to k_BT. When k_BT is greater than the energy of domain wall formation, we expect thermal fluctuations to destroy them. Both of these expectations are supported by the computer simulations shown in Figs. 24.23 and 24.24. The traveling cluster waves described represent an entirely different class of excitations than do the ordinary vibrational excitations—they arise from the strong nonlinearity (strong anharmonicity) of the interaction. Their existence does not modify those averaged structural or dynamical aspects of the crystal–crystal transition discussed, but does explain other phenomena and, presumably, influences the mechanism of the crystal–crystal transition.

The preceding discussion has concerned the dynamics of the crystal–crystal transition and relied on information about the two structures involved, presumably obtained from experimental data. How can we understand, in terms of the molecular interactions, the structures of different crystalline phases of the same substance? The general theory of crystal–crystal structural transformations has yet to be developed, but a good physical picture of the relative influence of energetic and entropic effects, and their dependence on the molecular interactions, can be obtained from a simple model. Consider, for definiteness, a crystal of methane. The interaction between CH_4 molecules is angle-dependent, and we anticipate that there can exist different crystal phases with different orientational orderings of the CH_4's. The interaction between two CH_4's depends on the angles θ_1, ϕ_1, ψ_1 and θ_2, ϕ_2, ψ_2 that describe their orientations. In the lowest order of approximation we suppose that each CH_4 feels only the average field generated by all surrounding molecules. If $f_i(\omega)d\omega_i$ is the probability for finding a molecule with orientation between θ_i, ϕ_i, ψ_i, and $\theta_i+d\theta_i$, $\phi_i+d\phi_i$, $\psi_i+d\psi_i$ (which we denote as between $\boldsymbol{\omega}_i$ and $\boldsymbol{\omega}_i+d\boldsymbol{\omega}_i$, where $\boldsymbol{\omega}_i$ stands for the set of angles θ_i, ϕ_i, ψ_i), then the average interaction of molecule i with orientation $\boldsymbol{\omega}_i$ and all its neighbors is

$$V_i(\boldsymbol{\omega}_i)=\sum_i\int d\boldsymbol{\omega}_j\,u_{ij}(\boldsymbol{\omega}_i,\boldsymbol{\omega}_j)f_j(\boldsymbol{\omega}_j), \qquad (24.99)$$

where $u_{ij}(\boldsymbol{\omega}_i,\boldsymbol{\omega}_j)$ is the direct pair potential. Note that $V_i(\boldsymbol{\omega}_i)$ depends on the distribution of orientations of the neighbors, $f_j(\boldsymbol{\omega}_j)$. To be consistent we must require that each molecule feel the same average field, and this in turn requires that each distribution function $f_i(\boldsymbol{\omega}_i)$ be the same and be determined by

$$f_i(\boldsymbol{\omega}_i)=\frac{\exp[-V_i(\boldsymbol{\omega}_i)/k_BT]}{\int d\boldsymbol{\omega}_i\,\exp[-V_i(\boldsymbol{\omega}_i)/k_BT]}$$

$$=\frac{\exp\left[(1/k_BT)\int d\boldsymbol{\omega}_j\,u_{ij}(\boldsymbol{\omega}_i,\boldsymbol{\omega}_j)f_j(\boldsymbol{\omega}_j)\right]}{\int d\boldsymbol{\omega}_i\,\exp\left[(1/k_BT)\int d\boldsymbol{\omega}_j\,u_{ij}(\boldsymbol{\omega}_i,\boldsymbol{\omega}_j)f_j(\boldsymbol{\omega}_j)\right]}. \qquad (24.100)$$

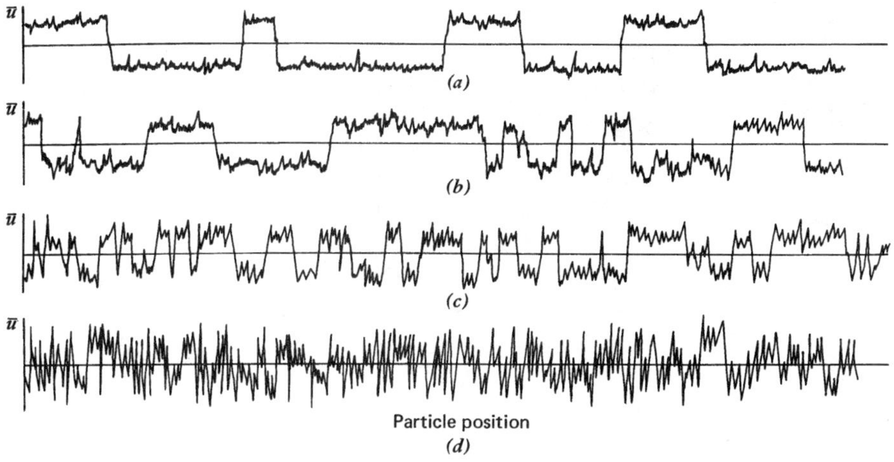

Figure 24.24 Particle displacements along a chain of double-well potentials. The model is described in the caption to Fig. 24.22. (*a*) $\bar{T} = 0.117$. (*b*) $\bar{T} = 0.19$. (*c*) $\bar{T} = 0.26$. (*d*) $\bar{T} = 1.68$. From T. R. Koehler, A. R. Bishop, J. A. Krumhansl, and J. R. Schrieffer, *Solid State Comm.* **17,** 1515 (1976).

That is, $f_i(\omega_i)$ is just the ordinary canonical distribution of orientation for a molecule subject to the effective field $V_i(\omega_i)$. Equation 24.97 is a *nonlinear* equation for $f_i(\omega_i)$.

Thus far we have treated the orientation of a CH_4 molecule by the methods of classical mechanics, and Eq. 24.97 is the proper expression for the distribution function using classical statistical mechanics. In the case of CH_4 one must use quantum mechanics to describe the orientations, since the moment of inertia is small and the temperatures of interest quite low. The formulation of the problem is the same, and the resulting quantum statistical equation for the distribution function is also the same except for summation over energy levels in place of integration over angles. For our purposes it is sufficient to use Eq. 24.97, since we seek only a qualitative picture of the possible phase transitions.

We now note again that Eq. 24.97 is a nonlinear equation for $f_i(\omega_i)$. It is a property of nonlinear equations of the type shown that the nature of the solution can change suddenly when the temperature passes a characteristic value. Just this happens for the potential between CH_4 molecules. The method used to solve Eq. 24.97 is complex because of the need to take into account quantization of rotation and the nature of the splittings in the energy levels induced by the field acting on a molecule. Fig. 24.25 displays the heat capacity of CH_4 as a function of temperature. The thermal anomaly at the transition temperature is relatively well described. The high-temperature phase is predicted to be orientationally disordered, and in the low-temperature phase there is partial ordering as shown in Fig. 24.26. These predictions concerning order have been confirmed by neutron diffraction experiments.

In fact there are three phases of solid CH_4 that appear in order of decreasing temperature at low pressure, and a fourth that is stable only at high pressure and low temperature. Although the theory described accurately predicts the properties of phase I (the disordered phase) and phase II (the partially ordered phase shown in Fig. 24.26), the predicted structure of phase III is incorrect. The theory described is a rigid lattice model (i.e., vibrational motion is not considered), and it neglects any correlation in the rotational motions of molecules on neighboring lattice sites [only the average over all neighbor orientations is used in defining the effective potential $V_i(\omega_i)$]. These omissions, and lack of complete knowledge of the intermolecular potential, are probably responsible for the erroneous predictions concerning the structure of phase III.

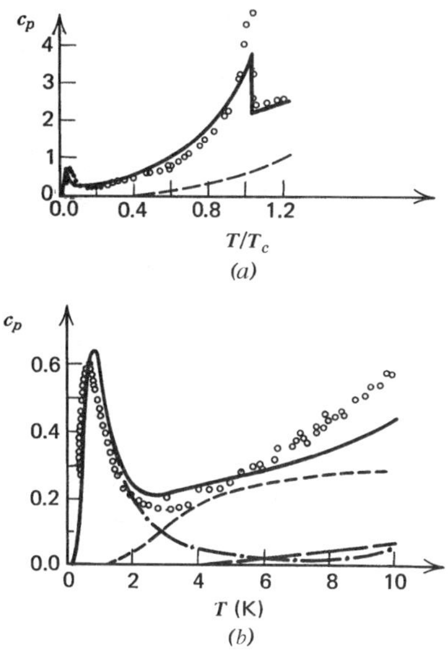

Figure 24.25 (*a*) Heat capacity of CH_4, versus T/T_c. The solid curve is the theoretical prediction derived from the quantum mechanical version of Eq. 24.97 with spin species equilibration. (*b*) Heat capacity of CH_4 versus T for $T < 10$ K. The solid curve is the theoretical prediction derived from the quantum mechanical version of Eq. 24.97; the dot-dash and dashed lines are the contributions from molecules in phase-I and phase-II environments, respectively. From T. Yamamoto, Y. Kataoka, and K. Okada, *J. Chem. Phys.* **66,** 2701 (1977).

We now examine the liquid–solid transition. This is a first-order transition (ΔV_{fus}, ΔS_{fus}, ΔH_{fus} nonzero) and is less well understood than are second-order transitions. One model of the melting process is based on the analog of the soft-mode instability discussed above. M. Born assumed that when the solid can no longer withstand a force that tends to make it flow, a transition to the liquid state occurs. This condition does not account for the thermodynamic condition of

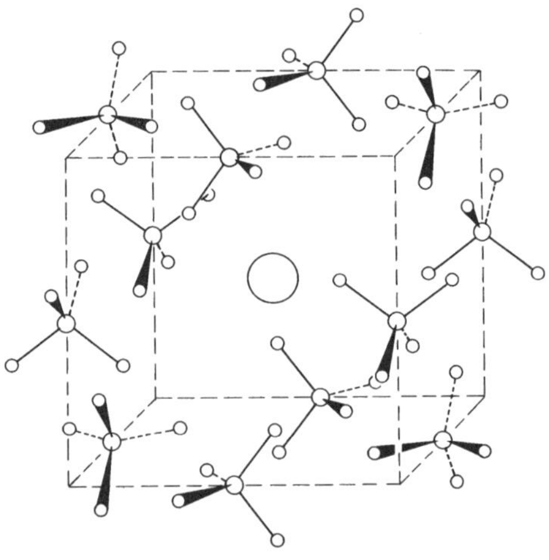

Figure 24.26 Phase relations between the orientations of CH_4 molecules in the partially ordered phase II. Dashed lines represent bonds in back of the plane of the C atom, doubled wedge lines bonds in front of the C atom, and solid lines bonds in the plane of the C atom parallel to the plane of this page. From T. Yamamoto, Y. Kataoka, and K. Okada, *J. Chem. Phys.* **66,** 2701 (1977).

equality of chemical potentials of the liquid and solid phases and leads to an overestimate of the stability of the solid. A different model of melting is based on the idea that, as the temperature of the solid increases, the concentrations of vacancies, interstitials, and other crystal imperfections increase, eventually becoming large enough that the solid is mechanically unstable. Just as in the case of the Born model of melting, the equality of chemical potentials of the liquid and solid phases is not accounted for in this model, and the stability of the solid is overestimated. One can also develop a description of the liquid–crystal phase transition from the point of view of instability of the liquid. One such model of the freezing process is based on the assumption that at some temperature the liquid becomes unstable to a density fluctuation that has a wavelength equal to the lattice separation characteristic of the crystalline solid. This condition also does not account for the thermodynamic condition of equality of chemical potentials of the liquid and solid phases and leads to an overestimate of the stability of the liquid. Although none of these "one-sided stability" theories is correct, each has an element that represents an important physical observation. To see this we now consider the results of some computer simulation studies.

J. G. Kirkwood and E. Monroe predicted[14] that a fluid of hard spheres becomes unstable with respect to a crystal of hard spheres when the density exceeds a critical value. This prediction was verified many years later by computer simulation studies. The hard sphere fluid → hard sphere crystal phase change is a first order transition; it occurs at a reduced liquid density $n\sigma^3 = 0.943$, where σ is the hard-sphere diameter and $n = N/V$ is the number density of molecules. The reduced density of the solid is 1.041, so there is a 10.3% change of volume when the fluid freezes. Since there is no direct attractive interaction between hard spheres, the fluid → solid transition must be a consequence of the advantages of the geometry of the crystal, and is driven by the difference in entropies of the two phases. It is easy to imagine (but has not yet been proven) that a disordered array of hard spheres will eventually jam together in a way that prevents any motion. This configuration, which has a lower density than that of a perfect close packed crystal of hard spheres in contact with each other, would have very small entropy. In contrast, if a regular array of hard spheres occupies a volume slightly larger than the volume corresponding to closest packing, small amplitude motion about the lattice positions is possible, and that hard sphere crystal configuration will have larger entropy than the jammed disordered array, hence be thermodynamically stable relative to it.

To what extent does the crystallization of a fluid of hard spheres mimic crystallization of a real liquid? To answer this question we display in Tables 24.2 and 24.3 the melting parameters for particles interacting with an inverse power repulsion of the form R^{-n}, and with a Lennard–Jones interaction, respectively. From the entries in Table 24.2 we see that as the exponent n in the inverse power repulsive potential decreases, so does the relative change of volume on freezing, whereas the entropy change on freezing is very weakly dependent on the nature of the repulsion between particles. The entries in Table 24.3 give the densities of coexisting solid and liquid phases of a Lennard–Jones system for several fixed temperatures. Along the solid–liquid coexistence line the densities of both phases increase as the transition temperature and pressure increase, but the difference in density between these phases slowly decreases. Consequently, both $\Delta V/V$ and $\Delta S/Nk_B$ slowly decrease as the transition temperature increases. Although the Lennard–Jones potential is a rather good representation of the Ar–Ar interaction, it is not an exact representation of that interaction.[15] Nevertheless, the computer-simulated melting parameters for a system of Lennard–Jones molecules are very close to those of Ar.

We now turn to a test of the idea that a mechanical instability of the solid triggers the transition to the liquid phase. An early version of this idea was proposed by F. Lindemann, who speculated that a crystal melts when the mean-square amplitude of vibration of the atoms exceeds some critical fraction of the atomic separation. We display in Table 24.2

[14] The basis for this prediction is the mean field equation discussed a little further on in this section.

[15] The Lennard–Jones potential determined from the best fit to, say, the second virial coefficient data for Ar, does not predict the correct lattice structure of crystalline Ar.

Table 24.2 Melting Characteristics of Inverse Power Potentials[a]

n	$\left(\left(\frac{\epsilon}{k_BT}\right)^{3/n} n\sigma^3\right)_L$	$\left(\left(\frac{\epsilon}{k_BT}\right)^{3/n} n\sigma^3\right)_S$	$\frac{\Delta V}{V_S}(\%)$	$\left(\frac{\epsilon}{k_BT}\right)^{3/n} \frac{p\sigma^3}{k_BT}$	$\frac{\Delta S}{Nk_B}$	$\frac{\beta \Delta U}{N}$	θ
∞	0.943	1.041	10.3	11.7	1.16	0	0.13
12	1.141	1.193	3.8	22.6	0.90	0.18	0.15
9	1.333	1.373	3.0	31.1	0.84	0.21	0.16
6	2.17	2.21	1.3	86.3	0.75	0.25	0.17
4	4.53	4.57	0.5	602	0.80	0.35	0.18

[a] From J. P. Hansen and I. R. McDonald, *Theory of Simple Liquids* (Academic Press, New York, 1976).

Table 24.3 Data on the Fluid–Solid Transition in the Lennard–Jones (LJ) System and in Argon[a]

	$\frac{k_BT}{\epsilon}$	$\frac{p\sigma^3}{\epsilon}$	$(n\sigma^3)_L$	$(n\sigma^3)_S$	$\frac{\Delta V}{V_S}(\%)$	$\frac{\Delta S}{Nk_B}$	θ
LJ	0.75	0.67	0.875	0.973	13.1	1.75	0.145
argon	0.75	0.59	0.856	0.967	12.9	1.64	
LJ	1.15	5.68	0.936	1.024	9.3	1.44	0.139
argon	1.15	6.09	0.947	1.028	8.4	1.25	
LJ	1.35	9.00	0.964	1.053	9.1	1.39	0.137
argon	1.35	9.27	0.982	1.056	7.6	1.20	
LJ	2.74	32.2	1.113	1.179	5.9	0.98	0.149
argon	2.74	37.4	1.167	1.230	5.3	1.04	

[a] From J. P. Hansen and I. R. McDonald, *Theory of Simple Liquids* (Academic Press, New York, 1976).

the Lindemann ratio θ, equal to the root-mean-square displacement of an atom from its lattice position, $\mathbf{R}_i$, divided by the nearest-neighbor distance a:

$$\theta = \frac{\langle u^2 \rangle^{1/2}}{a} = \frac{1}{a}\left(\frac{1}{N}\sum_{i=1}^{N}\left\langle |\mathbf{r}_i - \mathbf{R}_i|^2 \right\rangle\right)^{1/2}. \quad (24.101)$$

The value of θ can be calculated from the frequency spectrum of lattice vibrations. For example, using the Debye spectrum of lattice vibrations (see Chapter 22), one finds $\theta = (9\hbar^2/mk_BTa^2)^{1/2}(T/\Theta_D) \approx 0.13$. The entries in Table 22.3 show that θ is practically independent of the nature of the repulsive force between the atoms, and those in Table 22.4 show that θ is scarcely affected by adding an attractive force.

Turning to the criterion for instability of a liquid, computer simulations show that the amplitude of the first peak of the function

$$S(\mathbf{k}) = 1 + n\int d\mathbf{R} e^{-i\mathbf{k}\cdot\mathbf{R}} g_2(R) \quad (24.102)$$

for a variety of model liquids with different interactions between the atoms is practically constant at 2.85 ± 0.1 everywhere along the line of equilibrium between liquid and crystal. This behavior is likely a consequence of the insensitivity of the structure of a dense liquid to the detailed character of the intermolecular potential. Indeed, the structure of a dense hard sphere liquid, as represented by the pair correlation function, is very similar to those of systems with very different intermolecular interactions, for example, the dense Lennard–Jones liquid and a typical liquid metal (see Chapter 23).

Although both of the instability criteria cited are informative, we again note that they do not provide a basis for a molecular theory of melting since each deals with the limit of stability of a particular phase and neither requires equality of the chemical potentials of the liquid and solid phases at the point of instability. The outstanding characteristic of a second-order phase transition is the appearance at the same T and p of both a mechanical instability (e.g., $\omega_{soft} \to 0$) and the thermodynamic condition $\mu^{I} = \mu^{II}$. It appears to be the case that in a first-order phase transition, such as melting, the condition $\mu^{I} = \mu^{II}$ is reached before any purely mechanical instability occurs. Why this happens is not at present known, but the fact that it does prevents us from interpreting the information on the behavior of $S(\mathbf{k})$ or θ along the melting line in any fundamental fashion.

The thermodynamic condition of equality of chemical potential along the melting line can be incorporated into a mean field theory by using density functional theory (see Appendix 23B). The basis for this approach is a theorem, due to N. D. Mermin,[16] which states that there is a one-to-one correspondence between the local density, $n(\mathbf{R})$, and the external potential acting on the system, $\Psi(\mathbf{R})$ (see Appendix 23B). The Helmholtz free energy of the system is both a function of the ordinary thermodynamic variables and a functional of $\Psi(\mathbf{R})$, as is also the case for the Gibbs free

[16] N. D. Mermin, *Phys. Rev.* **137,** 1441 (1965).

energy and the grand potential $\Omega \equiv -k_BT \ln \Xi = -pV$ (see Chapter 19). The key physical idea utilized in the application of density functional theory to the freezing transition is that a crystal can be viewed as a strongly inhomogeneous liquid, in which the spatial variation of the density has the periodicity of the crystal lattice. To implement this idea we proceed as follows.

Consider a fluid of noninteracting molecules in an external field $\Psi(\mathbf{R})$. The chemical potential of this fluid can be represented in the form (Eq. 21A.31)

$$\mu_{id} = k_BT \ln \Lambda^3 n(\mathbf{R}) + \Psi(\mathbf{R}), \tag{24.103}$$

where the subscript *id* indicates that because of the absence of molecular interactions the fluid is an ideal gas. For a fluid of interacting molecules in an external field $\Psi(\mathbf{R})$, we write

$$\mu = \mu_{id}(n,\mathbf{R}) + \mu^E(n,\mathbf{R}) \equiv \mu_{id}(n,\mathbf{R}) - k_BTC_1(n,\mathbf{R}). \tag{24.104}$$

Clearly, $\mu^E \equiv -k_BTC_1$ is the excess chemical potential relative to that of an ideal gas of the same species in the same external field at the same number density and temperature. Eq. 24.104 expresses the constancy of the chemical potential throughout a nonuniform fluid, achieved by balancing the work required to create the spatial variation of the density against the external and internal potential energies. When $\Psi(\mathbf{R}) \equiv 0$ the fluid must have a position independent constant density n, and Eq. 24.104 reduces to $\mu(n) = \mu_{id}(n) - k_BTC_1(n)$. Eq. 24.104 implies that $-k_BTC_1$ acts as an effective one-particle potential energy that self-consistently accounts for the interactions between the molecules. From the combination of Eqs. 24.103 and 24.104, we find

$$n(\mathbf{R}) = \Lambda^{-3} \exp[(\mu - \Psi)/k_BT]\exp[C_1] \equiv z\exp[C_1], \tag{24.105}$$

with $z \equiv \Lambda^{-3} \exp[(\mu - \Psi)/k_BT]$. The fluid of interacting molecules can, of course, be a liquid.

We now consider equilibrium between the liquid and solid phases of a system of interacting molecules. In a fluid we know that $n = N/V$ is constant everywhere, hence, clearly not a function of $\mathbf{R}$. However, in a crystal the probability of finding a molecule in the volume element $d\mathbf{R}$ depends on whether or not that element contains a lattice point, hence, n is a function of $\mathbf{R}$. In an oriented and fixed crystalline phase, the one-particle density distribution can be represented in the form

$$n(\mathbf{R}) = n_L + \Delta n(\mathbf{R}) = n_L\left(1 + \sum_{\{\mathbf{G}_\alpha\}} \phi_{\mathbf{G}_\alpha} \exp(i\mathbf{G}_\alpha \cdot \mathbf{R})\right),$$

$$\phi_{\mathbf{G}_\alpha} = \frac{1}{w}\int d\mathbf{R}\frac{\Delta n(\mathbf{R})}{n_L}\exp(-i\mathbf{G}_\alpha \cdot \mathbf{R}), \tag{24.106}$$

where $n_S = \lim_{V\to\infty} V^{-1}\int d\mathbf{R}n(\mathbf{R})$ is the bulk density of the solid, w is the volume of a unit cell, and the $\{\mathbf{G}_a\}$ are reciprocal lattice vectors of the crystalline solid.[17] Using the method of analysis described in Appendix 23B it can be shown that (Appendix 24B)

$$C_2(\mathbf{R}_1,\mathbf{R}_2) = \frac{\delta C_1(\mathbf{R}_1)}{\delta n(\mathbf{R}_2)};$$

$$C_n(\mathbf{R}_1,\ldots,\mathbf{R}_n) = \frac{\delta C_{n-1}(\mathbf{R}_1,\ldots,\mathbf{R}_{n-1})}{\delta n(\mathbf{R}_n)}, \tag{24.107}$$

where $C_2(\mathbf{R}_1, \mathbf{R}_2)$ is the pair direct correlation function introduced in Chapter 23, and $C_n(\mathbf{R}_1, \ldots, \mathbf{R}_n)$ is the n-particle direct correlation function. Then, if we expand $C_1(n_L + \Delta n)$ in a functional Taylor series in which the expansion parameter is Δn, we can identify the coefficient of $(\Delta n)^m$ with $C_m(\mathbf{R}_1, \ldots, \mathbf{R}_m)$. If the functional Taylor series is truncated after the term with C_2 and we use

$$C_1(\mathbf{R}_1,n_L) = \ln(n_L/z_L);$$

$$C_1(\mathbf{R}_1,n_S) = \ln(n_S/z_S), \tag{24.108}$$

Eq. 24.106 can be rewritten in the form

$$\frac{n_L + \Delta n(\mathbf{R}_1)}{z_S} = \frac{n_L}{z_L}\exp\left[\int d\mathbf{R}_2 C_2(\mathbf{R}_{12},n_L)\Delta n(\mathbf{R}_2)\right]. \tag{24.109}$$

Given that $\mu_L = \mu_S$ at the freezing point, we must also have $z_L = z_S$ at the freezing point. Then, by combination of Eqs. 24.106 and 24.109, we obtain a relationship between the structure of the crystalline phase in equilibrium with the liquid phase, namely,

$$\left(1 + \sum_{\{\mathbf{G}_\alpha\}} \phi_{\mathbf{G}_\alpha} \exp(i\mathbf{G}_\alpha \cdot \mathbf{R})\right) = \exp\left[\int d\mathbf{R}_2 C_2(\mathbf{R}_{12},n_L)(n(\mathbf{R}_2) - n_L)\right]. \tag{24.110}$$

In principle, given the pair direct correlation function of the liquid and its density dependence, and the lattice vectors $\{\mathbf{G}_\alpha\}$, Eq. 24.110 can be solved for the coefficients $\{\phi_{\mathbf{G}_\alpha}\}$,

---

[17] Let **a**, **b**, and **c** be the elementary translations of a space lattice (the *direct lattice*). A second lattice, the *reciprocal lattice,* is defined by the translations $\mathbf{G}_1$, $\mathbf{G}_2$, $\mathbf{G}_3$, which satisfy the following two conditions:

$$\mathbf{G}_1 \cdot \mathbf{b} = \mathbf{G}_1 \cdot \mathbf{c} = \mathbf{G}_2 \cdot \mathbf{a} = \mathbf{G}_2 \cdot \mathbf{c} = \mathbf{G}_3 \cdot \mathbf{a} = \mathbf{G}_3 \cdot \mathbf{b} = 0$$
$$\mathbf{G}_1 \cdot \mathbf{a} = \mathbf{G}_2 \cdot \mathbf{b} = \mathbf{G}_3 \cdot \mathbf{c} = 1$$

Thus, $\mathbf{G}_1$ is normal to the plane generated by **b** and **c**, $\mathbf{G}_2$ to the plane generated by **a** and **c**, and $\mathbf{G}_3$ to the plane generated by **a** and **b**. Using the conditions relating the direct and recirpocal lattice vectors, we find

$$\mathbf{G}_1 = \frac{\mathbf{b}\times\mathbf{c}}{w};\quad \mathbf{G}_2 = \frac{\mathbf{c}\times\mathbf{a}}{w};\quad \mathbf{G}_3 = \frac{\mathbf{a}\times\mathbf{b}}{w}$$

$$\mathbf{a} = \frac{\mathbf{G}_2\times\mathbf{G}_3}{w^*};\quad \mathbf{b} = \frac{\mathbf{G}_3\times\mathbf{G}_1}{w^*};\quad \mathbf{c} = \frac{\mathbf{G}_1\times\mathbf{G}_2}{w^*}$$

$$w = \mathbf{a}\cdot\mathbf{b}\times\mathbf{c} = \frac{1}{w^*}.$$

and thereby the values of the densities of the coexisting crystal and liquid phases.

Before testing the accuracy of the prediction of the location and properties of the liquid-to-crystal phase transition obtained from Eq. 24.110, we must examine some its qualitative features. These include the following:

1. The theory of the solid–liquid phase transition that leads to Eq. 24.110 is based on the separate construction for each phase of a free energy versus density relationship and the use of the thermodynamic criterion for equilibrium to define the phase transition. Therefore, this theory does not automatically select the crystal phase with lowest free energy as the one to which the liquid freezes. To find the crystal phase with lowest free energy, independent calculations must be carried out for crystal phases with different symmetries, from which the crystal phase with lowest free energy can be selected.

2. The theoretical approach assumes that the free energy is an analytic function of the density and that a perturbation theory representation of the free energy difference, in terms of the density difference between the crystalline and liquid phases, is valid. As seen earlier in this chapter, the assumption concerning the analyticity of the free energy is not valid in the vicinity of the critical point of a second-order transition. It appears to be the case, from comparison of predictions made by the theory with empirical evidence, that the assumption of analyticity is valid for first-order transitions that do not have a critical point, but this has not been proven from first principles.

3. Equation 24.110 is nonlinear; for fixed temperature it has qualitatively different solutions at different densities and, correspondingly, for fixed density it has qualitatively different solutions at different temperatures. Consider the former case first; because of the nonlinearity of Eq. 24.110, density variations with periodic character are unstable relative to n = constant for high T but become stable when T reaches T_{fus} and remain stable for $T < T_{\text{fus}}$. The onset of this change in $n(\mathbf{R}_1)$ is sudden because of the feedback inherent to the nonlinear form. That is, a small-amplitude fluctuation that is favored is exponentially amplified and dominates the behavior of the solution to the equation. Similarly, for a fixed temperature, at low densities the only solution to 24.110 is $n(\mathbf{r}) = n_L$, corresponding to a stable liquid phase. As the density is increased a value is reached at which spatially periodic solutions to Eq. 24.110 exist, and these solutions are stable for all higher densities. At the transition point the densities $n(\mathbf{r})$ and n_L describe crystalline solid and liquid phases with the same temperature and chemical potential.

4. Equation 24.110 is not exact. One obvious source of approximation is the truncation of the functional Taylor series expansion of $C_1(n_L + \Delta n)$ after the pair direct correlation function term. A more subtle approximation is embodied in the assumption that the pair direct correlation function of the crystal can be obtained from that of the liquid. It is reasonable to expect, on physical grounds, that the angle averaged pair direct correlation function of the crystal will be qualitatively similar to that of the liquid, but the two functions are not the same in all details. These approximations have the consequence that for a specified temperature more than one pair of densities $n(\mathbf{r})$ and n_L can satisfy Eq. 24.110, much as an isotherm of the van der Waals equation in the liquid–gas coexistence region exhibits three values of the volume for a specified pressure (Fig. 24.17). The choice of the correct pair of densities is made by imposing the subsidiary condition that the pressures of the liquid and crystal phases must be the same. Since the theory described is formulated in the grand canonical ensemble (see Chapter 19) the total volume of the system is fixed and the numbers of molecules in the liquid and crystal phases are allowed to vary. The requirement that the pressures of the liquid and crystal phases be the same is then translated into the requirement that the difference in grand potential between those phases be zero. In Appendix 24B it is shown that this difference in grand potential is

$$\Delta\Omega = -k_BT\int d\mathbf{r}_1\,[n(\mathbf{r}_1)-n_L]+\frac{k_BT}{2}\int d\mathbf{r}_1 \times\int d\mathbf{r}_2 C_2(r_{12})[n(\mathbf{r}_2)-n_L][n(\mathbf{r}_2)+n_L]=0. \quad (24.111)$$

We show in Fig. 24.27 the phase diagram for a system of Lennard–Jones molecules calculated on the basis of the theory outlined and obtained from computer simulation data. Table 24.4 displays the densities of coexisting face centered cubic solid and liquid for a system in which the molecules interact with a potential of the form R^{-n}, and Table 24.5 displays similar data for a system in which the molecules interact with a Lennard–Jones potential. The overall agreement between the predicted behavior and that obtained from computer simulations is satisfactory but not perfect.

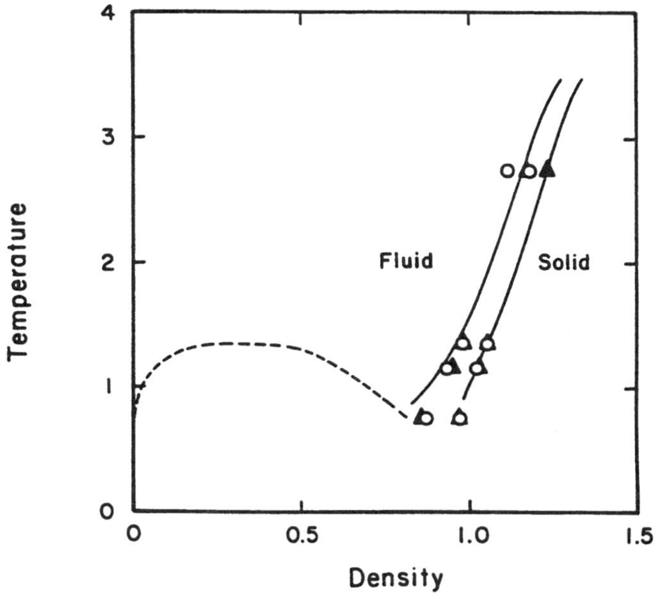

Figure 24.27 The phase diagram for a system of Lennard–Jones molecules. The solid lines are the densities of the coexisting liquid and face centered cubic crystal phases calculated using the theory discussed in the text, the circles are the results obtained from computer simulation studies of this system, and the triangles are experimental data for Ar. From C. Marshall, B. B. Laird, and A. D. J. Haymet, *Chem. Phys. Lett.* **122,** 320 (1985).

Table 24.4 Comparison of Predicted and Simulated Densities of Coexisting Face Centered Cubic Crystal and Liquid (Inverse Power Potentials)[a]

n	$\left(\left(\frac{\epsilon}{k_BT}\right)^{3/n} n\sigma^3\right)_L^{\text{calc}}$	$\left(\left(\frac{\epsilon}{k_BT}\right)^{3/n} n\sigma^3\right)_L^{\text{sim'n}}$	$\left(\left(\frac{\epsilon}{k_BT}\right)^{3/n} n\sigma^3\right)_S^{\text{calc}}$	$\left(\left(\frac{\epsilon}{k_BT}\right)^{3/n} n\sigma^3\right)_S^{\text{sim'n}}$	$(\Delta n)_{\text{calc}}$	$(\Delta n)_{\text{sim'n}}$
∞	0.967	0.939–0.948	1.147	1.036–1.045	0.180	0.09
12	1.28	1.15	1.37	1.19	0.09	0.04
6	3.43	2.18	3.52	2.21	0.09	0.03
4	12.30	5.54	12.47	5.57	0.17	0.03

[a] Adapted from Table 5 in A. D. J. Haymet, *Fundamentals of Inhomogeneous Fluids,* Ed. D. Henderson (Marcel Dekker, New York, 1992), p. 363.

Table 24.5 Comparison of Predicted and Simulated Densities of Coexisting Face Centered Cubic Crystal and Liquid (Lennard–Jones Potential)[a]

k_BT/ϵ	$(n_L\sigma^3)_{\text{calc}}$	$(n_L\sigma^3)_{\text{sim'n}}$	$(n_S\sigma^3)_{\text{calc}}$	$(n_S\sigma^3)_{\text{sim'n}}$	$(\Delta n)_{\text{calc}}$	$(\Delta n)_{\text{sim'n}}$
1.15	0.922	0.936	1.026	1.024	0.104	0.088
1.35	0.964	0.964	1.061	1.053	0.097	0.089
2.75	1.166	1.113	1.240	1.179	0.074	0.066

[a] Interpolated from data reported by C. Marshall, B. B. Laird and A. D. J. Haymet, *Chem. Phys. Lett.* **122,** 320 (1985).

The theory of liquid–crystal equilibrium sketched above has been improved by the introduction of better treatments of the expansion of C_1 in a functional Taylor series than the truncation approximation used to derive Eq. 24.109; details can be found in the literature.[18] The theory has also been extended to account for the equilibrium concentration of vacancies in the crystal along the liquid–crystal coexistence line,[19] with results that are consistent with experimental data for rare gas solids.

Finally, we examine briefly the theory of the first-order vapor–liquid transition. In this case there is an enormous difference between the densities of the two phases, the consequence of which is that the local structures of these fluids are very different. It is found that none of the considerations used in the analysis of crystal–crystal transitions, or even of melting, can be usefully applied. There is not any theoretical indication yet available of the nature of the microscopic motion that leads to the instability that generates the liquid–vapor phase transition. From the point of view of evolution of a structural change, the mechanism of *condensation* in a fluid system appears to involve the following. In any distribution of the molecules in the system (e.g., distribution of occupied and unoccupied lattice sites in the lattice gas, or positions in space in a real gas), there are isolated molecules, pairs of molecules close together, groups of three molecules close together, and so forth. These clusters are in mutual equilibrium associating and disassociating, and even fairly large clusters resembling droplets of liquid have some small but nonzero probability of occurring. The energy and entropy of each cluster depend on both the number of molecules composing the cluster and the amount of "surface" the cluster has. The existence of an excess free energy associated with the surface tends to minimize the amount of surface, hence to contract the cluster. This tendency is opposed by the increase in the free energy of the bulk that accompanies compression. For example, the bulk contribution to the entropy of a cluster will be larger if the molecules have more internal volume in which to move. At low temperature the minimization of the surface free energy dominates the behavior of the system, and it is advantageous to combine clusters to form ever larger groupings, since in this way the surface-to-volume ratio of the system is diminished, and thereby the total free energy of the system minimized. Indeed, if conditions are sufficiently favorable, we can imagine the clusters growing to macroscopic size. A macroscopic droplet represents the liquid, and its presence indicates that condensation has occurred. In contrast, if the temperature is high enough all the clusters dissociate, because the decrease in bulk free energy from the gain in translational entropy more than counterbalances the increase in "surface free energy." In this limit the system is completely dispersed as single molecules and no condensation can occur. The highest temperature at which a balance in the competing factors leads to some condensation, that is, the highest temperature at which some condensation can occur, is infinitesimally below the critical temperature. The critical temperature T_c is itself the limit above which, and at which, condensation cannot occur. The description of condensation

[18] A. D. J. Haymet in *Fundamentals of Inhomogeneous Fluids,* Ed. D. Henderson (Dekker, New York, 1992).
[19] R. McRae, J. D. McCoy and A. D. J. Haymet, *J. Chem. Phys.* **93,** 4281 (1990).

just given can be made consistent with what is now known of the peculiar approach to the critical region, that is, the values of α, β, γ, and δ.

All approximate theories of the equation of state of a fluid that are known predict isotherms in the two-phase region that are similar to those of the van der Waals equation (they have loops), hence contain an inherent error. Nevertheless, if an equation of state is accurate in the one-phase regions where liquid or gas are stable, an accurate prediction of the coexistence curve can be made by the *Maxwell equal areas construction.* Consider again the isotherm displayed in Fig. 24.17*b*. As discussed earlier, the isotherm should look like *AEE′D* and not have loops. Imagine that the fluid is carried around the closed cycle *EBCE′E*. For this closed cycle $\Delta u = 0$ and $\Delta s = 0$, and since only pV work is allowed we conclude that $\oint p\, dv = 0$; hence the two areas marked are equal.[20] Approximate theories of the liquid state, such as are discussed in Chapter 23, when supplemented with this equal areas construction, give a tolerably accurate description of the vapor–liquid coexistence curve. This is shown in Fig. 24.28.

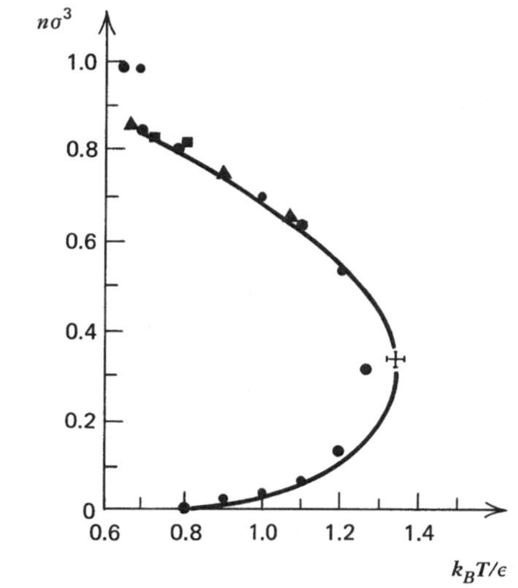

Figure 24.28 Densities of coexisting phases of a Lennard–Jones 6–12 fluid, calculated using the PY approximation mentioned in Chapter 23. The points represent the results of computer simulation calculations and experimental data for Ar, and the cross the critical point estimated from the simulations. From R. O. Watts and I. J. McGee, *Liquid State Chemical Physics* (Wiley, New York, 1976).

The several molecular models of phase transitions described are useful, but each contains an inherent approximation and hence the predictions must be viewed with some reserve. This is certainly the case for the theory of critical behavior, which we have seen is peculiar. What can be said about critical phenomena as a result of studies of models that can be treated exactly?

1. As the critical point is approached, the pair-correlation function becomes long-range. A hint that this must occur was given in Section 24.3 in the description of the coupling of density fluctuations near T_c. It is found that in all systems that exhibit a critical point the range of pair correlations becomes very large relative to the range of the intermolecular potential as T_c is approached, no matter what the form of the molecular interaction. For example, in this limit the correlation between directions of magnetic moments in a magnet behaves like the correlation between positions of atoms in a fluid. The general form found is

$$g_2(R) - 1 \propto \frac{e^{-R/\xi}}{R^{d-2+\eta}}, \tag{24.112}$$

where ξ is the scale length of the correlations, d is the number of dimensions (three for real space), and η is a critical exponent like α, β, γ, and δ. The scale length of correlations, ξ, diverges as $T \to T_c$.

2. Near T_c, ξ is very much greater than the range of the pair potential energy $u_2(R)$; hence the properties of the system become independent of $u_2(R)$. For this reason the behavior of magnets, fluids, and liquid mixtures is expected to be the same near T_c.

3. Provided that condition **2** is valid, the true approach to the critical region is very different from that predicted by the van der Waals and other classical equations of state. In particular, these classical theories predict that the critical exponents have the following values: $\alpha = 0$ and c_V has a jump discontinuity at T_c, $\beta = \frac{1}{2}$, $\gamma = 1$, $\delta = 3$, and $\eta = 0$. The values cited are not in agreement with those derived from experiment. As already mentioned (see Eqs. 24.33–24.37 and Table 24.6), the theoretical values of α, β, γ, and δ derived from renormalization group theory are in agreement with the experimental values, and the former values are considered to be more reliable and subject to less error. The best available data suggest that η is small and positive, but definitely not zero.

4. When condition **2** is not valid, that is, when the range of the interaction exceeds ξ, the behavior in the critical region is classical with $\beta = \frac{1}{2}$, $\gamma = 1$, $\delta = 3$, and $\eta = 0$. For example, a gas for which $u_2(R)$ becomes indefinitely long-range but weak, so that $\int d\mathbf{R} u_2(R)$ remains finite, satisfies the van der Waals equation of state supplemented by the Maxwell equal areas construction. This statement means that the supplementary condition is part of the equation of state and it determines the shape of an isotherm in the two-phase region. One example of a real system where $u_2(R)$ is very long-range is a solution of hydrogen in a transition metal. The insertion of the H atom causes displacements that strain the metal lattice, and the elastic energy associated with these displacements falls off only as R^{-1}. Two H atoms interact when the displacements they generate overlap, which occurs at very great distances. Consequently, the range of the interaction exceeds ξ for the hydrogen–hydrogen positional

[20] Note that this argument assumes it is legitimate to use the laws of thermodynamics in the region of absolute instability, i.e., along the hypothetical curve *CB*, where $(\partial p/\partial v)_T > 0$.

Table 24.6 Critical Exponents Predicted by Some Statistical Models

Property	Exponent	Classical Theory Mean Field Theory van der Waals Equation	Two-Dimensional Lattice Gas	Three-Dimensional Lattice Gas	Renormalization Group Theory
Heat capacity	α	0, finite discontinuity at $T = T_c$	0, logarithmic divergence at $T = T_c$	0.119±0.006	0.110±0.003
Coexistence curve	β	1/2	1/8	0.326±0.004	0.326±0.002
Compressibility	γ	1	7/4	1.239±0.003	1.239±0.002
Critical isotherm	δ	3	15	4.80±0.05	4.814±0.015

correlations. The hydrogen–metal system has a critical temperature for phase separation. It is found that the approach to the critical region is classical.

5. The Widom hypothesis has a basis in renormalization group theory, which is a mathematical and computational model by K. Wilson based on a physical idea suggested by L. Kadanoff. Kadanoff suggested that it is possible to systematically reduce the total number of degrees of freedom contained in a volume ξ^3 (or ξ^d for d dimensions) to a smaller number of effective degrees of freedom by iteratively removing fluctuations with wavelengths smaller than ξ. Starting with the original system Hamiltonian, each stage of transformation introduces an effective Hamiltonian which depends on a set of parameters, say $a_1, \ldots, a_n$, which can be thought of as the coordinates in a finite-dimensional space of Hamiltonians. In this space each set of values of the parameters corresponds to a particular Hamiltonian. We can also think of each transformation which reduces the number of degrees of freedom as a mapping in the space of Hamiltonians, so the successive transformations generate a trajectory in that space. In general, the space will also have one or more fixed points for which the transformation leaves the effective Hamiltonian unchanged. The particular fixed point of interest is the critical fixed point, which lies on some surface in the space of Hamiltonians that corresponds to $T = T_c$. Put another way, since $\xi \to \infty$ as $T \to T_c$, the successive transformations that reduce the number of degrees of freedom transform the Hamiltonian of the system into one that, eventually, is invariant to further transformations. Then the relations between critical exponents we have described earlier are a consequence of the form these transformations take, and the theory permits calculation of their values.[21]

Notwithstanding the successes of the modern theory of critical points, it should be clear from the discussion of, for example, the melting transition, that much remains to be done before a completely satisfactory molecular theory of phase transitions is created.

[21] A thorough discussion is given by K. G. Wilson and J. B. Kogut, *Physics Reports* **12C,** 2 (1974).

APPENDIX 24A

The Scaling Hypothesis for Thermodynamic Functions

In this appendix we examine the consequences of the assumption that near a critical point the thermodynamic potentials, such as the Helmholtz and Gibbs free energies, are generalized homogeneous functions of their independent variables. A generalized homogeneous function of x_1, x_2, . . . , x_n has the form

$$f(\lambda^{\zeta_1} x_1, \lambda^{\zeta_2} x_1, \ldots, \lambda^{\zeta_n} x_n) = \lambda f(x_1, x_2, \ldots), \qquad (24A.1)$$

for arbitrary λ. The parameters $\{\zeta_i\}$ appearing in Eq. 24A.1 are arbitrary numbers that do not simultaneously vanish. The geometric implications of the scale transformation defined by Eq. 24A.1 are displayed in Fig. 24.29 for the simple case $f(\lambda x_1, \lambda x_2) = \lambda f(x_1, x_2)$.

Because the thermodynamic functions of a fluid depend on only two independent variables, we now restrict attention to a function of two variables, namely,

$$f(\lambda^{\zeta_1} x_1, \lambda^{\zeta_2} x_2) = \lambda f(x_1, x_2). \qquad (24A.2)$$

Since Eq. 24A.1 is valid for arbitrary λ, we are free to choose a convenient value. Let $\lambda = \exp[\zeta_2^{-1} \ln(1/x_2)]$ with $\zeta_2 \neq 0$. Substitution of this value, followed by some algebraic manipulation, leads to

$$f(x_1, x_2) = x_2^{-1/\zeta_2} f(x_1 x_2^{-\zeta_1/\zeta_2}, 1). \qquad (24A.3)$$

If, instead, we let $\lambda = \exp[\zeta_1^{-1} \ln(1/x_1)]$ with $\zeta_1 \neq 0$, we find

$$f(x_1, x_2) = x_2^{-1/\zeta_1} f(1, x_2 x_1^{-\zeta_2/\zeta_1}). \qquad (24A.4)$$

These relations will be useful later.

Let the critical temperature T_c be the reference point for measuring the temperature, which we implement by defining the dimensionless reduced temperature $\epsilon \equiv (T - T_c)/T$. We assume that $T > T_c$, so $\epsilon \geq 0$. For a fluid, the molar Gibbs free energy of the system is a function of the temperature and pressure. We now assume that any nonsingular terms in the molar Gibbs free energy have been subtracted off and that the singular part of the molar Gibbs free energy, $g(\epsilon, p)$, is a generalized homogeneous function with respect to ϵ and p,

$$g(\lambda^{\zeta_\epsilon} \epsilon, \lambda^{\zeta_p} p) = \lambda g(\epsilon, p), \qquad (24A.5)$$

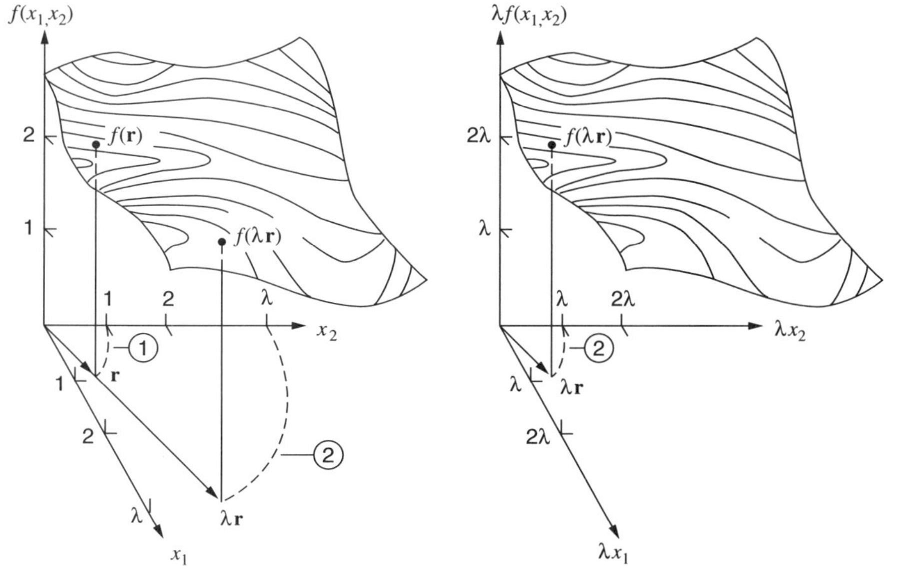

Figure 24.29 The scale transformation $f(\lambda x_1, \lambda x_2) = \lambda f(x_1, x_2)$ is equivalent to relabeling the x_1 and x_2 axes as shown on passage from (*a*) to (*b*). Then, knowledge of $f(\mathbf{r}) = f(x_1, x_2)$ on path 1 in (*a*) suffices to determine $f(\mathbf{r})$ on path 2 in (*b*). Because λ is an arbitrary number, knowledge of $f(\mathbf{r})$ along any path encircling the origin of coordinates suffices to determine the function everywhere. Figure adapted from H. E. Stanley, *Introduction to Phase Transitions and Critical Phemomena* (Oxford University Press, London, 1971).

for all values of λ. This assumption, known as the static scaling hypothesis, does not determine ζ_ϵ or ζ_p, but all of the critical point exponents can be expressed in terms of these parameters. By differentiation of both sides of Eq. 24A.5 with respect to p we find

$$\lambda^{\zeta_p} \frac{\partial g(\lambda^{\zeta_\epsilon}\epsilon, \lambda^{\zeta_p} p)}{\partial(\lambda^{\zeta_p} p)} = \lambda \frac{\partial g(\epsilon, p)}{\partial p}. \tag{24A.6}$$

The right-hand side of Eq. 24A.6 is the molar volume of the system, the reciprocal of the molar density ($\upsilon \equiv \rho^{-1}$). We now recall that the critical exponent β refers to the functional dependence of $(\rho_L - \rho_G)$ as $\epsilon \to 0$ (see Eq. 24.33), and the critical exponent δ refers to the functional dependence of $(p \to p_c)$ on $(\rho - \rho_c)$ as $\rho \to \rho_c$ (see Eq. 24.35). For the former case 24A.6 becomes

$$\upsilon(\epsilon, p_c) = \lambda^{\zeta_p - 1}\upsilon(\lambda^{\zeta_\epsilon}\epsilon, p_c). \tag{24A.7}$$

Since Eq. 24A.7 is valid for all values of λ, it must certainly be valid for the particular choice $\lambda = (-\epsilon)^{1/\zeta_\epsilon}$, so that

$$\upsilon(\epsilon, p_c) = (-\epsilon)^{(1-\zeta_p)/\zeta_\epsilon}\, \upsilon(-1, p_c). \tag{24A.8}$$

When $\epsilon \to 0^-$ the density difference between the liquid and gas behaves as $\Delta\rho \sim (-\epsilon)^\beta$ (Eq. 24.33) so that

$$\beta = \frac{\zeta_p - 1}{\zeta_\epsilon}. \tag{24A.9}$$

Similarly, setting $\epsilon = 0$ in Eq. 24A.7 and letting $(p - p_c) \to 0$ leads to

$$\upsilon(0, p) = \lambda^{\zeta_p - 1}\upsilon(0, \lambda^{\zeta_p} p). \tag{24A.10}$$

Then if we take the particular value $\lambda = p^{-1/\zeta_p}$, we find

$$\upsilon(0, p) = p^{(1-\zeta_p)/\zeta_p}\, \upsilon(0,1) \tag{24A.11}$$

When $(p - p_c) \to 0$, the difference between the critical pressure and the system pressure behaves as $|\rho - \rho_c|^{\delta-1}(\rho - \rho_c)$ (Eq. 24.35) so that

$$\delta = \frac{\zeta_p}{\zeta_p - 1}. \tag{24A.12}$$

Solving the simultaneous Eqs. 24A.9 and 24A.12 for ζ_ϵ and ζ_p gives

$$\zeta_\epsilon = -\frac{1}{\beta(\delta+1)}; \qquad \zeta_p = \frac{\delta}{\delta+1}. \tag{24A.13}$$

We can obtain additional relations involving the critical exponents by differentiation of Eq. 24A.6 with respect to p. We find

$$\lambda^{2\zeta_p} \chi_T(\lambda^{\zeta_\epsilon}\epsilon, \lambda^{\zeta_p} p) = \lambda \chi_T(\epsilon, p) \tag{24A.14}$$

where $\chi_T \equiv \kappa_T^{-1}$. If we consider the critical isobar and choose $\lambda = (-\epsilon)^{-1/\zeta_\epsilon}$, Eq. 24A.14 becomes

$$\chi_T(\epsilon, p_c) = (-\epsilon)^{-(2\zeta_p - 1)/\zeta_\epsilon}\, \chi_T(-1, 0). \tag{24A.15}$$

Then recalling that in the vicinity of the critical point $\kappa_T^{-1} \propto |T - T_c|^{\gamma}$, we obtain

$$\gamma = \frac{1 - 2\zeta_p}{\zeta_\epsilon}. \tag{24A.16}$$

Since there are only two scaling parameters, namely ζ_ϵ and ζ_p, we can find the value of γ in terms of β and δ. The result, obtained by substitution of Eq. 24A.13 in Eq. 24A.16, is

$$\gamma = \beta(\delta - 1), \tag{24A.17}$$

which is the Widom equality (see Eq. 24.81).

APPENDIX 24B

Aspects of Density Functional Theory

Using Eqs. 19.122 and 19.126 in the semiclassical limit (Appendix 21C), the normalized probability density for finding a particular molecular configuration is

$$\mathcal{P} = \Xi^{-1} \exp[-(KE + \mathcal{V}_N - \mu N) / k_B T], \tag{24B.1}$$

where KE is the kinetic energy and $\mathcal{V}_N$ the potential energy of the molecules in the system. By Mermin's theorem, Ξ is uniquely determined by $n(\mathbf{R})$, and $n(\mathbf{R})$ is in one-to-one correspondence with $\Psi(\mathbf{R})$, which then determines $\mathcal{P}$. Since $\mathcal{P}$ is a unique functional of $n(\mathbf{R})$ so is

$$\mathcal{F}[n] \equiv \langle KE + \mathcal{V}_N + k_B T \ln \mathcal{P} \rangle, \tag{24B.2}$$

where $\langle . . . \rangle$ denotes the average of the function enclosed. If the form Eq. 24B.1 and the relation $pV = k_BT \ln \Xi$ are substituted in Eq. 24B.2, we find that $\mathcal{F}$ is, sensibly, the Helmholtz free energy of the system. From $\mathcal{F}$ we construct another functional by use of the Legendre transform

$$\Omega_V[n] = \mathcal{F}[n] - \int d\mathbf{R}[\mu - \Psi(\mathbf{R})] n(\mathbf{R}). \tag{24B.3}$$

Imposing the condition that the variation of Ω_V with respect to the density distribution must vanish for the equilibrium density distribution, i.e.,

$$\frac{\delta \Omega_V[n]}{\delta n(\mathbf{R})} = 0, \tag{24B.4}$$

we have $\Omega_V[n] = \Omega = -pV$, the grand potential. The total Helmholtz free energy of the system can be written in the form

$$\begin{aligned} A &= N\mu - pV \\ &= \Omega + \mu \int d\mathbf{R}\, n(\mathbf{R}) = \mathcal{F}[n] + \int d\mathbf{R}\, n(\mathbf{R}) \Psi(\mathbf{R}), \end{aligned} \tag{24B.5}$$

since $N = \int d\mathbf{R} n(\mathbf{R})$. Using Eq. 24B.3 and the variational principle that determines the equilibrium density distribution, we find

$$\mu = \Psi(\mathbf{R}) + \frac{\delta \mathcal{F}[n]}{\delta n(\mathbf{R})}, \tag{24B.6}$$

which is another formal representation of the constancy of the chemical potential throughout a nonuniform fluid in an equilibrium state. We also have

$$\frac{\delta \mathcal{F}[n]}{\delta n(\mathbf{R})} = \mu - \Psi(\mathbf{R}) \equiv u(\mathbf{R}), \tag{24B.7}$$

$$-k_B T \frac{\delta \Omega}{\delta u(\mathbf{R})} = n(\mathbf{R}). \tag{24B.8}$$

It is convenient to represent $\mathcal{F}$ as the sum of an ideal gas contribution and an excess contribution resulting from intermolecular interactions:

$$\begin{aligned} \mathcal{F}[n] &= \mathcal{F}_{\text{id}}[n] + \mathcal{F}^E[n]; \\ \mathcal{F}_{\text{id}}[n] &= \int d\mathbf{R}[n k_B T(\ln(n\Lambda^3) - 1)]. \end{aligned} \tag{24B.9}$$

Comparison of Eqs. 24B.9 and 24B.7 with the separation of the chemical potential into a sum of an ideal gas contribution and an excess resulting from intermolecular interactions, displayed in Eq. 21.104, leads to the identification

$$-k_B T C_1[n] = \frac{\delta \mathcal{F}^E[n]}{\delta n(\mathbf{R})}. \tag{24B.10}$$

Finally, recognizing that $\mathcal{F}^E$ is effectively the excess Helmholtz free energy, the arguments of Appendix 23B lead to the definition

$$\begin{aligned} k_B T C_2(\mathbf{R}_1, \mathbf{R}_2) &= -\frac{\delta^2 \mathcal{F}^E[n]}{\delta n(\mathbf{R}_2) \delta n(\mathbf{R}_1)} \\ &= k_B T \frac{\delta C_1(\mathbf{R}_1)}{\delta n(\mathbf{R}_2)}, \end{aligned} \tag{24B.11}$$

and thereby to

$$C_n(\mathbf{R}_1, \dots, \mathbf{R}_n) = \frac{\delta C_{n-1}(\mathbf{R}_1, \dots, \mathbf{R}_{n-1})}{\delta n(\mathbf{R}_n)}. \tag{24B.12}$$

We now rewrite

$$\Omega_V[\tilde{n}] = \mathcal{F}[\tilde{n}] - \int d\mathbf{R}[\mu - \Psi(\mathbf{R})] \tilde{n}(\mathbf{R}) \tag{24B.13}$$

in the form

$$\Omega_V[n] = \mathcal{F}[n] - \mathcal{F}_{\text{id}} + \mathcal{F}_{\text{id}} - \int d\mathbf{R}[\mu - \Psi(\mathbf{R})]n(\mathbf{R})$$
$$= -\mathcal{F}^E + \int d\mathbf{R}\, n(\mathbf{R})[C_1(\mathbf{R}) - 1]. \qquad (24B.14)$$

The excess free energy of the system can be expanded using the excess free energy of the liquid as the reference state. The form of this expansion is

$$\mathcal{F}^E = \mathcal{F}^E_L + \int d\mathbf{R}_1 C_{1L}(n(\mathbf{R}_1) - n_L)$$
$$+ \int d\mathbf{R}_1 d\mathbf{R}_2 C_{2L}(\mathbf{R}_1, \mathbf{R}_2)[n(\mathbf{R}_1) - n_L][n(\mathbf{R}_2) - n_L] + \cdots. \qquad (24B.15)$$

The difference in the grand potentials between the state of interest and the liquid state is then, to terms of order the pair direct correlation function,

$$\Delta\Omega \equiv \Omega - \Omega_L = -k_B T \int d\mathbf{R}_1 [n(\mathbf{R}_1) - n_L]$$
$$+ \tfrac{1}{2} \int d\mathbf{R}_1\, d\mathbf{R}_2 C_{2L}(\mathbf{R}_1, \mathbf{R}_2)[n(\mathbf{R}_2) - n_L][n(\mathbf{R}_1) + n_L]. \qquad (24B.16)$$

• FURTHER READING

Ansimov, M. A., *Critical Phenomena in Liquids and Liquid Crystals* (Gordon and Breach, Philadelphia, 1991).

Binney, J. J., Dowrick, N. J., Fisher, A. J., and Newman, M. E. J., *The Theory of Critical Phenomena* (Clarendon Press, Oxford, 1993).

Brey, J., and Jones, R. B. (Eds.), *Critical Phenomena* (Springer-Verlag, Berlin, 1976).

Brout, R., *Phase Transitions* (W. A. Benjamin, Reading, Mass., 1965).

Cardy, J., *Scaling and Renormalization in Statistical Physics* (Cambridge University Press, Cambridge, 1996).

Epstein, P. S., *Textbook of Thermodynamics* (Wiley, New York, 1937), Chapters 6 and 7.

Fujimoto, M., *The Physics of Structural Phase Transitions* (Springer-Verlag, New York, 1997).

Kestin, J., *A Course in Thermodynamics,* Vol. II (Blaisdell, Waltham, Mass., 1968), Chapter 22.

Pfeuty, P., and Toulouse, G., *Introduction to the Renormalization Group and to Critical Phenomena* (Wiley, New York, 1977).

Rao, C. N. R., and Rao, K. J., *Phase Transitions in Solids* (McGraw-Hill, New York, 1978).

Scott, J. F., *Rev. Mod. Phys.* **46,** 83 (1974).

Sengers, J. V., and Levelt-Sengers, J. M. H., in *Progress in Liquid Physics,* C. A. Croxtoon, ed. (Wiley, New York, 1978), pp. 103–174.

Sengers, J. V., and Levelt-Sengers, J. M. H., *Ann. Rev. Phys. Chem.* **37,** 189 (1986).

Stanley, H. E., *Introduction to Phase Transitions and Critical Phenomena* (Oxford University Press, London, 1971).

Stephenson, J., in H. Eyring, D. Henderson, and W. Jost (Eds.), *Physical Chemistry: An Advanced Treatise,* Vol. VIII B, Chapter 10.

• PROBLEMS

1. At the normal boiling point of isopropanol, 82.3°C, its heat of vaporization is 665 J/g. Estimate its vapor pressure at 27°C.

2. Calculate the pressure at which water will boil at 95°C. The heat of vaporization of water is 40.6 kJ/mol.

3. The vapor pressure of supercooled liquid water is 2.149 torr at –10.0°C. The vapor pressure of ice at that temperature is 1.950 torr. The triple point of water is at 0.01°C. Calculate the heat of fusion of ice.

4. Calculate the change in freezing point per atmosphere of benzene at its normal freezing point, 5.5°C. The density of liquid benzene is 0.894 g/cm^3 and that of solid benzene is 1.014 g/cm^3 at 5.5°C. The heat of fusion of benzene is 9791 J/mol. Repeat this calculation for water, whose heat of fusion is 6000 J/mol. The density of liquid water at 0°C is 0.998 g/cm^3 and of ice, 0.9168 g/cm^3. Would a lake of frozen benzene be suitable for "ice" skating?

5. Prove that

$$c_\sigma = c_p - \alpha v \frac{\Delta h}{\Delta v}$$

where c_σ is defined in Eq. 24.19 and

$$\alpha \equiv \frac{1}{v}\left(\frac{\partial v}{\partial T}\right)_p.$$

6. The melting point of iodine (I_2) is 114°C. At this temperature, the vapor pressure of the solid increases 4.34 torr for a 1° rise in the temperature, and the value of the vapor pressure is 88.9 torr. Calculate the heat of sublimation of iodine.

7. The vapor pressure of liquid naphthalene is 10 torr at 85.8°C, and 40 torr at 119.3°C. Calculate the heat of vaporization and the boiling point. Calculate the vapor pressure at the melting point, 80°C. The vapor pressure of the solid is 1 torr at 52.6°C. Assume that the triple point is also 80°C, and calculate the heat of sublimation and the heat of fusion of the solid.

8. The vapor pressure of liquid arsenic depends on temperature as follows:

$$\log \frac{p_L}{\text{atm}} = -\frac{2460\ \text{K}}{T} + 3.81.$$

For solid arsenic, the vapor pressure is given by

$$\log \frac{p_S}{\text{atm}} = -\frac{6947\,\text{K}}{T} + 7.92.$$

Calculate the triple-point pressure and temperature.

9. Trouton's rule (see Section 14.4) states that the entropy of vaporization of most liquids is close to 85 J/K mol. Obtain a formula for the approximate vapor pressure of any liquid as a function of temperature and the boiling point of the liquid.

10. The vapor pressure of liquid iodine at 389.7 K is 100 torr and its boiling point is 456.2 K. For solid iodine, the vapor pressure is 1 torr at 311.9 K and the heat of fusion is 15.6 kJ/mol. Calculate the triple-point temperature and pressure of iodine.

11. At the normal melting point of mercury, 234.28 K, the density of the liquid is 13.69 g/cm^3 and the density of the solid is 14.19 g/cm^3. If the heat of fusion is 9.75 J/g, calculate the melting point of mercury at 20 atm.

12. The heat of vaporization of water is 44.8 kJ/mol and the heat of fusion of ice is 6.0 kJ/mol. If the vapor pressure of ice at 0°C is 4.58 torr, calculate its vapor pressure at –20°C.

13. It is found that the vapor pressure of a particular liquid varies with temperature by the equation

$$\log \frac{p_L}{\text{atm}} = -\frac{4849\,\text{K}}{T} - 14.70 \log \frac{T}{\text{K}} + 50.24.$$

How does the heat of vaporization vary with temperature? What is the heat of vaporization at 20°C?

14. Can a system containing three components in six phases be in equilibrium?

15. The discussion of the Gibbs phase rule in the text assumes that each component of the mixture occurs in each phase. How does the rule change if one or more components are absent from one or more phases?

16. Find the critical temperature, pressure, and volume of 1 mol of a van der Waals gas.

17. Show that

$$\frac{1}{v}\left(\frac{\partial v}{\partial T}\right)_p \quad \text{and} \quad -\frac{1}{v}\left(\frac{\partial v}{\partial p}\right)_T$$

for a van der Waals gas become infinite at the critical point.

18. Suppose that the virial equation of state,

$$\frac{p}{k_B T} = n + \sum_{l=2}^{\infty} B_l n^l ,$$

accurately describes a hard-sphere gas. For this case the virial coefficients B_l are independent of T. It is believed that all of the B_l are positive. Assume that this is correct and show that a consequence is that a hard-sphere gas cannot have a critical point. How do you interpret this result?

19. Find the critical temperature, pressure, and volume of a gas for which

$$\frac{p}{k_B T} = n + B_2 n^2 + B_3 n^3 ,$$

where B_2 and B_3 depend on the temperature but not on the density.

The following comments provide background material for the next eight problems.

The classical theory of the critical region assumes that near the critical point the following expansions are valid:

$$\tilde{A} = \sum_{m=0}^{\infty}\sum_{l=0}^{\infty} (m!l!)^{-1} A_{ml} (\Delta\tilde{n})^m (\Delta\tilde{T})^l ,$$

$$\tilde{\mu} = \sum_{m=0}^{\infty}\sum_{l=0}^{\infty} (m!l!)^{-1} \mu_{ml} (\Delta\tilde{n})^m (\Delta\tilde{T})^l ,$$

$$\tilde{p} = \sum_{m=0}^{\infty}\sum_{l=0}^{\infty} (m!l!)^{-1} p_{ml} (\Delta\tilde{n})^m (\Delta\tilde{T})^l ,$$

where $\tilde{T} \equiv T/T_c$, $\tilde{n} \equiv n/n_c$, and $\tilde{p} \equiv p/p_c$ are the reduced temperature, number density, and pressure, respectively. The reduced Helmholtz free energy is defined by $\tilde{A} \equiv A/Vp_c$ and the reduced chemical potential by $\tilde{\mu} \equiv \mu n_c/p_c$, whereas the expansion variables $\Delta\tilde{n}$ and $\Delta\tilde{T}$ are defined by $\Delta\tilde{n} \equiv (n - n_c)/n_c$ and $\Delta\tilde{T} \equiv (T - T_c)/T$. We shall also need the reduced functions $\tilde{\kappa}_T \equiv \kappa_T p_c/n_c^2$, $\tilde{S} \equiv ST_c/Vp_c$, $\tilde{C}_V \equiv C_V T_c/Vp_c$, and $\Delta\tilde{\mu} \equiv (n_c/p_c)[\mu(n, T) - \mu(n_c, T)]$. Note that because $\tilde{\mu} = (\partial\tilde{A}/\partial\tilde{n})_T$ and $\tilde{p} = \tilde{\mu}\tilde{n} - \tilde{A}$, the expansion coefficients A_{ml}, μ_{ml}, and p_{ml} are related by

$$\mu_{ml} = A_{m+1,l}, \qquad m \geq 0;$$

$$p_{ml} = A_{m+1,l} + (m-1)A_{ml}, \qquad m \geq 0;$$

$$p_{ml} = \mu_{ml} + (m-1)\mu_{m-1,l}, \qquad m \geq 1.$$

The usually stated conditions for a critical point, namely, $(\partial p/\partial V)_T = 0$ and $(\partial^2 p/\partial V^2)_T = 0$, imply that $A_{20} = 0$, $A_{30} = 0$, $\mu_{10} = 0$, $\mu_{20} = 0$, $p_{10} = 0$, and $p_{20} = 0$. It is, furthermore, usually assumed that $p_{30} = \mu_{30} = A_{30} \neq 0$.

20. Show that the condition of mechanical stability requires that for $T > T_c$

$$p_{30} = \mu_{30} = A_{40} > 0 \quad \text{and} \quad p_{11} > 0.$$

21. Show that the condition of thermal stability requires that $A_{02} < 0$ when $T > T_c$.

22. Show that the classical value of the exponent in Eq. 24.35 is

$$\delta_{\text{classical}} = 3.$$

23. Show that the classical value of the exponent in Eq. 24.36 is, for $\Delta\tilde{T} > 0$,

$$\alpha_{\text{classical}} = 0.$$

24. Show that the classical value of the exponent in Eq. 24.37 is, for $\Delta\tilde{T} > 0$,

$$\gamma_{\text{classical}} = 1.$$

25. Show that the coexistence curve near T_c satisfies

$$\frac{(\Delta\tilde{n}_L + \Delta\tilde{n}_G)^2 - (\Delta\tilde{n}_L)(\Delta\tilde{n}_G)}{\Delta\tilde{T}} = -\frac{6p_{11}}{p_{30}}.$$

26. From the result stated in Problem 25, derive the law of rectilinear diameters,

$$\frac{\Delta\tilde{n}_L + \Delta\tilde{n}_G}{2} = B_c\left|\Delta\tilde{T}\right| + \cdots,$$

where

$$B_c = \frac{\mu_{21}}{\mu_{30}} - \frac{3\mu_{11}\mu_{40}}{5\mu_{30}^2}$$
$$= \frac{p_{21}}{p_{30}} + \frac{4p_{11}}{5p_{30}} - \frac{3p_{11}p_{40}}{5p_{30}^2}.$$

27. A different theory of the critical region is generated if it is assumed that the first *four* isothermal derivatives of p with respect to V vanish at $T = T_c$. This theory is still classical in that the power series expansions set out above are assumed to be valid. In this case we have

$$A_{20} = A_{30} = A_{40} = A_{50} = 0,$$
$$\mu_{10} = \mu_{20} = \mu_{30} = \mu_{40} = 0,$$
$$p_{10} = p_{20} = p_{30} = p_{40} = 0,$$
$$p_{50} = \mu_{50} = A_{60} > 0,$$
$$p_{11} = \mu_{11} > 0.$$

(a) Show that under these assumptions the exponents δ and γ, defined in Eqs. 24.35 and 24.37, have the values 5 and 1, respectively.
(b) Show that under these assumptions the law of rectilinear diameters is no longer valid.

28. Use Eq. 24.50 to derive the Maxwell equal areas construction.

The following problems consider some of the thermodynamic implications of the free-energy function (Eq. 24.89).

29. Given Eq. 24.89 applied to a system that can undergo a second-order transition,

$$G(p,T) = G_0 + A\eta^2 + C\eta^4,$$

suppose that $A = a(T - T_c)$. Show that:
(a) This form for the coefficient A implies that a changes sign at $T = T_c$.
(b) There are two minima of G for $T < T_c$, located at

$$\eta = \pm\left(\frac{|a|}{2C}\right)^{1/2},$$

and one maximum at $\eta = 0$.
(c) The entropy has the form

$$S = S_0 + \frac{a^2}{2C}(T - T_c).$$

(d) The specific heat at constant pressure has the form

$$C_p = T\left(\frac{\partial S_0}{\partial T}\right)_p + T\left(\frac{a^2}{2C}\right).$$

(e) S is continuous and C_p discontinuous at $T = T_c$.

30. Suppose that the free energy of a system is described by

$$G = G_0 + A\eta^2 + B\eta^3 + C\eta^4,$$

with $B > 0$. In the case that $B = 0$, as in Problem 29, the free energy describes a second-order transition. When $B > 0$, as here, the free-energy function describes a different kind of transition.
(a) Draw several curves of G versus η, some for $T > T_c$, one for $T = T_c$, and some for $T < T_c$.
(b) Show that at $T = T_c$ there is one minimum at

$$\eta = \frac{1}{4|a|}[3B + (9B^2 + 32|a|C)^{1/2}],$$

at lower temperatures there are two minima at

$$\eta = \frac{-3 \pm (9B^2 - 32aC)^{1/2}}{4a},$$

and at high temperature there is only one minimum at $\eta = 0$.

(c) Show that $(\partial G/\partial T)_p$ is discontinuous at $T = T_c$.

(d) Trace the behavior of the system around a closed cycle starting from $T > T_c$ and going through $T < T_c$. Show that the system exhibits hysteresis.

Hint: Plot the entropy of the system along the cyclic path chosen.

31. Plot the extrema of G found in Problem 29 as a function of $-T$ for the following cases:
(a) $a > 0$ and $C > 0$.
(b) $a < 0$ and $C > 0$.
(c) Show that in case (a) the system is stable when $\eta = 0$, and that in case (b) the system is unstable when $\eta = 0$.

32. Suppose that some substance undergoes a solid–solid transition from a low-temperature form α to a high-temperature form β at $T = T_{\text{trans}}$. Construct a simple molecular description of the transition by assuming that solids α and β can be described by the Einstein model (see Chapter 22); the frequencies and binding energies of α and β are ν_α, $-\epsilon_\alpha$ and ν_β, $-\epsilon_\beta$, respectively. Show that

$$T_{\text{trans}} = \frac{\epsilon_\beta - \epsilon_\alpha}{3k_B \ln(\nu_\beta/\nu_\alpha)},$$

and that there is only one temperature at which α and β are in equilibrium.

33. Imagine a solid in equilibrium with its vapor, which is an ideal gas of atoms. Imagine, for simplicity, that the solid can be described by the Einstein model (Chapter 22), with binding energy per atom ϵ, frequency ν, and atomic mass m. Find an expression for the equilibrium vapor pressure of the solid.

34. Consider the same solid as in Problem 33, but suppose now that the vapor is an ideal gas mixture of atoms and diatomic molecules. If the binding energy of the diatomic molecule is D, its frequency ν', its moment of inertia I, and its mass $2m$, derive an expression for the vapor pressure of the solid.

35. The vapor pressure of a liquid at 300.00 K is equal to 0.100 atm. Using the Clausius–Clapeyron equation (*i.e.,* assume the vapor behaves ideally), compute the vapor pressure of the liquid at 400.00 K for two different situations: (a) the heat of vaporization of the liquid has a constant value (independent of the temperature) of 25.0 kJ/mol, (b) the heat of vaporization of the liquid is given by:

$$\Delta h_{\text{vap}} / (\text{kJ / mol}) = 25.0 \times e^{-T/6000}$$

Use a computer-based mathematical package such as MAPLE, MATHEMATICA, MACSYMA, or DERIVE to evaluate the required integral for part (b).

CHAPTER 25

Solutions of Nonelectrolytes

A great many physical and chemical phenomena are associated with the properties of solutions. For example, the vapor pressure of a mixture varies with its composition, the solubility of a compound may vary greatly from solvent to solvent, reaction equilibria may depend on the solvent medium in which the reaction occurs, and so forth. It is these and related properties that we examine in this chapter. As will be seen, much of the information available about the properties of solutions is derived from studies of the nature of the equilibria between a solution and other phases with which it may be in contact, say, a pure solid, vapor, or an electrode.

The study of the thermodynamic properties of solutions has two complementary aspects, reflecting alternatives in the analysis. If we assume that we know the dependences on temperature, pressure and concentration of the chemical potentials of all components in all the phases involved, the purpose of the thermodynamic analysis is to predict the characteristics of the equilibrium state. For example, knowing the forms of the chemical potentials in both liquid and vapor phases of a binary solution permits us to predict the vapor composition for a given composition of the liquid. This kind of information is of importance in the discussion of the practicality of a distillation procedure for the separation of the two components in the liquid. The alternative analysis uses the measured compositions of the liquid and vapor phases, the temperature and the pressure to deduce the chemical potentials of the components. The deviations of the chemical potentials from ideal behavior can then be interpreted in terms of molecular interactions. As in previous chapters, whenever possible we shall examine both the thermodynamic and the molecular interpretation of solution properties.

25.1 The Chemical Potential of a Component in an Ideal Solution

It has been shown, in Chapter 19, that knowledge of a substance's chemical potential is a necessary prerequisite to developing a quantitative description of the properties characteristic of phase equilibrium. Furthermore, it has been shown that the equilibrium composition of a reacting gas mixture can be predicted once the chemical potentials of all reacting species are known. It should, therefore, be obvious that the first step in studying the thermodynamics of solutions is to determine the functional dependence of the chemical potential of each species present on the composition, temperature, and pressure.

An examination of the procedure employed in Section 21.2 to determine this dependence for a component in a gaseous mixture reveals that knowledge of the equation of state of the gas plays an important role in the analysis. Starting from the general thermodynamic relationship connecting a change in chemical potential with changes in entropy and pressure, we used the fact that the real gas approaches perfect gas behavior in the zero-pressure limit, together with Dalton's law defining the partial pressure of a component in the gas mixture, to provide a representation of the chemical potential of the component in terms of composition and pressure. Once the chemical potentials of all components of the mixture were specified, the properties of the mixture and the relationships of these properties to those of the pure components could be easily analyzed.

The specific procedure used successfully for gaseous mixtures cannot now be used to analyze the behavior of liquid or solid mixtures. The reason is simple: At present there is not available any limiting form of the equation of state of a solid or liquid suitable for use in the same way as the perfect gas limit is used in the study of gases. For this reason we must modify the procedure used previously, so as to avoid requiring specific knowledge of the equation of state of the condensed phase. As already shown in Chapter 19 (see Eqs. 19.88–19.93), the composition dependence of the chemical potential can be deduced without knowledge of the equation of state, but determination of the temperature and pressure dependence of the chemical potential does require such knowledge.

Consider a mixture in which no chemical reactions occur. It is observed that when two liquids are mixed at constant temperature and pressure, there is usually a change in volume and evolution or absorption of heat. There are some mixtures for which these enthalpy and volume changes are very small, occasionally zero within the precision of measurement. Clearly, the more nearly alike the molecules of the several components, the smaller will be the changes in enthalpy and volume on mixing. For example, a mixture of perprotobenzene (C_6H_6) and perdeuterobenzene (C_6D_6) can be formed at constant temperature and pressure with negligibly small volume and enthalpy changes. We define an ideal solution, as an abstraction from observations, by the conditions

$$(\Delta H_{\text{mixing}})_{T,p} = (H^E)_{T,p} = 0 \tag{25.1}$$

and

$$(\Delta V_{\text{mixing}})_{T,p} = (V^E)_{T,p} = 0, \tag{25.2}$$

as a result of which we have

$$(\Delta U_{\text{mixing}})_{T,p} = (U^E)_{T,p} = 0 \tag{25.3}$$

and

$$(\Delta G_{\text{mixing}})_{T,p} = -T(\Delta S_{\text{mixing}})_{T,p}. \tag{25.4}$$

We shall now use these conditions to determine the composition dependence of the chemical potential for a component of an ideal solution. Note that this is the reverse of the procedure we used to study the properties of gaseous mixtures (Section 21.2). In that study we first established the composition dependence of the chemical potentials, and then deduced the limiting properties of a perfect gas mixture. It is the lack of knowledge of the equation of state of a liquid (or solid) that forces us to reverse the procedure in the study of condensed-phase mixtures.

Let the superscript zero (^{0}) refer to pure components. The change in Gibbs free energy on formation of a solution is related to the chemical potentials of the components as follows:

$$\Delta G_{\text{mixing}} \equiv G - G^0 = \sum_{i=1}^{r} n_i(\mu_i - \mu_i^0), \tag{25.5}$$

where n_i is the number of moles of component i. Equation 25.2 can be rewritten to read

$$V^E \equiv V - V^0 = \left(\frac{\partial \Delta G_{\text{mixing}}}{\partial p}\right)_{T,n} = 0, \tag{25.6}$$

and Eq. 25.1 can be rewritten to read

$$H^E \equiv H - H^0 = -T^2\left[\frac{\partial}{\partial T}\left(\frac{\Delta G_{\text{mixing}}}{T}\right)\right]_{p,n} = 0. \tag{25.7}$$

We now assume that the conditions on the right-hand sides of Eqs. 25.6 and 25.7, which pertain to $\Sigma_{i=1}^{r} n_i(\mu_i - \mu_i^0)$, are equally valid for each of the $\mu_i - \mu_i^0$. Then, term by term,

$$\left[\frac{\partial(\mu_i - \mu_i^0)}{\partial p}\right]_{T,n} = 0, \tag{25.8}$$

and

$$\left[\frac{\partial}{\partial T}\left(\frac{\mu_i - \mu_i^0}{T}\right)\right]_{p,n} = 0. \tag{25.9}$$

Equations 25.8 and 25.9 may be thought of as differential equations for the variable $(\mu_i - \mu_i^0)$; together they imply that $\mu_i - \mu_i^0$ can only be a function of the form

$$\mu_i - \mu_i^0 = Tw_i(n_1, \ldots, n_r), \tag{25.10}$$

where $w_i(n_1, \ldots, n_r)$ is a function of only the composition variables $n_1, \ldots, n_r$ and μ_i^0 is a function only of T and p. Substitution of Eq. 25.10 into Eqs. 25.8 and 25.9 verifies that it is a solution of the differential equations. Our remaining task is to determine the form of the composition function $w_i(n_1, \ldots, n_r)$.

It has been shown in Chapter 19 (Eq. 19.69) that for an ideal solution

$$w_i(n_1, \ldots, n_r) = R \ln x_i, \tag{25.11}$$

so that Eq. 25.10 becomes[1]

$$\mu_i = \mu_i^0 + RT \ln x_i. \tag{25.12}$$

The form 25.12 arises when it is asserted that the chemical potentials of all species in solution have the same analytical form. Although Eq. 25.12 is written with the mole

[1] We can also obtain Eq. 25.12 from the following argument. From Eqs. 25.4, 25.5, and 25.10 we can write

$$(\Delta S_{\text{mixing}})_{T,p} = -\sum_i n_i w_i(n_1, \ldots, n_r)$$

for the entropy of mixing. Note that nothing in our argument thus far has assumed any particular form of the equation of state, or any particular kind of phase. The equation written for the entropy of mixing is completely general. Since w_i is a function of composition only, an ideal solution of given composition must have the same value of ΔS_{mixing} for any T and p, independent of the state of aggregation. At sufficiently high temperature and low pressure any such mixture will approach the perfect gas limit, where we already know from Eq. 21.49 that

$$(\Delta S_{\text{mixing}})_{T,p} = -R\sum_i n_i \ln x_i,$$

x_i being the mole fraction of component i. A perfect gas mixture is ideal by our present definition ($V^E = 0$, $H^E = 0$). The two forms in which we have written ΔS_{mixing} are then identical if

$$w_i(n_1, \ldots, n_r) = R \ln x_i.$$

The chemical potential of a component in an ideal solution is then given by Eq. 25.12.

fraction x_i as the concentration unit, other choices are possible. For mixtures of small molecules the mole fraction is most often chosen, but for solutions of polymers the volume fraction, ϕ_i, is a far better choice. Finally, the most important aspect of the form 25.12 is its relationship to the properties of the pure component, displayed explicitly in the term $\mu_i^0(T, p)$. The function $\mu_i^0(T, p)$ plays a role in the theory of liquid mixtures analogous to the hypothetical perfect gas chemical potential, $\mu_i^*(T)$, in the theory of gaseous mixtures.

The procedure used to derive Eq. (25.12) emphasizes the lack of a generally valid equation of state for condensed systems and starts with the definition of an ideal mixture, embodied in Eqs. 25.1–25.4. Alternatively, noting that the chemical potential of a component i in an ideal gas mixture is proportional to $RT \ln x_i$, one can by analogy define the chemical potential of i in an ideal liquid or solid mixture by Eq. 25.12 and then deduce as consequences the conditions 25.1–25.4. The difference between these approaches is, obviously, a matter of emphasis. The one we have used takes as fundamental the (putative) observed properties of an ideal liquid or solid mixture, whereas the alternative mentioned takes as fundamental the definition formulated by analogy with the behavior of an ideal gas mixture.

25.2 The Chemical Potential of a Component in a Real Solution

As already noted, when liquids are mixed there is ordinarily some volume change and some heat absorbed or released. Thus, however useful the concept of an ideal solution as an abstraction from and extrapolation of observed behavior, it is necessary to analyze the properties of the chemical potential further if we wish to describe most real solutions.

There are many different formal ways of extending the representation of the chemical potential. For example, if the chemical potential of component i is written in the form

$$\mu_i = \mu_i^0(T, p) + RT \ln x_i + \mu_i^E, \tag{25.13}$$

then μ_i^E, known as the *excess chemical potential* of *i*, is directly related to the observable heat of mixing and volume change of mixing. (See Eqs. 25.6 and 25.7.) In fact, from the fundamental thermodynamic relations connecting the enthalpy and volume changes with the temperature and pressure derivatives of the corresponding Gibbs free energy change, we find

$$\begin{aligned} V^E &= \left(\frac{\partial \Delta G_{\text{mixing}}}{\partial p}\right)_{T,n_1,\ldots,n_r} \\ &= \sum_{i=1}^{r} n_i \left(\frac{\partial \mu_i^E}{\partial p}\right)_{T,n_1,\ldots,n_r}, \end{aligned} \tag{25.14}$$

$$\begin{aligned} H^E &= -T^2 \left(\frac{\partial (\Delta G_{\text{mixing}}/T)}{\partial T}\right)_{p,n_1,\ldots,n_r} \\ &= -T^2 \sum_{i=1}^{r} n_i \left(\frac{\partial (\mu_i^E/T)}{\partial T}\right)_{p,n_1,\ldots,n_r}. \end{aligned} \tag{25.15}$$

The first two terms on the right-hand side of Eq. 25.13, which correspond to the "ideal contribution" to the chemical potential of component *i*, make no contribution to the heat of mixing and volume change on mixing. As might be surmised from the use of the words *excess chemical potential* and the definition of an ideal solution, ΔV_{mixing} and ΔH_{mixing} are sometimes called the *excess volume* and *excess enthalpy* of the solution. It is for this reason that we introduced the notation V^E, H^E in Eqs. 25.2 and 25. 1. Clearly, if we have $\Delta V_{\text{mixing}} \neq 0$ and $\Delta H_{\text{mixing}} \neq 0$, the entropy of mixing will in general also deviate from the ideal form. Again using standard thermodynamic relations (see Table 19.2), we obtain

$$\begin{aligned} \Delta S_{\text{mixing}} &= -\left(\frac{\partial \Delta G_{\text{mixing}}}{\partial T}\right)_{p,n_1,\ldots,n_r} \\ &= -R \sum_{i=1}^{r} n_i \ln x_i - \sum_{i=1}^{r} n_i \left(\frac{\partial \mu_i^E}{\partial T}\right)_{p,n_1,\ldots,n_r}. \end{aligned} \tag{25.16}$$

The *excess entropy* of mixing is defined to be the difference $\Delta S_{\text{mixing}} - (-R \sum_i n_i \ln x_i)$. Thus, all the excess thermodynamic functions are referred to the properties of a hypothetical ideal solution with the same composition, temperature, and pressure. This definition of the excess entropy and the excess free energy differs from that introduced in Chapter 21. In Chapter 21 we defined the excess value of X as the X of mixing ($G^E = RT \sum_i n_i \ln x_i$) but in this chapter it is defined as the difference from the ideal solution value. For U, H and V the two definitions are equivalent, while for S and G they differ by the ideal entropy of mixing ($-R \sum_i n_i \ln x_i$) and ideal free energy of mixing ($RT \sum_i n_i \ln x_i$), respectively.

For convenience, instead of using the extensive excess functions V^E, H^E, . . . , we shall often describe a mixture in terms of deviations from its *mean molar properties*. For example, the mean molar volume is

$$v_m \equiv \frac{V}{\sum_i n_i}, \tag{25.17}$$

and the excess volume per mean mole of mixture is

$$v^E \equiv \frac{V}{\sum_i n_i} - \frac{\sum_i n_i v_i^0}{\sum_i n_i} = \frac{V}{\sum_i n_i} - \sum_i x_i v_i^0, \tag{25.18}$$

where v_i^0 is the molar volume of component i at the same temperature and pressure as is the mixture. Corresponding definitions hold for the mean molar excess enthalpy, h^E, entropy, s^E, and other properties.

In the discussion of the properties of imperfect gases we introduced the fugacity of the gas. When written in terms of the fugacity instead of the pressure, the algebraic forms of the chemical potentials of a perfect gas and an imperfect gas are the same. Similarly, the chemical potential of a component of a real solution can be written in forms that are different from Eq. 25.13. One form, much used in the chemical literature, consists of writing the chemical potential of component i as

$$\mu_i = \mu_i^0(T,p) + RT\ln\gamma_i x_i\,. \tag{25.19}$$

Then

$$\gamma_i x_i \equiv a_i \tag{25.20}$$

is known as the *activity* of component i; γ_i is called the *activity coefficient*. This language probably stems from the notion that the "effective concentration" of a component determines the extent to which it influences, or *acts* in determining, the equilibrium properties of the solution. The product $\gamma_i x_i$ replaces the mole fraction x_i in the analytical form of the chemical potential of i in an ideal solution, and therefore may be thought of as the effective concentration of i. Clearly γ_i must depend on temperature, pressure, and composition. Also, comparison of Eqs. 25.13 and 25.19 shows that

$$\mu_i^E = RT\ln\gamma_i\,, \tag{25.21}$$

so that ΔH_{mixing}, ΔV_{mixing}, ΔS_{mixing} may all be expressed in terms of derivatives of γ_i.

The two formulations displayed in Eqs. 25.13 and 25.19 do not exhaust the possibilities for formal analysis of the chemical potential,[2] but we shall not discuss any of the other representations of nonideality in this book. In this chapter we discuss the properties of solutions mostly in terms of the excess functions μ^E, h^E, s^E, and v^E, whereas in Chapter 26 we mostly use the activity and activity coefficient representation. Different choices of representation are made in the two chapters to conform to the respective common usages in studies of nonelectrolyte and electrolyte solutions.

Before we proceed further, it is necessary to make some comments about the term $\mu_i^0(T, p)$ in Eqs. 25.13 and 25.19. This term, usually called the chemical potential of the standard state, defines the reference level with respect to which changes in the chemical potential of component i are measured. There are two common conventions used to choose the standard state of a component in a solution. Since $\mu_i^0(T, p)$ is independent of composition, one obvious choice is the chemical potential of the pure liquid or solid at the same temperature and pressure,

$$\mu_i^0(T,p) = \mu_i^L(T,p) \tag{25.22a}$$

or

$$\mu_i^0(T,p) = \mu_i^S(T,p). \tag{25.22b}$$

If the pure phase chosen as the standard state for a given T and p is not stable at that T and p, then an extrapolation of the known properties of the liquid (or solid) is necessary to define the standard state. Although this seems an unnecessarily complicated way to choose a standard state, it is actually quite convenient in practical computation and is widely used. When this standard state is employed we denote it by the superscript (^{0}). A second convention used to define a standard state distinguishes between the component present in largest amount (solvent, labeled 1) and all other components. The standard state for component 1 is selected to be the pure liquid (or solid),

$$\mu_1^0(T,p) = \mu_1^L(T,p) \quad \text{(solvent)}, \tag{25.23a}$$

whereas for all the other components a *hypothetical* standard state is chosen: We imagine that there exists a fictitious state of component i with the properties that pure i would have if its limiting low-concentration properties in solution were to be retained in a pure substance. In mathematical terms this is equivalent to writing

$$\begin{aligned}\mu_i^0(T,p) &\equiv \mu_i^{\ominus}(T,p)\\ &= \lim_{x_i\to 1}(\mu_i - RT\ln x_i) \quad \text{(solute)}.\end{aligned} \tag{25.23b}$$

Note that the standard state in this convention, which we denote by the superscript ($^{\ominus}$), depends on temperature, pressure, *and* the nature of the solvent (component 1). How does this hypothetical standard state differ from the state of the true pure component? Consider a very dilute solution, one in which each solute molecule of species i is surrounded only by solvent molecules, and in which the average separation between molecules of i is so large that they do not interact with one another. Further dilution will not then alter the nature of the interaction of a molecule of i and the solvent. If the interaction between molecules of i and solvent is different from that between molecules of i alone, the motion and properties of an i molecule surrounded by solvent molecules must be different from those of an i molecule surrounded by only other i molecules. A hypothetical pure liquid constructed of particles imagined to interact as do i and solvent will, clearly, be different from pure liquid i. In macroscopic terms we expect that, for example, in a very

[2] For example, the fugacity can be extended to condensed phases, being defined in relation to the activity by $a_i = f_i/f_i^0$, where f_i^0 is the value in the standard state.

dilute solution the partial molar volume $\bar{v}_i \equiv (\partial V/\partial n_i)_{T,p,n'_j}$ becomes independent of composition, but not equal to the molar volume v_i^0 of pure component *i*. Extrapolation of the limiting low-concentration solution behavior then leads to a fictitious state with component *i* occupying the volume $v_i^\ominus$ and not the molar volume v_i^0. Similar discrepancies may exist for all of the thermodynamic functions. The standard state defined by the second convention is most useful in the description of dilute solutions. It is often called, misleadingly, the standard state of infinite dilution. The concentration of component *i* in this standard state is not infinitesimally small. Rather, we repeat, there is defined a fictitious state of matter that has the same properties per mole as does component *i* when present in dilute solution.

One of the principal goals of the description of real gases and real gas mixtures was the representation of the chemical potential in terms of the equation of state and specific heat (see Chapter 21). The form of the chemical potential used to describe solid and liquid mixtures, Eq. 25.13 or Eq. 25.19, has not yet been reduced to comparable form. Of course, from the definition of the excess properties of the solution we know that[3]

$$\left[\frac{\partial(\mu_i^E/T)}{\partial p}\right]_{T,n_1,\ldots,n_r} = R\left(\frac{\partial \ln\gamma_i}{\partial p}\right)_{T,n_1,\ldots,n_r} = \frac{\bar{v}_i - v_i^0}{T},$$

$$\left[\frac{\partial(\mu_i^E/T)}{\partial T}\right]_{p,n_1,\ldots,n_r} = R\left(\frac{\partial \ln\gamma_i}{\partial T}\right)_{p,n_1,\ldots,n_r} = \frac{\bar{h}_i - h_i^0}{T^2}, \tag{25.24}$$

which can be regarded as differential equations determining the behavior of the excess chemical potential at constant composition. From the Gibbs–Duhem equation (Eq. 19.68), at constant temperature and pressure we have

$$\left(\sum_{i=1}^{r} n_i\, d\mu_i\right)_{T,p} = 0, \tag{25.25}$$

which yields an equation defining the composition dependence of the excess chemical potential at constant T and p:

$$\left(\sum_{i=1}^{r} n_i\, d\mu_i^E\right)_{T,p} = RT\left(\sum_{i=1}^{r} n_i\, d\ln\gamma_i\right)_{T,p} = 0. \tag{25.26}$$

Equation 25.26 follows from Eq. 25.25 directly after substitution of Eqs. 25.13 and 25.21 into Eq. 25.25 and evaluation of the differentials. For a two-component system, Eq. 25.26 can be rewritten in the form

$$n_1(d\ln\gamma_1)_{T,p} + n_2(d\ln\gamma_2)_{T,p} = 0,$$

$$(1-x_2)\left(\frac{\partial \ln\gamma_1}{\partial x_2}\right)_{T,p} + x_2\left(\frac{\partial \ln\gamma_2}{\partial x_2}\right)_{T,p} = 0. \tag{25.27}$$

Therefore, at constant temperature and pressure, knowledge of μ_1^E enables us to compute μ_2^E:

$$\mu_2^E = RT\ln\gamma_2 = RT\int_{x_2}^{1} \frac{1-x'_2}{x'_2}\frac{\partial \ln\gamma_1}{\partial x'_2}\,dx'_2$$

$$= \int_{x_2}^{1} \frac{1-x'_2}{x'_2}\frac{\partial \mu_1^E}{\partial x'_2}\,dx'_2. \tag{25.28}$$

Simultaneous solution of the three differential equations in Eqs. 25.24 and 25.27 completely determines μ_i^E.

It is clear that actual evaluation of Eqs. 25.24 and 25.28 requires knowledge of the composition dependence of the equation of state, the solution heat capacity, and the activity coefficient. This requirement is analogous to the need to know the composition dependence of the equation of state of a gaseous mixture in order to evaluate Eq. 21.29. In that case it proved advantageous to use a model equation of state, such as the van der Waals equation, or the virial equation of state for more general results. What analogous relations exist for liquid and solid solutions?

A model for the behavior of solutions that is analogous to the van der Waals model of the gas is defined by the following two statements:

1. The entropy of mixing has the ideal form $-R\sum_{i=1}^{r} n_i \ln x_i$.
2. The excess chemical potentials have the form (binary mixture)

$$\mu_1^E = RT\ln\gamma_1 = wx_2^2, \tag{25.29}$$

$$\mu_2^E = RT\ln\gamma_2 = wx_1^2 = w(1-x_2)^2. \tag{25.30}$$

For simplicity we shall assume that w is an energy that is independent of temperature, pressure, and composition. From these statements defining the model mixture, known as a *regular mixture,* we deduce that the heat of mixing has the form

$$\Delta H_{\text{mixing}} = -T^2\left[n_1\left(\frac{\partial}{\partial T}\frac{\mu_1^E}{T}\right)_{p,n_2} + n_2\left(\frac{\partial}{\partial T}\frac{\mu_2^E}{T}\right)_{p,n_1}\right]$$

$$= n_1 w x_2^2 + n_2 w x_1^2$$

$$= n_1 w x_2(1-x_1) + n_2 w x_1(1-x_2)$$

$$= w(n_1+n_2)x_2(1-x_2). \tag{25.31}$$

Notice that ΔH_{mixing} is symmetrical in the mole fractions x_1 and x_2.

We shall show later that the regular mixture model predicts that if w/RT is sufficiently large a one-phase mixture separates

[3] As written, Eqs. 25.24 use pure liquid or solid of component *i* as the standard state. If, instead, the standard state is defined by Eq. 25.23*b*, v_i^0 and h_i^0 are to be replaced by $v_i^\ominus$ and $h_i^\ominus$.

into two phases with different compositions. Also, just as in the case of condensation of a gas, there exists a critical temperature above which phase separation does not occur.

Instead of using a specific model, such as a regular solution, we can employ a representation of solution properties more closely related to the concept of the virial expansion. It is obtained by noticing that, at constant temperature and pressure, the general solution of Eq. 25.27 is

$$\ln \gamma_1 = \int x_2 \varphi(x_2)\, dx_2, \tag{25.32a}$$

$$\ln \gamma_2 = -\int (1-x_2)\varphi(x_2)\, dx_2, \tag{25.32b}$$

where $\varphi(x_2)$ is some function of x_2. That Eqs. 25.32*a* and 25.32*b* are indeed a solution of Eq. 25.27 can be verified by direct substitution. In the spirit of the virial expansion we now write for a binary solution

$$\varphi(x_2) = \sum_j \sum_k a_{jk}(T,p) x_2^j (1-x_2)^k, \tag{25.33}$$

where the coefficients $a_{jk}(T, p)$ are independent of composition. Consider the lowest-order approximation, which retains only the term $a_{00}(T, p)$ in the series expansion. Then we have

$$\ln \gamma_1 = \int x_2 a_{00}(T,p)\, dx_2 = \tfrac{1}{2} a_{00} x_2^2, \tag{25.34a}$$

$$\ln \gamma_2 = -\int (1-x_2) a_{00}(T,p)\, dx_2 = \tfrac{1}{2} a_{00}(1-x_2)^2. \tag{25.34b}$$

Comparison with Eq. 25.30 shows that, to this order, if $a_{00}(T, p)$ is independent of temperature and pressure, retention of only the term a_{00} in the series generates what we have called regular mixture theory. Clearly, just as in the virial equation representation, higher-order terms can be retained to provide a better and better description of the excess chemical potential.

25.3 Partial Molar Quantities

It is natural, when examining the properties of a mixture, to try to assess the contributions made by the several components. If the excess volume, excess enthalpy, . . . of the mixture are not zero, then the total volume, enthalpy, . . . are not simply the sums of the volumes, enthalpies, . . . of the pure components. Let Z be some extensive thermodynamic function of the mixture. The partial molar function $\bar{z}_i$, defined by

$$\bar{z}_i \equiv \left(\frac{\partial Z}{\partial n_i}\right)_{T,p,n_{j\neq i}}, \tag{25.35}$$

can be used to provide a systematic subdivision of the extensive function Z into a sum of component contributions. Suppose that the amount of each component in the mixture is increased, at constant temperature and pressure, by a factor ξ. Then, because Z is an extensive function,

$$Z(\xi n_1, \xi n_2, \ldots, T, p) = \xi Z(n_1, n_2, \ldots, T, p). \tag{25.36}$$

Differentiation of $Z(\xi n_1, \xi n_2, \ldots, T, p)$ with respect to ξ, followed by setting $\xi = 1$, leads to[4]

$$Z = \sum_{i=1}^{r} n_i \bar{z}_i. \tag{25.37}$$

Partial molar quantities are intensive properties of the mixture; they depend on T, p, and the relative composition of the mixture. To illustrate how Eq. 25.37 provides the wanted subdivision into component contributions, consider forming a binary solution of n_1 mol of 1 and n_2 of 2 in a total volume V. We imagine that, at constant temperature and pressure, we mix δn_1 mol of 1 with δn_2 mol of 2, and repeat this procedure as many times as needed while keeping the ratio $\delta n_1/\delta n_2 = n_1/n_2$ the same for each addition. Then, since the partial molar volumes $\bar{v}_1$ and $\bar{v}_2$ depend only on the relative composition for fixed T and p, adding up all the increments in volume gives

$$V = n_1 \bar{v}_1 + n_2 \bar{v}_2, \tag{25.38}$$

which is just a special case of Eq. 25.37.

A differential change in the extensive property Z, at constant temperature and pressure, is given by

$$(dZ)_{T,p} = \left(\sum_{i=1}^{r} n_i\, d\bar{z}_i\right)_{T,p} + \left(\sum_{i=1}^{r} \bar{z}_i\, dn_i\right)_{T,p}. \tag{25.39}$$

Comparison of Eqs. 25.39 and 25.35 leads to a constraint on the possible changes of the several $\bar{z}_i$, namely,

$$\left(\sum_{i=1}^{r} n_i\, d\bar{z}_i\right)_{T,p} = 0. \tag{25.40}$$

[4] We have

$$\left[\frac{\partial Z(\xi n_1, \ldots, T, p)}{\partial \xi}\right]_{T,p,n_i} = Z(n_1, n_2, \ldots, T, p)$$
$$= \sum_{i=1}^{r} \left(\frac{\partial Z}{\partial(\xi n_i)}\right)_{T,p,n_j} \left(\frac{\partial(\xi n_i)}{\partial \xi}\right)_{T,p,n_j}$$
$$= \sum_{i=1}^{r} \left(\frac{\partial Z}{\partial(\xi n_i)}\right)_{T,p,n_j} n_i,$$

which becomes, for $\xi = 1$,

$$Z(n_1, \ldots, n_r, T, p) = \sum_{i=1}^{r} n_i \bar{z}_i.$$

When $\bar{z}_i = \mu_i$, Eq. 25.40 becomes the Gibbs–Duhem relation (see Eqs. 19.68 and 25.25). In general, then, the possible changes in the partial molar properties of a mixture satisfy, individually, Gibbs–Duhem type equations of constraint.

The experimental determination of partial molar properties is based on the following transformation of Eq. 25.35. Let z_m be the mean molar value defined by

$$z_m \equiv \frac{Z}{\sum_i n_i} = \sum_i x_i \bar{z}_i. \tag{25.41}$$

Consider a binary mixture, and differentiate z_m with respect to the mole fraction of component 1:

$$\begin{aligned}\left(\frac{\partial z_m}{\partial x_1}\right)_{T,p} &= \left(\frac{\partial}{\partial x_1}[x_1\bar{z}_1 + (1-x_1)\bar{z}_2]\right)_{T,p} \\ &= \bar{z}_1 - \bar{z}_2 + x_1\left(\frac{\partial \bar{z}_1}{\partial x_1}\right)_{T,p} + x_2\left(\frac{\partial \bar{z}_2}{\partial x_1}\right)_{T,p} \\ &= \bar{z}_1 - \bar{z}_2,\end{aligned} \tag{25.42}$$

the last line following from an application of Eq. 25.40. Combining Eqs. 25.42 and 25.41 leads to

$$\begin{aligned}\bar{z}_1 &= z_m - x_2\left(\frac{\partial z_m}{\partial x_2}\right)_{T,p}, \\ \bar{z}_2 &= z_m - x_1\left(\frac{\partial z_m}{\partial x_1}\right)_{T,p}.\end{aligned} \tag{25.43}$$

Similar but algebraically more complex expressions relate the partial molar and mean molar properties of multicomponent mixtures.

To use Eq. 25.41 we plot the mean molar property z_m as a function of mole fraction of, say, component 1. Then, as shown in Fig. 25.1, the tangent to the curve at $z_m(x_1')$ has intercepts $\bar{z}_2(x_1')$ at $x_1 = 0$ and $\bar{z}_1(x_1')$ at $x_1 = 1$.

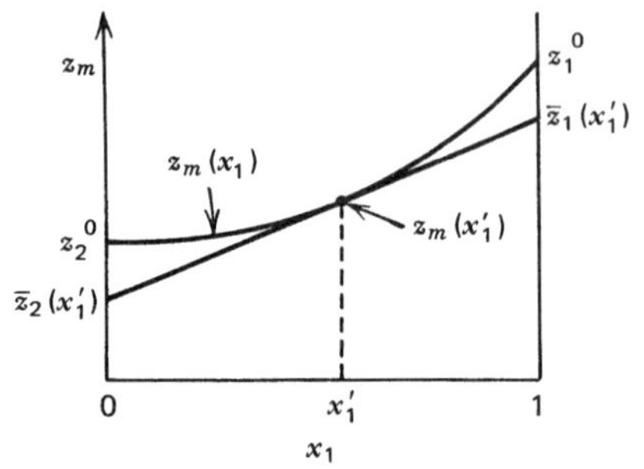

Figure 25.1 The mean molar function $z_m = Z/\Sigma n_i$ as a function of the mole fraction of component 1 in a binary mixture. The tangent to $z_m(x_1)$ at $x_1 = x_1'$ has intercepts $\bar{z}_2(x_1')$ at $x_1 = 0$ and $\bar{z}_1(x_1')$ at $x_1 = 1$.

25.4 Liquid–Vapor Equilibrium

We begin our study of nonelectrolyte mixtures with an examination of some of the characteristic features of liquid–vapor equilibrium. For each component of the mixture we must have

$$\mu_i\,(\text{liquid}) = \mu_i\,(\text{vapor}), \tag{25.44}$$

and, therefore, choosing the pure component as reference state (and recalling that the hypothetical perfect gas chemical potential is denoted $\mu_i^*(T)$),

$$\mu_i^0 + RT\ln\gamma_i x_i = \mu_i^*(T) + RT\ln\frac{f_i}{\mathfrak{p}}. \tag{25.45}$$

A simple rearrangement now gives

$$\gamma_i x_i = \frac{f_i}{\mathfrak{p}}\exp\left[\frac{1}{RT}(\mu_i^* - \mu_i^0)\right]. \tag{25.46}$$

But when $x_i = 1$ we have $\mu_i = \mu_i^0$ and $\gamma_i = 1$, since the pure component has been chosen as the reference state even for solutes, and for $x_i = 1$, $\mu_i = \mu_i^0$ no matter what the reference state is. Therefore, putting $x_i = 1$ in Eq. 25.46, we obtain

$$1 = \frac{f_i^0}{\mathfrak{p}}\exp\left[\frac{1}{RT}(\mu_i^* - \mu_i^0)\right]. \tag{25.47}$$

A combination of Eqs. 25.46 and 25.47 gives the simple result

$$\gamma_i x_i = \frac{f_i}{f_i^0}. \tag{25.48}$$

In the particular case that the solution is ideal ($\gamma_i = 1$) and the vapor phase is ideal ($f_i = p_i$), Eq. 25.48 reduces to

$$p_i = x_i p_i^0 \quad \begin{pmatrix}\text{solution ideal,} \\ \text{gas phase ideal}\end{pmatrix}, \tag{25.49}$$

which is known as *Raoult's law*. Equation 25.48 can be used in two ways: The activity coefficient of component *i* in the condensed phase can be deduced from measurements of the fugacity of component *i* in the vapor in equilibrium with the solution; on the other hand, if γ_i is known the vapor composition can be predicted.

Consider, first, equilibrium between an ideal liquid (or solid) mixture and an ideal gas mixture. Let y_i be the mole fraction of component *i* in the vapor phase. Then Dalton's law gives

$$p_i = y_i p \quad (\text{gas phase}), \tag{25.50}$$

where *p* is the total pressure of the gaseous mixture. By hypothesis the liquid mixture considered is ideal, and

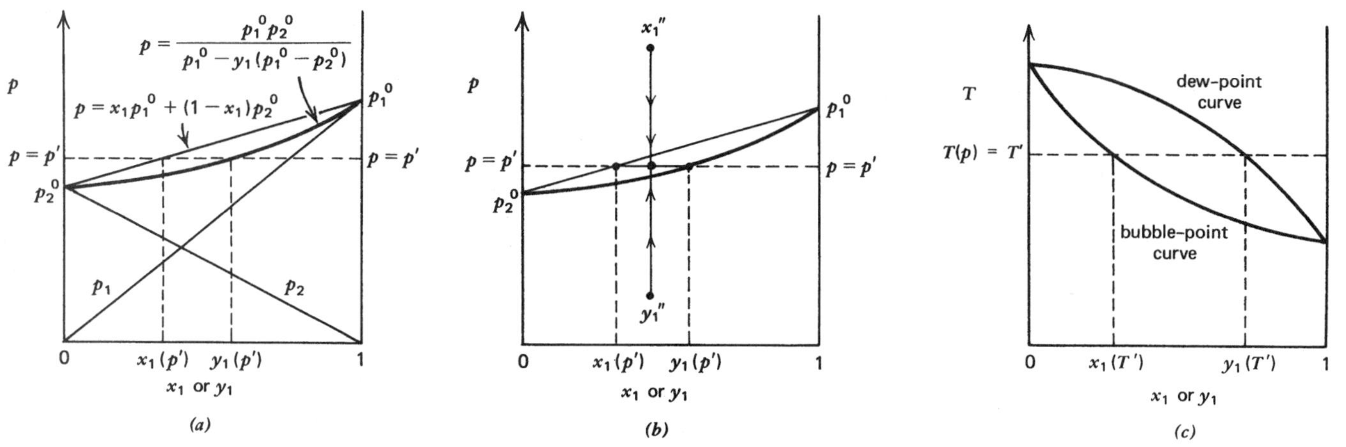

Figure 25.2 (*a*) Schematic representation of the isothermal dependence of the total pressure of an ideal binary mixture on the mole fraction of component 1 in the liquid, x_1, and on the mole fraction of component 1 in the vapor, y_1. (*b*) Schematic representation of the separation of a mixture into vapor and liquid phases when compressed to $p = p'$ isothermally from a state where p is less than the pressure along the dew-point curve, or when expanded isothermally from a state where p is greater than the pressure along the bubble-point curve. (*c*) Schematic diagram of the bubble-point and dew-point curves on a temperature-composition diagram at constant pressure (ideal binary mixture).

Eq. 25.49 thus describes the relation between the partial pressure of component i and its concentration in solution. Combining Eqs. 25.49 and 25.50, we have

$$y_i p = x_i p_i^0,$$
$$y_i = x_i \frac{p_i^0}{p} \tag{25.51}$$

The vapor and liquid compositions can be very different if the vapor pressures of the several pure components differ considerably. For example, at 80°C the vapor pressures of the monomethyl and monoethyl ethers of ethylene glycol are 154.4 torr and 102.2 torr, respectively. When mixed, these substances form an ideal solution. When the solution mole fractions are $x_1 = x_2 = 0.50$, the vapor mole fractions at 80°C are $y_1 = 0.602$, $y_2 = 0.398$, component 1 being the higher-vapor-pressure monomethyl ether. That is, the vapor is richer in the more volatile component.

It is useful to examine further liquid–vapor equilibrium in an ideal binary mixture. The total vapor pressure

$$p = p_1 + p_2 = x_1 p_1^0 + x_2 p_2^0 = x_1(p_1^0 - p_2^0) + p_2^0 \tag{25.52}$$

is, at constant temperature, a linear function of x_1 and/or x_2, as shown in Fig. 25.2*a*. On the other hand, because of the difference in composition between vapor and liquid, p is not a linear function of y_1. In fact, from Eq. 25.51 we find, for the concentration ratios in vapor and liquid,

$$\frac{(y_1/y_2)}{(x_1/x_2)} \equiv \alpha_{12} = \frac{y_1(1-x_1)}{x_1(1-y_1)} = \frac{p_1^0}{p_2^0}. \tag{25.53}$$

For an ideal mixture α_{12}, which measures the relative volatility of components 1 and 2, is independent of composition. If we now eliminate x_1 between Eqs. 25.53 and 25.52 we find for the total pressure as a function of y_1

$$p = \frac{p_1^0}{\alpha_{12} - y_1(\alpha_{12} - 1)} = \frac{p_1^0 p_2^0}{p_1^0 - y_1(p_1^0 - p_2^0)}, \tag{25.54}$$

which is a section of a rectangular hyperbola that is concave upward (see Fig. 25.2). Since vapor and liquid have the same pressure, the difference in their compositions at, say, $p = p'$, is obtained from the intersection of the line $p = p'$ with the two curves representing p as a function of x_1 and p as a function of y_1. As shown in Fig. 25.2, these intersections are $x_1(p')$ and $y_1(p')$, respectively. The line $p = x_1 p_1^0 + (1 - x_1)p_2^0$ is called the *bubble-point curve*, and $p = p_1^0 p_2^0/[p_1^0 - y_1(p_1^0 - p_2^0)]$ is called the *dew-point curve*. If a mixture is subjected to a pressure greater than that on its bubble-point curve, only the liquid phase will be stable; if it is subjected to a pressure less than that on its dew-point curve, only the vapor phase will be stable. If the pressure of the mixture falls between the dew-point and bubble-point pressures, it will separate into coexisting liquid and vapor phases. For example, if we imagine taking a gas mixture of composition y'', $(1 - y'')$ at a pressure below the dew-point pressure for that composition, and compressing that mixture isothermally until the pressure is p', then as shown in Fig. 25.2*b* the mixture separates into a liquid phase with composition $x_1(p')$, $[1 - x_1(p')]$ and a vapor phase with composition $y_1(p')$, $[1 - y_1(p')]$. The ratio of the number of moles of liquid to vapor is $[y_1(p') - y_1'']/[y_1'' - x_1(p')]$.[5] Liquid first appears when the vertical line representing the fixed composition y'' intersects the dew-point curve. Had we started with a liquid mixture with

[5] See the discussion of the lever rule, Chapter 24, Eq. 24.4.

composition x_1'', $(1 - x_1'')$, at a pressure greater than its bubble-point pressure, and lowered the pressure to p', vapor would first appear when the vertical line through x_1'' intersects the bubble-point curve.

We now note that, because p_1^0 and p_2^0 are, in general, different functions of T, the boiling point of the mixture is not a linear function of the liquid composition. The composition of the vapor changes with temperature according to

$$y_1(T) = \frac{x_1 \alpha_{12}(T)}{1 + x_1[\alpha_{12}(T) - 1]}, \tag{25.55}$$

where for an ideal mixture $\alpha_{12}(T) \equiv p_1^0(T)/p_2^0(T)$. Figure 25.2*c* displays the bubble-point and dew-point curves on a temperature-composition diagram. Since the vapor and liquid are at the same temperature, the difference in their compositions when the liquid is boiled at pressure p is obtained from the intersections of $T(p)$ with the bubble-point and dew-point curves, as shown in Fig. 25.2*c*. Of course, as the composition of the liquid changes, so does the boiling temperature, hence also the enrichment of the vapor in one component.

As an introduction to the study of nonideality in mixtures, we now consider a special case, but one that is very interesting. Specifically, we shall examine the nature of the equilibrium between a gas and a liquid in which the gas is only slightly soluble. When the gas and liquid do not react, for example, $H_2(g)$ and $H_2O(l)$, it is generally observed that the solubility of the gas is proportional to the gas pressure (strictly, to the gas fugacity) and often increases as the temperature is lowered. To make a thermodynamic analysis of this gas-liquid system we start with the observation that in the phase equilibrium between liquid and gas the boundary (the surface of the liquid) is permeable to transfer of the solute and the solvent. Consider a binary solution. Since both components are present in both phases, we know that

$$\begin{aligned} \mu_1(\text{gas phase}) &= \mu_1(\text{liquid phase}), \\ \mu_2(\text{gas phase}) &= \mu_2(\text{liquid phase}). \end{aligned} \tag{25.56}$$

We shall use the convention that the component normally a liquid is labeled 1, and the component normally a gas is labeled 2. Equation 21.39 gives the chemical potentials of the components in the gas phase. In the liquid phase we have, for a small range of concentration near $x_1 = 1$,

$$\begin{aligned} \mu_1(\text{liquid phase}) &= \mu_1^0 + RT\ln x_1 \\ \mu_2(\text{liquid phase}) &= \mu_2^{\ominus} + RT\ln x_2 . \end{aligned} \tag{25.57}$$

Note that in writing Eqs. 25.57 for the liquid phase we have assumed that the solution is ideal. We have chosen as the standard state for the liquid solvent, component 1, the pure liquid at the same temperature and pressure, and for the solute, component 2, the hypothetical state with the same properties per mole as an infinitely dilute solution ($\mu_2^{\ominus}$ given by Eq. 25.23*b*). Using Eqs. 25.56, 25.57, and 21.39, we obtain

$$\mu_2^*(T) + RT\ln\frac{f_2}{\mathfrak{p}} = \mu_2^{\ominus}(T, p, \text{solvent}) + RT\ln x_2, \tag{25.58}$$

which can be rearranged to read

$$\frac{f_2}{\mathfrak{p}} = x_2 \exp\left(\frac{\mu_2^{\ominus} - \mu_2^*}{RT}\right) = x_2 k_2(T, p, \text{solvent}). \tag{25.59}$$

Equation 25.59 (or more often Eq. 25.60 just below) is known as *Henry's law,* and $k_2(T, p, \text{solvent})$ as the *Henry's law constant.* If the vapor phase is at sufficiently low pressure to be described as an ideal gas mixture, f_i can be replaced by p_i, the partial pressure of component i. In that limit we have

$$p_2 = x_2 \mathfrak{p} k_2(T, p, \text{solvent}) \quad (\text{gas phase ideal}). \tag{25.60}$$

Equation 25.60 is in agreement with the observation that the solubility of the gas is proportional to its partial pressure.

Corresponding to Eq. 25.59, we find for the solvent the result

$$\frac{f_1}{\mathfrak{p}} = x_1 \exp\left(\frac{\mu_1^0 - \mu_1^*}{RT}\right), \tag{25.61}$$

or, when the vapor phase is ideal,

$$\frac{p_1}{\mathfrak{p}} = x_1 \exp\left(\frac{\mu_1^0 - \mu_1^*}{RT}\right). \tag{25.62}$$

Equations 25.59 and 25.61 differ in an important way despite their superficial similarity. The difference arises from the choices of standard states for solute and solvent, and thereby illustrates an important point. We shall now show that the partial pressure of the solvent is proportional to the pressure a hypothetical pure phase would have. To show this we need merely evaluate the right-hand side of Eq. 25.61 or 25.62. The pure solvent vapor–liquid equilibrium is described by

$$\mu_1^*(T) + RT\ln\frac{f_1^0}{\mathfrak{p}} = \mu_1^0(T, p), \tag{25.63}$$

(see Eq. 21.15) or, on rearrangement,

$$\frac{f_1^0}{\mathfrak{p}} = \exp\left(\frac{\mu_1^0 - \mu_1^*}{RT}\right). \tag{25.64}$$

But we have chosen the standard state for the solvent to be the pure liquid. Therefore, the exponentials on the right-hand sides of Eqs. 25.64 and 25.61 are equal, and we have

$$f_1 = x_1 f_1^0, \tag{25.65}$$

or, when the gas phase is ideal,

$$p_1 = x_1 p_1^0, \tag{25.66}$$

with p_1^0 the vapor pressure of pure component 1. In contrast, $k_2(T,\ p,$ solvent) is not the vapor pressure of pure component 2. We illustrate in Fig. 25.3 how $k_2(T,\ p,$ solvent) defines the vapor pressure (fugacity) of a hypothetical pure phase which has the same properties per mole as does the dilute solution.

Why does f_2^0 differ greatly from the vapor pressure the pure component would have at the same temperature and pressure when the concentration dependence of the chemical potential is of ideal form? The answer to this question becomes evident once it is realized that in a very dilute solution, such as the one we have described by Eqs. 25.57, every solute molecule is surrounded by solvent molecules only. Moreover, the solute molecules are far enough apart that solute–solute interaction is negligibly small. We expect that the solute molecule in the liquid can move only in an effective volume much smaller than that available in the gas, and therefore that the entropy of solution will be negative. However, the effective volume per molecule in solution need not be, and in general is not, the same as the effective volume per molecule in the pure liquid solute. In fact, if the interaction between two isolated solute molecules differs from that between an isolated solute–solvent molecule pair, there is no reason to expect the properties of the hypothetical state defined by extrapolating the low-concentration properties to be like those of the pure solute. For example, the solute–solvent attractive interaction will ordinarily be intermediate in magnitude between the solvent–solvent and solute–solute attractive interactions. Thus, there should be some excess or deficiency of binding of the solute to the solvent relative to the binding in the pure solute; that is, the enthalpy per solute molecule will be more negative or positive in the solution than in the pure solute.[6] Moreover, in addition to the very large loss of translational entropy when a gas molecule enters the liquid phase, and the likelihood that the effective volume per solute molecule in solution differs from that in the pure

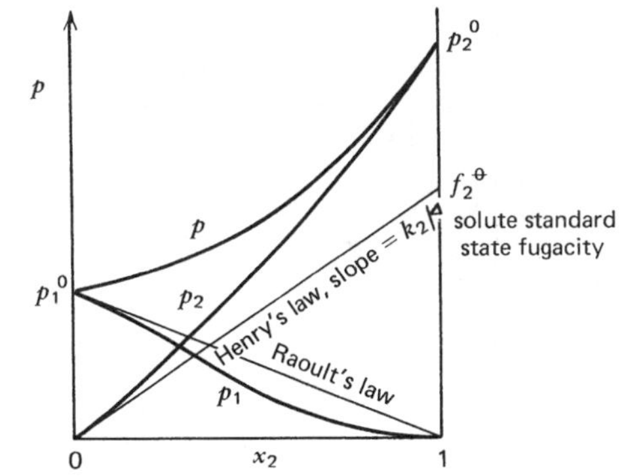

Figure 25.3 Schematic display of the Henry's law representation and the extrapolation to the standard state of the solute.

solute, differences in the solute–solute and solute–solvent attractive and repulsive interactions can lead to changes in the range of local ordering around a solute molecule, and thereby to further changes in the entropy per molecule. The balance between the enthalpy and entropy differences per molecule is rarely exactly the same in the hypothetical standard state and the true pure component, so that in general we have $k_2(T,\ p,$ solvent) $\neq p_2^0/\mathfrak{p}$. As indicated in the argument, the very existence of a difference between solute–solute and solute–solvent interactions also implies that k_2 and the chemical potential of the hypothetical standard state of the solute must depend on the nature of the solvent. Finally, the reason that the concentration dependence of the chemical potential, and of the entropy of mixing, is of the ideal form is that each solute molecule is effectively isolated in an environment of solvent. Changes in solute concentration in the range considered do not alter this environment. Therefore, decreases or increases of the solute concentration are equivalent to concentration changes in an ideal mixture.

We now turn to the study of nonideality in mixtures that are not dilute. In a nonideal mixture of volatile liquids (or solids), the partial pressures may be greater or less than those predicted by Raoult's law (see Fig. 25.4). Which of the two kinds of deviation occurs depends on molecular interactions in the condensed phase. To get a feeling for what happens, assume the entropy of mixing to be ideal. If unlike molecules attract one another less than do the respective pure-component molecular pairs, the mixture has a higher vapor pressure than predicted from Raoult's law; if unlike molecules are more strongly attracted to one another than like molecules, the mixture has a lower vapor pressure than predicted from Raoult's law. Deviations from the ideal form of the entropy of mixing also lead to deviations from Raoult's law, but these are often of less importance than the energetic changes mentioned, except when the unlike molecules are very different in size. Great differences in molecular size lead to considerable deviations from the mole-fraction form of the ideal entropy of mixing, and hence from Raoult's law.

---

[6] Indeed, at constant partial pressure of the solute gas, the solubility is defined by $\ln x_2 = \ln(p_2/\mathfrak{p}) - \ln k_2$, yielding for the temperature dependence

$$\left(\frac{\partial \ln x_2}{\partial T}\right)_{p_2} = -\left(\frac{\partial \ln k_2}{\partial T}\right)_{p_2} = -\left[\frac{\partial}{\partial T}\left(\frac{\mu_2^\ominus - \mu_2^*}{RT}\right)_{p_2}\right] = \frac{h_2^\ominus - h_2^*}{RT^2},$$

using Eqs. 25.59 and 25.7. The quantity $h_2^\ominus - h_2^*$ is the heat of solution of the gas in the liquid, that is, the difference in enthalpy between the hypothetical solution reference state and the perfect gas reference state. In slightly different words, $h_2^\ominus - h_2^*$ is the latent heat per mole required to drive a dissolved gas out of dilute solution. The sign of $h_2^* - h_2^\ominus$ is positive, so that the gas solubility decreases as the temperature increases. In general, the cohesive energy of the gaseous solute in the liquid decreases rapidly as the temperature increases. Since $\mu_2^* - \mu_2^\ominus = -RT \ln k_2$, the entropy of solution is easily found from $s_2^\ominus - s_2^* = -(\mu_2^\ominus - \mu_2^* - h_2^\ominus + h_2^*)/T$.

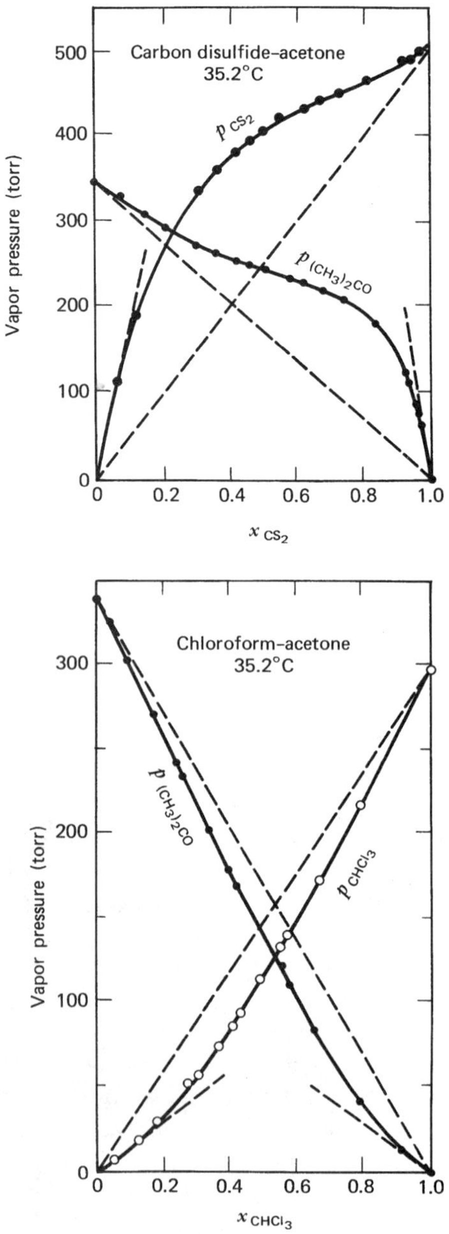

Figure 25.4 Partial pressure as a function of composition for carbon disulfide–acetone and chloroform–acetone solutions. From R. H. Cole and J. S. Coles, *Physical Principles of Chemistry* (Freeman, San Francisco, 1964).

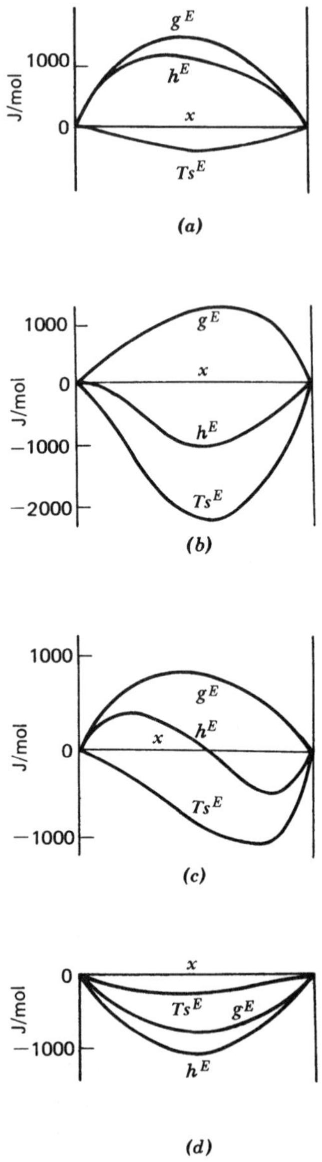

Figure 25.5 (*a*) The excess functions of the system acetonitrile + carbon tetrachloride at 45°C as a function of the mole fraction of acetonitrile. (*b*) The excess functions of water + pyridine at 80°C as a function of the mole fraction of water. (*c*) The excess functions of water + dioxane at 25°C as a function of the mole fraction of water. (*d*) The excess functions of water + hydrogen peroxide at 25°C as a function of the mole fraction of water. From J. S. Rowlinson, *Liquids and Liquid Mixtures,* 2nd ed. (Butterworths, London, 1969).

Some examples of the concentration dependence of the excess enthalpy and entropy are shown in Fig. 25.5.

There are a number of different ways of determining the activity coefficients (hence the excess chemical potentials by Eq. 25.21) of the components of a mixture. Sticking to the use of liquid–vapor equilibrium for the present, we see that if enough is known about the properties of the vapor, Eq. 25.48 can be employed to determine the several γ_i. The measurements required are the total pressure p and composition y_1, $(1 - y_1)$ of the vapor in equilibrium with a liquid of known composition x_1, $(1 - x_1)$ at a given temperature, and the equation of state of the vapor. Regarding the latter, it is usually sufficiently accurate to retain only the second virial coefficient (see Chapter 21) because the vapor pressure is not large. Although this prescription for measurements sounds simple, it is actually difficult to carry through with high precision. For this reason much effort has gone into reworking the analysis of activity coefficients into forms that take advantage of those measurements that can be made accurately.

In a binary solution the activity coefficients of species 1 and 2 are related by Eqs. 25.27. Suppose that the vapor is ideal, so that Eq. 25.48 can be written

$$p_1 = \gamma_1 x_1 p_1^0, \qquad p_2 = \gamma_2 x_2 p_2^0. \tag{25.67}$$

Substitution of Eq. 25.67 into Eq. 25.48 gives

$$(1-x_2)\left(\frac{\partial \ln p_1}{\partial x_2}\right)_{T,p} + x_2\left(\frac{\partial \ln p_2}{\partial x_2}\right)_T = 0. \tag{25.68}$$

We have shown, in Section 23.1, that the dependence of the vapor pressure of a liquid on the applied pressure is very small because of the small molar volume of the liquid. Therefore, although p_1 and p_2 in principle depend on p, that dependence is negligibly small and Eq. 25.68 can be accurately approximated by

$$(1-x_2)\left(\frac{\partial \ln p_1}{\partial x_2}\right)_T + x_2\left(\frac{\partial \ln p_2}{\partial x_2}\right)_T = 0, \tag{25.69}$$

in which the derivatives are at constant temperature only. Equation 25.69, known as the *Duhem–Margules equation,* is valid for arbitrary deviations from ideality in the liquid or solid mixture, provided that the vapor behaves as an ideal gas mixture and that the pressure dependences of p_1 and p_2 are indeed negligibly small. If we now use Eq. 25.52 to eliminate p_2 from Eq. 25.69, we find

$$\left(\frac{\partial p_1}{\partial x_2}\right)_T = \frac{(1-x_1)p_1}{p_1 - x_1 p}\left(\frac{\partial p}{\partial x_2}\right)_T, \tag{25.70}$$

which enables us to calculate p_1, p_2, γ_1, and γ_2 from a knowledge of the total vapor pressure as a function of liquid composition.[7]

We have seen how activity coefficients can be calculated from measurements of the composition dependence of the liquid–vapor equilibrium of a mixture. Consider, now, the inverse problem, namely, the calculation of the bubble-point and dew-point curves of a binary mixture from the known composition dependence of the chemical potentials, including the effects of nonideality. We start, as usual, with the condition for phase equilibrium. For equilibrium at temperature T and pressure p we have, from Eq. 25.63,

$$\begin{aligned}\mu_i^*(T) - \mu_i^0(T,p) + RT\ln\frac{f_i/\mathfrak{p}}{\gamma_i x_i} &= \mu_i^* - \mu_i^0(T,p) + RT\ln\frac{p}{\mathfrak{p}} + RT\ln\frac{f_i}{p}\\ &\quad - RT\ln\gamma_i x_i = 0.\end{aligned} \tag{25.71}$$

Suppose that we consider equilibrium at the adjacent temperature $T + dT$ and pressure $p + dp$. Then

$$\mu_i^V(T+dT, p+dp) = \mu_i^L(T+dT, p+dp), \tag{25.72}$$

which yields on expansion, using Eq. 25.71,

$$\begin{aligned}-R\,d\ln\frac{f_i/p}{\gamma_i x_i} &= \left[\frac{\partial}{\partial T}\frac{\mu_i^*(T) - \mu_i^0(T,p)}{T}\right]_p dT\\ &\quad + \left[\frac{\partial}{\partial p}\left(\frac{\mu_i^*(T) - \mu_i^0(T,p)}{T}\right)_T + R\ln\frac{p}{\mathfrak{p}}\right] dp\\ &= \frac{h_i^0 - h_i^*}{T^2}\,dT - \frac{v_i^0(T,p) - RT/p}{T}\,dp.\end{aligned} \tag{25.73}$$

Equation 25.73 must be integrated to yield predictions concerning the bubble-point and dew-point curves.

Consider first the composition dependence of isothermal liquid–vapor equilibrium. From Eq. 25.73,

$$\ln\left(\frac{\gamma_1^V y_1}{\gamma_1 x_1}\right) = -\frac{1}{RT}\int_{p_1^0}^{p} \Delta v_1^0\,dp \tag{25.74a}$$

and

$$\ln\left(\frac{\gamma_2^V y_2}{\gamma_2 x_2}\right) = -\frac{1}{RT}\int_{p_2^0}^{p} \Delta v_2^0\,dp, \tag{25.74b}$$

where $\gamma_i^V \equiv f_i/y_i p$ is the activity coefficient of i in the vapor. Since the molar volume of the liquid is very small relative to that of the gas, Δv_1^0 and Δv_2^0 become effectively the molar volumes of 1 and 2 in the gas phase. If we now make the approximation that the vapor phase is an ideal gas mixture,

$$\gamma_i^V \equiv \frac{f_i}{y_i p} = \frac{y_i p}{y_i p} = 1 \quad (i = 1,2),$$

---

[7] The following example illustrates the calculation of the activity coefficient for the case that the form of the activity coefficient given in Eq. 25.29 is valid. Then

$$p = p_1 + p_2 = \gamma_1 x_1 p_1^{\circ} + \gamma_2 x_2 p_2^{\circ} = x_1 p_1^{\circ}\exp\left(\frac{wx_2^2}{RT}\right) + x_2 p_2^{\circ}\exp\left(\frac{wx_1^2}{RT}\right).$$

We use the assumption that γ_1 and γ_2 are not far from unity to expand the exponentials to order x_2^2 and x_1^2, respectively:

$$p = x_1 p_1^{\circ}\left(1 + \frac{wx_2^2}{RT}\right) + x_2 p_2^{\circ}\left(1 + \frac{wx_1^2}{RT}\right).$$

Solving for w/RT gives

$$\frac{w}{RT} = \frac{p - (x_1 p_1^{\circ} + x_2 p_2^{\circ})}{x_1 x_2[p_1^{\circ} + (p_2^{\circ} - p_1^{\circ})x_1]}.$$

Since the total vapor pressure over an ideal mixture is just $x_1 p_1^0 + x_2 p_2^0 = p^{\text{ideal}}$, we can also write

$$\ln\gamma_1 = \frac{(p - p^{\text{ideal}})x_2^2}{x_1 x_2[p_1^{\circ} + (p_2^{\circ} - p_1^{\circ})x_1]},$$

permitting easy evaluation from experimental measurements of p as a function of x_1.

and set

$$\Delta v_1^0 = \Delta v_2^0 = \frac{RT}{p},$$

direct integration gives

$$\ln\left(\frac{y_1}{\gamma_1 x_1}\right) = \ln\left(\frac{p_1^0}{p}\right) \tag{25.75a}$$

and

$$\ln\left(\frac{y_2}{\gamma_2 x_2}\right) = \ln\left(\frac{p_2^0}{p}\right), \tag{25.75b}$$

which can be solved to give

$$x_2 = \frac{p_1^0\gamma_1 - p}{p_1^0\gamma_1 - p_2^0\gamma_2} \quad \text{(bubble-point curve)} \tag{25.76}$$

and

$$y_2 = \frac{p_1^0 p_2^0 \gamma_1\gamma_2 - p p_2^0\gamma_2}{p p_1^0\gamma_1 - p p_2^0\gamma_2} \quad \text{(dew-point curve)}. \tag{25.77}$$

Note that Eq. 25.76 is just a rearrangement of the familiar $p = p_1 + p_2 = \gamma_1 x_1 p_1^0 + \gamma_2 x_2 p_2^0$.

To calculate the composition dependence of the constant pressure liquid–vapor equilibrium we transform Eq. 25.73 to read

$$\ln\left(\frac{\gamma_1^V y_1}{\gamma_1 x_1}\right) = \int_{T_{1b}}^{T} \frac{h_1^* - h_1^0}{RT'^2}\,dT' \tag{25.78a}$$

and

$$\ln\left(\frac{\gamma_2^V y_2}{\gamma_2 x_2}\right) = \int_{T_{2b}}^{T} \frac{h_2^* - h_2^0}{RT'^2}\,dT'. \tag{25.78b}$$

Note that $h_1^* - h_1^0$ is just the latent heat of vaporization of component 1 and T_{1b} its boiling point at the given fixed pressure. Consider the approximation that $h_1^* - h_1^0$ and $h_2^* - h_2^0$ are independent of temperature over the range T_{1b}, $T_{2b} \to T$. Then the integrals in Eqs. 25.78 are immediately found to be

$$\int_{T_{1b}}^{T} \frac{h_1^* - h_1^0}{RT'^2}\,dT' = \frac{h_1^* - h_1^0}{R}\left(\frac{1}{T_{1b}} - \frac{1}{T}\right) \equiv \Lambda_{1b} \tag{25.79a}$$

and

$$\int_{T_{2b}}^{T} \frac{h_2^* - h_2^0}{RT'^2}\,dT' = \frac{h_2^* - h_2^0}{R}\left(\frac{1}{T_{2b}} - \frac{1}{T}\right) \equiv \Lambda_{2b}. \tag{25.79b}$$

With the additional approximation that the vapor is an ideal gas mixture, $\gamma_1^V y_1 \to y_1$, $\gamma_2^V y_2 \to y_2$, and we find[8]

$$y_2 = \frac{\exp(\Lambda_{1b})\gamma_2 - \gamma_1\gamma_2}{\exp(\Lambda_{1b})\gamma_2 - \exp(-\Lambda_{2b})\gamma_1}, \tag{25.80}$$

$$x_2 = \frac{\exp(\Lambda_{1b}) - \gamma_1}{\exp(\Lambda_{1b})\exp(\Lambda_{2b})\gamma_2 - \gamma_1}. \tag{25.81}$$

25.5 Liquid–Solid Equilibrium

Henry's law describes the relationship between the solubility and the partial pressure or fugacity of a gas. It is also possible to use measurements of the solubility of a solid or liquid component in a solid or liquid solution to obtain information about the thermodynamic properties of the solution. Suppose that some pure solid, component 2, is in equilibrium with a liquid solution in which component 2 is a solute. The equilibrium is described by

$$\mu_2^S(T,p) = \mu_2^\ominus(T,p,\text{ solvent}) + RT\ln\gamma_2 x_2. \tag{25.82}$$

Consider now the solution–solid equilibrium at the same pressure but at a different temperature, T_0. We shall assume that at the temperature T_0 component 2 has a solubility x_{20}. Then we have

$$\mu_2^S(T_0,p) = \mu_2^\ominus(T_0,p,\text{ solvent}) + RT\ln\gamma_{20}x_{20}. \tag{25.83}$$

If Eq. 25.82 is divided by T and Eq. 25.83 by T_0, and the resultant equations are subtracted, we find that

$$R\ln\gamma_2 x_2 - R\ln\gamma_{20}x_{20} = \frac{\mu_2^S(T) - \mu_2^\ominus(T)}{T} - \frac{\mu_2^S(T_0) - \mu_2^\ominus(T_0)}{T_0}. \tag{25.84}$$

---

[8] Eq. 25.81 can be used to test the accuracy of the regular solution model. For a regular solution

$$RT\ln\gamma_1 = wx_2^2,$$
$$RT\ln\gamma_2 = wx_1^2.$$

In this case, Eq. 25.81 can be written

$$x_2 = \frac{\exp(\Lambda_{1b}) - \exp(wx_2^2/RT)}{\exp(\Lambda_{1b})\exp(\Lambda_{2b})\exp(wx_1^2/RT) - \exp(-wx_2^2/RT)},$$

which can be solved for x_2 by iteration from an initial guess. For the mixture acetone (1)–ether (2) at 30°C, $w = 1.88$ kJ/mol, $h_1^* - h_1^0 = 33.97$ kJ/mol, $T_{1b} = 329.8$ K, $h_2^* - h_2^0 = 26.19$ kJ/mol, $T_{2b} = 308.0$ K. The calculated and observed values of x_2 are as follows:

T(K)	x_2 (calc.)	x_2 (obs.)
308.0	1	1
313.0	0.45	0.45
317.0	0.28	0.28
321.0	0.16	0.16
325.0	0.007	0.007
329.8	0	0

The right-hand side of Eq. 25.84 is easily evaluated. To do so we use the fundamental relation connecting the partial molar enthalpy and the chemical potential:

$$\left[\frac{\partial(\mu_2/T)}{\partial T}\right]_p = -\frac{\bar{h}_2}{T^2}. \tag{25.85}$$

The integral form of Eq. 25.85 is

$$\frac{\mu_2(T)}{T} - \frac{\mu_2(T_0)}{T_0} = -\int_{T_0}^{T} \frac{\bar{h}_2}{T'^2}\,dT'. \tag{25.86}$$

The use of Eq. 25.86 on the right-hand side of Eq. 25.84 finally yields

$$\ln\gamma_2 x_2 - \ln\gamma_{20} x_{20} = \int_{T_0}^{T} \frac{h_2^{\ominus} - h_2^S}{RT'^2}\,dT' = \int_{T_0}^{T} \frac{L_{2f}^{\ominus}\,dT'}{RT'^2}. \tag{25.87}$$

If the solubility at T_0 is sufficiently small that $\gamma_{20} = 1$, then

$$\ln\gamma_2 x_2 - \ln x_{20} = \int_{T_0}^{T} \frac{L_{2f}^{\ominus}}{RT'^2}\,dT'. \tag{25.88}$$

The quantity $L_{2f}^{\ominus} \equiv h_2^{\ominus} - h_2^S$ is the molar heat of fusion of pure solid 2 into the hypothetical state with the same properties per mole as exhibited by the infinitely dilute solution of component 2. Equation 25.87 enables us to compute the ratio of the activity coefficients of component 2 at two temperatures from solubility measurements at those temperatures. If one of the temperatures is sufficiently low that the chemical potential of the solute has the ideal form, then Eq. 25.88 enables us to calculate the absolute value of $\gamma_2(T)$.[9]

Note that when we set T_0 equal to T_{2f}, the melting temperature of pure solid 2, Eq. 25.87 becomes

$$\ln\gamma_2 x_2 = \int_{T_{2f}}^{T} \frac{L_{2f}^{\ominus}\,dT'}{RT'^2}, \tag{25.89}$$

which describes the temperature dependence of the coexistence curve between pure solid 2 and the liquid mixture with which it is in equilibrium.

How does the solubility of a pure solid in contact with a liquid mixture change when the applied pressure is changed at constant temperature? By virtue of the phase equilibrium at the two pressures, Eq. 25.82 is valid at both p and $p + dp$. We have, then, for this isothermal change,

$$\begin{aligned}&\mu_2^S(T, p+dp) - \mu_2^{\ominus}(T, p+dp,\ \text{solvent})\\ &\quad - \mu_2^S(T,p) + \mu_2^{\ominus}(T,p,\ \text{solvent})\\ &\quad = RT\ln(\gamma_2 x_2)_{p+dp} - RT\ln(\gamma_2 x_2)_p.\end{aligned} \tag{25.90}$$

Expansion of Eq. 25.90 as a Taylor series gives

$$\begin{aligned}&\frac{\partial}{\partial p}[\mu_2^S(T,p) - \mu_2^{\ominus}(T,p,\ \text{solvent})]_T\,dp\\ &\quad = RT\left[\left(\frac{\partial\ln\gamma_2}{\partial p}\right)_{T,x_2} dp + \left(\frac{\partial\ln\gamma_2 x_2}{\partial x_2}\right)_{T,p} dx_2\right],\end{aligned} \tag{25.91}$$

since γ_2 is a function of p and x_2. The left-hand side of Eq. 25.91 is the difference between the molar volume of the solid and the molar volume in the standard state, $v_2^S - v_2^{\ominus}$. The term $RT(\partial \ln \gamma_2/\partial p)_{T,x_2}$ on the right-hand side is, from Eq. 25.24, just $\bar{v}_2 - v_2^{\ominus}$. Introducing these terms and rearranging Eq. 25.91 leads to

$$(v_2^S - \bar{v}_2)_T\,dp + RT\left(\frac{\partial\ln\gamma_2 x_2}{\partial x_2}\right)_{T,p} dx_2 = 0 \tag{25.92}$$

and

$$\left(\frac{\partial x_2}{\partial p}\right)_T = \frac{\bar{v}_2 - v_2^S}{RT\left(\dfrac{\partial\ln\gamma_2 x_2}{\partial x_2}\right)_{T,p}}. \tag{25.93}$$

The denominator must be positive (see Eq. 19.86), so $(\partial x_2/\partial p)_T$ is positive or negative according to the sign of $\bar{v}_2 - v_2^S$. For an ideal mixture $\bar{v}_2 = v_2^S$, and the solubility is independent of the applied pressure.

Thus far we have examined the case that a pure solid is in equilibrium with a liquid mixture. In general we must expect that both the solid and liquid phases will be mixtures. What does the liquid–solid equilibrium coexistence curve look like in the more general case?

Suppose that both the liquid and solid mixtures exist over the entire composition range. The coexistence line at constant pressure can be calculated exactly as in the case of

---

[9] When a mixture is ideal, the standard states of the solute species are identical with the states of the pure liquids at the same temperature and pressure. Thus, for an ideal solution $L_{2f}^{\ominus}$ as defined in Eq. 25.87 is the molar heat of fusion of the solid to the supercooled liquid at the same temperature and pressure. This quantity must be calculated from the heat of fusion at the melting point and the difference in molar heat capacity between solid and liquid. We have

$$L_{2f}^{\ominus}(T) = \Delta h_{2f}(T_{2f}) + \int_{T_{2f}}^{T} (c_p^L - c_p^S)\,dT'.$$

Returning to Eq. 25.87, we set $\gamma_2 = 1$ for an ideal solution. If we take the temperature T_0 to be the melting temperature T_{2f}, we have $\gamma_{20}x_{20} = 1$; this is easily seen from Eq. 25.83, since $\mu_2^{\ominus} = \mu_2^L$ for an ideal solution and at the melting point $\mu_2^S = \mu_2^L$. Finally, substituting the expansion of $L_{2f}^{\ominus}$ into Eq. 25.87 with $T_0 = T_{2f}$ gives

$$\ln x_2 = \int_{T_{2f}}^{T} \frac{1}{RT'^2}\left[\Delta h_{2f}(T_{2f}) + \int_{T_{2f}}^{T'} \Delta c_p\,dT''\right] dT'.$$

When Δc_p is independent of temperature, this reduces to

$$\ln x_2 = \frac{(\Delta h_{2f} - T_{2f}\Delta c_p)(T - T_{2f})}{RTT_{2f}} + \frac{\Delta c_p}{R}\ln\frac{T}{T_{2f}}.$$

As already noted, when the solution is ideal we have $\mu_2^{\ominus} = \mu_2^L$ so that the calculated solubility at a given temperature is independent of the solvent. When the solution is not ideal this statement is not valid. The fact that the solubility of a component is the same in all ideal solutions is easily rationalized in terms of molecular interactions, because in an ideal solution solute–solute and solute–solvent interactions are identical.

liquid–vapor equilibrium. All that need be done is to replace $\mu_i^*(T)$ and $RT\ln(f_i/\mathfrak{p})$ in Eq. 25.71 with the corresponding quantities for a solid mixture, namely, $\mu_i^{0S}(T, p)$ and $RT\ln\gamma_i^S x_i^S$. Equation 25.78 then becomes

$$\ln\frac{\gamma_1^S x_1^S}{\gamma_1^L x_1^L} = \int_{T_{1f}}^{T} \frac{h_1^{0S} - h_1^{0L}}{RT'^2}\, dT' \qquad (25.94a)$$

and

$$\ln\frac{\gamma_2^S x_2^S}{\gamma_2^L x_2^L} = \int_{T_{2f}}^{T} \frac{h_2^{0S} - h_2^{0L}}{RT'^2}\, dT', \qquad (25.94b)$$

and $h_1^{0S} - h_1^{0L}$ is the latent heat of crystallization of component 1, with a similar definition for component 2. If $h_1^{0S} - h_1^{0L}$ and $h_2^{0S} - h_2^{0L}$ are independent of temperature over the range T_{1f}, $T_{2f} \to T$ and we define, analogous to Eq. 25.79, the following variables,

$$\frac{h_1^{0S} - h_1^{0L}}{R}\left(\frac{1}{T_{1f}} - \frac{1}{T}\right) \equiv \Lambda_{1f}, \qquad (25.95a)$$

$$\frac{h_2^{0S} - h_2^{0L}}{R}\left(\frac{1}{T_{2f}} - \frac{1}{T}\right) \equiv \Lambda_{2f}, \qquad (25.95b)$$

then

$$x_2^L = \frac{\exp(\Lambda_{1f})\gamma_1^L\gamma_2^S - \gamma_2^S\gamma_1^L}{\exp(\Lambda_{1f})\gamma_1^L\gamma_2^S - \exp(-\Lambda_{2f})\gamma_2^L\gamma_1^S} \qquad (25.96)$$

and

$$x_2^S = \frac{\exp(\Lambda_{1f})\gamma_1^L\gamma_2^L - \gamma_2^L\gamma_1^L}{\exp(\Lambda_{1f})\exp(\Lambda_{2f})\gamma_1^L\gamma_2^S - \gamma_2^L\gamma_1^S}. \qquad (25.97)$$

The line $T(x_2^L)$ is called the *liquidus;* it is analogous to the dew-point curve. The line $T(x_2^S)$ is called the *solidus;* it is analogous to the bubble-point curve. An example of a T, x_2 diagram is displayed in Fig. 25.6. Note that if both the liquid and solid mixtures are ideal, and all activity coefficients are unity, then Eqs. 25.96 and 25.97 reduce to

$$x_2^L = \frac{\exp(\Lambda_{1f}) - 1}{\exp(\Lambda_{1f}) - \exp(-\Lambda_{2f})}, \qquad (25.98)$$

and

$$x_2^S = \frac{\exp(\Lambda_{1f}) - 1}{\exp(\Lambda_{1f})\exp(\Lambda_{2f}) - 1}. \qquad (25.99)$$

Usually there are gross deviations from the ideal mixture behavior described by Eqs. 25.98 and 25.99.

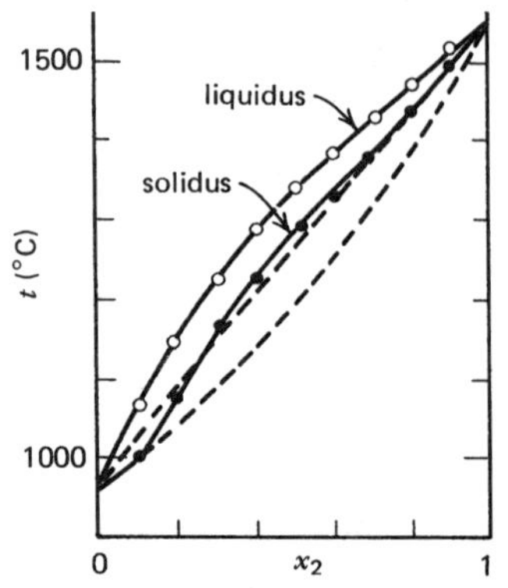

Figure 25.6 Phase diagram for liquid–solid equilibrium in the system Ag(1)–Pd(2). The dashed curves are the calculated liquidus and solidus lines for ideal solutions (Eqs. 25.98 and 25.99).

25.6 The Colligative Properties of Solutions: Boiling-Point Elevation, Freezing-Point Depression, and Osmotic Pressure

Thus far we have considered the thermodynamic properties of mixtures in which both components are volatile. If one of the two components of a binary mixture is nonvolatile, for example, sugar in water, special simplifications of the analysis are possible. Consider, for example, the equilibrium between the vapor and solution phases when only the solvent has appreciable vapor pressure. Equation 25.44 describes the solvent equilibrium. However, by assumption the vapor phase now contains only the solvent species. Proceeding as in Eqs. 25.45–25.48, we see that the net effect of adding a nonvolatile solute to a volatile solvent is to lower the vapor pressure. Put the other way around, because the vapor pressure is proportional to the mole fraction of the solvent and increases with temperature, the temperature at which the solution boils must be higher than the temperature at which the pure solvent boils. This boiling-point elevation can be used to determine the molecular weight of the nonvolatile solute or, if that is known, the activity coefficient of the solvent.

As usual, let component 1 be the solvent and component 2 the nonvolatile solute. As stated, Eq. 25.44 describes the gas–liquid phase equilibrium for the solvent in the solution. When the solvent is pure, gas–liquid equilibrium at the boiling point is described by

$$\mu_1^V(T_{1b}, p) = \mu_1^L(T_{1b}, p). \qquad (25.100)$$

Proceeding as in Eq. 25.82 *et seq.,* we divide Eq. 25.44 by T, Eq. 25.100 by T_{1b}, and subtract to find

$$\frac{\mu_1^V(T,p)}{T} - \frac{\mu_1^V(T_{1b},p)}{T_{1b}} = \frac{\mu_1^L(T,p)}{T} - \frac{\mu_1^L(T_{1b},p)}{T_{1b}} + R\ln\gamma_1 x_1. \qquad (25.101)$$

Using Eq. 25.86, we can rewrite Eq. 25.101 in the form

$$\ln \gamma_1 x_1 = -\int_{T_{1b}}^{T} \frac{h_1^V - h_1^L}{RT'^2}\, dT' = -\int_{T_{1b}}^{T} \frac{L_{1v}\, dT'}{RT'^2}, \tag{25.102}$$

where it is understood that the integration is along a path at constant pressure, with $L_{1v} \equiv h_1^V - h_1^L$ the latent heat of vaporization of pure solvent. Of course, the latent heat of vaporization depends on the temperature because the heat capacities of the vapor and liquid are different. Setting

$$L_{1v} = \Delta h_{1v} + [c_p^V - c_p^L](T - T_{1b}), \tag{25.103}$$

we find

$$\begin{aligned}\ln \gamma_1 x_1 &= -\int_{T_{1b}}^{T} \frac{\Delta h_{1v} + \Delta c_p (T' - T_{1b})}{RT'^2}\, dT' \\ &= -\int_0^{\theta} \frac{\Delta h_{1v} + \Delta c_p \theta'}{R(T_{1b} + \theta')^2}\, d\theta',\end{aligned} \tag{25.104}$$

where $\theta \equiv T - T_{1b}$ is the boiling-point difference between the solution and the solvent. Ordinarily the boiling-point change is small compared to the boiling point, so that we have $(\theta/T_{1b}) << 1$ and

$$\frac{1}{(T_{1b} + \theta)^2} \approx \frac{1}{T_{1b}^2}\left(1 - \frac{2\theta}{T_{1b}} + \cdots\right) \tag{25.105}$$

can be used in the integrand of Eq. 25.104. Finally, by substituting Eq. 25.105 in Eq. 25.104 and carrying out the integration, we obtain

$$\ln \gamma_1 x_1 = -\frac{\Delta h_{1v}\theta}{RT_{1b}^2} + \left(\frac{\Delta c_p}{2R} + \frac{\Delta h_{1v}}{RT_{1b}}\right)\left(\frac{\theta}{T_{1b}}\right)^2 + \mathcal{O}(\theta^3) \tag{25.106}$$

(where $\mathcal{O}$ means "order of"). The second term on the right-hand side of Eq. 25.106 is ordinarily small relative to the first term. Note that all the variables on the right-hand side except θ are functions of the solvent only. Thus, all nonvolatile solutes elevate the boiling point of a volatile solvent by the same amount when their mole fractions are the same.

Suppose that the solution we are studying is ideal and dilute so that, with only the term linear in θ retained, Eq. 25.106 reduces to

$$\ln x_1 = -\frac{\Delta h_{1v}\theta}{RT_{1b}^2}. \tag{25.107}$$

But in a dilute solution we can write

$$\ln x_1 = \ln(1 - x_2) \approx -x_2, \tag{25.108}$$

so that

$$x_2 = \frac{\Delta h_{1v}\theta}{RT_{1b}^2}. \tag{25.109}$$

We also have, when $x_2 << 1$,

$$x_2 = \frac{w_2/M_2}{w_1/M_1 + w_2/M_2} \approx \frac{w_2/M_2}{w_1/M_1}, \tag{25.110}$$

so that the molecular weight of the solute can be determined from

$$M_2 = \frac{w_2}{(w_1/M_1)} \cdot \frac{RT_{1b}^2}{\Delta h_{1v}\theta} = 1000\frac{w_2}{w_1} \cdot \frac{K_b}{\theta}. \tag{25.111}$$

For convenience, it is usual to tabulate, for various solvents, the quantity $K_b \equiv (M_1/1000)(RT_{1b}^2/\Delta h_{1v})$. This constant, known as the *ebullioscopic constant,* gives the boiling-point increase when 1 mol of nonvolatile solute is dissolved in 1000 g of solvent. This concentration can be described by saying that the *molality* (number of moles in 1000 g of solvent) is unity. For the case of CCl_4, the ebullioscopic constant is 5.03°C kg/mol. Values of K_b for several other solvents are given in Table 25.1.

The reason that the boiling-point elevation of a solution of a nonvolatile solute in a volatile solvent depends directly on the activity of the solvent is that there is no solute in the vapor. Similarly, if we consider the equilibrium between a solution and a crystalline phase of the solvent in which the solute is insoluble, the difference in freezing point between the solution and the pure solvent is related directly to the activity of the solvent. The analysis of this freezing-point depression parallels that for the boiling-point elevation. In fact, if in Eq. 25.100 *et seq.* we everywhere replace functions referring to the solvent in the gas phase with functions referring to the solvent in the solid phase, we obtain

$$\begin{aligned}\ln \gamma_1 x_1 &= -\frac{\Delta h_{1f}\theta}{RT_{1f}} \cdot \frac{1}{T_{1f}} \\ &\quad - \left(\frac{\Delta c_p}{2R} + \frac{\Delta h_{1f}}{RT_{1f}}\right)\left(\frac{\theta}{T_{1f}}\right)^2 + \mathcal{O}(\theta^3),\end{aligned} \tag{25.112}$$

Table 25.1 Ebullioscopic Constants

	T_b (°C)	K_b (°C kg/mol)
Benzene	80.15	2.53
Ethanol	78.26	1.22
Ethyl iodide	72.5	5.05
n-Heptane	98.42	3.43
Naphthalene	218.0	5.65
Water	100.0	0.512

Table 25.2 Cryoscopic Constants

	T_f (°C)	K_f (°C kg/mol)
Benzene	5.5	5.12
Bromoform	7.8	14.4
Diphenylamine	52.9	8.6
Naphthalene	80.2	6.9
Stannic bromide	31	28.0
Water	0	1.86

where the subscript f refers to fusion (or freezing) and here $\theta \equiv T_{1f} - T$ is the freezing-point depression. As before, θ measures the activity of the solvent in the presence of a given amount of solute.

When the solute concentration is low and $(\theta/T_{1f}) << 1$, we can write

$$\ln \gamma_1 x_1 = -\frac{\Delta h_{1f}\theta}{RT_{1f}^2} = -\frac{\theta}{K_f}\cdot\frac{M_1}{1000}, \qquad (25.113)$$

where, when the solute concentration is expressed in molality, the *cryoscopic constant, K_f,* is defined by the combination of constants $K_f \equiv (M_1/1000)(RT_{1f}^2/\Delta h_{1f})$. Then

$$\theta = K_f m_2 . \qquad (25.114)$$

For an ideal solution, we have $x_2 = \Delta h_{1f}\theta/RT_{1f}^2$, analogous to Eq. 25.109 for the boiling-point elevation. The values of K_f for a few solvents are given in Table 25.2.

There is one other simple method that can be used to exclude a component from one of two phases in equilibrium. Suppose that two phases are separated by a membrane permeable to solvent but not to solute (see Fig. 25.7). At equilibrium, assuming that the membrane is rigid and therefore the confined system cannot do mechanical work, but that heat can flow across the membrane, we have

$$\begin{aligned} p^{\mathrm{I}} &\neq p^{\mathrm{II}} && \text{(rigid membrane)},\\ T^{\mathrm{I}} &= T^{\mathrm{II}} && \text{(membrane permits flow of heat)},\\ \mu_1^{\mathrm{I}} &= \mu_1^{\mathrm{II}} && \text{(membrane permeable to solvent)}. \end{aligned} \qquad (25.115)$$

In general, the chemical potential can be related to the pressure in a system by (see Table 19.2)

$$\left(\frac{\partial \mu_i}{\partial p}\right)_{T,x_i} = \bar{v}_i . \qquad (25.116)$$

Integration of Eq. 25.116 between p_0 and $p_0 + \pi$ gives

$$\mu_i(T, p_0+\pi, x) - \mu_i(T, p_0, x) = \int_{p_0}^{p_0+\pi} \bar{v}_i \, dp. \qquad (25.117)$$

In the case under consideration the solvent chemical potential on side I, where there is no solute, is $\mu_1^L(T, p_0)$, whereas

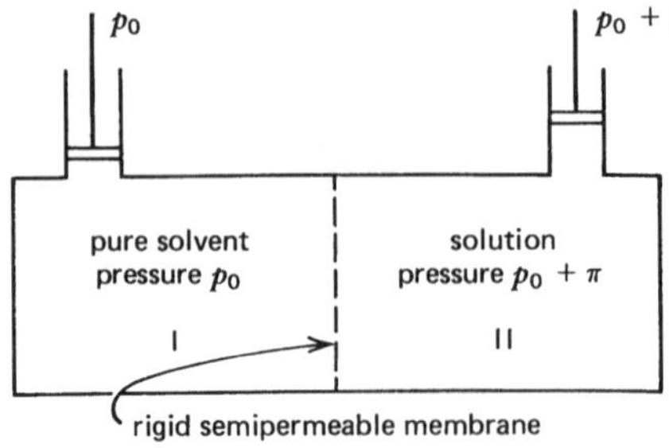

Figure 25.7 Schematic representation of an osmotic pressure experiment.

on side II the solvent chemical potential is $\mu_1^L(T, p_0 + \pi) + RT \ln \gamma_1 x_1$. The pressure π is known as the *osmotic pressure.*[10] Applying Eq. 25.117 to the particular case of the solvent we find, using Eq. 25.115,

$$\begin{aligned} RT \ln \gamma_1 x_1 &= \mu_1^L(T, p_0) - \mu_1^L(T, p_0+\pi)\\ &= -\int_{p_0}^{p_0+\pi} \bar{v}_1 \, dp. \end{aligned} \qquad (25.118)$$

Suppose that the partial molar volume of solvent varies smoothly with pressure, so that we can write

$$\bar{v}_1(p) = \bar{v}_1(p_0) + (p - p_0)\left(\frac{\partial \bar{v}_1}{\partial p}\right)_{T,x,p_0} + \cdots. \qquad (25.119)$$

The compressibilities of most liquids and the osmotic pressures generated in dilute solutions are so small that under most circumstances $\bar{v}_1(p_0) >> (p - p_0)(\partial \bar{v}_1/\partial p)_{T,x,p_0}$. In this limit we have

$$RT \ln \gamma_1 x_1 = -\pi \bar{v}_1(p_0). \qquad (25.120)$$

Once again consider the case that $x_2 << x_1$ and the solution is ideal. Then we obtain

$$RT \ln x_1 \approx -RTx_2 = -\pi \bar{v}_1 . \qquad (25.121)$$

But in a dilute solution we have

$$x_2 \approx \frac{w_2/M_2}{w_1/M_1} = \frac{n_2}{n_1} \qquad (25.122)$$

and

$$\frac{w_1}{M_1}\bar{v}_1 = n_1 \bar{v}_1 \approx V, \qquad (25.123)$$

[10] The osmotic pressure is interpretable as the incremental external pressure needed to raise the vapor pressure of the solvent in the solution in volume II to be equal to that of the pure solvent in volume I when both the solution and the pure solvent have the same temperature.

where V is the total volume of the solution. Combining Eqs. 25.121, 25.122, and 25.123, we find for an ideal dilute solution

$$\pi V = n_2 RT, \tag{25.124}$$

which closely resembles the perfect gas equation of state. In fact, the resemblance can be made even clearer if we use a different representation of the effects of nonideality in the solution. Suppose that we write

$$-\pi\bar{v}_1(p_0) = RT\ln\gamma_1 x_1 \equiv g_1 RT\ln x_1. \tag{25.125}$$

Equation 25.125 defines the quantity g_1 in terms of γ_1. It is equally valid to use γ_1 or g_1 to describe deviations from ideality, although in osmotic pressure experiments g_1, known as the *osmotic coefficient,* is more useful.[11] Now suppose that g_1 varies with the concentration, $c_2 = n_2/V$, as

$$g_1 = 1 + Bc_2 + Cc_2^2 + \cdots. \tag{25.126}$$

Then Eq. 25.125 becomes, in a dilute solution,

$$\pi = RTc_2(1 + Bc_2 + \cdots), \tag{25.127}$$

which closely resembles the virial equation of state. The solution virial coefficients can be calculated from the molecular theory of solutions. It is found that the solution virial coefficients depend on the effective interaction potential (potential of mean force) in the solution in the same way as do the gas virial coefficients on the true intermolecular potential.

Clearly, just as the freezing-point depression and boiling-point elevation can be used to determine solvent activity, solute molecular weight, and so on, so can the osmotic pressure of the solution be used. These three phenomena are known as the *colligative properties* of a solution.

[11] The osmotic coefficient representation of the activity of the solvent is very useful when the solution is dilute, since then the activity coefficient of the solvent is very close to unity. For a dilute solution, it is usually more convenient to use the molality concentration scale in place of the mole fraction concentration scale. The so-called practical osmotic coefficient, ϕ, is defined by

$$\phi \equiv -\left(\frac{\mathfrak{m}_1}{\sum_i m_i}\right)\ln\gamma_1 x_1$$

where $\mathfrak{m}_1 = 1000/M_1$, and the sum includes all solute species. The osmotic coefficients g_1 and ϕ can be interconverted by using the definition $x_1 = \left[1 + \left(\sum_i m_i/\mathfrak{m}_1\right)\right]^{-1}$ and expanding $\ln x_1$. The result is

$$\phi = g_1\left(1 - \frac{\sum_i m_i}{2\mathfrak{m}_1} + \cdots\right),$$

so that the chemical potential of the solvent is represented by

$$\mu_1 = \mu_1^0 - \frac{RT\phi}{\mathfrak{m}_1}\sum_i m_i,$$

and the osmotic pressure is

$$\pi = \phi\left(\frac{RT}{\mathfrak{m}_1 v_1^0}\right)\sum_i m_i.$$

25.7 Chemical Reactions in Nonelectrolyte Solutions

Given the concentration dependences and the standard states for the chemical potentials of all components of a solution, the treatment of chemical equilibrium is very simple. The general conditions describing chemical equilibrium in a mixture were derived in Section 19.5. There it was shown that at chemical equilibrium we have

$$\sum_{i=1}^{r} \nu_i\mu_i = 0, \tag{25.128}$$

where the ν_i are the stoichiometric coefficients for the reaction under consideration. Proceeding as in Section 19.5, but using for the chemical potentials the forms appropriate to a condensed phase, we find that

$$\Delta G^0 = -RT\ln K(T, p, \text{solvent}), \tag{25.129}$$

where

$$\Delta G^0 \equiv \sum_{i=1}^{r} \nu_i\mu_i^0 \tag{25.130}$$

and

$$K(T, p, \text{solvent}) \equiv \prod_{i=1}^{r} (\gamma_i x_i)^{\nu_i}. \tag{25.131}$$

When the standard state for each component is chosen as the pure liquid or solid, the equilibrium constant is a function only of the temperature and pressure. However, when the standard state is the hypothetical state with the same properties per unit mass as an infinitely dilute solution, the equilibrium constant also depends on the solvent. Clearly, the temperature and pressure derivatives of $K(T, p,$ solvent$)$ depend on the enthalpy change and the volume change in the reaction, just as in the gaseous reactions discussed in Section 21.3.

For an ideal solution we can write

$$K_{\text{ideal solution}} = \prod_{i=1}^{r} x_i^{\nu_i}. \tag{25.132}$$

Note that the equilibrium constant in Eq. 25.132 is dimensionless, as is the corresponding equilibrium constant in a gas (Eqs. 21.65 and 21.66).

There is a commonly used, but misleading, convention in which the equilibrium constant appears to have units.

This convention arises as follows. Suppose that, instead of mole fraction units, we use some other concentration variable to describe the composition dependence of the chemical potential. When the molality is chosen as the concentration unit, the only standard state that may be used is the one referring to the hypothetical substance with the same properties per unit mass as the solute at infinite dilution. The reason for this is that the molality scale presumes that one component is singled out as the solvent. The mole fraction of a component is

$$x_i = \frac{n_i}{n_1 + \sum_{j=2}^{r} n_j} = \frac{w_i/M_i}{(w_1/M_1) + \sum_{j=2}^{r}(w_j/M_j)}. \quad (25.133)$$

Recalling the definition of the molality of component j (the number of moles of j in 1000 g of solvent) we fix $w_1 =$ 1000 g; the value of w_j/M_j, is then the molality (m_j) of j in the solution. Thus, Eq. 25.133 becomes

$$x_i = \frac{m_i}{(1000/M_1) + \sum_{j=2}^{r} m_j}. \quad (25.134)$$

We now define the activity coefficients on the molality scale from the equation

$$\begin{aligned}\mu_i &= \mu_i^0(T,p) + RT\ln\left\{\frac{\gamma_i m_i}{(1000/M_1)\left[1+(M_1/1000)\sum_{j=2}^{r} m_j\right]}\right\}\\ &= \mu_i^0(T,p) + RT\ln\left\{\left[\frac{\gamma_i}{1+(M_1/1000)\sum_{j=2}^{r} m_j}\right]\left(\frac{m_i}{1000/M_1}\right)\right\}\\ &= \mu_i^0(T,p) - RT\ln \mathfrak{m}_1 + RT\ln\gamma_i' m_i,\end{aligned} \quad (25.135)$$

where

$$\gamma_i' \equiv \frac{\gamma_i}{1+(M_1/1000)\sum_{j=2}^{r} m_j} = \frac{\gamma_i}{1+\sum_{j=2}^{r}(m_j/\mathfrak{m}_1)} \quad (25.136)$$

is the activity coefficient of component i and

$$\mathfrak{m}_1 \equiv \frac{1000}{M_1}. \quad (25.137)$$

Note that, despite the change in concentration units, we have retained the dimensionless form of the argument of the logarithm, because the molality is a dimensionless number referred to a standard mass of solvent. Clearly, $\mathfrak{m}_1$ is the molality of the solvent, a different number for every different solvent but the same number for all solutions in the same solvent ($\mathfrak{m}_1 = 55.51$ for H_2O). It is usual for the factor $RT \ln \mathfrak{m}_1$ to be absorbed into the chemical potential of the standard state, thereby defining a new standard-state chemical potential. With reference to the molality concentration scale and the new standard state for which $\mu_i^{0m}(T, p, \text{solvent}) = \mu_i^0(T, p) - RT \ln \mathfrak{m}_1$, the equilibrium constant becomes

$$K_m(T,p,\text{ solvent}) = \prod_{i=1}^{r}(\gamma_i' m_i)^{\nu_i} \quad (25.138)$$

and is a dimensionless number.

A similar analysis can be used to write $K(T, p, \text{solvent})$ in terms of the molar concentrations, c_i (*molarity c*, number of moles per liter of solution). In this case the new standard-state chemical potential differs from $\mu_i^0(T, p)$ by the added term $RT \ln(v_1^0/1000)$ where v_1^0 is the molar volume of pure solvent, and the 1000 has units cm^3 and refers to the standard 1 liter of solution for which the molarity is defined. Of course, the activity coefficients γ_i'' on the molarity scale differ from γ_i and γ_i', as K and K_m differ from

$$K_c(T,p,\text{ solvent}) = \prod_{i=1}^{r}(\gamma_i'' c_i)^{\nu_i}. \quad (25.139)$$

The important point is that, whatever concentration unit is chosen, the argument of the logarithm in the chemical potential must be dimensionless. Then, strictly, $K(T, p, \text{solvent})$ will be dimensionless. However, because different standard states correspond to different choices of concentration units, it is common (as a convenience) to attach to $K(T, p, \text{solvent})$ units indicating the concentration scale in which it is defined.

Brushing aside questions of which concentration units are most convenient, we note that all the features of chemical equilibrium previously described are also relevant to the description of reactions in liquid or solid solution. Thus, an arbitrary reaction can be subdivided into different sequences of reactions, and from suitable tables of data the changes in free energy, enthalpy, and entropy for the reaction of interest can be computed as described in Chapters 14, 19, and 21.

25.8 More about Phase Equilibrium in Mixtures

The phenomena associated with phase equilibrium in mixtures are surprisingly numerous and diverse in character. And, of course, the more components there are in the mixture, the greater the range of possible behavior. Consequently, the study of phase equilibrium in mixtures is a very large subject, and we shall be able only to touch on a few points and illustrate a few important principles of interpretation.

Although we have examined liquid–vapor and liquid–solid equilibria for mixtures, in each case the focus of the argument was on exploitation of the features of the phase equilibrium in a fashion that permitted extraction of information about the properties of the components of the mixture. We now consider the stability of a mixture against phase separation. Unlike a pure component, even at constant temperature and pressure a mixture may be unstable with respect to separation into two phases with different compositions. We start our analysis with the observation that within one phase the Gibbs free energy at constant temperature and pressure satisfies the equation (see Table 19.2)

$$dG = \mu_1\, dn_1 + \mu_2\, dn_2 \quad (T, p, \text{constant}), \tag{25.140}$$

or, in terms of the mole fractions,

$$\begin{aligned} dG &= (n_1 + n_2)\mu_1\, dx_1 + (n_1 + n_2)\mu_2\, d(1 - x_1) \\ &= (n_1 + n_2)(\mu_1 - \mu_2)\, dx_1 \\ &= (n_1 + n_2)(\mu_2 - \mu_1)\, dx_2 . \end{aligned} \tag{25.141}$$

Thus, a plot of G versus x_2 has slope proportional to $\mu_2 - \mu_1$ at every point. Provided that phase separation does not occur, G versus x_2 is a continuous curve (see Fig. 25.8*a*). If phase separation occurs and phases I and II with different compositions are in equilibrium, we have

$$\mu_1^{\mathrm{I}} = \mu_1^{\mathrm{II}} \tag{25.142a}$$

and

$$\mu_2^{\mathrm{I}} = \mu_2^{\mathrm{II}}, \tag{25.142b}$$

or

$$\mu_1^{\mathrm{I}} - \mu_2^{\mathrm{I}} = \mu_1^{\mathrm{II}} - \mu_2^{\mathrm{II}} . \tag{25.143}$$

In a plot of G versus x_2 for a system of two phases, there will be separate lines for each phase (see Fig. 25.8*b*). Equation 25.143 implies that the slopes of the two curves are equal when the phases are in equilibrium. Hence the compositions of the two phases can be determined by drawing the common tangent to the two curves. For the example shown in Fig. 25.8, the two phases are in equilibrium when their compositions are those at points A and B, respectively.

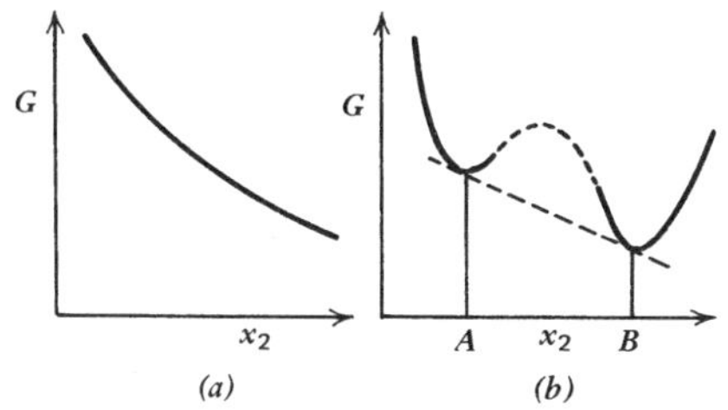

Figure 25.8 Schematic representation of the Gibbs free energy as a function of composition for a binary mixture. (*a*) One-phase region. (*b*) Two-phase region.

Consider a mixture for which the Gibbs free energy resembles the curve in Fig. 25.8*a* for temperatures $T > T_c$, and resembles the curve in Fig. 25.8*b* for temperatures $T < T_c$. The temperature T_c, called the *critical temperature of mixing,* characterizes that one isotherm for which the change from curve *a* behavior to curve *b* behavior occurs. Clearly, T_c plays a role in the thermodynamics of mixtures analogous to that of the critical temperature in the thermodynamics of pure fluids.

Everywhere along curve *a* of Fig. 25.8 we have $(\partial^2 G/\partial x_2^2) > 0$ (G is concave upward). If the separate parts of curve *b* are joined, then there is a region for which $(\partial^2 G/\partial x_2^2) < 0$ (G is concave downward) connecting two regions in which $(\partial^2 G/\partial x_2^2) > 0$. Therefore, there must be two points between those labeled A and B for which $(\partial^2 G/\partial x_2^2) = 0$. These are points of inflection of the free-energy curve. At T_c the points just defined merge into one point at which

$$\left(\frac{\partial^2 G}{\partial x_2^2}\right)_{T_c,p} = 0 \tag{25.144a}$$

and

$$\left(\frac{\partial^3 G}{\partial x_2^3}\right)_{T_c,p} = 0. \tag{25.144b}$$

Using 25.141 in Eqs. 25.144*a* and 25.144*b*, we find that at the critical temperature

$$\left(\frac{\partial \mu_1}{\partial x_2}\right)_{T_c,p} = \left(\frac{\partial \mu_2}{\partial x_2}\right)_{T_c,p} \tag{25.145a}$$

and

$$\left(\frac{\partial^2 \mu_1}{\partial x_2^2}\right)_{T_c,p} = \left(\frac{\partial^2 \mu_2}{\partial x_2^2}\right)_{T_c,p} . \tag{25.145b}$$

We also have, from the Gibbs–Duhem relation at constant temperature and pressure,

$$(1 - x_2)\left(\frac{\partial \mu_1}{\partial x_2}\right)_{T,p} + x_2\left(\frac{\partial \mu_2}{\partial x_2}\right)_{T,p} = 0. \tag{25.146}$$

After differentiation with respect to x_2, this yields

$$\begin{aligned} &(1 - x_2)\left(\frac{\partial^2 \mu_1}{\partial x_2^2}\right)_{T,p} + x_2\left(\frac{\partial^2 \mu_2}{\partial x_2^2}\right)_{T,p} \\ &\quad - \left(\frac{\partial \mu_1}{\partial x_2}\right)_{T,p} + \left(\frac{\partial \mu_2}{\partial x_2}\right)_{T,p} = 0. \end{aligned} \tag{25.147}$$

It follows from Eqs. 25.145–25.147 that at T_c we must have

$$\left(\frac{\partial \mu_1}{\partial x_2}\right)_{T_c,p} = \left(\frac{\partial \mu_2}{\partial x_2}\right)_{T_c,p} = 0 \tag{25.148a}$$

and

$$\left(\frac{\partial^2 \mu_1}{\partial x_2^2}\right)_{T_c,p} = \left(\frac{\partial^2 \mu_2}{\partial x_2^2}\right)_{T_c,p} = 0. \quad (25.148b)$$

When represented in terms of activity coefficients, the conditions 25.148 assume the form

$$\left(\frac{\partial \ln \gamma_1}{\partial x_2}\right)_{T_c,p} - \frac{1}{1-x_2} = 0, \quad (25.149a)$$

$$\left(\frac{\partial \ln \gamma_2}{\partial x_2}\right)_{T_c,p} + \frac{1}{x_2} = 0, \quad (25.149b)$$

$$\left(\frac{\partial^2 \ln \gamma_1}{\partial x_2^2}\right)_{T_c,p} - \frac{1}{(1-x_2)^2} = 0, \quad (25.149c)$$

$$\left(\frac{\partial^2 \ln \gamma_2}{\partial x_2^2}\right)_{T_c,p} - \frac{1}{x_2^2} = 0. \quad (25.149d)$$

To illustrate the preceding analysis of phase separation in a mixture, we evaluate T_c for the regular mixture model. For a regular mixture we have

$$RT \ln \gamma_1 = w x_2^2, \quad (25.29)$$

so that, using Eqs. 25.149*a* and 25.149*c*, we find

$$\frac{w}{RT_c}(2x_2)_c = \frac{1}{(1-x_2)_c} \quad (25.150a)$$

and

$$\frac{w}{RT_c} \cdot 2 = \frac{1}{(1-x_2)_c^2}. \quad (25.150b)$$

The solution to these simultaneous equations is

$$(x_2)_c = \frac{1}{2}, \qquad \frac{w}{RT_c} = 2. \quad (25.151)$$

Thus, in a regular mixture, if $w/RT > 2$ (i.e., if $T < T_c$), there is incomplete mixing.

Consider some temperature and pressure at which a binary liquid mixture separates into two liquid phases with compositions x_1^{I}, x_2^{I} and x_1^{II}, x_2^{II}. How do the compositions of these phases change when the temperature and pressure are altered? This problem is a variant of the calculation of the bubble-point and dew-point curves, or of the calculation of the solidus and liquidus curves (see Eqs. 25.71–25.78 and 25.94–25.97). The key point is that $\mu_1^{\mathrm{I}} = \mu_1^{\mathrm{II}}$ and $\mu_2^{\mathrm{I}} = \mu_2^{\mathrm{II}}$ at both T, p and $T + dT$, $p + dp$, so along the coexistence line $d\mu_1^{\mathrm{I}} = d\mu_1^{\mathrm{II}}$ and $d\mu_2^{\mathrm{I}} = d\mu_2^{\mathrm{II}}$. By expansion we then find

$$\left(\frac{\partial \mu_1}{\partial p}\right)_{T,x}^{\mathrm{I}} dp + \left(\frac{\partial \mu_1}{\partial T}\right)_{p,x}^{\mathrm{I}} dT + \left(\frac{\partial \mu_1}{\partial x_2}\right)_{T,p}^{\mathrm{I}} dx_2 = \left(\frac{\partial \mu_1}{\partial p}\right)_{T,x}^{\mathrm{II}} dp + \left(\frac{\partial \mu_1}{\partial T}\right)_{p,x}^{\mathrm{II}} dT + \left(\frac{\partial \mu_1}{\partial x_2}\right)_{T,p}^{\mathrm{II}} dx_2, \quad (25.152)$$

and similarly for component 2. Equation 25.152 is the analog of Eq. 25.73. Evaluation of the derivatives in Eq. 25.152, with use of Eq. 25.43 with $z_m = g_m$, followed by rearrangement leads to

$$(\bar{v}_1^{\mathrm{I}} - \bar{v}_1^{\mathrm{II}})\, dp - (\bar{s}_1^{\mathrm{I}} - \bar{s}_1^{\mathrm{II}})\, dT - x_2^{\mathrm{I}}\left(\frac{\partial^2 g_m}{\partial x_2^2}\right)_{T,p}^{\mathrm{I}} dx_2^{\mathrm{I}} + x_2^{\mathrm{II}}\left(\frac{\partial^2 g_m}{\partial x_2^2}\right)_{T,p}^{\mathrm{II}} dx_2^{\mathrm{II}} = 0, \quad (25.153)$$

and

$$(\bar{v}_2^{\mathrm{I}} - \bar{v}_2^{\mathrm{II}})\, dp - (\bar{s}_2^{\mathrm{I}} - \bar{s}_2^{\mathrm{II}})\, dT + x_1^{\mathrm{I}}\left(\frac{\partial^2 g_m}{\partial x_2^2}\right)_{T,p}^{\mathrm{I}} dx_2^{\mathrm{I}} - x_1^{\mathrm{II}}\left(\frac{\partial^2 g_m}{\partial x_2^2}\right)^{\mathrm{II}} dx_2^{\mathrm{II}} = 0. \quad (25.154)$$

Consider the equilibrium states that lie along the constant-pressure coexistence line. In this case we find

$$\left(\frac{\partial x_2}{\partial T}\right)_p^{\mathrm{I}} = -\frac{(\bar{h}_1^{\mathrm{I}} - \bar{h}_1^{\mathrm{II}})x_1^{\mathrm{II}} + (\bar{h}_2^{\mathrm{I}} - \bar{h}_2^{\mathrm{II}})x_2^{\mathrm{II}}}{T(\partial^2 g_m/\partial x_2^2)_{T,p}^{\mathrm{I}}(x_2^{\mathrm{I}} - x_2^{\mathrm{II}})} \quad (25.155)$$

and

$$\left(\frac{\partial x_2}{\partial T}\right)_p^{\mathrm{II}} = -\frac{(\bar{h}_1^{\mathrm{I}} - \bar{h}_1^{\mathrm{II}})x_1^{\mathrm{I}} + (\bar{h}_2^{\mathrm{I}} - \bar{h}_2^{\mathrm{II}})x_2^{\mathrm{I}}}{T(\partial^2 g_m/\partial x_2^2)_{T,p}^{\mathrm{II}}(x_2^{\mathrm{I}} - x_2^{\mathrm{II}})}. \quad (25.156)$$

Within the stable phases $(\partial^2 g_m/\partial x_2^2) > 0$, so the sign of $(\partial x_2/\partial T)_p$ is determined by the balance between $x_2^{\mathrm{I}} - x_2^{\mathrm{II}}$ and the sign of the numerator. The interpretation of the numerator is straightforward; it is, in the case of Eq. 25.155, the molar heat of solution of an infinitesimal amount of phase II in phase I at constant temperature and pressure, and similarly for Eq. 25.156.

If instead of the constant-pressure coexistence line we examine the constant-temperature coexistence line, we find

$$\left(\frac{\partial x_2}{\partial p}\right)_T^{\mathrm{I}} = \frac{x_2^{\mathrm{II}}(\bar{v}_2^{\mathrm{II}} - \bar{v}_2^{\mathrm{I}}) + x_1^{\mathrm{II}}(\bar{v}_1^{\mathrm{II}} - \bar{v}_1^{\mathrm{I}})}{(x_2^{\mathrm{II}} - x_2^{\mathrm{I}})(\partial^2 g_m/\partial x_2^2)_{T,p}^{\mathrm{I}}} \quad (25.157)$$

and

$$\left(\frac{\partial x_2}{\partial p}\right)_T^{\mathrm{II}} = \frac{x_2^{\mathrm{I}}(\bar{v}_2^{\mathrm{II}} - \bar{v}_2^{\mathrm{I}}) + x_1^{\mathrm{I}}(\bar{v}_1^{\mathrm{II}} - \bar{v}_1^{\mathrm{I}})}{(x_2^{\mathrm{II}} - x_2^{\mathrm{I}})(\partial^2 g_m/\partial x_2^2)_{T,p}^{\mathrm{II}}}. \quad (25.158)$$

Just as for $(\partial x_2/\partial T)_p$, the sign of $(\partial x_2/\partial p)_T$ depends on a balance between the composition difference $x_2^{\mathrm{II}} - x_2^{\mathrm{I}}$ and the sign of the numerator. The numerator of Eq. 25.157 is the volume change on solution of an infinitesimal amount of phase I in phase II, and similarly for Eq. 25.158.

Somewhat different conditions are satisfied along the constant-composition coexistence lines. If phase I has constant composition, so that $dx_1^{\mathrm{I}} = 0$, $dx_2^{\mathrm{I}} = 0$, we find

$$\left(\frac{\partial p}{\partial T}\right)_{x_2^{\mathrm{I}}} = \frac{x_2^{\mathrm{II}}(\bar{h}_2^{\mathrm{II}} - \bar{h}_2^{\mathrm{I}}) + x_1^{\mathrm{II}}(\bar{h}_1^{\mathrm{II}} - \bar{h}_1^{\mathrm{I}})}{T[x_2^{\mathrm{II}}(\bar{v}_2^{\mathrm{II}} - \bar{v}_2^{\mathrm{I}}) + x_1^{\mathrm{II}}(\bar{v}_1^{\mathrm{II}} - \bar{v}_1^{\mathrm{I}})]} \tag{25.159}$$

and

$$\left(\frac{\partial p}{\partial T}\right)^{\mathrm{II}}_{x_2^{\mathrm{I}}} = \frac{(\bar{v}_1^{\mathrm{II}} - \bar{v}_1^{\mathrm{I}})(\bar{h}_2^{\mathrm{II}} - \bar{h}_2^{\mathrm{I}}) - (\bar{v}_2^{\mathrm{II}} - \bar{v}_2^{\mathrm{I}})(\bar{h}_1^{\mathrm{II}} - \bar{h}_1^{\mathrm{I}})}{T(\partial^2 g_m/\partial x_2^2)^{\mathrm{II}}_{T,p}[x_2^{\mathrm{II}}(\bar{v}_2^{\mathrm{II}} - \bar{v}_2^{\mathrm{I}}) + x_1^{\mathrm{II}}(\bar{v}_1^{\mathrm{II}} - \bar{v}_1^{\mathrm{I}})]}. \tag{25.160}$$

Equation 25.159 is the analog of the Clausius–Clapeyron equation for this system in that it gives the variation of the total pressure with temperature of a phase of fixed composition in equilibrium with another phase.

The results just obtained lead to two important conclusions:

1. If $x_1^{\mathrm{I}} = x_1^{\mathrm{II}}$, $x_2^{\mathrm{I}} = x_2^{\mathrm{II}}$, then from Eqs. 25.155 and 25.156 we find that

$$\left(\frac{\partial T}{\partial x_2}\right)^{\mathrm{I}}_p = 0, \quad \left(\frac{\partial T}{\partial x_2}\right)^{\mathrm{II}}_p = 0. \tag{25.161}$$

   Therefore, if at constant pressure the temperature along the coexistence line between two phases of a binary mixture has a maximum, a minimum, or a point of inflection with horizontal tangent, the two phases have the same composition, and *vice versa.*
2. If in a series of isothermal equilibrium states of a two-phase binary system the composition of the phases becomes the same, Eq. 25.157 shows that the pressure of the system must pass through an extremum, and *vice versa.*

These conclusions are known as the *Gibbs–Konovalov theorems.* They are important in the study of azeotropy—at an *azeotropic point* the compositions of a binary liquid and its vapor are the same. More generally, an azeotropic transformation is said to occur if, in a closed system with several phases in equilibrium, the mass of one of the phases increases at the expense of the others without any change in the composition of any of the phases. Figure 25.9 shows T, x_2 diagrams for typical azeotropic systems. Systems with a maximum in the total vapor pressure curve have a positive excess Gibbs free energy, those with a minimum a negative excess Gibbs free energy.

We now consider again the stability of a mixture, but from a more general point of view than expressed at the beginning

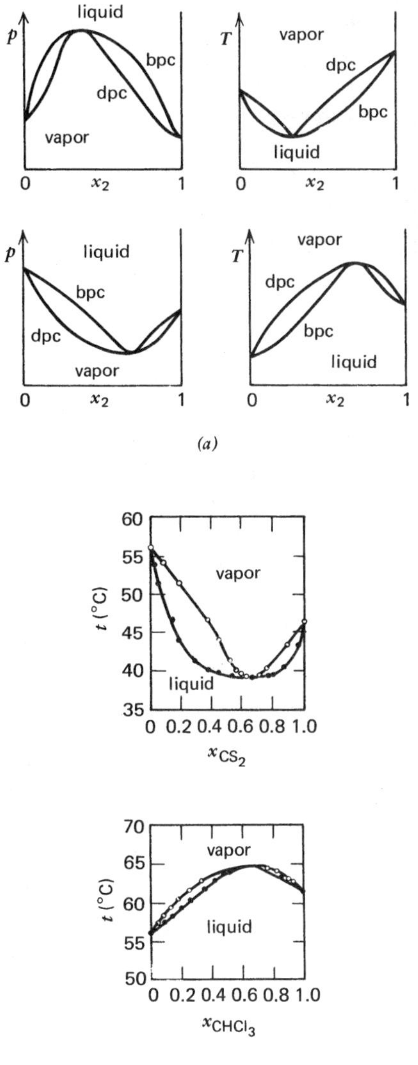

Figure 25.9 (*a*) Minimum- and maximum-boiling-point binary system phase diagrams. The bubble-point curve, labeled bpc, represents the liquid composition; the dew-point curve, labeled dpc, represents the vapor composition. (*b*) Phase diagrams for the binary systems acetone–carbon disulfide and acetone–chloroform. The partial pressures of components in these two systems are shown, at 35.2°C, in Fig. 25.4. From J. Hildebrand and R. Scott, *Regular Solutions* (Van Nostrand Reinhold, New York, 1950).

of this section. In Section 24.3 we introduced the energy surface $u(s, v)$ to aid in the visualization of the behavior of a one-component system in the region of phase separation. Alternatively, we could have used the Gibbs free energy surface $g(T, p)$ for that purpose. Since the properties of mixtures are usually studied via projections of the surface $g(T, p, x_1, \ldots)$, we start by examining that surface for a one-component system.

A schematic representation of the surface $g(T, p)$ for a one-component system is presented in Fig. 25.10. A point on that surface, say Γ_0, represents the Gibbs free energy of one mole of substance at pressure p_0 and temperature T_0.

The plane tangent to Γ_0 has slope $(\partial g/\partial T)_{p=p_0} = -s_0$ along the line of intersection with the plane $p = p_0$, and slope $(\partial g/\partial p)_{T=T_0} = v_0$ along the line of intersection with the plane $T = T_0$, hence has the equation

$$g - g_0 = -s_0(T - T_0) + v_0(p - p_0). \qquad (25.162)$$

This tangent plane cuts the g axis at

$$g = g_0 + T_0 s_0 - p_0 v_0 = u_0. \qquad (25.163)$$

We conclude that the surfaces $u(s, v)$ and $g(T, p)$ for the same system have a reciprocal relationship. That is, the quantities represented by the tangent plane to $g(T, p)$ are on the surface $u(s, v)$ and *vice versa*. This reciprocal relationship is fundamentally a consequence of the Legendre transformation used in Section 17.1 to define $g(T, p)$.

There are differences in the representations afforded by $u(s, v)$ and $g(T, p)$. Whereas there is only one surface $u(s, v)$, with any gap bridged by a derived surface, it is convenient to think of $g(T, p)$ as having several sheets, one for each homogeneous phase of the substance. Strictly, if we omit consideration of metastable states, $g(T, p)$ is one surface formed from the intersection and breaking off of pieces from three surfaces. That is, the intersection of $g^L(T, p)$ and $g^S(T, p)$ is a curve defining the melting line for the equilibrium between liquid and solid, and $g^S(T, p)$ does not extend beyond that intersection into the domain where one would have $g^S(T, p) > g^L(T, p)$. A corresponding statement can be made about $g^L(T, p)$. If, however, we include one-phase metastable states just as we did in the $u(s, v)$ representation,[12] then the surfaces do penetrate somewhat beyond their several intersections. In Fig. 25.10 the projections of the intersections *EM, MA,* and *MD* onto the (T, p) plane are the stable solid–vapor, solid–liquid, and liquid–vapor coexistence lines, respectively, and the projection of M is the triple point. The continuations $M'F'$, $M'B'$, and $M'C'$ are projections of the intersections of a metastable part of one surface with a metastable part of another surface. For example, $M'F'$ is the extension of the sublimation curve into the region where the liquid is the stable phase. Note that the extended curve describes coexistence between two phases each of which is unstable relative to the third phase in the region of the extension. In terms of the geometry of Fig. 25.10, above the space $E'M'B'$ $g^V < g^S < g^L$; above the space $B'M'D'$ $g^V < g^L < g^S$; above the space $D'M'F$ $g^L < g^V < g^S$; above the space $F'M'A'$ $g^L < g^S < g^V$; above the space $A'M'C'$ $g^S < g^L < g^V$, and above the space $C'M'E'$ $g^S < g^V < g^L$.

Given that every point on the surface $g(T, p)$ represents a stable state of the system, this surface is concave downward in its two principal directions, as can be verified by noting that

$$\left(\frac{\partial^2 g}{\partial T^2}\right)_p = -\left(\frac{\partial s}{\partial T}\right)_p = -\frac{c_p}{T} < 0 \qquad (25.164a)$$

and

$$\left(\frac{\partial^2 g}{\partial p^2}\right)_T = -\left(\frac{\partial v}{\partial p}\right)_T = -v\kappa_T < 0, \qquad (25.164b)$$

since $c_p > 0$, $T > 0$, $v > 0$, and $\kappa_T > 0$ for a stable system. Furthermore, each point on the surface $g(T, p)$ must be a minimum with respect to virtual variations (Section 19.3); these virtual variations generate different surfaces $g'(T, p)$, $g''(T, p)$, . . . , that are everywhere above $g(T, p)$ and above the tangent plane to $g(T, p)$ at the particular T, p point.

We now examine the surface $G(T, p, n_1, \ldots, n_r)$ for an r-component fluid. Let the total mass of the mixture be fixed. The system will then be completely specified if we are given the total mass and the masses of $r - 1$ components, as well as the temperature and pressure, since then

$$G = \sum_{i=1}^{r} n_i \mu_i \qquad (25.165)$$

can be written

$$g_m \equiv \frac{G}{\sum_i n_i} = \sum_{i=1}^{r} x_i \mu_i = \left(1 - \sum_{i=2}^{r} x_i\right)\mu_1 + \sum_{i=2}^{r} x_i \mu_i. \qquad (25.166)$$

If $T, p, x_2, \ldots, x_r$ are considered to be orthogonal axes, $g_m(T, p, x_2, \ldots, x_r)$ is a surface in an $(r + 2)$-dimensional space. Although it is difficult to visualize this surface because of the large number of dimensions required for the representation, it is a useful construct in the analysis of the restrictions imposed by the conditions that define a stable equilibrium state. As shown in Section 19.4, an r-component system will be stable at given fixed temperature and pressure if $g_m(T, p, x_2, \ldots, x_r)$ is a minimum.

Let $\Gamma_0 \equiv (T_0, p_0, x_2^0, \ldots, x_r^0)$ be a point on the surface $g_m(T, p, x_2, \ldots, x_r)$. Given that this point corresponds to a stable equilibrium state, virtual variations that generate new states in the vicinity of Γ_0 have Gibbs free energies that lie on surfaces g'_m, g''_m, . . . , each of which is above $g_m(T_0, p_0, x_2^0, \ldots, x_r^0)$ and also the tangent plane at Γ_0, measured parallel to the g_m-axis. Therefore, for small displacements $\Delta x_2, \ldots, \Delta x_r$ at fixed temperature and pressure,

$$\Delta g_m - (\Delta g_m)_{\text{tangent plane}} > 0 \qquad (\text{constant } T, p), \qquad (25.167)$$

where $(\Delta g_m)_{\text{tangent plane}}$ is given by the extension of Eq. 25.162. If Δg_m is expanded in a Taylor series about Γ_0, then $(\Delta g_m)_{\text{tangent plane}}$ is equal to the collection of terms linear in the Δx_i. Keeping terms quadratic in the Δx_i, we find that Eq. 25.167 is equivalent to

[12] The states on the continuation of $u(s, v)$ above the derived surface are metastable.

$$\left[\left(\frac{\partial^2 g_m}{\partial x_2^2}\right)_{T,p,n}(\Delta x_2)^2+\left(\frac{\partial^2 g_m}{\partial x_2\,\partial x_3}\right)_{T,p,n}\Delta x_2\Delta x_3+\cdots+\left(\frac{\partial^2 g_m}{\partial x_2\,\partial x_r}\right)_{T,p,n}\Delta x_2\Delta x_r\right.$$
$$+\left(\frac{\partial^2 g_m}{\partial x_3\,\partial x_2}\right)_{T,p,n}\Delta x_3\Delta x_2+\cdots\qquad+\left(\frac{\partial^2 g_m}{\partial x_3\,\partial x_r}\right)_{T,p,n}\Delta x_3\Delta x_r$$
$$\vdots$$
$$\left.+\left(\frac{\partial^2 g_m}{\partial x_r\,\partial x_2}\right)_{T,p,n}\Delta x_r\Delta x_2+\cdots\qquad+\left(\frac{\partial^2 g_m}{\partial x_r^2}\right)_{T,p,n}(\Delta x_r)^2\right]>0. \tag{25.168}$$

A sufficient condition that the left-hand side of Eq. 25.168 be positive is that the determinant of the coefficients in the quadratic form be positive definite. This requires that

$$\Delta_{r-1}\equiv\begin{vmatrix}\left(\frac{\partial^2 g_m}{\partial x_2^2}\right)_{T,p,n} & \cdots\left(\frac{\partial^2 g_m}{\partial x_2\partial x_r}\right)_{T,p,n}\\ \vdots & \\ \left(\frac{\partial^2 g_m}{\partial x_r\,\partial x_2}\right)_{T,p,n} & \cdots\left(\frac{\partial^2 g_m}{\partial x_r^2}\right)_{T,p,n}\end{vmatrix}>0 \tag{25.169}$$

and that all the minors of Δ_{r-1} be positive.[13] Any system that is in a state for which these conditions are satisfied is stable

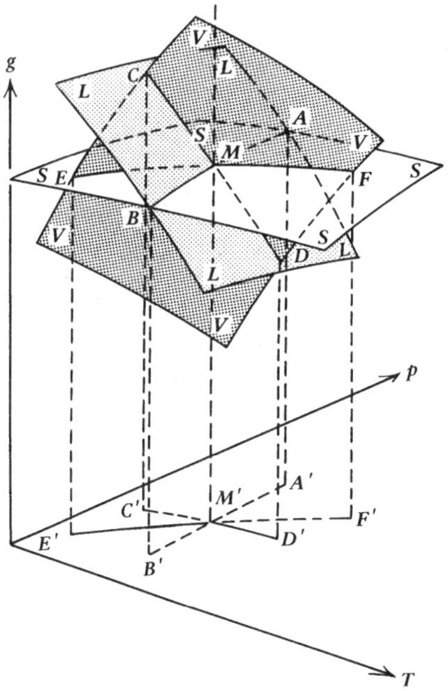

Figure 25.10 Free energy-temperature-pressure surface for a substance. The surfaces labeled *V*, *L*, and *S* refer to the vapor, liquid, and solid, respectively. (Courtesy of Prof. J. A. Beattie.)

[13] The minors of a determinant are the smaller determinants formed when a row and column are deleted.

with respect to so-called adjacent states, namely, those whose free energies are spanned by Δg_m of Eq. 25.167.

We now examine the surface $g_m(T, p, x_2)$ which represents the equilibrium states of a binary mixture. Even in this simple case we need a four-dimensional space to represent $g_m(T, p, x_2)$. But if we examine the free-energy surface for fixed pressure, $g_m(T, p, x_2)$ can be represented as a surface in a three-dimensional space defined by orthogonal g_m, T, and x_2 axes. Projection of the surface $g_m(T, p, x_2)$ onto the (T, x_2) plane generates a temperature-composition diagram such as shown in Fig. 25.11*a*. The intersections of $g_m(T, p, x_2)$ with the several planes on which $T = T_1$, $T = T_2$, . . . generate free energy-composition diagrams for T_1, T_2, . . . , as shown in Figs. 25.11*a*–25.11*j*. All of the diagrams 25.11*a*–25.11*j* refer to the same constant pressure.

For the case of a binary mixture, $\Delta_{r-1} = \Delta_1$ reduces to a single term. We find that the binary mixture will be stable if

$$\left(\frac{\partial^2 g_m}{\partial x_2^2}\right)_{T,p}>0 \tag{25.170}$$

and unstable if

$$\left(\frac{\partial^2 g_m}{\partial x_2^2}\right)_{T,p}<0. \tag{25.171}$$

The limit of stability occurs when

$$\left(\frac{\partial^2 g_m}{\partial x_2^2}\right)_{T,p}=0. \tag{25.172}$$

Condition 25.170 can be expressed in terms of the chemical potential. Using Eq. 25.165 and the Gibbs–Duhem equation (25.25) in the form

$$x_1\left(\frac{\partial \mu_2}{\partial x_2}\right)_{T,p}+x_2\left(\frac{\partial \mu_2}{\partial x_2}\right)_{T,p}=0, \tag{25.173}$$

we readily find that condition 25.170 requires

$$\left(\frac{\partial \mu_2}{\partial x_2}\right)_{T,p}>0, \tag{25.174}$$

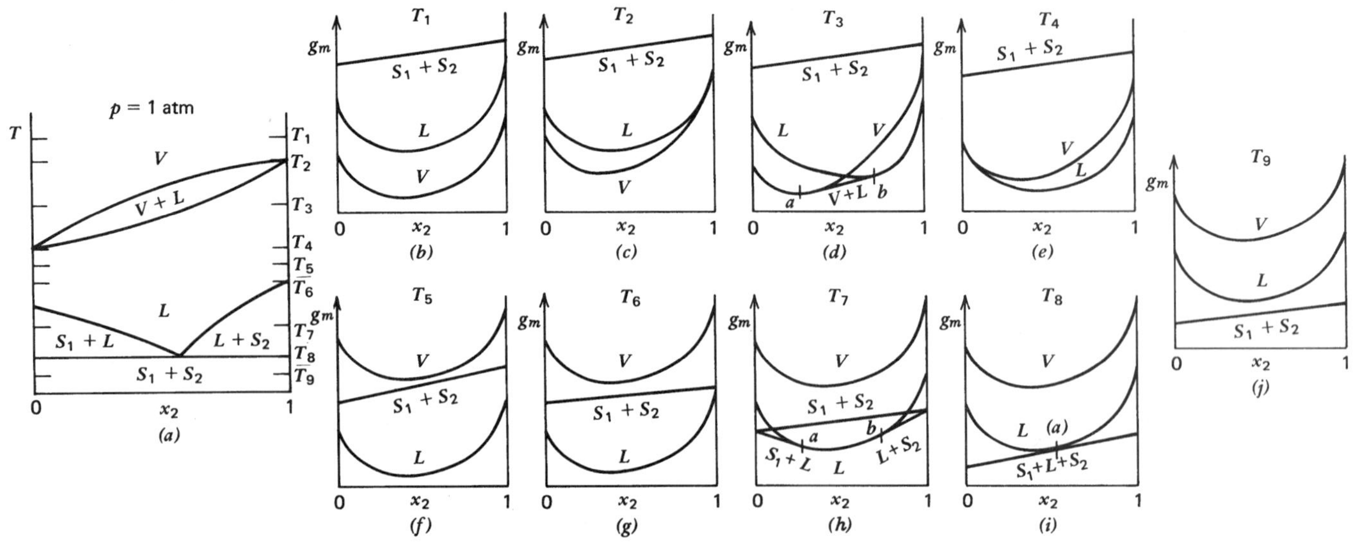

Typical temperature composition diagram for various temperatures

Figure 25.11 (*a*) Temperature-composition diagram for a binary mixture in which neither compounds nor solid solutions are formed, there is no miscibility gap, and there are no minimum- or maximum-boiling mixtures. (*b*) T_1 is greater than the boiling points of both components, hence the vapor is the stable state for all compositions. (*c*) The temperature is, in this case, equal to the boiling temperature of component 2, hence $g_m^L = g_m^V$ at $x_2 = 1$. Everywhere else the vapor is the most stable state. (*d*) T_3 intersects the region where vapor and liquid coexist. For large x_2 the liquid phase is most stable, for small x_2 the vapor phase is most stable. The common tangent is the line along which g_m is least, hence describes the equilibrium situation of coexisting liquid and vapor between domains where either only liquid ($x_2 \geq b$) or only vapor ($x_2 \leq a$) is stable. The phase coexistence domain is $a < x_2 < b$. (*e*) T_4 coincides with the boiling point of component 1, so this case is the analog of that in Fig. 25.11*c* and $g_m^L = g_m^V$ at $x_2 = 0$. (*f, g*) The analogs of Figs. 25.11*b* and 25.11*c* with liquid and solid replacing vapor and liquid in the descriptions. (*h*) T_7 intersects the coexistence lines for liquid–solid 1 and liquid–solid 2, hence two common tangents need to be drawn in the g_m, x_2 diagram. (*i*) T_8 touches the liquid–solid coexistence curve at only one point, where solids 1 and 2 and liquid are in equilibrium. This is called the eutectic temperature. (*j*) T_9 lies below the melting points of both solids, hence the two solids are coexisting phases. All of the preceding also apply to solid solutions; only the free energy-composition diagram needs relabeling for the appropriate phases. (Courtesy of Prof. J. A. Beattie.)

which also implies that

$$\left(\frac{\partial \mu_1}{\partial x_2}\right)_{T,p} < 0 \qquad (25.175)$$

and

$$\left(\frac{\partial \mu_2}{\partial x_1}\right)_{T,p} < 0. \qquad (25.176)$$

Thus, in a stable mixture μ_2 must increase when x_2 is increased and T and p are held constant.

We also note that at constant temperature and pressure the slope of the free energy-composition curve is (see also Eq. 25.42)

$$\left(\frac{\partial g_m}{\partial x_1}\right)_{T,p} = \mu_1 - \mu_2 = -\left(\frac{\partial g_m}{\partial x_2}\right)_{T,p}. \qquad (25.177)$$

When $x_1 \to 0$, $g_m \to \mu_2^0$ and $(\partial g_m/\partial x_1)_{T,p,n} \to -\infty$, whereas when $x_1 \to 1$, $g_m \to \mu_1^0$ and $(\partial g_m/\partial x_1)_{T,p,n} \to \infty$. The behavior of the derivatives follows from the dependence of the chemical potential on the logarithm of the mole fraction (see Eq. 25.12).

From the preceding we deduce that the free energy-composition diagram has the following properties:

1. For a homogeneous system g_m is a simple curve extending from $x_2 = 0$ to $x_2 = 1$; it is tangent to the axes $x_2 = 0$ and $x_2 = 1$, the slopes there being such that g_m is concave upward.
2. Corresponding to the existence of three phases of a homogeneous system there will be at least three lines on the g_m, x_2 diagram, one each for the vapor, liquid, and solid.
3. As shown by Eq. 25.143, points on the g_m, x_2 diagram that represent two coexisting phases have a common tangent.

The free energy-composition diagrams for a system are, of course, different at different temperatures. Figures 25.11*a*, 25.12*a*, and 25.13*a*–25.16*a* display a variety of temperature-composition diagrams. The simplest of these, Fig. 25.11*a*, represents a binary system in which neither compounds nor solid solutions are formed, in which there is not any region of liquid–liquid insolubility (miscibility gap), and in which there are not either minimum- or maximum-boiling solutions (azeotropic mixtures). Even in this case the

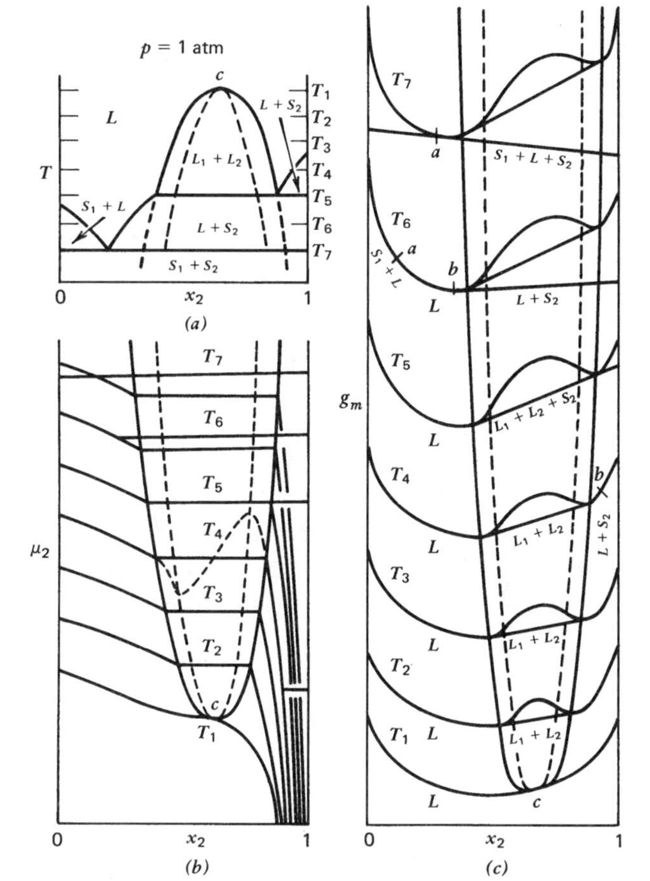

Figure 25.12 Temperature-composition diagram at constant pressure and the corresponding molar free energy-composition diagrams at constant pressure and temperature. (Courtesy of Prof. J. A. Beattie.)

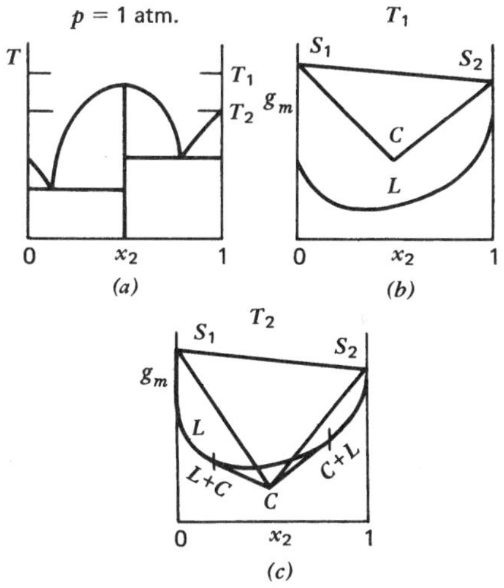

Figure 25.13 The components form a compound C that melts congruently. (Courtesy of Prof. J. A. Beattie.)

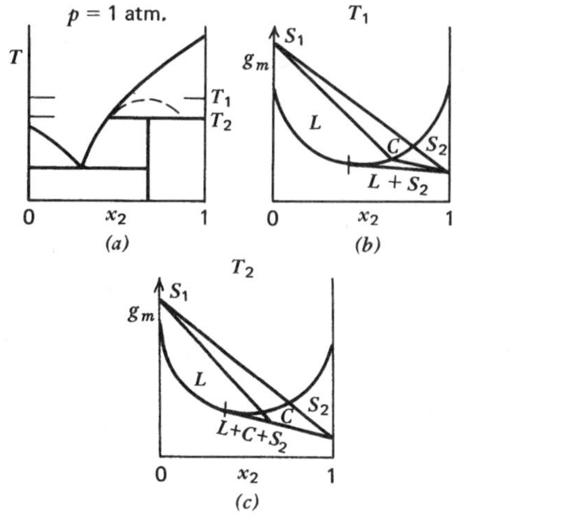

Figure 25.14 The components form a compound C that melts incongruently. (Courtesy of Prof. J. A. Beattie.)

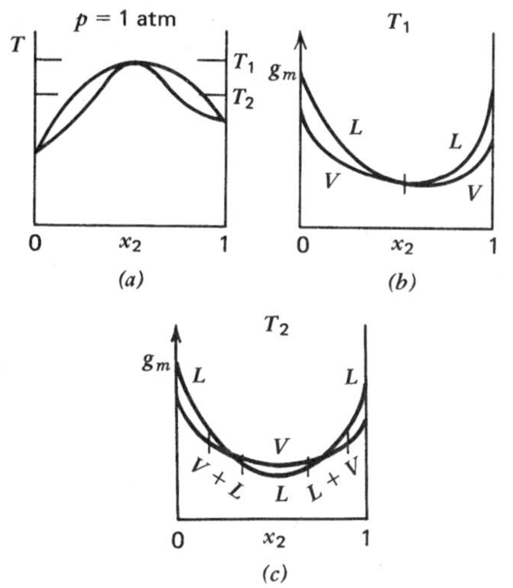

Figure 25.15 The components form a maximum-boiling liquid. (Courtesy of Prof. J. A. Beattie.)

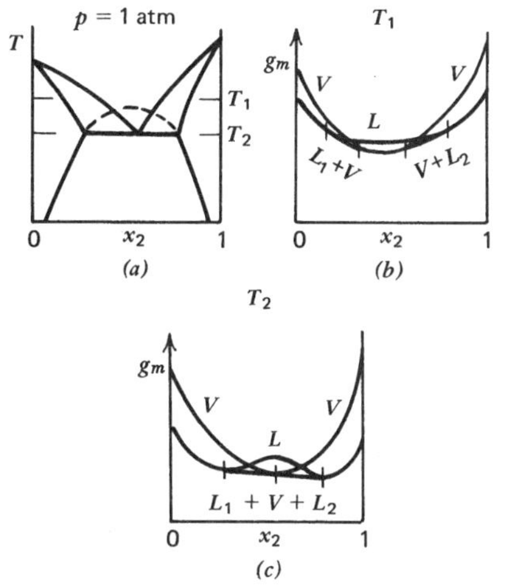

Figure 25.16 The liquid components are not miscible in all proportions. (Courtesy of Prof. J. A. Beattie.)

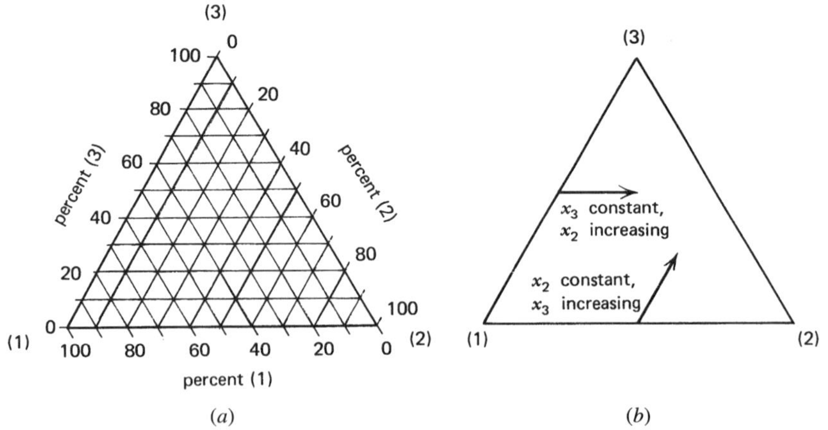

Figure 25.17 Triangular coordinates.

several free energy-composition diagrams for different temperatures, Figs. 25.11*b*–25.11*j*, look very different. The examples displayed are described in the caption, so only one will be cited here. Figure 25.11*d* illustrates a g_m, x_2 diagram when T_3 intersects the region where vapor and liquid coexist. For large x_2, namely, $x_2 \geq b$, the liquid phase is most stable; for small x_2, namely, $x_2 \leq a$, the vapor phase is most stable. The common tangent to g_m at $x_2 = a$ and $x_2 = b$ is the line along which g_m is least, hence describes the equilibrium between coexisting liquid and vapor when $a < x_2 < b$.

Figure 25.12 illustrates possible free energy-composition diagrams corresponding to a temperature-composition diagram for a binary mixture with a miscibility gap in the liquid region, and Figs. 25.13–25.16 illustrate the effects on the free energy-composition diagram of formation of a compound that melts at constant composition (congruently), formation of a compound that melts with a change of composition (incongruently), formation of a maximum-boiling liquid, and the case that the liquid components are not miscible in all proportions, respectively. In each case two temperatures are selected for display of the corresponding free energy-composition diagrams. A few comments about the diagrams are to be found in the respective figure captions.

We now consider, briefly, some properties of three-component mixtures. For this case $\Delta_{r-1} = \Delta_2$ reduces to (see Eq. 25.169)[14]

$$\left(\frac{\partial^2 g_m}{\partial x_1^2}\right)_{T,p,n}\left(\frac{\partial^2 g_m}{\partial x_2^2}\right)_{T,p,n} - \left[\left(\frac{\partial^2 g_m}{\partial x_1 \partial x_2}\right)_{T,p,n}\right]^2 > 0, \tag{25.178}$$

and the condition on the minors of Δ_2 to

$$\left(\frac{\partial^2 g_m}{\partial x_1^2}\right)_{T,p,n} > 0,$$
$$\left(\frac{\partial^2 g_m}{\partial x_2^2}\right)_{T,p,n} > 0. \tag{25.179}$$

[14] The labeling of the components is arbitrary.

The ternary mixture is stable when Eqs. 25.178 and 25.179 are satisfied. These conditions can, of course, be rewritten in terms of derivatives of the chemical potentials of the components.

To find the analog of Eq. 25.177, we write

$$g_m = x_1\mu_1 + x_2\mu_2 + (1 - x_1 - x_2)\mu_3 \tag{25.180}$$

and calculate (at constant temperature and pressure)

$$\left(\frac{\partial g_m}{\partial x_1}\right)_{T,p,x_2} = \mu_1 - \mu_2,$$
$$\left(\frac{\partial g_m}{\partial x_2}\right)_{T,p,x_1} = \mu_2 - \mu_3. \tag{25.181}$$

The subscripts 1, 2, 3 can be interchanged to find relations for other derivatives of $g_m(T, p, x_1, x_2)$. Then when $x_1 \to 1$, $g_m \to \mu_1^0$ and $(\partial g_m/\partial x_2)_{T,p,x_3} \to -\infty$, $(\partial g_m/\partial x_1)_{T,p,x_2} \to -\infty$; when $x_2 \to 1$, $g_m \to \mu_2^0$ and $(\partial g_m/\partial x_3)_{T,p,x_1} \to -\infty$, $(\partial g_m/\partial x_1)_{T,p,x_3} \to -\infty$; when $x_3 \to 1$, $g_m \to \mu_3^0$ and $(\partial g_m/\partial x_1)_{T,p,x_2} \to -\infty$, $(\partial g_m/\partial x_2)_{T,p,x_1} \to -\infty$. Also, when

$$x_1 \to 0: \quad g_m = x_2\mu_2 + (1 - x_2)\mu_3,$$
$$\left(\frac{\partial g_m}{\partial x_2}\right)_{T,p,x_3} \to \infty, \quad \left(\frac{\partial g_m}{\partial x_3}\right)_{T,p,x_2} \to \infty;$$
$$x_2 \to 0: \quad g_m = x_1\mu_1 + (1 - x_1)\mu_3,$$
$$\left(\frac{\partial g_m}{\partial x_3}\right)_{T,p,x_1} \to \infty, \quad \left(\frac{\partial g_m}{\partial x_1}\right)_{T,p,x_3} \to \infty;$$
$$x_3 \to 0: \quad g_m = x_1\mu_1 + x_2\mu_2,$$
$$\left(\frac{\partial g_m}{\partial x_1}\right)_{T,p,x_2} \to \infty, \quad \left(\frac{\partial g_m}{\partial x_2}\right)_{T,p,x_1} \to \infty. \tag{25.182}$$

The equilibrium states of a ternary system can be represented using a triangular prism. The base of this prism is formed by the coordinate axes x_1, x_2, and x_3 arranged as an equilateral triangle (see Fig. 25.17), and the height above the base is g_m, of course all for fixed temperature and pressure. The various limiting forms of g_m given in Eqs. 25.182 represent the free energy-composition diagrams for the three

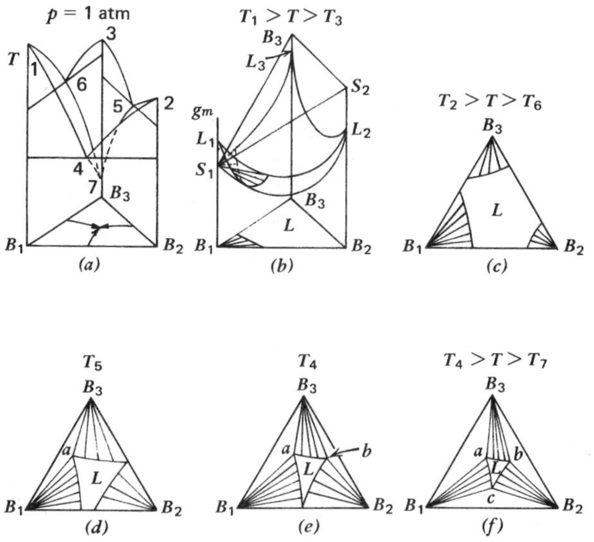

Figure 25.18 (a) The temperature-composition surface at $p = 1$ atm for the condensed phases of a ternary system with no compound formation, no solid solutions, and no miscibility gap in the liquid phase. Points in the surfaces (1, 4, 7, 6), (2, 5, 7, 4),(3, 6, 7, 5) give the compositions of the liquid phase in equilibrium at various temperatures with the solid components 1, 2, 3, respectively. Points on the curves (4, 7), (5, 7), (6, 7) give the composition of the liquid in equilibrium at various temperatures with the two solids 1 and 2, 2 and 3, 1 and 3, respectively. The point 7 gives the temperature and composition of the liquid in equilibrium with the three solids 1, 2, 3. Points 4, 5, and 6 are binary eutectic points, 7 is a ternary eutectic point. When the curves (4, 7), (5, 7), and (6, 7) are projected onto the composition triangle, an arrow is appended to show the direction of decreasing temperature. (b) Molar free energy-composition surface corresponding to (a). (c), (d), (e), (f) Projections of (b) on the composition triangle for different temperatures. The labels on the temperatures correspond to the numbers in part (a). (Courtesy of Prof. J. A. Beattie.)

binary systems into which a ternary system can be subdivided. These limiting curves appear on the appropriate faces of the triangular prism of coordinates just described. In general, the surface $g_m(T, p, x_1, x_2)$ extends over the entire composition triangle, touching the edges $x_1 = 1$, $x_2 = 1$, and $x_3 = 1$ at μ_1^0, μ_2^0, and μ_3^0, respectively. Moreover, $g_m(T, p, x_1, x_2)$ is tangent to the three edges and three sides of the coordinate prism, being concave upward in its two principle curvatures. There will be, in general, three or more sheets of $g_m(T, p, x_1, x_2)$, one each for the solid, liquid, and vapor phases, each covering the composition triangle. As before, points representing coexisting phases have a common tangent plane.

The preceding observations on the properties of $g_m(T, p, x_1, x_2)$ are seen to parallel those for a binary system. Ternary system diagrams, even in very simple cases, present a complicated appearance. Some examples are displayed in Fig. 25.18. The caption describes the features of interest.

25.9 Critical Phenomena in Mixtures

We saw in Chapter 24 that at the critical point of a pure fluid the differences between the vapor and liquid phases vanish. For a pure fluid the critical point lies on both the coexistence curve representing the vapor–liquid equilibrium and the curve of the limit of stability. The one point where both conditions are met occurs where $(\partial p/\partial v)_T = 0$ and $(\partial^2 p/\partial v^2)_T = 0$ simultaneously.

In general, we expect a critical state to occur in a system when the differences between two coexisting phases vanish. Consider the case of equilibrium between two phases, each of which has the same r components. The requirements of thermal and mechanical equilibrium ensure that the temperatures and pressures of the two phases are equal. To specify the concentrations of the r components in each phase (a total of $2r$ concentrations), we must specify r variables.[15] Suppose that T, p, and $r - 1$ concentrations are fixed at values characteristic of a critical state, and that the rth is varied. The set of states so generated consists of pairs of coexisting phases whose properties become identical when the rth concentration achieves the value characteristic of the critical state. Put another way, for fixed T, p, n_1/V, . . . , n_{r-1}/V, variation of n_r/V defines a coexistence line for two-phase equilibrium that terminates in the critical state. Suppose now that T, p, and n_1/V, . . . , n_{r-1}/V are varied infinitesimally, then fixed, and n_r/V is again varied. This procedure generates a new set of pairs of coexisting phases, each of which differs infinitesimally from the corresponding pairs of coexisting phases in the first variation of n_r/V, hence the new two-phase coexistence curve must terminate in a critical state differing infinitesimally from the first critical state. This argument shows that two-phase equilibrium in an r-component system leads to a critical state with $r - 1$ degrees of freedom. For a one-component system there is only a critical point ($r - 1 = 0$), but for a binary mixture there is a critical line ($r - 1 = 1$), and so forth. In the case of the binary mixture, in which at a fixed pressure two liquids do not mix at some temperatures but do at other temperatures, the critical solution temperature T_c is a function of the pressure; this function defines the critical line.

Just as for a one-component system, the critical state of a mixture lies on both the coexistence curve for two-phase equilibrium *and* the limit-of-stability curve, yet is itself a stable state of the system. For a binary mixture the limit of stability occurs when $\Delta_1 = 0$, that is, when

$$\left(\frac{\partial^2 g_m}{\partial x_2^2}\right)_{T,p} = \left[\frac{\partial(\mu_2 - \mu_1)}{\partial x_2}\right]_{T,p} = 0. \qquad (25.172)$$

Furthermore, again just as for a one-component system, Eq. 25.172 defines the stability of a homogeneous phase against small displacements not including formation of a new phase. Therefore, for portions of the coexistence region away from the critical point, phase separation occurs before Eq. 25.172 is satisfied, and there is a portion of the g_m, x_2 curve that lies inside the coexistence curve and describes

[15] The condition $\mu_i^{\mathrm{I}} = \mu_i^{\mathrm{II}}$, $i = 1$, . . . , r, fixes the ratio of the concentrations in the two phases.

metastable states (see Fig. 25.12 and, for the one-component analog, the discussion following Fig. 24.15). There is a gap in the free energy-composition surface where $\Delta_1 < 0$; this gap is spanned by the derived surface generated from the common tangent to the two portions of g_m representing stable phases. The gap disappears and the points of tangency on the two portions of $g_m(T, p, x_2)$ coincide at the critical point $T_c(p)$. Therefore, a critical state exists when both Eq. 25.172 and

$$\left(\frac{\partial^3 g_m}{\partial x_2^3}\right)_{T=T_c,p} = \left[\frac{\partial^2 (\mu_2 - \mu_1)}{\partial x_2^2}\right]_{T=T_c,p} = 0 \qquad (25.183)$$

hold. The extension of this definition of the critical state to r-component mixtures is algebraically tedious, but involves no new principles.

Because of the analogy between Eqs. 25.172 and 25.183 and the corresponding conditions for a pure fluid, $(\partial p/\partial v)_T = -(\partial^2 a/\partial v^2)_T = 0$ and $(\partial^2 p/\partial v^2)_T = (\partial^3 a/\partial v^3)_T = 0$, respectively, we expect that the behavior of a mixture in the region near its critical state will be very much like that of a pure fluid near its critical point. This expectation is strengthened when we recognize that the scale length for positional correlations between pairs of molecules diverges as the critical temperature is approached (Eq. 24.102). When the scale length for positional correlations between pairs of molecules exceeds the range of the molecular interactions of all species in the mixture, the differences between species in the mixture are masked. Indeed, it can be shown that the critical behavior of a pure fluid and of a fluid mixture are the same, and the values of the critical exponents quoted in Chapter 24 are averages that include data from studies of the critical points in both pure fluids and liquid mixtures.

However, the added complexity introduced by the increase in the number of degrees of freedom of a system when it has more than one component is such that the level of sophistication of the description of the critical behavior of a mixture is very much less than that of the description of the critical behavior of a one-component system.[16] Just as for the one-component system, the assumption that $g_m(T, p, x_2)$ can be expressed in a Taylor series expansion about the critical point (that is, about a point on the critical line) and the use of mean field-type equations of state predict incorrect behavior in the critical region. In summary form, it is found that:

1. At constant pressure,

$$x_2^{\mathrm{II}} - x_2^{\mathrm{I}} \propto |T_c - T|^{\beta}, \qquad (25.184)$$

with $\beta = 0.326 \pm 0.002$; the classical theory predicts that $\beta = 0.50$ (see Fig. 25.19).

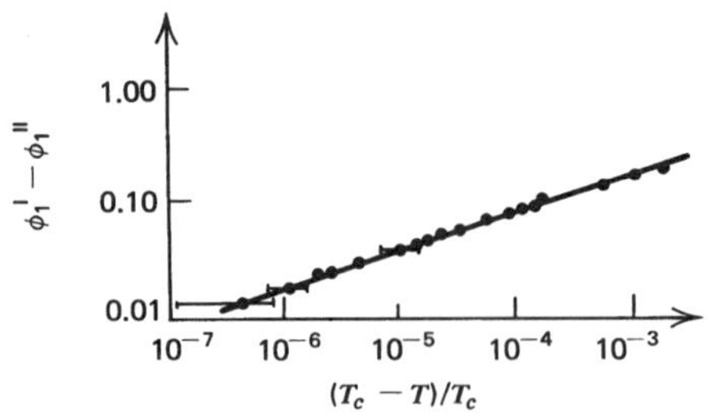

Figure 25.19 Coexistence curve for the liquid mixture C_7H_{14}–CCl_4. The critical exponent in

$$\phi_1^{\mathrm{I}} - \phi_1^{\mathrm{II}} \propto |T_c - T|^{\beta}$$

is $\beta = 0.335 \pm 0.02$. ϕ_1 is the volume fraction of component 1. Either the volume fraction or the mole fraction can be used to define the approach to T_c. Only the coefficient, but not the exponent, is changed when $x \to \phi$ in Eq. 25.184. From P. Heller, *Rept. Prog. Phys.* **30,** 731 (1967).

2. Along a line of constant temperature $T = T_c$ and constant pressure $p = p_c$,

$$(\mu_1 - \mu_2) - (\mu_1 - \mu_2)_c \propto (x_2 - x_{2c})^{\delta}, \qquad (25.185)$$

with $\delta = 4.80 \pm 0.02$; the classical theory predicts that $\delta = 3$.

3. Along a line of constant pressure $p = p_c$ and constant composition $x_2 = x_{2c}$,

$$\left[\frac{\partial(\mu_2 - \mu_1)}{\partial x_2}\right]_{T,p_c,n} \propto (T - T_c)^{\gamma}, \qquad (25.186)$$

with $\gamma = 1.239 \pm 0.002$; the classical theory predicts that $\gamma = 1$.

4. Along a line of constant pressure $p = p_c$ and constant composition $x_2 = x_{2c}$,

$$c_{px_2} \propto |T - T_c|^{-\alpha}, \qquad (25.187)$$

with $\alpha = 0.110 \pm 0.003$; the classical theory predicts that $\alpha = 0$ and that there is a jump discontinuity in c_{px_2}.

As in the case of pure fluids, the values quoted above for the critical exponents are those calculated from renormalization group theory (see Section 24.4); the available experimental values of the critical exponents agree with the theoretical values to within the precision of the experimental data.

The forms displayed in Eqs. 25.184–25.187 do not distinguish between the values of the critical exponents for approaches to the critical state from $T > T_c$ and $T < T_c$. Just as in the case of the pure fluid, the information available suggests that the values for the two approaches are the same, and just this is suggested by model calculations and also by the assumption that $g_m(T, p, x_2)$ is a generalized homoge-

[16] For a review see J. V. Sengers and J. M. H. Levelt-Sengers, *Ann. Rev. Phys. Chem.* **37,** 189 (1986).

neous function of $(T - T_c)$, $(p - p_c)$, and $(x_2 - x_{2c})$ very near the critical state (see also the discussion in Chapter 24).

Clearly, just as is the case for a pure fluid, the classical predictions of α, β, γ, and δ are incorrect. We showed in Chapter 24 that the assumption that the singular part of the chemical potential near the critical point is a homogeneous function of $T - T_c$ and $T_c - \tau(\rho)$ led to a universal equation of state near the critical point (Fig. 24.19). For a binary mixture the assumption that the singular part of $g_m(T, p, x_2)$ is a homogeneous function of $T - T_c$, $p - p_c$, and $x_2 - x_{2c}$ very near the critical state leads to an analogous universal equation of state. The available tests of this equation show, within the uncertainties of the data, about as good agreement as does the test for pure fluids (Fig. 24.19).

This description of the failure of the classical theory of the critical region should not be allowed to obscure the fact that far away from the critical region, where the range of correlations is not very much larger than the range of the molecular interactions, simple equations of state give very useful approximate descriptions of the phase equilibria and describe correctly all the qualitative features of such equilibria. It is only in the critical region that there is a qualitative breakdown of the classical analysis. Since a very large part of the interpretation of phase equilibria is unrelated to critical phenomena, there is ample justification for development of molecular models without the refinements necessary for the analysis of the critical region. Moreover, a very significant fraction of the effort in the study of mixtures is concerned with the role of molecular interactions in the one-phase region of complete miscibility, which further reinforces the need for model molecular theories of mixtures.

25.10 The Molecular Theory of Solutions of Nonelectrolytes

The molecular theory of mixtures has four principal goals. These are (1) development of a formalism that represents the thermodynamic properties of the mixture in terms of the distributions of the molecules, (2) accurate prediction of the thermodynamic properties of the homogeneous phases of a mixture, (3) accurate prediction of the location of phase boundaries and the compositions of coexisting phases, and (4) interpretation of the generic properties of mixtures in terms of the nature of the interactions between the molecules. Clearly, the existence of different interactions between pairs of like and unlike molecules makes the molecular theory of mixtures more difficult than the corresponding theory for the pure components. The many theoretical investigations that have been made range from elementary treatments of very crude models to sophisticated generalizations of the interpretation of the relationships between thermodynamic functions, fluctuations in composition, and the character of molecular distribution functions in different thermodynamic states. As in Chapter 23, we shall discuss the important ideas in the theoretical analyses but not work out the mathematical details; a few of the ideas will be illustrated by model calculations. We consider first the general characteristics of the representation of thermodynamic functions in terms of molecular distribution functions, then the molecular theory of crystalline mixtures and, finally, the molecular theory of liquid mixtures.

A. The Statistical Mechanics of Mixtures

In Chapter 19 we showed that a system at equilibrium is subject to a number of stability conditions. The argument used to establish those stability conditions was based on examination of the change in a thermodynamic potential, such as the internal energy, associated with a virtual variation of the state of the system. We now choose to invert the argument in the following sense. The molecular theory of equilibrium states shows that there are fluctuations in thermodynamic functions, which we can think of as defining virtual variations of the state of the system. It then should be possible to use the theory of fluctuations in a mixture to obtain relationships between its thermodynamic properties and the molecular distribution functions. Kirkwood and Buff[17] exploited this approach to show that in a binary mixture the derivative of the osmotic pressure with respect to the concentration of solute, and the derivatives of the chemical potentials of the components with respect to their own concentrations, are

$$\begin{aligned} \left(\frac{\partial \pi}{\partial n_2}\right)_{T,\mu_1} &= \frac{k_B T}{1+G_{22}n_2}, \\ \frac{1}{k_B T}\left(\frac{\partial \mu_2}{\partial n_2}\right)_{T,p} &= \frac{1}{n_2}+\frac{G_{12}-G_{22}}{1+n_2(G_{22}-G_{12})}, \\ \frac{1}{k_B T}\left(\frac{\partial \mu_1}{\partial n_1}\right)_{T,p} &= \frac{1}{n_1}+\frac{G_{12}-G_{11}}{1+n_1(G_{11}-G_{12})}, \end{aligned} \tag{25.188}$$

where n_1 and n_2 are the average numbers of molecules of species 1 and 2 per unit volume, and the functions G_{ij} are defined in terms of the pair correlation function for species i and j, denoted g_{ij}, by

$$G_{ij} \equiv \int d\mathbf{R}(g_{ij}-1). \tag{25.189}$$

In addition to Eq. 25.188, the Kirkwood–Buff theory of solutions leads to many other *exact* relationships between the thermodynamic properties of a solution and the molecular distribution functions.

Equation 25.188 exhibits the dependences of the chemical potentials of the species in a binary solution on the differences between the local structures about each of the species in the solution; the local structures are characterized

[17] J. G. Kirkwood and F. P. Buff, *J. Chem. Phys.* **19**, 774 (1951).

by the pair correlation functions g_{ij}. Note that if all the G_{ij} ($i, j = 1, 2$) are equal, we recover the ideal solution forms for the chemical potentials of both species; the second terms on the right-hand sides of the last two of Eq. 25.188 are, then, related to the excess chemical potentials of species 1 and 2.

Equations 25.188 provide a formal basis for the calculation of the thermodynamic properties of a solution, but that calculation requires as input all the independent pair correlation functions of the system. The Kirkwood–Buff theory of solutions, although formally exact, is incomplete in the sense that information about the needed pair correlation functions must come from some other molecular theory or from experimental data. However, the formalism of the theory can be used to analyze experimental data. For example, in a dilute solution in which $n_2G_{22} << 1$, the first of Eqs. 25.188 becomes

$$\frac{1}{k_BT}\left(\frac{\partial \pi}{\partial n_2}\right)_{T,\mu_1} = 1 - G_{22}n_2^2 + \cdots. \qquad (25.190)$$

If the osmotic pressure is fit to a power series in the solute concentration, G_{22} can be determined from the fit. Similarly, measurements of the excess chemical potentials of the components of the solution can be used to evaluate the combinations of G_{ij} on the right hand sides of Eq. 25.188. The values of the G_{ij}, and their temperature dependences, can be used to infer some information about the distribution of molecules in the solution.

A different theory of solutions, also formally exact, can be based on an extension of the ideas used to develop the relationship between thermodynamic properties of an open system and the distribution of molecules in the grand canonical ensemble (Chapter 19). The development of this theory starts with the observation that the grand potential $\Omega \equiv -pV$ is related to the grand partition function by Eqs. 19.125 and 19.126, which we write here in the form

$$\exp\left(-\frac{\Omega}{k_BT}\right) = \Xi = \sum_{\mathbf{N}\geq 0}\sum_{k} g_k \exp\left(\frac{\mathbf{N}\cdot\boldsymbol{\mu} - E_k}{k_BT}\right), \qquad (25.191)$$

where $\mathbf{N}$ and $\boldsymbol{\mu}$ are vectors whose components are the numbers and chemical potentials of species $1, 2, \ldots, r$, so that $\mathbf{N}\cdot\boldsymbol{\mu} = \sum_{i=1}^{r} N_i\mu_i$. We refer to $\mathbf{N}$ and $\boldsymbol{\mu}$ as the molecular number set and the chemical potential set, respectively. Eq. 25.191 clearly shows that the pressure of the system is a function of the chemical potential set of the system. Mayer[18] first showed for the one component case, and McMillan and Mayer[19] for the multicomponent case, that the formalism that leads to 25.190 can be extended to provide a representation of the n-particle distribution function in an r-component system with a specified chemical potential set and the n-particle distribution function in another r-component system with a different chemical potential set. One of the most important results obtained from this formalism is a relation for the difference in pressure between two r-component systems with different chemical potential sets.

Let $z_1 = \Lambda_i^{-3}\exp(\mu_i/k_BT)$ be the fugacity of component i, $\mathbf{z}$ a vector with components $z_1, \ldots, z_r$, and $\boldsymbol{\gamma}$ the vector whose components are the activity coefficients $\gamma_1 = z_1/n_1, \ldots, \gamma_r = z_r/n_r$. It should be noted that, although we have used the same notation, these activity coefficients differ from those we introduced in the earlier part of this chapter in that they are defined to be unity in the limit that the system becomes a perfect gas mixture. McMillan and Mayer showed that the difference in pressure between two r-component systems with fugacity sets $\mathbf{z}$ and $\mathbf{z}^*$ is

$$\exp\left[-\frac{\Omega(\mathbf{z})-\Omega(\mathbf{z}^*)}{k_BT}\right] = \sum_{\mathbf{m}\geq 0}\frac{1}{\mathbf{m}!}\frac{(\mathbf{z}-\mathbf{z}^*)^{\mathbf{m}}}{\boldsymbol{\gamma}^{\mathbf{m}}}\int d\{\mathbf{m}\}\exp\left(-\frac{w_{\mathbf{m}}(\mathbf{z}^*)}{k_BT}\right), \qquad (25.192)$$

where $d\{\mathbf{m}\}$ is shorthand notation for the volume elements associated with the coordinates of the $\mathbf{m}$ molecule subset and $w_m(\mathbf{z}^*)$ is the potential of mean force acting on the $\mathbf{m}$ molecules (see Chapter 23). The $\mathbf{m}$-particle potential of mean force is related to the $\mathbf{m}$-particle correlation function $g_{\mathbf{m}}$ by the definition (see Eq. 23.34)

$$\frac{n^{(\mathbf{m})}}{n^{\mathbf{m}}} \equiv g_{\mathbf{m}} \equiv \exp\left(-\frac{w_{\mathbf{m}}}{k_BT}\right), \qquad (25.193)$$

where $n^{(\mathbf{m})}$ is the $\mathbf{m}$-particle distribution function and $n^{\mathbf{m}} \equiv \prod_{i=1}^{\mathbf{m}} n_i$. For the case of osmotic equilibrium, one system consists of pure solvent and the other a solution; the two systems are separated by a membrane permeable only to the solvent. Then the fugacity set for the solution side of the semipermeable membrane is $z_1, \ldots, z_r \neq 0$ and for the solvent side is $z_1^* \neq 0$, $z_j^* = 0$ ($j = 2, \ldots, r$). From Eq. 25.192 and the condition of osmotic equilibrium, $z_1 = z_1^*$, the osmotic pressure is found to be

$$\exp\left(\frac{\pi V}{k_BT}\right) = \sum_{\mathbf{m}\geq 0}\frac{\mathbf{z}'^{\mathbf{m}'}}{\boldsymbol{\gamma}^{*\mathbf{m}'}\,\mathbf{m}'!}\int d\{\mathbf{m}\}\exp\left(-\frac{w_{\mathbf{m}}(z_1^*,0)}{k_BT}\right), \qquad (25.194)$$

where the prime (′) on $\mathbf{z}'$ and $\mathbf{m}'$ corresponds to the restriction that the sum excludes the solvent molecules. Note that $\boldsymbol{\gamma}^*$ is the set of activity coefficients of the solute species in a solution that is infinitely dilute with respect to the solutes but not the solvent. Although the usual activity coefficients are unity in such a case, $\boldsymbol{\gamma}^*$ has been defined in a fashion such that it is unity for a perfect gas. A power series expansion of the right-hand side of Eq. 25.194 yields, after some algebraic manipulation,

$$\frac{\pi}{k_BT} = n_2[1 + B_2^*n_2 + B_3^*n_2^2 + \cdots], \qquad (25.195)$$

[18] J. E. Mayer, *J. Chem. Phys.* **10,** 629 (1942).
[19] W. G. McMillan and J. E. Mayer, *J. Chem. Phys.* **13,** 276 (1945).

where the B_j^* have the same structure as the corresponding coefficients in the virial expansion for the equation of state of an imperfect gas except that the direct interaction potential in the gas phase virial coefficients is replaced by the potential of mean force at infinite dilution of the solute species in the solution phase virial coefficients. For example (see Eq. 21.164),

$$B_2^* = \frac{1}{2V}\int d\mathbf{R}_1\, d\mathbf{R}_2\left[1-\exp\left(-\frac{w_2}{k_B T}\right)\right]. \tag{25.196}$$

Note that the properties of the solvent nowhere enter explicitly in the expression for B_2^* or any of the other solution virial coefficients. Rather, the properties of the solvent lay hidden in the potential of mean force at infinite dilution of the solute.

Not enough is known about the potentials of mean force in mixtures of nonelectroytes to make *a priori* calculations of the solution virial coefficients, hence, of the thermodynamic properties of those solutions. The solution virial coefficients can be calculated by adopting models for the potentials of mean force. In essence, this is the approach adopted in the theory of dilute electrolyte solutions (Chapter 26). To illustrate the utility of models of the solution virial coefficients, consider a solution of long rod-shaped molecules in some solvent composed of small molecules. Examples of systems of this type are tobacco mosaic virus in water or a dilute electrolyte solution, and helical polypeptides in organic solvents. As a simple model of the potential of mean force between rodlike molecules, we shall treat the solvent as a continuum and the solute molecule as a "hard" right circular cylinder. Thus, by analogy with the behavior of hard spheres in a gas (see Eq. 21.168), the second solution virial coefficient is determined by the volume excluded to the center of one hard rod by another hard rod. Clearly, this depends on the length of the rod. The calculation of the virial coefficients for even this model is difficult and tedious. In the limit that the length-to-diameter ratio is very large, $(l/d) >> 1$, the second solution virial coefficient is proportional to l^2d, and higher coefficients to higher powers of (l^2d). Indeed, the concentration dependence of the osmotic pressure is represented by a series of terms of the form $(cl^2d)^n$ with various numerical coefficients (where $c \equiv N/V$). It is found that the solute–solute interactions are so effective that above a critical concentration the solution becomes unstable with respect to separation into two phases, each a solution, one concentrated and one dilute with respect to solute. The calculated solution virial coefficients lead to the prediction that the phase transition is first-order, so is accompanied by a density change.

How does calculating the set of solution virial coefficients help us to understand the predicted phase transition? What is the nature of that transition? As the concentration of solute is increased, solute–solute interactions become more and more important. The key to the following argument is the observation that a long rodlike molecule sweeps out a large volume when it rotates about its center of mass. At a relatively low concentration, the number of solute molecules multiplied by the volume swept out per molecule equals or exceeds the volume of the solution. When this concentration is exceeded, the ends of the rods tend to get in each other's way and solute–solute interaction results in partial parallelism of the solute molecules, since when the volumes swept by two rods intersect, collision can still be avoided by correctly orienting the molecules. However, as the orientations of the molecules become more and more parallel, the entropy per solute molecule decreases. Of course, the energy per solute molecule also decreases, so that the free energy is minimized. The important point is that the loss of entropy per particle eventually becomes so large that the chemical potential of a dilute solution with no hindrance of orientation becomes equal to that of the concentrated solution with partial parallel orientation of the rodlike molecules. At that concentration the solution separates into two phases, one with a small concentration of solute and no restriction on molecular orientation and one with a large concentration of solute and all solute molecules partially parallel (see the discussion of paraffin hydrocarbon liquids in Chapter 23). This predicted phase separation has been observed in tobacco mosaic virus solutions in water and in certain polypeptide solutions in organic solvents.

Finally, we note that the virial expansion for the osmotic pressure (or some other thermodynamic property of the solution) can be used to analyze experimental data. For example, the concentration dependence of the osmotic pressure can be fit to 25.195; B_2^* is obtained from the initial slope of a plot of π/n_2 versus n_2. The dependence of B_2^* on temperature can then be used to infer properties of the local environment of a solute molecule.

Despite the formal elegance of the Kirkwood–Buff and McMillan–Mayer theories, they are so difficult to use for practical calculations that much of the analysis of the properties of nonelectrolyte mixtures is based on approximate theories, which we discuss in the next two subsections. The Kirkwood–Buff and McMillan–Mayer theories of mixtures are, however, useful in providing exact results that can be used to measure the accuracy of the results obtained from approximate theories of mixtures.

B. Molecular Theory of Crystalline Mixtures

We now examine the molecular theory of the equilibrium properties of crystalline mixtures.

We have shown in Chapter 22 that, if the binding energy at 0 K, the vibrational spectrum, and the anharmonicity of vibration of a one-component crystalline solid are known, its free energy and all other thermodynamic properties can be computed. Construction of a similar description of the thermodynamic states of a crystalline mixture requires consideration of the following.

1. The several components are distributed in some way over the total number of lattice sites. Consider, for example, a binary mixture of A and B. If the AA, BB, and AB interactions are all the same, then in the nonvibrating lattice A and B will be randomly distributed. On the other hand, if AB interactions are more attractive than AA or BB interactions, then in the static lattice there will be preferential clustering and a larger-than-random fraction of AB near neighbors. Thus, depending on the nature of the intermolecular interactions, there will be some contribution to the free energy of the mixed crystal arising from the static distribution of components.

2. In general, irrespective of the nature of the components and their distribution over the lattice sites, the vibrational spectrum of the mixed crystal differs from the vibrational spectra of the pure components. When the AB interaction differs from the AA and BB interactions, it is easy to see why this is so, since in this case it is likely that the curvatures of the pair potentials differ at their respective minima, hence so do the force constants that determine the vibrational spectra. It is also likely that the anharmonicities are different in the mixture and the pure components.

3. Even when the pair potentials AA, AB, and BB are the same,[20] if the masses of A and B differ then the vibrational spectra of pure A, pure B, and the AB mixture differ. The reason for this is also easy to see. The vibrational frequency of a mass on a spring is proportional to $m^{-1/2}$, and the vibrational spectrum of a collection of masses also depends on the magnitudes of the masses, although in a more complicated way. For example, if only one light particle is substituted in a heavy-particle lattice and all the force constants remain the same, the frequency spectrum changes as follows: First, one normal mode and the corresponding normal frequency are removed from the dense set of frequencies characteristic of the pure crystal. Second, there appears a new frequency, larger than $\nu_{\max}$ (the maximum frequency for the pure crystal), which corresponds to the symmetric ("breathing") motion of the light particle and the surrounding nearest-neighbor particles in antiphase motion. For this motion the effective mass is dominated by the smallest mass, which in the case under discussion is the mass of the substituent atom. Then, because the force constant is the same as in the pure lattice, the corresponding frequency must be larger than that in the pure lattice (which would be $\nu_{\max}$). Finally, the frequencies of all the other normal modes are changed very little, in fact by a fraction of order N^{-1}, where N is the number of particles in the crystal. Clearly, when there are many particles with different masses in the crystal, and in addition the force constants differ and the distribution of atoms over the lattice sites is not as regular as is the crystal lattice itself, the distribution of vibrational frequencies in the mixed crystal will differ considerably from that in a pure crystal. It is to be expected that the thermal properties of the mixed crystal will then differ somewhat from those of the pure components.

4. The distribution of A and B atoms over the lattice sites is not independent of the spectrum of vibrations. As shown in Chapter 22, a typical normal mode in a one-component crystal involves coherent motion of many atoms (see Fig. 22.3). In a mixture the corresponding group of many atoms can have a range of compositions, and even for fixed composition many different arrangements of A and B within the group. Each of these arrangements has, in general, a different influence on the vibrational spectrum. At equilibrium that distribution of atoms is favored which, other things being equal, minimizes the vibrational contribution to the free energy. The effect of this dependence of vibrational spectrum on atomic arrangement is to generate an effective interaction between atoms. This effective interaction is temperature-dependent and is in addition to the ordinary interaction between stationary atoms.

5. An AB mixed crystal can have a different density of imperfections than pure A or B crystals.

6. Although conditions **1–5** have been phrased in terms of the distribution of A and B over lattice sites, it must not be assumed that the lattice sites of the mixed crystal are the same as those of pure A or B. Put another way, we expect that when the mixed crystal is formed there will be some change in lattice structure. For a dilute solution of A and B this will take the form of displacement of the B atoms near an A atom from the positions they would have occupied in a pure B crystal. In an equimolar AB mixed crystal, all atoms are displaced from their nominal positions in pure A or B crystals. This effect, which depends mostly on the repulsive part of the AA, AB, and BB interactions, leads to changes in the energy of the static lattice and the vibrational spectrum of the lattice.

Many theories of the thermodynamic properties of crystalline mixtures completely neglect points **2**, **3**, **4**, **5**, and **6**. When point **6** is neglected, the complete partition function can be separated into the product of a vibrational partition function and a configurational partition function. It is usually the case that the effective interaction described in point **4** is negligibly small compared to other interactions, except in some special instances,[21] so factorization of the mixed-crystal partition function is usually a good approximation. However, the neglect of points **2** and **6** leads to inaccurate predictions of the properties of the mixture. As we shall see, the structure imposed by short-range repulsive forces is very important in the determination of the excess thermodynamic properties of the mixed crystal. As an illustration of the effects to be expected, we consider the calculation of the entropy of mixing from a very simple model.

Consider a crystal lattice with N_A A atoms and N_B B atoms, and suppose that the AA, AB, and BB interactions,

[20] As, for example, in a mixture of isotopes.

[21] If the vacancies in a crystal are thought of as a component, and a crystal with vacancies as a mixture, then the effective interaction induced by vibrational motion leads to clustering of vacancies at low temperature. This phenomenon occurs, for example, in metals exposed for long periods to radiation from nuclear reactors, such as fuel rod casings.

both repulsive and attractive, are identical. Then, neglecting effects of types **3** and **4**, it is clear that the energy and the volume of the mixture are independent of composition, hence there is no heat of mixing and no volume change on mixing. Clearly, the model described is an ideal mixture, the thermodynamics of which is completely described by the entropy of mixing.

Because all compositions have the same energy, we can use the simple Boltzmann definition of the entropy,

$$S = k_B \ln \Omega, \tag{25.197}$$

to compute the entropy of mixing. If the $N = N_A + N_B$ particles occupy all the sites of the lattice, the number of distinguishable arrangements is just

$$\Omega = \frac{N!}{N_A!N_B!}, \tag{25.198}$$

because the particles can be permuted among the N sites in $N! \equiv (N_A + N_B)!$ ways, but $N_A!$ permutations of the A's among themselves and $N_B!$ permutations of the B's among themselves lead to indistinguishable distributions of A and B over the lattice. The numbers of distinguishable arrangements of A and B in their pure lattices are, of course, just $N_A!/N_A! = 1$ and $N_B!/N_B! = 1$. Therefore, in this static crystal model the entropy of the mixed crystal is just the entropy of mixing. Using Stirling's approximation, $\ln N! \approx N \ln N - N$, Eq. 25.198 is readily reduced to

$$\begin{aligned}\Delta S_{\text{mixing}} &= -N_A k_B \ln \frac{N_A}{N_A + N_B} - N_B k_B \ln \frac{N_B}{N_A + N_B} \\ &= -N_A k_B \ln x_A - N_B k_B \ln x_B .\end{aligned} \tag{25.199}$$

As expected, the entropy of the model crystal considered is just the ideal entropy of mixing.

In the model just examined, each lattice site could be occupied by only one molecule, and each molecule could occupy only one lattice site, so there are no geometric constraints on the occupation of the sites neighboring a given lattice site just because that site is occupied. What happens to the distribution of molecules when A and B have different volumes and, say, a B molecule occupies more than one lattice site? So as to avoid unnecessary complications, we consider a model crystal in which there are no attractive interactions between AA, AB, and BB pairs of molecules. Even with this simplification, counting exactly the number of distinguishable configurations available to the system is very difficult. It is fairly easy, however, to evaluate the entropy of mixing when the concentration of the component with larger molecular volume is very small. The argument goes as follows: Suppose that component B, with molecular volume v_B^0, occupies many lattice sites per molecule, and component A, with molecular volume v_A^0, occupies only one lattice site per molecule (see Fig. 25.20).

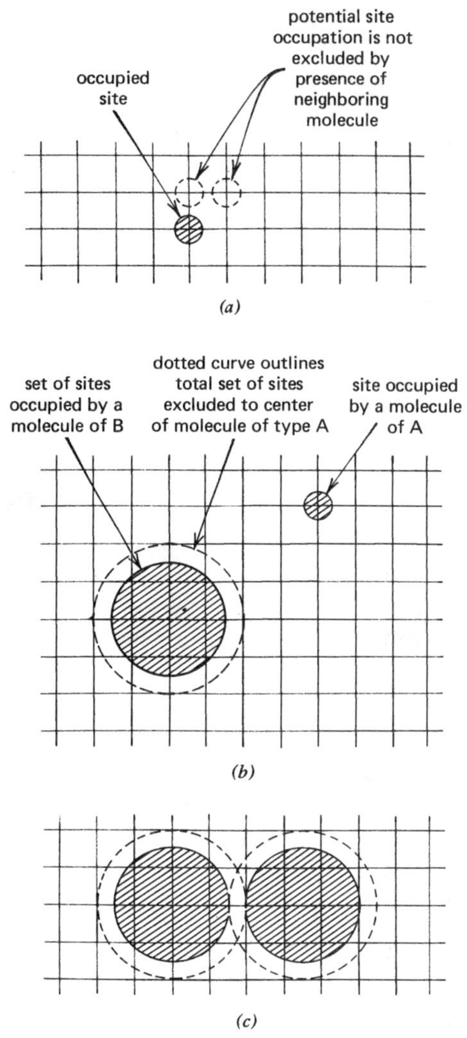

Figure 25.20 Schematic representation of excluded volume effects in a static lattice. (*a*) Each molecule occupies only one site and does not exclude occupation of neighboring sites. (*b*) A molecule of species B occupies more than one lattice site and blocks the center of an A molecule from neighboring sites that are closer to the center of B than the sum of the radii of an A and a B atom. (*c*) The volume excluded by two nearest-neighbor particles is less than the sum of the excluded volumes of each would be if they were far apart.

Neglecting edge effects, we find that the number of ways in which the center of mass of a B molecule can be put into an empty lattice of N sites is just

$$\omega_1 = N = \left(\frac{1}{\Delta}\right)V, \tag{25.200}$$

where $\Delta \equiv V/N$ is the volume of a lattice cell. As shown in Fig. 25.20, the presence of a molecule of B results in the exclusion of A from possible occupation of a number of lattice sites, say z of them. The volume excluded thereby is $z\Delta$. The number of ways of introducing a second molecule of species B, given that the first one is somewhere in the lattice, is

$$\omega_2 = N - z = \left(\frac{1}{\Delta}\right)(V - b), \qquad (25.201)$$

where $b \equiv z\Delta$. Similarly, the third molecule will be excluded from $N - 2z$ sites, and we have

$$\omega_3 = N - 2z = \left(\frac{1}{\Delta}\right)(V - 2b). \qquad (25.202)$$

Equation 25.202 involves the use of an approximation. When the two molecules already present are as close together as possible, they exclude from occupation fewer than $2z$ sites, because the exclusion zones overlap in the region between the molecules. (See Fig. 25.20.) In writing Eq. 25.202 we are assuming that the number of configurations in which two B molecules are adjacent is negligibly small relative to the number of configurations in which they are far apart. If this approximation is made at each step as a B particle is added to the lattice, the number of ways the ith B molecule can be placed in a lattice already occupied by $i - 1$ other B molecules is

$$\omega_i = N - (i-1)z = \left(\frac{1}{\Delta}\right)[v - (i-1)b]. \qquad (25.203)$$

This approximate calculation of the excluded-volume factors is only accurate for very dilute solutions, for which $N_B v_B^0 << V$. We shall assume that this restriction is met.

After placing N_B B molecules on the lattice we have left, in our approximation, $N - zN_B$ empty lattice sites. These can be filled in $N_A!$ ways with A molecules. The total number of distinguishable ways of arranging N_A A molecules and N_B B molecules on the lattice is thus

$$\Omega = \frac{N_A!}{N_A!}\frac{1}{N_B!}\prod_{i=1}^{N_B}\omega_i = \frac{1}{N_B!}\left(\frac{1}{\Delta}\right)^{N_B}\prod_{i=1}^{N_B}[V - (i-1)b]. \qquad (25.204)$$

As in the case of Eq. 25.198, the factors $1/N_A!$ and $1/N_B!$ account for the indistinguishability of A and B molecules—these may be interchanged among themselves $N_A!$ and $N_B!$ ways, respectively, without altering the state of the system.

In a dilute solution we have $N_B b << V$, so that

$$\begin{aligned}\ln\left\{\prod_{i=1}^{N_B}[V - (i-1)b]\right\} &= \ln\left[\prod_{j=0}^{N_B-1}(V - jb)\right] \\ &= \ln\left[\prod_{j=0}^{N_B-1}V\left(1 - \frac{jb}{V}\right)\right] \\ &\approx N_B \ln V - \frac{b}{V}\sum_{j=0}^{N_B-1} j. \end{aligned} \qquad (25.205)$$

Therefore, the entropy of the mixed crystal is just

$$\begin{aligned}S_{\text{mixture}} &= k_B \ln\Omega \\ &= N_B k_B \ln\frac{V}{\Delta} - k_B \ln N_B! - \frac{k_B b}{V}\sum_{j=0}^{N_B-1} j. \end{aligned} \qquad (25.206)$$

The sum in the last term of Eq. 25.206 is equal to the number of terms, N_B, multiplied by the average value of the terms, $(N_B - 1)/2$. Expressing the total volume in terms of the volumes of the components,

$$V = N_A v_A^0 + N_B v_B^0, \qquad (25.207)$$

and assuming that $N_B >> 1$, we convert Eq. 25.206 to the form

$$\begin{aligned}S_{\text{mixture}} &= N_B k_B \ln\left(\frac{N_A v_A^0 + N_B v_B^0}{\Delta}\right) - k_B \ln N_B! \\ &\quad - \frac{N_B^2 k_B b}{2(N_A v_A^0 + N_B v_B^0)}. \end{aligned} \qquad (25.208)$$

Equation 25.208 represents, for the model considered, the entropy of the mixture. We can obtain the entropy of pure A and pure B on the basis of this same model by setting $N_B = 0$ and $N_A = 0$, respectively.[22] We then find for the entropy of mixing

$$\begin{aligned}\Delta S_{\text{mixing}} &= N_B k_B \ln\left(\frac{N_A v_A^0 + N_B v_B^0}{N_B v_B^0}\right) \\ &\quad - \frac{N_B^2 k_B b}{2(N_A v_A^0 + N_B v_B^0)} + \frac{N_B k_B b}{2 v_B^0}. \end{aligned} \qquad (25.209)$$

[22] It is necessary to add a few words to the description of the calculation of the entropy of mixing. For the model under discussion the vibrational contributions to the entropies of the mixture and the pure components have been neglected. We then expect to find $S_A = 0$ and $S_B = 0$, respectively, when N_B is set equal to zero and when N_A is set equal to zero in the expression for the entropy of the mixture. Eq. 25.199 gives the correct value for S_A when $N_B = 0$ because in that case the number of configurations available to the system has been correctly evaluated. However, Eq. 25.199 does not give $S_B = 0$ when $N_A = 0$ because of the approximate treatment of the number of configurations of the system when one molecule blocks the occupancy of z lattice sites. To make the analysis more consistent we require that $S_B = 0$, which implies that b has the value

$$b = z v_B^0\left(1 + \ln\frac{v_B{}^0}{\Delta}\right).$$

Since $b = z\Delta$, we can substitute $\Delta = b/z$ in the argument of the logarithm and iterate the procedure to successively remove Δ from the expression for b. This leads to the approximate formula

$$b \approx z v_B^0\left(1 + \ln\frac{z}{2}\right),$$

which varies very slowly with z and is consistent with the simple idea that the excluded volume per sphere is larger than the volume of one sphere. For the purposes of this discussion we assume that the excluded volume parameter b is constrained to have the value that leads to $S_B = 0$ when $N_A = 0$.

Equation 25.209 differs in two important ways from Eq. 25.190. First, the entropy of mixing depends on the volume fraction $\phi_B = N_B v_B^0/(N_A v_A^0 + N_B v_B^0)$, and not on the mole fraction. Second, because of the disparity in sizes of A and B molecules, there is a nonideality contribution to ΔS_{mixing}. These points are best illustrated when Eq. 25.200 is written in the form

$$\begin{aligned}\Delta S_{\text{mixing}} &= -N_B k_B \ln \phi_B - \frac{1}{2} N_B k_B \phi_B \frac{b}{v_B^0} + \frac{1}{2} N_B k_B \frac{b}{v_B^0} \\ &= -N_B k_B \ln \phi_B + \frac{N_B k_B b}{2 v_B^0} \phi_A \qquad (25.210)\end{aligned}$$

and compared with Eq. 25.199 for a solution dilute in component B:

$$\begin{aligned}\Delta S_{\text{mixing}}^{\text{ideal}} &= -N_B k_B \ln x_B - N_A k_B \ln(1 - x_B) \\ &\approx -N_B k_B \ln x_B + N_A k_B x_B \\ &= -N_B k_B \ln x_B + N_B k_B x_A . \qquad (25.211)\end{aligned}$$

We may now ask which of the features characteristic of these simple models are of general validity. First, note that the absence of a heat of mixing does not imply the absence of intermolecular interactions, but only that the AA, AB, and BB interactions are the same; conversely, even if this is true there can still be a nonzero heat of mixing if $\Delta V_{\text{mixing}} \neq 0$. Second, when the sizes of the two component molecules are very different, the entropy of mixing depends on the volume fractions of the components, and not on the mole fractions. For molecules of the same size, of course, the volume fraction and mole fraction are identical. Third, mixtures in which one component is much larger than the other are intrinsically nonideal because of the volume from which the center of mass of the smaller particle is excluded by the larger particle. In the particular case we have examined, the nonideality increases with the excluded volume. Although our deductions are derived from a crude static lattice model, it is found that in real solutions all of the conclusions cited remain valid.

We have thus far not considered explicitly the influence of molecular interactions on the distributions of molecules over the lattice sites, except for the influence of dissimilar molecular volumes. In a sense, the formulation of the excluded volume used in Eqs. 25.200 *et seq.* is a crude representation of the effect of dissimilar repulsive interactions between AA, AB, and BB molecular pairs. In general, the AA, AB, and BB pair interactions differ in both their repulsive and attractive components, and the distribution of molecules over the lattice sites is a complicated function of the interaction energies. To simplify our description, let us now assume that the molecular volumes of A and B are the same, so that attention can be focused on differences in the attractive interactions. The crudest approach to the evaluation of the lattice partition function is to assume that the distribution of components over the lattice is random despite the differences in interactions. When carried through, this assumption leads to the theory of regular mixtures, the thermodynamics of which we have outlined in Sections 25.2 and 25.8. In this crude model the entropy of mixing is of ideal form because the distribution of components is defined to be random and other possible contributions to the entropy of mixing are ignored. The energy difference $\epsilon_{AB} - (\epsilon_{AA} + \epsilon_{BB})/2$ determines the heat of mixing.

Three different approaches have been used to obtain a more exact description of the molecular distribution for this model. First, there are a variety of approximations designed to account for the major effects of dissimilar molecular interaction without being so mathematically clumsy as to make the evaluation of the partition function very difficult. For example, if we think of the AA, AB, and BB near-neighbor pairs of a binary solution as diatomic molecules in a mixed gas, their relative concentrations can be related through an "equilibrium constant." In the nonvibrating lattice model we are considering, there is no entropy difference between the several "diatomic molecules," so that the "equilibrium constant" $N_{AB}^2/N_{AA}N_{BB}$ is determined by the near-neighbor energy difference $\epsilon_{AB} - (\epsilon_{AA} + \epsilon_{BB})/2$. The distribution of A and B molecules over the lattice is then calculated, in this approximation, subject to the subsidiary constraint that the number of near-neighbor AB pairs satisfy the equilibrium $N_{AB}^2 = KN_{AA}N_{BB}$ at each temperature.[23] This approximation leads to some minor changes in the thermodynamics of the mixture, but in most cases the predicted behavior is close to that of the regular solution with the same interaction energies. In other words, although the entropy of mixing differs from the ideal form, for reasonable values of ϵ_{AB}, ϵ_{AA}, and ϵ_{BB} the deviation is not large.

A second approach is based on an analysis similar to the virial expansion used in Chapter 21 to describe a gas. In place of an integration over the translational space in which gas molecules move, for this model there is a summation over the lattice points at which molecules may be placed. The results of the analysis are a set of virial expansions (the concentration replaces the gas density) for the thermodynamic functions, with virial coefficients defined by lattice sums instead of integrations. Just as in the theory of the gas, this calculation is useful for dilute solutions of, say, A in B, for which only terms of lowest order in the mole fraction need be retained. The lattice virial coefficients are difficult to calculate, and this theory is not much used.

A third scheme is based on term-by-term evaluation of the partition function after the exponential term is expanded as a series in powers of $(1/T)$. This procedure is tedious, and requires the evaluation of many terms to get an accurate evaluation of the partition function. If enough terms are calculated, and if the properties of the full partition function can be guessed from the properties of the terms calculated, valuable information about the approach to phase separation can be obtained. In Section 24.4 we described a lattice

[23] This is known as the quasi-chemical approximation.

gas, which can be thought of as a mixture of occupied and unoccupied lattice sites, or of molecules and "holes." The results quoted there are typical of what can be learned from this type of model and this scheme of calculation. The qualitative results confirm the picture given by the quasi-chemical equilibrium approximation to the molecular distribution: When unlike molecules repel more strongly than like molecules, AB near-neighbor pairs occur with less than random frequency, there is heat absorbed on mixing, and the total vapor pressure is greater than that of an ideal mixture. On the other hand, when unlike molecules attract more strongly than like molecules, AB near-neighbor pairs occur with greater than random frequency, heat is evolved on mixing, and the vapor pressure is less than for an ideal mixture. Corresponding to the decreased or increased frequency of AB pairs relative to the random distribution, the entropy of mixing deviates from the ideal form. In this crude model, in which all vibrational motion is neglected, the entropy of mixing is less than the ideal entropy of mixing in both cases. In real mixtures vibrational effects alter this result, and the deviation from the ideal entropy of mixing can be in either direction. Comparison of the several calculations shows that, although differences in attractive interaction do not change the entropy of mixing much, differences in molecular volume have a large effect on the entropy of mixing. This result is similar to that obtained for the effects of attractive (a-term) and repulsive (b-term) interactions on the entropy of a van der Waals gas, and are confirmed by the general virial expansion results. Finally, all theories agree that in the limit $T \to \infty$ the molecular distribution approaches a random distribution. This is a simple consequence of the thermal energy k_BT becoming large relative to $\epsilon_{AB} - (\epsilon_{AA} + \epsilon_{BB})/2$, so that the Boltzmann factor approaches unity and all configurations of A and B on the lattice have equal weight.

All of the preceding discussion was based on the analysis of a model mixture in which vibrational motion was neglected. In any real mixture the difference in vibrational spectrum from the spectra of the pure components can lead to considerable alteration of the thermodynamic functions. The calculation of a mixture's vibrational spectrum is extremely complicated because the disorderly distribution of molecules over the lattice prevents the use of many of the simplifications of the theory of pure-crystal lattices. There are also insufficient experimental data relevant to the spectra of mixed crystals, so that simplified analyses have not yet emerged in response to experimental findings. In general, we expect the thermodynamic functions to be affected most when new low-frequency vibrations appear in the mixed crystal (say, when very heavy atoms are substituted in the crystal), or when AB pairs are so tightly bound that many crystal vibrations appear as new high-frequency "quasi-molecular" vibrations. The total number of vibrational modes is determined by the number of atoms, so that the appearance of new modes must mean the loss of modes elsewhere in the frequency spectrum if the number of atoms is the same. When the mixed crystal has more low-frequency components in its spectrum than do the pure crystals the vibrational entropy of the mixture will be greater than that of the pure crystals, and *vice versa.* Corresponding to these entropy differences will be differences in other thermodynamic functions, for example, the heat capacities. Although the alterations in vibrational spectra described are also accompanied by energy differences between the pure and mixed crystals greater than those expected in the absence of vibration, in most cases the principal influence of the changed vibrational spectrum will be on the entropy of mixing.

To illustrate and test the ideas about the thermodynamic properties of crystalline mixtures just described, we now examine alkali halide solid solutions. These are, of course, ionic crystals. We showed in Section 14.10 how the lattice energy of a pure ionic crystal can be computed from a knowledge of the ion–ion potential energy function; the extension of the calculation to treat mixed crystals is straightforward.

The lattice energy of an ionic crystal is very large compared to its zero-point and thermal vibrational energies, and very large compared to the heat of mixing. Also, the vibrational spectra of the mixture and the pure components are expected to be only slightly different, because of the similarity in the ion–ion potentials. In this case it is likely that the static lattice energy of the mixture is very nearly the same for all distributions of, say, Li^+ and Na^+ in the mixed crystal LiF–NaF, hence to a very good approximation the distribution of Li^+ and Na^+ can be taken to be random. Of course the static lattice contribution to the entropy of mixing is then of the ideal solution form, but the full entropy of mixing, which includes a contribution from the small change in the vibrational spectrum, is not the ideal solution entropy of mixing.

Although the Coulomb interaction is the same for every ion pair with the same charges, the other contributions to the ion–ion potential energy are different for each pair of ions. When one ion is substituted for another in an ionic lattice there are three major components of the change in energy:

1. The ion–ion repulsion, of the form $\lambda \exp(-R/\rho)$, with λ and ρ constants, is different for each pair of ions. This, of course, corresponds to the observation that the ionic diameters are all different, hence so must be the possible distances of closest approach. We expect the change in repulsive interaction on substitution to be an important contributor to the force that leads to a change in ionic positions.
2. When ions are displaced from the lattice positions characteristic of the parent pure component, there is a change in the electrostatic field acting on each ion. Because the displacements are small, the change in electrostatic energy per ion can be calculated in terms of the interactions between all of the effective dipole moments created by the displacements of the ions from their lattice sites.

3. The ion polarizabilities are all different. Since the electric field of one ion distorts the charge cloud of a neighboring ion, the interaction energy being proportional to the induced dipole moment, this contribution to the energy will change when the ion polarizabilities and positions change.

If we consider an isolated substitution of one ion for another, so that the environment of the solute ion is entirely defined by the host, the change in energy is obtained by just summing contributions **1**, **2**, and **3** for the one geometry of substitution. If we consider a mixed crystal in the "middle concentration range," an ion of one type can be found in many different environments. For the case that there is random mixing in the static lattice, the concentration of each type of environment is constructed from the known bulk concentrations. For example, let x_A and x_B be the mole fractions of the two cations in the cubic mixed crystal $A_{x_A} B_{x_B} X$, and note that in the cubic crystal the six nearest neighbors of an A or B cation are always X anions. Then the differences between the environments of different A's and B's arise from differences in the makeup of the 12 second-neighbor ions. To take one configuration, the concentration of sites each surrounded by a second-neighbor shell with seven A's and five B's is proportional to $x_A^7 x_B^5$. The energy associated with each such configuration must be evaluated, then weighted with the appropriate concentration, and all terms summed to give the change in energy. The energy must then be minimized with respect to the ion displacements.

Calculations of the type described have been made both for dilute and concentrated alkali halide mixtures. Consider the dilute solutions first. Table 25.3 shows the calculated heats of mixing (or formation) of 1% mixtures. The results illustrate the common-sense idea that the energy of formation of the mixture increases with increasing degree of deformation required. For example, KBr–Na has a smaller heat of mixing than does NaBr–K because it is easier to put a small ion in a large site than *vice versa.* Also, the series of mixtures KCl–Na, KBr–Na, KI–Na shows a decreasing heat of formation, as expected, since the cation site increases in volume steadily as the anion size increases.

Table 25.3*a* Heat of Mixing for 1% Cation Defect Concentration (in J/mol)[a]

LiF–Na	206	NaF–Li	556	LiF–K	2130	KF–Li	1540
LiCl–Na	156	NaCl–Li	120	LiCl–K	1100	KCl–Li	498
LiBr–Na	126.4	NaBr–Li	100	LiBr–K	812	KBr–Li	427
LiI–Na	92.5	NaI–Li	73	LiI–K	590	KI–Li	315
LiF–Rb	3490	RbF–Li	1940	NaF–K	494	KF–Na	305
LiCl–Rb	1720	RbCl–Li	686	NaCl–K	266	KCl–Na	165
LiBr–Rb	1260	RbBr–Li	598	NaBr–K	197	KBr–Na	143
LiI–Rb	1020	RbI–Li	460	NaI–K	143	KI–Na	103
NaF–Rb	1020	RbF–Na	531	KF–Rb	55.6	RbF–K	46.8
NaCl–Rb	536	RbCl–Na	296	KCl–Rb	39.1	RbCl–K	19.2
NaBr–Rb	401	RbBr–Na	265	KBr–Rb	23.0	RbBr–K	27.0
NaI–Rb	324	RbI–Na	202	KI–Rb	23.8	RbI–K	21.6

Table 25.3*b* Heat of Mixing for 1% Anion Defect Concentration (in J/mol)[a]

LiF–Cl	933	LiCl–F	858	LiF–Br	1950	LiBr–F	1080
NaF–Cl	946	NaCl–F	485	NaF–Br	1850	NaBr–F	736
KF–Cl	686	KCl–F	351	KF–Br	1230	KBr–F	561
RbF–Cl	540	RbCl–F	289	RbF–Br	946	RbBr–F	481
LiF–I	4070	LiI–F	1070	LiCl–Br	60.6	LiBr–Cl	43.9
NaF–I	3550	NaI–F	1070	NaCl–Br	52.7	NaBr–Cl	44.3
KF–I	2590	KI–F	870	KCl–Br	43.9	KBr–Cl	30.2
RbF–I	2220	RbI–F	774	RbCl–Br	36.6	RbBr–Cl	28.9
LiCl–I	402	LiI–Cl	150	LiBr–I	102	LiI–Br	59.8
NaCl–I	347	NaI–Cl	194	NaBr–I	101	NaI–Br	69.9
KCl–I	284	KI–Cl	158	KBr–I	79	KI–Br	61.1
RbCl–I	264	RbI–Cl	151	RbBr–I	71	RbI–Br	61.1

[a] From D. L. Fancher and G. R. Barsch, *J. Phys. Chem. Solids* **30,** 2503 (1969).

Table 25.4 contains a comparison of the observed and calculated heats of mixing, and we show in Fig. 25.21 a comparison of the observed and calculated concentration dependences of the heats of mixing. The agreement is seen to be good but not perfect.

Heat capacity studies of 1:1 KC1:KBr mixtures reveal a residual entropy (see Chapter 18) of 5.77 J/K mol, which is the same, within experimental error, as the entropy of a random mixture. This observation confirms the hypothesis that the static distribution of ions is accurately described as random. On the other hand, for $T > 0$ the same experiments show that the entropy of mixing deviates from the ideal value, so there must be a vibrational contribution. For $KCl_{0.5}Br_{0.5}$ at 300 K, calorimetric experiments reveal there is a contribution of 0.46 J/K mol to the entropy of mixing from the change in the vibrational spectrum. This is actually a rather large effect, as can be seen from the change in solubility from that predicted if the entropy of mixing were ideal. We saw, earlier, that at equilibrium between a mixture and pure components the chemical potential of each species must be the same in the mixture and the pure component. The difference in Gibbs free energy between the solid solution and the pure components is, of course, $\Delta G_{\text{mix}} = \Delta H_{\text{mix}} - T\Delta S_{\text{mix}}$. The heat of mixing of the solid solution can be calculated from the ionic lattice model as described. Then separating ΔS_{mix} into an ideal mixture component and a vibrational component, and assuming these to be independent, $\Delta S_{\text{mix}} = \Delta S_{\text{mix}}^{\text{ideal}} + \Delta S_{\text{mix}}^{\text{vib}}$, which allows the solubility of one alkali halide in another to be computed if $\Delta S_{\text{mix}}^{\text{vib}}$ can be computed. As shown in Chapter 22, the entropy of a crystal is rather well approximated by the value predicted from a Debye spectrum of vibrations. The Debye spectrum of a mixture differs from that of its parent components because of changes in the elastic constants. When such changes are taken into account, and $\Delta S_{\text{mix}}^{\text{vib}}$ calculated and combined with ΔH_{mix} obtained earlier, one predicts the solubility curves shown in Fig. 25.22. It is clear that the agreement with the observed solubility is better when the vibrational contribution to ΔS_{mix} is taken into account. The estimated magnitudes of the vibrational contribution to ΔS_{mix} are given in Table 25.5.

We conclude that a proper theoretical accounting for the thermodynamic properties of a mixture requires (1) very accurate pair potentials, (2) knowledge of how the structure changes when the mixture is formed, and (3) inclusion in the entropy of mixing of contributions arising from the change in the vibrational spectrum of the solid.

We shall soon see that each of these requirements has a counterpart in the theory of liquid mixtures.

Table 25.4 Comparison of Observed and Computed Heats of Formation of 1% Mixtures of Alkali Halides (all in J/mol)[a]

x		$Na_xK_{1-x}Cl$	$Na_xK_{1-x}Br$	$K_xRb_{1-x}Cl$	$K_xRb_{1-x}I$
0.99	Exp.	192.5–230.1	192.5	108.8	37.6
	Theory	266.1	197.5	142.7	39.3
0.01	Exp.	163.2–192.5	144.3	92.1	30.5
	Theory	164.8	143.1	102.9	19.2
x		$NaCl_xBr_{1-x}$	$NaBr_xI_{1-x}$	KCl_xBr_{1-x}	KBr_xI_{1-x}
0.99	Exp.	45.2–59.8	92.0	35.6	62.7
	Theory	52.7	100.8	43.9	79.0
0.01	Exp.	43.9–52.3	75.3	34.7	69.0
	Theory	44.3	69.8	30.1	61.1

[a] From D. L. Fancher and G. R. Barsch, *J. Phys. Chem. Solids* **30,** 2503 (1969).

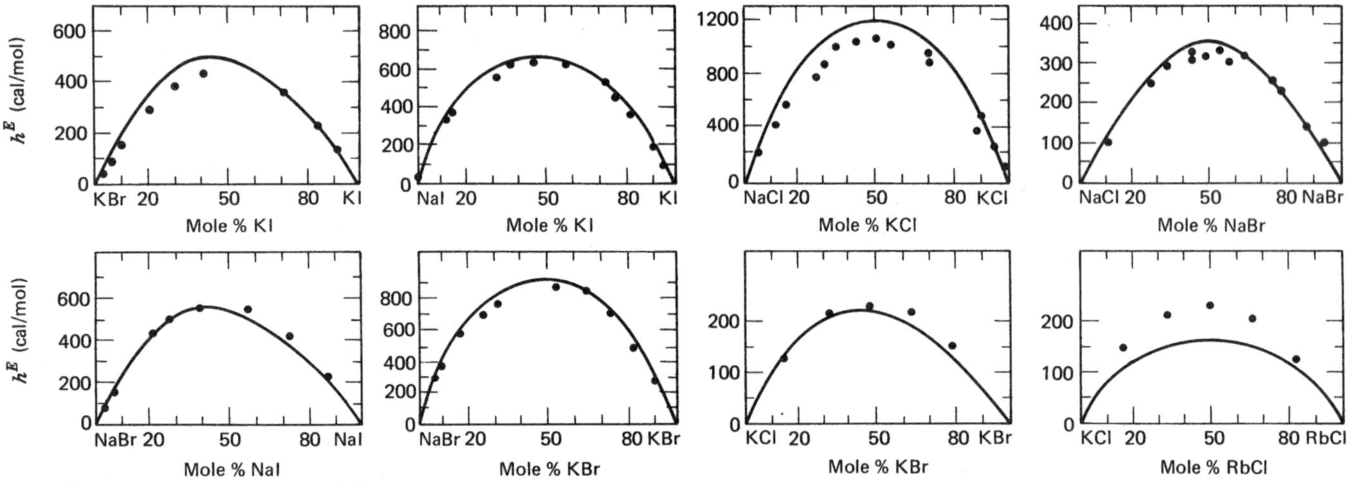

Figure 25.21 Comparison of theoretical and observed heats of mixing for alkali halide systems: ——— Theory; • experiment. From D. L. Fancher and G. R. Barsch, *J. Phys. Chem. Solids* **30,** 2503 (1969).

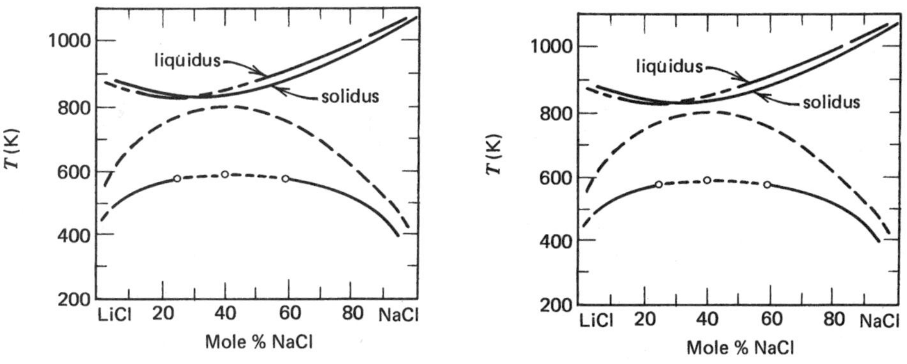

Figure 25.22 Solid solubility curves for two alkali halide systems. – – – Theoretical curve calculated using $\Delta S^{\text{ideal}}_{\text{mix}}$ only; ——— Theoretical curve calculated using $\Delta S^{\text{ideal}}_{\text{mix}} + \Delta S^{\text{vib}}_{\text{mix}}$; – ··· – ··· Semitheoretical curve calculated from the experimental ΔH_{mix} and $\Delta S^{\text{ideal}}_{\text{mix}} + \Delta S^{\text{vib}}_{\text{mix}}$; ooo and ··· experimental data. From D. L. Fancher and G. R. Barsch, *J. Phys. Chem. Solids* **30,** 2517 (1969).

Table 25.5 Vibrational Contribution to the Entropy of Mixing (in J/K mol)—1:1 Mixtures[a]

	LiCl–NaCl	NaCl–KCl	NaBr–KBr	NaCl–NaBr	KCl–KBr	KBr–KI
$\Delta S^{\text{ideal}}_{\text{mix}}$	5.77	5.77	5.77	5.77	5.77	5.77
$\Delta S^{\text{vib}}_{\text{mix(calc.)}}$[b]	1.17	1.42	1.04	1.33	0.83	0.83
$\Delta S^{\text{vib}}_{\text{mix(exp.)}}$	–	–	–	–	0.46	–

[a] From D. L. Fancher and G. R. Barsch, *J. Phys. Chem. Solids* **30,** 2517 (1969).

[b] From application of the Debye model.

C. Molecular Theory of Liquid Mixtures

Examination of the contents of Chapters 22 and 23 reveals that the concept of structure is employed somewhat differently in the theoretical descriptions of crystals and liquids. In the case of a crystalline solid the regularity of the atomic arrangement permits straightforward calculation of both the static lattice energy and the energy of vibrational motion. To decide whether one or another lattice is more stable, the Gibbs free energy of each is computed and the one with the lowest Gibbs free energy chosen. The point is that the lattice structure is known, the distribution of molecules is therefore also known, and the focus of the theoretical analysis is the influence on the thermodynamic properties of displacements of the atoms from lattice sites. In the case of a liquid the distribution of atoms is not *a priori* known, and one of the functions of the theory is to determine that distribution. The same difference in use of the concept of structure separates the theory of crystalline mixtures from the theory of liquid mixtures. In this section we outline a few of the main ideas used in the theory of liquid mixtures.

We start our discussion by considering a very crude model, since it provides a pictorial representation of some important characteristics of a mixture. It must be emphasized that this model is not an accurate representation of real liquid mixtures yet has features that real liquid mixtures also have.

The essence of the difference between liquids and solids can be traced to the nature of the molecular motions in the two classes of condensed matter. The simplest approximation to a liquid mixture that retains some of the features distinguishing a liquid from a solid is called the *cell model.* It is based on the following ideas: Although molecular displacement in a liquid is not difficult, the rate at which a molecule diffuses is small compared to the rate at which a free molecule would move. At the density characteristic of the liquid phase, each molecule is surrounded by a "cage" of other molecules. If the "cage" molecules confine the central molecule to a small volume element for a long time compared to the time the molecule takes to move across the "cage," we can think of the central molecule as restricted to move only within a cell centered about some point. Of course the "cage" molecules move also, the "cage" decays in time, and the center of the cell defined by the "cage" moves. The cell model of liquids adopts the idea that a "cage" exists, but neglects the decay of the "cage" and the drift of the cell center in time. In other words, every molecule is assumed to be confined to a cell in which its motion is defined by the intermolecular potential of the "caging" molecules. In this model a molecule cannot move throughout the entire liquid, but only in a small volume about some point in space. The distribution of component molecules, say, A and B, throughout the liquid is now calculated on the basis of two ideas: First, it is imagined that some regular lattice can be passed through the centers of the cells occupied by the molecules. Clearly, this approximation overestimates the long-range order in the liquid (see Chapter 23) and makes the model too solid-like. The distribution of molecules over the cell centers (lattice sites) is calculated as if the lattice were rigid. Second, a molecule can move about its cell in an irregular way. The trajectory in the cell is weighted at each point by a Boltzmann factor involving the instantaneous energy of interaction of the molecule with its surroundings. That is, vibrational motion (as in a crystal) is replaced by a thermal motion weighted according to the instantaneous interaction energy. Each molecule is assumed to move in its cell independently of all others, except insofar as the near neighbors (assumed to be stationary for the calculation of the field in the cell) provide the field in which it moves. Because of the assumed independence of motion, the model is an analogue of the Einstein model of a crystal. Like the Einstein model, it errs by neglecting correlation in the motions of neighboring particles.

The canonical partition function for a binary liquid mixture has the form

$$Q_{N_A N_B} = (N_A!N_B!h^{3(N_A+N_B)})^{-1}\int_V \cdots \int dx_1 \cdots dp_{zN} \times \exp\left[-\frac{1}{k_BT}\left(\sum_i \frac{p_i^2}{2m_i} + \mathcal{V}_N\right)\right], \tag{25.212}$$

where $N = N_A + N_B$ molecules are contained in the volume V. Equation 25.212 should be compared with Eq. 21C.3 of Appendix 21C, which is the canonical partition function of a pure fluid. Just as there, Eq. 25.212 can be reduced to

$$Q_{N_A N_B} = \frac{1}{N_A!N_B!}\left(\frac{2\pi m_A k_BT}{h^2}\right)^{3N_A/2}\left(\frac{2\pi m_B k_BT}{h^2}\right)^{3N_B/2} \times \int_V \cdots \int dx_1 \cdots dz_N \exp\left(-\frac{\mathcal{V}_N}{k_BT}\right). \tag{25.213}$$

The cell model we have described represents the total potential energy, $\mathcal{V}_N$, in the form

$$\mathcal{V}_N = N_A\epsilon_A(0) + N_B\epsilon_B(0) + \sum_{i=1}^{N}[\epsilon_i(r) - \epsilon_1(0)]. \tag{25.214}$$

That is, the total potential energy is thought of as the sum of (1) the lattice energy that would exist if each molecule were stationary at the center of its cell, $N_A\epsilon_A(0) + N_B\epsilon_B(0)$, and (2) the energy associated with the deviation of a molecule from its lattice position in the field generated by the near-neighbor molecules at the centers of their cells, $\epsilon_i(r) - \epsilon_i(0)$, summed over all molecules. Using Eq. 25.214 in Eq. 25.213, we find that

$$Q_{N_A N_B} = \Lambda_A^{-3N_A}\Lambda_B^{-3N_B}\frac{(N_A+N_B)!}{N_A!N_B!} \times J_A^{N_A} J_B^{N_B}\exp\left\{-\frac{1}{k_BT}[N_A\epsilon_A(0) + \epsilon_B(0)]\right\} \tag{25.215}$$

because there are, neglecting near-neighbor correlations, $(N_A + N_B)!$ ways of distributing $N_A + N_B$ molecules over $N_A + N_B$ sites. The cell partition functions are

$$J_B = 4\pi\int_\Delta \left\{-\frac{1}{k_BT}[\epsilon_A(r) - \epsilon_A(0)]\right\} r^2\,dr \tag{25.216}$$

$$J_B = 4\pi\int_\Delta \left\{-\frac{1}{k_BT}[\epsilon_B(r) - \epsilon_B(0)]\right\} r^2\,dr \tag{25.217}$$

where Δ is the cell volume and $N\Delta = V$. Note that the cell partition functions appear to the powers N_A and N_B because each molecule is considered to move independently of the others. From $Q_{N_A N_B}$ and cell model approximations to Q_{NA}^0 and Q_{NB}^0, the canonical partition functions of pure liquid A and B, one easily finds the excess entropy of mixing to be

$$\begin{aligned} s^E &\equiv \Delta s_{\text{mixing}} - \Delta s_{\text{ideal mixing}} \\ &= x_A k_B\left[\ln\left(\frac{J_A}{J_A^0}\right) + T\frac{\partial}{\partial T}\ln\left(\frac{J_A}{J_A^0}\right)\right] \\ &\quad + x_B k_B\left[\ln\left(\frac{J_B}{J_B^0}\right) + T\frac{\partial}{\partial T}\ln\left(\frac{J_B}{J_B^0}\right)\right], \end{aligned} \tag{25.218}$$

where the superscript zero refers to the pure component. We have already assumed that the assignment of molecules to cells is random, hence s^E should vanish for a static lattice. When the molecular volumes are the same and no displacement from the lattice sites is allowed, we have $J_A = J_A^0$, $J_B = J_B^0$, and Eq. 25.218 gives $s^E = 0$ as required. The cell partition functions depend on the composition because the field in which a molecule moves is determined by the composition of the surrounding medium. Without making a detailed calculation of the J's, we can still deduce the following interesting information:

1. Suppose that the average composition-dependent potential is such that the molecule in the cell in the solution can move over a larger fraction of the cell volume than in either pure component. Then we have $J_A > J_A^0$, $J_B > J_B^0$, and there is a positive contribution to the excess entropy of mixing. If the fraction of the cell volume available in solution is less than in the pure components, the contribution to s^E is negative. Note how this entropy change is related to the ratio of volumes available for center-of-mass motion, a result that can be compared with the entropy change when a gas is expanded isothermally.
2. Examination of Eqs. 25.216 and 25.217 shows that the temperature derivative of the cell partition function is related to the average energy of a particle in its cell. Since, in thermodynamic terms, the temperature derivative of the average energy is related to the specific heat, the integral of c_p/T to the entropy, and finally the entropy to the density of states for motion in the cell, differences in average energy per particle should be expected to contribute to an excess entropy of mixing.

In sum, part of the entropy change on mixing comes from the change in "available volume," and part from the way a molecule moves in the "available volume."

The cell model provides a qualitative understanding of some of the principal contributions to the deviations of the properties of a mixture from ideal behavior, but is much too crude to provide a quantitative description. As mentioned earlier, the approximations used to simplify the calculation of the partition function make the liquid too much like a crystal in the sense of assuming too much long-range order, as well as neglecting important correlations in the motion of near-neighbor molecules. As we shall see, the cell model also does not make explicit the role of structure change in the determination of the properties of the mixture, although it does implicitly account crudely for some of the effects of structure change (e.g., by the change in cell volume).

In addition to the exact Kirkwood–Buff and McMillan–Mayer theories of solutions, an exact theory of mixtures can be based on generalization of the arguments used in Chapter 23 to relate the internal energy and pressure of a liquid to the distribution of molecules and the intermolecular potential. In a binary mixture of A and B there are three pair distribution functions (for AA, AB, and BB pairs), six triplet distribution functions (for AAA, ABA and AAB, ABB, BAA, BBA and BAB, and BBB triplets, but only four of these are independent functions), and so forth. This theory provides a correct description of the liquid mixture for all compositions, provided only that the densities of A and B and the interaction potentials are known, from which we must calculate all the pair-correlation functions $g_{ij}(R)$. These pair-correlation functions are then used to calculate the equation of state and internal energy of the mixture from the exact relations

$$\frac{p}{k_B T} = n - \frac{1}{6}\sum_{i=1}^{2}\sum_{j=1}^{2} n_i n_j \int \mathbf{R}\cdot\nabla\left(\frac{u_{ij}}{k_B T}\right) g_{ij}(R)\, d\mathbf{R} \qquad (25.219)$$

and

$$\frac{U}{k_B T V} = \frac{3}{2}n + \frac{1}{2}\sum_{i=1}^{2}\sum_{j=1}^{2} n_i n_j \int\left(\frac{u_{ij}}{k_B T}\right) g_{ij}(R)\, d\mathbf{R}, \qquad (25.220)$$

which are obvious extensions of Eqs. 23.21 and 23.17 to a two-component liquid. In practice the pair-correlation functions $g_{ij}(R)$ can be calculated only after introduction of approximations.

Just as in the case of pure liquids, computer simulation of the properties of mixtures provides invaluable information obtainable in no other fashion. Figure 25.23 shows the three pair-correlation functions that characterize a binary equimolar mixture of molecules that interact with a Lennard–Jones 6–12 potential. The parameters chosen are appropriate to the Ar–Kr mixture at $T = 118$ K and $v_m = 33.3$ cm^3/mol. Note how the peaks are separated even though the molecular diameters are only 7% different. Figure 25.24 shows the effect of changes in the ratio of molecular diameters on the excess Gibbs free energy for equimolar binary mixtures of Lennard–Jones molecules. The parameters chosen correspond to $p = 0$, $T = 117$ K, $\epsilon_{BB}/\epsilon_{AA} = 1.11$, and $\epsilon_{AB} = (\epsilon_{AA}\epsilon_{BB})^{1/2}$. The results of computer simulation studies of

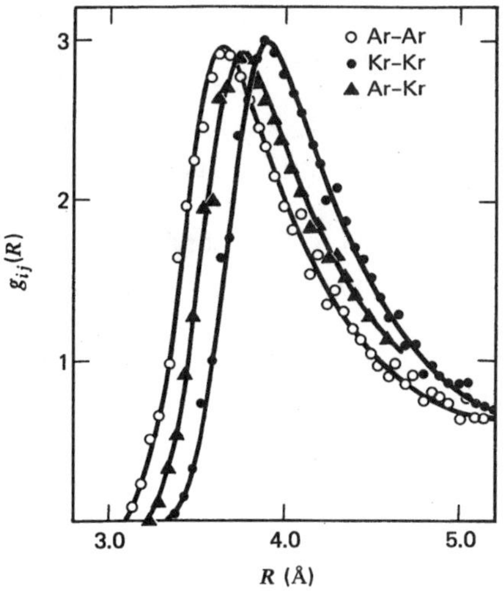

Figure 25.23 Pair-correlation functions from a molecular dynamics simulation of an equimolar liquid mixture of argon and krypton ($T = 118$ K, $v_m = 33.3$ cm^3/mol). The three main peaks are well separated. From I. R. McDonald and K. Singer, *Chemistry in Britain* **9**, 56 (1973).

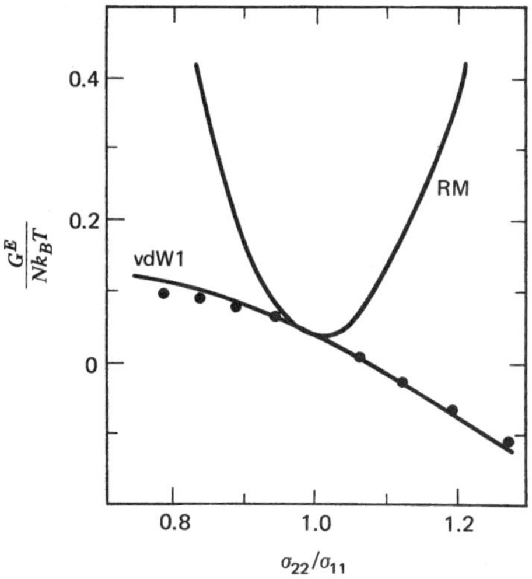

Figure 25.24 The effect of changes in the molecular diameter ratio on the excess Gibbs free energy of binary mixtures of Lennard–Jones liquids ($T = 117$ K, $p = 0$) for the case when $\epsilon_{22}/\epsilon_{11} = 1.11$. The points give the results of Monte Carlo calculations and the curves show the predictions of various theories: random mixing (RM) and one-fluid (vdW1) versions of the van der Waals model. From I. R. McDonald and K. Singer, *Chemistry in Britain* **9**, 56 (1973).

mixtures of hard spheres and mixtures of Lennard–Jones molecules lead to the following conclusions:

1. Just as for pure fluids, the repulsive part of the molecular interaction determines the structure of the liquid. The addition of an attractive component to the interaction has little influence on the liquid's structure.
2. Because of conclusion **1**, the packing geometry in a liquid mixture will be very like that in a mixture of hard spheres with the same ratio of diameters. When that ratio differs much from unity, there is considerable deviation from random mixing.
3. The assumption that the AB attractive interaction scales as

$$\epsilon_{AB} = \xi(\epsilon_{AA}\epsilon_{BB})^{1/2}$$

with $\xi = 1$ is not accurate enough to be useful. Although ξ deviates from unity by only a few percent, the excess thermodynamic properties of the mixture are themselves only a few percent of the total values, so inaccurate predictions result when ξ is set equal to 1. The consequence of this observation is that prediction of the properties of a mixture requires very accurate interaction potentials for all pairs of molecules, and one cannot calculate the mixture properties from knowledge of only the properties of the pure components.

Although Eqs. 25.219 and 25.220 are exact relations, they cannot be used to predict the properties of a liquid mixture without knowledge of the independent pair-correlation functions of the mixture. At present, those pair-correlation functions can only be obtained by use of the same class of approximations as used to develop a theory of pure liquids (see Chapter 23). Before examining an example of such a theory of liquid mixtures, we use the results of the computer simulation studies to construct a simple analysis of the properties of a mixture.

First, what mileage can we get out of the concept of random mixing that is so useful in the case of crystalline mixtures? The random mixing assumption, applied to a liquid mixture, amounts to assuming that all $g_{ij}(R)$ have the same dependence on R. Then, for a binary mixture,

$$g_{11}(R) = g_{22}(R) = g_{12}(R) \equiv g_2^x(R) \qquad (25.221)$$

and

$$u_2^x(R) = \sum_{i,j} x_i x_j u_{ij}(R), \qquad (25.222)$$

so that the energy, Eq. 25.220, becomes

$$\frac{U}{k_B TV} = \frac{3}{2}n + \frac{2\pi n^2}{k_B T}\int u_2^x(R) g_2^x(R) R^2\, dR. \qquad (25.223)$$

In the random mixing approximation $g_2^x(R)$ is the pair-correlation function of a hypothetical fluid with the composition-dependent potential $u_2^x(R)$. We expect this approximation to be inaccurate because it neglects the change in structure that occurs on mixing. Put another way, if the molecular diameters are unequal, Eq. 25.221 is very inaccurate. Since it is the repulsive forces that determine the structure of the liquid, the first peak in $g_2(R)$ is always close to the distance of closest approach of two molecules. This distance is different in the pure components and the mixture if the molecular diameters are different, whereas the approximation Eq. 25.221 would make the first peak of $g_2(R)$ occur at the same R for both pure components and mixture regardless of the molecular sizes. Consequently, the predicted thermodynamic properties of the mixture are inaccurate.

Why, then, is the random mixing approximation useful to describe a crystalline mixture? Simply put, if the two components to be mixed in the solid phase have very different molecular sizes, they will be immiscible. A stable crystalline mixture can be formed only when the two components have nearly equal sizes, and then deviations from random mixing, although they do occur, are usually not large. In contrast, a liquid binary mixture is stable over a considerable range of variation of the ratio of species diameters, and the difference in sizes of the species does lead to deviations of the entropy of mixing from the ideal form. However, in the absence of stabilizing attractive interactions, when the ratio of species diameters becomes very large a liquid binary mixture separates into two phases. For example, when the packing fractions of the two species are comparable, a binary liquid mixture of hard sphere molecules with $\sigma_{22}/\sigma_{11} \geq 4$ separates into two phases.

A much better way of representing the consequences **1** and **2** of the computer simulations is to assume that

$$g_{11}(R/\sigma_{11}) = g_{12}(R/\sigma_{12}) = g_{22}(R/\sigma_{22}) \equiv g_2^x(R/\sigma_x). \qquad (25.224)$$

This form for $g_2^x(R/\sigma_x)$ preserves the key feature that its first peak occurs near to the distance of closest approach. In other words, once scaled by the distance of closest approach, we assume the form of $g_2(R)$ is the same for all pairs of species, as it is in a hard-sphere fluid. Using Eq. 25.215 in Eq. 25.211, we find that

$$\begin{aligned}\frac{U}{k_B TV} &= \frac{3}{2}n + \frac{2\pi n^2}{k_B T}\sum_{i,j} x_i x_j \epsilon_{ij}\sigma_{ij}^3 \\ &\quad \times \int F(R/\sigma_{ij}) g_{ij}(R/\sigma_{ij}) \left(\frac{R}{\sigma_{ij}}\right)^2 d\left(\frac{R}{\sigma_{ij}}\right) \\ &= \frac{3}{2}n + \frac{2\pi n^2}{k_B T}\epsilon_x \sigma_x^3 \int F(\tilde{R}) g_2^x(\tilde{R})\tilde{R}^2\, d\tilde{R},\end{aligned} \qquad (25.225)$$

where $u_2(R) = \epsilon F(R/\sigma) = \epsilon F(\tilde{R})$ is the form assumed for all pair interactions and

$$\epsilon_x \sigma_x^3 = \sum_{i,j} x_i x_j \epsilon_{ij}\sigma_{ij}^3. \qquad (25.226)$$

Table 25.6 Comparison of Experimental and Theoretical Excess Thermodynamic Functions for Some Liquid Mixtures

$T(K)$	Ar + Kr 116	Ar + N_2 84	Ar + O_2 84	Ar + CO 84	Ar + CH_4 91	Xe + Kr 161	N_2 + O_2 84	CO + CH_4 91
(a) g^E (J/mol)								
Exp.	84	34	37	57	74	115	39	115
Comp.	45	35	0	27	–10	28	38	76
RM	203	98	—	99	277	—	143	137
VDW1	46	39	—	29	–17	—	43	83
Pert'n	39	32	—	25	–12	—	36	69
(b) h^E (J/mol)								
Exp.	—	51	60	—	103	—	42	105
Comp.	–18	40	0	34	–35	–52	48	13
RM	254	142	—	149	406	—	220	115
VDW1	–30	43	—	35	–55	—	52	26
Pert'n	–28	35	—	30	–30	—	42	24
(c) v^E (cm³/mol)								
Exp.	–0.52	–0.18	0.14	0.10	0.17	—	–0.31	–0.32
Comp.	–0.60	–0.23	0.00	–0.17	–0.13	–0.62	–0.25	–0.68
RM	0.42	0.11	—	0.23	1.34	—	0.35	–0.32
VDW1	–0.68	–0.25	—	–0.19	–0.23	—	–0.28	–0.71
Pert'n	–0.62	–0.23	—	–0.17	–0.12	—	–0.25	–0.63

In this approximation $g_2^x(\tilde{R})$ is the pair-correlation function for a fluid whose depth and range potential parameters satisfy Eq. 25.226. The theory of mixtures just described is known as a van der Waals one-fluid theory (vdWl). As shown in Fig. 25.24, it agrees rather well with the observed excess free energy as a function of molecular diameter ratio, whereas the random mixing assumption agrees only at the one point $\sigma_{11} = \sigma_{22} = \sigma_{12}$. In Table 25.6 we compare the observed and calculated excess thermodynamic functions of some simple mixtures. The random mixing assumption is again shown to be very poor. The van der Waals one-fluid theory is in good agreement with the computer simulation results, and both of these are in modest agreement with experimental data, with some glaring inaccuracies. It is likely that a principal source of the errors in the predictions of the properties of real liquid mixtures is the uncertainty in the pair potentials for unlike molecules, since the approximate theory agrees well with the computer experiments based on the same interactions.

We have also included in Table 25.6 the predictions of a perturbation theory of liquid mixtures just like the perturbation theory of pure liquids described briefly in Chapter 23. The perturbation theory is seen to be about as accurate as the van der Waals one-fluid theory.

In summary, the structure of a liquid mixture must be known accurately before its excess thermodynamic properties can be predicted. That structure is determined by short-range repulsive forces, hence the use of models that accurately represent the equation of state of mixed hard spheres is likely to be fruitful. It is also necessary to know accurately the mixed-pair interaction—this cannot be adequately approximated by the geometric mean of the pure component interactions.

A very powerful analysis of the properties of liquid mixtures is based on an extension[24] of the density functional theory of liquids sketched in Chapter 23. Eq. 24.104 related the excess chemical potential of a pure fluid to an effective one particle potential energy, $-k_BTC_1$, which self-consistently accounts for the interactions between the molecules, and in Appendix 23B it was shown that successive functional derivatives of C_1 with respect to the density define the pair (C_2) and higher direct correlation functions that describe the structure of the liquid. This analysis can be extended to the description of multicomponent fluids. It is found that

$$\ln g_{ij}(R) = -\frac{u_{ij}(R)}{k_BT} + \frac{n_iB_{ij}(R) + n_jB_{ji}(R)}{n_i + n_j} + \sum_j n_j \int d\mathbf{R}' C_{ji}(|\mathbf{R} - \mathbf{R}'|)[g_{ij}(R') - 1] \tag{25.227}$$

where B's are bridge functions (see Eq. 23.53). In the analogous case of pure liquids, it is known that replacement of the bridge function of the system under study by the bridge

[24] G. Kahl, B. Bildstein, and Y. Rosenfeld, *Phys. Rev.* **E54,** 5391 (1996).

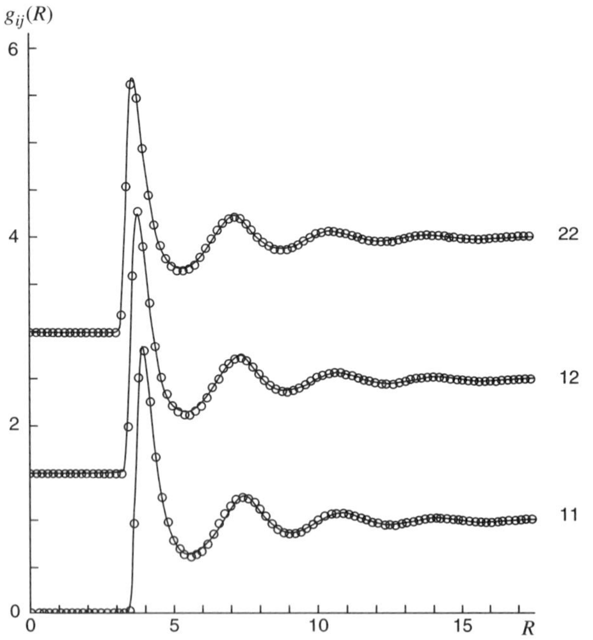

Figure 25.25 Partial pair-correlation functions for an equimolar binary mixture of Lennard–Jones molecules chosen to represent Ar and Kr. The lines represent the results of calculations based on Eq. 25.227, and the circles are the results of molecular dynamics simulations of the liquid mixture. The temperature and density of the mixture are $T = 115.8$ K and $n = 0.01834$ Å^{-3} = 1.834×10^{22} molecules/cm^3. From G. Kahl, B. Bildstein, and Y. Rosenthal, *Phys. Rev.* **E54,** 5391 (1996).

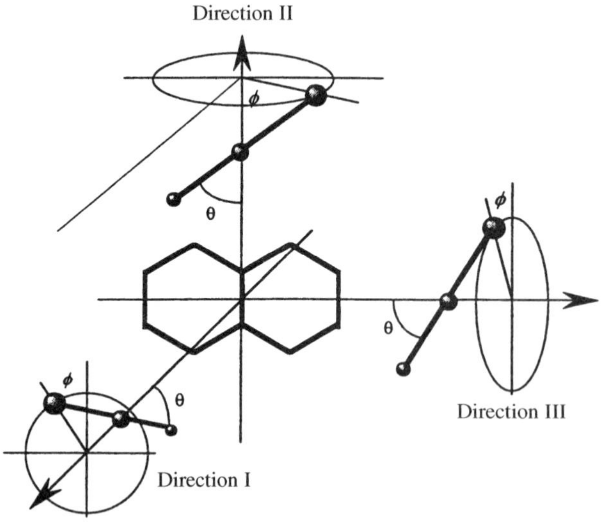

Figure 25.26 Directions along which the naphthalene-carbon dioxide orientational correlations are examined. The orientation of a CO_2 molecule about a naphthalene axis is specified by the polar angle θ and the azimuthal angle ϕ. From K. Koga, H. Tanaka, and X. C. Zeng, *J. Phys. Chem.* **100,** 16711 (1996).

function of a hard sphere fluid is a very accurate approximation. The success of this approximation can be formalized in the hypothesis that the bridge function has a universal form. For the case of the liquid mixture, the assumption made is that universality occurs at the level of the bridge functional, which is obtained via functional differentiation from the part of the free energy functional which represents the excess free energy attributable to terms in the functional series expansion involving triplet and higher order direct correlation functions. This part of the free energy functional can be constructed for a fluid mixture of hard spheres,[25] which then becomes the reference system of choice. In the mixture case the bridge functions contain (via the free energy functional) the structure of both the reference system and the system under study. A comparison of the results of numerical calculations of the g_{ij} for a mixture of Lennard–Jones molecules chosen to mimic Ar and Kr and computer simulation data is shown in Fig. 25.25. Clearly, the theory of mixtures based on Eq. 25.227 is very accurate when the species are simple molecules.

In this section we have surveyed only the simplest features of the statistical molecular theory of solutions, focusing attention on those aspects of molecular interaction connected with the translational motion of the molecules. Of course, in real systems the internal degrees of freedom of the molecules can be different in solution and in the pure component, different solvents may stabilize different molecular geometries of the solute molecule, local directional bonding of the solute to the solvent can occur, and so forth. Many of the subtler aspects of the chemistry of condensed phases are connected with just such phenomena. The theory of solutions of polyatomic molecules with directional and nondirectional interactions is so complicated that little progress has been made in deducing generally valid descriptions. An example of the level of detail possible with modern theories of liquid structure is provided by a study of the environment an isolated naphthalene molecule dissolved in supercritical carbon dioxide.[26] The mixture is described using the RISM theory (see Section 23.6) of polyatomic liquids, extended to the case of mixtures.[27] Fig. 25.26 shows the three directions along which the naphthalene-carbon dioxide orientational correlations were examined, Fig. 25.27 displays several of the orientation dependent naphthalene-carbon dioxide pair distribution functions along the directions specified in Fig. 25.26, and Fig. 25.28 illustrates the three most favorable conformations for a naphthalene-carbon dioxide molecular pair. The complexity of the structure of a solution of naphthalene in

[25] Y. Rosenfeld, *Phys. Rev. Lett.* **63,** 980 (1989); *J. Chem. Phys.* **98,** 8126 (1993).

[26] K. Koga, H. Tanaka, and X. C. Zeng, *J. Phys. Chem.* **100,** 16711 (1996).

[27] D. Chandler and H. C. Anderson, *J. Chem. Phys.* **57,** 1930 (1972); F. Hirata and P. J. Rossky, *Chem. Phys. Lett.* **83,** 329 (1981); F. Hirata, B. M. Pettit, and P. J. Rosky, *J. Chem. Phys.* **77,** 509 (1982); S. W. Chen and P. J. Rossky, *J. Phys. Chem.* **97,** 6078 (1993); M. Matsumoto, H. Tanaka, and K. Nakanishi, *J. Chem. Phys.* **99,** 6935 (1993); S. J. Singer and D. Chandler, *Mol. Phys.* **55,** 621 (1985); D. A. Zichi and P. J. Rossky, *J. Chem. Phys.* **84,** 1712 (1986).

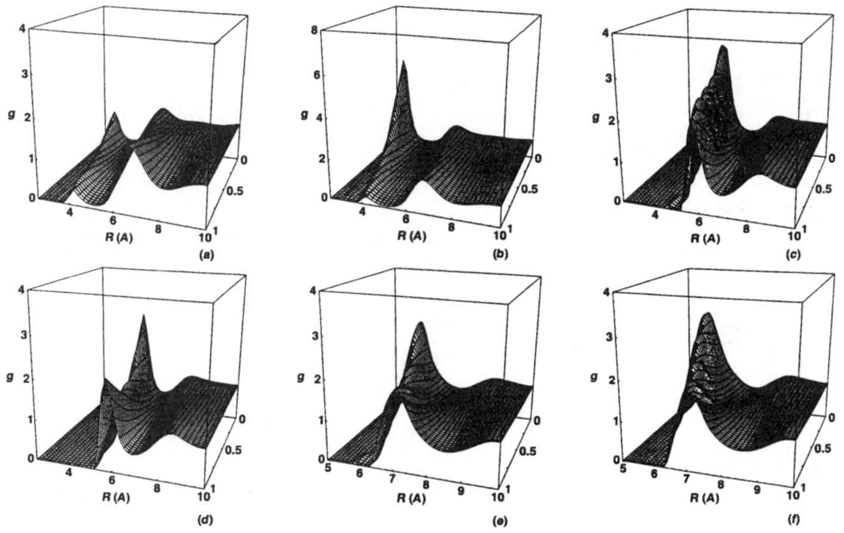

Figure 25.27 Orientation dependent naphthalene-carbon dioxide pair-correlation functions for (*a*) $\phi = 0°$ and (*b*) $\phi = 90°$ in direction I (see Fig. 25.26), (*c*) $\phi = 0°$ and (*d*) $\phi = 90°$ in direction II, and (*e*) $\phi = 0°$ and (*f*) $\phi = 90°$ in direction III. The temperature and density of the mixture are $T = 400$ K and $\rho = 1.5$ g/cm^3. From K. Koga, H. Tanaka, and X. C. Zeng, *J. Phys. Chem.* **100,** 16711 (1996).

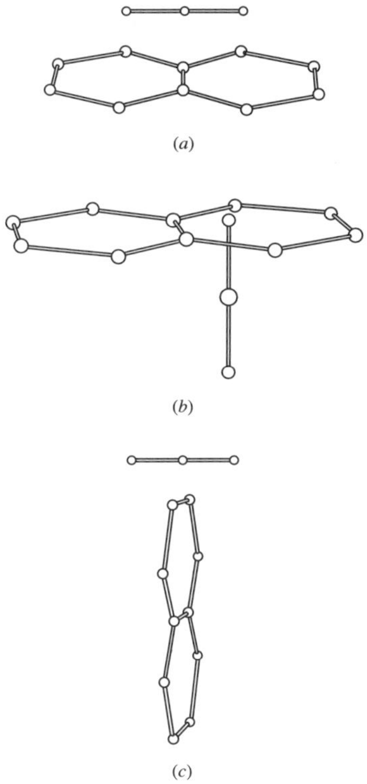

Figure 25.28 The three most-favored configurations of a naphthalene-carbon dioxide pair in directions I, II, and III. From K. Koga, H. Tanaka, and X. C. Zeng, *J. Phys. Chem.* **100,** 16711 (1996).

supercritical carbon dioxide is representative of that in mixtures in which much of chemistry is carried out. Clearly, the local orientation effects illustrated influence the excess chemical potentials of the solute and solvent, and thereby the molecular conformations and chemical equilibria.

As of this date the molecular interpretation of chemical effects arising from specific interactions in solution remains difficult because of the complexity of the theory of mixtures of polyatomic molecules. Consequently, it is still common to make use of the predictions obtained from very simple models, even though these predictions are often not thermodynamically self consistent. What is needed is a set of new concepts into which the complications of the detailed molecular theory of solutions can be swept and on the basis of which a simple and incisive theoretical description of the thermodynamic properties of liquid mixtures can be based.

FURTHER READING

Davis, H. Ted, *Statistical Mechanics of Phases, Interfaces, and Thin Films* (VCH Publishers, New York, 1996), Chapter 6.

Findlay, A., *The Phase Rule and Its Applications,* 9th ed., rev. by A. N. Campbell and N. O. Smith (Dover, New York, 1951).

Hildebrand, J. H., Prausnitz, J. M., and Scott, R. L., *Regular and Related Solutions* (Van Nostrand Reinhold, New York, 1970).

Pitzer, K. S., *Thermodynamics,* 3d ed. (McGraw-Hill, New York, 1995), Chapters 10–14.

Prausnitz, J. M., Lichtenthaler, R. N., and Gomes de Azevado, E., *Molecular Thermodynamics of Fluid Phase Equilibria* (Prentice Hall, Englewood Cliffs, 1986).

Prigogine, I., *The Molecular Theory of Solutions* (North Holland, Amsterdam, 1957).

Rushbrooke, G. S., *Introduction to Statistical Mechanics* (Oxford University Press, London, 1949), Chapters 13, 14, 17, and 18.

Advanced Treatments

Henderson, D., and Leonard, P. J., in D. J. Henderson (Ed.), *Physical Chemistry; An Advanced Treatise,* Vol. VIIIB, p. 414 (Academic Press, New York, 1971).

McDonald, I. R., *Specialist Periodical Report, Statistical Mechanics* **1,** 134 (1973), The Chemical Society, London.

Münster, A., *Statistical Thermodynamics,* Vol. II (Springer-Verlag, Berlin, 1974), Chapter 17.

Palatnik, L. S., and Landau, A. I., *Phase Equilibria in Multicomponent Systems* (Holt, Rinehart and Winston, New York, 1964).

Rowlinson, J. S., and Swinton, F. L., *Liquids and Liquid Mixtures,* 3d ed. (Butterworths, London, 1982).

PROBLEMS

1. A mixture of $C_6H_5CH_3$ and C_6H_6 contains 30% by weight of $C_6H_5CH_3$. At 30°C the vapor pressure of pure $C_6H_5CH_3$ is 36.7 torr and that of pure C_6H_6 is 118.2 torr. Assuming ideality, calculate the total pressure and the partial pressure of each component above the solution at 30°C.

2. Consider a system that contains two components in two phases, say, liquid and vapor. From the Gibbs phase rule, the system has two degrees of freedom, so that p and n_2^V may be taken as independent variables. The temperature can therefore be written $T = T(p, n_2^V)$. If the system is azeotropic, however, there is an additional constraint $x_2^L = x_2^V$ that results in the loss of one degree of freedom. The temperature of the system then depends only on the pressure. Evaluate dT/dp for the azeotropic mixture.

3. Mixtures of C_6H_5Cl and C_6H_5Br form nearly ideal solutions. At 136.7°C the vapor pressure of pure C_6H_5Cl is 863 torr and that of C_6H_5Br is 453 torr. Find the composition of a mixture whose total vapor pressure is 760 torr. Find the composition of a mixture whose vapor is equimolar in C_6H_5Br and C_6H_5Cl.

4. Assuming that methane forms ideal solutions in C_8H_{18}, calculate the change in chemical potential at 298 K when 3 mol of methane at 1 atm pressure is dissolved in an infinite amount of a CH_4–C_8H_{18} solution, in which the partial pressure of CH_4 is 27 torr.

5. Calculate the change in free energy per gram-atom when monoclinic sulfur is changed to rhombic sulfur at 298 K. Sulfur forms ideal solutions of S_8 molecules in benzene, and the solubilities of monoclinic and rhombic sulfur in 1000 g of benzene are 27.1 g and 21.2 g, respectively.

6. The partial vapor pressure of chloroform in a mixture with acetone at 300 K is 20.0 torr when its mole fraction is 0.12, and 220 torr when its mole fraction is 0.80. How does the chemical potential of the chloroform in the second solution differ from the chemical potential in the first solution? What would your answer be if the solutions were ideal?

7. Consider a binary, nonideal solution defined by the equations (see Eqs. 25.29–25.30)

$$\mu_1 = \mu_1^0 + RT\ln x_1 + wT^2(1-x_1)^2,$$

$$\mu_2 = \mu_2^0 + RT\ln(1-x_1) + wT^2x_1^2,$$

where x_1 is the mole fraction of component 1 and w is a constant. Calculate Δh_{mixing}, Δs_{mixing}, and Δv_{mixing}. The reader is cautioned that the form wT^2 is not a realistic description of the temperature dependence of the interactions in any real solution.

8. A mixture contains 1 mol each of liquid benzene and liquid toluene. At 300 K the vapor pressure of pure toluene is 32.06 torr and that of pure benzene is 103.01 torr.
   (a) If the pressure over the mixture is reduced at 300 K, at what pressure does boiling start?
   (b) What is the composition of the first trace of vapor boiled off?
   (c) If the pressure is lowered further, at what pressure does the last trace of liquid disappear?
   (d) What is the composition of the last trace of liquid?

9. For nondilute solutions, show that Eq. 25.107 should be replaced by

$$\ln x_1 = -\frac{\Delta h_{1v}\theta}{RT_{1b}T}$$

if Δh_{1v} is considered constant. An analogous expression is found for freezing-point depression. If the heat of fusion of water is 6.01 kJ/mol, calculate the freezing point in solutions where the mole fraction of water is 0.8, 0.6, 0.4, 0.2.

10. Two grams of benzoic acid dissolved in 25 g of benzene (K_f = 4.90°C kg/mol) produce a freezing point depression of 1.62°C. Calculate the molecular weight and compare your answer with the molecular weight obtained from the formula C_6H_5COOH.

11. At 0°C the Henry's law constant for argon in water is 2.17×10^4 and at 30°C the constant is 3.97×10^4 for $\mathfrak{p}$ = 1 atm. Find the approximate heat of solution over this temperature range.

12. 10.0 g of an unknown nonvolatile solid was dissolved in 100.0 g of H_2O. At 26°C, the vapor pressure of the solution was observed to be 24.75 torr. At that same temperature, pure water has a vapor pressure of 25.20 torr. What is the molecular weight of the solid?

13. At 30°C a 1M solution of a sugar has a vapor pressure of 31.207 torr, whereas pure water has a vapor pressure of 31.824 torr. Calculate the osmotic pressure of the solution using Eq. 25.124 and Eq. 25.120. In the latter case, approximate the partial molar volume of the water by the molar volume of pure water.

14. The Henry's law constant for N_2 in H_2O at 25°C is 6.43×10^7, and that for O_2 is 3.30×10^7 for $\mathfrak{p}$ = 1 atm. If air contains 80% N_2 and 20% O_2 by volume, calculate the percentage composition of air dissolved in water.

15. Verify that the chemical potentials for an ideal two-component solution (Eq. 25.12) satisfy the Gibbs–Duhem relation.

16. Suppose that the excess chemical potentials of a binary solution are given by

$$\mu_1^E = a_1 x_2 + b_1 x_2^2 + c_1 x_2^3,$$
$$\mu_2^E = a_2 x_1 + b_2 x_1^2 + c_2 x_1^3.$$

Prove that $a_1 = 0$, $a_2 = 0$, $c_2 = -c_1$, $b_2 = b_1 + \frac{3}{2}c_1$.

17. Consider an ideal solution of two volatile components. The partial pressures of the components in the vapor phase are p_1 and p_2, and the vapor occupies a volume V. If the partial molar volumes of the gases are $\bar{v}_1 = RT/p_1^0$ and $\bar{v}_2 = RT/p_2^0$, show that Raoult's law requires that $V = n_1\bar{v}_1 + n_2\bar{v}_2$, where n_1 and n_2 are the number of moles of the components in the vapor.

18. Calculate the partial molar volumes of methanol and water in a water–methanol solution that is 12.5% methanol by weight, from the following data obtained at 15°C and 1 atm: A 12.0% methanol solution has a density of 0.97942 g/cm^3, and a 13.0% methanol solution has a density of 0.97799 g/cm^3.

19. The vapor pressure of pure ethanol at 35°C is 102.8 torr and that of pure chloroform is 295.1 torr. In an ethanol–chloroform solution in which the mole fraction of ethanol is 0.20, the total vapor pressure is 304.2 torr, and the mole fraction of ethanol in the vapor is 0.138. Calculate the activity coefficients of the ethanol and the chloroform, and the excess free energy of mixing.

20. When phenol (C_6H_5OH) is dissolved in a particular solvent, it is partially dimerized. If 2.58 g of phenol is dissolved in 100 g of the solvent, the freezing point of the solvent (K_f = 14.1°C kg/mol) is depressed by 2.37°C. Find how much of the phenol is dimerized in solution.

21. Use the Gibbs phase rule to show that Fig. 25.11a cannot have a eutectic line or area.

22. Is it possible for one component of a binary solution to behave ideally but not the other component?

23. The volume of an aqueous solution of NaCl at 25°C in which m mol of NaCl are dissolved in 1 kg of water is found to be

$$V = 1003 + 16.6m + 1.77m^{3/2} + 0.119m^2,$$

where the volume is given in cubic centimeters. Find the partial molal volumes of NaCl and H_2O when $m = 0.1$.

24. In an acetone–chloroform solution where the mole fraction of acetone is 0.531, the partial molar volume of acetone is 74.2 cm^3/mol and that of chloroform is 80.2 cm^3/mol. What volume of solution weighs 100 g? The densities of the pure components are 1.351×10^{-2} mol/cm^3 for acetone and 1.240×10^{-2} mol/cm^3 for chloroform. What is the total volume of the unmixed components of the 100-g solution?

25. Suppose that you are given a binary solution with components A and B, and the partial molar volume of A is known as a function of x_A : $\bar{v}_A = f(x_A)$. Obtain an equation for $\bar{v}_B$ in terms of x_A.

26. Polybenzyl glutamate (PBG) is a polypeptide that has an internally hydrogen-bonded rigid helical conformation in chloroform, and a nonhydrogen-bonded, flexible, randomly coiled conformation in trichloracetic acid. If trichloracetic acid is added to a solution of PBG in chloroform, the PBG can be made to undergo the transformation helix → random coil. The entropy change in this transformation is found to be negative. But the random coil conformation of PBG has many more configurations accessible for energy between E and $E + dE$ than does the helix conformation, so we expect the entropy change for the transformation helix → random coil to be positive. Provide a molecular model-type explanation for the observation.

27. Consider a mixed liquid solvent containing components 1 and 2, in equilibrium with a gas labeled component 3. Suppose that the solubility of the gas is proportional to the gas fugacity. For the case discussed in Section 25.4, where the solvent has only one component, the change in solvent chemical potential when the concentration of solute is changed can be obtained directly by use of Henry's law. Show that for the case that the solvent is a mixture, knowledge of the Henry's law constant is insufficient to determine the separate changes of the chemical potentials of components 1 and 2 when the solute concentration is changed.

28. Consider a system of two immiscible liquids, in each of which there is dissolved some third component. An example of this situation is the two-phase system consisting of alcohol in water in contact with alcohol in benzene. The Nernst distribution law asserts that, in a system such as that described, the ratio of the concentrations of solute in the two phases is a constant.

(a) Derive the Nernst distribution law for the case that the solution in each of the immiscible liquid phases is ideal.

(b) Derive the generalization of the Nernst distribution law when the solutions in each of the phases are not ideal. Is the ratio of solute concentrations in the two phases necessarily a constant?

(c) Describe the relationship between the Nernst distribution law as formulated above and Henry's law.

29. In Section 25.6 we discussed the freezing-point depression of a solution when the solid formed is pure (has no solute). Consider, now, the case that an ideal dilute liquid solution freezes to form an ideal dilute solid solution of different concentration. Let the ratio of the concentrations in the solid and liquid phases be K. Show that the rate of change of freezing point with solute concentration is

$$\frac{dT}{dx_2} = (K-1)\frac{RT_{1f}^2}{\Delta h_{1f}}$$

for $\theta \equiv T_{1f} - T$ very small, which corresponds to $x_2 << 1$.

30. Suppose that a two-component liquid mixture is in equilibrium with a two-component solid mixture, neither mixture being ideal. Show that

$$RT_{1f}\ln\frac{\gamma_1^L x_1^L}{\gamma_1^S x_1^S} = -\frac{\Delta\bar{h}_{1f}\theta}{T_{1f}} - \left(\frac{\Delta\bar{h}_{1f}}{T_{1f}^2} + \frac{\Delta\bar{c}_p}{2T_{1f}}\right)\theta^2 + \cdots,$$

where, as usual, solvent ≡ 1 and solute ≡ 2. What is the meaning of $\Delta\bar{h}_{1f}$ and $\Delta\bar{c}_p$ in this situation?

31. Consider a multicomponent mixture in which there are many solutes, denoted 2, 3, . . . , r, each at very low concentration. It is found that the activity coefficients of components 2, . . . , r can be represented in the form

$$\ln\gamma_2 = \ln\gamma_2^0 + k_{22}x_2 + k_{23}x_3 + \cdots + k_{2r}x_r,$$
$$\vdots$$
$$\ln\gamma_r = \ln\gamma_r^0 + k_{r2}x_2 + k_{r3}x_3 + \cdots + k_{rr}x_r.$$

The coefficients k_{ij} represent the influence of interactions between the different solute species. Show that

$$k_{ij} = k_{ji}.$$

32. In the case of a binary solution, knowledge of the chemical potential of one component for all concentrations can be used to determine completely the chemical potential of the otherµ component (see Eq. 25.28). Consider now a ternary solution. Show that the knowledge of the chemical potential of one component for all concentrations, both of that component and the other two components, determines the chemical potentials of the other components. *Hint:* Fix the ratio of concentrations of two components and treat the mixture as a pseudo-binary system. Use the Gibbs–Duhem equation to integrate along different pseudo-binary system lines.

33. A slight generalization of the regular mixture model allows the quantity w defined by Eqs. 25.29 and 25.30 to be temperature-dependent. Show that, when w depends on T,

$$S^E = -(n_1+n_2)x_1x_2\frac{dw}{dT},$$
$$H^E = (n_1+n_2)x_1x_2\left(w - T\frac{dw}{dT}\right),$$
$$C_p^E = -(n_1+n_2)x_1x_2T\frac{d^2w}{dT^2}.$$

For the equimolar benzene–cyclohexane mixture at 293 K, $w = 1270$ J/mol, $dw/dT = -7.0$ J/K mol, $d^2w/dT^2 = 0.046$ J/K^2 mol. Calculate $\ln\gamma_1$, H^E, S^E, and C_p^E, and compare these with the observed values, which are 0.130, 836 J/mol, 1.76 J/K mol, and –3.3 J/K mol, respectively.

34. The simplest useful description of a polymer solution (solvent ≡ 1, polymer solute ≡ 2) is the Flory–Huggins model, which gives for the Gibbs free energy of mixing

$$\Delta G_{\text{mixing}} = RT(n_1\ln\phi_1 + n_2\ln\phi_2) + wn_1\phi_2,$$

with ϕ_1 and ϕ_2 the volume fractions of solvent and solute, and w a temperature-dependent interaction energy.

(a) Show that the heat of mixing is

$$\Delta H_{\text{mixing}} = \left(w - T\frac{dw}{dT}\right)n_1\phi_2.$$

(b) Show that the solvent activity is

$$\ln a_1 = \ln(1-\phi_2) + \phi_2\left(1 - \frac{v_1^0}{v_2^0}\right) + \phi_2^2\left(\frac{w}{RT}\right),$$

with v_1^0, v_2^0 the molar volumes of solvent and solute.

(c) Calculate the partial molar entropy of the solvent.

(d) For the case $v_2^0 = 1000v_1^0$, $M_2 = 1000\,M_1$, plot on the same graph $\bar{s}_1 - s_1^0$ as a function of ϕ_2 using the Flory–Huggins model and also as a function of x_2 using Raoult's law. *Hint:* Convert x_2 to ϕ_2 assuming additivity of molecular volumes.

35. Imagine a crystal of N_A A atoms and N_B B atoms, where $N_B << N_A$. Suppose that the following model can be used: (a) The interaction between molecules is limited to nearest neighbors, of which there are z;

(b) the mass of a B atom is much smaller than the mass of an A atom, so that the vibrational spectrum of the dilute solid solution has one high-frequency mode ν', which lies above all other crystal frequencies; (c) all other crystal frequencies are essentially unchanged by the substitution of a B atom for an A atom; (d) the vibrational spectrum of the "A part" of the mixed crystal is adequately described by the Einstein model (Chapter 22).

(a) Calculate the chemical potential, partial molecular enthalpy, and partial molecular entropy of the solute B.

(b) Suppose that the mixed crystal just described is in equilibrium with an ideal gas of B atoms. Calculate the Henry's law constant for the system.

36. In a classical mechanical description the only difference between two isotopes of the same element, say, ^{3}He and ^{4}He, is the difference in their masses; the pair interaction, $u_2(R)$, is the same for both isotopic species. In a quantum mechanical description this is no longer true, because the effective interaction must include the consequences of the lack of commutativity of the kinetic and potential energy operators (Section 3.4). It can be shown that the chemical potential of a monatomic species has the form

$$\mu = \mu_{\text{classical}} + \frac{\hbar^2}{24mk_BT}\langle(\nabla u_2)^2\rangle$$

to order $\hbar^2$. The term $\langle(\nabla u_2)^2\rangle$ is the mean square force on one atom arising from all the other atoms in the system.

(a) Show that in the classical limit the difference in chemical potentials of isotopic species of masses m_1 and m_2 is

$$\mu_1 - \mu_2 = \frac{3}{2}k_BT\ln\frac{m_2}{m_1}.$$

(b) Show that, to order $\hbar^2$, the difference between the vapor pressures of the pure isotopic liquids (or solids) is

$$p_1 - p_2 = p_{\text{classical}}\frac{\hbar^2\langle(\nabla u_2)^2\rangle}{24(k_BT)^2}\left(\frac{1}{m_1} - \frac{1}{m_2}\right),$$

where $p_{\text{classical}}$ is the common limiting value of the vapor pressure of each isotopic species.

(c) Assuming that the form for the quantum correction to the chemical potential is valid to arbitrarily low temperature, predict whether or not an isotopic mixture will remain one homogeneous phase in the limit $T \to 0$. Justify your conclusion.

37. Suppose that hydrogen is slightly soluble in some crystal in which it occupies sites that prohibit molecular rotation. Estimate the ratio of solubilities of o-H_2 and p-H_2 in this crystal at 60 K, assuming that conversion of o-H_2 to p-H_2 and *vice versa* does not occur.

38. Consider a liquid mixture of monatomic species B in monatomic host A, with $N_B << N_A$. Suppose that this mixture is in equilibrium with a perfect gas of B atoms. Using the cell model to describe the liquid phase, derive a relationship for the Henry's law constant. Compare your result with that of Problem 35(b); what are the differences and similarities?

39. The principle of corresponding states is a powerful tool in the study of mixtures. Consider some pure reference substance, denoted 0, in which the pair potential is

$$u_{00}(R) = \epsilon_{00}F\left(\frac{R}{\sigma_{00}}\right),$$

and some other pure substance, denoted 1, with pair potential

$$u_{11}(R) = \epsilon_{11}F\left(\frac{R}{\sigma_{11}}\right).$$

(a) Show that the potential energy contribution (superscript c for "configurational") to the Helmholtz free energy of substance 1 is given by

$$A_1^c(T,V) = f_{11}A_0^c\left(\frac{T}{f_{11}}, \frac{V}{g_{11}^3}\right) - 3Nk_BT\ln g_{11},$$

where

$$f_{11} \equiv \frac{\epsilon_{11}}{\epsilon_{00}},$$

$$g_{11} \equiv \frac{\sigma_{11}}{\sigma_{00}}.$$

(b) Show also that the equation of state of substance 1 can be expressed in the form

$$p_1^c(T,V) = \frac{f_{11}}{g_{11}^3}p_0^c\left(\frac{T}{f_{11}}, \frac{V}{g_{11}^3}\right).$$

(c) If $f_{11} - 1 << 1$ and $g_{11} - 1 << 1$, show that

$$A_1^c(T,V) - A_0^c(T,V)$$
$$= U_0^c(f_{11} - 1) - 3(Nk_BT - p_0^cV)(g_{11} - 1),$$

where U_0^c is the potential energy contribution to the internal energy of the reference substance.

40. Imagine a hypothetical pure substance in which the pair potential is the average, with random mixing, of the pair potentials in a mixture. Explicitly, define

$$\langle u(R_{ij})\rangle = \sum_{\alpha}\sum_{\beta} x_\alpha x_\beta u_{\alpha\beta}(R_{ij}).$$

Then the potential energy of a mixture, averaged over all assignments of molecules to positions with random distribution, is

$$\langle \mathcal{V}\rangle = \sum_{i>j}\sum_{\alpha}\sum_{\beta} x_\alpha x_\beta u_{\alpha\beta}(R_{ij}) = \sum_{i>j}\langle u(R_{ij})\rangle.$$

Substitution of $\langle \mathcal{V}\rangle$ into the configurational partition function

$$Z = \frac{1}{\Pi N_\alpha!}\int\cdots\int \exp\left(-\frac{\mathcal{V}}{k_B T}\right) d\mathbf{R}_1\cdots d\mathbf{R}_N,$$

defines the hypothetical pure substance for the particular composition. Call it the equivalent substance, and note that its properties depend on the composition.

(a) Suppose that the molecules in the mixture interact via a Lennard–Jones potential. Show that

$$A_x^c(T,V) = f_x A_0^c\left(\frac{T}{f_x}, \frac{V}{g_x^3}\right) - 3Nk_BT\ln g_x,$$

where the zero refers to a reference substance as in Problem 38 and

$$f_x = \left(\sum_\alpha\sum_\beta x_\alpha x_\beta f_{\alpha\beta} g_{\alpha\beta}{}^6\right)^2 \times \left(\sum_\alpha\sum_\beta x_\alpha x_\beta f_{\alpha\beta} g_{\alpha\beta}{}^{12}\right)^{-1},$$

$$g_x = \left(\sum_\alpha\sum_\beta x_\alpha x_\beta f_{\alpha\beta} g_{\alpha\beta}{}^6\right)^{-1/6} \times \left(\sum_\alpha\sum_\beta x_\alpha x_\beta f_{\alpha\beta} g_{\alpha\beta}{}^{12}\right)^{1/6},$$

$$f_{\alpha\beta} = \left(\frac{\epsilon_{\alpha\beta}\sigma_{\alpha\beta}{}^6}{\epsilon_{00}\sigma_{00}{}^6}\right)^2\left(\frac{\epsilon_{\alpha\beta}\sigma_{\alpha\beta}{}^{12}}{\epsilon_{00}\sigma_{00}{}^{12}}\right)^{-1},$$

$$g_{\alpha\beta} = \left(\frac{\epsilon_{\alpha\beta}\sigma_{\alpha\beta}{}^6}{\epsilon_{00}\sigma_{00}{}^6}\right)^{-1/6}\left(\frac{\epsilon_{\alpha\beta}\sigma_{\alpha\beta}}{\epsilon_{00}\sigma_{00}}\right)^{1/6}.$$

(b) Show that the equation of state is

$$p_x^c(T,V) = \frac{f_x}{g_x^3} p_0^c\left(\frac{T}{f_x}, \frac{V}{g_x^3}\right).$$

(c) Show that the excess Gibbs free energy of the mixture is

$$G^E(T,p,x) = f_x G_0^c\left(\frac{T}{f_x}, \frac{pg_x^3}{f_x}\right) - \sum_\alpha x_\alpha f_{\alpha\alpha} G_0^c\left(\frac{T}{f_{\alpha\alpha}}, p\frac{g_{\alpha\alpha}{}^3}{f_{\alpha\alpha}}\right) - 3Nk_BT\sum_\alpha x_\alpha \ln\left(\frac{g_x}{g_{\alpha\alpha}}\right).$$

(d) For the particularly simple case that $g_{\alpha\beta} = 1$ for all α, β, show that to first order in $f_x - \Sigma_\alpha x_\alpha f_{\alpha\alpha}$ the excess thermodynamic functions of the mixture are given by

$$G^E = U_0^c\left(f_x - \sum_\alpha x_\alpha f_{\alpha\alpha}\right) = U_0\left[\sum\sum_{\alpha>\beta} x_\alpha x_\beta(2f_{\alpha\beta} - f_{\alpha\alpha} - f_{\beta\beta})\right],$$

$$H^E = (U^c - TC_p^c)_0\left(f_x - \sum_\alpha x_\alpha f_{\alpha\alpha}\right),$$

$$S^E = -(C_p^c)_0\left(f_x - \sum_\alpha x_\alpha f_{\alpha\alpha}\right),$$

$$V^E = (-VT\alpha^c)_0\left(f_x - \sum_\alpha x_\alpha f_{\alpha\alpha}\right).$$

These formulas describe a *conformal solution.*

41. The characteristic feature of a gravitational field is that the work required to move a mass M from a position where the gravitational potential is ϕ^{I} to another position where it is ϕ^{II} is simply

$$w = M(\phi^{\mathrm{II}} - \phi^{\mathrm{I}}),$$

and is independent of the chemical nature of the matter. This leads us to define the chemical potential of species i in a solution in the gravitational potential ϕ by

$$\mu_i = \mu_i^0 + RT\ln\gamma_i x_i + M_i\phi$$

(see also Chapter 26).

(a) Show that at constant temperature and pressure

$$RT\,d\ln\gamma_i x_i + \left(M_i - \bar{v}_i\frac{M_m}{v_m}\right)d\phi = 0,$$

where M_m and v_m are the mean molar mass and mean molar volume, respectively.

(b) Show that for an ideal binary mixture

$$\frac{x_2^{\mathrm{II}}}{x_2^{\mathrm{I}}} = \exp\left[-(M_2 - \rho\bar{v}_2)\frac{\phi^{\mathrm{II}} - \phi^{\mathrm{I}}}{RT}\right],$$

where ρ is the mass density.

CHAPTER 26

Equilibrium Properties of Solutions of Electrolytes

In Chapter 25 we studied the equilibrium properties of nonelectrolyte mixtures. These mixtures are operationally distinguishable from those containing electrolytes—the latter conduct electricity whereas the former do not. The most elementary implication of this observation is that some of the constituents of an electrolyte solution are charged. Because the Coulomb interaction between two charges is of very long range and is very strong, it differs profoundly from the short-range attraction and repulsion typical of the van der Waals-type interaction between uncharged molecules. It is then reasonable to expect that the excess chemical potential of a component in an electrolyte solution will have a different concentration dependence than does the excess chemical potential of a component in a nonelectrolyte mixture. Of course, the interaction between two charged molecules, or between a charged and a neutral molecule, also has a short-range van der Waals-type component. Nevertheless, the effect of the Coulomb interactions on the properties of the solution is the feature that distinguishes electrolyte from nonelectrolyte solutions.

In this chapter we examine the equilibrium properties of electrolyte solutions. Although most of the analysis will be devoted to dilute solutions, we shall also consider briefly mixtures of molten salts. Since water is the most frequently used solvent for electrolytes, special attention will be paid to the nature of aqueous solutions.

26.1 The Chemical Potential

Just as in the study of nonelectrolyte mixtures, the first step in a thermodynamic analysis of electrolyte solutions is the determination of the form of the chemical potential. We adopt the convention that in solutions of strong electrolytes the solute is completely dissociated into ions. From the point of view of thermodynamic analysis, it is not necessary to adopt this convention. However, the complete-dissociation model, although an oversimplification, is reasonably accurate as a description of the state of ionization of strong electrolytes. When the properties of a real electrolyte solution are compared with the properties of a hypothetical ideal solution containing independent ions, the deviations are found to be relatively small. In contrast, if we chose to describe the strong electrolyte solution in terms of the concentration of neutral molecules, it would be found that deviations from the hypothetical ideal solution behavior were much larger. The thermodynamic analysis can be formulated either way, but the availability of a molecular interpretation and the possibility of minimizing "nonideal" effects lead us to choose the convention that the solute exists in a completely dissociated form in solution.

Despite the fact that we shall use concentrations of ions in the thermodynamic formalism, it is not possible to make a solution of ions of one charge type. All we can do is make an electroneutral solution in which the amount of positive charge equals the amount of negative charge. Thus, the chemical potential of the "electroneutral" electrolyte can be measured, but not that of the positive ion or the negative ion separately. For example, consider a salt—say, NaCl—dissolved in water. We cannot determine

$$\mu_{Na^+} = \left(\frac{\partial G}{\partial n_{Na^+}}\right)_{T,p,n_{Cl^-},n_{H_2O}}, \tag{26.1}$$

because we cannot vary n_{Na^+} and n_{Cl^-} independently. However, if we add to the solution dn mol of NaCl, the change of Gibbs free energy at constant T and p is

$$dG = \left(\frac{\partial G}{\partial n_{NaCl}}\right)_{T,p,n_{H_2O}} dn_{NaCl} \equiv \mu_{NaCl}\, dn_{NaCl}. \tag{26.2}$$

It is often convenient to define individual ion chemical potentials by Eq. 26.1, but it must be kept in mind that only

$$\mu_{NaCl} \equiv \mu_{Na^+} + \mu_{Cl^-} \tag{26.3}$$

is operationally definable and experimentally measurable.

For an electrolyte of the $CaCl_2$ type, we have, similarly,

$$\begin{aligned} dG &= \mu_{CaCl_2}\, dn_{CaCl_2} \\ &= (\mu_{Ca^{2+}} + 2\mu_{Cl^-})\, dn, \\ dn_{CaCl_2} &= dn_{Ca^{2+}} = \tfrac{1}{2}\, dn_{Cl^-}\, . \end{aligned} \tag{26.4}$$

In general, for an electrolyte of the type $A_{\nu_+}B_{\nu_-}$, the definition corresponding to Eq. 26.3 is

$$\mu_{A_{\nu_+}B_{\nu_-}} = \nu_+\mu_+ + \nu_-\mu_-\,, \tag{26.5}$$

because dissolution of 1 mol of $A_{\nu_+}B_{\nu_-}$ creates ν_+ mol of A^{Z_+} and ν_- mol of B^{Z_-}, where Z_+ and Z_- are the charges on the A and B ions. The electroneutrality condition for this electrolyte is

$$\nu_+Z_+ + \nu_-Z_- = 0. \tag{26.6}$$

We now restrict attention to solutions in which the concentration of electrolyte is small compared with the concentration of solvent. For this reason, it is convenient to adopt the molality concentration scale, and to use as the reference state for the electrolyte the hypothetical state in which the properties per mole are the same as in the infinitely dilute electrolyte solution (see Chapter 25). Recalling that μ_{Na^+} and μ_{Cl^-} are not independently definable, but only the sum $\mu_{Na^+} + \mu_{Cl^-}$, we write for the chemical potential of NaCl

$$\begin{aligned} \mu_{NaCl} &= \mu^{\theta}_{NaCl} + RT\ln a_{Na^+}a_{Cl^-} \\ &\equiv \mu^{\theta}_{NaCl} + RT\ln a_{\pm}{}^2 \\ &= \mu^{\theta}_{NaCl} + 2RT\ln \gamma_{\pm}m_{\pm}\, . \end{aligned} \tag{26.7}$$

Note that we have used the convention defined in Eq. 25.23*b* to represent the chemical potential. The product $a_{Na^+}a_{Cl^-}$ is defined to be the activity of NaCl, and its square root $a_{\pm} = \gamma_{\pm}m_{\pm}$ is the *mean ionic activity.* In terms of activity coefficients and molalities, we have

$$\begin{aligned} a_{Na^+} &= \gamma_{Na^+}m_{Na^+}\,, \\ a_{Cl^-} &= \gamma_{Cl^-}m_{Cl^-}\,, \\ \gamma_{\pm}^2 &\equiv \gamma_+\gamma_-\,, \\ m_{\pm}^2 &\equiv m_+m_-\,, \\ m_+ &= m_- = m_{NaCl}\,. \end{aligned} \tag{26.8}$$

Similarly, the chemical potential of $CaCl_2$ is

$$\begin{aligned} \mu_{CaCl_2} &= \mu_{Ca^{2+}} + 2\mu_{Cl^-} \\ &= \mu^{\theta}_{CaCl_2} + RT\ln a_{Ca^{2+}}a^2_{Cl^-} \\ &\equiv \mu^{\theta}_{CaCl_2} + RT\ln a_{\pm}^3 \\ &= \mu^{\theta}_{CaCl_2} + RT\ln \gamma_{\pm}^3 m_{\pm}^3\,, \end{aligned} \tag{26.9}$$

where, in this case,

$$\begin{aligned} \gamma_{\pm}^3 &= \gamma_{Ca^{2+}}\gamma^2_{Cl^-} \\ m_{\pm}^3 &= m_{Ca^{2+}}m^2_{Cl^-}\,. \end{aligned} \tag{26.10}$$

It is easily seen that for the general salt the chemical potential is

$$\begin{aligned} \mu_{salt} &= \nu_+\mu_+ + \nu_-\mu_- \\ &= \mu^{\theta}_{salt} + RT\ln a_+^{\nu_+}a_-^{\nu_-} \\ &\equiv \mu^{\theta}_{salt} + RT\ln a_{\pm}^{(\nu_+ + \nu_-)}\,, \end{aligned} \tag{26.11}$$

with the corresponding mean activity coefficients and mean ionic molality:

$$\begin{aligned} \gamma_{\pm}^{(\nu_+ + \nu_-)} &= \gamma_+^{\nu_+}\gamma_-^{\nu_-}\,, \\ m_{\pm}^{(\nu_+ + \nu_-)} &= m_+^{\nu_+}m_-^{\nu_-} = \nu_+^{\nu_+}\nu_-^{\nu_-}m_{salt}^{(\nu_+ + \nu_-)}\,. \end{aligned} \tag{26.12}$$

The reason we have introduced these extra complications in the activity coefficient and concentration scales is that they are necessary if the properties of μ^{θ}_{salt} are to be obtainable by extrapolation of experimental data. For example, in the case of NaCl-type electrolytes,

$$\lim_{m_{NaCl}\to 0} (\mu_{NaCl} - RT\ln m_{NaCl}) \tag{26.13}$$

does not exist, whereas

$$\lim_{m_{\pm}\to 0} (\mu_{NaCl} - 2RT\ln m_{\pm}) \tag{26.14}$$

does exist. Clearly, if μ^{θ}_{NaCl} is to be determined by extrapolation to zero concentration of the properties of a NaCl solution, a limit such as Eq. 26.13 or 26.14 must be used. Why does Eq. 26.14 exist as a limit when Eq. 26.13 diverges? In simple terms, the reason that Eq. 26.14 is finite whereas Eq. 26.13 diverges is that the electrolyte is dissociated into ions. To determine the thermodynamic functions of an electrolyte solution in the hypothetical limit of ideal behavior (the so-called infinitely dilute solution), it is necessary to count the number of solute particles correctly. But Eq. 26.14 is nothing more than ($\mu_{NaCl} - RT\ln m_+ - RT\ln m_-$), which accounts for all solute particles, whereas Eq. 26.13 uses a concentration scale that is the same whether NaCl gives one, two, . . . , ten particles in solution. Thus, ($\mu_{NaCl} - RT\ln m_{NaCl}$) does not give the correct concentration dependence for the chemical potential of a solution of independent positive ions of concentration $m_+ = \nu_+m$ and negative ions of concentration $m_- = \nu_-m$. In an ideal solution each solute particle makes the same contribution to the chemical potential. Hence, by omitting some of the particles from the particle count, Eq. 26.13 in effect neglects a term of the form, say, $RT\ln m_+$, and this term diverges as $m \to 0$.

Thus far we have developed the description of an electrolyte solution under the assumption that the solute is completely dissociated into ions, the so-called strong electrolyte convention. This is the description of choice for many solutes but not for all solutes. There is a very large class of compounds that, when present as a solute, confer electrical conductivity on the solution, but that are poorly described under the strong electrolyte convention. These weak electrolytes are distinguishable from strong electrolytes by the following criterion. The electrical conductivity of a solution is conveniently represented as the sum over the ionic species present of the product of the ion concentration and the ion mobility (the ratio of the ion velocity to the applied electric field; see Chapter 29). In zero-th approximation, neglecting the influence of ion–ion interactions, the electrical mobility of an ion is independent of its concentration. This condition is satisfied by solutions of strong electrolytes, for which the ratio of the electrical conductivity to the solute concentration (called the equivalent conductivity) varies very little over a large concentration range in the dilute solution regime. However, there are many solutes, such as acetic acid (CH_3COOH), for which the equivalent conductivity increases strongly as the concentration of solute decreases. Under the assumption that the ion mobility is a constant, this observation implies that the number of ions relative to the number of unionized solute molecules must increase as the solute concentration decreases, i.e., that the weak electrolyte is only partially ionized at any nonzero concentration, and increasingly ionized as the solute concentration decreases. A quantitative representation of the behavior described is based on the requirement that at infinite dilution the weak electrolyte is completely dissociated and has the limiting equivalent conductivity Λ^0. If the ion mobility is independent of concentration, Λ/Λ^0 gives the extent of dissociation at each concentration. Suppose the molality of the solution of the weak electrolyte is m. Then $m(1 - \Lambda/\Lambda^0)$ is the concentration of the undissociated solute and the ratio $m(\Lambda/\Lambda^0)^2[1 - (\Lambda/\Lambda^0)]^{-1} = K_\Lambda$ is a pseudo-equilibrium constant, since it is defined under the assumptions of constant ionic mobility and proportionality between the activity and the concentration of the ions, both of which are approximations. The true equilibrium constant for the weak electrolyte dissociation can be measured via construction of an electrochemical cell (see Section 26.2B), without appeal to conductivity measurements. In favorable cases, in the dilute solution regime, the corrections for solute nonideality and for the concentration dependence of the equivalent conductivity are small and tend to cancel so that the pseudo-equilibrium constant is very nearly equal to the true equilibrium constant. This is the case for typical carboxylic acids, such as acetic acid, and for many other weak electrolytes.

26.2 Cells, Chemical Reactions, and Activity Coefficients

All the colligative properties of solutions discussed in Chapter 25 can be used to study the properties of solutions of strong electrolytes. The major new property of electrolyte solutions, the fact that they conduct an electric current, can be used to derive electrical work from a chemical reaction. To describe the operation of electrochemical cells, we must extend somewhat our previous considerations.

Thus far we have described how a phase containing ions can be treated as if it were a phase containing only electrically neutral particles. The only significant difference between the analysis of electrolyte and nonelectrolyte solutions is the imposition of the electroneutrality condition

$$\sum_i n_i Z_i = 0. \qquad (26.15)$$

What happens if the electroneutrality condition is relaxed?

We have remarked many times that electrostatic interactions are very strong. The charge on the electron is 4.803×10^{-10} esu, or 1.602×10^{-19} C. Thus, 1 mol of charge is just $6.022 \times 10^{23} \times 1.602 \times 10^{-19} = 0.9648 \times 10^5$ C. This quantity of charge is known as the *Faraday,* and denoted $\mathscr{F}$. Suppose that a sphere consisting of 1 mol of matter deviates from electroneutrality by one part in 10^{10}. The excess charge Q in the sphere will collect on the surface, and in vacuum will generate a potential of magnitude $Q/4\pi\epsilon_0 R$ at a distance R from the center of the sphere. In our case, an amount of charge equal to $10^{-10}\,\mathscr{F}$ generates a potential of 0.87×10^6 V at a point 10 cm from the center of the sphere. This example shows that minuscule deviations from electroneutrality generate very large electric potentials and fields. In particular, a departure from electroneutrality that cannot be detected chemically leads to significant electrical effects: The mass balance and charge balance implied by Eq. 26.15 can be met to one part in 10^{15} and sufficient charge imbalance still exist to give potentials in the volt range.[1]

We conclude that it is meaningful to use as shorthand notation the statement that two phases can have the same chemical composition but different electrical potentials. It is true that the origin of the potential differences is charge imbalance, but the corresponding composition differences are not important chemically except insofar as they generate electrical potentials.

Given the above definition, we see that the chemical potentials of all solutes in a solution must depend on the electrical potential. The difference in chemical potential of component i between two phases of identical chemical composition is proportional to the charge on the ions of component i, and is independent of all other properties of the ion. For two phases, say I and II, we write

$$\mu_i^{\mathrm{II}} - \mu_i^{\mathrm{I}} = Z_i \mathscr{F}(\psi^{\mathrm{II}} - \psi^{\mathrm{I}}), \qquad (26.16)$$

[1] To measure their spectra, experiments have been carried out in which ions of a single charge are trapped in a low-temperature matrix. When the concentration of the trapped ion becomes too large, such matrices explode.

where the ψ's are the electrical potentials of the phases of identical chemical composition. The quantity $\mathscr{E} \equiv \psi^{\mathrm{II}} - \psi^{\mathrm{I}}$ is the electrical potential difference between the two phases.[2]

Following our previous analysis of phase equilibrium, one can clearly see that if the boundary between phases I and II is permeable to charge, at thermodynamic equilibrium $\psi^{\mathrm{I}} = \psi^{\mathrm{II}}$. On the other hand, if the boundary is *not* permeable to charge, it is *not* necessary that the electrical potentials of the two phases be equal. This observation about phase equilibrium suggests that, by suitable selection of the boundaries between phases, measurements of the electrical potential difference between phases can provide information about the equilibrium state. Finally, we note that only potential differences such as $\psi^{\mathrm{II}} - \psi^{\mathrm{I}}$ are measurable. Therefore, separation of the chemical potential into a "chemical" contribution and an "electrical" contribution, however useful for thinking, must be considered to be arbitrary. There is, in fact, one particular decomposition of the chemical potential in solution that simplifies the treatment of equilibria in electrochemical cells so much that it is extensively employed. We write formally

$$\mu_i = \mu_i^{\theta}(T,p) + RT \ln \gamma_i m_i + Z_i \mathscr{F}\psi. \qquad (26.17)$$

We shall use the form of Eq. 26.17, but again remind the reader that it is the chemical potential that is fundamental, that ψ cannot be measured, and that the right-hand side of Eq. 26.17, although eminently reasonable in form, is arbitrary in its subdivision into chemical and electrical terms.

A galvanic cell is a device, such as that displayed in Fig. 26.1, in which a chemical reaction can be made to result directly in the performance of electrical work. We seek to relate the voltage difference across the cell, $\psi^{\mathrm{VI}} - \psi^{\mathrm{I}}$, measured between two wires of the *same metal* attached to the two electrodes, to the properties of the chemical reaction in the cell. First, we note that the boundary between two metals is permeable to electrons, so that, when two different metals are placed in contact electrons flow until the chemical potential of the electron is the same in both metals. Although this results in a net excess or deficiency of electrons in each piece of metal, this concentration change is negligibly small. Similarly, when a Zn electrode is dipped into a solution containing ions of Zn^{2+}, the boundary between electrode and solution must be considered permeable to Zn^{2+}. Thus, Zn^{2+} ions dissolve from the electrode and enter the solution, or deposit onto the electrode from the solution, until the chemical potential of Zn^{2+} is the same in the two phases. Again, the charge transfer is sufficiently small to cause negligible change in the electron/ion ratio of the Zn electrode. In the Zn metal we have

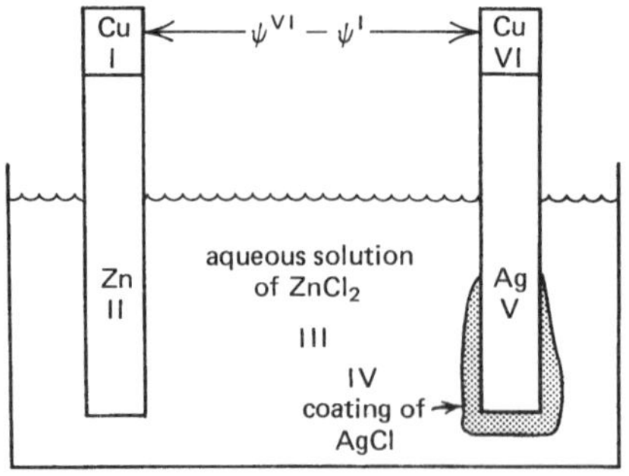

Figure 26.1 Schematic representation of a typical galvanic cell without liquid junctions.

$$2\mu_{e^-}^{\mathrm{II}} + \mu_{\mathrm{Zn}^{2+}}^{\mathrm{II}} = \mu_{\mathrm{Zn}}^{\mathrm{II}}, \qquad (26.18)$$

using the strong-electrolyte convention. That is, we regard the interior of the Zn metal as dissociated into Zn^{2+} ions and electrons. For the electrode-solution equilibrium we have

$$\mu_{\mathrm{Zn}^{2+}}^{\mathrm{II}} = \mu_{\mathrm{Zn}^{2+}}^{\mathrm{III}}. \qquad (26.19)$$

Combining Eqs. 26.18 and 26.19, we find that

$$\mu_{e^-}^{\mathrm{II}} = \tfrac{1}{2}(\mu_{\mathrm{Zn}}^{\mathrm{II}} - \mu_{\mathrm{Zn}^{2+}}^{\mathrm{III}}). \qquad (26.20)$$

This describes the "half-cell" of Zn plus $ZnCl_2$. Consider now the full cell of Fig. 26.1. At the Ag electrode the following equations are valid:

$$\mu_{\mathrm{Cl}^-}^{\mathrm{IV}} = \mu_{\mathrm{Cl}^-}^{\mathrm{III}}, \qquad (26.21)$$

$$\mu_{\mathrm{Ag}^+}^{\mathrm{V}} = \mu_{\mathrm{Ag}^+}^{\mathrm{IV}}, \qquad (26.22)$$

$$\mu_{\mathrm{Ag}^+}^{\mathrm{IV}} + \mu_{\mathrm{Cl}^-}^{\mathrm{IV}} = \mu_{\mathrm{AgCl}}^{\mathrm{IV}}, \qquad (26.23)$$

$$\mu_{\mathrm{Ag}^+}^{\mathrm{V}} + \mu_{e^-}^{\mathrm{V}} = \mu_{\mathrm{Ag}}^{\mathrm{V}}. \qquad (26.24)$$

These can be combined to yield

$$\mu_{e^-}^{\mathrm{V}} = \mu_{\mathrm{Ag}}^{\mathrm{V}} - \mu_{\mathrm{AgCl}}^{\mathrm{IV}} + \mu_{\mathrm{Cl}^-}^{\mathrm{III}}. \qquad (26.25)$$

Finally, combining Eqs. 26.20 and 26.25, we obtain

$$\begin{aligned}\mu_{e^-}^{\mathrm{V}} - \mu_{e^-}^{\mathrm{II}} &= \mu_{\mathrm{Ag}}^{\mathrm{V}} - \mu_{\mathrm{AgCl}}^{\mathrm{IV}} + \mu_{\mathrm{Cl}^-}^{\mathrm{III}} - \tfrac{1}{2}\mu_{\mathrm{Zn}}^{\mathrm{II}} + \tfrac{1}{2}\mu_{\mathrm{Zn}^{2+}}^{\mathrm{III}}\\ &= \mu_{\mathrm{Ag}}^{\mathrm{V}} + \tfrac{1}{2}\mu_{\mathrm{ZnCl}_2}^{\mathrm{III}} - \tfrac{1}{2}\mu_{\mathrm{Zn}}^{\mathrm{II}} - \mu_{\mathrm{AgCl}}^{\mathrm{IV}}. \end{aligned} \qquad (26.26)$$

Clearly, the last line of Eq. 26.26 shows that the difference in electron chemical potential between phases II and V is determined by the Gibbs free energy change in the reaction

$$\tfrac{1}{2}\mathrm{Zn(II)} + \mathrm{AgCl(IV)} = \tfrac{1}{2}\mathrm{Zn}^{2+}\mathrm{(III)} + \mathrm{Cl}^-\mathrm{(III)} + \mathrm{Ag(V)}. \qquad (26.27)$$

[2] We note that the ions of opposite charge in an electrolyte solution need not have the same solubilities in the liquid and solid phases of the solvent, so that when an electrolyte solution freezes there can be an electrostatic potential built up at the interface between the solid and liquid phases (the freezing potential).

Each of the phase boundaries in the cell under consideration is only semipermeable: I–II to electrons but not ions, II–III to Zn^{2+} but not Cl^-, III–IV to Cl^- but not Zn^{2+} or Ag^+, IV–V to Ag^+ but not Cl^-, and V–VI to electrons but not ions. None of the potential differences I–II, . . . , V–VI is measurable, because the phases have different chemical compositions. However, the potential $\psi^{VI} - \psi^{I}$ is between two Cu wires and hence measurable.

Let us derive the chemical information contained in Eq. 26.27 by working backwards. Consider what happens if we connect the two Cu wires through some device that can do work, such as a motor. Suppose that a current is made to flow through the cell and motor. At the Zn-solution boundary there occurs the reaction

$$\mathrm{Zn(II)} \rightarrow \mathrm{Zn}^{2+}\,\mathrm{(III)} + 2e^-, \tag{26.28}$$

and the resulting electrons move from the Zn in phase II to the Cu in phase I, through the motor to the Cu in phase VI, and then to the Ag of phase V. The two extra electrons then react with AgCl to form Ag,

$$2e^-\,\mathrm{(V)} + 2\mathrm{AgCl(IV)} \rightarrow 2\mathrm{Ag(V)} + 2\mathrm{Cl}^-\mathrm{(III)}, \tag{26.29}$$

so that the net change is just

$$\mathrm{Zn(II)} + 2\mathrm{AgCl(IV)} \rightarrow \mathrm{Zn}^{2+}\mathrm{(III)} + 2\mathrm{Cl}^-\mathrm{(III)} + 2\mathrm{Ag(V)}, \tag{26.30}$$

which is the same as Eq. 26.27 written for the transfer of two electrons.

Finally, let us use the representation Eq. 26.16 of the chemical potential to derive a relationship between the cell reaction and the electrical potential. Noting that the chemical potentials of the electrons in phases I and VI satisfy the equations

$$\mu_{e^-}^{\mathrm{I}} = \mu_{e^-}^{\mathrm{II}}, \tag{26.31}$$

$$\mu_{e^-}^{\mathrm{VI}} = \mu_{e^-}^{\mathrm{V}}, \tag{26.32}$$

and using Eq. 26.16, we find that

$$\psi^{\mathrm{IV}} - \psi^{\mathrm{I}} = -\frac{1}{\mathcal{F}}(\mu_{e^-}^{\mathrm{V}} - \mu_{e^-}^{\mathrm{II}}), \tag{26.33}$$

since for electrons $Z = -1$. With Eqs. 26.19, 26.21, and 26.22, this becomes

$$\mathcal{E} = \psi^{\mathrm{VI}} - \psi^{\mathrm{I}} - \frac{1}{\mathcal{F}}(\mu_{e^-}^{\mathrm{V}} + \mu_{\mathrm{Ag}^+}^{\mathrm{V}}) + \frac{1}{\mathcal{F}}(\mu_{\mathrm{Ag}^+}^{\mathrm{IV}} + \mu_{\mathrm{Cl}^-}^{\mathrm{V}})$$
$$-\frac{1}{\mathcal{F}}\left(\mu_{\mathrm{Cl}^-}^{\mathrm{III}} + \frac{1}{2}\mu_{\mathrm{Zn}^{2+}}^{\mathrm{III}}\right) + \frac{1}{\mathcal{F}}\left(\frac{1}{2}\mu_{\mathrm{Zn}^{2+}}^{\mathrm{II}} + \mu_{e^-}^{\mathrm{II}}\right). \tag{26.34}$$

Only the third term on the right-hand side of Eq. 26.34 refers to the solution phase. All the other terms can be lumped into a constant, say, $\mathcal{E}'$. Then we have

$$\mathcal{E} = \mathcal{E}' - \frac{1}{2\mathcal{F}}(2\mu_{\mathrm{Cl}^-}^{\mathrm{III}} + \mu_{\mathrm{Zn}^{2+}}^{\mathrm{III}})$$
$$= \mathcal{E}' - \frac{1}{2\mathcal{F}}\mu_{\mathrm{ZnCl}_2}^{\theta} - \frac{RT}{2\mathcal{F}}\ln(a_{\mathrm{Zn}^{2+}}a_{\mathrm{Cl}^-}^2). \tag{26.35}$$

The term $\mathcal{E}'$ is nothing but

$$\mathcal{E}' = -\frac{1}{\mathcal{F}}(\mu_{\mathrm{Ag}}^{\mathrm{V}} - \mu_{\mathrm{AgCl}}^{\mathrm{IV}} - \frac{1}{2}\mu_{\mathrm{Zn}}^{\mathrm{II}}), \tag{26.36}$$

so that

$$\mathcal{E}^{\theta} \equiv \mathcal{E}' - \frac{\mu_{\mathrm{ZnCl}_2}^{\theta}}{2\mathcal{F}}$$
$$= -\frac{1}{2\mathcal{F}}(2\mu_{\mathrm{Ag}}^0 - 2\mu_{\mathrm{AgCl}}^0 - \mu_{\mathrm{Zn}}^0 + \mu_{\mathrm{ZnCl}_2}^{\theta}), \tag{26.37}$$

where we have used the fact that the standard-state chemical potential of the Zn is just that of the solid, and so on. Note that, in the cell considered, $ZnCl_2$ is in solution and all other chemical components are in the solid state. The quantity $\mathcal{E}^{\theta}$ defined in Eq. 26.37 is clearly $-\Delta G^{\theta}/2\mathcal{F}$ for the reaction 26.30. Thus, the electrical potential of the cell is

$$\mathcal{E} = \mathcal{E}^{\theta} - \frac{RT}{2\mathcal{F}}\ln(a_{\mathrm{Zn}^{2+}}a_{\mathrm{Cl}^-}^2). \tag{26.38}$$

Equation 26.38 is known as the *Nernst equation.*

In general, if the reaction

$$a\mathrm{A} + b\mathrm{B} + \cdots = x\mathrm{X} + y\mathrm{Y} + \cdots \tag{26.39}$$

can be made to occur in a cell that has no interfaces where two solutions of different composition are in contact, the cell potential is

$$\mathcal{E} = \mathcal{E}^{\theta} - \frac{RT}{z\mathcal{F}}\ln\frac{a_{\mathrm{X}}^x a_{\mathrm{Y}}^y \cdots}{a_{\mathrm{A}}^a a_{\mathrm{B}}^b \cdots}, \tag{26.40}$$

where z is the number of electrons transferred in the chemical reaction. Following the procedure used to define the reference state for a component in solution, it is conventional to set equal to unity the activity of a pure solid or a pure liquid. Thus, if pure solids or liquids are either reactants or products in the cell reaction, their activities do not appear in the second term on the right-hand side of Eq. 26.40.

Having deduced the relationships among cell potential, activity, and chemical equilibrium, we can use suitable cells to determine free energies of reaction, activity coefficients, and so on. We shall give a few examples here, leaving others for the problems. In all of the following we shall use the convention that the symbol | denotes a solid/solution interface. Thus, the cell we have discussed would be represented by

$$\mathrm{Cu, Zn}\,|\mathrm{ZnCl}_2\,(aq)|\,\mathrm{AgCl, Ag, Cu.}$$

The usual convention is that the cell potential is positive if the left-hand electrode (as written) supplies electrons to the outside circuit when the cell acts as a source of energy. With this convention, when the cell potential is positive the left-hand electrode is negative.

A. Determination of Solute Activities

Consider the cell

$$\text{Cu, Pt, } H_2(g)\ |HX(aq)|\ \text{AgX, Ag, Cu},$$

where X represents any of the halogens. The Pt electrode is usually coated with colloidal Pt so as to catalyze the reaction

$$\tfrac{1}{2}H_2(g) = H^+ + e^-.$$

This is the reaction at the left-hand electrode, which serves as a source of electrons for the outside circuit. At the right-hand electrode the reaction is

$$e^- + \text{AgX} = \text{Ag} + X^-,$$

which serves as a sink for electrons. The overall chemical reaction in the cell is

$$\tfrac{1}{2}H_2(g) + \text{AgX} = \text{Ag} + H^+ + X^-.$$

Thus, the cell potential (electromotive force) is, from Eq. 26.40,

$$\begin{aligned}\mathscr{E} &= \mathscr{E}^\ominus - \frac{RT}{\mathscr{F}}\ln\frac{a_\pm^2}{a_{H_2}^{1/2}}\\ &= \mathscr{E}^\ominus - \frac{RT}{2\mathscr{F}}\ln a_{H_2} - \frac{2RT}{\mathscr{F}}(\ln\gamma_\pm + \ln m_\pm).\end{aligned} \tag{26.41}$$

In this reaction H_2 is gaseous. Examination of the mathematical form of the chemical potential of a real gas, Eq. 21.39, shows that we may identify $f_i/\mathfrak{p}$ with the activity. In general, when gases are involved in the cell reaction, their activities are $a_i = f_i/\mathfrak{p}$.

The constant $\mathscr{E}^\ominus$ can be deduced from measurements of $\mathscr{E}$ versus m and a_{H_2} by the extrapolation

$$\mathscr{E}^\ominus = \lim_{m_\pm\to 0}\left(\mathscr{E} - \frac{RT}{2\mathscr{F}}\ln a_{H_2} + \frac{2RT}{\mathscr{F}}\ln m_\pm\right). \tag{26.42}$$

Of course, the activity of gaseous H_2 is determined separately. Once $\mathscr{E}^\ominus$ and a_{H_2} are known, measurements of $\mathscr{E}$ for given concentrations provide a direct means of evaluating the solute activity coefficients. Measurements of this type have been made for many electrolyte solutions. The method is limited only by the availability of suitable reversible electrodes.

Instead of using a chemical reaction, we can generate a potential from the free energy of dilution, if matters are arranged properly. Consider the cell

$$\text{Ag, AgCl}\ |KCl(m_1)|K,\ \text{Hg(amalgam)}|KCl(m_2)|\text{AgCl, Ag}.$$

In this cell the two solution phases have different concentrations. They are separated from one another by a solution of K in Hg that will permit transfer of K^+ but not of Cl^-. At the left-hand electrode the reaction is

$$\text{Ag}(s) + Cl^-(aq, m_1) + K^+(aq, m_1) = \text{AgCl}(s) + \text{K(in Hg)}.$$

At the right-hand electrode we have

$$\text{AgCl}(s) + \text{K(in Hg)} = \text{Ag}(s) + Cl^-(aq, m_2) + K^+(aq, m_2),$$

which is the same reaction in reverse, but creates ions at a different concentration. The net "reaction" is[3]

$$K^+(aq, m_1) + Cl^-(aq, m_1) = K^+(aq, m_2) + Cl^-(aq, m_2).$$

Since there is no chemical change, we have $\mathscr{E}^\ominus = 0$, and Eq. 26.40 gives

$$\mathscr{E} = -\frac{2RT}{\mathscr{F}}\ln\frac{a_\pm(m_2)}{a_\pm(m_1)}. \tag{26.43}$$

If the concentration m_1 is held fixed while m_2 is varied, we can calculate $a_\pm(m_1)$ from

$$\ln a_\pm(m_1) = \lim_{m_\pm\to 0}\left(\frac{\mathscr{F}\mathscr{E}}{2RT} + \ln m_2\right). \tag{26.44}$$

A cell of the type just considered is called a *concentration cell.* Another example of a concentration cell is shown in Fig. 26.2.

B. Determination of an Equilibrium Constant

Consider the reaction

$$H_2O = H^+ + OH^-,$$

for which the equilibrium constant is

$$K_w \equiv \frac{a_{H^+}a_{OH^-}}{a_{H_2O}}. \tag{26.45}$$

---

[3] Alternatively, we can write the following set of reactions: Left-hand electrode: $Ag(s) + Cl^-(aq, m_1) \to AgCl + e^-$; Right-hand electrode: $AgCl(s) + e^- \to Ag(s) + Cl^-(aq, m_2)$; K/Hg interface: $K^+(aq, m_1) + K(\text{in Hg}) \to K(\text{in Hg}) + K^+(aq, m_2)$; Net reaction: $K^+(aq, m_1) + Cl^-(aq, m_1) \to K^+(aq, m_2) + Cl^-(aq, m_2)$.

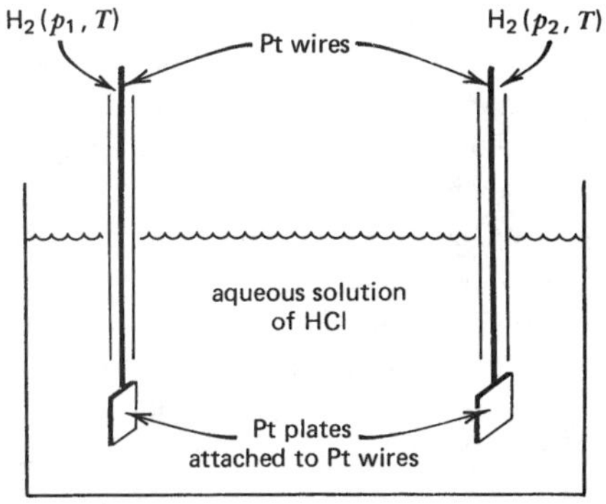

Figure 26.2 Schematic representation of a concentration cell. The hydrogen electrodes operate under different pressures. If oxidation occurs at the left electrode the reaction is

$$H_2(p_1) \rightarrow H_2(p_2)$$

and the cell potential is

$$\mathscr{E} = -\frac{RT}{2\mathscr{F}} \ln \frac{p_2}{p_1}.$$

We wish to construct a cell with which K_w can be measured. In the cell

$$\text{Pt, } H_2(g)\ |\text{NaOH}(m_1), \text{NaCl}(m_2)|\ \text{AgCl, Ag, Pt}$$

the cell reaction is

$$\tfrac{1}{2}H_2(g) + \text{AgCl} = \text{Ag} + H^+ + Cl^-,$$

although the H^+ concentration is controlled by the supplementary equilibrium 26.45. In this cell NaOH is deliberately added so as to vary m_{OH^-}. Suppose that conditions are arranged so that $a_{H_2} = 1$. Then Eq. 26.40 gives

$$\mathscr{E} = \mathscr{E}^\ominus - \frac{RT}{\mathscr{F}} \ln a_{H^+} a_{Cl^-} \qquad (26.46)$$

or

$$\mathscr{E} = \mathscr{E}^\ominus - \frac{RT}{\mathscr{F}} \ln K_w \cdot \frac{a_{H_2O} a_{Cl^-}}{a_{OH^-}}, \qquad (26.47)$$

where we have used Eq. 26.45 to eliminate a_{H^+} in Eq. 26.46. Thus, K_w can be evaluated from

$$\ln K_w = \lim_{\substack{m_{OH^-} \to 0 \\ m_{Cl^-} \to 0}} \left[\frac{\mathscr{F}}{RT}(\mathscr{E}^\ominus - \mathscr{E}) - \ln \frac{m_{Cl^-}}{m_{OH^-}} \right]. \qquad (26.48)$$

By making measurements with $(m_{Cl^-}/m_{OH^-}) = 1$, the last term in Eq. 26.48 can be dropped. It is found that $K_w = 1.008 \times 10^{-14}$ at 25°C and 1 atm.

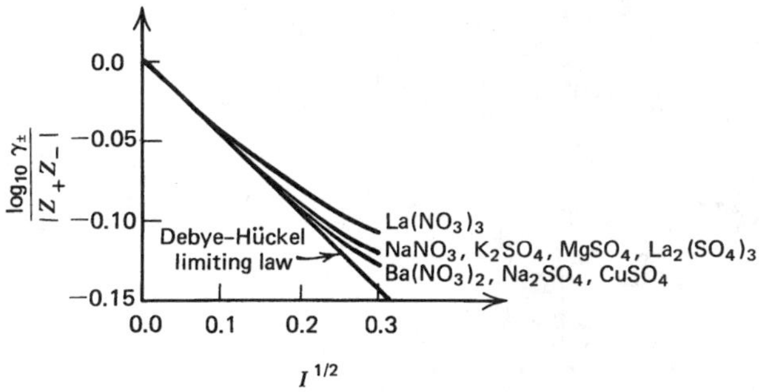

Figure 26.3 Schematic representation of the variation of the mean activity coefficients of some electrolytes with ionic strength.

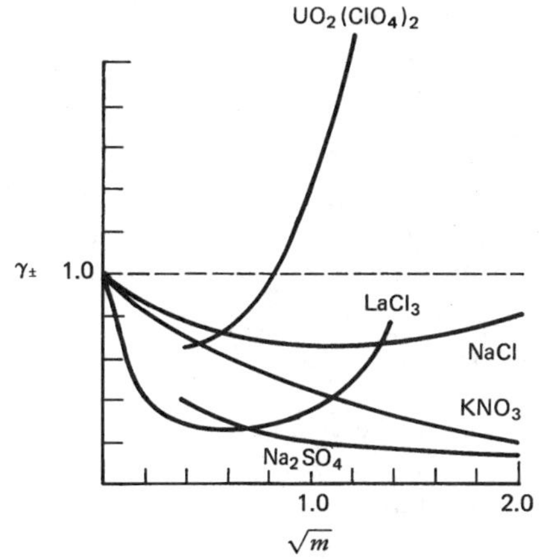

Figure 26.4 Schematic representation of the variation of the mean activity coefficients of several electrolytes at high concentration.

We close this section with a brief description of what has been learned about the thermodynamic properties of electrolyte solutions. All of the techniques described, as well as several others, have been used to study electrolyte solutions. Figures 26.3, 26.4, and 26.5 show the variation with concentration of the activity coefficients of a few electrolytes. We note that:

1. In dilute solution the activity coefficient decreases with increasing concentration.

2. In very dilute solution $-\ln \gamma_\pm$ is proportional to the square root of the ionic concentration (see Fig. 26.3). More specifically, $-\ln \gamma_\pm$ is proportional to the square root of the *ionic strength I*, defined by

$$I \equiv \tfrac{1}{2} \sum_i Z_i^2 m_i, \qquad (26.49)$$

which weights the contribution of each ionic species by the square of its charge.

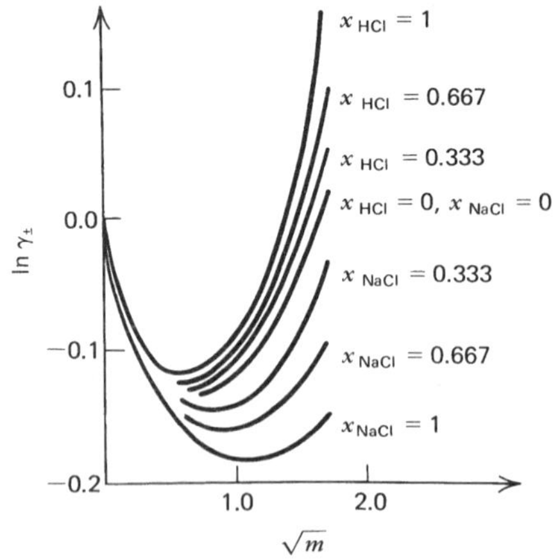

Figure 26.5 Schematic representation of the mean activity coefficients of HCl and NaCl in mixed electrolyte solutions. Along a vertical line m is fixed, but the proportions of NaCl and HCl vary as shown.

3. Although the excess free energy of the electrolyte solution is proportional to the square root of the ionic strength in very dilute solutions, the concentration range over which that proportionality is valid is very small (see Fig. 26.4), and among several electrolytes is smallest for the highest-valence types.

4. For some electrolytes the activity coefficient plotted against concentration shows a minimum, and at high solute concentration the value of $\gamma_\pm$ may be very large. For example, in uranyl perchlorate we have $\gamma_\pm = 1510$ at 5.5 m. An extreme example in the opposite direction is CdI_2, for which $\gamma_\pm = 0.0168$ at 2.5 m.

5. For unsymmetrical electrolytes part of the deviation of the activity coefficient from the limiting law behavior arises from incomplete inclusion of the Coulomb interactions which lead to higher order terms such as $I \ln I$, which are not always negligible relative to the $I^{1/2}$ even in very dilute solutions. Nevertheless, the preceding observations suggest that non-Coulombic interactions are also important in electrolyte solutions, and that they become dominant when the solute concentration is large. Thus, very low activity coefficients can often be interpreted in terms of the formation of complex ions, new species in which non-Coulombic binding plays an important role. On the other hand, very large activity coefficients can often be interpreted in terms of short-range ion–ion repulsion, much as in a fluid of rigid spheres. Intermediate cases, that is, moderately low activity coefficients, are sometimes interpreted as arising from ion–pair formation. By ion–pair formation is meant the existence of an equilibrium of the type

$$A^+ + B^- = (A^+ \cdots B^-)$$

and sometimes higher multiple ions,

$$(B^- \cdots A^+ \cdots B^-),\ (A^+ \cdots B^- \cdots A^+),$$

and so on. In a few instances there is spectroscopic evidence for such ion pairing, but often the concept is used as a cloak for our ignorance of the true interactions. An example of short-range non-Coulombic contributions to the excess free energy is displayed in Fig. 26.5. Were there no non-Coulombic interactions, all the curves in Fig. 26.5 would fall on top of one another.

6. Electrolytes with polyvalent cations usually have much larger activity coefficients than electrolytes of analogous valency type containing a polyvalent anion. Lanthanum chloride and potassium ferricyanide are a contrasting pair of this type. If, as discussed later in this chapter, polyvalent cations form complexes with water and thereby are "large charged hard spheres," whereas polyvalent anions do not form such complexes, the observation can be rationalized, for the repulsion between hard-sphere ions increases the activity, the more so the larger the radius of the hard sphere.

7. Numerous empirical formulas have been used to describe the concentration dependence of the activity coefficient. All have in common:

(a) An approach to the limiting form

$$\lim_{m \to 0} \ln \gamma_\pm = -AI^{1/2} = -Bm^{1/2}. \tag{26.50}$$

We shall see, in Section 26.5, that this limiting law is the form predicted by the theory of Debye and Hückel and by more formal statistical mechanical analysis.

(b) The addition of some representation of specific ion interactions to the limiting-law expression. Non-Coulombic short-range interactions lead to changes in the shielded Coulomb contribution to $\ln \gamma_\pm$, and also to "nonelectrolyte"-type contributions that depend on m, m^2, and so on, but not on $m^{1/2}$.

(c) At higher concentrations there are Coulombic contributions to the excess free energy that depend on $m \ln m$, m, $m^{3/2}$, and products of these functions of the solute concentration.

8. The temperature and pressure derivatives of $\ln \gamma_\pm$ also confirm the observations and deductions listed above.

26.3 Comments on the Structure of Water

In Chapter 23 we very briefly characterized the structure of liquid water as revealed by x-ray and neutron diffraction studies. The data available are consistent with a picture in which each water molecule is surrounded by four others, to which it is hydrogen-bonded, with approximately tetrahedral coordination. There is, of course, dispersion in the distribution of molecular separations, and also in the distribution of relative molecular orientations. Indeed, translational ordering extends only about two molecular diameters from a given molecule, and orientational ordering does not extend beyond the first shell of neighbors.

As of the date of this text, despite intensive study, our understanding of how the molecular interactions determine the structure and properties of water remains incomplete. One of the difficulties encountered in generating a fully quantitative molecular description of water is that the representation of the potential energy of the liquid as a sum of pair interactions is much less accurate for water than for, say, Ar. There is considerable evidence, from quantum mechanical calculations, that some configurations of isolated groups of water molecules have energies significantly lower than determined by the sum of the (isolated) pair interactions of the molecules,[4] and it is likely that configurations resembling these are present in the liquid. Nevertheless, it has proven possible to define several different empirical *effective* pair potentials to represent the interaction between water molecules in the liquid. These effective pair potentials partially account for the contributions arising from the excess multimolecule interactions (see Eq. 21.159), but in an unknown fashion; they are inherently density and temperature dependent. A second difficulty is attributable to the strength and directional character of the effective pair interaction between water molecules. At room temperature, at the typical spacing of water molecules in the liquid, the molecular interaction is dominated by hydrogen bonding with an energy of about 10 k_BT per bond, and this dominant water–water interaction saturates when four other molecules surround the central molecule. The latter difficulty is sufficiently great that none of the integral equation descriptions of liquid structure outlined in Chapter 23 provides a satisfactory representation of the structure of liquid water. Finally, there is much evidence that the translational, librational, and vibrational motions of liquid water molecules cannot be quantitatively described without use of quantum mechanical methods, so that the use of classical mechanics in a simulation introduces some error.

Despite the difficulties cited, both Monte Carlo and molecular dynamics simulations of liquid water with the several proposed effective potentials have yielded valuable information. It is important to note that the majority of the effective potentials used in the simulations assume the water molecule is a rigid body with a specified charge distribution; the interaction between water molecules is constructed as the sum of a central potential, representing the van der Waals type interaction, and an electrostatic interaction that generates the directional hydrogen bonding. These effective potentials are not equally accurate for all possible configurations of a pair of molecules (or of larger groups of molecules). For example, none of the available empirical effective water–water potentials accounts for both the densities and the molecular packing in the structures of the proton ordered ices II, VII, and IX;[5] an ab initio pair potential derived from electronic structure calculations succeeds better than any of the empirical effective pair potentials in predicting the hydrogen bond topology and the OOO angles characteristic of the various ice lattices but consistently underestimates the ice densities by about 21%. Since the ab initio pair potential is free of any contributions from many molecule interactions, we infer that the many-molecule component of the potential energy of condensed phases of water acts, in zero-th order, to reduce all lengths in the system by a uniform factor ~1.07. Presumably, this is the kind of multimolecule interaction effect that is captured in the construction of an empirical effective pair potential between water molecules.

It is found that simulations using the best of the effective potentials generate most of the structural features of water that are inferred from the x-ray and neutron diffraction experiments. The structure of water found in all of the simulations exhibits near tetrahedral local ordering along with considerable dispersion in the molecular separations and relative orientations. Given that the range of separations over which hydrogen bonding is effective in water is about 2.7–3.0 Å, it is also possible to trace a continuous network of hydrogen bonds among the molecules in an instantaneous configuration of molecules in the liquid. Because of the inherent positional and orientational disorder we shall refer to this kind of connectivity as a continuous random network. Clearly, the hydrogen bonds that make up the network are not all equally strong—in addition to the obvious dependence on the O · · · O separation, the further the OH · · · O angle deviates from 180°, the weaker the hydrogen bond. It is probably only a semantic distinction to assert that there are broken and unbroken hydrogen bonds, since there is a continuous range of variation in the hydrogen-bond energy with O · · · O separation and OH · · · O angle. Another way of stating this conclusion about the hydrogen bonding is that the interaction of water molecules is described by a continuous potential energy surface, which does not provide a natural distinction between "broken" and "intact" hydrogen bonds; rather, we expect there to be a smooth distribution of hydrogen bond strength as a function of O · · · O separation and OH · · · O angle. Nevertheless, it is useful for some purposes to adopt a point of view that categorizes the molecules according to the numbers of hydrogen bonds they form. If a threshold energy for the existence of a hydrogen bond is adopted, the number of hydrogen bonds per molecule is a function of that threshold value; this number is small when the threshold has a value corresponding to very strong bonding and large when the threshold corresponds to weak bonding. For any reasonable value of the threshold, computer simulation studies of water show that there is a continuous network of hydrogen bonding even when not all water molecules are hydrogen bonded to four other molecules.[6] A molecular dynamics study of both the angular and distance dependences of the pair-correlation function in water, $g_{OO}(R, \Omega)$, supports the existence of a continuous distorted fully

[4] J. E. Del Bene and J. A. Pople, *J. Chem. Phys.* **58,** 3605 (1973).
[5] M. D. Morse and S. A. Rice, *J. Chem. Phys.* **76,** 650 (1982).

[6] A. Geiger and P. Mausbach, in *Hydrogen Bonded Liquids,* eds. J. C. Dore and J. Teixeira, Kluwer Academic Publishers, Dordrecht (1991), p. 171.

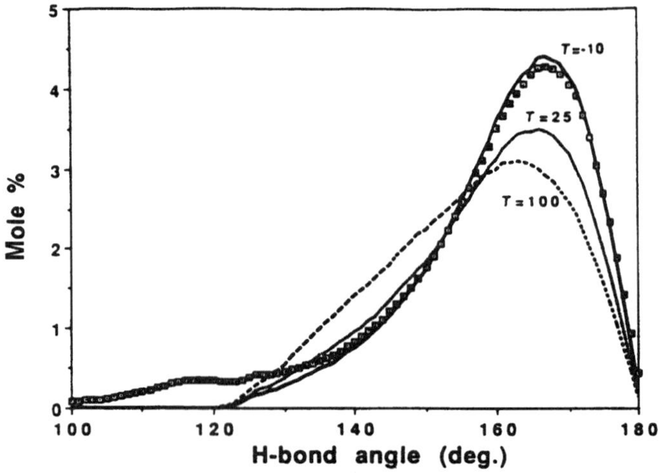

Figure 26.6 The distribution of hydrogen bond angles of tetrahedrally coordinated water molecules in a computer simulated model of water. The solid, dashed, and dotted lines are the distributions at −10, 25, and 100°C, respectively. In this model, in addition to the tetrahedrally coordinated water molecules, there are some interstitial water molecules. The squares show the distribution of hydrogen bond angles for all hydrogen bonded pairs of molecules at −10°C. From I. M. Svischev and P. G. Kusalik, *J. Chem. Phys.* **99,** 3049 (1993).

hydrogen bonded tetrahedral network of molecules in coexistence with additional interstitial molecules.[7] Figure 26.6 displays the distribution of hydrogen bond angles in the tetrahedral network component, and the distribution of hydrogen bond angles for all hydrogen bonded pairs of molecules; the small excess number of hydrogen bond angles in the range 100° to 120° corresponds to the interstitial molecules. The tetrahedral network component is found to have a nearest neighbor coordination number, 4, that is independent of temperature, while the concentration of interstitial molecules increases with increasing temperature. The interstitial molecules interact strongly with the central water molecule, the average energy of interaction being only slightly smaller in magnitude than the energy of the typical hydrogen bond. We note that many of the models proposed for water assume that the liquid can be considered to be a mixture of two components, the concentrations of which vary with temperature and density; it is usually assumed that one of the components is fully hydrogen bonded with a local structure like that of ice Ih, and that the other component is only partially hydrogen bonded. Because our intent is to convey how the concepts used in a model of water can be employed in the interpretation of the structure and properties of aqueous solutions we adopt, as a vehicle, the random network model of water, to which we now turn.

As implied by the name, and as inferred from molecular configurations generated in computer simulations, the random network model[8] assumes that liquid water retains a continuous distorted network of hydrogen bonded water molecules. In the immediate vicinity of a water molecule the temporary local order resembles that in ice Ih, but at greater distance from the central molecule the arrangement and connectivity can be quite different from that in ice Ih because of the stretching and bending of the hydrogen bonds between water molecules. A key aspect of the analysis is the assumption that the thermodynamic and structural properties of water can be thought of as having two major contributions, namely, one arising from vibrations about temporary positions in the locally ordered structure and the other arising from the arrangement and deformations of the hydrogen bonds that define the temporary local order. These contributions are not completely independent, since the internal vibrations of a water molecule are influenced by the strength of its hydrogen bonding to neighbors, as are its hindered translational and librational motions. The characteristic features of the hydrogen bond network are represented by three structural parameters: the mean and standard deviation of the hydrogen bond length, and the root mean square hydrogen bond angle. The values of these parameters are obtained self consistently at any given temperature by minimizing the total free energy with respect to their variations. The effects on the vibrational spectrum of the liquid are also accounted for using the same parameters.

The random network model generates a very good (but not perfect) description of the distance dependence of the oxygen–oxygen pair-correlation function, $g_{OO}(R)$; the positions and widths of the several peaks in the pair-correlation function are defined by the allowed large fluctuations in hydrogen bond length and angle (Fig. 26.7*a*).[9] The model also provides a good account of the free energy, enthalpy and entropy of water over a considerable temperature range (Fig. 26.7*b*). However, the model does not account for the change in sign of the isothermal compressibility in water at 4°C (corresponding to a maximum in the liquid density at that temperature), nor for the pronounced decrease in the

---

[7] I. M. Svishchev and P. G. Kusalik, *J. Chem. Phys.* **99,** 3049 (1993).

[8] A simple version of the random network model was first introduced by J. A. Pople (Proc. Roy. Soc. (London) **A205,** 163 (1951)) and qualitatively devloped by G. S. Kell (Chapter 9 in *Water and Aqueous Solutions: Structure, Thermodynamics and Transport Properties,* ed. R. A. Horne, Wiley (1972)) and by B. Kamb (p. 507 in *Structural Chemistry and Molecular Biology,* eds. A. Rich and N. Davidson, Freeman, San Francisco (1968)). M. G. Sceats and S. A. Rice greatly extended the conceptual basis of the random network model and developed applications to the calculation of the structural, spectral and thermodynamic properties of liquid water and the many ice phases (for a review see Chapter 2, Volume 7 of *Water: A Comprehensive Treatise,* ed. F. Franks, Plenum Press, New York (1982)). A. R. Henn and W. Kauzmann (*J. Phys. Chem.* **93,** 3770 (1989)) extended the Sceats–Rice analysis to include volume as an independent variable.

[9] The average nearest neighbor coordination number in liquid water is larger than 4; it is defined by integration of the pair-correlation function over the first peak, and has the typical value 4.4–4.5. There is no direct evidence which establishes that this coordination number reflects fluctuations in the distribution of second, third, . . . neighbor water molecules, or the presence of nontetrahedrally bonded water molecules in a distribution of local environments. The latter interpretation is implied by the results of the simulations cited in footnote 5.

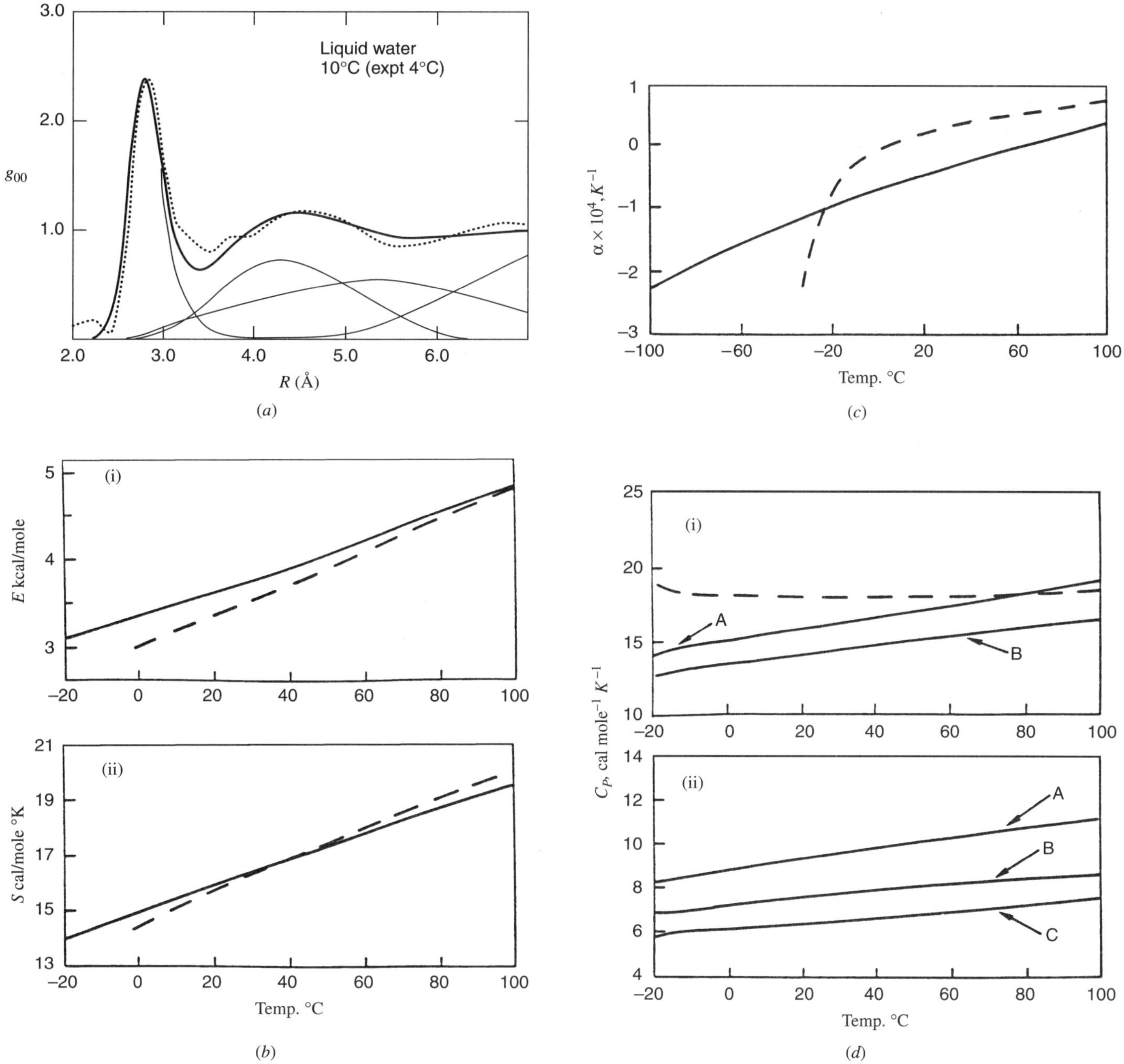

Figure 26.7 (*a*) The OO pair-correlation function of water at 10°C predicted by the Random Network Model (heavy solid line) compared with the observed OO pair-correlation function at 4°C (dots). The light solid lines show the contributions to the structure of the OO pair-correlation function from hydrogen bonded molecules in the first, second, and third shells of water molecules. From M. G. Sceats, M. Stavola, and S. A. Rice, *J. Chem. Phys.* **70,** 3927 (1979). (*b*) Comparison of the temperature dependence of the energy (i) and entropy (ii) of water at zero pressure predicted by the Random Network Model (solid lines) and that observed (dashed lines). From A. R. Henn and W. Kauzmann, *J. Phys. Chem.* **93,** 3770 (1989). (*c*) Comparison of the temperature dependence of the coefficient of thermal expansion of water at zero pressure predicted by the random network model (solid line) and that observed (dashed line). From A. R. Henn and W. Kauzmann, *J. Phys. Chem.* **93,** 3770 (1989). (*d*) Comparison of the constant pressure heat capacity of water at zero pressure predicted by the random network model (solid lines) and that observed (dashed lines). (i) Curve A, $C_p = -T^2(\partial^2 A/\partial T^2)_p$; curve B, $C_p = (\partial U/\partial T)_p$. (ii) The vibrational and configurational contributions to the constant pressure heat capacity. Curve A, $C_{p,\text{vib}} = -T^2(\partial^2 A_{\text{vib}}/\partial T^2)_p$; curve B, $C_{p,\text{vib}} = (\partial U_{\text{vib}}/\partial T)_p$; curve C, $C_{p,\text{config}} = (\partial U_{\text{config}}/\partial T)_p$. From A. R. Henn and W. Kauzmann, *J. Phys. Chem.* **93,** 3770 (1989).

isothermal compressibility in the supercooled liquid region; it does predict a change in sign of the isothermal compressibility at 58°C (Fig. 26.7*c*). Also, the calculated heat capacity is much less accurate than the calculated energy and entropy and does not exhibit the experimentally observed dramatic increase in the supercooled liquid region (Fig. 26.7*d*). On balance, the random network model of water provides a useful set of concepts with which to frame a discussion of the properties of aqueous solutions. From this point of view, its most important attribute is that it formally incorporates the notion, derived from the results of computer simulations, that there are very many environments generated by local differences in hydrogen bonding. The observed range of variation can be taken as indicative of the extent to which a continuous hydrogen-bonded network can be deformed and still retain connectivity. This characteristic of the network of hydrogen bonding must in some way influence the interaction between water and solute molecules, hence also the thermodynamic properties of aqueous solutions. It is this subject that we now examine.

26.4 The Influence of Solutes on the Structure of Water

When a substance is dissolved in water, the arrangement of molecules near the solute is different from that far away. Of course, the alteration of local structure induced by the solute depends on the nature and strength of its interaction with water. We have shown that in a simple liquid, whether pure or mixed, the repulsive interaction between molecules plays the dominant role in determining their spatial distribution (see Chapters 23 and 25). Although the repulsive force between water molecules does not play so direct a role in determining the structure, that structure is influenced locally by the size and shape of a solute molecule. Moreover, if the solute is charged, the combination of a strong electric field acting on the water molecules with the difference in sizes and shapes of the solute and water molecules can produce a striking difference between the local and distant arrangements of the water molecules in the liquid. In some cases the solute ion and its near-neighbor shell of water molecules can be considered to be a single chemical species, for example, the hexaaquo complex of Ni^{++}, which has a lifetime for water exchange of 30 μs. The electric field of the ion must also generate a subtle change in the liquid water structure that extends some distance from the ion, since that field induces a density change in the surrounding liquid (electrostriction) that decreases slowly with increasing distance from the ion.

To examine how solutes influence the structure of water, we first consider the properties of aqueous solutions of gases. The thermodynamic data for such systems are usually derived from a study of the temperature dependence of the Henry's law constant (see Section 25.4). Suppose that an infinitesimal quantity of substance is transferred from the gas to the solution phase. Accompanying this transfer are changes in enthalpy, entropy, and volume, which we denote $\Delta\bar{h}_2$, $\Delta\bar{s}_2$, and $\Delta\bar{v}_2$, respectively. For the dilute solutions we consider the solute–solute interaction is negligibly small and the derived thermodynamic functions are properly interpreted solely in terms of the effects of solute–solvent interaction. The natural reference state is then the hypothetical pure substance which has the same properties as the solute in an infinitely dilute solution.

We list in Table 26.1 values of the enthalpy, entropy, and volume changes that accompany the transfer of 1 mol of solute from the hypothetical pure state defined above to a gas at 1 atm. These are denoted $\Delta\bar{h}_2^{\ominus}$, $\Delta\bar{s}_2^{\ominus}$, and $\Delta\bar{v}_2^{\ominus}$, respectively. From the data presented we deduce the following:

1. The increase in entropy associated with transfer of the solute from solution to the gas phase is greater than the entropy of vaporization of the pure liquid solute at its normal boiling point.

Table 26.1 Thermodynamic Properties of Solutions of Simple Gases in Water

	PROPERTIES OF SOLUTE IN REFERENCE STATE									PROPERTIES OF SOLUTE AS PURE LIQUID		
	$\Delta\bar{h}_2^{\ominus}$ (J/mol)			$\Delta\bar{s}_2^{\ominus}$ (J/K mol)			$\Delta\bar{v}_2^{\ominus}$ (cm³/mol)					
Solute	25°C	50°C	80°C	25°C	50°C	80°C	25°C	50°C	80°C	$\Delta h_{2(vap)}$ (J/mol)	$\Delta s_{2(vap)}$ (J/K mol)	v_2^0 (cm³/mol)
He	3,510	2,790	2,300	111	109	107	(15.5)[a]					32
Ne	7,870	5,360	2,430	120	111	103				1,740	64.0	17
Ar	11,420	7,700	2,970	126	114	98				6,530	74.9	29
Kr	14,850	9,830	3,100	135	116	97				9,030	75.3	34
Xe	18,790	11,210	3,010	141	117	92				12,640	76.6	43
Rn	21,130	12,790	3,010	144	118	88				16,400	77.8	50
H_2	5,360	1,780[b]	711	109	99[b]	91	24	26	24			28
N_2	8,950	8,330[b]	753	125	123[b]	100	44	40	38	5,560	72.0	35
CO	16,360	8,240[b]	42	125	119[b]	95	37	36	32	6,040	74.1	35
O_2	12,510	8,660[b]		131	126[b]		31	31	32	6,820	75.7	28
CH_4	13,310			133			36	37	38	8,180	72.8	39

[a] Probably too small. An estimate based on extrapolation of the error in other (similar) measurements would indicate a value of 20 cm³/mol.
[b] At 40°C.

2. The enthalpy change associated with transfer of the solute from solution to the gas phase is greater than the enthalpy of vaporization of the pure liquid solute at its normal boiling point.
3. The enthalpy and entropy of transfer from solution to gas phase are strongly temperature-dependent.
4. The volume of the hypothetical pure-component reference state is about equal to the molar volume of the real pure liquid solute at its normal boiling point. Furthermore, the volume of the reference state is much less sensitive to variation of the temperature than are $\Delta \bar{h}_2^{\ominus}$ and $\Delta \bar{s}_2^{\ominus}$.

Interpretation of the listed inferences is, with present knowledge, necessarily speculative. The following arguments are intended to illustrate the reasoning used to convert macroscopic observations into structural concepts that can be incorporated into a molecular model of aqueous solutions.

Because the entropy of vaporization of the pure liquid solute is less than the entropy of transfer from the hypothetical standard state of the solute to the gas, and since in both cases the gas is taken to be in the same thermodynamic state, we must conclude that the entropy of the aqueous solution is less than the entropy of pure water. But in the hypothetical standard state of the solute there are no solute–solute interactions, so the difference in entropies must be attributed to the influence of the solute on the behavior of the solvent water. Then the decrease in entropy implies that the number of configurations of equal energy available to the water molecules must be smaller in solution than in the pure liquid, i.e., there is a region of enhanced order in the water surrounding each solute molecule. An extreme version of this observation is the assertion that introduction of a solute molecule of the type under discussion creates a well-defined structure in the solvent proximate to the solute molecule. Indeed, there does exist a class of compounds known as *clathrates,* examples of which are simple crystalline hydrates of Xe, Cl_2, $CHCl_3$, and other similar small molecules.[10] A clathrate compound can be characterized as a complex cage of host water molecules enclosing a cavity in which a guest molecule resides. The host water structures of clathrates involve hydrogen bonding patterns not used in any of the known forms of ice, e.g., there are hydrogen bonded rings of five water molecules. Clathrate compounds are usually less dense than ice, and are unstable with respect to ice if the guest molecule is removed. The clathrate compound is rendered stable by virtue of nonspecific, nonsaturable dispersion interactions between the guest and the surrounding water molecules. It is tempting to think that the structure of water near a given solute molecule resembles that in the corresponding clathrate compound.

Computer simulation studies of the properties of an inert gas atom in water provide information concerning the structure of water around the solute.

In one simulation[11] the pair potential energy hypersurface of Ar–H_2O was obtained from self consistent field quantum mechanical calculations of the energy, and a similar quality water–water pair interaction was employed. It is important to note that this water–water potential does not exhibit strong tetrahedral character, yet does generate a good pair-correlation function for the liquid. The calculated Ar–O and Ar–H pair-correlation functions show the existence of a first solvation shell which contains approximately 17 water molecules (Fig. 26.8*a*), and a second broad shell of water molecules separated by a distance typical of ordinary hydrogen bonding. From the near equality of the separations corresponding to the first peaks in the Ar–O and Ar–H pair-correlation functions it is possible to infer that, on average, the OH bonds of the waters in the first solvation shell do not point toward the Ar atom. With respect to the fundamental unit of a central oxygen with four (more or less) tetrahedrally distributed hydrogens, on average the Ar atom sits inside the umbrella formed by three of the OH bonds, with the fourth OH bond pointing radially outward. A comparison of the distribution of water molecule orientations in the first solvation shell with those in pure water and in a clathrate (Fig. 26.8*b*) shows that the structure of the first hydration shell has similarities and differences with the structures of both of the latter. In particular, the broad maximum in the ArOH angle distribution in the first solvation shell is coincident with the maximum in the narrow distribution of that angle in the clathrate, while the breadth and amplitude of the angle distribution, except for very small angles, is like that in bulk water. Radial orientation of the OH bond, which is very rare in the first hydration shell, occurs frequently at greater distances from the Ar atom. Thus the presence of the Ar atom increases the orientational order in the neighboring water, the locally ordered water shares some structural characteristics with the structure of a clathrate, but there is no evidence for formation of a complete clathrate cage of water molecules as opposed to a distorted polyhedral cage derived by bending and stretching the ordinary water hydrogen bonds.

In another simulation[12] the model inert gas atom and the model rigid water molecule were assigned the same diameter, and their interaction was taken to be of the Lennard–Jones form (Eq. 21.173) with a well depth intermediate between those expected for Ne–H_2O and Ar–H_2O. The effective water–water pair potential used in these simulations has very strong tetrahedral character. The results of the simulations show that the solute is surrounded by a water

[10] See M. Falk and O. Knop, Chapter 2 in *Water: A Comprehensive Treatise,* Vol. 2, Ed. F. Franks, Plenum Press, New York (1973); D. W. Davidson, Chapter 3 in *Water: A Comprehensive Treatise,* Vol. 2, Ed. F. Franks, Plenum Press, New York (1973).

[11] G. Algona and A. Tani, *J. Chem. Phys.* **72,** 580 (1980).

[12] A. Geiger, A. Rahman, and F. H. Stillinger, *J. Chem. Phys.* **70,** 263 (1979).

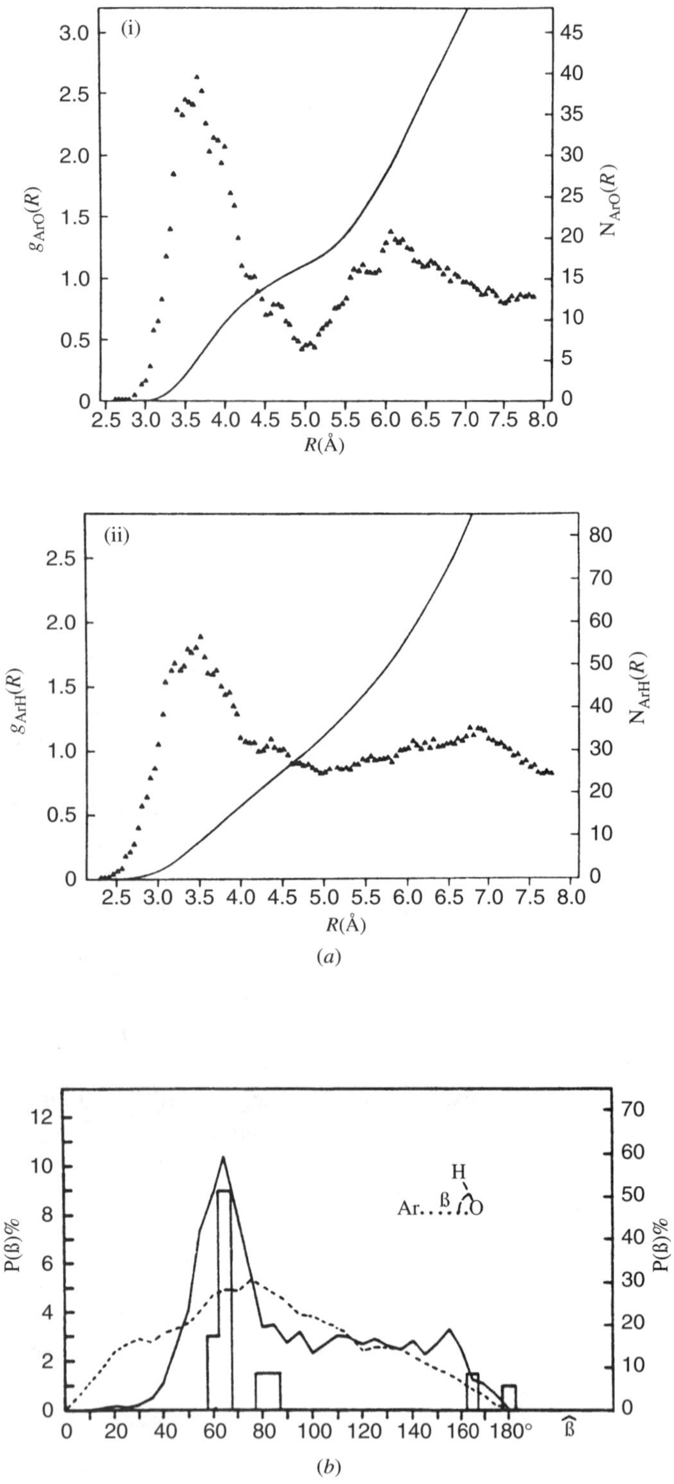

Figure 26.8 (*a*) The results of Monte Carlo simulations of the ArO (i) and ArH (ii) pair-correlation functions (triangles) and the cumlative number of water molecules (lines). (*b*) The probability density for finding the angle between an Ar atom, an O atom, and each of the four hydrogens closest to the Ar atom. The histograms show the same distribution for the crystalline clathrate hydrate (right-hand scale). The full line refers to first shell water molecules and the dashed line to bulk water (with Ar . . . O separations greater than 5 Å). From G. Alagona and A. Tani, *J. Chem. Phys.* **72,** 580 (1980).

cage whose structure differs from that of pure water. The nearest-neighbor hydration shell resembles a highly strained net of six hexagonal rings of hydrogen-bonded waters with some clathrate like character. The orientational structure of this shell persists to some extent, but only weakly, in the surrounding water molecules of the second hydration shell. The differences between the water structure around the solute and that in pure water are everywhere small, the hydrogen bonding between a water molecule and its four nearest neighbors is preserved, and the hydration structure blends continuously into that of the surrounding liquid. Even though some of the preferred water orientations in the hydration shell resemble those in a clathrate, there are profound differences between the two structures. The most important of these differences, which is also the one responsible for the continuous blending of the hydration shell into the surrounding water, is that the hydration-shell water is not immobilized as it is in a crystalline clathrate. The computer simulations indicate that there is only a small difference between the mobilities of water molecules in the hydration shell and in pure water. In summary, the computer simulations show that introduction of an inert gas atom into water alters the structure of the liquid immediately around the solute, but that the local environment of the solute is neither rigid nor precisely ordered as in a crystalline arrangement of water molecules.

Even the limited local structure modifications found in the computer simulations described can be expected to lower the entropy per water molecule, as is observed in real solutions. Moreover, because of the continuous blending of the hydration structure into the liquid structure it is reasonable to expect $\Delta\bar{v}_2^{\ominus}$ to be much less sensitive to temperature than $\Delta\bar{h}_2^{\ominus}$ and $\Delta\bar{s}_2^{\ominus}$, both of which depend on the thermal excitation of quasi-vibrational motions and on the extent of the difference between the local and the bulk liquid structures. The data of Table 26.1 support this view.

Given the extreme sensitivity of the structure of a liquid to repulsive interactions between the molecules, we must expect some changes from the picture presented by the computer simulations when the solute and water molecule have substantially different diameters. Although there are no computer simulations of this class of solutions available, the dependence of the free energy of hydration on solute size has been studied using a variant of the RISM integral equation theory[13] described in Chapters 23 and 25. Moreover, the data of Table 26.1 provide some clues which can be used for the construction of crude structural models.

The results of the calculations of the change in chemical potential associated with inserting a Lennard–Jones atom in water show that it can be represented as the sum of two terms, one of which scales with the solute surface area and the other of which scales with the solute volume; the solute

[13] J. Perkyns and B. M. Pettitt, *J. Phys. Chem.* **100,** 1323 (1996).

volume term dominates the excess chemical potential in the limit of large solute size, but both terms are important when the ratio of solute to water diameters is in the range typical of most solutes. The relative importance of the surface area term increases as the solute–water interaction strength increases. When decomposed into its energetic and entropic components, it is found that, for large solute atom diameters, the excess chemical potential of solvation is dominated by the entropy change on solvation.

Proceeding with our illustrative arguments, we can understand the data in Table 26.1 qualitatively within the framework of the continuous-random-network, distorted-hydrogen-bond model of water. Insertion of a He atom leads to a reference state with $\Delta\bar{v}_2^{\theta} < v_2^0$ because of the existence of a distribution of O · · · O separations in water. Geometric constraints imposed by the hydrogen bonding often leave voids too small for other water molecules or solutes as large or larger, but of sufficient size to accommodate a He atom.[14] These voids are not "vacuum"; there is, in each void, an intermolecular potential field generated by the surrounding water molecules. The field is not uniform. All we are asserting is that a He atom interacts so weakly with the molecules surrounding a void that occupancy of the void is possible, whereas the interaction of those molecules with a larger atom is sufficiently repulsive to make occupation of the void unlikely. In both cases our conception includes rearrangement of the water. We expect the equilibrium configuration of an occupied void to be different from that of an unoccupied void, and the change in configuration to reflect the effect of He–H_2O repulsion. The values of $\Delta\bar{h}_2^{\theta}$, observed for He as a solute greatly exceed the expected heat of vaporization of a He atom from liquid Ne (which is isoelectronic with H_2O). Thus, the combination of the dispersion attraction of He for H_2O with the relaxation of the water about a He atom is energetically favored. Presumably, since $|\Delta\bar{s}_2^{\theta}| > |\Delta s_{2(\text{vap})}^0|$ the water around a He atom is more ordered than in the bulk liquid. This picture of the structure of a He-atom-in-water solution requires that $\Delta\bar{h}_2^{\theta}$ and $\Delta\bar{s}_2^{\theta}$ be insensitive to temperature because preexisting voids are the initial configurations from which relaxation starts, a conclusion which is supported by the data in Table 26.1. The values of $\Delta\bar{h}_2^{\theta}$ and $\Delta\bar{s}_2^{\theta}$ for He change much less with increasing temperature than do those for other gaseous solutes.

Insertion of solutes larger than He into the water structure requires much more extensive alteration of that structure, primarily because of the lack of suitable voids for placement of a solute molecule. In the simplest picture the process of inserting a solute molecule can be thought of as proceeding in two steps. First, a cavity of suitable size must be created in the liquid. In general, it requires work to create such a cavity; the amount of work required depends on both the volume and the surface area of the cavity. Typical model calculations for water at 298 K indicate that creation of a cavity to hold an Ar atom requires about 2.9 kJ/mol and leads to an entropy change of −52 J/K mol. Second, the solute molecule is inserted in the cavity and the liquid allowed to relax in the new potential field. Calculation of the changes in thermodynamic functions due to this relaxation is difficult, and the results are rather dependent on the model chosen. The first step described (cavity formation) is an approximate way of accounting for the effect on liquid structure of repulsion between solute and solvent molecules. We have already indicated that such repulsive interactions play a role in He/H_2O solutions, and have shown how they determine the local structure in simple liquid mixtures, so we expect them to be very important, possibly dominant, in the cases now under discussion. If this view is taken seriously, and only enthalpy changes considered in the second step (solvent relaxation around the inserted solute molecule), reasonable agreement between calculation and experiment is obtained. For example, for Ar–H_2O at 298 K one finds:

$$\Delta\bar{h}_2^{\theta}(\text{calc.}) = 10.2 \text{ kJ / mol}; \quad \Delta\bar{h}_2^{\theta}(\text{exp.}) = 11.4 \text{ kJ / mol}$$

$$\Delta\bar{s}_2^{\theta}(\text{calc.}) = 120 \text{ J / K mol}; \quad \Delta\bar{s}_2^{\theta}(\text{exp.}) = 126 \text{ J / K mol.}$$

The model described does not provide any information about the nature of changes in hydrogen bonding in water, or other details—it is semimacroscopic in character. There is not, at present, a satisfactory microscopic theory of structure changes in aqueous solutions of uncharged species.

Ionic solutes interact very strongly with water, and we therefore expect these to generate a greater relative alteration of the local water structure than do neutral solutes. If electrostatic effects in the interaction are of dominant importance, we expect there to be a regular variation of the thermodynamic properties of individual ions with ionic radius. If nonelectrostatic effects in the interaction are of importance, we expect there to be differences between the properties of ions with different electronic structures but the same charge and size.

Following the method used in our discussion of solutions of simple gases, we first examine the available experimental data. As noted in Section 14.6, it is not possible to determine the properties of individual ions, only those of neutral salts. Following the convention introduced in Section 14.6, Table 26.2 displays thermodynamic functions for individual ions computed under the *ad hoc* assumption that the Gibbs free energy of formation, the enthalpy of formation, and the entropy of formation of the hydrogen ion at 25°C and 1 atm in water are *all* taken to be zero:

$$\Delta g_f^{\theta}(\text{H}^+, aq) \equiv 0, \tag{26.51}$$

$$\Delta h_f^{\theta}(\text{H}^+, aq) \equiv 0, \tag{26.52}$$

$$\Delta s_f^{\theta}(\text{H}^+, aq) \equiv 0. \tag{26.53}$$

[14] Helium diffuses through glass for the same reason. The He atom is sufficiently small to be able to migrate between atoms that would strongly repel a larger atom.

Table 26.2 Gibbs Free Energies, Enthalpies, and Entropies of Formation of Some Ions in Water at 25°C

Ion	$\Delta g_f^{\ominus}$(kJ/mol)	$\Delta h_f^{\ominus}$(kJ/mol)	$\Delta s_f^{\ominus}$(J/K mol)
H^+	0(convention)	0(convention)	0(convention)
OH^-	−157.3	−230.0	−10.5
F^-	−276.5	−329.1	−9.6
Cl^-	−131.2	−167.4	+55.2
Br^-	−102.8	−120.9	+80.7
I^-	−51.7	−55.9	+109.4
HS^-	+12.6	−17.2	+62.8
S^{2-}	+8.6	+32.6	−17.
SO_4^{2-}	−742.0	−907.5	+17.2
HSO_4^-	−752.9	−885.8	+127.7
NO_3^-	−110.6	−206.6	+125.
NH_4^-	−79.5	−132.8	+112.8
Li^+	−293.8	−278.4	+14.
Na^+	−261.9	−239.7	+60.2
K^+	−282.3	−251.2	+103.
Rb^+	−283.0	−248.5	+120.
Cs^+	−296.0	−261.9	+133.
Tl^+	−32.5	+5.8	+127.
Ag^+	+77.1	+105.9	+73.93
Sn^{2+}	−24.3	−10.0	+21.
Pb^{2+}	−24.3	+1.6	+21.
Zn^{2+}	−147.2	−152.4	−106.5
Cd^{2+}	−77.7	−72.4	−61.1
Hg^{2+}	+164.8	+174.0	−23.
Fe^{2+}	−84.9	−87.9	−113.
Fe^{3+}	−1060.	−47.7	−293.
Mg^{2+}	−456.0	−462.0	−118.
Ca^{2+}	−553.0	−543.0	−55.2
Sr^{2+}	−557.3	−545.5	−39.
Ba^{2+}	−560.7	−538.4	+13.

Table 26.3 Some Enthalpies and Entropies of Hydration at 25°C

	$\Delta h^0_{\text{cryst}}$ [a] (kJ/mol)	$\Delta h^{\ominus}_{\text{hydration}}$ [b] (kJ/mol)	$\Delta s^{\ominus}_{\text{hydration}}$ [b] (J/mol K)
LiF	−1032	−1005	−276
NaCl	−779	−760	−185
KF	−813	−819	−207
$SrCl_2$	−2110	−2161	−400
AlF_3	−6140	−6161	−930
$AlCl_3$	−5452	−5774	−757
AgCl	−916	−851	−191

[a]For the reaction $A^+(g) + B^-(g) = AB\ (crystal)$.
[b]For the reaction $A^+(g) + B^-(g) = A^+(aq) + B^-(aq)$.

Table 26.2 contains data also included in Table 14.5; these are repeated here for convenience. By combination of these quantities, as illustrated in Sections 14.6 and 18.1, we can deduce the thermodynamics of reactions. Note that for the purpose of computing changes that occur in a chemical reaction the arbitrary choice of standard state (Eqs. 26.51–26.53) is irrelevant.

A quantitative measure of the absolute magnitude of the ion–solvent interaction is provided by the heat of the reaction

$$A^+(g) + B^-(g) = A^+(aq) + B^-(aq). \quad (26.54)$$

When the final state in the reaction is the pure crystal instead of the ionic solution, the enthalpy change is just the heat of formation of the crystal from the ions. This is, of course, also just the enthalpy required to separate the crystal into a set of infinitely separated ions. Using tables of energies of ionization, heats of sublimation, and so on, one can easily compute the values given in Table 26.3.

In an ionic crystal the energy of stabilization comes primarily from the ion–ion Coulomb interactions. These interactions are sensitive to the sizes of the ions, since it is the ionic size that determines their minimum separation. In turn, the ionic size is determined by the strong repulsions generated by the overlap between closed-shell charge distributions. An examination of the data in Table 26.3 shows that $\Delta h^0_{\text{cryst}}$ and $\Delta h^{\ominus}_{\text{hyd}}$ are nearly equal in magnitude, and that they change in a similar fashion when the ionic size is changed. This suggests that it might be useful to decompose $\Delta h^{\ominus}_{\text{hyd}}$ and $\Delta h_f^{\ominus}$ into contributions from separate ions. We have already made such a decomposition for $\Delta h_f^{\ominus}$ using the convention cited in Eqs. 26.51–26.53. These are the values in Table 26.2. What we now seek are "absolute" ionic enthalpies, that is, the assignment of "realistic values" for the enthalpies of hydration of ions. From the point of view of thermodynamics, all choices of convention are equally valid, and there is no set of observations that can be used, without theoretical modeling, to obtain single-ion properties. However, from the point of view of a molecular theory, it is clearly desirable to choose a convention that, for a simple model, gives sensible results. In the earliest attempts to construct tables of ionic properties, the enthalpy of hydration of KF was equally divided into an enthalpy of hydration of K^+ and an enthalpy of hydration of F^-. This choice was based on the observation that the ionic radii of K^+ and F^- are almost the same, hence the heat of reaction of each with water should be the same. Of course, given this new convention, the tables of thermodynamic data based on Eqs. 26.51–26.53 can easily be converted, since once one value has been defined all others can be worked out by hypothetical reaction sequences.

Other schemes besides the one based on equality of enthalpies of hydration of K^+ and F^- have been proposed. We select for discussion just one of these schemes.

A set of values that represent a close approach to absolute single-ion free energies of hydration can be obtained from the concept of *absolute half-cell potential.* The idea is as follows. Consider two half-cells, the left one consisting of an electrode M_1 immersed in a solution of M_1^+ and X^- ions, and the right one consisting of an electrode M_2 immersed in a solution of M_2^+ and X^- ions. Electrical contact between these two half-cells is via a liquid junction, that is, a boundary region in which the concentrations of M_1^+, M_2^+, and X^- vary continuously from those characteristic of the left-hand half-cell to those characteristic of the right-hand half-cell. In general, cells that have a liquid junction have a potential that depends on the nature of that junction. Conditions can be found under which the potential difference attributed to the liquid junction is very small. We shall assume that in the case under discussion it is zero, recognizing that this assumption is one of the inherent approximations in the scheme of determining absolute half-cell potentials. Now, in the left-hand cell the reaction is $M_1(s) \rightarrow M_1^+(sol'n) + e^-(M_1)$, and in the right-hand cell it is $M_2^+(sol'n) + e^-(M_2) \rightarrow M_2(s)$. Let the Gibbs free energy change for the left-hand cell reaction be separated into the following steps:

(i) $M_1(s) \rightarrow M_1(g);$ $\quad \Delta g = \Delta g^0_{1(\text{vap})}$

(ii) $M_1(g) \rightarrow M_1^+(g) + e^-(M_1);$ $\quad \Delta g = \mathcal{F}(I_1 - \phi_1)$

(iii) $M_1^+(g) \rightarrow M_1^+$ $(sol'n\ adjacent\ to\ M_1);$ $\quad \Delta g = \Delta g^\theta_{1(\text{hyd})}$

(iv) $M_1^+(sol'n\ adjacent\ to\ M_1)$ $\rightarrow M_1^+(bulk\ sol'n);$ $\quad \Delta g = \mathcal{F} V_{M_1 s}$

where $\Delta g^0_{1(\text{vap})}$ is the Gibbs free energy of vaporization of neutral M_1, I_1, and ϕ_1 the ionization potential of the metal and the work function[15] of the metal in contact with the solution, respectively, $\Delta g^\theta_{\text{hyd}}$ the Gibbs free energy of hydration of M_1^+, and $V_{M_1 s}$ the electrostatic potential difference between M_1 and the solution. Similarly, for the right-hand half-cell:

(i) $M_2^+(bulk\ sol'n) \rightarrow M_2^+$ $(sol'n\ adjacent\ to\ M_2);$ $\quad \Delta g = -\mathcal{F} V_{M_2 s}$

(ii) $M_2^+(sol'n\ adjacent\ to\ M_2)$ $\rightarrow M_2^+(g);$ $\quad \Delta g = -\Delta g^\theta_{2(\text{hyd})}$

(iii) $M_2^+(g) + e^-(M_2) \rightarrow M_2(g);$ $\quad \Delta g = -\mathcal{F}(I_2 - \phi_2)$

(iv) $M_2(g) \rightarrow M_2(s);$ $\quad \Delta g = -\Delta g^0_{2(\text{vap})}$

The total changes in Gibbs free energy for the left and right half-cells are obtained by summation of all contributions.

[15] The work function is the free-energy change on transferring an electron from the inside of the metal to the outside; it depends on the nature of the medium in contact with the metal.

For the overall cell reaction $M_1 + M_2^+ \rightarrow M_1^+ + M_2$, we have, at equilibrium,

$$(\Delta g^0_{1(\text{vap})} - \Delta g^0_{2(\text{vap})}) + (\Delta g^\theta_{1(\text{hyd})} - \Delta g^\theta_{2(\text{hyd})}) + \mathcal{F}(I_1 - I_2) = -[\mathcal{F}(V_{M_1 s} - \phi_1) - \mathcal{F}(V_{M_2 s} - \phi_2)]. \quad (26.55)$$

The left-hand side of Eq. 26.55 is just the Gibbs free energy change for the cell reaction carried out nonelectrically, so the right-hand side is $-\mathcal{F}\mathcal{E}^\theta$. Note that because M_1 and M_2 are different metals, the cell potential is the difference of the electrostatic potentials $V_{M_1 s} - V_{M_2 s}$ minus the work function difference $\phi_1 - \phi_2$. We now define a half-cell potential as

$$\mathcal{E}_{1/2} \equiv V_{\text{Ms}} - \phi = \frac{(\Delta g^0_{\text{vap}} + \Delta g^\theta_{\text{hyd}} + \mathcal{F} I)}{\mathcal{F}}, \quad (26.56a)$$

which is the difference between the mean electrostatic energy of an electron in solution (but not hydrated) and the chemical potential of the electrons in the metal in contact with the solution.

How can one measure $\mathcal{E}_{1/2}$? Suppose that another electrode, say R, is connected by a wire to M_1. The electrode R is in contact with air, and does not come into contact with the solution. Since M_1 and R are connected with a wire, the chemical potential of the electrons in M_1 must be the same as that of the electrons in R. Now consider the reaction $M_1 \rightarrow M_1^+(bulk\ sol'n) + e^-(M_1)$ to consist of the following steps:

(i) $M_1(s) \rightarrow M_1(g,\ at\ sol'n\text{–}air\ interface)$

(ii) $M_1(g) \rightarrow M_1^+ + e^-(both\ in\ gas\ at\ the\ sol'n\text{–}gas\ interface)$

(iii) $e^-(gas\ at\ sol'n\text{–}gas\ interface) \rightarrow e^-(M_1)$

(iv) $M_1^+(gas\ at\ sol'n\text{–}gas\ interface) \rightarrow M_1^+(bulk\ sol'n)$

At equilibrium the sum of the Gibbs free energy changes for these processes is

$$\Delta g^0_{1(\text{vap})} + \Delta g^\theta_{1(\text{hyd})} + \mathcal{F} I_1 - (-\mathcal{F}\mathcal{E}_{1/2}) = 0, \quad (26.56b)$$

since the transfer of an electron from a region of electrostatic potential corresponding to that just outside the solution's free surface to the metal M_1 is given by $-(-\mathcal{F}\mathcal{E}_{1/2})$. Note that $\Delta g^\theta_{1(\text{hyd})}$ corresponds to the transfer of an ion to the interior of the solution from a gas phase with the same electrostatic potential as the mean electrostatic potential in the bulk solution.

Since R and M_1 are connected by a wire and the chemical potentials of the electrons in them are equal, we have

$$V_{M_1 s} - \phi_1 = V_{Rs} - \phi_R = \mathcal{E}_{1/2}, \quad (26.56c)$$

with ϕ_R the work function of R in contact with air. Then, if ϕ_R is known and the electrostatic potential is the same in the bulk and surface, we need only measure the potential difference between the solution and an electrode R outside it (say, just over the solution surface) in order to obtain $\mathscr{E}_{1/2}$ for M_1 immersed in a solution of M_1^+ and X^- ions. In fact, that potential difference can be readily measured. However, we must expect that orientation of molecules at the solution surface will lead to a potential difference between the bulk and surface of the solution. This potential difference depends on temperature, electrolyte concentration, and other variables. Its sign and magnitude can be reasonably estimated from theoretical and experimental studies, hence the measured potential difference between R and the solution surface corrected, and $\mathscr{E}_{1/2}$ thereby determined. The use of this correction is a second approximation in this method of determining absolute half-cell potentials.

R. Gomer has used the method described and found that $\mathscr{E}_{1/2}$ for the reaction $H_2 \rightarrow 2H^+ + 2e^-$ is -4.73 ± 0.05 V. Hence absolute half-cell potentials can be obtained from tabulations of standard half-cell potentials by adding -4.73 V. With these values of $\mathscr{E}_{1/2}$, absolute values of the Gibbs free energy of hydration can be obtained from Eq. 26.56*b;* some values are shown in Table 26.4. The values obtained in this way are seen to be rather different from those based on Eq. 26.51.

Reviews of the uses to which single ion thermodynamic data can be put are presented in the books by Y. Marcus and N. S. Hush listed at the end of this chapter. Our present interest in these data is limited to examination of how they illuminate the nature of ion hydration and its variation from ion to ion. For this purpose we have included in Table 26.4 thermodynamic data for ions obtained by methods other than the one just discussed.

In Figs. 26.9 and 26.10 are plotted the variation of the "absolute" enthalpy and entropy of hydration of individual ions as a function of the reciprocal of the ionic radius. Note that there is near proportionality of $\Delta h^{\ominus}_{\text{hydration}}$ and $\Delta s^{\ominus}_{\text{hydration}}$ to r_{ion}^{-1}, but there are also significant deviations from linearity. The data for transition metal ions fall so far off the curves for the corresponding closed-shell simple ions that for these ions the ion–water interaction must contain a significant contribution from other than electrostatic interactions. A similar conclusion can be drawn for the case of simple ions: The data displayed show a clear dependence of $\Delta h^{\ominus}_{\text{hydration}}$ and $\Delta s^{\ominus}_{\text{hydration}}$ on the ionic charge, but the differences in properties between oppositely charged ions of like size show that more than just electrostatic interaction between an ion and water is involved in determining the thermodynamic properties of the ion. Thus the simple electrostatic model of hydration does not adequately account for the dependence of the thermodynamic properties of ions on the diameter, electronic structure, or sign of the charge. Incidentally, the fact that the thermodynamic functions for an ion depend on the sign of its charge points up another difficulty in defining a satisfactory convention for the decomposition of observable changes into separate ionic contributions.

Table 26.4 Absolute Thermodynamic Data for Aqueous Ions (298 K)

Ion	r_{ion} Å	$\Delta h^{\ominus}_{\text{hyd}}$ (kJ/mol)[a]	$\Delta s^{\ominus}_{\text{hyd}}$ (J/mol K)[a]	$\Delta g^{\ominus}_{\text{hyd}}$ (kJ/mol)[a]	$\Delta g^{\ominus}_{\text{hyd}}$ (kJ/mol)[b]
H^+		−1121	−123	−1084	−1059
F^-	1.33	−474	−143	−431	−463
Cl^-	1.81	−338	−85	−313	−341
Br^-	1.96	−304	−68	−284	−318
I^-	2.19	−261	−47	−247	
Zn^{2+}	0.70	−2105	−295	−2017	
Cu^2	0.69	−2161	−302	−2071	2030
Ni^{2+}	0.72	−2174	−369	−2056	
Co^{2+}	0.74	−2110	−366	−2004	
Fe^{2+}	0.76	−1981	−321	−1885	−1846
Fe^{3+}	0.64	−4343	−506	−4192	
Mn^{2+}	0.80	−1905	−286	−1820	
Cr^{2+}	0.84	−1910	−183	−1856	
Cr^{3+}	0.69	−4368	−531	−4209	
Al^{3+}	0.45	−4750	−507	−4599	−4533
La^{3+}	1.15	−3374	−404	−3253	
Mg^{2+}	0.65	−1982	−294	−1895	−1850
Ca^{2+}	0.94	−1653	−238	−1582	−1514
Sr^{2+}	1.10	−1482	−222	−1416	−1376
Ba^{2+}	1.29	−1364	−187	−1308	−1259
Li^+	0.68	−545	−133	−505	−494
Na^+	0.98	−436	−100	−406	−379
K^+	1.33	−351	−66	−332	−306
Rb^+	1.48	−326	−54	−310	
Cs^+	1.67	−293	−51	−278	

[a] From R. M. Noyes, *J. Am. Chem. Soc.* **84,** 513 (1962); **86,** 971 (1964).
[b] Determined by R. Gomer and G. Tryson, *J. Chem. Phys.* **66,** 4413 (1977); W. G. Madden, R. Gomer, and M. J. Mandell, *J. Phys. Chem.* **81,** 2652 (1977).

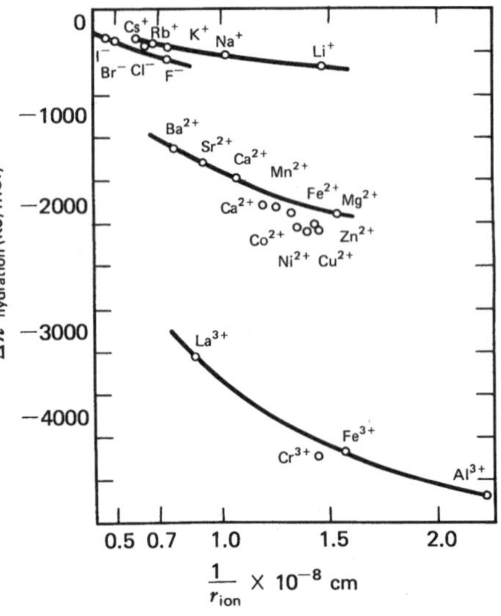

Figure 26.9 The computed heat of hydration of ions as a function of ionic size.

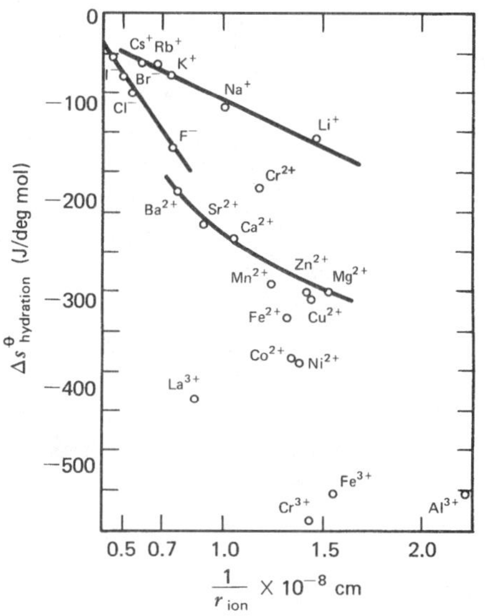

Figure 26.10 The computed entropy of hydration of ions as a function of ionic size.

The roles of the sign of the electrostatic interaction and of non-Coulombic interactions within water are particularly obvious when the entropies of hydration of the ions are compared with the entropy change associated with the solution of an inert gas of the same radius. To make this comparison we must use the same standard states. The data in Table 26.1 use as standard state the hypothetical pure substance retaining the same properties as the solute in dilute solution. The data in Table 26.4 use as standard state the hypothetical 1 *m* solution retaining the same properties as the solute in dilute solution. Thus the entropies of these standard states differ by $R \ln(18/1000)$, or 33 J/K mol. We must subtract 33 J/K mol from the entries in Table 26.1 to compare them with those in Table 26.4. With reference to a hypothetical perfect 1 *m* solution, the entropy change associated with solution of 1 mol of inert gas is about −92 J/K mol. The data in Table 26.4 show that for small ions the entropy of hydration is more negative, and for large ions less negative, than would correspond to the solution of the same number of uncharged inert gas atoms. For example, we have $\Delta s^\theta_{\text{hyd}}(\text{LiF}) = -275$ J/K mol, which is more negative than $2 \times (-92) = -184$ J/K mol, and $\Delta s^\theta_{\text{hyd}}(\text{RbI}) = -101$ J/K mol, which is more positive than −184 J/K mol. There are at least two effects that compete in influencing the enthalpy and entropy of hydration. First, the ion–dipole and ion–quadrupole interactions become stronger with decreasing ion–water molecule separation. It is then to be expected that the orientation of the water dipoles in the field of a small ion will be more nearly complete than in the field of a large ion. Second, insertion of the ion into the hydrogen-bond network of water alters that network. Using the data of Tables 26.1 and 26.4, we see that the first effect tends to produce larger negative entropies of hydration for smaller ions, and the second effect to produce larger negative entropies of hydration for larger ions.

More detailed interpretation of the thermodynamics of hydration and other ionic properties requires consideration of the structure of the water in the immediate neighborhood of the ion. The information currently available about the structure of the water surrounding an ion comes from neutron and x-ray diffraction experiments and computer simulations. The qualitative basis for the following arguments is derived from the observations that an ion tends to orient the surrounding water molecules and that the ion–water interaction energy is large. It is then tempting to regard an ion plus some number of water molecules as a molecular complex. The idea that a fixed number of water molecules is associated with a given ion is very old. However, determinations of the number of water molecules per ion, the hydration number, when derived from macroscopic measurements, give very different results depending on the nature of the measurement. This is not surprising, because the definition of an ion–water complex has some inherent ambiguities. For example, inferring the hydration number from the effective hydrodynamic radius determined from a conductance measurement and a hydrodynamic interpretation of the molecular friction coefficient (see Chapter 29) involves assumptions about the nature of the poorly defined boundary between the moving ion and the host water. Similarly, inferring the hydration number from a partitioning of the partial molar volume of an ion into an intrinsic ("hard core") contribution and a hydration shell contribution depends on the convention used to subdivide the solute partial molar volume between the ions, and is subject to the assumption that the intrinsic component for each ion is equal to its value in the crystalline phase of the solute. A similar procedure can be used to infer hydration numbers from other thermodynamic data, e.g., partial molar heat capacities, but because of the conventions and assumptions used to reduce the data it can rarely be asserted that no other explanation of the observations is possible and that ionic hydration must be present. Despite these uncertainties, there is good evidence for ionic hydration in some instances. In general, the hydration numbers of univalent ions are small, particularly those of anions. For these ions the concept of a solvent–ion complex is useful only in particular cases (e.g., Li^+). In the case of multivalent ions (such as Cr^{3+}), the rate of exchange of a water molecule adjacent to an ion for a water molecule from the outer solvent is very small (half-life of many hours). Also, the spectra of some multivalent rare earth ions, as well as their heats of hydration, suggest local field effects that alter the electronic structure of the ion (see later and also the discussion in Chapter 9). In these cases it is useful to consider the hydrated ion as a specific complex with well-defined structure.

Neutron and x-ray diffraction studies of ionic solutions[16] provide quantitative data supporting the statements made in

---

[16] For reviews see G. W. Neilson and R. H. Tromp, *Ann. Rept. Chem. Soc. (R. Soc. Chem.)*, **1991,** 45; J. E. Enderby, *Chem. Soc. Rev.* **1995,** 159.

the preceding paragraph.[17] Fig. 26.11*a* displays the CrO and CrH pair-correlation functions in a solution of the trivalent ion Cr^{3+}. The narrow first peaks in both the CrO and CrH pair-correlation functions, which correspond to six O and 12 H nearest neighbors, respectively, show that the hydration shell of six water molecules is very well defined; the broader but still well-defined second neighbor peaks in these functions show the persistence of orientational ordering of the water molecules in the second shell. Similar hydration structure is found in solutions of Fe^{2+} and Fe^{3+} (with a 0.1 Å shift in the Fe–O separation accompanying the electron exchange $Fe^{2+} \rightarrow Fe^{3+}$), and in an aqueous solution of Ni^{2+}. However, the first hydration shell in an aqueous solution of the divalent ion Sr^{2+} is very disordered, unlike the first hydration shells of Fe^{2+} and Ni^{2+}. The first hydration shell of the monovalent aqueous ion Li^+ is well defined in dilute solution, corresponding to a hydration number of six, but the hydration number decreases as the Li^+ concentration increases, indicating that the nearest neighbor shell of water molecules is much less stable than in Ni^{2+}. In contrast, aqueous K^+ is only weakly hydrated and the hydration shell is strongly disordered. There is no evidence for persistence of water orientation beyond the first shell. In this case no dynamical significance can be attributed to the presence of water molecules in the first shell; the hydration number just expresses the geometrical fact that there is a high probability of finding water molecules around the ion.

How does the hydration structure of an anion differ from that of a cation? To answer this question, we examine, as an example, the environment of aqueous Cl^-. The ClO and ClH pair-correlation functions obtained from neutron diffraction experiments are displayed in Fig. 26.11*b*. It is found that there are approximately six O atoms and six H atoms in the first hydration shell of Cl^-; when combined with the measured ClO and ClH separations, these data are consistent with a local structure around Cl^- with water molecules oriented as shown in Fig. 26.11*c*.

There have been a considerable number of molecular dynamics simulations of the structure of water surrounding an ion in an electrolyte solution. The results of these studies, carried out with a variety of water–water and ion–water potentials, are consistent with respect to the major features of the hydration structures of cations and anions, but differ with respect to details of those structures.

Figure 26.12 displays the cation–oxygen pair-correlation functions for several univalent and divalent ions; these results are typical of those obtained in many other simulation studies. The heights of the first peaks of these pair-correlation functions decrease as the ion radius increases, and are typically larger for divalent ions than for univalent ions. The first hydration shell of Li^+ is well defined (the distribution of Li–O separations is narrow and goes to zero between the first and second peaks), with a hydration number of six, in agreement with the experimental value. In contrast, the first hydration shell of Cs^+ is poorly defined. Despite the decrease in the electrostatic component of the ion–water interaction as the cation radius increases, which leads to a corresponding decrease in the stability of the first hydration shell, the coordination number of Cs^+ is larger than that of Li^+, namely 10. This behavior is a simple consequence of the increase in the volume of the first hydration

---

[17] As noted in Chapters 23 and 25, specification of the structure of a liquid requires determination of all of the independent pair-correlation functions. For example, to characterize the structure of an aqueous solution of the electrolyte MX_n we must determine the functions g_{OO}, g_{OH}, and g_{HH} associated with the water molecules, g_{MM}, g_{MX}, and g_{XX} associated with the electrolyte, and g_{MO}, g_{MH}, g_{XO}, and g_{XH} associated with the local structure of the water surrounding the ions. Although it is in principle possible to determine all ten pair-correlation functions by carrying out ten independent diffraction studies of the same solution, in practice that is not possible. Thus, although the scattering cross section for x-rays varies with wavelength, and that variation is strong near an x-ray absorption edge (Section 2.3), the combination of the nonzero width of the incident x-ray energy distribution and the (generally) small change in scattering cross section with change in wavelength make it very difficult to carry out measurements of the scattered x-ray intensity at many different x-ray wavelengths with sufficient precision to determine the independent pair-correlation functions. This method of determining the independent pair-correlation functions has been exploited in only a few favorable cases. The situation is dramatically different when we consider neutron scattering. Because the scattering cross section for neutrons is different for each isotope of an element, and the differences between the scattering cross sections are typically large, neutron diffraction with isotopic substitution can be used to determine the independent pair-correlation functions in a liquid mixture, subject only to the limit imposed by the availability of a suitable number of isotopes of the atoms in the solution. If neutrons are incident on a liquid containing several nuclear species the intensity of the scattered neutrons is given by a relation analogous to Eq. 23A.6 (or the equivalent Eq. 23A.7), namely,

$$I(s) = \sum_{\alpha}\sum_{\beta} b_\alpha b_\beta \left\langle \sum_{i\alpha}\sum_{j\beta} \exp[is\cdot(\mathbf{R}_j(\beta) - \mathbf{R}_i(\alpha))] \right\rangle$$

$$= N\left[\sum_{\alpha} c_\alpha b_\alpha^2 + \sum_{\beta} c_\alpha c_\beta b_\alpha b_\beta [S_{\alpha\beta}(s) - 1]\right],$$

$$g_{\alpha\beta}(R) = 1 + \frac{V}{2\pi^2 NR}\int ds[S_{\alpha\beta}(s) - 1]s\sin(sR),$$

where $c_\alpha = N_\alpha/N$ and b_α, the neutron scattering length of nucleus α, measures the intensity of neutron scattering from that nucleus. Suppose $I(s)$ is measured for two electrolyte solutions which are identical except for the isotopic character of the cation. We denote the two isotopes α and α'. The two functions $I(s)$ and $I'(s)$ will not be the same; the difference between them will contain only those pair-correlation functions associated with the cation, since those due to the water will have been subtracted out. The Fourier transform of the difference $I(s) - I'(s)$ is

$$\Delta G_\alpha(R) = c_\alpha \Delta b_\alpha \sum_{\beta\neq\alpha} 2c_\beta b_\beta [g_{\alpha\beta}(R) - 1] + c_\alpha^2 \Delta b_\alpha^2 [g_{\alpha\alpha}(R) - 1],$$

with $\Delta b_\alpha = b_\alpha - b_{\alpha'}$ and $\Delta b_\alpha^2 = b_\alpha^2 - b_{\alpha'}^2$. Fortunately, in a not too concentrated solution (say 1 molar), only the MO and MH pair-correlation functions contribute significantly to $\Delta G_\alpha(R)$. If, in addition, $\Delta G_\alpha(R)$ is determined for two solutions of the same electrolyte concentration in protonated and deuterated water, the individual pair-correlation functions g_{MO} and g_{MH} can be determined from the experimental data.

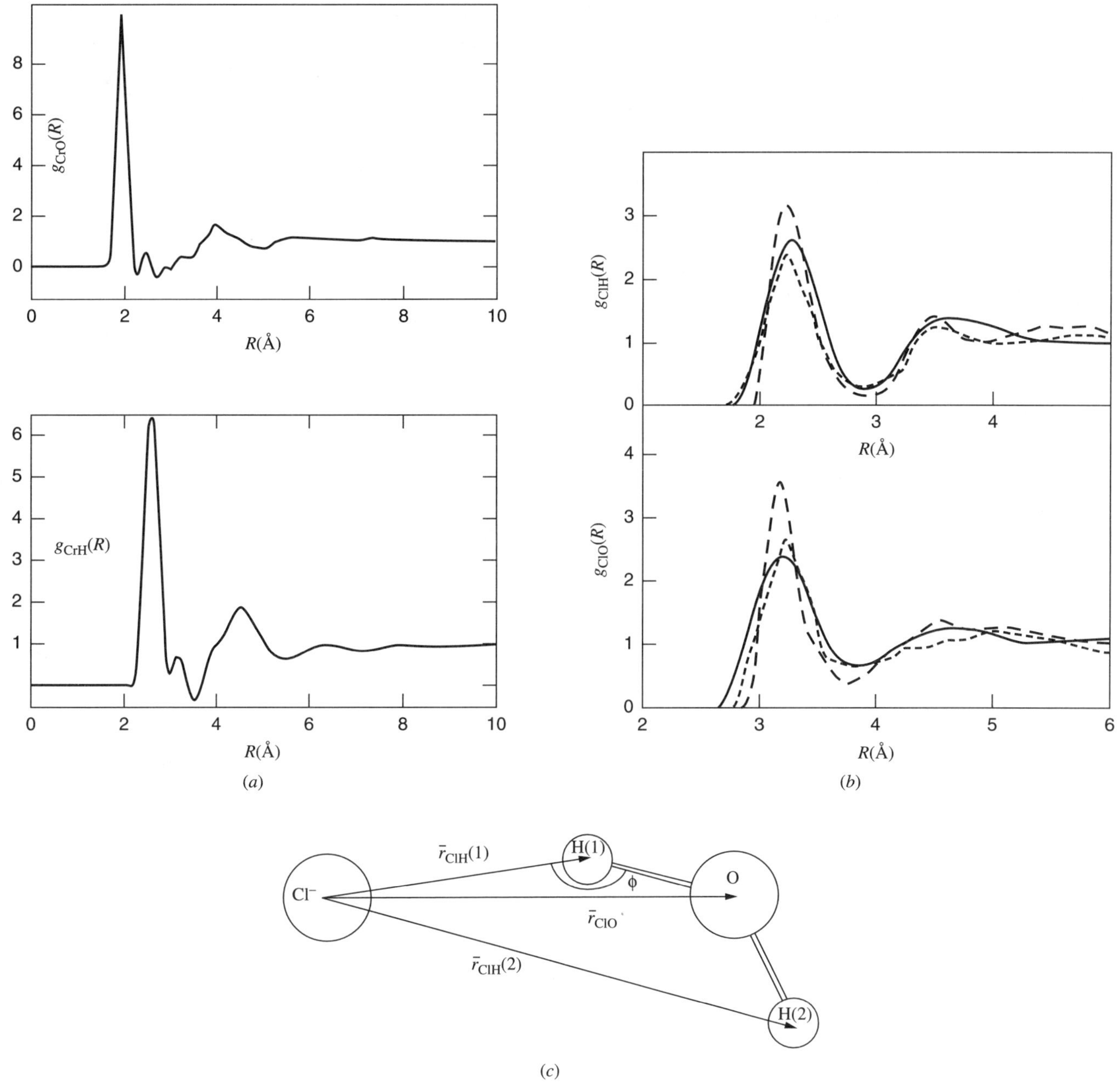

Figure 26.11 (*a*) The pair CrO and CrH correlation functions determined by neutron diffraction experiments. From R. D. Broadbent, G. W. Neilson, and M. Sandstrom, *J. Phys. Condens. Matter* **4,** 1992 (1992). (*b*) The ClO and ClH pair-correlation functions in a 2 *m* $NiCl_2$ solution. The experimental data are shown by the full line (from D. H. Powell, G. W. Neilson, and J. E. Enderby, *J. Phys. Condens. Matter* **5,** 5723 (1993)), and the results of simulations by the short dashed line (from M. Sprik, M. L. Klein, and K. Watanabe, *J. Phys. Chem.* **94,** 6483 (1990)) and the long dashed line (from L. X. Dang, J. E. Rice, J. Caldwell, and P. A. Kollman, *J. Am. Chem. Soc.* **113,** 2481 (1991)). (*c*) The instantaneous arrangement of a water molecule near to a chloride ion (from J. E. Enderby, *Chem. Soc. Rev.* **159** (1995)).

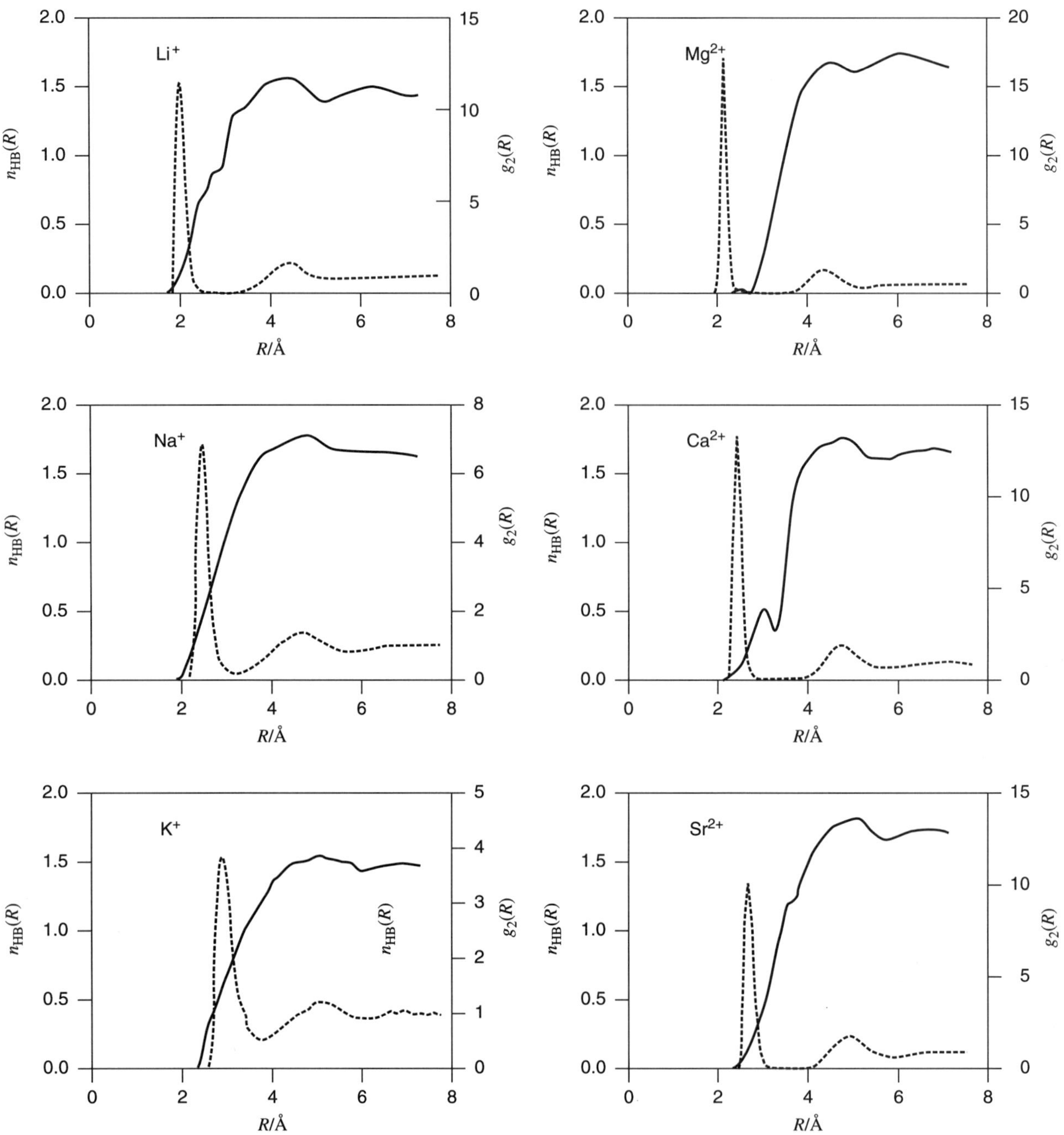

Figure 26.12 Cation–oxygen pair-correlation functions for several univalent and divalent cations obtained from molecular dynamics simulations. The cumulative number of waters surrounding each cation is also shown. From S. Obst and H. Bradaczek, *J. Phys. Chem.* **100,** 15677 (1996).

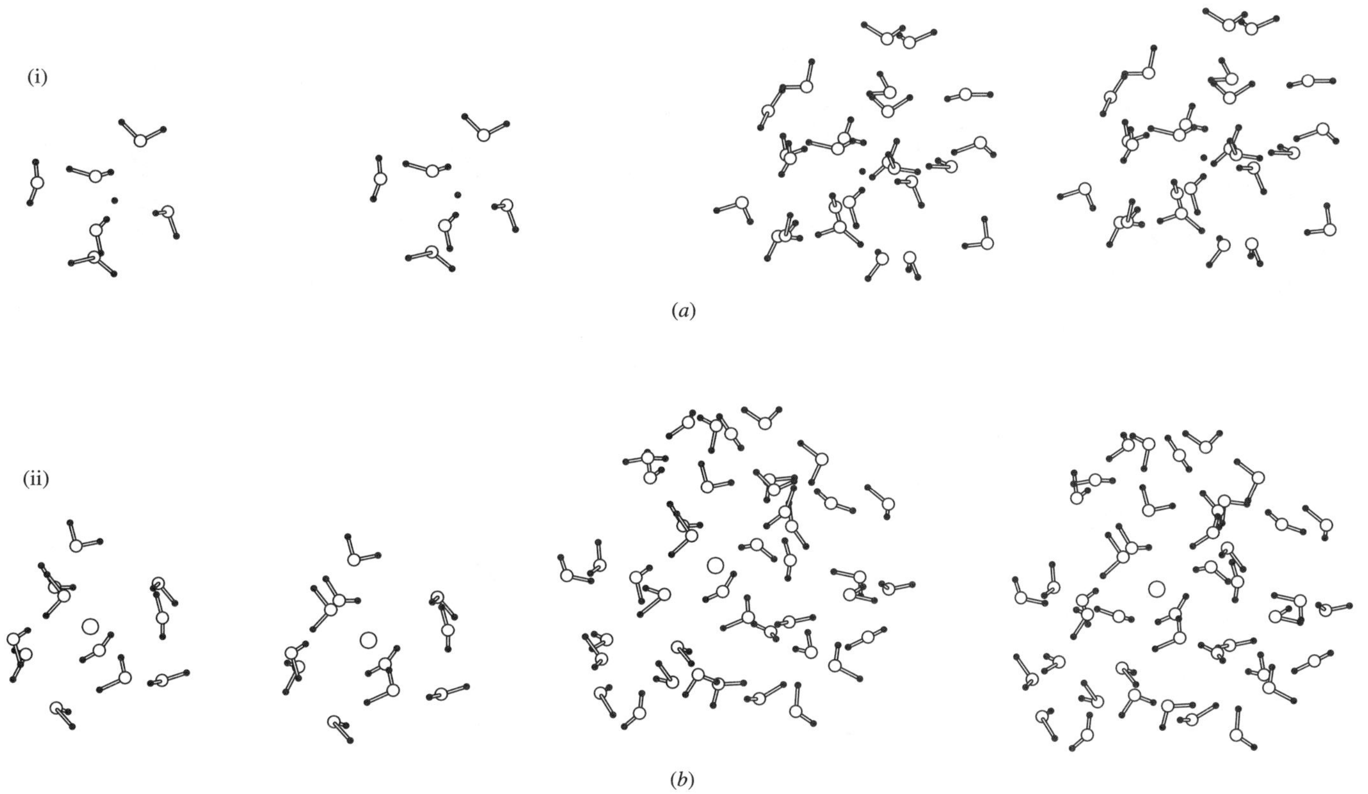

Figure 26.13 Instantaneous first (*a*) and first and second (*b*) hydration shells of (i) Li^+ and (ii) Cs^+. From S. H. Lee and J. C. Rasaiah, *J. Chem. Phys.* **101,** 6964 (1994).

shell with increasing ion radius which, at constant density, fills with more water molecules. These features are evident in the instantaneous first and second hydration shell structures of these ions displayed in Figs. 26.13*a* and 26.13*b*, respectively. Figure 26.14*a* shows the distribution of the angle between the water molecule dipole and the line between the cation and the oxygen atom in the first hydration shells of Li^+, Na^+, K^+, Rb^+, and Cs^+, and Fig. 26.14*b* shows the distribution of the angle between the direction of the OH bond of the water molecule and the line between the cation and the oxygen atom in the first and second hydration shells of the same cations. Clearly, the most favorable orientation of the dipole moment of a water molecule in the first hydration shell is not parallel to the line between the cation and the oxygen atom; it is about 22° for Li^+ and increases monotonically as the cation radius increases. The distribution of this angle also broadens monotonically as the cation radius increases. The deviation from parallelism of the water dipole moment and the cation–water line is a signature of the competition between the electrostatic cation–water interaction and the interactions that maintain the hydrogen bond network of water.

As expected, the hydration structure of divalent ions is, in most cases, much better defined than is that of monovalent ions; typical instantaneous configurations of the water molecules in the first hydration shells of Mg^{2+}, Ca^{2+}, and Sr^{2+} are displayed in Fig. 26.15. The ordered octahedral structure of the first hydration shell of Mg^{2+} and of the distorted hexahedral structure of the first hydration shell of Ca^{2+} agree with the neutron scattering results for these ions. Although the reported molecular dynamics simulations yield about the right hydration number for Sr^{2+}, that shell appears to be more ordered than is inferred from the neutron diffraction data.

What is the structure of water adjacent to a closely spaced anion and cation? Some early computer simulations by Watts address this question. Figures 26.16 and 26.17 show, respectively, the distribution of oxygen atoms and hydrogen atoms around Li^+ and F^- ions separated by 4 Å. Clearly, both the Li^+ and F^- ions have a strongly bound nearest-neighbor shell of oxygens, with some evidence for a diffuse second-neighbor shell of oxygens. The center of the $Li^+ \cdots O$ distance distribution is about 2 Å, that of the $F^- \cdots O$ distribution about 2.7 Å, but each has considerable

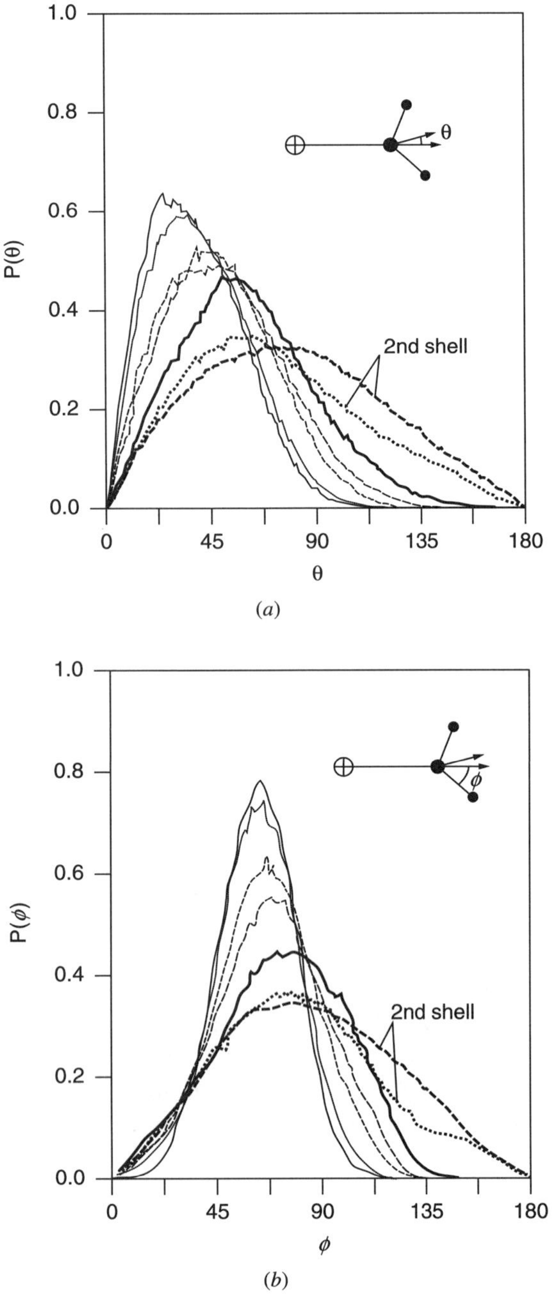

Figure 26.14 (*a*) Distribution of the angle between the water molecule dipole and the line between the cation and the oxygen in the first hydration shell of several cations (from top to bottom Li^+, Na^+, K^+, Rb^+, Cs^+). (*b*) Distribution of the angle between the direction of the OH bond of the water molecule and the line between the cation and the oxygen in the first and second hydration shells of several cations (from top to bottom Li^+, Na^+, K^+, Cs^+, Rb^+). From S. H. Lee and J. C. Rasaiah, *J. Chem. Phys.* **101,** 6964 (1994).

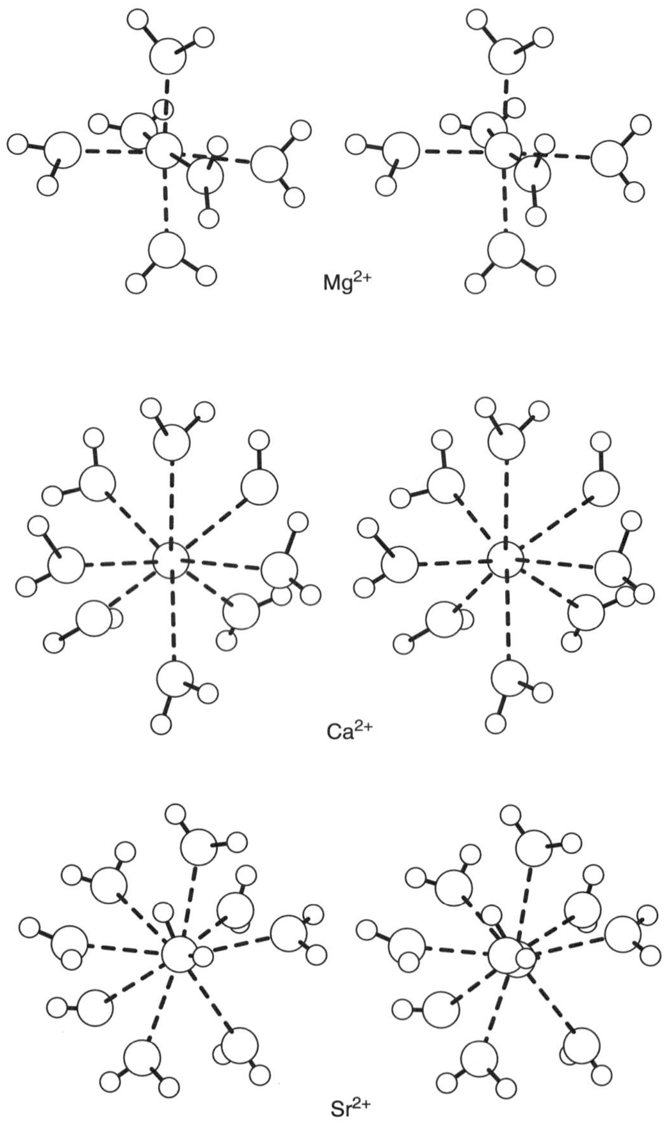

Figure 26.15 Snapshots of the hydration structure of several alkaline metal cations. From S. Obst and H. Bradaczek, *J. Phys. Chem.* **100,** 15677 (1996).

dispersion. The hydrogen atom distributions show a complementary pattern—there is a strongly bound nearest-neighbor shell around each ion and a second-neighbor shell about the F^- ion. The H atoms are, on average, closer to the F^- (~1.7 Å) than to the Li^+ (~3.0 Å). Obviously, the differences in atom locations shown in Figs. 26.16 and 26.17 reflect the influence of the charge distribution of the water molecule on the directionality of the ion–water interaction. Any given H_2O will orient so that the O atom is toward the cation and at least one of the H atoms is toward the anion. In fact, each of the four water molecules around the anion has only one H pointed toward it, and the second-neighbor distribution of H's arises from the other H on each water molecule. These second-neighbor H atoms are oriented so that hydrogen bonding to the surrounding water is possible. Beyond two coordination layers the water structure is no longer altered by the presence of the ion pair, although a charge–dipole interaction still contributes to the energy. In general, when two ions are close together, as at high solute concentration, their respective solvation shells are different than when the ions are far apart. Put another way, the hydra-

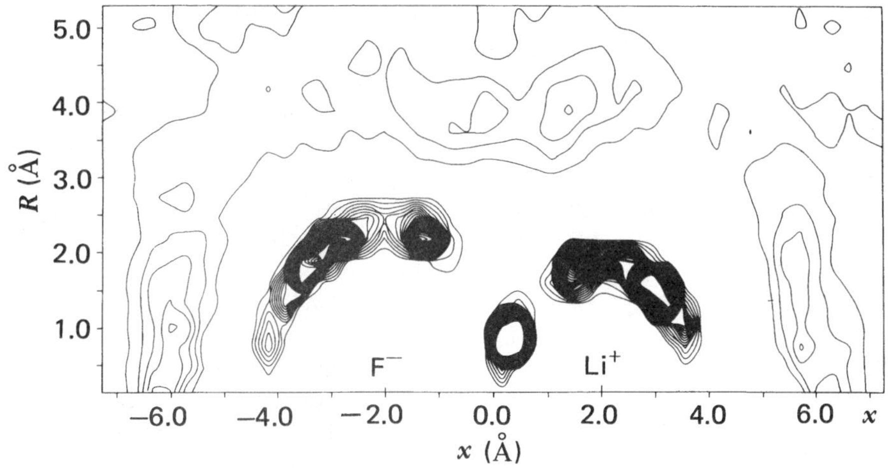

Figure 26.16 Distribution of oxygen atoms around the ion pair LiF for an ionic separation of 4.0 Å at 298 K. From R. O. Watts, *J. Chem. Phys.* **61,** 2550 (1974).

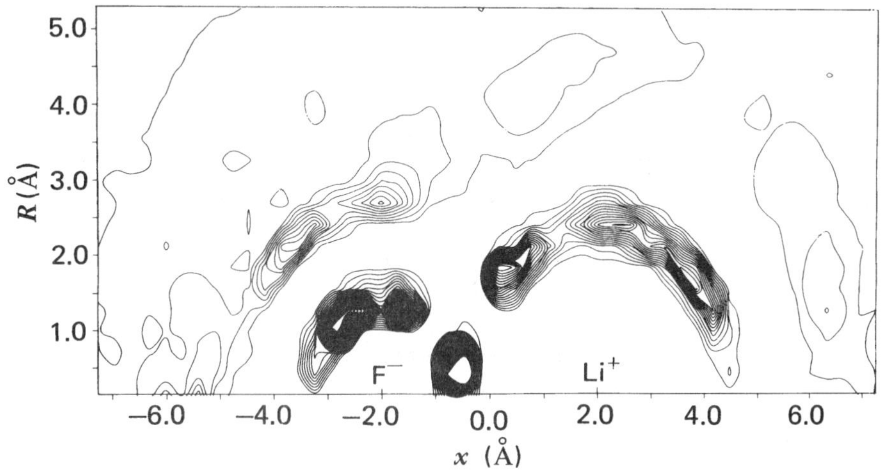

Figure 26.17 Distribution of hydrogen atoms around the ion pair LiF for an ionic separation of 4.0 Å at 298 K. From R. O. Watts, *J. Chem. Phys.* **61,** 2550 (1974).

tion structures of the ions in an electrolyte solution are not always independent of one another.

Prior to the availability of information about the hydration of ions from neutron diffraction experiments and computer simulations, several simple theories of the influence of the ion–solvent interaction on the thermodynamic properties of the electrolyte solution had been proposed. Although these can now be seen to be deficient in many respects, they are still worthy of consideration because of the fashion in which different important ionic properties are singled out as contributory to the thermodynamic properties of the solution. Moreover, these simple models permit categorization of ions in terms of such properties as diameter and charge, and give order-of-magnitude or better estimates of the thermodynamic properties of the ions in solution.

A very simple molecular model of the ion–solvent interaction was proposed by Born. The ion is regarded as a charged sphere, and the solvent as a dielectric continuum. Although this model cannot be an exact representation of the solution, and the effects of molecular structure in the solvent are not accounted for, the model is useful in providing a measure of the influence of the very strong electrostatic ion–solvent interaction. The potential at a distance R from the center of a charged sphere in vacuum is just $q/4\pi\epsilon_0 R$. The work required to bring up an element of charge dq from infinity to a distance R is then $q\,dq/4\pi\epsilon_0 R$. Thus, the total work required to charge up a sphere of radius a from zero to Ze is just

$$\int_0^{Ze} \frac{q\,dq}{4\pi\epsilon_0 a} = \frac{Z^2e^2}{8\pi\epsilon_0 a}, \tag{26.57}$$

since the potential at the surface of the sphere is $q/4\pi\epsilon_0 a$. Now let the sphere be placed in a continuous dielectric. We shall assume that there is no change in the structure of the dielectric because of insertion of the ion, and that the dielectric constant of the medium, D, is constant everywhere. Then, since in the dielectric the potential at a distance R

from a sphere with charge q is $q/4\pi\epsilon_0 DR$, the work of charging up the ion at constant temperature and pressure is

$$\int_0^{Ze} \frac{q\,dq}{4\pi\epsilon_0 Da} = \frac{Z^2e^2}{8\pi\epsilon_0 Da}. \qquad (26.58)$$

The dielectric constant of a liquid depends on the temperature and pressure, which was the reason for our stating the constancy of T and p. The difference between Eqs. 26.57 and 26.58 is the difference between the amounts of work for creating an ion in vacuum and in solution, at least according to this model. This difference is, then, the Gibbs free energy of hydration. Thus, we can write

$$\Delta g^{\ominus}_{\text{hyd}} = -\frac{N_A Z^2 e^2}{8\pi\epsilon_0 a}\left(1-\frac{1}{D}\right), \qquad (26.59)$$

where N_A is Avogadro's number. Equation 26.59 predicts an inverse dependence of the free energy of hydration on ionic radius, in qualitative agreement with experimental data. However, quantitative predictions based on Eq. 26.59 are not very good. For example, the free energies of hydration of Na^+, Ca^{2+}, Al^{3+}, and F^- in H_2O are predicted to be −699, −2920, −12340, and −515 kJ/mol, whereas the observed values are −406, −1582, −4599, and −431 kJ/mol, respectively. Although the sign and the order of magnitude are correct, discrepancies by a factor of 2 or 3 between prediction and observation are common.

It is not hard to point out a number of gross oversimplifications in the Born model: the use of a structureless dielectric continuum in place of the true structure of the solvent, the neglect of all bonding effects except the electrostatic ion–solvent continuum interaction, the assumption that the dielectric properties of the solvent are independent of distance from the ion, and so on. It is hard, however, to improve the model. It has been suggested that the ionic radii be adjusted empirically (for the case of water as solvent, by adding 0.75 Å to the radii of cations and 0.1 Å to the radii of anions) on the basis of the rationalization that the effective radius should be some combination of ionic radius and the width of the solvated layer. However, this adjustment accounts poorly for the very strong ion–bound layer interaction, and so cannot be considered satisfactory. Reversing the procedure, one can use empirical values of $\Delta g^{\ominus}_{\text{hyd}}$ to derive an effective dielectric constant, always bearing in mind that the local-structure-disrupting effects of ions are too complex to be summed up into a simple effective dielectric constant. The effective dielectric constants so derived are less than the bulk dielectric constant and vary with ion size in the expected manner, i.e., they increase toward the bulk value as the ion size increases. These effective dielectric constants are independent of charge for closed-shell cations, but for open-shell cations and anions new effects appear, specifically, indirect evidence of covalent bonding in the case of the open-shell cations.

There are a number of more detailed models of the local structure of the water surrounding an ion that have been used to calculate ionic enthalpies and entropies. In these it is assumed that the ion and the first-neighbor shell of water molecules form a chemical species with well-defined geometry.[18] Both tetrahedral and octahedral complexes have been suggested. Although the several models emphasize different aspects of the local ion–water structure, all agree that charge–dipole and charge–quadrupole interactions are important and all agree that the dipole moment of a water molecule bound to a monovalent ion is considerably larger (by as much as a factor of 2) than the dipole moment of the free water molecule. The agreement between the calculated and experimental thermodynamic properties of a hydrated ion obtained with the more complicated models is better than that given by the simple Born model, but it is still not completely satisfactory.

The best currently available calculations of the free energies of hydration of ions are obtained from computer simulations. The results of these simulations are sensitive to the assumed form of the ion–water and water–water interactions and to the accuracy with which the difficult calculation of the long range electrostatic interactions in the system is handled. The results obtained from several simulations are compared with experimental data in Table 26.5. The data displayed show both the uncertainties in the experimental data, the differences between the results of different simulations, and the discrepancies between current theory and the experimental data.

We previously noted that the properties of the transition metal ions differ considerably from those of simple closed-shell ions of the same size, that is, ions that have the ground-state term ${}^1S_0$. In all the models considered so far, it has been assumed that the field of the water molecules does not alter the electronic state of the ion. Somewhat better stated, it has been assumed that no degeneracies of the ionic electronic states are altered by the field of the water molecules, and that the lowest excited states of the ion are so high in energy that they do not influence the thermodynamic properties of the solution. But the transition metal ions have d^n electron configurations, and the degeneracy of the d-orbital states of the free ion can be lifted by a field of suitable symmetry. In fact, the lattice energies of the transition metal halides do

[18] The ions H_3O^+ and OH^- require special consideration because of the possibility of hydrogen bonding to other water molecules. M. Eigen has proposed a model in which the principal species present in these solutions are $H_9O_4^+$ and $H_7O_4^-$. In each case there is assumed to be approximate tetrahedral coordination about the O atom, as in ice, except that only three hydrogen bonds are formed. In this model the species H_3O^+ hydrogen-bonds only through its own OH bonds, whereas the species OH^- hydrogen-bonds only through the hydrogens of the other water molecules.

Table 26.5* Calculated and Experimental Free Energies of Ionic Hydration (kJ/mol)**

ion	calc[a]	calc[b]	calc[b]	calc[c]	expt[d]	expt[e]	expt[f]
Na^+	−398	−508	−431	−459	−365	−371	−372
K^+	−271	−425	−349	−321	−295	−298	−298
Ca^{2+}	−1306	−1623	−1394		−1505	−1553	
F^-	−580	−497	−421	−418	−465	−394	−441
Cl^-	−371	−315	−239	−237	−340	−277	−324
Br^-	−358	−304	−228		−315	−263	−310

* Adapted from G. Hummer, L. R. Pratt, and A. E. Garcia, *J. Phys. Chem.* **100,** 1206 (1996).

** The experimental data are adjusted to give the free energy of hydration for H^+ to be −1050 kJ/mol, so the values tabulated differ from those in Table 26.4.

[a] G. Hummer, L. R. Pratt, and A. E. Garcia, *J. Phys. Chem.* **100,** 1206 (1996).

[b] T. P. Straatsma and H. J. C. Berendsen, *J. Chem. Phys.* **89,** 5876 (1988).

[c] M. Migliore, G. Corongiu, E. Clementi and G. C. Lie, *J. Chem. Phys.* **88,** 7766 (1988).

[d] Y. Marcus, *J. Chem. Soc. Faraday Trans.* **87,** 2995 (1991).

[e] H. L. Friedman and C. V. Krishnan, in *Water: A Comprehensive Treatise,* ed. F. Franks, Plenum Press, New York, Vol. 3, p. 1.

[f] B. E. Conway, *J. Solution Chem.* **7,** 721 (1978).

Table 26.6 Ligand–Field Stabilization Energies in an Octahedral Field

Configuration	Stabilization[a]
d^0	0
d^1	$\frac{2}{5}\Delta_{oct}$
d^2	$\frac{4}{5}\Delta_{oct}$
d^3	$\frac{6}{5}\Delta_{oct}$
d^4	$\frac{3}{5}\Delta_{oct}$
d^5	0
d^6	$\frac{2}{5}\Delta_{oct}$
d^7	$\frac{4}{5}\Delta_{oct}$
d^8	$\frac{6}{5}\Delta_{oct}$
d^9	$\frac{3}{5}\Delta_{oct}$
d^{10}	0

[a]Δ_{oct} is the total splitting in an octahedral field of the degenerate d orbitals into orbitals with symmetries e_g and t_{2g}.

Table 26.7 Values of Octahedral-Field Stabilization Energies of Hexaaquo Complexes of Transition Metal Ions

Configuration	Ion	Δ_{oct}(kJ/mol)[a]
d^1	Ti^{3+}	245
d^2	V^{3+}	215
d^3	V^{2+}	146
	Cr^{3+}	211
d^4	Cr^{2+}	167
	Mn^{3+}	251
d^5	Mn^{2+}	90
	Fe^{3+}	167
d^6	Fe^{2+}	120
	Co^{3+}	—
d^7	Co^{2+}	120
d^8	Ni^{2+}	103
d^9	Cu^{2+}	160
d^{10}	Zn^{2+}	0

[a]Δ_{oct} is the total splitting, in an octahedral field, of the degenerate free-ion d orbitals into two groups of orbitals with symmetries t_{2g} and e_g.

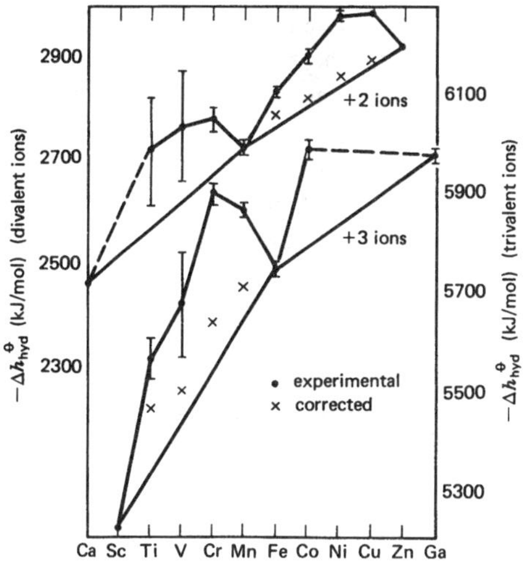

Figure 26.18 The heats of hydration of transition metal ions, showing the influence of ligand–field stabilization on the bulk thermodynamic properties. From D. S. McClure, in T. M. Dunn, D. S. McClure, and R. G. Pearson (Eds.), *Aspects of Crystal Field Theory* (Harper & Row, New York, 1965).

deviate from those expected on the basis of the simple Born–Mayer ionic model (see Chapters 9 and 14).

Spectroscopic studies of transition metal ions in solution are consistent with the existence of hexaaquo octahedral complexes of the ion and water. In these cases there appears to be ample evidence that on solution a reaction occurs and a well-defined compound is formed. The ligand–field stabilization energies in an octahedral field are tabulated in Table 26.6. Typical values of Δ_{oct} are 85–200 kJ/mol (see Table 26.7).

To analyze the enthalpy of hydration of transition metal ions we assume that, except for the existence of the ligand–field splitting, there would be a smooth variation of $\Delta h^{\ominus}_{hyd}$ with atomic number as the ionic size varies. As shown in Fig. 26.18, when the extra stabilization arising from the lowering of the energies of some *d* orbitals is subtracted, the heats of hydration of the divalent transition metal ions vary smoothly across the series. The properties of the trivalent ions still deviate from the expected behavior. Probably some covalent bonding is important in these cases. This simply means that the elementary electrostatic interpretation of the properties of the hexaaquo ion complex must be replaced by a more refined model.

To sum up, all of the interpretations of the alteration of the structure of water caused by inserting a simple closed-shell ion agree that the factors leading to a large negative enthalpy of hydration also lead to a large negative entropy of hydration. These factors are, primarily, small ionic radius and high ionic charge. Deductions from all the theories also suggest that ions cause an alteration of the structure of the water, over and above that describable as dipole orientation or restriction of rotational motion. In solutions of small ions the orienting effects of high field outweigh those associated with altering the hydrogen-bond network of water. In solutions of large ions the solvent orientation is diminished relative to that in a solution of small ions because the field is weaker; however, the effects on the hydrogen-bond network are increased because of the greater size of the ion. None of the theories available adequately describes nonelectrostatic interactions, except in very special cases. In one such case, the properties of transition metal ions in solution are qualitatively described by a model in which the ion reacts with water to form a hexaaquo complex. Clearly, this reaction leads to exaggeration of the alteration of the structure of water arising from electrostatic and nonbonding forces.

26.5 The Statistical Mechanics of Electrolyte Solutions

We have already argued that the existence of a Coulomb potential between the ions of an electrolyte solution grossly differentiates that solution from a nonelectrolyte solution. The counterpart of this macroscopic difference occurs in the molecular theory; the long range of the ion–ion interaction prevents use of the kind of analysis developed for nonelectrolyte solutions. In fact, because the ions in a solution are mobile and the Coulomb interaction at small separations is large relative to the thermal energy k_BT, positive ions tend to be surrounded by negative ions, and *vice versa.* When the ions are arranged as described, the effective interaction between them is altered and becomes short-range. To represent the properties of electrolyte solutions we shall first discuss a primitive model, introduced by P. Debye and E. Hückel in 1923. This model is a mixture of microscopic and macroscopic concepts, based on a picture of ions as impenetrable charged spheres in a solvent treated as a continuous and structureless dielectric. The Debye–Hückel analysis of the properties of this model is based on easily stated physical ideas and an approximation that is valid only in the very low electrolyte concentration regime; it has provided the basis for our understanding of the influence of coulomb interactions on the thermodynamic and transport properties (Chapter 29) of dilute electrolyte solutions. To connect the ideas introduced in the Debye–Hückel theory with those used in the modern theory of liquids we then discuss the same model using the kind of analysis employed in Chapters 23 and 25. Finally, we shall indicate how the structure of the modern theory of electrolyte solutions can be used to develop an accurate semiempirical theory of the thermodynamic properties of those solutions.

A. Debye–Hückel Theory

Because the interaction between a pair of ions depends on their separation, it is clear that our first step must be to examine the pair distribution function (see Chapter 23). Consider an electrolyte solution occupying volume V and containing $N_1, N_2, \ldots N_r$ ions of species $1, 2, \ldots r$. Suppose that an ion of species i is at the origin of a coordinate system. The probability per unit volume of finding an i ion at the origin and, simultaneously, an ion of species j at the point $\mathbf{R}_j$ is

$$\frac{N_i}{V}\frac{N_j}{V}\exp\left(-\frac{w_{ij}}{k_BT}\right), \tag{26.60}$$

where w_{ij} is the potential of mean force (see Section 23.3) acting between ions i and j. Recall that w_{ij} is also the work required to bring i and j from infinite separation to the required configuration, in the presence of all the other ions and solvent. Because of the nature of interionic forces, w_{ij} must have some short-range repulsive part and some longer-range modified Coulombic part. The analysis assumes that:

1. The short-range repulsive part of w_{ij} can be treated as if the ions were impenetrable spheres of radius a.
2. The long-range modified Coulombic part of w_{ij} can be represented as $eZ_j\Phi_i(\mathbf{R}_j)$ where $\Phi_i(\mathbf{R}_j)$ is the average electrostatic potential at $\mathbf{R}_j$ arising from the ion i at the origin and all the other ions in the system.
3. The electrostatic potential Φ_i satisfies the Poisson equation when $|\mathbf{R}| \geq a$,

$$\nabla^2\Phi_i = -\frac{1}{\epsilon_0 D}q_i(\mathbf{R}), \quad (|\mathbf{R}| \geq a), \tag{26.61}$$

and the Laplace equation when $|\mathbf{R}| < a$,

$$\nabla^2\Phi_i = 0, \quad (|\mathbf{R}| < a). \tag{26.62}$$

In Eq. 26.61, D is the dielectric constant of the medium and $q_i(\mathbf{R})$ is the charge density at $\mathbf{R}$, given that ion i is at the origin of coordinates.

4. The charge density results from the distribution of the charges in the medium. We do not yet know this distribution, but whatever it is it must be consistent with Eq. 26.60. We assume that, when $|\mathbf{R}| \geq a$,

$$q_i(\mathbf{R}) = \sum_\alpha eZ_\alpha \frac{N_\alpha}{V} \exp\left[-\frac{eZ_\alpha \Phi_i}{k_B T}\right], \quad (|\mathbf{R}| \geq a). \tag{26.63}$$

Note that this description of the charge distribution neglects all short-range ion–ion interactions, since it is obtained from Eq. 26.60 by using just the long-range component of w_{ij}.

Combining Eqs. 26.61 and 26.63, we obtain the *Poisson–Boltzmann equation,*

$$\nabla^2 \Phi_i = -\frac{1}{\epsilon_0 D} \sum_\alpha eZ_\alpha \frac{N_\alpha}{V} \exp\left[-\frac{eZ_\alpha \Phi_i}{k_B T}\right], \quad (|\mathbf{R}| \geq a). \tag{26.64}$$

The solution of Eq. 26.64, in principle, permits calculation of the ion–ion distribution function and then the thermodynamic properties of the electrolyte solution. However, Eq. 26.64 cannot be solved analytically, so numerical methods must be used.

There is one case, fortunately of considerable interest, for which Eq. 26.64 can be reduced to a simpler form and solved. Moreover, that simplified equation is consistent in its treatment of w_{ij}, whereas Eq. 26.64 is not.[19] Suppose that the ionic solution is sufficiently dilute that the average interionic spacing corresponds to $\Phi_i/k_B T << 1$. In this limit, expanding the exponential in Eq. 26.64, we find that

$$\nabla^2 \Phi_i(R) = -\frac{e}{\epsilon_0 D} \sum_\alpha \frac{N_\alpha Z_\alpha}{V} + \left[\frac{e^2}{\epsilon_0 D k_B T} \sum_\alpha Z_\alpha^2 \frac{N_\alpha}{V}\right] \Phi_i(R), \quad (|\mathbf{R}| \geq a). \tag{26.65}$$

The condition that the solution be electrically neutral requires that

$$\sum_\alpha eZ_\alpha \frac{N_\alpha}{V} = 0. \tag{26.66}$$

Thus, using the definition

$$\kappa^2 \equiv \frac{e^2}{\epsilon_0 D k_B T} \sum_\alpha Z_\alpha^2 \frac{N_\alpha}{V}, \tag{26.67}$$

we can reduce Eq. 26.65 to

$$\nabla^2 \Phi_i(R) = \kappa^2 \Phi_i(R), \quad (|\mathbf{R}| > a). \tag{26.68}$$

In our model the ions are spheres, and the only distance dependence is through the ionic separation R. In this case, Eq. 26.68 becomes

$$\frac{1}{R} \frac{d^2(R\Phi_i)}{dR^2} = \kappa^2 \Phi_i. \tag{26.69}$$

To solve Eq. 26.69 we rewrite it in the form

$$\frac{d^2}{dR^2}(R\Phi_i) = \kappa^2 (R\Phi_i), \quad (|\mathbf{R}| \geq a). \tag{26.70}$$

Direct substitution shows that

$$R\Phi_i = C_1 e^{-\kappa R} + C_2 e^{\kappa R} \tag{26.71}$$

is a solution. But the potential must approach zero as the distance from the central ion gets indefinitely large,

$$\lim_{R \to \infty} \Phi_i = 0, \tag{26.72}$$

so that $C_2 = 0$ and we find that

$$\Phi_i = \frac{C_1 e^{-\kappa R}}{R}, \quad (|\mathbf{R}| \geq a) \tag{26.73}$$

is the proper solution.

To complete the solution we must evaluate the constant C_1. This can be done by solving Eq. 26.62 and requiring that the electrostatic potential and the electrostatic field be continuous at the surface of the ion. We use a different method here. Our method is based on the conservation of charge. If a positive ion is at the origin of coordinates, in any spherical shell at distance R from it there will, on average, be more negative ions than positive ions; the spherical shell will therefore have net negative charge. The totality of all shells surrounding the cation at the origin must carry a negative charge just equal to the positive charge of the ion. This condition of overall electroneutrality may be written in the form

$$\int_a^\infty 4\pi R^2 q_i(R)\, dR = -Z_i e. \tag{26.74}$$

The charge density is, of course, related to the electrostatic potential by Eq. 26.63. However, we are solving Eq. 26.61

[19] See later comments.

in the domain in which the electrostatic interaction is small compared with k_BT (see Eq. 26.68). Comparison of Eqs. 26.61, 26.68, and 26.73 shows that

$$q_i(R) = -\kappa^2 \epsilon_0 D \Phi_i = -C_1 \kappa^2 \epsilon_0 D \frac{e^{-\kappa R}}{R}. \quad (26.75)$$

Using Eq. 26.75 in Eq. 26.74 leads to

$$4\pi C_1 \kappa^2 \epsilon_0 D \int_a^\infty e^{-\kappa R} R \, dR = Z_i e, \quad (26.76)$$

or

$$C_1 = \frac{Z_i e}{4\pi\epsilon_0 D} \frac{e^{\kappa a}}{1+\kappa a}, \quad (26.77)$$

so that the potential Φ_i is given by

$$\Phi_i = \frac{Z_i e}{4\pi\epsilon_0 D} \frac{e^{\kappa a}}{1+\kappa a} \frac{e^{-\kappa R}}{R}. \quad (26.78)$$

One of the fundamental principles of electrostatics is that of linear superposition of potentials. This principle states that the total electric potential at some point may be treated as a sum of the potentials arising from all sources in the medium. In our case, Eq. 26.78 represents the total electrostatic potential at a distance R from an ion with charge Z_ie. We now treat this total electrostatic potential as the sum of the direct potential of the ion at the origin,

$$\Phi_i'' = \frac{Z_i e}{4\pi\epsilon_0 D R}, \quad (26.79)$$

and the potential due to all the other ions in the solution,

$$\Phi_i' = \frac{Z_i e}{4\pi\epsilon_0 D R}\left[\frac{e^{\kappa(a-R)}}{1+\kappa a} - 1\right]. \quad (26.80)$$

Equation 26.80 is obtained by using the superposition principle

$$\Phi_i = \Phi_i' + \Phi_i'' \quad (26.81)$$

and subtracting Eq. 26.79 from Eq. 26.78. In our model Eq. 26.80 is valid for all R down to $R = a$. No ions can penetrate the region $R < a$, and for all $R < a$ the potential due to the spherical distribution of ions outside the central ion is constant and equal to its value at $R = a$. We write, then,

$$\Phi_i'(a) \equiv \Phi_i^{\text{atm}} = -\frac{Z_i e}{4\pi\epsilon_0 D} \frac{\kappa}{1+\kappa a}. \quad (26.82)$$

Thus, the electrostatic energy of the central ion itself is reduced by the product of its charge Z_ie and the potential 26.82 because of the interaction with all other ions in the solution. Corresponding to the existence of ionic electrostatic interaction energy, the chemical potential of the ions is different from that of molecules in an ideal solution. We now calculate the activity coefficients of the ions under the assumption that the solution would be ideal if there were no electrostatic interactions, that is, that the short-range ion–solvent and ion–ion forces do not lead to nonideality.

From the discussion of Section 19.4 we can infer that μ_i is the Gibbs free energy change when 1 mol (or 1 molecule, depending on units) of solute is added to the solution at constant temperature and pressure. In the standard representation of the chemical potential of a solute, $\mu^\ominus + RT \ln m_i$ is the part of μ_i that does not involve solute–solute interaction and $RT \ln \gamma_i$ is the part of the chemical potential arising from solute–solute interaction. In other words, $RT \ln \gamma_i$ is the work (at constant T and p) done against intermolecular forces when 1 mol of solute is added to the solution.

Suppose that it were possible to discharge an ion continuously. We assume that the uncharged ion, which we call a ghost ion, does not interact with the other ions. If the ghost ion is recharged, the work done in charging up the ion, allowing the other ions to adjust to the new equilibrium at every point along the charging process, is just $RT \ln \gamma_i$. This follows because, under our assumptions, addition of a ghost ion to a solution does not contribute to the interaction energy. But the final state obtained when either a fully charged ion is added to the solution, or a ghost ion is added and then charged, is the same. Hence the Gibbs free energy changes for the two processes must be the same.

Suppose that it were possible, as just suggested, to discharge an ion. If the fractional charge on ion i is ξ_i, then the atmosphere contribution to the potential is

$$\Phi_i^{\text{atm}}(\xi_i) = -\frac{\xi_i Z_i e \kappa}{4\pi\epsilon_0 D(1+\kappa a)}. \quad (26.83)$$

To bring up an amount of charge $Z_ie \, d\xi_i$ to the ion i, the work done is just the product of the charge and the ion–atmosphere potential, $\Phi_i^{\text{atm}} Z_ie \, d\xi_i$. Therefore, we obtain

$$\begin{aligned} k_B T \ln \gamma_i &= \int_0^1 \Phi_i^{\text{atm}} Z_i e \, d\xi_i \\ &= -\int_0^1 \frac{\xi_i Z_i^2 e^2 \kappa}{4\pi\epsilon_0 D(1+\kappa a)} d\xi_i \\ &= -\frac{Z_i^2 e^2 \kappa}{8\pi\epsilon_0 D(1+\kappa a)}. \end{aligned} \quad (26.84)$$

Here γ_i is, by definition, the activity coefficient of the ion i. The single–ion activity coefficients must be combined as described earlier in this chapter before comparison with experiment is possible. The result is, using Eq. 26.12 and $|\nu_+ Z_+| = |\nu_- Z_-|$,

$$\ln \gamma_\pm = -\frac{e^2 \kappa}{8\pi\epsilon_0 D k_B T}\left(\frac{\nu_+ Z_+^2 + \nu_- Z_-^2}{\nu_+ + \nu_-}\right)$$
$$= -|Z_+ Z_-| \frac{e^2 \kappa}{8\pi\epsilon_0 D k_B T(1+\kappa a)}. \tag{26.85}$$

Equation 26.84 is the principal result of the Debye–Hückel theory of electrolyte solutions. From it we deduce that:

1. Since κ is proportional to the square root of the concentration (see Eq. 26.67), we recover the empirically discovered rules concerning ionic strength.
2. In the limit of very small concentration, we obtain

$$\lim_{\kappa a \to 0} k_B T \ln \gamma_i = \frac{Z_i^2 e^2 \kappa}{8\pi\epsilon_0 D}, \tag{26.86}$$

which is known as the *limiting law value.* It is found that the limiting law form for the activity coefficient (and other derived quantities) is exact for the model considered. On the other hand, Eq. 26.84 and similar extensions are not completely correct because of the approximations made to w_{ij}. Only in the limit of infinite dilution are the cited assumptions valid.

3. We can verify the qualitative picture of the screening given earlier by the following arguments. First, expansion of the exponential of Eq. 26.60 shows that the distribution of pairs of ions is of the form

$$\frac{N_i N_j}{V^2}\left(1 - \frac{Z_i Z_j e^2}{4\pi\epsilon_0 k_B T}\frac{e^{-\kappa R}}{R}\right), \tag{26.87}$$

so that the correlation between two ions extends over a distance of the order of $(1/\kappa)$, the *Debye screening length.* When $\kappa R >> 1$, the ions are independent, since then their interaction is screened out. Moreover, as anticipated, the screening arises because the ions distribute themselves in such a way that a cloud of charge is interposed between any two ions. To see how the screening charge arises, we calculate the total charge $Q_i(B)$ outside the ion i but inside a sphere of radius B. This is, in the limiting-law case,

$$Q_i(B) = -\int_0^B d\mathbf{R} \sum_\alpha \frac{N_\alpha}{V} e^3 Z_i Z_\alpha^2 \frac{e^{-\kappa R}}{4\pi\epsilon_0 D R}$$
$$= -eZ_i \kappa^2 \int_0^B dR R e^{-\kappa R}$$
$$= -eZ_i[1 - e^{-\kappa B}(1+\kappa B)]. \tag{26.88}$$

Clearly, as $\kappa B \to \infty$, we find that $Q_i(B) \to -Z_i e$, just equal and opposite to the charge of ion i. For any given value of B, an ion "outside" B sees a net charge less than $Z_i e$, a result that shows how interposition of other ions reduces the effective charge from $Z_i e$ to some smaller value.

A comparison of some predicted and observed thermodynamic properties of dilute electrolyte solutions is displayed in Fig. 26.19. Note the good agreement near zero concentration.

We have remarked that Eq. 26.64 is not self-consistent. The inconsistency arises from two approximations in the derivation. The first of these is easy to see. As shown in Chapter 23, the pair-correlation function for simple spherical molecules has oscillatory structure for a range of separations of the order of a few molecular diameters. This structure in the pair-correlation function is suppressed by the assumption that the potential of mean force is $eZ_j\Phi_i(R_j)$ for all $|\mathbf{R}| \geq a$, hence the influence of short-range interactions on the ionic distribution is omitted except for the region $|\mathbf{R}| < a$. The second approximation is more subtle. The mean electrostatic potential depends on the distribution of pairs of ions. Just as in the theory of simple liquids, the determination of the pair-correlation function requires knowledge of the distribution of triplets of ions, and so on (see Eq. 23.23). The assumption that the pair potential of mean force $w_2 = w_{ij}$ is $eZ_j\Phi_i(R_j)$ is equivalent to the assumption that the potential of mean force between three molecules is the sum of the three pair potentials of mean force between them: $w_3 = w_2(1, 2) + w_2(1, 3) + w_2(2, 3)$. This approximation neglects interference effects in the distribution of three molecules relative to the superposed distributions of pairs of molecules. It is exact in the limit that the density of molecules becomes vanishingly small, but otherwise in error.

There are a variety of ways of showing that the Debye–Hückel approximation becomes exact in the limit of zero electrolyte concentration. Although the detailed analysis is complicated, it is possible to see that this is likely the case by noting that both of the approximations used to derive Eq. 26.64 become exact in the limit of zero concentration, the first because the pair correlations induced by short-range interactions vanish, and the second because interference between the configurations of widely separated pairs of molecules vanishes.

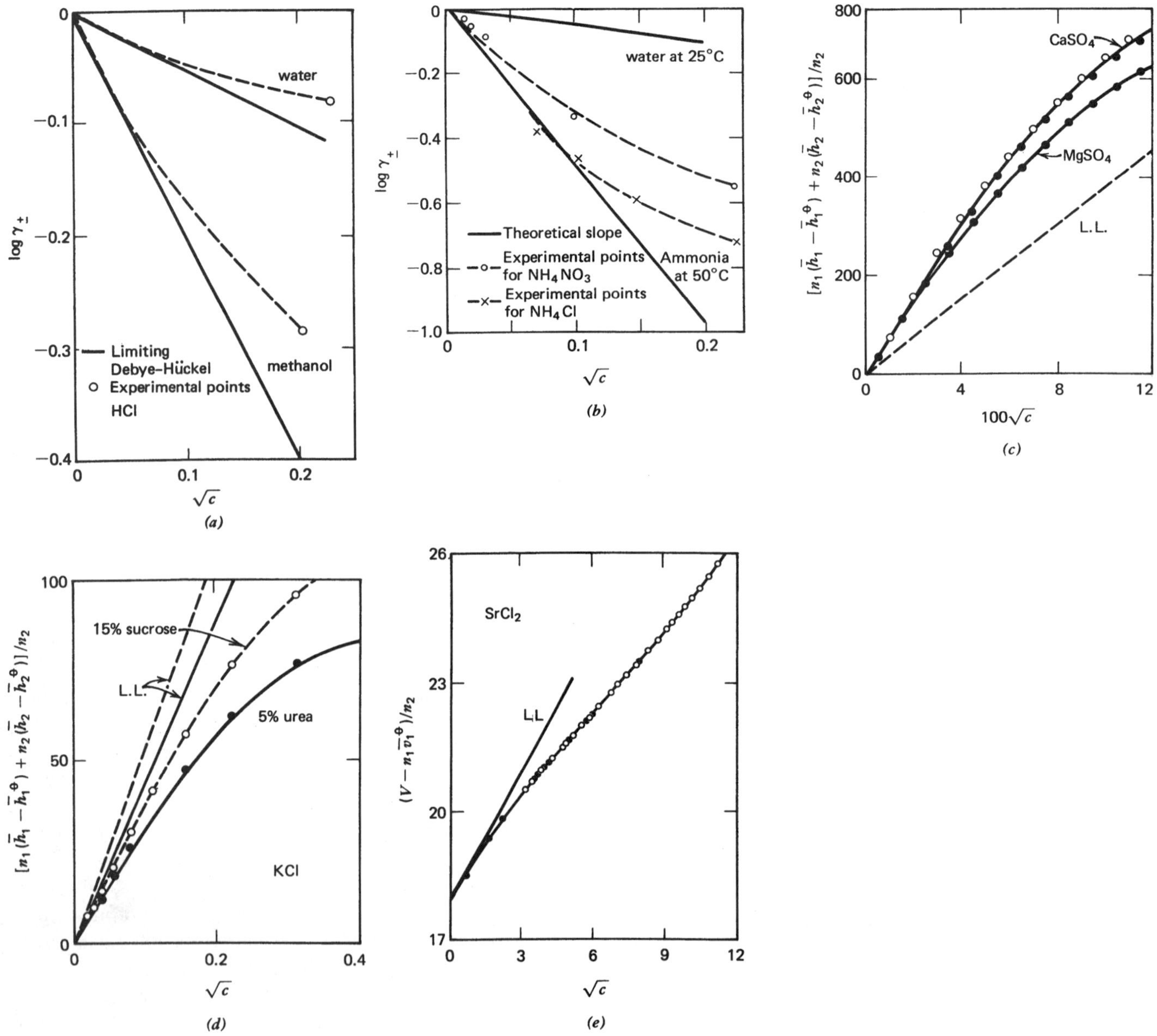

Figure 26.19 (*a*) Activity coefficients of hydrochloric acid in water (upper curves) and methanol (lower curves) at 25°C. (*b*) Activity coefficients of ammonium chloride and nitrate in ammonia at −50°C. (*c*) Relative apparent molal enthalpies of calcium sulfate and magnesium sulfate in water at 25°C. L.L. denotes the Debye–Hückel limiting law behavior. (*d*) Relative apparent molal enthalpy of KCl in mixed solvents at 25°C; the straight lines represent the theoretical limiting law. (*e*) Apparent molal volume of strontium chloride in water at 25°C; note the departure from the theoretical limiting law at moderate dilutions. In all these figures *c* is the molality of the solution.
(*a*) and (*b*) from R. A. Robinson and R. H. Stokes, *Electrolyte Solutions* (Butterworths, London, 1955); (*c*), (*d*), and (*e*) from H. Harned and B. B. Owen, *Physical Chemistry of Electrolytic Solutions* (Van Nostrand Reinhold, New York, 1958).

B. An Alternative View of the Debye–Hückel Approximation

It is interesting to connect the original Debye–Hückel analysis of the properties of an electrolyte solution, described in the last section, with an analysis using the methods developed to describe liquids (Chapter 23). Recall that one exact representation of the pair correlation function of a liquid is the Ornstein–Zernike relation, Eq. 23.24, which we rewrite, using the definition $h(R) \equiv g_2(R) - 1$, in the form

$$h(R_{ij}) = C(R_{ij}) + n\int \mathbf{dR}_k C(R_{jk})h(R_{jk}). \qquad (26.89)$$

The Fourier transform of Eq. 26.89 is

$$(1+n\tilde{h})(1-n\tilde{C})=1, \tag{26.90}$$

or

$$\tilde{h}=\tilde{C}+\tilde{C}n\tilde{C}+\tilde{C}(n\tilde{C})^2+\cdots. \tag{26.91}$$

When the liquid is a multicomponent system, as is the case for an electrolyte solution, we must generalize the functions appearing in Eq. 26.89. For components α and β we define

$$\begin{aligned}\mathsf{h}_{\alpha\beta}(\mathbf{R}_i,\mathbf{R}_j)&\equiv(n_\alpha n_\beta)^{1/2}h_{\alpha\beta}(\mathbf{R}_i,\mathbf{R}_j),\\ \mathsf{C}_{\alpha\beta}(\mathbf{R}_i,\mathbf{R}_j)&\equiv(n_\alpha n_\beta)^{1/2}C_{\alpha\beta}(\mathbf{R}_i,\mathbf{R}_j),\end{aligned} \tag{26.92}$$

with $n_\alpha = N_\alpha/V$, etc. The definitions 26.92 can be thought of as the $\alpha\beta$ elements of matrices **h** and **C**. Then the Fourier transform of the multicomponent Ornstein–Zernike equation can be written

$$(1+\tilde{\mathbf{h}})\cdot(1-\tilde{\mathbf{C}})=\mathbf{I}, \tag{26.93}$$

and the expansion equivalent to 26.91 reads

$$\tilde{\mathbf{h}}=\tilde{\mathbf{C}}+\tilde{\mathbf{C}}\cdot\tilde{\mathbf{C}}+\tilde{\mathbf{C}}\cdot\tilde{\mathbf{C}}\cdot\tilde{\mathbf{C}}+\cdots. \tag{26.94}$$

In Eqs. 26.93 and 26.94 the tilde indicates the function is a Fourier transform. We remarked in Chapter 23 that in the limit $R\to\infty$ the direct correlation function, $C(R)$, has the form

$$\lim_{R\to\infty} C(R)=-\frac{u_2(R)}{k_BT}, \tag{26.95}$$

where $u_2(R)$ is the interaction potential between a pair of particles. We now assume that this asymptotic form for $C(R)$ is valid for all R, and that the interaction between a pair of ions has the form

$$\begin{aligned}u_{ij}&=\Phi_i(R)Z_je,\\ \nabla^2\Phi_i(|\mathbf{R}|)&=-\frac{1}{\epsilon_0 D}Z_ie\delta(\mathbf{R}).\end{aligned} \tag{26.96}$$

A convenient way to solve Eq. 26.96 is to Fourier transform both sides of the differential equation; the result is

$$\tilde{\Phi}_i(k)=\frac{Z_ie}{4\pi\epsilon_0 Dk^2}, \tag{26.97}$$

where k is the transform variable. The corresponding forms of the $\alpha\beta$ component (matrix element) of the direct correlation function, and its Fourier transform, are

$$\begin{aligned}\mathsf{C}_{\alpha\beta}&=-(n_\alpha n_\beta)^{1/2}\frac{Z_\alpha Z_\beta e^2}{4\pi\epsilon_0 Dk_BTR},\\ \tilde{\mathsf{C}}_{\alpha\beta}&=-(n_\alpha n_\beta)^{1/2}\frac{Z_\alpha Z_\beta e^2}{\epsilon_0 Dk_BTk^2}.\end{aligned} \tag{26.98}$$

Proceeding in a similar fashion, we find for the matrix elements of the products $\tilde{\mathbf{C}}\cdot\tilde{\mathbf{C}}$, $\tilde{\mathbf{C}}\cdot\tilde{\mathbf{C}}\cdot\tilde{\mathbf{C}}$, etc.,

$$\begin{aligned}(\tilde{\mathbf{C}}\cdot\tilde{\mathbf{C}})_{\alpha\beta}&=-\tilde{\mathsf{C}}_{\alpha\beta}\frac{\kappa^2}{k^2},\\ (\tilde{\mathbf{C}}\cdot\tilde{\mathbf{C}}\cdot\tilde{\mathbf{C}})_{\alpha\beta}&=-\tilde{\mathsf{C}}_{\alpha\beta}\left(\frac{\kappa^2}{k^2}\right)^2, \text{ etc.}\end{aligned} \tag{26.99}$$

Substituting Eq. 26.99 into Eq. 26.94 generates a simple geometric series which can be summed; the result is

$$\begin{aligned}\tilde{\mathsf{h}}_{\alpha\beta}(k)&=\frac{k^2\tilde{\mathsf{C}}_{\alpha\beta}}{k^2+\kappa^2},\\ \tilde{h}_{\alpha\beta}(k)&=-\frac{Z_\alpha Z_\beta e^2}{\epsilon_0 Dk_BT(k^2+\kappa^2)},\end{aligned} \tag{26.100}$$

or, in real space,

$$h_{\alpha\beta}(R)=-\frac{Z_\alpha Z_\beta e^2 e^{-\kappa R}}{4\pi\epsilon_0 DR}. \tag{26.101}$$

Equation 26.101, with the definition $h_{\alpha\beta}(R)\equiv g_{\alpha\beta}(R)-1$, yields the Debye–Hückel pair-correlation function and, of course, the Debye–Hückel potential of mean force. The thermodynamic properties of the electrolyte solution are then calculated using Eq. 25.195 for the osmotic pressure and keeping only the second virial coefficient term displayed in Eq. 25.196. From the osmotic coefficient one can obtain the solute activity coefficient (Eq. 25.125). The result obtained is the Debye–Hückel limiting law relation for the activity coefficient of the electrolyte solution.

C. Beyond the Debye–Hückel Approximation

The Debye–Hückel theory becomes inaccurate for concentrations in excess of about 10^{-3} *M*. There is a vast literature that deals with extensions of the theory of electrolytes to more concentrated solutions. The different approaches can be sorted into four categories.

1. First, and simplest, are theories that retain the essential structure of the linearized Poisson–Boltzmann equation and use models of one sort or another to extend the range of its validity. The best known of these is the Bjerrum theory of ion association. In this theory it is assumed that some of the ions form neutral pairs and thereby do not participate in generating the mean electrostatic potential in the solution. The concentrations of ion pairs and free ions are related by an equilibrium constant, hence vary as the electrolyte concentration varies. The simplest form of this theory considers as paired only those ions in contact; other versions broaden the ion–pair concept to include oppositely charged ions separated by only one solvent molecule, or all oppositely charged ions that have mutual attraction larger than $2k_BT$.

It is useful to summarize the main features of the Bjerrum ion pairing theory because the concepts introduced will reappear in other analyses of the properties of electrolyte solutions. As before, we consider the ions to be charged hard spheres of radius a in a structureless dielectric continuum with dielectric constant D. The association constant for the equilibrium between the ion pair C^+A^- and the free cation C^+ and anion A^-, $C^+ + A^- \leftrightarrow C^+A^-$, is

$$K = \frac{\alpha m \gamma_\pm^2}{(1-\alpha)^2} = \text{constant} \times \int_a^q e^{2q/R} R^2 \, dR, \qquad (26.102)$$

where α is the fraction of ions that are paired. The ion pairs are assumed not to contribute to the electrostatic interactions in the solution, and the Debye screening length is now determined entirely by the free ion concentration, so that $\kappa^B = (1 - \alpha)^{1/2}\kappa$. The expression for the association constant assumes that the activity coefficient of the ion pairs is taken to be unity and that the free ions have activity coefficients of the Debye–Hückel form. In the integral expression for the association constant, $q \equiv |Z_+Z_-e^2/8\pi\epsilon_0 Dk_BT|$ has the dimensions of length. In fact, when a pair of singly charged anions and cations are separated a distance q their electrostatic interaction energy is just $2k_BT$. The integrand of 26.102 has a minimum at q, so this value for the upper limit of integration is chosen to define an ion pair as a cation and anion separated by a distance in the range $a \le R \le q$. As shown in Fig. 26.19*c*, the thermodynamic properties of aqueous 2:2 electrolyte solutions deviate from the Debye–Hückel limiting law behavior at the lowest concentrations for which accurate measurements are feasible. In contrast, the simple Bjerrum theory modification of the Debye–Hückel analysis agrees rather well with the experimental data (Fig. 26.20).

The theory of ion pairing has some *ad hoc* features, so that it appears that the improved representation of experimental data has more of the character of a parametric fit than the character of a prediction. If viewed as a scheme to generate a parametric fit to experimental data it is the case that, by adjustment of the ionic radii, which are the parameters that determine the ion–pair equilibrium constant, the activity coefficients and other thermodynamic properties of many electrolyte solutions can be fitted over a much larger concentration range than by Debye–Hückel theory. However, this view of the ion pairing theory is too restrictive, since there are observations, for example, absorption spectra, that are reasonably interpreted as indicating that ion pairing occurs for high-valency electrolytes even at room temperature in water, for lower-valency electrolytes at high temperature in aqueous solution, and for many electrolytes in low-dielectric–constant solvents. It is also found that ion pairing plays a major role in determining the critical behavior of an electrolyte solution,[20] and it has been incorporated into integral equation theories of electrolyte solution behavior.[21]

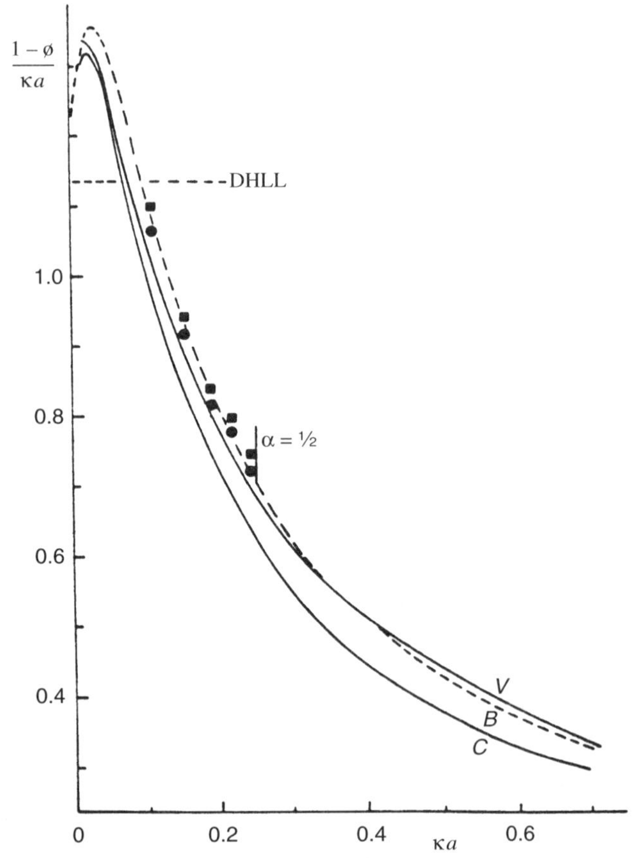

Figure 26.20 Practical osmotic coefficient of a 2:2 aqueous electrolyte solution as a function of ionic concentration. The ion core diameter was taken to be 4.2 Å; the reduced temperature is $8\pi a\epsilon_0 Dk_BT/|Z_+Z_-e^2| = 0.1468$. The solid lines are obtained from the HNC integral equation representation of the primitive model of the electrolyte solution; the curve labeled V is calculated from the virial equation and that labelled C is calculated from the compressibility equation (see Chapter 23). The dashed curve labelled B is obtained from the Bjerrum ion pairing approximation. The experimental data are for aqueous solutions of $ZnSO_4$ at 0°C. The point on the Bjerrum curve where the ions are 50% paired is indicated as $\alpha = 1/2$. From H. L. Friedman and B. Larsen, *J. Chem. Phys.* **70,** 92 (1979). The practical osmotic coefficient is defined by the equation $-\pi v_1^0 = \phi RT\,(m_1/1000)\,\sum_i m_i$ in place of Eq. 25.125.

2. Second, the full uncorrected Poisson–Boltzmann equation can be integrated numerically, thereby avoiding the linearization approximation that leads to Eq. 26.65. However, the use of the full nonlinear Poisson–Boltzmann equation leads to a thermodynamic inconsistency, in the sense that different equally valid procedures used to calculate the activity coefficient lead to (slightly) different results, just as different procedures used to calculate the pressure of a liquid from an approximate integral equation theory of the distribution function lead to different results (Chapter 23).

[20] See, for example, Y. Levin and M. E. Fisher, *Physica* **A225,** 164 (1996); M. E. Fisher, *J. Phys.: Condens. Matter* **8,** 9103 (1996).

[21] W. Ebeling and M. Grigo, *Ann. Phys.* **37,** 21 (1980); *J. Solution Chem.* **11,** 151 (1982); S. Yeh, Y. Zhou, and G. Stell, *J. Phys. Chem.* **100,** 1415 (1996).

Table 26.8 The Osmotic Coefficent Given by the Various Theories Discussed for the 1:1 Primitive Electrolyte Model (Hard-Core Ions in a Continuous Dielectric Fluid) Compared with Numerical Simulations for the Same Model

M[a]	DH[b]	PB[c]	MPB[d]	PY(p)[e]	PY(C)[f]	AC[g]	Simulation[h]
0.00911	0.968	0.970	0.970	0.9703	0.9705	0.9707	0.9701 ± 0.0008
0.10376	0.933	0.946	0.945	0.94542	0.9461	0.9452	0.9445 ± 0.0012
0.42502	0.949	0.981	0.980	0.9765	0.9844	0.9787	0.9774 ± 0.0046
1.0001	1.03	1.08	1.09	1.079	1.108	1.091	1.094 ± 0.005
1.9676	1.20	1.26	1.34	1.311	1.386	1.342	1.346 ± 0.009

[a] Moles per liter.
[b] Debye–Hückel theory.
[c] Full Poisson–Boltzmann (26.64).
[d] Modified Poisson–Boltzmann equation with correction for short-range correlations and deviations from additivity of pair potential of mean force.
[e] See caption to Fig. 23.20.
[f] See caption to Fig. 23.20.
[g] Andersen–Chandler optimized virial expansion theory.
[h] Numerical simulation results of D. N. Card and J. P. Valleau, *J. Chem. Phys.* **52,** 6232 (1970); **56,** 248 (1972).

Alternatively, the Poisson–Boltzmann equation can be improved by including corrections that account for the omissions or the approximations used in its derivation.[22]

Despite its lack of thermodynamic consistency, useful information can be obtained from numerical solutions of the full Poisson–Boltzmann equation. Calculations have been carried out[23] for a variety of types of electrolytes that have the common ionic diameter $2a \approx 4$ Å and for a range of ionic strength up to 0.1 mol/kg. It is found that: (1) ion pairing is negligible for 1:1 electrolytes but important for 2:2 electrolytes; (2) the activity coefficients of 1:1 electrolytes are practically identical with those predicted by Debye–Hückel theory; (3) for 1:2 and 2:1 electrolytes the values of γ predicted differ from the Debye–Hückel values by no more than 2%; (4) for 2:2 electrolytes the values of γ agree with those predicted by the ion–pairing theory to within 2%, but differ from those predicted by Debye–Hückel theory by up to 20%.

Complementing the results just cited, the improved Poisson–Boltzmann equation approach to calculating the influence of ionic interactions leads to predictions of the ionic activity coefficient that are superior to those of Debye–Hückel theory and the original Poisson–Boltzmann equation 26.64, as shown in Table 26.8. Note that in this table the nonideality contribution is represented by the osmotic coefficient (see Chapter 25, Eq. 25.125).

3. Third, the electrolyte solution can be described using the generalization of the virial expansion mentioned in Chapter 25. This theory, originally due to J. E. Mayer,[24,25] is mathematically complex because of the need to take special measures to account for the influence of the long-range Coulomb potential between ions.

The starting point for Mayer's theory of electrolyte solutions is Eqs. 25.194 and 25.195. The interactions that determine the values of the generalized virial coefficients that appear in this expansion for the osmotic pressure are the potentials of mean force at infinite dilution of the solute (see Eq. 25.196). If the electrolyte solution is modeled as hard sphere ions in a continuum dielectric, at infinite dilution the pair potential of mean force is just the Coulomb interaction $Z_iZ_je^2/4\pi\epsilon_0 DR$. But the integrals that define the generalized virial coefficients diverge when the interaction has the form of a coulomb potential. This fact implies that the properties of the solution cannot be constructed sequentially in terms of the contributions from single ions, pairs of ions, triplets of ions, etc.; rather the interactions of all of the ions must be treated simultaneously. Indeed, in a gas of ions (called a plasma) there is a well defined signature of the collective effects induced by the simultaneous interactions of the ions, namely, charge density oscillations that play the role of elementary excitations of the system. In electrolyte solutions the friction introduced by the presence of the solvent heavily damps those collective oscillations. The Poisson–Boltzmann equation accounts for the collective nature of the interactions in an electrolyte solution by

[22] The improved Poisson–Boltzmann equations that have been studied all have the form

$$\nabla^2\Phi_i(R) = -(\epsilon_0 D)^{-1}\sum_\alpha Z_\alpha e n_\alpha \xi_{i\alpha}\exp[-(Z_\alpha e\Phi_i(R)+\phi_{i\alpha}(R))/k_BT],$$

where $\xi_{i\alpha}$ is a short-ranged function which approaches unity rapidly as $R \to \infty$ and $\phi_{i\alpha}(R)$ has the form of a third order fluctuation representing, mainly, the departure of the potential of mean force from the mean potential. The original form of this improved Poisson–Boltzmann equation was derived by J. G. Kirkwood, *J. Chem. Phys.* **2,** 767 (1934). For a review of more recent versions see C. W. Outhwaite, *Specialist Periodical Reports: Statistical Mechanics* **2,** 188 (1975), The Chemical Society, London.

[23] E.A. Guggenheim and R. H. Stokes, *Equilibrium Properties of Aqueous Solutions of Single Strong Electrolytes,* The International Encyclopedia of Physical Chemistry and Chemical Physics, Topic 15, Pergamon Press, London (1969).

[24] J. E. Mayer, *J. Chem. Phys.* **18,** 1426 (1950).

[25] The Mayer theory has been extensively further developed by H. L. Friedman: H. L. Friedman, *Ionic Solution Theory,* Wiley, New York (1962); H. L. Friedman in *Modern Aspects of Electrochemistry,* No. 6, eds. J. M. O'Bockris and B. E. Conway, Butterworths, London (1971).

including the influence of the entire charge distribution in the system on the mean electrostatic potential. In the Mayer theory of electrolytes, it is the infinite sum of the terms which are divergent in the limit that the potential is of Coluomb form that represents the collective behavior which gives rise to a screening of the Coulomb interaction. The mathematical procedure introduced by Mayer amounts to systematically reordering the expansion 25.195 to identify the most strongly divergent terms, the next most strongly divergent terms, etc., and then successively summing each category of divergent terms. The infinite sum of the most divergent terms generates the Debye–Hückel screened Coulomb potential and, of course, also the limiting law forms for the thermodynamic properties of the electrolyte solution with the characteristic $m^{1/2}$ concentration dependence. For a symmetric electrolyte the infinite sums of the next two most divergent terms lead to additional contributions to the activity coefficient which have the concentration dependences m and $m^2 \ln m$; for an unsymmetric electrolyte those contributions have concentration dependences m, $m \ln m$ and $m^{3/2}$. The calculation of further terms in the expansion becomes increasingly difficult (and tedious) as the order of the terms increases.

Mayer's theory permits accurate calculation of the activity coefficient or osmotic coefficient up to about 0.1 *M*. An optimized virial expansion theory, due to H. C. Andersen and D. Chandler,[26] extends the range of accurate prediction considerably (see Table 26.8).

We note that the Kirkwood–Buff theory of solutions, briefly described in Chapter 25, can also be used to describe the properties of electrolyte solutions, but is not much used for that purpose.[27]

4. Fourth, the methods used in the development of the theory of liquids (see Chapter 23) can be extended to models of electrolyte solutions. The so-called method of integral equations has been largely developed by H. Friedman and his co-workers. The problems discussed in Chapter 23 with respect to finding an accurate and tractable description of the distribution of triplets of molecules, or of the direct correlation function, occur again in the theory of electrolyte solutions, and the methods of solution used are fundamentally the same. As shown in Table 26.8, the predicted osmotic coefficients are in much better agreement with numerical simulations of an electrolyte solution than are those from the Debye–Hückel, Poisson–Boltzmann, or modified Poisson–Boltzmann theories; the optimized virial expansion theory is about as good.

Despite the success of these extended theories in describing the primitive model of an electrolyte solution (hard-core ions in a continuous dielectric fluid), they do less well in predicting the properties of real electrolyte solutions (see Fig. 26.21). Given the agreement between

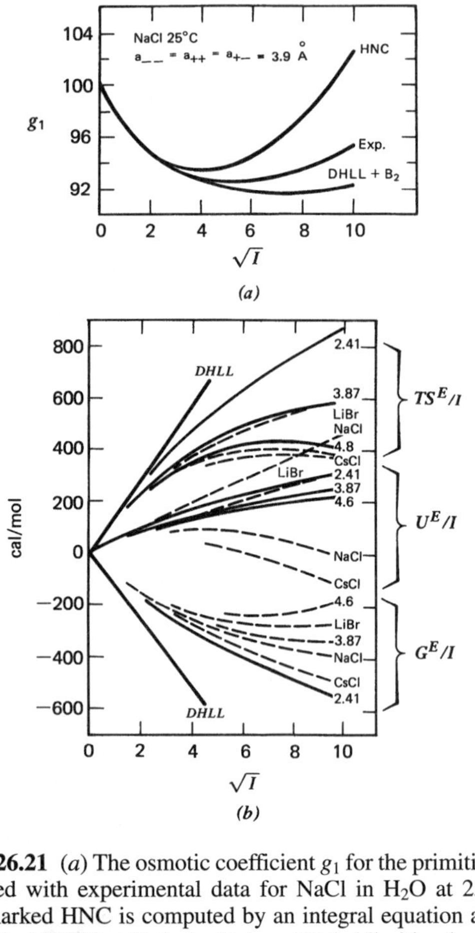

Figure 26.21 (*a*) The osmotic coefficient g_1 for the primitive model compared with experimental data for NaCl in H_2O at 25°C. The curve marked HNC is computed by an integral equation approach, that marked DHLL + B_2 is the Debye–Hückel limiting law plus the first cluster correction (analogous to the second virial coefficient of a gas). (*b*) The thermodynamic functions G^E/I, U^E/I and TS^E/I for the primitive model with equal anion and cation sizes, compared with experimental data for aqueous LiBr, NaCl, and CsCl. The numbers entered on the right hand side of the figure are the ion size parameters. From J. C. Rasaiah, *J. Chem. Phys.* **52,** 704 (1970).

---

[26] H. C. Anderson and D. Chandler, *J. Chem. Phys.* **53,** 547 (1970); D. Chandler and H. C. Anderson, *J. Chem. Phys.* **54,** 26 (1971); H. C. Anderson and D. Chandler, *J. Chem. Phys.* **55,** 1497 (1971).

[27] In Chapter 19 we displayed the relationships between the several pair-correlation functions in a mixture and the thermodynamic properties of that mixture. Specifically, for a binary mixture

$$\left(\frac{\partial \pi}{\partial n_2}\right)_{T,\mu_1} = \frac{k_B T}{1+G_{22}n_2},$$

$$\frac{1}{k_B T}\left(\frac{\partial \mu_2}{\partial n_2}\right)_{T,p} = \frac{1}{n_2} + \frac{G_{12}-G_{22}}{1+n_2(G_{22}-G_{12})},$$

$$\frac{1}{k_B T}\left(\frac{\partial \mu_1}{\partial n_1}\right)_{T,p} = \frac{1}{n_1} + \frac{G_{12}-G_{11}}{1+n_1(G_{11}-G_{12})}, \qquad (25.188)$$

where n_1 and n_2 are the average numbers of molecules of species 1 and 2 per unit volume, and the functions G_{ij} are defined in terms of the pair-correlation function for species i and j, denoted g_{ij}, by

$$G_{ij} \equiv \int d\mathbf{R}(g_{ij}-1). \qquad (25.189)$$

Consider a solution of the electrolyte MX_p in water. If the solution is described in terms of the pair-correlation functions of all species, ten such functions are required (G_{MM}, G_{MX}, G_{XX}, G_{OO}, G_{HO}, G_{HH}, G_{MO}, G_{MH}, G_{XO} and G_{XH}). However, M, X, O and H cannot be varied independently because of the molecular character of water; in this case it can be shown that there are only three independent terms $G_{\alpha\beta}$. If we choose these to be G_{OO}, G_{MM} and G_{OM}, then it can be shown that

$$\left(\frac{\partial \mu_2}{\partial x_2}\right)_{T,p} = \frac{k_B T n_1 n_2 v_1^2}{x_2^2(n_2 G_{MM}+1-\kappa_T k_B T n_2)} = \frac{k_B T x_1 n v_1 v_2}{x_2(1-x_2)(k_B T\kappa_T - G_{MO})},$$

from which $\gamma_\pm$ can be calculated if the concentration dependences of G_{MM} and G_{MO} are known.

theory and the primitive model, we conclude that a more refined model of the electrolyte solution is needed. There is at least one glaring omission in the primitive mode—the van der Waals-type interactions between the ion and the solvent have been neglected. Of course, so have the individual ion–dipole interactions, but these are at least crudely accounted for by the introduction of the dielectric constant of the fluid. In fact, when any particle is inserted in a dielectric, whether charged or not, the electric field in the medium is altered because the dielectric is no longer homogeneous.[28] Moreover, we have seen that in the immediate neighborhood of an ion the solvent structure is altered from that characteristic of the bulk. These phenomena contribute to the effective interaction (the potential of mean force) between the ions. Some extensions of the integral equation theory to include short-range attractions have been studied. As shown in Fig. 26.22, inclusion of a square-well attraction does improve the agreement between theory and experiment for simple electrolytes. A more complicated multiparameter potential is needed for salts such as the tetraalkylammonium halides. That a consistent set of these parameters can be found is shown in Fig. 26.23.

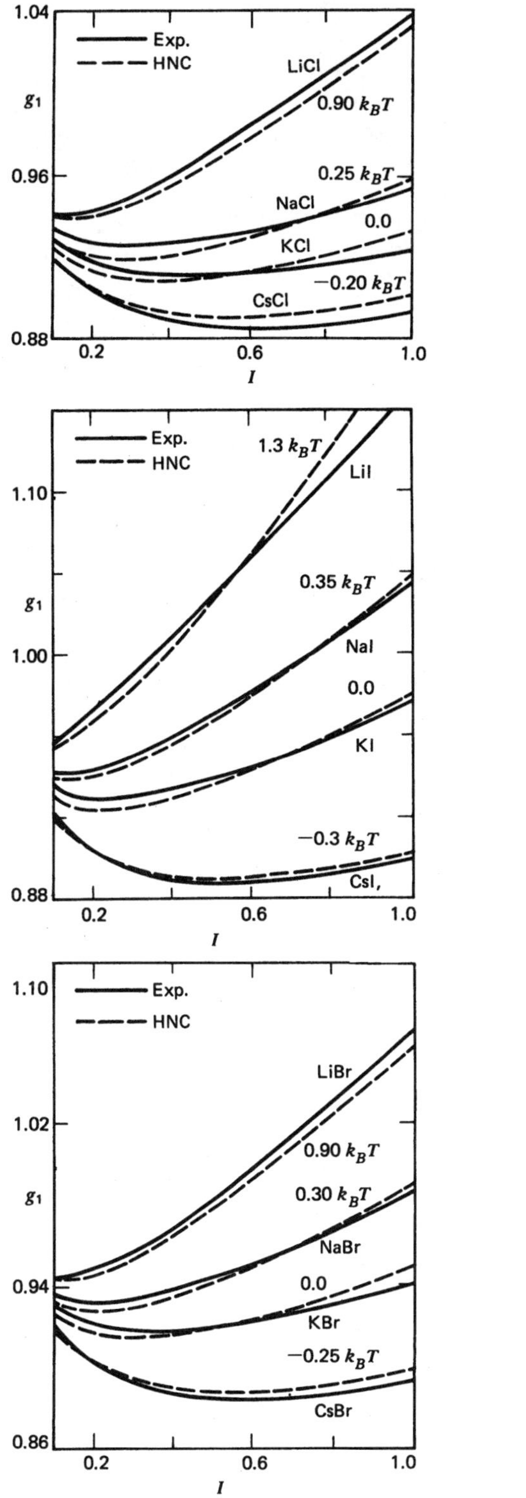

Figure 26.22 Theoretical osmotic coefficients for a model of the electrolyte solution in which there is a square-well interaction between the ions—otherwise the model is the same as the primitive model. The well depths used to fit the data are shown next to each curve. From J. C. Rasaiah, *J. Chem. Phys.* **52,** 704 (1970).

[28] The influence of the discrete structure of the host dielectric on the properties of the electrolyte solution has been investigated by P. G. Kusalik and G. N. Patey, *J. Chem. Phys.* **88,** 7715 (1988); **89,** 5843 (1988); **89,** 7478 (1988). Kusalik and Patey studied, by solving an integral equation (related to the HNC equation described in Chapter 23), the thermodynamic, dielectric, and structural properties of a model system consisting of hard sphere ions immersed in a solvent of hard spheres, each of which has the polarizability, dipole moment, and quadrupole moment of the water molecule. To compare the various properties obtained for this model with those obtained for the model in which the hard core ions are immersed in a continuum dielectric, they calculated the solvent averaged pair potential of mean force, and represented it in the form $w_{ij}(R) = u_{ij}^{HS} + (Z_iZ_je^2/4\pi\epsilon_0 DR) + w_{ij}^x$, the last term of which is a short-range interaction arising from the molecular nature of the solvent. They then solved the HNC integral equation (Chapter 23) with this solvent averaged pair potential of mean force. A comparison of the predicted thermodynamic properties from the two approaches can be considered a test of the accuracy with which the infinite dilution *m*-ion potential of mean force can be represented as the sum of infinite dilution pair potentials of mean force. Kusalik and Patey report calculations for systems with parameters that represent several different 1:1 electrolytes, distinguished by their different ionic diameters. It is found that at low electrolyte concentration (say, up to 0.1 *m*) the pair-correlation functions calculated by the two theoretical schemes are in very good agreement, but that agreement deteriorates as the concentration is increased. The agreement between the calculated activity coefficients showed similar behavior as shown below:

Values of $-\ln \gamma_\pm$ Obtained from Calculations Using the Discrete Solvent Model (I) and the Solvent Averaged Dielectric Constant Model (II)*,**

	KCl		NaCl		MBr	
c	I	II	I	II	I	II
0.025	0.142	0.142	0.141	0.142	0.143	0.145
0.100	0.279	0.280	0.275	0.288	0.291	0.297
0.250	0.425	0.426	0.430	0.486	0.482	0.468

*Adapted from P. G. Kusalik and G. N. Patey, *J. Chem. Phys.* **89,** 7478 (1988).

**The solvent hard sphere diameter, a_s, was taken to be 2.8 Å, and the reduced ionic diameters, a_i/a_s, were taken to be 0.84, 1.08, 1.16, and 1.96 for Na^+, K^+, Cl^-, Br^-, and M^+, respectively.

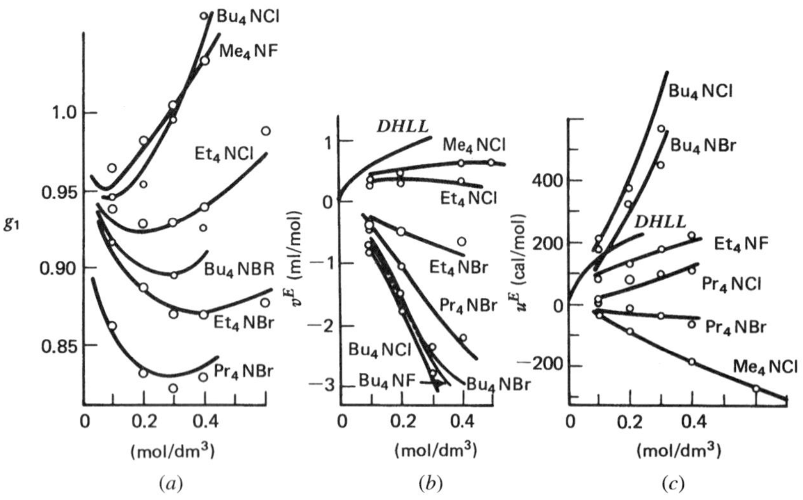

Figure 26.23 (*a*) Comparison of model osmotic coefficients with experiment in typical cases. The curves are experimental values of g_1, whereas the model values are shown as discrete data. (*b*) Comparison of model excess volumes with experiment in typical cases. The curves are experimental values of v^E, whereas the model values are shown as discrete data. (*c*) Comparison of model excess energies with experiment in typical cases. The curves are experimental values of u^E, whereas the model values are shown as discrete data. From P. S. Ramanathan, C. V. Krishnan, and H. L. Friedman, *J. Solution Chem.* **1,** 237 (1972).

D. Semiempirical Representation of the Properties of Electrolyte Solutions

Clearly, the molecular theory of electrolyte solutions needs further development before it can be trusted to give a realistic and quantitative description of the influence of ions on the structure of water and on the thermodynamic properties of the solution for the entire range of composition. Since knowledge of the properties of electrolyte solutions is important for the understanding of many chemical, geochemical, and biochemical processes, it is worthwhile examining how the existing theoretical base can be augmented with empirical information to provide an accurate description over the entire concentration range. Pitzer[29] has stated the development objective as "the accurate, compact, and convenient representation of the experimentally determined properties of electrolyte solutions. The equations must be consistent with the basic principles of thermodynamics and statistical mechanics. A second and equally important objective is the prediction of properties of systems not directly measured but related to those measured. This predictive capacity must be based on the underlying statistical mechanical theory." Building on earlier work by Bronsted[30] and Guggenheim,[31] Pitzer has developed a semiempirical representation of the thermodynamic properties of electrolyte solutions which is firmly based on the Mayer theory up through the B_2^* terms. This theory employs some astute approximations for the ionic strength dependences of the several terms and it correctly allows for specific cation–anion, cation–cation and anion–anion interactions (including ion pairing).

The starting point for the Pitzer representation is the *rearranged Mayer series* for the excess Helmholtz free energy,

$$\frac{A^{ex}}{Vk_BT} = -\frac{\kappa^3}{12\pi} + \sum_i \sum_j c_i c_j B_{ij}(\kappa) + \sum_i \sum_j \sum_k c_i c_j c_k C_{ijk}(\kappa) + \cdots. \tag{26.103}$$

The first term on the right hand side of 26.103 is the Debye–Hückel limiting law contribution. The virial coefficients $B_{ij}(\kappa)$, $C_{ijk}(\kappa)$, . . . differ from those in the straightforward McMillan–Mayer expansion by the omission of the terms which, when rearranged, generate the Debye–Hückel limiting law contribution; they depend on the ionic strength as well as the ion–ion excluded volume and the character of the solvent. Because the virial coefficients can be calculated only for simple models of the electrolyte solution, Pitzer accepts the structure of Eq. 26.103 but regards the $B_{ij}(\kappa)$, $C_{ijk}(\kappa)$, . . . as empirical quantities to be evaluated from experimental data, using theoretical guidance as to their ionic strength dependences. After transformation to the molality scale for concentration, the mean activity coefficient of a solution of the single electrolyte $M_{\nu_M}^{Z_M} X_{\nu_X}^{Z_X}$, which dissociates into $\nu = \nu_M + \nu_X$ ions, is

$$\ln \gamma_\pm = |Z_M Z_X| f^\gamma + m\left(\frac{2\nu_M \nu_X}{\nu}\right) B_{MX}^\gamma + m^2 \left[\frac{2(\nu_M \nu_X)^{3/2}}{\nu}\right] C_{MX}^\gamma \tag{26.104}$$

where[32]

[29] K. S. Pitzer, Chapter 3 in *Activity Coefficients in Electrolyte Solutions,* 2nd ed., ed. K. S. Pitzer, CRC Press, Boca Raton, Florida (1991).
[30] J. N. Bronsted, *J. Am. Chem. Soc.* **44,** 877 (1922); **45,** 2898 (1923).
[31] E. A. Guggenheim, *Phil. Mag.* **19,** 558 (1935).

[32] The coefficient of the ionic strength dependent term includes Avogadro's number, N_A, the mass density of water, d_w, and the mass of solvent water, 1000 gms, because of the conversion of concentration units from number density, used in the original Mayer theory, to molality in Eq. 26.105.

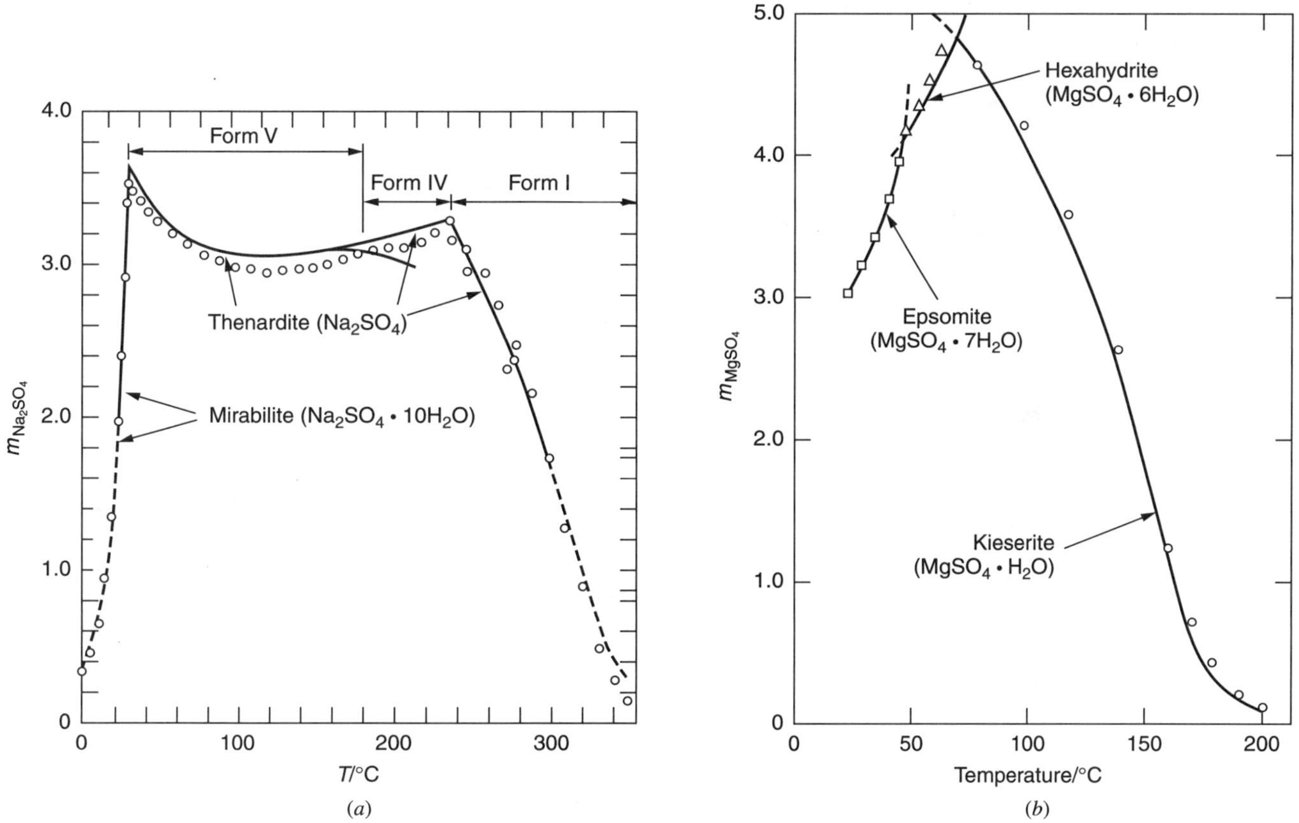

Figure 26.24 Solubilities, as a function of temperature, of various forms of crystalline (*a*) Na_2SO_4 and (*b*) $MgSO_4$. The experimental data are shown as circles and the theoretical predictions based on the Pitzer equations as lines. From K. S. Pitzer, *Thermodynamics,* 3rd ed., McGraw-Hill, New York (1995).

$$
\begin{aligned}
f^{\gamma} &= -\frac{1}{3}\left(\frac{2\pi N_A d_w}{1000}\right)^{1/2}\left(\frac{e^2}{4\pi\epsilon_0 D k_B T}\right)^{3/2} \\
&\quad \times \left(\frac{I^{1/2}}{1+bI^{1/2}}+\frac{2}{b}\ln(1+bI^{1/2})\right), \\
B^{\gamma}_{MX} &= 2\beta^{(0)}_{MX}+\beta^{(1)}_{MX} \\
&\quad \left(\exp(-\alpha_1 I^{1/2}) + \frac{2[1-(1+\alpha_1 I^{1/2})\exp(-\alpha_1 I^{1/2})]}{\alpha_1^2 I}\right) \\
&\quad + \beta^{(2)}_{MX}\left(\exp(-\alpha_2 I^{1/2}) + \frac{2[1-(1+\alpha_2 I^{1/2})\exp(-\alpha_2 I^{1/2})]}{\alpha_2^2 I}\right), \\
C^{\gamma}_{MX} &= 3|Z_M Z_X|^{1/2}\, C_{MX}.
\end{aligned}
\qquad (26.105)
$$

In Eq. 26.105, b has the universal value 1.2 $kg^{1/2}mol^{-1/2}$, α_1 = 2.0 $kg^{1/2}mol^{-1/2}$ and α_2 = 0 for electrolytes that do not exhibit ion pairing, but α_1 = 1.4 $kg^{1/2}mol^{-1/2}$ and α_2 = 12 $kg^{1/2}mol^{-1/2}$ for 2:2 electrolytes which exhibit ion pairing. The parameters $\beta^{(0)}_{MX}$, $\beta^{(1)}_{MX}$, $\beta^{(2)}_{MX}$, and C_{MX} have different values, specific for each solute; $\beta^{(2)}_{MX}$ is related to the ion pairing association constant by $\beta^{(2)} = -K/2$. A similar expression can be written for the osmotic coefficient, and the semiempirical representation can be extended to the description of mixed electrolyte solutions. Extensive tables of the coefficients in Eq. 26.105 can be found in K. S. Pitzer, *Thermodynamics,* McGraw-Hill, New York (1995). An example of the accuracy of the Pitzer equations is displayed in Fig. 26.24, which shows the predicted and observed solubilities of several minerals in concentrated mixed electrolyte solutions.

26.6 Molten Salts and Molten Salt Mixtures[33]

Molten salts and molten salt mixtures are representative of fluids in which all of the particles are ions. It is to be expected, and is found, that their properties differ from those of a dilute mixture of ions and uncharged particles (as in a dilute electrolyte solution). Yet in each class of systems the competition between the ion–ion Coulomb interaction and the short range ion–ion interaction determines the thermodynamic properties.

[33] The properties of molten salts are reviewed by M. Rovere and M. P. Tosi, *Repts. Prog. Phys.* **49,** 1001 (1986).

For this reason it is instructive to examine how the properties of a molten salt depend on the ionic size, charge, temperature, and other variables.

In a molten salt, as in all liquids, there is positional disorder with respect to the locations of the ions. Now, in an ionic crystal every positive ion is surrounded by nearest-neighbor negative ions and *vice versa,* all of the ions being located at lattice sites. In a dilute electrolyte solution, Debye–Hückel theory establishes the existence of a diffuse cloud of negative ions about each positive ion, and *vice versa.* It should not be surprising, then, that the local structure of a molten salt is very much like that of a simple liquid, such as Ar, with respect to the distribution of ionic centers of mass, but that every positive ion is surrounded by nearest-neighbor negative ions, and *vice versa* (see Fig. 23.13). Put another way, the pair-correlation function for particles of any type a distance R from a central anion or cation, $g_m(R) = (1/2)[g_{++}(R) + g_{+-}(R)] = (1/2)[g_{-+}(R) + g_{--}(R)]$, has significant variation only in the immediate vicinity of the ion at the origin, whereas the ion–ion pair-correlation functions display fluctuations in the anion and cation number densities, relative to each other, even at fairly large distances from the central ion.

It is worthwhile examining the alternating charge distribution in ionic systems a little further before proceeding to an analysis of the behavior of molten salts. In particular, it is interesting to survey how the charge distribution in an assembly of ions changes as the density of ions increases. Of course, as the ion density increases the importance of the short-range ion–ion interactions, the excluded volume contribution, increases, as also does the electrostatic energy of the system. The interplay between these two interactions leads to a dramatic change in the character of the distribution of ions at a specific ion density (which depends on temperature).

We have seen that in a dilute electrolyte solution the potential of mean force has the form calculated from Debye–Hückel theory, namely, a shielded Coulomb potential with spatial dependence $\exp(-\kappa R)/R$. The shielding arises from the interposition of other ions between the two that are separated by the distance R (Eq. 26.88), but that ionic cloud is diffuse. Consequently, the ion–ion pair-correlation function decays exponentially to its asymptotic (large R) value. In contrast, in a crystalline salt the pair-correlation function is oscillatory for large R, corresponding to the regularity of the arrangement of the ions. Kirkwood[34,35] was the first to show, using a version of the improved Poisson–Boltzmann equation displayed in footnote 22, that as the concentration of an electrolyte solution is increased the asymptotic form of the ion–ion pair-correlation function changes from exponential decay to oscillatory; with the approximations employed by Kirkwood, at room temperature the transition from exponential decay to oscillatory behavior occurs discontinuously

[34] J. G. Kirkwood, *Chem. Rev.* **19,** 275 (1936).
[35] J. G. Kirkwood and J. C. Poirier, *J. Phys. Chem.* **58,** 591 (1954).

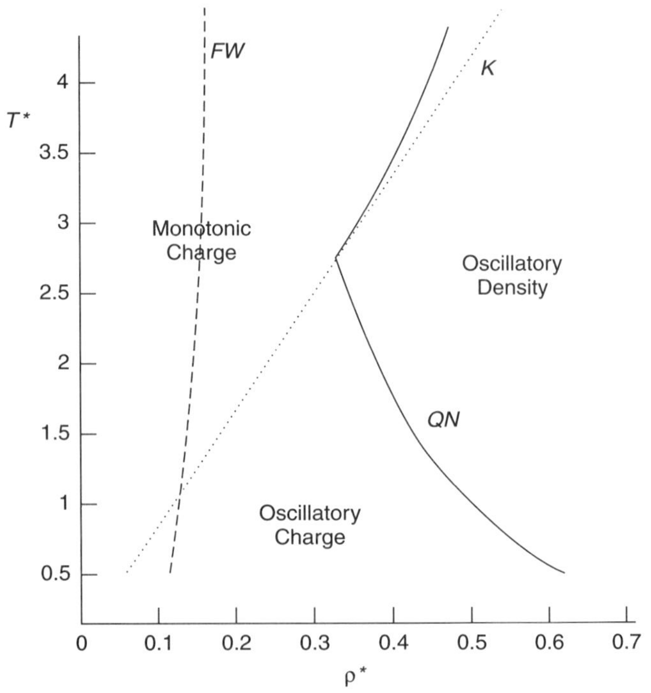

Figure 26.25 Generalized phase diagram for an ionic fluid, showing the boundaries where the asymptotic behavior of the pair-correlation function changes from an exponential decay to oscillatory, calculated using a generalization of the mean spherical approximation integral equation (Chapter 23). The line marked K is the boundary between a system with exponentially decaying pair-correlation function and a system with oscillatory pair-correlation function, while the line marked FW is the boundary between a system in which the total density correlation function decays exponentially and one in which the total density correlation function oscillates. The line marked QN is the boundary between regions with charge dominated and with oscillatory density dominated decay of $R[g_{ij}(R) - 1]$ at longest range. The calculations are for a Z:Z electrolyte, with $T^* = k_BTDa/(Ze)^2$, $\rho^* = na^3$. From R. J. F. Leote de Carvalho and R. Evans, *Molec. Phys.* **83,** 619 (1994).

when $\kappa a \approx 1.03$. Other calculations, based on inclusion of ion pairing in the Poisson–Boltzmann representation,[36] different versions of the improved Poisson–Boltzmann equation,[37] or integral equation approximations,[38] lead to values of κa at which the transition to oscillatory behavior occurs in the range 1.178–1.241. Figure 26.25 shows a phase diagram for a 1:1 electrolyte, calculated using a generalization of the mean spherical approximation integral equation (Chapter 23); the line marked K is the boundary between a system with exponentially decaying pair-correlation function and a system with oscillatory pair-correlation function, while the line marked FW is the boundary between a system in which the total density correlation function decays exponentially and

[36] B. P. Lee and M. E. Fisher, *Phys. Rev. Lett.* **76,** 2906 (1996).
[37] C. W. Outhwaite, *Specialist Periodical Reports: Statistical Mechanics* **2,** 188 (1975), The Chemical Society, London.
[38] R. J. F. Leote de Carvalho and R. Evans, *Molec. Phys.* **83,** 619 (1994).

one in which the total density correlation function oscillates. Clearly, the competition between short-range repulsive interactions and long-range Coulomb interactions generates a complex phase diagram, the common feature of which is charge ordering, with the ordering weak at low ion density and increasingly stronger as the ion density increases.

A complete molecular theory of molten salts involves subtle considerations related to the way the long-range Coulomb potential generates the charge ordering observed in the liquid state. That class of theory is not suitable for description in this text. Instead, we shall restrict attention to an extension of the theory of corresponding states. This approach provides a satisfying unification and overview of many properties of molten salts and molten salt mixtures.

Consider, as a simple case, a pure binary molten salt, such as NaCl. Recall that a positive ion has only negative-ion nearest neighbors, and *vice versa.* It is to be expected, then, that two like-charged ions are always far enough apart that any short-range repulsion between them can be neglected, and that the dominant part of their interaction is the Coulomb repulsion. In contrast, the short-range repulsion between unlike ions that are nearest neighbors plays an important role in defining the structure of the liquid. As in the case of simple liquids, it is expected that a qualitatively correct model of that short-range repulsion is the hard-core interaction (see Chapter 23). We now assume that the following simple model potential[39] will adequately describe the principal properties of a molten salt:

$$\begin{aligned} u_2^{++}(R) &= u_2^{--}(R) = \frac{(Ze)^2}{R} && (R>0), \\ u_2^{+-}(R) &= \infty && (R\le\sigma), \\ u_2^{+-}(R) &= \frac{Z_+Z_-e^2}{R} && (R>\sigma). \end{aligned} \qquad (26.106)$$

In Eq. 26.106, σ is the sum of unlike ion radii. The model potential 26.106 treats the ions as unpolarizable rigid spheres without any form of multipole interaction or dispersion interaction. We must expect the neglect of these interactions and the use of a hard-core repulsion to lead to some error in the predicted thermodynamic properties. However, drawing on the results of our study of ionic crystals, we expect that error to be small enough that qualitative properties of the molten salt are accounted for.

To introduce a principle of corresponding states, we define the reduced distance

$$R^* \equiv \frac{R}{\sigma} \qquad (26.107)$$

[39] To be consistent with the literature on the subject, we use cgs units in the following discussion.

and rewrite Eqs. 26.106 in the form

$$\begin{aligned} u_2^{++}(R^*) &= u_2^{--}(R^*) = \frac{(Ze)^2}{R^*\sigma} \\ &= \frac{(Ze)^2}{\sigma} u_2^*(R^*) && (R^*>0), \\ u_2^{+-}(R^*) &= \infty && (R^*\le 1), \\ u_2^{+-}(R^*) &= -\frac{(Ze)^2}{\sigma} u_2^*(R^*) && (R^*>1). \end{aligned} \qquad (26.108)$$

The potential energy of interaction of all pairs of ions can then be written as

$$\begin{aligned} \mathcal{V}_{2N} &= \sum_{i>j} u_{ij}^{++} + \sum_{i>j} u_{ij}^{--} + \sum_{i>j} u_{ij}^{+-} \\ &= \frac{(Ze)^2}{\sigma} \mathcal{V}_{2N}^*(\mathbf{R}_1^* \ldots, \mathbf{R}_{2N}^*), \end{aligned} \qquad (26.109)$$

where we have assumed that there are a total of $2N$ ions in the system, with $N_- = N_+ = N$. The canonical partition function for the ionic melt is just

$$\begin{aligned} Q_{2N} &= \frac{1}{N!}\left(\frac{2\pi m_+ k_B T}{h^2}\right)^{3N/2} \frac{1}{N!}\left(\frac{2\pi m_- k_B T}{h^2}\right)^{3N/2} \\ &\quad \times \int_V \exp\left(-\frac{\mathcal{V}_{2N}}{k_B T}\right) d\mathbf{R}_1 \cdots d\mathbf{R}_{2N}, \end{aligned} \qquad (26.110)$$

where V is the volume of the melt. Introduction of the reduced distance of Eq. 26.107 then gives

$$\begin{aligned} Q_{2N} &= \frac{\sigma^{6N}}{(N!\Lambda_+^{3N/2})(N!\Lambda^{3N/2})} && (26.111) \\ &\quad \times \int_{V/\sigma^3} \exp\left(-\frac{\mathcal{V}_{2N}^*}{k_B T}\right) d\mathbf{R}_1^* \cdots d\mathbf{R}_{2N}^* \\ &= \frac{\sigma^{6N}}{(N!\Lambda_+^{3N/2})(N!\Lambda_-^{3N/2})} \mathcal{J}(T^*, V^*, N), && (26.112) \end{aligned}$$

where $N!\Lambda_+^{3N/2}$ and $N!\Lambda_-^{3N/2}$ are the kinetic energy contributions to Q_{2N} for positive and negative ions and $\mathcal{J}$ depends only on the number of ions in the reduced volume V^* and the reduced temperature T^*, defined by

$$V^* \equiv \frac{V}{\sigma^3}, \qquad (26.113)$$

$$T^* \equiv \frac{\sigma k_B T}{(Ze)^2}. \qquad (26.114)$$

From the fundamental relation between the Helmholtz free energy and the canonical partition function,

$$A_{2N} = -k_B T \ln Q_{2N},$$

we readily find for the pressure

$$p = k_B T\left[\frac{\partial \ln \mathcal{J}(T^*,V^*,N)}{\partial V}\right]_{T,N}$$
$$= \left(\frac{Z^2e^2}{\sigma^4}\right)T^*\left(\frac{\partial \ln \mathcal{J}}{\partial V^*}\right)_{T^*,N}$$
$$= \left(\frac{Z^2e^2}{\sigma^4}\right)p^*(T^*,V^*,N), \qquad (26.115)$$

using the definition

$$p^*(T^*,V^*,N) \equiv T^*\left(\frac{\partial \ln \mathcal{J}}{\partial V^*}\right)_{T^*,N}. \qquad (26.116)$$

The reduced pressure p^* is simply calculated from the system pressure p by inverting Eq. 26.105, that is,

$$p^* \equiv \left(\frac{\sigma^4}{Z^2e^2}\right)p. \qquad (26.117)$$

The reduced equation of state

$$p^* = p^*(T^*,V^*) \qquad (26.118)$$

should be a universal function for all molten salts with pair interactions of the form Eq. 26.106. A test of the validity of this approach is shown in Fig. 26.26, which is a plot of the reduced vapor pressure as a function of reduced temperature for the fluorides, chlorides, bromides, and iodides of Na, K, Rb, and Cs. Although at a given value of the absolute temperature T the vapor pressures of these salts are very different, all of the values conform to one reduced equation of state. The ideas sketched above can be, and have been, extended to treat unsymmetrical electrolytes, such as CaF_2. The reduced equation of state found is a universal function within a particular valence class, but is not necessarily the same for different valence classes. For example, using the methods described, it is possible to compare the properties of two different unidivalent salts, but not those of a unidivalent salt with those of a unitrivalent salt.

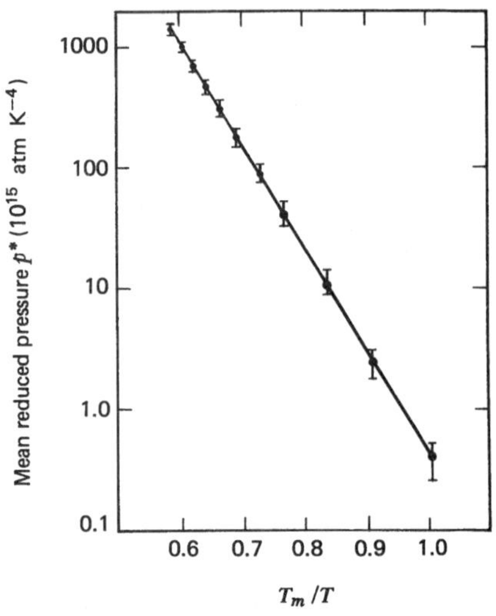

Figure 26.26 Mean reduced vapor pressure curve for the halides of Na, K, Rb, and Cs. Average deviation from the mean is shown by the vertical lines. T_m is the melting temperature of an alkali halide. From M. Blander, *Adv. Chem. Phys.* **11,** 83 (1967).

Many different tests of the applicability of the theorem of corresponding states to molten salts have been made, all rather successful. We shall, therefore, use this approach to describe the excess thermodynamic properties of molten salt mixtures.

The extension of the corresponding-states principle to the description of molten salt mixtures, although straightforward, involves some new considerations. First, since there are more than two cations and anions in the mixture, a single distance of anion–cation closest approach cannot provide the necessary parameterization for a reduced equation of state. Instead, several distances of closest approach, one for each unlike anion–cation pair, are required. Second, it is necessary to make some assumption concerning the distribution of different cations amongst the positions occupied by all cations, and similarly for the anions. A plausible zeroth-order approximation is that the cations mix randomly among themselves, and the anions mix randomly among themselves. A test of this assumption can be made by the method of computer simulation; it will be discussed later. Third, the excess energy, entropy, and volume of molten salt mixtures are very small relative to the internal energy, entropy, and volume of either the pure molten salts or the molten mixture. It is not obvious, then, that the neglect of dispersion interactions between the ions is acceptably accurate. A better model, still within the framework of the corresponding-states approach, must include these interactions.

To apply these ideas to the description of a molten salt mixture we write for the anion–cation interaction

$$u_2^{+-}(R) = u_{+-}^{\text{Coulomb}} + \xi_{+-}u_{+-}^{\text{non-Coulomb}} \qquad (R>\sigma),$$
$$u_2^{+-}(R) = \infty \qquad (R\le\sigma), \qquad (26.119)$$

and for the anion–anion and cation–cation interactions

$$u_2^{--}(R) = u_{--}^{\text{Coulomb}} + \xi_{--}u_{--}^{\text{non-Coulomb}},$$
$$u_2^{++}(R) = u_{++}^{\text{Coulomb}} + \xi_{++}u_{++}^{\text{non-Coulomb}}, \qquad (26.120)$$

where ξ is a measure of the strength of the non-Coulomb potential. When we use the form $\xi u^{\text{non-Coulomb}}$ we imply that $u^{\text{non-Coulomb}}$ is a function of R only and all the coefficients are contained in ξ. For example, if $\xi u^{\text{non-Coulomb}}$ were to be a Lennard–Jones attraction, we would have $u^{\text{non-Coulomb}} = -R^{-6}$ and $\xi = 4\epsilon\sigma^6$ (see Eq. 21.173).

Following the spirit of the analysis of a pure molten salt, Eqs. 26.119 and 26.120 should be reduced by introduction of the anion–cation contact separation as a reference length. Instead of using several such contact separations, one for each unlike anion–cation pair in the mixture, we suppose that differences in ionic size are sufficiently small that the properties of a melt can be expanded in a Taylor series about the properties of a reference melt with anion–cation contact separation σ_0, the expansion parameter being $(\sigma_{+-} - \sigma_0)$. In the case of a mixture, there must be terms in this expansion for all possible ion combinations. The non-Coulomb part of the interaction is handled similarly. Let ξ_0 be a single value of ξ for the reference melt. Then the effect of differences among ξ_{++}, ξ_{+-}, and ξ_{--} is described by a Taylor series expansion about the properties of the reference melt, the expansion parameter being $(\xi_{ij} - \xi_0)$, where *i, j* run over all ion species. Writing down the Taylor series expansion of the partition function is now very simple, and evaluation of the many derivatives that appear is straightforward but tedious. For a molten salt mixture with a common anion and the same charge on both cations (call the cations 1 and 2), and for which all non-Coulomb interactions are the same in the pure melts and the mixture ($\xi_{++} = \xi_{+-} = \xi_{--}$), the excess Helmholtz free energy of mixing is

$$A^E = N_1 N_2 \left(\frac{k_B T}{2}\right)\left(\frac{\sigma_1 - \sigma_2}{\sigma_0}\right)^2 \times \text{function dependent on the two, three, and four particle distribution functions of the melt.} \quad (26.121)$$

It is not necessary, for our purposes, to specify the nature of the several distribution functions that appear in Eq. 26.121. The crux of the matter is that a difference in ionic size generates local structural changes that lead to a nonzero excess free energy of mixing. If the non-Coulombic terms in the interaction are not all the same, there is an additional contribution to the excess Helmholtz free energy of mixing of the form

$$-N_1 N_2 \left[\left(\frac{\xi_{12} - \xi_0}{\xi_0}\right) a_{12} - \frac{1}{2}\left(\frac{\xi_{11} - \xi_0}{\xi_0}\right) a_{11} - \frac{1}{2}\left(\frac{\xi_{22} - \xi_0}{\xi_0}\right) a_{22}\right], \quad (26.122)$$

where the functions a_{ij} are the averages of the *R*-dependent part of the non-Coulomb interaction, $u_{ij}^{\text{non-Coulomb}}$, over the distribution of ions. The form 26.122 implies that in the first approximation the non-Coulombic part of the excess free energy of mixing arises from the difference in averaged non-Coulombic energies of ions with the configurations M_1XM_2, M_1XM_1, and M_2XM_2, where X is the anion and M_1, M_2 the two cations.

In molten salt mixtures the excess entropy of mixing is sufficiently small that a^E is very nearly the same as h^E. Figure 26.27 shows experimental data for mixtures of alkali nitrates plotted versus $(\sigma_1 - \sigma_2)^2/(\sigma_1^2\sigma_2^2)$. For the values of

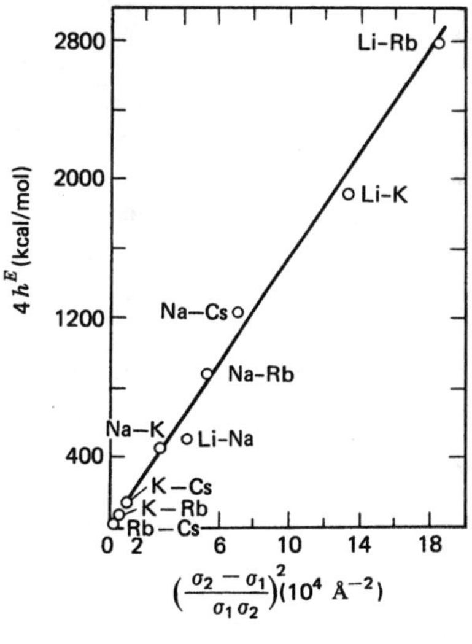

Figure 26.27 Plot of $4h^E$ for the alkali nitrates versus the quantity $[(\sigma_2 - \sigma_1)/\sigma_1\sigma_2]^2$. The cation pair to which each point corresponds is shown next to each point. From M. Blander, *Adv. Chem. Phys.* **11,** 83 (1967).

σ_1 and σ_2 characteristic of these ions, the product $\sigma_1\sigma_2$ varies very little compared to $\sigma_1 - \sigma_2$, so the functional form predicted in Eq. 26.121 is verified over a wide range by the observations. In fact, the fit of the observations to theory is even better when the non-Coulombic contribution 26.122 is included. That contribution is small for the alkali nitrates, but much larger in mixtures with highly polarizable ions, for example, Ag^+ or Tl^+. In such systems there can also occur partial covalent bonding, which simply means that a description of the system as simple polarizable ions is not completely accurate.

It is interesting to compare the simple ideas of the corresponding-states representation of molten salt properties with the results of computer simulations. Consider, first, pure salts. It was shown in Chapter 14 that the lattice energy of alkali halide crystals is well accounted for if the ion–ion interaction is of the Born–Mayer form (Eq. 14.33). This potential is semiempirical in nature; it is an extreme test to require that it account for material properties at temperatures far removed from that where it was fitted. Nevertheless, since the Born–Mayer potential or slight variants of it are the best available ion–ion potentials, we must make just that test. Tables 26.9 and 26.10 show calculated and observed properties of alkali halide crystals, at 298 K and near the melting points of the several substances. The computer simulations are based on a version of the Born–Mayer potential like Eq. 14.33 that neglects the charge-induced dipole part of the interaction but does include the dispersion interactions. Other simulations including the ion-induced dipole term show that it contributes about 0.2% of the lattice energy at 298 K. These calculations also show that inclusion of the charge-induced dipole interaction considerably improves the values of the predicted lattice vibration frequencies. Examination of the entries in Tables 26.9 and

Table 26.9 Thermodynamic Properties of Alkali Halide Crystals at 298 K and Zero Pressure[a,b]

	R_0(Å)		u(kJ/mol)		κ_T(10^{-12} cm²/dyn)		α(10^{-4}/K)		C_p(J/K mol)	
Salt	**exp.**	MC	**exp.**	MC	**exp.**	MC	**exp.**	MC	**exp.**	MC
NaF	2.317	2.322	897.0	910.3	2.15	2.88	0.98	1.37	46.8	51.6
KF	2.674	2.681	794.1	798.9	3.28	3.99	1.00	1.20	50.0	49.0
RbF	2.815	2.840	759.0	773.2	3.81	4.41	0.95	1.17	50.5	50.0
LiCl	2.570	2.573	832.2	837.0	3.36	3.67	1.22	1.41	50.3	52.5
NaCl	2.820	2.842	764.0	762.6	4.17	4.36	1.10	1.20	50.8	52.2
KCl	3.147	3.151	693.7	693.1	5.73	6.42	1.01	1.43	51.5	53.5
RbCl	3.291	3.297	666.5	673.3	6.40	6.32	0.99	1.07	51.2	49.9
NaBr	2.989	2.997	726.3	730.5	5.02	5.51	1.19	1.14	52.3	50.1
KBr	3.298	3.314	663.2	663.8	6.75	7.34	1.10	1.30	52.5	50.1
RbBr	3.445	3.444	638.5	651.7	7.69	7.69	1.04	1.01	51.7	46.6

[a]exp. = experimental results; MC = results of Monte Carlo computations.
[b]From D. J. Adams and I. R. McDonald, *J. Phys. C (Solid State Phys.)* **7,** 2761 (1974).

Table 26.10 Thermodynamic Properties of Alkali Halides at Zero Pressure and Temperature Near the Experimental Melting Points[a,b]

		v(cm³/mol)				$-u$(kJ/mol)			
		crystal		liquid		crystal		liquid	
Salt	T(K)	**exp.**	MC	**exp.**	MC	**exp.**	MC	**exp.**	MC
LiF	1120	11.1		14.3		990.2		963.4	
KF	1130	26.0	27.91	30.4	36.09	768.8	775.3	740.6	758.3
CsF	976	31.8		41.6		701.6		679.8	
LiCl	883	22.4	22.78	28.2	30.35	814.9	819.0	795.0	801.1
NaCl	1073	30.0	31.37	37.6	39.72	739.0	744.2	711.0	717.3
KCl	1045	41.6	42.52	48.8	53.10	669.7	676.3	643.1	649.5
RbCl	995	47.2	47.91	53.9	58.66	645.7	658.4	627.3	629.9
KBr	1007	48.0	49.12	56.0	60.26	639.9	648.3	614.4	622.0
RbBr	953	53.6	54.68	60.9	65.85	618.6	629.8	603.1	603.9

[a] exp. = experimental results; MC = results of Monte Carlo computations.
[b] From D. J. Adams and I. R. McDonald, *J. Phys. C (Solid State Phys.)* **7,** 2761 (1974).

26.10 shows that agreement between computer simulation and experiment is generally better at 298 K than near 1000 K. Note that the volume of the crystal is systematically overestimated. Also, the rather good agreement between computed and observed lattice energies masks the fact that the remaining error is confined to the non-Coulombic contribution. This is so because, for a given lattice, the Coulombic contribution to the lattice energy can be computed exactly. Since it is very much larger than the non-Coulombic contribution, the relative error in the latter is larger than meets the eye on examining the table entries. All in all, the Born–Mayer potential provides a fair description of the thermodynamic properties of the solid, but one that worsens as T increases.

Table 26.10 also has entries for liquid alkali halides. It is seen that the predicted properties are less accurate than was the case for the crystals. The estimated contribution of the charge-induced dipole interaction is 0.5–0.8% of the internal energy of the liquid. Despite the small contribution of the charge-induced dipole interaction to the energy, it does influence the distribution of ions quite a bit, as shown in Fig. 26.28.

We now turn to the case of molten salt mixtures. Given the documented inaccuracy of the Born–Mayer potential, we cannot expect a completely quantitative description to emerge from a computer simulation. Nevertheless, that potential is adequate to test the qualitative notions used in formulating the corresponding-states analysis of molten salt

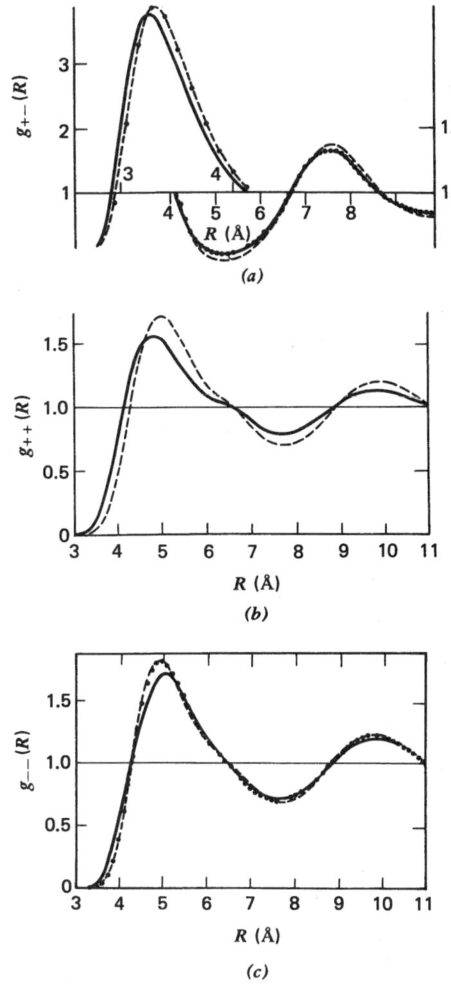

Figure 26.28 The pair-correlation functions for the three pairs of ions in a model of liquid KI, (*a*) $g_{+-}(R)$: -- rigid-ion model; — polarizable-ion model. (*b*) $g_{++}(R)$: -- rigid-ion model;— polarizable-ion model. (*c*) $g_{--}(R)$: -- rigid-ion model; — polarizable-ion model. From G. Jacucci, I. R. McDonald, and A. Rahman, *Phys. Rev.* **A13,** 1581 (1976).

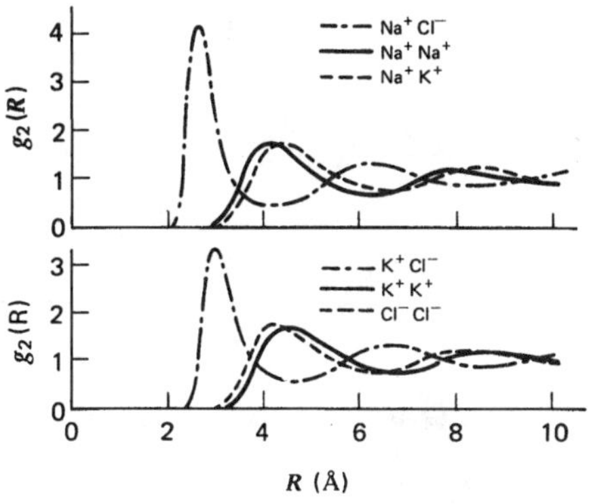

Figure 26.29 Pair-correlation functions for NaCl-KCl(*l*) at T = 1083 K and zero pressure. Composition of the mixture: $x_{NaCl} = x_{KCl} = 0.5$. From B. Larsen, T. Førland, and K. Singer, *Mol. Phys.* **26,** 1521 (1973).

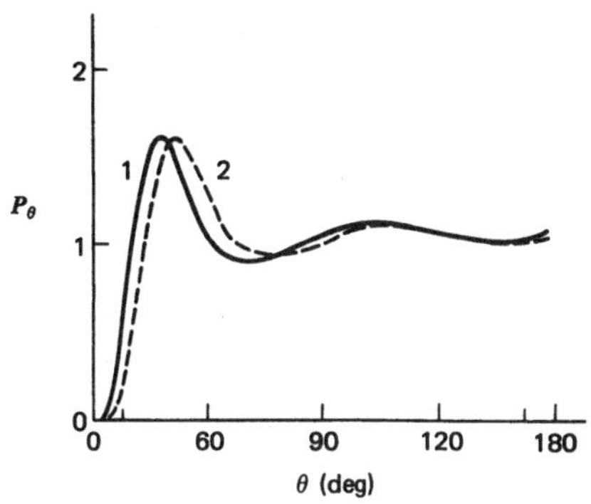

Figure 26.30 Selected angular distribution functions for NaCl-KCl (*l*) at $T = 1083$ K and zero pressure. Composition of the mixture: $x_{NaCl} = x_{KCl} = 0.5$. 1: $Na^+K^+Na^+$ distribution, K^+ as central ion; 2: $K^+Na^+K^+$ distribution, Na^+ as central ion. From B. Larsen, T. Førland, and K. Singer, *Mol. Phys.* **26,** 1521 (1973).

mixtures. Consider, for example, the hypothesis that the cations mix randomly amongst themselves in a mixture with a common anion. If this is correct, the pair-correlation functions $g_{Na^+K^+}$, $g_{Na^+Na^+}$, and $g_{K^+K^+}$ (see Chapter 23) should be the same for all compositions of the mixture. As shown in Fig. 26.29, this is not the case; the size difference between Na^+ and K^+ necessarily generates some deviation from random mixing of the Na^+ and K^+ among themselves. The same effect of ionic size can be seen in the distribution of angles between triplets of ions. Let $P_\theta(\theta, R)2\pi R^2 \sin\theta\, dR\, d\theta$ be the probability of finding a third ion, say k, in the volume element $2\pi R^2 \sin\theta\, dR\, d\theta$ around the pair of ions ij when their separation is in the range R to $R + dR$ and the angle between $\mathbf{R}_{ij}$ and $\mathbf{R}_{ik}$ is between θ and $\theta + d\theta$. Figure 26.30 shows, for a simulation of an equimolar KCl/NaCl melt, the $Na^+K^+Na^+$ and $K^+Na^+K^+$ angular distributions. If mixing among the K^+ and Na^+ ions were random, these distributions should peak at the same angle, which they obviously do not.

We must conclude, then, that the cations in a molten salt mixture with common anion do not mix randomly among themselves, although the deviation from random mixing is not large.

What can be learned from the computer simulations about the energetics of molten salt mixtures? Because of the inadequacy of the ion–ion potential, only qualitative conclusions can be drawn from the calculated results. These do show that dispersion-energy contributions to the excess enthalpy of mixing are important, and that nearest- and next-nearest-neighbor interactions contribute about equally to h^E because of the delicate balance between the changes in the Coulomb and non-Coulomb components of the energy. The contributions to the entropy by each ionic species are very nearly the same in the pure melt and the mixture, so that, as already hinted, deviations from random mixing among ions of the same charge must be a small effect. Finally, there is order-of-magnitude agreement

Table 26.11 Mixing Properties of Equimolar Alkali Chloride Melts at $T = 1073$ *K*, $p = 0$[a]

	h^E(kJ/mol)		v^E(cm^3/mol)	
System	**MC**	**exp.**	**MC**	**exp.**
LiCl/KCl	–3.5	–4.4[b]	0.31	0.30
LiCl/RbCl	–3.9	–5.3[b]	0.50	0.36
NaCl/KCl	0	–0.5[c]	0.54	0.25

[a] From D. J. Adams and I. R. McDonald, *Mol. Phys.* **34,** 287 (1977).
[b] At 1013 K.
[c] At 1083 K.

between computed and observed excess heats and volumes of mixing even though the ion–ion potential energy function is inexact (see Table 26.11).

26.7 The Structure of an Electrolyte Solution Near an Electrode

The properties of electrolyte solutions adjacent to solid and/or liquid surfaces influence a number of important phenomena, for example, the passage of current in an electrochemical cell. Many of the phenomena of interest depend on both the solution-surface interface structure and the kinetics of processes that occur in that interface. In this section we briefly examine the structures of the interfaces between an electrolyte solution and various solid surfaces, leaving the discussion of the kinetic processes to Part III.

A. The Structure of Water Near a Surface

We start with an examination of how the contact between a smooth wall and a nonmetallic atomic liquid, such as Ar, alters the structure of the liquid in the immediate region of contact. The presence of the contact between the wall and the liquid can be thought of as the imposition of an external field on the liquid. Since the Hamiltonian of an infinite liquid has no dependence on the location of a particular origin of coordinates (the liquid has translational invariance), the equilibrium states of an infinite liquid must also be translationally invariant and therefore have the same density at all points in space. When the external field representing a wall is imposed on a liquid the translational invariance of the liquid is broken, and a density distribution that varies with distance from the wall is induced. In particular, at the average densities characteristic of most liquids, the density distribution adjacent to the wall exhibits several oscillations as the distance from the wall into the liquid increases. The theoretical analysis of this response of the liquid to the presence of a smooth wall is based on Eq. 23B16, which represents the spatial dependence of the one particle density, $n(\mathbf{R})$, as the functional derivative of the free energy, $-k_BT \ln Q_N[\psi]$, with respect to the external field, $\psi(\mathbf{R})$. This equation is reduced to a form suitable for determination of $n(\mathbf{R})$ by use of arguments and approximations similar to those employed in the development of the theory of the pair-correlation function in a uniform liquid. For example, when the Percus–Yevick approximation (Eq. 23.45) is used the spatial dependence of the density distribution is determined from

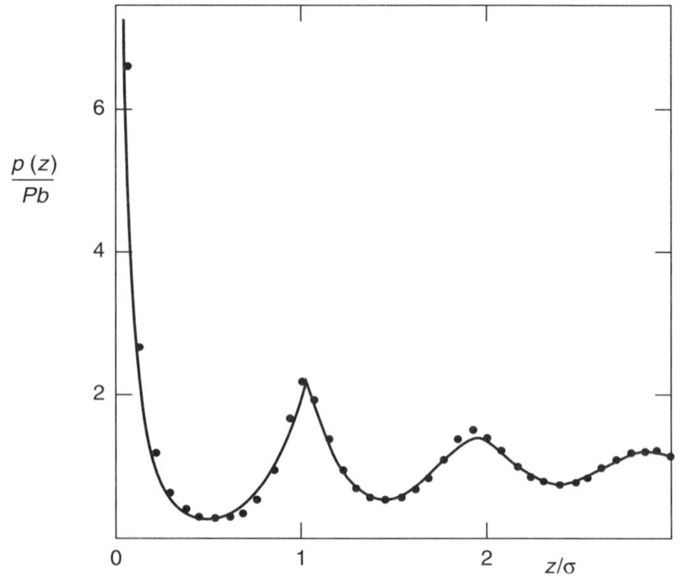

Figure 26.31 Density profile of a hard sphere liquid (reduced density $\rho\sigma^3 = 0.878$) close to a smooth hard wall. From R. J. Evans, Chapter 3 in *Fundamentals of Inhomogeneous Fluids,* ed. D. Henderson, Marcel Dekker, New York (1992).

$$n(\mathbf{R}_1) = n\exp[-\psi(\mathbf{R}_1)/k_BT] \times [1+\int d^3R_2 C(n;R_{12})(n(\mathbf{R}_2)-n)], \quad (26.123)$$

where n is the bulk number density of the liquid. Just as in the case of the theory of the pair-correlation function in a uniform liquid, different approximations for the direct correlation function lead to different integral equations for $n(\mathrm{R})$. The results of Monte Carlo simulations of a hard sphere liquid adjacent to a smooth hard wall are shown in Fig. 26.31, as is the predicted density variation near the wall obtained from an integral equation similar to, but more accurate than, Eq. 26.123. Clearly, the oscillations in the liquid density adjacent to the wall are quite pronounced.

It is reasonable to expect that the gross features of the density distribution of water adjacent to an electrode will resemble those of a hard sphere liquid adjacent to a smooth hard wall, specifically, the existence of oscillations in the density distribution. However, a hard sphere liquid is not a good model of water, and a smooth hard wall is not a good model of an electrode surface, so we must expect there to be important features of the density distribution which are unique to the water–electrode interface.

The surface of a real electrode can be, in rare instances, a single crystal plane, but is usually not. The interaction between a water molecule and a surface can involve hydrogen bonding, or it can involve only nonspecific, nondirectional interactions. The nonspecific interactions between a water molecule and a metal surface that is a single crystal plane, and between a water molecule and a dielectric surface that is a single crystal plane, are rather different, and both of these interactions are at least somewhat changed when the surfaces become rough. Each of the many possibilities for water–surface interactions has a signature in the structure of the water adjacent to the surface. And, because the water molecule is polar, application of a nonzero voltage to the electrode with which liquid water is in contact can alter the interface structure.

To learn how the structure of water adjacent to a solid depends on the competition between water–water and water–surface interactions, we examine the results of molecular dynamics simulations. In all of the simulations we consider the water–water interaction was represented by one or another of the empirical potentials mentioned in Section 26.3, and the water–surface interaction by some other potential. We focus attention on the cases in which the solid surface has groups to which water can hydrogen bond, or it is a salt or a metal.[40]

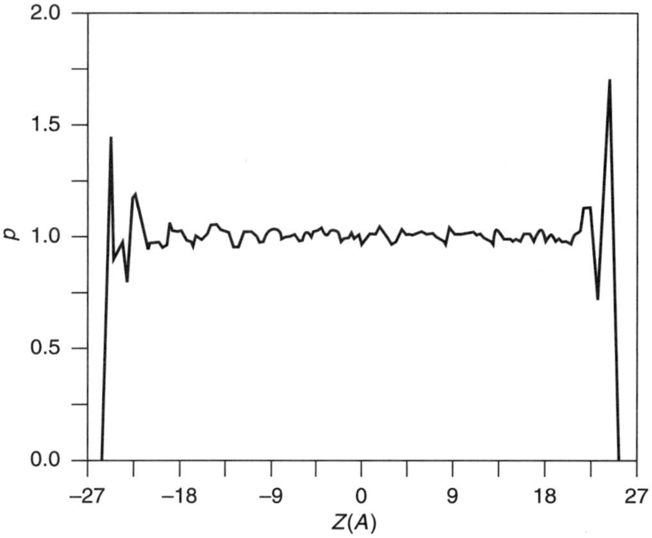

Figure 26.32 Oxygen number density profile of liquid water between hydroxylated silica surfaces. From S. H. Lee and P. J. Rossky, *J. Chem. Phys.* **100,** 3334 (1994).

Molecular dynamics simulations show that when water is in contact with a surface to which it can hydrogen bond—for example, a fully hydroxylated silica surface,[41] the oxygen density distribution in the interfacial region has two narrow peaks corresponding to well defined positions for the first two layers of water adjacent to the surface (Fig. 26.32); the local ordering induced in the interface by the surface does not extend beyond the second layer and the surface–water OO pair-correlation function strongly resembles that for the bulk liquid. In this case, the structure of the interface is primarily determined by the favorable energetics of hydrogen bonding to the surface, which leads to a major reorganization of the water structure in the layer closest to the surface and, by virtue of layer-to-layer hydrogen bonding, somewhat less reorganization in the second layer (Fig. 26.33). A typical water molecule in the layer adjacent to the surface forms up to three hydrogen bonds with silanol groups.

Drastic reorganization of water adjacent to a surface does not require hydrogen bonding to that surface. Figure 26.34 displays the results of a molecular dynamics simulation of the water density adjacent to a 001 plane of MgO, and Fig. 26.35 displays a typical snapshot of the molecular distribution.[42] The layer adjacent to the MgO surface is densely packed and well ordered, with one water molecule per Mg atom; the planes of the water molecules are oriented, with sensibly equal numbers at average angles with respect to the surface plane of –17° (both hydrogens pointing toward the surface) and +30° (one hydrogen angled up and one parallel

---

[40] It is also interesting to examine the structure of the interfacial region between water and a hydrophobic surface. A hydrophobic interaction between water and a surface can be simulated by use of a Lennard–Jones form for the water molecule–surface potential. It is found (C. Y. Lee, J. A. McCammon and P. J. Rossky, *J. Chem. Phys.* **80,** 4448 (1984)) that when water is in contact with a smooth plane or with a close packed atomic plane, with which the water molecules interact via a Lennard–Jones potential, the first and second layers of liquid water near the surfaces are distinguishably different from water further away. Specifically, the simulation results show that: (i) the oxygen atom density (effectively the same as the water density) along a line perpendicular to the wall shows a few oscillations, (ii) the peak heights of the density oscillations are smaller, and the peak widths greater, than in the hard sphere system or in the equivalent system of Lennard–Jones atoms interacting with the same walls, (iii) the first two peaks in the water density distribution adjacent to the close packed plane of atoms are closer to the wall and the peak heights are larger than for the corresponding distribution of water adjacent to the smooth plane, (iv) the separation of the first two peaks is greater than the O–O separation in the bulk liquid, and (v) about 75% of the hydrogen bonding typical of the bulk liquid is maintained at the hydrophobic surface. Observation (i) is a consequence of the molecular packing near the wall, just as in a hard sphere fluid in contact with a smooth hard wall. All the other observations depend on the particular properties of water and the wall. The small amplitude of the first peak of the interface water density distribution (observation (ii)) is a consequence of the lack of "wetting" by water of the Lennard–Jones surface, and the shift in separation from the wall of the first two peaks of the water density distribution when the wall is changed from smooth to atomically rough (observation (iii)) is to be expected given the increased attractive interaction in the hollows of the plane of close packed atoms relative to that of the smooth plane. These and the remainder of the observations are consistent with optimization of hydrogen bonding interactions in the water phase, which generates substantial structural reorganization adjacent to the hydrophobic surface. A compromise between increasing the density of the water layer adjacent to the surface and maintaining the largest possible number of hydrogen bonds between water molecules can be achieved by orienting a water molecule adjacent to the surface so as to direct one hydrogen bond away from the bulk liquid, while retaining the remaining three hydrogen bonds. Of course, the water molecules to which it is bonded cannot lie in the same plane, although all such molecules can lie in a stratum about 0.9 Å thick, consistent with the increase in O–O separation near the wall to 3.7 Å from the 2.8 Å separation characteristic of the bulk liquid. In this picture, in the limit of maximum hydrogen bonding, those molecules adjacent to the wall have only 75% of their possible hydrogen bonds, as inferred from the simulations.

[41] S. H. Lee and P. J. Rossky, *J. Chem. Phys.* **100,** 3334 (1994).

[42] M. I. McCarthy, G. K. Schenter, C. A. Scamehorn, and J. B. Nicholas, *J. Phys. Chem.* **100,** 16989 (1996).

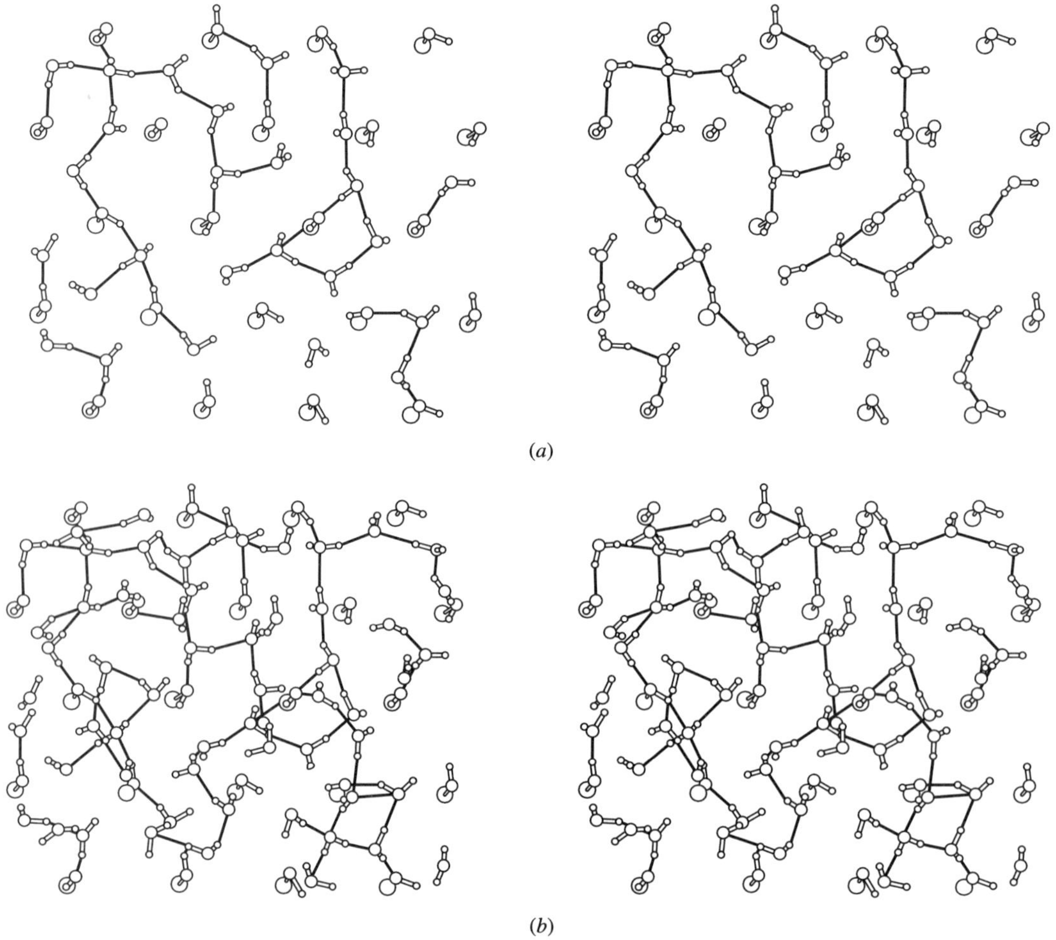

Figure 26.33 Instantaneous views of (*a*) the first layer of water adjacent to a hydroxylated silical surface and (*b*) the first and second layers of water adjacent to the same surface. The hydrogen bonding is indicated by solid lines and the Si atoms by large circles. From S. H. Lee and P. J. Rossky, *J. Chem. Phys.* **100,** 3334 (1994).

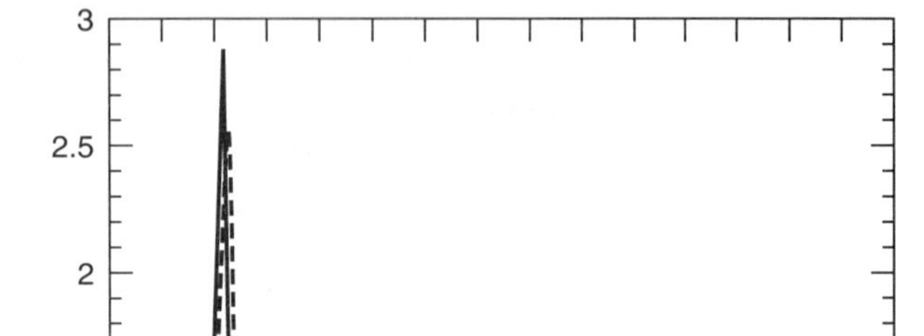

Figure 26.34 Density of oxygens in water molecules adjacent to the 001 plane of MgO. From M. I. McCarthy, G. K. Schenter, C. A. Scamehorn, and J. B. Nicholas, *J. Phys. Chem.* **100,** 16989 (1996).

to, or slightly tilted toward, the surface). The second layer is less well ordered, and ordinary liquid order is found in the third and subsequent layers.

The interaction between a water molecule and a metal surface is substantially different from that between a water molecule and a dielectric surface. Since the metal has very high electrical conductivity and the water molecule has a dipole moment, a macroscopic description of the water–metal surface interaction leads to the conclusion that an image dipole is induced in the metal, and that part of the overall interaction energy arises from the water dipole–image dipole interaction. Although useful as a conceptualization of the polarization of the electron distribution of the metal by the water molecule, at the water–surface separations of interest to us that description is oversimplified and inaccurate. A better description of the water molecule–surface interaction energy is obtained from quantum mechanical calculations of the total energy, as a function of separation, of the water molecule–metal surface system (see Chapters 9 and 10); it is this approach that has been used to define an interaction potential for input into the computer simulation studies.

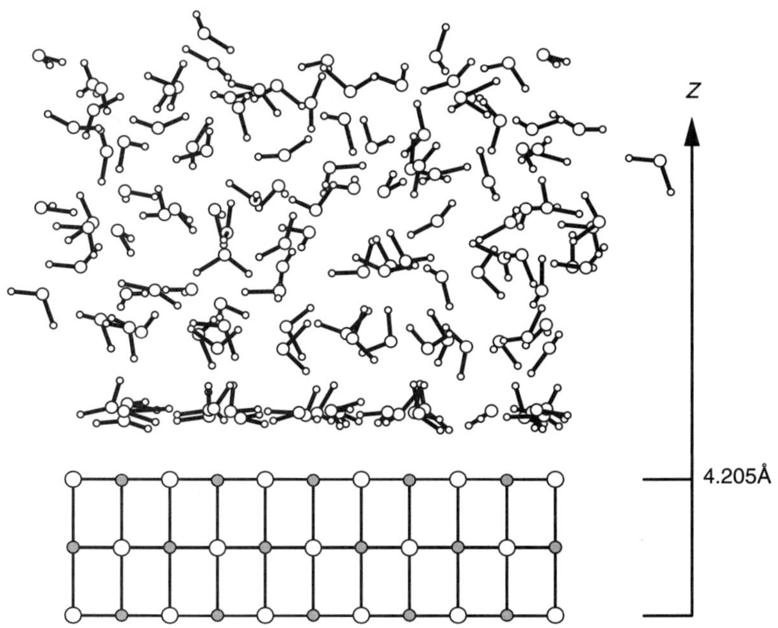

Figure 26.35 Instantaneous configuration of water molecules adjacent to the 001 plane of MgO. From M. I. McCarthy, G. K. Schenter, C. A. Scamehorn, and J. B. Nicholas, *J. Phys. Chem.* **100,** 16989 (1996).

Molecular dynamics simulations reveal that the generic character of the distribution of water in the water–metal surface interface is like that already described. That is, the water molecules form layers with diminishing definition as the distance from the surface increases; typically two, and sometimes three, layers have densities and ordering which are distinguishably different from that in the bulk liquid (Fig. 26.36). For the case of water immediately adjacent to the 111 surface of Ag, the water dipole moment and the OH bonds lie nearly parallel to the surface.[43] Because of the tetrahedrality of hydrogen bonding, these molecules form a sublayer (the small separation half of the first peak in Fig. 26.36) which is bonded to the adjacent sublayer (the larger separation half of the first peak in Fig. 26.36) that completes the first layer of water; the water molecules in the second sublayer have two preferred orientations, one with the dipole moment pointing slightly toward the metal surface and the other with the dipole moment pointing slightly away from the metal surface. This molecular arrangement is like that in ice Ih. The structure of the interface between water and liquid Hg is similar.[44] Fig. 26.36 also displays the density of water adjacent to the Ag 111 surface when the applied field strength is nonzero. At the lower of the applied field strengths, 1.0 V/Å, the separations of the O atoms of the water molecules from the positively and negatively charged Ag surface remain the same as in the absence of the field, but there are changes in the hydrogen densities near the positively and negatively charged surfaces, implying that the water molecules reorient in the presence of the field. At the higher of the applied field strengths, 3.0 V/Å, the layering of the water molecules extends further into the bulk liquid than in the case of no applied field, and the increased separation between the peaks of the oxygen and hydrogen density distributions implies that the water molecules are strongly oriented by the field. Associated with the field induced orientation of the water molecules is a reorganization of the liquid structure, shown by the simultaneous decrease in the height of the first peak of the O atom distribution and the appearance of extra peaks in that distribution.[45] The density profile of water adjacent to an Ag electrode has been measured;[46] there is qualitative agreement between the results of the molecular dynamics simulations and experiment (Fig. 26.37).

---

[43] K. J. Schweighofer, X. Xia, and M. L. Berkowitz, *Langmuir* **12,** 3747 (1996).

[44] J. Böcker, Z. Gurskii, and K. Heinzinger, *J. Phys. Chem.* **100,** 14969 (1996).

[45] A more detailed analysis of the water orientation near the Ag surface reveals the nature of the interplay between (i) hydrogen bonding, (ii) adsorption of the oxygen of a water molecule to a Ag atom, (iii) repulsion of the hydrogen atoms of a water molecule by a Ag atom, and (iv) interaction of the water dipole moment with the electric field. In the absence of an electric field (ii) and (iii) favor a configuration with the plane of the water molecule nearly parallel to the plane of the surface, with the hydrogens pointing slightly away from the surface. Optimization of the hydrogen bonding between adsorbed water molecules and the other water molecules adjacent to the surface, which maintains the network of hydrogen bonds to the greatest possible extent, then requires the dipole moments of the latter to point slightly toward the metal surface. In the presence of a nonzero electric field the water adsorption and hydrogen atom-Ag atom repulsion are increased at the positively charged surface and decreased at the negatively charged surface. This effect is relatively small at 1.0 V/Å, but rather large at 3.0 V/Å, which accounts for the propagation of order in the water adjacent to the Ag surface to more distant layers in the presence of the stronger electric field. At 3.0 V/Å some of the field reoriented water near the positively charged surface is desorbed so as to better maintain the hydrogen bonding in the liquid, and at the negatively charged surface there is desorption because of the repulsion between the surface and the oxygen atom of the water molecule.

[46] M. Toney, J. Howard, J. Richer, G. Borges, J. Gordon, O. Melroy, D. Weisler, D. Yee, and L. Sorenson, *Nature* **368,** 444 (1994).

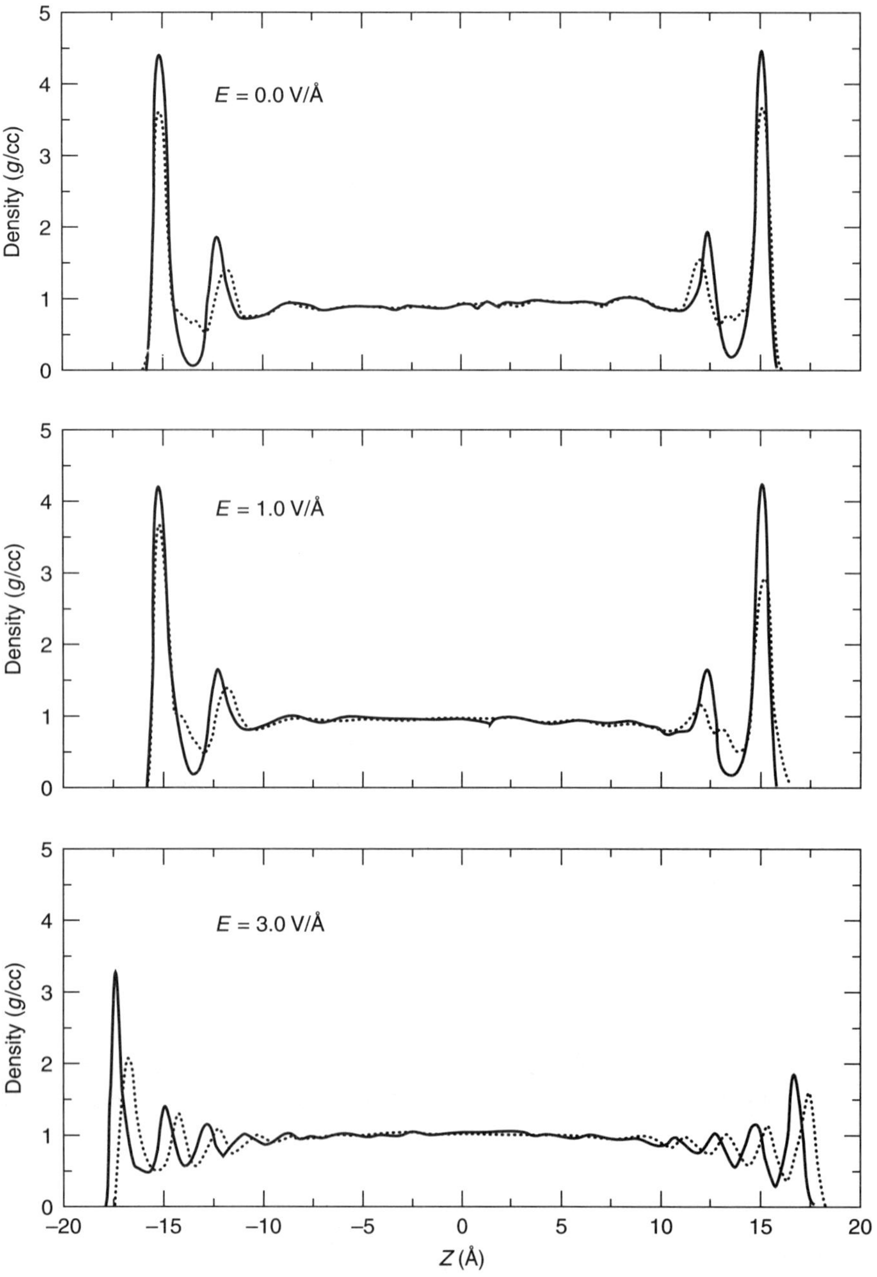

Figure 26.36 Density of water adjacent to the 111 surface of Ag. The solid lines display the oxygen atom distribution and the dotted lines the hydrogen atom distribution. From K. J. Schweighofer, X. Xia, and M. L. Berkowitz, *Langmuir* **12,** 3747 (1996).

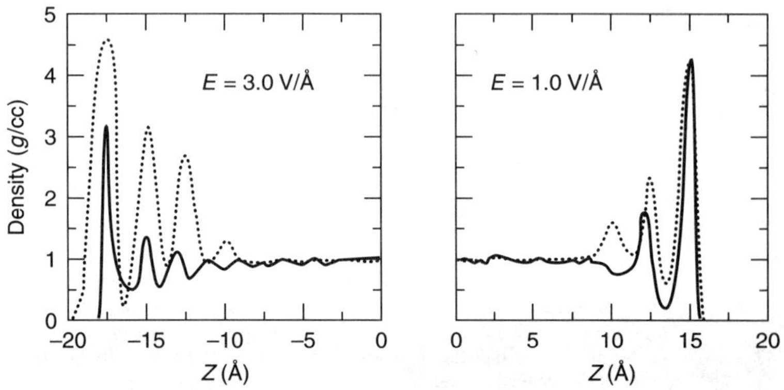

Figure 26.37 Comparison of the measured density distributions of water adjacent to an Ag electrode (· · · ·) with that obtained from a molecular dynamics simulation (——). From K. J. Schweighofer, X. Xia, and M. L. Berkowitz, *Langmuir* **12,** 3747 (1996).

B. The Electrolyte Solution–Electrode Interface

The region over which the structure of water is influenced by the presence of a surface overlaps the region in which transfer of electrons and ions between a solution and an electrode occurs. It might seem, then, that understanding the structure of the electrolyte solution–electrode interface, and its dependence on applied electric field, requires an analysis that includes simultaneous consideration of all of the particles, both the ions and the water molecules in the electrolyte solution, and also the atoms in the electrode surface. Clearly, such an approach will yield the most accurate and most complete description of the structure of the interface. Nevertheless, when examining the structure of a comparably complex electrolyte solution we found that a primitive model, consisting of charged hard spheres in a continuum dielectric, provided the key elements with which the most important qualitative features of the structure and properties of the electrolyte solution can be understood. We consider the analogue of that model for the electrolyte solution–electrode interface first, the Gouy–Chapman theory,[47] and then the changes introduced when the molecular character of the dielectric medium is accounted for.

Suppose the electrolyte solution contains a mixture of ions with densities n_α, charges $Z_\alpha e$, and diameters $\sigma_\alpha \equiv 2a_\alpha$. We denote by $n_\alpha(z)$ the number density of ions of species α a distance z from the flat electrode. If the solvent is represented as a continuous dielectric and each of the ionic species has the distribution $n_\alpha(z) = n_\alpha \exp[-Z_\alpha e\Phi/k_BT]$, with $\Phi(z)$ the mean electrostatic potential at z, then arguments similar to those used to derive Eq. 26.64 lead to

$$\frac{d^2\Phi(z)}{dz^2} = -\frac{1}{\epsilon_0 D}\sum_\alpha n_\alpha Z_\alpha e \exp\left[-\frac{Z_\alpha e\Phi(z)}{k_BT}\right], \qquad (26.124)$$

which is the one-dimensional Poisson–Boltzmann equation. For the special case of a symmetric simple electrolyte, with $Z_+e = -Z_-e = |Ze|$, Eq. 26.124 can be solved for $\Phi(z)$. In this case

$$\frac{d^2\Phi(z)}{dz^2} = -\frac{n|Ze|}{\epsilon_0 D}\left[e^{-|Ze|\Phi(z)/k_BT} - e^{|Ze|\Phi(z)/k_BT}\right]. \qquad (26.125)$$

Multiplying both sides of Eq. 26.125 by $2(d\Phi/dz)$ and the use of the definition of the hyperbolic sine function then leads to

$$\frac{d}{dz}\left(\frac{d\Phi(z)}{dz}\right)^2 = \frac{4k_BTn}{\epsilon_0 D}\sinh\left(\frac{|Ze|\Phi(z)}{k_BT}\right)\left(\frac{d\Phi(z)}{dz}\right), \qquad (26.126)$$

[47] G. Gouy, *J. Phys.* **9**, 457 (1910); D. L. Chapman, *Philos. Mag.* **25**, 475 (1913).

which, using the boundary condition $(d\Phi(z)/dz) \to 0$ as $z \to \infty$, can be integrated immediately to yield

$$\left(\frac{d\Phi(z)}{dz}\right)^2 = \frac{4k_BTn}{\epsilon_0 D}\left[\cosh\left(\frac{|Ze|\Phi(z)}{k_BT}\right) - 1\right],$$

$$\frac{d\Phi(z)}{dz} = \pm\left(\frac{8k_BTn}{\epsilon_0 D}\right)^{1/2}\sinh\left(\frac{|Ze|\Phi(z)}{2k_BT}\right). \qquad (26.127)$$

Dividing both sides of the second of Eqs. 26.127 by $\sinh(|Ze|\Phi(z)/2k_BT)$ and integrating again leads to

$$\tanh\left(\frac{|Ze|\Phi(z)}{4k_BT}\right) = \exp\left[\pm\left(\frac{2|Ze|^2 n}{\epsilon_0 Dk_BT}\right)^{1/2} z + \frac{C|Ze|}{2k_BT}\right], \qquad (26.128)$$

where C is a constant of integration. Since $\Phi(z) \to 0$ as $z \to \infty$ only the solution with the negative sign in the exponent on the right-hand side of Eq. 26.128 is acceptable. If the change in electrostatic potential from $z = 0$ to $z = \infty$ is denoted $\Delta\Phi$, then $\Phi(z = 0) = \Delta\Phi$,[48] and the integration constant can be eliminated by forming the ratio

$$\frac{\tanh(|Ze|\Phi/4k_BT)}{\tanh(|Ze|\Delta\Phi/4k_BT)} = \exp(-\kappa z), \qquad (26.129)$$

where κ is the Debye–Hückel inverse screening length (Eq. 26.67). Equation 26.129 defines an implicit relation between the mean electrostatic potential, the potential at the electrode, and the ionic concentration. After some algebraic manipulation the ionic density profile is given by

$$g_\alpha(z) \equiv \frac{n_\alpha(z)}{n_\alpha} = \left(\frac{1 + Z_\alpha \tanh(|Ze|\Delta\Phi/4k_BT)\exp(-\kappa z)}{1 - Z_\alpha \tanh(|Ze|\Delta\Phi/4k_BT)\exp(-\kappa z)}\right)^2. \qquad (26.130)$$

Figure 26.38 compares the density profile for hard sphere ions near an electrode predicted by the Gouy–Chapman theory with the density profile obtained from computer simulations.

[48] We can relate the charge density on the electrode to the potential at $z = 0$ as follows. If there is, say, a positive surface charge density q^{E+} on one electrode, the condition of electroneutrality requires that there be a distribution of anions which sastisfies the condition

$$q^{A-} = \int_0^\infty dz\, n(z) = -\epsilon_0 D\int_0^\infty dz\left(\frac{d^2\Phi}{dz^2}\right) = \epsilon_0 D\left(\frac{d\Phi}{dz}\right)_{z=0} = -q^{E+}.$$

Of course, at the other electrode there will be a negative surface charge density and an equivalent distribution of cations, so that overall electroneutrality is maintained. Using the second of Eqs. 26.128, evaluated at $z = 0$, we find

$$q^{E+} = (8k_BTn\epsilon_0 D)^{1/2}\sinh\left(\frac{|Ze|\Delta\Phi}{2k_BT}\right).$$

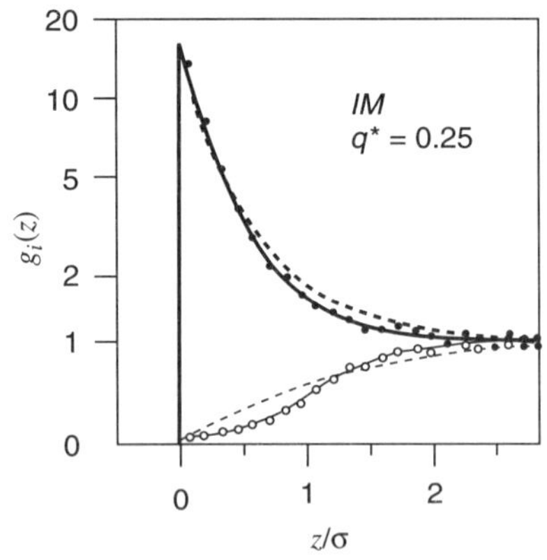

Figure 26.38 Density profile of hard charged spheres near a smooth hard charged electrode for a 1 *m* 1:1 electrolyte solution. The reduced charge density $q^* = q^E\ \sigma^2/e = 0.25$, $\sigma = 4.21$ Å, $T = 298$ K, $D = 78.5$. The points show the results of simulations and the solid and dashed curves are the predictions obtained from an integral equation theory combining features of the HNC and MSA approximations (see Chapter 23) and of the Gouy–Chapman theory, respectively. From L. Blum and D. Henderson, Chapter 6 in *Fundamentals of Inhomogeneous Fluids,* ed. D. Henderson, Marcel Dekker, New York (1992).

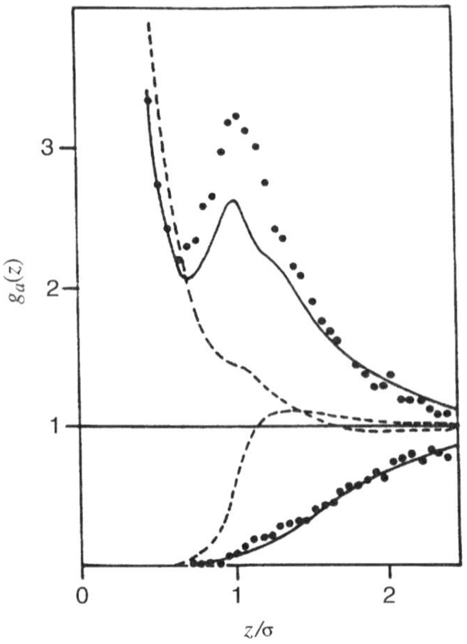

Figure 26.39 Density profile of hard charged spheres near a smooth hard charged electrode for a 1 *m* 1:1 electrolyte solution. The reduced charge density (defined in the caption to Fig. 26.38) is 0.7; all other parameters are the same as for Fig. 26.38. The solid and the dashed curves are the predictions obtained from an integral equation theory of the HNC type with and without inclusion of an approximation for the bridge function, respectively. From L. Blum and D. Henderson, Chapter 6 in *Fundamentals of Inhomogeneous Fluids,* ed. D. Henderson, Marcel Dekker, New York (1992).

Many of the faults of the Gouy–Chapman theory are similar to those of the Poisson–Boltzmann equation based theory of electrolyte solutions. In particular, the potential of mean force is approximated as the mean electrostatic potential, the effect of excluded volume on the distribution of ions is treated crudely, the difference between the solvent structure adjacent to the electrode and in the bulk solution is ignored and the molecular character of the solvent is completely ignored. Nevertheless, as shown in Fig. 26.38, the predicted ion density distribution near the electrode is in reasonable agreement with the simulation data. The source of this agreement can be traced to two features of the Gouy–Chapman theory: it satisfies the constraint of electroneutrality and it gives a reasonable value for the pressure of the liquid in contact with the electrode.[49] Since the density profile predicted by the Gouy–Chapman theory is monotone decreasing with increasing z, and the integral over the density profile is fixed by the constraint of electroneutrality, the prediction of a reasonable value for the pressure at the electrode implies that the predicted density distribution cannot deviate grossly from the correct distribution. This argument fails when the pressure at the electrode is poorly accounted for, and the Gouy–Chapman theory predictions are inaccurate in those cases.

One class of theories intended to be more accurate than the Gouy–Chapman theory is concerned with improving the treatment of the ion core–ion core interactions within the framework of the primitive model of charged hard spheres in a continuum dielectric solvent.[50] As might be expected, these theories use the same methodology and the same categories of approximations as used in the theory of liquids. In particular, there are integral equation representations of the ion density profile near an electrode that have features in common with most of the integral equations mentioned in Chapter 23. Figure 26.39 displays a comparison of the ion density profiles near an electrode calculated from computer simulations and from two integral equations, one related to the ordinary HNC equation (Eq. 23.48) and the other to a modified HNC equation which includes an approximation to the bridge function (Eq. 23.53). The system studied has very high surface charge density, and

[49] The electrostatic forces in a continuous dielectric medium with charges contribute to the pressure. For the primitive model of charged hard spheres in a continuum dielectric it can be shown that

$$k_BT\sum_\alpha n_\alpha(0) = p + \frac{DE_0^2}{8\pi},$$

where E_0 is the applied field at $z = 0$. The Poisson–Boltzmann equation leads to the pressure

$$k_BT\sum_\alpha n_\alpha(0) = k_BT\sum_\alpha n_\alpha + \frac{DE_0^2}{8\pi},$$

which is equivalent to replacing the true pressure with the ideal gas pressure.

[50] L. Blum and D. Henderson in *Fundamentals of Inhomogeneous Fluids,* ed. D. Henderson, Marcel Dekker, New York (1992), Chapter 6.

the ion density distribution shows oscillations because of the excluded volume generated from the ion core–ion core interactions. It is clear that the more advanced of the integral equation theories captures the essential features of the computer simulation results.

Another class of theories intended to be more accurate than the Gouy–Chapman theory takes account of the molecular nature of the solvent as well as providing a more accurate treatment of the effects of ion core–ion core interactions. In these theories, the polarization of the medium associated with the charges present is accounted for by the interaction between the ions and the multipole charge distribution of the solvent molecules. One well studied model consists of a mixture of charged hard spheres, with diameters σ_α, and dipolar hard spheres, with diameter σ_s. The analogue of the mean spherical approximation integral equation (Eq. 23.52) for this system can be solved analytically. In the dilute solution limit it is found that $\Delta\Phi$ can be expanded in a power series in the reciprocal Debye–Hückel screening length κ:

$$\Delta\Phi = \frac{E_0}{D\kappa} + \frac{E_0}{2D}\left(\sigma_\alpha + \frac{D-1}{\lambda}\sigma_s\right) + \cdots, \qquad (26.131)$$

with $\lambda(1+\lambda)^2 = 4D^{1/2}$. The first term on the right-hand side of Eq. 26.131 is the same as obtained from the linearized version of the Gouy–Chapman theory. Note that the second term on the right-hand side of 26.131 is independent of the concentration.

One of the observable properties of the electrolyte solution–electrode interface is its differential capacitance, defined by

$$C_d \equiv \left(\frac{\partial q^E}{\partial(\Delta\Phi)}\right)_{p,T,n_\alpha}. \qquad (26.132)$$

The differential capacitance corresponding to the terms displayed in Eq. 26.131 is

$$\frac{1}{C} = \frac{1}{C_{GC}} + \frac{1}{C_H}, \qquad (26.133)$$

where C_{GC} is the differential capacitance obtained from the Gouy–Chapman theory; it is a function of the ionic concentration. The second term on the right-hand side of 26.133 is independent of concentration and is often called the Helmholtz capacitance.[51] In the model under consideration C_H is determined almost entirely by the properties of the solvent because $(D-1)/\lambda >> 1$; this observation and the values calculated are consistent with the experimental data. A test of Eq. 26.133 against experimental data is displayed in Fig. 26.40.

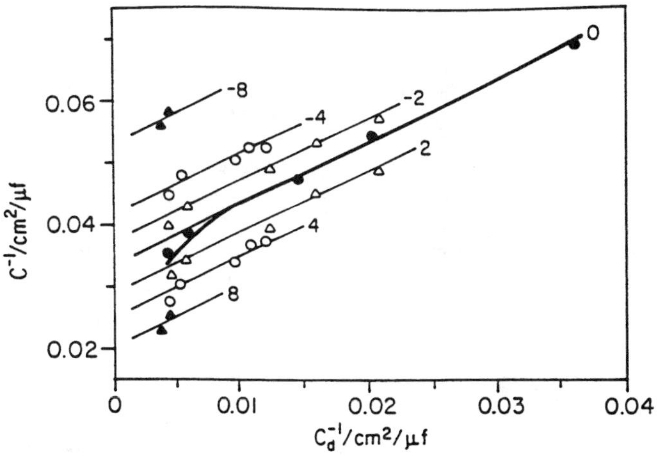

Figure 26.40 Inverse capacitance of an aqueous solution of NaH_2PO_4 at 25°C near a mercury electrode as a function of the inverse Gouy–Chapman capacitance. The light straight lines are obtained from Eq. 26.133. The heavy line is the prediction obtained from an integral equation theory of the MSA type using $\sigma_\alpha = \sigma_s =$ 2.76 Å. From L. Blum and D. Henderson, Chapter 6 in *Fundamentals of Inhomogeneous Fluids,* ed. D. Henderson, Marcel Dekker, New York (1992).

None of the theoretical approaches just described takes account of the influence of the electrolyte on the properties of the electrode, for example the alteration of the electron distribution at the surface of a metal by virtue of contact with an electrolyte solution.[52] Self-consistent inclusion of the effects of the electrolyte solution on the electrode are necessary before the many different phenomena associated with electrode processes can be fully understood.

• FURTHER READING

Ben-Naim, A., *Water and Aqueous Solutions* (Plenum, New York, 1974).

Friedman, H. L., *Ionic Solution Theory* (Wiley, New York, 1962).

Gordon, J. E., *The Organic Chemistry of Electrolyte Solutions* (Wiley, New York, 1975).

Gurney, R. W., *Ionic Processes in Solution* (Dover, New York, 1953).

Harned, H. S., and Owen, B. B., *The Physical Chemistry of Electrolytic Solutions,* 3rd ed. (Van Nostrand Reinhold, New York, 1958).

Hush, N. S., Ed., *Reactions of Molecules at Electrodes* (Wiley, New York, 1971), Chapter 2.

McQuarrie, D. A., *Statistical Mechanics* (Harper & Row, New York, 1976), Chapter 15.

[51] It is often also called the inner layer capacitance, because it is often regarded as the capacitance of a layer of water assumed to be tightly bound to the electrode. However, the structure of the layer of water adjacent to, or bound to, an electrode depends on many features of the interaction with the surface and the applied field, as discussed above, so this interpretation is not satisfactory. The integral equation based theory shows that the additional capacitance arises from the nonzero size of a solvent molecule, and the fact that the solvent polarization is spread over the entire diffuse ion density profile.

[52] See, for example, J. Goodisman, *Electrochemistry: Theoretical Foundations,* Wiley, New York (1987); D. L. Price and J. W. Halley, *J. Chem. Phys.* **102,** 6603 (1995).

Marcus, Y., *Introduction to Liquid State Chemistry* (Wiley, New York, 1977), Chapters 6 and 7.
Outhwaite, C. W., *Specialist Periodical Report, Statistical Mechanics* **2,** 188 (1975), The Chemical Society, London.
Pitzer, K. S., *Thermodynamics,* 3rd ed. (McGraw-Hill, New York, 1995), Chapters 15–19.
Resibois, P., *Electrolyte Theory* (Harper & Row, New York, 1968).
Robinson, R. A., and Stokes, R. H., *Electrolyte Solutions,* 2nd ed. (Butterworths, London, 1959).
Watts, R. O., and McGee, I. J., *Liquid State Chemical Physics* (Wiley, New York, 1976), Chapters 9 and 10.

• PROBLEMS

1. Describe how the concentration cell shown in Fig. 26.2 might be used to measure the fugacity of hydrogen at various pressures.

2. Below are given the molalities of a saturated TlCl solution at two temperatures, and the corresponding activity coefficients. Calculate the heat of solution of TlCl.

T(K)	m_{TlCl}	$\gamma_\pm$
298.1	0.0161	0.81
312.8	0.0252	0.77

3. The dissociation pressure of oxygen over silver oxide, Ag_2O, at 298 K is 5 × 10^{-4} atm. Calculate $\mathscr{E}$ for the cell

$$\text{Ag}|\text{Ag}_2\text{O}(sat.)|\text{Pt, O}_2\,(0.2\text{ atm}).$$

4. Calculate the solubility of silver iodide, AgI, in pure water at 298 K. The following cell generates 0.807 V:

$$\text{Ag}|\text{AgI}(sat.),\text{KI}(0.1\ m)\,|\text{AgNO}_3\,(0.1\ m)|\text{Ag}.$$

Regard the solutions as ideal.

5. A cell

$$\text{Cu}(amalgam,\ a')\,|\text{CuSO}_4\,(aq.)|\text{Cu}(amalgam,\ a'')$$

generates 0.124 V at 298 K. In the more dilute amalgam, which may be regarded as an ideal solution, the concentration of copper is 1.664 × 10^{-5}% by weight, and in the other amalgam the copper concentration is 0.04472% by weight. Calculate the activity coefficient of Cu in the more concentrated amalgam.

6. Calculate the cell potential of the cell

$$\text{Cl}_2\,(g,\ 1\text{ atm}),\ \text{Pt}|\ \text{HCl}(10\ m)\,|\text{Pt, O}_2\,(g,\ 1\text{ atm})$$

at 298 K. The equilibrium constant of the gas phase reaction

$$4\text{HCl}+\text{O}_2 \rightarrow 2\text{H}_2\text{O}+2\text{Cl}_2$$

is $K_p = 10^{13}$ atm^{-1}, and the partial vapor pressures over the HCl solution in the cell are p_{H_2O} = 9.4 torr and p_{HCl} = 4.2 torr.

7. The cell potential of a particular cell has the temperature dependence

$$\mathscr{E}(\text{volts}) = 0.96466 + 1.74\times10^{-4}(t-25) + 3.8\times10^{-7}(t-25)^2,$$

where t is the Celsius temperature. Find the heat of reaction at 25°C.

8. How does the potential of a cell depend on the external pressure?

9. A solution of KI has an ionic strength of 0.32. Calculate its concentration. What concentrations of K_2SO_4 and $MgSO_4$ have the same ionic strength?

10. Using the Debye–Hückel limiting law, Eq. 26.86, calculate the ionic activity coefficients γ_+ and γ_- and the mean activity coefficient $\gamma_\pm$ for 10^{-4} M and 10^{-3} M solutions of NaCl and $CaCl_2$.

11. Obtain a limiting expression for the solubility of a salt (say, AgCl) in the presence of a soluble electrolyte, in terms of the sparingly soluble salt's solubility–product constant and the ionic strength.

12. The value of K_{sp} for AgCl at 25°C is 1.71 × 10^{-10}. Calculate the solubility of AgCl in pure water and in 10^{-3} M $ZnSO_4$.

13. For a reaction such as Eq. 26.39, which can occur in a cell at constant temperature and pressure, obtain a relation between the cell potential $\mathscr{E}$ and the Gibbs free energy change for the reaction.

14. Consider a membrane that is permeable to K^+ but not to Cl^- or H_2O. The membrane separates two solutions of KCl, one of which is 0.1 m and the other is 0.01 m. A voltage is found to arise spontaneously across the membrane.
(a) What is the source of the voltage?
(b) Calculate the magnitude of the voltage. (*Hint:* Consider Eq. 26.17 and disregard activity coefficients.)
(c) The charge separation Q necessary to produce a potential difference across an electric capacitor is given by $Q = CV$, where C is called the capacitance.

For a biological membrane whose area is 1 cm^2, the value of C is of the order 1 μF. (If C is given in F (farads), then V is in volts and Q in coulombs.) Calculate the number of ions that must pass through a 1-cm^2 membrane to produce the voltage found in part (b).

(d) Suppose that the two KCl solutions have volumes of 1 cm^3 each. How does the number of ions found in part (c) compare to the total number of ions present? Is the charge imbalance implied by the potential difference detectable by chemical means?

15. The following values of $\gamma_\pm$ have been found for NaCl at 25°C:

m	0.001	0.002	0.005	0.10	0.20
$\gamma_\pm$	0.9649	0.9519	0.9275	0.9024	0.8712

Show that the above data are consistent with the limiting law, Eq. 26.86.

16. For a particular reaction occurring in an electrochemical cell, the cell potential was found to vary with temperature according to

$$\mathscr{E} = 0.134 - 2.94\times10^{-4}\,t - 4.20\times10^{-6}\,t^2,$$

where $\mathscr{E}$ is in volts and t is the Celsius temperature. Find ΔH, ΔS, and ΔG for the reaction at 25°C.

17. The potential of the cell

$$\text{Pt, } H_2\,(1\text{ atm})\,|\,\text{HCl}(0.01\ m)\,|\,\text{Cl}_2\,(p\text{ atm}),\ \text{Pt}$$

was measured at 25°C for various values of p, the pressure of chlorine. The results, with $\mathscr{E}$ in volts, were

p	1	50	100
$\mathscr{E}$	1.5962	1.6419	1.6451

What is the fugacity of Cl_2 at each pressure? What assumption must you make about the hydrogen fugacity to do this problem? How might you test this assumption?

18. What is the potential of the cell

$$\text{Pt, } H_2\,(1\text{ atm})|\text{HCl}(0.1\ m)|\text{HCl}(0.2\ m)|H_2\,(10\text{ atm}),\ \text{Pt?}$$

19. What is the mean ionic activity of K_2SO_4 in terms of the molality of the salt?

20. Calculate the ionic strength (Eq. 26.49) of the following solutions: 0.1 m KCl, 0.1 m K_2SO_4, 0.1 m $MgSO_4$.

21. The potential of the following cell was measured at various HCI molalities at 20°C:

$$\text{Pt, } H_2\,(\text{unit activity})|\,\text{HCl}(sol'n)\,|\text{AgCl}\,|\text{Ag}.$$

The voltages obtained were

m	0.005314	0.008715	0.013407	0.021028	4.0875
$\mathscr{E}$	0.49395	0.46987	0.44899	0.42726	0.12307

Find $\mathscr{E}^\ominus$ for this cell and $\gamma_\pm$ for HCl at $m = 4.0875$.

22. Show that $\mathscr{E}^\ominus$ for the cell described in Eq. 26.42 can be obtained more accurately by the extrapolation

$$\mathscr{E}^\ominus = \lim_{m_\pm\to 0}\left(\mathscr{E} - \frac{RT}{2\mathscr{F}}\ln a_{H_2} + \frac{2RT}{\mathscr{F}}\ln m_\pm - \text{constant}\, m_\pm^{1/2}\right).$$

23. Use the data of Problem 21 to calculate K_{sp} for AgCl.

24. Suppose that two electrolyte solutions, I and II, are separated by a rigid semipermeable membrane that allows passage of some, but not all, ions. At equilibrium we expect both a pressure difference and an electric potential difference between I and II. Using the method introduced to discuss osmotic pressure, show that

$$\ln\frac{(m_i\gamma_i)^{\text{II}}}{(m_i\gamma_i)^{\text{I}}} + \frac{Z_i\mathscr{F}}{RT}(\psi^{\text{II}} - \psi^{\text{I}}) = \frac{M_1}{1000}\frac{\bar{v}_i}{\bar{v}_1}\left(g_1^{\text{I}}\sum_i m_i^{\text{I}} - g_1^{\text{II}}\sum_i m_i^{\text{II}}\right).$$

25. Show that, at constant temperature and pressure, the osmotic coefficient and ionic activity coefficients are related by

$$-(\nu_+ + \nu_-)m\ dg_1 + (1-g_1)(\nu_+ + \nu_-)dm + \nu_+ m\ d\ln\gamma_+ \nu_- m\ d\ln\gamma_- = 0,$$

where m is the molality of the salt $M_{\nu+}X_{\nu-}$. Then show that

$$-\ln\gamma_\pm = (1-g_1) + \int_0^m (1-g_1)\ d\ln m.$$

Hint: Use the Gibbs–Duhem equation.

26. Given that the osmotic coefficient for an electrolyte solution has the form

$$1 - g_1 = a_1 m^{1/2} + a_2 m + a_3 m^{3/2},$$

what is the mean activity coefficient of the solute?

27. Suppose that a 1:1 ionic salt is accurately described by the Born–Mayer model, and that the Born–Mayer model for ionic hydration accurately describes the changes that occur when the salt dissolves. Show that the expected Gibbs free energy of solution of a series of 1:1 ionic salts with common anion has a maximum with respect to cation radius. What are the consequences of this result for ionic solutions? *Hint:* Treat the ionic solution as ideal.

28. Consider a dilute gas of positive and negative ions (a plasma). Show that the second virial coefficient for such a gas does not exist (diverges). The second virial coefficient of a gas is defined by (see Chapter 21)

$$B(T) = -2\pi \int_0^\infty \left[\exp\left(-\frac{u_2(R)}{k_B T} \right) - 1 \right] R^2 \, dR.$$

29. The McMillan–Mayer theory of solutions leads to a virial expansion for the osmotic pressure. The virial coefficients found have the same functional form as for a gas, but the direct intermolecular potential is replaced by the potential of mean force. Show that when the Debye–Hückel approximation to the potential of mean force is used in

$$B(T) = -2\pi \int_0^\infty \left[\exp\left(-\frac{w_2(R)}{k_B T} \right) - 1 \right] R^2 \, dR,$$

the Debye–Hückel limiting law is recovered.

30. The charge density around the ith ion is, according to Debye–Hückel theory,

$$q_i(R) = -\frac{Z_i e \kappa^2}{4\pi R} \cdot \frac{e^{-\kappa(R-a)}}{(1+\kappa a)}.$$

Find the most probable value of R, and the mean value of R, for this distribution of charge. What physical insight into the importance of κ does this calculation give?

31. The pair-correlation function in Debye–Hückel theory is, for very low concentration where $\kappa a \to 0$,

$$g_{ij}(R) = 1 - \frac{Z_i Z_j e^2}{4\pi\epsilon_0 D k_B T} \cdot \frac{e^{-\kappa R}}{R}.$$

Use this result in the general equation

$$\frac{U}{V} = \frac{3}{2} n k_B T + 2\pi \sum_{i=1}^{2} n_i n_j \int_a^\infty g_{ij}(R) u_{ij}(R) R^2 \, dR$$

to derive the limiting-law properties of the ionic solution. Note that $u_{ij} \to \infty$ for $R \le a$, and u_{ij} is the direct Coulomb interaction $Z_i Z_j e^2 / 4\pi\epsilon_0 DR$ for $R > a$.

32. Show that the osmotic coefficient, g_1, for a 1:1 electrolyte solution is, according to Debye–Hückel theory,

$$1 - g_1 = \frac{\sum n_i Z_i^2 e^2 \kappa}{24\pi\epsilon_0 D k_B T \sum n_i}.$$

33. Show that the total electrical work required to charge simultaneously all of the ions in an electrolyte solution is

$$w = -\sum \frac{N_i Z_i^2 e^2 \kappa}{12\pi\epsilon_0 D} = -\frac{V k_B T}{12\pi} \kappa^3.$$

Use the Gibbs–Helmholtz equation to calculate, from w, the integral heat of dilution of the electrolyte solution. On what solution parameters does the integral heat of dilution depend? Is it altered if the dielectric constant D depends on temperature?

34. The Bjerrum–Fuoss theory of ion association was originally introduced to extend the domain of validity of the analysis of electrolyte solutions. The idea is as follows: The number of ions j a distance R from one at the origin is $n_j g_{ij}\, d\mathbf{R}_j$ or $4\pi R^2 n_j \exp(-w_{ij}/k_B T)\, dR_j$. For ions that are very close together, Bjerrum and Fuoss make the approximation $w_{ij} = Z_i Z_j e^2 / 4\pi\epsilon_0 DR$. The number of ions j surrounding i goes through a minimum at $q_0 \equiv |Z_i Z_j| / 8\pi\epsilon_0 D k_B T$. It is now asserted that all ions separated by less than q_0 are "associated," and all others free. The associated ions are treated as ideal neutral solute particles, whereas the free ions interact as in Debye–Hückel theory.

(a) Let $1 - \alpha$ be the degree of association of ions. Show that for a 1:1 electrolyte

$$1 - \alpha = \frac{4\pi N}{V} \left(\frac{e^2}{4\pi\epsilon_0 D k_B T} \right)^3 Q(b),$$

where

$$Q(b) = \int_2^b e^y y^{-4} \, dy,$$

$$b \equiv \frac{e^2}{4\pi\epsilon_0 D a k_B T},$$

$$y \equiv \frac{e^2}{4\pi\epsilon_0 D R k_B T}.$$

(b) Take the Debye–Hückel ionic size parameter a to be equal to q_0, and treat the relation between free and associated ions as an equilibrium. Show that

$$\frac{1-\alpha}{\alpha^2}=\gamma_\pm{}^2\frac{4\pi N}{V}\left(\frac{e^2}{4\pi\epsilon_0 Dk_BT}\right)^3 Q(b).$$

(c) Given the following values for $1-\alpha$ with a = 1.76 Å, calculate $\gamma_\pm$ for the electrolyte solution and compare these values with those deduced from Debye–Hückel theory.

m	0.001	0.005	0.01	0.05	1.10	0.50
$1-\alpha$	0.001	0.007	0.012	0.046	0.072	0.204

The theory of association has been very considerably extended beyond the simple model considered in this problem.

35. The arguments used in this chapter suggest that assuming

$$g_{ij}(R)=g_{ij}^{HS}(R)\exp\left(-\frac{w_{ij}^{DH}(R)}{k_BT}\right)$$

is probably a good approximation for the primitive model of an electrolyte solution. Here $g_{ij}^{HS}(R)$ is the pair-correlation function for hard spheres of diameter $2a$, and

$$w_{ij}^{DH}(R)=\frac{Z_iZ_je^2e^{-\kappa(R-\alpha)}}{4\pi\epsilon_0 DR(1+\kappa a)}.$$

Assume that all ions have the same size, and calculate, for a 1:1 electrolyte, the excess energy and free energy of the solution. For the concentrations considered (say, 0.1 m or less), it is adequate to use the approximation

$$\begin{aligned} g_{ij}^{HS}(R) &= 0 \quad \text{for } R\le 2a,\\ &= 1+\frac{4\pi}{3}(8na^3)\left[1-\frac{3}{8}\left(\frac{R}{a}\right)+\frac{1}{128}\left(\frac{R}{a}\right)^3\right] \quad \text{for } 2a<R\le 4a,\\ &= 1 \quad \text{for } R\ge 4a. \end{aligned}$$

Appendices

APPENDIX I

Systems of Units

The measurement of physical and chemical properties requires the definition of units. Frequently the chosen units in a given investigation are first the commonly used ones, such as inches, pounds, and so on. The complexity and multiplicity of common units for weights and measures is analogous to that of any other aspect of culture, such as language. The need for easy comparison of measurements requires a common "language," an agreed-upon set of units. Just as an agreed-upon language evolves in time, so do agreed-upon units. The currently recommended units are called the *International System of Units* (abbreviated *SI*, for *Système International d'Unités*). This system is the result of much discussion and compromise. For a fuller description of the process of development of units, see the NIST (National Institute of Science and Technology) Web page "Fundamental Physical Constants" http://physics.nist.gov/cuu/index.html. M. L. McGlashan, *Physicochemical Quantities and Units,* 2nd ed. (Royal Institute of Chemistry, London, 1971), and *Ann. Rev. Phys. Chem.* **24,** 51 (1973); F. D. Rossini, *Fundamental Measures and Constants for Science and Technology* (CRC Press, Cleveland, 1974); *The International System of Units (SI)*, National Bureau of Standards (U.S.) Special Publication 330 (Washington, D.C., 1977); M. A. Paul, *J. Chem. Soc.* **11,** 3 (1971). The following tables list the essentials of the SI units, and give their relationship to some other commonly used units (including the older cgs and mks systems).

The International System of Units (SI)

BASE UNITS

Quantity	**Unit**	**Symbol**	**Definition (abridged)**
Length	meter	m	The length of the path travelled by light in a vacuum during a time interval of 1/299 792 458 of a second.
Mass	kilogram	kg	The mass of the international prototype of the kilogram (a piece of platinum-iridium kept in Sèvres, France).
Time	second	s	9 192 631 770 periods of the radiation corresponding to the transition between the hyperfine levels of the ^{133}Cs ground state.
Electric current	ampere	A	The current that produces a force of 2×10^{-7} newton per meter of length between two infinitely long, negligibly thick parallel conductors 1 meter apart in a vacuum.
Thermodynamic temperature	kelvin	K	The fraction 1/273.16 of the thermodynamic temperature of the triple point of water.
Amount of substance	mole	mol	The amount of substance of a system containing as many specified elementary entities (atoms, molecules, ions, etc.) as there are atoms in 0.012 kilograms of ^{12}C.
Luminous intensity	candela	cd	The perpendicular luminous intensity of a surface that emits monochromatic radiation of 540×10^{12} hertz with a radiant intensity of 1/683 watt per steradian.

PREFIXES

Fraction	**Prefix**	**Symbol**	**Multiple**	**Prefix**	**Symbol**
10^{-1}	deci	d	10	deca	da
10^{-2}	centi	c	10^{2}	hecto	h
10^{-3}	milli	m	10^{3}	kilo	k
10^{-6}	micro	μ	10^{6}	mega	M
10^{-9}	nano	n	10^{9}	giga	G
10^{-12}	pico	p	10^{12}	tera	T
10^{-15}	femto	f			
10^{-18}	atto	a			

DERIVED UNITS

Quantity	Unit	Symbol	Definition
Frequency	hertz	Hz	s^{-1}
Energy	joule	J	kg m^2/s^2
Force	newton	N	kg m/s^2 = J/m
Power	watt	W	kg m^2/s^3 = J/s
Pressure	pascal	Pa	kg/m s^2 = N/m^2 = J/m^3
Electric charge	coulomb	C	A s
Electric potential difference	volt	V	kg m^2/s^3 A = J/A s = J/C
Electric resistance	ohm	Ω	kg m^2/s^3A^2 = V/A
Electric conductance	siemens	S	s^3A^2/kg m^2 = Ω^{-1}
Electric capacitance	farad	F	A_2s^4/kg m^2 = A s/V = C/V
Magnetic flux	weber	Wb	kg m^2/s^2A = V s
Inductance	henry	H	kg m^2/s^2A^2 = V s/A = Wb/A
Magnetic flux density (magnetic induction)	tesla	T	kg/s^2A = V s/m^2

The SI is an extension of the *meter-kilogram-second* (*mks*) *system.* A parallel system of units formerly predominant in scientific work is the *centimeter-gram-second* (*cgs*) *system,* in which the indicated three units (cm, g, s) are used as base units. The cgs mechanical units with special names include:

Quantity	Unit	Symbol	Definition	Relation to SI Unit
Energy	erg	erg	g cm^2/s^2	10^{-7} J
Force	dyne	dyn	g cm/s^2	10^{-5} N
Viscosity	poise	P	g/cm s	10^{-1} Pa s
Kinematic viscosity	stokes	St	cm^2/s	10^{-4} m^2/s

The following named units are decimal multiples of SI or cgs units:

Quantity	Unit	Symbol	SI Definition	cgs Definition
Length	angstrom	Å	10^{-10} m	10^{-8} cm
Length	micron	μ	10^{-6} m	10^{-4} cm
Volume	liter	l	1 dm^3 (=10^{-3} m^3)	10^3 cm^3
Mass	tonne	t	10^3 kg	10^6 g
Pressure	bar	bar	10^5 Pa	10^6 dyn/cm^2
Concentration	(molarity)	*M*	mol/dm^3	10^{-3} mol/cm^3 (=mol/liter)

The degree Celsius (°C) is equal in magnitude to the kelvin, and may be used in either system for temperature *differences* or for the *Celsius temperature,* defined as $T - 273.15$ K, where T is the thermodynamic temperature (cf. Section 12.5).

Electric and magnetic units present a problem, since several inconsistent systems have been used, with even the basic equations (and thus the unit dimensions) differing between them. The SI is a so-called four-quantity system, with the ampere (on which other electromagnetic units are based) defined independently of the mechanical units; a four-quantity cgs system is also possible, for example, with the franklin (Fr), equal to the esu of charge, as the fourth unit. However, the usual practice was to use a three-quantity cgs system, defining electric and magnetic units in terms of mechanical units by suppressing the constants in either Coulomb's law (*electrostatic system*) or Ampere's law (*electromagnetic system*); the *Gaussian system* uses electrostatic units for electric quantities and electromagnetic units for magnetic quantities. Some of the units in these systems include:

Quantity	Unit	Definition	Relation to SI Unit
Electrostatic Units			
Electric charge	statcoulomb ("esu of charge")	$g^{1/2}$ $cm^{3/2}/s$	3.33564×10^{-10} C
Electric current	statampere	statcoulomb/s	3.33564×10^{-10} A
Electric potential	statvolt	erg/statcoulomb	299.793 V
Electric resistance	statohm	statvolt/statampere	8.98758×10^{11} Ω
Dipole moment	debye (D)	10^{-18} statcoulomb cm	3.33564×10^{-30} C m
Electromagnetic Units			
Electric current	abampere ("emu of current")	$g^{1/2}$ $cm^{1/2}/s$	2.99793×10^{9} A
Magnetic field strength	oersted (Oe)	abampere/cm	79.5775 A/m
Magnetic flux density	gauss (G)	abampere/cm	10^{-4} T
Magnetic flux	maxwell (Mx)	abampere cm	10^{-8} Wb

The following units belong to neither the SI nor any variation of the cgs system, but have been in sufficiently common use to require definition (a few of the major English-system units are included):

Quantity	Unit	Symbol	Value	Other Equivalents
Length	inch	in.	2.54 cm	
Length	bohr	(a_0)	5.291772×10^{-11} m	0.5292 Å $(\equiv \epsilon_0 h^2/\pi m_e e^2)$
Mass	pound	lb	0.45359237 kg	
Mass	atomic mass unit	amu	1.660539×10^{-27} kg	$(\equiv 1\ g/N_A)$
Time	minute	min	60 s	
Time	hour	h	3600 s	60 min
Energy	calorie[a]	cal	4.184 J	
Energy	electron volt	eV	1.602176×10^{-19} J	$(\equiv e \times 1\ V)$
Energy	hartree	$(R_\infty hc)$	4.35974×10^{-18} J	27.2114 eV $(\equiv m_e e^4/8\epsilon_0^2 h^2)$
Energy/*hc* (wavenumber)	—	cm^{-1}	1.98648×10^{-23} J/*hc*	1.23985×10^{-4} eV/*hc*
Pressure	torr	Torr	133.3224 Pa	(1/760) atm ≈ 1 mmHg
Pressure	atmosphere	atm	1.01325×10^{5} Pa	1.01325 bar = 760 Torr
Celsius temperature	degree Fahrenheit	°F	(5/9)°C	$(\theta/°F = \frac{9}{5}\ \theta/°C + 32)$
Thermodynamic temperature	degree Rankine	°R	(5/9) K	$(T/°R = \theta/°F + 459.688)$

[a] This is the defined thermochemical calorie (cf. Section 13.7); note that several other "calories," with slightly different values, are also in use.

APPENDIX II

Partial Derivatives

Suppose that a given function $u(x, y)$ depends on the values of two independent variables, x and y, and that we are interested in just how u varies with x and y. Such a problem can be conveniently analyzed by separating the effects of the variables, that is, varying only one at a time. By a generalization of the process used to define the derivative in elementary calculus, we can define the *partial derivative* of u with respect to x at constant y:

$$\left(\frac{\partial u}{\partial x}\right)_y = \lim_{\Delta x \to 0} \frac{u(x+\Delta x, y) - u(x, y)}{\Delta x}. \qquad \text{(II.1)}$$

The symbol ∂ (rather than d) is used for partial differentiation; a subscript (here y) is used to designate a variable held constant. Similarly, we have

$$\left(\frac{\partial u}{\partial y}\right)_x = \lim_{\Delta y \to 0} \frac{u(x, y+\Delta y) - u(x, y)}{\Delta y}. \qquad \text{(II.2)}$$

The derivatives of various functions have the same form as in ordinary differentiation; for example, if $u(x, y) = x^2 y$, the partial derivatives are

$$\left(\frac{\partial u}{\partial x}\right)_y = y\frac{d}{dx}(x^2) = 2xy \text{ and } \left(\frac{\partial u}{\partial y}\right)_x = x^2\frac{d}{dy}(y) = x^2.$$

The extension to functions of three or more variables is obvious: The function $v(x, y, z)$ has the partial derivatives $(\partial v/\partial x)_{y,z}$, $(\partial v/\partial y)_{x,z}$, $(\partial v/\partial z)_{x,y}$. Second and higher derivatives can be taken in the usual way; one can easily show that

$$\left[\frac{\partial}{\partial x}\left(\frac{\partial u}{\partial y}\right)_x\right]_y = \left[\frac{\partial}{\partial y}\left(\frac{\partial u}{\partial x}\right)_y\right]_x = \frac{\partial^2 u}{\partial x \partial y}, \qquad \text{(II.3)}$$

that is, that the order of differentiation is immaterial, as long as u and its derivatives are continuous.

As with ordinary differentiation, we can obtain a simple geometric interpretation of the partial derivative. A function of only one variable can be represented by a simple curve in a plane; this curve, so long as it is smooth, has at any point a single well-defined slope, equivalent to the derivative of the function. To plot all the possible values of a function like $u(x, y)$ we need a three-dimensional surface, as shown in Fig. II.1. Such a surface has an infinite number of slopes at a given point, corresponding to the infinite number of directions in which the slope can be taken. As can be seen from the figure, the derivatives defined in Eqs. II.1 and II.2 correspond to the slopes parallel to the x and y axes. More generally, for a function of n variables, the partial derivatives are equivalent to slopes on an $(n + 1)$-dimensional surface.

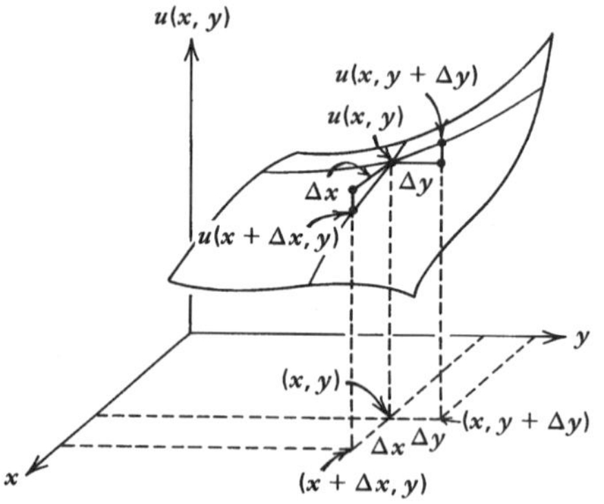

Figure II.1 Geometric interpretation of partial differentiation. The figure shows a portion of the surface $u(x, y)$. The partial derivatives $(\partial u/\partial x)_y$ and $(\partial u/\partial y)_x$ for given values of x and y are the slopes of this surface parallel to the x and y axes, respectively, at the point $u(x, y)$.

The partial derivatives we have defined correspond to slopes parallel to the principal axes representing the independent variables. However, it is possible and sometimes important to take slopes in other directions. For example, one might want to know how $u(x, y)$ varies when we impose the condition that $x = \alpha y^2$ at all times; this will be given by the slope along the curve $x = \alpha y^2$. The derivative of u in the direction corresponding to the steepest slope is called the *gradient* of u; it equals the magnitude of a vector (in the xy plane) whose x and y components are $(\partial u/\partial x)_y$ and $(\partial u/\partial y)_x$, respectively.[1]

How does $u(x, y)$ change when x and y vary simultaneously? For infinitesimal changes in x and y, the resulting infinitesimal change in u is clearly

$$du = \left(\frac{\partial u}{\partial x}\right)_y dx + \left(\frac{\partial u}{\partial y}\right)_x dy; \qquad \text{(II.4)}$$

[1] In three dimensions the gradient corresponds to the vector **grad** u or ∇u defined in footnote 9 on page 65.

this expression is called the *total differential* of u. Suppose we vary x and y together in such a way that u remains constant; then $du = 0$ and we have

$$\left(\frac{\partial u}{\partial x}\right)_y dx + \left(\frac{\partial u}{\partial y}\right)_x dy = 0 \quad (u = \text{const.}). \tag{II.5}$$

The ratio of dx to dy at constant u is simply $(\partial x/\partial y)_u$, so we can rearrange Eq. II.5 in the forms

$$\left(\frac{\partial x}{\partial y}\right)_u = -\frac{(\partial u/\partial y)_x}{(\partial u/\partial x)_y} \quad \text{or}$$

$$\left(\frac{\partial u}{\partial x}\right)_y \left(\frac{\partial x}{\partial y}\right)_u \left(\frac{\partial y}{\partial u}\right)_x = -1. \tag{II.6}$$

(The second form is easy to remember because the variables appear in cyclic order.) Note that u is as legitimate an independent variable as x or y: If we have an equation $f(u, x, y) = 0$ connecting the values of u, x, and y, we can solve for any one in terms of the other two.

Suppose that there is a set of related variables $x, y, z, u, \ldots$ of which only two—*any* two—are independent. (An example would be the pressure, temperature, density, energy, . . . of a fluid.) We can then, for example, express the variable u as a function of x and y, of x and z, of y and z, and so forth. By differentiating Eq. II.4 we can obtain

$$\left(\frac{\partial u}{\partial x}\right)_z = \left(\frac{\partial u}{\partial x}\right)_y + \left(\frac{\partial u}{\partial y}\right)_x \left(\frac{\partial y}{\partial x}\right)_z \tag{II.7}$$

(since $\partial x/\partial x = 1$, no matter what is held constant),

$$\left(\frac{\partial u}{\partial z}\right)_v = \left(\frac{\partial u}{\partial x}\right)_y \left(\frac{\partial x}{\partial z}\right)_v + \left(\frac{\partial u}{\partial y}\right)_x \left(\frac{\partial y}{\partial z}\right)_v, \tag{II.8}$$

and many other such relationships. Generalization of all this to more than two independent variables is straightforward.

Two other equations that hold for any combination of variables are the "chain rule,"

$$\left(\frac{\partial u}{\partial y}\right)_z = \left(\frac{\partial u}{\partial x}\right)_z \left(\frac{\partial x}{\partial y}\right)_z, \tag{II.9}$$

and the inverse relationship,

$$\left(\frac{\partial x}{\partial y}\right)_z = \frac{1}{(\partial y/\partial x)_z}; \tag{II.10}$$

we used the latter in deriving Eq. II.6.

Consider a general differential expression

$$df = M(x, y)\, dx + N(x, y)\, dy, \tag{II.11}$$

where M and N are some functions of x and y. If this expression is actually the differential of some function f, it is called an *exact differential.* According to Eqs. II.3 and II.4, a differential of the form II.1 is exact if

$$\left(\frac{\partial M}{\partial y}\right)_x = \left(\frac{\partial N}{\partial x}\right)_y. \tag{II.12}$$

For the two-variable case, an inexact differential can always be converted to an exact differential by multiplying by some *integrating factor.* For example, $2y\, dx + x\, dy$ is inexact; multiplication by x converts it to $2xy\, dx + x^2\, dy$, which is the exact differential of x^2y.

We often have occasion to transform from one set of variables to another—to introduce new coordinates in mechanics, or new independent variables in thermodynamics. Given the equations relating the new to the old variables, it is a straightforward process to transform an algebraic expression; for an example, see the derivation of Eq. 3.125. Things are more difficult when one has to differentiate or integrate with respect to the variables in question.

Suppose that we want to transform from the variables $x, y, z, \ldots$ to the variables $u, v, w, \ldots$. Derivatives with respect to one of the old variables can be transformed by an equation of the form

$$\left(\frac{\partial}{\partial x}\right)_{y,z,\ldots} = \left(\frac{\partial u}{\partial x}\right)_{y,z,\ldots} \left(\frac{\partial}{\partial u}\right)_{v,w,\ldots} + \left(\frac{\partial v}{\partial x}\right)_{y,z,\ldots} \left(\frac{\partial}{\partial v}\right)_{u,w,\ldots} + \cdots, \tag{II.13}$$

which is simply the generalization of Eq. II.8 in operator form (cf. Section 3.4). For an example, consider the transformation from Cartesian coordinates x, y to circular polar coordinates r, ϕ. We must evaluate the operators

$$\left(\frac{\partial}{\partial x}\right)_y = \left(\frac{\partial r}{\partial x}\right)_y \left(\frac{\partial}{\partial r}\right)_\phi + \left(\frac{\partial \phi}{\partial x}\right)_y \left(\frac{\partial}{\partial \phi}\right)_r,$$

$$\left(\frac{\partial}{\partial y}\right)_x = \left(\frac{\partial r}{\partial y}\right)_x \left(\frac{\partial}{\partial r}\right)_\phi + \left(\frac{\partial \phi}{\partial y}\right)_x \left(\frac{\partial}{\partial \phi}\right)_r. \tag{II.14}$$

From the transformation equations given with Fig. 3.9, we obtain

$$\left(\frac{\partial r}{\partial x}\right)_y = \frac{x}{(x^2 + y^2)^{1/2}} = \frac{x}{r} = \cos\phi,$$

$$\left(\frac{\partial r}{\partial y}\right)_x = \frac{y}{(x^2 + y^2)^{1/2}} = \frac{y}{r} = \sin\phi,$$

$$\left(\frac{\partial \phi}{\partial x}\right)_y = \frac{-y}{x^2 + y^2} = \frac{y}{r^2} = -\frac{\sin\phi}{r},$$

$$\left(\frac{\partial \phi}{\partial y}\right)_x = \frac{x}{x^2 + y^2} = \frac{x}{r^2} = \frac{\cos\phi}{r}. \tag{II.15}$$

Substitution in Eqs. II.14 then yields

$$\left(\frac{\partial}{\partial x}\right)_y = \cos\phi\left(\frac{\partial}{\partial r}\right)_\phi - \frac{\sin\phi}{r}\left(\frac{\partial}{\partial \phi}\right)_r,$$

$$\left(\frac{\partial}{\partial y}\right)_x = \sin\phi\left(\frac{\partial}{\partial r}\right)_\phi + \frac{\cos\phi}{r}\left(\frac{\partial}{\partial \phi}\right)_r; \qquad \text{(II.16)}$$

these expressions are used in Appendix 3B to obtain the Hamiltonian operator in circular coordinates.

When transforming integrals, one needs to know the "volume element" in the new variables. That is, for any quantity f that depends on the variables, one must know the factor J for which

$$\int_V f(x,y,\ldots)\, dx\, dy \ldots = \int_V f(u,v,\ldots) J\, du\, dv \ldots \qquad \text{(II.17)}$$

when the two integrals are taken over corresponding ranges of coordinates. The expression $J\, du\, dv\ \ldots$ has the same dimensions as $dx\, dy\ \ldots$, which for Cartesian coordinates is a volume element. We state without proof the rule giving the value of J:

$$J\left(\frac{x,y,\ldots}{u,v,\ldots}\right) = \begin{vmatrix} \left(\frac{\partial x}{\partial u}\right)_{v,w,\ldots} & \left(\frac{\partial x}{\partial v}\right)_{u,w,\ldots} & \cdots \\ \left(\frac{\partial y}{\partial u}\right)_{v,w,\ldots} & \left(\frac{\partial y}{\partial v}\right)_{u,w,\ldots} & \cdots \\ \cdots & \cdots & \cdots \end{vmatrix}; \qquad \text{(II.18)}$$

the determinant J is called the *Jacobian* of the transformation. For two-variable systems the Jacobian has the simple form

$$J(x,y/u,v) = \left(\frac{\partial x}{\partial u}\right)_v \left(\frac{\partial y}{\partial v}\right)_u - \left(\frac{\partial x}{\partial v}\right)_u \left(\frac{\partial y}{\partial u}\right)_v. \qquad \text{(II.19)}$$

For the transformation from Cartesian to circular coordinates, we have (again using Fig. 3.9)

$$\begin{aligned} J(x,y/r,\phi) &= \left(\frac{\partial x}{\partial r}\right)_\phi \left(\frac{\partial y}{\partial \phi}\right)_r - \left(\frac{\partial x}{\partial \phi}\right)_r \left(\frac{\partial y}{\partial r}\right)_\phi \\ &= (\cos\phi)(r\cos\phi) - (-r\sin\phi)(\sin\phi) \\ &= r(\cos^2\phi + \sin^2\phi) = r, \end{aligned} \qquad \text{(3A.20)}$$

and the area element corresponding to $dx\, dy$ is $r\, dr\, d\theta$. One can also deduce this result directly from a diagram, as shown in Fig. II.2. On the other hand, the volume element of Fig. 2A.2 could have been obtained by calculating

$$J(x,y,z/r,\theta,\phi) = r^2 \sin\theta \qquad \text{(II.21)}$$

from the transformation relations.

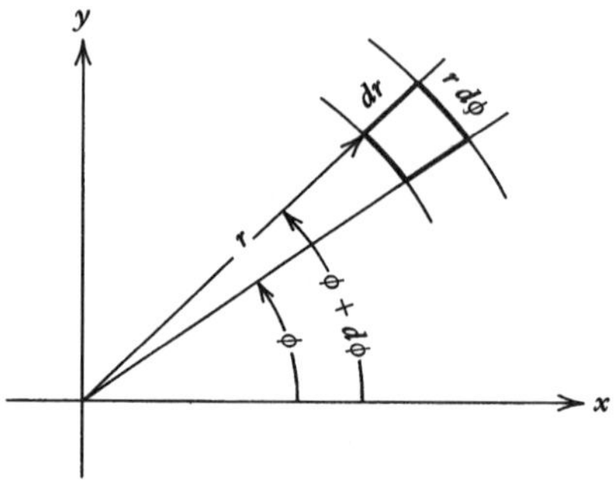

Figure II.2 Area element in circular polar coordinates: $dA = (dr)(r\, d\phi) = r\, dr\, d\phi$.

Searching the Scientific Literature

by Andrea Twiss-Brooks
and Fritz Whitcomb

Introduction

Finding chemical and physical data is one of the most difficult tasks in searching the literature, at times frustrating even experienced researchers and information professionals. Chemical and physical data appear in the journal articles, data handbooks and compilations, technical reports, commercial databases available online or on CD-ROM and diskette, and more recently in sites on the World Wide Web. Navigating this vast landscape of resources requires at least a basic understanding of these various resources and how to use them. This appendix is intended to start the physical chemistry student on the search for chemical and physical data by providing references to information guides and printed data resources, general guidelines for effective searching, specific examples of data searches and results, and a selected list of on-line and Web resources that report chemical and physical data.

The print literature remains a valuable source of chemical and physical data, even in this era of desktop computers and widespread access to the Internet. Effective searching of the printed literature requires access to a research library and a significant investment of time. Guidance for searching the printed literature may be found from a research library's science reference staff as well as from a number of guides devoted to the chemical and physical literature, a few of which are listed later.

In addition to the print literature, there are numerous physical and chemical databases available for searching through commercial database sources. Need for familiarity with the search interface and the often high cost involved means that these resources are often available to students only through a science library's search service or other supervised access. However, some providers offer special academic discounts, and many universities have taken advantage of these discounts to offer online searching to their students and research staff. Despite the costs involved, the sophisticated search interfaces offer faster retrieval of high-quality data and sometimes may be the only source of needed data.

In recent years, the growth of the World Wide Web has provided some new and exciting resources for chemical and physical data. In some cases, the Web has provided a simplified search interface and delivery vehicle for commercial sources. In others, it has provided a more effective distribution mechanism for data sources at little or no cost to the user. These factors have allowed many universities and their libraries to offer the student direct access to chemical and physical data resources. The examples in this appendix are taken from both commercial and noncommercial applications on the Web and are intended to provide some idea of the wealth of data available as well as a few starting points in a search for chemical and physical data. There are many excellent guides to searching the chemical and physical print, commercial on-line and Internet-based literature. The selected list of titles below provide overviews of data producing agencies, descriptions and guides to use of print format standard data reference sets, discussions of scope and currency of resources, lists of on-line databases for physical and chemical data, and in some cases, specific search strategies and examples. Many of these titles may be found at any research or university library.

Selected Guides to the Literature of Chemical and Physical Data

Arny, Linda Ray, *The Search for Data in the Physical and Chemical Sciences* (Special Libraries Association, New York, 1984).

Information Sources in Chemistry, 4th ed., R. T. Bottle and J. F. B. Rowland, eds. (Bowker-Saur, London, 1993).

Information Sources in Physics, 3rd ed., Dennis F. Shaw, ed. (Bowker-Saur, London, 1994).

The Internet: A Guide for Chemists, Steven M. Bachrach, ed. (American Chemical Society, Washington, D.C., 1996).

Maizell, Robert E. (Robert Edward), *How to Find Chemical Information: A Guide for Practicing Chemists, Educators, and Students,* 3rd ed. (Wiley, New York, 1998).

Wiggins, Gary, *Chemical Information Sources* (McGraw-Hill, New York, 1991).

www.nist.gov/srd/#begin.htm

General Guidelines for Chemical and Physical Data Searching

Perhaps the most important part of the search for chemical and physical data is the formulation of the question. Clearly, one first must know what is the question before it is certain that the answer has been found. Always review what information is known already that might be helpful with a search. Some of the questions to ask before going to the library or the computer workstation are:

- Will the data be recognized when found? What are the names or labels of the data? Are there synonyms (in English and/or other languages), alternate spellings, standard symbols that are used by different sources?
- What units will the desired data have? Standard SI or other system? What conversions might be necessary?
- What conditions of data measurement are desired? What temperature or pressure? Are the data sought independent of some conditions but not others? Are actual laboratory measurements required, or will extrapolated or calculated values suffice?
- How many data are likely to be found? Will the data be for a common substance with thousands of literature references? Is a Chemical Abstracts Registry Number (CAS RN) available? Can a systematic name be found, or a list of common synonyms for each compound? Is the molecular formula or structure known?
- When the data are found, how are they known to be accurate? What are the approximate values expected? What institutions and/or research groups are reliable sources of the data? Will older values be accurate, or have the methods of measurement changed significantly over time?
- Where should the search for data begin? Would the data most likely be found in the older printed literature? Are there guides or references that might point to a likely resource? Have some of the standard handbooks been consulted to see what kind of data is already compiled in those resources?

Examples of Searches for Chemical and Physical Data

Since there are many excellent guides to searching the printed literature for chemical and physical data, the examples given here will focus on data searches in Web-based resources. However, it is important to remember to search the printed literature thoroughly for data suspected to have been measured some years ago. Many values that have been previously reported in journals, handbooks and other resources will not be repeated in online resources. This is just a small sample of the many printed resources available that contain collections of physical and chemical data:

Beilstein, Friedrich Konrad, *Beilstein Handbook of Organic Chemistry,* 4th ed., 5th suppl. series (Springer-Verlag, Berlin, 1984–).

Chemical Rubber Company, *CRC Handbook of Chemistry and Physics,* 79th ed., David R. Lide, ed. (CRC Press, Cleveland, Ohio, 1998).

Journal of Physical and Chemical Reference Data, Malcolm W. Chase, ed. (American Chemical Society, Washington, D.C., 1972–).

Landolt, Hans, Richard Börnstein, and K. H. Hellwege, *Zahlenwerte und Funktionen aus Naturwissenschaften und Technik. Neue Serie (Numerical Data and Functional Relationships in Science and Technology. New Series)* (Springer, Berlin, 1961–).

TRC Thermodynamic Tables. Hydrocarbons (Thermodynamic Research Center, Texas A & M University, College Station, TX, 1985–).

The three examples shown next are not intended as a substitute for more formal search instruction, database documentation, or consultation with an information specialist trained in chemical and physical literature searching. The examples are intended rather to illustrate a few of the possibilities available to students and researchers, and to demonstrate some general principles of data searching. They are also intended to point out some pitfalls for searchers to avoid. Consulting a local expert (instructor, librarian or other experienced person) for help in planning searchers for coursework or research is strongly encouraged. The URLs (uniform resource locators) for Web resources mentioned in the examples are included in the list at the end of the appendix.

Example 1: Electronic Spectra of Aromatic Hydrocarbons

An area of significant research is the investigation of theoretical models of aromatic hydrocarbons and the measurement of molecular spectra to test these models. Studies of spectra obtained at extremely low temperatures are an important contribution to this research. Identification of an article reporting a high-resolution, low-temperature spectrum of naphthalene using a bibliographic database is presented. There are many excellent chemical and physical databases to choose from, including INSPEC (Physics Abstracts) and SciFinder Scholar (Chemical Abstracts), to name two. The search shown here was performed using a Web-based version of the INSPEC database. The search terms are entered via a Web form (Fig. IV.1) to construct a search strategy.

All citations, abstracts and subject keywords for articles published from 1969 to the present are searchable. In the search shown, the words *naphthalene* and *high resolution* are combined using a logical operator *and* to give a set of articles containing the word in the title, abstract, or subject fields in the record (Fig. IV.2).

An example of a citation retrieved by the above search strategy is shown in Fig. IV.3. The search resulted in about 45 citations. More citations may be retrieved by broadening the scope of the search by using synonyms, truncation to allow different endings, or using alternate spellings. If too many articles are retrieved by a particular search, it may be

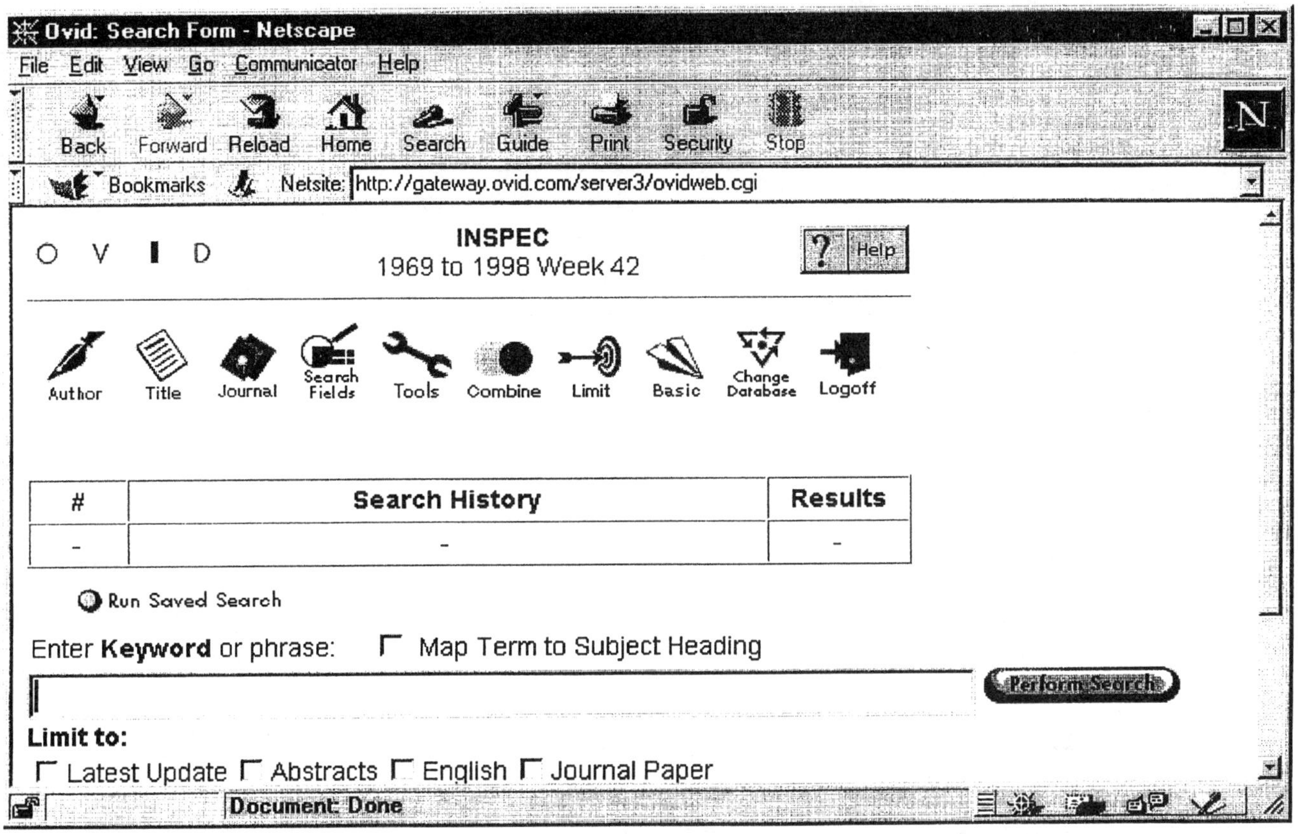

Figure IV.1 Search screen for *INSPEC via OVID* database.

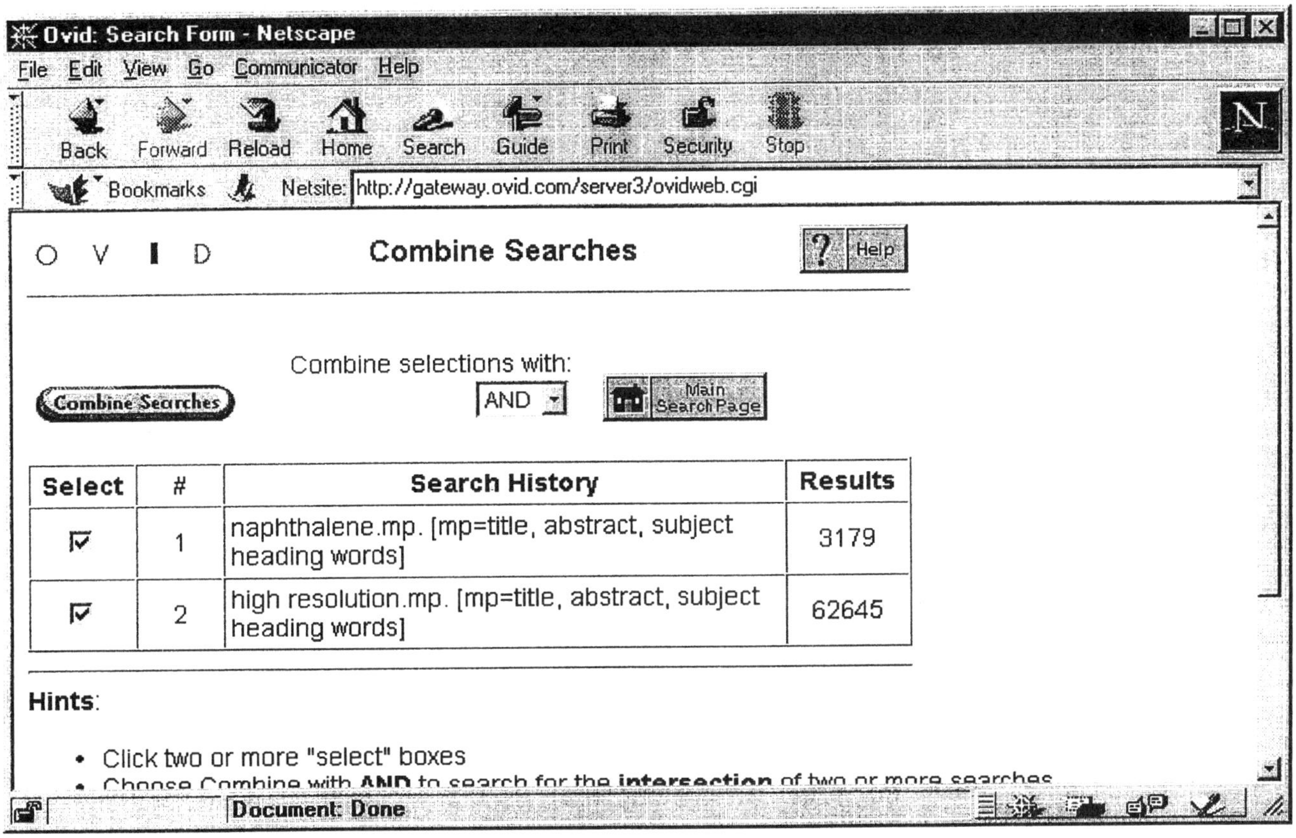

Figure IV.2 Combining search terms in *INSPEC via OVID* database.

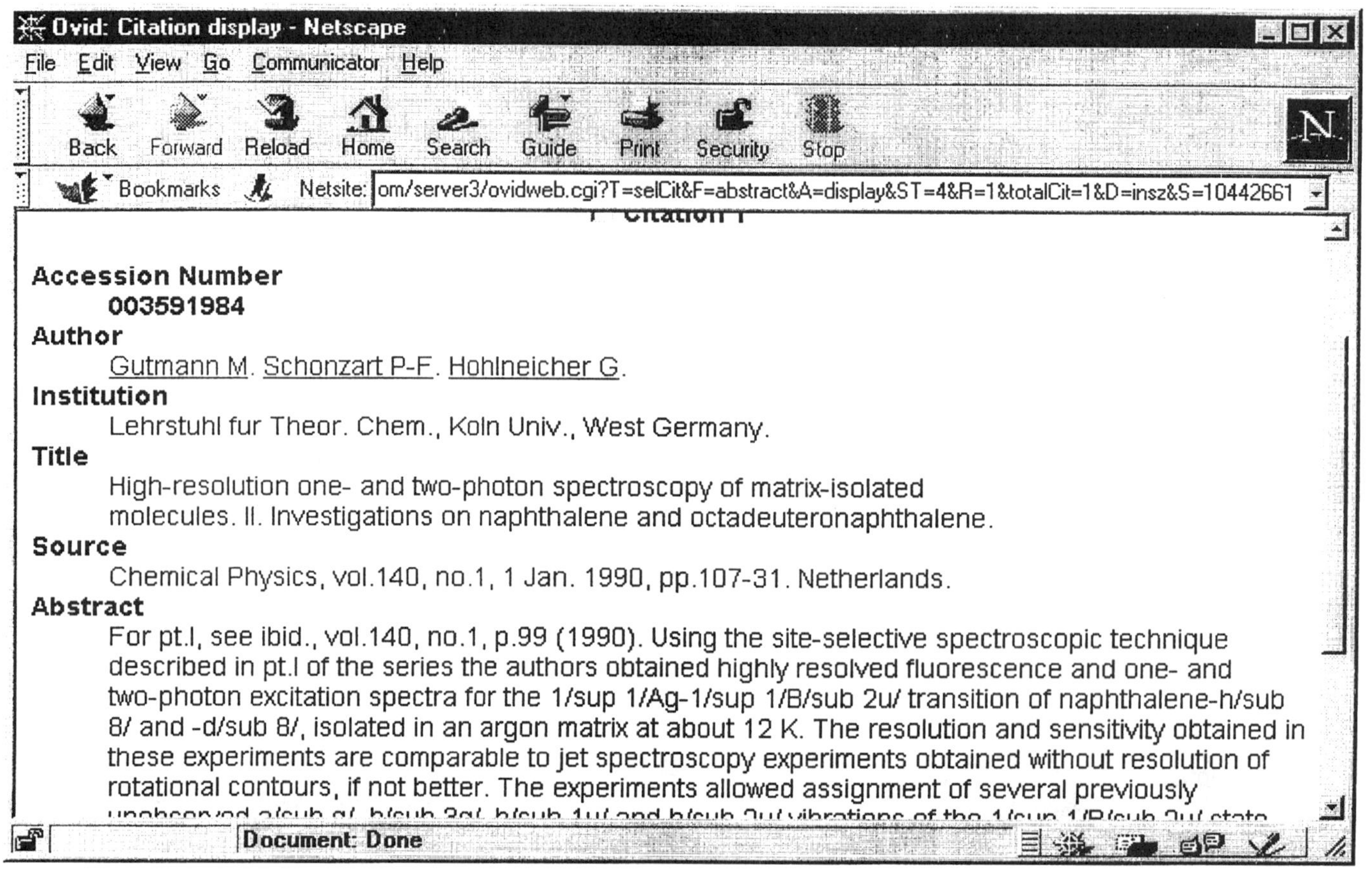

Figure IV.3 Sample citation and abstract retrieved from *INSPEC via OVID* database.

necessary to limit the search by specifying a span of years, combining the subject words with particular author names, or including additional search terms.

Often, an article abstract will contain important data values from the original article, although it is usually advisable to consult the article itself for complete experimental details and conditions. Full-text on-line journals are currently available (including titles produced by the American Institute of Physics, the Institute of Physics, and several commercial publishers), and it may be possible to retrieve the article from the Web or other networked sites. However, if this option is not available, users may still go to the library to obtain the original report for full details of the complete data available.

Example 2: Thermodynamic Properties for NH_3 (Ammonia)

There are very simple substances of economic and research importance that have a wealth of chemical and physical data available. Chemical manufacturing and purifying process industries rely on these data in their plant designs. Researchers doing precise theoretical calculations or interpreting experimental measurements also need reliable data. In many cases, this means identifying sources of evaluated data from highly reliable sources. The next example demonstrates the use of one source of evaluated thermophysical and thermochemical data for ammonia over a range of temperatures.

It is possible to examine a number of reliable print sources for data on ammonia. Electronic data sources are also available. One organization that produces a number of databases is the National Institute of Standards Technology (NIST) through their Standard Reference Data program. The *Standard Reference Data Series* includes about 50 databases, including materials properties, surface data, atomic and molecular physics, and chemical kinetics titles. NIST also produces a Web resource called the *NIST Chemistry WebBook*. The *NIST Chemistry WebBook* contains thermochemical data for 5000 organic and small inorganic compounds, thermophysical property data for 16 fluids and a variety of other data on a number of compounds, including spectra, ion energetics, and reaction thermochemistry. After connecting to the *NIST Chemistry WebBook,* it is possible to search for a particular substance by chemical name, molecular formula, partial formula, or other identifier (e.g., Chemical Abstracts Service Registry Number). In this example, the molecular formula, NH_3, is used to retrieve the appropriate record from the database (Fig. IV.4). In some search interfaces, it is necessary to enter the molecular formula in a particular order (e.g., Hill order), or with special syntax for the numeric subscripts, so always consult the help files, search examples, or other available docu-

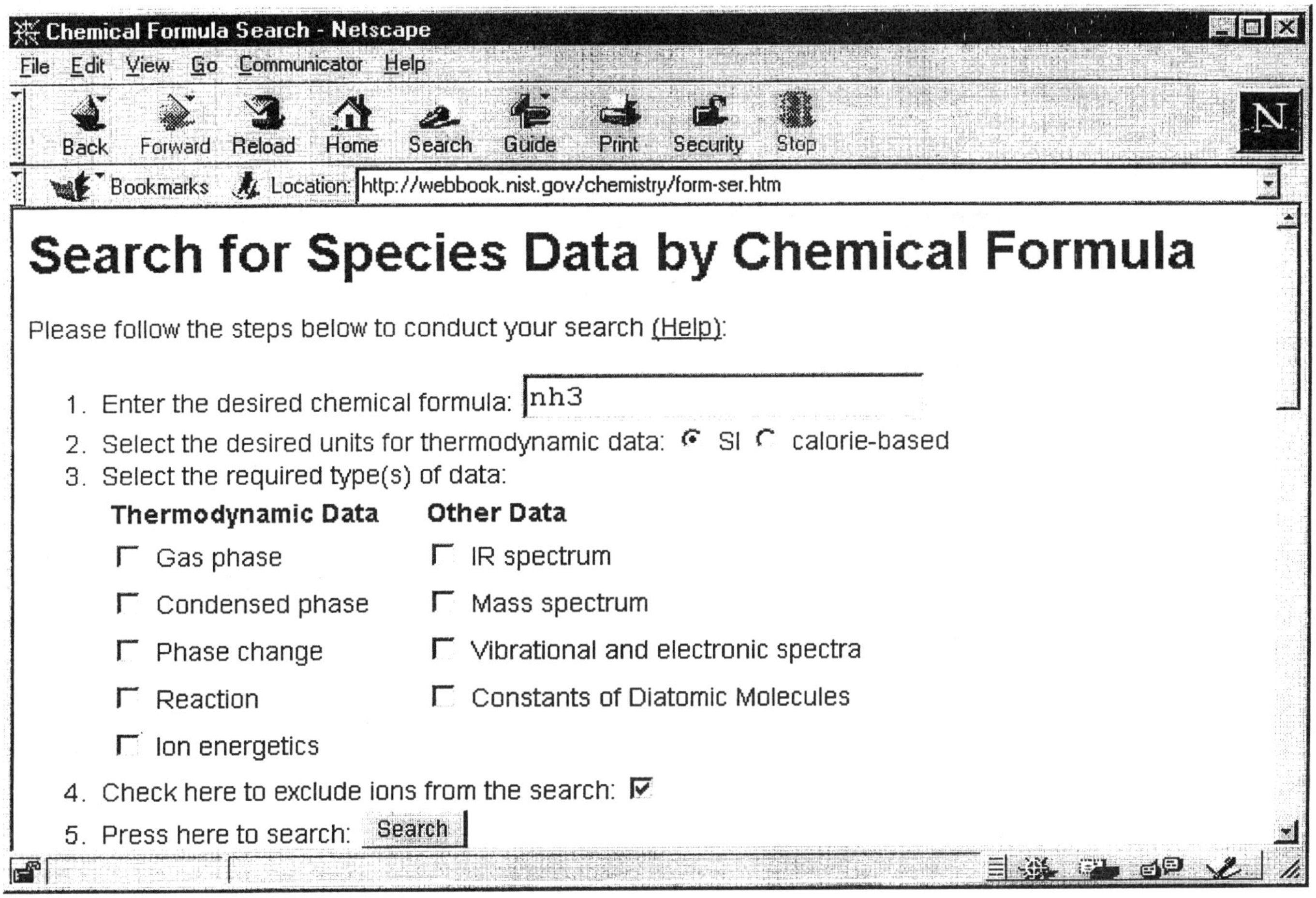

Figure IV.4 Search form for *NIST Chemistry WebBook,* molecular formula option.

mentation for guidance if zero answers in a search are found.

After completion of the search, the results screen indicates structure, alternate names, and available data for ammonia. The Chemical Abstracts Registry number (CAS RN) and molecular weight are also given. Note the data available include thermodynamic data, reaction data, fluid properties, ion energetics and spectral data. Choosing fluid properties will lead to another input screen (Fig. IV.5) for specifying the parameters of the fluid properties data table.

Let's examine the isobaric fluid properties data for ammonia at 200 MPa and over a range of temperatures of 200–500 °K. When searching for chemical and physical data, it may be necessary to make conversions in units for temperature, pressure or other criteria to obtain the desired results. The isobaric data retrieved may be viewed in tabular or graphical format. The isobaric fluid data for ammonia is shown in chart format in Fig. IV.6.

References for the source(s) of the data in the *NIST Chemistry WebBook* are given for all available data. The availability of references for any data compilation is an important feature that should be consulted to judge the quality of the data. Another indication of the quality is acceptance of the data by other scientific or research organizations. For example, the Committee on Data for Science and Technology (CODATA) is an organization dedicated to the production and distribution of critically evaluated numeric data. Many of the *NIST Chemistry WebBook* data values are marked as CODATA accepted values. CODATA also has a list of approved fundamental physical constants available on the Web (see the list of Web sites listed at the end of this appendix for the URL).

Example 3: Kinetic Data for Ozone Reactions with Chlorinated Atmospheric Pollutants

Chlorofluorocarbons, or CFCs, have been implicated in the destruction of the ozone layer surrounding the Earth. A hole in the ozone layer discovered over Antarctica has spurred an increase in scientific and public interest in the reaction of various chemical pollutants with atmospheric gases. This example asks for articles reporting recent studies on the kinetics of atmospheric reactions involving chlorinated species and ozone. Since this is a topic familiar even to the nonscientist, there is a lot of information to be found on the World Wide Web. However, starting a review of ozone reaction chemistry and kinetics by doing a simple text word search on *ozone* and *CFCs* using a Web search engine would

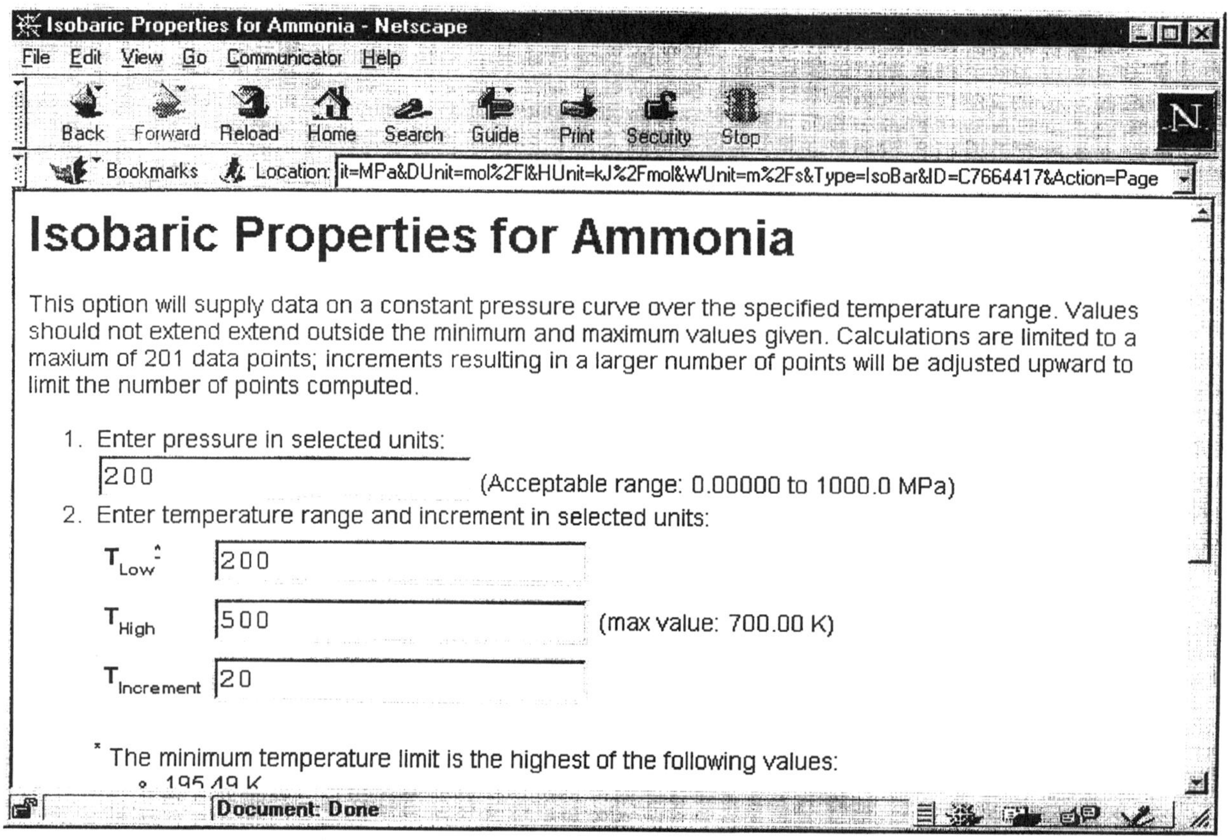
Isobaric Properties for Ammonia - Netscape

File Edit View Go Communicator Help

Back Forward Reload Home Search Guide Print Security Stop

Bookmarks Location: it=MPa&DUnit=mol%2Fl&HUnit=kJ%2Fmol&WUnit=m%2Fs&Type=IsoBar&ID=C7664417&Action=Page

Isobaric Properties for Ammonia

This option will supply data on a constant pressure curve over the specified temperature range. Values should not extend extend outside the minimum and maximum values given. Calculations are limited to a maxium of 201 data points; increments resulting in a larger number of points will be adjusted upward to limit the number of points computed.

1. Enter pressure in selected units:
   200 (Acceptable range: 0.00000 to 1000.0 MPa)
2. Enter temperature range and increment in selected units:
   T_{Low}* 200
   T_{High} 500 (max value: 700.00 K)
   $T_{Increment}$ 20

* The minimum temperature limit is the highest of the following values:
  ∘ 195.49 K

Document: Done

Figure IV.5 Form for defining parameters of isobaric properties of ammonia data display in *NIST Chemistry WebBook*.

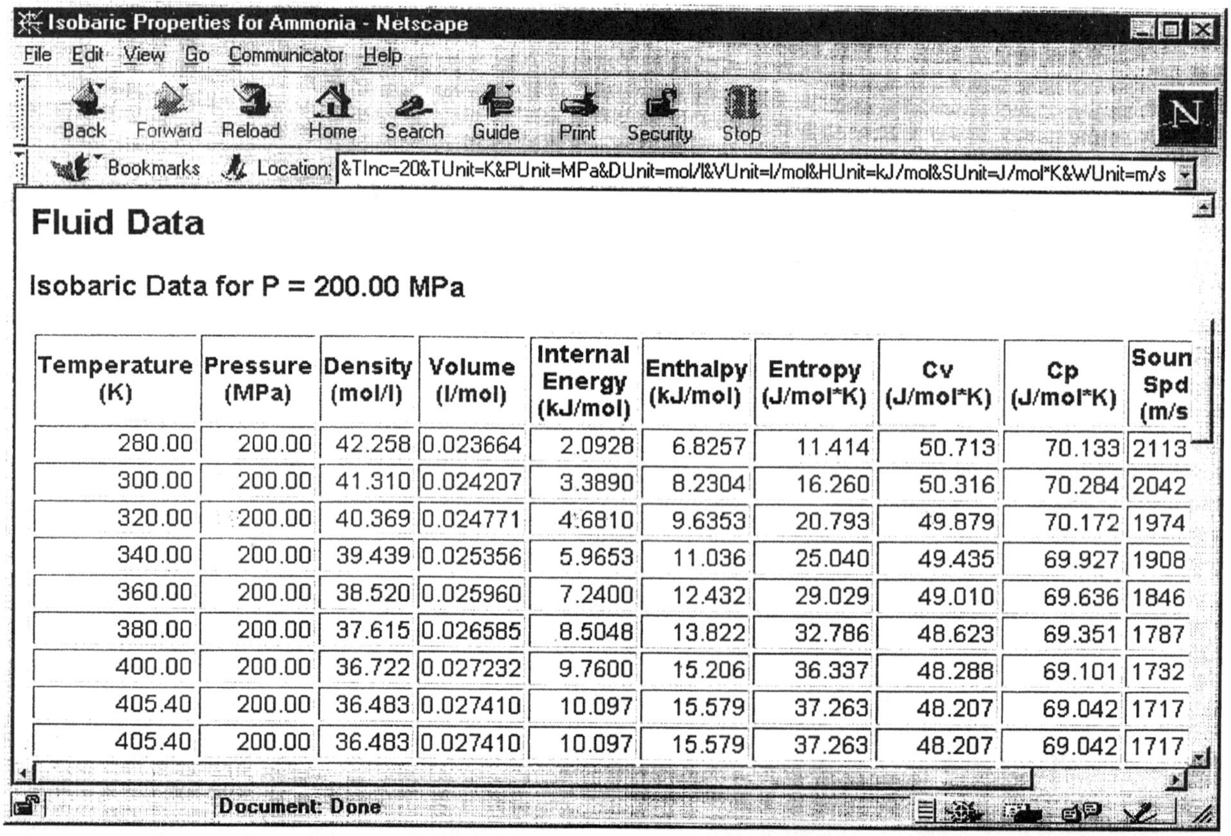
Isobaric Properties for Ammonia - Netscape

File Edit View Go Communicator Help

Back Forward Reload Home Search Guide Print Security Stop

Bookmarks Location: &TInc=20&TUnit=K&PUnit=MPa&DUnit=mol/l&VUnit=l/mol&HUnit=kJ/mol&SUnit=J/mol*K&WUnit=m/s

Fluid Data

Isobaric Data for P = 200.00 MPa

Temperature (K)	Pressure (MPa)	Density (mol/l)	Volume (l/mol)	Internal Energy (kJ/mol)	Enthalpy (kJ/mol)	Entropy (J/mol*K)	Cv (J/mol*K)	Cp (J/mol*K)	Sound Spd (m/s
280.00	200.00	42.258	0.023664	2.0928	6.8257	11.414	50.713	70.133	2113
300.00	200.00	41.310	0.024207	3.3890	8.2304	16.260	50.316	70.284	2042
320.00	200.00	40.369	0.024771	4.6810	9.6353	20.793	49.879	70.172	1974
340.00	200.00	39.439	0.025356	5.9653	11.036	25.040	49.435	69.927	1908
360.00	200.00	38.520	0.025960	7.2400	12.432	29.029	49.010	69.636	1846
380.00	200.00	37.615	0.026585	8.5048	13.822	32.786	48.623	69.351	1787
400.00	200.00	36.722	0.027232	9.7600	15.206	36.337	48.288	69.101	1732
405.40	200.00	36.483	0.027410	10.097	15.579	37.263	48.207	69.042	1717
405.40	200.00	36.483	0.027410	10.097	15.579	37.263	48.207	69.042	1717

Document: Done

Figure IV.6 Isobaric fluid data for ammonia (displayed as HTML table) found in *NIST Chemistry WebBook*.

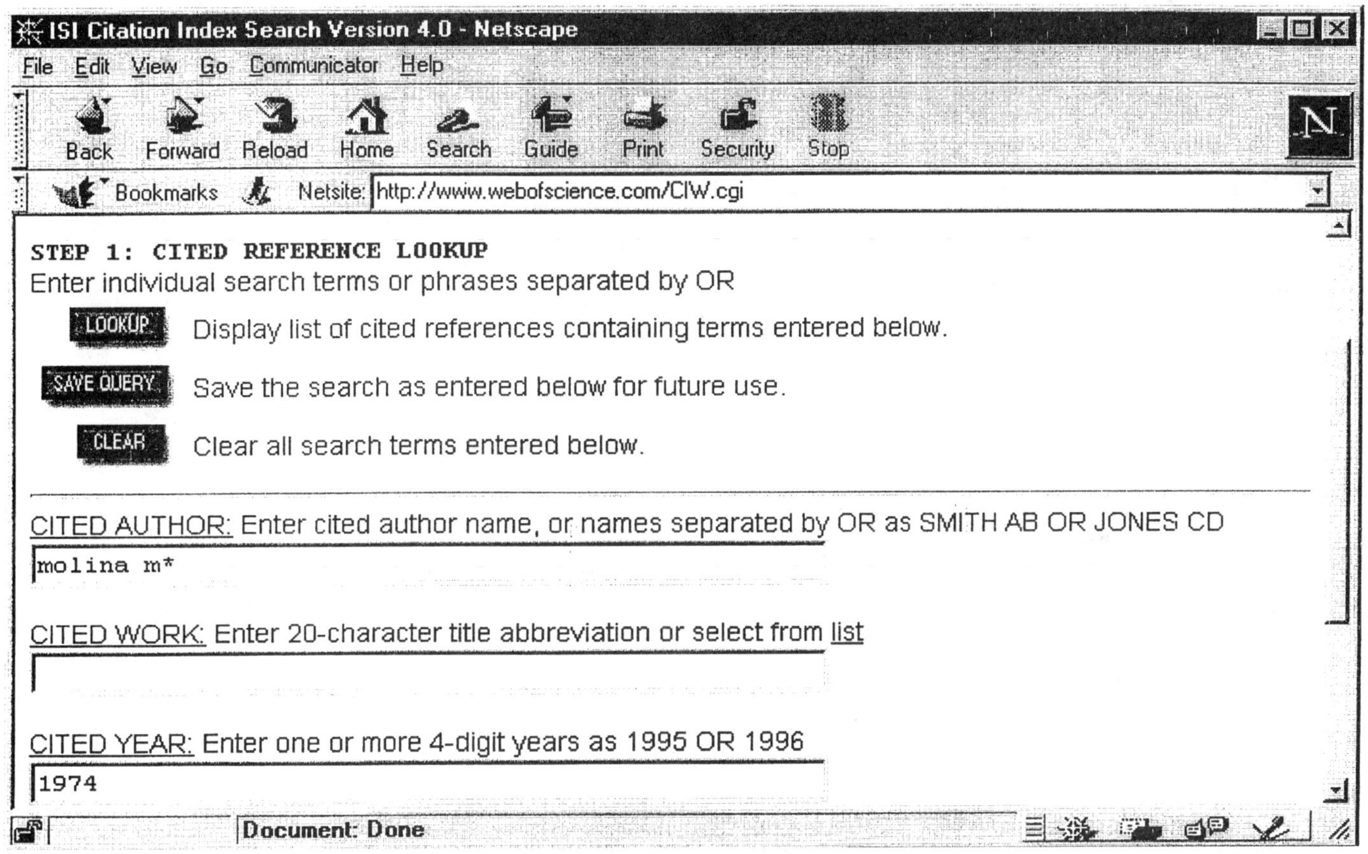

Figure IV.7 Cited reference search form for *Web of Science (Science Citation Index).*

retrieve hundreds or even thousands of hits, many of which are of dubious value. Adding the word *kinetics* to the search might help decrease the number of answers, but there are other, more direct ways of focusing a search.

One approach which might be more effective and cover the most appropriate range of the available scientific literature is a citation search. This approach relies on identifying a "classic" paper in a field and searching for subsequent articles that list the "classic" paper in their references. In this case, the classic paper is fairly easy to identify, once one remembers that there was a Nobel Prize awarded to some scientists working in the area of atmospheric ozone chemistry. Mario Molina and F. Sherwood Rowland were two of the corecipients of the Nobel Prize for their work on ozone and CFCs. By tracing the citation history of a particular paper they published, later research efforts that published kinetic and other reaction studies on ozone and various species can be identified. To trace this history, use the Web version of the *Science Citation Index* called *Web of Science.* This database is available only at sites that have licensed it; its on-line counterpart is available from a number of database providers. There is also a printed version available.

In *Web of Science,* search terms are entered into a Web form. Molina's name is truncated after the first initial, since his middle initial may not be known, or he may not use it in publications. Since there are many, many M. Molina's listed in the cited references in this database, narrowing the lookup down to the year in which the paper was published will retrieve manageable results (Fig. IV.7). By looking up some biographical information on Molina (on the Nobel Prize web site) before starting a citation search, it can be found that Molina and Rowland published their landmark paper in 1974.

Once the lookup is performed, see that there is one particular paper published by Molina in 1974 that has been very heavily cited since then (Fig. IV.8). This paper appeared in volume 249 of the journal *Nature.* To verify that it is the correct paper, you probably could go to the library and look at this reference. However, performing the cited reference search at this point should give evidence from the results as to whether or not the papers deal with atmospheric ozone reactions.

The results retrieved are for research articles on atmospheric ozone chemistry. In fact, a number of articles retrieved might have been missed if the search had involved using species or compound names or formula, since many of the articles discuss compounds related to ozone and CFC's. For the particular reference shown below, some of the rate constants for various reactions of interest are given in the abstract (Fig. IV.9).

The cited references approach is an excellent way to bring together related papers without constructing a complex search statement including many formula or name search terms. This type of approach is also useful in research

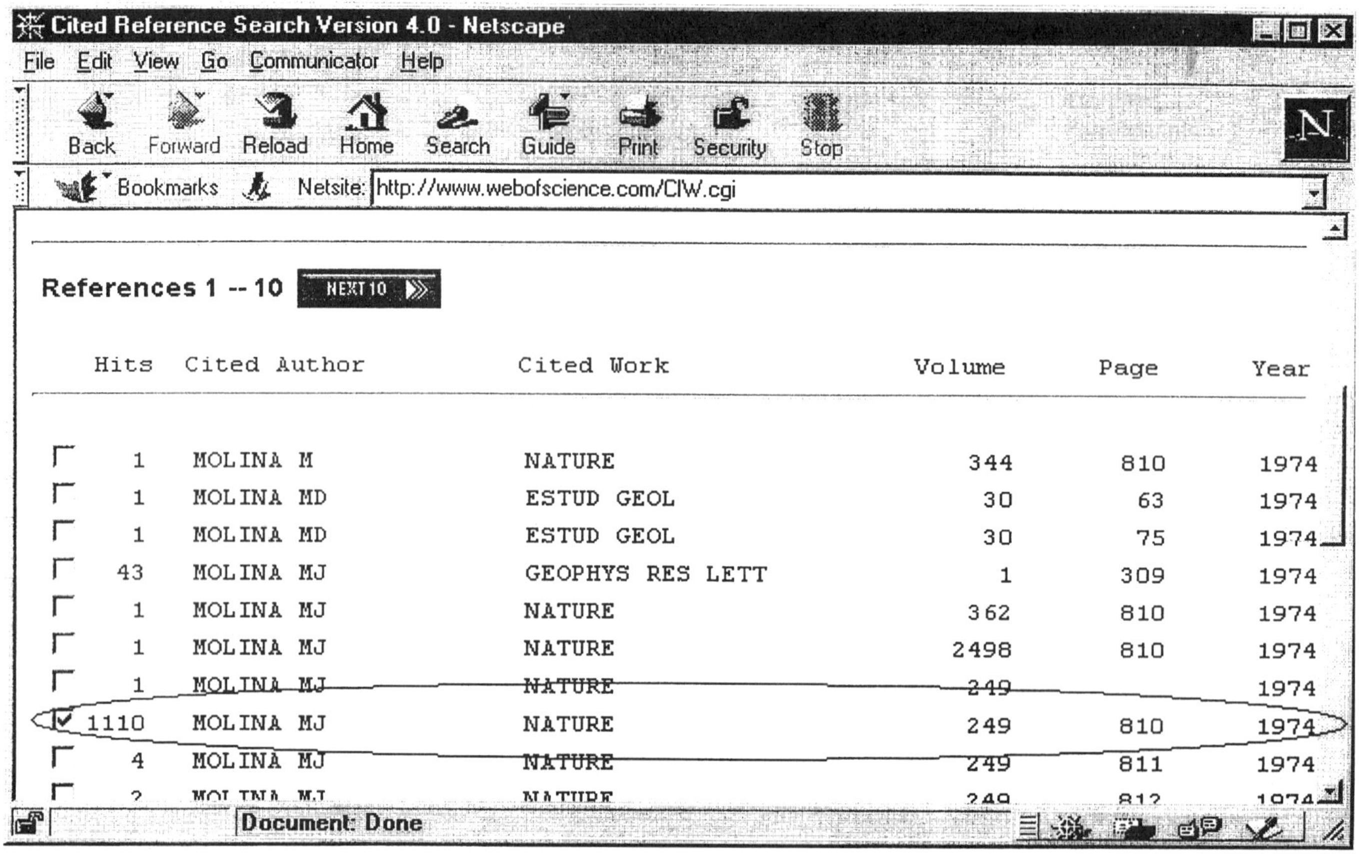

Hits	Cited Author	Cited Work	Volume	Page	Year
1	MOLINA M	NATURE	344	810	1974
1	MOLINA MD	ESTUD GEOL	30	63	1974
1	MOLINA MD	ESTUD GEOL	30	75	1974
43	MOLINA MJ	GEOPHYS RES LETT	1	309	1974
1	MOLINA MJ	NATURE	362	810	1974
1	MOLINA MJ	NATURE	2498	810	1974
1	MOLINA MJ	NATURE	249		1974
1110	MOLINA MJ	NATURE	249	810	1974
4	MOLINA MJ	NATURE	249	811	1974
2	MOLINA MJ	NATURE	249	812	1974

Figure IV.8 List of papers written by M. Molina in 1974 that are cited by later papers as reported in *Web of Science (Science Citation Index)*.

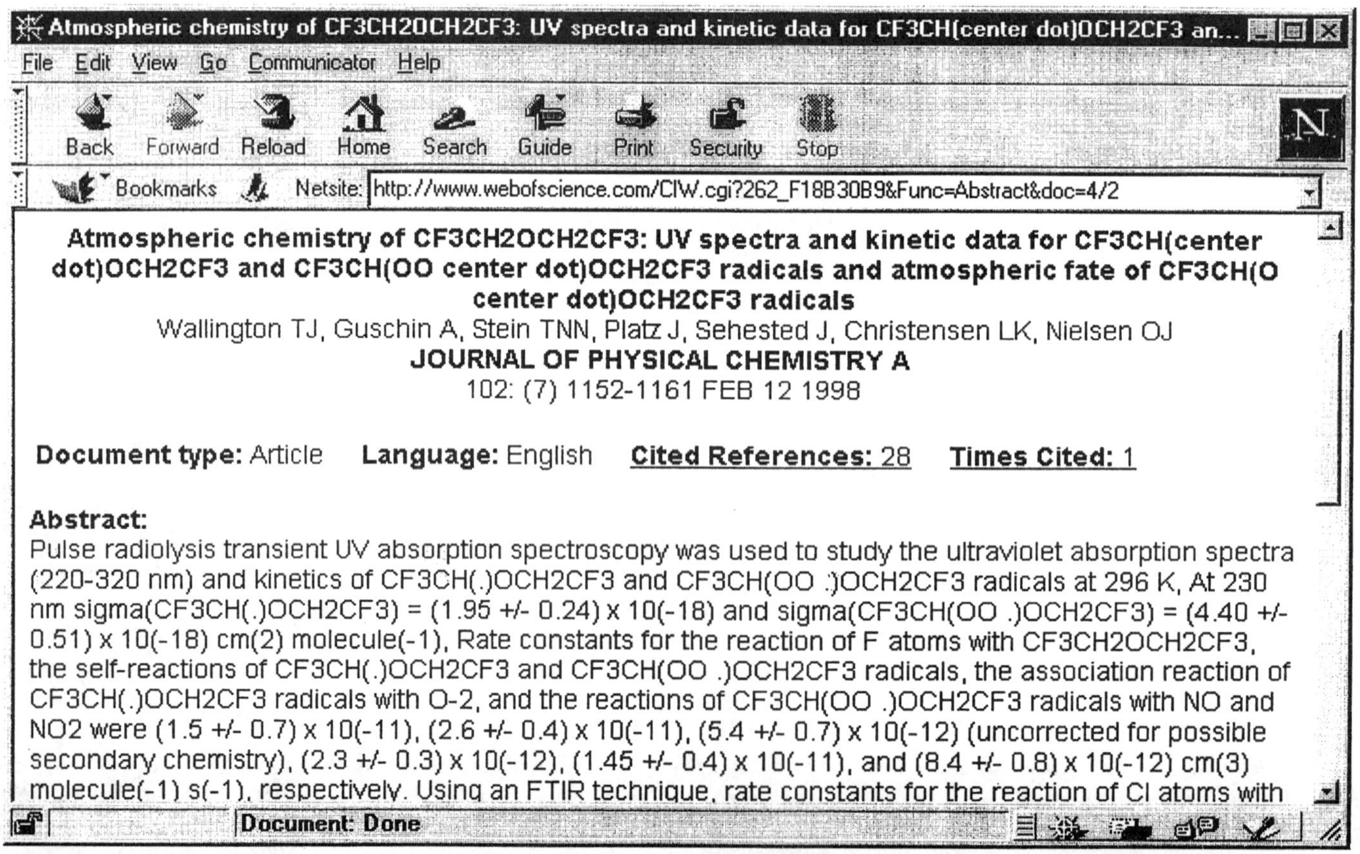

Figure IV.9 Sample citation and abstract for article citing Molina, M. J., and Rowland, F. S., *Nature* 249:810–814, 1974.

areas where compilations of data have not been assembled, or in areas where publication of data is in a broad range of journals in a variety of fields like chemistry, physics, atmospheric science, and environmental studies.

Conclusion

The range of resources available to the searcher for data in the chemical and physical sciences should now be known. Successful searches require planning and preparation, as well as practice. The following list of resources on the World Wide Web are just a few of the many places to begin the search for chemical and physical data:

Andrea Twiss-Brooks, Chemistry Librarian/Chemical Information Specialist
Fritz Whitcomb, Bibliographer for Physics, Astronomy, and Technology
The John Crerar Library, The University of Chicago
http://www.lib.uchicago.edu/LibInfo/Libraries/Science/

Some Web Sites of Interest

http://www.lib.uchicago.edu/LibInfo/SourcesBySubject/Chemistry/
data_bibliography.html
This is an annotated bibliography of print and Web resources for data collections maintained at the University of Chicago.

http://www.nist.gov/srd/jpcrd.htm
Cumulative index to the *Journal of Physical and Chemical Reference Data.*

http://physics.nist.gov/cuu/Constants/
CODATA recommended values of fundamental physical constants.

http://ojps.aip.org/journals/doc/JCPSA6-home/top.html
Journal of Chemical Physics home page.

http://webbook.nist.gov/
NIST Chemistry WebBook.

http://www.isinet.com/
Institute for Scientific Information, producer of *Web of Science.*

http://www.nobel.se/prize/index.html
The Electronic Nobel Museum Project, including a searchable index of Nobel laureates.

Index

Page references of tables and figures are in italics

Table of Atomic Weights

Element	Symbol	Atomic Number	Atomic Weight
Actinium	Ac	89	(227)[a]
Aluminum	Al	13	26.98154
Americium	Am	95	(243)
Antimony	Sb	51	121.75
Argon	Ar	18	39.948
Arsenic	As	33	74.9216
Astatine	At	85	(210)
Barium	Ba	56	137.34
Berkelium	Bk	97	(247)
Beryllium	Be	4	9.01218
Bismuth	Bi	83	208.9804
Boron	B	5	10.81
Bromine	Br	35	79.904
Cadmium	Cd	48	112.40
Calcium	Ca	20	40.08
Californium	Cf	98	(251)
Carbon	C	6	12.01115
Cerium	Ce	58	140.12
Cesium	Cs	55	132.9054
Chlorine	Cl	17	35.453
Chromium	Cr	24	51.996
Cobalt	Co	27	58.9332
Copper	Cu	29	63.546
Curium	Cm	96	(247)
Dysprosium	Dy	66	162.50
Einsteinium	Es	99	(254)
Erbium	Er	68	167.26
Europium	Eu	63	151.96
Fermium	Fm	100	(257)
Fluorine	F	9	18.99840
Francium	Fr	87	(223)
Gadolinium	Gd	64	157.25
Gallium	Ga	31	69.72
Germanium	Ge	32	72.59
Gold	Au	79	196.9665
Hafnium	Hf	72	178.49
Hahnium[b]	Ha	105	(260)
Helium	He	2	4.00260
Holmium	Ho	67	164.9304
Hydrogen	H	1	1.00797
Indium	In	49	114.82
Iodine	I	53	126.9045
Iridium	Ir	77	192.22
Iron	Fe	26	55.847
Krypton	Kr	36	83.80
Kurchatovium[c]	Ku	104	(260)
Lanthanum	La	57	138.9055
Lawrencium	Lr	103	(256)
Lead	Pb	82	207.19
Lithium	Li	3	6.941
Lutetium	Lu	71	174.97
Magnesium	Mg	12	24.305
Manganese	Mn	25	54.9380
Mendelevium	Md	101	(258)
Mercury	Hg	80	200.59
Molybdenum	Mo	42	95.94
Neodymium	Nd	60	144.24
Neon	Ne	10	20.179
Neptunium	Np	93	237.0482
Nickel	Ni	28	58.71
Niobium	Nb	41	92.9064
Nitrogen	N	7	14.0067
Nobelium	No	102	(255)
Osmium	Os	76	190.2
Oxygen	O	8	15.9994
Palladium	Pd	46	106.4
Phosphorus	P	15	30.97376
Platinum	Pt	78	195.09
Plutonium	Pu	94	(244)
Polonium	Po	84	(210)
Potassium	K	19	39.098
Praseodymium	Pr	59	140.9077
Promethium	Pm	61	(147)
Protactinium	Pa	91	231.0359
Radium	Ra	88	226.0254
Radon	Rn	86	(222)
Rhenium	Re	75	186.2
Rhodium	Rh	45	102.9055
Rubidium	Rb	37	85.4678
Ruthenium	Ru	44	101.07
Samarium	Sm	62	150.4
Scandium	Sc	21	44.9559
Selenium	Se	34	78.96
Silicon	Si	14	28.086
Silver	Ag	47	107.868
Sodium	Na	11	22.98977
Strontium	Sr	38	87.62
Sulfur	S	16	32.06
Tantalum	Ta	73	180.9479
Technetium	Tc	43	98.9062
Tellurium	Te	52	127.60
Terbium	Tb	65	158.9254
Thallium	Tl	81	204.37
Thorium	Th	90	232.0281
Thulium	Tm	69	168.9342
Tin	Sn	50	118.69
Titanium	Ti	22	47.90
Tungsten	W	74	183.85
Uranium	U	92	238.029
Vanadium	V	23	50.9414
Xenon	Xe	54	131.30
Ytterbium	Yb	70	173.04
Yttrium	Y	39	88.9059
Zinc	Zn	30	65.38
Zirconium	Zr	40	91.22

[a] Value in parentheses is the mass number of the most stable or best-known isotope.
[b] Suggested by American workers but not yet accepted internationally.
[c] Suggested by Russian workers. American workers have suggested the name Rutherfordium.